ENCYCLOPEDIA OF INDUSTRIAL CHEMICAL ANALYSIS

VOLUME 20

Index
to
Volumes 4–19

ENCYCLOPEDIA
OF INDUSTRIAL
CHEMICAL
ANALYSIS

Edited by Foster Dee Snell *and* Leslie S. Ettre

VOLUME 20

Index
to
Volumes 4–19

Interscience Publishers
division of John Wiley & Sons
New York • London • Sydney • Toronto

CONTENTS

Volume 20

CONTENTS OF THE ENCYCLOPEDIA

The first three volumes of the Encyclopedia cover methods and techniques common in analytical chemistry. The contents of Volumes 1–3 are included in this listing. The material in these volumes is not included in the subject index. A subject index for Volumes 1–3 is included in Volume 3.

VOLUME 1

Title	Authors	Pages
ABSORPTION AND EMISSION SPECTROSCOPY	Clifford L. Hilton	1–29
ACID–BASE TITRATIONS	David C. Wimer	30–52
ACTIVATION ANALYSIS	Vincent P. Guinn	52–77
ADSORPTION CHROMATOGRAPHY	Lloyd R. Snyder	78–98
AIR POLLUTION AND ATMOSPHERIC ANALYSIS	Morris B. Jacobs	99–121
AMPEROMETRIC TITRATIONS	Peter Kabasakalian and M. D. Yudis	122–141
AQUAMETRY	John Mitchell, Jr.	142–159
ATOMIC ABSORPTION SPECTROPHOTOMETRY	Radu Mavrodineanu	160–186
AZEOTROPY	W. S. Tamplin	186–209
BALANCES AND WEIGHING	John Flood and Joan L. Battle	209–228
BOILING POINT DETERMINATION	John R. Anderson	229–242
BUFFER SOLUTIONS	Clifford L. Hilton	242–247
CALORIMETRY	C. B. Murphy	248–263
CAPACITY AND DIELECTRIC CONSTANT	E. N. Shawhan and J. West Loveland	263–302
CHROMATOGRAPHIC METHODS	Lloyd R. Snyder	302–314
COLOR DESIGNATION AND SPECIFICATION	Dean B. Judd	315–351
COLORIMETRIC METHODS OF ANALYSIS	Foster Dee Snell and Cornelia T. Snell	352–376
COMPLEXATION IN AQUEOUS MEDIA	H. A. Flascha and A. J. Barnard, Jr.	376–428
CONDUCTOMETRY AND OSCILLOMETRY	J. West Loveland	429–489
COULOMETRY AND COULOMETRIC TITRATIONS	John W. Miller	489–506
COUNTERCURRENT DISTRIBUTION	Lyman C. Craig and Te Piao King	506–534
CRITICAL-CONSTANT DETERMINATIONS	W. S. Tamplin	534–545
DENSITY AND SPECIFIC GRAVITY	J. K. Taylor	546–560
DIFFERENTIAL REACTION RATE ANALYSIS	J. Gordon Hanna and Sidney Siggia	561–574

VOLUME 1 (*continued*)

Title	*Authors*	*Pages*
DIFFERENTIAL THERMAL ANALYSIS	C. B. Murphy	574–598
DISTILLATION	Frederick E. Williams	599–633
ELECTRICAL METHODS OF ANALYSIS	Clifford L. Hilton	634–637
ELECTRON DIFFRACTION	Robert G. Scott	638–648
ELECTRON MICROSCOPY	W. A. Ladd and M. W. Ladd	649–685
ELECTRON PROBE MICROANALYSIS	Bani R. Banerjee	685–702
EMISSION SPECTROSCOPY	Ford R. Bryan	702–745
EVALUATION OF ANALYTICAL DATA	W. J. Youden	746–763

VOLUME 2

FIELD EMISSION MICROSCOPY	Erwin W. Müller	1–10
FLAME PHOTOMETRY	Radu Mavrodineanu	11–88
FLAMMABILITY DETERMINATIONS	W. S. Tamplin	89–94
FLUOROMETRY AND PHOSPHORIMETRY	Alfred Weissler	94–113
FREEZING POINT DETERMINATION	John R. Anderson	113–138
GAS ANALYSIS	Anson P. Hobbs	139–157
GAS CHROMATOGRAPHY	C. Merritt, Jr.	157–195
GRAVIMETRIC METHODS OF ANALYSIS	David H. Klein	195–225
HEAT OF FUSION	W. S. Tamplin	226–238
HEAT OF VAPORIZATION	W. S. Tamplin	238–246
INDICATORS	Clifford L. Hilton	246–253
INFRARED SPECTROMETRY	George L. Covert	253–304
INTERFERENCE MICROSCOPY	Peter Bartels	304–334
INTERFEROMETRY	W. E. Degenhard	334–346
ION EXCHANGE METHODS	Robert Kunin	346–369
IONIZATION CONSTANTS	W. S. Tamplin	369–375
LABORATORY DESIGN AND EQUIPMENT	Leslie Albert	375–389
LABORATORY SAFETY	F. A. Van Atta	389–412
LIGHT SCATTERING	L. H. Tung	412–427
LITERATURE OF ANALYTICAL CHEMISTRY	David R. Gaskill	427–444
MAGNETIC SUSCEPTIBILITY	L. N. Mulay	445–464
MASS SPECTROMETRY	Maurice L. Bazinet and C. Merritt, Jr.	464–510
MELTING POINT DETERMINATION	Clifford L. Hilton	511–516
MICRO- AND SEMIMICROANALYSIS	Clifford L. Hilton	516–561
MICROSCOPY	T. G. Rochow	561–600
MICROWAVE SPECTROSCOPY	D. R. Lide	600–611
MOLECULAR WEIGHT DETERMINATION	George L. Beyer	611–648
NEAR-INFRARED SPECTROPHOTOMETRY	Kermit B. Whetsel	648–691
NEPHELOMETRY AND TURBIDIMETRY	Frank P. Hochgesang	692–707
NUCLEAR MAGNETIC AND ELECTRON PARAMAGNETIC RESONANCE	E. G. Brame, Jr.	707–734
OPTICAL ACTIVITY	Pierre Crabbé	735–769

VOLUME 3

Title	Authors	Pages
PAPER CHROMATOGRAPHY	H. J. Pazdera	1–40
PARTICLE SIZE AND SURFACE AREA	Riyad R. Irani	40–77
PHASE EQUILIBRIA AND ITS ANALYTICAL APPLICATIONS	G. Richard Atwood	77–146
pH DETERMINATION	Roger G. Bates	146–161
POLAROGRAPHY AND VOLTAMMETRY	Peter Kabaskalian and M. D. Yudis	161–199
POTENTIOMETRY AND POTENTIOMETRIC TITRATIONS	W. M. Banick, Jr.	200–216
PRECIPITATION AND CRYSTALLIZATION	Clifford L. Hilton	216–221
PRESSURE MEASUREMENT AND CONTROL	Anson P. Hobbs	222–229
QUALITATIVE ANALYSIS, INORGANIC	David R. Gaskill	229–251
QUALITATIVE ANALYSIS, ORGANIC	John E. Zarembo	251–279
RADIOACTIVE ISOTOPE METHODS	V. J. Molinski	279–305
RAMAN SPECTROSCOPY	R. F. Stamm	305–333
RECORDERS	W. R. Clark	334–359
REDOX TITRATIONS	Robert Strickman	359–376
REFLECTANCE SPECTROPHOTOMETRY	Deane B. Judd	376–392
REFRACTOMETRY	W. E. Degenhard	392–407
RHEOLOGICAL MEASUREMENTS, FLUID	J. G. Savins	408–463
SAMPLING AND SAMPLE PREPARATION	Clifford L. Hilton	464–475
SEPARATION BY DIFFUSION	S. B. Tuwiner	475–486
SEPARATION BY ELECTRICAL METHODS	Philip A. Crispino	487–502
SEPARATION BY EXTRACTION	G. S. Somekh	502–533
SEPARATION BY MECHANICAL METHODS	Francis B. Hutto and Frederick W. Best	534–548
SOLIDIFICATION POINT DETERMINATION	Clifford L. Hilton	548–550
SOLUBILITY	Clifford L. Hilton	550–555
SPECIFICATIONS, STANDARDS, AND STANDARD-SETTING ORGANIZATIONS	Edwin A. Wynne and Clifford A. Flanders	555–565
STANDARDIZATION OF SOLUTIONS	Joan L. Battle	566–571
SUBLIMATION	E. C. Kuehner and R. T. Leslie	572–584
SURFACE AND INTERFACIAL MEASUREMENTS	Lloyd L. Osipow	584–598
TEMPERATURE MEASUREMENT AND CONTROL	J. C. Melcher	599–642
THERMAL METHODS OF ANALYSIS	Clifford L. Hilton	642–643
THERMOGRAVIMETRIC ANALYSIS	C. B. Murphy	644–665
THERMOMETRIC ANALYSIS	C. B. Murphy	666–672
THERMOMETRIC TITRATIONS	C. B. Murphy	672–685
THIN-LAYER CHROMATOGRAPHY	H. J. Pazdera and A. E. Kaptionak	686–725
ULTRAVIOLET AND VISIBLE ABSORPTION SPECTROPHOTOMETRY	Clifford L. Hilton	726–754
VACUUM ULTRAVIOLET SPECTROSCOPY	Clifford L. Hilton	755–768

VOLUME 3 (*continued*)

Title	*Authors*	*Pages*
VISCOSITY	V. L. Burger	768–774
VOLUME MEASUREMENT: LIQUIDS	J. J. Moran and E. M. Cook	774–795
X-RAY METHODS OF ANALYSIS	William J. Croft	796–817

VOLUME 4

ABLATIVE MATERIALS	C. A. Gaulin and W. T. Barry	1–30
ABRASIVES	W. M. Hazel	31–61
ACETALS AND KETALS	Frank S. Wagner, Jr.	62–74
ACETATE AND TRIACETATE FIBERS	B. S. Sprague and C. L. Kolb	74–93
ACETIC ACID	E. F. Joy and A. J. Barnard, Jr.	93–101
ACETIC ANHYDRIDE	E. F. Joy and A. J. Barnard, Jr.	102–107
ACETONE	D. K. Banerjee and C. C. Budke	107–121
ACETOPHENONE	V. D. Johnston	122–126
ACETYLENE	W. F. Tiedge and F. E. Caskey	126–135
ACETYLENIC CHEMICALS	W. F. Tiedge and F. E. Caskey	136–147
ACROLEIN	E. D. Peters and G. C. Hood, Jr.	148–159
ACRYLAMIDE	M. V. Norris	160–168
ACRYLAMIDE POLYMERS	M. V. Norris	168–181
ACRYLIC AND METHACRYLIC ACIDS AND ESTERS	Leo S. Luskin	181–218
ACRYLIC AND MODACRYLIC FIBERS	R. H. Heidner and M. E. Gibson, Jr.	219–367
ACRYLONITRILE	M. V. Norris	368–381
ACRYLONITRILE POLYMERS	M. V. Norris	382–390
ADHESION	R. F. Blomquist	391–398
ADHESIVES	Eino Moks	398–407
ADIPIC ACID	Robert E. Keller	408–423
ADIPIC ACID DERIVATIVES	Robert E. Keller	423–431
ADSORBENTS AND ABSORBENTS	T. D. Oulton	431–452
AEROSOLS	Maurice Sage	452–461
ALCOHOLIC BEVERAGES, DISTILLED	George I. deBecze	462–494
ALCOHOLS	D. K. Banerjee and C. C. Budke	495–586
ALKALOIDS	G. J. Papariello and Frederick Tishler	587–618

VOLUME 5

ALKANOLAMINES	John A. Riddick and Carl J. Wassink	1–22
ALKYD RESINS	William E. Link	22–56
ALLOYS	Clifford L. Hilton	57–75
ALLYL COMPOUNDS	Wolf Karo	75–109
ALUMINUM		
Aluminum and aluminum alloys	George P. Koch	110–158
Aluminum compounds	Evan Morgan	158–199

VOLUME 5 (continued)

Title	Authors	Pages
Aluminum determination	Clifford L. Hilton	199–258
AMINES	H. J. Stolten	259–275
AMINO RESINS	Seymour Friedman	276–290
AMMONIA	D. E. Scheirer, K. Theurer, and E. L. Nelson	290–324
AMMONIUM COMPOUNDS	D. E. Scheirer, K. Theurer, and E. L. Nelson	325–355
ANESTHETICS	Clark A. Kelly	355–421
ANILINE AND ITS DERIVATIVES	John W. McDowell and Jean Northcott	421–459
ANTIBIOTICS	G. A. Brewer, Jr., and T. B. Platt	460–633

VOLUME 6

Title	Authors	Pages
ANTIFREEZES	E. Beynon	1–23
ANTIHISTAMINES	Clark A. Kelly	24–38
ANTIMONY	George J. Abel, Jr.	39–70
ANTIOXIDANTS AND ANTIOZONANTS	R. W. Wise and R. H. Campbell	71–122
ANTIPYRETICS AND ANALGESICS	J. G. Rutgers and Clifford L. Hilton	123–222
ARSENIC	George J. Abel, Jr.	223–254
ASBESTOS	C. N. Thompson and T. Sopoci	254–273
ASPHALTS, TARS, AND PITCHES	Ward K. Parr	274–350
AUTOMOTIVE CHEMICAL SPECIALTIES	Eric Beynon	350–380
AZINE AND RELATED DYES	Raymond B. Crawford and W. Allan Fisher	381–400
AZO DYES	Roger L. Rounds	401–424
BAKERY CHEMICALS	W. B. Chess	425–438
BAKERY PRODUCTS	W. A. Baer and M. J. Horn	438–467
BARBITURATES	J. G. Rutgers, J. N. Bathish, and G. J. Papariello	467–532
BARIUM	A. Procko	532–582
BEAD AND FLAME TESTS	Clifford L. Hilton	583–593
BENZALDEHYDE AND ITS DERIVATIVES	Editorial staff	594–639

VOLUME 7

Title	Authors	Pages
BENZENE	Editorial staff	1–58
BENZOIC ACID	Editorial staff	58–88
BENZYL AND PHENYLETHYL ALCOHOLS	Editorial staff	89–102
BERYLLIUM		
Beryllium and beryllium alloys	Paul C. Kempchinsky	103–117
Beryllium compounds	Editorial staff	118–129
Beryllium determination	Paul C. Kempchinsky	129–141
BILE ACIDS AND SALTS	Editorial staff	141–154
BIOASSAY	W. M. Hoskins	155–186

VOLUME 7 (*continued*)

Title	*Authors*	*Pages*
BISMUTH	E. A. Wynne	187–220
BLEACHING AGENTS	C. H. Schmiege	221–293
BONDED ABRASIVES	Daniel E. Thomas and Robert P. Anibal	293–312
BORON		
Boron and boron alloys	Robert S. Braman	312–329
Boron compounds	Robert S. Braman	329–358
Boron compounds	Editorial staff	358–384
Boron determination	Robert S. Braman	384–423
BRAKE FLUIDS	R. W. Shiffler	423–440
BRAKE LININGS	W. R. Jensen, J. T. Trainor, A. J. Lacroix, and R. R. Moalli	440–474
BREWERY PRODUCTS	Harold E. Weissler	475–679

VOLUME 8

Title	Authors	Pages
BROMINE	V. A. Stenger	1–29
CADMIUM	David R. Gaskill	30–55
CAFFEINE	George F. McCutcheon	55–71
CALCIUM	Joseph B. Bodkin and the Editorial staff	72–114
CAPROLACTAM	John H. Bonfield and Jean Northcott	114–129
CARBIDES	J. W. Tereshko	130–139
CARBON, ACTIVATED	G. H. Scheffler	139–161
CARBONATED BEVERAGES	Charles P. Grier	161–178
CARBON BLACK	B. Schubert, F. P. Ford, and F. Lyon	179–243
CARBON DIOXIDE	George E. Schmauch	244–267
CARBON DISULFIDE	Andrew Procko	268–279
CARBON, HYDROGEN, AND NITROGEN MICROANALYSIS	Orville Hinsvark	280–298
CARBON MONOXIDE	George E. Schmauch	299–319
CARBON TETRACHLORIDE	Marion E. Mitten, Kenneth S. Dress, William G. Krochta, and Bernard J. De Witt	320–330
CARBONYL COMPOUNDS	Gerald R. Umbreit	331–367
CARBOXYLIC ACIDS	Gabor Karoly, Allen E. Klein, and Gerald R. Umbreit	368–409
CARBOXYLIC ACIDS, AMINO	Graham A. Stoner and Fred Fischer	410–556
CARBOXYLIC ACIDS, DERIVATIVES	Leslie S. Ettre	557–612
CARRIER GAS AND VACUUM FUSION METHODS	Wayne E. Dallmann	613–693

VOLUME 9

Title	*Authors*	*Pages*
CASEIN	Carolyn N. Bye	1–15
CAULKING COMPOUNDS	V. W. Wells	16–27
CELLULOSE	W. J. Alexander and G. C. Daul	27–59
CELLULOSE DERIVATIVES	W. J. Alexander and G. C. Daul	59–94
CEMENTS, MORTARS, AND CONCRETE	W. G. Hime	94–155
CENTRIFUGAL SEPARATION	Frederick W. Keith, Jr.	156–188
CERAMICS	J. P. Williams and D. E. Campbell	189–232
CEREALS	James Grove	232–245
CERIUM	R. E. Harris and M. M. Woyski	246–277
CESIUM	J. A. Hill	278–303
CHLORINE	G. Oplinger	304–333
CHLORINE COMPOUNDS, INORGANIC	George Oplinger	333–368
CHLOROCARBONS	Marion E. Mitten, Kenneth S. Dress, William G. Krochta, F. P. Ewald, and Bernard J. De Witt	369–510
CHLOROPHENOLS	Norman E. Skelly	511–528
CHLOROSULFONIC ACID	E. F. Joy and A. J. Barnard, Jr.	528–534
CHOCOLATE AND COCOA PRODUCTS	Elwood M. Meyers, A. S. Graham, E. J. Finnegan, and W. R. Kreiser	534–611
CHOLINE AND ITS DERIVATIVES	Emil Lorz	611–632
CHROMIUM		
Chromium metal	Eugene D. Olsen and Craig C. Foreback	632–660
Chromium alloys	Eugene D. Olsen and Craig C. Foreback	660–680
Chromium compounds	Winslow H. Hartford	680–709

VOLUME 10

CIGARETTES	James T. Dobbins, Jr.	1–42
CINNAMALDEHYDE	Martin H. Infeld and Kenneth A. Connors	42–55
CINNAMIC ACID AND ITS DERIVATIVES	Kenneth A. Connors	56–76
CINNAMYL ALCOHOL	Kenneth A. Connors	76–85
CITRIC ACID	R. M. Speights and A. J. Barnard, Jr.	85–100
CLAYS AND CLAY MINERALS	T. D. Oulton	100–138
COAGULANTS AND ANTICOAGULANTS	Clark A. Kelly	138–209
COAL AND COKE	R. F. Abernethy, S. Ergun, R. A. Friedel, J. T. McCartney, and I. Wender	209–262

VOLUME 10 (*continued*)

Title	*Authors*	*Pages*
COATED ABRASIVES	Paul F. Forsyth	263–283
COBALT		
Cobalt and cobalt alloys	Roland S. Young	283–326
Cobalt compounds	F. R. Morral	327–347
Cobalt determination	Roland S. Young	348–373
COFFEE	Pierre Navellier	373–447
COLORS FOR FOODS, DRUGS, AND COSMETICS	Daniel M. Marmion	447–547
CONFECTIONERY PRODUCTS	M. B. Sherman	548–567
CONTRACEPTIVES	Neil H. Brown and Clark A. Kelly	568–597
COPPER		
Copper metal	Roland S. Young	598–630
Copper alloys	Roland S. Young	630–650
Copper compounds	Carol Freedenthal	651–680

VOLUME 11

CORN PRODUCTS	Robert J. Smith	1–63
COSMETICS	Charles M. Fairchild and Irwin Margolin	64–116
COTTON	Robert T. O'Connor	117–159
COUMARINS	F. Tishler and J. Mollica	159–173
CROTONALDEHYDE	Frank S. Wagner, Jr.	174–183
CYCLIC ALCOHOLS	Frank S. Wagner, Jr.	184–197
CYCLOHEXANONE	Frank S. Wagner, Jr.	197–208
CYCLOHEXYLAMINE	James G. Theivagt, Preston F. Helgren, and Delbert R. Leubke	209–219
CYCLOHEXYLAMINE DERIVATIVES	James G. Theivagt, Preston F. Helgren, and Delbert R. Luebke	220–255
DENTAL MATERIALS	Frank H. Freeman	256–321
DETERGENTS	Robert M. Kelley	322–406
DIAMINES AND HIGHER AMINES, ALIPHATIC	James M. Weber	407–431
DIAMINES AND HIGHER AMINES, AROMATIC	John S. Heil	432–462
DIAMONDS, INDUSTRIAL	Richard P. Hight	462–484
DIATOMACEOUS EARTH	C. N. Thompson and T. Sopoci	484–500
DIENES	Ralph C. Vollmar	500–515
DIOXANE	Quentin E. Quick	516–524
DIPHENYLS AND TERPHENYLS	Charles G. Smith	524–537
DISINFECTANTS AND ANTISEPTICS	Robert S. Browning, Jr.	537–572
DITHIOCARBAMATES	Quentin E. Thompson	573–602
DRYCLEANING AGENTS	Albert R. Martin	603–617

VOLUME 12

Title	Authors	Pages
DYES	Kurt A. Dellian and Robert E. Tellis	1–22
EDTA AND RELATED COMPOUNDS	E. A. Pinzer and B. B. Stewart	23–56
EGGS AND EGG PRODUCTS	Glenn W. Froning	57–81
ELASTOMERS, SYNTHETIC	W. Cooper, F. C. J. Poulton, and P. R. Sewell	81–161
EMBALMING CHEMICALS	Leandro Rendon	162–169
ENAMELS	L. B. Odell	170–192
EPOXY COMPOUNDS	Milton Lapkin	192–232
EPOXY RESINS	Milton Lapkin and E. Dale Weir	232–259
ESSENTIAL OILS	Robert J. Eiserle	259–290
ETHANOLAMINES	F. W. Neumann	291–295
ETHERS	Martha Fabuss	296–316
ETHYLENE OXIDE	Robert L. Anderson	317–340
ETHYLENE AND PROPYLENE POLYMERS	O. Fuchs, L. Bohn, M. Fleissner, G. Gann, H. H. Suhr, and H. D. R. Schüddemage	341–395
EUROPIUM	M. M. Woyski and R. E. Harris	395–404
EXPLOSIVES	Frank Pristera	405–471
FATS AND FATTY OILS	C. Y. Hopkins	472–493
FELT	R. H. Armitage	494–514
FERRITES	R. C. Hughes, R. C. Sweet, and R. Mavodineanu	514–538
FILMS, PHOTOGRAPHIC	Gerald Russell	539–599
FLOUR	Cho C. Tsen and G. D. Miller	599–618

VOLUME 13

Title	Authors	Pages
FLUORINE	A. M. G. Macdonald and Joan L. Battle	1–5
FLUORINE COMPOUNDS, INORGANIC	A. M. G. Macdonald	6–36
FLUORO ACIDS, INORGANIC	E. F. Joy and A. J. Barnard, Jr.	37–58
FLUOROCARBONS	Joan L. Battle	58–73
FLUOROCARBONS, POLYMERS		
Tetrafluoroethylene polymers	John A. Blair	73–93
Other fluorine-containing polymers	Joan L. Battle and Robert P. Bringer	93–100
FORMALDEHYDE	R. A. Bleidt	100–117
FORMIC ACID	R. M. Speights and A. J. Barnard, Jr.	117–138
FOURIER TRANSFORM SPECTROSCOPY	M. J. D. Low	139–176
FUMIGANTS	B. G. Tweedy, J. A. Ross, and G. A. Bozarth	176–197
FUNGICIDES	B. G. Tweedy and Carol Loeppky	198–227
FURAN AND FURAN DERIVATIVES	John W. Madden	227–250

VOLUME 13 (*continued*)

Title	*Authors*	*Pages*
GALLIUM	Hiroshi Onishi	251–285
GASES, NOBLE	George E. Schmauch	286–309
GAS, NATURAL	Ralph C. Vollmar	310–339
GASOLINE	David L. Camin	339–361
GELATIN	C. H. Barnstein	361–375
GERMANIUM	Frederick A. Lowenheim	375–402
GLASS AND GLASS-CERAMICS	J. P. Williams and D. E. Campbell	403–487
GLUES OF ANIMAL ORIGIN	C. L. Pearson	487–494
GLYCEROL	John E. Going	494–532
GLYCOLS AND POLYHYDRIC ALCOHOLS	C. C. Budke and D. K. Banerjee	533–607

VOLUME 14

Title	Authors	Pages
GOLD	J. C. Van Loon and F. E. Beamish	1–51
GRAPHITE AND RELATED CARBONS	W. E. Parker and M. L. Pearce	52–87
GYPSUM AND GYPSUM PRODUCTS	D. A. Randolph	87–102
HAFNIUM AND ZIRCONIUM	Richard T. vanSanten, James H. Schlewitz, Gary L. Beck, and Ronald E. Walsh	103–152
HALOHYDRINS	Gregor H. Riesser and Joanna G. Riesser	153–178
HORMONES	S. Ahuja	179–231
HYDRAZINE	Edward Kuchar	232–241
HYDRIDES	A. P. Scanzillo	242–260
HYDROCARBON RESINS	Paul O. Powers	261–297
HYDROCARBONS	Ralph C. Vollmar	298–354
HYDROCHLORIC ACID	Andrew Procko	354–376
HYDROGEN	David M. Johnson	377–389
HYDROGEN CHLORIDE	Andrew Procko	389–403
HYDROGEN CYANIDE	Grady L. Roberts	403–426
HYDROGEN PEROXIDE	Owen B. Mathre	427–439
HYDROXYLAMINE AND THEIR SALTS	Carl J. Wassink	440–460
IMINES	James M. Weber	460–495
INDIGOID DYES	Hansjoerg Vollmann, Douglas W. Allan, Theresa A. Moretti, and Nicholas M. Scarpa	495–518
INDIUM	J. M. Ramaradhya, R. C. Bell, S. Brownlow, and C. J. Mitchell	518–565
INSULIN	H. Stelmach	565–583
IODINE	Vernon A. Stenger and Leslie S. Ettre	584–616

VOLUME 15

Title	*Authors*	*Pages*
ION EXCHANGE RESINS	Robert Kunin	1–20
IRON	Frederick A. Lowenheim	21–93
ISOCYANIC ACID ESTERS	D. J. David	94–124
LACTIC ACID AND ITS DERIVATIVES	G. D. Smith, Rose Mary Fritz, George R. King, and Robert L. Maute	125–149
LANTHANUM	Ray E. Harris and Roland Christensen	149–161
LEAD	M. K. Weiss	161–201
LIME AND LIMESTONE	D. R. Janninck	202–233
LIQUEFIED PETROLEUM GAS	J. F. Hickerson	234–260
LITHIUM	Frederick A. Lowenheim	260–289
LSD	David C. Fenimore	290–300
LUBRICANTS	A. E. Baker	301–355
MAGNESIUM	H. W. Emmel, R. O. Olson, and C. E. Van Hall	356–420
MALONIC ACID AND DERIVATIVES	R. L. Erickson and George Kesslin	420–447
MANGANESE METAL, ALLOYS, AND COMPOUNDS	Frederick A. Lowenheim	447–496
MARGARINE	R. L. Bradley, Jr.	496–504
MATCHES	Isadore Kowarsky	504–551
MERCAPTANS	John W. Miller	551–574

VOLUME 16

MERCURY	Frederick A. Lowenheim	1–67
METHACRYLATE POLYMERS	R. Kretz and D. H. Melchior	67–102
METHANOL	Robert L. Maute	102–122
MILK AND MILK PRODUCTS	R. L. Bradley, Jr.	123–149
MOLYBDENUM	John W. Mann	150–191
MORPHOLINE	Joseph R. Mann	192–203
NAPHTHALENE AND NAPHTHALENE DERIVATIVES	Hans George Treibl	204–242
NARCOTICS	J. A. Mollica and B. J. Kline	243–284
NICKEL	Frederick A. Lowenheim and Roland S. Young	284–348
NIOBIUM METAL, ALLOYS, AND COMPOUNDS	Frederick A. Lowenheim	348–395
NITRILES	J. Gordon Hanna	396–411
NITRO COMPOUNDS, ORGANIC	Carl J. Wassink and J. T. Allen	412–448
NITROGEN AND NITROGEN COMPOUNDS, INORGANIC	Frederick A. Lowenheim	449–512
OXYGEN	Frederick A. Lowenheim	514–538
OZONE	Frederick A. Lowenheim	538–551
PENICILLINS AND RELATED COMPOUNDS	Ronald P. Upton and William B. Hardie	552–583

VOLUME 17

Title	Authors	Pages
PHENOLS	Leslie S. Ettre and Edward Obermiller	1–50
PHOSPHORUS	Frederick A. Lowenheim	51–150
PHTHALIC ACID AND DERIVATIVES	Jon J. Harper and Hans Pohlmann	150–175
PIGMENTS	Winslow H. Hartford	176–213
PLATINUM GROUP METALS	Frederick A. Lowenheim	214–275
POLYAMIDES		
Polyamides, general	Anthony Anton	275–305
Polyamide fibers	Orville E. Snider	305–329
POLYCARBONATES	A. Horbach, W. Stuck, and H. Wunderlich	329–352
POTASSIUM AND POTASSIUM COMPOUNDS	Frederick A. Lowenheim	353–411
PYRIDINE AND PYRIDINE DERIVATIVES	C. S. Giam	412–442
RADIUM	Frederick A. Lowenheim	443–469
RARE EARTHS	Ray E. Harris	470–515
RHENIUM AND TECHNETIUM	Frederick A. Lowenheim	515–543
RUBIDIUM AND RUBIDIUM COMPOUNDS	Frederick A. Lowenheim	543–559
SCANDIUM	Frederick A. Lowenheim	560–580
SELENIUM AND TELLURIUM	Frederick A. Lowenheim	580–608
SENSORY TESTING METHODS	William H. Stahl and Margery A. Einstein	608–644

VOLUME 18

Title	Authors	Pages
SILICON	Frederick A. Lowenheim	1–48
SILICON COMPOUNDS, ORGANIC	Charles R. Thrash and G. Beall	48–155
SILVER	Frederick A. Lowenheim	155–178
SOAPS	Robert M. Kelley	179–207
SODIUM	Frederick A. Lowenheim	207–260
STEEL	Frederick A. Lowenheim	261–263
STRONTIUM	Frederick A. Lowenheim	264–285
STYRENE	Kent Hodges and James McLean	285–313
STYRENE POLYMERS	John G. Cobler and Gerald E. Stobby	314–336
SUGAR	Emma J. McDonald	337–359
SULFUR	Frederick A. Lowenheim	360–440
TANTALUM	Frederick A. Lowenheim	440–454
TEA	P. J. Hilton	455–518
THALLIUM	Frederick A. Lowenheim	519–536
THIOPHENE	L. S. Bark and F. Robinson	537–545

VOLUME 19

Title	Authors	Pages
THORIUM	Frederick A. Lowenheim	1–21
TIN	Frederick A. Lowenheim	21–80

VOLUME 19 (*continued*)

Title	Authors	Pages
TITANIUM	Frederick A. Lowenheim	81–132
TOBACCO	T. C. Tso and R. A. Andersen	133–160
TRIPHENYLMETHANE AND RELATED DYES	Louis J. Papa	161–189
TUNGSTEN	Frederick A. Lowenheim	190–224
URANIUM AND THE TRANSURANIUM ELEMENTS	Frederick A. Lowenheim	224–262
URETHANE POLYMERS	Robert L. Sandridge and Ronald F. Bargiband	262–319
VANADIUM	Frederick A. Lowenheim	320–347
VANILLIN	James W. Barr	347–351
WATER	Martha A. Fabuss and Bela M. Fabuss	352–396
WINE AND MUST	Maynard A. Amerine and Cornelius S. Ough	397–514
YEAST	George I. deBecze	514–543
ZINC	Frederick A. Lowenheim	544–577
PESTICIDES AND INSECTICIDES	Louis Lykken	578–619

CONTRIBUTORS TO THE ENCYCLOPEDIA

George J. Abel, Jr., *American Smelting & Refining Company,* South Plainfield, N. J., Antimony, Vol. 6, pp. 39–70; Arsenic, Vol. 6, pp. 223–254

R. J. Abernethy, *U. S. Bureau of Mines, Pittsburgh Energy Research Center,* Pittsburgh, Pa., Coal and coke, Vol. 10, pp. 209–262

S. Ahuja, *Geigy Pharmaceuticals Division, CIBA-Geigy Corporation,* Suffern, N. Y., Hormones, Vol. 14, pp. 179–231

W. J. Alexander, *ITT Rayonier Incorporated,* Whippany, N. J., Cellulose, Vol. 9, pp. 27–59; Cellulose derivatives, Vol. 9, pp. 59–94

Douglas W. Allan, *American Hoechst Corporation,* Coventry, R. I., Indigoid dyes, Vol. 14, pp. 495–518

J. T. Allen, *Commercial Solvents Corporation,* Terre Haute, Ind., Nitro compounds, organic, Vol. 16, pp. 412–448

Maynard A. Amerine, *University of California, Davis,* Davis, Calif., Wine and must, Vol. 19, pp. 397–514

R. A. Andersen, *United States Department of Agriculture,* Lexington, Ky., Tobacco, Vol. 19, pp. 133–160

Robert L. Anderson, *Union Carbide Corporation,* South Charleston, West Va., Ethylene oxide, Vol. 12, pp. 317–340

Robert P. Anibal, *Carborundum Company,* Niagara Falls, N. Y., Bonded abrasives, Vol. 7, pp. 293–312

Anthony Anton, *E. I. du Pont de Nemours & Co., Inc.,* Wilmington, Del., Polyamides, general under Polyamides, Vol. 17, pp. 275–305

R. H. Armitage, *United States Testing Company, Inc.,* Hoboken, N. J., Felt, Vol. 12, pp. 494–514

W. A. Baer, *Ward Foods, Inc.,* Bronx, N. Y., Bakery products, Vol. 6, pp. 438–467

A. E. Baker, *General Electric Company,* Philadelphia, Pa., Lubricants, Vol. 15, pp. 301–355

D. K. Banerjee, *U. S. Industrial Chemicals Company, Division of National Distillers and Chemical Corporation,* Cincinnati, Ohio, Acetone, Vol. 4, pp. 107–121; Alcohols, Vol. 4, pp. 495–586; Glycols and polyhydric alcohols, Vol. 13, pp. 533–607

Ronald F. Bargiband, *Mobay Chemical Company,* New Martinsville, West Va., Urethane polymers, Vol. 19, pp. 262–319

L. S. Bark, *University of Salford,* Salford, M5 4WT, England, Thiophene, Vol. 18, pp. 537–545

A. J. Barnard, Jr., *J. T. Baker Chemical Co.,* Phillipsburg, N. J., Acetic acid, Vol. 4, pp. 93–101; Acetic anhydride, Vol. 4, pp. 102–107; Chlorosulfonic acid, Vol. 9, pp. 528–534; Citric acid, Vol. 10, pp. 85–100; Fluoro acids, inorganic, Vol. 13, pp. 37–58; Formic acid, Vol. 13, pp. 117–138

C. H. Barnstein, *National Formulary,* Washington, D. C., Gelatin, Vol. 13, pp. 361–375

James W. Barr, *Sterwin Chemicals, Inc., Subsidiary of Sterling Drug, Inc.,* Rothschild, Wis., Vanillin, Vol. 19, pp. 347–351

W. T. Barry, *Aerospace Corporation,* Los Angeles, Calif., Ablative materials, Vol. 4, pp. 1–30

J. N. Bathish, *Wyeth Laboratories, Inc.*, Philadelphia, Pa., Barbiturates, Vol. 6, pp. 467–532

Joan L. Battle, *Encyclopedia of Industrial Chemical Analysis*, Fluorine, Vol. 13, pp. 1–5; Fluorocarbons, Vol. 13, pp. 58–73; Fluorocarbon polymers, Vol. 13, pp. 93–100

G. Beall, *Union Carbide Corp.*, Sisterville, West Va., Silicon compounds, organic, Vol. 18, pp. 48–155

F. E. Beamish, *University of Toronto*, Toronto, Canada, Gold, Vol. 14, pp. 1–51

Gary L. Beck, *Teledyne Wah Chang Albany*, Albany, Ore., Hafnium and zirconium, Vol. 14, pp. 103–152

R. C. Bell, *Cominco Ltd.*, Trail, British Columbia, Canada, Indium, Vol. 14, pp. 518–565

Eric Beynon, *Union Carbide Corporation, Consumer Products Division*, Tarrytown, N. Y., Antifreezes, Vol. 6, pp. 1–23; Automotive chemical specialties, Vol. 6, pp. 350–380

John Blair, *E. I. du Pont de Nemours & Co. Inc.*, Wilmington, Del., Fluorocarbon polymers, Vol. 13, pp. 73–93

R. A. Bleidt, *Union Carbide Corporation*, South Charleston, West Va., Formaldehyde, Vol. 13, pp. 93–117

R. F. Blomquist, *Forest Products Laboratory, Forest Service, U. S. Department of Agriculture*, Athens, Ga., Vol. 4, pp. 391–398

Joseph B. Bodkin, *Mineral Constitution Laboratories, The Pennsylvania State University*, University Park, Pa., Calcium, Vol. 8, pp. 72–114

L. Bohn, *Farbwerke Hoechst A. G.*, Frankfurt am Main, German Federal Republic, Ethylene and propylene polymers, Vol. 12, pp. 341–395

John H. Bonfield, *Allied Chemical Corporation, Plastics Division*, Morristown, N. J., Caprolactam, Vol. 8, pp. 114–129

G. A. Bozarth, *University of Missouri*, Columbia, Mo., Fumigants, Vol. 13, pp. 176–197

R. L. Bradley, Jr., *University of Wisconsin*, Madison, Wis., Margarine, Vol. 15, pp. 496–504; Milk and milk products, Vol. 16, pp. 123–149

Robert S. Braman, *University of South Florida*, Tampa, Fla., Boron and boron alloys under Boron, Vol. 7, pp. 312–329; Boron compounds (introduction) under Boron, Vol. 7, pp. 329–358; Boron determination under Boron, Vol. 7, pp. 384–423

G. A. Brewer, Jr., *The Squibb Institute for Medical Research*, New Brunswick, N. J., Antibiotics, Vol. 5, pp. 460–633

Robert P. Bringer, *3M Company*, St. Paul, Minn., Fluorocarbons, polymers, Vol. 13, pp. 93–100

Neil H. Brown, *Sterling-Winthrop Research Institute, a division of Sterling Drug, Inc.*, Rensselaer, N. Y., Contraceptives, Vol. 10, pp. 568–597

Robert S. Browning, Jr., *Sterling-Winthrop Research Institute, a division of Sterling Drug, Inc.*, Rensselaer, N. Y., Disinfectants and antiseptics, Vol. 11, pp. 527–572

S. Brownlow, *Cominco, Ltd.*, Trail, British Columbia, Canada, Indium, Vol. 14, pp. 518–565

C. C. Budke, *U. S. Industrial Chemicals Company, Division of National Distillers & Chemical Corp.*, Cincinnati, Ohio, Acetone, Vol. 4, pp. 107–121; Alcohols, Vol. 4, pp. 495–586; Glycols and polyhydric alcohols, Vol. 13, pp. 533–607

Carolyn N. Bye, *National Casein of New Jersey*, Riverton, N. J., Casein, Vol. 9, pp. 1–15

David L. Camin, *Sun Oil Company*, Marcus Hook, Pa., Gasoline, Vol. 13, pp. 339–361

D. E. Campbell, *Corning Glass Works, Research and Development Laboratories,* Corning, N. Y., Ceramics, Vol. 9, pp. 189–232; Glass and glass-ceramics, Vol. 13, pp. 403–487

R. H. Campbell, *Monsanto Company, Organic Chemicals Division,* Nitro, West Va., Antioxidants and antiozonants, Vol. 6, pp. 71–121

F. E. Caskey, *Process Engineering Equipment Company,* Summit, N. J., Acetylene, Vol. 4, pp. 126–135; Acetylenic chemicals, Vol. 4, pp. 136–147

W. B. Chess, *Stauffer Chemical Co.,* Dobbs Ferry, N. Y., Bakery chemicals, Vol. 6, pp. 425–438

Roland Christensen, *Kerr-McGee Chemical Corporation,* Oklahoma City, Okla., Lanthanum, Vol. 15, pp. 141–161

John G. Cobler, *The Dow Chemical Company,* Midland, Mich., Styrene polymers, Vol. 18, pp. 314–336

Kenneth A. Connors, *University of Wisconsin, School of Pharmacy,* Madison, Wis., Cinnamaldehyde, Vol. 10, pp. 42–55; Cinnamic acid and its derivatives, Vol. 10, pp. 56–76; Cinnamyl alcohol, Vol. 10, pp. 76–85

W. Cooper, *Dunlop Research Centre,* Birmingham, England, Elastomers, synthetic, Vol. 12, pp. 81–161

Raymond B. Crawford, *Allied Chemical Corporation, Industrial Chemicals Division,* Buffalo, N. Y., Azine and related dyes, Vol. 6, pp. 381–400

Wayne E. Dallman, *Institute for Atomic Research and Department of Chemistry, Iowa State University,* Ames, Iowa, Carrier gas and vacuum fusion methods, Vol. 8, pp. 613–693

G. C. Daul, *ITT Rayonier Incorporated,* Whippany, N. J., Cellulose, Vol. 9, pp. 27–59; Cellulose derivatives, Vol. 9, pp. 59–94

D. J. David, *Columbia Scientific Industries,* Austin, Texas, Isocyanic acid esters, Vol. 15, pp. 94–124

George I. deBecze, *Schenley Distillers, Inc.,* Lawrenceburg, Ind., Alcoholic beverages, distilled, Vol. 4, pp. 462–494; *Industrial Consultant,* Lawrenceburg, Ind., Yeast, Vol. 19, pp. 514–543

Kurt A. Dellian, *CIBA-GEIGY Corporation,* Ardsley, New York, Dyes, Vol. 12, pp. 1–22

Bernard J. DeWitt, *PPG Industries,* Barberton, Ohio, Carbon tetrachloride, Vol. 8, pp. 320–330; Chlorocarbons, Vol. 9, pp. 369–510

James T. Dobbins, Jr., *R. J. Reynolds Tobacco Company,* Winston-Salem, N. C., Cigarettes, Vol. 10, pp. 1–42

Kenneth S. Dress, *PPG Industries,* Barberton, Ohio, Carbon tetrachloride, Vol. 8, pp. 320–330; Chlorocarbons, Vol. 9, pp. 369–510

Margery A. Einstein, *McCormick & Company, Inc.,* Cockeysville, Md., Sensory testing methods, Vol. 17, pp. 608–644

Robert J. Eiserle, *Fritzche-Dodge & Olcott, Inc.,* New York, Essential oils, Vol. 12, pp. 259–290

H. W. Emmel, *The Dow Chemical Company,* Midland, Mich., Magnesium, Vol. 15, pp. 356–420

S. Ergun, *U. S. Bureau of Mines, Pittsburgh Energy Research Center,* Pittsburgh, Pa., Coal and coke, Vol. 10, pp. 209–262

R. L. Erickson, *Kay-Fries Chemicals, Inc.,* West Haverstraw, N. Y., Malonic acid and derivatives, Vol. 15, pp. 420–447

Leslie S. Ettre, *Encyclopedia of Industrial Chemical Analysis,* Carboxylic acids, derivatives, Vol. 8, pp. 557–612; Iodine, Vol. 14, pp. 584–616; *The Perkin-Elmer Corp.,* Norwalk, Conn., Phenols, Vol. 17, pp. 1–50

F. P. Ewald, *PPG Industries, Inc.,* Barberton, Ohio, Chlorocarbons, Vol. 9, pp. 369–510

Bela M. Fabuss, *Raytheon Service Co.,* Burlington, Mass., Water, Vol. 19, pp. 352–396

Martha Fabuss, *Consultant,* Winchester, Mass., Ethers, Vol. 12, pp. 296–316; Water, Vol. 19, pp. 352–396

Charles M. Fairchild, *Revlon Research Center,* Bronx, N. Y., Cosmetics, Vol. 11, pp. 64–116

David C. Fenimore, *Texas Research Institute of Mental Sciences,* Houston, Texas, LSD, Vol. 15, pp. 290–300

E. J. Finnegan, *Hershey Foods Corporation,* Hershey, Pa., Chocolate and cocoa products, Vol. 9, pp. 534–611

Fred Fischer, *International Minerals & Chemical Corporation,* Libertyville, Ill., Carboxylic acids, amino, Vol. 8, pp. 410–556

W. Allan Fisher, *Allied Chemical Corporation, Industrial Chemicals Division,* Buffalo, N. Y., Azine and related dyes, Vol. 6, pp. 381–400

M. Fleissner, *Farbwerke Hoechst A. G.,* Frankfurt am Main, German Federal Republic, Ethylene and propylene polymers, Vol. 12, pp. 341–395

F. P. Ford, *Columbian Carbon Company,* Princeton, N. J., Carbon black, Vol. 8, pp. 179–243

Craig C. Foreback, *University of South Florida,* Tampa, Fla., Chromium metal under Chromium, Vol. 9, pp. 632–660; Chromium alloys under Chromium, Vol. 9, pp. 660–680

Paul F. Forsyth, *The Carborundum Company,* Niagara Falls, N. Y., Coated Abrasives, Vol. 10, pp. 263–283

Carol Freedenthal, *Kennecott Copper Corporation, Special Products Division,* Houston, Texas, Copper compounds under Copper, Vol. 10, pp. 651–680

Frank H. Freeman, *Kerr Manufacturing Company,* Romulus, Mich., Dental materials, Vol. 11, pp. 256–321

R. A. Friedel, *U. S. Bureau of Mines, Pittsburgh Energy Research Center,* Pittsburgh, Pa., Coal and coke, Vol. 10, pp. 209–262

Seymour Friedman, *Catalin Corporation, Division of Ashland Oil & Refining Company,* Fords, N. J., Amino resins, Vol. 5, pp. 276–290

Rose Mary Fritz, *Monsanto Company,* Texas City, Texas, Lactic acid and its derivatives, Vol. 15, pp. 125–149

Glenn W. Froning, *University of Nebraska,* Lincoln, Neb., Eggs and egg products, Vol. 12, pp. 57–81

O. Fuchs, *Farbwerke Hoechst A. G.,* Frankfurt am Main, German Federal Republic, Ethylene and propylene polymers, Vol. 12, pp. 341–395

G. Gann, *Farbwerke Hoechst A. G.,* Frankfurt am Main, German Federal Republic, Ethylene and propylene polymers, Vol. 12, pp. 341–395

David R. Gaskill, *Mobil Chemical Company,* Metuchen, N. J., Cadmium, Vol. 8, pp. 30–55

C. A. Gaulin, *Aerospace Corporation,* Los Angeles, Calif., Ablative materials, Vol. 4, pp. 1–30

C. S. Giam, *Texas A&M University,* College Station, Texas, Pyridine and pyridine derivatives, Vol. 17, pp. 412–422

Myron E. Gibson, Jr., *Chemstrand Research Center, Inc., a subsidiary of Monsanto Company,* Durham, N. C., Acrylic and modacrylic fibers, Vol. 4, pp. 219–367

John E. Going, *University of Wisconsin,* Milwaukee, Wis., Glycerol, Vol. 13, pp. 494–532

A. S. Graham, *Hershey Foods Corporation,* Hershey Pa., Chocolate and cocoa products, Vol. 9, pp. 534–611

Charles P. Grier, *Dr. Pepper Company,* Dallas, Texas, Carbonated beverages, Vol. 8, pp. 161–178

James Grove, *General Foods Corporation,* Battle Creek, Mich., Cereals, Vol. 9, pp. 232–245

J. Gordon Hanna, *The Connecticut Agricultural Experiment Station,* New Haven, Conn., Nitriles, Vol. 16, pp. 369–411

William B. Hardie, *Pfizer, Inc.,* Brooklyn, N. Y., Penicillins and related compounds, Vol. 16, pp. 552–583

Jon J. Harper, *Amoco Chemicals Corporation,* Naperville, Ill., Phthalic acids and derivatives, Vol. 17, pp. 151–175

R. E. Harris, *Kerr-McGee Corporation,* Oklahoma City, Okla., Cerium, Vol. 9, pp. 246–277; Europium, Vol. 12, pp. 395–404; Lanthanum, Vol. 15, pp. 149–161; Rare earths, Vol. 17, pp. 470–515

Winslow H. Hartford, *Chemical Consultant,* Charlotte, N. C., Chromium compounds under Chromium, Vol. 9, pp. 680–709; *Belmont Abbey College,* Charlotte, N. C., Pigments, Vol. 17, pp. 176–213

W. M. Hazel, *Norton Company,* Chippawa, Ontario, Canada, Abrasives, Vol. 4, pp. 31–61

R. H. Heidner, *Chemstrand Research Center, Inc., a subsidiary of Monsanto Company,* Durham, N. C., Acrylic and modacrylic fibers, Vol. 4, pp. 219–367

John S. Heil, *First Chemical Corporation, A Subsidiary of First Mississippi Corporation,* Pascagoula, Miss., Diamines and higher amines, aromatic, Vol. 11, pp. 432–462

Preston F. Helgren, *Abbott Laboratories,* North Chicago, Ill., Cyclohexylamine, Vol. 11, pp. 209–219; Cyclohexylamine derivatives, Vol. 11, pp. 220–255

J. F. Hickerson, *Humble Oil & Refining Company,* Baytown, Texas, Liquefied petroleum gas, Vol. 15, pp. 234–260

Richard P. Hight, *Norton Company,* Worcester, Mass., Diamonds, industrial, Vol. 11, pp. 462–484

J. A. Hill, *Kerr-McGee Corporation,* Oklahoma City, Okla., Cesium, Vol. 9, pp. 278–303

Clifford L. Hilton, *Interscience Publishers, a division of John Wiley & Sons, Inc.,* Alloys, Vol. 5, pp. 57–75; Aluminum determination under Aluminum, Vol. 5, pp. 199–258; Antipyretics and analgesics, Vol. 6, pp. 123–222; Bead and flame tests, Vol. 6, pp. 583–593

P. J. Hilton, *The Tea Research Foundation of Central Africa,* Mulanje, Malawi, Africa, Tea, Vol. 18, pp. 455–518

W. G. Hime, *Portland Cement Association,* Skokie, Ill., Ceramics, mortars, and concrete, Vol. 9, pp. 94–155

Orville Hinsvark, *Pennwalt Corporation, Pharmaceutical Division*, Rochester, N. Y., Carbon, hydrogen, and nitrogen microanalysis, Vol. 8, pp. 280–298

Kent Hodges, *The Dow Chemical Company*, Midland, Mich., Styrene, Vol. 18, pp. 285–313

G. C. Hood, Jr., *Shell Chemical Company*, Emeryville, Calif., Acrolein, Vol. 4, pp. 148–159

C. Y. Hopkins, *National Research Council of Canada*, Ottawa, Canada, Fats and fatty oils, Vol. 12, pp. 472–493

A. Horbach, *Farbenfabriken Bayer AG*, Werk Uerdingen, Krefeld-Uerdingen, West Germany, Polycarbonates, Vol. 17, pp. 329–352

M. J. Horn, *Ward Foods, Inc.*, Bronx, N. Y., Bakery products, Vol. 6, pp. 438–467

W. M. Hoskins, *University of California*, Lafayette, Calif., Bioassay, Vol. 7, pp. 155–186

R. C. Hughes, *Philips Laboratories, A Division of North American Philips Corporation*, Briarcliff Manor, N. Y., Ferrites, Vol. 12, pp. 514–538

Martin H. Infeld, *University of Wisconsin, School of Pharmacy*, Madison, Wis., Cinnamaldehyde, Vol. 10, pp. 42–55

D. R. Janninck, *United States Gypsum Company*, Chicago, Ill., Lime and limestone, Vol. 15, pp. 202–233

W. R. Jensen, *Raybestos Division, Raybestos-Manhattan, Inc.*, Bridgeport, Conn., Brake linings, Vol. 7, pp. 440–474

David M. Johnson, *Glass Containers Corporation*, Indianapolis, Ind., Hydrogen, Vol. 14, pp. 377–389

V. D. Johnston, *The Givaudan Corporation*, Delawanna, N. J., Acetophenone, Vol. 4, pp. 122–126

E. F. Joy, *J. T. Baker Chemical Company*, Phillipsburg, N. J., Acetic acid, Vol. 4, pp. 93–101; Acetic anhydride, Vol. 4, pp. 102–107; Chlorosulfonic acid, Vol. 9, pp. 528–534; Fluoro acids, inorganic, Vol. 13, pp. 37–58

Wolf Karo, *The Borden Chemical Company, A Division of the Borden Company*, Philadelphia, Pa., Allyl compounds, Vol. 5, pp. 75–109

Gabor Karoly, *M&T Chemicals, Inc.*, Rahway, N. J., Carboxylic acids, Vol. 8, pp. 368–409

Frederick W. Keith, Jr., *Pennwalt Corporation*, Warminster, Pa., Centrifugal separation, Vol. 9, pp. 156–188

Robert E. Keller, *Monsanto Company*, St. Louis, Mo., Adipic acid, Vol. 4, pp. 408–423; Adipic acid derivatives, Vol. 4, pp. 423–431

Robert M. Kelley, *Colgate-Palmolive Company*, Piscataway, N. J., Detergents, Vol. 11, pp. 322–406; Soaps, Vol. 18, pp. 179–207

Clark A. Kelly, *Sterling-Winthrop Research Institute, a division of Sterling Drug Inc.*, Rensselaer, N. Y., Anesthetics, Vol. 5, pp. 355–421; Antihistamines, Vol. 6, pp. 24–38; Coagulants and anticoagulants, Vol. 10, pp. 138–209; Contraceptives, Vol. 10, pp. 568–597

Paul C. Kempchinsky, *The Beryllium Corporation*, Hazleton, Pa., Beryllium and beryllium alloys under Beryllium, Vol. 7, pp. 103–117; Beryllium determination under Beryllium, Vol. 7, pp. 129–141

George Kesslin, *Kay-Fries Chemicals, Inc.*, West Haverstraw, N. Y., Malonic acid and derivatives, Vol. 15, pp. 420–447

George R. King, *Monsanto Company*, Texas City, Texas, Lactic acid and its derivatives, Vol. 15, pp. 125–149

Allen E. Klein, *Tenneco Chemicals, Inc., Heyden Division*, Garfield, N. J., Carboxylic acids, Vol. 8, pp. 368–409

B. J. Kline, *CIBA Pharmaceutical Company, a division of CIBA-Geigy Corporation*, Summit, N. J., Narcotics, Vol. 16, pp. 243–284

George P. Koch, *Reynolds Metals Company*, Richmond, Va., Aluminum alloys under Aluminum, Vol. 5, pp. 110–158

C. L. Kolb, *Celanese Research Company*, Summit, N. J., Acetate and triacetate fibers, Vol. 4, pp. 74–93

Isadore Kowarsky, *Universal Match Industries, Inc.*, St. Louis, Mo., Matches, Vol. 15, pp. 504–551

W. R. Kreiser, *Hershey Foods Corporation*, Hershey, Pa., Chocolate and cocoa products, Vol. 9, pp. 534–611

R. Kretz, *Rhöm GmbH Chemische Fabrik*, Darmstadt, West Germany, Methacrylate polymers, Vol. 16, pp. 67–102

William G. Krochta, *PPG Industries, Inc.*, Barberton, Ohio, Carbon tetrachloride, Vol. 8, pp. 320–350; Chlorocarbons, Vol. 9, pp. 369–510

Edward Kuchar, *Olin Corporation*, New Haven, Conn., Hydrazine, Vol. 14, pp. 232–241

Robert Kunin, *Rohm and Haas Company*, Philadelphia, Pa., Ion exchange resins, Vol. 15, pp. 1–20

A. J. Lacroix, *Raybestos Division, Raybestos-Manhattan, Inc.*, Bridgeport, Conn., Brake linings, Vol. 7, pp. 440–474

Milton Lapkin, *Olin Corporation, Thompson Plastics*, Assonet, Mass., Epoxy compounds, Vol. 12, pp. 192–232; Epoxy resins, Vol. 12, pp. 232–259

William E. Link, *ADM Chemicals, Division of Ashland Oil & Refining Company*, Minneapolis, Minn., Alkyd resins, Vol. 5, pp. 22–56

Carol Loeppky, *University of Missouri*, Columbia, Mo., Fungicides, Vol. 13, pp. 198–227

Emil Lorz, *Hoffman-Taff, Inc.*, Springfield, Mo., Choline and its derivatives, Vol. 9, pp. 611–632

M. J. D. Low, *New York University*, New York, Fourier transform spectroscopy, Vol. 13, pp. 139–176

Frederick A. Lowenheim, *Consultant*, Plainfield, N. J., Germanium, Vol. 13, pp. 375–402; Iron, Vol. 15, pp. 21–93; Lithium, Vol. 15, pp. 260–289; Manganese metal, alloys, and compounds, Vol. 15, pp. 447–496; Mercury, Vol. 16, pp. 1–67; Nickel, Vol. 16, pp. 284–348; Niobium metal, alloys, and compounds, Vol. 16, pp. 348–395; Nitrogen and nitrogen compounds, inorganic, Vol. 16, pp. 449–512; Oxygen, Vol. 16, pp. 514–538; Ozone, Vol. 16, pp. 538–551; Phosphorus, Vol. 17, pp. 51–150; Platinum group metals, Vol. 17, pp. 214–275; Potassium and potassium compounds, Vol. 17, pp. 353–411; Radium, Vol. 17, pp. 443–469; Rhenium and technetium, Vol. 17, pp. 515–543; Rubidium and rubidium compounds, Vol. 17, pp. 543–559; Scandium, Vol. 17, pp. 560–580; Selenium and tellurium, Vol. 17, pp. 580–608; Silicon, Vol. 18, pp. 1–48; Silver, Vol. 18, pp. 155–178; Sodium, Vol. 18, pp. 207–260; Steel, Vol. 18, pp. 261–263; Strontium, Vol. 18, pp. 264–285; Sulfur, Vol. 18, pp. 360–440; Tantalum, Vol. 18, pp. 440–454; Thallium, Vol. 18, pp. 519–536; Thorium, Vol. 19, pp. 1–21; Tin, Vol. 19, pp. 21–80; Titanium, Vol. 19, pp. 81–132; Tungsten, Vol.

19, pp. 120–224; Uranium and the transuranium elements, Vol. 19, pp. 224–262; Vanadium, Vol. 19, pp. 320–347; Zinc, Vol. 19, pp. 544–577

Delbert R. Luebke, *Abbott Laboratories*, North Chicago, Ill., Cyclohexylamine, Vol. 11, pp. 209–219; Cyclohexylamine derivatives, Vol. 11, pp. 220–255

Leo S. Luskin, *Rohm and Haas Company*, Philadelphia, Pa., Acrylic and methacrylic acids and esters, Vol. 4, pp. 181–218

Louis Lykken, *Consultant*, Richmond, Calif., Pesticides and insecticides, Vol. 19, pp. 578–619

F. Lyon, *Columbian Carbon International*, New York, Carbon black, Vol. 8, pp. 179–243

A. M. G. Macdonald, *University of Birmingham*, Birmingham, England, Fluorine, Vol. 13, pp. 1–5; Fluorine compounds, inorganic, Vol. 13, pp. 6–36

John W. Madden, *The Quaker Oats Company*, Barrington, Ill., Furan and furan derivatives, Vol. 13, pp. 227–250

John W. Mann, *Climax Molybdenum Company*, Ann Arbor, Mich., Molybdenum, Vol. 16, pp. 192–203

Joseph R. Mann, *The Dow Chemical Company*, Midland, Mich., Morpholine, Vol. 16, pp. 150–191

Irwin Margolin, *Revlon Research Center*, Bronx, N. Y., Cosmetics, Vol. 11, pp. 64–116

Daniel M. Marmion, *Allied Chemical Corporation, Specialty Chemicals Division*, Buffalo, N. Y., Colors for food, drugs, and cosmetics, Vol. 10, pp. 447–547

Albert R. Martin, *National Institute of Drycleaning*, Silver Spring, Md., Drycleaning agents, Vol. 11, pp. 603–617

Owen B. Mathre, *E. I. du Pont de Nemours & Co., Inc., Electrochemicals Department*, Wilmington, Del., Hydrogen peroxide, Vol. 14, pp. 427–439

Robert L. Maute, *Monsanto Polymers and Petrochemicals Co.*, Texas City, Texas, Lactic acid and its derivatives, Vol. 15, pp. 125–149; Methanol, Vol. 16, pp. 102–122

R. Mavrodineanu, *Philips Laboratories*, Briarcliff Manor, New York, Ferrites, Vol. 12, pp. 514–538

J. T. McCartney, *U. S. Bureau of Mines, Pittsburgh Energy Research Center*, Pittsburgh, Pa., Coal and coke, Vol. 10, pp. 209–262

George F. McCutcheon, *S. B. Penick & Company*, Lyndhurst, N. J., Caffeine, Vol. 8, pp. 30–71

Emma J. McDonald, Madison, Wis., Sugar, Vol. 18, pp. 337–359

John W. McDowell, *Allied Chemical Corporation, Industrial Chemicals Division*, Buffalo, N. Y., Aniline and its derivatives, Vol. 5, pp. 421–459

James McLean, *The Dow Chemical Company*, Midland, Mich., Styrene, Vol. 18, pp. 285–313

D. H. Melchior, *Röhm GmbH Chemische Fabrik*, Darmstadt, West Germany, Methacrylate polymers, Vol. 16, pp. 67–102

Elwood M. Meyers, *Hershey Foods Corporation*, Hershey, Pa., Chocolate and cocoa products, Vol. 9, pp. 534–611

G. D. Miller, *Kansas State University*, Manhattan, Kan., Flour, Vol. 12, pp. 559–618

John W. Miller, *Phillips Petroleum Company*, Bartlesville, Okla., Mercaptans, Vol. 15, pp. 551–574

C. J. Mitchell, *Cominco, Ltd.*, Trail, British Columbia, Canada, Indium, Vol. 14, pp. 518–565

Marion E. Mitten, *PPG Industries, Inc.,* Barberton, Ohio, Carbon tetrachloride, Vol. 8, pp. 320–330; Chlorocarbons, Vol. 9, pp. 369–510

R. R. Moalli, *Raybestos Division, Raybestos-Manhattan, Inc.,* Bridgeport, Conn., Brake linings, Vol. 7, pp. 440–474

Eino Moks, *Weyerhaeuser Company,* Seattle, Wash., Adhesives, Vol. 4, pp. 398–407

J. A. Mollica, *CIBA Pharmaceutical Company, Division of CIBA-Geigy Corporation,* Summit, N. J., Coumarins, Vol. 11, pp. 159–173; Narcotics, Vol. 16, pp. 243–284

Theresa A. Moretti, *American Hoechst Corporation,* Coventry, R. I., Indigoid dyes, Vol. 14, pp. 495–518

Evan Morgan, *Reynolds Metals Company,* Richmond, Va., Aluminum compounds under Aluminum, Vol. 5, pp. 158–199

F. R. Morral, *Cobalt Information Center, Battelle Memorial Institute,* Columbus, Ohio, Cobalt compounds under Cobalt, Vol. 10, pp. 327–347

Pierre Navellier, *Laboratoire Central, Préfecture de Police de Paris,* Paris, France, Coffee, Vol. 10, pp. 373–447

E. L. Nelson, *Allied Chemical Corporation, Agricultural Division,* Hopewell, Va., Ammonia, Vol. 5, pp. 290–324; Ammonium compounds, Vol. 5, pp. 325–355

F. W. Neumann, *The Dow Chemical Company,* Midland, Mich., Ethanolamines, Vol. 12, pp. 291–295

M. V. Norris, *American Cyanamid Company,* Stamford, Conn., Acrylamide, Vol. 4, pp. 160–168; Acrylamide polymers, Vol. 4, pp. 168–181; Acrylonitrile, Vol. 4, pp. 368–381; Acrylonitrile polymers, Vol. 4, pp. 382–390

Jean Northcott, *Allied Chemical Corporation, Specialty Chemicals Division,* Buffalo, N. Y., Aniline and its derivatives, Vol. 5, pp. 421–459; Caprolactam, Vol. 8, pp. 114–129

Edward Obermiller, *Consolidation Coal Company,* Library, Pa., Phenols, Vol. 17, pp. 1–50

Robert T. O'Connor, *United States Department of Agriculture, Agricultural Research Service, Southern Utilization Research and Development Division,* New Orleans, La., Cotton, Vol. 11, pp. 117–159

L. B. Odell, *Napko Corporation,* Houston, Texas, Enamels, Vol. 12, pp. 170–192

Eugene D. Olsen, *University of South Florida,* Tampa, Fla., Chromium metal under Chromium, Vol. 9, pp. 632–660; Chromium alloys under Chromium, Vol. 9, pp. 660–680

R. O. Olsen, *The Dow Chemical Company,* Midland, Mich., Magnesium, Vol. 15, pp. 356–420

Hiroshi Onishi, *Japan Atomic Energy Research Institute,* Tokai-Mura, Ibaraki-ken, Japan, Gallium, Vol. 13, pp. 251–285

G. Oplinger, *Allied Chemical Corporation, Industrial Chemicals Division,* Solvay, N. Y., Chlorine, Vol. 9, pp. 304–333; Chlorine compounds, inorganic, Vol. 9, pp. 333–368

Cornelius S. Ough, *University of California, Davis,* Davis, Calif., Wine and must, Vol. 19, pp. 397–514 ·

T. D. Oulton, *Minerals & Chemicals Phillip Corporation,* Menlo Park, N. J., Adsorbents and absorbents, Vol. 4, pp. 431–452; *Engelhard Minerals & Chemicals Corp., Minerals & Chemicals Division,* Edison, N. J., Clays and clay minerals, Vol. 10, pp. 100–138

Louis J. Papa, *E. I. du Pont de Nemours & Co., Inc.,* Wilmington, Del., Triphenylmethane and related dyes, Vol. 19, pp. 161–189

G. J. Papariello, *Wyeth Laboratories, Inc.,* Philadelphia, Pa., Alkaloids, Vol. 4, pp. 587–618; Barbiturates, Vol. 6, pp. 467–532

W. E. Parker, *Airco Speer, Division of Air Reduction Company,* Niagara Falls, N. Y., Graphite and related carbons, Vol. 14, pp. 52–87

Ward K. Parr, *Chicago Testing Laboratory, Inc.,* Chicago, Ill., Asphalts, tars, and pitches, Vol. 6, pp. 274–350

M. L. Pearce, *Airco Speer, Division of Air Reduction Company,* Niagara Falls, N. Y., Graphite and related carbons, Vol. 14, pp. 52–87

C. L. Pearson, *Peter Cooper Corporation,* Gowanda, N. Y., Glues of animal origin, Vol. 13, pp. 487–494

E. D. Peters, *Shell Chemical Company,* Emeryville, Calif., Acrolein, Vol. 4, pp. 148–159

E. A. Pinzer, *CIBA-GEIGY Corporation,* Ardsley, N. Y., EDTA and related compounds, Vol. 12, pp. 23–56

T. B. Platt, *The Squibb Institute for Medical Research,* New Brunswick, N. J., Antibiotics, Vol. 5, pp. 460–633

Hans P. Pohlmann, *Amoco Chemicals Corporation,* Naperville, Ill., Phthalic acids and derivatives, Vol. 17, pp. 151–175

F. C. J. Poulton, *Dunlop Research Centre,* Birmingham, England, Elastomers, synthetic, Vol. 12, pp. 81–161

Paul O. Powers, *AMOCO Chemicals Corporation,* North Plainfield, N. J., Hydrocarbon resins, Vol. 14, pp. 261–297

Frank Pristera, *U. S. Army, Picatinny Arsenal,* Dover, N. J., Explosives, Vol. 12, pp. 405–471

Andrew Procko, *PPG Industries,* New Martinsville, West Va., Barium, Vol. 6, pp. 532–582; Carbon disulfide, Vol. 8, pp. 268–279; Hydrochloric acid, Vol. 14, pp. 354–376; Hydrogen chloride, Vol. 14, pp. 389–403

Quentin E. Quick, *Union Carbide Corporation,* South Charleston, West Va., Dioxane, Vol. 11, pp. 516–524

J. M. Ramaradhya, *Cominco Ltd.,* British Columbia, Canada, Indium, Vol. 14, pp. 518–565

D. A. Randolph, *United States Gypsum Company,* Des Plaines, Ill., Gypsum and gypsum products, Vol. 14, pp. 87–102

Leandro Rendon, *Champion Company,* Springfield, Ohio, Embalming chemicals, Vol. 12, pp. 162–169

John A. Riddick, *Commercial Solvents Corporation,* Terre Haute, Ind., Alkanolamines, Vol. 5, pp. 1–22

Gregor H. Riesser, *Shell Chemical Company,* Houston, Texas, Halohydrins, Vol. 14, pp. 153–178

Joanna G. Riesser, Houston, Texas, Halohydrins, Vol. 14, pp. 153–178

Grady L. Roberts, *Monsanto Company,* Texas City, Texas, Hydrogen cyanide, Vol. 14, pp. 403–426

F. Robinson, *West Cumberland College of Science and Technology,* Whitehaven, England, Thiophene, Vol. 18, pp. 537–545

J. A. Ross, *University of Missouri,* Columbia, Mo., Fumigants, Vol. 13, pp. 176–197

Roger L. Rounds, *General Aniline & Film Corporation*, Rensselaer, N. Y., Azo dyes, Vol. 6, pp. 401–424

Gerald Russell, *Picker Chemicals*, Rainham, Essex, England, Films, photographic, Vol. 12, pp. 539–599

J. G. Rutgers, *Wyeth Laboratories, Inc.*, Philadelphia, Pa., Antipyretics and analgesics, Vol. 6, pp. 132–222; Barbiturates, Vol. 6, pp. 467–532

Maurice Sage, *Sage Laboratories, Inc.*, New York, Aerosols, Vol. 4, pp. 452–461

Robert L. Sandridge, *Mobay Chemical Company*, New Martinsville, West Va., Urethane polymers, Vol. 19, pp. 262–319

A. P. Scanzillo, *Ventron Corporation, Chemicals Division*, Beverley, Mass., Hydrides, Vol. 14, pp. 242–260

Nicholas M. Scarpa, *American Hoechst Corporation*, Coventry, R. I., Indigoid dyes, Vol. 14, pp. 495–518

G. H. Scheffler, *Atlas Chemical Industries, Inc.*, Wilmington, Del., Carbon, activated, Vol. 8, pp. 139–161

D. E. Scheirer, *Allied Chemical Corporation, Agricultural Division*, Hopewell, Va., Ammonia, Vol. 5, pp. 290–324; Ammonium compounds, Vol. 5, pp. 325–355

James H. Schlewitz, *Teledyne Wah Chang Albany*, Albany, Ore., Hafnium and zirconium, Vol. 14, pp. 103–152

George E. Schmauch, *Air Products and Chemicals, Inc.*, Allentown, Pa., Carbon dioxide, Vol. 8, pp. 244–267; Carbon monoxide, Vol. 8, pp. 299–319; Gases, noble, Vol. 13, pp. 286–309

C. H. Schmiege, *Olin Mathieson Chemical Corporation*, Niagara Falls, N. Y., Bleaching agents, Vol. 7, pp. 221–293

B. Schubert, *Columbian Carbon Company*, Princeton, N. J., Carbon black, Vol. 8, pp. 179–243

H. D. R. Schüddemage, *Farbwerke Hoechst A. G.*, Frankfurt am Main, German Federal Republic, Ethylene and propylene polymers, Vol. 12, pp. 341–395

P. R. Sewell, *Dunlop Research Centre*, Birmingham, England, Elastomers, synthetic, Vol. 12, pp. 81–161

M. B. Sherman, *Consultant*, Brooklyn, N. Y., Confectionery products, Vol. 10, pp. 548–567

R. W. Shiffler, *Union Carbide Corporation, Chemicals and Plastics*, Tarrytown, N. Y., Brake fluids, Vol. 7, pp. 423–440

Norman E. Skelly, *The Dow Chemical Company*, Midland, Mich., Chlorophenols, Vol. 9, pp. 511–528

Charles G. Smith, *The Dow Chemical Company*, Midland, Mich., Diphenyls and terphenyls, Vol. 11, pp. 524–537

G. D. Smith, *Monsanto Company*, Texas City, Texas, Lactic acid and its derivatives, Vol. 15, pp. 125–149

Robert J. Smith, *CPC International, Inc.*, Argo, Ill., Corn products, Vol. 11, pp. 1–63

Orvill E. Snider, *Chemtex, Inc.*, New York, Polyamide fibers under Polyamides, Vol. 17, pp. 305–329

T. Sopoci, *Johns-Manville Research and Engineering Center*, Manville, N. J., Asbestos, Vol. 6, pp. 254–269; Asbestos–cement products, Vol. 6, pp. 270–273; Diatomaceous earth, Vol. 11, pp. 484–500

R. M. Speights, *J. T. Baker Chemical Company*, Phillipsburg, N. J., Citric acid, Vol. 10, pp. 85–100; *Geochemical Systems, Inc.*, Orange, Calif., Formic acid, Vol. 13, pp. 117–138

B. S. Sprague, *Celanese Research Company*, Summit, N. J., Acetate and triacetate fibers, Vol. 4, pp. 74–93

William H. Stahl, *McCormick & Company, Inc*, Cockeysville, Md., Sensory testing methods, Vol. 17, pp. 608–644

H. Stelmach, *Abbott Laboratories*, Chicago, Ill., Insulin, Vol. 14, pp. 565–583

Vernon A. Stenger, *The Dow Chemical Company*, Midland, Mich., Bromine, Vol. 8, pp. 1–29; Iodine, Vol. 14, pp. 584–616

B. B. Stewart, *CIBA-GEIGY Corporation*, Ardsley, N. Y., EDTA and related compounds, Vol. 12, pp. 23–56

Gerald E. Stobby, *The Dow Chemical Company*, Midland, Mich., Styrene polymers, Vol. 18, pp. 314–336

H. J. Stolten, *General Aniline & Film Corporation*, Easton, Pa., Amines, Vol. 5, pp. 259–275

Graham A. Stoner, *Infotronics Corporation, Special Products Division*, Houston, Texas, Carboxylic acids, amino, Vol. 8, pp. 410–556

W. Stuck, *Farbenfabriken Bayer AG*, Werk Uerdingen, Krefeld-Uerdingen, West Germany, Polycarbonates, Vol. 17, pp. 329–352

H. H. Suhr, *Farbwerke Hoechst AG*, Frankfurt am Main, German Federal Republic, Ethylene and propylene polymers, Vol. 12, pp. 341–395

R. C. Sweet, *Philips Laboratories, A Division of North American Philips Corporation*, Briarcliff Manor, N. Y., Ferrites, Vol. 12, pp. 514–538

Robert E. Tellis, *CIBA-GEIGY Corporation*, Ardsley, N. Y., Dyes, Vol. 12, pp. 1–22

J. W. Tereshko, *Union Carbide Corporation, Carbide Products Division*, Cleveland, Ohio, Metal borides under Boron compounds, Vol. 7, pp. 358–368; Carbides, Vol. 8, pp. 130–139

James G. Theivagt, *Abbott Laboratories*, North Chicago, Ill., Cyclohexylamine, Vol. 11, pp. 209–219; Cyclohexylamine derivatives, Vol. 11, pp. 220–255

K. Theurer, *Allied Chemical Corporation, Agricultural Division*, Hopewell, Va., Ammonia, Vol. 5, pp. 290–324; Ammonium compounds, Vol. 5, pp. 325–355

Daniel E. Thomas, *Carborundum Company*, Niagara Falls, N. Y., Bonded abrasives, Vol. 7, pp. 293–312

C. N. Thompson, *Johns-Manville Research and Engineering Center*, Manville, N. J., Asbestos, Vol. 6, pp. 254–269; Asbestos–cement products, Vol. 6, pp. 270–273; Diatomaceous earth, Vol. 11, pp. 484–500

Quentin E. Thompson, *Monsanto Company*, St. Louis, Mo., Dithiocarbamates, Vol. 11, pp. 573–602

Charles R. Thrash, *Union Carbide Corp.*, Sistersville, West Va., Silicon compounds, organic, Vol. 18, pp. 48–155

W. F. Tiedge, *Airco Chemicals and Plastics Company, A Divison of Air Reduction Company, Inc.*, Middlesex, N. J., Acetylene, Vol. 4, pp. 126–135; Acetylenic chemicals, Vol. 4, pp. 136–147

Frederick Tishler, *Ciba Pharmaceutical Company*, Summit, N. J., Alkaloids, Vol. 4, pp. 587–618; *Geigy Parmaceuticals*, Ardsley, N. Y., Coumarins, Vol. 11, pp. 159–173

J. J. Trainor, *Raybestos Division, Raybestos-Manhattan, Inc.*, Bridgeport, Conn., Brake linings, Vol. 7, pp. 440–474

Hans George Treibl, *Standard Naphthalene Products Co., Inc.,* Kearny, N. J., Naphthalene and naphthalene derivatives, Vol. 16, pp. 204–242

Cho C. Tsen, *Kansas State University,* Manhattan, Kan., Flour, Vol. 12, pp. 599–618

T. C. Tso, *United States Department of Agriculture,* Beltsville, Md., Tobacco, Vol. 19, pp. 133–160

B. G. Tweedy, *University of Missouri,* Columbia, Mo., Fumigants, Vol. 13, pp. 176–197; Fungicides, Vol. 13, pp. 198–227

Gerald R. Umbreit, *Greenwood Laboratories, Inc.,* Chaddsford, Pa., Carbonyl compounds, Vol. 8, pp. 331–367; Carboxylic acids, Vol. 8, pp. 368–409

Ronald Philip Upton, *Pfizer, Inc.,* Groton, Conn., Penicillins and related compounds, Vol. 16, pp. 552–583

C. E. Van Hall, *The Dow Chemical Company,* Midland, Mich., Magnesium, Vol. 15, pp. 356–420

J. C. Van Loon, *University of Toronto,* Toronto, Canada, Gold, Vol. 14, pp. 1–51

Richard T. vanSanten, *Teledyne Wah Chang Albany,* Albany, Ore., Hafnium and zirconium, Vol. 14, pp. 103–152

Hansjoerg W. Vollmann, *American Hoechst Corporation,* Coventry, R. I., Indigoid dyes, Vol. 14, pp. 495–518

Ralph C. Vollmar, *Consultant,* Berkeley, Calif., Dienes and polyenes, Vol. 11, pp. 500–515; Gas, natural, Vol. 13, pp. 310–339; Hydrocarbons, Vol. 14, pp. 298–354

Frank S. Wagner, Jr., *Celanese Chemical Company, Technical Center,* Corpus Christi, Texas, Acetals and ketals, Vol. 4, pp. 62–74; Crotonaldehyde, Vol. 11, pp. 174–183; Cyclic alcohols, Vol. 11, pp. 184–197; Cyclohexanone, Vol. 11, pp. 197–208

Ronald E. Walsh, *Teledyne Wah Chang Albany,* Albany, Ore., Hafnium and zirconium, Vol. 14, pp. 103–152

Carl J. Wassink, *Commercial Solvents Corporation,* Terre Haute, Ind., Alkanolamines, Vol. 5, pp. 1–22; Hydroxylamines and their salts, Vol. 14, pp. 440–460; Nitro compounds, organic, Vol. 16, pp. 412–448

James M. Weber, *The Dow Chemical Company,* Freeport, Texas, Diamines and higher amines, aliphatic, Vol. 11, pp. 407–431; Imines, Vol. 14, pp. 460–495

E. Dale Weir, *CIBA Products Company,* Toms River, N. J., Epoxy resins, Vol. 12, pp. 232–259

M. K. Weiss, *The Bunker Hill Company,* Kellog, Idaho, Lead, Vol. 15, pp. 161–201

Harold E. Weissler, *Falstaff Brewing Corporation,* St. Louis, Mo., Brewery products, Vol. 7, pp. 475–679

V. W. Wells, *Presstite, A Division of Interchemical Corporation,* St. Louis, Mo., Caulking compounds, Vol. 9, pp. 16–27

I. Wender, *U. S. Bureau of Mines, Pittsburgh Energy Research Center,* Pittsburgh, Pa., Coal and coke, Vol. 10, pp. 209–262

J. P. Williams, *Corning Glass Works, Research and Development Laboratories,* Corning, N. Y., Ceramics, Vol. 9, pp. 189–232; Glass and glass-ceramics, Vol. 13, pp. 403–487

R. W. Wise, *Monsanto Chemical Company, Organic Chemicals Division,* Akron, Ohio, Antioxidants and antiozonants, Vol. 6, pp. 71–122

M. M. Woyski, *Kerr-McGee Corporation,* Oklahoma City, Okla., Cerium, Vol. 9, pp. 246–277; Europium, Vol. 12, pp. 395–404

H. Wunderlich, *Farbenfabriken Bayer AG,* Werk Uerdingen, Krefeld-Uerdingen, West Germany, Polycarbonates, Vol. 17, pp. 329–352

E. A. Wynne, *Fisher Scientific Company*, Fairlawn, N. J., Bismuth, Vol. 7, pp. 187–220

Roland S. Young, *Department of Mines and Petroleum Resources*, Victoria, B. C., Canada, Cobalt alloys under Cobalt, Vol. 10, pp. 283–326; Cobalt determination under Cobalt, Vol. 10, pp. 348–373; Copper alloys under Copper, Vol. 10, pp. 598–630; Copper metal under Copper, Vol. 10, pp. 630–650; Nickel, Vol. 16, pp. 284–348

FINIS

Somewhat over 10 years ago, agreement was reached to publish this Encyclopedia. Since it started with a preface and an introduction, the above title seems appropriate to introduce the index.

Over the course of 10 years, contributions by 370 authors have gone through the essential steps of technical editing, copy editing, typesetting, and printing. Hopefully we planned to discuss basic techniques for analysis with stress on those which in many fields have superceded gravimetry and titrimetry. The degree of acceptance of the first three volumes of the Encyclopedia is indicative of a substantial degree of success.

Representatives of many industries have contributed the methods which they have found practical for their purpose. It is hoped that the student will refer to these as a contrast to the standard texts on analytical chemistry, not as a substitute for, but as a supplement which shows how the methods taught in the university are modified in industry. Over the years authors have contributed 400 articles on 12,423 pages which have now reached print and are indexed herein.

The term analysis has been interpreted, we believe properly, to extend beyond determination of ingredients to evaluation of the performance of the finished product. Of course, the Encyclopedia being completed is not perfect. It could have covered many more products, covered many more methods of analysis and evaluation.

So in bringing the effort to an end, we can only hope that it is a worthwhile contribution to the tools of the analytical laboratory as broadly defined above.

FOSTER DEE SNELL
LESLIE S. ETTRE

PREFACE TO THE GENERAL INDEX

The following listing contains the index to Volumes 4–19. The index of the first three volumes dealing with general analytical techniques can be found at the end of the third volume. However, the five general analytical technique articles included in Volumes 6–13 are indexed here.

Entries in the present compilation are indexed by volume and page number. A number before the colon (:) indicates volume number; a number after the colon indicates page number. If in the same volume the particular subject is discussed on different pages, the page numbers are divided by a comma (,). If the same subject is discussed in more than one volume, the volume:page entries are divided by a semicolon (;). For example:

> Novobiocin 5:461, 494
> Infrared spectroscopy
>   of carboxylic acids 4:416; 7:65; 8:378; 12:576

Titles of main articles are in italics and inclusive page numbers are given. For example:

> *Ablative materials* 4:1–30

An entry such as "Aromatic acids. See *Carboxylic acids.*" means that all entries for aromatic acids will be found under *Carboxylic acids.* On the other hand, an entry such as:

> Mevinphos. (See also *Pesticides and insecticides.*)

means that specific materials will be found in the subentries and related material under the cross reference.

In the alphabetization of the subentries words such as, in, for, of, as, etc, are not considered. For example:

> Pentadiene
>   in cigarette smoke 10:31
>   gas chromatography 10:31
> Reduction methods
>   for azo dyes 6:404
>   identification test for dyes 6:390, 391
>   for nitrate 19:138

Methods for determination using a particular reagent are usually subheaded by the main reagent. For example, the determination of silver by precipitation with sodium chloride is listed as:

> *Silver*
>   sodium chloride gravimetric method 10:621; 18:164

and the same reference is also repeated under sodium chloride as:

> Sodium chloride
>   gravimetric method for silver 10:621; 18:164

Analysis of a reagent and its use in the determination of other elements or substances is under one main heading representing the reagent.

Abbreviations are used only in connection with chemical names. For example: n for normal, t and *tert* for tertiary, etc. For chemical names and indication of substitutions the usual organic chemical nomenclature is followed.

A

Abderhalden drying apparatus 7:531
Abietic acid 8:407
Ablation 4:1–2
Ablative materials 4:1–30
  carbon 4:7
  charring 4:15
  graphite 4:7
  noncharring 4:4
  silica 4:2
Abrasion number
  of activated carbon 8:155
Abrasion resistance
  of activated carbon 8:157
Abrasion test
  of dental porcelains 11:276
  of dental restorative materials, composite
    11:315
Abrasives. (See also Aluminum oxide abrasives;
    Bonded abrasives; *Ceramics;* Coated abrasives;
    Diamonds, industrial; and Silicon carbide
    abrasives.) 4:31–61; 7:293; 9:189
  analysis 4:34; 7:295, 303; 10:264; 15:536
  in brake linings 7:441
  bulk density 7:311
  capillarity 10:267
  classification 4:31
  crystal size 7:308
  density 10:267
  in detergents 11:349
  grain size 7:303
  grain type 7:295
  hardness 7:309
  identification 7:295
  magnetic content 7:311
  manufactured 4:33
  in match strikers 15:536
  natural 4:32
  particle size 15:536
  particle size distribution 7:303; 10:264
  in rubbing compounds 6:353
  sampling and sample preparation 4:35
  in soaps 18:194
  specific gravity 7:295
  in waxes and polishes 6:363
Abrasive wheel. (See also Grinding wheels.) 7:293
Absolute viscosity 4:307
  of asphalt cements 6:300
  of bituminous materials 6:300
  of silicone fluids 18:113
  temperature susceptibility 6:305
Absom method

recovery of asphalt from bituminous paving
    mixtures 6:334
Absorbents. See *Adsorbents and absorbents.*
Absorption methods
  for carbon dioxide 8:248, 260, 263
  for nitrogen 16:454
  for oxygen 16:522
Absorption spectra
  of dysprosium(III) perchlorate 17:484
  of erbium(III) perchlorate 17:485
  of gadolinium(III) perchlorate 17:483
  of holmium(III) perchlorate 17:485
  of neodymium(III) perchlorate 17:482
  of praseodymium(III) perchlorate 17:481
  of samarium(III) perchlorate 17:482
  of terbium(III) perchlorate 17:484
  of thulium(III) perchlorate 17:485
  of ytterbium(III) perchlorate 17:486
Absorption spectroscopy
  of acetaldehyde 9:564
  of acetone 9:564
  of acrolein 9:564
  of benzaldehyde 9:564
  of benzophenone 9:564
  of 2-butanone 9:564
  of butyraldehyde 9:564
  of caffeic acid 10:64
  of carbonyls in cocoa beans 9:562
  of cerium 17:483
  of chalcone 9:564
  of chlorogenic acid 10:64
  of cinnamaldehyde 9:564
  of crotonaldehyde 9:564
  of 2,4-dimethylacetophenone 9:564
  of *p*-dimethylaminobenzaldehyde 9:564
  of 1,3-diphenyl-2-propanone 9:564
  of dysprosium 17:483
  of erbium 17:483
  of europium 17:483
  of formaldehyde 9:564
  of 2-furaldehyde 9:564
  of 2-furanacrylaldehyde 9:564
  of β-2-furylacrylophenone 9:564
  of 4-(2-furyl)-3-buten-2-one 9:564
  of gadolinium 17:483
  of 2-heptadecanone 9:564
  of 4-heptanone 9:564
  of hexanal 9:564
  of holmium 17:483
  of hydrocinnamaldehyde 9:564
  of *o*-hydroxybenzaldehyde 9:564

1

of 4-hydroxy-4-methyl-2-pentanone 9:564
of isobutyraldehyde 9:564
of mesityl oxide 9:564
of *p*-methoxybenzaldehyde 9:564
of *p*-methylacetophenone 9:564
of 3-methyl-2-butanone 9:564
of 5-methyl-2-furaldehyde 9:564
of neodymium 17:483
of 2-nonadecanone 9:564
of 2-nonanone 9:564
of 3-pentanone 9:564
of phenylacetaldehyde 9:564
of 4-phenyl-3-buten-2-one 9:564
of praseodymium 17:483
of propionaldehyde 9:564
of rare earths 17:480
of samarium 17:483
of terbium 17:483
of tiglaldehyde 9:564
of titanium 9:666
of 2-tridecanone 9:564
of 2-undecanone 9:564
of ytterbium 17:483
ABS polymer. See Acrylonitrile–butadiene–
    styrene polymer.
7-ACA. See 7-Aminocephalosporanic acid.
Acaricides. See *Pesticides and insecticides.*
Accelerator test
  for amino resins 5:280
Acenaphthene
  gas chromatography 16:217
Acenocoumarol. (See also *Coumarins.*) 10:153;
    11:160
  identification 10:151
  infrared spectrum 10:153
  paper chromatography 10:152
  polarography 10:154
  properties 10:153
  solubility 10:153
  thin-layer chromatography 10:151
  ultraviolet spectroscopy 10:151, 153
  uses 11:159
Acetal. (See also *Acetals and ketals.*)
  of benzaldehydes 6:625, 630
  gas chromatography 16:109
  in methanol 16:109, 114
Acetaldehyde 8:331
  absorption spectroscopy 9:564
  in acrolein 4:154
  in alcohols 4:544
  in black tea 18:510
  in 1,3-butadiene 11:504
  in cigarette smoke 10:27, 31
  in cocoa bean vapor 9:570
  commercial specifications 8:340
  in crotonaldehyde 11:179
  derivatives for identification 8:343

2,4-dinitrophenylhydrazine colorimetric method
    8:363
  in embalming fluids 12:165
  gas chromatography 4:154, 212; 10:28, 31; 11:
    181; 12:204, 300, 339; 16:108, 109, 425; 18:
    510; 19:455
  identification test using morpholine and sodium
    nitroprusside 8:344
  infrared spectrum 8:345
  iodometric titration 19:453
  mass spectroscopy 9:570
  in methanol 16:108, 109, 114
  3-methylbenzothiol-2-one hydrazone colori-
    metric method 19:454
  in oxyethylene–oxypropylene copolymers 19:
    308
  piperidine–sodium nitroprusside colorimetric
    method 19:455
  polarography of 4:71
  in production of acrolein 4:150
  properties 8:334
  sodium nitroprusside–diethanolamine colori-
    metric method 19:308
  sodium nitroprusside–morpholine colorimetric
    method 12:237
  ultraviolet spectroscopy 11:180
  use in preparation of peracetic acid 8:373
  uses 8:332
  in vinyl acetate 8:589
  in wines 19:452
Acetaldol
  commercial specifications 8:340
Acetal polymers 4:64
  hydrolysis of 4:70
  identification by gas chromatography 4:66
Acetals and ketals 4:62–74
  in alcohols 4:545
  analysis 4:64
  azeotropes 4:63
  commercial grades and specifications 4:64
  hydrolysis of 4:67
  hydroxylamine hydrochloride method 4:70
  identification 4:65
  manufacture 4:64
  mixtures, determination in 4:68
  properties 4:63
  separation 4:65
  uses 4:63
Acetamide
  derivative of amines 5:265
  derivatives for identification 8:601
  infrared spectrum 8:602
  properties 8:558, 597
Acetaminophen 6:177
  analysis 6:181
  colorimetric method using *N*-(1-naphthyl)-*N*-
    diethylpropylenediamine 6:195

column chromatography 6:154
commercial grades and specifications 6:181
diazotization method 6:195
impurities, determination of 6:196
infrared spectrum 6:183
ion exchange chromatography 6:182
manufacture 6:179
properties 6:178
separation 6:182
ultraviolet spectroscopy 6:156, 182, 183, 184, 196
Acetanilide 5:421; 6:177
  acid–base titration 5:444
  analysis 6:181
  colorimetric method using hydroxamic acid 6:193
  colorimetric method using α-naphthol 6:194
  colorimetric methods 5:452
  commercial grades and specifications 6:180
  derivatives for identification 8:601
  diazotization method 6:194
  gas chromatography 6:187; 8:602
  gravimetric method 6:189
  identification 6:181
  impurities, determination of 6:196
  infrared spectrum 5:438; 6:182
  manufacture 5:426; 6:178
  in phenacetin 6:196
  properties 5:422; 6:178; 8:597
  separation 6:181
  standard for carbon, hydrogen, and nitrogen analysis 8:290, 291, 295, 297
  ultraviolet spectroscopy 5:433
  ultraviolet spectrum 6:181
  uses 5:428
  volumetric method using sodium nitrite 6:190
Acetate and triacetate fibers 4:74–93
  acetyl value determination 4:80
  additives, determination of 4:91
  analysis 4:78
  identification 4:78
  impurities, determination of 4:89
  infrared spectroscopy 4:78
  manufacture 4:78
  mechanical properties, determination of 4:89
  mixtures, determination in 4:91
  physical properties, determination of 4:85
  polymer composition, determination of 4:80
  properties and composition 4:75
  uses 4:74
Acetate fibers. (See also Acetate and triacetate fibers.)
  microscopy 4:264
  mixtures, determination in 4:353
  solubility 4:248, 250, 353
Acetates
  in aluminum acetate 5:162

in cobalt acetate 10:329
ferric chloride color test 14:581
identification 5:161
in insulin 14:581
Acetic acid 4:93–101; 8:395
  in acetic anhydride 4:104
  acid–base titration 4:98; 13:127; 15:145; 16:437
  in acid salts 8:596
  analysis 4:97
  azeotropes 4:95
  in beer 7:579, 581
  in black tea 18:507, 511
  in calcium acetate 8:78
  in ceramics 9:224
  in cocoa beans 9:544
  combined, in cellulose acetate polymers 4:75, 80
  commercial grades and specifications 4:96
  density 7:37
  derivatives for identification 8:376
  esterification of tertiary alcohols with 4:540
  flammability 4:94
  in formic acid 13:127
  freezing point method 4:97
  freezing points of aqueous solutions 4:95
  gas chromatography 8:383, 384; 11:181; 13:127; 18:511; 19:449
  gravimetric method for titanium 19:93
  identification 4:97
  impurities, determination of 4:98
  in lactic acid and its derivatives 15:145
  manufacture 4:95
  in methanol 16:114
  microscopy 9:224
  near–infrared spectroscopy 4:104
  in nitroethane 16:437
  in nitromethane 16:437
  in nitroparaffins 16:437
  in 1-nitropropane 16:438
  in 2-nitropropane 16:438
  paper chromatography 15:133
  properties 4:94; 8:396, 558
  properties, determination of 4:98
  refractive index 7:37
  sampling 4:97
  separation 4:97
  specific gravity of aqueous solutions 4:96
  temperature-rise method, reaction with triethylamine 4:104
  thin-layer chromatography 8:381
  in tobacco 19:148
  uses 4:93
  in wines 19:405, 412, 413
Acetic acid dimer 8:368
Acetic acid–hydroxylamine chloride
  colorimetric method for ethers 12:299
Acetic acid insolubles

in cadmium acetate 8:37
Acetic acid–sulfuric acid
  colorimetric method for medroxyprogesterone
    acetate 10:582
Acetic anhydride 4:102–107
  acetic acid in 8:566
  acetylation reagent 4:141, 529, 568
  acid–base titration 4:103, 105
  acylation of hydroxyl groups in coal 10:212
  analysis 4:103
  aniline-reaction method 4:103
  colorimetric method for ethynodiol diacetate
    10:587
  colorimetric method for lanolin 11:82
  colorimetric method for malonic acid 15:424
  commercial grades and specifications 4:103;
    8:560
  impurities, determination of 4:106
  infrared spectroscopy 8:562, 563
  manufacture 4:103
  morpholine-reaction method 4:103
  morpholine titration 8:566
  properties 4:102; 8:558, 559
  properties, determination of 4:106
  temperature-rise method, reaction with triethyl-
    amine 4:103
  temperature-rise method, reaction with water 4:
    103, 104, 106
  p-toluidine derivative 8:561
  use in manufacture of acid anhydrides 8:560
  uses 4:102
  volumetric method for 1,3-dichloro-2-propanol
    14:161
  volumetric method for ethylene chlorohydrin
    14:161
Acetic anhydride–perchloric acid
  acylation of cinnamyl alcohol 10:82
  volumetric method for nitrogen 11:427
Acetic anhydride–pyridine
  volumetric method for pentaerythritol 13:593
Acetic anhydride–sodium acetate
  acylation of cinnamyl alcohol 10:81
Acetin
  volumetric method for glycerol 13:508
Acetoacetanilide
  conductometric titration 5:445
  manufacture 5:426
  properties 5:422
  ultraviolet spectroscopy 5:433
  use in manufacture of azo dyes 6:403
Acetoacetic acid 8:395
  properties 8:395, 396
Acetoacetic ester condensation 8:580
Aceto DIPP (PAN, PBN, and POD) 6:72
Acetoin
  acidic steam distillation 19:458
  gas chromatography 19:449, 457

α-naphthol–creatine colorimetric method 19:
    457
  structure 19:446, 456
  in wines 19:446, 452, 456
2-Acetonaphthone
  hydroxylamine hydrochloride method 8:357
Acetone 4:107–121; 8:331
  absorption spectroscopy 9:564
  in acetylenic chemicals 4:145
  in acrolein 4:154
  in alcoholic beverages 4:489
  analysis 4:110
  azeotropes 4:108
  in black tea 18:510
  in chloroform 9:435
  in cigarette smoke 10:27, 31
  in cocoa bean vapor 9:570
  commercial grades and specifications 4:110
  density 7:38
  derivatives for identification 8:343
  2,4-dinitrophenylhydrazine colorimetric method
    8:363
  2,4-dinitrophenylhydrazone method 4:116
  in enamels 12:175, 188
  gas chromatography 4:118, 154, 212, 214; 8:
    351, 353; 11:73, 362; 10:28, 31; 12:204, 300,
    302, 332; 16:109, 425, 432; 18:510; 19:458
  hydroxylamine hydrochloride method 4:112,
    145; 8:357
  2-hydroxy-1-naphthaldehyde colorimetric
    method 11:204
  identification 4:110
  impurities, determination of 4:113
  indigo test 4:112
  infrared spectrum 8:347
  iodine–lime water gravimetric method 12:188
  iodoform method 4:111
  manufacture 4:109
  mass spectroscopy 7:31; 9:570
  in methanol 16:109, 112, 114
  mixtures, determination of 4:116
  Nessler reagent color test 16:112
  in nail enamels 11:73
  polarography 4:119
  properties 4:108; 12:175
  properties, determination of 4:113
  refractive index 7:38
  sodium nitroprusside test 4:111
  solubility 13:539, 552
  sulfur content 18:371, 376
  ultraviolet spectroscopy 4:116
  use in manufacture of BLE 6:110
  uses 4:107, 109, 185
Acetone–carbon disulfide
  colorimetric method for sulfur 18:371
Acetone cyanohydrin process
  manufacture of methacrylic acid

and methacrylates 4:185
Acetone diethyl acetal. (See also *Acetals and
    ketals.*) 4:62
  melting point of derivatives 4:68
  properties 4:63
Acetone-2,4-dinitrophenylhydrazone
  thin-layer chromatography 8:381
Acetone–diphenylamine reaction products 6:73,
    110
  analysis 6:111
  colorimetric method based on formation of
    Wurster salts 6:101, 103
  colorimetric method using diazotized *p*-nitro-
    aniline 6:99
  color tests 6:90
  column chromatography 6:89
  gas chromatography 6:111
  identification 6:111
  manufacture 6:110
  properties 6:110
  separation 6:111
  ultraviolet spectroscopy 6:97
Acetone extractables
  in brake linings 7:448
  in carbon black 8:235
Acetonitrile. (See also *Nitriles.*)
  in acrylonitrile 4:371, 375
  azeotropes 16:403
  in cigarette smoke 10:31
  derivatives for identification 16:401, 402
  gas chromatography 10:31; 16:404, 425
  in hydrogen cyanide 14:422
  mass spectroscopy 14:422
  polarography 16:398
  properties 16:396
  solubility 16:397
Acetonylacetone
  properties 8:335
3-(α-Acetonylbenzyl)-4-hydroxycoumarin. See
    Warfarin.
3-(α-Acetonylbenzyl)-4-hydroxycoumarin potas-
    sium. See Warfarin potassium.
3-(α-Acetonylbenzyl)-4-hydroxycoumarin
    sodium. See Warfarin sodium.
3-(α-Acetonyl-4-nitrobenzyl)-4-hydroxycou-
    marin. See Acenocoumarol.
Acetophenetidine. See Phenacetin.
Acetophenone 4:122–126
  in acetylenic chemicals 4:145
  acidity 8:358
  analysis 4:123
  azeotropes 8:336
  commercial grades and specifications 4:123
  commercial specifications 8:341
  derivatives for identification 8:343
  gas chromatography 4:124
  hydroxylamine hydrochloride method

    4:124, 145; 8:357
  identification 4:123
  impurities, determination of 4:124
  infrared spectrum 8:347
  manufacture 4:122
  mixtures, determination of 4:124
  properties 4:122; 8:335
  properties, determination of 4:124
  separation 4:123
  uses 4:122
Acetovanillone
  gas chromatography 11:167
21-Acetoxy-3-hydroxy-5-pregnane-11,20-diones
  liquid chromatography 14:208
Acetylacetone (2,4-Pentanedione)
  acidity 8:359
  azeotropes 8:337
  colorimetric method for formaldehyde 11:73
  colorimetric method for titanium 19:99
  commercial specifications 8:341
  Karl Fischer titration 8:360
  properties 8:335
  water in 8:360
Acetylacetone–ammonium acetate
  colorimetric method for formaldehyde 13:106,
    107
Acetylacetone derivatives
  color reagents for boric acid 7:388
Acetylacetone, fluorinated
  uses 8:332
N-Acetyl-*p*-aminophenol. See Acetaminophen.
Acetylation method
  for acetylenic glycols 4:141
  for alcohols 4:529, 568
  for allyl alcohols 5:78
  for amines 5:270
  for molecular weight of polytetramethylene
    ether glycols 13:246
  for β-phenylethyl alcohol 7:99
  for trialkanolamines 5:20
  for trimethylolpropane allyl ethers 5:100
Acetylation tests
  for cellulose 9:55
N-Acetylaziridine. (See also *Imines.*)
  properties 14:469
  structure 14:469
Acetyl benzene. See *Acetophenone*.
Acetyl bromide
  properties 8:570
Acetyl chloride
  commercial specifications 8:571
  properties 8:558, 570
Acetyl chloride–dimethylaniline acylation
  of tertiary alcohols 12:279
Acetyl chloride–sodium hydroxide
  volumetric method for hydroxyl groups 12:248
Acetylcholine. (See also *Choline and its derivatives.*)

uses 9:611
Acetylcholine bromide. (See also *Choline and its derivatives.*)
  properties 9:613
Acetylcholine chloride. (See also *Choline and its derivatives.*)
  dipicrylamine colorimetric method 9:617
  Dragendorff's colorimetric method 9:617
  iodine colorimetric method 9:617
  iodoplatinate colorimetric method 9:617
  properties 9:613
  thin-layer chromatography 9:616
Acetyl codeine. (See also Codeine.)
  in aspirin 16:270
  ion exchange methods 16:270
Acetyl content
  of cellulose acetate 9:70
  of cellulose acetate butyrate 9:74
  of cellulose acetate propionate 9:74
  of starch derivatives 11:28
Acetylene. (See also *Hydrocarbons.*) 4:126–135; 14:317
  in air 16:485
  analysis 4:128
  bromine-absorption method 4:128
  in carbon monoxide 8:310, 313
  commercial grades and specifications 4:127
  in ethylene 5:362; 14:327
  in ethylene oxide 12:223, 333, 338
  gas chromatography 4:130, 212; 7:53; 8:310; 12:338; 14:327; 16:458, 485
  in hydrogen 14:381
  Ilosvay's reagent color test 16:532
  impurities, determination of 4:130
  infrared spectroscopy 8:313; 13:294
  manufacture 4:127
  in noble gases 13:294
  in oxygen 16:532
  properties 4:127
  safety precautions 4:128
  silver nitrate colorimetric method 12:223, 333
  in trichloroethylene 5:367
  uses 4:126, 184
Acetylene black 8:179
  commercial grades and specifications 8:196
  composition 8:191
  manufacture 8:194
Acetylene dichloride. See 1,2–Dichloroethylene.
Acetylenes
  in 1,3-butadiene 11:504, 507
  in isoprene 11:510
  silver nitrate absorption method 11:507
Acetylene tetrabromide
  commercial specifications 8:12
  properties 8:12
Acetylene tetrachloride. See 1,1,2,2-Tetrachloro-ethane.

Acetylenic alcohols 4:136, 143, 575
  analysis 4:139, 143, 577
  azeotropes 4:139
  commercial grades and specifications 4:139
  impurities, determination of 4:143
  manufacture 4:139, 575
  properties 4:136, 576
  silver acetylide method 4:139, 577
  solubility 4:137
  uses 4:137, 576
Acetylenic chemicals. (See also Acetylenic alcohols; Acetylenic glycols.) 4:136–147
  analysis 4:139
  azeotropes 4:138, 193
  commercial grades and specifications 4:139
  impurities, determination of 4:143
  manufacture 4:139
  mixtures, determination in 4:145
  properties 4:136
  uses 4:139
Acetylenic glycols
  acetylation method 4:141
  alkaline decomposition method 4:140
  analysis 4:140
  commercial grades and specifications 4:139
  impurities, determination of 4:143
  manufacture 4:139, 575
  properties 4:137
  uses 4:137
Acetyl fluoride
  properties 8:570
Acetyl groups
  in acrylonitrile–vinyl acetate copolymers 4:387
Acetyl halides
  derivatives for identification 8:572
Acetyl iodide
  properties 8:570
Acetyl isocyanate. (See also *Isocyanic acid esters.*)
  infrared spectroscopy 15:106
N-Acetyl-*p*-phenetidine. See Phenacetin.
Acetylpropionyl
  in black tea 18:510
  gas chromatography 18:510
3-Acetylpyridine. (See also *Pyridine and pyridine derivatives.*)
  ionization constants 17:417
  in tobacco 19:136
Acetylsalicylic acid (Aspirin) 6:124; 8:401
  acid–base titration, nonaqueous 6:166, 170
  analysis 6:127
  colorimetric method using ferric ion 6:148
  column chromatography 6:36, 128, 134, 137, 139, 150, 152, 154
  commercial grades and specifications 6:127
  extraction 6:128, 133, 158, 166
  gas chromatography 6:135, 138, 156, 158

hydrolysis method 6:136
identification 6:128, 136
impurities determination of 6:138
infrared spectroscopy 6:128, 131
liquid chromatography 8:59, 60
manufacture 6:126
nuclear magnetic resonance spectroscopy 6:172, 174
paper chromatography 6:183
properties 6:124; 8:402
salicylic acid in 6:138
separation 6:128, 183
solvent extraction 8:61
thin-layer chromatography 8:59
ultraviolet spectroscopy 6:130, 132, 136, 140, 147, 149, 150, 153, 159, 162; 8:67
Acetyl tributyl citrate. (See also Citrate esters.)
properties 10:97
Acetyl triethyl citrate. (See also Citrate esters.)
properties 10:97
Acetyl value
of cellulose acetate polymers 4:75, 79
Achroic R value
of reducing sugars 7:556
Acid absorption
of hydrocarbons 14:340
Acid anhydrides 8:559
acid–base titration 8:563
amide formation 8:563
analysis 8:561
commercial grades and specifications 8:561
derivatives for identification 8:561
ferric hydroxamate colorimetric method 8:561, 569, 590
formation from carboxylic acids 8:372
gas chromatography 8:562
identification 8:561
impurities, determination of 8:567
infrared spectroscopy 8:562
manufacture 8:560
properties 8:558, 559
use in preparation of carboxylic acids 8:372
uses 8:557
Acid azoic dyes
ammonium hydroxide colorimetric method 10:538
in meat 10:538
Acid–base titration
of acetic acid 4:98; 13:127; 15:145; 16:437
of acetic anhydride 4:103, 105
of acid in soaps 18:200
of acrylic acid 4:197
of alkali carbonate 15:402
of alkali in soaps 18:200
of alkaloids 16:257, 273
of alkaloids in tobacco 19:135
of alkanolamines 5:11

of aluminum 5:219
of amines 4:607; 5:268, 270; 7:428
of amino acids 19:474
of aminoalkyl esters (neutralization equivalent) 4:198
of ammonia 11:351; 14:558; 15:252; 16:189, 438, 509; 17:439; 18:85; 19:138, 471, 473
of ammonium bicarbonate 5:329
of ammonium chloride 9:453
of ammonium hydroxide 5:317
of amolanone hydrochloride 5:378
of anisindione 10:177
of apomorphine hydrochloride 16:276
of ash content of cocoa beans 9:554
of barium hydroxide 6:549
of beer 7:508, 579
of benzoic acid 7:65, 68
of beryllium 7:133
of bithionol 11:547
of borates 7:406
of borax 7:428
of boric acid 7:371, 402
of boric oxide 13:417
of butacaine sulfate 5:381
of calcium carbonate 10:38; 19:360
of carbonate 17:401
of carbon dioxide 15:285; 19:361, 501
of carbonyls in 1,3-butadiene 11:507
of carboxylic acid (free) in acid chlorides 7:76
of carboxylic acids 8:389
of carboxyl groups in starch derivatives 11:34
of cephalosporins 16:563
of cesium carbonate 9:286
of chloride 18:79, 88, 124
of chlorophenols 9:514
of chlorosulfonic acid 9:530
of cholic acid 7:149
of cinchophen 6:203
of cinnamic acid 10:61
of citric acid 8:64; 10:92; 19:149
of cocaine 16:273
of cocaine hydrochloride 5:385
of codeine 16:267
of codeine phosphate 16:269
of codeine sulfate 16:269
of crystal violet 19:175
of cyanoacetic acid 15:439, 442
of cyclohexanone 11:206
of cyclohexylamine 11:213
of dehydrocholic acid 7:149
of deoxycholic acid 7:149
of dichlorophene 11:547
of dicyclohexylamine 11:217
of diethylenetriamine 11:424, 425
of emetine hydrochloride 4:614
of epoxides 11:568
of esters 4:201
of ethanolamines 12:295

of ethyl biscoumacetate 10:163
of ethyleneamines 11:422, 425
of ethylenediamine 11:423, 425
of ethylenimine 14:476
of ethylenimine derivatives 14:476
of ethynodiol diacetate 10:581
of fats and fatty oils 12:477
of fatty acids 11:57; 15:498; 18:189, 205
of fluoride 19:75
of fluoroboric acid 13:39
of fluorosilicate 13:33
of fluorosilicic acid 13:40, 48; 18:7
of formic acid 13:125
of formaldehyde 11:566
of formol nitrogen in beer 7:615
of glutamic acid 8:474
of heroin 16:271
of hexachlorophene 11:547
of hydrazoic acid 16:472
of hydrochloric acid 14:361, 366, 368
of hydrofluorosilicic acid 13:34
of hydrogen chloride 9:330; 18:63
of imidazole alkaloids 4:608
of ion exchange resins 15:14
of isophthalic acid 17:158
of lactic acid 15:133, 136
of lidocaine 5:395
of lidocaine hydrochloride 5:395
of lime 9:109
of lime and limestone 15:229
of lithium carbonate 15:279
of lithium hydroxide 15:285
of lubricants 15:314, 322
of magnesium carbonate 15:401
of magnesium hydroxide 15:363
of magnesium oxide 15:363
of malic acid 19:149
of malonic acid 15:429
of meperidine hydrochloride 16:274
of meprylcaine hydrochloride 5:398
of mestranol 10:577
of metabutoxycaine hydrochloride 5:400
of metaphosphoric acid 17:131
of methacrylic acid 4:197
of mixed acids 16:498
of morphine 16:259, 261, 267
of morphine hydrochloride 16:260
of morphine sulfate 16:260
of morpholine 16:198, 201
of musts 19:411
of nalorphine hydrobromide 16:268
of narcotics 16:255
of nicotinamide 17:435
of nicotinic acid 17:435
of nitric acid 14:367; 16:497
of 2-nitro-2-methyl-1-propanol 16:445
of noscopine 16:272

of noscopine hydrochloride 16:271
of organic acid in elastomers 12:147
of organosilicon copolymers 18:147
of orthophosphates 17:139
of orthophosphoric acid 17:120, 128
of oxalates 7:616
of oxalic acid 19:149
of papaverine hydrochloride 16:271
of penicillins 16:563
of phenindione 10:182
of phenobarbital 6:521
of phenprocoumon 10:165
of phenylenediamines 11:447
of phosphates 17:139
of phosphorus 17:64, 65, 74, 80
of phosphorus pentoxide 17:111
of phthalic acid 17:158
of phthalic anhydride 17:154
of picric acid 12:419
of polyamides 17:293
of polyols 19:278
of polyphosphates 17:139
of potassium hydrogen phthalate 17:401
of potassium hydroxide 17:401; 18:250
of procaine hydrochloride 5:408
of propionic acid 14:457
of pyridine and its derivatives 17:434
of pyridine bases in phenols 17:38
of pyridoxine hydrochloride 17:435
of silicates 18:44
of silicone fluids 18:114
of silver 11:289
of sodium benzoate 7:69; 8:63
of sodium bicarbonate 18:246
of sodium carbonate 6:359; 18:245
of sodium dehydrocholate 7:149
of sodium hexobarbital 5:370
of sodium hydroxide 14:253; 18:250
of sodium oxide 18:122, 231
of sodium silicates 18:44
of sparteine 4:610
of strontium carbonate 18:275
of strontium oxide 18:275
of strychnine 4:602
of styphnic acid 12:419
of succinic acid 19:424
of sulfur dioxide 19:480
of sulfuric acid 13:40; 16:499; 18:416
of sulfuryl fluoride 13:195
of tartaric acid 19:420
of terephthalic acid 17:158, 161
of thiuram disulfides 11:595
of triethanolamine 12:292
of triethylenetetraamine 11:424
of trimellitic anhydride 17:172
of trimethylamine 9:626
of uranium 14:366

of vinbarbital 6:529
of water 19:360
of wines 19:411
of zirconyl nitrate 14:148
Acid–base titration, nonaqueous
of acetylsalicylic acid 6:166, 170
of acid halides 8:574
of acid salts 8:592
of acrylamide 4:164
of acrylic acid copolymers 4:275
of acrylic fibers and groups 4:288
of alkaloids 4:593
of alkanolamines 5:12
of N-alkylhydroxylamines 14:449
of amides 8:605
of amines 5:88, 269, 271, 273
of amines in polymers 4:273, 386
of ammonium benzoate 7:86
of amobarbital 6:503
of ampicillin 16:577
of aniline and its derivatives 5:443, 445, 447
of antihistamines 6:25
of aprobarbital 6:512
of barbiturates 6:492
of benzoic anhydride 7:73
of beryllium acetate 7:120
of bile acids 7:150
of bishydroxycoumarin 10:157
of butyl aminobenzoate 5:382
of caffeine 6:170; 8:63, 65
of carboxylic acids 8:390
of choline derivatives 9:621
of cinchona alkaloids 4:606
of decaborane 7:346
of dicyclohexylamine 11:240
of N-1,3-dimethylbutyl-N'-phenyl-p-phenylene-
    diamine 6:114
of N,N'-di-3-(5-methylheptyl)-p-phenylene-
    diamine 6:115
of 3,5-dinitrobenzoate derivatives of alcohols
    7:93
of erythromycin 5:531
of ethyl-p-aminobenzoate 5:388
of ethyleneamines 11:425
of fats and fatty oils 12:478
of fatty acids 11:365
of glycine 8:482
of isobucaine hydrochloride 5:394
of hydrogen chloride 7:76
of lauryl cetyl dimethylamine oxide 11:348
of lauryl dimethylamine oxide 11:348
of leucine 8:494
of levorphanol tartrate 16:280
of mephobarbital 6:517
of mepivacaine hydrochloride 5:396
of methadone hydrochloride 16:277
of monomer mixtures 4:341

of morphine 16:260
of nicotine and nornicotine 4:610
of nitrilotrispropionamide in acrylamide 4:166
of nystatin 5:578
of phenacetin 6:166, 170
of phenindamine tartrate 6:170
of phenobarbital 6:171
of phenols 17:27
of pilocarpine 4:608
of phenylalanine 8:511
of poly(methacrylic acid) 16:91
of pramoxine hydrochloride 5:402
of pyrrocaine hydrochloride 5:414
of secobarbital 6:524
of sodium acid malonate 15:431
of sodium formate 13:135
solvent systems 8:390
of trialkanolamines 5:20
of tyrosine 8:538
of vinylpyridine polymers 4:272
Acid casein. See Casein.
Acid chlorides
preparation 8:374
reaction of amides with 8:608
volumetric method using m-chloroaniline 7:76
Acid chrome dark blue
colorimetric method for tantalum 16:369
Acid chrome violet K
colorimetric method for niobium 16:369, 374
Acid colors. (See also Certified colors; and individ-
    ual compounds.)
liquid chromatography 10:527
Acid-consuming capacity
of aluminum hydroxide 5:188
Acid content. (See also Acidity; Acid number;
    and pH.)
of acetone 4:113
of adsorbents 4:437
of aerosols 4:461
of alcoholic beverages 4:480, 485, 490
of alcohols 4:543
of ammonium sulfate 5:352
of anhydride curing agents 4:24
of chloroform 5:359
of cigarette smoke 10:17
of felt 12:509
Acid cross staining test
for reactive dyes 12:19
Acid-detergent fiber
in tobacco 19:143
Acid dichromate
colorimetric method for chromium 9:647
Acid dissolution
of antimonides 9:215
of arsenides 9:215
of basic refractories 9:215
of cements 9:214

of ceramics 9:214
of chalcogenides 9:215
of chrome ore 9:215
of clay 9:214
of enamels 9:215
of ferrites 9:215
of glass ceramics 9:215
of glasses 9:214
of glazes 9:214
of high-silica ceramics 9:215
of iron oxides 9:215
of lime 9:215
of mortars 9:215
of niobates 9:215
of phosphides 9:215
of porcelain 9:215
of silica brick 9:214
of tantalates 9:215
of whiteware 9:215
Acid-dye method
  for nitrogenous base 4:592
  for cinchona alkaloids 4:606
Acid–dye response
  of polyamides 17:299
Acid dyes 6:402; 12:2
Acid fluoride fusion
  of titanium 19:93
Acid-forming substances
  in hexachloroethane 9:451
Acid, free
  acid–base titration 18:200
  in ammonium sulfate 5:352
  in cadmium chloride 8:42
  in casein 9:11
  in cellulose acetate 9:70
  in cellulose acetate butyrate 9:74
  in cellulose acetate propionate 9:74
  in citrate esters 10:98
  in copper(II) acetate 10:654
  in essential oils 12:274
  in ethers 12:308
  in ethyl ether 12:308
  in ethyl formate 13:133
  in fluorocarbons 13:69
  in formic acid esters 13:133
  in glycerol 13:524
  in hydrogen peroxide 14:436
  in methyl formate 13:133
  in nitroethane 16:437
  in nitromethane 16:437
  in nitroparaffins 16:437
  in 1-nitropropane 16:438
  in 2-nitropropane 16:438
  in pesticides and insecticides 19:615
  in R salt 16:238
  in silicone fluids 18:114
  in silver nitrate 18:175
  in soaps 18:200
  in sodium fluoride 13:33
  in triethyl citrate 10:98
  in victoria blue BO 19:187
Acid fuchsin D. See D & C Red No. 33.
Acid green
  in enamels 12:187
  identification 12:187
Acid halides
  acid–base titration, nonaqueous 8:574
  analysis 8:571
  commercial specifications 8:571
  derivatives for identification 7:63; 8:572
  ferric hydroxamate colorimetric method 8:561
  formation from carboxylic acids 8:372
  identification 8:571
  impurities, determination of 8:574
  infrared spectroscopy 8:573
  manufacture 8:571
  preparation 7:63; 8:375
  properties 8:558, 570
  use in preparation of carboxylic acids 8:372
Acid hydrolysis
  of amides 8:606
  of caprolactam 8:121
  of cysteine 8:467
  of formamide 8:606
  of protein 8:411
  of starch in chocolate liquor 9:585
  of sugar 18:346
Acid hydrolysis method
  for carbohydrate determination 4:474
  for dithiocarbamate esters 11:589
  for dithiocarbamic acid salts 11:574
  for fat 6:449
  for thiuram disulfides 11:593
Acidic copolymers
  acid–base titration of 4:387
Acidic steam distillation
  of acetoin 19:458
Acidimetric methods
  for fluoroboric acid 13:39
Acidimetric titration
  of sulfide 18:397
Acid-insoluble matter
  in gypsum and gypsum products 14:94
  in hydrated alumina 5:188
  in lime and limestone 15:206
Acid, ionized
  in carbon black 8:230
Acidity. (See also Acid content; Acid number;
    and pH.)
  of aluminum silicate 5:190
  of ammonium nitrate solutions 5:342
  of ammonium oxalate solutions 5:345
  of ammonium persulfate solutions 5:347
  of barium sulfate 6:567, 568

of beer 7:508, 579
of benzene 7:41
of benzoic acid esters 7:84
of bromohydrocarbons 8:13
of caprolactam 8:126
of carbonated beverages 8:164, 165
of carbon black 8:192, 230
of carbon dioxide 8:259
of carbon tetrachloride 8:327
of carbonyl compounds 8:358
of confectionery products 10:564
of corn products 11:7
of crotonaldehyde 11:182
of cyclopropane 5:360
of diazodinitrophenol 12:416
of diethyl malonate 15:434
of dimethyl malonate 15:434
of dioxane 11:521
of essential oils and related substances 7:84
of ethylene 5:362
of ethyl ether 5:361
of ethylene oxide 12:221, 333
of fermented milk products 16:135
of fluorocarbons 13:68
of formaldehyde 13:114
of furfural 13:239
of glycerol 13:523
of hydrocarbons 14:344
of hydrogen cyanide 14:421
of hydrogen peroxide 14:436
of isocyanates 19:277
of isocyanic acid esters 15:118
of mercuric oxide 16:62
of mercury fulminate 12:409
of methanol 16:112
of musts 19:411
of nitrous oxide 5:366
of phenprocoumon 10:166
of phosphorus sesquisulfide 15:531
of pigments 17:177
of propylene oxide 12:225
of red phosphorus 15:536
of Stoddard solvent 11:606
of sulfuric acid 18:416
of tolylene diisocyanates 15:118
of trichloroethylene 5:367
of water 19:360
of wine ash 19:489
of wines 19:411
Acidity, aprotonic
  in aluminum oxide 5:188
Acid magenta. (See also *Triphenylmethane and
    related dyes.*)
  titanous chloride titration 19:177
Acid mixtures
  hydrochloric acid content 14:367
  nitric acid content 14:367

phosphoric acid content 14:367
sulfuric acid content 14:367
Acid number. (See also Acid content; Acidity;
    and pH.) 8:392
of acetophenone 4:124
of alkyl resins 5:50
of chlorinated paraffins 9:500
of dimethyl terephthalate 17:165
of drycleaning detergents 11:612
of drycleaning solvents, used 11:609
of essential oils 12:274
of fatty alcohols 4:572
of glycols 13:587
of hydrocarbon resins 14:284
of isophthalic acid 17:168
of molybdenum disulfide 16:181
of organic coating materials 8:393
of organosilicon copolymers 18:147
of plasticizer 8:393
of polyols 19:278
of polytetramethylene ether glycols 13:247
of rosin 8:393
of silicone fluids 18:114
of terephthalic acid 17:161
Acid number stability
of dimethyl terephthalate 17:165
Acids
  in acid halides 8:574
  in chlorobenzene 9:478
  in chloroform 9:434
  in 1,2-dichloroethane 9:440
  in diethylethoxymethylene malonate 15:438
  in ethyl cyanoacetate 15:442
  in formic acid 13:127
  in hexachloroethane 9:451
  in iodoform 14:615
  in lactic acid and its derivatives 15:145
  in methyl chloride 9:425
  in methyl cyanoacetate 15:442
  in methylene chloride 9:429
  tetrabutylammonium hydroxide titration 10:19
  in 1,1,2,2-tetrachloroethane 9:448
  in tetrachloroethylene 9:464
  in 1,1,1-trichloroethane 9:443
  in trichloroethylene 9:457
  turbidimetric method for thiosulfate 18:431
α-Acids
  in beer 7:604
  ultraviolet spectroscopy 7:605
Acid salts 8:590
  acid–base titration, nonaqueous 8:592
  analysis 8:592
  commercial grades and specifications 8:592
  formation from carboxylic acids 8:372
  identification 8:592
  impurities, determination of 8:593
  infrared spectroscopy 8:592

manufacture 8:591
properties 8:590
Acids, organic. See *Carboxylic acids.*
Acids, volatile
in beer 7:579
microdiffusion method 7:579
Acidulants
in confectionery products 10:552
Acid value. (See also Acid number.)
of benzaldehydes 6:623
of cinnamic acid and its derivatives 10:70
of cinnamyl esters 10:84
of fats and fatty oils 12:477
of fatty acids 18:184
of formic acid esters 13:133
Acid violet. (See also F D & C Violet No. 1.)
in enamels 12:187
identification 12:187
Acid wash color
of aromatic hydrocarbons 14:345
of benzene 7:42
cis-Aconitic acid
thin-layer chromatography 8:381
cis-Aconitic anhydride
colorimetric method for lidocaine 5:395
colorimetric method for tetracaine 5:415
Acribel
cross section 4:224, 263
solubility 4:249
Acridine
gravimetric method for tantalum 16:367
Acridine dyes 6:606; 12:7
Acrilan. (See also *Acrylic and modacrylic
fibers.*) 4:220
birefringence 4:258
cross section 4:224, 263
infrared spectrum 4:241
melting point 4:262
microscopy 4:264
mixtures, determination in 4:353
moisture regain 4:333
properties 4:222
solubility 4:247–252, 353
staining tests 4:252
Acrolein 4:148–159
absorption spectroscopy 9:564
in acrylonitrile 4:371
analysis 4:151
azeotropes 4:149
bromination method 4:151, 156
in cigarette smoke 10:27, 31
commercial grades and specifications 4:150
commercial specifications 8:340
gas chromatography 4:371; 10:28, 31; 16:108
in glycerol 13:527
hydroxylamine hydrochloride method 4:152,
157

impurities 4:151
impurities, determination of 4:154
infrared spectrum 4:153
manufacture 4:149
in methanol 16:108
polarography 4:153
properties 4:148
safety precautions 4:151
uses 4:148
Acrolein-2,7-dihydroxynaphthalene
colorimetric method for glycerol 13:501
Acrolein dimer
in acrolein 4:154, 157
Acrolein polymer
in acrolein 4:157
Acrylaldehyde. See *Acrolein.*
Acrylamide 4:160–168
acid–base titration, nonaqueous 4:164
analysis 4:163
bromate–bromide method 4:163, 175
chemical reactions 4:161
commercial grades and specifications 4:162
differential reaction rate analysis 4:167
identification 4:163
impurities, determination of 4:164
infrared spectrum 4:163
manufacture 4:162
mixtures, determination in 4:167
morpholine-addition method 4:164
polarography 4:164, 175, 337, 341
polymerization 4:168
properties 4:161
residual monomer in polymers 4:175, 337
separation 4:163
toxicity and handling precautions 4:162
ultraviolet spectroscopic method 4:176, 337
Acrylamide–acrylic acid polymers
analysis 4:171
carboxyl content 4:172
molecular weight 4:173
residual acrylamide monomer 4:175
viscosity 4:173
Acrylamide–ethylene–propylene terpolymer. (See
also *Ethylene and propylene polymers.*) 12:
345
manufacture 12:345
properties 12:345
Acrylamide polymers. (See also Polyacrylamide.)
4:168–181
analysis 4:171
carboxyl content 4:172
copolymers 4:171
fractionation 4:174
hydrolysis 4:173
molecular weight 4:173
polymer composition, determination of 4:171
preparation 4:168

properties, 4:169
refractive index 4:177
residual acrylamide monomer 4:175
terpolymers 4:171
viscosity, determination of 4:173
Acrylate elastomers. See Acrylate rubbers.
Acrylate resins
  in finishing agents 11:126
  solubility 11:126
Acrylate rubbers. (See also *Elastomers, synthetic.*)
    12:82
  commercial elastomers 12:86
  composition 12:82
  curing agents 12:82
  density 12:91
  infrared spectroscopy 12:94
  uses 12:82
Acrylates. See *Acrylic and methacrylic acids and
    esters.*
Acrylic acid. (See also *Acrylic and methacrylic
    acids and esters.*) 8:398
  acid–base titration 4:197, 341
  bromine–bromide method 4:192
  commercial grades and specifications 4:187
  dimer acid in 4:197
  freezing point 4:199
  manufacture 4:184
  melting point 4:199
  in methanol 16:114
  polymer in 4:200
  properties 4:182; 8:636
  solubility 4:181
Acrylic acid–acrylamide polymers. See Acrylamide-
    acrylic acid polymers.
Acrylic acid–acrylonitrile polymers. See Acrylo-
    nitrile–acrylic acid polymers.
Acrylic acid–polyethylene copolymers. (See also
    Ethylene and propylene polymers.) 12:345
  manufacture 12:345
Acrylic acid polymers
  acrylic acid content 4:275
Acrylic acid–polypropylene copolymers. (See
    also *Ethylene and propylene polymers.*)
    12:345
  manufacture 12:345
  properties 12:345
Acrylic and methacrylic acids and esters. (See
    also individual compounds.) 4:181–218
  acidity 4:197–201
  acrylate content of acrylamide 4:165
  acrylate content of alkyd resins by spectroscopy
    5:45
  in air 4:215
  analysis 4:187
  azeotropes 4:181, 187
  bromination methods 4:190
  commercial grades and specifications 4:187

dodecyl mercaptan addition method 4:192
gas chromatography 4:211
heating test 12:588
impurities, determination of 4:200
infrared spectroscopy 4:215
inhibitors, determination of 4:201
manufacture 4:182
mercuric acetate method 4:193
morpholine-addition method 4:194
neutralization equivalent 4:198
in photographic films 12:588
polarography 4:216
properties 4:181
properties, determination of 4:199
residual monomers in acrylic polymers 4:214
sampling and sample preparation 4:187
saponification method 4:196
sodium bisulfite addition method 4:195
solubility 4:181, 184
uninhibited monomers, preparation of 4:201
Acrylic and modacrylic fibers. (See also *Acrylo-
    nitrile polymers.*) 4:219–367
  additives, determination of 4:329, 343
  analysis 4:229
  basic dye capacity 4:328
  birefringence 4:257
  color properties, determination of 4:321
  composition 4:222
  consumption of 4:219, 222
  cross sections 4:224
  crystallinity 4:315
  end groups 4:227, 288
  heat, effect of 4:260
  identification 4:229
  impurities, determination of 4:332
  infrared spectroscopy 4:229
  manufacture 4:226
  microscopy 4:263
  mixtures, determination in 4:353
  molecular weight characterization 4:297
  polymer composition, determination of 4:267
  properties 4:222
  refractive index 4:257
  solubility 4:224, 247
  staining tests 4:252
  structure 4:316
  trade marks 4:219
  ultraviolet spectroscopy 4:267
Acrylic automotive lacquer. (See also *Methacry-
    late polymers.*)
  pyrolysis–gas chromatogram 16:89
Acrylic coatings. (See also *Methacrylate polymers.*)
  pyrolysis–gas chromatogram 16:88
Acrylic elastomers. See Acrylate rubbers.
Acrylic fibers. (See also *Acrylic and modacrylic
    fibers; Acrylonitrile polymers.*) 4:219, 383
  attenuated total reflectance 4:246

basic dye capacity 4:328
crystallinity 4:315
end groups, determination of 4:288
heat and flame test, identification by 4:261
infrared spectra 4:240 385
microscopy 4:264
mixtures, determination in 4:353
residual monomers in 4:333
structure 4:316
Acrylic monomers. See *Acrylic and methacrylic
    acids and esters;* and individual compounds.
Acrylic polymers. (See also *Acrylic and modacrylic
    fibers; Acrylonitrile polymers.*)
crystallinity 4:315
in embalming fluids 12:166
gas chromatography 4:286
infrared spectroscopy 4:238, 240
molecular weight characterization 4:297
osmotic pressure measurement 4:310
residual monomers 4:214, 333
structure 4:316
sulfur determination 4:280
viscosity 4:302
Acrylic resin teeth. (See also Methyl methacrylate
    dental materials.) 11:306
bonding strength 11:307
color stability 11:308
commercial grades and specifications 11:257,
    306
craze resistance 11:308
hardness 11:307
indentation resistance 11:308
specification tests 11:306
surface finishing 11:307
Acrylics
in chrome protectors 6:354
Acrylonitrile. (See also *Nitriles.*) 4:368–381;
    13:193
in acrylonitrile–butadiene–styrene polymer 18:
    334
in air 4:378; 16:408
in alkyd resins by infrared spectroscopy 5:45
analysis 4:370; 16:407
azeotropes 4:369; 16:403
bisulfite-addition method 4:341, 373
in Buna-N copolymers 16:411
commercial grades and specifications 4:370
composition of acrylic and modacrylic fibers
    4:219
derivatives for identification 16:402, 407
n-dodecyl mercaptan volumetric method 16:408
dodecyl mercaptan addition method 4:335
flammability 13:193
gas chromatography 4:212, 342, 371, 388; 12:
    105; 16:404; 18:335
identification 4:372; 16:407
impurities, determination of 4:375

infrared spectroscopy 4:372, 378; 16:411
iodometric methods 16:408
Kjeldahl volumetric method 18:334
manufacture 4:370
mercaptan-addition method 4:373, 380
modifier for alkyd resins 5:24
morpholine-addition method 4:374
morpholine volumetric method 16:409
paper chromatography 4:380
piperidine, reaction with 4:335
polarography 4:333, 374, 379
polymerization 4:226, 382
properties 4:368; 13:193; 16:396
properties, determination of 4:375
residual monomer in polymers 4:214, 333, 342,
    388
separation 4:372
sodium bisulfite volumetric method 16:409
solubility 4:368; 13:193
solvent extraction 18:335
in styrene–acrylonitrile copolymer 18:334
toxicity 4:368, 370; 13:193
ultraviolet spectroscopy 16:411
use in production of acrylamide 4:162
uses 4:368; 13:193
volumetric determination 16:407
Acrylonitrile–acrylic acid polymers. (See also
    *Acrylic and modacrylic fibers; Acrylonitrile
    polymers.*)
acrylic acid content 4:275
acrylic ester content 4:276
acrylonitrile content 4:270
infrared spectroscopy 4:238
Acrylonitrile–butadiene copolymers.
    (See also *Elastomers, synthetic; NBR.*) 12:83
commercial elastomers 12:86
density 12:91
identification 12:90
uses 12:83
Acrylonitrile–butadiene polymers. See Butadiene-
    acrylonitrile polymers.
Acrylonitrile–butadiene–styrene polymers. (See
    also Styrene–acrylonitrile–butadiene polymers;
    Styrene polymers.)
acrylonitrile content 18:334
analysis 18:334
composition, determination of 18:334
impurities, determination of 18:335
infrared spectroscopy 18:319
manufacture 18:314
properties 18:314
styrene content 18:334
Acrylonitrile–isoprene copolymer. See Isoprene-
    acrylonitrile copolymer.
Acrylonitrile–methyl methacrylate polymers. (See
    also *Acrylic and modacrylic fibers; Acrylonitrile
    polymers;* and Orlon.)

in acrylic fibers 4:223
infrared spectroscopy 4:238, 241, 245, 384
precipitation fractionation methods 4:299
Acrylonitrile-methylvinylpyridine polymers
  acrylonitrile content 4:270
  ultraviolet spectroscopy 4:267
Acrylonitrile-methylvinylpyridine-vinyl acetate
    polymers
  precipitation fractionation methods 4:299
Acrylonitrile-polyethylene copolymers. (See also
    Ethylene and propylene polymers.) 12: 345
  manufacture 12:345
  properties 12:345
  solubility 12:345
Acrylonitrile polymers. (See also *Acrylic and
    modacrylic fibers;* Polyacrylonitrile.) 4:382–
    390
  acid–base titrations 4:386
  acrylonitrile content 4:237, 270
  amine content 4:273
  analysis 4:383
  basic dye capacity of fibers 4:328
  colorimetric test 4:388
  end group determination 4:288
  fractionation 4:298, 389
  gas chromatography 4:286, 387
  infrared spectroscopy 4:230, 296, 384
  microgel, measurement of 4:320
  molecular weight 4:298, 388
  nitrogen, determination of 4:278, 384
  osmotic pressure measurements 4:310
  preparation 4:382
  properties 4:222, 382
  properties, determination of 4:390
  structure by polarized infrared 4:318
  ultraviolet spectroscopy 4:267
  uses 4:223, 383
  viscosity 4:389
Acrylonitrile-styrene polymers. See Styrene–
    acrylonitrile polymers.
Acrylonitrile–vinyl acetate polymers. (See also
    *Acrylic and modacrylic fibers; Acrylonitrile
    polymers.*)
  acetyl groups in 4:387
  in acrylic fibers 4:223
  acrylonitrile content 4:271
  infrared spectroscopy 4:240
  precipitation fractionation methods 4:299
  vinyl acetate content 4:274
Acrylonitrile–vinyl chloride polymers. (See also
    *Acrylic and modacrylic fibers; Acrylonitrile
    polymers;* and Dynel.)
  infrared spectroscopy 4:240, 244
  properties 4:222
  solubility 4:330
Acrylonitrile–vinyl pyridine polymers
  acrylonitrile content 4:270

residual vinyl pyridine monomers 4:341
structure by polarized infrared 4:318
ultraviolet spectroscopy 4:267
vinyl pyridine content 4:272
ACTH. See Corticotropin.
Actidione. See Cycloheximide.
Actinide fluorides
  carrier gas distillation fusion methods 8:636
  oxygen in 8:636, 637
Actinide metals
  nitrogen in 8:631
  oxygen in 8:631
Actinide series. (See also individual elements.)
    19:258
  manufacture 19:259
  uses 19:260
Actinomycin 5:468, 549
  analysis 5:553, 558, 562
  identification 5:553
  microbiological assay 5:558
  properties 5:550
  separation 5:553
  solubility 5:551
  structure 5:468
  uses 5:549
Actinomycin C 5:468, 549
  structures 5:468
Actinomycin D 5:549
  infrared spectrum 5:553
Actinomycins 5:468, 549
  analysis 5:553, 558, 562
  bioautography 5:553
  column chromatography 5:553
  countercurrent distribution 5:554
  identification 5:553
  microbiological assay 5:558
  paper chromatography 5:553
  properties 5:550
  separation 5:553
  solubility 5:551
  thin-layer chromatography 5:554
  uses 5:549
Activated carbon. See *Carbon, activated.*
Activation analysis. See Neutron activation
    analysis.
Active amyl alcohol. See Amyl alcohol, active.
Active dried baker's yeast. (See also *Yeast.*) 19:
    517
  manufacture 19:517
  properties 19:517
  species 19:516
  uses 19:516
Acylation methods
  for alcohols in essential oils 12:277
Additive content
  of polyurethanes 19:311
Adenine

in gelatin 12:552
ion exchange chromatography 12:552
ultraviolet spectroscopy 12:553
Adhesion. (See also *Adhesives.*) 4:391–398
of bituminous paving mixtures 6:341
of enamels 12:180
of match strikers 15:548
physical testing 4:392
Adhesives. (See also *Adhesion.*) 4:398–407
analysis 10:269; 15:535
ash content 10:276
binders in match strikers 15:535
in coated abrasives 10:281
evaluation of 4:398
foam content 10:275
foam stability 15:535
gel characteristics 10:273
gel time 10:274
grease content 10:275
identification 4:405
jelly strength 10:273
in match strikers 15:520, 535
moisture content 10:275
pH 10:274
physical testing 4:392
separation 4:405
strength tests 4:393
viscosity 10:269; 15:535
volatile content 10:275
water tolerance 10:274
Adipamic acid
properties 8:396
Adipamide 4:423, 427
Adipates. See Adipic acid esters; Adipic acid salts.
Adipic acid 4:408–423; 5:22; 8:400
in adipic acid 4:426
in alkyd resins 5:37
analysis 4:411
chemical properties 4:408, 410
commercial grades and specifications 4:411
derivatives for identification 4:412
freezing point depression method 4:412
gas chromatography 5:25; 17:290; 19:294
identification 4:412
impurities, determination of 4:413
infrared spectra 4:414
manufacture 4:410
mixtures, determination in 4:419
occurrence 4:410
paper chromatography 15:133
properties 4:408; 8:396
properties, determination of 4:413
purification 4:411
radioactive isotope method 4:419
separation 4:411
solubility 4:410

thin-layer chromatography 17:292
titration methods 4:412
toxicity 4:410
uses 4:408; 8:399
Adipic acid derivatives. (See also Adipamide; Adipic acid esters; Adipic acid salts; Adipic anhydride; Adiponitrile; Adipoyl chloride; Hexamethylenediamine.) 4:423
Adipic acid esters 4:423
analysis 4:426
commercial grades and specifications 4:425
gas chromatography 4:426
impurities, determination of 4:426
manufacture 4:425
mixtures, determination in 4:426
properties 4:424
properties, determination of 4:426
saponification method 4:426
Adipic acid salts 4:428
Adipic anhydride 4:428
Adiponitrile. (See also *Nitriles.*) 4:423, 428
gas chromatography 16:404
Adipoyl chloride 4:428
Adiprene 12:86
Adrenal hormones. See also Steroids; and individual compounds.) 14:183
analysis 14:216
composition 14:184
properties 14:193
sources 14:193
structure 14:184
uses 14:184, 193
Adrenalin. See Epinephrine.
Adrenocorticoids. (See also Steroids.)
analysis 14:216
fluorometric methods 14:219
gas chromatography 14:216
thin-layer chromatography 14:216
Adrenocorticotropic hormone. See Corticotropin.
Adrenotropic hormone. (See also Pituitary hormones.)
analysis 14:206
assay 14:206
Adsorbents and absorbents 4:431–452
chemical analysis 4:439
classification 4:433
composition, test of 4:435
efficiency 4:447
evaluation 4:435
filtration rate and oil retention 4:450
particle size 4:439
properties 4:434
properties, determination of 4:443
testing 4:432, 435
uses 4:433
Adsorption 4:431
Adsorption chromatography. See Column

chromatography; Gas chromatography; and Liquid chromatography.
Adsorption isotherm technique 8:145
Adsorption methods
  for organic matter in water 19:379
Adsorption theory of adhesion 4:391
Adsorptive capacity
  of activated carbon 8:142
  of gas-phase carbons 8:143
  of liquid-phase carbons 8:144
Aerating agents
  in confectionery products 10:551
Aerosol OT titration
  of cationic surfactants 11:79
Aerosols 4:452–461; 11:95
  air, determination of 4:460
  analysis 4:457; 11:96
  container safety test 11:98
  containers and valves 4:457
  for cosmetics 11:95
  net weight delivery 11:97
  nonfood production 4:452
  particle size, determination of 4:458
  physical tests 11:96
  pressure, determination of 4:457
  propellants in 11:95, 96, 98
  properties 4:453
  spray test 11:97
  water, determination of 4:460
Aesculetin. See 6,7-Dihydroxycoumarin.
Agar
  composition 12:583
  gravimetric determination 12:581
  identification 6:362
  infrared spectroscopy 12:583
  iodine colorimetric method 12:582
  in photographic films 12:582
Agar-based impression materials. See Dental impression materials, agar-based.
Agar diffusion method
  of microbiological assay 5:483
Agebest 1293–22A 6:72
Ageing tests
  for antioxidants 6:78
  for silicone rubbers 18:140
AgeRite Alba 6:72
  colorimetric method using diazotized *p*-nitroaniline 6:106
AgeRite DPPD 6:72
AgeRite Excel
  colorimetric method using diazotized *p*-nitroaniline 6:99
AgeRite Gel 6:72
  colorimetric method using diazotized *p*-nitroaniline 6:99
AgeRite Geltrol 6:72

AgeRite HP 6:72
  colorimetric method based on formation of Wurster salts 6:101, 103
  colorimetric method using diazotized *p*-nitroaniline 6:99
AgeRite Hipar 6:72
  colorimetric method based on formation of Wurster salts 6:101, 103
  colorimetric method using diazotized *p*-nitroaniline 6:99
AgeRite Iso 6:72
AgeRite Powder 6:72
  colorimetric method using diazotized *p*-nitroaniline 6:99
AgeRite Resin D 6:72
  in brake fluids 7:425, 428
  colorimetric method using diazotized *p*-nitroaniline 6:99
  color test 6:94
AgeRite Spar 6:72
  colorimetric method using diazotized *p*-nitroaniline 6:106
AgeRite Stalite 6:72
  column chromatography 6:89
  colorimetric method using diazotized *p*-nitroaniline 6:99
  color test 6:93
AgeRite Stalite S 6:72
AgeRite Superflex 6:72
AgeRite Superlite 6:72
  in brake fluids 7:428
  colorimetric method using diazotized *p*-nitroaniline 6:106
AgeRite Superlite Solid 6:72
AgeRite Syrup
  color test 6:95
AgeRite White 6:72
  colorimetric method based on formation of Wurster salts 6:101, 103
  colorimetric method using diazotized *p*-nitroaniline 6:98
AgeRite White D 6:72
Ageusia 17:612
Agglomerating index
  of coal and coke 10:246
Aggregate content
  of concrete 9:142
Aggregate gradation
  in concrete 9:142
Agitation haze
  in beer 7:520
AHEW
  of ethyleneamines 11:430
  phenyl glycidyl ether volumetric method 11:430
Air
  acrylonitrile content 4:378; 16:408

in aerosols 4:460
in beer 7:620
in beer package headspace 7:629
benzene in 7:18, 35, 39
beryllium in 7:138
boron hydrides in 7:349, 411
in carbonated beverages 8:166
in carbon dioxide 8:251
in carbon disulfide 8:274
carbon disulfide in 8:277
in carbon monoxide 8:310, 313
carbon monoxide in 8:319
carbon tetrachloride in 8:328
in cements 9:141
in chlorine 9:316
chlorine content 9:331
in concrete 9: 143
epichlorohydrin content 14:176
ethylene oxide content 12:223; 13:191
ethylene chlorohydrin content 14:155
ethylenimine content 14:478
fluoride content 13:26, 29
fluorine content 13:4
freeze-out method 4:461
gas chromatography 8:262, 263, 274; 12:324,
339; 14:328, 329, 393, 415; 16:425, 485
halogenated hydrocarbon content 13:182
hydrogen chloride content 14:395
hydrogen cyanide content 14:411, 412
hydrogen sulfide content 18:394
impurities, determination of 16:485
lead content 15:192
mercury content 16:22, 42, 44, 45, 46
methyl bromide content 13:183
nitric oxide content 16:484
nitrogen dioxide content 16:479, 484
nitrous oxide content 16:477
oxygen difluoride content 13:4
ozone content 16:484, 546
perchloroethylene absorption method 4:460
m-phenylenediamine content 11:449
in styrene 18:291
styrene content 18:311
sulfur dioxide content 18:408
sulfuryl fluoride content 13:195
tetraethyllead content 15:193
thallium content 18:524
2,4-toluenediamine content 11:455
triethylene glycol content 13:599
Air elutriation
particle size distribution of adsorbents 4:442
Air permeability
of felt 12:512
Akermanite 8:75
Akroflex 6:72
Akroflex C
colorimetric method based on formation of

Wurster salts 6:101, 103
colorimetric method using diazotized *p*-nitro-
aniline 6:99
Akroflex CD 6:72
colorimetric method based on formation of
Wurster salts 6:101, 103
colorimetric method using diazotized *p*-nitro-
aniline 6:99
Alabaster 14:88
Alanine. (See also *Carboxylic acids, amino.*) 8:
412, 445
analysis 8:450
in cocoa beans 9:556
commercial specifications 8:449
composition 8:281, 412
crystalline forms 8:447
enzymatic method 8:450
gas chromatography 8:429
identification 8:450
infrared spectra 8:448
ion exchange chromatography 8:415, 421, 450
manufacture 8:449
ninhydrin test 8:450
paper chromatography 8:422
potassium permanganate test 8:450
properties 8:446
in protein 8:414
solubility 8:447
structure 8:443, 447
in yeast 19:529
β-Alanine
gas chromatography 17:290
Albasan 6:72
colorimetric method using diazotized *p*-nitro-
aniline 6:99
Albite
composition 10:104
Albumen
in finishing agents 11:126, 132
lead acetate gravimetric method 11:127, 132
Millon's reagent gravimetric test 10:435
in roasted coffee 10:435
solubility 11:126
ALCA method. See Schorlemmer method.
Alcoholic beverages
potassium content 17:363, 390
Alcoholic beverages, distilled 4:462–494
analysis 4:468
definitions 4:462
quality control 4:466
Alcohol polyether sulfates
infrared spectroscopy 11:331
Alcohols. (See also Acetylenic alcohols; Alcohols,
higher; Alcohols, lower; Ethylenic alcohols;
Fatty alcohols; and individual compounds.)
4:495–586
acetylation method 8:360

acylation methods 12:277
analysis 6:5
antifreezes 6:1
in beer 7:541
by boiling point difference 7:549
in brake fluids 7:423, 424
in brandies 19:439
in carbonyl compounds 8:360
in carburetor cleaners 6:356
cooling system flushes 6:352
in creams and lotions 11:94
derivatives for identification 4:67, 516
for disinfectants and antiseptics 11:539
in essential oils 12:277
esterification procedures 4:529
gas chromatography 4:212, 214; 7:93; 18:75,
    77
nuclear magnetic resonance spectroscopy 19:
    297
in organosilane esters 18:75
phthalation method 8:360
qualitative tests 4:521
by specific gravity 7:544
use in manufacture of acrylates and meth-
    acrylates 4:186
windshield washer solvents 6:354
in wines 19:404, 425
Alcohols, higher
ultraviolet spectroscopy 4:545
Alcohols, higher, fatty. See Fatty alcohols.
Alcohols, higher, synthetic. (See also individual
    compounds.) 4:551
analysis 4:558
commercial grades 4:555
gas chromatography 4:524, 558
manufacture 4:552
phthalic anhydride method 4:535, 559
properties 4:551
uses 4:554
Alcohols, lower. (See also individual compounds.)
    4:496
acetylation methods 4:529
acid chloride method 4:537
analysis 4:516
azeotropes 4:498
bromination method 4:541, 542
column chromatography 4:528
commercial grades and specifications 4:512
in de-icers 6:356
derivatives for identification 4:516
3,5-dinitrobenzoyl chloride method 4:547
esterification procedures 4:529, 540
gas chromatography 4:523; 6:369
identification 4:516
impurities, determination of 4:543, 546
infrared spectroscopy 4:523, 524
manufacture 4:501

mass spectroscopy 4:526
mixtures, determination in 4:545
p-nitrobenzoyl chloride method 4:548
paper chromatography 4:527
phthalic anhydride acylation method 4:535
polarography 4:546
properties 4:496
properties, determination of 4:542
pyromellitic dianhydride acylation method
    4:535, 537
separation 4:516
solubility 4:497
stearic anhydride method 4:537
thin-layer chromatography 4:528
ultraviolet spectroscopy 4:545
uses 4:507
in windshield washer antifreezes 6:356
x-ray diffraction 4:526
Alcohols, primary
analysis 4:529
secondary alcohols in the presence of 4:548
use in preparation of carboxylic acids 8:372
Alcohols, secondary
analysis 4:529
primary alcohols in the presence of 4:548
Alcohols, tertiary
analysis 4:541
Alcohols, unsaturated. See Acetylenic alcohols;
    Ethylenic alcohols.
Alcoholysis
preparation of alkyd resins 5:23
Aldehydes. (See also Carbonyl compounds;
    Formaldehyde; Glutaraldehyde; and individual
    compounds.) 8:331
in acetone 4:114
in acetylenic chemicals 4:145
in acrylonitrile 4:375
in alcoholic beverages 4:484, 490
in alcohols 4:543
in amino resins 5:276
analysis 8:339; 11:566
assay 11:566
basic fuchsin method 4:484
in benzyl alcohol 5:379
in chloroform 5:359; 9:435
in cinnamyl alcohol 10:83
in cocoa beans 9:543
colorimetric determination 17:298
o-dianisidine colorimetric method 17:298
derivatives for identification 4:67, 111
dimethylcyclohexanedione turbidimetric reagent
    12:309
2,4-dinitrophenylhydrazine colorimetric method
    17:298
2,4-dinitrophenylhydrazone colorimetric method
    4:116
in disinfectants and antiseptics 11:565

in essential oils 12:280
in ethanol 4:544
in ethers 12:309
in ethylene 5:362
in ethylene oxide 12:222, 334
in ethyl ether 5:362; 12:309
gas chromatography 4:490; 8:354
in gelatin 12:550
hydroxylamine hydrochloride method 4:124, 145, 375
hydroxylamine hydrochloride volumetric method 10:83; 12:281; 18:306
identification 8:344
iodometric methods 12:334
iodometric titration 12:222, 226
ion exchange chromatography 4:116
in lactic acid 15:145
manufacture 8:339
in methanol 16:112, 114
Nessler's reagent color test 16:112
Nessler's reagent–sodium chloride turbidimetric method 12:309
nomenclature 8:331
paper chromatography 4:116
in polyamides 17:298
potassium hydroxide colorimetric method 12:309
properties 8:332
in propylene oxide 12:225
Schiff's reagent method 4:375
sodium bisulfite method 4:484
sodium bisulfite volumetric method 12:280
sodium metabisulfite volumetric method 12:334
sodium sulfite volumetric method 12:281
Stillman-Reed volumetric method 12:283
in styrene 18:291, 306
thin-layer chromatography 8:354
Tollens reagent color test 15:145
in 1,1,1-trichloroethane 9:443
ultraviolet spectroscopy 12:335
use in preparation of carboxylic acids 8:372
uses 8:332
in vanillin 19:351
in vinyl ether 5:367
Aldicarb. (See also *Pesticides and insecticides.*)
commercial grades and specifications 19:595
composition 19:599
manufacture 19:599
properties 19:590
solubility 19:587
structure 19:584
toxicity 19:592
Aldol
in crotonaldehyde 11:179
Aldol-α-naphthylamine 6:72
Aldosterone. (See also Steroids.)
analysis 14:219

radiochemical methods 14:219
structure 14:184
Aldridge method
color test for hydrogen cyanide 14:418
Aldrin. (See also *Pesticides and insecticides.*)
commercial grades and specifications 19:593
composition 19:588, 596
diazotized dinitroaniline colorimetric method 19:611
gas chromatography 19:389
manufacture 19:596
properties 19:588
in raw milk 16:129
solubility 19:587
structure 19:580
toxicity 19:592
in water 19:389
Alfol process
manufacture of higher alcohols 4:552
Algae meal
composition 10:530
Alginate
in beer 7:639
carbazole colorimetric method 12:584
composition 12:583
ferric hydroxide–sulfuric acid colorimetric method 9:589
ferric oxide–sulfuric acid colorimetric method 12:582
gravimetric determination 12:581
orcinol colorimetric method 12:584
in photographic films 12:582
Alginate-based impression materials. See Dental impression materials, alginate-based.
Alginate content
of cocoa products 9:589
Algoflon. See Polytetrafluoroethylene.
Aliphatic quaternary ammonium chlorides
infrared spectroscopy 11:331
Alite. See Tricalcium silicate.
Alizarin
colorimetric method for aluminum 11:140
colorimetric method for indium 14:526
Alizarin Blue-S
colorimetric method for boric acid 7:408
Alizarin cyanine green. See D & C Green No. 5.
Alizarine lakes
in enamels 12:182
identification 12:182
Alizarin fluorine blue
colorimetric method for fluoride 13:28
Alizarine red B. (See also *Pigments.*)
structure 17:182
Alizarin red S
colorimetric method for fluorine 15:227
colorimetric method for niobium 16:368, 375
colorimetric method for rare earths 19:19

colorimetric method for scandium 17:571
colorimetric method for tantalum 16:368
colorimetric method for zirconium 14:110,
    118; 15:396
Alizarin S. (See also Sodium alizarin sulfonate.)
colorimetric method for boric acid 7:386, 408
Alizarin sulfonic acid
colorimetric method for scandium 17:574
Alizarin–zirconium oxide
colorimetric method for fluorine 14:155
Alizurol purple. See D & C Violet No. 2.
Alkacid paper
color test for cyclohexylamine 11:227
Alkali barium silicate glass. (See also Silicate
    glasses.)
barium oxide content 9:218
emission spectroscopy 9:218
impurities, determination of 9:218
silica content 9:218
Alkali blue. (See also *Pigments.*)
in enamels 12:187
identification 12:187
structure 17:186
Alkali borates
flux for ceramics 9:216
Alkali carbonate
acid–base titration 15:402
in barium carbonate 6:544
flux for ceramics 9:216
in magnesium carbonate 15:402
Alkali chlorides
in calcium chloride 8:88
Alkali chromate
in barium chromate 6:547
Alkali content, combined
of soaps 18:188
Alkali fluidity
of starch and dextrin adhesives 4:404
Alkali fluorides
carrier gas distillation fusion methods 8:636
oxygen in 8:636, 637
Alkali, free
acid–base titration 18:200
in copper(II) oxide 10:674
in glycerol 13:524
in pesticides and insecticides 19:615
in potassium stannate 19:72
in soaps 18:200
in sodium fluoride 13:33
in sodium stannate 19:72
Alkali fusion
of platinum group metals 17:231
Alkali metals
in aluminum oxide 5:186
in cesium 9:283
in cesium alum 9:293
in polymers 4:24

in zinc chromate 9:707
Alkaline chromate
colorimetric method for chromium 9:646
Alkaline decomposition method
for acetylenic glycols 4:140
Alkaline earths
in aluminum chloride 5:174
in aluminum sulfate 5:193
in alums 5:193
in bismuth nitrate 7:202
in bismuth oxide 7:205
in bismuth subcarbonate 7:196
in bismuth subnitrate 7:203
in cadmium bromide 8:38
in cadmium chloride 8:42
in cobalt(II) acetate 10:330
in cobalt carbonate 10:333
in cobalt chloride 10:336
in cobalt nitrate 10:340
in cobalt(II) sulfate 10:343, 344
ion exchange chromatography 18:268
Alkaline ferricyanide
etchant for ceramics 9:224
Alkaline hydrolysis
of protein 8:411
Alkaline metal alkyls
use in preparation of carboxylic acids 8:372
Alkaline metal fluorides
carrier gas distillation fusion methods 8:636
Alkaline metal fluorides
oxygen in 8:636, 637
Alkalinity. (See also Basicity.)
of black tea ash 18:488
of cocoa beans 9:548, 549
of detergents 11:385
of glycerol 13:523
of glycols 13:587
of hypochlorite bleaches 9:354
of mercuric oxide 16:62
of methanol 16:112
of pigments 17:177
in polyols 19:278
of potassium hydroxide 18:250
of potassium iodide 14:605
of silicates 18:44
of sodium carbonate 18:245
of sodium hydroxide 18:250
of sodium iodide 14:605
of sodium tungstate 19:221
of water 19:361
of wine ash 19:489
Alkalis
in acid salts 8:595
in aluminates 5:159, 160
in aluminum chloride 5:174
in aluminum sulfate 5:193
in alums 5:193

in ammonium phosphate, dibasic 5:350
in asbestos 6:266
in barium hydroxide 6:551
in barium peroxide 6:554
in bismuth nitrate 7:202
in bismuth oxide 7:205
in bismuth subcarbonate 7:196
in bismuth subnitrate 7:203
in cadmium chloride 8:42
in calcium acetate 8:79
in chrysotile 6:266
in cobalt(II) acetate 10:330
in cobalt carbonate 10:333
in cobalt chloride 10:336
in cobalt nitrate 10:340
in cobalt(II) sulfate 10:343, 344
flame photometric method 6:266
in iodoform 14:615
ion exchange chromatography 18:268
in methycellulose 9:80
in phenolic resin adhesives 4:402
in potassium stannate 19:72
in sodium stannate 19:72
in tetrachloroethylene 9:464
in trichloroethylene 9:457
Alkali salts
in ammonium chromate 9:698
in ammonium dichromate 9:698
in calcium lactate 15:144
in chromium compounds 9:698
in strontium nitrate 18:284
in zinc chromates 17:199
Alkali silicate glasses. (See also Silicate
    glasses.) 9:197
Alkaloids. (See also individual compounds.)
    4:587–618
  acid–base titration 16:257, 273; 19:135
  analysis 4:592
  in black tea 18:488
  classification 4:587
  in coca leaves 16:273
  colorimetric methods 8:66
  cyanogen bromide colorimetric method 19:141
  ion exchange chromatography 8:62
  solvent extraction 19:135
  in tobacco 19:135, 141
Alkanes.
  (See also *Hydrocarbons;* and individual
    compounds.) 14:299
  commercial specifications 14:303
  occurrence 14:303
  uses 14:303
Alkanethiols. See *Mercaptans.*
Alkanoic acids 8:369
Alkanolamines. (See also individual compounds.)
    5:1–21
  acid–base titration 5:11

amine function determination 5:11
analysis 5:4
azeotropes 5:3
in brake fluids 7:427
colorimetric method using methyl orange 5:392
in creams and lotions 11:94
commercial grades and specifications 5:3
gas chromatography 5:4
impurities, determination of 5:13
infrared spectroscopy 5:7
ion exchange chromatography 11:82
manufacture 5:1
mixtures, determination of 5:15
nitrogen content of 5:8
properties 5:1
properties, determination of 5:13
in shampoos 11:82
Alkarb
  composition 17:547
Alkenes. See Olefins.
Alkoxide content
  of aluminum alkoxides 5:165
  of aluminum alkyls 5:171
Alkoxy content
  of organofunctional silanes 18:85
  of organosilane esters 18:75
  of silicone resins 18:123
  of starch derivatives 11:29
Alkoxy determination
  for allyl alcohol 5:86
Alkoxyl groups
  hydriodic acid volumetric method 16:92
  iodometric method 18:75
  in methacrylate polymers 16:92
  transesterification 18:77
  Zeisel method 18:75
Alkyd resins 5:22–56
  analysis 5:25
  benzoic acid in 7:67
  in chrome protectors 6:354
  composition 5:22
  composition, determination of 5:25
  in finishing agents 11:126
  gas chromatography 5:25
  impurities, determination of 5:50
  infrared spectroscopy 5:40
  ion exchange chromatography 5:48
  Kappelmeier separation 5:33
  manufacture 5:23
  nuclear magnetic resonance spectroscopy 5:50
  paper chromatography 5:49
  phenol colorimetric reagent 11:127
  polarography 5:49
  properties 5:22
  properties, determination of 5:52
  solubility 11:126
  thin-layer chromatography 5:49

ultraviolet spectroscopy 5:50
Alkylamines
  acetylation method 5:270
  acid–base titration 5:268, 271
  analysis 5:264
  azeotropes 5:261
  commercial grades and specifications 5:264
  derivatives for identification 5:264
  gas chromatography 5:267
  identification 5:264
  manufacture 5:262
  mixtures, determination of 5:271
  properties 5:260
  separation 5:264
  solubility 5:261
  toxicity 5:262
  uses 5:263
Alkylaryl polyether alcohol content
  of shampoos 11:82
Alkylaryl sulfates
  paper chromatography 12:575
  ultraviolet spectroscopy 12:575
Alkylaryl sulfonates. (See also Anionic
    detergents.) 11:322
  biodegradability 11:399
  infrared spectroscopy 12:576
  manufacture 11:323
  solvent extraction 11:345
  ultraviolet spectroscopy 11:343
  uses 11:323
Alkylbenzenes. (See also Hydrocarbons; and
    individual compounds.) 14:324
  commercial grades and specifications 14:324
  gas chromatography 14:334
  in styrene 18:291
  uses 14:324
Alkylbenzene sulfonates
  infrared spectroscopy 11:330
Alkylbenzylammonium halides
  ultraviolet spectroscopy 12:575
Alkylbismuth halides 7:208
Alkyl content
  of aluminum alkyls 5:168
Alkylene oxides. (See also Epoxy compounds.)
  use in manufacture of alkanolamines 5:1
Alkyl halides. See Chlorocarbons; C$_1$, C$_2$, C$_3$, and
    C$_4$ Chlorinated hydrocarbons; and individual
    compounds.
N-Alkylhydroxylamines. (See also Hydroxylamines
    and their salts; and individual compounds.)
  acid–base titration, nonaqueous 14:449
  assay 14:449
  colorimetric determination 14:453
  1,2-dinitrobenzene colorimetric method 14:453
  manufacture 14:444
  properties 14:442
  uses 14:443

volumetric determination 14:449
O-Alkylhydroxylamines. (See also Hydroxyl-
    amines and their salts.)
  manufacture 14:444
  properties 14:443
  uses 14:443
Alkyl iodides
  aniline–sodium methoxide volumetric method
    16:93
  gas chromatography 16:92
  iodometric titration 16:93
  thin-layer chromatography 16:93
Alkylnaphthalenes. (See also individual compounds.)
    16:216–219
  analysis 16:216
  gas chromatography 16:216
  impurities, determination of 16:219
  occurrence 16:216
Alkyl nitriles
  use in preparation of carboxylic acids 8:372
Alkyl nitrite method
  for higher alcohols 4:545
Alkylolamines
  infrared spectroscopy 12:576
Alkylphenols. (See also Phenols; and individual
    compounds.)
  commercial grades and specifications 17:15
  manufacture 17:10
  properties 17:4
Alkylpyridines. (See also Pyridine and pyridine
    derivatives.)
  manufacture 17:420
N-Alkyl pyridinium halides
  paper chromatography 12:575
O-Alkyl saccharin derivative
  of higher alcohols 4:559
  of lower alcohols 4:519
Alkyl sulfates
  in creams and lotions 11:93
  hydrolysis 11:83
  infrared spectroscopy 12:576
  ion exchange chromatography 11:85
  methylene blue colorimetric method 11:93
  paper chromatography 12:575
  in shampoos 11:83
Alkyl sulfonates
  infrared spectroscopy 12:576
Alkyltrimethyl pyridinium halides
  paper chromatography 12:575
Allanite 17:470
Allene
  in cyclopropane 5:361
Allobarbital. See Diallylbarbituric acid.
Allophanates
  infrared spectroscopy 15:94
Allotetrahydrocortisol
  gas chromatography 14:218

Alloxan
  colorimetric method for thiophene 18:541
Alloys. (See also Alloys, nonferrous; Ferro alloys;
    Ferrous alloys; and individual metals.) 5:57–
    75
  aluminum in 5:200, 204, 247
  analysis 5:72
  antimony in 6:57
  boron in 7:407, 410, 417, 418
  chromium content 9:652
  definition 5:57
  sample preparation 5:72; 6:236; 7:210
  sodium peroxide fusion 7:400
Alloys, acid-insoluble
  aluminum in 5:204
Alloys, acid-soluble
  aluminum in 5:200
Alloys, dental. See Dental alloys.
Alloys, ferrous. See Ferro alloys; Ferrous alloys.
Alloys, nonferrous 5:57, 65
Alloy steels. (See also *Steel.*)
  aluminum in 5:227
  composition 5:59
All-purpose detergents. (See also *Detergents.*)
    11:325
  composition 11:325
Allyl acetate
  boiling point 5:93
  bromate–bromide method 5:97
  residual monomer in polymer 5:97
Allylacetone 5:106
N-Allylacrylamide 5:106
Allyl alcohol. (See also Ethylenic alcohols.)
    13:193
  acetylation 4:530; 5:78
  in acrolein 4:154
  alkoxy determination 5:86
  analysis 5:78
  azeotropes 5:77
  bromination methods 5:82
  coulometric titration 5:80
  gas chromatography 4:154, 578; 5:79
  impurities, determination of 5:86
  iodine monochloride method 5:84
  manufacture 4:575; 5:76
  mercurimetric methods 5:85
  oxidation methods 5:86
  potentiometric titration with *N*-bromosuccinimide
    5:82
  properties 4:576; 5:77; 13:193
  pyromellitic dianhydride method 4:537
  solubility 5:76; 13:193
  specifications 4:577
  sulfonation method 5:86
  toxicity 5:78; 13:193
  uses 4:576; 13:193
Allyl alcohols. (See also Allyl alcohol; *Allyl*

compounds; and Methallyl alcohol.) 5:76
  analysis 5:78
Allyl (*N*-allyl)carbamate 5:106
Allyl (2-allyloxy)propionate 5:106
Allylamine
  acid–base titration 5:89
  derivatives for identification 5:87
  gas chromatography 5:88
  identification 5:87
  paper chromatography 5:89
  potentiometric titration with *N*-bromosuccinimide
    5:89
  properties 5:87, 260
  uses 5:86
Allylamines. (See also *Allyl compounds;* and
    individual compounds.) 5:86
  acid–base titration 5:88
  analysis 5:87
  colorimetric method using dipicrylamine 5:89
  gas chromatography 5:88
  identification 5:87
  manufacture 5:88, 262
  mixtures, determination of 5:88
  paper chromatography 5:89
  properties 5:87
  uses 5:263
N-Allylaniline 5:87
Allyl *p*-anisate
  bromination methods 5:96
Allyl aromatic compounds. (See also *Allyl com-
    pounds;* and individual compounds.) 5:90
  analysis 5:90
  boiling points 5:90
  bromination methods 5:90
  melting points 5:90
  mercurimetric methods 5:90
Allylbarbituric acids. (See also *Allyl compounds;*
    and individual compounds.) 5:91; 6:468
  analysis 5:92
  bromide–bromate titration 6:497
  N-bromosuccinimide titration 5:92
  derivatives for identification 6:477
  gas chromatography 6:483
  melting points 5:91
  properties 6:470
  solubility 6:476
  thin-layer chromatography 6:481
Allylbenzene
  analysis 5:90
  boiling point 5:90
  bromination methods 5:90
  mercurimetric methods 5:91
Allyl bromide 5:101
5-Allyl-5-*sec*-butylbarbituric acid 5:91
Allyl carbamate 5:106
Allyl chloride. (See also 3-Chloropropene.) 9:468
  azeotropes 9:468

Beilstein test 9:382
colorimetric method using phloroglucinol 5:102
composition 9:411
derivatives for identification 5:101
gas chromatography 5:102
infrared spectroscopy 5:102; 9:394
manufacture 9:468
properties 5:101; 9:468
solubility 9:468
toxicity 9:468
uses 9:468
Allyl compounds 5:75–109
  analysis 5:76
  bromination methods 5:82
  iodine monochloride method 5:84
  mercurimetric methods 5:85
  occurrence 5:75
  oxidation methods 5:86
  thiocyanation methods 5:85
  uses 5:75
  Wijs method 5:84
Allyl crotonate
  boiling point 5:93
  dodecyl mercaptan addition method 5:97
Allyl cyanide
  analysis 5:92
  commercial specifications 5:92
  properties 5:92
Allyl cyanides. (See also *Allyl compounds;*
    Allyl cyanide; and Methallyl cyanide.) 5:92
Allyl cyanurates. (See also *Allyl compounds.*)
    5:93
Allyl cyclohexanepropionate
  saponification method 5:95
Allylcyclopentenyl barbituric acid. See Cyclopal.
N-Allyldiisopropylamine 5:87
Allyl esters. (See also *Allyl compounds;* and
    individual compounds.) 5:93
  analysis 5:94
  boiling points 5:93
  bromate–bromide method 5:97
  bromination methods 5:82
  dodecyl mercaptan addition method 5:97
  halogenation methods 5:96
  mercurimetric methods 5:85, 96
  polarography 5:98
  residual monomer in polymers 5:97
  saponification methods 5:94
  uses 5:93
Allyl ethers. (See also *Allyl compounds; Ethers;*
    and individual compounds.) 5:98
  analysis 5:98
  boiling points 5:98
  bromination method 5:99
  gas chromatography 5:100
  halogenation methods 5:98
  mercurimetric methods 5:85, 100

Allyl ethyl ether. (See also *Ethers.*)
  properties 12:297
Allyl fluoride 5:101
Allyl glycidyl ether. (See also *Ethers.*)
  properties 12:297
Allyl halides. (See also *Allyl compounds;* and
    individual compounds.) 5:101
  gas chromatography 5:102
  halogenation methods 5:102
  hexamethylenetetraamine method 5:102
  identification 5:101
Allylidene diacetate 5:106
Allyl iodide. (See also Iodine compounds, organic.)
    5:101
  properties 14:612
5-Allyl-5-isobutylbarbituric acid
  analysis 5:92
  melting point 5:91
Allyl isocyanate. (See also *Allyl compounds.*) 5:103
5-Allyl-5-isopropylbarbituric acid
  analysis 5:92
  melting point 5:91
  paper chromatography 6:479
  thin-layer chromatography 6:480
Allyl isothiocyanate
  analysis 5:105
  boiling point 5:104
Allyl mercaptan. (See also *Mercaptans.*)
  occurrence 15:551
Allylmercaptomethylpenicillin. See Penicillin O.
N-Allylmethacrylamide 5:106
Allyl methacrylate
  boiling point 5:93
  polarography 5:98
  residual monomer in polymer 5:98
  thin-layer chromatography 16:91
5-Allyl-5-(1-methylbutyl)barbituric acid 5:91
N-Allylmorphine. (See also Morphine.)
  potassium ferricyanide fluorometric method
    16:263
2-Allylnaphthalene
  analysis 5:90
Allylnaphthols. See Allyl aromatic compounds.
N-Allylnormorphine. See Nalorphine.
Allyl olefins. (See also *Allyl compounds;* and
    individual compounds.) 5:103
  analysis 5:103
  gas chromatography 5:103
  halogenation methods 5:104
  properties 5:103
2-Allyloxypyridine 5:87, 106
2-Allylphenol. (See also *Phenols.*)
  properties 17:4
4-Allylphenol. (See also *Phenols.*)
  properties 17:4
Allylphenols. See Allyl aromatic compounds;
    Phenols; and individual compounds.

2-Allylphenyl methacrylate 5:106
N-Allylpiperidine 5:87
Allyl propionate
  boiling point 5:93
  bromination methods 5:96
Allyl propyl disulfide
  boiling point 5:104
  gas chromatography 5:105
Allyl starch
  halogenation methods 5:96, 99
Allyl sucrose
  halogenation methods 5:96, 99
Allyl sulfides
  gas chromatography 5:105
  halogenation methods 5:105
Allyl sulfur compounds. (See also *Allyl com-*
    pounds; Allyl sulfides; and individual com-
    pounds.) 5:104
  analysis 5:105
  boiling points 5:104
  gas chromatography 5:105
Allylvinyl. See Divinylmethane.
Almond oil, bitter. (See also *Essential oils.*)
  benzaldehyde content 12:282
  refractive index 12:271
Alnico alloys. (See also Cobalt alloys.) 10:285
  composition 10:285
Alpha olefins. (See also *Hydrocarbons;* and in-
    dividual compounds.) 14:313
  commercial grades and specifications 14:314,
    315
  manufacture 14:313
  properties 14:313
  uses 14:313
Alpha-particle spectroscopy
  of radium 17:453, 456, 459, 462
Alphaprodine. (See also *Narcotics.*)
  ammonium vanadate colorimetric method 16:
    246
  color tests 16:246
  formaldehyde–sulfuric acid colorimetric method
    16:246
  gas chromatography 16:255
  microchemical tests 16:249
  picric acid microchemical method 16:249
  structure 16:245
  ultraviolet spectroscopy 16:251
Alphaprodine hydrochloride. (See also *Narcotics.*)
  ultraviolet spectroscopy 16:251
Alphasem
  extraction 6:491
Alphazurine. See D & C Blue No. 4.
Alphenal
  gas chromatography 6:483
Alsterberg titration
  of oxygen 16:525
Alterungsschutzmittel AH (AP, DDA, DNP, DOD,

    PAN, PBN, and 4010) 6:72
Alum
  color test for natural coloring materials 10:537
Alumel. (See also Nickel alloys.)
  composition 16:290
Alumina. (See also Aluminum oxide.) 5:183
  analysis 5:116
  as adsorbent 4:434
  calcium in 5:116
  chromium in 5:116
  commercial specifications 10:534
  composition 10:534
  fluorine in 5:116
  iron in 5:116
  manganese in 5:116
  oxygen content of 5:116
  phosphorus in 5:116
  in pollucite 9:295
  properties 9:211
  silica in 5:116
  silicon in 5:116
  sodium in 5:116
  sulfate in 5:116
  titanium in 5:116
  in titanium dioxide pigments 19:125
  zinc in 5:116
  zinc sulfate–EDTA volumetric method 19:125
Alumina, activated 5:183
Alumina, adsorbent 5:183
Alumina, calcined 5:183
  analysis 5:116
Alumina catalysts 5:183
Alumina cream. (See also Aluminum hydroxide.)
  clarifying reagent for sugar 18:345
Alumina, fused 5:184
Alumina gel. (See also Aluminum hydroxide.)
  5:182
Alumina, sintered 5:184
Aluminates 5:158
  aluminum content of 5:159
  analysis 5:159
  impurities, determination of 5:160
  sample preparation 6:59, 236
Aluminoborosilicate glass. (See also Silicate
    glasses.)
  composition 9:199
  uses 9:199
Aluminon
  colorimetric method for aluminum 7:652; 10:
    618; 11:496
  colorimetric method for tantalum 16:368
  color reagent for boric acid 7:388
  reagent for identification of aluminum 5:212,
    223
Aluminosilicate. See Aluminum silicate.
Aluminosilicate glass. (See also *Glass and glass-*
    ceramics; Silicate glasses.) 9:200

composition 9:198; 13:407, 409,
    410
uses 9:198; 13:410
Aluminothermic process
  for chromium metal 9:634
Aluminous cements. (See also Hydraulic
    cements.)
  hardening 9:97
Aluminum. (See also Aluminum alloys;
    Aluminum metal.) 5:110–258
acid–base titration 5:217, 219
in acid-insoluble alloys 5:204
in acid-soluble alloys 5:200
alizarin colorimetric method 11:140
in alloys 5:247
in alloy steels 5:227
in aluminates 5:159
aluminon colorimetric method 5:223; 7:652;
    10:618; 11:496
aluminon test 5:212
in aluminum acetate 5:162
in aluminum alkoxides 5:164
in aluminum alkyls 5:168
in aluminum ammonium sulfate 5:254
in aluminum and aluminum alloys 5:110, 114
in aluminum fluoride 5:178
in aluminum hydroxide 5:185
in aluminum nitrate 5:181
in aluminum oxide 5:185
in aluminum silicate 5:189
in aluminum sulfate 5:192
in alums 5:192
ammonium benzoate–8-hydroxyquinoline
    gravimetric method 15:376
ammonium chloride gravimetric method
    16:323
ammonium hydroxide gravimetric method
    9:665; 10:307, 632; 12:151
amperometric titration 5:251
in antimony–aluminum–gallium alloys 13:269
in antimony metal 5:225
in asbestos 6:263, 267
in asbestos–cement products 6:270
atomic absorption spectroscopy 5:212, 247;
    9:346; 15:196
in bakery chemicals 6:432
in bauxite 5:115
in beer 7:650, 652
benzoate, precipitation as 5:216
bismuth in 7:215, 216
in boiler water condensates 5:241
borax bead test 6:584
in bronzes 5:239, 246, 248
in calcium metal 8:76
in carbon 14:66, 74
in cesium alum 9:292
in cesium sulfate 9:293

in chrome alum 12:584
in chrome ores 9:701, 704
in chromic acid 9:699
in chromium 9:635
in chromium alloys 9:663, 665
in chromium compounds 9:699
in chromium–molybdenum–aluminum steel 5:
    248
in chrysotile 6:263, 267
in clays and clay minerals 10:136
in cobalt alloys 10:307
in cobalt compounds 10:327
in cobalt metal 10:293
in cocoa products 9:595
colorimetric methods 5:223; 15:376
in copper alloys 5:228; 10:632
in copper metal 10:618
in cotton 11:140, 152, 157
cupferron extraction 5:204; 15:286
1,2-cyclohexylenedinitrilotetraacetic acid
    volumetric method 5:218
in diatomaceous earth 11:490, 496
diethyldithiocarbamate extraction 5:205
displacement of copper 10:604
EDTA volumetric method 5:218, 221; 6:432;
    12:154; 13:269; 14:141, 257
in elastomers, synthetic 12:151
in electrical resistance alloys 10:290
electrolytic separation 5:200
emission spectroscopy 4:3; 5:211, 242; 7:650;
    8:76; 9:346; 11:152, 493; 13:277; 14:34, 48,
    74, 122, 540; 15:155, 375; 16:160, 387
in enamels 12:173
Eriochrome Cyanine R colorimetric method
    5:226
Eriochrome Cyanine R test 5:213
in explosive priming mixtures 12:418, 430
extraction procedures 5:204
in ferrochromium 9:661
in ferroniobium 16:387
flame photometry 5:212, 244; 15:287
flame tests 6:589
fluorometry 5:236
in gallium arsenide 13:276
gallium content 13:262
in gallium metal 13:266, 267
gas chromatography 13:264
in gold 14:45
in gold alloys 14:34
in graphite 14:66, 74
gravimetric methods 5:214; 12:430; 15:375
in hafnium 14:122
hematoxylin test 5:213
in high-temperature alloys 10:286
4-hydroxy-3-nitrophenyl phosphoric acid
    volumetric method 5:218
8-hydroxyquinolate, precipitation as 5:217

8-hydroxyquinoline colorimetric method 5:228; 10:293; 15:377
8-hydroxyquinoline extraction 5:204
8-hydroxyquinoline gravimetric method 5:217; 9:664, 704; 14:256; 16:324
8-hydroxyquinoline volumetric method 5:222
identification 5:161, 199, 211
in indium 14:540
in indium alloys 14:553
ion exchange chromatography 13:254; 14:113, 140
ion exchange separation 5:208, 230
in iron 15:59
in iron alloys 5:228
in iron ores 5:228
in lanthanum oxide 15:155
in lead alloys 5:226; 15:195
in lead metal 5:225
in lithium aluminum hydride 14:256
in lithium hydroxide 15:286
in magnesium 15:375
in magnesium alloys 5:245; 15:375
microcosmic salt bead test 6:587
in minerals 5:239, 246
in molybdenum 16:161
in molybdenum trioxide 16:161
morin colorimetric method 11:140
morin fluorometric method 5:237, 240
in nickel 16:323, 324
in nickel alloys 5:228; 16:323, 324
in niobium 16:387
occurrence 5:110
oxide, precipitation as 5:215
in plutonium alloys 5:230
polarography 5:250
Pontachrome Blue Black R fluorometric method 5:236, 238; 15:376
Pontachrome Blue Black R test 5:214
Pontachrome Violet SW fluorometric method 5:237, 240
Pontachrome Violet SW polarographic method 5:250
Pontachrome Violet SW voltammetric method 5:251, 254
in potash alum 12:584
potassium fluoride volumetric method 5:217
precipitation methods 5:207, 214
powder, analysis of 5:152; 17:203
powder, commercial specifications 17:203
powder impurities, determination of 17:203
powder, oxygen content of 5:153
pyrocatecholsulfonphthalein colorimetric method 7:652
quinalizarin colorimetric method 11:140
2-quinizarin sulfonic acid colorimetric method 14:66
8-quinolinol colorimetric reagent 13:267

8-quinolinol gravimetric reagent 9:292
in rare earth fluorides 17:508
N-salicylidene-2-amino-3-hydroxyfluorene fluorometric method 5:237, 241
salicylidene-o-aminophenol fluorometric method 5:237
scandium content 17:577
separation 5:199
separation from gallium and indium 5:210
separation from plutonium 5:230
separation from zirconium 5:208
in silica ablators 4:3
in silicon 18:39
in silicon carbide abrasives 4:56
sodium alizarin sulfonate colorimetric method 5:233
sodium alizarin sulfonate test 5:214
sodium ammonium phosphate gravimetric method 14:554
in sodium chloride 9:346
in sodium dichromate 9:684
in sodium fluoaluminate 5:178
sodium fluoride volumetric method 5:217
in solders 5:226
in spelters 5:228
in steels 5:227, 239, 246, 248
5-sulfo-4'-diethylamino-2,2'-dihydroxyazobenzene colorimetric method 5:235
Superchrome Garnet Y polarographic method 5:250, 251
in tanning compounds 9:690
thioacetamide separation 5:207
in thorium compounds 5:250, 251
in tin alloys 5:226
in tin metal 5:226; 19:66, 67
in titanium metal 19:120
in trialkylaluminums 5:249
tri-n-octylphosphine oxide extraction 5:206
voltammetry 5:251, 254
volumetric methods 5:217; 15:376
x-ray fluorescence spectroscopy 10:136; 11:157
x-ray diffraction 12:418
in zinc alloys 5:246, 248, 253
in zircon 14:139
in zirconium 14:122
zirconium content 14:120
Aluminum acetate 5:160
acetate content of 5:162
aluminum content of 5:162
analysis 5:161
commercial grades and specifications 5:161
heavy metals in 5:163
manufacture 5:161
properties 5:160
uses 5:160
Aluminum alkoxides 5:163
alkoxide content of 5:165

aluminum content of 5:164
analysis 5:164
formation for identification of formals 4:68
properties 5:164
uses 5:163
Aluminum alkyls 5:165
    alkyl content of 5:168
    aluminum content of 5:168, 249
    analysis 5:167
    chloride in 5:171
    commercial specifications 5:167
    hydride content of 5:171
    hydrolysis 5:168
    manufacture 5:166
    properties 5:166
    sampling 5:167
    titanium in 5:171
    uses 5:165
Aluminum–alizarin
    colorimetric method for mercury 16:19, 21
Aluminum alloy corrosion test
    of trichloroethane 9:443
Aluminum alloys. (See also Alumina; Aluminum
        metal; Aluminum powder; Bauxite.) 5:65,
        111, 112
    aluminum content of 5:110, 114, 248
    analysis 5:114
    antimony in 5:118
    beryllium in 5:118
    bismuth in 5:121; 7:215, 216
    boron in 5:122; 7:410, 418
    cadmium in 5:123
    calcium in 5:124
    carrier gas fusion methods 8:682
    cerium in 5:125
    chromium in 5:125; 9:654
    cobalt in 5:127
    composition of 5:67, 111, 112
    copper in 5:128; 10:617
    gallium in 5:129
    identification 5:117
    indium in 5:131
    iron in 5:131
    lead in 5:133
    lithium in 5:134
    magnesium in 5:135; 15:371
    manganese in 5:136; 15:462, 477
    molybdenum in 5:138
    neodymium in 5:152
    nickel in 5:138; 16:310, 319
    niobium in 5:139
    oxygen in 5:140; 8:682
    potassium in 5:152
    praseodymium in 5:152
    properties 5:110, 113, 114
    samarium in 5:152
    sampling 5:117
    silicon in 5:141
    silver in 5:144
    sodium in 5:144
    sulfur in 5:144
    thallium in 5:152
    thorium in 5:152
    tin in 5:144; 19:53, 64
    titanium in 5:145; 19:115, 119, 120
    tungsten in 5:152
    uses 5:110
    vanadium in 5:147; 19:339, 342
    zinc in 5:148
    zirconium in 5:151
Aluminum amalgam
    gravimetric reagent for tellurium 17:591
Aluminum ammonium sulfate 5:191
    aluminum content of 5:254
    analysis 5:192
    commercial grades and specifications 5:192
    manufacture 5:191
    properties 5:191
    uses 5:191
Aluminum antimonide 5:194
    analysis 5:195
Aluminum antimonide semiconductors. (See also
        Ceramics.)
    properties 9:213
    uses 9:213
Aluminum benzylpenicillin. (See also Benzyl-
        penicillin.)
    iodometric methods 16:562
    structure 16:567
Aluminum bromide 5:171
    analysis 5:172
    commercial specifications 8:9
    properties 5:171; 8:9
Aluminum bronze 5:68
Aluminum t-butoxide. (See also Aluminum
        alkoxides.) 5:163
Aluminum chloride 5:172
    analysis 5:173
    chloride content by Volhard titration 5:174
    commercial grades and specifications 5:173
    condensing capacity 5:176
    impurities, determination of 5:174
    manufacture 5:172
    properties 5:172; 9:334
    sample preparation 5:173
    solubility 9:336
    uses 5:172
Aluminum compounds. (See also individual com-
        pounds.) 5:158–199
Aluminum distearate
    in enamels 12:176
Aluminum enamel. (See also Enamels.)
    composition 9:195
Aluminum ethoxide. (See also Aluminum

alkoxides.) 5:164
Aluminum flake
  cooling system sealer 6:353
Aluminum fluoride 5:177
  analysis 5:178
  assay methods 5:178
  carrier gas distillation fusion methods 8:636
  commercial grades 5:177
  impurities, determination of 5:180
  manufacture 5:177
  properties 5:177; 13:11
  uses 5:177
Aluminum formate
  uses 13:120, 135
Aluminum hydroxide 5:181
  acid-consuming capacity 5:188
  analysis 5:184
  assay methods 5:185
  commercial grades and specifications 5:182
  impurities, determination of 5:188, 189
  properties 5:182
  sample preparation 5:184
  uses 5:182
Aluminum hydroxychloride
  in deodorants 11:88
Aluminum iodide 5:180
Aluminum isopropoxide. (See also Aluminum
      alkoxides.) 5:163
Aluminum lactate. (See also Lactic acid salts.)
  uses 15:130
Aluminum metal. (See also Aluminum alloys.)
  in aluminum chloride 5:174
  aluminum content of 5:110, 114
  analysis 5:114
  arsenic in 5:118
  barium in 5:118
  calcium in 5:124
  carbon in 5:124
  carrier gas fusion methods 8:624, 626, 682
  cerium in 5:125
  chromium in 5:125
  composition 5:111
  copper in 5:128; 10:617
  dysprosium in 5:129
  gadolinium in 5:152
  gallium in 5:129
  gold in 5:152
  hot extraction methods 8:684
  hydrogen in 8:613, 616, 617, 618, 619, 638,
      681, 684
  iron in 5:131
  lanthanum in 5:133
  lead in 5:133
  lithium in 5:134
  magnesium in 5:135
  manganese in 5:136
  manufacture 5:111

  mercury in 5:152
  neodymium in 5:152
  nickel in 5:138
  niobium in 5:139, 152
  nitrogen in 5:140; 8:616, 618
  occurrence 5:110
  oxygen in 8:616, 617, 618, 619, 624, 626, 681,
      682
  potassium in 5:152
  praseodymium in 5:152
  properties 5:110
  samarium in 5:152
  sampling 5:117
  scandium in 5:141
  silicon in 5:141; 18:34
  silver content 18:172
  sodium in 5:144; 18:218, 226
  solubility in gold 14:9
  strontium in 5:152
  sulfur in 5:144; 18:385
  tantalum in 5:152
  tellurium in 5:152
  thallium in 5:152
  thorium in 5:152
  tin in 5:144; 19:53, 64
  titanium in 5:145; 19:115, 119, 120
  tungsten in 5:152
  uranium in 5:152
  uses 5:110
  vacuum fusion methods 8:616, 617, 618, 619,
      681
  vacuum hot extraction methods 8:638
  vanadium in 5:147
  zinc in 5:148; 19:568
Aluminum monostearate
  in enamels 12:176
Aluminum–nickel alloys. (See also Nickel alloys.)
  uses 16:289
Aluminum nitrate 5:181
Aluminum nitride 5:194
  analysis 5:195
  carrier gas fusion methods 8:624, 682
  nitrogen in 8:624
  oxygen in 8:624, 682
Aluminum oxide 5:181
  abrasive 7:294
  in alkali barium silicate glass 9:218
  ammonium chloride gravimetric method 13:420
  ammonium hydroxide gravimetric method 9:
      112; 14:94
  analysis 5:184; 7:295
  assay methods 5:185
  in calcium oxide 8:94
  carrier gas fusion methods 8:626
  in cements 9:112, 113, 126
  in ceramics 9:194, 224
  in clays and clay minerals 10:115

commercial grades and specifications 5:184
crystal size 7:308
cupferron colorimetric method 9:126
EDTA titration 13:445, 446
emission spectroscopy 9:218
in enamels 9:195
friability, 7:308, 309
in glass and glass-ceramics 13:413, 415, 418,
    445, 446
in glazes 9:196
in gypsum and gypsum products 14:94, 96
hardness 7:309
in hydrated lime 9:191
in hydraulic cements 9:96, 100
identification 7:295
impurities, determination of 5:186, 188
in lime and limestone 9:191; 15:215
manufacture 5:183
microscopy 9:224
nitrogen in 8:616
oxygen in 8:616, 626, 681
particle diameter values 7:304
properties 5:183; 9:204
properties, determination of 5:188
in quicklime 9:191
in refractories 9:206
sample preparation 5:184
in silicate glasses 9:198
specific gravity 7:296
uses 5:183
vacuum fusion methods 8:616, 681
Aluminum oxide abrasives 4:31, 33, 35
    aluminum oxide content 4:37
    analysis 4:35
    decomposition of samples 4:36
    emission spectroscopy 4:51
    impurities, determination of 4:37
    x-ray emission spectroscopy 4:49
Aluminum penicillin G. See Aluminum
    benzylpenicillin.
Aluminum phosphide 5:194
    analysis 5:195
Aluminum potassium sulfate 5:192
    analysis 5:192
    commercial grades and specifications 5:192
    insolubles in 5:194
    occurrence 5:192
    properties 5:192
    sodium in 5:194
    uses 5:192
Aluminum silicate. (See also Silicates.) 5:189
    aluminum content of 5:189
    analysis 5:189
    commercial grades and specifications 5:189
    in enamels 12:174
    impurities, determination of 5:190
    manufacture 5:189

properties 5:189
sample preparation 5:189
silicon content of 5:190
uses 18:44
Aluminum silicate pigment
    silicon content 18:29
Aluminum soap. (See also Soaps.)
    uses 18:180
Aluminum–sodium hydroxide
    color test for nitrogen compounds 14:605
Aluminum sodium sulfate 5:192
Aluminum sulfate 5:190
    aluminum content of 5:192
    analysis 5:192
    commercial grades and specifications 5:191
    impurities, determination of 5:193
    manufacture 5:191
    properties 5:190
    sulfate content of 5:193
    uses 5:191
Aluminum tristearate
    in enamels 12:176
Aluminum–uranium alloys. (See also Uranium
    alloys.)
    decomposition 19:235
Alums. (See also individual compounds.) 5:191
    analysis 5:192
Alunite 14:519
    composition 10:104
Alvite 14:104
Amalgamation time
    of dental amalgams 11:290
Amalgams. (See also Mercury alloys; Mercury
    metal.)
    properties 16:4
    sample decomposition 16:14
    uses 16:10
Amalgams, dental. See Dental amalgams.
Amaranth. See F D & C Red No. 2.
Amaranth lakes
    in enamels 12:183
    identification 12:183
Amatol. (See also High-explosive mixtures.)
    analysis 12:446
    composition 12:446
    uses 12:446
Amazonite
    composition 17:544
    occurrence 17:544
Amberlite resins. (See also Ion exchange resins.)
    15:2
Ambifunctional silanes. See Organofunctional
    silanes.
Amblygonite 13:2; 17:71
AM-9 chemical grout 4:160, 167
American process
    for chocolate and cocoa products 9:545

Americium. (See also Transuranium elements.)
  manufacture 19:259
  separation 19:259
Amerine method
  for nitrogen in wines 19:471
Ameripol 12:86
  intrinsic viscosity 12:119
Amesite
  composition 10:103
Amide derivative
  of adipic acid 4:412
  of amines 5:265
Amides 8:596
  acid–base titration, nonaqueous 8:605
  in acrylamide polymers 4:172
  analysis 8:600
  commercial grades and specifications 8:600
  copper sulfate identification test 8:600
  derivatives for identification 8:601
  derivative for identification of benzoic acid 7:63
  derivatives of acid halides 8:572
  derivatives of carboxylic acids 8:374
  differential rates of hydrolysis 4:167
  in embalming fluids 12:166
  ferric hydroxamate colorimetric method 8:601
  ferric hydroxamate test 8:561
  formation from acid anhydrides 8:564
  formation from carboxylic acids 8:372
  gas chromatography 8:605
  hydrolysis 8:606
  identification 8:600
  impurities, determination of 8:609
  infrared spectroscopy 8:602, 609; 12:576
  ion exchange chromatography 8:610
  manufacture 8:599
  nitrogen content 8:607
  phenol–hypochlorite identification test 8:600
  properties 8:558, 598
  reaction with acid chlorides 8:608
  reduction to amines 8:607
  use in preparation of carboxylic acids 8:372
Amides, primary
  reaction with acid chlorides 8:608
Aminco C and H analyzer 8:284
  operation details 8:284, 286
  schematic diagram 8:284
Amine antioxidants. (See also *Antioxidants and antiozonants*.) 6:78
  in brake fluids 7:424, 425, 428
  colorimetric method based on formation of Wurster salts 6:100
  colorimetric method using diazotized *p*-nitroaniline 6:98
  gas chromatography 6:81
  paper chromatography 6:85
Amine curing agents
  nitrogen content of 4:24

Amine derivatives
  color reagents for boric acid 7:389
Amine end groups
  in polyamides 17:293
Amine equivalent
  of isocyanates 19:269, 271, 272
  of isocyanic acid esters 15:112
  of tolylene diisocyanates 14:112
Amine hydrogen equivalent weight. See AHEW.
Amines. (See also *Alkanolamines*; Alkylamines; Allylamines; *Aniline and its derivatives*; and individual compounds.) 5:259–275
  acid–base titration 7:428
  in alkanolamines 5:11
  in amino resins 5:276
  analysis 5:264
  in antifreezes 6:15
  in beer 7:613
  in brake fluids 7:425, 427
  classification 5:259
  colorimetric method using dipicrylamine 5:89
  in copolymers 4:273, 386
  in cyclohexylamine 11:217
  determination in dithiocarbamic acid salts 11:582
  gas chromatography 6:404; 19:305
  in grapes 19:469
  identification 19:302
  ion exchange methods 19:292
  in methanol 16:114, 117
  mixtures, determination of 5:271
  nuclear magnetic resonance spectroscopy 19:302
  picric acid colorimetric method 16:117
  of polyurethanes 19:291, 302
  reaction of carbonyl compounds with 8:333
  reduction of amides to 8:607
  solvent extraction 19:292
  steam distillation 19:138
  thin-layer chromatography 19:140, 302, 305
  in tobacco 19:138
  use in manufacture of azo dyes 6:402
  in wines 19:469
  in yeast 19:469
Amine soaps. (See also *Soaps*.)
  manufacture 18:179
  uses 18:179
Amines, primary
  acetylation method 5:270
  acid–base titration 5:88, 270, 271, 448
  colorimetric method using methyl orange 5:273
  derivatives for identification 5:88, 265
  mixtures, determination in 5:271
Amines, primary aliphatic
  titration of 4:607
Amines, secondary
  acetylation method 5:270
  acid–base titration 5:88, 270, 271, 448

colorimetric method for 4:607
colorimetric method using bromcresol green
5:273
colorimetric method using methyl orange 5:273
derivatives for identification 5:88, 265
mixtures, determination in 5:271
polarography 5:89
Amines, secondary aliphatic
titration of 4:607
Amines, tertiary
acid–base titration 5:88, 271, 273, 448
colorimetric method using methyl orange 5:273
derivatives for identification 5:266
mixtures, determination in 5:291
titrant for free acid in anhydrides 8:567
Amine value 11:425
of diethylenetriamine 11:425
of ethyleneamines 11:425
of ethylenediamine 11:425
of pentaethylenehexamine 11:425
of tetraethylenepentamine 11:425
of triethylenetetraamine 11:425
Amino acids. (See also *Carboxylic acids, amino.*)
acid–base titration 19:474
bioassay 7:182
in cocoa beans 9:556
formol titration 19:474
gas chromatography 19:473
in gelatin 12:552
in green tea leaf 18:476
ion exchange chromatography 9:557; 12:552
ion exchange method 19:474
in musts 19:469, 473
ninhydrin colorimetric method 19:474
paper chromatography 9:557
separation 19:473
solvent extraction 18:477; 19:140
Strecker degradation 9:542
in tobacco 19:140
ultraviolet spectroscopy 12:553
in wines 19:469, 473
Amino alcohols. See *Alkanolamines.*
Aminoalkyl esters
neutralization equivalent of 4:198
Aminoantipyrine
colorimetric method for 4-*tert*-butylphenol
17:18
colorimetric method for cresols 17:18
colorimetric method for 2,6-di-*tert*-butylphenol
17:18
colorimetric method for 2,6-diisopropylphenol
17:18
colorimetric method for dimethylphenols 17:18
colorimetric method for *n*-heptyl *p*-hydroxy-
benzoate 7:602
colorimetric method for 4-methyl 2,6-di-*tert*-
butylphenol 17:18

colorimetric method for methylisopropylphenols
17:18
colorimetric method for phenol 8:147; 12:310;
17:18
colorimetric method for phenols 17:17, 30; 19:
386
colorimetric method for pyrocatechol 17:18
colorimetric method for resorcinol 17:18
4-Aminoantipyrine–potassium ferricyanide
colorimetric method for phenols 11:544
Aminoazobenzene
in D & C Red No. 17 10:498, 509
visible spectroscopy 10:509
o-Aminobenzaldehyde 6:596
commercial specifications 6:607
derivatives for identification 6:612
manufacture 6:600
properties 6:598
p-Aminobenzaldehyde 6:596
derivatives for identification 6:612
manufacture 6:600
properties 6:598
Aminobenzaldehydes
manufacture 6:600
properties 6:598
uses 6:606
2-Amino-1,4-benzenedisulfonic acid
properties 5:423
solubility 5:421
4-Amino-1,3-benzenedisulfonic acid
in sulfanilic acid 5:446
m-Aminobenzoic acid
paper chromatography 8:379, 404
o-Aminobenzoic acid. See Anthranilic acid.
p-Aminobenzoic acid 8:404
paper chromatography 8:379, 404
properties 8:402
uses 8:404
D-α-(-)-Aminobenzylpenicillin. See Ampicillin.
2-Amino-1-butanol
gas chromatography 5:5
infrared spectrum 5:9
properties 5:2
specifications 5:5
γ-Aminobutyric acid
gas chromatography 17:290
ion exchange chromatography 8:421
paper chromatography 8:422
Aminocaproic acid
polarography 17:299
in polyamides 17:299
thin-layer chromatography 17:292
ε-Amino-*n*-caproic acid
gas chromatography 17:290
η-Amino-*n*-caprylic acid
gas chromatography 17:290
7-(D-5-Amino-5-carboxyvaleramido)cephalo-

sporanic acid. See Cephalosporin C.
7-Aminocephalosporanic acid
  in cephaloglycin 16:579
4-Amino-6-chloro-*m*-benzenedisulfonamide
  colorimetric method for morphine 16:262
Aminocyclohexane. See Cyclohexylamine.
p-Aminodimethylaniline
  use in preparation of methylene blue 6:386
p-Aminodimethylaniline–ferric chloride
  colorimetric method for hydrogen sulfide 18:
    394, 398
  colorimetric method for sulfide 19:365
p-Aminodimethylaniline–ferric sulfate
  colorimetric method for sulfur 14:545
p-Aminodimethylaniline sulfate
  colorimetric method for hydrogen sulfide 7:641
  colorimetric method for sulfur dioxide 7:645
2-Amino-4,6-dinitrophenols
  in hair dyes 11:103
  thin-layer chromatography 11:103
Aminodiphenylamines. (See also *Diamines and
    higher amines, aromatic.*)
  derivatives for identification 11:438
  properties 11:434
  solubility 11:435
Aminoethane 5:1
Aminoethanol 5:1
2-Aminoethanol. See Ethanolamine; Monoethanol-
    amine.
2-(2-Aminoethylamino)ethanol
  infrared spectrum 5:8
  properties 5:2
N-β-(Aminoethyl)-γ-aminopropyltrimethoxy-
    silane. (See also Organofunctional silanes.)
  commercial specifications 18:82
  gas chromatography 18:85
  properties 18:80
N-(2-Aminoethyl)aziridine. (See also *Imines.*)
  infrared spectrum 14:492
  properties 14:469
  structure 14:469
Aminoethylethanolamine 5:4
4-(2-Aminoethyl)morpholine
  gas chromatography 16:200
  in morpholine 16:200
2-Amino-2-ethyl-1,3-propanediol
  gas chromatography 5:5
  infrared spectrum 5:9
  properties 5:2
  specifications 5:5
2-Aminoheptane
  properties 5:260
ε-Aminohexanoic acid
  derivatives for identification 8:122
1-Amino-4-hydroxyanthraquinone
  fluorescence reagent for boric acid 7:387, 411
2-Amino-2-(hydroxymethyl)-1,3-propanediol.

See Tris(hydroxymethyl)aminomethane.
4-Amino-5-hydroxy-1,3-naphthalenedisulfonic
    acid
  colorimetric method for aniline 5:451
ε-Aminohexanoic acid
  derivatives for identification 8:122
  formation from caprolactam 8:121
4-Amino-5-hydroxy-2,7-naphthalenedisulfonic acid.
    (See also Aminonaphtholsulfonic acids.)
  commercial grades and specifications 16:240
  properties 16:240
  solubility 16:240
Aminoketone dyes 12:9
λ-Amino-*n*-lauric acid
  gas chromatography 17:290
2-Amino-4-methylhexane 5:260
2-Amino-2-methyl-1,3-propanediol
  gas chromatography 5:5
  infrared spectrum 5:10
  properties 5:2
  specifications 5:5
2-Amino-2-methyl-1-propanol
  gas chromatography 5:5
  infrared spectrum 5:10
  manufacture 5:3
  properties 5:2
  specifications 5:5
2-Aminomethylpyridine. (See also Picolylamine.)
  ionization constants 17:417
3-Amino-1,5-naphthalenedisulfonic acid. See 2-
    Naphthylamine-4,8-disulfonic acid.
4-Amino-1-naphthalenesulfonic acid. See 1-
    Naphthylamine-4-sulfonic acid.
4-Amino-2-naphthalenesulfonic acid. See 1-
    Naphthylamine-3-sulfonic acid.
5-Amino-1-naphthalenesulfonic acid. See 1-
    Naphthylamine-5-sulfonic acid.
5-Amino-2-naphthalenesulfonic acid. See 1-
    Naphthylamine-6-sulfonic acid.
6-Amino-2-napthalenesulfonic acid. See 2-
    Naphthylamine-6-sulfonic acid.
8-Amino-1-naphthalenesulfonic acid. See 1-
    Naphthylamine-8-sulfonic acid.
8-Amino-2-naphthalenesulfonic acid. See 1-
    Naphthylamine-7-sulfonic acid.
Aminonaphthalenesulfonic acids. (See also individ-
    ual compounds.) 16:227
  analysis 16:228
  impurities, determination of 16:228
  sodium nitrate titration 16:228
8-Amino-1,3,6-naphthalenetrisulfonic acid. See 1-
    Naphthylamine-3,6,8-trisulfonic acid.
8-Amino-1-naphthol-3,6-disulfonic acid
  use in manufacture of azo dyes 6:403
Aminonaphthols 16:240
  manufacture 16:240
  properties 16:240

Aminonaphtholsulfonic acids. (See also individual
    compounds.)
  manufacture 16:240
Amino nitrogen
  in yeast 19:529
α-Amino nitrogen
  automated analysis 7:665
  in beer 7:614, 665
  colorimetric method using 2,4,6-trinitrobenzene-
    1-sulfonic acid 7:614
  in wines 19:474
Aminonitrophenols
  in hair dyes 11:103
  thin-layer chromatography 11:103
6-Aminopenicillanic acid
  manufacture 16:553
  structure 16:553
o-Aminophenol
  in aniline 5:446
p-Aminophenol
  in acetaminophen 6:197
  colorimetric method using 1-naphthol 6:156
  colorimetric method using N-(1-naphthyl)-N-
    diethylpropylenediamine 6:195
  column chromatography 6:154
  diazotization method 6:195
  ion exchange chromatography 6:182
  separation 6:182
p-Aminophenol derivatives. (See also individual
    compounds.) 6:177
Aminophenols
  in hair dyes 11:103
  thin-layer chromatography 11:103
7-(D-α-Aminophenylacetamido)cephalosporanic
    acid. See Cephaloglycin.
7-(D-α-Aminophenylacetamido)-3-methyl-3-
    cephem-4-carboxylic acid. See Cephalexin.
Aminophenylpyridines. (See also Pyridine
    derivatives.)
  ionization constants 17:416
Aminophylline
  extraction 6:497
  ion exchange chromatography 6:499
2-Amino-1,3-propanediol 5:5
1-Amino-2-propanol 5:9
2-Amino-1-propanol 5:5
N-(3-Aminopropyl)cyclohexylamine 11:248
  gas chromatography 11:249
  infrared spectroscopy 11:248
  manufacture 11:248
  properties 11:244
  solubility 11:249
  toxicity 11:249
  uses 11:248
γ-Aminopropyltriethoxysilane. (See also Organo-
    functional silanes.)
  commercial specifications 18:82

  infrared spectroscopy 18:83
  properties 18:80
  solubility 18:89
  specification tests 18:89
γ-Aminopropyltrimethoxysilane. (See also Organo-
    functional silanes.)
  commercial specifications 18:82
  properties 18:80
Aminopyridines. (See also Pyridine and pyridine
    derivatives.)
  commercial grades and specifications 17:423
  manufacture 17:422
  trisodium pentacyanoammino ferrate colorimetric
    method 17:433
Aminopyrine 6:206
  acid–base titration 6:216
  analysis 6:207
  bromate titration 6:217
  colorimetric method using diazotized p-nitro-
    aniline 6:217
  colorimetric method using ferric chloride 6:217
  colorimetric method for nitrate 16:504
  commercial grades and specifications 6:207, 215
  EDTA titration 6:216
  extraction 6:213, 215, 219
  gas chromatography 6:210, 211
  gravimetric methods 6:215, 219
  identification 6:213
  impurities, determination of 6:218
  infrared spectrum 6:213
  manufacture 6:207
  oxidative-cleavage method 6:217
  properties 6:206
  separation 6:212
  ultraviolet spectroscopy 6:212
Amino resin adhesives 4:399
Amino resins 5:276–290
  analysis 5:280
  composition 5:278
  identification 5:282
  infrared spectroscopy 5:282
  properties 5:277
  properties, determination of 5:280
  uses 5:280
p-Aminosalicylic acid 8:404
  properties 8:402, 404
4-Amino-3,5,6-trichloropicolinic acid. See Tordan.
κ-Amino-n-undecanoic acid
  gas chromatography 17:290
δ-Amino-n-valeric acid
  gas chromatography 17:290
Aminox 6:72
  color test 6:94
  colorimetric method using diazotized p-nitro-
    aniline 6:99
Ammonal. (See also High-explosive mixtures.)
  analysis 12:446

composition 12:446
uses 12:446
Ammonia. (See also Ammonium hydroxide.)
5:290–324; 16:471
in acetone 4:114
in acetylene 4:131
acid–base titration 11:351; 14:458; 15:252;
16:189, 438, 509; 17:439; 18:85; 19:138,
471, 473
in air 16:485
in aluminum nitrate 5:181
in amines 5:88
in ammonium dichromate 9:685
analysis 5:298
assay methods 5:305
azeotropes 5:294
in beer 7:610
in bismuth subnitrate 7:203
in 1,3-butadiene 11:504
in cadmium carbonate 8:39
in cadmium chloride 8:42
in cadmium nitrate 8:45
in calcium carbonate 8:83
caustic distillation 9:685
chemical methods 5:293
in chromium compounds 9:699
in cobalt chloride 10:336
in cobalt nitrate 10:340
in cobalt(II) sulfate 10:343, 344
in cocoa beans 9:556
commercial grades and specifications 5:298
Conway-O'Brink diffusion method 11:99
in detergents 11:350
distillation 14:457
evolution method 19:138
gas chromatography 5:268; 14:239; 16:485
in glycerol 13:527
gravimetric method for nickel 16:296
in hair dyes 11:99
in hydrazines 14:239
in hydroxylammonium sulfate 14:457
identification 5:302
impurities, determination of 5:305
infrared spectroscopy 14:458
infrared spectrum 5:304
ion exchange chromatography 8:415, 421;
19:473
Kjeldahl method 19:382
in liquefied petroleum gas 15:252
manufacture, methods of 5:296
in molybdates 16:189
in musts 19:469, 471, 472
Nessler's reagent color test 10:336; 19:383
in nitrogen fertilizers 16:509
in nitroparaffins 16:438
occurrence 5:295
properties 5:291

in protein 8:414
in pyridine 17:439
sampling 5:298
separation 5:302
in soaps 18:194
sodium hydroxide distillation 11:350
solubility 5:292
solvent properties 5:192, 193
in tobacco 19:138
toxicity 5:295
uses 5:290
volumetric method for lithium amide 15:285
volumetric method for nitrogen 14:580
in water 19:382, 384
in wines 19:469, 471, 472
Ammoniacal cadmium chloride
colorimetric method for sulfide 9:119
Ammoniacal nitrogen content
of acrylamide polymers 4:172
Ammonia–iodide
colorimetric method for mercury 16:42
Ammoniated mercury. (See also Mercurials;
Mercury compounds.) 16:53
assay 11:563
commercial grades and specifications 11:562;
16:53, 54
manufacture 16:53
uses 16:53
zinc–potassium permanganate volumetric method
11:563
Ammoniated mercury ointment
mercury content 16:28
Ammoniobrite 7:314
Ammonium acetate
analysis 5:325
assay 5:325
colorimetric method for lead styphnate 12:429
commercial grades and specifications 5:325
extraction of calcium sulfate from gypsum
products 14:91
impurities, determination of 5:325
infrared spectrum 5:326
properties 5:325
Ammonium acetate–acetylacetone
colorimetric method for formaldehyde 11:567
Ammonium acetate–sodium arsenite
colorimetric method for epichlorohydrin 14:176
Ammonium acetate–thiocyanate–pyridine
colorimetric method for copper 10:337
Ammonium acid carbonate. See Ammonium bicarbonate.
Ammonium adipate 4:429
Ammonium alum. See Aluminum ammonium
sulfate.
Ammonium aurin tricarboxylate. See Aluminon.
Ammonium benzeneseleninate
gravimetric method for niobium 16:367

gravimetric method for tantalum 16:367
Ammonium benzoate. (See also Benzoic acid
  salts.) 7:85
  acid–base titration, nonaqueous 7:86
  analysis 7:86
  properties 7:85
  separation 7:62
Ammonium benzoate-8-hydroxyquinoline
  gravimetric method for aluminum 15:376
Ammonium bicarbonate 5:327, 331
  acid–base titration 5:329
  analysis 5:328
  assay methods 5:328
  commercial grades 5:328
  distillation method 5:329
  impurities, determination of 5:330
  infrared spectrum 5:328
  properties 5:327
  uses 5:327
Ammonium borates 7:379
Ammonium bromide 12:542
  commercial specifications 8:9, 10; 12:542
  impurities, determination of 12:542
  properties 8:9
Ammonium carbamate 5:331
Ammonium carbonate 5:331
  analysis 5:332
  assay 5:332
  commercial grades and specifications 5:331, 332
  gravimetric method for strontium 18:272, 274
  impurities, determination of 5:332
  infrared spectrum 5:332
  uses 5:331
Ammonium carbonatozirconate. See Zirconyl
  ammonium carbonate.
Ammonium chlorate. (See also Chlorates.)
  properties 9:358
  solubility 9:359
Ammonium chloride 12:543
  acid–base titration 9:453
  analysis 5:335
  assay methods 5:335
  commercial grades and specifications 5:335;
    12:542
  etchant for ceramics 9:224
  formaldehyde method 5:336
  gravimetric method for aluminum 16:323
  gravimetric method for aluminum oxide 13:420
  gravimetric method for beryllium 16:324
  gravimetric method for platinum 17:258
  in hexachloroethane 9:453
  impurities, determination of 5:337; 12:543
  infrared spectrum 5:334
  ion exchange method 5:336
  manufacture 5:333
  properties 5:333; 9:334
  solubility 5:333, 334; 9:336

volumetric method for formaldehyde 13:111
  uses 5:333
Ammonium chloride–diammonium phosphate
  gravimetric method for manganese 15:457, 460
Ammonium chloride–hydrogen peroxide
  gravimetric method for iron 16:238
Ammonium chloroplatinate
  platium recovery from 17:406
Ammonium chromate
  analysis 9:693
  assay 9:693
  commercial grades and specifications 9:694
Ammonium compounds. (See also individual compounds.) 5:325–355
  identification 5:325
Ammonium content
  in ammonium formate 13:136
  in ammonium nitrate 12:451
  in ammonium perchlorate 12:452
  in ammonium picrate 12:451
  in hydrochloric acid 9:340, 351; 14:371
  in hydrogen peroxide 14:437
  in magnesium chloride 15:404
  in magnesium nitrate 15:407
  in magnesium sulfate 15:412
  Nessler's reagent color test 9:351; 14:372, 437;
    15:404, 407, 412; 17:111
  in phosphorus pentoxide 17:111
  sodium hydroxide volumetric method 13:136
Ammonium cyanide. (See also Cyanides.) 14:424
  properties 14:424
  solubility 14:424
  toxicity 14:424
Ammonium dichromate. (See also Chromium
  compounds.) 9:681
  ammonia content 9:685
  analysis 9:683, 693
  assay 9:684, 693
  chromate content 9:684
  commercial grades and specifications 9:683, 694
  ferrous sulfate titration 9:685
  gravimetric method for barium 18:274
  gravimetric method for thallium 18:526
  impurities, determination of 9:687
  iodometric titration 9:684
  properties 9:682
  uses 9:681
Ammonium fluoberyllate 7:122
Ammonium fluoborate 7:337
Ammonium fluoride
  commercial grades and specifications 13:12
  etchant for ceramics 9:224
  gravimetric method for rare earths 17:477
  toxicity 13:13
Ammonium formate
  ammonium content 13:136
  analysis 13:136

assay 13:136
formate content 13:137
uses 13:120, 134
Ammonium hexachloroiridate
properties 17:226
uses 17:226
Ammonium hexachloropalladate
manufacture 17:225
Ammonium hexachloroplatinate
gravimetric method for platinum 17:254
properties 17:227
uses 17:227
Ammonium hexanitratocerate. (See also Ceric
ammonium nitrate.)
titration of nitrogen dioxide 16:479
volumetric method for hydrazoic acid 16:472
Ammonium hexathiocyanatocobaltate
colorimetric method for nonionic detergents
11:328; 18:183
Ammonium hydrogen carbonate. See Ammonium bicarbonate.
Ammonium hydrogen fluoride
etchant for ceramics 9:224
Ammonium hydroxide. (See also *Ammonia.*) 5:
290
acid–base titration 5:317
analysis 5:316
assay methods 5:317
chemical properties 5:293
colorimetric method for acid azoic dyes 10:538
colorimetric method for cochineal 10:538
colorimetric method for enocianin 10:538
colorimetric method for tetryl 12:421
colorimetric method for TNT 12:421
commercial grades and specifications 5:298
gravimetric method for aluminum 5:162; 10:
307, 632; 12:151
gravimetric method for aluminum oxide 9:112,
295, 665
gravimetric method for beryllium 16:324
gravimetric method for copper 10:604
gravimetric method for ferric oxide 9:112
gravimetric method for gallium 13:253, 257
gravimetric method for indium 14:528
gravimetric method for iron 9:670; 10:313, 639;
12:151; 16:164, 180
gravimetric method for magnesium 12:186
gravimetric method for mercury 16:6, 7
gravimetric method for nickel 16:292, 298
gravimetric method for rare earths 17:476
gravimetric method for silver 18:164
gravimetric method for thorium 19:9
gravimetric method for tin 19:33, 39
gravimetric method for titanium 16:174, 176;
19:94, 103, 111
gravimetric method for tungsten 19:200, 219
gravimetric method for uranium 19:243

gravimetric method for zinc 14:552
gravimetric method for zirconium 14:111; 16:
173, 174
identification 5:317
impurities, determination of 5:320
microchemical method for *l*-narcotine 16:250
sampling 5:316
separation 5:317
turbidimetric method for barium 10:150
uses 5:291
Ammonium hydroxide–barium chloride
gravimetric method for citric acid 19:423
Ammonium hydroxide group
ammonium oxalate gravimetric method 14:606
in asbestos 6:261
in barite 6:559
in cements 9:112
in chrysotile 6:261
in clays and clay minerals 10:115
in lime and limestone 15:208
in magnesium oxide 15:409
in potassium chlorate 9:359, 361
in potassium chloride 9:339
in potassium iodide 14:606
potassium pyrosulfate fusion 15:208
in sodium chloride 9:337, 349
in sodium iodide 14:606
Ammonium hydroxide insolubles
in casein 9:13
Ammonium hydroxide precipitation 8:100
in acid salts 8:594
of beryllium 7:129, 131
in calcium acetate 8:77
in calcium carbonate 8:83
in copper(II) oxide 10:675
Ammonium hydroxide, substances not precipitated
by
in thorium nitrate 19:19
Ammonium iodide. (See also Iodides.) 12:544;
14:595
commercial grades and specifications 12:542;
14:595
impurities, determination of 12:544; 14:604
properties 14:595
uses 14:595
Ammonium iodide–sodium bisulfite
colorimetric method for phosphate 14:604
Ammonium mercaptoacetate. See Ammonium
thioglycolate.
Ammonium mercuric thiocyanate
gravimetric method for zinc 5:149
Ammonium metavanadate. (See also Metavanadates.)
analysis 19:326
dissolution 19:326
manufacture 19:325
uses 19:343

Ammonium molybdate. (See also Molybdates.)
colorimetric method 13:281
colorimetric method for arsenic 6:244, 245
colorimetric method for benzylmorphine 16:
246
colorimetric method for codeine 16:246
colorimetric method for cotarnine 16:246
colorimetric method for dihydrocodeine 16:246
colorimetric method for dihydromorphine 16:
246
colorimetric method for ethylmorphine 16:246
colorimetric method for fentanyl 16:246
colorimetric method for heroin 16:246
colorimetric method for hydrocodone 16:246
colorimetric method for o-hydroxyphenols 17:
20
colorimetric method for levomethorphan 16:
246
colorimetric method for levorphanol 16:246
colorimetric method for metazocine 16:246
colorimetric method for metopon 16:247
colorimetric method for morphine 16:247
colorimetric method for nalorphine 16:247
colorimetric method for narceine 16:247
colorimetric method for l-narcotine 16:247
colorimetric method for norlevorphanol 16:247
colorimetric method for orthophosphate 17:61
colorimetric method for oxycodone 16:247
colorimetric method for oxymorphone 16:247
colorimetric method for papaverine 16:247
colorimetric method for phenazocine 16:247
colorimetric method for pholcodine 16:247
colorimetric method for phosphorus 9:242, 296
colorimetric method for phosphorus pentoxide
9:130
colorimetric method for racemethorphan 16:
247
colorimetric method for racemorphan 16:247
colorimetric method for silica 7:430
colorimetric method for silicate 11:142
colorimetric method for silicon 14:64; 15:274;
18:31
colorimetric method for thebaine 16:247
colorimetric method for tin 19:37
gravimetric method for ferrocyanides 15:89
gravimetric method for lead 15:179
gravimetric method for menadiol sodium diphosphate 10:203
gravimetric method for phosphate 11:75, 132,
335; 18:183
properties 16:184
titration of lead 15:180
uses 16:184
volumetric method for phosphate 17:507
volumetric method for phosphorus 9:295; 15:
75; 17:74, 80, 86, 96, 97, 145
volumetric method for phosphorus

pentoxide 15:214
Ammonium molybdate–acetone
colorimetric method for germanium 13:401
Ammonium molybdate–aminonaphthol sulfonic
acid
colorimetric method for phosphate 17:102; 19:
498
Ammonium molybdate–ammonium thiocyanate
colorimetric method for tin 19:37
Ammonium molybdate–ammonium vanadate
colorimetric method for phosphorus 17:146
Ammonium molybdate–benzidine
colorimetric method for phosphate 11:142
colorimetric method for silicon 18:16
Ammonium molybdate–hydrazine sulfate
colorimetric method for phosphorus 11:369;
15:60; 17:83
Ammonium molybdate–hydroquinone
colorimetric method for phosphorus 17:98
Ammonium molybdate–magnesia
gravimetric method for phosphorus 17:78, 95,
101
Ammonium molybdate–p-methylaminophenol
sulfate
colorimetric method for phosphorus 9:348; 13:
34; 14:437; 15:401, 404; 16:536; 18:223
Ammonium molybdate–oxine
gravimetric method for silicon 18:78
Ammonium molybdate–stannous chloride
colorimetric method for phosphate 14:604; 17:
105
colorimetric method for silicon 16:334
Ammonium molybdate–sulfuric acid
colorimetric method for thiosulfate 18:431
Ammonium molybdate tetrahydrate
gravimetric method for phosphorus pentoxide
13:432
Ammonium muriate. See Ammonium chloride.
Ammonium naphthaleneseleninate
gravimetric method for niobium 16:367
gravimetric method for tantalum 16:367
Ammonium nitrate. (See also High explosives.)
5:338; 12:451
ammonium content 12:451
analysis 5:340
assay methods 5:340
commercial grades and specifications 5:339;
16:508
gravimetric method for beryllium 10:308
identification 5:340; 12:451
impurities, determination of 5:342
infrared spectrum 5:339; 12:459
manufacture 5:338
melting point 12:461
nitrate content 12:451
nitrogen content 5:340
properties 5:338; 12:434, 451; 16:497, 508

solubility 12:451
uses 5:339; 12:451; 16:508
Ammonium nitrogen content
of yeast 19:529
Ammonium nitromolybdate
volumetric method for phosphorus 17:85
Ammonium oxalate 5:344
analysis 5:344
commercial grades and specifications 5:344
gravimetric method for ammonium hydroxide
group 14:606
gravimetric method for calcium 10:296; 12:
186; 14:606
gravimetric method for calcium oxide 9:116;
13:421; 15:215
gravimetric method for magnesium 14:606
gravimetric method for strontium 18:272
gravimetric method for thorium 19:12
impurities, determination of 5:345
infrared spectrum 5:344
oxalate content 5:344
precipitation reagent for calcium 8:82, 102
properties 5:344
Ammonium oxalate–potassium permanganate
volumetric method for calcium 14:68; 17:194
volumetric method for calcium oxide 17:201
Ammonium perchlorate. (See also High explosives;
Perchlorates.) 12:451
ammonium content 12:452
properties 9:363; 12:451
solubility 9:364
uses 9:362; 12:451
Ammonium perchloratocerate
volumetric method for glycerol 13:512
Ammonium perrhenate
manufacture 17:519, 520
Ammonium persulfate 5:346
analysis 5:347
assay 5:347
catalyst for production of AM-9 chemical grout
4:160, 167
colorimetric method for manganese 9:49
commercial grades and specifications 5:346
impurities, determination of 5:347
infrared spectrum 5:346
properties 5:346
Ammonium persulfate–ammonium thiocyanate
colorimetric method for iron 9:349, 362, 629;
10:95, 330; 13:34; 14:373; 15:399, 402, 405,
408, 410, 491; 16:60, 344, 537; 17:403; 18:
176, 177, 234; 19:19, 74
Ammonium phosphate
gravimetric method for beryllium 7:133
gravimetric method for magnesium 12:153;
15:360
gravimetric method for titanium 19:94
gravimetric method for zirconium 14:109,

114; 16:173, 174
manufacture 17:143
solubility 17:141
uses 17:141
Ammonium phosphate, dibasic 5:348
analysis 5:349
assay methods 5:349
commercial grades and specifications 5:348
commercial specifications 17:127
impurities, determination of 5:349
infrared spectrum 5:348
phosphate content 5:349
properties 5:348
uses 5:348
Ammonium phosphate, monobasic
commercial specifications 17:127
Ammonium phosphomolybdate
gravimetric method for phosphorus 17:59, 62,
101
volumetric method for phosphorus 17:64, 189
Ammonium picrate. (See also High explosives.)
12:451
ammonium content 12:451
identification 12:451
infrared spectrum 12:460
manufacture 12:451
melting picrate 12:461
properties 12:434, 438, 451
solubility 12:451
uses 12:438
Ammonium purpurate. See Murexide.
Ammonium pyrrolidinedithiocarbamate
gravimetric method for niobium 16:366
Ammonium reineckate
colorimetric method for antihistamines 6:33
colorimetric method for lidocaine 5:395
Ammonium salicylate
gravimetric method for niobium 16:367
Ammonium salts
in acrylamide 4:165
in aluminum sulfate 5:193
in alums 5:193
in amino acids 8:435
Ammonium sebacate–oxalic acid
gravimetric method for rare earths 15:389
Ammonium soaps. (See also *Soaps.*)
manufacture 18:179
uses 18:179
Ammonium stearate
gravimetric method for lithium 15:265
Ammonium sulfamate
in finishing agents 11:126
mercuric chloride gravimetric method 11:127
solubility 11:126
Ammonium sulfate. (See also Sulfates.)
analysis 5:351
assay methods 5:351

in caprolactam 8:127
commercial grades and specifications 5:351;
    18:422
gravimetric method for barium sulfate 17:187
gravimetric method for cesium 17:554
gravimetric method for potassium 17:554
gravimetric method for rubidium 17:554
impurities, determination of 5:352
infrared spectrum 5:351
ion exchange chromatography 11:84
manufacture 5:351; 18:422
in manufacture of ammonium chloride 5:335
properties 5:350
in shampoos 11:84
uses 5:351; 18:422
Ammonium sulfate–formic acid
    colorimetric method for peroxides 17:298
Ammonium sulfate test
    for protease activity 7:594
Ammonium sulfide. (See also Sulfides.)
    colorimetric method for benzoic acid 19:485
    colorimetric method for picric acid 12:419
    gravimetric method for mercury 16:27
    gravimetric method for molybdenum 16:153
    gravimetric method for nickel 16:292
    gravimetric method for thallium 18:522
    gravimetric method for tungsten 19:205
    gravimetric method for zinc 19:554
    manufacture 18:393
Ammonium sulfide–hydrochloric acid
    colorimetric method for antimony 12:186
Ammonium sulfide, substances not precipitated by
    in manganese sulfate 15:490
    in nickel sulfate 16:345
Ammonium sulfide–tartrate
    gravimetric method for iron 15:33
Ammonium sulfite
    in manufacture of ammonium chloride 5:335
Ammonium sulfosalicylate
    colorimetric method for niobium 16:368
    gravimetric method for niobium 16:367
Ammonium tartrate
    gravimetric method for scandium 17:570
Ammonium thiocyanate. (See also Thiocyanates.)
    analysis 5:353
    assay methods 5:353
    colorimetric method for cobalt 10:348, 366
    colorimetric method for copper 17:439
    colorimetric method for iron 8:440; 9:533;
        10:655; 13:130
    colorimetric method for niobium 16:378
    colorimetric method for tungsten 16:162
    colorimetric method for uranium 19:248
    commercial grades and specifications 5:353;
        18:434
    gravimetric method for copper 10:607
    impurities, determination of 5:354

infrared spectrum 5:353
manufacture 18:433
microchemical method for benzylmorphine 16:
    249
properties 5:352; 18:434
thiocyanate content 5:353
titration of mercuric oxide 11:563
titration of mercury 11:564
titration of phenyl mercuric acetate 16:63
uses 18:433
volumetric method for ferric oxide 17:192
volumetric method for merbromin 11:564
volumetric method for phenylmercuric acetate
    11:564
volumetric method for phenylmercuric nitrate
    11:564
volumetric method for silver 15:392
volumetric method for thimerosal 11:564
Ammonium thiocyanate–acetone
    colorimetric method for cobalt 11:140
Ammonium thiocyanate–ammonium persulfate
    colorimetric method for hydrogen peroxide
    14:438
Ammonium thiocyanate–cobalt chloride
    colorimetric method for polyethers 17:300
Ammonium thiocyanate method. (See also Thio-
    cyanate method for iron.) 4:114
Ammonium thiocyanate–tetraphenylarsonium
    chloride
    colorimetric method for cobalt 10:370
Ammonium thioglycolate
    colorimetric method for iron 15:51
Ammonium trisulfatocerate. See Ceric ammonium
    sulfate.
Ammonium vanadate (See also Vanadates.)
    colorimetric method for alphaprodine 16:246
    colorimetric method for cotarnine 16:246
    colorimetric method for dihydrocodeine 16:246
    colorimetric method for dihydromorphine 16:
        246
    colorimetric method for heroin 16:246
    colorimetric method for isomethadone 16:246
    colorimetric method for metazocine 16:246
    colorimetric method for methadone 16:246
    colorimetric method for metopon 16:247
    colorimetric method for morphine 16:247
    colorimetric method for narceine 16:247
    colorimetric method for papaverine 16:247
    colorimetric method for peroxides 11:522
    colorimetric method for thebaine 16:247
    solubility 19:345
Ammonium vanadate–ammonium molybdate
    colorimetric method for orthophosphate 17:61
    colorimetric method for phosphate 19:369
    colorimetric method for phosphorus 11:36, 377;
        17:66, 70, 71, 76, 91, 94
    colorimetric method for phosphorus

pentoxide 15:213
Ammonium yttrium tartrate
  gravimetric method for scandium 17:568, 572
Ammonolysis 8:578
Amobarbital 6:468, 502
  acid–base titration, nonaqueous 6:503
  analysis 6:503, 504
  commercial grades and specifications 6:502
  derivatives for identification 6:477
  extraction 6:491
  gas chromatography 6:210, 211, 483, 484
  gravimetric method 6:493
  impurities, determination of 6:504
  infrared spectroscopy 6:496
  paper chromatography 6:183, 478, 479
  properties 6:470, 504
  in sodium amobarbital 6:510
  solubility 6:476
  thin-layer chromatography 6:480, 481, 482
  ultraviolet spectroscopy 6:501
Amoco 6:533, 72
Amolanone hydrochloride 5:374
  acid–base titration 5:378
  analysis 5:378
  anesthetic 5:378
  assay 5:378
  properties 5:378
  ultraviolet spectroscopy 5:378
Amorphous silica
  composition 10:104
Amosite. (See also Chrysotile.) 6:254
  analysis 6:267
  in asbestos–cement products 6:272
  composite 6:257
  electron micrograph 6:258
  properties 6:256
Amperometric titration
  of aluminum 5:251
  of chromium 9:655
  of cocaine hydrochloride 5:385
  of lead 15:189
Amphetamine
  gas chromatography 6:209
Amphibole fibers. (See also Amosite; Crocidolite.)
  6:254
Amphiboles 15:358
Amphomycin 5:468, 549
  analysis 5:555, 558, 562
  bioautography 5:555
  identification 5:555
  infrared spectrum 5:554
  microbiological assay 5:558
  paper chromatography 5:555
  properties 5:551
  separation 5:555
  solubility 5:552
  uses 5:549

Amphoteric detergents. (See also *Detergents.*)
  11:324
  uses 11:324
Amphotericin A 5:472
  colorimetric method 5:578
  infrared spectrum 5:568
Amphotericin B 5:472
  colorimetric method 5:577
  infrared spectrum 5:568
Amphotericins 5:472, 564
  analysis 5:568, 571, 577
  bioautography 5:568
  identification 5:568
  infrared spectrum 5:568
  microbiological assay 5:571
  paper chromatography 5:568
  properties 5:565
  separation 5:568
  solubility 5:566
  uses 5:564
Amphoterics
  upholstery cleaners 6:354
Ampicillin. (See also *Penicillins;* Sodium
    ampicillin.) 16:575
  acid–base titration, nonaqueous 16:577
  analysis 16:575
  assay 16:575
  bioassay 16:575
  in biological fluids 16:576
  commercial grades and specifications 16:575
  impurities, determination of 16:577
  iodometric methods 16:562
  lithium methoxide titration 16:577
  manufacture 16:575
  microbiological methods 16:575
  properties 16:556, 575
  solubility 16:557
  specification tests 16:576
  structure 16:554
  ultraviolet spectroscopy 16:565, 576, 577
  uses 16:575
Ampicillin penicilloic acid
  in ampicillin 16:577
Amyl acetate
  commercial specifications 8:578
  in enamels 12:175, 189
  in essential oils 12:276
  gas chromatography 11:73
  infrared spectrum 8:585
  molecular weight 12:276
  in nail enamels 11:73
  potassium hydroxide volumetric method 12:189
  properties 8:577; 12:175
  saponification method 8:586
  solubility 13:540
Amyl alcohol
  gas chromatography 11:73

in nail enamels 11:73
n-Amyl alcohol. (See also Alcohols, lower;
    Amyl alcohols.)
    3,5-dinitrobenzoyl chloride method 4:547
    gas chromatography 16:107; 19:442, 446, 449
    in methanol 16:107, 114
    pyromellitic dianhydride method 4:537
Amyl alcohol, active
    in brandies 19:440
    gas chromatography 19:442
    in wines 19:425, 440, 442
Amyl alcohols. (See also Alcohols, lower.) 4:496
    α-amylase content of malt 4:476
    azeotropes 4:501, 504
    commercial grades and specifications 4:514
    derivatives for identification 4:518
    manufacture 4:506
    properties 4:498
    uses 4:512
Amylamines
    in brake fluids 7:425, 427
    commercial grades and specifications 5:264
    manufacture, methods of 5:262
    properties 5:260
    uses 5:263
Amylase
    iodine color test 7:556
α-Amylase
    in flour 12:607
    limit-dextrin colorimetric method 12:607
    volumetric method for damaged starch in flour
        12:606
Amylbenzenes. (See also Hydrocarbons; and
        individual compounds.) 14:323
    commercial grades and specifications 14:323
    uses 14:323
Amyl butyrate
    in essential oils 12:276
    molecular weight 12:276
sec-Amylchlorobenzene
    mass spectroscopy 9:405
α-n-Amyl cinnamaldehyde
    commercial specifications 10:44
    infrared spectroscopy 6:616
α-Amyl cinnamaldehyde dimethyl acetal
    commercial specifications 10:44
α-Amyl cinnamyl alcohol
    commercial specifications 10:78
Amyl-p-N,N-dimethylaminobenzoate
    in sunscreens 11:92
    ultraviolet spectroscopy 11:92
n-Amyl ether. (See also Ethers.)
    properties 12:297
sec-Amyl ether. (See also Ethers.)
    properties 12:297
Amyl formate. See Pentyl formate.
2-Amylfuran

in black tea 18:510
    gas chromatography 18:510
n-Amyl iodide. (See also Iodine compounds,
    organic.)
    properties 14:612
n-Amyl isocyanate. (See also Isocyanic acid
        esters.)
    infrared spectroscopy 15:106
    near-infrared spectroscopy 15:105
n-Amyl lactate. (See also Lactic acid esters.)
    properties 15:131
Amyl mercaptan. (See also Mercaptans.)
    derivatives for identification 15:559
    properties 15:552
Amyl nitrite
    color method for amine antioxidants 6:93
Amylodextrin
    iodine color test 7:556
Amylograph 11:18
Amylograph method
    for viscosity of flour paste 12:611
Amyl salicylate
    in essential oils 12:276
    molecular weight 12:276
Amyltrichlorosilane. (See also Organo-chloro-
        silanes.)
    commercial specifications 18:56
    properties 18:54
Amyltriethoxysilane. (See also Organosilane
        esters.)
    commercial specifications 18:71
    properties 18:70
Anabasine 4:591
    gas chromatography 19:137
    paper chromatography 19:136
    in tobacco 19:136
L-Anabasine. (See also Pyridine and pyridine
        derivatives.)
    trisodium pentacyanoammino ferrate colorimetric
        method 17:433
Anagyrine 4:590, 609
Analgesics. See Antipyretics and analgesics.
Anatabine
    gas chromatography 19:137
    in tobacco 19:137
Anatase. (See also Titanium dioxide.) 19:81, 123
    composition 10:104
Anchorage strength
    of dental porcelains 11:273
Andradite 8:75
Androgenic hormones. See Androgens.
Androgens. (See also Steroids; and individual
        compounds.) 14:186
    analysis 14:224
    properties 14:203
    sources 14:203
    structure 14:185

uses 14:203

Androsta-3,5-dien-17-one
  liquid chromatography 14:209

5-Androstane-3,17-diones
  liquid chromatography 14:208

5α-Androstan-17-one
  gas chromatography 14:224

Δ^4-Androstenedione
  gas chromatography 14:224

4-Androsten-17β-ol-3-one. See Testosterone.

5-Androst-2-en-17-one
  liquid chromatography 14:209

Androsterone
  gas chromatography 14:216
  liquid chromatography 14:209

Anesthetics. (See also individual compounds.)
  5:355–421
  general 5:356
  local 5:373
  spinal 5:373

Anethole
  in manufacture of anisaldehyde 6:600

Angelicin. (See also *Coumarins.*) 11:160
  gas chromatography 11:168, 170

Anglesite. (See also Lead sulfate.) 15:166

Anhydride content
  of trimellitic anhydride 17:173

Anhydride curing agents
  acid and anhydride content of 4:24

Anhydrides. See Acid anhydrides.

Anhydrite 8:75, 97; 14:88; 18:360
  properties 14:88

Anhydrite cement. See Keene's cement.

Anhydrous borax. See Borax, anhydrous.

Anhydrous gypsum cement. See Keene's cement.

Anhydrous soap content
  of soaps 18:188

Anhydroxyprogesterone. See 17-Ethyltestosterone.

Anileridine. (See also *Narcotics.*)
  color tests 16:246
  p-dimethylaminobenzaldehyde colorimetric
    method 16:246
  gas chromatography 6:210
  structure 16:245
  ultraviolet spectroscopy 16:251

Anileridene hydrochloride. (See also *Narcotics.*)
  ultravioiet spectroscopy 16:251

Anilide derivatives
  of acid halides 8:572
  of adipic acid 4:412
  of carboxylic acids 8:376

Anilides
  analysis 5:445
  derivative for identification of benzoic acid
    7:63

Aniline. (See also *Aniline and its derivatives.*)
  5:276, 421

acid–base titration 5:443, 447
analysis 5:428
azeotropes 5:424
azo dye colorimetric method 5:445, 450
azo dye volumetric method 5:448
azo dyes for identification 5:430
bromination methods 5:444
colorimetric methods 5:450
colorimetric method using β-naphthol 6:409
colorimetric method for vanadium 19:331, 334
color reactions for identification 5:431
cyanogen bromide method 6:35
in cyclohexylamine 11:217
derivatives for identification 5:429
in diphenylamine 11:459
gas chromatography 5:268, 436, 445; 11:212,
  213, 241, 247, 459
gravimetric method for isocyanic acid esters 15:
  111
gravimetric methods 5:448
in hydrazines 14:239
infrared spectroscopy 5:434, 452
iron in 5:446
manufacture 5:424
near-infrared spectroscopy 5:435, 443, 453
nitrobenzene in 5:446
phenolic compounds 5:446
picryl chloride method 5:447
in preparation of Safranine T 6:384
properties 5:421
reagent for amide formation with acid anhydrides
  8:564
reagent for formation of derivatives of acid halides
  8:572
redox titration 5:445
sampling 5:428
separation 5:435
toxicity 5:429
ultraviolet spectroscopy 5:432, 449; 11:217;
  14:239
uses 5:426
vapor in air 5:455
x-ray diffraction 5:435

Aniline and its derivatives. (See also Aniline; and
  individual compounds.) 5:421–459
acid–base titration, nonaqueous 5:443
analysis 5:428
assay methods 5:443
azeotropes 5:424
azo dye colorimetric method 5:445, 450
azo dyes for identification 5:430
bromination method 5:444
chronopotentiometry 5:454
colorimetric methods 5:445, 450
color reactions for identification 5:430
derivatives for identification 5:429
gas chromatography 5:268, 436, 445

identification 5:429
impurities, determination of 5:446
infrared spectroscopy 5:433, 453
manufacture, methods of 5:424
mixtures, determination of 5:447
near-infrared spectroscopy 5:434, 453
polarography 5:454
properties 5:421
refractive index methods 5:454
separation 5:435
ultraviolet spectroscopy 5:432, 449
uses 5:426
vapors in air 5:455
x-ray diffraction 5:435
Aniline Black 6:381
Aniline derivatives. (See also individual compounds.) 6:177
in cinchophen 6:205
Aniline–formaldehyde resins 5:426
properties 5:277, 281
Aniline hydrochloride 5:421
acid–base titration 5:444
azeotropes 11:436
manufacture 5:424
properties 5:422
solubility 5:421
ultraviolet spectroscopy 5:432
uses 5:427
Aniline point
of hydrocarbon resins 14:285
of hydrocarbons 14:348
rubber, relation to 15:320
Aniline-reaction method
for acetic anhydride 4:103
Aniline salts
uses 5:427
Aniline–sodium methoxide
volumetric method for alkyl iodides 16:93
Aniline–sulfanilamide
gravimetric method for mercury 16:22
Aniline sulfate 5:421
acid–base titration 5:444
manufacture 5:424
properties 5:422
solubility 5:421
Animal feed
calcium in 8:107
iodide content 14:601
Animal glue. (See also Glues of animal origin.) 4:399
Biuret test 10:282
foam content 10:275
grease content 10:275
jelly strength 10:273
moisture content 10:275
pH 10:274
tannic acid gravimetric method 10:282

viscosity 10:269
Animal grease
in gelatin 12:547
solvent extraction 12:546
Animal tissue
benomyl content 13:206
benzylpenicillin content 16:568
chloroneb content 13:218
mercury content 16:15
organomercury content 16:64
Animal units
in bioassay 7:156
Animal waste products
in cocoa beans 9:551
Anion exchange capacity
of ion exchange resins 15:19
Anion exchange resins. See Ion exchange resins.
Anionic detergents. (See also Detergents.) 11:322
benzethonium chloride titration 11:337
benzidine gravimetric method 11:336
carbon adsorption method 11:340
methylene blue colorimetric method 11:328, 339; 18:183
methylene blue titration 11:337
1-methylheptylamine spectroscopic method 11:340
molecular weight, determination of 11:336
phosphoric acid desulfonation method 11:329
in soaps 18:183, 204
p-toluidine hydrochloride volumetric method 11:338
Anionic materials
electrodeposition method 11:131
in finishing agents 11:126, 131
solubility 11:126, 131
Anionic surfactants. (See also Surfactants.)
disodium phosphate spectroscopic method 12:575
Hyamine 1622 titration 11:78
ion exchange chromatography 11:81
paper chromatography 12:575
in shampoos 11:77
Anisaldehyde (4-Methoxybenzaldehyde) 6:597
column chromatography 6:618
commercial grades and specifications 6:607, 609
derivatives for identification 6:612
fluorometry 6:626
gas chromatography 8:355
hydroxylamine hydrochloride method 6:620
infrared spectroscopy 6:616
manufacture 6:600
near-infrared spectroscopy 6:629
properties 6:598
ultraviolet spectroscopy 6:633
uses 6:606
Anise oil. (See also Essential oils.)
refractive index 12:271

Anisic acid. See *p*-Methoxybenzoic acid.
m-Anisidine
  infrared spectrum 5:439
  manufacture 5:425
  properties 5:422
  ultraviolet spectroscopy 5:433
o-Anisidine 5:421
  derivatives for identification 5:429
  infrared spectrum 5:439
  manufacture 5:425
  near-infrared spectroscopy 5:435
  picryl chloride method 5:447
  properties 5:422
  ultraviolet spectroscopy 5:433
p-Anisidine
  acid–base titration 5:447
  azo dyes for identification 5:430
  color reactions for identification 5:431
  derivatives for identification 5:429
  infrared spectrum 5:439
  manufacture 5:425
  near-infrared spectroscopy 5:435
  picryl chloride method 5:447
  properties 5:422
  ultraviolet spectroscopy 5:433
Anisidines 5:421
  in hair dyes 11:103
  thin-layer chromatography 11:103
Anisindione 10:176
  acid–base titration 10:177
  analysis 10:177
  identification 10:177
  infrared spectrum 10:176
  melting point 10:177
  properties 10:176
  solubility 10:176
  ultraviolet spectroscopy 10:176
Anisole
  in methanol 16:115
Anisomycin 5:476, 598
  analysis 5:602, 606, 613
  identification 5:602
  infrared spectrum 5:603
  microbiological assay 5:606
  paper chromatography 5:602
  properties 5:600
  separation 5:602
  solubility 5:601, 602
  ultraviolet spectrum 5:604
  uses 5:598
Ankerite 8:75
Annabergite
  composition 16:286
Annatto
  assay 10:539
  in cheese 10:544
  commercial specifications 10:530
  composition 10:530
  identification 10:536, 537
  impurities, determination of 10:541
  paper chromatography 10:541
  stannous chloride colorimetric method 10:536
  structure 10:539
  visible spectroscopy 10:539
Anodic-stripping voltammetry
  of mercury 16:53
Anorthite
  composition 10:104
Anosmia 17:612
Anprolene. See *Ethylene oxide.*
Antazoline
  gas chromatography 6:210, 211
Antazoline phosphate. (See also *Antihistamines.*)
    6:26
  properties 6:32
Antergan 6:24
Anthocyanins
  bisulfite bleach test 19:464
  pH shift 19:464
  structure 19:461
  ultraviolet spectroscopy 7:647
  visible spectroscopy 19:464
  in wines 19:460, 464
Anthocyanogens
  in beer 7:551
  colorimetric determination 19:467
  ferrous sulfate colorimetric method 19:468
  fractionation 7:647
  separation 19:468
  in wines 19:461, 467
Anthoxanthin pigments
  in wines 19:461
Anthracene
  derivatives for identification 16:208
Anthracene chrome violet B
  colorimetric method for niobium 16:369, 374
9-Anthraldehyde
  fluorometry 6:625, 627
  ultraviolet spectroscopy 6:632
Anthranilaldehyde. See 2-Aminobenzaldehyde.
Anthranilic acid (*o*-Aminobenzoic acid) 8:401
  paper chromatography 8:379, 404
  properties 8:401, 402
  uses 8:401
Anthraquinone dyes. (See also Certified colors.)
    12:9
  paper chromatography 10:475
Anthrazinedione
  color test for malonic acid 15:424
Anthrocyanidin
  in cacao 9:536
Anthrocyanin
  in cacao 9:536
Anthrone

colorimetric method for carbohydrates 7:557
colorimetric method for galactose 12:550
colorimetric method for glycerol 12:587
colorimetric method for glyoxal 12:585
colorimetric method for melamine–formaldehyde resin 11:134
colorimetric method for sodium heparin 10:149
for lactose in casein 9:10
Anti-Age 33 (and 34) 6:72
Anti-Age 44 (55 and 66) 6:73
Antibiotic K-288. See Vancomycin.
Antibiotics 5:460–633
  analysis 5:478
  assay methods 5:478
  biochemical determination 16:128
  classification 5:460
  identification 5:478
  impurities, determination of 5:491
  microbiological assay 5:479
  in raw milk 16:128
  separation 5:478
  standards 5:478
Anticogulants (See also Acenocoumarol; Anisin-
    dione; Bishydroxycoumarin; *Coumarins;* Di-
    phenadione; Ethyl biscoumacetate; Phenin-
    dione; Phenprocoumon; Sodium heparin;
    Warfarin; Warfarin potassium; and Warfarin
    sodium.) 10:140–184
  classification 10:140
Antidegradants. (See also *Antioxidants and
    antiozonants.*)
  colorimetric determination 12:142
  in elastomers, synthetic 12:140
  gas chromatography 12:142
  infrared spectroscopy 12:141
  thin-layer chromatography 12:140
  ultraviolet spectroscopy 12:141, 142
  volumetric determination 12:143
Antifoaming agents
  in confectionery products 10:552
Antifogging agents
  in gelatin 12:553
Antifreezes 6:1–23
  additives, determination of 6:9
  analysis 6:3
  gas chromatography 6:5
  mixtures, determination in 6:17
  performance tests 6:18
  properties 6:2, 3, 5
  propylene glycol content 13:599
  sampling 6:3
  specific gravity 6:5
Antifreezes, gas line 6:355
  analysis 6:368
  performance tests 6:378
Antifreezes, windshield washer 6:355
  analysis 6:369

performance tests 6:378
Antigen A (BC and C) 6:73
Antigorite
  composition 10:103
Antihistamines 6:24–38
  acid–base titration, nonaqueous 6:25
  analysis 6:24
  codeine content 16:270
  column chromatography 6:36
  cyanogen bromide colorimetric method 6:35
  hydrocodone content 16:272
  identification 6:24
  infrared spectroscopy 6:25
  ion exchange chromatography 6:34
  ninhydrin colorimetric method 6:35
  paper chromatography 6:35
  properties 6:24, 32
  reineckate colorimetric method 6:33
  separation 6:25
  thin-layer chromatography 6:35
  ultraviolet spectroscopy 6:31
Anti-icing agents
  in gasoline 13:359
Antimonial lead
  antimony in 6:58, 61
  sample preparation 6:61
Antimonides
  acid dissolution 9:215
  fluxes for 9:216
Antimonites 6:52
Antimony. (See also Antimony alloys; Antimony
    metal.) 6:39–70
  in alloys 6:57
  in aluminum alloys 5:118
  ammonium sulfide–hydrochloric acid colorimetric
    method 12:186
  in antimonial lead 6:58
  in antimony fluoborate 6:54
  in antimony halides 6:49
  in antimony metal and alloys 6:42
  in antimony oxides 6:53
  in antimony sulfides 6:54
  in arsenic metal 6:225
  in arsenic(III) oxide 6:231
  atomic absorption spectroscopy 6:57; 15:195,
    196
  in bismuth metal 7:193
  borax bead test 6:584
  bromate volumetric method 6:64
  cerate volumetric method 6:64
  cerium(IV) sulfate colorimetric method 13:393
  in chromium 9:635
  in cobalt alloys 10:308
  in cobalt metal 10:294
  colorimetric method 6:67
  colorimetric method using rhodamine B 7:193
  in copper 6:60

in copper alloys 6:59; 10:633
in copper metal 10:619
in cotton 11:141, 156
detection 6:57
in dross 6:58
in elastomers, synthetic 12:151
emission spectroscopy 6:57; 14:34, 542; 15:197; 16:387
in enamels 12:186
in ferrochromium 9:661
in ferroniobium 16:387
flame tests 6:589, 592
in flue dusts 6:58
in gallium metal 13:266
in germanium 13:393
in gold alloys 14:34
in hydrochloric acid 14:374
hydrogen sulfide gravimetric method 12:151, 420
in indium 14:542
in indium alloys 14:551
iodine volumetric method 6:65
iodoantimonite colorimetric method 6:67
ion exchange chromatography 13:254
in iron 15:59
in lead 15:195, 196, 197
in lead alloys 6:58, 60; 15:195
manganese dioxide gravimetric method 6:63; 10:294
in matte 6:58
9-methyl-2,3,7-trihydroxy-6-fluorone colorimetric method 11:141
microcosmic salt bead test 6:587
neutron activation analysis 13:265
in nickel 16:323
in nickel alloys 16:323
in niobium 16:387
occurrence 6:39
in ores 6:58
in organic medicinals 6:58
permanganate volumetric method 6:66
phenylfluorone colorimetric method 6:67
polarography 6:58
potassium bromate titration 10:633; 14:551
in potassium pyroantimonate 6:56
properties 6:39
rhodamine B colorimetric method 6:68, 225, 231; 10:294, 619; 11:141, 144; 15:197; 16:323
separation 6:62
in sodium arsenite 6:235
sulfide precipitation 6:62
thiosulfate volumetric method 6:66
in tin alloys 6:60
in tin metal 19:66, 68
x-ray fluorescence spectroscopy 6:57; 11:156
volumetric method 6:64

Antimony alloys. (See also Antimony metal.) 6:39
antimony content 6:42
carrier gas fusion methods 8:682
composition 6:39
oxygen in 8:682
uses 6:39
Antimony–aluminum–gallium alloys 13:269
aluminum content 13:269
analysis 13:269
gallium content 13:269
Antimony bromide 6:48
antimony content 6:49
Antimony chloride 6:48
antimony content 6:49
copper in 6:51
iron in 6:51
sample preparation 6:61, 62
sulfate in 6:51, 52
Antimony compounds. (See also individual compounds.) 6:46–56
sample preparation 6:61
Antimony fluoborate 6:54
analysis 6:54
antimony content of 6:54
boric acid in 6:56
boron content of 6:55
fluoboric acid in 6:55
fluorine content of 6:54
hydrofluoric acid in 6:56
Antimony fluoride 6:48
antimony content 6:49
fluorine content 6:50
sample preparation 6:61
Antimony halides 6:48
analysis 6:49
antimony content 6:49
halogen content 6:50
impurities, determination of 6:50
Antimony–indium alloys. (See also Indium alloys.)
indium content 14:548
Antimony iodide 6:48
antimony content 6:49
Antimony lactate. (See also Lactic acid salts.)
uses 15:130
Antimony–lead alloys
tin content 19:64
Antimony metal. (See also Antimony alloys.) 6:39
aluminum in 5:225
antimony content of 6:42
arsenic in 6:42
commercial grades and specifications 6:41
copper in 6:42, 46
iron in 6:43
lead in 6:43, 46
metallurgy 6:40

nickel in 6:44, 46
properties 6:39
sample preparation 6:58, 236
silver in 6:44, 46; 18:172
sulfur in 6:45; 18:382
tin in 6:46
uses 6:39
Antimony oxide 6:52
  analysis 6:53; 17:190
  antimony content of 6:53
  commercial specifications 17:190
  in enamels 12:173
  impurities, determination of 6:53
  metallurgy 6:40
  permanganate titration 17:190
  sample preparation 6:61, 62
  uses 17:190
Antimony, pentavalent 6:47
Antimony potassium tartrate 6:54
  analysis 6:54
Antimony salt 6:49
Antimony sulfide 6:39, 53, 54
  analysis 6:54
  commercial grades and specifications 15:522
  in explosive priming mixtures 12:418, 420, 424,
    425, 426, 428, 429, 430
  gravimetric determination 12:420
  hydrochloric acid gravimetric method 12:420,
    424
  in match strikers 15:522, 536
  metallurgy 6:40
  particle size 15:536
  properties 6:47, 53, 54
  solubility 12:420
  solvent extraction 12:424, 426, 428
  x-ray diffraction 12:418
Antimony tetroxide 6:47
Antimony trichloride
  colorimetric method for codeine 16:246
  properties 9:334
  solubility 9:336
Antimony trihydride 6:47
Antimony trioxide
  in alkali barium silicate glass 9:218
  emission spectroscopy 9:218
  in enamels 9:195
  in glass 9:200
  in glass and glass-ceramics 13:413, 415, 428
  hydrobromic acid distillation 13:430
  hydrogen sulfide–sulfuric acid gravimetric
    method 13:430
  potassium bromate titration 13:430
  rhodamine B colorimetric method 13:431
Antimony, trivalent 6:47
Antimycin 5:477, 598
  analysis 5:602, 607, 613
  bioautography 5:603

countercurrent distribution 5:602
fluorometry 5:613
fish-kill method 5:613
identification 5:602
microbiological assay 5:607
paper chromatography 5:602
properties 5:600
separation 5:602
solubility 5:601, 602
structure 5:477
thin-layer chromatography 5:613
uses 5:598
Antioxidant 5
  colorimetric method using diazotized *p*-nitro-
    aniline 6:106
Antioxidant 108 (116, 123, 184, and 237) 6:73
Antioxidant 425 6:73
  colorimetric method using diazotized *p*-nitro-
    aniline 6:106
Antioxidant 2246 6:73
  colorimetric method using diazotized *p*-nitro-
    aniline 6:106
Antioxidant 4010
  colorimetric method using diazotized *p*-nitro-
    aniline 6:98
Antioxidants and antiozonants. (See also Amine
  antioxidants; Phenolic antioxidants; and
  individual compounds and trade names.) 6:
  71–122
analysis 6:81; 7:428
in brake fluids 7:424, 425, 428
in cereals 9:242
colorimetric methods 6:97
color tests 6:91
column chromatography 6:87
composition 6:72
in confectionery products 10:552
in engine starting fluids 6:356
extraction from rubber 6:108
evaluation 6:78
gas chromatography 6:81, 108
in gasoline 13:359
identification 6:81
infrared spectroscopy 6:97
in methacrylate polymers 16:94
paper chromatography 6:84
in photographic films 12:592
in polyamides 17:301
in polyethylene films 12:592
in polyurethanes 19:312
separation 6:81
solvent extraction 16:94; 19:312
in styrene polymers 18:321
testing 6:78
thin-layer chromatography 6:89; 16:94; 19:312
trade names 6:72
ultraviolet spectroscopy 6:97; 16:94

uses 6:71
Antioxygene (A, AFL, AN, BN, CAS, DIP,
    MC, and RES) 6:73
Antiozonants. See *Antioxidants and antiozonants.*
Antiperspirants. See Deodorants.
Antipyretics and analgesics. (See also *p*-Amino-
    phenol derivatives; Aniline derivatives;
    Pyrozolone derivatives; Quinoline derivatives;
    Salicylic acid; Salicylic acid derivatives; and
    individual compounds.) 6:123–222
Antipyrine 6:206
    acid–base titration 6:215
    in aminopyrine 6:218
    analysis 6:207
    colorimetric method using *p*-dimethylamino-
        benzaldehyde 6:214
    colorimetric method using α-naphthol 6:214
    column chromatography 6:218
    commercial grades and specifications 6:207
    gas chromatography 6:209, 210, 211
    gravimetric method 6:214
    identification 6:209
    impurities, determination of 6:218
    infrared spectrum 6:208
    iodometric method 6:215
    ion exchange chromatography 6:218; 8:62
    manufacture 6:206
    polarography 6:215
    properties 6:206
    ultraviolet spectroscopy 6:208, 209, 214
Antirusts 6:351
    analysis 6:357
    in gasoline 13:359
    performance tests 6:372
Antiseptics. (See also *Disinfectants and anti-
    septics.*)
    definition 11:538
Anti-skinning agents
    in enamels 12:177
Antox 6:73
    colorimetric method using diazotized *p*-nitro-
        aniline 6:99
6-APA. See 6-Aminopenicillanic acid.
APANS. See Thorin.
Apatites 8:75; 17:51, 143
    in silicate rocks 17:71
APHA color standards 4:199
API gravity
    of gasoline 13:360
Aplite
    titanium content 19:110
β-Apo-8'-carotenal
    assay 10:540
    commercial specifications 10:530
    structure 10:538
    visible spectroscopy 10:540
Apomorphine. (See also Apomorphine hydro-

chloride; and *Narcotics.*) 16:276
    analysis 16:276
    in biological materials 16:276
    colorimetric determination 16:277
    color tests 16:246
    p-dimethylaminobenzaldehyde colorimetric
        method 16:277
    formaldehyde–sulfuric acid colorimetric method
        16:246
    gas chromatography 16:254
    gold chloride microchemical method 16:249
    hydrochloric acid microchemical method 16:249
    manufacture 16:276
    microchemical tests 16:249
    nitric acid colorimetric method 16:246
    paper chromatography 16:255
    polarography 16:268
    potassium iodide microchemical method 16:249
    properties 16:276
    solvent extraction 16:276
    structure 16:245
    thin-layer chromatography 16:256
    ultraviolet spectroscopy 16:251, 276
    uses 16:276
Apomorphine hydrochloride. (See also Apomor-
    phine.)
    acid–base titration 16:276
    phosphorimetry 16:264
    solvent extraction 16:276
    ultraviolet spectroscopy 16:276
Apple acid 8:399
Aprobarbital 6:468, 512
    acid–base titration, nonaqueous 6:512
    analysis 6:512
    bromide-bromate titration 6:497
    commercial specifications 6:512
    derivatives for identification 6:477
    extraction 6:491
    gas chromatography 6:483, 484
    impurities, determination of 6:512
    properties 6:470
    solubility 6:476
    thin-layer chromatography 6:481
Aprotonic acidity
    in aluminum oxide 5:188
Aqua fortis. See Nitric acid.
Aqua regia
    composition 16:496
    properties 16:496
Arabinitol
    liquid chromatography 13:503
Arabinose
    gas chromatography 13:507; 19:300
    in green tea leaf 18:478
    paper chromatography 18:478
Arachidonic acid. (See also Fatty acids.)
    properties 12:473

Aragonite 8:75
Aranox 6:73
  colorimetric method using diazotized p-nitro-
    aniline 6:99
Arasan. See Thiram.
Arathane. See Dinocap.
Ardealite 17:121
Areal shrinkage of fabrics 4:25
Arecoline 4:591
Argentite 18:155
Argentometric method
  for epoxy compounds 12:212
  for halogens 14:156
  for sodium borohydride 7:347
  for warfarin sodium 10:169
Argentometric titration
  of barbiturates 6:493
  of chlorosulfonic acid 9:530
Argillaceous limestone 15:203
Arginine. (See also *Carboxylic acids, amino.*)
    8:412, 452
  analysis 8:455
  in cocoa beans 9:556
  color tests 8:455
  commercial specifications 8:455
  composition 8:412
  crystalline form 8:453
  diacetyl-α-naphthol color test 8:455
  enzymatic decarboxylation 8:455
  gas chromatography 8:429
  infrared spectrum 8:454
  ion exchange chromatography 8:415, 421
  ninhydrin test 8:455
  paper chromatography 8:422
  properties 8:452
  in protein 8:414
  Sakaguchi reaction 8:455
  structure 8:452
  uses 8:454
  in yeast 19:529
Argon. (See also *Gases, noble.*)
  in air 16:485
  in carbon dioxide 8:252
  in carbon monoxide 8:310, 312, 315
  in cigarette smoke 10:13
  commercial specifications 13:287
  gas chromatography 8:252, 310, 312; 13:297,
    307; 16:485, 523
  in helium 13:298
  in hydrogen 14:380
  impurities, determination of 13:299
  isotopic abundance 13:289
  in krypton 13:300
  manufacture 13:290
  mass spectroscopy 8:315; 13:292, 301, 304
  in neon 13:299
  occurrence 13:290

  in oxygen 16:531
  properties 13:288
  purification of carrier gas 8:673
  solubility 13:289
  thermal conductivity 14:380
  in xenon 13:300
Argyrodite
  occurrence 13:376
Argyrol. (See also Silver compounds.)
  uses 18:174
Aristocort. See 9-Fluoro-16-hydroxy predniso-
    lone.
Arnow's reagent
  colorimetric method for chlorogenic acids 19:
    146
  colorimetric method for polyphenols 19:146
  colorimetric method for rutin 19:146
Aroma
  of black tea 18:507
  of cocoa beans 9:567
  of cocoa products 9:592
  of roasted coffee 10:436
Aromatic acids. See *Carboxylic acids.*
Aromatic aldehydes
  acetal formation 6:630
  fluorometry 6:625
  gas chromatography 6:619
  hydroxylamine hydrochloride method 6:620
  infrared spectroscopy 6:615
  near-infrared spectroscopy 6:628
  ultraviolet spectroscopy 6:630
Aromatic antibiotics 5:461, 494
  analysis 5:496, 500, 506
  properties 5:495
Aromatic diamines. (See also *Diamines and
    higher amines, aromatic;* and individual
    compounds.)
  properties 11:432
  uses 11:432
Aromatic petroleum resins. (See also *Hydro-
    carbon resins.*) 14:263
  commercial grades and specifications 14:276, 279
  composition 14:263
  polymer distribution 14:269
  uses 14:261
Arrhenal. See Disodium methylarsenate.
Arsenates 6:229
Arsenazo
  colorimetric method for rare earths 17:479
  colorimetric method for scandium 17:574
Arsenazo I
  colorimetric method for niobium 16:369, 374
  colorimetric method for tantalum 16:369; 18:
    449
  colorimetric method for thorium 19:13
  structure 19:13
Arsenazo II

colorimetric method for thorium 19:13
structure 19:13
Arsenazo III
  colorimetric method for tantalum 16:369
  colorimetric method for thorium 19:13, 14
  colorimetric method for uranium 19:249
  structure 19:14
Arsenic. (See also Arsenic alloys; Arsenic
    metal.) 6:223, 254
  in aluminum chloride 5:175
  in aluminum hydroxide 5:188
  in aluminum metal 5:118
  in aluminum sulfate 5:193
  in alums 5:193
  in amino acids 8:436
  in ammonium phosphate, dibasic 5:350
  in ammonium sulfate 5:352
  in antimony halides 6:50
  in antimony metal 6:42
  in arsenic compounds 6:241
  in arsenic metal 6:225
  arsenic trichloride distillation 16:187
  arsenomolybdate extraction method 6:244
  by arsine evolution 5:175; 8:41
  atomic absorption spectroscopy 15:196
  in barium sulfate 6:567, 568
  in bismuth chloride 7:199
  in bismuth metal 7:193
  in bismuth nitrate 7:202
  in bismuth subcarbonate 7:196
  in bismuth subnitrate 7:203
  bromate volumetric method 6:241
  in cadmium chloride 8:41
  in calcium acetate 8:78
  in calcium carbonate 8:84
  in certified colors 10:493
  in cinnamic acid and its derivatives 10:70
  in citrate esters 10:99
  in citric acid and salts 10:95
  in cobalt alloys 10:308
  in cobalt metal 10:294
  in cocoa products 9:595, 598
  in copper alloys 10:634
  in copper(I) chloride 10:661
  in copper metal 10:619
  in cotton 11:140
  in cyclamates 11:231
  in dental restorative materials, composite 11:316
  in dental silicate cements 11:261
  in dental zinc phosphate cements 11:263
  detection 6:237
  distillation 6:237
  in eggs and egg products 12:66
  emission spectroscopy 6:237, 243; 14:34, 542;
    15:144, 197; 16:387
  extraction 6:240
  ferric hydroxide gravimetric method 10:294

in ferrochromium 9:661
in ferroniobium 16:387
flame tests 6:589, 592
in fluorosilicic acid 13:50
in formic acid esters 13:134
in gallium metal 13:266
in gelatin 13:369
in germanium 13:391
in germanium dioxide 13:391
in glycerol 13:526
in gold alloys 14:34
gold chloride colorimetric method 11:140
gravimetric determination 16:20
Gutzeit method 6:246
in gypsum 14:98
in hydrochloric acid 9:340, 351; 14:373, 374
identification 6:237
in indium 14:542
in indium alloys 14:550
iodometric titration 6:241; 14:550; 16:188
in iron 15:59
in lactic acid and its derivatives 15:143, 144
in lead 15:196
in lead alloys 15:195
in lime and limestone 15:223
in magnesium carbonate 15:402
in magnesium oxide 15:411
in magnesium sulfate 15:412
mercuric bromide colorimetric method 13:526
in mercury 16:20
molybdenum blue colorimetric method 6:244,
    245; 10:294, 619; 13:391
in molybdoarsenic acid 16:187
in nickel 16:323
in nickel alloys 16:323
in niobium 16:387
in nitric acid 16:497
permanganate volumetric method 6:242
in phosphate rocks 17:71
in phosphorus pentoxide 17:111
potassium bromate titration 10:634
in potassium chlorate 9:359, 361
separation 6:237
in silicon 18:39
silver diethyldithiocarbamate colorimetric method
    6:250; 8:436; 10:95; 13:369; 14:373; 15:143,
    223; 18:419
in sodium tungstate 19:221
solvent extraction 16:20
stannous chloride colorimetric method 11:140
stannous chloride gasometric method 15:223
sulfide precipitation 6:240
in sulfuric acid 18:419
thiosulfate volumetric method 6:243
in tin metal 19:66, 68
in uranium 19:238
x-ray fluorescence spectroscopy 6:243

Arsenic acid 6:228
Arsenic alloys. (See also Arsenic metal.) 6:223
   analysis 6:225
   arsenic content of 6:225
   carrier gas fusion methods 8:682
   impurities, determination of 6:225
   oxygen in 8:682
Arsenicals. (See also individual compounds.)
      19:580
   commercial grades and specifications 19:593
   manufacture 19:591, 595
   properties 19:586
   solubility 19:587
   toxicity 19:592
Arsenic bromide
   manufacture 6:229
   properties 6:229
Arsenic chloride 6:229
   analysis 6:234
   commercial specifications 6:230
   distillation 6:237
   impurities, determination of 6:234
   manufacture 6:229
   properties 6:229, 230
Arsenic compounds. (See also individual com-
      pounds.) 6:228, 235
   analysis 6:230
   arsenic content of 6:241
   properties 6:228, 229
   sample preparation 6:236
   toxicity 6:224, 228
Arsenic compounds, pentavalent
   arsenic content of 6:237
   sample preparation 6:237
Arsenic compounds, trivalent
   arsenic content of 6:241
   sample preparation 6:236
Arsenic fluoride 6:229
   properties 6:230
Arsenic iodide
   manufacture 6:229
   properties 6:230
Arsenic metal. (See also Arsenic alloys.) 6:223
   analysis 6:225
   arsenic content of 6:225
   impurities, determination of 6:225
   metallurgy 6:224
   occurrence 6:223
   properties 6:224
   sample preparation 6:236
   uses 6:223
Arsenic oxide 6:223, 228, 229
   in alkali barium silicate glass 9:218
   analysis 6:230, 233
   arsenic(III) oxide in arsenic(V) oxide 6:233
   commercial specification 6:230
   emission spectroscopy 9:218

   in enamels 9:195
   in glass 9:200
   impurities, determination of 6:230, 233
   metallurgy 6:224, 229, 230
   properties 6:229, 230
   in sodium arsenate 6:234
   uses 6:229
Arsenic sulfide
   properties 6:229, 230
   toxicity 6:224
Arsenic trichloride
   distillation of arsenic 16:187
   properties 9:334; 19:96
   solubility 9:336
Arsenic trioxide
   boric acid distillation 13:429
   in glass and glass-ceramics 13:413, 415, 428
   potassium bromate titration 13:430
Arsenic, white. See Arsenic(III) oxide.
Arsenides
   acid dissolution 9:215
   fluxes for 9:216
   sample preparation 15:31
   solubility 15:31
Arsenite absorption
   of chlorine 9:317
Arsenites 6:229
   in antifreezes 6:10
Arsenite titration
   of chloroisocyanuric acids and salts 7:229
Arsenomolybdate extraction method
   for arsenic 6:244
Arsenomolybdate method
   for carbomycin 5:531
Arsenous oxide. See Arsenic oxide.
Arsine 6:223, 243
   evolution of 6:10, 246, 250
   Gutzeit method 6:246
Arsine generator 15:224; 18:420
3-(2-Arsonophenylazo)-4,5-dihydroxy-2,7-naph-
      thalenedisulfonic acid disodium salt. See
      Arsenazo.
1-(*o*-Arsonophenylazo)-2-naphthol-3,6-disulfonic
      acid disodium salt. See Thorin.
Arylamines
   colorimetric method using β-naphthol 6:409
   colorimetric method using R-salt 6:409
   identification 6:408
Arylbismuth halides 7:208
Aryl hydrazines. (See also *Hydrazine.*)
   assay 14:236
   manufacture 14:234
   potassium iodate titration 14:236
   solubility 14:232
   uses 14:232
N-Arylhydroxylamines. (See also *Hydroxylamines
      and their salts.*)

manufacture 14:445
properties 14:443
uses 14:443
Arylide maroon. (See also *Pigments.*)
structure 17:179
Arylpyridines. (See also *Pyridine and pyridine
derivatives.*)
manufacture 17:422
Asarcolo alloys 8:32
Asbestos. (See also Amosite; Chrysotile; and
Crocidolite.) 6:254, 269
in asbestos–cement products 6:270, 271, 272
analysis 6:258; 7:446
in brake linings 7:440, 441, 446
composition 6:257
in cooling system sealers 6:353, 361
mining 6:254
properties 6:255
Asbestos–cement products 6:270, 273
analysis 6:270
composition, determination of 6:270
manufacture 6:270
uses 6:255, 270
Asbestos fibers. See *Asbestos.*
Asbolite 10:284
Ascarite
absorbent for carbon dioxide 8:262
Ascharite camsellite 7:315
Ascorbic acid
in carbonated beverages 8:166
in chocolate and cocoa products 9:538
colorimetric method for niobium 16:368, 375
colorimetric method for selenium 14:99
colorimetric method for titanium 19:107
2,6-dichlorophenolindophenol titration 19:424
2,4-dinitrophenylhydrazine colorimetric method
19:425
in green tea leaf 18:458
iodometric method 19:425
in musts 19:424
oxidation methods 19:425
phosphotungstic acid colorimetric method 19:
424
potassium permanganate titration 18:459
potassium titanium oxalate colorimetric method
18:458
properties 8:396
solvent extraction 19:424
thin-layer chromatography 8:381; 19:425
volumetric method for gold 14:22
in wines 19:424
Ascorbic acid–ammonium molybdate
colorimetric method for phosphorus 17:72
Ascorbic acid–benzidine
colorimetric method for peroxidase 18:485
Ash content
of activated carbon 8:158

of adhesives 10:276
of adsorbents 4:436
of antifreezes 6:10
of beer 7:553
of carbon black 8:234
of carbon fabrics 4:26
of carbon materials 14:62
of casein 9:9
of cellulose 9:48
of cellulose acetate 9:69
of cellulose nitrate 9:64
of cereals 9:236
of chocolate 9:540, 550
of coal and coke 10:236
of cocoa beans 9:548, 549, 553
of cocoa products 9:573
of confectionery products 10:561
of copper stearate 10:679
of corn products 11:11
of cotton 11:143, 148, 157
of creams and lotions 11:92
of cyanoacetic acid 15:440
of dental restorative materials, composite 11:
310
of diamonds, industrial 11:481
of eggs and egg products 12:64
of felt 12:509
of flour 12:602
of formaldehyde 13:115
of glues of animal origin 13:493
of glycerol 13:521
of graphite 4:14; 14:62
of green coffee 10:374, 391
of high-explosive mixtures 12:467
of hydrocarbon resins 14:284
of hydrochloric acid 9:340, 350
of lactic acid and its derivatives 15:141
of methylcellulose 9:79
of musts 19:488
of perchloric acid 9:364
of phenols 17:38
of photographic films 12:592
of pigments 17:177
of polytetramethylene ether glycols 13:249
of roasted coffee 10:375, 417, 423
of shampoos 11:75
Skala and Swanson method 12:64
of sodium carboxymethylcellulose 9:88
of sodium heparin 10:150
of sugar 8:175; 18:357
of tea 18:487
of terephthalic acid 17:162
of tobacco 19:153
of wines 19:488
of yarns 4:91
of yeast 19:530
Ashing procedures

for acrylic and modacrylic fibers 4:283, 344
for beer 7:553, 648
for boron compounds, organic 7:399
for plant material 7:400, 416
Aspartic acid. (See also *Carboxylic acids, amino.*)
    8:412, 457
    analysis 8:461
    in cocoa beans 9:556
    colorimetric method using 2,4-dinitrophenyl-
        hydrazine 8:461
    commercial specifications 8:460
    composition 8:412
    crystalline form 8:458
    enzymatic decarboxylation 8:461
    gas chromatography 8:429
    ion exchange chromatography 8:415, 421
    metal complexes 8:459
    occurrence 8:460
    paper chromatography 8:421
    polarography 8:461
    properties 8:457
    in protein 8:414
    solubility 8:457, 459
    uses 8:457
    in yeast 19:529
Asphalt cements
    commercial grades and specifications 6:280
    specific gravity 6:319
    temperature susceptibility 6:306
    viscosity 6:300
Asphaltic lubricants. See Naphthenic lubricants.
Asphalts. (See also Bituminous materials.) 6:274
    in brake linings 7:441
    commercial grades and specifications 6:280
    composition 6:274
    durability 6:306
    manufacture 6:277
    occurrence 6:274, 277
    properties 6:274
    properties, determination of 6:289
    purity 6:311
    recovery from bituminous paving mixtures
        6:334
    solubility 6:311
    temperature–viscosity relationships 6:282, 284
Asphalts, blown 6:275
    commercial grades and specifications 6:282
    manufacture 6:279
Asphalts, cracked 6:275
    manufacture 6:278
Asphalts, cut-back
    commercial grades and specifications 6:282
    distillation test 6:313
    flash point 6:320
Asphalts, emulsified
    analysis 6:324
    commercial grades and specifications 6:285

demulsibility tests 6:327
    properties, determination of 6:326
    water content of 6:324
Asphalts, oxidized. See Asphalts, blown.
Asphalts, residual
    identification 6:318
    manufacture 6:278
    specific gravity 6:318
Asphalts, tars, and pitches 6:274, 350
Aspirin. (See also Acetylsalicylic acid.)
    acetyl codeine content 16:270
    codeine content 16:270
    codeine phosphate content 16:269
ASRC 12:86
Astrazone pink FG
    color test for sulfur dioxide 18:403
Atacamite
    copper content 10:600
Atactic content
    of polypropylene 12:361
Aterrimin 5:477, 598
    analysis 5:603, 608, 614
    composition 5:477
    identification 5:603
    microbiological assay 5:608
    paper chromatography 5:603
    properties 5:600
    separation 5:603
    solubility 5:601, 602
    uses 5:598
Atomic absorption spectra
    of europium oxide 12:401
    of europium(III) perchlorate 12:401
    of rare earth perchlorates 12:401
Atomic absorption spectroscopy
    of aluminum 5:212, 247; 9:346; 15:196
    of antimony 6:57; 15:195, 196
    of arsenic 15:196
    of beryllium 7:140
    of bismuth 7:217; 15:195, 196; 16:167
    of cadmium 12:571; 15:186, 196
    of calcium 8:109; 9:346; 10:37, 133; 11:157;
        13:136; 15:157, 380; 19:493
    of ceramics 9:217
    of cerium 17:494
    of chloride 14:128
    of chromium 9:346, 636, 649; 14:125
    of clays and clay minerals 10:130
    of cobalt 9:346
    of copper 7:200; 9:346; 10:132, 167; 11:157;
        14:49, 125, 126; 15:157, 186, 195, 196, 382;
        19:495
    of cotton 11:146, 156
    of dimethyldiethyllead 13:358
    of dysprosium 17:494
    of erbium 17:494
    of europium 17:494

of gadolinium 17:494
of gallium 13:260
of glass and glass-ceramics 13:450
of gold 14:26, 30, 41, 43, 47
of hafnium 14:108, 120, 124
of holmium 17:494
of impurities in antimony metal 6:46
of indium 14:535
of iron 7:200; 9:346; 10:131, 655, 661;
    11:157; 12:572; 14:49, 125; 15:56, 195,
    196; 16:117
of lanthanum 15:159; 17:494
of lead 7:200; 9:346; 10:668; 11:157; 12:572;
    14:49; 15:185, 196; 16:167
of lead additives 13:358
of lithium 15:266, 386
of lutetium 17:494
of magnesium 9:346, 667; 10:133; 11:157;
    14:126; 15:157, 359, 370
of manganese 9:346, 671; 11:157; 15:386, 478
of mercury 16:22, 43
of molybdenum 9:346; 14:126; 16:160
of neodymium 17:494
of nickel 9:346; 10:132; 14:125; 15:196; 16:
    321
of nitrosyl peroxide 16:476
of potassium 10:37, 131; 11:157; 15:157;
    16:165; 17:372, 390; 19:492
of praseodymium 17:494
of rare earths 17:491
of rhenium 17:534
of rubidium 17:552
of samarium 17:494
of scandium 15:390; 17:494, 577
of silicon 9:346; 18:37
of silver 10:622; 14:49; 15:195, 196; 18:172
of silver halides 12:564
of sodium 10:37, 131; 11:157; 14:125; 15:157,
    393; 16:164; 18:220
of strontium 9:346; 18:270, 280
of technetium 17:534
of terbium 17:494
of tetraethyllead 13:358
of tetramethyllead 13:358
of thallium 12:572
of thulium 17:494
of tin 14:125; 15:196; 19:64
of titanium 9:346; 19:119
of triethylmethyllead 13:358
of trimethylethyllead 13:358
of tungsten 19:218
of uranium 19:255
of vanadium 9:346; 10:132
of ytterbium 17:494
of yttrium 17:494
of zinc 7:663; 10:131; 11:157; 14:49; 15:157,
    186, 195, 196, 395; 19:568

of zirconium 14:108, 120, 124
Atranal. See 2,6-Dihydroxy-4-methylbenzaldehyde.
Atropine 4:587
    analysis 4:594
    gas chromatography 6:210, 211
Atropine sulfate tablets
    morphine content 16:261
Attapulgite. (See also *Clays and clay minerals*.)
    composition 10:103
Aubepine Liquid. (See also Anisaldehyde.) 6:607,
    609
Aubepine Prime. (See also Anisaldehyde.) 6:607,
    609
Augite 8:75
Auramine. (See also *Triphenylmethane and
    related dyes*.)
    colorimetric method for tantalum 16:370
    indigo carmine titration 19:175
Auramine G. (See also *Triphenylmethane and
    related dyes*.)
    paper chromatography 19:172
Auramine O. (See also *Triphenylmethane and
    related dyes*.)
    paper chromatography 19:172
Aureothricin 5:462, 508
    analysis 5:510, 511, 515
    identification 5:510
    infrared spectroscopy 5:515
    microbiological assay 5:511
    paper chromatography 5:510
    properties 5:509
    separation 5:510
    solubility 5:509, 510
    structure 5:463
    uses 5:508
Aurin
    structure 19:393
Aurintricarboxylic acid. See Aluminon.
Automobile brake system road test 7:455, 465
Automobile exhaust
    gas chromatography 7:7
    nitrogen oxide content 16:481
Automotive chemical specialities. (See also *Anti-
    freezes*.) 6:350, 380
    analysis 6:357
    performance tests 6:371
Autunite
    composition 19:225
    decomposition 19:235
    occurrence 19:225
Auxochromes. See Indole–thionaphthene dyes.
Available chlorine. (See also Chlorine content.)
    9:352
    in detergents 11:355
    in lithium hypochlorite 15:287
    sodium arsenite titration 9:353
    in water 19:377

Available halogen content
  of halogen donors 11:553
Available lime index 15:229
Avicennite 18:519
Avogadrite 7:315
Awaruite 15:21; 16:285
Axinite 7:316; 8:75
Azaserine 5:477, 599
  analysis 5:604, 608, 614
  microbiological assay 5:608
  properties 5:600
  solubility 5:601, 602
  structure 5:477
  uses 5:599
Azelaic acid 5:22; 8:400
  gas chromatography 5:25; 17:290; 19:294
  properties 8:396, 400
Azeotropic distillation
  of ethylene glycol in glycerol 13:507
  of glycerol 13:507
  of water in glycerol 13:507
Azides. (See also individual compounds.) 16:472
  analysis 16:472
  gravimetric method for silver 18:164
  in lead azide 12:411
Azine 6:381
Azine and related dyes. (See also Azine dyes;
    Oxazine dyes; and Thiazine dyes.) 6:381–
    400
  analysis 6:388
  application methods 6:397
  commercial grades and specifications 6:388
  identification 6:389
  impurities, determination of 6:399
  manufacture 6:383
  separation 6:392
  solubility 6:383
  visible spectroscopy 6:393
  volumetric method using titanous chloride
    6:394, 395
Azine dyes 6:381, 382; 12:8
  analysis 6:388
  identification 6:388
  manufacture 6:384
  solubility 6:383
Azinphosmethyl. (See also *Pesticides and
    insecticides.*)

colorimetric determination 19:611
commercial grades and specifications 19:
  594
composition 19:597
manufacture 19:597
potassium hydroxide–N-1-naphthylethylene-
  diamine colorimetric method 19:611
properties 19:590
solubility 19:587
structure 19:581
toxicity 19:592
Aziridines. See *Imines.*
4-Azobenzenediazonium fluoborate
  as colorimetric reagent 5:451
Azo Bordeaux
  in enamels 12:186
  identification 12:186
Azocoll–cysteine mixture
  colorimetric method for protease activity
    7:589
Azo compounds
  Kjeldahl method 16:458
Azo dyes 6:401–424; 12:6
  analysis 6:403
  application method 6:414
  colorimetric determination of aniline and its
    derivatives 5:445, 450
  column chromatography 6:414
  hydrolysis methods 6:407
  identification 6:403; 12:20
  for identification of aniline and its derivatives
    5:430
  infrared spectroscopy 6:404
  manufacture 6:402
  reaction with nitric acid 6:407
  reduction methods 6:404, 416
  salt test 6:415
  separation 6:413
  titanous chloride titration 6:416
  uses 6:401
  visible spectroscopy 6:404, 421
Azoic dyes 12:5
  structure 12:6
Azored N-resorcinol
  colorimetric method for boron 7:410
Azurite 10:657
  copper content 10:600

B

Babbitt metal. (See also Copper alloys; Tin
  alloys.) 5:72
  composition 10:631; 19:24
Babcock test
  for cheese 16:141
  for chocolate milk 16:132
  for cream 16:132
  for frozen desserts 16:143
  for homogenized milk 16:132
  for ice cream 16:143
  for raw milk 16:127
  for skim milk 16:132
Bacitracin 5:466, 533
  analysis 5:537, 542, 547
  bioautography 5:538
  gravimetric method using phosphotungstic acid
    5:547
  identification 5:537
  infrared spectrum 5:537
  microbiological assay 5:542
  paper chromatography 5:538
  properties 5:534
  separation 5:537
  solubility 5:535, 536
  thin-layer chromatography 5:537
  ultraviolet spectrum 5:537
  uses 5:533
Bacitracin A
  structure 5:466
Backing
  for coated abrasives 10:276
Bacteria
  in carbonated beverages 8:167
  in casein 9:13
  in cocoa products 9:599
  in penicillins 16:566
Bacteriological methods
  for disinfectants and antiseptics 11:538
Bacteriologic examinations
  of water 19:390
Bacteriostats. (See also Antiseptics.)
  in soaps 18:199
Baddeleyite 14:104
  composition 9:203
Bagginess
  of fabrics 4:25
Bahnmetall. (See also Sodium alloys.)
  composition 18:210
  uses 18:210
Bailey and Andrew method
  for caffeine 10:431

  for caffeine in tea 18:489
Bakerite 7:316
Baker's yeast. (See also *Yeast.*) 19:516
  species 19:516
  uses 19:516
Bakery chemicals. (See also *Bakery products* and
    individual chemicals.) 6:425–438
  analysis 6:430
  sample preparation 6:431
Bakery products 6:438–467
  analysis 6:448
  chemically leavened 6:444
  filth tests 6:461
  loss in processing 6:459
  microbiological testing 6:455, 460
  quality control 6:448, 462
  unleavened 6:447
  yeast-raised 6:438
Baking powders
  in cakes 6:445
  composition 6:430
  neutralizing value 6:426
  potassium content 17:388
  sodium content 17:388
Baking soda. See Sodium bicarbonate.
Baking tests
  for flour 12:615
Balance of grinding wheels 7:300
Balling 4:462
Ball-drop viscosity test
  for cellulose derivatives 9:66
Balzers gas analyzer Exhalograph 8:659
Bandylite 7:315
Baratol. (See also High-explosive mixtures.)
  analysis 12:446
  composition 12:446
  uses 12:446
Barbital 6:468
  colorimetric method 6:491
  derivatives for identification 6:477
  extraction 6:491
  gas chromatography 6:209, 211, 483, 484
  paper chromatography 6:479
  properties 6:470
  solubility 6:476
  thin-layer chromatography 6:480, 481
  ultraviolet spectroscopy 6:501
Barbiturates. (See also individual compounds.)
    6:467–532
  acid–base titration, nonaqueous 6:492
  analysis 6:475

argentometric titration 6:493
bromate titration 6:495
colorimetric methods 6:491
color tests 6:477
derivatives for identification 6:477
extraction 6:475, 491, 497
gas chromatography 6:482
gravimetric methods 6:493
identification 6:477
infrared spectroscopy 6:484, 492, 496
ion exchange chromatography 6:499
manufacture 6:470
paper chromatography 6:478
properties 6:469
separation 6:475
solubility 6:476
thin-layer chromatography 6:479
ultraviolet spectroscopy 6:483, 484,
    500
Barbituric acid 6:468
    properties 6:469
Barbituric acid-*p*-toluidine
    colorimetric method for hydroxymethyl-
        furfural 19:459
Barite. (See also Barium sulfate.) 6:533,
    555
    analysis 6:556
    barium content of 6:556, 557, 561
    x-ray fluorescence spectroscopy 6:
        564
Barium. (See also Barium alloys; Barium metal.)
    6:532–582
    in acid salts 8:595
    in aluminum metal 5:118
    ammonium dichromate gravimetric method
        18:274
    ammonium hydroxide turbidimetric method
        10:150
    in ammonium iodide 14:605
    in barite 6:556, 557, 561
    in barium acetate 6:536
    in barium bromate 6:537
    in barium chloride 6:545
    in barium compounds 6:536
    in barium hydroxide 6:548
    in barium nitrate 12:420
    in barium oxide 6:553
    in barium sulfate 6:556
    barium sulfate gravimetric method 9:293
    in barium titanate 19:126
    in black ash 6:569
    borax bead test 6:584
    in calcium acetate 8:79
    in calcium carbonate 8:84
    in cements 9:131
    in cesium sulfate 9:293
    chromate gravimetric method 6:536, 553,

    556, 576
    in chromic acid 9:684, 699
    in chromium compounds 9:684
    colorimetric determination 18:270
    conductometry 6:579
    in cotton 11:140
    in dehydrocholic acid 7:150
    differential thermal analysis 6:578
    EDTA titration 6:577; 12:154; 13:445
    in elastomers, synthetic 12:151
    emission spectroscopy 6:579; 15:156, 187;
        16:160
    in enamels 12:184
    in europium oxide 12:403
    flame photometry 6:579
    flame tests 6:583, 589, 592; 12:184
    in glass and glass-ceramics 13:445
    gravimetric methods 6:572, 575
    in hydrogen peroxide 14:437
    identification 6:572
    ion exchange chromatography 6:573; 18:268
    in lanthanum oxide 15:156
    in magnesium acetate 15:399
    in magnesium carbonate 15:402
    in magnesium chloride 15:404
    in magnesium nitrate 15:407
    in magnesium oxide 15:410
    microcosmic salt bead test 6:587
    in molybdenum 16:161
    in molybdenum trioxide 16:161
    neutron activation 6:580
    occurrence 6:533
    in organic salts 6:571
    oxalate gravimetric method 6:572
    paper chromatography 6:574; 18:270
    polarography 6:578
    in potassium chloride 9:339
    potassium chromate turbidimetric method 15:
        402
    potassium dichromate gravimetric method 12:
        186
    potassium dichromate turbidimetric method
        14:606; 15:399, 404, 410
    in potassium iodide 14:606
    potassium sulfate turbidimetric method 14:605
    precipitation as chromate 8:100
    precipitation as nitrate 8:101
    precipitation as sulfate 8:100
    in rocks 15:187
    separation 6:572; 18:267
    sodium acetate–potassium dichromate turbidi-
        metric method 18:283
    in sodium chloride 9:337, 349
    in sodium dichromate 9:684
    in sodium heparin 10:150
    in sodium iodide 14:606
    sodium rhodizonate colorimetric

method 11:140, 144
sodium sulfate–potassium permanganate
   colorimetric method 11:140
spot tests 6:574
in strontium 18:267, 273
in strontium chloride 18:283
in strontium nitrate 18:283
sulfate gravimetric method 6:539, 572, 575
sulfonazo III colorimetric method 18:276
sulfuric acid gravimetric method 19:126
sulfuric acid turbidimetric method 14:437, 606
volumetric methods 6:577
x-ray fluorescence spectroscopy 6:578
Barium acetate 6:535
   analysis 6:536
   barium content of 6:536
   chloride in 8:593
   commercial specifications 6:535; 8:591
   gravimetric method for sulfate 11:75
   gravimetric method for terephthalic acid 17:160
   impurities, determination of 6:536
   properties 6:535; 8:591
Barium alloys. (See also Barium metal.) 6:532
   analysis 6:534
   uses 6:532
Barium benzoate
   properties 8:591
Barium borotartrate gravimetric method
   for boron 7:401
Barium bromate 6:537
   barium content of 6:537
   properties 8:7
Barium bromide 6:537
   barium content of 6:538
   commercial specifications 8:10
   properties 8:9
Barium carbonate 6:534, 538
   analysis 6:539
   barium content of 6:539
   in black ash 6:569
   commercial grades and specifications 6:538
   in enamels 12:177
   impurities, determination of 6:539
   properties 12:177
   solubility 18:267
   uses 6:538
Barium carbonate–phenolphthalein
   color test for sulfate 18:423
Barium chloranilate
   colorimetric method for boron 7:410
   colorimetric method for sodium sulfate 11:84
   colorimetric method for sulfate 18:427
Barium chlorate. (See also Chlorates.) 6:545
   properties 9:358
   solubility 9:359
   volumetric method for sulfate 18:425
Barium chloride 6:545

analysis 6:545
barium content of 6:545
commercial specifications 6:545
gravimetric method for selenium 17:591
gravimetric method for sodium sulfate 10:492;
   18:247
gravimetric method for sulfate 9:117, 292, 344,
   348, 365, 686; 10:252, 330, 333, 340, 654,
   661, 664, 672; 11:48, 228, 335, 381; 12:186;
   13:56; 14:146, 371, 604; 15:399, 404, 407,
   431, 480, 492; 16:239, 536; 17:188, 197, 508;
   18:184, 194, 258, 283, 386, 423, 424; 19:499
gravimetric method for sulfite 18:402, 403
gravimetric method for sulfonated alcohol 11:
   126, 131
gravimetric method for sulfur 13:437; 14:72,
   514; 15:62, 76, 223, 329; 16:179, 213, 335;
   18:374
gravimetric method for sulfuric acid 14:456
gravimetric method for sulfurous acid 18:406
gravimetric method for sulfur trioxide 14:96;
   15:222
gravimetric method for tellurium 17:591
gravimetric method for thiophene 18:543
impurities, determination of 6:545
properties 9:334
solubility 9:336
turbidimetric method for sulfate 9:129, 350;
   11:228; 13:33, 34, 129, 525; 14:371, 437,
   455, 605; 15:143; 16:62; 17:438; 18:176,
   233, 426; 19:74, 222, 364
turbidimetric method for sulfur 15:253, 280,
   283
turbidimetric method for sulfur compounds
   15:401, 410; 17:403
volumetric method for sodium sulfate 10:492
volumetric method for sulfate 18:387, 425
Barium chloride, dihydrate. See Barium chloride.
Barium chloride–hydrochloric acid
   gravimetric method for ethylene oxide 12:574
Barium chloride–methanol
   turbidimetric method for sulfate 10:94
Barium chromate 6:546
   analysis 6:547; 9:693
   assay 9:693
   commercial grades and specifications 6:547;
      9:694, 696
   gravimetric method for barium 6:536, 553, 556,
      575, 576
   gravimetric method for chromium 9:639
   impurities, determination of 6:547
   solubility 9:696; 18:267
   volumetric method for sulfate 18:425
Barium citrate complex
   stability 10:89
Barium complexes 8:98
Barium compounds. (See also individual

compounds.) 6:534–571
occurrence 6:533
Barium dioxide. See Barium peroxide.
Barium ferrite
in ceramics 9:224
microscopy 9:224
Barium fluoride 6:548
carrier gas distillation fusion methods 8:636
properties 13:11
Barium fluosilicate 6:548
Barium glass. (See also *Glass and glass-ceramics.*)
composition 13:407
Barium hydroxide 6:548
analysis 6:549
barium content of 6:549
commercial grades and specifications 6:548
gravimetric method for carbon dioxide 8:263
hydrolysis of choline 9:620
impurities, determination of 6:549
volumetric method for carbon dioxide 8:261
Barium iodate
volumetric method for sulfate 18:425
Barium iodide 6:551
Barium metal. (See also Barium alloys.) 6:532
analysis 6:534
manufacture 6:533
properties 6:532
uses 6:532
Barium nitrate 6:551
analysis 6:552
barium content 12:420
commercial specifications 6:552
in explosive priming mixtures 12:417, 420,
423, 427, 429, 430
impurities, determination of 6:552
microscopy 12:417
nitrate content 12:420
refractive index 12:417
solubility 12:420
solvent extraction 12:423, 429
sulfuric acid gravimetric method 12:427
uses 6:551
x-ray diffraction 12:418
Barium nitrate–sodium chloride
turbidimetric method for sulfate 18:426
Barium organic salts 6:571
Barium oxalate
gravimetric method for barium 6:572
properties 8:591
solubility 18:267
Barium oxide 6:553
in alkali barium silicate glass 9:218
barium content of 6:553
emission spectroscopy 9:218
in enamels 9:195
in glass and glass-ceramics 13:413, 415, 418
in glazes 9:196

in silicate glasses 9:198
sulfuric acid gravimetric method 13:419
Barium perchlorate. (See also Perchlorates.)
properties 9:363
solubility 9:364
volumetric method for sulfate 18:386
volumetric method for sulfur 12:117
Barium perchlorate–sodium sulfate
volumetric method for sulfuric acid 14:367
Barium peroxide 6:553
analysis 6:554
commercial grades and specifications 6:554
impurities, determination of 6:554
uses 16:535
Barium rhodizonate
structure 17:456
Barium salt of 1-(4-chloro-*o*-sulfo-5-tolylazo)-2-
naphthol. See D & C Red No. 9.
Barium salt of 2-(2-hydroxy-1-napthylazo)-1-
napthalenesulfonic acid. See D & C Red No. 12.
Barium stearate 6:555
in enamels 12:176
Barium sulfate. (See also Barite.) 6:534, 555
ammonium sulfate gravimetric method 17:187
analysis 6:556; 17:192
barium content of 6:556
in black ash 6:570
in brake linings 7:441
in cadmium yellow 17:200
commercial grades and specifications 6:556;
17:192
decomposition method for hydrochloric acid
14:368
gravimetric method for barium 6:539, 572, 575;
9:293
gravimetric method for cyclamates 11:224
gravimetric method for sulfur 17:189
impurities, determination of 6:566; 17:192
solubility 18:267
in titanium–barium pigment 17:187
uses 17:192
in zinc sulfide pigments 17:190
Barium sulfide 6:534, 569
analysis 6:569
in black ash 6:569
Barium sulfite 6:570
Barium sulfonates 6:570
Barium titanate. (See also Titanates.) 6:571
analysis 19:126
barium content 19:126
in ceramics 9:224
microscopy 9:224
properties 9:212
titanium content 19:126
Barley
composition 9:233
mineral content 9:233

vitamin content 9:233
Barley malt
  analysis 4:469
  quality control 4:467
Baronal. (See also High-explosive mixtures.)
  analysis 12:446
  composition 12:446
  uses 12:446
Barytes. (See also Barite.)
  in brake linings 7:441
  composition 12:174
  in enamels 12:174
Basalts 19:225
Base exchange capacity
  of clays and clay minerals 10:109
Basic blue 7. See Victoria blue BO.
Basic brick. (See also Refractories.)
  composition 9:207
Basic chromic chloride. (See also Chromium
    compounds.)
  analysis 9:689
  basicity of 9:690
Basic colors. (See also Certified colors; and
    individual compounds.)
  liquid chromatography 10:527
Basic copolymers
  acid–base titration of 4:273, 386
Basic copper sulfate. (See also Copper sulfate.)
    10:676
  solubility 10:676
Basic dye capacity
  of acrylonitrile polymer fibers 4:328
Basic dyes 6:402; 12:3
Basic fuchsin. See Schiff's reagent.
Basicity. (See also Alkalinity; pH.)
  of acetone 4:114
  of adsorbents 4:437
  of aluminum nitrate 5:181
  of antifreezes 6:9
  of automotive chemical specialties 6:357
  of barium carbonate 6:544
  of barium sulfate 6:567, 568
  of basic chromium salts 9:690
  of calcium carbonate 8:83
  of calcium chloride 8:88
  of caprolactam 8:126
  of chrome tanning compounds 9:690
  of cyclopropane 5:360
  of ethylene 5:362
  of nitroparaffins 16:438
  of nitrous oxide 5:366
  Schorlemmer method 9:691
  of sodium benzoate 7:87
  of water for carbonated beverages 8:177
Basic refractories. (See also Refractories.)
  acid dissolution 9:215
  fluxes for 9:216

Basic violet 4. See Ethyl violet.
Basic violet 14. See Fuchsine.
Basic weight
  of paper matchboard 15:534
  of pulp handsheets 7:254
Basket centrifuges 9:171
  characteristics 9:160
  settling curves 9:161
Bastnasite 9:246; 12:395; 13:2; 15:149; 17:470,
    499
Bathochromic shifts
  of flavonol glycosides 18:474
Bathocuproine
  colorimetric method for copper 4:479; 13:267
  colorimetric method for iron 4:480
Bathophenanthroline
  colorimetric method for iron 8:175; 13:267;
    15:54
Bathophenanthrolinedisulfonic acid
  colorimetric method for ferric oxide 15:209
Batrilex. See Terraclor.
Baumé gravity
  of sulfuric acid 18:411, 421
Baumé scale 18:411
Bauxite 13:251
  aluminum content of 5:115
  analysis 5:114
  beryllium in 5:116
  calcium in 5:115
  chromium in 5:116
  commercial specifications 17:194
  composition 9:203
  gallium in 5:115
  iron in 5:115
  magnesium in 5:116
  molybdenum in 5:116
  nickel in 5:116
  phosphorus in 5:115
  sample preparation 5:184
  silicon in 5:115; 18:37
  titanium in 5:115; 19:120
  uranium in 5:115
  vanadium in 5:116
  zinc in 5:116
  zirconium in 5:116
Bauxite, adsorbent
  chemical analysis 4:439
  efficiency for percolation processes 4:449
  hardness 4:446
  uses 4:434
Bauxite ore
  silica in 4:37
Bayer 4010 6:73
Bay leaves, oil of. (See also *Essential oils.*)
  refractive index 12:271
Bazzite 17:561
BCOD. See Bis(cyclohexanone oxalyl dihydrazone).

Bead and flame tests 6:583–593
Bead tests 6:583
  for chromium 12:14
  for cobalt 10:348; 12:14
  for copper 10:602; 12:14
  for iron 12:14
  for lead 15:172
  for metal complex dyes 12:14
  for nickel 12:14; 16:300
Bearing alloys 14:547
  manufacture 14:547
Bearing bronze 5:68
Bearing metals
  bismuth in 7:211
  sample preparation 7:210
Beef tallow. (See also *Fats and fatty oils.*)
  properties 12:473
Beer
  acidity 7:508, 579
  α-acids 7:604
  acids, volatile 7:579
  air in 7:620
  alcohol content 7:541
  alginate in 7:639
  aluminum in 7:650, 652
  α-amino nitrogen in 7:614, 665
  ammonia 7:610
  analysis 7:480
  anthocyanogens 7:551
  ash content 7:553
  bitterness test 7:603
  boiling point 7:503, 543
  boron in 7:653
  buffer capacity 7:508
  cadmium in 7:651
  calcium in 7:652, 656
  calcium oxalate in 7:616
  carbohydrates in 7:554, 665
  carbon dioxide in 7:570, 620
  carboxylic acids in 7:579
  chelating agents in 7:584
  chlorides in 7:581
  chromium in 7:657
  cobalt in 7:651, 657
  color 7:517
  copper in 7:659
  coproporphyrin in 7:638
  dextrins in 7:555, 564
  diacetyl in 7:582, 608
  distillation 7:543
  EDTA and related compounds in 7:584; 12:47
  enzyme activity in 7:586
  ethanol content 7:541
  extracts 7:495
  foaming tendency of 7:535
  freezing point 7:504
  gushing 7:538

haze in 7:519
n-heptyl *p*-hydroxybenzoate in 7:594
hops in 7:603
hydrogen sulfide in 7:640
invertase activity in 7:586
iron in 7:650, 660
isohumulones in 7:603
lead in 7:651
lithium in 7:652
Lundin protein fractions 7:612
magnesium in 7:656
maltose in 7:555, 563, 566
mercaptans in 7:642
metals in 7:648
molecular weight, apparent 7:505
molybdenum in 7:651
nickel in 7:651, 661
nitrogen in 7:608
oxalates in 7:616
oxidizable substances in 7:618
oxygen in 7:620
package headspace volume 7:633
pasteurization, evaluation of 7:586
pH 7:507
phosphorus in 7:633
polyethylene oxide in 7:634
polyphenols in 7:552, 604, 646
polyvinylpyrrolidone in 7:636
porphyrinogen in 7:638
potassium in 7:652
preservatives in 7:597
production 7:478
properties 7:477
properties, determination of 7:485
propylene glycol alginate in 7:639
protease activity in 7:588
protein in 7:608, 611, 612, 665
reducing substances in 7:618
reducing sugars in 7:563
sample preparation 7:481, 553, 648
sediment in 7:529
silica in 7:639
sodium in 7:652; 18:220
specific gravity 7:485, 544
specific heat 7:510
stability, gaseous 7:538
stability, physical 7:532
starch in 7:555
strontium in 7:652
sulfhydryl in 7:620, 642
sulfur compounds in 7:640
sulfur dioxide in 7:581, 643; 18:406
surface tension 7:511, 595
suspended particles in 7:529
tannins in 7:604, 646
tin in 7:650, 662
turbidity of 7:519

vanadium in 7:651
viscosity 7:513
zinc in 7:651, 663
Beer sediments. (See also *Yeast.*) 19:519
  composition 19:519
  species 19:516
  uses 19:516, 519
Beeswax
  in automobile waxes 6:353
Beeswax–borax emulsions
  identification 11:93
Beet powder. See Beets, dehydrated.
Beets, dehydrated
  commercial specifications 10:530
  composition 10:530
Beidellite. (See also *Clays and clay
    minerals.*)
  composition 10:102
Beilstein test 8:13; 9:381–384
  of allyl chloride 9:382
  of benzyl chloride 9:382
  of carbon tetrachloride 8:324; 9:382
  of chlorocarbons 9:381
  of chloroform 9:382
  of dichlorobenzenes 9:382
  of dichloroethanes 9:382
  of 1,2-dichloroethylene 9:382
  of dichloromethane 9:382
  of 1,2-dichloropropane 9:382
  of ethyl chloride 9:382
  for halogens in essential oils 12:
    289
  of methyl chloride 9:382
  of 1,1,2,2-tetrachloroethane 9:382
  of tetrachloroethylene 9:382
  of 1,1,1-trichloroethane 9:382
  of trichloroethylene 9:382
  of vinyl chloride 9:382
Bendix method
  volumetric method for tin on tinplate 19:
    55
Benedict's test
  for aldehydes 8:344
Benlate. See Benomyl.
Benomyl 13:206
  analysis 13:206
  in animal tissues 13:206
  colorimetric determination 13:207
  fluorometric determination 13:207
  in plant tissues 13:206
  properties 13:206
  in soil 13:206
  solubility 13:206
  solvent extraction 13:206
  uses 13:206
Benoxinate hydrochloride 5:374
  analysis 5:378

properties 5:378
Bentonites. (See also *Clays and clay minerals.*)
    10:106
  determination of pH 4:437
  uses 10:106
Benzalacetone
  commercial specifications 10:44
Benzalacetophenone oxides. (See also *Epoxy
    compounds.*)
  ultraviolet spectroscopy 12:203
Benzal chloride. (See also Chlorinated toluenes.)
    9:484
  azeotropes 9:485
  composition 9:411
  Fujiwara colorimetric method 9:386
  gas chromatography 9:488
  infrared spectroscopy 9:488
  manufacture 9:486
  properties 9:485
  solubility 9:486
  uses 9:484
Benzaldehyde 6:594
  absorption spectroscopy 9:564
  analysis 6:611
  in benzyl alcohol 7:89, 99
  in bitter almond oil 12:282
  in black tea 18:511
  color method for deoxycholic acid 7:147
  column chromatography 6:617
  commercial grades and specifications 6:607, 608
  derivatives for identification 6:611; 8:343
  in embalming fluids 12:165
  gas chromatography 6:619; 8:355; 9:378; 18:
    571
  gravimetric method for pentaerythritol 13:592
  hydroxylamine hydrochloride volumetric method
    6:620; 8:357; 12:282
  impurities, determination of 6:623
  infrared spectroscopy 6:614, 615; 9:394
  infrared spectrum 8:346
  manufacture 6:599; 8:339
  near-infrared spectroscopy 6:628
  properties 6:594, 598; 8:334
  solubility 13:540
  in styrene 18:306
  ultraviolet spectroscopy 6:632; 7:99
  uses 6:594, 605; 8:332
Benzaldehyde and its derivatives. (See also individ-
    ual compounds.) 6:594–639
  acetal formation 6:630
  analysis 6:611
  column chromatography 6:617
  commercial grades and specifications 6:607
  derivatives for identification 6:611
  fluorometry 6:625
  gas chromatography 6:619
  hydroxylamine hydrochloride method 6:620

impurities, determination 6:623
infrared spectroscopy 6:615
manufacture 6:599
near-infrared spectroscopy 6:628
paper chromatography 6:620
properties 6:594
thin-layer chromatography 6:620
ultraviolet spectroscopy 6:630
uses 6:605
Benzaldehyde–modified phenol formaldehyde resin
    pyrolysis–gas chromatography 7:446
Benzalkonium chloride. (See also Quaternary am-
    monium compounds.)
  assay 11:557
  bromphenol blue colorimetric method 11:559
  commercial specifications 11:555
  composition 11:555
  in disinfectants and antiseptics 11:555
  potassium iodate titration 11:557
  ultraviolet spectroscopy 11:556
Benzamide
  derivative of amines 5:265
  derivative of aniline and its derivatives 5:429
  derivatives for identification 8:601
  properties 8:597
Benzanilide
  properties 8:597
Benzanthrone
  fluorometric method for glycerol 13:508
Benzathine benzylpenicillin. (See also Benzyl-
    penicillin.)
  iodometric methods 16:562
  properties 16:556
  solubility 16:557
  structure 16:567
Benzathine penicillin G. See Benzathine benzyl-
    penicillin.
Benzathine penicillin V. See Benzathine phenoxy-
    methylpenicillin.
Benzathine phenoxymethylpenicillin. (See also
    Phenoxymethylpenicillin.)
  iodometric methods 16:562
  solubility 16:557
Benzene. (See also Hydrocarbons.) 7:1–58;
    14:317
  in air 7:18, 39
  analysis 7:7
  in automotive exhaust gases 7:7
  azeotropes 7:4
  azeotropic distillation 7:12
  carbon disulfide content 18:543
  in cigarette smoke 10:31
  in cocoa bean vapor 9:570
  colorimetric method using butanone 7:17
  commercial grades and specifications 7:5, 6
  density 7:36
  distillation 7:12, 32

in ethanol 4:544
gas chromatography of 4:212; 7:7; 10:31; 11:73,
    98; 13:350; 14:330, 333, 335, 336, 337, 375;
    16:220; 18:63, 299
in gasoline 13:350, 353
gravimetric determination with ammoniacal
    nickel cyanide 7:25
heat of reaction with nitrating acid 7:25
hydrogenation 7:25
identification 7:7
impurities, determination of 7:41
infrared spectroscopy 7:28
liquid chromatography 7:14
mass spectroscopy 7:31; 9:570
in nail enamels 11:73
oscillometry 7:33
in paraffins 14:336
polarography 7:34
properties 7:3; 13:350
properties, determination of 7:41
Raman spectroscopy 7:36
refractive index 7:36
separation 7:7
solubility 13:539, 552
specific gravity 7:36
in styrene 18:299
thiophene content 18:543
ultraviolet spectroscopy 7:39
uses 7:1
Benzene-2-arsonic acid-(1-azo-7)-1,8-dihydroxy-
    naphthalene-3,6-disulfonic acid. See Arsenazo
    I.
Benzene-2-arsonic acid-(1-azo-1)-2-hydroxynaph-
    thalene-3,6-disulfonic acid. See Thoron.
m-Benzenedicarboxylic acid. See Isophthalic acid.
o-Benzenedicarboxylic acid. See Phthalic acid.
p-Benzenedicarboxylic acid. See Terephthalic acid.
Benzene extractables
  in carbon black 8:235
Benzene miscibility
  of acid halides 8:576
Benzene sulfonamide
  derivatives of amines 5:265
  derivative of aniline and its derivatives 5:429
Benzenesulfonyl chloride
  properties 8:570
Benzenethiol. See Phenyl mercaptan.
Benzethonium chloride. (See also Quaternary am-
    monium compounds.)
  assay 11:557
  commercial specifications 11:555
  in disinfectants and antiseptics 11:555
  properties 11:555
  sodium tetraphenylboron gravimetric method
    11:558
  sodium tetraphenylboron titration 11:557
  ultraviolet spectroscopy 11:556

volumetric method for anionic detergents 11:
337
volumetric method for sodium lauryl sulfate
12:65
Benzidine. (See also *Diamines and higher amines,
aromatic.*) 11:433
colorimetric method for chromium 11:140, 144
colorimetric method for lead 15:172
colorimetric method for manganese 11:141; 15:
457
colorimetric method for vanadium 19:331
derivatives for identification 11:438
gravimetric method for anionic detergents 11:
336
properties 11:434
solubility 11:435
Benzidine chloride
gravimetric method for tungsten 19:202
Benzidine–copper acetate
color test for hydrogen cyanide 14:411
Benzidine hydrochloride
titration of sulfate 18:425
Benzidine hydrochloride–pyridine
colorimetric method for thiocyanate 18:437
Benzidine orange. (See also *Pigments.*)
structure 17:180
Benzidine yellow GR. (See also *Pigments.*)
structure 17:181
Benzidine yellow toner
composition 12:173
in enamels 12:173, 184
identification 12:184
Benzilic acid
gravimetric method for scandium 17:570
Benzoates
in antifreezes 6:16
Benzocaine. See Ethyl-*p*-aminobenzoate.
Benzofuran
boiling point 16:209
Benzoguanamine
alkyd resins by infrared spectroscopy 5:44
Benzohydroxamic acid
gravimetric method for niobium 16:367
gravimetric method for tantalum 16:367
Benzoic acid 7:58–88; 8:400
acid–base titration 7:65, 68
in alkyd resins 7:67
ammonium sulfide colorimetric method 19:485
analysis 7:61
in antifreezes 6:16
commercial grades and specifications 7:61
content of alkyd resins 5:44
derivatives for identification 7:62; 8:376
ferric chloride colorimetric method 19:485
in food products 7:68
gas chromatography 5:25; 7:72; 8:385, 388;
17:154, 162

gravimetric method for thorium 19:9, 12
hydroxylamine hydrochloride colorimetric method 19:485
identification 7:61
impurities, determination of 7:66
infrared spectroscopy 5:44; 7:65
ion exchange chromatography 6:144
manufacture 7:58, 60
modifier for alkyd resins 5:24
occurrence 7:60
paper chromatography 8:379
in phthalic anhydride 7:70; 17:154, 156
preparation from benzaldehyde 6:594
preparation of benzoyl chloride 7:62
properties 7:59; 8:402
in salicylic acid 6:143
separation 7:62, 67
solubility 7:59
solvent extraction 19:485
specification tests 7:66
standard sample for carbon and hydrogen analysis
8:285, 286
in terephthalic acid 17:162, 163
ultraviolet spectroscopy 7:67, 69, 70
uses 7:58; 8:400
in wines 19:485
Benzoic acid derivatives. (See also individual compounds.) 7:72
of benzaldehydes 6:612, 613
Benzoic acid esters. (See also individual compounds.)
7:78
derivatives for identification 7:80
identification 7:61, 80
impurities, determination of 7:84
saponification method 7:83
separation 7:62
Benzoic acid–oxalic acid
gravimetric method for thorium 15:393
Benzoic acid salts
identification 7:61
Benzoic anhydride 7:72
acid–base titration, nonaqueous 7:73
analysis 7:73
anilide formation 7:73
bromination method 7:74
derivatives for identification 7:73
2,4-dichloroaniline reaction 7:74
identification 7:73
infrared spectroscopy 8:562
properties 7:72; 8:559
Benzoin
fluorescence method for boron 7:411
hydroxylamine hydrochloride method 8:357
preparation from benzaldehyde 6:595
Benzoin anti-oxime
gravimetric method for molybdenum 16:155
gravimetric method for molybdenum

trioxide 16:181
Benzoin–lead acetate
  colorimetric method for sulfur 18:371
Benzoinoxime. (See also Benzoin anti-oxime.)
  colorimetric method for copper 11:140
  colorimetric method for vanadium 19:331, 333
  gravimetric method for molybdenum 10:316
  gravimetric method for tungsten 19:200
Benzonitrile. (See also *Nitriles.*)
  azeotropes 16:403
  derivatives for identification 16:402
  properties 16:396
Benzophenone
  absorption spectroscopy 9:564
  derivatives for identification 8:343
  hydroxylamine hydrochloride method 8:357
  properties 8:335
Benzopyran
  bioassay 7:182
Benzo[α]pyrene
  in cigarette smoke 10:19
  fluorometry 10:20
  gas chromatography 10:20
  ultraviolet spectroscopy 10:19
1,2-Benzopyrone. See Coumarin.
2-Benzothiazolyl diethyldithiocarbamate. See
    Ethylac.
Benzothiazyl disulfide. (See also *Mercaptans.*)
  iodometric method 15:571
Benzothiophene
  boiling point 16:209
Benzotriazole
  in antifreeze 13:600
  in glycols 13:600
1,2,3-Benzotriazole
  gravimetric method for silver 18:162, 166
Benzotrichloride. (See also Chlorinated toluenes.)
    9:484
  azeotropes 9:485
  color method for amine antioxidants 6:93
  composition 9:411
  Fujiwara colorimetric method 9:386
  manufacture 9:486
  properties 9:485
  use in manufacture of benzoic acid 7:60
  uses 9:484
n-Benzoyl-*dl*-arginine-*p*-nitroanilide hydrolysis
    method
  for protease activity 7:591
o-Benzoyl benzoic acid 8:405
  properties 8:402, 405
  uses 8:405
Benzoyl chloride 7:75
  analysis 7:76
  commercial specifications 7:75; 8:571
  derivatives for identification 7:63
  gravimetric method for thiophene 18:540

impurities, determination of 7:6
preparation from benzaldehyde 6:594
preparation from benzoic acid 7:63
properties 7:75; 8:570
volumetric method using *m*-chloraniline 7:76
N-Benzoyl glycine. See Hippuric acid.
Benzoyl halides
  derivatives for identification 8:572
Benzoyl peroxide
  colorimetric method for antioxidants 6:92, 94
  colorimetric method for N,N'-diphenyl-*p*-phen-
      ylenediamine 4:210
  polarography 16:99
N-Benzoylphenylhydroxylamine
  gravimetric method for niobium 16:366
  gravimetric method for tantalum 16:366
N-Benzoylphenylhydroxylamine–thiocyanate
  colorimetric method for niobium 16:370, 375
N-Benzoyl-*o*-tolylhydroxylamine
  colorimetric method for vanadium 19:342
Benzyl acetate
  in essential oils 12:276
  molecular weight 12:276
  properties 8:577
Benzyl alcohol 5:374; 7:89
  analysis 5:379; 7:91
  azeotropes 7:92
  benzaldehyde in 7:89, 99
  in black tea 18:507, 511
  commercial grades and specifications 7:91, 94
  derivatives for identification 7:93, 95
  gas chromatography 7:93, 96, 99; 16:98; 18:
      511
  identification 7:93
  impurities, determination of 7:99
  infrared spectroscopy 7:97, 100
  manufacture 7:90
  mass spectroscopy 7:95, 96
  near-infrared spectrum 7:98
  nuclear magnetic resonance spectrum 7:98
  phthalic anhydride method 4:535
  preparation from benzaldehyde 6:594, 595
  properties 5:379; 7:90
  properties, determination of 7:99
  separation 7:93
  solubility 13:539
  turbidimetric method 7:100
  ultraviolet spectroscopy 7:99
  uses 7:89
Benzylamine derivatives
  of acid halides 8:572
  of esters 8:582
Benzyl and phenylethyl alcohols. (See also Benzyl
    alcohol; β-Phenylethyl alcohol.) 7:89–102
Benzyl benzoate. (See also Benzoic acid esters.)
    7:78
  commercial grades and specifications 7:78, 80

3,5-dinitrobenzoate derivative 7:83
occurrence 7:60
properties 7:77
saponification method 7:84
ultraviolet spectroscopy 7:85
Benzylbenzoic acid–sulfuric acid
colorimetric method for naphthalene 16:208
Benzylcarbinyl cinnamate. See Phenethyl
cinnamate.
Benzyl chloride. (See also Chlorinated toluenes.)
9:484
azeotropes 9:485
Beilstein test 9:382
composition 9:411
gas chromatography 9:488
manufacture 9:487
in manufacture of benzaldehyde 6:599
mass spectroscopy 9:400
properties 9:485
solubility 9:486
stabilizers in 9:484
toxicity 9:484
use in manufacture of benzyl alcohol 7:90
uses 9:484
Benzyl cinnamate. (See also *Cinnamic acid and
its derivatives.*)
commercial specifications 10:66
properties 10:65
saponification methods 10:67
in sunscreens 11:92
ultraviolet spectroscopy 11:92
Benzyldimethyl[2-[2[*p*-(1,1,3,3-tetramethyl-
butyl)phenoxy]ethoxy]ethyl]ammonium
chloride. See Benzethonium chloride; Hyamine
1622.
Benzyldimethyl[2-[2-[*p*-(1,1,3,3-tetramethyl-
butyl)tolyloxy]ethoxy]ethyl]ammonium
chloride. See Methylbenzethonium chloride.
Benzyl ether. (See also *Ethers.*)
properties 12:297
Benzyl ethyl ether. (See also *Ethers.*)
properties 12:297
N-Benzylhydroxylamine (See also *Hydroxylamines
and their salts.*)
properties 14:443
Benzylic acid
properties 8:402
Benzyl isocyanate. (See also *Isocyanic acid esters.*)
infrared spectroscopy 15:106
Benzyl methyl ether. (See also *Ethers.*)
properties 12:297
Benzylmorphine. (See also *Narcotics.*)
ammonium molybdate colorimetric method
16:246
ammonium thiocyanate microchemical method
16:249
color tests 16:246

formaldehyde–sulfuric acid colorimetric method
16:246
hydrochloric acid microchemical method 16:249
microchemical tests 16:249
potassium iodide microchemical method 16:249
structure 16:244
Benzylpenicillenic acid
in penicillins 16:561, 566
ultraviolet spectroscopy 16:561, 566
Benzylpenicillic acid
in penicillins 16:561
ultraviolet spectroscopy 16:561, 564
Benzylpenicillin. (See also Aluminum benzylpenicillin; Benzathine benzylpenicillin; Calcium
benzylpenicillin; Dibenzylamine benzylpeni-
cillin; 1-Ephenamine benzylpenicillin; *N*-Ethyl-
piperidine benzylpenicillin; Hydrabamine benzyl-
penicillin; Penicillins; Potassium benzylpeni-
cillin; Procaine benzylpenicillin; and Sodium
benzylpenicillin.) 16:566–571
analysis 16:568
assay 16:568
bioassay 16:568
bioautography 16:575
in biological fluids 16:568
bromometric method 16:571
in carbenicillin 16:566, 575
N-ethylpiperidine gravimetric method 16:570
in foodstuff 16:568
impurities, determination of 16:566, 570
iodometric method 16:571
manufacture 16:553, 566
microbiological methods 16:568
paper chromatography 16:558, 575
penicillin content 16:570
in penicillin O 16:566, 571
properties 16:566
solvent extraction 16:570, 572
specification tests 16:569
structure 16:553, 554, 567
thin-layer chromatography 16:559
in tissues 16:568
ultraviolet spectroscopy 16:560, 564, 571, 572
uses 16:566
Benzylpenicillin salts. (See also Benzylpenicillin;
and individual compounds.) 16:567
uses 16:568
N-Benzyl-*N*-phenyl-*N*′,*N*′-dimethylethylenediamine.
See Antergan.
2-Benzylpyridine. (See also *Pyridine and pyridine
derivatives.*)
ionization constants 17:416
manufacture 17:422
4-Benzylpyridine. (See also *Pyridine and pyridine
derivatives.*)
manufacture 17:422
Benzyl salicylate

in sunscreens 11:92
ultraviolet spectroscopy 11:92
S-Benzylthiuronium
gravimetric method for cinnamic acid 10:59
S-Benzylthiuronium chloride
derivative for identification of alcohols 4:517
Benzyltrimethylammonium periodate
volumetric method for α-glycol groups 12:249
Benzyl violet 4B. (See also *Triphenylmethane and related dyes.*)
liquid chromatography 19:173
Bergamot oil. (See also *Essential oils.*)
refractive index 12:271
residue on evaporation 12:273
Bergapten. (See also *Coumarins.*) 11:161
gas chromatography 11:168, 170
Berkelium. (See also Transuranium elements.)
manufacture 19:259
Beryl 7:103, 108, 118; 9:278; 15:261
analysis 7:118
composition 7:118
extraction of beryllium from 7:108
properties 7:118
sample fusion 7:118
Beryllia. (See also Beryllium oxide.)
composition 9:207
properties 9:211
Beryllium. (See also Beryllium alloys; Beryllium metal.) 7:103–141
acid–base titration 7:133
in air 7:138
in aluminum alloys 5:118
ammonium chloride gravimetric method 16:324
ammonium hydroxide gravimetric method 16:324
ammonium nitrate gravimetric method 10:308
atomic absorption spectroscopy 7:140
in bauxite 5:116
in beryllium alloys 7:135
in beryllium copper 10:289, 308
in beryllium metal 7:112
in biological samples 7:138
in cobalt alloys 10:308
colorimetric determination 15:378
colorimetric method using *p*-nitrophenyl-azoorcinol 7:136
in copper alloys 10:634
in copper metal 10:620
EDTA gravimetric method 10:634
emission spectroscopy 7:138; 13:277; 14:542; 15:375
fluorometric method using morin 5:118; 7:138
in gallium arsenide 13:276
gold content 14:42
gravimetric determination as pyrophosphate 7:132
gravimetric method using ammonium hydroxide

7:131
in high-temperature alloys 10:286
in hydrochloric acid 14:375
in indium 14:542
iodometric titration 7:134
ion exchange chromatography 18:268
in magnesium 15:375, 378
in magnesium alloys 15:375, 378
morin fluorometric method 15:378
in nickel 16:324
in nickel alloys 16:324
p-nitrobenzeneazoorcinol colorimetric method 10:309
p-nitrophenylazoorcinol colorimetric method 10:620
precipitation with ammonium hydroxide 7:129
radiochemical determination 7:139
rare earth content 17:473
separation 7:129
solvent extraction 15:379
in spring steels 10:290
Beryllium acetate 7:119
acid–base titration, nonaqueous 7:120
analysis 7:120
Beryllium alloys. (See also Beryllium–copper alloys; and Beryllium metal.) 7:103
beryllium content of 7:135
production 7:111
Beryllium carbide 7:120
Beryllium carbonate 7:121
Beryllium chloride 7:121
Beryllium compounds. (See also individual compounds.) 7:118
toxicity 7:107
Beryllium copper. (See also Cobalt alloys.) 10:289
beryllium content 10:308
composition 10:289
Beryllium–copper alloys. (See also Copper alloys.) 7:103
composition 7:104
production 7:111
properties 7:105, 106; 10:631
uses 7:104
Beryllium fluoride 7:122
analysis 7:123
preparation from beryl 7:108
Beryllium hydroxide 7:123
composition 7:124
preparation from beryl 7:108, 123
Beryllium metal. (See also Beryllium alloys.) 7:103
analysis 7:112
beryllium content of 7:112
beryllium nitride in 7:117
beryllium oxide in 7:113
carbon in 7:114
carrier gas fusion methods 8:625, 682

chromium in 7:115
commercial grades and specifications 7:111
copper in 7:116
hot extraction methods 8:684
hydrogen in 8:622, 638, 680
iron in 7:114
manganese in 7:115
metallurgy 7:108
nickel in 7:116
nitrogen in 7:117; 8:622, 625
oxygen in 8:617, 621, 622, 623, 625, 630, 633, 635, 636, 680, 682
production 7:109
properties 7:104
toxicity 7:107
uses 7:103
vacuum distillation fusion methods 8:635
vacuum fusion methods 8:617, 621, 622, 623, 680
vacuum hot extraction methods 8:638
Beryllium nitrate 7:124
analysis 7:124
nitrate content of 7:125
Beryllium nitride 7:125
in beryllium metal 7:117
Beryllium oxide 7:126
in alkali barium silicate glass 9:218
in beryl 7:108
in beryllium metal 7:113
boron in 7:410
carrier gas fusion methods 8:626
in ceramics 9:224
composition 7:126
emission spectroscopy 9:218
microscopy 9:224
oxygen in 8:626
properties 7:107, 127; 9:204
in refractories 9:207
Beryllium oxyfluoride 7:122
Beryllium sulfate 7:128
Betaine
ion exchange chromatography 9:618
Betamethasone. (See also Steroids.)
manufacture 14:194
properties 14:194
solubility 14:194
uses 14:194
Betanox
colorimetric method using diazotized *p*-nitro-aniline 6:99
color test 6:94
Beta OH method
for water in glass 13:462
Beta-ratio
spectroscopic method for iron oxide 13:464
Bethanechol chloride
dipicrylamine colorimetric method 9:617

Dragendorff's colorimetric method 9:617
iodine colorimetric method 9:617
iodoplatinate colorimetric method 9:617
thin-layer chromatography 9:616
BET technique 8:187, 214
apparatus 8:216, 218, 220, 221
for surface area of carbonaceous materials 4:14
Betterton-Knoll process
for recovery of bismuth from lead ores 7:189
Betts process
for recovery of bismuth from lead ores 7:189
Beverages
caramel content 10:538
Beverages, carbonated. See *Carbonated beverages*.
Beverages, nonalcoholic
caffeine in 8:66
BHA
2,2'-bipyridine colorimetric method 9:243
in cereals 9:242
dichloroquinonechloroimide colorimetric method 9:243
BHC
gas chromatography 19:389
in raw milk 16:129
in water 19:389
BHT. (See also Butylated hydroxytoluene.)
2,2'-bipyridine colorimetric method 9:243
in cereals 9:242
Biaberite 10:343
Biallyl
gas chromatography 5:104
halogenation methods 5:104
properties 5:103
thiocyanation methods 5:104
Bicarbonate
carbon dioxide absorption method 11:353
in detergents 11:335, 353
hydrochloric acid effervescence test 18:183
in soaps 18:183, 192
in water 19:362
Bicarbonate of soda. See Sodium bicarbonate.
Bile 7:141
Bile acids and salts. (See also individual compounds.) 7:141–154
acid–base titration 7:149
analysis 7:146
colorimetric methods 7:151
color tests 7:146
fluorometry 7:152
gas chromatography 7:148
identification 7:146
liquid chromatography 7:148
separation 7:148
specification tests 7:150
thin-layer chromatography 7:148
Binders
adhesive properties 15:535

analysis 7:444; 15:535
in brake linings 7:441, 444
in match strikers 15:520, 535
Binox M 6:73
Bioassay 7:155–186
applications 7:182
biological factors 7:179
classification 7:156
of cycloheximide 13:223
definition 7:155
of gonadotropic hormones 14:205
of growth hormone 14:205
of insulin 14:206, 573, 575
mathematical procedures for data 7:171
of pesticides and insecticides 19:614
precision 7:179
selection of procedure 7:181
sources of error 7:158
Bioautography
of actinomycins 5:533
of amphomycin 5:555
of amphotericins 5:568
of antimycin 5:603
of bacitracin 5:538
of benzylpenicillin 16:575
of candicidin 5:569
of capreomycin 5:555
of carbomycin 5:521
of cephalosporins 16:559
of cephalothin 16:579
of chloramphenicol 5:469
of chromomycin 5:498
of colistin 5:539
of cycloheximide 5:510
of erythromycin 5:522
of etamycin 5:555
of filipin 5:569
of leucomycin 5:524
of lincomycin 5:585
of mikamycin 5:555
of neomycins 5:586
of novobiocin 5:500
of nystatin 5:570
of oleandomycin 5:524
of penicillins 16:559
of pimaricin 5:571
of polymyxin 5:541
of rifamycins 5:500
of spiramycins 5:525
of staphylomycin 5:556
of streptomycin 5:588
of thiazolidomycin 5:511
of thiostrepton 5:556
of viomycin 5:557
Biochemical methods
for antibiotics in raw milk 16:128
for enzymes in detergents 11:356

for nitrate 19:138
for salmonella 9:603
Biochemical oxygen demand
of water 19:373
Biodegradability
of alkyl aryl sulfonates 11:399
Biological assay methods. See *Bioassay.*
Biological factors
in bioassay 7:179
Biological fluids
ampicillin content 16:576
benzylpenicillin content 16:568
EDTA and related compounds in 12:47
lactic acid content 15:137
Biological materials
apomorphine content 16:276
beryllium in 7:138
boron in 7:416, 417
bromine 8:25
chromium content 9:651
iron content 15:53
lead content 15:191
manganese content 15:472, 477
meperidine content 16:275
mercury content 16:39
methadone content 16:278
molybdenum content 19:217
nickel content 16:320
potassium content 17:383
tin content 19:62
titanium content 19:96
tungsten content 19:213, 217
vanadium content 19:340, 342
Biological methods
for thiuram disulfides 11:598
Biological samples
radium content 17:454, 465
thallium content 18:524
Biological substances
cyclamate content 11:233, 238
Biological systems
bishydroxycoumarin content 11:171
warfarin content 11:171
Biological tests
for penicillins 16:566
Biotin
in eggs and egg products 12:66
Biotite
composition 10:102; 11:464
2,2'-Biphenaldehyde
near-infrared spectroscopy 6:629
Biphenyl. See Diphenyl.
2,2'-Biphenylcarboxylic acid. See Diphenic acid.
o-Biphenyl isocyanate. (See also *Isocyanic acid
esters.*)
near-infrared spectroscopy 15:105
2,2'-Bipyridine colorimetric method

for iron 4:376
2,2'-Biquinoline. (See also Cuproine.)
  colorimetric method for copper 10:602, 613;
    13:370
Birch oil, sweet. (See also *Essential oils.*)
  refractive index 12:271
Birefringence
  of acrylic and modacrylic fibers 4:257
  of polyamide fibers 17:314
β-Bisabolene
  gas chromatography 12:268
  in majoram, oil of 12:268
Bis(2-aminoethyl) disulfide hydrochloride. (See
    also *Imines.*)
  properties 14:471
  structure 14:471
Bis(2-aminoethyl) sulfide. (See also *Imines.*)
  properties 14:471
  structure 14:471
Bis(2-bromoethyl) ether. (See also *Ethers.*)
  properties 12:297
4,4'-Bis(2,6-di-*t*-butylphenol). (See also *Phenols.*)
  structure 17:2
Bis(*N*,*N'*-carballyloxy)hydrazine 5:106
Bis(4-chlorobutyl) ether. (See also *Ethers.*)
  properties 12:297
Bis(1-chloroethyl) ether. (See also *Ethers.*)
  properties 12:297
Bis(2-chloroethyl) ether. (See also *Ethers.*)
  properties 12:297
Bis(2-chloroisopropyl) ether. (See also *Ethers.*)
  properties 12:297
Bis(chloromethyl) ether. (See also *Ethers.*)
  properties 12:297
1,3-Bis(chloromethyl)tetramethyldisilazane. (See
    also Silylation reagents.)
  properties 18:90
Bis-*p*-chlorophenylmethane
  mass spectroscopy 9:407
1,1-Bis(4-chlorophenyl)-2,2,2-trichloroethanol.
    See Kelthane.
Bis(2-chloropropyl) ether. (See also *Ethers.*)
  properties 12:297
Bis(3-chloropropyl) ether. (See also *Ethers.*)
  properties 12:297
Bis(cyclohexanone oxalyl dihydrazone)
  colorimetric method for copper 12:555
Biscyclohexylsulfamide
  in cyclamates 11:226
  infrared spectroscopy 11:226
Biscyclopentadienylmanganese. (See also Manganese
    compounds.)
  uses 15:489
Bis(di-*n*-butylthiocarbamoyl)disulfide. See
    Tetrabutylthiuram disulfide.
2,2'-Bis[di(carboxymethyl)imino] diethyl ether.
    See Chel ME.

1,2-Bis {[2-di(carboxymethyl)imino]ethoxy}
    ethane. See EGTA.
Bis(diethylthiocarbamoyl)disulfide. See Tetra-
    ethylthiuram disulfide.
4,4'-Bis(dimethylamino)diphenylamine
  colorimetric method for mercury 16:42
4,4'-Bis-dimethylaminothiobenzophenone
  colorimetric method for halogens 11:142
Bis(1,3-dimethylbutyl)amine 5:260
N,*O*-Bis(dimethylsilyl)acetamide. (See also
    Silylation reagents.)
  properties 18:90
Bis(dimethylthiocarbamoyl)disulfide. See Tetra-
    methylthiuram disulfide; Thiram
Bis(dipentaerythritol) monoformal
  gas chromatography 13:578, 594
Bisepigallocatechin
  in black tea 18:493
  structure 18:493
  thin-layer chromatography 18:495, 496
Bis(3,4-epoxy-6-methylcyclohexylmethyl) adipate.
    (See also *Epoxy resins.*)
  infrared spectrum 12:240
Bis(2-ethylhexyl)amine 5:260
Bis(1-ethyl-3-methylpentyl)amine 5:260
N,*N'*-Bis(1-ethyl-3-methylpentyl)-*p*-phenylene-
    diamine 6:74
Bis(2-hydroxyanil)glyoxal
  colorimetric method for calcium 8:98, 108
Bishydroxycoumarin. (See also *Coumarins.*) 10:
    154; 11:159
  acid–base titrations, nonaqueous 10:157
  analysis 10:156
  assay 10:157; 11:169
  in biological materials 10:159
  in biological systems 11:171
  bromometric titration 10:158
  crystallography 10:155
  diazotized dianisidine colorimetric method 10:
    161
  Folin-Ciocalteu colorimetric method 10:160
  gravimetric determination 10:156, 157
  identification 10:151, 156
  infrared spectroscopy 10:157, 159; 11:162
  lithium methoxide volumetric method 10:159
  paper chromatography 10:152, 156, 158, 161
  properties 10:154
  separation 10:156
  solubility 10:154
  thin-layer chromatography 10:151, 156; 11:165
  ultraviolet spectroscopy 10:151, 157, 158, 160
  uses 11:159
  x-ray diffraction methods 10:155
N,*N*-(Bis-2-hydroxyethyl)cyclohexylamine
  gas chromatography 11:247
2,2-Bis(hydroxymethyl)-1,3-propanediol. See
    Pentaerythritol.

4-Bis(*p*-hydroxyphenyl)methylene-2,5-cyclo-
  hexadien-1-one. See Aurin.
2,2-Bis(4-hydroxyphenyl)propane. See Bisphenol
  A.
2,2′-Bisindole indigo. See Indigo.
2,3′-Bisindole indigo. See Indirubin.
Bis(isocyanatocyclohexyl)methane
  gas chromatography 19:315
  in polyurethanes 19:315
Bis isocyanatocyclohexyl methane diisocyanate.
  (See also Isocyanates.)
  structure 19:271
  trade names 19:271
1,3-Bis(isocyanatomethyl)benzol. (See also
  Isocyanic acid esters.)
  infrared spectroscopy 15:106
1,1-Bis(4-isocyanatophenyl)cyclohexane. (See
  also *Isocyanic acid esters.*)
  infrared spectroscopy 15:106
Bis(4-isocyanatophenyl)methane. (See also
  Isocyanic acid esters.)
  infrared spectroscopy 15:106
2,2-Bis(4-isocyanatophenyl)propane. (See also
  Isocyanic acid esters.)
  infrared spectroscopy 15:106
Bis(2-methallyl) ether. (See also *Ethers.*)
  properties 12:297
N,N′-Bis(1-methylheptyl)-*p*-phenylenediamine
  6:74
Bis(methylmercuric) sulfate. (See also Mercury-
  containing compounds.) 13:198
  manufacture 13:198
  properties 13:198
  solubility 13:198
  toxicity 13:198
  uses 13:198
Bismuth. (See also Bismuth alloys; Bismuth
  metal.) 7:187–220
  in aluminum alloys 5:121
  analysis 7:202
  atomic absorption spectroscopy 7:217; 15:195,
    196; 16:167
  basic nitrate precipitation method 7:212
  in bismuth alloys 7:211
  borax bead test 6:584
  cinchonine–potassium iodide colorimetric method
    11:140, 144
  of cobalt alloys 10:309
  in cobalt metal 10:295
  commercial specifications 7:201
  in copper alloys 10:634
  in copper metal 10:620
  in cotton 11:140, 152
  cupferron extraction method 15:198
  diethyldithiocarbamate colorimetric method 7:
    215
  dithizone colorimetric method 7:215; 15:177

EDTA titration 7:213; 13:445
electrodeposition 7:213
emission spectroscopy 7:209, 216; 11:152; 13:
  277; 14:48, 540; 15:187, 197; 16:387
ferric hydroxide gravimetric method 10:295
in ferrochromium 9:661
in ferroniobium 16:387
flame photometry 7:218
flame tests 6:589, 592
in gallium arsenide 13:276, 279
in gallium metal 13:266
in glass and glass-ceramics 13:445
in gold 14:45
gravimetric methods 7:211; 16:20
in hydrochloric acid 14:374
identification 7:208
impurities, determination of 7:202
in indium 14:540
in indium alloys 14:551
iodide colorimetric method 7:215
ion exchange chromatography 13:255
in iron 15:59
in lead 15:177, 195, 196, 198
in lead alloys 15:195
in mercury 16:20
microcosmic salt bead test 6:587
in molybdenum 16:167
in molybdenum disulfide 16:167, 178
in molybdenum trioxide 16:167
in nickel 16:324
in nickel alloys 16:324
in niobium 16:387
occurrence 7:187
oxybromide precipitation method 7:210
oxychloride gravimetric method 5:121
oxychloride precipitation method 7:211
perchloric acid gravimetric method 15:199
phosphate precipitation method 7:211
polarography 7:217; 13:279
potassium chromithiocyanate colorimetric method
  11:140
potassium iodide colorimetric method 10:295,
  620; 15:198
radiochemical methods 7:218
in rocks 15:187
separation 7:210
in silver 18:173
solutions 7:208
solvent extraction 16:167
thiourea colorimetric method 7:214; 10:635;
  15:198
in tin metal 19:66, 67
visible spectroscopy 7:216
voltammetry 7:217
volumetric methods 7:213
x-ray methods 7:218
Bismuth alloys. (See also Bismuth metal.) 5:65; 7:190

analysis 7:193
bismuth content of 7:211
composition 5:68; 7:190
melting points 5:68
properties, determination of 7:193
sample preparation 7:209, 210
uses 7:190
Bismuth bromide 7:197
properties 7:197
Bismuth carbonate, basic 7:195
analysis 7:195
commercial specifications 7:195
gravimetric determination as oxide 7:195
gravimetric determination as oxychloride 7:195
impurities, determination of 7:196
Bismuth chloride 7:198
commercial specifications 7:198
impurities, determination of 7:199
properties 7:197, 198; 9:334
solubility 9:336
Bismuth compounds. (See also individual compounds.) 7:194
Bismuth compounds, organic 7:207
sample preparation 7:209
Bismuth(III) dimethyldithiocarbamate. (See also Dithiocarbamic acid salts.)
uses 11:576
Bismuth fluoride 7:198
properties 7:197, 198; 13:11
Bismuth halides. (See also individual compounds.) 7:197
analysis 7:199
Bismuthides 7:201
Bismuthine 7:201
Bismuth iodide 7:199
properties 7:197, 199
Bismuth metal. (See also Bismuth alloys.) 7:187
analysis 7:193
commercial grades and specifications 7:192
impurities, determination of 7:193
metallurgy 7:189
occurrence 7:187
properties 7:187
properties, determination of 7:193
recovery from copper ores 7:189
recovery from lead ores 7:189
recovery from tin concentrates 7:190
refining 7:190
sample preparation 7:209
Bismuth nitrate 7:201
colorimetric method for o-phenylenediamine 11:446
Bismuth nitrate, basic. See Bismuth subnitrate.
Bismuth oxide 7:204
analysis 7:205
commercial specifications 7:204
impurities, determination of 7:205

properties 7:204
Bismuth oxycarbonate. See Bismuth carbonate, basic.
Bismuth oxychloride
EDTA titration 11:72
in nail enamels 11:72
Bismuth oxyhalides 7:199
Bismuth oxynitrate. See Bismuth subnitrate.
Bismuth potassium iodide
gravimetric method for morpholine 16:195
Bismuth–pyrocatechol violet complex
colorimetric method for EDTA 12:40
Bismuth solder 5:65
Bismuth stannate. (See also Stannates.)
uses 19:72
Bismuth subcarbonate. See Bismuth carbonate, basic.
Bismuth subnitrate 7:202
analysis 7:202
commercial specifications 7:202
impurities, determination of 7:203
Bismuth sulfate 7:206
Bismuth sulfide 7:206
Bismuth telluride semiconductors. (See also Ceramics.)
properties 9:213
uses 9:213
Bismuth trihydride 7:201
Bis(pentaerythritol) monoformal
gas chromatography 13:578
Bispentaerythritol monoformal hexaacetate
gas chromatography 13:578
Bisphenol A. (See also Phenols.)
in brake fluids 7:425, 428
gas chromatography 17:338
manufacture 17:3
in polycarbonates 17:348
structure 17:3
thin-layer chromatography 17:338
ultraviolet spectroscopy 7:428; 17:348
uses 17:3
Bisphenol A epoxy resins. (See also Epoxy resins.) 12:233
infrared spectra 12:238
manufacture 12:233, 234
properties 12:234
viscosity 12:235
Bisphenol A polycarbonate. (See also Polycarbonates.)
infrared spectrum 17:332
nuclear magnetic resonance spectroscopy 17:333
properties 17:330
uses 17:330
Bisphenol B. (See also Phenols.)
manufacture 17:3
structure 17:3
uses 17:3

Bisphenols. (See also Bithionol; Dichlorophene;
    and Hexachlorophene.)
  analysis 11:546
  in disinfectants and antiseptics 11:545
  gas chromatography 11:550
Bisphenol Z polycarbonate. (See also *Poly-*
    carbonates.)
  infrared spectrum 17:332
p-Bisphenyl isocyanate. (See also *Isocyanic*
    acid esters.)
  near-infrared spectroscopy 15:105
Bisphthalimidomethyl derivative of allylamine
    5:87
2,7-Bis(2'-sulfonophenylazo)-1,8-dihydroxy-3,6-
    naphthalene-disulfonic acid. See Sulfonazo
    III.
Bis(thiocarbamoyl)sulfides. See Thiuram di-
    sulfides.
2,2'-Bisthionaphthene indigo. See Thioindigo.
1,4-Bis(*p*-toluino)anthraquinone. See D & C
    Green No. 6.
Bistrichlorosilylethane. (See also Organo-
    chlorosilanes.)
  gas chromatography 18:62
Bistrichlorosilylethylene (See also Organo-
    chlorosilane.)
  gas chromatography 18:62
N,O-Bis(trimethylsilyl)acetamide. (See also
    Silylation reagents.)
  gas chromatography 18:94
  impurities, determination of 18:94
  infrared spectroscopy 18:92
  manufacture 18:91
  properties 18:90
Bis(trimethylsilyl)ethylenediamine. (See also
    Silylation reagents.)
  properties 18:90
N,O-Bis(trimethylsilyl)trifluoroacetamide. (See
    also Silylation reagents.)
  properties 18:90
Bisulfate fusion
  of silicon 18:14
Bisulfite-addition method. (See also Sodium
    bisulfite addition method.)
  for acetals 4:69
  for acrylonitrile 4:373
Bisulfite bleach test
  for anthocyanins 19:464
Bisulfites. (See also Sulfites.)
  iodometric titration 19:416
  manufacture 18:401
  properties 18:402
  solubility 18:402
  in wines 19:416
Bithionol. (See also Bisphenols.)
  acid–base titration 11:547
  analysis 11:546

assay 11:547
identification 11:546
loss on drying 11:547
melting range 11:547
in pharmaceuticals 11:548
properties 11:546
residue on ignition 11:547
solubility 11:546
specification tests 11:547
ultraviolet spectroscopy 11:548, 549
uses 11:546
3,3'-Bitolylene-4,4'-diamine
  in polyurethanes 19:304
  structure 19:304
Bitolylene diisocyanate. (See also Isocyanates.)
  properties 15:95
  structure 19:270
  trade names 15:102; 19:270
Bitterness test
  for beer 7:603
Bittersweet chocolate. See Chocolate.
Bitumens. See Bituminous materials.
Bituminous materials 6:274
  analysis 6:289
  in bituminous paving mixtures 6:330
  chemical composition 6:323
  commercial grades and specifications 6:280
  durability 6:306
  identification 6:323
  manufacture 6:277
  occurrence 6:277
  properties 6:274, 289
  purity 6:311
  separation 6:323
  solubility 6:311
  temperature susceptibility factors 6:304
Bituminous paving mixtures
  analysis 6:330
  bitumen content of 6:330
  properties, determination of 6:335
  recovery of asphalt from 6:334
  volatile content of 6:333
  water in 6:317
Bituminous shale 19:225
Biurets
  infrared spectroscopy 15:96
  manufacture 15:96
Biuret test
  for animal glue 10:282
  for protein 11:75; 19:476
Bixin. See Annatto.
Black ash. (See also Barium sulfide.) 6:534, 569
  analysis 6:569
Black iron oxide
  analysis 15:531
  assay 15:531
  commercial grades and specifications 15:515

in matches 15:515
in match strikers 15:522
particle size 15:531
Blackness
of carbon black 8:181, 211
Black phosphorus. See Phosphorus, black.
Black pigments. (See also *Pigments*; and individual compounds.) 17:202
Black powder. (See also High-explosive mixtures.)
analysis 12:446
composition 12:446
uses 12:446
Black tea
alkaloid content 18:488
analysis 18:485
aroma 18:507
ash, alkalinity of 18:488
ash content 18:487
briskness 18:486 '
color 18:486, 494, 499
commercial grades and specifications 18:485
composition, determination of 18:486
cream content 18:506
cream index 18:506
creaming powder content 18:506
flavor 18:486
gas chromatography 18:508
headspace analysis 18:508
infusion tests 18:486
insoluble ash content 18:487
manufacture 18:455
mesh size 18:485
moisture content 18:486
polyphenol content 18:492
properties, determination of 18:485
quality 18:486
soluble ash content 18:487
solvent extraction 18:487, 489, 492, 494
standard brew 18:485
standard infusion 18:494
steam distillation 18:512
strength 18:486
tannin content 18:505
Black tin oxide. See Stannous oxide.
Blanc fixe. (See also Barium sulfate.) 6:555
Blancophor G titration
of cationic dyes 12:13
BLE. See Acetone–diphenylamine reaction
products.
BLE-25 6:73
Bleach demand tests
for cellulose 9:52
Bleaching 7:221
chemical 7:221
mechanism 7:223
paper 7:222
strength 7:224

textile 7:221
Bleaching agents 7:221–293
analysis 7:225
identification 7:225
impurities, determination of 7:230
performance tests 7:232
Bleaching powder 7:221
available chlorine 7:224
Bleaching strength 7:223
Bleach liquor
available chlorine 7:225
iodometric titration 7:225
Bleach, tropical. See Tropical bleach.
Blended whisky 4:465
analysis 4:480
quality control 4:470
Blending agents
for alkyd resins 5:22
Blood
glycerol content 13:520
hexachlorophene content 11:550
lead content 15:186, 189
nitrous oxide content 16:477
zinc content 19:568
Blood glues, evaluation 4:400
Blood plasma
cyanoacetic acid content 15:439
nalorphine content 16:268
Blood serum
bromine in 8:24
calcium content 8:110; 18:222
magnesium content 15:371; 18:222
potassium content 18:222
sodium content 18:222
Bloom ratings
of gelatin 13:368
Bloom retardants
in confectionery products 10:552
Blowout method
for dye mixtures 6:388, 413
for triphenylmethane and related dyes 19:164
Blue basic lead sulfate. See Blue lead.
Blue casse
in wines 19:494
Blue lead
analysis 17:202
commercial specifications 17:202
composition 17:202
lead content 17:202
uses 17:202
Blue organic pigments
in enamels 12:187
identification 12:187
Blue pigments. (See also *Pigments*; and individual
compounds.) 17:201
Blue reflectance 7:233
of bleached fabrics 7:233

Blue ring test
  for thiosulfate 18:431
Bluestone. See Copper sulfate.
Blue tetrazolium
  colorimetric method for norethindrone 14:207
  colorimetric method for norethindrone acetate
    14:207
  colorimetric method for norgestrel 14:207
  colorimetric method for steroids 14:207
Blue vitriol. See Copper sulfate.
Blue VRS. (See also *Triphenylmethane and
    related dyes.*)
  paper chromatography 19:172
Boart, crushed (See also *Diamonds, Industrial.*)
    11:468
Boehmite
  composition 10:104
Boiler water condensates
  aluminum in 5:241
Boiling point
  of antifreezes 6:5
  of beer 7:503, 543
  of brake fluids 7:429
  of 1,3-butadiene 11:503
  Cottrell-type apparatus 7:503
  of ethanol 19:427
  of ethylenediamine 11:414
  of fluorocarbons 13:66
Boiling point elevation method
  for molecular weight of polyethylenimine 14:
    486
Boiling range
  of benzene 7:43
  of carbon disulfide 8:274
  of carbon tetrachloride 8:327
  of fluorocarbons 13:66
  of methyl chloride 9:425
Bois de rose oil. (See also *Essential oils.*)
  refractive index 12:271
Bomb method
  for sulfur in lubricants 15:328
Bond content
  of coated abrasives 10:279
Bonded abrasives. (See also Grinding wheels.)
    7:293–312
Bonding. See *Adhesion.*
Bonding agents
  in grinding wheels 7:294
  in vitrified wheels 7:297
Bonding strength
  of acrylic resin teeth 11:307
Bone
  calcium in 8:110
  radium content 17:465
  strontium content 18:269
Bone char
  ash determination 4:436

carbon and acid-insoluble matter 4:437
chemical analysis 4:439
efficiency for percolation processes 4:449
pH 4:437
uses 4:434
water-soluble materials 4:438
BON maroon. (See also *Pigments.*)
  structure 17:179
BON reds
  composition 12:173
  in enamels 12:173
Boracite 7:315
Boranes. See Boron hydrides.
Borate esters 7:334
  gas chromatography 7:353
Borates. (See also Metal borates; and individual
    compounds.) 7:368
  acid–base titration 7:406
  analysis 7:380
  in antifreezes 6:11
  antitrusts 6:351
  in cotton 11:141
  in creams and lotions 11:92
  EDTA volumetric method 15:160
  electrolytic reduction to elemental boron 7:321
  flame photometry 11:75
  flame test 18:184
  identification 5:282
  ion exchange chromatography 7:396
  in lanthanum glass 15:160
  in manufacture of ferroboron 7:323
  p-nitrobenzene azochromotropic acid colori-
    metric method 11:141
  occurrence 7:313
  polarimetric methods 7:412
  precipitation methods 7:394
  in shampoos 11:75
  in soaps 18:182, 184
  in starch 15:527
  turmeric colorimetric method 15:527
Borax. (See also Sodium borate.) 18:208
  acid–base titration 7:428
  analysis 7:380, 428
  boron content of 7:416
  in brake fluids 7:424, 425, 428
  commercial grades and specifications 7:374, 375
  detection 7:391
  in embalming fluids 12:166
  impurities, determination of 7:382
  infrared spectrum 7:380
  mannitol volumetric method 18:192
  in manufacture of boric acid 7:371
  occurrence 7:313, 314, 373
  primary standard 7:415
  properties 7:373
  screen analysis 7:375
  in soaps 18:192

solubility 7:373, 374
uses 7:373
Borax, anhydrous 7:377
  commercial grades and specifications 7:377, 378
  manufacture 7:377
  properties 7:377
  screen analysis 7:378
  uses 7:374, 377
Borax bead test 6:583
Borax fusion
  of silicon 18:13
  of titanium minerals 19:93
Borazine 7:344
Borazole. See Borazine.
Borazoles, substituted
  gas chromatography 7:353
Bordeaux lakes
  in enamels 12:183
  identification 12:183
Boric acid 7:312, 370
  acid–base titration 7:371, 402
  alizarin-S colorimetric method 7:386, 408
  in aluminum acetate solutions 5:163
  analysis 7:380
  on antimony fluoborate 6:55
  carmine colorimetric method 7:387, 409
  colorimetric methods 7:407
  color tests 7:385
  commercial grades and specifications 7:372,
    373
  curcumin colorimetric method 7:388, 409
  detection 7:385
  1,1′-dianthrimide colorimetric method 7:389,
    407, 410
  diols, complexes with 7:402
  distillation of antimony trioxide 13:429
  extraction 7:395
  fluorescence tests 7:385, 410
  in fluoroboric acid 13:39
  flux for ceramics 9:216
  gravimetric method for tantalum 16:388
  identification 7:385
  impurities, determination of 7:381
  infrared spectrum 7:380
  ion exchange methods 7:396
  manufacture 7:371
  occurrence 7:313, 314
  pH change methods 7:390
  polarimetric methods 7:412
  polyols, complexes with 7:402
  primary standards 7:416
  properties 7:370, 371
  from pyrohydrolysis of boron 7:325
  quinalizarin colorimetric method 7:386, 407
  sample handling 7:398
  solubility 7:371, 372
  turmeric colorimetric method 7:388, 409

  uses 7:372, 374
  volatility 7:398
  volumetric method for boron 15:83
Boric acid, anhydrous. See Boric oxide.
Boric acid esters
  analysis 7:349
Boric anhydride. See Boric oxide.
Boric oxide 7:369
  acid–base titration 13:417
  color 13:461
  colorimetric method for fluoride 13:414, 433
  commercial grades and specifications 7:370, 371
  fusion of silicon 18:13
  in glass and glass-ceramics 13:413, 415, 417, 459
  infrared spectroscopy 13:459
  manufacture 7:369
  primary standard 7:416
  properties 7:369
  pyrohydrolysis 7:394
  reduction to elemental boron 7:319
  screen analysis 7:370
  solubility 7:370
  in titanium boride 7:367
  uses 7:370
Borides. (See also Metal borides.) 7:323, 358
  fluxes for 9:216
  pyrohydrolysis 7:325
Borinic acid 7:330
  analysis 7:349
Borker and Sloman spectroscopic method
  for caffeine 10:407
Bornite
  copper content 10:600
Bornyl acetate
  in essential oils 12:276
  molecular weight 12:276
Borofluorides 7:315
Borohydrides 7:330, 343
  iodine titration 7:346
  manufacture 7:344
Boron. (See also Boron alloys; Boron, elemental.)
  7:312–423
  alizarin-S colorimetric method 7:386, 408
  in aluminum alloys 5:122
  analysis 7:384
  in antimony fluoborate 6:55
  barium borotartrate gravimetric method 7:401
  in beer 7:653
  in borax 7:416
  boric acid volumetric method 15:83
  in boron alloys 7:323
  in boron carbide 8:135
  in boron compounds, organic 7:349
  in boron, elemental 7:323
  in boron hydrides 7:345, 408
  in carbon 14:70
  carmine colorimetric method 5:122; 7:387, 409

in cements 9:131
chromatographic methods 7:397
in cobalt alloys 10:309
colorimetric methods 7:407
color tests 7:385
in copper alloys 10:635
curcumin colorimetric method 7:388, 409, 654;
   10:309, 635; 14:77
detection 7:385
1,1'-dianthrimide colorimetric method 7:407,
   410
distillation as boron trichloride 7:393
distillation as boron trifluoride 7:394
distillation method 16:325
distillation as methyl borate 7:391
by emission spectroscopy 4:4, 345; 7:411; 14:
   34, 48, 122, 542; 15:155, 187; 16:387
extraction 7:395
in ferroboron 15:83
in ferrochromium 9:661
in ferroniobium 16:387
flame photometry 7:351, 352, 390, 411
flame tests 6:583, 592; 7:389
fluorescence tests 7:385, 410
in germanium 13:397
in germanium dioxide 13:397
in glass 7:394, 396, 400
in gold 14:45
in gold alloys 14:34
in graphite 14:70
gravimetric methods 7:401
in hafnium 14:122
in high-temperature alloys 10:286
identification 7:385
in indium 14:542
ion exchange chromatography 7:396; 15:84
in iron 15:59
isotope abundance 7:415
in lanthanum oxide 15:155
mannitol titration method 7:361, 402; 14:144;
   16:325
manufacture 7:318
in metal borides 7:361
methanol gravimetric method 18:13
neutron absorption 7:351, 413
in nickel 16:325
in nickel alloys 16:325
in niobium 16:387
nitron tetrafluoborate gravimetric method 7:
   401
occurrence 7:313
polarimetric methods 7:412
polarography 7:412
in polymers 4:345
precipitation methods 7:394, 401
primary standards 7:415
properties 7:316

pyrohydrolysis 7:325, 394
quinalizarin colorimetric method 7:386, 407
in rocks 15:187
sample handling 7:398
separation 7:391
in silica ablators 4:4
in silicon 18:13, 23
in sodium borohydride 14:254
in specific materials 7:416
titrimetric methods 7:402
turmeric colorimetric method 7:388, 409
uses 7:321
volumetric determination 14:254
in zirconium 14:122
in zirconium diboride 14:144
Boron alloys 5:59, 61; 7:321, 323
   boron content of 7:323
   composition 7:322
   sample decomposition 7:323
Boron carbide. (See also *Carbides*.) 7:331
   analysis 7:332
   boron content of 8:134
   carbon content of 8:134
   composition 7:322
   formation 8:130
   free carbon in 8:134
   manufacture 7:331
   properties 7:331; 8:131, 132
   in titanium boride 7:367
   uses 7:323; 8:132
Boron compounds. (See also individual compounds.) 7:312, 329–384
   reduction to elemental boron 7:319
   sampling 7:331
   toxicity 7:331
Boron compounds, heterocyclic 7:330
   analysis 7:349
Boron compounds, organic 7:330, 349
   analysis 7:349
   carbon in 7:352
   flame photometry 7:351
   gas chromatography 7:352
   hydrogen in 7:352
   infrared spectroscopy 7:353, 354
   mass spectroscopy 7:355
   neutron absorption 7:351, 415
   nitrogen in 7:352
   oxidation methods 7:349
   sample preparation 7:399
Boron compounds, refractory. (See also Boron
   carbide; Boron nitride; and Metal borides.)
   7:330, 331, 358
Boron-containing minerals 7:314
Boron, elemental. (See also Boron alloys.) 7:312
   analysis 7:323
   boron content of 7:323
   carrier gas fusion methods 8:624, 682

chemical properties 7:317, 320
crystalline modifications 7:317
hydrogen in 8:618, 681
impurities, determination of 7:326
isotope abundance methods 7:326
isotopes 7:317
manufacture 7:319
mechanical properties 7:318
nitrogen in 8:624
oxygen in 8:616, 618, 624, 681, 682
properties 7:316
sample decomposition 7:323
thermal properties 7:319
use in manufacture of metal borides 7:358
uses 7:321
vacuum fusion methods 8:616, 618, 681
Boron, free
oxidation methods 7:324
Boron halides. (See also individual compounds.)
. 7:332
analysis 7:337
reduction to elemental boron 7:321
Boron hydrides 7:330, 339, 344
in air 7:349
analysis 7:345
boron content of 7:345, 408
Carius oxidation 7:400
colorimetric methods 7:347
flame photometry 7:351, 411
fractional distillation 7:352
gas chromatography 7:352
iodine titration 7:346
manufacture 7:343
mass spectroscopy 7:346
monitors for detection in air 7:349
neutron absorption 7:352
nitrogen in 7:352
nomenclature 7:339
properties 7:342
toxicity 7:331
uses 7:342
Boronic acid 7:330
analysis 7:349
Boron monoxide 7:368
Boron nitride 7:331
analysis 7:332
carrier gas fusion methods 8:624
manufacture 7:332
nitrogen in 8:624
oxygen in 8:624
properties 7:331
in titanium borides 7:367
Boron oxide 7:368
analysis 7:381
carrier gas fusion methods 8:624, 682
oxygen in 8:624, 682
Boron subhalides 7:337

analysis 7:339
Boron tribromide 7:332
analysis 7:337
manufacture 7:333
properties 7:332
Boron trichloride 7:333
analysis 7:337
chemical reactions 7:334
commercial grades and specifications 7:335
distillation 7:393
manufacture 7:334
properties 7:333; 9:334
solubility 9:336
uses 7:333
Boron trifluoride 7:335
analysis 7:337
commercial grades and specifications 7:336
distillation 7:394
manufacture 7:336
mass spectroscopy 7:338
methylation reagent for carboxylic acids 8:386
properties 7:335
toxicity 13:13
uses 7:336
Boron trifluoride–methanol
esterification reagent for fatty acids 11:364
Boron triiodide 7:336
analysis 7:339
properties 7:336
Boron trioxide. (See also Boric oxide.)
in alkali barium silicate glass 9:218
emission spectroscopy 9:218
in enamels 9:195
in glazes 9:196
in silicate glasses 9:198
Borosilicate glass. (See also *Glass and glass-ceramics*;
Silicate glasses.)
composition 9:198; 13:407, 409, 410
uses 9:198; 13:410
Borosilicate minerals 7:316
Botanicals. (See also *Pesticides and insecticides.*)
19:580
commercial grades and specifications 19:593
manufacture 19:596
properties 19:586
solubility 19:587
toxicity 19:592
Botran. See Dicloran.
Bottle centrifuges 9:162, 163, 165
characteristics 9:160
settling curves 9:161, 169
Bound rubber content
of SBR 12:132
Bourbonal. See Ethyl vanillin.
Bourbon whisky 4:464
analysis 4:480
congeneric content 4:467

quality control 4:470
Bourdon-type pressure gauge 7:281
BPHA. See *N*-Benzoylphenylhydroxylamine.
Brabender oven method 4:471
Braggite
  composition 17:215
Brake drum polishing tests 7:455, 457, 470
Brake dynamometer 7:453
Brake fluids 7:423–440
  analysis 7:426
  commercial grades and specifications 7:426
  compatibility test fluid 7:433
  composition 7:424
  composition, determination of 7:426
  evaluation methods 7:428
  performance test 7:435
  properties 7:424
Brake linings 7:440–474
  analysis 7:443–449
  asbestos in 7:440
  composition 7:441, 443
  dry mold type 7:441
  dynamometer testing 7:452
  extruded type 7:442
  finished product testing 7:449
  identification 7:449
  in-process controls 7:448
  master batch testing 7:447
  millboard type 7:440, 443
  performance tests 7:452
  pyrolysis–gas chromatography 7:449
  raw materials testing 7:444
  road tests 7:465
  sheeter type 7:442
  types 7:441
  vehicle testing 7:465
Brake system road tests 7:455, 465
Brake systems, hydraulic 7:423
Brandy 4:466
  alcohol content 19:493
  analysis 4:480
  congeneric content 4:467
  quality control 4:470
Brannerite
  composition 19:225
  occurrence 19:225
Brass. (See also Copper alloys; Tin alloys; and
    Zinc alloys.) 10:630
  composition 10:630
  manganese content 15:462, 466
  phosphorus content 17:89
  properties 10:630
  sample decomposition 18:25
  silicon content 18:25
  sulfur content 18:382, 390
  tin content 19:51
  uses 19:547

zinc content 19:558
Brass corrosion test
  for lubricants 15:337
Brassicol. See Terraclor.
Bratton-Marshall reaction
  p-aminobenzoic acid esters analysis 5:389
  atropine analysis 4:596
  chloramphenicol analysis 5:306
  propoxycaine hydrochloride analysis 5:412
  tropane alkaloids analysis 4:595
Braunite 15:448
Bravoite
  composition 16:286; 19:321
  occurrence 19:321
Brazing alloys
  phosphorus content 17:93
Bread 6:439
  analysis 6:448
  iron content 15:32
  manufacture 6:440
  quality control 6:448
Breaking length
  of pulp handsheets 7:256
Breaking strength
  of fabrics 4:25
  of felt 12:510
Break-in tests
  of brake linings 7:454, 469
Breithauptite
  composition 16:286
Breon 12:86
Brewery products. (See also Beer.) 7:475–679
  analysis 7:480
  definitions 7:476
  production 7:479
  properties 7:477
Brightness
  of cellulose 9:52
  of paper matchboard 15:534
  of pulp 7:257, 262
Brightness reversion
  of paper 7:263
Brightness testing
  master instrument 7:258
  of pulp handsheets 7:262
  reflectance standards 7:262
Brilliant blue. See F D & C Blue No. 1.
Brilliant blue FCF. (See also *Triphenylmethane
    and related dyes.*)
  liquid chromatography 19:173
Brilliant green. (See also *Triphenylmethane and
    related dyes.*)
  colorimetric method for boron 7:410
  colorimetric method for tantalum 16:370
  colorimetric method for thallium 18:531
  in enamels 12:187
  identification 12:187

orange II titration 19:175
visible spectroscopy 19:167
Brilliant lake red. See D & C Red No. 31.
Brilliant red
in enamels 12:185
identification 12:185
Brilliant yellow
colorimetric method for magnesium 15:368
Brimstone. (See also Sulfur, elemental.) 18:360,
364
Brine 18:239
iodide content 14:600
mercury content 16:44, 45, 46
Briskness
of black tea 18:486
Bristol glaze. (See also Glazes.)
composition 9:196
Britannia metal. (See also Copper alloys.) 5:72
composition 10:631
British Standards method for analysis of fiber
mixtures 4:356
Brix hydrometers 18:341
Brochantite
copper content 10:600
Brocide. See Ethylene dichloride.
Brodie solution 8:474
Broenner's acid. See 2-Naphthylamine-6-sulfonic
acid.
Bromapatite
manufacture 17:121
Bromate–bromide method
for acrylamide 4:163
for acrylic monomers 4:191
for allyl esters 5:97
for butacaine sulfate 5:381
for residual acrylamide in polymers 4:175
Bromate–bromide titration
of barbiturates 6:495
of calcium cyclobarbital 6:515
of monosulfonated dyes 10:519
Bromates 8:7
in ammonium bromide 12:543
analysis 8:8
in cesium bromate 9:285
commercial grades and specifications 8:7
in cotton 11:141
formation from bromine 8:3, 7
impurities, determination of 8:8
iodometric method 8:8; 9:361
manganous sulfate–sulfuric acid colorimetric
method 11:141
polarography 12:543
in potassium chlorate 9:359, 361
potassium iodide–starch colorimetric method
12:543
properties 8:7
sodium thiosulfate–potassium iodide

titration 9:285
uses 8:4, 7
Bromate volumetric method
for antimony 6:64
of aminopyrine 6:217
for arsenic(III) 6:241
of barbiturates 6:495
Bromcresol green
colorimetric method for amines 5:273
Bromeline
in beer 7:588
Bromides. (See also *Bromine.*) 8:8
in ammonium iodide 12:544; 14:604
analysis 8:10, 13
in bromates 8:8
bromine content of 8:10, 13
in cadmium iodide 8:44
in cesium bromide 9:285
chloramine-T colorimetric method 9:350; 14:370
in chlorides 9:340
chromic acid extraction 9:347
commercial grades and specifications 8:8, 9, 10,
12
formation from bromine 8:3
in glass 9:200
in glass and glass-ceramics 13:413, 415, 435
in hydriodic acid 14:603
in hydrochloric acid 9:340, 350; 14:370
impurities, determination of 8:11, 13
in iodides 14:597
in iodine 14:592
in nitrates 16:505
polarography 12:545
in potassium chloride 9:339; 12:545
in potassium iodate 14:608
in potassium iodide 12:545; 14:605
potassium persulfate–sulfuric acid redox method
9:340
properties 8:8, 9, 12
silver nitrate titration 13:436
silver nitrate turbidimetric method 14:592, 604
in sodium chloride 9:337, 347
in sodium iodide 14:605
sodium thiosulfate–Chloramine-T colorimetric
method 19:496
solvent extraction 19:496
uses 8:4
by Volhard method 9:285
in wines 19:495
Brominated compounds
in D & C Red No. 21. 10:523
paper chromatography 10:523
Brominated salicylanilides. (See also Salicylanilides.)
ion exchange chromatography 11:552
thin-layer chromatography 11:552
Bromination method

for acrolein 4:151, 156
for acrylic monomers 4:190
for alcohols, tertiary 4:542
for allyl alcohol 5:82
for allyl aromatic compounds 5:90
for allylbenzene 5:90
for allyl esters 5:96
for allyl ethers 5:98
for allyl halides 5:102
for allyl olefins 5:104
for aniline and its derivatives 5:444, 448
for benzoic anhydride 7:74
for 2-butyne-1,4-diol 4:145
for cinnamic acid 10:61
for crotonaldehyde 11:178
for diphenylamine 5:447
for olefins 5:82, 90
for phenindione 10:181
for rubber, natural 12:122
for sodium hexobarbital 5:369
Bromination titration
  of chlorophenols 9:514
Bromine. (See also Bromine, elemental.) 8:1–29
  in biological materials 8:25
  in blood serum 8:24
  in bromohydrocarbons 8:13
  in cadmium bromide 8:37
  in chlorine 9:309, 321
  in chlorocarbons 9:409
  colorimetric method for amine antioxidants 6:95
  colorimetric method for histidine 8:487
  colorimetric method using phenol red 8:23
  in cotton 11:154, 156
  coulometric determination 8:26
  distillation method 8:16, 24
  in elastomers, synthetic 12:111
  in elemental bromine 8:5
  fluorescein identification test 8:15
  in foods 8:25
  gravimetric method for phenols 16:212
  in hydrocarbon resins 14:286
  in hydrocarbons 14:336, 339
  identification 8:14
  mercuric nitrate titration 12:111
  method using catalytic oxidation of iodine 8:23
  neutron activation analysis 8:27; 14:155
  occurrence 8:1
  in organic compounds 8:18, 22
  peroxide bomb method 8:18
  polarography 8:27
  potentiometric titration 8:21
  Schöniger method 8:22
  separation 8:14, 16
  separation from iodine 8:15
  silver bromide gravimetric method 8:17
  silver nitrate identification test 8:15
  in sodium chloride 9:344

sodium thiosulfate titration 9:322, 344
thin-layer chromatography 8:16
uses 8:1, 4
van der Meulen method 8:19, 25
visible spectroscopy 8:21
Volhard titration 8:17
x-ray fluorescence spectroscopy 8:26; 11:156
x-ray spectroscopy 9:409
Bromine–absorption method
  for acetylene 4:128
Bromine–ammonium hydroxide
  gravimetric method for manganese 16:297
Bromine–benzidine
  colorimetric method for hydrogen cyanide 14:
    418
Bromine–bromide method
  for acrylic monomers 4:192
Bromine compounds, (See also Bromates; Bro-
    mides; Bromohydrocarbons; and individual
    compounds.) 8:1, 6–14
  uses 8:4
Bromine–dimethylglyoxime
  colorimetric method for nickel 10:330
Bromine, elemental 8:1–6
  analysis 8:5
  bromine content of 8:5
  chemical properties 8:2
  commercial grades and specifications 8:4
  impurities, determination of 8:5
  manufacture 8:3
  properties 8:1, 2
  solubility 8:2
  toxicity 8:3
  uses 8:1, 4
Bromine monochloride method
  for diallylbarbituric acid 5:92
Bromine–nitric acid
  extraction method for tetraethyllead 15:194
Bromine–nitric acid oxidation
  of sulfur 18:384
Bromine pentafluoride
  manufacture 13:9
  properties 13:8
  uses 13:9
Bromine–potassium cyanide
  colorimetric method for pyridine and its deriva-
    tives 17:424
Bromine–sodium thiosulfate
  volumetric method for iodide 14:600
Bromine trifluoride
  manufacture 13:9
  properties 13:8
  uses 13:9
Bromoacetic acid
  properties 8:396
Bromoacetone
  properties 8:335

p-Bromoaniline
  acid–base titration 5:447
3-Bromobenzaldehyde
  infrared spectroscopy 6:617
Bromobenzamide
  derivatives for identification 8:601
p-Bromobenzonitrile. (See also *Nitriles.*)
  derivatives for identification 16:402
1-Bromo-2-butanol. (See also Halohydrins.)
  derivatives for identification 14:160
2-Bromo-1-butanol. (See also Halohydrins.)
  derivatives for identification 14:160
1-Bromo-2-chloroethane. See Ethylene
  chlorobromide.
Bromochloropropene 13:181
  manufacture 13:181
  properties 13:181
  toxicity 13:181
  uses 13:181
Bromocinnamic acid. (See also *Cinnamic acid
  and its derivatives.*)
  properties 10:63
Bromocresol purple
  colorimetric method for dodine 13:205
  colorimetric reagent for glyodin 13:208
Bromocyanide. See Cyanogen bromide.
Bromodiphenhydramine
  gas chromatography 6:210
Bromoethane. See Ethyl bromide.
Bromofluoresceins
  in D & C Orange No. 5 10:520
  paper chromatography 10:520
Bromoform
  commercial specifications 8:12
  Fujiwara colorimetric method 9:386
  properties 8:12
Bromohydrocarbons. (See also Bromides; and
  individual compounds.) 8:11
  in air 8:13
  analysis 8:13
  bromine content of 8:13
  commercial specifications 8:12
  impurities, determination of 8:13
  properties 8:12
Bromomethane. See Methyl bromide.
Bromometric methods
  for benzylpenicillin 16:571
  for penicillin O 16:571
Bromometric titration
  of bishydroxycoumarin 10:158
  of furfural 13:236
  of G salt 16:237, 239
  of hydrocarbons 14:336
  of morpholine 16:200
  of oxidizable sulfur 13:325
  of phenols 17:26
  of resorcinol 17:46

  of R salt 16:234, 237, 239
  of Schaeffer's salt 16:234, 237, 239
  of thiophene 18:540
p-Bromophenacyl derivative
  of benzoic acid 7:64
p-Bromophenacyl ester
  derivative of adipic acid 4:412
  derivatives for identification of carboxylic acids
    8:376
Brompheniramine maleate. (See also *Antihistamines.*) 6:26
  properties 6:32
p-Bromophenyl isocyanate. (See also *Isocyanic
  acid esters.*)
  near-infrared spectroscopy 15:105
2-Bromo-1,3-propanediol. (See also Glycerol
  halohydrins.)
  structure 14:167
1-Bromo-2-propanol. (See also Halohydrins.)
  derivatives for identification 14:160
  structure 14:153
2-Bromo-1-propanol. (See also Halohydrins.)
  derivatives for identification 14:160
  infrared spectrum 14:159
  structure 14:153
Bromopyridines. (See also *Pyridine and pyridine
  derivatives.*)
  ionization constants 17:416
  manufacture 17:422
Bromopyrogallol red
  colorimetric method for niobium 16:368, 373,
    384
  colorimetric method for tantalum 16:368
4'-Bromosalicylanilide. (See also Salicylanilides.)
  ultraviolet spectroscopy 11:552
5-Bromosalicylanilide. (See also Salicylanilides.)
  ultraviolet spectroscopy 11:552
Bromosilane
  properties 18:40
N-Bromosuccinimide
  qualitative test for alcohols 4:523
  titrant for allyl alcohol 5:82
  titrant for allylamine 5:89
  titrant for allylbarbituric acids 5:92
Bromothallic acid. (See also *Thallium compounds.*)
  solvent extraction 18:524
Bromothymol blue
  colorimetric method for tetracaine 5:417
Bromotrifluoromethane
  gas chromatography 13:64
Bromphenol blue
  colorimetric method for cationic surfactants
    12:578
  colorimetric method for benzalkonium chloride
    11:559
  colorimetric method for cationic detergents 11:
    328

colorimetric method for quaternary ammonium
  compounds 11:82, 93
color test for emulsifiers 15:98
volumetric method for quaternary ammonium
  compounds 11:560
Bronze. (See also Copper alloys; Tin alloys.)
  5:68; 10:631
  aluminum in 5:239, 246, 248
  chromium content 9:648
  composition 10:631; 19:24
  manganese content 15:462, 466
  phosphorus content 17:89
  properties 10:631
  sample decomposition 18:25
  silicon content 18:25
  sulfur content 18:382, 390
Bronze powder
  analysis 17:203
  commercial specifications 17:203
Brookite. (See also Titanium dioxide.) 19:81, 123
Brown pigments. (See also *Pigments*; and individual
  compounds.) 17:192
Brown ring test
  for nitrates 16:501
  for nitric oxide 16:477
Brucine 4:589, 602
  colorimetric method for nitrates 16:503
Brucine–sulfanilic acid
  colorimetric method for nitrates 16:504
Brucine sulfate
  colorimetric method for chlorate 9:348
  colorimetric method for nitrate 9:348; 10:329;
    15:404, 410
Brucine–sulfuric acid
  colorimetric method for nitrate 11:142
Brucite 15:357
Brunauer, Emmett, Teller method. See BET
  technique.
BSA. See *N,O*-Bis(trimethylsilyl)acetamide.
BTTN. (See also High explosives.) 12:451
  infrared spectrum 12:455
  manufacture 12:451
  properties 12:432, 451
  solubility 12:451
  uses 12:451
Bubble method
  for viscosity of epoxy resins 12:251
Bubble rise method
  for viscosity of hydrocarbon resins 14:295
Budene 12:86
Buffer capacity
  of beer 7:508
Building materials. See *Asbestos-cement
  products.*
Building plasters. See also *Gypsum and
  gypsum products.*)
  commercial grades and specifications 14:90

Bulk density
  of abrasive grains 7:311
  of activated carbon 8:152
  of adsorbents 4:443
  of carbon black 8:230
  of diamonds, industrial 11:477
  of high-explosive mixtures 12:467
  of molding compounds 4:23
Bulk volume
  of pulp handsheets 7:256
Bullets
  silver content 18:172
Buna-N copolymers
  acrylonitrile content 16:411
Burkeite 17:405
Burning rate
  of matches 15:547
Burning rate, static
  of cigarettes 10:4
Burnish tests
  of brake linings 7:454, 468
Burnt sienna
  commercial specifications 17:193
  impurities, determination of 17:194
Burnt umber
  commercial specifications 17:193
  impurities, determination of 17:194
Burow's solution. (See also Aluminum acetate.)
  5:161
  boric acid content 5:163
Bursting strength
  of felt 12:511
  of paper 7:280
  of pulp handsheets 7:254
Butabarbital 6:468, 512
  gas chromatography 6:483
  paper chromatography 6:478, 479
  properties 6:470, 515
  in sodium butabarbital 6:515
  thin-layer chromatography 6:480
Butacaine sulfate 5:374
  analysis 5:380
  properties 5:380
1,3-Butadiene 11:501–507
  analysis 11:503
  assay 11:505
  azeotropes 11:502
  boiling point 11:503
  1,3-butadiene content 11:504
  commercial grades and specifications 11:
    502
  gas chromatography 4:212; 11:505; 12:103,
    104, 105; 14:329
  impurities 11:503
  impurities, determination of 11:504, 506
  inhibitors 11:503
  maleic anhydride absorption method 11:505

manufacture 11:501
near-ultraviolet spectroscopy 11:505
properties 11:501
solubility 11:501
ultraviolet spectroscopy 11:505
uses 11:501
Butadiene–acrylic acid polymers
  acrylic acid content 4:275
Butadiene–acrylonitrile polymers. (See also
  Acrylic and modacrylic fibers; Acrylic
  polymers; and *Acrylonitrile polymers.*)
  4:383
  acrylonitrile content by infrared spectroscopy
    4:237, 386
  gas chromatography 4:286
Butadiene–acrylonitrile–styrene polymers. See
  Styrene–acrylonitrile–butadiene polymers.
Butadiene copolymer content
  of impact polystyrene 18:330
Butadiene dimer
  in 1,3-butadiene 11:506
  gas chromatography 11:506
Butadiene dioxide. (See also *Epoxy compounds.*)
  infrared spectroscopy 12:200
Butadiene–methacrylic acid copolymer. (See also
  Elastomers, synthetic.)
  infrared spectroscopy 12:96
  uses 12:84
Butadiene monoxide. (See also *Epoxy compounds.*)
  hydrochlorination methods 12:208
  near-infrared spectroscopy 12:217, 218
Butadiene polymers. (See also Polybutadienes.)
  gas chromatography 4:286
Butadiene–styrene–vinyl-2-pyridine terpolymer.
  (See also *Elastomers, synthetic.*)
  infrared spectroscopy 12:97
  uses 12:83
Butalbital
  gas chromatography 6:483
Butallylonal 6:468, 487
  derivatives for identification 6:477
  gas chromatography 6:484
  infrared spectrum 6:487
  properties 6:470
  solubility 6:476
  thin-layer chromatography 6:481
  ultraviolet spectroscopy 6:501
Butanal
  in black tea 18:510
  gas chromatography 18:510
Butane. (See also *Liquefied petroleum gas.*)
  commercial grades and specifications 15:235
  gas chromatography 4:212; 15:240
  in propane 15:241
  propane content 15:242
  specific gravity 15:245
  vapor pressure 15:249

n-Butane. (See also Butanes.)
  in aerosols 11:95
  calorific value 13:331
  commercial specifications 14:301
  gas chromatography 11:98; 13:317, 318, 350;
    14:329; 18:52
  in gasoline 13:350, 353
  mass spectrum 16:418
  in methyltrichlorosilane 18:52
  in natural gas 13:320
  properties 11:95; 13:350; 14:301
  specific gravity 13:329
1,4-Butanediamine. See Tetramethylenediamine.
1,2-Butanediol. (See also Glycols.)
  azeotropes 13:541
  gas chromatography 13:504
  liquid chromatography 13:580
1,3-Butanediol. (See also Glycols.)
  azeotropes 13:541
  commercial grades and specifications 13:546
  derivatives for identification 13:568
  gas chromatography 13:504, 576; 19:298, 300
  manufacture 13:543
  paper chromatography 13:581
  in polyurethanes 19:297
  properties 13:536
  structure 13:534
  thin-layer chromatography 13:582
  uses 13:544
1,4-Butanediol. (See also Glycols.) 5:31
  commercial grades and specifications 13:546
  derivatives for identification 13:568
  gas chromatography 13:504, 507, 576, 577; 19:
    298, 300, 301
  manufacture 13:544
  paper chromatography 13:581
  in polyurethanes, 19:297
  properties 13:536
  structure 13:534
  thin-layer chromatography 13:582
  uses 13:544
2,3-Butanediol. (See also Glycols.)
  colorimetric determination 19:447
  commercial grades and specifications 13:546
  derivatives for identification 13:568
  gas chromatography 13:504, 576, 577; 19:300,
    446, 447
  liquid chromatography 13:580
  manufacture 13:544; 19:446
  oxidation methods 19:447
  paper chromatography 13:581
  piperidine colorimetric method 19:447
  in polyurethanes 19:297
  properties 13:536
  separation 19:444
  structure 13:534; 19:446
  thin-layer chromatography 13:502, 582
  uses 13:544

in wines 19:425, 444, 446
1,4-Butanediol dimethacrylate
nuclear magnetic resonance spectroscopy 16:86
1,4-Butanediol dipropylene glycol
gas chromatography 13:579
2,3-Butanedione. See Diacetyl.
Butanes. (See also Alkanes; and individual compounds.) 14:301
occurrence 14:301
uses 14:301
1-Butanethiol. See *n*-Butyl mercaptan.
2-Butanethiol. See *sec*-Butyl mercaptan.
1,2,4-Butanetriol. (See also Polyhydric alcohols.)
commercial grades and specifications 13:566
manufacture 13:563
paper chromatography 13:581
properties 13:562
solubility 13:560
structure 13:561
uses 13:564
1,2,4-Butanetriol trinitrate. See BTTN.
Butanol
in brake fluids 7:423, 424
in enamels 12:175, 189
gas chromatography 9:475; 11:73; 16:98
in nail enamels 11:73
properties 12:175
1-Butanol. See *n*-Butyl alcohol.
2-Butanol. See *sec*-Butyl alcohol.
Butanol insolubles
in acrylamide 4:165
Butanols. See Butyl alcohols.
Butanone. See Methyl ethyl ketone.
2-Butanone
absorption spectroscopy 9:564
in black tea 18:510
gas chromatography 18:510
2-hydroxy-1-naphthaldehyde colorimetric method 11:204
reaction rate analysis 8:364
2-Butenal. See *Crotonaldehyde*.
1-Butene. (See also *Hydrocarbons*.)
in 1,3-butadiene 11:504, 506
in butane 15:240
commercial grades and specifications 14:312, 313
gas chromatography 11:506; 14:329; 15:240
in propane 15:241
properties 14:312
2-Butene. (See also *Hydrocarbons*.)
in 1,3-butadiene 11:506
in butane 15:240
commercial grades and specifications 14:312, 313
gas chromatography 11:506; 14:329; 15:240
properties 14:312
1,3-Butenediol

gas chromatography 13:576
2-Butene-1,4-diol
gas chromatography 13:576
3-Butenenitrile. (See also Allyl cyanide; *Nitriles*.)
morpholine volumetric method 16:410
Butenes. (See also Butylenes; *Hydrocarbons*.)
gas chromatography 13:350
in gasoline 13:350, 353
properties 13:350
Butethal 6:468, 487
derivatives for identification 6:477
gas chromatography 6:483, 484
infrared spectrum 6:487
properties 6:470
solubility 6:476
thin-layer chromatography 6:481
Butethamine hydrochloride 5:374
analysis 5:382
properties 5:382
Butoxy triglycol
gas chromatography 7:427
Butter 16:136
analysis 16:137
color content 10:545
commercial specifications 16:136
composition, determination of 16:137
grading 16:137
impurities, determination of 16:138
Kohman procedure 16:137
liquid chromatography 10:545; 16:138
manufacture 16:136
milkfat content 16:138
moisture content 16:138
potassium content 17:363
salt content 16:138
sampling 16:137
specification tests 16:138
thin-layer chromatography 10:545
Buttermilk. (See also Fermented milk products.)
manufacture 16:134
N-Butyl acetanilide
ultraviolet spectroscopy 7:85
Butyl acetate
in enamels 12:175, 189
extraction of iron 9:645
gas chromatography 11:73
in nail enamels 11:73
properties 12:175
n-Butyl acetate
commercial specifications 8:578
gas chromatography 16:108
in methanol 16:108
properties 8:577
saponification method 8:586
sec-Butyl acetate
commercial specifications 8:578
gas chromatography 8:351; 16:108

in methanol 16:108
properties 8:577
saponification method 8:586
Butyl acrylate. (See also *Acrylic and methacrylic acids and esters.*)
in air 4:216
azeotropes 4:184
commercial grades and specifications 4:187
gas chromatography 4:212
mercuric acetate method 4:193
polymer in 4:200
properties 4:182
saponification method 4:196
sodium sulfite addition method 4:341
solubility 4:184
n-Butyl alcohol. (See also Alcohols, lower;
Butyl alcohols.) 5:5
in brandies 19:440
in *n*-butyl mercaptan 15:565
3,5-dinitrobenzoyl chloride method 4:547
gas chromatography 7:96; 12:204; 15:565; 16:
107; 19:442
in methanol 16:107, 114
phthalic anhydride method 4:535
pyromellitic dianhydride method 4:537
in wines 19:425, 440
sec-Butyl alcohol. (See also Alcohols, lower;
Butyl alcohols.)
in black tea 18:510
3,5-dinitrobenzoyl chloride method 4:547
2,4-dinitrophenylhydrazone colorimetric method
4:549
gas chromatography 7:96; 9:475; 12:204; 16:
107; 18:510
in methanol 16:107
tert-Butyl alcohol. (See also Alcohols, lower;
Butyl alcohols.)
in distilled alcoholic beverages and intermediates
4:489
gas chromatography 12:204
in methanol 16:114
Butyl alcohols. (See also Alcohols, lower; and
individual compounds.) 4:496
azeotropes 4:500, 504
commercial grades and specifications 4:514
derivatives for identification 4:518
manufacture 4:506
properties 4:497
uses 4:512
n-Butylamine
azeotropes 5:261
manufacture 5:262
in methanol 16:114
properties 5:260
sec-Butylamine
in methanol 16:114
properties 5:260
uses 5:263

tert-Butylamine
manufacture 5:262
properties 5:260
uses 5:263
n-Butylamine method
for hydroquinone 4:204
Butylamines (See also individual compounds.) 5:263
in brake fluids 7:427
manufacture 5:262
Butyl aminobenzoate 5:374
acid–base titration 5:382
analysis 5:382
anesthetic 5:382
properties 5:382
tert-Butylaminoethyl methacrylate. (See also
Acrylic and methacrylic acids and esters.)
commercial grades and specifications 4:187
polymer in 4:200
properties 4:182
saponification method 4:196
Butylated hydroxytoluene (BHT) 6:73, 115
analysis 6:116
column chromatography 6:87
extraction 6:116
gas chromatography 6:116
identification 6:116
manufacture 6:116
properties 6:115
separation 6:116
thin-layer chromatography 6:89
ultraviolet spectroscopy 6:116
ultraviolet spectrum 6:87
Butylated urea 5:276
Butylated urea–formaldehyde resin
infrared spectrum 5:283
N-n-Butylaziridine. (See also *Imines.*)
properties 14:468
structure 14:468
N-t-Butylaziridine. (See also *Imines.*)
properties 14:468
structure 14:468
Butylbenzene
gas chromatography 7:7, 8
liquid chromatography 7:13, 14, 15, 16
Butylbenzenes. (See also *Hydrocarbons*; and
individual compounds.) 14:323
commercial grades and specifications 14:322
gas chromatography 13:352
in gasoline 13:352, 354
properties 13:352; 14:323
solubility 14:323
uses 14:323
Butyl benzoate. (See also Benzoic acid esters.)
7:78
3,5-dinitrobenzoate derivative 7:83
properties 7:77
saponification method 7:84

p-tert-Butylbenzoic acid
  content of alkyd resins 5:44
  gas chromatography 5:27
  infrared spectroscopy 5:44
  modifier for alkyd resins 5:24
  ultraviolet spectroscopy 5:50
Butyl benzyl phthalate
  gas chromatography 16:97
n-Butyl α-butoxypropionate. (See also Lactic
    acid esters.)
  properties 15:131
Butyl carbitol
  sulfur content 18:377
p-t-Butylcatechol
  sodium hydroxide colorimetric method 18:
    304
  solvent extraction 18:304
  in styrene 18:290, 303
Butyl Cellosolve
  in carburetor cleaners 6:356
  gas chromatography 11:73
  in nail enamels 11:73
n-Butyl chloride. See 1-Chlorobutane.
sec-Butyl chloride. See 2-Chlorobutane.
t-Butyl chloride. See 3-Chloro-2-methylpro-
    pane.
t-Butylchlorobenzene
  mass spectroscopy 9:405
t-Butyl cinnamate. (See also *Cinnamic acid and
    its derivatives.*)
  ultraviolet spectroscopy 10:69
Butyl 3,5-dinitrobenzoate
  thin-layer chromatography 16:94
4,4′-Butylenebis(6-*tert*-butyl-3-methylphenol)
  paper chromatography 6:85
1,3-Butylene dimethacrylate. (See also *Acrylic and
    methacrylic acids and esters.*)
  polymer in 4:200
  saponification method 4:196
β-Butylene glycol. See 1,3-Butanediol.
1,3-Butylene glycol. See 1,3-Butanediol.
1,4-Butylene glycol. See 1,4-Butanediol.
2,3-Butylene glycol. See 2,3-Butanediol.
Butylenes. (See also *Hydrocarbons*; and individual
    compounds.) 14:312
  uses 14:312
N-Butylethanolamine 5:2
n-Butyl ether. (See also *Ethers.*)
  properties 12:297
sec-Butyl ether. (See also *Ethers.*)
  properties 12:297
t-Butyl ether. (See also *Ethers.*)
  properties 12:297
n-Butyl formate
  gas chromatography 16:108
  in methanol 16:108
  uses 13:120

n-Butyl glycidyl ether. (See also *Epoxy com-
    pounds.*)
  infrared spectroscopy 12:200
2-and 3-*tert*-Butyl-4-hydroxyanisol. See BHA.
4,4′-Butylidenebis-6-*tert*-butyl-*m*-cresol 6:76
  colorimetric method using diazotized *p*-nitro-
    aniline 6:106
  column chromatography 6:87
  ultraviolet spectrum 6:87
p,p′-sec-Butylidenediphenol. See Bisphenol B.
n-Butyl iodide. (See also Iodine compounds,
    iodine.)
  gas chromatography 16:92
  properties 14:612
Butyl isocyanate. (See also *Isocyanic acid esters.*)
  commercial sources 15:102
  infrared spectroscopy 15:106
  near-infrared spectroscopy 15:105
  properties 15:95
Butyl ketoxime
  in enamels 12:177
n-Butyl lactate acetate. (See also Lactic acid esters.)
  properties 15:131
Butyl lactates. (See also Lactic acid esters.)
  properties 15:131
  solubility 15:130
n-Butyllithium. (See also Organolithium com-
    pounds.)
  assay 15:288
n-Butyl mercaptan. (See also *Mercaptans.*)
  assay 15:565
  derivatives for identification 15:559
  gas chromatography 15:558, 565
  impurities, determination of 15:565
  occurrence 15:551
  properties 15:552
sec-Butyl mercaptan. (See also *Mercaptans.*)
  in *n*-butyl mercaptan 15:565
  gas chromatography 15:558, 565
  properties 15:552
tert-Butyl mercaptan. (See also *Mercaptans.*)
  in *n*-butyl mercaptan 15:565
  gas chromatography 15:557, 565
  manufacture 15:554
  properties 15:552
  uses 15:555
Butyl methacrylate. (See also *Acrylic and meth-
    acrylic acids and esters.*)
  commercial grades and specifications 4:187
  gas chromatography 4:212
  polymer in 4:200
  properties 4:182
  saponification method 4:196
n-Butyl methacrylate
  nuclear magnetic resonance spectroscopy 16:86
  thin-layer chromatography 16:90
sec-Butyl methacrylate

nuclear magnetic resonance spectroscopy
16:86
tert-Butyl methacrylate
nuclear magnetic resonance spectroscopy
16:86
2-*n*-Butylphenol. (See also *Phenols.*)
derivatives for identification 17:21
properties 17:4
3-*n*-Butylphenol. (See also *Phenols.*)
properties 17:4
4-*n*-Butylphenol. (See also *Phenols.*)
derivatives for identification 17:21
properties 17:4
2-*sec*-Butylphenol. (See also *Phenols.*)
commercial grades and specifications 17:15
4-*sec*-Butylphenol. (See also *Phenols.*)
properties 17:4
2-*tert*-Butylphenol. (See also *Phenols.*)
commercial grades and specifications 17:15
4-*tert*-Butylphenol. (See also *Phenols.*)
4-aminoantipyrine colorimetric method 17:
18
commercial grades and specifications 17:15
derivatives for identification 17:21
Millon's reagent color test 17:18
Butylphenols. (See also individual compounds.)
uses 17:2
4-*tert*-Butylphenyl groups
in polycarbonates 17:339
ultraviolet spectroscopy 17:339
4-*tert*-Butylphenyl salicylate
thin-layer chromatography 16:96
ultraviolet spectroscopy 16:95
tert-Butylpyridines. (See also *Pyridine and pyridine derivatives.*)
ionization constants 17:416
Butylrhodamine B
colorimetric method for tantalum 16:369
Butyl rubber. (See also *Elastomers, synthetic; Polyisobutenes.*)
colorimetric determination 12:91
gravimetric determination 12:92
infrared spectroscopy 12:93, 97
pyrolysis–gas chromatography 12:102, 105
uses 12:84
n-Butyl thiocyanate. (See also *Isocyanic acid esters.*)
gas chromatography 15:110
n-Butyltrichlorosilane. (See also Organo-chlorosilanes.)
manufacture 18:55
properties 18:54
1-Butyne
gas chromatography 4:144; 11:510
in isoprene 11:510
properties 4:137
2-Butyne

gas chromatography 4:144; 11:510
in isoprene 11:510
properties 4:137
Butynediol
acetylation method 4:141
bromination method 4:145
commercial grades and specifications 4:139
manufacture 4:139
properties 4:137
Butyraldehyde
absorption spectroscopy 9:564
in acetylenic chemicals 4:145
commercial specifications 8:340
derivatives for identification 8:343
2,4-dinitrophenylhydrazine colorimetric
method 8:363
gas chromatography 16:108, 109
hydroxylamine hydrochloride method 8:357
infrared spectroscopy 6:616
in methanol 16:108, 109, 114
properties 8:334
ultraviolet spectroscopy 6:632
n-Butyraldehyde
in cocoa bean vapor 9:570
gas chromatography 12:204
mass spectroscopy 9:570
Butyrals. See *Acetals and ketals.*
n-Butyramide
derivatives for identification 8:601
properties 8:558, 597
Butyric acid
in black tea 18:511
derivatives for identification 8:376
gas chromatography 8:383; 18:511
in methanol 16:114
paper chromatography 15:133
properties 8:396, 398, 558
thin-layer chromatography 8:381
Butyric anhydride
butyric acid in 4:104
morpholine-reaction method 4:103
properties 8:558, 559
Butyric chloride
properties 8:558, 570
Butyrobetaine
ion exchange chromatography 9:618
γ-Butyrolactone
in black tea 18:511
gas chromatography 18:511
Butyryl content
of cellulose acetate butyrate 9:74
Butyryl halides
derivatives for identification 8:572
n-Butyronitrile. (See also *Nitriles.*)
azeotropes 16:403
derivatives for identification 16:401, 402
gas chromatography 16:404

C

Cacao. (See also Cocoa.) 9:534
  agronomy 9:535
  anthrocyanidin content 9:536
  anthrocyanin content 9:536
  catechin content 9:536
  in chocolate 9:540
  color 9:536
  cyanin content 9:536
  fermentation 9:535
  phobophene content 9:536
  tannin content 9:536
Cacotheline
  colorimetric method for tin 11:141, 144
  colorimetric method for vanadium 19:331, 332
Cadmium. (See also Cadmium alloys; Cadmium
    metal.) 8:30–55
  in aluminum alloys 5:123
  atomic absorption spectroscopy 12:571; 15:186,
    196
  in beer 7:651
  borax bead test 6:584
  in cadmium carbonate 8:39
  in cadmium chloride 8:40
  in cadmium metal 8:34
  in cadmium yellow 17:200
  in cements 9:131
  colorimetric method 8:53
  complexes with ligands 8:35
  in cotton 11:152, 154, 156
  in copper alloys 10:635
  dithizone colorimetric method 8:53
  EDTA titration 8:50; 10:635; 13:445
  in elastomers, synthetic 12:151
  electrolytic deposition method 8:39, 40, 50, 51
  emission activation analysis 8:54
  emission spectroscopy 7:651; 11:152; 13:277;
    14:48, 122, 538; 15:187, 197; 16:387; 19:570
  in ferroniobium 16:387
  flame tests 6:589
  in gallium arsenide 13:276, 279
  in gallium metal 13:266
  in gallium phosphide 13:282
  in glass and glass-ceramics 13:445
  in gold 14:45
  gravimetric determination 8:34, 40, 50, 51; 16:
    20
  in hafnium 14:122
  in hydrochloric acid 14:374
  hydrogen sulfide gravimetric method 12:151
  2-(o-hydroxyphenyl)benzoxazole precipitation
    method 8:51
  identification 8:49
  in indium 14:538
  in indium alloys 14:551
  ion exchange chromatography 13:255; 17:364
  in iron 15:59
  in lead 15:196
  in lead alloys 15:195
  in mercury 16:20
  microcosmic salt bead test 6:587
  neutron activation analysis 8:54; 13:265
  in niobium 16:387
  occurrence 8:30
  in photographic films 12:571
  polarography 5:123; 8:52; 12:571; 13:279
  in rocks 15:187
  separation 8:34, 48, 50
  in soil 15:186
  solvent extraction 12:571
  sulfide precipitation 8:34, 48
  in tin metal 19:67
  x-ray fluorescence spectroscopy 11:156
  in zinc metal 19:570
  in zinc or zinc alloys 8:52
  in zinc powder 17:204
  in zirconium 14:122
Cadmium acetate 8:36
  analysis 8:36
  commercial grades and specifications 8:36
  impurities, determination of 8:36
  properties 8:36
Cadmium acetate–α-naphthylamine
  colorimetirc method for nitrate 19:476
Cadmium alloys 8:31, 32, 33
  analysis 8:34
  composition 8:32, 33
  properties 8:32, 33
  sample preparation 8:34
Cadmium antimonide semiconductors. (See also
    Ceramics.)
  properties 9:213
  uses 9:213
Cadmium bromide 8:37
  analysis 8:37
  commercial specifications 8:37
  impurities, determination of 8:38
  properties 8:37
Cadmium carbonate 8:38
  analysis 8:39
  cadmium content of 8:39
  commercial grades and specifications 8:39
  impurities, determination of 8:39

properties 8:38
Cadmium chloride 8:40
  analysis 8:40
  cadmium content of 8:40
  commercial specifications 8:40
  gravimetric method for choline chloride 9:623
  gravimetric method for dicholine citrate 9:623
  impurities, determination of 8:41
  properties 8:40; 9:334
  solubility 9:336
Cadmium citrate complex
  stability 10:89
Cadmium complexes 8:35
Cadmium compounds. (See also individual compounds.) 8:35-48
  solubility 8:35
  toxicity 8:33, 36
Cadmium fluoride 8:42
  properties 8:42; 13:11
Cadmium hydroxide 8:43
  commercial specifications 8:43
  properties 8:43
Cadmium iodide 8:43
  analysis 8:43
  commercial specifications 8:43
  impurities, determination of 8:44
  iodide content of 8:43
  properties 8:43
Cadmium lithopones 8:31, 44
  commercial grades and specifications 17:195
Cadmium manganese ferrite
  in ceramics 9:224
  microscopy 9:224
Cadmium-mercury lithopones
  commercial grades and specifications 17:195
Cadmium-mercury reds 17:194
  analysis 17:195
  cadmium sulfate content 17:195
  commercial grades and specifications 17:194
Cadmium metal
  analysis 8:34
  carrier gas distillation fusion methods 8:636
  carrier gas fusion methods 8:682
  commercial grades and specifications 8:34
  metallurgy 8:33
  oxygen in 8:636, 682
  properties 8:31
  sample preparation 8:34
  silver content 18:172
  solubility in gold 14:9
  thallium content 18:524
  tin content 19:64
  toxicity 8:33
Cadmium niobate
  in ceramics 9:224
  microscopy 9:224
Cadmium nitrate 8:44

analysis 8:45
commercial grades and specifications 8:45
impurities, determination of 8:45
properties 8:44
volumetric method for EGTA 12:47
Cadmium ores
  sample decomposition 8:34
Cadmium organometallic compounds 8:45
Cadmium oxide 8:46
  commercial specifications 8:46
  in glass and glass-ceramics 13:413, 415
  properties 8:46
Cadmium red
  commercial specifications 12:173
  composition 12:173
  in enamels 12:173
Cadmium selenide
  in glass and glass-ceramics 13:461
Cadmium soap. (See also Soaps.)
  uses 18:180
Cadmium sulfate 8:46
  analysis 8:47
  in cadmium-mercury reds 17:195
  commercial grades and specifications 8:47
  gravimetric method for hydrogen sulfide 15:568
  impurities, determination of 8:47
  properties 8:46
Cadmium sulfide 8:47
  adsorption method for mercury 16:16
  commercial specifications 8:48
  in glass 9:200
  gravimetric method for iron 15:48
  properties 8:47
Cadmium sulfide semiconductors. (See also
    Ceramics.)
  properties 9:213
  uses 9:213
Cadmium sulfoselenides. See Cadmium-mercury
    reds.
Cadmium telluride semiconductors. (See also
    Ceramics.)
  properties 9:213
  uses 9:213
Cadmium yellow
  analysis 17:200
  barium sulfate content 17:200
  cadmium content 17:200
  commercial grades and specifications 17:200
Caffeetannic acid. See Chlorogenic acid.
Caffeic acid. (See also Cinnamic acid and its
    derivatives.)
  absorption spectroscopy 10:64
  gas chromatography 19:146
  occurrence 10:63
  paper chromatography 8:379
  in roasted coffee 10:375, 396
  solvent extraction 19:146

in tobacco 19:146
Caffeine 8:55-71
  acid–base titration, nonaqueous 6:170; 8:63, 65
  amperometric titration 8:68
  analysis 8:57
  Bailey and Andrew method 10:431; 18:489
  in beverages, nonalcoholic 8:66
  bioassay 7:181
  in black tea 18:488
  Borkar and Sloman method 10:407; 18:490
  in carbonated beverages 8:167
  in cocoa beans 9:548
  in coffee 8:59, 62, 67, 68
  column chromatography 6:36, 128, 134, 150,
    152, 154
  colorimetric methods 8:66
  commercial grades and specifications 8:56
  in decaffeinated coffee 10:409
  extraction 6:128, 133
  gas chromatography 6:135, 156, 210, 211; 8:59,
    603; 18:491
  gravimetric determination 8:63, 68, 167
  in green coffee 10:374, 404, 433
  identification 8:58
  impurities, determination of 8:64
  infrared spectroscopy 8:68
  infrared spectrum 6:129
  in instant coffee 10:409
  iodine titration 6:166; 8:65
  ion exchange chromatography 6:182; 8:62; 18:
    489, 490
  Kjeldahl method 18:490
  Lee spectroscopic method 10:433
  liquid chromatography 8:59; 10:407
  nephelometry 8:68
  nuclear magnetic resonance spectroscopy 6:172,
    174; 8:68
  paper chromatography 6:183; 8:59
  in pharmaceuticals 8:63
  phosphomolybdic acid colorimetric method 8:
    66, 167
  Power and Chestnut method 10:431
  properties 8:55; 10:404
  in roasted coffee 10:375, 409, 417, 431
  separation 6:128, 133; 8:59
  Smith and Rees spectroscopic method 10:407
  solubility 8:56; 10:405; 18:489
  solvent extraction 8:61, 63; 18:489
  structure 18:488
  sublimation 8:62
  thin-layer chromatography 8:59; 18:492
  toxicity 8:55
  ultraviolet spectroscopy 6:133, 147, 149, 150,
    154, 156, 159, 162; 8:67; 10:407, 408, 412,
    433; 18:491
  uses 8:55
Caffeine tablets

codeine phosphate content 16:269
3-Caffeoylquinic acid. See Chlorogenic acid.
Cahn electrobalance 7:599
Cahnite 7:315
Cake batter. (See also *Eggs and egg products.*)
  specific gravity 12:74
Cakes. (See also *Eggs and egg products.*) 6:444
  analysis 6:448
  compressibility 12:74
  manufacture 6:446
  quality control 6:448
  rape seed displacement 12:71
  tenderness 12:73
  tensile strength 12:74
  volume 12:71
Calamine
  composition 19:545
Calamite 14:519
Calcein
  fluorometric method for calcium 15:272
Calcined clay 4:33
Calcined magnesia
  composition 9:203
Calcined magnesium oxide–sodium carbonate
    fusion
  of sulfur, inorganic 18:380
Calcined soda. See Sodium carbonate.
Calcines
  sulfur content 18:383
Calcite
  composition 10:104
Calcium. (See also Calcium alloys; Calcium metal.)
    8:72-114
  in acid salts 8:594, 595
  in aluminates 5:160
  in aluminum alloys 5:116, 124
  in aluminum metal 5:124
  in aluminum oxide 5:186
  in aluminum oxide abrasives 4:40, 51
  in ammonium chloride 5:337
  ammonium oxalate gravimetric method 10:296;
    12:186; 14:606
  ammonium oxalate–potassium permanganate
    volumetric method 14:68; 17:194
  in amosite 6:267
  in animal feeds 8:107
  in asbestos 6:263, 267
  in asbestos–cement products 6:270, 271
  atomic absorption spectroscopy 8:109; 9:346;
    10:37, 133; 11:157; 13:136; 15:157, 380;
    19:493
  in bakery chemicals 6:432
  in barite 6:557, 561
  in barium acetate 6:536
  in barium carbonate 6:541, 544
  in barium chloride 6:546
  in barium hydroxide 6:550, 551

in barium nitrate 6:552
in barium peroxide 6:554
in bauxite 5:115
in beer 7:652, 656
bis(2-hydroxyanil)glyoxal colorimetric method
    8:98, 108
in blood serum 8:110; 18:222
in bone 8:110
in borax 7:382
borax bead test 6:584
in boric acid 7:381
calcein fluorometric method 15:272
in calcium acetate 8:77
in calcium alloys 8:76
in calcium boride 7:361
in calcium carbide 8:80, 135
in calcium carbonate 8:82
in calcium chloride 8:87
in calcium cyclamate 11:225
in calcium fluoride 8:90
in calcium hydride 14:251
in calcium metal 8:76
in calcium oxide 8:93
in calcium phosphate 8:95
in carbon 14:68, 74
in cellulose 9:50
in cement 8:110
in cereals 9:233, 241
in cerous chloride 9:260
chloranilic acid method 8:108
in chocolate and cocoa products 9:595
in chromium alloys 9:666
in chromium compounds 9:699
in chrysotile 6:263, 267
in cigarette paper 10:36
in citric acid 10:96
in clays and clay minerals 10:133, 136
in cobalt compounds 10:327
in cobalt metal 10:293, 296
colorimetric methods 8:108; 18:270
in copper metal 10:620
in cotton 11:140, 154, 157
in diatomaceous earth 11:490, 494, 495, 498
dihydroxytartaric acid osazone identification
    test 8:99
in distilled alcoholic beverages and intermediates
    4:481, 489
EDTA titration 6:264; 7:656; 8:82, 87, 90, 93,
    95, 104; 9:343; 10:96; 11:44, 225, 490, 495;
    12:153, 154; 19:363, 492
EGTA titration 8:105
in elastomers, synthetic 12:151
emission spectroscopy 8:112; 9:346; 11:493;
    14:48, 74, 122, 542; 15:144, 155, 375; 16:
    160, 387
in enamel 12:186
eriochrome Black T colorimetric method 15:369

in europium oxide 12:403
in ferroniobium 16:387
flame photometry 7:381, 382, 652; 8:111, 595;
    9:666; 11:494, 498; 15:280, 399
flame tests 6:583, 589, 592; 8:98; 18:270
glyoxal–bis(2-hydroxyanil) colorimetric method
    11:140, 145, 491
in gold 14:45
in graphite 14:68, 74
gravimetric determination 8:102
in hafnium 14:122
identification 8:98
in indium 14:542
ion exchange chromatography 8:101; 13:255;
    18:268
ion exchange method 19:492
ion selective electrodes 8:106
in iron oxide pigments 17:194
in lactic acid and its derivatives 15:144
in lanthanum oxide 15:155
liquid chromatography 10:37
in lithium 15:272
in lithium carbonate 15:280
in lithium chloride 9:345
in lithium hydride 14:251
in lithium hydroxide 15:286
in magnesium 15:375, 380
in magnesium acetate 15:399
in magnesium alloys 8:111; 15:375, 380
in magnesium carbonate 15:402
in magnesium chloride 15:405
in magnesium nitrate 15:408
in magnesium oxide 15:411
in magnesium sulfate 15:413
in manganese 15:479
in manganese oxide pigments 17:194
microcosmic salt bead test 6:587
in molybdenum 16:161
in molybdenum trioxide 16:161
in niobium 16:387
oxalate gravimetric method 6:263; 7:361; 8:82,
    102; 9:50; 12:151
permanganate titration 8:106
in plants 8:110; 18:221
in potassium chlorate 9:359, 361
in potassium chloride 9:339
in potassium iodide 14:606
potassium permanganate titration 9:241; 10:296
in rare earth chlorides 17:506
in rare earth fluorides 17:508
separation 8:99; 18:267
in silica ablators 4:3
in silicon carbide abrasives 4:56, 59
in sodium chloride 9:337, 338, 343, 346, 349
in sodium dichromate 9:684
in sodium formate 13:136
in sodium iodide 14:606

sodium rhodizonate colorimetric method 11:140

in soil 8:110; 18:220

in starch hydrolysates 11:44

in strontium 18:267, 273

in strontium chloride 18:283

in strontium nitrate 18:283

sulfuric acid gravimetric method 15:402

in terephthalic acid 17:163

in tissue 8:110

in tobacco 19:152

in urine 8:110; 18:222

volumetric methods 8:104

in water 18:221; 19:362

in wines 19:491, 492

x-ray fluorescence spectroscopy 10:136; 11: 154, 157

in zinc powder 17:204

in zirconium 14:122

Calcium acetate 8:77

analysis 8:77

chloride in 8:593

commercial specifications 8:77, 591

properties 8:77, 591

specification tests 8:77

turbidimetric method for oxalic acid 10:543

uses 8:77

Calcium adipate 4:428

Calcium alloys. (See also Calcium metal.) 8:72

analysis 8:76

calcium content of 8:76

uses 8:72

Calcium aluminate

light microscopy 9:103

in portland cement 9:103

Calcium aluminate cement. (See also Refractories.)

composition 9:203

in hydraulic cements 9:107

Calcium aluminum oxide

in hydraulic cements 9:100

Calcium aluminum silicate content

of hydraulic cements 9:107

Calcium amphomycin

infrared spectrum 5:554

Calcium arsenate. (See also *Pesticides and insecticides.*) 6:229

arsenic content 19:595

commercial grades and specifications 19:593

composition 19:580, 586

manufacture 19:595

solubility 19:587

toxicity 19:592

Calcium arsenite 6:229

Calcium benzoate

properties 8:591

Calcium benzylpenicillin. (See also Benzylpenicillin.)

structure 16:567

Calcium borates 7:313

Calcium boride. (See also Metal borides.) 7:358

calcium content of 7:361

composition 7:322

manufacture 7:358

properties 7:359, 360

uses 7:323

Calcium bromide 8:79

commercial specifications 8:10

properties 8:9

Calcium carbide. (See also Carbides.)

analysis 8:80

calcium content 8:80, 135

carbon content 8:134

formation 8:130

generation of acetylene 8:79

properties 8:79, 131, 132

use in preparation of acetylene 4:126

uses 8:80, 132

Calcium carbide reactor

for production of acetylene from water 7:52

Calcium carbonate 8:80

acid–base titration 10:38; 19:360

analysis 8:82; 15:530

in asbestos–cement products 6:272

analysis 17:191

assay 15:530

in baking powders 6:430

calcium content of 8:82

in cigarette paper 10:36, 38

in coated abrasives 10:281

commercial grades and specifications 15:514; 17:190

commercial specifications 8:81; 10:534

composition 10:534

in matches 15:514

particle size 15:530

properties 8:80

solubility 18:267

specification tests 8:82

titration of EDTA and related compounds 12:32

uses 8:81; 17:190

in Venetian red 17:194

in water 19:360

volumetric method for EDTA 12:32

Calcium carbonate equivalent

of lime and limestone 15:230

Calcium carbonate fusion

of silicon 18:12

Calcium chelates

DTPA content 12:50

EDTA content 12:50

ferric chloride titration 12:50

thorium nitrate titration 12:50

Calcium chlorate. (See also Chlorates.)

properties 9:358

solubility 9:359
Calcium chloride 6:1; 8:85
    analysis 8:87
    calcium content 8:87
    commercial grades and specifications 8:85, 86
    titration of EDTA and related compounds 12:31, 51
    gravimetric method for fluoride 13:42
    properties 8:85; 9:334
    solubility 9:336
    specification tests 8:87
    uses 8:85
    volumetric method for NTA 12:33
Calcium chloride–ferric chloride
    volumetric method for EDTA 12:51
Calcium chromate
    commercial specifications 9:696
    solubility 9:696; 18:267
Calcium citrate complex
    stability 10:89
Calcium citrate tetrahydrate. (See also Citric acid and salts.)
    commercial grades and specifications 10:91
    uses 10:86
Calcium complexes 8:98
Calcium compounds. (See also individual compounds.) 8:77–97
    solubility 8:97
Calcium cyanide. (See also Cyanides.) 13:176; 14:424
    commercial specifications 14:425
    impurities in 14:425
    manufacture 13:176
    properties 14:424
    solubility 14:425
    uses 13:176; 14:425
Calcium cyclamate. (See also Cyclamates.) 11:220
    calcium content 11:225
    commercial grades and specifications 11:223
    properties 11:221
    solubility 11:221
Calcium cyclobarbital. See Cyclobarbital calcium.
Calcium cyclohexylsulfamate. See Calcium cyclamate.
Calcium dichromate
    commercial specifications 9:697
    solubility 9:697
Calcium EDTA. See EDTA; and *EDTA and related compounds.*
Calcium ferrocyanide. (See also Ferrocyanides.)
    uses 15:85
Calcium fluoride 8:88
    analysis 8:90
    calcium content of 8:90
    carrier gas distillation fusion methods 8:636
    in ceramics 9:224

commercial grades and specifications 8:89
    in glass 9:200
    microscopy 9:224
    properties 8:88; 13:11
    specification tests 8:90
    uses 8:89
Calcium fluoride–sulfuric acid
    gravimetric method for silicate 18:16
Calcium hydride. (See also *Hydrides.*) 8:91
    analysis 14:247, 251
    assay 14:247, 251
    calcium content 14:251
    commercial grades and specifications 14:246
    composition, determination of 14:247, 251
    hydrogen content 14:247
    manufacture 14:245
    properties 14:243, 244
    uses 14:242
Calcium hydroxide 8:91
    commercial specifications 8:91
    volumetric method for glycerol dichlorohydrins 14:172
    volumetric method for glycerol monochlorohydrins 14:172
Calcium hydroxide–salicylate dental products 11:298
    analysis 11:298
    composition 11:298
    consistency 11:298
    crushing strength 11:298
    setting time 11:298
    uses 11:295
Calcium hypochlorite. (See also Hypochlorites.) 8:91; 9:352
    available chlorine 7:224, 225
    bleaching agent 7:222
    iodometric titration 7:225
    properties 9:352
    uses 9:352
    water in 7:230
Calcium hypophosphite. (See also Hypophosphites.)
    manufacture 17:112
Calcium iodide 8:92
    analysis 8:92
    iodide content of 8:92
    properties 8:92
Calcium lactate. (See also Lactic acid salts.)
    assay 15:134
    in baking powders 6:430
    commercial grades and specifications 15:130
    EDTA titration 15:134
    impurities, determination of 15:144
    manufacture 15:128, 129
    properties 8:591
    solubility 15:129
    uses 15:130

Calcium–manganese–silicon alloy. (See also
    Silicon alloys.) 5:61
  composition 5:64; 18:2
Calcium metal. (See also Calcium alloys.) 8:72
  analysis 8:76
  calcium content of 8:76
  commercial grades and specifications 8:76
  emission spectroscopy 8:76
  hydrogen in 8:617, 681
  impurities, determination of 8:76
  manufacture 8:74
  occurrence 8:72, 74
  oxygen in 8:681
  properties 8:73
  uses 8:73
  vacuum fusion methods 8:617, 681
Calcium naphthanate
  in enamels 12:176
Calcium nitrate
  volumetric method for glycine 12:37
  volumetric method for IDA 12:37
  volumetric method for NTA 12:37
Calcium orthophosphate 17:120
  commercial specifications 17:128
  manufacture 17:121, 125
Calcium oxalate
  in beer 7:616
  solubility 18:267
  turbidimetric method for oxalate 10:93
  volumetric method for calcium sulfate 17:187
Calcium oxide 8:92
  in alkali barium silicate glass 9:218
  ammonium oxalate gravimetric method 9:116;
    13:421; 15:215
  ammonium oxalate–potassium permanganate
    volumetric method 17:201
  analysis 8:93
  calcium content of 8:93
  in cements 9:115, 128
  in ceramics 9:194, 224
  in chrome greens 17:201
  in clays and clay minerals 10:119
  in cobalt(II) orthophosphate 10:343
  in cobalt oxides 10:342
  in cobalt(II) sulfate 10:344
  commercial specifications 8:93
  EDTA titration 9:128; 10:119, 121; 13:445,
    447
  EGTA titration 15:216
  emission spectroscopy 9:218
  in enamels 9:195
  in glass and glass-ceramics 13:413, 415, 418,
    445, 447
  in glazes 9:196
  in gypsum and gypsum products 14:95
  light microscopy 9:103
  in lime and limestone 15:215

  in hydrated lime 9:191
  in hydraulic cements 9:96, 107
  in limestone 9:191
  in magnesium carbonate 15:402
  microscopy 9:224
  oxalic acid–ammonium hydroxide gravimetric
    method 14:95
  oxalic acid gravimetric method 10:119
  in portland cements 9:103
  potassium permanganate titration 10:121; 15:
    216
  properties 9:204
  in quicklime 9:191
  in refractories 9:206
  in silica ablators 4:3
  in silicate glasses 9:198
  specification test 8:93
  uses 8:93
Calcium pectate
  solvent extraction 19:143
  in tobacco 19:143
Calcium penicillin G. See Calcium benzylpenicillin.
Calcium perchlorate. (See also Perchlorates.)
  properties 9:363
  solubility 9:364
Calcium permanganate. (See also Manganese com-
    pounds.)
  uses 15:489
Calcium phosphate
  baking acid 6:426
  in baking powders 6:429, 430
  commercial specifications 6:428
  in glass 9:200
  uses 17:141
Calcium phosphates 8:94
  analysis 8:95
  calcium content of 8:95
  commercial grades and specifications 8:95
  phosphate content of 8:96
  specification tests 8:96
Calcium phosphite. (See also Phosphites.)
  manufacture 17:112
Calcium salt of 2-(2-hydroxy-1-naphthylazo)-1-
    naphthalenesulfonic acid. See D & C Red No.
    11.
Calcium salt of 3-hydroxy-4-phenylazo-2-naphthoic
    acid. See D & C Red No. 31.
Calcium salt of 4-(1-sulfo-2-naphthylazo)-3-hydroxy-
    2-naphthoic acid. See D & C Red No. 34.
Calcium salt of 4-(o-sulfo-p-tolylazo)-3-hydroxy-
    2-naphthoic acid. See D & C Red No. 7.
Calcium silicate. (See also Silicates.) 8:97
  in enamels 12:174
  uses 18:44
Calcium silicate cement. (See also Refractories.)
  microscopy 9:227
Calcium silicide

in explosive priming mixtures 12:418, 421, 424
hydrochloric acid gravimetric method 12:421, 424
solubility 12:421
x-ray diffraction 12:418
Calcium–silicon alloy. (See also Silicon alloys.)
composition 18:2
Calcium sodium fluoride
in ceramics 9:224
microscopy 9:224
Calcium stearate
in enamels 12:176
Calcium stearyl-2-lactate. (See also *Lactic acid and its derivatives.*)
uses 15:125
Calcium sulfate 8:97
ammonium acetate extraction 14:91
in ammonium dichromate 9:687
in baking powders 6:429, 430
calcium oxalate volumetric method 17:187
commercial specifications 17:191
in gypsum and gypsum products 14:91
solubility 18:267
in titanium–calcium pigments 17:187; 19:125
uses 17:191
Calcium sulfate dihydrate. See Gypsum.
Calcium sulfate hemihydrate. (See also Densite; Hydrocal; and Plaster of Paris.) 11:263
California mastitis test
for raw milk 16:125
Californium. (See also Transuranium elements.)
manufacture 19:259
Caliper
of paper matchboard 15:534
Calogerasite. See Simpsonite.
Calomel. (See also Mercurous chloride.) 16:1
Caloric content. (See also Food energy content.)
of beer 7:502
of chocolate 9:540
Calorific value
calculation from composition 13:332
of coal and coke 10:241
continous reading calorimeter method 13:337
of natural gas 13:331
water-flow calorimeter method 13:332
Calverite 14:1
Campesterol
gas chromatography 19:150
in tobacco 19:150
Camphene
gas chromatography 12:268
in majoram, oil of 12:268
structure 14:266
Camphene oxide. (See also *Epoxy compounds.*)
zinc bromide–2,4-dinitrophenylhydrazine gravimetric method 12:214
Camphoric acid 8:407

gravimetric method for gallium 13:253, 258
Camphor oil, white. (See also *Essential oils.*)
refractive index 12:271
Canadian whisky 4:465
analysis 4:480
congeneric content 4:467
quality control 4:470
Cananaga oil. (See also *Essential oils.*)
refractive index 12:271
Candelilla wax 6:353
Candicidins 5:472, 564
analysis 5:568, 572, 578
bioautography 5:569
countercurrent distribution 5:568
identification 5:568
microbiological assay 5:572
paper chromatography 5:568
properties 5:565
separation 5:568
solubility 5:566, 567
ultraviolet spectroscopy 5:578
uses 5:564
Candy test 10:565
Cane sugar
Fehling's reagent color test 11:127, 132
in finishing agents 11:126
solubility 11:126
Canfieldite 19:22
occurrence 13:376
Canned food
lactic acid content 15:140
Canthaxanthin
commercial specifications 10:530
composition 10:530
Capillarity
of abrasive grains 10:267
Capillary test
for triphenylmethane and related dyes 19:164
Capillary tube method
for melting range of margarine 15:503
Cappelenite 7:316
Capreomycin 5:468, 549
analysis 5:555, 559, 562
bioautography 5:555
column chromatography 5:555
identification 5:555
microbiological assay 5:559
paper chromatography 5:555
properties 5:551
separation 5:555
solubility 5:552
uses 5:549
Capric acid
gas chromatography 8:382, 388
properties 8:558
thin-layer chromatography 8:381
Capric anhydride

properties 8:558

n-Caprinitrile. See *n*-Decanonitrile.

m-Caproamide
derivatives for identification 8:601

Caproic acid
in adipic acid 4:416
gas chromatography 8:383
properties 8:396, 558

n-Caproic acid
in black tea 18:507

n-Caproic aldehyde
properties 8:334

Caproic anhydride
infrared spectroscopy 8:562
properties 8:558

Caprolactam 8:114-129
acid hydrolysis 8:121
ε-aminohexanoic acid formation 8:122
analysis 8:120
commercial grades and specifications 8:119
gas chromatography 8:121; 17:297
infrared spectroscopy 8:122; 17:296, 322
impurities, determination of 8:122
manufacture 8:117
in polyamide fibers 17:322
in polyamides 17:296
potassium dichromate volumetric method 17:322
properties 8:114
purification 8:119, 120
separation 8:120
solidification point 8:123
solvent extraction 17:296, 322
solubility 8:116

Caprolactam oligomers
in polyamide fibers 17:322
potassium chloride gravimetric method 17:322
potassium dichromate volumetric method 17:322

γ-Caprolactone
in black tea 18:511
gas chromatography 18:511

n-Capronitrile. See *n*-Hexanonitrile.

Capryl alcohol. See 2-Octanol.

Caprylic acid
gas chromatography 8:388
properties 8:558

n-Caprylic aldehyde
properties 8:334

Caprylic anhydride
properties 8:558

n-Caprylonitrile. See *n*-Octanonitrile.

Capsules
weight variation test 6:510

Captan 13:209
analysis 13:210
gas chromatography 13:212
liquid chromatography 13:211
properties 13:209

resorcinol colorimetric method 13:210
solubility 13:209
thin-layer chromatography 13:211
toxicity 13:209
uses 13:209

Captan epoxide
gas chromatography 13:212
thin-layer chromatography 13:211

Cap tests 11:114

Caramel. (See also *Confectionery products.*) 10:
548
in beverages 10:538
commercial specifications 10:530
composition 10:530
in fruit juices 10:538
identification 10:536, 537
in lemonades 10:538
liquid chromatography 10:544
manufacture 10:553
in maple syrup 10:544
paraldehyde gravimetric method 10:536
phenylhydrazine hydrochloride–sodium acetate gravimetric method 10:538
phloroglucinol colorimetric method 10:538
resorcinol–fuming hydrochloric acid colorimetric method 10:538
in salt solutions 10:544
in sugar solutions 10:544

Caramelized sugar content
of roasted coffee 10:375

Carats 14:3

Caraway oil. (See also *Essential oils.*)
carvone content 12:282
refractive index 12:271

Carbamate insecticides. (See also individual compounds.) 19:584
colorimetric determination 19:611
commercial grades and specifications 19:594
manufacture 19:599
properties 19:590
solubility 19:587
toxicity 19:579, 592
uses 19:579

Carbamate pesticides. (See also Carbamate insecticides.)
solvent extraction 19:157
in tobacco 19:157

Carbamide. See Urea.

4-Carbamidobenzaldehyde
infrared spectroscopy 6:616

2-Carbamoylpyridine. See Picolinamide.

3-Carbamoylpyridine. See Nicotinamide.

4-Carbamoylpyridine. See Isonicotinamide.

Carbamylcholine chloride. (See also *Choline and its derivatives.*)
dipicrylamine colorimetric method 9:617
Dragendorff's colorimetric reagent 9:617

iodine colorimetric method 9:617
iodoplatinate colorimetric method 9:617
thin-layer chromatography 9:616
Carbanilides. (See also individual compounds.)
    analysis 11:551
    in disinfectants and antiseptics 11:550
    nuclear magnetic resonance spectroscopy 11:
        551
    properties 8:597
Carbanthrene blue. See D & C Blue No. 9.
Carbaryl. (See also *Pesticides and insecticides.*)
    colorimetric determination 19:611
    commercial grades and specifications 19:594
    composition 19:590, 599
    manufacture 19:599
    properties 19:590
    solubility 19:587
    structure 19:585
    toxicity 19:592
Carbathene. (See also *Ethylene and propylene
        polymers.*) 12:344
    properties 12:344
Carbazole
    colorimetric method for alginate 12:584
    colorimetric method for synthetic resins 10:282
Carbenicillin. (See also Penicillins.) 16:574
    analysis 16:574
    assay 16:574
    bioassay 16:574
    commercial specifications 16:574
    impurities, determination of 16:566, 575
    microbiological methods 16:574
    paper chromatography 16:575
    penicillin content 16:575
    properties 16:574
    specification tests 16:574
    structure 16:554
    in urine 16:574
    uses 16:574
Carbenicillin penicilloic acid
    in carbenicillin 16:575
Carbides. (See also individual compounds.) 8:
        130–139
    analysis 8:133
    carbon content 8:133
    fluxes for 9:216
    formation 8:130
    free carbon in 8:134
    inert gas fusion 8:135
    metal content 8:133, 135
    nitrogen in 8:135
    oxygen in 8:135
    uses 8:130, 132
Carbinoxamine maleate. (See also *Antihist-
        amines.*) 6:26
    properties 6:32
Carbitol

in brake fluids 7:423
gas chromatography 11:73
in nail enamels 11:73
Carbodiimides
    infrared spectroscopy 15:98
    manufacture 15:98
Carbofuran. (See also *Pesticides and insecticides.*)
    commercial grades and specifications 19:594
    composition 19:599
    manufacture 19:599
    solubility 19:587
    structure 19:585
    toxicity 19:592
Carbohydrates
    automated analysis 7:665
    in beer 7:554, 665
    in chocolate and cocoa products 9:538, 540
    chromatographic methods 7:565
    colorimetric method using anthrone 7:557
    in flour 12:606
    in gelatin 12:550
    in grains for distilled alcoholic beverages 4:469,
        474
    in green coffee 10:374, 391
    in green tea leaf 18:457, 477
    in roasted coffee 10:375
    in undistilled intermediates for distilled alcoholic
        beverages 4:480
Carbohydrates, soluble
    in green coffee 10:391
Carbolic acid. See Phenol.
Carbomycin 5:463, 517
    analysis 5:521, 525, 531
    bioautography 5:521
    colorimetric methods 5:531
    countercurrent distribution 5:521
    identification 5:521
    infrared spectrum 5:521
    microbiological assay 5:525
    paper chromatography 5:521
    properties 5:518
    separation 5:521
    solubility 5:519, 520
    structure 5:463
    ultraviolet spectrum 5:522
Carbon. (See also *Carbon, activated; Carbon black;*
        Carbon, free; *Carbon, hydrogen, and nitrogen
        microanalysis;* and Graphite.)
    ablator 4:7
    in aluminum metal 5:124
    in beryllium metal 7:114
    in black ash 6:570
    in boron 7:326
    in boron compounds, organic 7:352
    in brake linings 7:447
    in carbides 8:133
    in carbon adsorbents 4:437

carbon dioxide gasometric methods 14:126; 16:181
carbon dioxide gravimetric method 15:69
in carbon fibers 4:21
in carbon or graphite fabrics 4:26
in ceramics 9:204
in cesium metal 9:283
in chrome steel 9:662
in chromic oxide 9:692
in chromium alloys 9:668
in chromium metal 9:635
in coal and coke 10:236
in cobalt alloys 10:293, 297, 310
combustion method 7:363; 16:325; 18:232
commercial grades and specifications 14:52
in copper alloys 10:636
conductometric method 16:168
crystalline structure 14:53
in cutting alloys 10:289
in dental alloys 10:290
in elastomers, synthetic 12:109
in ferrochromium 9:661; 15:80
in ferromanganese 15:78
in ferrosilicon 15:74
in hafnium 14:126
in hard facing alloys 10:289
heat of sublimation of 4:11
in high-speed steels 10:289
in high-temperature alloys 10:286
in iron 15:59, 69
in lithium hydride 14:251
in manganese 15:482
manufacture 14:54
in metal borides 7:363
microscopy 9:204
in molybdenum 16:168
in molybdenum disulfide 16:181
in nickel 16:325
in nickel alloys 16:325
in organic materials 8:280
in organofunctional silanes 18:87
oxidation methods 19:379
properties 9:204; 14:52
pyrolysis–gas chromatography 18:53
in rubidium 17:555
in silicon carbide abrasives 4:55
in sodium metal 18:232
in sodium–potassium alloy 18:232
in spring steels 10:290
in steel 15:69
in surgical alloys 10:290
in titanium boride 7:367
in trichloromethylsilane 18:53
in wear-resistant alloys 10:289
in zirconium 14:126
in zirconium carbide 14:143
Carbonaceous material

in sulfur 18:377
Carbon, activated 8:139–161
adsorptive capacity 8:142
analysis 8:140
classification 8:139
crystallite structure 8:140, 141
manufacture 8:139
properties, determination of 8:140, 151
specification tests 8:151
structure 8:140, 141
uses 8:140
Carbonados. (See also *Diamonds, industrial.*) 11:469
properties 11:463, 469
uses 11:469
Carbon, adsorbent
acid-insoluble matter in 4:437
ash determination 4:436
carbon content of 4:437
chemical analysis 4:439
decolorizing efficiency 4:448
hardness 4:446
pH 4:437
uses 4:434
water-soluble material 4:438
Carbon adsorption method
for anionic detergents 11:340
Carbon and hydrogen. (See also *Carbon, hydrogen, and nitrogen microanalysis.*)
Pregl method 8:281, 292
Carbonate aggregates
in concrete 9:143
Carbonate–apatite 17:121
Carbonated ash content
of sugar 18:358
Carbonated beverages 8:161–178
analysis 8:165
composition 8:162
ingredients 8:162
manufacture 8:163
package tests 8:175
packaging 8:163
properties 8:162
Carbonate hardness
of water 19:362
Carbonate rocks
impurities, determination of 17:72
sample decomposition 18:27
silicon content 18:27
sulfur content 18:380
titanium content 19:120
Carbonates
acid–base titration 9:286; 17:401
in asbestos 6:265
in asbestos–cement products 6:272
in barite 6:558, 562
in barium hydroxide 6:549

in cadmium carbonate 8:39
carbon dioxide absorption method 11:351
carbon dioxide evolution method 18:252
in cesium carbonate 9:286
in chrysotile 6:265
in clays and clay minerals 10:124
in coal and coke 10:254
in cotton 11:141
in creams and lotions 11:92
in detergents 11:335, 351
in glycerol 13:525
gravimetric method for silver 18:164
hydrochloric acid effervescence test 18:183
oxidation methods 19:379
perchloric acid gravimetric method 18:255
phenolphthalein–sodium carbonate colorimetric
    method 11:141
in polycarbonates 17:340
in potassium hydroxide 17:401; 18:252
potassium hydroxide volumetric method 17:340
in rare earth carbonates 17:508
sample preparation 15:31
silicon content 18:37
in soaps 18:182, 183, 192
in sodium hydroxide 18:252
in water 19:362, 379
Carbonate white lead
    analysis 17:188
    commercial specifications 17:188
    impurities, determination of 17:188
    lead carbonate content 17:188
Carbonation
    of carbonated beverages 8:164, 172
Carbon bisulfide. See Carbon disulfide.
Carbon black 8:178–243
    acidity 8:192, 230
    analysis 8:197; 17:202
    blackness 8:181, 211
    in brake linings 7:441
    in coatings 8:241
    commercial grades and specifications 8:196;
        17:202
    composition 8:191
    compressibility 8:227
    crystallinity 8:188
    density 8:230
    dibutyl phthalate absorption 8:225
    in elastomers, synthetic 12:148
    electron microscopy 8:180, 202
    in enamels 12:173
    in explosive priming mixtures 12:421, 424
    gas adsorption 8:186, 214
    gas chromatography 8:229
    gravimetric determination 12:148
    identification 8:197; 12:44
    impurities, determination of 8:234; 17:202
    in ink 8:241

iodine adsorption 8:187, 224
    manufacture 8:193
    masstone 8:181, 211
    microstructure 8:187
    oil absorption 8:226
    in paint 8:241
    in paper 8:241
    particle size 8:180, 202
    particle size distribution 8:184; 12:150
    pH 8:192, 229
    in plastic 8:238
    properties 8:180
    properties, determination of 8:202
    resistivity 8:193, 233
    in rubber 8:197, 238
    separation 8:197
    solubility 12:421
    structure 8:179, 187, 225
    surface area 8:186, 214
    tinting strength 8:186, 213
    types 8:179; 17:202
    uses 8:179
    visible spectroscopy 12:150
    volatile content 8:228
Carbon–carbon composites. See Graphite com-
    posites, fiber reinforced.
Carbon compounds
    in chromium compounds 9:699
    in copper(II) oxide 10:674
Carbon, decolorizing. See Carbon, activated.
Carbon dioxide 8:244–267
    absorption method 7:577
    absorption method for bicarbonates 11:353
    absorption method for carbonates 11:351
    in acetylene 4:130
    acid–base titration 15:285; 19:361, 501
    in air 16:485
    analysis 8:248
    in asbestos 6:265
    in beer 7:570, 620
    in carbonated beverages 8:172
    in carbon monoxide 8:306, 309, 310, 313, 315
    caustic absorption method 8:248, 260
    in cesium hydroxide 9:290
    in chlorine 9:309, 316, 320
    in chrysotile 6:265
    in cigarette smoke 10:13
    commercial grades and specifications 8:247
    in cyclopropane 5:360
    enzymatic method 19:501
    in ethylene 5:362
    in ethylene oxide 12:338
    evolution method for carbonate 18:252
    in fluorine, elemental 13:4
    gas chromatography 8:251, 262, 263, 310, 313;
        12:324, 338; 13:297, 317, 318; 14:384, 392,
        416; 15:557; 16:456, 485; 19:502

in gas mixtures 8:260
gasometric methods 6:435, 436; 15:220;
  19:501
gasometric method for carbon 14:126, 133;
  16:181
Gehaltemeter 7:571
gravimetric method 7:577; 8:262
in gypsum and gypsum products 14:93
in hydrated lime 9:191
in hydrogen 14:380, 381, 384
identification 8:248
impurities, determination of 8:249, 251
infrared spectroscopy 8:252, 264, 313; 13:294
in krypton 13:300
in lime and limestone 9:191; 15:220
in lithium hydroxide 15:285
manufacture 8:246
mass spectroscopy 8:255, 266, 315; 13:302,
  306
in natural gas 13:320
in nitrous oxide 5:366
in noble gases 13:294, 297, 302
occurrence 8:244
Orsat technique 14:381
in oxygen 16:531, 532
potassium hydroxide absorption method 8:306
pressure methods 7:575
properties 8:244
in quicklime 9:191
radiochemical method 19:502
in rare earth oxides 17:510
in roasted coffee 10:375
solubility 8:245
specification tests 8:259
specific gravity 13:329
table of gas volumes 8:168
thermal conductivity 14:380
titrimetric methods 8:261
toxicity 8:246
uses 8:244
volumetric determination 19:501
in water 19:361
in wines 19:415, 500
in xenon 13:301
in zirconyl ammonium carbonate 14:147
Carbon disulfide 8:268–279; 13:193
  analysis 8:271
  azeotropes 8:269, 271
  in benzene 7:43; 18:543
  in carbon tetrachloride 8:325
  colorimetric method using cupric acetate–di-
    ethylamine 8:277
  colorimetric method for morpholine 16:200
  commercial grades and specifications 8:271
  cupric dithiocarbamate method 7:43
  gas chromatography 8:274, 325
  identification 8:272

impurities, determination of 8:274
infrared spectroscopy 8:325; 18:543
infrared spectrum 8:272
iodine titration 8:273
manufacture 8:269; 18:365
mass spectroscopy 8:272
polarography 8:273
properties 8:268; 13:193
Raman spectroscopy 8:272
safety 18:367
separation 8:272
solubility 8:269
solubility of bituminous materials 6:311
toxicity 13:194; 18:367
ultraviolet spectroscopy 7:45; 8:272, 278; 18:
  398
uses 8:268; 13:193
xanthate reaction 8:273
Carbon, elemental
  properties 18:4
  silicon content 18:30, 37
Carbon fibers
  ablator 4:21
  carbon content of 4:21
  differential thermal analysis 4:21
  elemental analysis 4:21
  quality control tests 4:25
  surface area 4:14
  water content of 4:21
Carbon, free
  in carbides 8:134
  in metal borides 7:363
  in silicon carbide abrasives 4:53
  in titanium boride 7:367
  in vanadium carbide 8:137
Carbon, hydrogen, and nitrogen microanalysis
    8:280–298
  gas chromatography 8:283
  instruments 8:281
Carbonic acid 8:244, 248
Carbonizable substances
  in acetylsalicylic acid 6:140
  in amobarbital 6:508
  in aprobarbital 6:512
  in chloroform 5:359
  in cinchophen 6:205
  in mephobarbital 6:518
  in phenobarbital 6:523
  in secobarbital 6:525
  in sodium pentobarbital 6:520
  in sodium phenobarbital 6:523
  in sodium secobarbital 6:525
Carbon monoxide 8:299–319
  in acetylene 4:130
  in air 8:319; 16:485
  analysis 8:304
  in carbon dioxide 8:251, 260

in chlorine 9:320
in cigarette smoke 10:13, 26
commercial grades and specifications 8:303
cuprous chloride absorption method 8:306
cuprous sulfate absorption method 8:306, 308
in ethylene 5:362
in fluorocarbons 13:70
gas absorption method 8:306
gas chromatography 8:251, 310, 318; 10:26;
    13:71; 14:384, 392, 415; 16:454, 485
gas detector kits 8:305
hemoglobin identification test 8:304
in hydrogen 14:380, 381, 384
identification 8:304
impurities, determination of 8:306, 309
infrared spectroscopy 8:313, 318; 13:294
iodine pentoxide method 8:316
liquefaction 8:303
manufacture 8:302
mass spectroscopy 8:315; 13:302
in noble gases 13:294, 302
occurrence 8:299
Orsat technique 14:381
oxidation indicator 8:319
in oxygen 16:532
palladous chloride identification test 8:304
properties 8:299
solubility 8:299, 300
tetraphenylbenzidine colorimetric method 16:
    532
thermal conductivity 14:380
toxicity 8:301
uses 8:299
Carbon oxysulfide
    in carbon disulfide 8:274
    gas chromatography 8:274
Carbon refractories. (See also Refractories.)
    composition 9:207
Carbon residue
    Conradson method 15:325
    of lubricants 15:323
    Ramsbottom method 15:323
Carbon steel. (See also Chrome steels; *Steel.*)
    composition 5:59
Carbon support films
    for electron microscopy 8:202
Carbon tetrachloride 8:320–330; 13:182
    adsorption by activated carbon 8:143
    in air 8:328
    analysis 8:324
    azeotropes 8:322
    Beilstein identification test 8:324; 9:382
    carbon disulfide in 8:279
    in chlorine 9:321
    in chloroform 9:433, 434
    commercial grades and specifications
        8:323

composition 9:411
cyclopentanol–sodium hydroxide colorimetric
    method 9:387
density 7:37
2,7-dihydroxynaphthalene colorimetric method
    9:387
distillation of water in explosives 12:467
Fujiwara colorimetric method 9:385, 386, 434
gas chromatography 8:324; 9:374, 375, 376,
    377, 380, 381, 442
in hydrochloric acid 9:392
identification 8:324
impurities, determination of 8:324
infrared spectroscopy 8:324, 325, 328; 9:390,
    392, 394
liquid chromatography 7:15
manufacture 8:322
mass spectroscopy 9:403
α-naphthol colorimetric method 9:387
phenolphthalein colorimetric method 9:387
properties 8:320
refractive index 7:37
separation 8:324
solubility 8:320; 13:539, 552
solubility of bituminous materials 6:311
soluble matter in felt 12:494
specification tests 8:326
toxicity 8:321; 13:182
in 1,1,1-trichloroethane 9:442
uses 8:320; 13:182
Carbon tetrafluoride. (See also *Fluorocarbons.*)
    gas chromatography 13:64
    infrared spectroscopy 13:294
    in noble gases 13:294
    properties 13:60
Carbonylation of acetylene
    manufacture of acrylic acid and acrylates 4:184
Carbonyl chloride. See Phosgene.
Carbonyl compounds. (See also Aldehydes; *Car-
    boxylic acids; Carboxylic acids, derivatives;*
    Ketones; and individual compounds.) 8:331–
    367
    in acetylenic chemicals 4:145
    in acrolein 4:154
    in acrylonitrile 4:357
    analysis 8:339
    in cigarette smoke 10:13
    commercial grades and specifications 8:340
    2,4-dinitrophenylhydrazine colorimetric method
        8:364
    gas chromatography 4:154; 8:351, 366
    hydroxylamine hydrochloric method 4:145, 157,
        375; 8:356
    identification 8:339
    infrared spectroscopy 8:345
    kinetic analysis 8:364
    manufacture 8:339

nuclear magnetic resonance spectroscopy 8:348
oximation methods 8:356
polarography 8:365
properties 8:332
separation 8:348
sodium bisulfite addition 4:155
specification tests 8:358
thin-layer chromatography 8:348
ultraviolet spectroscopy 8:347
uses 8:332
Carbonyl iron. (See also Iron metal.)
commercial grades and specifications 15:28
uses 15:28
Carbonyls
absorption spectroscopy 9:562
acid–base titration 11:507
in 1,3-butadiene 11:504, 507
in cellulose 9:45
in chocolate liquor 9:562
in cocoa beans 9:562
column chromatography 9:562
2,4-dinitrophenylhydrazine colorimetric method
11:521
in dioxane 11:521
in glycols 13:600
in polyhydric alcohols 13:600
thin-layer chromatography 9:565
in wines 19:452
Carbonyl sulfide
in carbon dioxide 8:255
gas chromatography 14:392
mass spectroscopy 8:255
Carboranes 7:330
Carborundum. (See also Silicon carbide.) 18:39
in explosive priming mixtures 12:417, 421, 424,
426
microscopy 12:417, 421, 424
solubility 12:421
x-ray diffraction 12:418
Carbowax. (See also Polyethylene glycols.)
Rosen color test 13:573
Carboxide. See *Ethylene oxide*.
Carboxin 13:212
analysis 13:213
in cottonseed meal 13:213
p-dimethylaminobenzaldehyde colorimetric
method 13:213
manufacture 13:213
in plants 13:213
in potatoes 13:213
properties 13:213
solubility 13:213
in sugar beets 13:213
toxicity 13:213
uses 13:212
4-Carboxybenzaldehyde
polarography 17:162

in terephthalic acid 17:162, 163
p-Carboxybenzyl alcohol
in terephthalic acid 17:163
α-Carboxybenzylpenicillin. See Carbenicillin.
Carboxyl end groups
in polyamides 17:293
Carboxyl groups 8:368
acid–base titration 11:34
in cellulose 9:44
in coal and coke 10:211
in methacrylate polymers 16:91
in starch derivatives 11:34
2-Carboxy-2'-hydroxy-5'-sulfoformazyl benzene.
See Zincon.
Carboxylic acid polymers
carboxylic acid content 4:275
Carboxylic acids. (See also individual compounds.)
8:368–409
in acid anhydrides 8:565, 567
acid–base titrations 8:389
in acid halides 8:574
in acid salts 8:596
in acrylamide polymers 4:172
in adipic acid 4:415
in amides 8:609
analysis 8:374
in beer 7:579
classes 8:369
in cocoa beans 9:543, 544
column chromatography 4:415
conductometric titration 4:172
derivatives for identification 7:62; 8:374
esterification method 8:394
in esters 8:587
gas chromatography 4:416; 5:25; 8:383
identification 8:374
infrared spectroscopy 7:65; 8:378; 12:576
ion exchange chromatography 8:382
methylation 8:385
paper chromatography 8:380
polarography 8:395
preparation 8:372
preparation of acid halides 7:63
properties 8:371
separation 8:378
thin-layer chromatography 8:380
use in manufacture of anhydrides 8:560
volumetric methods 8:389
Carboxylic acids, alicyclic 8:407
Carboxylic acids, aliphatic 8:369, 395
Carboxylic acids, amino. (See also individual compounds.) 8:370, 410–556
analysis 8:411
colorimetric methods 8:431
copper complex colorimetric method 8:433
dissociation constant 8:445
electrophoresis 8:427

gas chromatography 8:428
impurities, determination of 8:434
ion exchange chromatography 8:414
isoelectric point 8:445
isotope dilution methods 8:433
ninhydrin colorimetric method 8:416, 432
nitrogen content of 8:434
optical rotation 8:443
paper chromatography 8:422, 435
paper electrophoresis 8:427
specification tests 8:434
Carboxylic acids, aromatic 8:369, 400
Carboxylic acids, derivatives. (See also Acid
    anhydrides; Acid halides; Acid salts; Amides;
    Esters; and individual compounds.) 8:557–
    612
Carboxylic acids, free
acid–base titration 7:76
in acid chlorides 7:76
Carboxylic acids, halogenated 8:370
Carboxylic acids, heterocyclic 8:407
Carboxylic acids, unsaturated 8:371, 374
Carboxymethylcellulose 11:353
colorimetric determination 12:584
composition 12:583
in detergents 11:353
2,7-dihydroxynaphthalene colorimetric method
    11:354
gravimetric determination 12:581
in matches 15:509
paper chromatography 12:582
3,3′-Carboxymethylene-bis (4-hydroxycoumarin)
    ethyl ester. See Ethyl biscoumacetate.
{[(Carboxymethyl)imino] bis{(ethylenedinitrilo)}
    tetraacetic acid. See DTPA.
p-Carboxyphenylhydrazine derivative of acetals
    4:67
5-(2-Carboxyphenyl)-1-(2-hydroxy-5-sulfophenyl)-
    3-phenyl formazon. See Zincon.
4-Carboxypiperidine. See Isonipecotic acid.
Carburetor cleaners 6:356
analysis 6:370
Car cure
of brake linings 7:472
Cardanol content, residual
in brake linings 7:449
Cardolites
in brake linings 7:441
Cariflex 12:86
gel permeation chromatography 12:121
intrinsic viscosity 12:119
Carius method
for boron compounds, organic 7:399
for boron hydrides 7:400
for chlorine content of chlorocarbons
    9:411
for iodine in organic iodine

compounds 14:613
for sulfur compounds, organic 18:378
Carmine. (See also Carminic acid.)
assay 10:540
colorimetric method for boric acid 7:387, 409
commercial specifications 10:530
composition 10:530
visible spectroscopy 10:540
Carmine red
colorimetric method for boric acid 7:387, 409
Carminic acid. (See also Carmine.)
colorimetric method for boric acid 7:387, 409
colorimetric method for boron 5:122
structure 10:539
Carnallite 9:278, 304, 338; 15:357, 403; 17:353,
    544, 547
Carnauba wax 6:353
Carnitine
ion exchange chromatography 9:618
Carnotite 19:225
composition 19:225, 321
occurrence 19:225, 321
uranium content 19:325
vanadium content 19:325
Carotene
in cheese 10:544
colorimetric determination 12:612
in feed products 11:52
in flour 12:612
liquid chromatography 11:53
solvent extraction 19:152
in tobacco 19:152
visible spectroscopy 11:55; 19:152
α-Carotene
in β-carotene 10:542
in green tea leaf 18:480
liquid chromatography 10:542
thin-layer chromatography 18:480
β-Carotene
assay 10:539
commercial specifications 10:530
in dried food 10:545
enzyme method 10:545
in green tea leaf 18:480
identification 10:537
impurities, determination of 10:542
in juices 10:545
liquid chromatography 10:542, 545
structure 10:538
thin-layer chromatography 18:480
visible spectroscopy 10:539, 545
γ-Carotene
in green tea leaf 18:480
thin-layer chromatography 18:480
β-Carotene equivalent
of color in egg products 12:68
Carotenoid content

of green tea leaf 18:479
Carotenoids, synthetic
 in juices 10:545
 liquid chromatography 10:545
Carpenter alloys. (See also Nickel alloys.)
 composition 16:290, 351
Carrageenin
 composition 12:583
 ferric chloride gravimetric method 12:582
 gravimetric determination 12:581
 infrared spectroscopy 12:583
 in photographic films 12:582
 thorium nitrate gravimetric method 12:582
Carrier distillation methods
 for rare earths 17:479
Carrier gas and vacuum fusion methods 8:613–693
 apparatus 8:653
 applications 8:683
 calibration 8:677
 crucible assemblies 8:640, 642, 645, 671
 distillation–reduction techniques 8:634
 fusion techniques 8:614
 heating sources 8:641
 hot extraction techniques 8:638
 for nitrogen 16:469
 operating blanks 8:639, 671
 operating procedures 8:668
 sample preparation 8:641, 670
 synthetic standards 8:679
 technology 8:639
Carrier gas fusion methods. (See also *Carrier gas and vacuum fusion methods.*)
 apparatus 8:661
 applications 8:624, 682
 distillation 8:635
 gas collection traps 8:675
 gas purification 8:672
 hot extraction 8:638
 for nitrogen in metal borides 7:365
 for oxygen 16:169
 for oxygen in metal borides 7:365
 reaction media 8:629
 theory 8:614
Carrollite 10:284
Carrot oil
 commercial specifications 10:531
 composition 10:531
Carvacrol. See 2-Methyl-5-isopropylphenol.
Carvone
 in caraway oil 12:282
 in dill oil 12:282
 hydroxylamine hydrochloride volumetric
  method 12:282
 in spearmint oil 12:282
Car washes 6:354
 analysis 6:364
 performance tests 6:377

Caryocerite 7:316
Caryophylene
 gas chromatography 12:266, 268
 in majoram, oil of 12:268
 in peppermint oil 12:266
Carzinophilin 5:477, 599
 analysis 5:604, 609, 614
 composition 5:477
 microbiological assay 5:609
 properties 5:600
 solubility 5:601, 602
 ultraviolet spectroscopy 5:614
Casein 9:1–15
 amino acid composition 9:4
 analysis 9:7
 commercial grades and specifications 9:5
 fat content 9:9
 ferric chloride colorimetric method 11:127, 132
 in finishing agents 11:126, 132
 flame test 12:588
 identification 9:7
 impurities, determination of 9:8
 infrared spectrum 9:7
 lactose content 9:10
 manufacture 9:1
 in photographic films 12:588
 precipitation 9:8
 properties 9:2
 protein content 9:9
 separation 9:8
 solubility 9:3; 11:126
 specification tests 9:8
 stability 9:5
 uses 9:1
 viscosity 9:11
Casein digestion method
 for protease activity 7:590
Casein fibers 9:1
Casein–formaldehyde resin
 ferric chloride colorimetric method 11:127, 133
 in finishing agents 11:126, 133
 solubility 11:126
 sulfuric acid–carbazole colorimetric method 11:133
Casein glues 4:400
Casein plastics 9:2
Cashew nut shell liquid
 analysis 7:447
 in brake linings 7:441, 447, 450
 pyrolysis–gas chromatography 7:450
Cashmilon. (See also Acrylic fibers; *Acrylic and modacrylic fibers.*) 4:220, 249
Cassella's acid. See Naphthylamine-4,8-disulfonic acid.
Cassia oil. (See also *Essential oils.*)
 cinnamic aldehyde content 12:282
 refractive index 12:271

solubility 13:540

Cassiterite. (See also Stannic oxide.) 14:519; 17:561; 19:22
decomposition 19:235
scandium content 17:577

Cast gold alloys, dental. See Dental cast gold alloys.

Cast iron. (See also Iron metal.)
commercial grades and specifications 15:29
composition 15:26
impurities, determination of 15:58
magnesium content 15:371
manganese content 15:467
manufacture 15:26
uses 15:27

Cast iron enamels. (See also *Enamels.*)
composition 9:195

Castor oil. (See also *Fats and fatty oils.*)
in alkyd resins 5:23
analysis 7:427
in brake fluids 7:423, 424, 427
in enamels 12:176
hydroxyl value 12:478
properties 12:176, 473
in rubber, natural 12:145
solubility 13:539, 552
thin-layer chromatography 12:145

Castor oil derivatives
analysis 7:427
in brake fluids 7:424, 427

Catalase
hydrogen peroxide gasometric method 16:124
in raw milk 16:124

Catalytic hydrogenation method
for sulfur compounds, organic 18:379

Catalytic methods
for silver 18:172

Catechin
in black tea 18:495
in cacao 9:536
colorimetric determination 19:467
gas chromatography 18:471
in green tea leaf 18:457, 465
liquid chromatography 18:465
paper chromatography 18:468
structure 18:461
thin-layer chromatography 18:463, 470, 495, 496
ultraviolet spectroscopy 18:465
vanillin colorimetric method 19:467
in wines 19:461, 467

Catechin gallate
in green tea leaf 18:461
structure 18:461
thin-layer chromatography 18:463

Catechol
in 1,3-butadiene 11:507

colorimetric method for erythritol 13:572
colorimetric method for glycerol 13:572
colorimetric method for mannitol 13:572
colorimetric method for pentaerythritol 13:572
colorimetric method for propylene glycol 11:74
ferric chloride colorimetric method 11:507
in hair dyes 11:103
thin-layer chromatography 11:103

Catenary of raw roving 4:25

Cation content
of tobacco for cigarettes 10:5

Cation exchange capacity
of ion exchange resins 15:16

Cation exchange resins. See *Ion exchange resins.*

Cationic detergents. (See also *Detergents.*) 11:323
bromphenol blue colorimetric method 11:328
lauryl sulfate titration 11:347
methylene blue titration 11:347
uses 11:323

Cationic dyes 12:3
Blancophor G titration 12:13
identification 12:13

Cationic materials
electrodeposition method 11:131
in finishing agents 11:126, 131
solubility 11:126, 131
turbidimetric determination 11:126, 131

Cationic surfactants. (See also Surfactants.)
Aerosol OT titration 11:79
bromphenol blue colorimetric method 12:578
in cotton 11:135
ion exchange chromatography 11:81
Orange II colorimetric method 11:135
paper chromatography 12:575
in shampoos 11:78
ultraviolet spectroscopy 12:575

Caulking compounds. 9:16–27
analysis 9:18
classification 9:16
composition 9:17
manufacture 9:17
properties, determination of 9:18
specification tests 9:18
uses 9:16

Caustic
mercury content 16:44

Caustic color method
for hydroquinone 4:202

Caustic distillation
of ammonia 9:685

Caustic potash. (See also Potassium hydroxide.)
sampling 17:359

Caustic soda. (See also Sodium hydroxide.)
commercial grades and specifications 18:242
manufacture 18:241

C$_1$ Chlorinated hydrocarbons. (See also

Chlorocarbons; and individual compounds.)
9:422–435
C₂ Chlorinated hydrocarbons. (See also *Chlorocarbons*; and individual compounds.) 9:435–
465
C₃ Chlorinated hydrocarbons. (See also *Chlorocarbons*; Chlorinated propanes; and individual
compounds.) 9:465–472
  analysis 9:470
  gas chromatography 9:470
  methoxymercuration 9:470
C₄ Chlorinated hydrocarbons. (See also *Chlorocarbons*; and individual compounds.) 9:472–
475
  analysis 9:474
  gas chromatography 9:474
CDTA. See DCYTA.
Cedar leaf oil. (See also *Essential oils.*)
  thujone content 12:282
Cedarwood oil. (See also *Essential oils.*)
  refractive index 12:271
Cedryl acetate
  in essential oils 12:276
  molecular weight 12:276
Celestite. (See also Strontium sulfate.) 18:264,
281
Celite
  in brake linings 7:441
Cell constituents
  of yeast 19:526
Cell count
  of yeast 19:522, 523
Cellosolve
  in brake fluids 7:423
Cell structure
  of polyurethane foams 19:280
Cellular polyurethanes. (See also Polyurethane
  foams.)
  analysis 19:281
  cell structure 19:280
  composition 19:264
  manufacture 19:263
  properties 19:281
  types 19:265
  uses 19:265
Cellulose 4:75; 9:27–59
  acetylation 4:78
  analysis 9:33
  composition 9:28
  composition, determination of 9:34
  crystalline content 9:53
  cupriethylenediamine disperse viscosity of 9:34
  end-use tests 9:55
  flame test 12:588
  fluidity of dispersions 7:235
  history 9:27
  impurities, determination of 9:47

  manufacture 9:31
  nitric acid gravimetric method 19:142
  in photographic films 12:588
  properties 9:28
  properties, determination of 9:52
  regeneration 4:80
  in shampoos 11:76
  sodium content 18:220
  solubility 9:30, 42, 44
  of tobacco 19:142
  viscosity of dispersions 7:265
  x-ray diffraction 9:53
Cellulose acetate. (See also *Acetate and triacetate
  fibers*; *Cellulose derivatives.*) 4:74; 9:69
  acetyl content 9:70
  analysis 9:69
  commercial grades 9:69
  in finishing agents 11:126, 132
  flame test 12:588
  heat stability 9:72
  impurities, determination of 9:69
  infrared spectroscopy 12:590
  lanthanum nitrate–iodine colorimetric method
    11:127, 132
  mixtures, determination in 4:359
  in photographic films 12:588
  properties 9:69
  solubility 4:330, 359; 11:126; 13:539, 552
  uses 9:69
  viscosity 9:62, 71
Cellulose acetate butyrate. (See also *Cellulose
  derivatives.*) 9:72
  acetyl content 9:74
  analysis 9:73
  butyryl content 9:74
  impurities, determination of 9:73
  properties 9:73
Cellulose acetate filters
  for cigarettes 10:2
Cellulose acetate phthalate
  uses 9:60
Cellulose acetate propionate. (See also *Cellulose
  derivatives.*) 9:72
  acetyl content 9:74
  analysis 9:73
  impurities, determination of 9:73
  properties 9:73
  propionyl content 9:74
Cellulose acetate sulfate
  uses 9:60
Cellulose acetobutyrate
  flame test 12:588
  infrared spectroscopy 12:590
  in photographic films 12:588
  solubility 12:589
Cellulose butyrate. (See also Cellulose acetate butyrate.)
  infrared spectroscopy 12:590

in photographic films 12:590
Cellulose derivatives. (See also individual com-
    pounds.) 9:59–94
analysis 9:60
heating test 12:588
infrared spectroscopy 12:590
1-naphthol-sulfuric acid color test 12:588
paper chromatography 12:582
in photographic films 12:588
solubility 12:589
uses 9:59
viscosity 9:66
Cellulose diacetate. (See also Cellulose acetate.)
infrared spectroscopy 12:590
in photographic films 12:589
solubility 12:589
Cellulose ester films
in photographic films 12:559
Cellulose esters
    hydroxylamine hydrochloride–ferric chloride
        colorimetric method 12:588
    infrared spectroscopy 12:590
    lanthanum nitrate–iodine colorimetric method
        12:589
    in photographic films 12:588
    solubility 12:589
Cellulose fibers. See *Cellulose.*
Cellulose fibers, bleached
    fluidity of dispersions 7:235
Cellulose, modified
    infrared spectrum 4:244
Cellulose nitrate. (See also Cellulose derivatives;
    High explosives.) 9:62; 12:443
in alkyd resins 5:24
analysis 9:64
colorimetric determination 12:421
commercial specifications 12:443
diphenylamine colorimetric method 11:127,
    132
diphenylamine–sulfuric acid color test 12:588
in explosive priming mixtures 12:418, 421, 422
in finishing agents 11:126, 132
flame test 12:588
heat stability 9:64
history 9:62
impurities, determination of 9:64
infrared spectroscopy 12:455, 590
manufacture 9:63; 12:443
nitrogen content 9:66
in photographic films 12:588
properties 9:63; 12:432, 443
solubility 9:63; 11:126; 12:443, 589
solvent extraction 12:422
uses 9:62; 12:443
viscosity 9:62, 66
x-ray diffraction 12:418
Cellulose propionate. (See also Cellulose

acetate propionate.)
infrared spectroscopy 12:590
in photographic films 12:590
Cellulose, regenerated
ablator 4:21
intrinsic viscosity 4:82
Cellulose triacetate. (See also Cellulose acetate.)
    4:74
infrared spectroscopy 12:590
in photographic films 12:589
solubility 12:589
Cellulose trinitrate
viscosity 9:62
Cemented carbides. (See also Cobalt alloys.) 10:
    291
composition 10:291
decomposition 10:349
uses 10:291
Cementite
in iron 15:70
Cement mixing test
for emulsified asphalt 6:328
Cements. (See also *Ceramics*; Concretes; Hydraulic
    cements; Mortars; and Portland cement.) 9:94,
    190
acid dissolution 9:214
air content 9:141
aluminum oxide content 9:112, 113, 126
ammonium hydroxide group content 9:112
analysis 9:102; 17:361
calcium oxide content 9:115, 128
in cement-treated soils 9:151
commercial specifications 9:98
composition, determination of 9:110, 124
compressive strength 9:136
consistency 9:133
electron microscopy 9:146
false set 9:133
ferric oxide content 9:112, 114, 126
fluxes for 9:216
Gillmore time-of-set test 9:136
heat of hydration 9:141
impurities, determination of 9:131
iron content 15:56
magnesium content 15:371
magnesium oxide content 9:116, 128
manganic oxide content 9:119
particle size 9:131
phosphorus content 17:97
phosphorus pentoxide content 9:130
potassium oxide content 9:120
properties, determination of 9:131
sample preparation 17:361
silica content 9:111, 126
silicon content 18:38
sodium hydroxide fusion 9:125
sodium oxide content 9:120

sulfate content 9:113, 129
tensile strength 9:136
time of setting 9:134
titanium dioxide content 9:122
unsoundness tests 9:140
Vicat time-of-set test 9:135
water content 9:133
Cements, mortars, and concrete. (See also
    Cements; Concrete; and Mortars.) 9:94–155
Cement-treated soils 9:150
  analysis 9:150
  cement content 9:151
Centrifugal extractor 6:331
Centrifugal force method
  for specific gravity of natural gas 13:330
Centrifugal separation. (See also Centrifuges.)
    9:156–188
  drainage 9:158
  particle size distribution 9:169, 178
  sedimentation 9:156
  settling curves 9:161
  Stokes' law 9:156
  theory 9:156
Centrifuges
  basket 9:171
  bottle 9:163
  comparative performance 9:161
  continous conveyor discharge 9:172
  disc 9:173
  laboratory 9:159
  tubular 9:176
  types 9:160
  ultracentrifuges 9:179
Centrifuge test
  for water in lubricants 15:329
Cephalexin. (See also Cephalosporins.)
  structure 16:555
Cephaloglycin. (See also Cephalosporins.)
  assay 16:579
  impurities, determination of 16:579
  properties 16:556
  structure 16:555
  uses 16:578
Cephaloridine. (See also Cephalosporins.)
  iodometric methods 16:562
  properties 16:556
  solubility 16:557
  specification tests 16:579
  structure 16:555
  ultraviolet spectroscopy 16:560
Cephalosporidine. (See also Cephalosporins.)
  uses 16:578
Cephalosporin C. (See also Cephalosporins.)
  assay 16:579
  bioassay 16:578
  paper chromatography 16:559
  properties 16:556

structure 16:555
ultraviolet spectroscopy 16:579
Cephalosporins. (See also *Penicillins and related
    compounds*; and individual compounds.) 5:
    464; 16:578
  acid–base titration 16:563
  analysis 16:558, 578
  assay 16:560, 578
  bioassay 16:578
  hydroxylamine–ferric ammonium sulfate colori-
    metric method 16:563
  identification 16:558
  infrared spectroscopy 16:559
  iodometric methods 16:562
  manufacture 16:578
  microbiological methods 16:578
  nuclear magnetic resonance spectroscopy 16:560
  paper chromatography 16:558
  properties 16:553, 578
  separation 16:558
  solubility 16:557
  structure 16:553, 555
  thin-layer chromatography 16:559
  ultraviolet spectroscopy 16:560, 579
  uses 16:578
Cephalothin. (See also Cephalosporins; Sodium
  cephalothin.)
  assay 16:579
  bioassay 16:579
  bioautography 16:579
  paper chromatography 16:559
  specification tests 16:579
  structure 16:555
  uses 16:578
Ceramics. (See also *Abrasives*; Cements; Clay pro-
    ducts; *Enamels*; Glass; Glass ceramics; Glazes;
    Gypsum; Lime; Refractories; Porcelains; White-
    wares; and individual compounds.) 9:189–232
  acid dissolution 9:214
  analysis 9:210, 216
  atomic absorption spectroscopy 9:217
  boron in 7:417
  cerium content of 9:260
  chemical etchants 9:224
  combustion 9:217
  composition 9:194
  for electronics 9:209, 212
  electron microscopy 9:223
  emission spectroscopy 9:217
  flame emission spectroscopy 9:217
  fluxes for 9:215
  fusion methods 9:214
  in glass industry 9:208
  inert gas fusion 9:217
  infrared spectroscopy 9:217
  impurities, determination of 9:217, 218
  microscopy 9:222

optical emission spectroscopy 9:217
polarography 9:217
pyrohydrolysis 9:217
sampling 9:212
silicon content 18:30
sodium carbonate fusion 7:400
solids mass spectroscopy 9:217
x-ray diffraction 9:219
x-ray fluorescence spectroscopy 9:217
Ceramics, dental. See Dental ceramics.
Cerargyrite 18:155
Cerate volumetric method
   for antimony 6:64
Cereals 9:232–245
   analysis 9:235
   antioxidants in 9:242
   ash content of 9:236
   color 9:235
   composition of 9:233
   crude fiber in 9:237
   fat in 9:237
   food dyes in 9:244
   manufacture 9:234
   mineral content 9:233
   moisture in 9:235
   pH of 9:236
   protein content of 9:236
   sampling 9:235
   starch content of 9:238
   sugar content of 9:238
   vitamin content 9:233
Ceresan. See 2-Methoxyethyl mercuric chloride.
Ceric ammonium nitrate 9:250
   analysis 9:251
   cerium content 9:251
   color test for silver 18:164
   commercial grades and specifications 9:250
   identification 9:251
   impurities, determination of 9:251
   manufacture 9:250
   titration of hydrogen peroxide 11:107
   volumetric method for hypophosphites 17:113
   volumetric method for lead azide 12:411, 419
   volumetric method for oxalates 7:618
   volumetric method for phosphites 17:113
Ceric ammonium sulfate 9:252
   analysis 9:253
   cerium content of 9:253
   commercial specifications 9:252
   impurities, determination of 9:253
   manufacture 9:252
   properties 9:252
Ceric ammonium sulfate–arsenious acid
   colorimetric method for iodide 14:602
Ceric hydroxide 9:254
   analysis 9:255
   cerium content of 9:255

commercial grades and specifications 9:255
impurities, determination of 9:255
uses 9:254
Ceric nitrate
   colorimetric method for halohydrins 14:157
   color reagent for antioxidants 6:92
   test for alcohols 4:522
Ceric oxide. (See also Cerium oxide.) 9:254
   analysis 9:255
   cerium content of 9:255
   in cerium stannate 9:263
   in cerium titanate 9:261
   in cerium zirconate 9:261
   commercial grades and specifications 9:254
   emission spectroscopy 9:256, 258
   in glass and glass-ceramics 13:415, 442
   impurities, determination of 9:255
   manufacture 9:254
   potassium permanganate titration 9:261
   potassium persulfate–silver sulfate colorimetric
      method 13:442
   sodium peroxide fusion 9:257
   uses 9:254
Ceric sulfate 9:253
   titration of crystal violet 19:175
   titration of erioglaucine O 19:175
   titration of eriogreen B 19:175
   titration of ethyl violet 19:175
   titration of magenta 19:175
   titration of setocyanine supra 19:175
   titration of setoglaucine O 19:175
   titration of uranium 19:246
   titration of xylenecyanol FF 19:175
   uses 9:253
   volumetric method for aniline 5:445
   volumetric method for antimony 6:64
   volumetric method for copper 10:611
   volumetric method for ferrous gluconate 10:540
   volumetric method for hydroquinone 17:47
   volumetric method for iron 12:523, 526; 15:38
   volumetric method for menadiol sodium diphos-
      phate 10:204
   volumetric method for menadione 10:198
   volumetric method for menadione sodium bi-
      sulfite 10:201
   volumetric method for sodium chlorate 9:353
   volumetric method for tellurium 17:602
Ceric sulfate–nitric acid
   colorimetric method for diphenyl 11:526
   colorimetric method for terphenyls 11:526
Ceric titration
   of hydrogen peroxide 14:433
Cerium. (See also Cerium alloys; Cerium metal;
      and *Rare earths.*) 9:246–277
   absorption spectroscopy 17:483
   in aluminum and aluminum alloys 5:125
   atomic absorption spectroscopy 17:494

borax bead test 6:584
in ceric ammonium nitrate 9:251
in ceric ammonium sulfate 9:253
in ceric hydroxide 9:255
in ceric oxide 9:255
in cerium metal 9:247
colorimetric methods 9:274; 15:381
emission spectroscopy 9:275; 15:153, 155,
   375; 17:488, 501
in europium oxide 17:502
ferric ammonium sulfate titration 9:271
fluoride precipitation 9:268, 269
in glass 9:200, 260
gravimetric methods 9:272
hydroxide precipitation 9:266
8-hydroxyquinoline colorimetric method 9:275
8-hydroxyquinoline extraction 9:270
identification 9:264
iodate gravimetric method 9:272
ion exchange chromatography 15:157
in iron 15:59
in lanthanum oxide 15:152, 153, 155; 17:501
in magnesium 15:375, 380
in magnesium alloys 15:375, 380
microcosmic salt bead test 6:587
in Misch metal 9:247
oxalate precipitation 9:264
o-phenanthroline–ferrous sulfate colorimetric
   method 274
potassium periodate colorimetric method 15:
   152
in praseodymium oxide 17:501
in samarium oxide 17:502
separation 9:264
thenoyltrifluoroacetone colorimetric method 9:
   275
thenoyltrifluoroacetone extraction 9:269
titration of uranium 13:274
in uranium 19:238
ultraviolet spectroscopy 9:273
volumetric methods 9:270
x-ray fluorescent spectroscopy 17:490, 505
in yttrium oxide 17:501, 505
Cerium alloys. (See also Misch metal.) 9:246
analysis 9:246
uses 9:247
Cerium compounds. (See also individual compounds.) 9:250
in ceramics 9:260
emission spectroscopy 9:256
in glass 9:260
Cerium earth group. (See also *Rare earths.*) 17:
   470
Cerium fluorides
carrier gas distillation fusion methods 8:636
oxygen in 8:636
Cerium hydroxide. (See also Ceric hydroxide;

Rare earth hydroxides.)
properties 17:510
Cerium metal. (See also Rare earth metals.)
analysis 9:246
carrier gas fusion methods 8:627, 682
cerium content of 9:247
impurities, determination of 9:249
manufacture 9:246
nitrogen in 8:623, 627
oxygen in 8:623, 627, 637, 682
properties 9:247; 17:471
rare earth content of 9:248
sample preparation 8:641
uses 17:470
vacuum fusion methods 8:623
Cerium naphthenate 9:253
analysis 9:253
Cerium nitrate
colorimetric method for thiophene 18:537
Cerium oxide. (See also Ceric oxide; Rare earth
   oxides.)
emission spectroscopy 17:501
in glass 9:200
impurities, determination of 17:501
properties 17:497, 509
Cerium perchlorate
ultraviolet spectrum 9:273
Cerium stannate
analysis 9:263
in ceramic dielectric materials 9:260
Cerium sulfate
colorimetric method for antimony 13:393
Cerium titanate
analysis 9:261
in ceramic dielectric materials 9:260
Cerium zirconate
analysis 9:261
in ceramic dielectric materials 9:260
Cerous chloride 9:259
analysis 9:259
identification 9:259
impurities, determination of 9:260
manufacture 9:259
Cerous nitrate
volumetric method for phosphoric acid 14:367
Cerrolow. (See also Indium alloys.)
composition 14:547
properties 14:547
uses 14:547
Cerro-seal. (See also Indium alloys.)
composition 14:547
properties 14:547
uses 14:547
Certified colors. (See also Colors; and individual
   compounds.) 10:462–529
analysis 10:471
assay 10:484

classification 10:462
commercial specifications 10:462
D & C colors 10:452, 462, 465
electrophoresis 10:482
Ext D & C colors 10:460, 462, 470
F D & C colors 10:448, 462, 463
gravimetric determination 10:487
identification 10:471
impurities, determination of 10:489
liquid chromatography 10:524
nitrogen content, organic 10:488
paper chromatography 10:475
separation 10:473
sulfur content, organic 10:488
titanous chloride titration 10:484
Cerussite. (See also Lead carbonate.) 15:166
Cervantite 6:39
Ceryl alcohol. (See also Fatty alcohols.)
  properties 4:561
Cesium. (See also Cesium metal.) 9:278–303
in alkali salts 15:267
ammonium sulfate gravimetric method 17:554
in cesium alum 9:292
in cesium bromate 9:284
in cesium bromide 9:285
in cesium carbonate 9:285
in cesium chlorate 9:287
in cesium chromate 9:288
in cesium dichromate 9:288
in cesium metal 9:282, 284
chloroplatinate gravimetric method 9:297;
  17:553
emission spectroscopy 9:301
flame emission spectroscopy 13:452; 17:552
flame photometry 9:284, 294, 300; 17:555
flame tests 6:592
gravimetric determination of 9:296
identification 9:296
iodometric method 17:555
ion exchange chromatography 9:300; 17:366,
  550
neutron activation analysis 9:300
perchlorate gravimetric method 9:298; 15:267
periodate gravimetric method 9:299
9-phosphomolybdic acid gravimetric method
  17:553
in pollucite ore 9:293
in polymers 4:24
in rubidium 17:555
separation 9:296
silicomolybdic acid gravimetric method 17:549
silicotungstic acid gravimetric method 17:549,
  553
sodium bismuth nitrite gravimetric method 17:
  549
sodium tetraphenylboron volumetric method 9:
  298

sulfate gravimetric method 9:297
volumetric determination 9:298
x-ray emission spectroscopy 9:300, 301
Cesium alum. (See also Cesium sulfate.) 9:292
analysis 9:292
cesium content 9:292
flame photometry 9:293
impurities, determination of 9:293
manufacture 9:292
properties 9:292
solubility 9:292
Cesium aluminum silicate. See Pollucite.
Cesium aluminum sulfate. See Cesium alum.
Cesium–bismuth nitrite
gravimetric method for sodium 18:229
Cesium bromate 9:284
analysis 9:284
bromate content 9:285
cesium content 9:284
manufacture 9:284
properties 9:284
Cesium bromide 9:285
analysis 9:285
bromide content 9:285
cesium content 9:285
manufacture 9:285
properties 9:285
uses 9:285
Cesium carbonate 9:286
analysis 9:286
carbonate content 9:286
cesium content 9:285
manufacture 9:286
properties 9:286
uses 9:286
Cesium chlorate 9:287
analysis 9:287
cesium content 9:287
chlorate content 9:287
manufacture 9:287
properties 9:287
Cesium chloride 9:287
analysis 9:287
commercial specifications 9:288
manufacture 9:287
properties 9:287, 334
solubility 9:336
uses 9:287
Cesium chromate 9:288
analysis 9:288
cesium content 9:288
chromate content 9:288
manufacture 9:288
properties 9:288
uses 9:288
Cesium compounds. (See also individual compounds.) 9:284–296

analysis 9:284
solubility 9:284
Cesium cyanide. (See also Cyanides.) 14:424
properties 14:424
solubility 14:424
Cesium dichromate 9:288
analysis 9:288
cesium content 9:288
dichromate content 9:288
manufacture 9:288
properties 9:288
uses 9:288
Cesium fluoride 9:288
analysis 9:289
fluoride content 9:289
manufacture 9:288
properties 9:288; 13:11
uses 9:289
Cesium hydroxide 9:289
analysis 9:289
impurities, determination of 9:290
manufacture 9:289
properties 9:289
uses 9:289
Cesium iodate 9:290
analysis 9:290
iodate content 9:290
manufacture 9:290
properties 9:290
Cesium iodide 9:290
analysis 9:290
manufacture 9:290
properties 9:290
Cesium metal 9:278–284
analysis 9:282
emission spectroscopy 9:281
flame photometry 9:283
impurities, determination of 9:283
isotopes 9:278
manufacture 9:280
occurrence 9:278
properties 9:278
sample preparation 9:282
uses 9:278
x-ray spectroscopy 9:281
Cesium nitrate 9:290
analysis 9:291
manufacture 9:291
nitrate content 9:291
properties 9:290
uses 9:291
Cesium ores. (See also Pollucite.) 9:246
Cesium oxide
in glass and glass-ceramics 13:415, 452
Cesium sulfate 9:291
analysis 9:292
impurities, determination of 9:293

manufacture 9:291
properties 9:291
sulfate content 9:292
uses 9:292
Cetane
analysis 14:334
gas chromatography 14:334
n-Cetane. (See also Alkanes.) 14:306
in cetane 14:334
commercial specifications 14:306
gas chromatography 14:330, 334
uses 14:306
Cetyl alcohol. (See also Fatty alcohols.)
commercial grades 4:563
properties 4:561
specifications 4:564
uses 4:562
Cetyl iodide. (See also Iodine compounds, organic.)
properties 14:612
Cetylpyridinium chloride. (See also Quaternary ammonium compounds.)
assay 11:557
commercial specifications 11:555
in disinfectants and antiseptics 11:555
properties 11:555
sodium tetraphenylboron titration 11:557
turbidimetric method for sodium heparin 10:149
ultraviolet spectroscopy 11:556
Cetyltrimethylammonium borohydride 7:344
Chabazite 8:75
Chain length
of polyphosphates 17:139
Chalcocite
copper content 10:600
Chalcogenides
acid dissolution 9:215
fluxes for 9:216
Chalcone
absorption spectroscopy 9:564
Chalcopyrite 14:519; 16:285; 18:360
copper content 10:600
Chalkone derivatives
colorimetric method for boric acid 7:387
Chamosite
composition 15:22
Channel carbon 8:179
commercial grades and specifications 8:196
composition 8:191
manufacture 8:193
particle size distribution 8:184
Charcoal. (See also Carbon, activated.)
in explosive priming mixtures 12:431
Charcoal adsorption
of gasoline 13:320
Charcoal filters
for cigarettes 10:3
Charring ablators 4:15

oxidation 4:19
quality control methods 4:22
thermogravimetry 4:18
transpiration 4:20
Chavicol. See 4-Allylphenol.
Cheelox Iron 12
  cupric nitrate titration 12:35
Cheese 16:139
  analysis 16:140
  annatto content 10:544
  carotene content 10:544
  commercial specifications 16:140
  grading 16:140
  impurities, determination of 16:142
  manufacture 16:139
  milkfat content 16:141
  moisture content 16:141
  potassium content 17:363
  salt content 16:142
  sampling 16:140
  specification tests 16:140
Chel 138. (See also *EDTA and related com-
    pounds.*)
  cupric nitrate titration 12:34
  structure 12:24
Chelating agents. (See also *EDTA and related
    compounds*; and individual compounds.)
  in beer 7:584
Chel CD. See DCYTA.
Chel ME. (See also *EDTA and related compounds.*)
  structure 12:24
Chelometric titration
  of chromium 9:640
  of glass and glass-ceramics 13:444
Chemical additives
  in wines 19:477
Chemical oxygen demand
  of water 19:375
Chemigum N 12:86
Chemstrand method for identification of acrylic
    and modacrylic fibers
  by solubility 4:249
  by staining tests 4:255
Chenodeoxycholic acid
  color tests 7:147
  fluorometry 7:152
  preparation 7:146
  properties 7:143
  structure 7:142
Cheralite
  composition 19:2
  thorium content 19:2
Cherries
  dibromochloropropane content 13:185
Chessylite 10:657
Chicago acid. See 4-Amino-5-hydroxy-1,3-
    naphthalenedisulfonic acid.

Chiddy method
  gravimetric method for gold 14:37
Chile saltpeter. (See also Sodium nitrate.) 18:207
Chill haze
  in beer 7:520, 527
China. (See also *Ceramics.*)
  composition 9:194
China clay. (See also Aluminum silicate.) 4:33
  commercial grades and specifications 17:191
  commercial specifications 5:189
  in elastomers synthetic 12:152
  infrared spectrum 12:152
China tea. (See also Black tea.) 18:455
Chinese blue. See Iron blues.
Chinomethionate. (See also *Pesticides and insecti-
    cides.*)
  commercial grades and specifications 19:594
  composition 19:591, 599
  manufacture 19:599
  properties 19:591
  solubility 19:587
  structure 19:585
  toxicity 19:592
Chi square test 7:165
Chloanthite
  composition 16:286
Chloral
  in chloroform 9:434
  Fujiwara colorimetric method 9:434
  infrared spectroscopy 6:616
  properties 8:334
Chloral hydrate
  Fujiwara colorimetric method 9:386
  properties 8:334
2-Chlorallyl alcohol
  in epichlorohydrin 14:175
  gas chromatography 14:175
Chloramine-T
  colorimetric method for bromide 9:350; 14:370
  colorimetric method for thiodiphenylamine 11:
    458
Chloramphenicol 5:461, 493
  analysis 5:496, 501, 506
  bioautography 5:496
  chlorine content 5:506
  colorimetric methods 5:508
  column chromatography 5:496
  countercurrent distribution 5:496
  identification 5:496
  infrared spectrum 5:497
  ion exchange separation 5:496
  microbiological assay 5:501
  paper chromatography 5:496
  properties 5:493
  separation 5:496
  solubility 5:494
  structure 5:461

thin-layer chromatography 5:496
titration methods 5:506
ultraviolet spectroscopy 5:507
uses 5:493
Chloramphenicol palmitate
infrared spectrum 5:497
solubility 5:494
Chloranil 13:200
analysis 13:200
colorimetric method for antioxidants 6:97
diphenyl-*p*-phenylenediamine colorimetric
method 13:200
in food crops 13:200
manufacture 13:200
properties 13:200
solubility 13:200
toxicity 13:200
use in preparation of dioxazine dyes 6:385
uses 13:200
Chloranilic acid
colorimetric method for calcium 8:108
Chlorapatite
manufacture 17:121
Chlorates. (See also individual compounds.)
9:357–362
in ammonium persulfate 5:347
analysis 9:359
brucine sulfate colorimetric method 9:348
in cesium chlorate 9:287
in cotton 11:141
diphenylamine–sulfuric acid color test 12:428
ferrous ammonium sulfate–ceric sulfate method
9:360
identification 9:359
manganous sulfate–diphenylcarbazide colorimetric
method 11:141
in potassium chlorate 12:420, 423
in potassium chloride 9:339
in potassium iodate 14:609
in potassium permanganate 15:492
silver nitrate–formaldehyde gravimetric method
12:423
in sodium chlorate 9:360
in sodium chloride 9:337, 348
sulfur dioxide reduction 9:287
Chlorcyclizine
gas chromatography 6:210, 211
Chlorcyclizine hydrochloride. (See also *Antihistamines.*) 6:26
properties 6:32
Chlordane. (See also *Pesticides and insecticides.*)
commercial grades and specifications 19:593
composition 19:588, 596
gas chromatography 19:389
manufacture 19:596
properties 19:588
solubility 19:587

structure 19:580
toxicity 19:592
in water 19:389
Chlorendic acid
in alkyd resins 5:36
infrared spectroscopy 5:46
2-Chlorethanol. See Ethylene chlorohydrin.
Chlorex. See Dichloroethyl ether.
Chlorhexidine. (See also Guanidine derivatives.)
in disinfectants and antiseptics 11:569
sodium hypobromite colorimetric method 11:
570
Chloride, hydrolyzable
in isocyanic acid esters 15:119
in tolylene diisocyanate 15:119
Chlorides. (See also *Chlorine*; *Chlorine compounds,
inorganic*; and individual compounds.) 9:336–
351
in acetic acid 4:100
in acetone 4:114
in acetophenone 4:125
in acid anhydrides 8:567
acid–base titration 18:79, 88, 124
in acid salt 8:593
in alcohols 4:545
in aluminum alkyls 5:171
in aluminum chloride 5:174
in aluminum hydroxide 5:188
in aluminum sulfate 5:193
in alums 5:193
in amino acids 8:438
in ammonium acetate 5:325
in ammonium bromide 12:542
in ammonium carbonate 5:332
in ammonium iodide 12:544; 14:604
in ammonium nitrate 5:342, 343
in ammonium oxalate 5:345
in ammonium persulfate 5:347
in ammonium phosphate, dibasic 5:350
in ammonium sulfate 5:352
in ammonium thiocyanate 5:354
in amosite 6:267
analysis 9:340
in antimony trioxide 6:53
in arsenic(III) oxide 6:231
in arsenic(V) oxide 6:233
in asbestos 6:265, 267
atomic absorption methods 14:128
in barium acetate 6:536
in barium carbonate 6:544
in barium chromate 6:547
in barium hydroxide 6:550, 551
in barium nitrate 6:552
in barium peroxide 6:554
in barium sulfate 6:567
in beer 7:581
in bismuth nitrate 7:203

in bismuth oxide 7:205
in bismuth subcarbonate 7:196
in bismuth subnitrate 7:203
in borax 7:383
in boric acid 7:381
in cadmium acetate 8:37
in cadmium bromide 8:38
in cadmium iodide 8:44
in cadmium nitrate 8:45
in cadmium sulfate 8:47
in calcium acetate 8:78
in calcium carbonate 8:84
in carbonyl compounds 8:362
in cereals 9:242
in ceric ammonium nitrate 9:251
chlorine content 9:340
in chloroform 5:359; 9:434
in chromium compounds 9:686, 692, 698
in chrysotile 6:265
in citric acid 10:93
in cobalt(II) acetate 10:329
in cobalt carbonate 10:333
in cobalt nitrate 10:340
in cobalt(II) sulfate 10:343, 344
combustion methods 18:308
in copper(II) acetate 10:654
in copper(II) bromide 10:656
in copper(II) carbonate, basic 10:658
in copper(II) nitrate 10:668
in copper(I) oxide 10:672
in cotton 11:141
coulometric titration 15:142
in creams and lotions 11:92
in crocidolite 6:267
in cuprous oxide 17:196
in diammonium citrate 10:93
in epoxy resin adhesives 4:401
in ethyl chloride 5:391
in ethyleneamines 11:429
in ethylene oxide 12:336
Fajans titration 9:342
in fluorocarbons 13:69
in fluorosilicic acid 13:50
in formic acid 13:129
in glass 9:200
in glass and glass-ceramics 13:413, 415, 435
in glycerol 13:525
in gypsum and gypsum products 14:96
in hafnium 14:127
in hydriodic acid 14:603
in hydrobromic acid 8:11
in hydrochloric acid 14:366
hydrofluoric acid titration 14:127
in hydrogen peroxide 14:436
8-hydroxyquinoline–silver nitrate gravimetric
    method 11:141
impurities, determination of 9:343

in iodides 14:597
in iodine 14:592
ion exchange method 19:497
in isocyanic acid esters 15:120
in lactic acid and its derivatives 15:142
in lithium 15:275
in lithium bromide 15:282
in lithium carbonate 15:279
in lithium hydroxide 15:286
in magnesium acetate 15:398
in magnesium carbonate 15:401
in magnesium nitrate 15:407
in magnesium oxide 15:410
in magnesium perchloriate 9:364, 366
in magnesium sulfate 15:412
in malonic acid 15:431
in manganese sulfate 15:490
mercuric acetate titration 13:69
mercuric nitrate titration 7:581
in mercuric oxide 16:61
mercurimetric titration 9:342
in mercury fulminate 12:410
in methanol 16:113
in methylcellulose 9:80
in methylene chloride 9:430
Mohr titration 9:342
neutron activation method 12:568
in nickel sulfate 16:344
in nicotinic acid 17:439
in nitric acid 16:498
in organo-chlorosilanes 18:63
in organofunctional silanes 18:87
in organosilane esters 18:78
in perchloric acid 9:364, 365
in phenyl mercuric acetate 16:63
in potassium chlorate 9:359, 361
in potassium hydroxide 18:257
in potassium iodate 14:608
in potassium iodide 12:545; 14:605
in potassium permanganate 15:492
in pyridine 17:438
in pyrrocaine hydrochloride 5:413
in rare earth chlorides 17:506
in silicone resins 18:124
silver chloride gravimetric method 9:341
silver chloride turbidimetric method 10:93, 329,
    654, 656, 658
in silver nitrate 18:175
silver nitrate gravimetric method 9:242; 10:672;
    11:92; 15:282
silver nitrate spectroscopic methods 14:127
silver nitrate titration 19:364, 497, 531
silver nitrate turbidimetric method 8:438; 9:251,
    361, 365; 12:337, 410, 542, 544; 13:33, 129,
    525; 14:436, 592, 604; 15:142, 275, 398, 401,
    407, 410, 490; 16:61, 63, 113, 344, 498, 536;
    17:438; 18:175, 233, 283; 19:19, 222

silver nitrate volumetric method 6:434; 7:582;
　　8:88; 9:686; 11:47, 429; 12:338; 13:436, 525;
　　14:96; 15:120, 122, 279, 431; 16:183; 18:88,
　　203
　in soaps 18:203
　in sodium arsenate 6:234
　in sodium arsenite 6:235
　in sodium bismuthate 7:207
　in sodium chlorate 9:359, 360
　in sodium fluoride 13:33
　in sodium hydroxide 18:257
　in sodium iodide 14:605
　in sodium metal 18:233
　in sodium peroxide 16:536
　in sodium tungstate 19:222
　in starch hydrolysates 11:47
　in strontium chromate 9:707; 17:200
　in strontium nitrate 18:283
　in styrene 18:308
　in thorium nitrate 19:19
　in titanium 19:120
　in 1,1,1-trichloroethane 9:444
　in trichloroethylene 5:367
　Volhard titration 6:265; 9:341; 18:257
　in water 19:364
　in wines 19:496
　x-ray methods 14:128
　in yeast 19:531
　in zinc chromates 9:707; 17:199
　in zinc powder 17:204
　in zirconium 14:127
Chloride titrator, automatic 9:417
Chlorinated benzenes. (See also *Chlorocarbons*;
　　and individual compounds.) 9:475–484; 13:
　　214–222
Chlorinated biphenyls. (See also *Chlorocarbons*.)
　　9:491–494
　commercial grades and specifications 9:493
　in enamels 12:176
　gas chromatography 9:494
　manufacture 9:494
　properties 9:492; 12:176
　toxicity 9:492
　uses 9:491
Chlorinated camphene. See Toxaphene.
Chlorinated compounds
　in benzaldehydes 6:623
　in benzoic acid 7:66
　in benzoic acid esters 7:84
　in caffeine–sodium benzoate powder 8:64
　in carbonyl compounds 8:362
　in cinnamyl alcohol 10:84
　flame photometry 10:84
　in sodium benzoate 7:87
Chlorinated hydantoins
　available chlorine 7:228
　bleaching agents 7:222

Chlorinated hydrocarbon insecticides. (See also
　　Chlorinated hydrocarbon pesticides.)
　in tobacco 19:155
Chlorinated hydrocarbon pesticides. (See also
　　individual compounds.) 19:580
　colorimetric determination 19:611
　commercial grades and specifications 19:593
　manufacture 19:596
　properties 19:588
　solubility 19:587
　toxicity 19:592
Chlorinated hydrocarbons. (See also *Chloro-*
　　carbons.) 9:369
Chlorinated isocyanuric acids
　available chlorine 7:228
　bleaching agents 7:222
　water in 7:232
Chlorinated naphthalenes. (See also *Chlorocarbons*.)
　　9:488–491; 16:220
　analysis 9:491; 16:220
　chlorine content 16:220
　dimethylaniline colorimetric method 9:491
　manufacture 9:490; 16:220
　properties 9:489
　solubility 9:489
　toxicity 9:490
　uses 9:489
Chlorinated *p*-nitroaniline
　composition 12:173
　in enamels 12:173
Chlorinated paraffins. (See also *Chlorocarbons*.)
　　9:494–501
　acid number 9:500
　analysis 9:498
　chlorine content 9:498
　commercial grades and specifications 9:497
　impurities, determination of 9:498
　manufacture 9:497
　properties 9:496
　sodium peroxide bomb ignition 9:498
　softening point 9:501
　solubility 9:496
　specification tests 9:498
　specific gravity 9:499
　toxicity 9:497
　uses 9:496
Chlorinated para red. (See also *Pigments*.)
　structure 17:178
Chlorinated polyethylene. (See also *Ethylene and*
　　propylene polymers.)
　infrared spectrum 12:367
　manufacture 12:346
Chlorinated polypropylene. (See also *Ethylene*
　　and propylene polymers.)
　infrared spectrum 12:367
　manufacture 12:346
Chlorinated propanes. (See also *Chlorocarbons*;

and individual compounds.) 9:465–468
solubility 9:468
Chlorinated solvents
  degreasers 6:355
  gas chromatography 6:367
Chlorinated toluenes. (See also *Chlorocarbons*;
    and individual compounds.) 9:484–488
  analysis 9:487
  gas chromatography 9:487
  infrared spectroscopy 9:487
Chlorinated trisodium phosphate
  available chlorine 7:224, 228
Chlorination–hydrolysis process
  manufacture of amyl alcohols 4:506
Chlorination methods
  for platinum group metals 17:245
Chlorine. (See also Chlorine, elemental.) 9:304–
    333
  in air 9:331
  arsenite absorption 9:317
  automatic chloride titrator 9:417
  bleaching agents 7:221, 223
  in 1,3-butadiene 11:504
  Carius combustion method 9:411
  in chloramphenicol 5:506
  in chlorides 9:340
  in chlorinated naphthalenes 16:220
  in chlorinated paraffins 9:498
  in chlorine dioxide absorption tower solution
    7:227
  in chlorocarbons 9:409, 410, 414, 416
  in chloroform 9:434
  in coal and coke 10:253
  combustion methods 9:410
  in cotton 11:154, 157
  coulometric titration 9:417
  in decaffeinated coffee 10:441
  in elastomers, synthetic 12:111
  in epichlorohydrin 12:227
  in epoxy resin adhesives 4:401
  in epoxy resins 12:250
  Eschka method 10:253
  in formaldehyde 13:114
  gas chromatography 13:64; 14:392, 416
  in halohydrins 14:157
  in hydrochloric acid 14:366, 370
  in hydrogen chloride 14:395
  in hypochlorite bleaches 9:354
  identification 9:340
  iodometric methods 14:370
  iodometric titration 9:330; 11:355; 14:370, 395
  ion exchange chromatography 12:251
  in isocyanates 19:277
  in lubricants 15:327
  mercuric nitrate volumetric method 12:111; 14:
    157
  mercuric perchlorate titration 12:251

mercury absorption 9:314
in molybdenum chlorides 16:182
in molybdenum oxychlorides 16:182
neutron activation analysis 14:155
orthotolidine dihydrochloride colorimetric
  method 19:377
orthotolidine–sodium arsenite colorimetric
  method 19:377
oxygen flask ignition 9:415
in pesticides and insecticides 19:608
redox reactions 9:333
in resins 4:24
in roasted coffee 10:424
silver chloride gravimetric method 9:412
silver chloride turbidimetric method 9:419
silver nitrate gravimetric method 10:424, 442;
  16:222
silver nitrate titration 12:250; 19:154
sodium arsenite titration 9:317
sodium peroxide bomb ignition 9:414
in tobacco 19:154
in vinyl chloride and vinylidene chloride polymers 4:280
in vinyl ether 5:368
Volhard titration 9:413, 418
in water 9:330; 19:376
x-ray fluorescence spectroscopy 11:157
x-ray spectrography 9:409
Chlorine–chloroform
  color test for iodides 14:596
Chlorine compounds
  in potassium hydrogen phthalate 17:403
  redox reactions 9:333
  silver nitrate turbidimetric method 17:403
Chlorine compounds, inorganic. (See also Chlo-
    rates; Chlorides; Chlorites; Hypochlorites;
    Perchlorates; and individual compounds.) 9:
    333–368
Chlorine-containing bleaches
  analysis 7:225
  available chlorine 7:223, 225
  identification 7:225
  iodometric titration 7:225
Chlorine demand
  of water 19:378
Chlorine dioxide 9:355
  absorption tower solution 7:227
  analysis 9:356
  available chlorine 7:224
  bleaching agents 7:222, 223
  in chlorine dioxide bleaches 9:356
  properties 9:356
  uses 9:355
Chlorine dioxide absorption tower solution
  analysis 7:227
Chlorine dioxide bleaches 9:356
  analysis 9:356

chlorine dioxide content 9:356
  iodometric titration 9:356
  sodium chlorite content 9:356
  sodium hypochlorite content 9:356
Chlorine, elemental 9:304–333
  analysis 9:309
  assay 9:314
  chlorine content 9:314
  commercial grades and specifications 9:309
  density 9:307
  gas chromatography 9:320
  history 9:304
  identification 9:314
  impurities, determination of 9:314, 320
  infrared spectroscopy 9:321
  isotopes 9:304
  manufacture 9:304, 308
  properties 9:305
  sampling 9:310
  solubility 9:307
  toxicity 9:308
  uses 9:304
  vapor pressure 9:307
Chlorine, free
  in carbon tetrachloride 8:327
  in carbonyl compounds 8:362
  in cellulose 9:51
  in chlorinated paraffins 9:498
  in chloroform 5:359
  in hydrochloric acid 9:340, 351
  silver nitrate titration 9:52
Chlorine germicidal equivalence test
  for disinfectants and antiseptics 11:539
Chlorine isotope peaks
  mass spectroscopy 9:395
Chlorine monofluoride
  manufacture 13:9
  properties 13:8
Chlorine number
  of pulp 9:52
Chlorine, residual
  in water 19:377
Chlorine–starch
  color test for iodides 14:596
Chlorine trifluoride
  gas chromatography 13:64
  manufacture 13:9
  properties 13:8
  toxicity 13:13
  uses 13:7
Chlorine values
  of water 19:376
Chlorine water 9:351
1-Chlor-6-isocyanato-*n*-hexane. (See also *Iso-cyanic acid esters.*)
  infrared spectroscopy 15:106
Chlorite 11:464; 15:358
Chlormadinone acetate. (See also Steroids.) 10:578

analysis 10:579, 594
  assay 10:579, 594
  gas chromatography 10:577
  identification 10:579
  manufacture 14:203
  in mestranol 10:577
  picric acid–sodium hydroxide colorimetric
    method 10:595
  properties 10:578; 14:203
  solubility 10:578; 14:203
  structure 14:202
  sulfuric acid odor test 10:579
  thin-layer chromatography 10:579
  ultraviolet spectroscopy 10:579, 594
  uses 14:203
p-Chlor-*o*-nitraniline red
  in enamels 12:185
  identification 12:185
Chloroacetamide
  derivatives for identification 8:601
p-Chloroacetanilide
  gas chromatography 6:187
  in phenacetin 6:196
Chloroacetic acid
  properties 8:396
Chloroacetone
  properties 8:335
Chloroacetyl chloride
  properties 8:570
Chloroacetylene. (See also C_2 Chlorinated hydro-carbons.)
  mass spectroscopy 9:397
  properties 9:466
Chloroacetyl halides
  derivatives for identification 8:572
β-Chloroallyl alcohol
  bromination method 5:83
3-Chloroallylbenzene
  mass spectroscopy 9:403
γ-Chloroallyl chloride. See 1,3-Dichloropropene.
2-Chloroallyl diethyldithiocarbamate. See Vegadex.
2-Chloroallylidene diacetate 5:106
Chloro-4-aminobenzoic acid
  in chloroprocaine hydrochloride 5:383
2-Chloro-5-aminotoluene-4-sulfonic acid. See
  Lake Red C amine.
3-Chloro-androst-5-ene-17-one
  liquid chromatography 14:209
m-Chloroaniline
  acid–base titration 5:443
  azo dyes for identification 5:430
  color reactions for identification 5:431
  infrared spectrum 5:440
  near-infrared spectroscopy 5:435
  properties 5:423
  ultraviolet spectroscopy 5:433
  volumetric method for acid chlorides 7:76
o-Chloroaniline

acid–base titration 5:443
azo dyes for identification 5:430
color reactions for identification 5:431
derivatives for identification 5:429
gas chromatography 5:438
infrared spectroscopy 5:440, 454
near-infrared spectroscopy 5:435
properties 5:423
ultraviolet spectroscopy 5:433
p-Chloroaniline
  acid–base titration 5:444, 447
  azo dye colorimetric method 5:445
  azo dyes for identification 5:430
  color reactions for identification 5:431
  derivatives for identification 5:429
  infrared spectroscopy 5:440, 454
  near-infrared spectroscopy 5:435
  properties 5:423
  ultraviolet spectroscopy 5:433
Chloroanilines
  gas chromatography 5:441
  manufacture 5:426
  uses 5:428
Chloroauric acid
  gravimetric method for pyridine 17:434
Chlorobenzaldehyde
  derivatives for identification 8:343
  properties 8:334
2-Chlorobenzaldehyde 6:596
  commercial specifications 6:607, 608
  derivatives for identification 6:612
  infrared spectroscopy 6:616
  properties 6:598
3-Chlorobenzaldehyde 6:596
  commercial specifications 6:607
  derivatives for identification 6:612
  infrared spectroscopy 6:617
  properties 6:598
4-Chlorobenzaldehyde 6:596
  commercial specifications 6:607, 608
  derivatives for identification 6:612
  infrared spectroscopy 6:617
  properties 6:598
Chlorobenzaldehydes
  manufacture 6:600
  uses 6:606
Chlorobenzamide
  derivatives for identification 8:601
Chlorobenzene 9:476
  analysis 9:478
  azeotropes 9:476
  Beilstein detection 9:382
  commercial grades and specifications 9:477
  composition 9:411
  distillation range 9:478
  gas chromatography 9:478; 14:375
  impurities, determination of 9:478

infrared spectroscopy 9:394
manufacture 9:477
mass spectroscopy 9:403
properties 9:476
solubility 9:477; 13:539, 552
specification tests 9:478
toxicity 9:477
uses 9:476
p-Chlorobenzenesulfonyl isocyanate. (See also
  Isocyanic acid esters.)
  commercial sources 15:102
Chlorobenzoic acids 8:400
  paper chromatography 8:379, 400
  properties 8:400, 402
p-Chlorobenzoyl chloride
  infrared spectrum 8:573
Chlorobenzyl chlorides
  gas chromatography 9:378, 488
2-Chlorobiphenyl
  mass spectroscopy 9:407
4-Chlorobiphenyl
  mass spectroscopy 9:407
Chlorobiphenyls. (See also Chlorinated biphenyls.)
  gas chromatography 9:495
  properties 9:492
1-Chloro-3-bromo-1-propene. See Bromochloropropene.
1-Chloro-1,3-butadiene. (See also C_4 Chlorinated
  hydrocarbons.)
  infrared spectroscopy 9:390
2-Chloro-1,3-butadiene. (See also Chloroprene.)
  gas chromatography 9:375
  infrared spectroscopy 9:390
  mass spectroscopy 9:396
1-Chlorobutane. (See also C_4 Chlorinated hydrocarbons.) 9:474
  azeotropes 9:474
  composition 9:411
  gas chromatography 9:376, 474, 475
  infrared spectroscopy 9:390
  manufacture 9:474
  mass spectroscopy 9:397
  properties 9:473
  sodium nitrite colorimetric method 9:388
  toxicity 9:474
  uses 9:474
2-Chlorobutane. (See also C_4 Chlorinated hydrocarbons.)
  gas chromatography 9:474, 475
  infrared spectroscopy 9:390
  mass spectroscopy 9:396
  properties 9:473
  sodium nitrite colorimetric method 9:388
1-Chloro-2-butanol. (See also Halohydrins.)
  derivatives for identification 14:160
2-Chloro-1-butanol. (See also Halohydrins.)
  derivatives for identification 14:160

1-Chloro-1-butene. (See also C$_4$ Chlorinated
  hydrocarbons.)
  gas chromatography 9:375
  infrared spectroscopy 9:390
3-Chloro-1-butene. (See also C$_4$ Chlorinated
  hydrocarbons.)
  gas chromatography 9:375
  infrared spectroscopy 9:390
1-Chloro-2-butene. (See also C$_4$ Chlorinated
  hydrocarbons.)
  gas chromatography 9:375
  infrared spectroscopy 9:390
2-Chloro-2-butene. (See also C$_4$ Chlorinated
  hydrocarbons.)
  gas chromatography 9:375
  infrared spectroscopy 9:390
  mass spectroscopy 9:397
Chlorobutyric acid
  properties 8:396
Chlorocarbonate. See Ethyl chloroformate.
Chlorocarbons. (See also C$_1$, C$_2$, C$_3$, and C$_4$
    Chlorinated hydrocarbons; Chlorinated
    benzenes; Chlorinated toluenes; Chlorinated
    naphthalenes; Chlorinated biphenyls; Chlori-
    nated paraffins; and individual compounds.)
    9:369–510
  analysis 9:373
  Beilstein detection 9:381
  Carius combustion method 9:411
  chlorine content 9:409, 410
  combustion methods 9:410
  flame photometry 9:382
  Fujiwara colorimetric method 9:384
  gas chromatography 9:374
  history 9:369
  impurities, determination of 9:409
  infrared spectroscopy 9:389
  manufacture 9:371
  mass spectroscopy 9:395, 408
  oxygen flask ignition 9:415
  properties 9:369
  sodium biphenyl hydrolysis 9:421
  sodium–ethanolamine hydrolysis 9:419
  sodium hydrolysis method 15:327
  sodium–liquid ammonia hydrolysis 9:418
  sodium peroxide bomb ignition 9:414
  toxicity 9:370
  uses 9:370
  x-ray spectroscopy 9:409
Chlorocinnamic acid. (See also *Cinnamic acid
    and its derivatives.*)
  properties 10:63
p-Chloro-*m*-cresol
  in embalming fluids 12:165
β-Chlorocumene
  mass spectroscopy 9:402
Chlorocyclohexane

mass spectroscopy 9:398
Chlorocyclopentane
  mass spectroscopy 9:398
1-Chlorodecane
  mass spectroscopy 9:401
  sodium nitrite colorimetric method 9:388
p-Chloro-*β*,*β*-dichlorocumene
  mass spectroscopy 9:404
1-Chloro-1,1-difluoroethane
  gas chromatography 11:98
Chlorodifluoromethane. (See also Chlorofluoro-
    carbons.)
  commercial specifications 13:63
  gas chromatography 13:64
  properties 13:61
4-Chloro-6-(2,4-dihydroxyphenylazo)-1-hydroxy-
    benzene-2-sulfonic acid. See Lumogallion.
4-Chloro-3,5-dimethylphenol. (See also *Chloro-
    phenols.*)
  manufacture 9:512
  properties 9:511
  uses 9:511
1-Chloro-2,4-dinitrobenzene
  gravimetric method for morphine 16:267
  colorimetric method for pyridine 17:424
1-Chlorododecane
  mass spectroscopy 9:401
1-Chloro-2,3-epoxypropane. See Epichlorohydrin.
Chloroethane. (See also Ethyl chloride.)
  mass spectroscopy 9:398
2-Chloro-1-ethanol. (See also Ethylene chloro-
    hydrin.) 5:1
α-Chloroethylbenzene
  mass spectroscopy 9:402
β-Chloroethylbenzene
  mass spectroscopy 9:400
1-Chloroethylbenzene
  mass spectroscopy 9:402
2-Chloroethylbenzene
  mass spectroscopy 9:401
p-Chloroethylbenzene
  mass spectroscopy 9:404
α-Chloroethyldimethylbenzene
  mass spectroscopy 9:404
Chloroethylene. See Vinyl chloride.
1-Chloroethyl ethyl ether. (See also *Ethers.*)
  properties 12:297
2-Chloroethyl ethyl ether. (See also *Ethers.*)
  properties 12:297
1-Chloro-2-ethylhexane
  mass spectroscopy 9:397
Chlorofluorocarbons
  analysis 13:62
  gas chromatography 13:62
  manufacture 13:62
  nomenclature 13:59
  properties 13:59

uses 13:58
Chlorofluoromethane. (See also Chlorofluorocarbons.)
properties 13:61
Chloroform. (See also *Chlorocarbons.*) 9:430;
13:182
analysis 5:356; 9:433
anesthetic 5:356
appearance 9:434
azeotropes 9:432
Beilstein detection 9:382
in carbon tetrachloride 8:325
in chlorine 9:321
colorimetric method for naphthylamines 5:274
commercial grades and specifications 9:432
composition 9:411
2,7-dihydroxynaphthalene colorimetric method
9:387
Fujiwara colorimetric method 9:385, 386, 433
gas chromatography 8:325; 9:375, 376, 377,
380, 381, 433, 442; 13:64; 14:375
in hydrochloric acid 9:392
impurities, determination of 5:358; 9:433, 434
infrared spectroscopy 9:390, 392, 394, 433
liquid chromatography 7:15
manufacture 8:323; 9:432
mass spectroscopy 9:399
α-naphthol colorimetric method 9:388
odor 9:434
properties 5:356, 357; 9:431
solubility 5:358; 9:431
specification tests 9:434
toxicity 9:431; 13:182
in 1,1,1-trichloroethane 9:442
uses 9:430; 13:182
Chloroformate groups
in polycarbonates 17:340
sodium hydroxide–silver nitrate volumetric
method 17:340
Chloroform, insoluble
in antimony halides 6:51
Chloroform miscibility
of acid halides 8:576
Chlorogenic acid. (See also *Cinnamic acid and its
derivatives.*)
absorption spectroscopy 10:64
Arnow's colorimetric method 19:146
Folin's colorimetric method 19:147
gas chromatography 10:401; 19:146
in green coffee 10:374, 396, 398, 413
in green tea leaf 18:462, 475
iodometric titration 10:413
lead acetate gravimetric method 10:398
liquid chromatography 10:400; 18:476
occurrence 10:63, 69
paper chromatography 19:146
in roasted coffee 10:375, 430

solvent extraction 19:146
structure 18:462
thin-layer chromatography 18:463, 475
in tobacco 19:146
ultraviolet spectroscopy 10:69, 398, 399, 400
1-Chloroheptane
mass spectroscopy 9:400
sodium nitrite colorimetric method 9:388
2-Chloroheptane
mass spectroscopy 9:397
sodium nitrite colorimetric method 9:388
3-Chloroheptane
sodium nitrite colorimetric method 9:388
4-Chloroheptane
sodium nitrite colorimetric method 9:388
1-Chlorohexadecane
mass spectroscopy 9:397
1-Chlorohexane
mass spectroscopy 9:400
sodium nitrite colorimetric method
9:388
Chlorohydrin
in halohydrin 14:155
potassium hydroxide–silver nitrate volumetric
method 14:156
Chlorohydrocarbons
gas chromatography 8:325
2-Chloro-5-hydroxybenzaldehyde
near-infrared spectroscopy 6:629
5-Chloro-2-hydroxybenzophenone
thin-layer chromatography 16:96
ultraviolet spectroscopy 16:95
6-Chloro-17-hydroxypregna-4,6-diene-3,20-dione
acetate. See Chlormadinone acetate.
1-Chloroindane
mass spectroscopy 9:403
β-Chloroisopropylbenzene
mass spectroscopy 9:402
Chloromandelic acid
gravimetric method for zirconium 5:152
p-Chloromercuribenzoic acid
colorimetric method for cysteine 15:561
colorimetric method for 2,3-dimercapto-1-
propanol 15:561
colorimetric method for glutathione 15:561
colorimetric method for mercaptans 15:561
Chloromethane. See Methyl chloride.
Chloromethylbenzene
mass spectroscopy 9:400
1-Chloro-3-methylbutane
gas chromatography 9:376
mass spectroscopy 9:396
2-Chloro-2-methylbutane
gas chromatography 9:376
mass spectroscopy 9:399
α-Chloromethylethylbenzene
mass spectroscopy 9:403

Chloromethyl-2-methylbenzene
  mass spectroscopy 9:402
Chloromethyl-3-methylbenzene
  mass spectroscopy 9:402
Chloromethyl-4-methylbenzene
  mass spectroscopy 9:402
Chloromethylnaphthalene
  mass spectroscopy 9:407
1-Chloromethylnaphthalene
  mass spectroscopy 9:405
3-Chloro-3-methylpentane
  mass spectroscopy 9:396
4-Chloro-2-methylphenol. (See also *Chloro-phenols.*)
  manufacture 9:512
  properties 9:511
  uses 9:511
1-Chloro-2-methylpropane. (See also Isobutyl chloride.)
  mass spectroscopy 9:396
1-Chloro-3-methylpropane. (See also C$_4$ Chlorinated hydrocarbons.)
  gas chromatography 9:376
2-Chloro-2-methylpropane. (See also C$_4$ Chlorinated hydrocarbons.)
  gas chromatography 9:475
  mass spectroscopy 9:397
3-Chloro-2-methylpropene. (See also C$_4$ Chlorinated hydrocarbons.)
  gas chromatography 9:376
  mass spectroscopy 9:396
  properties 9:473
α-Chloromethyl-2,4,6-triisopropylbenzene
  mass spectroscopy 9:408
1-Chloronaphthalene
  mass spectroscopy 9:406
2-Chloronaphthalene
  mass spectroscopy 9:406
Chloronaphthalenes. (See also Chlorinated naphthalenes.)
  properties 9:489
Chloroneb 13:218
  analysis 13:218
  in animal tissue 13:218
  gas chromatography 13:218
  in plant tissue 13:218
  properties 13:218
  solubility 13:218
  solvent extraction 13:218
  toxicity 13:218
  uses 13:218
Chloronitroalkanes. (See also individual compounds.) 13:186–190
2-Chloro-nitrobenzene–diazonium naphthalene sulfonate
  colorimetric method for cresols 17:20
1-Chloro-1-nitroethane

  gas chromatography 16:432
1-(*o*-Chloro-*p*-nitrophenylazo)-2-naphthol. See D & C Red No. 36.
1-Chloro-1-nitropropane
  gas chromatography 16:432
1-Chloro-2-nitropropane 13:186
  analysis 13:186
  in crops 13:186
  gas chromatography 13:186
  manufacture 13:186
  properties 13:186
  solubility 13:186
  toxicity 13:186
  uses 13:186
2-Chloro-1-nitropropane
  gas chromatography 16:432
2-Chloro-2-nitropropane
  gas chromatography 16:427, 432
3-Chloro-1-nitropropane
  gas chromatography 16:432
1-Chlorononane
  mass spectroscopy 9:401
1-Chlorooctadecane
  mass spectroscopy 9:397
  sodium nitrite colorimetric method 9:388
1-Chlorooctane
  mass spectroscopy 9:401
  sodium nitrite colorimetric method 9:388
2-Chlorooctane
  sodium nitrite colorimetric method 9:388
3-Chlorooctane
  mass spectroscopy 9:397
Chloropentafluoroethane. (See also Chlorofluorocarbons.)
  commercial specifications 13:63
  properties 13:61
1-Chloropentane 9:376
  gas chromatography 9:376
  mass spectroscopy 9:396, 399
  sodium nitrite colorimetric method 9:388
2-Chloropentane
  mass spectroscopy 9:399
  sodium nitrite colorimetric method 9:388
3-Chloropentane
  mass spectroscopy 9:399
m-Chloroperbenzoic acid
  volumetric method for hydrocarbon resins 14:288
p-Chlorophenacyl derivative
  of benzoic acid 7:65
m-Chlorophenol. (See also *Phenols.*)
  gas chromatography 17:36
o-Chlorophenol. (See also *Phenols.*)
  colorimetric determination 17:30
  gas chromatography 17:36
p-Chlorophenol. (See also *Phenols.*)
  colorimetric determination 17:30

commercial specifications 11:540
gas chromatography 11:544; 17:36
Chlorophenols. (See also individual compounds.)
    9:511–528
  analysis 9:513
  bromination titration 9:514
  color 9:526
  commercial grades and specifications 9:513
  distillation range 9:526
  freezing point 9:526
  gas chromatography 9:518; 19:386
  identification 9:526
  infrared spectroscopy 9:515
  ion exchange chromatography 9:521
  manufacture 9:512
  moisture content 9:527
  properties 9:511
  solubility 9:512
  specification tests 9:526
  specific gravity 9:527
  toxicity 9:512
  trimethylsilyl derivatives 9:520
  ultraviolet spectroscopy 9:521
  uses 9:511
p-Chlorophenylacetylene
  mass spectroscopy 9:404
4-*p*-Chlorophenylethynylpyridine. (See also
    Pyridine and pyridine derivatives.)
  ionization constants 17:416
Chlorophenyl isocyanate. (See also *Isocyanic
    acid esters.*)
  commercial sources 15:103
  infrared spectroscopy 15:106
  near-infrared spectroscopy 15:105
  properties 15:95
4-Chloro-2-phenylphenol. (See also *Phenols.*)
  gas chromatography 11:544
6-Chloro-2-phenylphenol. (See also *Phenols.*)
  gas chromatography 11:544
p-Chlorophenyl phenyl sulfone. See Sulphenone.
p-Chlorophenyl 2,4,5-trichlorophenyl sulfone. See
    Tetradifon.
Chlorophyll
  in green tea leaf 18:478
  identification 10:537
  phaeophytin colorimetric method 18:478
  potassium hydroxide colorimetric method 10:
    538
  solvent extraction 18:478
Chloropicrin 13:187
  analysis 13:187
  gas chromatography 8:13; 13:183; 16:427
  infrared spectroscopy 8:13
  in methyl bromide 8:13
  N-(1-naphthyl)ethylenediamine hydrochloride
    colorimetric method 13:187
  properties 13:187

  solubility 13:187
  toxicity 13:187
  uses 13:187
Chloroplatinate
  derivative of amines 5:267
  gravimetric method for cesium 9:297
Chloroplatinic acid 17:273
  assay 17:273
  commercial specifications 17:273
  electrolytic methods 17:273
  gravimetric method for cesium 9:298; 17:553
  gravimetric method for potassium 17:367, 372,
    384, 386, 388, 405; 18:190
  gravimetric method for potassium oxide 13:424
  gravimetric method for rubidium 17:553
  gravimetric method for thallium 18:526
  impurities, determination of 17:273
  manufacture 17:227, 407
  potassium determination, suitability for 17:273
  sodium content 18:229
  solubility 17:273
  specification tests 17:273
  uses 17:273
Chloroprene. (See also 2-Chloro-1,3-butadiene.)
  gas chromatography 9:375
  properties 9:473
Chloroprene dimer
  mass spectroscopy 9:400
Chloroprene polymers. See Polychloroprenes.
Chloroprocaine hydrochloride 5:375
  analysis 5:383
  properties 5:382
Chloroprocaine penicillin O. (See also Penicillin
    O.)
  iodometric methods 16:562
  properties 16:556
  solubility 16:557
1-Chloropropane. (See also Chlorinated propanes.)
  gas chromatography 9:376
  infrared spectroscopy 9:390, 394
  mass spectroscopy 9:396
  properties 9:471
  sodium nitrite colorimetric method 9:388
2-Chloropropane. (See also Isopropyl chloride.)
    9:465
  azeotropes 9:467
  gas chromatography 9:376
  infrared spectroscopy 9:390
  manufacture 9:468
  mass spectroscopy 9:396
  properties 9:467
  sodium nitrite colorimetric method 9:388
  solubility 9:468
  toxicity 9:468
  uses 9:465
2-Chloro-1,3-propanediol. (See also Glycerol 2-
    monochlorohydrin.)

properties 14:168
solubility 14:168
3-Chloro-1,2-propanediol. (See also Glycerol
   1-monochlorohydrin.)
   derivatives for identification 14:160
   infrared spectrum 14:160
   properties 14:167, 168
   solubility 14:167
   structure 14:167
1-Chloro-2-propanol. (See also Propylene
   chlorohydrin.)
   azeotropes 14:166
   commercial grades and specifications 14:166
   derivatives for identification 14:160
   infrared spectrum 14:159
   manufacture 14:165
   properties 14:165
2-Chloro-1-propanol. (See also Propylene
   chlorohydrins.)
   azeotropes 14:166
   derivatives for identification 14:160
   manufacture 14:165
   properties 14:165
3-Chloro-1-propanol. (See also Trimethylene
   chlorohydrin.)
   analysis 14:167
   infrared spectrum 14:159
   properties 14:167
1-Chloropropene. (See also C_3 Chlorinated
   hydrocarbons.)
   gas chromatography 9:376
   infrared spectroscopy 9:390, 393, 394
   mass spectroscopy 9:396
   properties 9:471
2-Chloropropene. (See also C_3 Chlorinated
   hydrocarbons.)
   in allyl chloride 5:102
   gas chromatography 5:102; 9:376
   infrared spectroscopy 9:390, 393, 394
3-Chloropropene (See also Allyl chloride.)
   gas chromatography 9:376
   infrared spectroscopy 9:390, 393
   mass spectroscopy 9:396
3-Chloropropenylbenzene
   mass spectroscopy 9:403
Chloropropionic acid
   properties 8:396
2-Chloro-n-propylbenzene
   mass spectroscopy 9:401
3-Chloropropylbenzene
   mass spectroscopy 9:401
β-Chloro-n-propylbenzene
   mass spectroscopy 9:401
3-Chloropropylene oxide. See Epichlorohydrin.
Chloropropyltrimethoxysilane. (See also Organofunctional silanes.)
   commercial specifications 18:82

3-Chloropropyne
   mass spectroscopy 9:396
Chloropyridines. (See also *Pyridine and pyridine
   derivatives.*)
   ionization constants 17:416
   manufacture 17:422
Chloroquin
   gas chromatography 6:210, 211
o-Chloro-p-red
   in enamels 12:182, 185
   identification 12:182, 185
Chlorosilanes. (See also Organochlorosilanes; and
   individual compounds.) 18:48–53
   analysis 18:49
   commercial grades and specifications 18:49
   composition, determination of 18:49
   gas chromatography 18:50
   identification 18:49
   impurities, determination of 18:51
   infrared spectroscopy 18:49
   manufacture 18:49
   mass spectroscopy 18:50
   properties 18:40, 48
   safety 18:48, 49
   uses 18:49
Chlorosilanes, organic. See Organo-chlorosilanes.
Chlorostannous acid
   colorimetric method for phosphorus 10:301
β-Chlorostyrene
   mass spectroscopy 9:401
p-Chlorostyrene
   mass spectroscopy 9:405
4-p-Chlorostyrylpyridine. (See also *Pyridine and
   pyridine derivatives.*)
   ionization constants 17:416
N-Chlorosuccinimide
   spot test for ferrocyanide 15:89
Chlorosulfonated ethylene–propylene copolymers.
   (See also *Ethylene and propylene polymers.*)
   12:344
   properties 12:344
Chlorosulfonated polyethylene. (See also *Elastomers,
   synthetic; Ethylene and propylene polymers;*
   and Hypalon.) 12:84
   commercial elastomers 12:86
   composition 12:85
   density 12:91
   identification 12:90
   infrared spectroscopy 12:99, 368
   manufacture 12:346
   properties 12:85
   pyrolysis–gas chromatography 12:107
   solubility 12:85
   uses 12:85
Chlorosulfonic acid. 9:528–534
   acid–base titration 9:530
   analysis 9:530

commercial grades and specifications 9:529
hydrogen chloride content 9:533
identification 9:530
impurities, determination of 9:533
properties 9:529
sampling 9:530
silver nitrate titration 9:530
sulfur trioxide content 9:531
uses 9:528
4-(4-Chloro-2-sulfo-5-tolylazo)-1-naphthol
in D & C Red No. 8 and No. 9 10:521
Chlorosulfuric acid. See *Chlorosulfonic acid.*
1-Chlorotetradecane
mass spectroscopy 9:397
1,1,1,2,2-Chlorotetrafluoroethane. (See also
Chlorofluorocarbons.)
properties 13:61
Chlorothen citrate. (See also *Antihistamines.*)
6:26
properties 6:32
Chlorothene
gas chromatography 11:73
in nail enamels 11:73
Chlorothiodicarboximides 13:209–212
analysis 13:210
Chlorothymol. (See also *Phenols.*)
commercial specifications 11:540
α-Chlorotoluene. See Benzyl chloride.
m-Chlorotoluene. See Monochlorotoluene.
o-Chlorotoluene. See Monochlorotoluene.
p-Chlorotoluene. See Monochlorotoluene.
2,1,1,1-Chlorotrifluoroethane. (See also Chlorofluorocarbons.)
properties 13:61
Chlorotrifluoroethylene
gas chromatography 12:64
uses 15:302
1-Chloro-1,2,2-trifluoroethylene
gas chromatography 11:98
Chlorotrifluoroethylene–vinylidene fluoride copolymer 13:94
properties 13:94
uses 13:94
Chlorotrifluoromethane. (See also Chlorofluorocarbons.)
gas chromatography 13:64
properties 13:61
1-Chloro-2,4,6-trinitrobenzene
colorimetric method for cyclohexylamine 11:
228
α-Chloro-*m*-xylene
mass spectroscopy 9:402
α-Chloro-*o*-xylene
mass spectroscopy 9:402
α-Chloro-*p*-xylene
mass spectroscopy 9:402
2-Chloro-*m*-xylene

mass spectroscopy 9:402
2-Chloro-*p*-xylene
mass spectroscopy 9:402
4-Chloro-*m*-xylene
mass spectroscopy 9:402
Chlorpheniramine
gas chromatography 6:210
Chlorpheniramine maleate. (See also *Antihistamines.*) 6:26
cyanogen bromide method 6:35
properties 6:32
thin-layer chromatography 6:35
Chlorpromazine
gas chromatography 6:210, 211
Chlorprophenpyridamine maleate
column chromatography 6:150
ultraviolet spectroscopy 6:150
Chocolate. (See also Cacao; Cocoa; Cocoa beans;
Cocoa products; Chocolate liquor; *Chocolate
and cocoa products*; and *Confectionery products.*) 9:534; 10:548
analysis 9:572
caloric content 9:540
in cocoa beans 9:555
in cocoa products 9:590
commercial grades and specifications 10:550
composition 9:540, 550
in confectionery products 10:550
manufacture 9:541, 545; 10:554
milk fat content 9:580
milk protein content 9:580
nutrient content 9:538
oxalic acid content 9:541
particle size 10:564
particle size distribution 9:571
salt content 9:537
sample preparation 9:572
sugar content 9:540, 545
vanillin content 9:537, 545
viscosity 9:572
Chocolate and cocoa products. (See also Cacao;
Cocoa; Cocoa beans; Cocoa products; Chocolate;
Chocolate liquor; and *Confectionery products.*)
9:534–611
analysis 9:550
composition 9:537
manufacture 9:541
microbiology 9:598
nutrient content 9:538
organoleptic factors 9:536
Chocolate candy
nutrient content 9:538
Chocolate coatings
commercial grades and specifications 10:550
specifications 10:550
viscosity 10:556
Chocolate-enrobed products. (See also

Confectionery products.) 10:548
Chocolate liquor. (See also Cocoa; Cocoa beans;
    Cocoa products; Chocolate; *Chocolate and
    cocoa products.*) 9:544
  carbonyl content 9:562
  in cocoa beans 9:555
  cocoa butter content 9:545
  starch content 9:585
  theobromine content 9:588
  uses 9:545
  wet screening methods 9:569
Chocolate milk. (See also Milk.)
  milkfat content 16:132
Chocolate syrup
  nutrient content 9:538
Cholanic acid 7:141
  structure 7:142
Cholesterol. (See also Steroids.) 7:141
  carbon and hydrogen analysis 8:290, 291
  in egg yolk 12:63
  gas chromatography 19:150
  liquid chromatography 14:209
  structure 14:192
  in tobacco 19:150
  Zlatkis' reagent color test 12:64
Cholic acid 7:141
  acid–base titration 7:149
  color tests 7:146, 147
  commercial specifications 7:145
  fluorometry 7:152
  identification 7:146
  Pettenkofer color reaction 7:146
  preparation 7:144
  properties 7:143
  structure 7:142
  uses 7:144
  visible spectroscopy 7:151
Choline. (See also *Choline and its derivatives.*)
    9:611
  assay 9:620
  barium hydroxide hydrolysis 9:620
  in green coffee 10:411
  ion exchange chromatography 9:617, 618
  methanol extraction 9:620
  paper chromatography 9:615
  thin-layer chromatography 9:616
  uses 9:611
Choline and its derivatives. (See also Choline;
    and individual compounds.) 9:611–632
  acid–base titration, nonaqueous 9:621
  analysis 9:614
  assay 9:620
  impurities, determination of 9:626
  ion exchange chromatography 9:617
  Lim and Schall method 9:629
  manufacture 9:614
  microbiological methods 9:624

  physiological methods 9:626
  reineckate colorimetric method 9:622
  solubility 9:613
  tetraphenylboron gravimetric method 9:623
  thin-layer chromatography 9:616
Choline ascorbate. (See also *Choline and its
    derivatives.*)
  properties 9:613
  uses 9:611
Choline base. (See also *Choline and its derivatives.*)
    9:611
  analysis 9:614
  assay 9:620
  commercial grades and specifications 9:614
  manufacture 9:614
  properties 9:612
  reineckate gravimetric method 9:620
  uses 9:611
Choline bicarbonate. (See also *Choline and its
    derivatives.*)
  properties 9:613
  uses 9:612
Choline bitartrate. (See also *Choline and its
    derivatives.*)
  commercial grades and specifications 9:615
  properties 9:613
  uses 9:611
Choline borate. (See also *Choline and its derivatives.*)
  properties 9:613
  uses 9:611
Choline bromide. (See also *Choline and its
    derivatives.*)
  commercial grades and specifications 9:615
  properties 9:613
  uses 9:611
Choline chloride. (See also *Choline and its
    derivatives.*) 9:611
  assay 9:621
  cadmium chloride gravimetric method 9:623
  commercial grades and specifications 9:615
  dipicrylamine colorimetric method 9:617
  Dragendorff's colorimetric method 9:617
  infrared spectra 9:612
  iodine colorimetric method 9:617
  iodoplatinate colorimetric method 9:617
  properties 9:612
  thin-layer chromatography 9:616
  toxicity 9:613
  uses 9:611
  x-ray crystallography 9:613
Choline dihydrogen citrate. (See also *Choline and
    its derivatives.*)
  assay 9:621
  commercial grades and specifications 9:615
  properties 9:613
  uses 9:611
Choline gluconate. (See also

Choline and its derivatives.)
  properties 9:613
  uses 9:611
Choline iodide. (See also *Choline and its
    derivatives.*)
  commercial grades and specifications 9:615
  properties 9:613
  uses 9:611
Choline theophyllinate. (See also *Choline and
    its derivatives.*)
  properties 9:613
  solubility 9:613
  uses 9:611
Chorionic gonadotropin. (See also Pituitary
    hormones.)
  occurrence 14:187
  uses 14:187
Chow and Beamish method,
  for gold ores 14:10
Chromated zinc chloride. (See also Zinc com-
    pounds.)
  uses 19:573
Chromate pigments. (See also individual com-
    pounds.) 17:197
Chromates
  in alkaline flushes 6:359
  in antirusts 6:351, 358
  in cesium chromate 9:288
  gravimetric method for barium 6:536, 553, 556,
    576
  gravimetric method for silver 18:164
  precipitation method for chromium 9:636
  solvent extraction 19:330
  in vanadium 19:330
  volumetric method using ferrous ammonium
    sulfate 6:358
  in zinc chromates 17:199
Chromaticity chart 4:324
Chromatographic suitability
  of aluminum oxide 5:188
Chromatography. See Column chromatography;
    Gas chromatography; Ion exchange chroma-
    tography; Liquid chromatography; Paper chro-
    matography; and Thin-layer chromatography.
Chrome alum
  aluminum content 12:584
  chromium content 12:584
  in photographic films 12:584
Chrome cake 18:244
Chrome cleaners 6:353
  analysis 6:364
  performance tests 6:375
Chrome green
  commercial grades and specifications 17:200
  composition 17:200
  in enamels 12:183
  extender content 17:200

  identification 12:183
  impurities, determination of 17:200
  lead chromate content 17:200
Chromel 5:72
Chromel alloys. (See also Nickel alloys.)
  composition 16:290
Chrome lignosulfonates. (See also Chromium
    compounds.)
  analysis 9:689
Chrome orange. (See also Chromium pigments.)
  commercial grades and specifications 9:706;
    17:198
  commercial specifications 12:173
  composition 12:173
  in enamels 12:173
  uses 9:705; 17:197
Chrome ores 9:700
  acid dissolution 9:215
  aluminum content 9:701, 704
  analysis 9:700
  chromium content 9:700
  classification 9:700
  commercial grades and specifications 9:701
  composition 9:203
  impurities, determination of 9:704
  iron content 9:701, 703
  magnesium content 9:701, 704
  properties 9:204
  sample preparation 9:700
  silicon content 9:701, 703
  sodium peroxide fusion 9:700
Chrome oxide green. (See also Chromium pig-
    ments.)
  chromium content 9:705
  commercial grades and specifications 9:706
  uses 9:705
Chrome protectors 6:354
  analysis 6:364
  performance tests 6:376
Chrome refractories. (See also Refractories.)
  composition 9:208
  fluxes for 9:216
  properties 9:208
Chrome steels 9:662
  classification 9:662
  commercial grades and specifications 9:662
  composition 9:662
  properties 9:664
  tensile strength 9:664
  uses 9:662
Chrome tanning compounds. (See also Chromium
    compounds.) 9:688
  aluminum content 9:690
  analysis 9:689
  basicity 9:690
  manufacture 9:688
Chrome yellow. (See also Chromium

pigments; Lead chromate.)
 analysis 9:705
 chromium content 9:705
 commercial grades and specifications 9:706;
  17:197, 198
 commercial specifications 12:173
 composition 12:173
 in enamels 12:173, 183
 identification 12:183
 lead content 9:705
 uses 9:705; 17:197
Chromic acetate
 commercial specifications 9:696
 properties 9:696
Chromic acetylacetonate
 commercial specifications 9:696
 properties 9:696
Chromic acid. (See also Chromium compounds.)
  9:681
 analysis 9:683
 assay 9:684, 693
 chromium content 9:684
 colorimetric method for *p*-phenetidine 6:195
 commercial grades and specifications 9:683, 695
 extraction of bromide 9:347
 ferrous sulfate titration 9:685
 iodometric titration 9:684
 impurities 9:684
 oxidation of methylhydroxylcellulose 9:81
 properties 9:682
 uses 9:681
 volumetric method for polyisoprene 12:138
Chromic chloride
 analysis 9:693
 assay 9:693, 696
 commercial grades and specifications 9:694
 properties 9:334
 solubility 9:336, 696
Chromic fluoride
 commercial specifications 9:696
 solubility 9:696
Chromic formate
 commercial specifications 9:696
 solubility 9:696
Chromic nitrate
 analysis 9:693
 assay 9:693
 commercial grades and specifications 9:694
Chromic oxide. (See also Chromium compounds.)
 analysis 9:689
 chromium content 9:690
 diphenyl carbazide colorimetric method 13:440
 ferrous sulfate titration 9:703
 in glass 9:200
 in glass and glass-ceramics 13:413, 415, 440,
  461
 impurities, determination of 9:692

 peroxide fusion 9:690
 in refractories 9:206
Chromic oxide green. See Chrome oxide green.
Chromic sulfate
 assay 9:693
 commercial grades and specifications 9:694
Chromite. (See also Chrome ores.) 9:633, 700
 composition 9:700
Chromium. (See also Chromium alloys; Chromium
  metal.) 9:632–709
 acid dichromate colorimetric method 9:647
 adsorption chromatography 9:639
 alkaline chromate colorimetric method 9:646
 in alloys 9:652
 in alumina 5:116
 in aluminum alloys 9:654
 in aluminum and aluminum alloys 5:125
 in aluminum oxide 5:187
 in aluminum oxide abrasives 4:36, 47
 in ammonium dichromate 9:684
 amperometric titration 9:655
 in asbestos 6:266, 267
 atomic absorption spectroscopy 9:346, 649;
  14:125
 barium chromate gravimetric method 9:639
 in bauxite 5:116
 bead tests 12:14
 in beer 7:657
 benzidine colorimetric method 11:140, 144
 in beryllium metal 7:115
 in biological materials 9:651
 borax bead test 6:584
 in bronze 9:648
 in carbon 14:74
 in cements 9:131
 chelometric titration 9:640
 chromate precipitation method 9:636
 in chrome alum 12:584
 in chrome oxide green 9:705
 in chrome steels 9:662
 in chromic acid 9:684
 in chromic oxide 9:690
 in chromium compounds 9:688, 689, 692
 in chromium ores 9:633, 700
 in chromium pigments 9:705
 chromyl chloride gravimetric method 10:352
 chromyl chloride volatilization 9:664
 in chrysotile 6:266, 267
 in cobalt alloys 10:310, 352
 in cobalt metal 10:293, 297
 colorimetric method using diphenylcarbazide
  5:126; 7:115, 657
 in constant-modulus alloys 10:290
 in copper alloys 9:654; 10:636
 in copper metal 10:620
 in cotton 11:140, 152, 154, 156
 coulometric methods 9:656

in cutting alloys 10:289
in dental alloys 10:290
2,7-diaminodiphenylene oxide colorimetric
    method 11:140, 144
diphenylcarbazide colorimetric method 9:636,
    642, 643, 644, 645, 646; 10:195, 297, 621;
    11:140; 14:129; 15:381; 16:326
EDTA colorimetric method 9:648
EDTA titration 9:638, 641; 13:445
in electrical resistance alloys 10:290
emission spectroscopy 9:346, 653; 11:152;
    13:277; 14:48, 74, 122, 542; 15:155, 187;
    16:160, 387
in F D & C Blue No. 1 10:494
in ferrochromium 9:661; 15:79
in ferroniobium 16:387
ferrous ammonium sulfate–potassium dichromate
    titration 9:640
ferrous ammonium sulfate–potassium permanga-
    nate titration 9:655; 10:636; 15:79
in ferrous metals 9:650
flame photometry 9:652
in gallium arsenide 13:276
in glass and glass-ceramics 13:445
in gold 14:45
in graphite 14:74
gravimetric determination 9:636, 639
in hafnium 14:122, 124, 129
in hard facing alloys 10:289
in high-speed steels 10:289
in high-temperature alloys 10:286
in hydrochloric acid 14:375
hydrogen peroxide colorimetric method 19:108
8-hydroxyquinoline method 9:639
identification test 6:391; 9:635
in indium 14:542
iodometric titration 9:641
ion exchange chromatography 9:637; 13:255
in iron 9:645, 647, 650; 15:59
in iron ore 9:644
in lanthanum oxide 15:155
in lead pigments 9:705
in lead silico chromate 9:705
in low-expansion alloys 10:290
in magnesium 15:381
in magnesium alloys 15:381
in menadione 10:195
in metal complex dyes 12:14
methyl isobutyl ketone extraction 9:652
microcosmic salt bead test 6:587
in molybdenum 16:161
in molybdenum trioxide 16:161
neutron activation analysis 13:265
in nichrome 9:663
in nickel 16:326
in nickel alloys 16:326
in niobium 16:387

oxidation–reduction titration 9:640
perchloric acid gravimetric method 9:654
perchloric acid–hydrochloric acid gravimetric
    method 16:298
persulfate method 9:689
in petroleum 9:655
in phosphate rocks 17:71
polarographic methods 9:653
in potassium dichromate 9:684
potassium permanganate titration 10:310
potentiometric titration 5:125
in rocks 15:187
separation 9:636
in silica–alumina samples 9:656
in silicate rocks 9:643, 646, 648, 653
in silver 18:173
sodium bicarbonate gravimetric method 9:647
in sodium chloride 9:346
in sodium chromate 9:684
in sodium dichromate 9:684
sodium hydroxide gravimetric method 9:688
sodium peroxide fusion 9:637, 653
in soft magnetic alloys 10:288
spectroscopic methods 9:641
in spring steels 10:290
in steel 9:645, 647, 648, 650, 653, 655; 19:207
in strontium chromate 9:705; 17:199
in surgical alloys 10:290
in terephthalic acid 17:162
in titanium 19:121
visible spectroscopy 10:494
volumetric determination 9:640
in wear-resistant alloys 10:289
wet peroxide method 9:689
x-ray fluorescence spectroscopy 11:154, 156;
    17:162
in zinc chromate 9:705
in zirconium 14:122, 124, 129
Chromium alloys. (See also Chrome steels; Ferro-
    chrome; and Ferrochromium.) 5:61; 9:660–
    680
analysis 9:663
carrier gas fusion methods 8:624
composition 9:661, 662
impurities, determination of 9:673
oxygen in 8:624
sample preparation 9:663
uses 9:662
vanadium content 19:339
Chromium ammonium sulfate. (See also Chromium
    compounds.)
analysis 9:689
Chromium–cobalt alloys, dental. See Dental
    chromium–cobalt alloys.
Chromium complexes. (See also Chromium com-
    pounds.) 9:681
analysis 9:689

Chromium compounds. (See also individual
  compounds.) 9:680–709
 analysis 9:684, 688, 693, 705
 chromium content 9:688, 692
 impurities, determination of 9:686, 691, 693,
  705
 manufacture 9:681, 688
 properties 9:680
 uses 9:681, 688, 693, 705
Chromium metal 9:632–660
 aluminothermic process 9:634
 analysis 9:634, 657
 atomic absorption spectroscopy 9:636
 carrier gas fusion methods 8:627, 682
 commercial grades and specifications 9:635
 elasticity 9:635
 electrolytic process 9:634
 emission spectroscopy 9:635
 history 9:632
 hydrogen in 8:618, 680
 impurities, determination of 9:636
 manufacture 9:633
 nitrogen in 8:618, 623, 627
 occurrence 9:633
 oxygen in 8:616, 618, 619, 620, 621, 622, 623,
  627, 629, 632, 633, 680, 682
 properties 9:633
 silicothermic process 9:634
 solubility 9:634
 solubility in gold 14:9
 tensile properties 9:635
 uses 9:632
 vacuum fusion methods 8:616, 618, 619, 620,
  621, 622, 623, 680
Chromium–molybdenum–aluminum steel
 aluminum in 5:248
Chromium–nickel–iron castings. See Chrome
  steel.
Chromium ores. See Chrome ores.
Chromium oxide–pyridine complex test
 for alcohols 4:522
Chromium oxides
 carrier gas fusion methods 8:626
 commercial specifications 12:173; 17:201
 in enamels 12:173
 oxygen in 8:626
Chromium pigments. (See also individual compounds.) 9:705
 analysis 9:705
 chromium content 9:705
 impurities, determination of 9:707
 uses 9:705
Chromium plating solutions
 sulfate content 18:427
Chromium potassium sulfate. (See also Chromium
  compounds.)
 analysis 9:689, 693

assay 9:693
 commercial grades and specifications 9:694
Chromium sesquioxide
 in ceramics 9:224
 microscopy 9:224
Chromium steel
 nitrogen in 8:613
Chromium trioxide. (See also Chromic acid.)
 colorimetric method for cocaine hydrochloride
  16:246
Chromium–vanadium steel. (See also Vanadium
  alloys.)
 decomposition 19:327
Chromomycin A₃ 5:461, 493
 analysis 5:497, 501, 507
 bioautography 5:498
 column chromatography 5:498
 countercurrent distribution 5:498
 fluorometry 5:507
 identification 5:497
 microbiological assay 5:501
 paper chromatography 5:497
 properties 5:493
 separation 5:497
 solubility 5:494
 structure 5:462
 uses 5:493
Chromomycins
 separation 5:498
Chromotrope 2B. (See also p-Nitrobenzene azochromotropic acid.)
 colorimetric method for boron 7:387,
  410
Chromotrope 2R
 in D & C Red No. 33 10:523
 liquid chromatography 10:523
Chromotropic acid
 colorimetric method for formaldehyde 11:134;
  12:585; 13:106, 107
 colorimetric method for methanol 4:488, 550;
  15:146; 19:437
 colorimetric method for niobium 16:369, 374
 colorimetric method for nitro alcohols 16:443
 colorimetric method for 2-nitro-1-butanol 16:
  444
 colorimetric method for 2-nitro-2-ethyl-1,3-
  propanediol 16:444
 colorimetric method for 2-nitro-2-methyl-1,3-
  propanediol 16:444
 colorimetric method for 2-nitro-2-methyl-1-
  propanol 16:444
 colorimetric method for penicillins 16:560
 colorimetric method for serine 8:523
 colorimetric method for tantalum 16:369; 18:
  449
 colorimetric method for titanium 19:98, 107,
  115

colorimetric method for tris(hydroxymethyl)-
  nitromethane 16:444
color test for formic acid 13:125
color test for HMX 12:460
color test for RDX 12:421, 460
identification reagent for formaldehyde 8:344
method for formals 4:68, 71
Chromous chloride
  commercial specifications 9:696
  properties 9:334
  solubility 9:336, 696
  uses 9:693
Chromous reduction method
  nitrogen content of ammonium nitrate 5:341
Chromyl chloride
  commercial specifications 9:697
  gravimetric method for chromium 10:352
  volatilization of chromium 9:664
Chronopotentiometry
  of aniline and its derivatives 5:454
Chrysoberyl 7:118
Chrysocolla
  copper content 10:600
Chrysotile
  alkali in 6:266
  aluminum in 6:263, 267
  ammonium hydroxide group, determination of
    6:261
  analysis 6:258
  in asbestos–cement products 6:270
  calcium in 6:263, 267
  carbonate in 6:265
  carbon dioxide in 6:265
  chloride in 6:265
  chromium in 6:266, 267
  composition 6:257; 10:103
  electron micrograph 6:258
  iron in 6:262, 267
  magnesium in 6:263, 267
  magnetite in 6:266
  manganese in 6:263, 267
  nickel in 6:266, 267
  organic matter in 6:266
  phosphorus in 6:261
  potassium in 6:266
  properties 6:256
  sample preparation 6:259
  silica in 6:259, 267
  sodium in 6:266, 267
  sulfate, acid-soluble in 6:265
  titanium in 6:261, 267
  water in 6:259
  zirconium in 6:267
Chrysotile asbestos
  analysis 7:446
  in brake linings 7:446
CI Acid Blue 74. See Indigo carmine.

CI Acid Green 1. See Ext D & C Green No. 1.
CI Acid Green 25. See D & C Green No. 5.
CI Acid Orange 7. See D & C Orange No. 4.
CI Acid Red 33. See D & C Red No. 33.
CI Acid Red 87. See D & C Red No. 22.
CI Acid Red 92. See D & C Red No. 28.
CI Acid Yellow 1. See Ext D & C Yellow No. 7.
CI Acid Yellow 3. See D & C Yellow No. 10.
CI Acid Yellow 36. See Ext D & C Yellow No. 1.
CI Acid Yellow 73. See D &C Yellow No. 7; and
  No. 8.
CI Basic Violet 10. See D & C Red No. 19.
CI Blacks
  colorimetric determination 12:15, 16, 17
CI Blues
  colorimetric determination 12:15, 16, 17
CI Browns
  colorimetric determination 12:15, 16, 17
CI Food Blue 1. See F D & C Blue No. 2; Indigo
  carmine.
CI Food Blue 2. See F D & C Blue No. 1.
CI Food Green 3. See F D & C Green No. 3.
CI Food Red 1. See F D & C Red No. 4.
CI Food Red 9. See F D & C Red No. 2.
CI Food Red 14. See F D & C Red No. 3.
CI Food Violet 2. See F D & C Violet No. 1.
CI Food Yellow 3. See F D & C Yellow No. 6.
CI Food Yellow 4. See F D & C Yellow No. 5.
Cigarette filters
  analysis 10:38
  cellulose acetate 10:2
  charcoal 10:3
  draft resistance 10:39
  efficiency 10:38
  polyolefin tow 10:2
  surface area 10:39
  triacetin content 10:40
Cigarette paper 10:2
  analysis 10:31
  chemical additives 10:36
  dry tensile strength 10:34
  opacity 10:31
  porosity 10:32
  wet tensile strength 10:35
Cigarettes. (See also Cigarette filters; Cigarette
  paper; Cigarette smoke; and Cigarette tobacco.)
  10:1–42
  analysis 10:4
  burning rate, static 10:4
  cellulose acetate filters 10:2
  charcoal filters 10:3
  commercial specifications 10:3
  draft resistance 10:4
  fire retention 10:4
  history 10:1
  humectants in 10:29
  moisture conditioning 10:10

paper for 10:2
polyolefin tow filters 10:2
smoking machines 10:11
tobacco blend 10:1
Cigarette smoke
  acetaldehyde content 10:27, 31
  acetone content 10:27, 31
  acetonitrile content 10:31
  acid content 10:17
  acrolein content 10:27, 31
  analysis 10:8
  benzene content 10:31
  benzo[α]pyrene content 10:19
  carbon monoxide content 10:26
  cigarette selection 10:8
  composition 10:13
  diethyl ketone content 10:31
  gas chromatography 10:30, 31
  isoprene content 10:31
  mass spectroscopy 10:30
  menthol content 10:21
  methanol content 10:31
  methyl ethyl ketone content 10:31
  nicotine content 10:16
  nitrogen oxide content 10:26
  particulate matter content 10:14
  pentadiene content 10:31
  pH 10:21, 24, 25
  phenols in 10:23
  propionaldehyde content 10:31
  sample collection 10:12
  toluene content 10:31
  water content 10:14
Cigarette tobacco 10:1
  analysis 10:5
  cation content 10:5
  moisture content 10:5
  potassium content 10:6
  sugar content 10:6
CI Greens
  colorimetric determination 12:15, 17
CI Mordant Black 17. See Pontachrom Blue
  Black R.
Cinchomeronic acid. (See also *Pyridine and
  pyridine derivatives.*) 17:413
Cinchona 4:589
  analysis 4:605
Cinchonidine 4:589
  analysis 4:605
  gas chromatography 6:210, 211
Cinchonine 4:589
  analysis 4:605
  gas chromatography 6:210, 211
  gravimetric method for tungsten 8:137; 10:322;
    16:338; 19:199, 203
Cinchonine–oxime
  gravimetric method for tungsten 15:81

Cinchonine–potassium iodide
  colorimetric method for bismuth 11:140, 144
Cinchonine sulfate
  gravimetric method for tannin phenols 19:466
Cinchophen
  acid–base titration 6:203
  analysis 6:199
  colorimetric method using α-naphthol 6:205
  commercial grades and specifications 6:198
  gravimetric methods 6:202
  impurities, determination of 6:205
  infrared spectrum 6:200
  iodometric method 6:203
  manufacture 6:198
  in neocinchophen 6:205
  properties 6:198
  ultraviolet spectroscopy 6:200
Cinders
  sample preparation 7:209
Cineole. (See also Eucalyptol.)
  o-cresol congealing point test 12:286
  in essential oils 12:285
Cinnabar 16:1; 18:360
Cinnamal. See *Cinnamaldehyde.*
Cinnamaldehyde 10:42–55
  absorption spectroscopy 9:564
  analysis 10:45
  assay 10:49
  commercial grades and specifications 10:44
  derivatives 10:44
  derivatives for identification 10:45
  unsym-dimethylhydrazine–hydrochloric acid
    titration 10:51
  2,4-dinitrophenylhydrazine gravimetric method
    10:45
  fluorometry 6:627
  gas chromatography 10:49
  hydroxylamine gravimetric method 10:45
  hydroxylamine hydrochloride–potassium hydrox-
    ide titration 10:50
  hydroxylamine hydrochloride–sodium hydroxide
    titration 10:50
  identification 10:45
  infrared spectroscopy 6:616; 10:46
  liquid chromatography 10:49
  manufacture 10:43
  p-nitrophenylhydrazine gravimetric method 10:
    45
  nuclear magnetic resonance spectroscopy 10:48
  phenylhydrazine gravimetric method 10:45
  polarographic methods 10:53
  properties 8:334; 10:42
  semicarbazide gravimetric method 10:45
  separation 10:45
  sodium bisulfite–sodium hydroxide titration
    10:52
  solubility 10:42

spectroscopic methods 10:52
thin-layer chromatography 10:48
toxicity 10:42
ultraviolet spectroscopy 6:631, 633;
    10:47
ultraviolet spectroscopy differential 10:52
uses 10:42
Cinnamaldehyde derivatives
    thin-layer chromatography 10:49
Cinnamein. See Benzyl cinnamate.
Cinnamene. See Styrene.
Cinnamic acid 8:406; 10:56
    acid–base titration 10:61
    analysis 10:58
    assay 10:61
    S-benzylthiuronium gravimetric method 10:59
    commercial grades and specifications 10:58
    derivatives for identification 10:58
    detection 10:62
    gas chromatography 8:388; 10:60
    identification 10:58
    infrared spectroscopy 10:60
    ion exchange chromatography 10:60
    manufacture 10:57
    Mohler color test 10:60
    p-nitrobenzyl chloride gravimetric method 10:
        59
    occurrence 10:57
    paper chromatography 8:379; 10:59, 64
    properties 8:402; 10:56
    separation 10:58
    in sunscreens 11:92
    thionyl chloride–ammonium hydroxide gravi-
        metric method 10:58
    thionyl chloride–aniline gravimetric method 10:
        58
    thionyl chloride–p-toluidine gravimetric method
        10:58
    ultraviolet spectroscopy 10:59; 11:92
    uses 10:56
cis-Cinnamic acid. See Cinnamic acid.
trans-Cinnamic acid. See Cinnamic acid.
Cinnamic acid and its derivatives. (See also
        Cinnamic acid; and individual compounds.)
        10:56–76
    acid value 10:70
    analysis 10:64, 66, 71
    assay 10:61
    bromination methods 10:61
    gas chromatography 10:69
    identification 10:60, 66
    impurities, determination of 10:70
    paper chromatography 10:64
    specification tests 10:70
    ultraviolet spectroscopy 10:69
    uses 10:66
Cinnamic alcohol. See Cinnamyl alcohol.

Cinnamic aldehyde. (See also Cinnamaldehyde.)
    in cassia oil 12:282
    in cinnamon oil 12:282
    hydroxylamine hydrochloride volumetric method
        12:282
    sodium nitroprusside test 4:111
Cinnamic anhydride. (See also Cinnamic acid and
        its derivatives.)
    analysis 10:71
    properties 10:70
    ultraviolet spectroscopy 10:71
    uses 10:71
Cinnamonitrile. (See also Nitriles.)
    polarography 16:398
Cinnamon oil. (See also Essential oils.)
    cinnamic aldehyde content 12:282
    refractive index 12:271
Cinnamoyl chloride. (See also Cinnamic acid and
        its derivatives.)
    analysis 10:71
    properties 10:70
    ultraviolet spectroscopy 10:72
    uses 10:71
N-Cinnamoylimidazole. (See also Cinnamic acid
        and its derivatives.)
    properties 10:70
    ultraviolet spectroscopy 10:72
    uses 10:71
N-Cinnamoylphenylhydroxylamine
    gravimetric method for niobium 16:366
    gravimetric method for tantalum 16:366
Cinnamyl acetate
    commercial specifications 10:78
    in essential oils 12:276
    molecular weight 12:276
    saponification methods 10:83
    ultraviolet spectroscopy 10:80
Cinnamyl alcohol 10:76–85
    acetic anhydride–perchloric acid acylation 10:82
    acetic anhydride–sodium acetate acylation 10:81
    analysis 10:78
    assay 10:81
    commercial grades and specifications 10:77
    derivatives for identification 10:78
    3,5-dinitrobenzoyl chloride gravimetric method
        10:78
    identification 10:78
    impurities, determination of 10:83
    manufacture 10:76
    α-naphthyl isocyanate gravimetric method 10:79
    p-nitrobenzyl chloride gravimetric method 10:78
    nuclear magnetic resonance spectroscopy 10:81
    occurrence 10:76
    phenylisocyanate gravimetric method 10:79
    properties 10:77
    separation 10:78
    specification tests 10:83

p-toluenesulfonyl chloride–*p*-nitroso-*N*,*N*-
    dimethylaniline colorimetric method 10:80
  ultraviolet spectroscopy 10:79
  uses 10:76
Cinnamyl anthranilate
  commercial specifications 10:78
  saponification methods 10:83
Cinnamyl butyrate
  commercial specifications 10:78
  saponification methods 10:83
Cinnamyl cinnamate. (See also *Cinnamic acid
    and its derivatives.*)
  ultraviolet spectroscopy 10:69, 80
Cinnamyl esters. (See also individual compounds.)
  acid value 10:84
  in cinnamyl alcohol 10:84
Cinnamyl formate. (See also Formic acid esters.)
  commercial specifications 10:78; 13:132
  properties 13:131
  saponification methods 10:83
  solubility 13:131, 134
Cinnamylhydroxamic acid
  gravimetric method for niobium 16:367
  gravimetric method for tantalum 16:367
Cinnamyl isobutyrate
  commercial specifications 10:78
  saponification methods 10:83
Cinnamyl isovalerate
  commercial specifications 10:78
  saponification methods 10:83
Cinnamyl propionate
  commercial specifications 10:78
  saponification methods 10:83
CI Oranges
  colorimetric determination 12:15, 16, 17
CI Pigment Red 4. See D & C No. 36.
CI Pigment Red 49. See D & C Red No. 10; No.
    11; No. 12; and No. 13.
CI Pigment Red 53. See D & C Red No. 8; and
    No. 9.
CI Pigment Red 57. See D & C Red No. 6; and
    No. 7.
CI Pigment Red 63. See D & C Red No. 34.
CI Pigment Red 64. See D & C Red No. 31.
CI Pigment Red 88. (See also Thioindigoid dyes.)
  infrared spectrum 14:507
  properties 14:497
  visible spectroscopy 14:508
CI Reactive Blacks
  colorimetric determination 12:20
CI Reds
  colorimetric determination 12:15, 16, 17
Cis-4 12:86
  gel permeation chromatography 12:120
  intrinsic viscosity 12:119
Cisdene 12:86
CI Solvent Green 3. See D & C Green No. 6.

CI Solvent Red 23. See D & C Red No. 17.
CI Solvent Red 43. See D & C Red No. 21.
CI Solvent Red 48. See D & C Red No. 27.
CI Solvent Red 49. See D & C Red No. 37.
CI Solvent Violet 13. See D & C Blue No. 3.
CI Solvent Yellow 33. See D & C Yellow No. 11.
Citraconic acid
  gas chromatography 17:154
  in phthalic anhydride 17:154
  preparation from citric acid 8:373
Citral
  hydroxylamine hydrochloride volumetric method
      12:282
  in lemongrass oil 12:282
  in lemon oils 12:282
  in lime oil 12:282
  solubility 13:540
  ultraviolet spectroscopy 6:632
Citramalic acid
  structure 19:423
  in wine 19:423
Citrate esters. (See also individual compounds.)
    10:97
  analysis 10:98
  assay 10:98
  impurities, determination of 10:98
  saponification method 10:98
  toxicity 10:97
  uses 10:97
Citrates
  metal complexes 10:89
Citric acid 8:399; 10:85–100
  acid–base titration 8:64; 10:92; 19:149
  ammonium hydroxide–barium chloride gravi-
      metric method 19:423
  assay 10:92
  in beer 7:585
  calcium hydroxide turbidimetric method 15:145
  colorimetric method for iodide 14:608
  commercial grades and specifications 10:90
  in confectionery products 10:552
  gas chromatography 19:149
  gravimetric determination 10:92
  in green coffee 10:374
  impurities, determination of 10:93
  in lactic acid and its derivatives 15:145
  lead tetraacetate colorimetric method 19:423
  liquid chromatography 15:133; 19:149
  manufacture 10:89
  in musts 19:412, 423
  occurrence 10:89
  paper chromatography 15:133; 19:418
  pentabromoacetone gravimetric method 10:92
  pH 10:88
  in pharmaceuticals 8:64
  preparation 8:373
  properties 8:396; 10:86

in roasted coffee 10:375
sample preparation 10:96
solubility 10:87, 88
solvent extraction 19:149
structure 19:423
thin-layer chromatography 8:381; 15:133;
    19:419
in tobacco 19:148
uses 10:85
in wines 19:412, 418, 422
Citric acid and salts. (See also *Citric Acid*; and
    individual compounds.) 10:85–97
analysis 10:91
identification 10:91
impurities, determination of 10:93
separation 10:91
uses 10:85
Citronellal
in citronella oil 12:282
hydroxylamine hydrochloride volumetric method
    12:282
Citronella oil. (See also *Essential oils.*)
citronellal content 12:282
refractive index 12:271
Citronellyl formate. (See also Formic acid esters.)
commercial specifications 13:132
properties 13:131
solubility 13:131, 134
Citropten. See 5,7-Dimethoxycoumarin.
Citrovorum factor
in cocoa beans 9:540
Citrus fruit
diphenyl content 11:527
morpholine content 16:202
Citrus oils. (See also individual oils.)
ultraviolet spectroscopy 12:287
Citrus Red No. 2. (See also Certified colors.)
classification 10:450
commercial specifications 10:464
impurities, determination of 10:500
liquid chromatography 10:500
manufacture 10:450
structure 10:450
visible spectroscopy 10:486
City traffic tests
of brake linings 7:470
CI Vat Black 1. (See also *Indigoid dyes.*)
infrared spectrum 14:507
properties 14:497
structure 14:497
visible spectroscopy 14:508
CI Vat Blue 1. See D & C Blue No. 6; Indigo.
CI Vat Blue 2. See 4,4'-Dichloro-5,5'-dibromo-
    indigo.
CI Vat Blue 5. (See also *Indigoid dyes.*)
infrared spectrum 14:505
properties 14:496

visible spectroscopy 14:508, 511
CI Vat Blue 6. See D & C Blue No. 9.
CI Vat Blue 35. See 5,5'-Dibromoindigo.
CI Vat Blue 48. See 4,5,7,4',5',7'-Hexabromo-
    indigo.
CI Vat Brown 5. (See also Thioindigoid dyes.)
infrared spectrum 14:505
properties 14:496
visible spectroscopy 14:508
CI Vat Brown 42. (See also *Indigoid dyes.*)
properties 14:497
structure 14:497
CI Vat dye. See 5,5'-Dibromothioindigo.
CI Vat Orange 5. (See also Thioindigoid dyes.)
infrared spectrum 14:507
properties 14:497
visible spectroscopy 14:508, 511
CI Vat Red 1. (See also D & C Red No. 30;
    Thioindigoid dyes.)
infrared spectrum 14:506
properties 14:497
visible spectroscopy 14:508, 511
CI Vat Red 41. See Thioindigo.
CI Vat Red 45. (See also *Indigoid dyes.*)
properties 14:497
structure 14:497
CI Vat Violet 2. (See also Thioindigoid dyes.)
infrared spectrum 14:506
properties 14:497
visible spectroscopy 14:508
CI Violets
colorimetric determination 12:15, 16, 17
CI Yellows
colorimetric determination 12:15, 16, 17
Clarke weight-loss method
for tin on tinplate 19:55
Clay. (See also Ceramics; *Clays and clay minerals.*)
    9:191
acid dissolution 9:214
in concrete 9:142
fluxes for 9:216
in rubbing compounds 6:353
uses 9:191
vanadium content 19:343
Clay products. (See also *Ceramics.*) 9:191
manufacture 9:191
Clays, adsorbent
chemical analysis 4:439
decolorizing efficiency 4:447
pH 4:437
uses 4:434
water-soluble materials 4:438
Clay and clay minerals 10:100–138
aluminum oxide 10:115
analysis 10:101
atomic absorption spectroscopy 10:130
base exchange capacity 10:109

calcium oxide content 10:119
carbonate content 10:124
classification 10:101, 102, 105
composition 10:100, 102
identification 10:107
impurities, determination of 10:124, 130
iron content 10:117
iron oxide content 10:116
magnesium oxide content 10:119
phosphorus content 10:123
potassium oxide content 10:123
preliminary tests 10:105
sampling 10:104
silica content 10:113
silicon dioxide content 10:112
sodium oxide content 10:123
sulfate content 10:124
thermal analysis 10:108
titanium dioxide content 10:118
uses 10:101, 105
volatile matter in 10:111
x-ray diffraction 10:107
x-ray fluorescence spectroscopy 10:134
Clayton's method
  identification of dyes 6:391
Cleaners. (See also Flushes; Polishes; and Waxes.)
  acid-type 6:351, 359, 373
  analysis 6:358, 362
  automobile 6:353, 362, 375
  carburetor 6:356
  chrome 6:353, 364, 375
  cooling system 6:351, 358, 373
  crankcase 6:356
  engine 6:355, 367, 378
  metal 6:355, 367
  performance tests 6:373
  top engine 6:356
  upholstery 6:354, 366
  whitewall tire 6:354, 367, 378
Clear glazes. (See also Glazes.)
  composition 9:196
Cleland method
  light scattering of acrylic polymers 4:312
Cleland-Stockmayer viscosity relationship 4:304
Clemizole hydrochloride. (See also Antihistamines.)
  6:27
Cleveland open cup method
  flash point of bituminous materials 6:321
  flash tester used for 6:321
Cleve's acid. See 1-Naphthylamine-3-sulfonic acid.
Clinochlore
  composition 10:103
Clinoptiolite 8:75
Cloth
  bleaching 7:221
Cloth backing
  for coated abrasives 10:277

filler content 10:276
Cloudiness
  of distilled alcoholic beverages 4:489
Cloud point
  of coumarone–indene resins 14:292
  of fatty alcohols 4:571
  of furfuryl alcohol 13:242
  of hydrocarbon resins 14:292
  of lubricants 15:331, 332
  of nonionic detergents 11:383
  of organosilicon copolymers 18:148
  of phenols 17:42
  of styrene resins 14:292
  of tetrachloroethylene 9:463
Clove oil. (See also Essential oils.)
  commercial specifications 12:261
  refractive index 12:271
  solubility 13:540
Cloxacillin. (See also Isoxazolylpenicillins; Sodium
    cloxacillin.)
  analysis 16:577
  assay 16:577
  bioassay 16:577
  manufacture 16:577
  microbiological methods 16:577
  properties 16:577
  specification tests 16:577
  structure 16:554
  ultraviolet spectroscopy 15:565
  uses 16:577
CMC. See Carboxymethylcellulose.
Coagulants. (See also Menadiol sodium diphos-
    phate; Menadione; Menadione sodium bisulfite;
    Phytonadione; and Protamine sulfate.) 10:184–
    204
Coagulants and anticoagulants. (See also Anticoag-
    ulants; Coagulants.) 10:138–209
Coal. (See also Coal and coke.)
  in concrete 9:142
  mercury content 16:41
  phosphorus content 17:96
Coal and coke 10:209–262
  absolute reflectance 10:222
  agglomerating index 10:246
  analysis 10:211
  ash fusibility 10:243
  caking 10:245
  calorific value 10:241
  carbon content 10:237
  carboxyl groups in 10:211
  classification 10:256, 257, 258
  coking 10:245
  composition 10:210, 211
  density 10:226
  electron microscopy 10:226
  electron paramagnetic resonance spectroscopy
    10:214

extinction coefficient 10:224
free-swelling index 10:245
Gray-King test 10:246
hydrogen content 10:237
hydroxyl groups in 10:211
impurities, determination of 10:234, 252
infrared spectroscopy 10:212
macerals in 10:222
mass spectroscopy 10:213
nitrogen content 10:239
nuclear magnetic resonance spectroscopy 10:
    214
oxygen content 10:241
petrographic analysis 10:215
plasticity 10:249
polished surface analysis 10:219
properties 10:210
properties, determination of 10:211
reflectance measurement 10:221
refractive index 10:224
Roga index 10:246
sampling 10:211
specification tests 10:233
sulfur content 10:240, 252
thin-section analysis 10:217
ultraviolet spectroscopy 10:213
uses 10:209
visible spectroscopy 10:213
x-ray diffraction 10:228
x-ray fluorescence 10:231
Coal ash
    scandium content 17:574, 577
    strontium content 18:280
Coal tar 14:263
    uses 14:263
Coal-tar color scale 14:293
Coal-tar disinfectants 11:540
Coal tar dyes. (See also Certified colors; and
        individual compounds.)
    in embalming fluids 12:167
    liquid chromatography 10:525
Coal tar pitch
    manufacture 14:55
    uses 14:55
Coal tar products
    analysis 14:282
Coal tars. (See also Bituminous materials.) 6:276
    commercial specifications 6:285
    manufacture 6:277
    properties 6:277
Coated abrasives 10:263–283
    adhesives 10:269, 281
    analysis 10:264, 278
    backing 10:276
    classification 10:263
    composition, determination of 10:279
    filler content 10:282

manufacture 10:263
    performance 10:283
Coating defects
    in photographic films 12:561
Coating test
    for bituminous materials 6:345
    for emulsified asphalts 6:329
Coatings
    on ammonium nitrate 5:343
    carbon black in 8:241
Coatings, organic
    acid number 8:393
Cobalt. (See also Cobalt alloys; Cobalt metals.)
        10:283–373
    in aluminum alloys 5:127
    ammonium thiocyanate colorimetric method
        10:348, 366; 11:140
    ammonium thiocyanate–tetraphenylarsonium
        chloride colorimetric method 10:370
    analytical standards 10:357
    atomic absorption spectroscopy 9:346
    bead tests 10:348; 12:14
    in beer 7:651, 657
    in beryllium copper 10:289
    borax bead test 6:583, 584
    in cemented carbides 10:292
    in cements 9:131
    in chromium 9:639
    in chromium alloys 9:668
    in cobalt alloy electroplates 10:291
    in cobalt alloys 10:307
    in cobalt carbonate 10:332
    in cobalt metal 10:293
    in cobalt(II) orthophosphate 10:343
    cobalt oxide gravimetric method 10:360
    in cobalt oxides 10:342
    colorimetric methods 10:366; 16:328
    colorimetric method using 3-methoxy-2-
        nitrosophenol 7:658
    colorimetric method using Nitroso-R-salt 5:127
    colorimetric method using quinoxalinedithiol
        7:658
    in constant-modulus alloys 10:290
    in copper alloys 10:637
    in copper metal 10:621
    in cotton 11:140, 152, 156
    in cutting alloys 10:289
    in dental alloys 10:290
    EDTA titration 9:639; 10:365; 13:445
    in electrical resistance alloys 10:290
    electrolysis 16:303
    electrolytic deposition method 10:353, 358
    emission spectroscopy 7:651; 9:346; 10:348, 371;
        11:152; 13:277; 14:48, 122, 542; 15:156, 187;
        16:160, 387; 18:451
    ferricyanide titration 16:344
    in ferrochromium 9:661

in ferroniobium 16:387
field tests 10:371
flame tests 6:589
in gallium arsenide 13:276
in gallium metal 13:266
in glass and glass-ceramics 13:445
in gold 14:45
gravimetric determination 10:350, 360
in hafnium 14:122
in hard facing alloys 10:289
in high-speed steels 10:289
in high-temperature alloys 10:286
in hydrochloric acid 14:374
hydrogen sulfide gravimetric method 19:554
identification 10:348
in indium 14:542
iodometric titration 10:332, 366
ion exchange chromatography 10:356; 13:255;
    17:526
in iron 15:59, 64
in lanthanum oxide 15:156
lead dioxide oxidation 10:370
in low-expansion alloys 10:290
in magnesium 15:381
in magnesium alloys 15:381
in manganese 15:479
mercury content 16:44
mercury thiocyanate colorimetric method 19:
    557
in metal complex dyes 12:14
microcosmic salt bead test 6:587
neutron activation analysis 13:265
nickel in 10:360; 16:297, 298, 303, 326
in nickel alloys 16:326
in nickel sulfate 16:344
in niobium 16:387
p-nitrophenylhydrazone of diacetylmonoxime
    colorimetric method 10:348
2-nitroso-1-napthol colorimetric method 10:369,
    370
1-nitroso-2-naphthol gravimetric method 10:351,
    637; 16:297, 327
nitroso-R-salt colorimetric method 10:348, 358,
    367, 621; 15:64, 381; 16:327
occurrence 10:284
in organic materials 10:349
PAN colorimetric method 10:370
phenylthiohydantoic acid colorimetric method
    10:358
polarographic methods 10:370
potassium cobaltinitrite gravimetric method 10:
    361
potassium cyanide titration 10:363
potassium ferricyanide titration 10:361
potassium nitrite gravimetric method 10:352;
    16:297
potassium permanganate oxidation 10:370

potassium thiocarbonate colorimetric method
    10:358
radiochemical methods 10:371
in rocks 10:369; 15:187
rubeanic colorimetric method 11:140, 144
sample preparation 10:359
separation 10:350
in sodium chloride 9:346
sodium perborate oxidation 10:370
sodium phosphate separation scheme 10:352
sodium pyrosulfate fusion 16:327
in soft magnetic alloys 10:288
in soil 10:369
solvent extraction 10:354; 15:157; 16:298
in spring steels 10:290
in stainless steel 10:369
in styrene 18:309
in surgical alloys 10:290
in tin metal 19:67
uses 10:283
volumetric determination 10:363
in wear-resistant alloys 10:289
x-ray fluorescence spectroscopy 10:348, 363;
    11:156
zinc oxide gravimetric method 10:350
in zirconium 14:122
Cobalt abietate. See Cobalt resinate.
Cobalt acetate. (See also Cobalt compounds.)
    10:328
acetate content 10:329
commercial grades and specifications 10:328
impurities, determination of 10:329
manufacture 10:328, 329
perchloric acid titration 10:329
properties 10:328
solubility 10:328, 329
specification tests 10:329
uses 10:329
Cobalt alloy electroplates. (See also Cobalt alloys.)
    10:291
composition 10:291
Cobalt alloys. (See also individual alloys.) 10:285
aluminum content 10:307
analysis 10:306
beryllium content 10:308
bismuth content 10:309
boron content 10:309
carbon content 10:310
chromium content 10:310
cobalt content 10:307
copper content 10:311
decomposition 10:349
impurities, determination of 10:308
iron content 10:313
manganese content 10:314
molybdenum content 10:315
nickel content 10:316

niobium content 10:318
sampling procedures 10:306
silicon content 10:320
tantalum content 10:318
titanium content 10:320
tungsten content 10:321
vanadium content 10:323; 19:340
zirconium content 10:324
Cobalt aluminate. (See also Cobalt compounds.)
    10:331
  manufacture 10:331; 17:202
  properties 10:331
  uses 10:331; 17:202
Cobalt alums 10:344
Cobalt ammines. See Cobalt complexes.
Cobalt ammonium sulfate
  manufacture 10:344
  properties 10:344
  solubility 10:344
  uses 10:344
Cobalt-base alloys. (See also Cobalt alloys.)
  composition 10:289
Cobalt blues
  uses 17:202
Cobalt–borax bead test 6:583
Cobalt bromide. (See also Cobalt compounds.)
    10:331
  color test for water 14:345; 15:251
  manufacture 10:331
  properties 10:331
  solubility 10:331
  uses 10:331
Cobalt carbonate. (See also Cobalt compounds.)
    10:331
  assay 10:332
  cobalt content 10:332
  commercial grades and specifications 10:332
  impurities, determination of 10:333
  manufacture 10:331
  occurrence 10:331
  solubility 10:332
  specification tests 10:332
  uses 10:332
Cobalt carbonyls. (See also Cobalt compounds;
    and individual compounds.) 10:333
  analysis 10:334
  toxicity 10:334
  uses 10:333
Cobalt chelates
  EDTA content 12:51
Cobalt chloride. (See also Cobalt compounds.)
    10:335
  analysis 10:336
  assay 10:336
  commercial grades and specifications 10:335,
    336
  impurities, determination of 10:336

manufacture 10:335
properties 10:335
specification tests 10:336
uses 10:336
Cobalt citrate complex
  stability 10:89
Cobalt complexes. (See also Cobalt compounds.)
    10:346
  history 10:346
  properties 10:346
  stability 10:346
  uses 10:346
Cobalt compounds. (See also individual compounds.)
    10:327–347
  analysis 10:327
  cupferron–chloroform extraction of cations
    10:356
  impurities, determination of 10:327
Cobalt-containing samples
  decomposition 10:349
Cobalt cyanide. See Cobalt complexes.
Cobalt disulfide
  manufacture 10:345
Cobalt driers 10:345
Cobalt ethyl hexoate. (See also Cobalt octoate.)
    10:345
Cobalt ferrite. (See also Cobalt alloys; Ferrites.)
  composition 10:288
  Curie temperature 9:212
  magnetic constants 9:212
  magnetic properties 12:515
Cobalt fluoride. (See also Cobalt compounds.)
    10:337
  manufacture 10:337
  properties 10:337; 13:11
  solubility 10:337
  uses 10:337
Cobalt fluorosilicate. (See also Cobalt compounds.)
    10:337
  solubility 10:338
  uses 10:338
Cobalt hydrate. See Cobalt hydroxide.
Cobalt hydrocarbonyl. (See also Cobalt carbonyls.)
  in air 10:334
  2-nitroso-1-naphthol-4-sulfonic acid colorimetric
    method 10:334
  toxicity 10:334
Cobalt hydroxide. (See also Cobalt compounds.)
    10:338
  commercial grades and specifications 10:338
  manufacture 10:338
  properties 10:338
  solubility 10:338
  uses 10:338
Cobaltiferous speisses
  decomposition 10:349
Cobalt iodide. (See also Cobalt compounds.) 10:339

manufacture 10:339
properties 10:339
solubility 10:339
uses 10:339
Cobaltite 10:284
Cobalt linoleate 10:345
Cobalt metal. (See also Cobalt alloys.)
  analysis 10:293
  carrier gas fusion methods 8:682
  cobalt content 10:293
  determination 10:293
  hydrogen in 8:618, 680, 681
  impurities in 10:293
  metallurgy 10:292
  nitrogen in 8:618
  oxygen in 8:618, 680, 681, 682
  properties 10:285, 291, 292
  reaction medium for fusion methods 8:632,
    633
  solubility in gold 14:9
  uses 10:284
  vacuum fusion methods 8:618, 680, 681
Cobalt-microcosmic salt bead test 6:587
Cobalt minerals
  decomposition 10:349
Cobalt monoxide. See Cobalt oxide.
Cobalt naphthenate. (See also Cobalt compounds.)
  in enamels 12:176
  properties 10:346
  uses 10:346
Cobalt nitrate. (See also Cobalt compounds.)
    10:339
  analysis 10:340
  colorimetric method for EDTA 12:41
  commercial grades and specifications 10:339,
    340
  impurities, determination of 10:340
  manufacture 10:339
  properties 10:339
  solubility 10:339
  specification tests 10:340
  uses 10:340
Cobalt nitrate-ammonium thiocyanate
  colorimetric method for nonionic detergents
    12:578
  colorimetric method for papaverine 16:272
Cobalto-cobaltic oxide. (See also Cobalt oxide.)
  manufacture 10:341
  properties 10:341
Cobalto-cobaltic sulfide
  manufacture 10:345
Cobalt octoate. (See also Cobalt compounds.)
  manufacture 10:345
Cobalt oleate. (See also Cobalt compounds.)
  manufacture 10:345
  solubility 10:345
  uses 10:345

Cobalt ores 10:284
Cobalt orthophosphate. (See also Cobalt com-
    pounds.)
  cobalt content 10:343
  commercial grades and specifications 10:343
  manufacture 10:342
  solubility 10:342
  uses 10:342
Cobaltous nitrate-ferric chloride
  colorimetric method for DCYTA 12:28
  colorimetric method for EDTA 12:28
  colorimetric method for PDTA 12:28
  colorimetric method for TTHA 12:28
Cobalt oxalate. (See also Cobalt compounds.)
    10:341
  properties 10:341
  solubility 10:341
  uses 10:341
Cobalt oxide. (See also Cobalt compounds.)
    10:341
  cobalt content 10:342
  commercial grades and specifications 10:342;
    12:519
  composition 12:519
  in glass 9:200
  in glass and glass-ceramics 13:413, 415, 441, 461
  gravimetric method for cobalt 10:360
  manufacture 10:341
  nitroso-R salt colorimetric method 13:441
  properties 10:341
  solubility 10:341
  solvent extraction 13:441
  uses 10:341, 342
Cobalt phosphate. (See also Cobalt orthophos-
    phate.) 10:342
Cobalt-platinum alloys. (See also Cobalt alloys.)
  platinum content 10:319
Cobalt resinate. (See also Cobalt compounds.)
  manufacture 10:346
Cobalt soaps 10:345
  manufacture 10:345
Cobalt-sodium azide
  color test for sulfite 18:403
Cobalt soyate. (See also Cobalt compounds.)
  manufacture 10:346
Cobalt steels. (See also Cobalt alloys.) 10:285
Cobalt sulfate. (See also Cobalt compounds.)
    10:343
  analysis 10:344
  cobalt content 10:343, 344
  commercial grades and specifications 10:343,
    344
  impurities, determination of 10:344
  manufacture 10:343, 344
  properties 10:343, 344
  solubility 10:343
  specification tests 10:344

uses 10:343
Cobalt sulfate–acetone
  colorimetric method for thiocyanate 18:435
Cobalt sulfide. (See also Cobalt compounds.)
    10:345
  manufacture 10:345
  properties 10:345
  solubility 10:345
Cobalt sulfite. (See also Cobalt compounds.)
    10:345
  manufacture 10:345
  properties 10:345
Cobalt tetracarbonyl. (See also Cobalt carbonyls.)
  manufacture 10:333
  properties 10:334
  solubility 10:334
  uses 10:333
Cobalt tricarbonyl. (See also Cobalt carbonyls.)
  properties 10:334
  solubility 10:334
Cobalt ultramarine. See Cobalt aluminate.
Cobenium. (See also Cobalt alloys.) 10:290
  composition 10:290
Cocaine. (See also Cocaine hydrochloride; Coca
    leaves; and Narcotics.) 4:587; 16:273
  acid–base titration 16:273
  analysis 4:595; 5:383; 16:273
  anesthetic 5:383
  in coca leaves 16:273
  colorimetric determination 16:274
  gas chromatography 16:254
  gold chloride microchemical method 16:249
  lead iodide microchemical method 16:249
  methyl orange dye complex method 5:384
  microchemical tests 16:249
  mixtures, determination in 5:385
  platinic chloride microchemical method 16:249
  polarography 16:274
  properties 5:383
  solvent extraction 16:273
  structure 16:244
  thin-layer chromatography 16:256
  ultraviolet spectroscopy 5:383; 16:251
  uses 16:273
  volumetric determination 16:273
Cocaine hydrochloride. (See also Cocaine.)
  acid–base titration 5:385
  amperometric titration 5:385
  analysis 5:383
  anesthetic 5:383
  chromium trioxide colorimetric method 16:246
  color tests 16:246
  column chromatography 5:384
  ion exchange method 5:383
  potassium permanganate colorimetric method
    16:246
  properties 5:383

ultraviolet spectroscopy 16:251
Coca leaves. (See also Cocaine.) 16:273
  alkaloid content 16:273
  analysis 16:273
  cocaine content 16:273
  composition 16:273
  occurrence 16:273
Cochineal
  ammonium hydroxide colorimetric method
    10:538
  colorimetric method for boric acid 7:387, 409
  identification 10:537
  in meat 10:538
  uranyl acetate colorimetric method 10:538
Cocoa. (See also Cacao; Cocoa beans; Cocoa pro-
    ducts; Chocolate; and Chocolate and cocoa
    products.) 9:534, 545
  caffeine in 8:67
  commercial grades and specifications 10:550
  monocarbonyl content 9:543
  nutrient content 9:538
  theobromine in 8:67
Cocoa beans. (See also Cacao; Cocoa; and Choco-
    late and cocoa products.)
  acetic acid content 9:544
  aldehyde content 9:543
  amino acid content 9:556
  analysis 9:550
  aroma 9:567
  carbonyl content 9:562
  carboxylic acid content 9:543, 544
  chocolate liquor content 9:555
  commercial grades and specifications 9:547
  composition 9:548, 549
  cut test 9:550
  fat content 9:555
  flavor 9:567
  fructose content 9:558
  furan aldehyde content 9:563
  gas chromatography 9:567, 569, 570
  glucose content 9:558
  grading 9:541
  headspace analysis 9:567
  HMF content 9:542
  iodine number 9:548
  impurities, determination of 9:551
  mass spectroscopy 9:567, 570
  melting point 9:548
  moisture content 9:552
  nitrogen content 9:554
  organic acid content 9:565
  pyrrole aldehyde content 9:563
  reducing sugar content 9:558
  refractive index 9:548
  roasting 9:541
  sampling procedures 9:550
  selection 9:541

sucrose content 9:558
sugar content 9:558
vitamin content 9:540
Cocoa butter. (See also Theobroma oil.)
  in chocolate 9:540
  in chocolate liquor 9:545
  in cocoa products 9:590
  refractive index 9:576
Cocoa dialysates
  thin-layer chromatography 9:593
  ultraviolet spectroscopy 9:593
Cocoa nibs. (See also Cocoa beans.) 9:544
  composition 9:548, 549
Cocoa powder. See Cocoa.
Cocoa products. (See also Cocoa; Chocolate;
    Chocolate liquor; and *Chocolate and cocoa
    products*.) 9:534, 545, 549, 568
  alginate content 9:589
  analysis 9:568, 591
  aroma 9:592
  bacteria in 9:599
  chocolate content 9:590
  cocoa butter content 9:590
  color 9:571
  composition 9:590
  crude fiber content 9:574
  dialysis methods 9:593
  fat content 575, 590
  flavor 9:592
  infrared spectroscopy 9:593
  impurities, determination of 9:573, 595
  lactose content 9:581
  lecithin content 9:587
  milk solids content 9:580
  moisture content 9:573, 590
  molds in 9:600
  nitrogen content 9:575
  particle size distribution 9:568
  pH 9:573
  phenolic glycoside content 9:593
  Ramachandran extractions 9:593
  salmonella in 9:601
  sample preparation 9:595
  shell content 9:573
  sucrose content 9:578, 590
  sugar content 9:578
  tannin content 9:593
  wet screening methods 9:569
  yeasts in 9:600
Cocoa shells. (See also Cocoa beans.)
  composition 9:548, 549
Coconut oil. (See also *Fats and fatty oils*.)
  in alkyd resins 5:23
  ethylene oxide in 12:330
  gas chromatography 5:29
  properties 12:473
  solubility 13:539, 552

Codeine. (See also Acetyl codeine; Codeine phos-
    phate; Codeine sulfate; and *Narcotics*.) 4:588;
    16:268-270
  acid-base titration 16:267
  ammonium molybdate colorimetric method
    16:246
  analysis 4:597; 16:269
  in antihistamines 16:270
  antimony trichloride colorimetric method 16:
    246
  in aspirin 16:270
  color test for ethylene glycol 13:570
  color test for propylene glycol 13:570
  color tests 16:246
  commercial grades and specifications 16:269
  p-dimethylaminobenzaldehyde colorimetric
    method 16:246, 270
  dioctyl sodium sulfosuccinate titration 16:279
  in ephedrine 16:270
  ferric chloride–sulfuric acid colorimetric method
    16:246
  fluorometry 16:265
  formaldehyde–sulfuric acid colorimetric method
    16:246
  gas chromatography 6:210; 16:254, 266
  gravimetric determination 16:270
  iodine–potassium iodide microchemical method
    16:249
  ion exchange methods 16:270
  manufacture 16:268
  mass spectroscopy 16:270
  microchemical tests 16:249
  nitric acid colorimetric method 16:246
  occurrence 16:269
  in opium 16:266
  paper chromatography 6:183
  in phenacetin 16:270
  in phenylephrine 16:270
  phosphomolybdic acid colorimetric method
    16:270
  phosphorimetry 16:264
  polarography 16:268, 270
  potassium–cadmium iodide microchemical method
    16:249
  properties 16:268
  solubility 16:268
  structure 16:243
  thin-layer chromatography 16:256
  ultraviolet spectroscopy 16:251, 270
  uses 16:268
Codeine phosphate. (See also Codeine.)
  acid-base titration 16:269
  in aspirin 16:269
  assay 16:269
  in caffeine tablets 16:269
  column chromatography 6:150
  gas chromatography 16:269

infrared spectroscopy 6:163
in phenacetin 16:269
solvent extraction 16:269
ultraviolet spectroscopy 6:150; 16:251, 270
Codeine sulfate. (See also Codeine.)
  acid–base titration 16:269
  column chromatography 6:150
  ultraviolet spectroscopy 6:150; 16:270
Coefficient of friction
  of brake linings 7:452
Co-Elinvar. (See also Cobalt alloys.) 10:290
  composition 10:290
Coesite. (See also Silica.) 18:5
Coffee. (See also Decaffeinated coffee; Green
    coffee; Instant coffee; and Roasted coffee.)
    10:373–447
  agronomy 10:374
  analysis 10:377
  caffeine in 8:59, 62, 67, 68
  curing 10:374
  history 10:373
  roasting 10:375
Coffee beans. See *Coffee*; Green coffee; and
    Roasted coffee.
Coffinite 19:225
  composition 19:225
  occurrence 19:225
Cohesion. See *Adhesion*
Cohesion loss
  of bituminous paving mixtures 6:344
Coin gold 5:68
Coin silver. (See also Silver alloys.)
  composition 18:156
Coke. (See also *Coal and coke.*)
  in brake linings 7:447
  phosphorus content 17:96
Colchicine 4:592
  analysis 4:611
Colchicum alkaloids 4:592
  analysis 611
Cold test
  for corn oil 11:58
Colemanite
  occurrence 7:313, 314
  use in manufacture of boric acid 7:371
Coliform bacteria
  membrane filter technique 19:391
  in water 19:391
Coliform count, total
  filter membrane technique 19:391
  of water 19:391
Colistin 5:466, 533
  analysis 5:538, 543, 548
  bioautography 5:539
  identification 5:538
  microbiological assay 5:543
  paper chromatography 5:538

separation 5:538
uses 5:533
Colistin A
  structure 5:466
Colistin sodium methanesulfonate
  infrared spectrum 5:538
  properties 5:534
  solubility 5:535
Colistin sulfate
  infrared spectrum 5:538
  properties 5:534
  solubility 5:535
Collagen. (See also Animal glues; *Glues of animal
    origin.*) 13:487
Collapsible tube tests 11:114
Collidines. See Trimethylpyridines.
Colloidal silicates. (See also Sodium silicates.)
    18:42
  commercial grades and specifications 18:42
Colloids
  in photographic films 12:558
Colloids, dental. See Dental colloids.
Color
  of acetate and triacetate fibers 4:76, 85
  of acetic acid 4:98
  of acetic anhydride 4:107
  of acetone 4:114
  of acetophenone 4:125
  of acrylamide 4:165
  of acrylic and modacrylic polymers and fibers
    4:321
  of acrylic monomers 4:199
  of acrylonitrile 4:376
  of adipic acid 4:417
  of alcoholic beverages, distilled 4:481, 487
  of alcohols 4:542, 572
  of alkanolamines 5:13
  of alkyd resins 5:51
  of ammonium hydroxide 5:323
  of aniline and its derivatives 5:428
  APHA method 4:199
  ASTM method 4:199
  of beer 7:517, 532
  of benzene 7:45
  of black tea 18:486, 494, 499
  of caprolactam 8:123
  of carbonated beverages 8:165, 172
  of carbon black 8:181
  of carbon disulfide 8:275
  of carbon tetrachloride 8:326
  of carbonyl compounds 8:360
  β-carotene equivalent 12:68
  of cereals 9:235
  of chlorophenols 9:526
  coal-tar color scale 14:293
  of cocoa 9:536
  of cocoa products 9:571

of corn oil 11:57
of corn products 11:12
of corn syrup 10:558
of coumarone–indene resins 14:293
of crotonaldehyde 11:182
of cyclohexylamine 11:216
of diatomaceous earth 15:529
of 1,2-dichloroethane 9:439
of drycleaning solvents, used 11:609
of eggs and egg products 12:67
of enamels 12:179
of epoxy resins 12:253
of ethyleneamines 11:430
of ethylene and propylene polymers 12:347
of ethylene oxide 12:223
of fats and fatty oils 12:477
of fatty alcohols 4:572
Gardner color scale 14:293
in glass 9:200
of glass and glass-ceramics 13:461
of glycerol 13:530
Hazen method 4:199
of hydrocarbon resins 14:266, 281, 293
of hydrocarbons 14:343
of hydrochloric acid 9:340
in hydrogen cyanide 14:423
of hydroxylamine salts 14:454
of iodoform 14:615
of lactic acid 15:147
of limestone 15:231
of liquids 6:622, 623
Lovibond method 15:502
of margarine 15:502
of methanol 16:113
of methyl chloride 9:425
of morpholine 16:199
of naphthalene 16:212
of nicotinamide 17:439
of nitro alcohols 16:447
of nitroparaffins 16:438
of pentaerythritol 13:601
of perchloric acid 9:364, 365
of pigments 17:205
platinum–cobalt method 13:601; 14:344, 423
of polytetrafluoroethylene 13:88
of polyurethane elastomers 19:282
of polyurethane foams 19:281
of resins 4:23
of roasted coffee 10:420
rosin color scale 14:294
Saybolt value 14:343
of styrene 18:309
of sugar 8:175; 10:558
of terpenes 14:284
of tobacco 19:157
of 1,1,1-trichloroethane 9:443
of water 19:353

of wines 19:504
Colored glass. (See also *Glass and glass-ceramics.*)
composition 13:407
Coloring agents
in embalming fluids 12:167
Colors. (See also Certified colors; and individual
compounds.)
in butter 10:545
in confectionery products 10:552
in cordials 10:527
in cosmetics 19:527
in drugs 10:527
in lipsticks 10:529
in macaroni 10:527
in oils 10:527
in oranges 10:527
in soft drinks 10:527
Colors for foods, drugs, and cosmetics. (See also
Certified colors; Colors; Uncertified colors;
and individual compounds.) 10:447–547
Color stability
of acrylic resin teeth 11:308
of cellulose acetate 9:72
of dental restorative materials, composite
11:313
of methyl methacrylate dental materials
11:304
Color standards
for phenols 17:44
Color tests
for antioxidants and antiozonants 6:90
for barbiturates 6:477
for boron 7:385, 407
for dimethyl terephthalate 17:166
for isophthalic acid 17:168
for phthalic anhydride 17:155
for terephthalic acid 17:163
for trimellitic anhydride 17:173
Columbite 18:440
composition 16:349; 18:441
rhenium content 17:529
Columbium
in ferrochromium 9:661
Column chromatography. (See also Liquid chromatography.)
of acetaminophen 6:154
of acetylsalicylic acid 6:128, 134, 137, 139, 150,
152, 154
of actinomycins 5:553
of alcohols 4:528
of atropine 4:596
of *p*-aminophenol 6:154
of antihistamines 6:36
of antioxidants and antiozonants 6:87
of antipyrine 6:218
of azo dyes 6:414
of benzaldehydes 6:617

of caffeine 6:128, 134, 150, 152, 154
of capreomycin 5:555
of carbonyls in cocoa beans 9:562
of carboxylic acids 4:415
of chloramphenicol 5:496
of chlorprophenpyridamine maleate 6:150
of chromomycin 5:498
of cocaine hydrochloride 5:384
of codeine phosphate 6:150
of codeine sulfate 6:150
cyclopentenylallyl barbituric acid 6:150
of doxylamine succinate 6:150
of ergot alkaloids 4:600
of griseofulvin 5:498
of itobarbital 6:152
of kanamycins 5:583
of leucomycin 5:523
of methapyrilene hydrochloride 6:150
of mikamycin 5:555
of mitomycin 5:500
of neomycins 5:585
of ostreogrycin 5:556
of paromomycin 5:586
of phenacetin 6:128, 134, 150, 152
of phenindamine tartrate 6:150
of phenobarbital 6:150, 495, 498
of phenothiazine derivatives 6:152
of plasticizers 12:592
of pyrilamine maleate 6:150
of ristocetin 5:605
of salicylic acid 6:139, 150, 154
of staphylomycin 5:556
of tetracaine 5:414
of theobromine 9:589
of thonzylamine hydrochloride 6:150
of trichomycins 5:471
of tyrocidine 5:542
Column density
  of ion exchange resins 15:9
Combustion methods
  for carbon 16:329; 18:232
  for ceramics 9:217
  for chloride 18:308
  for chlorocarbons 9:410
  for elastomers, synthetic 12:109
  for fluorocarbons 13:66
  for hydrides 14:249
  for sulfur 13:437; 18:308, 385
  for sulfur in natural gas 13:324
Combustion of organic materials 8:282
Comeback test
  for matches 15:545
Comol. (See also Cobalt alloys.) 10:285
Compatibility test
  for brake fluids 7:434
Complexin. (See also Cephalosporins.)
  structure 16:555

Complexometric titration (See also EDTA titra-
    tion.)
  for sulfur content of acrylic polymers 4:286
Composite ceramics. (See also *Ceramics.*)
  properties 9:210
  uses 9:210
Composite dental restorative materials. See Dental
    restorative materials, composite.
Compound E. See Cortisone.
Compound F. See Hydrocortisone.
Compound S
  gas chromatography 14:218
  thin-layer chromatography 14:217
Compressed baker's yeast. (See also *Yeast.*) 19:
    516
  composition 19:517
  properties 19:517
  species 19:516
  uses 19:516
Compressed yeast. (See also *Yeast.*) 19:516
  manufacture 19:516
  species 19:516
  uses 19:516
Compressibility
  of cakes 12:74
  of carbon black 8:227
Compression cell
  for measurement of fibers 4:322
Compression set
  of vinylidene fluoride–hexafluoropropylene co-
    polymer 13:98
Compression set tests
  for silicone rubbers 18:131
Compressive strength
  of bituminous paving mixtures 6:344
  of carbonaceous ablative materials 4:13
  of cements 9:136
  of concrete 9:136, 144
  of dental elastomers 11:286
  of dental gypsum products 11:266
  of dental impression materials, agar-based 11:280
  of dental impression materials, alginate-based
    11:283
  of dental investments 11:270
  of dental silicate cements 11:259
  of dental zinc phosphate cements 11:263
  of mortar 9:136
Conalbumin
  in egg white 12:58, 61
  liquid chromatography 12:60
  properties 12:58
Concrete. (See also Cements; Mortars.) 9:94
  aggregate content 9:142
  aggregate gradation 9:142
  air content 9:143
  analysis 9:144, 147, 149
  composition 9:142

compressive strength 9:136, 144
consistency 9:143
durability 9:143
electron microscopy 9:146
impurities, determination of 9:142
microscopic methods 9:145, 146
mix proportions test 9:149
mix water 9:143
slump test 9:143
tensile strength 9:136
Condensation
of silicone resins 18:123
Condensed milk 16:144
analysis 16:145
commercial specifications 16:144
manufacture 16:144
milkfat content 16:145
sampling 16:145
solids content 16:145
Condensing capacity
of aluminum chloride 5:176
Conductometric methods
of acetoacetanilide 5:445
of acrylate content of acrylamide 4:165
of barium 6:579
for carboxyl content of acrylamide polymers
4:172
of fluorosulfuric acid 13:52
for hydrochloric acid 14:365
for molybdenum 16:169
for oxygen 8:650
Cone penetration test
for caulking compounds 9:22
for dental gypsum products 11:265
Cone penetrometer 9:21
Confectionery products 10:548–567
acidity 10:564
analysis 10:554
ash content 10:561
candy test 10:565
classification 10:548
color 10:558
ingredients 10:548
manufacture 10:553
moisture content 10:558
protein content 10:564
reducing sugar content 10:562
specific gravity 10:555
sucrose content 10:561
sugar content 10:555, 561
viscosity 10:556
Congealing point
of benzaldehydes 6:623
in cineole 12:286
of safrole 12:286
Congeners 4:462
in alcoholic beverages, distilled 4:466, 484

Congo diamonds 11:466
properties 11:466
uses 11:466
Coniine 4:591
Conradson method
for carbon residue of lubricants 15:325
Consistency
of bituminous materials 6:290
of caulking compounds 9:19
of cements 9:133
Constantan. (See also Nickel alloys.) 5:68
composition 16:290
Constant-modulus alloys. (See also Cobalt alloys.)
10:290
composition 10:290
Container safety test
for aerosols 11:98
Continous reading calorimeter method
for calorific value of natural gas 13:337
Contraceptives. (See also Estrogens; Progestogens.)
10:568–597
analysis 10:585
classification 10:568
estrogens in 10:569
pharmaceutical tablets 10:585
progestogens in 10:578
Conversion, degree of
of epoxy resins 12:254
Conversion time of malt 4:478
Conveyor discharge centrifuges 9:172, 173
characteristics 9:160
settling curves 9:161
Conway microdiffusion cell 7:580
Conway-O'Brink diffusion method
for ammonia 11:99
Cooker mash
analysis 4:480
definition 4:462
quality control 4:468
Cookies 6:446
Cooley-Tukey algorithm 13:157
Cooling system additives. (See also Antifreezes.)
6:351
analysis 6:357
performance tests 6:378
Cooperite
composition 17:215
Copaiba oil. (See also Essential oils.)
refractive index 12:271
Copo 12:86
Copper. (See also Copper alloys; Copper metal.)
10:598–680
in acrylic polymers and fibers 4:345, 349
in alcoholic beverages 4:481, 489
in aluminum alloys 5:128
in aluminum metal 5:128; 10:617
in aluminum oxide 5:187

ammonium acetate–thiocyanate–pyridine
  colorimetric method 10:337
ammonium thiocyanate colorimetric method
  17:439
ammonium thiocyanate gravimetric method
  10:607
analysis of powder 17:203
antimony in 6:60
in antimony metal 6:42, 46
in antimony pentachloride 6:51
in antlerite 10:600
in arsenic(III) chloride 6:234
in arsenic metal 6:226
in atacamite 10:600
atomic absorption spectroscopy 6:46; 7:200;
  9:346; 10:132, 617; 11:157; 14:49, 125, 126;
  15:157, 186, 195, 196, 382; 19:495
in azurite 10:600
bathocuproine colorimetric method 13:267
bead tests 10:602; 12:14
in beer 7:659
benzoinoxime colorimetric method 11:140
in beryllium metal 7:116
2,2'-biquinoline colorimetric method 10:602,
  613; 13:370
bis(cyclohexanone oxalyl dihydrazone) colorimetric method 12:555
in bismuth chloride 7:199, 200
in bismuth metal 7:194
in bismuth nitrate 7:203
in bismuth subcarbonate 7:196
in bismuth subnitrate 7:203
borax bead test 6:584
in bornite 10:600
in brass 10:630
in brochantite 10:600
in cadmium chloride 8:42
in carbonated beverages 8:174
in cellulose 9:51
in cellulose acetate polymers 4:90
in cements 9:131
in cereals 9:233, 241
ceric sulfate titration 10:611
in chalcocite 10:600
in chalcopyrite 10:600
in chlorocarbons 9:409
in chromic acid 9:684
in chromium alloys 9:669; 10:311
in chromium metal 9:635
in chrysocolla 10:600
in citric acid 10:95
in clays and clay minerals 10:127, 130
in cobalt(II) acetate 10:330
in cobalt carbonate 10:333
in cobalt chloride 10:337
in cobalt compounds 10:328
in cobalt metal 10:293, 297

in cobalt nitrate 10:340, 341
in cobalt(II) orthophosphate 10:343
in cobalt oxides 10:342
in cobalt(II) sulfate 10:343, 344
in cocoa products 9:595, 598
colorimetric determination 10:612; 13:370;
  15:382
colorimetric method using cuprethol 7:659
colorimetric method using cuproine 8:174;
  10:127, 298
commercial specifications for powder 17:203
in copper(II) acetate 10:653
in copper alloys 10:631
in copper(II) bromide 10:656
in copper(II) carbonate, basic 10:658
in copper(I) chloride 10:660
in copper(II) chloride 10:662
in copper(I) cyanide 10:663
in copper metal 10:601
in copper naphthenate 10:679
in copper oleate 10:679
in copper oxide 12:534
in copper(I) oxide 10:670
in copper(II) oxide 10:674
corrosion test of trichloroethylene 9:458
in cotton 11:140, 153, 154, 156
in covellite 10:600
in cuprite 10:600
in cuprous oxide 17:195
in dental amalgams 11:289
in dental restorative materials, composite 11:315
in diamines and higher amines, aromatic 11:441
diethylammonium diethyldithiocarbamate
  colorimetric method 14:543
2,9-dimethyl-1,10-phenanthroline colorimetric
  method 10:614
2,2'-diquinolyl colorimetric method 11:140
displacement by aluminum 10:604
displacement by lead 10:604
displacement by zinc 10:604
dithizone colorimetric method 7:116; 10:613;
  15:177; 16:344
EDTA titration 10:611, 653; 13:445
in eggs and egg products 12:66
in elastomers, synthetic 12:156
electrodeposition 10:603, 606, 654, 663, 670;
  15:65; 17:195
electrolysis 10:311; 12:534; 16:301, 328
emission spectroscopy 4:345; 9:346; 10:602,
  616; 11:153; 14:48, 122, 536; 15:144, 155,
  187, 197, 375; 16:160, 387; 19:570⁻
in enargite 10:600
extraction methods 10:604
of ferrites 12:534
in ferrochromium 9:661
in ferroniobium 16:387
flame photometry 10:602, 617

flame tests 6:583, 589, 592
in formaldehyde 13:114
in gallium arsenide 13:279
in gallium metal 13:266, 267
in gallium phosphide 13:282
in gelatin 12:554; 13:370
in germanium 13:395
in glass 9:200
in glass and glass-ceramics 13:445
in gold 14:45, 49
gold content 14:42
gravimetric determination 10:603, 606; 16:20
gravimetric method using ammonium hydroxide
    10:604
in green tea leaf 18:481
in hafnium 14:122, 124
in high-temperature alloys 10:286
hydrobromic acid colorimetric method 10:615;
    15:382
in hydrochloric acid 14:374
hydrogen sulfide gravimetric method 10:603,
    608; 15:65; 16:302
in hypochlorite bleaches 9:355
identification 10:602
in indium 14:536, 542
in indium alloys 14:551
iodometric method 5:128; 10:312, 608, 658,
    660; 11:289, 315; 15:382
ion exchange chromatography 13:254; 17:526
in iron 15:59, 65
in lactic acid and its derivatives 15:144
in lanthanum oxide 15:155
in lead 10:617; 15:177, 195, 196
in lead alloys 15:195
in magnesium 15:375, 382
in magnesium alloys 15:375, 382
in malachite 10:600
in manganese 15:479
in mercury 16:20
in metal complex dyes 12:14
microcosmic salt bead test 6:587
in molybdenum 16:161
in molybdenum trioxide 16:161
neocuproine colorimetric method 6:42, 226;
    9:669; 13:395; 14:129; 15:383
neutron activation analysis 4:349; 13:265; 15:
    382
in nickel 16:301, 328
in nickel alloys 16:328
in nickel–copper alloys 16:305
in nickel sulfate 16:344
in niobium 16:387
oxidation potentials 10:651; 18:158
phosphorus content 17:91
in photographic films 12:571
in plants 18:221
polarography 10:297, 616; 12:541, 542,

    555, 571; 13:279, 370
potassium cyanide titration 10:611
potassium ferrocyanide colorimetric method
    7:196, 199; 10:330
potassium iodate titration of copper(I) thiocya-
    nate 10:610
in pyridine 17:439
radiochemical methods 10:618
redox titration 10:660
in reducing water 4:479
reduction method for iridium 17:253
reduction method for mercury 16:7
in rocks 15:187
rubeanic acid colorimetric method 11:140, 144
sample preparation 6:60
selenium content 17:593, 595
separation 10:603
in silver 18:173
in silver nitrate 12:542; 18:176
in soaps 18:202
in sodium chloride 9:338, 345, 346
in sodium dichromate 9:684
sodium diethyldithiocarbamate colorimetric
    method 4:90; 9:51, 241, 345; 10:95, 612;
    11:45; 13:114; 18:176, 202; 19:495
in soil 10:614; 15:186
solvent extraction 15:157; 19:495
in stannous fluoride 19:75
in starch hydrolysates 11:45
in steel 10:617
in stream sediment 10:614
in styrene 18:309
tellurium content 17:595, 601, 604
in tennantite 10:600
in tenorite 10:600
in tetrahedrite 10:600
in tin alloys 10:617
in tin metal 19:67
in titanium metal 19:121
in tobacco 19:152
uses 10:631
in vegetation 10:614
volumetric determination 10:608
in water 10:614
in wines 19:491, 494
x-ray fluorescence spectroscopy 10:602, 617;
    11:156
x-ray spectroscopy 9:409
in zinc 10:617
in zinc alloys 10:617
zinc diethyldithiocarbamate colorimetric method
    12:157
in zinc metal 19:569
in zircaloy 14:129
in zirconium 14:122, 124, 129
of zirconium–copper–molybdenum alloys 14:126
Copper acetate 10:653

analysis 10:653
commercial specifications 10:653
copper content 10:653
impurities, determination of 10:654
manufacture 10:653
properties 10:653
specifications tests 10:654
uses 10:653
volumetric method for dextrose 9:579
Copper acetate–benzidine acetate
colorimetric method for mercuric cyanide 16:23
colorimetric method for thiocyanate 18:435
color test for formic acid 13:125
Copper acetoarsenite 6:229
Copper alloys 10:630–650
aluminum in 5:228; 10:632
analysis 10:631
antimony in 6:59; 10:633
arsenic content 10:634
beryllium content 10:634
bismuth in 7:216; 10:634
boron content 10:635
cadmium content 10:635
carbon content 10:636
chromium content 9:654; 10:636
cobalt content 10:636
composition 5:68
copper content 10:631
gold content 10:638
hydrogen content 10:643
iron content 10:638
lead content 10:639
magnesium content 10:640
manganese content 10:640; 15:466
molybdenum content 10:641
nickel content 10:642
nitrogen content 10:643
oxygen content 10:643
phosphorus content 10:644; 17:91
platinum content 10:644
sample preparation 6:59
selenium content 10:644
silicon content 10:645
silver content 10:645
sulfur content 10:645
tellurium content 10:645
tin content 10:646; 19:65
titanium content 10:646
tungsten content 10:648
uses 10:630
zinc content 10:649
zirconium content 10:649
Copper ammonium sulfate
turbidimetric method for sulfur 18:377
Copper arsenite 6:229
Copper bromide 10:656
analysis 10:656

commercial specifications 10:656
copper content 10:656
impurities, determination of 10:656
manufacture 10:656
properties 10:656
solubility 10:656
specification tests 10:656
uses 10:656
x-ray diffraction 10:652
Copper–cadmium acetate
volumetric method for sulfur 18:375
Copper carbonate
commercial grades and specifications 12:519
composition 12:519
Copper carbonate, basic 10:657
analysis 10:658
commercial grades and specifications 10:657
copper content 10:658
impurities, determination of 10:658
iodometric titration 10:658
manufacture 10:657
solubility 10:657
specification tests 10:658
uses 10:657
x-ray diffraction 10:652
Copper chelates
EDTA content 12:49
Copper chloride 10:659, 661
analysis 10:660, 662
colorimetric method for glycerol 13:509
commercial grades and specifications 10:660
commercial specifications 10:662
copper content 10:660, 662
gravimetric method for gold 14:38
impurities, determination of 10:661, 663
iodometric titration 10:660
manufacture 10:659, 662
properties 10:659, 661
solubility 10:659, 662
specification tests 10:660, 662
uses 10:659, 662
volumetric method for glycerol 13:509
x-ray diffraction 10:652
Copper chloride–chlorine fusion
of platinum group metals 17:240
Copper chloride–chlorine–potassium chloride fusion
of platinum group metals 17:242
Copper chloride–silver nitrate
volumetric method for ferrocyanide 15:90
Copper–chromium alloys. (See also Copper alloys.)
uses 10:631
Copper citrate complex
stability 10:89
Copper complex method for amino acids 8:433
Copper compounds. (See also individual compounds.) 10:651–680
manufacture 10:651

uses 10:651
Copper corrosion test
  of aromatic hydrocarbons 14:347
  of benzene 7:45
  for carbon tetrachloride 8:326
  for greases 15:335
  of hydrocarbons 14:346
  for liquefied petroleum gas 15:257
  for lubricants 15:336
  of petroleum gases 14:347
  of petroleum solvents 14:347
  of terpenes 14:347
Copper cyanide 10:663
  analysis 10:663
  commercial grades and specifications 10:663
  copper content 10:663
  impurities, determination of 10:664
  manufacture 10:663
  properties 10:663
  solubility 10:663
  specifications tests 10:664
  x-ray diffraction 10:652
Copper dibutyldithiocarbamate. (See also Dithio-
    carbamic acid salts.)
  properties 11:575
Copper(II) diethyldithiocarbamate. (See also
    Dithiocarbamic acid salts.)
  properties 11:575
  ultraviolet spectroscopy 11:580
  visible spectroscopy 11:580
Copper(II) dimethyldithiocarbamate. (See also
    Dithiocarbamic acid salts.)
  properties 11:575
  uses 11:576
  visible spectroscopy 11:583
Copper dimorpholinodithiocarbamate
  colorimetric method for morpholine 16:202
Copper dithiocarbamate
  colorimetric method for thiuram disulfides 11:
    594
Copper enamel. (See also *Enamels.*)
  composition 9:195
Copper ethylacetoacetate-4,4'-tetramethyldi-
    aminodiphenylmethane
  colorimetric method for nitriles 16:399
Copper ferrite. (See also Ferrites.)
  Curie temperature 9:212
  magnetic constants 9:212
  magnetic properties 12:515
Copper ferrocyanide. (See also Ferrocyanides.)
    17:196
  uses 15:85
Copper-gallium alloys 13:270
  analysis 13:270
  gallium content 13:270
Copper hydroxide 10:664
  analysis 10:665

commercial grades and specifications 10:665
  identification 10:665
  impurities, determination of 10:666
  infrared spectrum 10:666
  manufacture 10:664
  oil absorption 10:666
  properties 10:664
  specification tests 10:666
  uses 10:665
  x-ray diffraction 10:652
Copper-8-hydroxyquinoline complex
  spectroscopic method for EDTA 12:41
Copper iodide
  colorimetric method for mercury 16:18, 21, 42
  colorimetric method for thiuram disulfides 11:
    593
Copper lactate. (See also Lactic acid salts.)
  uses 15:130
Copper metal 10:598-630
  analysis 10:618
  carrier gas fusion methods 8:627, 682
  commercial grades and specifications 10:600
  copper content 10:601
  in cuprous oxide 17:195
  density 10:627
  electrical conductivity 10:627
  grain structure 10:627
  gravimetric method for silver 18:165
  history 10:598
  hydrogen in 8:616, 618, 680
  impurities, determination of 10:618
  manufacture 10:600
  nitrogen in 8:616, 618
  occurrence 10:600
  oxygen in 8:616, 617, 618, 623, 627, 680, 682
  properties 10:598
  reaction medium for fusion methods 8:633
  silicon content 18:32
  softening temperature 10:627
  solubility in gold 14:9
  toxicity 10:599
  uses 10:598
  vacuum fusion methods 8:616, 617, 618, 623,
    680
  x-ray diffraction 10:652
Copper minerals 10:600
  sample preparation 10:601
Copper-molybdenum alloys
  composition 16:351
Copper naphthenate 10:678
  analysis 10:679
  commercial specifications 10:678
  copper content 10:679
  in enamels 12:176
  iodometric titration 10:679
  manufacture 10:678
Copper-nickel alloys. (See also Nickel alloys.)

composition 16:351
tin content 19:52
Copper–nickel–cobalt alloys. (See also Cobalt
    alloys.) 10:285
composition 10:285
Copper–nickel–zinc alloys. (See also Nickel
    alloys.)
    nickel content 16:300
    tin content 19:52
    zinc content 19:559
Copper nitrate 10:667
    analysis 10:667
    commercial grades and specifications 10:667
    identification 10:667
    manufacture 10:667
    properties 10:667
    solubility 10:667
    specification tests 10:668
    uses 10:667
    x-ray diffraction 10:652
    volumetric method for glycine 12:37
    volumetric method for IDA 12:37
    volumetric method for NTA 12:37
Copper oleate 10:678
    analysis 10:679
    copper content 10:679
    iodometric titration 10:679
    manufacture 10:678
    solubility 10:678
    uses 10:678
Copper ores
    recovery of bismuth from 7:189
    sample preparation 10:601
Copper oxide 10:669, 673
    in alkali barium silicate glass 9:218
    analysis 10:670, 674
    commercial grades and specifications 10:669,
        673; 12:519
    composition 12:519
    copper content 10:670; 12:534
    copper(I) oxide content 10:671
    emission spectroscopy 9:218
    ferric chloride reduction method 10:671
    in glass 9:200
    in glass and glass-ceramics 13:415, 461
    in glazes 9:196
    impurities, determination of 10:672, 674
    manufacture 10:669, 673
    particle size 10:672
    properties 10:669, 673
    solubility 10:673
    specification tests 10:672, 674
    stability 10:672
    uses 10:670, 673
    x-ray diffraction 10:652
Copper phthalocyanine 10:675
    manufacture 10:675

properties 10:675
uses 10:675
Copper phthalocyanine blue. (See also *Pigments.*)
    commercial specifications 17:204
    flame test 17:205
    identification 17:205
    impurities, determination of 17:205
    solubility 17:205
    specification tests 17:205
    structure 17:183
    sulfuric acid gravimetric method 17:205
Copper phthalocyanine green. (See also *Pigments.*)
    structure 17:183
Copper reds. (See also individual compounds.)
        17:195
Copper–salicylaldehyde method
    monoalkanolamines, determination of 5:18
Copper silicate. (See also Silicates.)
    uses 18:44
Copper soaps 10:678
    analysis 10:679
Copper stearate 10:679
    analysis 10:679
    ash content 10:679
    manufacture 10:679
    solubility 10:679
    uses 10:679
Copper sulfate. (See also Hydrocyanite.) 10:676
    analysis 10:678
    color reagent for methionine 8:505
    color test for glues of animal origin 13:491
    commercial grades and specifications 10:676
    copper content 10:678
    gold stamping test 11:115
    gravimetric method for gold 14:39
    gravimetric method for hypophosphites 17:113
    gravimetric method for thiocyanate 18:435
    identification method for amides 8:600
    impurities, determination of 10:678
    properties 10:676
    solubility 10:676
    specification tests 10:678
    spectroscopic method for EDTA 12:41
    uses 10:676
    volumetric method for EDTA 12:43
    volumetric method for ethylenediaminetetra-
        acetate 11:363
    volumetric method for tetramethylthiuram di-
        sulfide 11:595
    x-ray diffraction 10:652
Copper sulfate–carbonate–tartrate
    colorimetric method for reducing sugars 9:559
Copper sulfate–hydroquinone titration
    of thiuram disulfides 11:594
Copper sulfate–lactic acid
    colorimetric method for thiophene 18:538
Copper sulfide

volumetric method for sulfur 12:146
Copper sulfite–iodide
  colorimetric method for mercury 16:42
Copper–tellurium alloys
  tellurium content 17:602
Copper thiocyanate
  potassium iodate titration 10:610
Copra
  ethylene oxide in 12:330
Coproporphyrin
  in beer 7:638
  iodine colorimetric method 7:638
Cordials 4:466
  analysis 4:480
  colors in 10:527
Cordierite
  properties 9:211
  in whitewares 9:191
Coriander oil. (See also *Essential oils.*)
  refractive index 12:271
Cork
  in brake linings 7:441, 447
Corn. (See also *Corn products.*)
  composition 9:233
  1,1-dichloro-1-nitroethane content 13:189
  dry milling 11:2
  mineral content 9:233
  uses 11:1
  vitamin content 9:233
  wet milling 11:1
Corn endosperm oil
  commercial specifications 10:531
  composition 10:531
Corn, for distilled alcoholic beverages
  analysis 4:469
  quality control 4:467
Corn oil. (See also *Corn products*; *Fats and fatty oils.*)
  analysis 11:55
  cold test 11:58
  color 11:57
  composition 11:55
  impurities, determination of 11:56
  iodine value 11:55
  manufacture 11:2, 55
  peroxide value 11:56
  properties 12:473
  smoke point 11:58
Corn products. (See also Corn oil; Corn starch; Feed products; Starch derivatives; and Starch hydrolysates.) 11:1–63
  acidity 11:7
  analysis 11:2
  color 11:12
  density 11:15
  drying methods 11:2
  fat content 11:7

impurities, determination of 11:9
  manufacture 11:1
  moisture content 11:2
  particle size 11:14
  pH 11:8
  protein content 11:5
  uses 11:1
  visible spectroscopy 11:12
Corn starch. (See also *Corn products.*)
  analysis 11:16; 15:526
  in baking powders 6:430
  commercial grades and specifications 15:508
  composition, determination of 11:16
  fat content 11:17
  gel strength 11:25
  hydrolysis 11:17
  impurities, determination of 15:527
  iodine staining method 11:16
  in matches 15:508
  reducing value 11:27
  uses 15:508
  viscosity 15:526
  viscosity, hot paste 11:17
  viscosity, inherent 11:24
Corn syrup. (See also Starch hydrolysates.)
  color 10:558
  commercial grades and specifications 10:549
  in confectionery products 10:550
  properties 10:550
Corn whisky 4:464
  analysis 4:480
Corrodkote method
  performance test for chrome protectors 6:376
Corrosion inhibitors
  for cooling systems 6:351
  morpholine content 16:200
Corrosion inhibitors, alkaline
  in brake fluids 7:424, 425
Corrosion test
  for antifreezes 6:18
  for automotive chemicals 6:372
  for detergents 11:398
  for lubricants 15:333
  for silicone resins 18:124
  for Stoddard solvent 11:606
Corrosion test strips 7:430
Corrosiveness
  of brake fluids 7:429
Corrosive sublimate. See Mercuric chloride.
Cortate. See Desoxycorticosterone.
Cortef. See Hydrocortisone.
Cortexone
  liquid chromatography 14:210
Corticosteroids
  gas chromatography 14:215
  liquid chromatography 14:209, 210
  thin-layer chromatography 14:213

Corticosterone. (See also Steroids.)
  liquid chromatography 14:209, 210
  structure 14:184
  thin-layer chromatography 14:217
Corticotropin. (See also Pituitary hormones.)
  14:180
  commercial grades and specifications 14:187
  composition 14:180
  manufacture 14:187
  occurence 14:187
  uses 14:180, 187
Cortisol. (See also Hydrocortisone.)
  gas chromatography 14:216, 219
  liquid chromatography 14:209, 210
  thin-layer chromatography 14:217
Δ'-Cortisol. See Δ'-Prednisolone.
Cortisone. (See also Steroids.)
  gas chromatography 14:216, 219
  liquid chromatography 14:210
  manufacture 14:193
  preparation from deoxycholic acid 7:144
  thin-layer chromatography 14:217
  trade names 14:192
Δ-Cortisone. See Prednisone.
Cortisone acetate. (See also Steroids.)
  infrared spectrum 14:193
  manufacture 14:193
  properties 14:193
  solubility 14:193
  structure 14:193
  uses 14:193
Cortogen. See Cortisone.
Cortolones
  gas chromatography 14:218
Cortols
  gas chromatography 14:218
Cortone. See Cortisone.
Corundum 4:31
  composition 9:208
  fluxes for 9:216
  properties 9:208
Cosmetics. (See also Creams and lotions; Deodorants; Hair dyes; Hair sprays; Lipsticks; Nail
  enamels; Shampoo; and Sunscreens.) 11:64–
  116
  aerosols 11:95, 96
  analysis 11:64
  colors in 10:527
  history 11:64
  morpholine content 16:201
  packaging components 11:109
  propellants in 11:95, 96
Cotarnine. (See also Narcotics.)
  ammonium molybdate colorimetric method
  16:246
  ammonium vanadate colorimetric method 16:
  246

color tests 16:246
  mercuric chloride microchemical method 16:249
  microchemical tests 16:249
  platinic chloride microchemical method 16:249
  potassium ferrocyanide microchemical method
  16:249
  ultraviolet spectroscopy 16:251
Cottage cheese. (See also Fermented milk products.)
  commercial specifications 16:136
  gelatin content 13:364
  manufacture 16:135
  moisture content 16:136
Cotton. (See also Finishing agents.) 11:117–159
  agronomy 11:117
  analysis 11:118
  application of dyes to 6:398
  ash content 11:143, 148, 157
  atomic absorption spectroscopy 11:146, 156
  behavior with dyes 6:390, 391
  bleaching 7:222
  cationic surfactants in 11:135
  composition 11:117
  dimethylol urea in 11:138
  emission spectroscopy 11:145, 147
  finishing agents 11:126, 129
  FMIR 11:124
  formaldehyde content 11:133
  history 11:117
  infrared spectroscopy 11:118
  inorganic constituents, determination of 11:139
  manufacture 11:118
  melamine–formaldehyde resin content 11:134
  N-methylol derivatives in 11:138
  properties 11:117
  sample preparation 11:118
  spot tests 11:139
  urea–formaldehyde resin content 11:134
  x-ray fluorescence spectroscopy 11:146, 151
Cotton cellulose. See Cellulose; and Cotton.
Cotton cloth, bleached
  fluidity of cellulose dispersions 7:235
  properties, determination of 7:233
Cotton dyeing test
  for triphenylmethane and related dyes 19:165
Cotton fabrics. See Cotton.
Cotton fibers. (See also Cotton.)
  microscopy 4:265
  mixtures, determination in 4:353
  solubility 4:248, 250, 353
Cotton finishing agents. See Finishing agents.
Cotton hull ash
  potassium content 17:406
Cottonseed flour
  commercial specifications 10:531
  composition 10:531
  impurities, determination of 10:542

sample preparation 10:542
Cottonseed meal
  carboxin content 13:213
  oxycarboxin content 13:213
Cottonseed oil. (See also *Fats and fatty oils.*)
  in alkyd resins 5:23
  properties 12:473
  solubility 13:539, 552
Cottrell-type boiling point apparatus 7:503
Cotunnite. See Lead chloride.
Coulometry
  of allyl alcohol 5:80
  of ceramics 9:217
  of chloride 15:142
  of chlorine in chlorocarbons 9:417
  for chromium 9:656
  for *unsym*-dimethylhydrazine 14:238
  of glass and glass-ceramics 13:458
  for hydrazine 14:238
  for hydrochloric acid 14:365
  for mercaptans 15:569, 571
  for methylhydrazine 14:238
  for oxygen 8:649
  for papaverine 16:272
  for sulfur 15:254
  of water in ammonia 5:306
Coumadin sodium. See Warfarin sodium.
Coumaphos. (See also *Pesticides and insecticides.*)
  uses 19:586
Coumaric acid. (See also *Cinnamic acid and its
    derivatives.*)
  properties 10:63
o-Coumaric acid
  paper chromatography 8:379
Coumarin. (See also *Cinnamic acid and its
    derivatives*; *Coumarins.*) 10:63; 11:159
  diazonium colorimetric method 11:166
  in flavors 11:166
  gas chromatography 11:163, 166, 168, 170
  manufacture 11:159
  occurrence 11:159
  ultraviolet spectroscopy 6:635
Coumarin derivatives. See *Coumarins.*
Coumarins 10:151; 11:159–173
  analysis 10:151; 11:161
  assay 11:162
  colorimetric determination 11:161, 162
  fluorescence spectroscopy 11:162
  gas chromatography 11:163, 167, 170
  gravimetric determination 11:162
  identification 10:151; 11:161
  ion exchange chromatography 11:166
  liquid chromatography 11:164
  occurrence 11:160
  paper chromatography 10:152; 11:164
  in perfumes 11:167
  in plants 11:167

polarography 11:163
separation 10:151
thin-layer chromatography 10:151; 11:164
ultraviolet spectroscopy 11:163
volumetric determination 11:162
Coumarone
  boiling point 16:209
Coumarone–indene resins. (See also *Hydrocarbon
    resins.*) 14:262
  cloud point 14:292
  color 14:293
  commercial grades and specifications 14:272,
    278
  density 14:267
  manufacture 14:261, 262, 263, 271
  polymer distribution 14:270
  structure 14:267
p-Coumarylquinic acid
  in green tea leaf 18:462, 475
  liquid chromatography 18:476
  structure 18:462
  thin-layer chromatography 18:463, 475
Countercurrent distribution
  of actinomycins 5:554
  of aniline and its derivatives 5:435
  of antimycin 5:602
  of aterrimins 5:604
  of bacitracin 5:537
  of candicidin 5:568
  of carbomycin 5:521
  of chloramphenicol 5:496
  of chromomycin 5:498
  of colistin 5:539
  of etamycin 5:555
  of ferrimycin 5:604
  of fumagillin 5:569
  of lincomycin hydrochloride 5:584
  of mitomycin C 5:499
  of neomycin 5:585
  of nisin 5:540
  of nystatin 5:569
  of ostreogrycin 5:556
  of stendomycins 5:605
  of streptomycin 5:587
  of trichomycins 5:571
  of tyrocidine 5:542
Courtelle. (See also Acrylic fibers; *Acrylic and
    modacrylic fibers.*) 4:220, 249
  cross section 4:223, 263
  infrared spectroscopy 4:246
  microscopy 4:264
  properties 4:222
  staining tests 4:255
Covellite
  copper content 10:600
CPPD. See *N*-Cyclohexyl-*N*′-phenyl-*p*-phenylene-
    diamine.

Crag fruit fungicide. See Glyodin.
Crankcase cleaners 6:357
  analysis 6:370
Crazy water crystals 18:244
Cream. (See also Confectionery products; Milk.)
    16:131
  analysis 16:131
  manufacture 16:131
  milkfat content 16:131
  pasteurization 16:131
Cream index
  of black tea 18:506
Creaming powder content
  of black tea 18:506
Cream of tartar 6:425
Creams and lotions. (See also *Cosmetics.*) 11:91
  alkyl sulfates in 11:93
  analysis 11:91
  ash content 11:92
  borate in 11:92
  carbonate in 11:92
  chloride in 11:92
  composition, determination of 11:94
  emulsifiers in 11:93
  p-hydroxybenzoic acid esters in 11:93
  nonvolatile materials in 11:92
  phosphate in 11:92
  potassium in 11:92
  propylene glycol content 11:92
  quaternary ammonium compounds in 11:93
  silicones in 11:92
  sodium in 11:92
  solvent extraction 11:94
  sulfate in 11:92
  titanium dioxide in 11:92
  types 11:91
  water content 11:91
  zinc oxide in 11:93
Creep
  of polymers 6:79
Cresidine
  manufacture 5:425
  properties 5:422
Creslan. (See also Acrylic fibers; *Acrylic and modacrylic fibers.*) 4:220
  birefringence 4:258
  cross section 4:223, 263
  infrared spectrum 4:241
  microscopy 4:264
  moisture regain 4:333
  properties 4:222
  solubility 4:247, 248
  staining tests 4:252
m-Cresol. (See also Cresols; *Phenols.*)
  4-aminoantipyrine colorimetric method 17:18
  colorimetric determination 17:30
  derivatives for identification 17:21

ferric chloride colorimetric method 17:16
gas chromatography 14:580; 17:33; 19:386
infrared spectroscopy 17:29
in insulin 14:580
manufacture 17:10
Millon's reagent color test 17:18
paper chromatography 17:24
properties 17:4
Raschig gravimetric method 17:28
selenous acid colorimetric method 17:20
solubility 17:5
thin-layer chromatography 17:24
m,p-Cresol. (See also Cresols; *Phenols.*)
  commercial grades and specifications 17:11
  manufacture 17:9
  specific gravity 17:44
o-Cresol. (See also Cresols; *Phenols.*)
  4-aminoantipyrine colorimetric method 17:18
  2-chloro-nitrobenzene-diazonium naphthalene-
    2-sulfonate colorimetric method 17:20
  colorimetric determination 17:30
  commercial grades and specifications 17:11
  congealing point test for cineole 12:286
  derivatives for identification 17:21, 28
  ferric chloride colorimetric method 17:16
  gas chromatography 17:30
  infrared spectroscopy 17:29
  manufacture 17:10
  Millon's reagent color test 17:18
  paper chromatography 17:24
  phthalic anhydride colorimetric method 17:19
  properties 17:4
  solubility 17:5
  specific gravity 17:44
  thin-layer chromatography 17:24
  in 2,6-xylenol 17:33
p-Cresol. (See also Cresols; *Phenols.*)
  4-aminoantipyrine colorimetric method 17:18
  2-chloro-nitrobenzene-diazonium naphthalene
    sulfonate colorimetric method 17:20
  colorimetric determination 17:32
  derivatives for identification 17:21
  ferric chloride colorimetric method 17:16
  gas chromatography 17:33
  infrared spectroscopy 17:29
  liquid chromatography 17:32
  manufacture 17:10
  Millon's reagent color test 17:18
  paper chromatography 17:24
  properties 17:4
  sodium nitroprusside colorimetric method 17:20
  solubility 17:5
  sulfanilic acid colorimetric method 17:32
  thin-layer chromatography 17:24
Cresol–modified phenol–formaldehyde resin
  pyrolysis–gas chromatography 7:446
o-Cresol monosulfide 6:76

Cresols. (See also Cresol; *Phenols*; and
    individual compounds.) 17:1
  color tests 17:20
  commercial grades and specifications 17:11
  ferric chloride colorimetric method 11:126, 131
  in finishing agents 11:126, 131
  gas chromatography 6:367
  in green tea leaf 18:459
  manufacture 17:9
  in metal cleaners 6:355, 367
  occurrence 17:1, 9
  in pharmaceuticals 11:548
  phenol content 11:542
  potassium permanganate titration 18:459
  potassium titanium oxalate colorimetric method
    18:459
  properties 17:4
  saponated cresol solution method 11:541
  in soap 11:541
  solubility 11:126
  ultraviolet spectroscopy 11:548
  uses 17:2
o-Cresotic acid (Methylsalicylic acid) 8:401
  properties 8:401, 402
Cresylic acid. (See also Cresols; *Phenols.*) 17:1
  commercial grades and specifications 17:12
  specific gravity 17:44
Cresylic resins
  in brake linings 7:441
Cresylite. (See also High explosives.) 12:437
  manufacture 12:437
  melting point 12:461
  properties 12:434
  solubility 12:437
  uses 12:437
Cristobalite
  composition 10:104
Critchfield and Johnson
  colorimetric method for ethylene oxide 12:328
Crocidolite. (See also Chrysotile.) 6:254
  analysis 6:267
  in asbestos–cement products 6:272
  composition 6:257
  electron micrograph 6:258
  properties 6:256
Crocoite. See Lead chromate.
Crookesite 18:519
Crosslink density
  of elastomers, synthetic 12:124
  of ethylene and propylene polymers 12:380
Crosslinked gel
  solubility 18:316
Crosslinking, degree of
  of epoxy resins 12:257
Crosslinking reactions
  of ethylene and propylene polymers 12:346
Crossover technique

  in bioassay 7:180
Crotonaldehyde 11:174–183
  absorption spectroscopy 9:564
  acidity 11:182
  analysis 11:176, 178
  assay 11:177
  bromination method 11:178
  color 11:182
  commercial grades and specifications 11:176
  commercial specifications 8:340
  derivatives for identification 11:177
  2,4-dinitrophenylhydrazine colorimetric method
    8:363
  gas chromatography 11:181; 16:108
  4-hexylresorcinol colorimetric method 11:180
  hydroxylamine hydrochloride oximation 11:178
  identification 11:176
  impurities, determination of 11:179
  infrared spectrum 8:345
  manufacture 11:175
  in methanol 16:108, 114
  paper chromatography 11:180
  m-phenylenediamine dihydrochloride colorimetric
    method 11:179
  polarography 11:181
  properties 8:334; 11:174
  refractometry 11:178
  specification tests 11:182
  specific gravity 11:182
  toxicity 11:175
  ultraviolet spectroscopy 6:632; 11:180
  water content 11:183
Crotonic acid 8:398
  properties 8:396, 398
  uses 8:398
Crotonic aldehyde. See *Crotonaldehyde.*
Crotonic anhydride
  infrared spectroscopy 8:562
Crotonitrile. (See also *Nitriles.*)
  gas chromatography 16:404
  properties 16:396
Crotyl acetate
  gas chromatography 11:181
Crown and bridge cement. See Dental zinc phos-
    phate cements.
Crucibles
  for carrier gas and vacuum fusion systems 8:640,
    642, 645, 671
  degassing 8:669
  preparation for induction heating 8:670
Crude fiber
  in cereals 9:237
  in chocolate 9:540
  in cocoa beans 9:548, 549
  in cocoa products 9:574
  in feed products 11:49
  in flour 12:606

in tobacco 19:142
in yeast 19:532
Crushing strength
of calcium hydroxide–salicylate dental products
11:298
of dental restorative materials, composite 11:313
Crylor. (See also Acrylic fibers; *Acrylic and
modacrylic fibers.*) 4:220, 249
cross section 4:226, 263
moisture regain 4:333
staining tests 4:255
Cryolite. (See also Sodium fluoaluminate.) 13:2
toxicity 13:13
Cryoscopic methods
for pesticides and insecticides 19:609
Cryptoxanthin
in green tea leaf 18:480
thin-layer chromatography 18:480
Crystal glazes. (See also Glazes.)
composition 9:196
Crystallinity
of acrylic polymers and fibers 4:315
by electron diffraction 4:316
of polyamide fibers 17:314
of polyethylene 12:361
of polytetrafluoroethylene 13:80
of triacetate fibers 4:77, 86
by x-ray diffraction 4:77, 86
Crystallinity, degree of
of ethylene and propylene polymers 12:381
Crystallinity index
of polycarbonates 17:333
Crystallite size
of graphite 14:76
Crystallography
of bishydroxycoumarin 10:155
Crystal size
of abrasive grains 7:308
of diazodinitrophenol 12:416
Crystal violet. (See also *Triphenylmethane and
related dyes.*)
acid–base titration 19:175
ceric sulfate titration 19:175
colorimetric method for tantalum 16:370;
18:449
dye capacity of acrylonitrile copolymer fibers
4:328
impurities, determination of 19:185
paper chromatography 19:172
polarography 19:182
thin-layer chromatography 19:173
titanous chloride titration 19:181
visible spectroscopy 19:166, 167
voltammetry 19:169
Cuam
fluidity of bleached cellulose in 7:235, 238
preparation of 7:239

standardization 7:239
viscometer 7:238
viscosity conversion tables 7:264, 265
Cube-in-water method
for softening point of bituminous materials
6:298
Cuen
fluidity of bleached cellulose in 7:235
preparation of 7:267
standardization 7:268
storage bottle 7:265
viscosity conversion tables 7:264, 265
viscosity of pulp dispersions in 7:265
Cumaldehyde. See Cuminal.
Cumene. (See also *Hydrocarbons.*) 14:321
commercial grades and specifications 14:322
gas chromatography 7:9
manufacture 14:321
occurrence 14:321
properties 14:321, 322
toxicity 14:321
uses 14:309, 321
o-Cumenol. See *o*-Isopropylphenol.
Cuminal (4(1-Methylethyl)benzaldehyde) 6:597
commercial grades and specifications 6:608, 609
derivatives for identification 6:612
hydroxylamine hydrochloride method 6:620
infrared spectrum 6:614
manufacture 6:601
properties 6:598
uses 6:606
Cuminic aldehyde
in cumin oil 12:282
hydroxylamine hydrochloride volumetric method
12:282
Cumin oil. (See also *Essential oils.*)
cuminic aldehyde content 12:282
Cundiff's method
for triacetin in cigarette filters 10:40
Cunico alloys. (See also Cobalt alloys.) 10:285
Cupellation
of gold 14:13
Cupellation beads
gold content 14:30
Cupferron
colorimetric method for aluminum oxide 9:126
colorimetric method for ferric oxide 9:126
extraction method for aluminum 5:204;
15:286
extraction method for bismuth 15:198
gravimetric method for gallium 13:253, 258
gravimetric method for iron 15:46
gravimetric method for niobium 10:318; 16:333,
366, 378, 386, 392
gravimetric method for tantalum 10:318; 16:366,
378, 392; 18:445
gravimetric method for tin 19:34, 40

gravimetric method for titanium 9:665; 14:252;
15:83; 16:391; 19:95, 101, 111
gravimetric method for titanium dioxide 13:426
gravimetric method for vanadium 19:329, 335
gravimetric method for zirconium 9:665; 14:
111, 252; 16:340
Cupferron–chloroform extraction
of cations in cobalt compounds 10:356
of iron 9:645
of vanadium 9:645
Cuprammonium hydroxide solvent. See Cuam.
Cuprethol
colorimetric method for copper 7:659
Cupric acetate–diethylamine
colorimetric method for carbon disulfide 8:277
Cupric acetate titration
of EDTA 12:38
of NTA 12:38
Cupric chloride
electrophoretic method for NTA 12:44
properties 9:334
solubility 9:336
volumetric method for NTA 12:34
Cupric dithiocarbamate
colorimetric method for carbon disulfide 7:43
Cupric nitrate–sodium diethyl dithiocarbamate
colorimetric method for poly(methacrylic acid)
16:92
Cupric nitrate titration
of Cheelox Iron 12 12:35
of Chel 138 12:34
of DHEG 12:35
of EDTA 12:35
of NTA 12:35
Cupric sulfate. See Copper sulfate.
Cupric tartrate
colorimetric method for reducing sugars 10:543
Cupriethylenediamine solvent. See Cuen.
Cuprite 10:669
copper content 10:600
Cuproauride 14:1
Cuprodescloizite
composition 19:321
occurrence 19:321
Cuproin. (See also 2,2'-Diquinolyl.)
colorimetric method for copper 8:174; 10:127,
298
Cupro–nickel alloys. (See also Copper alloys;
Nickel alloys.) 10:631
properties 10:631
uses 16:289
Cuprous chloride
absorbing agent for carbon monoxide 8:306
properties 9:334
solubility 9:336
Cuprous oxide
analysis 17:195

assay 17:195
commercial specifications 17:195
copper content 17:195
ferric chloride–ceric ammonium nitrate volumetric
method 17:195
gravimetric method for lactose 9:581
impurities, determination of 17:195
Munson-Walker values 9:583
reducing power 17:195
stability 17:196
uses 17:195
Cuprous sulfate–β-naphthol
absorbing agent for carbon monoxide 8:306,
308
Curare-type alkaloids 4:592
analysis 4:613
Curcumin. (See also Turmeric.)
colorimetric method for boric acid 7:388, 409
colorimetric method for boron 7:654; 10:309,
635
structure 10:539
Curcumin–oxalic acid
colorimetric method for boron 14:70
Cure evaluation tests
for epoxy resins 12:253
Cure temperatures
of epoxy resins 12:255
Curie 17:445
Curie temperature
of barium titanate 9:212
of cobalt ferrite 9:212
of copper ferrite 9:212
of iron ferrite 9:212
of lead zirconate 9:212
of magnesium ferrite 9:212
of manganese ferrite 9:212
of nickel ferrite 9:212
Curing ability
of urea formaldehyde resins 15:536
Curing agents 4:24
for epoxy resins 12:255
Curium. (See also Transuranium elements.)
half-life 19:259
isotopes 19:259
manufacture 19:259
Curium oxide
carrier gas fusion methods 8:626
oxygen in 8:626
Curl test
for matchbook covers 15:537
Custards. (See also *Eggs and egg products.*)
penetrometer test 12:72
Custom alloys
composition 16:351
Cut test
for cocoa beans 9:551
Cutting alloys. (See also Cobalt alloys.) 10:289

composition 10:289
Cutting quality
  of paper matchboard 15:534
Cuvée 19:408
Cyanacryl 12:86
Cyanamide process
  for ammonia 5:296
Cyanates
  in chlorides 9:340
Cyanides. (See also *Hydrogen cyanide*; and individual compounds.) 13:176–179; 14:424; 16:473
  analysis 13:177; 14:425
  in chlorides 9:340
  colorimetric determination 13:177
  distillation 13:177
  ferrous sulfate–sodium hydroxide colorimetric method 19:483
  gravimetric method for silver 18:164
  pyridine–benzidine colorimetric method 13:179
  pyridine–pyrazolone colorimetric method 13:178
  silver nitrate titration 13:178
  volumetric determination 13:177
  in wines 19:482
Cyanide solutions
  gold content 14:37
Cyanidin
  structure 19:461
  in wines 19:461, 464
Cyanin
  in cacao 9:536
  structure 18:461
1-Cyan-5-isocyanato-*n*-pentane. (See also *Isocyanic acid esters.*)
  infrared spectroscopy 15:106
1-Cyan-3-isocyanatopropane. (See also *Isocyanic acid esters.*)
  infrared spectroscopy 15:106
Cyanoacetamide
  analysis 15:442
  assay 15:442
  commercial specifications 15:425
  derivatives for identification 8:601; 15:442
  hydrolytic methods 15:443
  identification 15:442
  infrared spectroscopy 15:442
  manufacture 15:424
  paper chromatography 15:442
  properties 15:421
  separation 15:442
  solubility 15:422
  tetrabutylammonium methoxide titration 15:442
Cyanoacetic acid 8:395
  acid–base titration 15:439, 442
  analysis 15:439

assay 15:439
  in blood plasma 15:439
  colorimetric determination 15:439
  commercial specifications 15:425
  derivatives for identification 15:439
  diazotized sulfanilic acid colorimetric method 15:439
  in ethyl cyanoacetate 15:442
  ferrous nitrate–sulfur colorimetric method 15:439
  identification 15:439
  impurities, determination of 15:439
  infrared spectroscopy 15:439
  liquid chromatography 15:439
  manufacture 15:424
  in methyl cyanoacetate 15:442
  paper chromatography 15:439
  properties 8:395, 396; 15:421, 422
  solubility 15:422
4-Cyanobenzaldehyde
  fluorometry 6:625, 627
  infrared spectroscopy 6:616
  ultraviolet spectroscopy 6:634
1-Cyanocyclohexene oxide. (See also *Epoxy compounds.*)
  infrared spectroscopy 12:200
N-(2-Cyanoethyl)aziridine. (See also *Imines.*)
  properties 14:469
  structure 14:469
Cyanoethyl cellulose
  uses 9:60
N-(2-Cyanoethyl)cyclohexylamine
  gas chromatography 11:249
  infrared spectroscopy 11:248
Cyanoethylene. See Acrylonitrile.
Cyanogen
  gas chromatography 14:416
Cyanogen bromide
  colorimetric method for antihistamines 6:35
  colorimetric method for niacin 6:458
  colorimetric method for nicotinic acid 17:432
  colorimetric method for pyridine and its derivatives 17:424, 432
  colorimetric method for sugar 19:141
  manufacture 18:434
  volumetric method for thiocyanate 18:436
5-Cyanopentanoic acid
  in adipic acid 4:415
Cyanoprene 12:86
Cyanopyridines. (See also *Pyridine and pyridine derivatives.*)
  commercial grades and specifications 17:423
  ionization constants 17:417
  trisodium pentacyanoammino ferrate colorimetric method 17:433
Cyanuric chloride
  gas chromatography 14:416

Cyclamates. (See also Calcium cyclamate;
      Cyclamic acid; and Sodium cyclamate.)
      11:220–238
   analysis 11:224
   assay 11:224
   barium sulfate gravimetric method 11:224
   in biological substances 11:233, 238
   in carbonated beverages 8:173
   colorimetric determination 11:233
   cyclohexylamine content 11:217
   ferrous phenanthroline colorimetric method
      11:234
   in fruits 11:232
   gas chromatography 11:235
   identification 11:224
   impurities, determination of 11:226
   infrared spectroscopy 11:236
   ion exchange chromatography 11:234
   manufacture 11:222
   nephelometry 11:238
   nitrous acid titration 11:231
   paper chromatography 11:234
   polarography 11:237
   radiochemical methods 11:238
   sodium nitrite titration 11:225
   in soft drinks 11:233
   thin-layer chromatography 11:235
   volumetric determination 11:238
   in wine 11:233
Cyclamen aldehyde
   commercial specifications 10:44
Cyclamic acid. (See also Cyclamates.) 11:220
   commercial grades and specifications 11:223
   infrared spectroscopy 11:237
   properties 11:220
   solubility 11:221
Cyclic alcohols 11:184–197
   analysis 11:188
   assay 11:193
   colorimetric determination 11:194
   derivatives for identification 11:189
   ferric hydroxamate colorimetric method 11:195
   gas chromatography 11:193
   identification 11:189
   infrared spectroscopy 11:190
   manufacture 11:187
   mass spectroscopy 11:190
   near-infrared spectroscopy 11:196
   properties 11:184
   proton magnetic resonance spectroscopy 11:191
Cyclic ketones. See Cyclohexanone.
Cycloaliphatic epoxy resins. (See also Epoxy
      resins.) 12:235
   manufacture 12:235
Cycloalkanes. (See also Hydrocarbons; and
      individual compounds.) 14:306
   gas chromatography 14:330

   occurrence 14:306
Cyclobarbital 6:468, 488, 515
   in calcium cyclobarbital 6:516
   derivatives for identification 6:477
   gas chromatography 6:483, 484
   infrared spectrum 6:488
   paper chromatography 6:479
   properties 6:470
   solubility 6:476
   thin-layer chromatography 6:480, 481
Cyclobarbital calcium 6:515
   analysis 6:515
   bromometric titration 6:515
   commercial grades and specifications 6:515
   impurities, determination of 6:516
Cyclobutanol. (See also Cyclic alcohols.)
   derivatives for identification 11:189
   mass spectroscopy 11:190
   properties 11:185
Cyclobutanone
   2-hydroxy-1-naphthaldehyde colorimetric method
      11:204
Cyclobutyl alcohol. See Cyclobutanol.
Cyclodecanol. (See also Cyclic alcohols.)
   derivatives for identification 11:189
   properties 11:185
Cyclodecene oxides. (See also Epoxy compounds.)
   infrared spectroscopy 12:200
Cyclodecyl alcohol. See Cyclodecanol.
Cyclododecane. (See also Hydrocarbons.)
   properties 14:307, 309
Cyclododecanol. (See also Cyclic alcohols.)
   derivatives for identification 11:189
   properties 11:185
Cyclododecene. (See also Hydrocarbons.)
   commercial specifications 14:316
Cyclododecyl alcohol. See Cyclododecanol.
Cycloeicosanol. (See also Cyclic alcohols.)
   derivatives for identification 11:189
   properties 11:185
Cycloeicosyl alcohol. See Cycloeicosanol.
Cycloheptadecanol. (See also Cyclic alcohols.)
   derivatives for identification 11:189
   properties 11:185
Cycloheptadecyl alcohol. See Cycloheptadecanol.
Cycloheptane
   gas chromatography 13:351
   in gasoline 13:351, 354
   properties 13:351
Cycloheptanol. (See also Cyclic alcohols.)
   derivatives for identification 11:189
   mass spectroscopy 11:190
   properties 11:185
Cycloheptanone
   2-hydroxy-1-naphthaldehyde colorimetric method
      11:204
Cycloheptyl alcohol. See Cycloheptanol.
Cyclohexadecanol. (See also Cyclic alcohols.)

derivatives for identification 11:189
properties 11:185
Cyclohexadecyl alcohol. See Cyclohexadecanol.
1,3-Cyclohexadiene
  liquid chromatography 7:13, 14, 15, 16
1,4-Cyclohexadiene
  liquid chromatography 7:13, 14, 15, 16
Cyclohexane. (See also *Hydrocarbons.*) 14:306
  azeotropes 14:307
  in benzene 7:11
  commercial grades and specifications 14:308
  gas chromatography 7:11; 11:73, 192, 212, 213;
    13:319, 350; 14:330
  in gasoline 13:350, 353
  infrared spectroscopy 7:30
  in nail enamels 11:73
  properties 13:350; 14:306
  solubility 14:306
  toxicity 14:307
  uses 14:307
Cyclohexanediol
  gas chromatography 13:579
1,2-Cyclohexanedione dioxime. See Nioxime.
Cyclohexanethiol. See Cyclohexyl mercaptan.
Cyclohexanol. (See also *Cyclic alcohols.*)
  commercial grades and specifications 11:188
  derivatives for identification 11:187, 189
  in enamels 12:175
  gas chromatography 11:192, 212, 213, 245
  infrared spectrum 11:189, 196
  manufacture 11:187
  mass spectroscopy 11:190
  phthalic anhydride method 4:535
  properties 11:184, 185, 186; 12:175
  pyromellitic dianhydride method 4:537
  solubility 11:184
  ultraviolet spectroscopy 11:196
  use in manufacture of fatty alcohols 4:560
  uses 11:184
Cyclohexanone 11:197–208
  in acetylenic chemicals 4:145
  acid–base titration 11:206
  analysis 11:201
  assay 11:206
  azeotropes 11:198
  in caprolactam 8:121
  commercial grades and specifications 11:200
  o-cyclohexylphenol colorimetric method 11:205
  derivatives for identification 8:343; 11:204
  gas chromatography 11:192, 206, 241
  H-acid diazonium salt colorimetric method 11:
    194, 205
  hydroxylamine hydrochloride method 8:357
  2-hydroxy-1-naphthaldehyde colorimetric
    method 11:204
  identification 11:201
  impurities, determination of 11:207

infrared spectroscopy 11:196
  manufacture 11:200
  mass spectroscopy 11:202
  12-molybdosilicic acid colorimetric method
    11:204
  nuclear magnetic resonance spectroscopy 11:201
  polarography 11:207
  properties 11:197
  ultraviolet spectroscopy 11:206
  use in manufacture of caprolactam 8:117
  uses 11:197
Cyclohexanone oxime
  in caprolactam 8:121, 125
Cyclohexene. (See also *Hydrocarbons.*)
  commercial grades and specifications 14:314,
    315
  gas chromatography 7:8; 11:212, 213
  liquid chromatography 7:13, 14, 15, 16
  toxicity 14:314
  uses 14:314
1,2-Cyclohexene oxide. (See also *Epoxy com-
    pounds.*)
  infrared spectroscopy 12:200
  properties 12:195
2-(1-Cyclohexenyl)cyclohexanone
  in cyclohexanone 11:207
  mass spectroscopy 11:202
  polarography 11:207
  properties 11:207
Cycloheximide 5:462, 508
  analysis 5:510, 512, 516; 13:223
  bioassay 13:223
  bioautography 5:510
  colorimetric methods 5:516
  in fruit 13:223
  hydroxylamine hydrochloride–ferric perchlorate
    colorimetric method 13:224
  identification 5:510
  microbiological assay 5:512
  paper chromatography 5:510
  properties 5:509; 13:223
  separation 5:510
  solubility 5:509, 510; 13:223
  solvent extraction 13:223
  structure 5:463
  toxicity 13:223
  uses 5:508; 13:223
Cyclohexyl acetate
  gas chromatography 11:192
N-Cyclohexyl-*β*-alanine
  properties 11:244
Cyclohexyl alcohol. See Cyclohexanol.
Cyclohexylamine 11:209–219
  acid–base titration 11:213
  Alkacid paper color test 11:227
  analysis 11:211
  assay 11:213

1-chloro-2,4,6-trinitrobenzene colorimetric method 11:228
color 11:216
commercial grades and specifications 11:210
in cyclamates 11:217, 227
derivatives for identification 11:211
in drugs 11:217
in food products 11:217
gas chromatography 11:212, 213, 214, 215, 245, 249
identification 11:211
impurities, determination of 11:214
infrared spectroscopy 11:211
manufacture 11:210
methyl orange–boric acid colorimetric method 11:228
methyl orange colorimetric method 11:217
paper chromatography 11:211
properties 5:260; 11:209
salicylaldehyde colorimetric method 11:227
separation 11:211
solubility 11:209
thin-layer chromatography 11:212
toxicity 11:210
in urine 11:212
uses 5:264; 11:209
Cyclohexylamine derivatives. (See also Cycla-mates; and individual compounds.) 11:220–255
Cyclohexylamines
manufacture 5:263
uses 5:264
2-Cyclohexylamino-1-phenylethanol 11:250
infrared spectroscopy 11:250
manufacture 11:250
properties 11:244
solubility 11:250
toxicity 11:250
uses 11:250
1-Cyclohexylamino-2-propanol
colorimetric method using methyl orange 5:392
N-Cyclohexyl-2-benzothiazole sulfenamide
in elastomers, synthetic 12:143
ultraviolet spectrum 12:143
Cyclohexyl cyanide. (See also *Nitriles.*)
properties 16:396
(1,2-Cyclohexylenedinitrilo)tetraacetic acid. (See also DCYTA.)
volumetric method for aluminum 5:218
Cyclohexylethylene oxide. (See also *Epoxy compounds.*)
infrared spectroscopy 12:200
N-Cyclohexylhydroxylamine. (See also *N*-Alkyl-hydroxylamines.)
manufacture 14:444
properties 14:443
uses 14:443

N-Cyclohexylidene aniline
gas chromatography 11:212, 241
2-Cyclohexylidenecyclohexanone
in cyclohexanone 11:207
manufacture 11:207
polarography 11:207
properties 11:207
N-Cyclohexylidenecyclohexylamine
gas chromatography 11:212, 241
Cyclohexyl isocyanate. (See also *Isocyanic acid esters.*)
commercial sources 15:102
infrared spectroscopy 15:106
properties 15:95
Cyclohexyl mercaptan. (See also *Mercaptans.*)
derivatives for identification 15:559
properties 15:552
o-Cyclohexylphenol
colorimetric method for cyclohexanone 11:205
N-Cyclohexyl-*N*′-phenyl-*p*-phenylenediamine (CPPD) 6:72
gas chromatography 6:81, 108
α-Cyclohexyl-α-phenyl-1-piperidinepropanol
gas chromatography 6:210, 211
Cyclohexylsulfamates
in carbonated beverages 8:173
Cyclohexylsulfamic acid. See Cyclamic acid.
Cyclonite. See RDX.
Cyclononadecanol. (See also *Cyclic alcohols.*)
derivatives for identification 11:189
properties 11:185
Cyclononadecyl alcohol. See Cyclononadecanol.
Cyclononanol. (See also *Cyclic alcohols.*)
derivatives for identification 11:189
properties 11:185
Cyclononanone
2-hydroxy-1-naphthaldehyde colorimetric method 11:204
Cyclononyl alcohol. See Cyclononanol.
Cyclooctadecanol. (See also *Cyclic alcohols.*)
derivatives for identification 11:189
properties 11:185
Cyclooctadecyl alcohol. See Cyclooctadecanol.
1,5-Cyclooctadiene 11:513
assay 11:514
commercial specifications 11:513
manufacture 11:513
properties 11:513
toxicity 11:514
uses 11:513
Cyclooctane. (See also *Hydrocarbons.*)
properties 14:307, 309
Cyclooctanol. (See also *Cyclic alcohols.*)
derivatives for identification 11:189
mass spectroscopy 11:190
properties 11:185
Cyclooctanone

2-hydroxy-1-naphthaldehyde colorimetric
  method 11:204
Cyclooctene. (See also *Hydrocarbons.*)
  commercial specifications 14:316
Cyclooctene oxides. (See also *Epoxy compounds.*)
  infrared spectroscopy 12:200
Cyclooctyl alcohol. See Cyclooctanol.
Cyclopal 6:468
  bromide–bromate titration 6:497
  derivatives for identification 6:477
  gas chromatography 6:483, 484
  properties 6:470
  solubility 6:476
  thin-layer chromatography 6:481
Cyclopal sodium 6:488
  infrared spectrum 6:488
Cycloparaffins. See Cycloalkanes.
Cyclopentadecanol. (See also *Cyclic alcohols.*)
  derivatives for identification 11:189
  properties 11:185
Cyclopentadecanone
  2-hydroxy-1-naphthaldehyde colorimetric method
    11:204
Cyclopentadecyl alcohol. See Cyclopentadecanol.
Cyclopentadiene 11:511
  analysis 11:512
  azeotropes 11:511
  in 1,3-butadiene 11:504
  fulvene colorimetric method 11:512
  gas chromatography 11:510
  in hydrocarbons 11:512
  in isoprene 11:510
  manufacture 11:511
  properties 11:511
  uses 11:511
Cyclopentane. (See also *Hydrocarbons.*)
  commercial grades and specifications 14:307,
    308
  gas chromatography 13:350; 14:330
  in gasoline 13:350, 353
  properties 13:350
  uses 14:307
Cyclopentanol. (See also *Cyclic alcohols.*)
  commercial grades and specifications 11:188
  derivatives for identification 11:189
  gas chromatography 11:192
  manufacture 11:188
  mass spectroscopy 11:190
  properties 11:185
  uses 11:184
Cyclopentanol–sodium hydroxide colorimetric
    reagent
  for carbon tetrachloride 9:387
  for methylene chloride 9:387
Cyclopentanone
  derivatives for identification 8:343
  gas chromatography 11:192

hydroxylamine hydrochloride method 8:357
2-hydroxy-1-naphthaldehyde colorimetric method
    11:204
  mass spectroscopy 7:31
Cyclopentene
  gas chromatography 13:350
  in gasoline 13:350, 353
  properties 13:350
1,2-Cyclopentene oxide. (See also *Epoxy compounds.*)
  infrared spectroscopy 12:200
  properties 12:195
2-Cyclopentenol
  acetylation with acetic anhydride 4:530
  phthalic anhydride method 4:536
Cyclopentenylallyl barbituric acid
  column chromatography 6:150
  ultraviolet spectroscopy 6:150
Cyclopentyl alcohol. See Cyclopentanol.
1-Cyclopentyl-2-aminopropane 5:260
Cyclopropane
  analysis 5:359
  anesthetic 5:359
  assay 5:360
  impurities, determination of 5:360
  properties 5:357, 359
Cyclopropane 1,1-dicarboxylic acid
  properties 8:396
Cyclopropanol. (See also *Cyclic alcohols.*)
  derivatives for identification 11:189
  properties 11:185
Cyclopropene groups
  in fats and fatty oils 12:484
  Halphen test 12:484
  sulfur colorimetric method 12:484
Cyclopropyl alcohol. See Cyclopropanol.
Cyclopropyl methyl ether. (See also *Ethers.*)
  properties 12:297
Cycloserine 5:508, 513
  analysis 5:510, 513, 516
  colorimetric methods 5:516
  infrared spectrum 5:511
  microbiological assay 5:513
  properties 5:509
  solubility 5:509, 510
  structure 5:463
  uses 5:508
Cyclotetradecanol. (See also *Cyclic alcohols.*)
  derivatives for identification 11:189
  properties 11:185
Cyclotetradecyl alcohol. See Cyclotetradecanol.
Cyclotetramethylenetetranitramine. See HMX.
Cyclotol. (See also High-explosive mixtures.)
  analysis 12:447
  composition 12:447
  uses 12:447
Cyclotridecanol. (See also *Cyclic alcohols.*)

derivatives for identification 11:189
properties 11:185
Cyclotridecyl alcohol. See Cyclotridecanol.
Cyclotrimethylenetrinitramine. See RDX.
Cycloundecanol. (See also *Cyclic alcohols.*)
  derivatives for identification 11:189
  properties 11:185
Cycloundecyl alcohol. See Cycloundecanol.
Cylindrite 14:519; 19:22
p-Cymene. (See also Terpenes.)
  commercial specifications 14:316
  gas chromatography 12:266, 268
  in majoram, oil of 12:268
  manufacture 14:316
  in peppermint oil 12:266
  properties 14:317
  uses 14:317
Cyprex. See Dodine.
Cyrtolite 14:104
Cysteic acid
  ion exchange chromatography 8:415, 421
  paper chromatography 8:422
Cysteine. (See also *Carboxylic acids, amino*; and
    Mercaptans.) 8:412, 463
  acid hydrolysis 8:467
  analysis 8:466
  p-chloromercuribenzoic acid colorimetric method
    15:561
  colorimetric method for menadione 10:196
  colorimetric method using dimethyl-*p*-phenylene-
    diamine 8:468
  crystalline form 8:464
  gas chromatography 8:429
  hydrazine hydrolysis 8:467

manufacture 8:466
occurrence 15:551
phospho-18-tungstic acid colorimetric method
  15:561
properties 8:463
Cystine. (See also *Carboxylic acids, amino.*)
    8:412, 462
  acid hydrolysis 8:467
  analysis 8:466
  colorimetric method using dimethyl-*p*-phenylene-
    diamine 8:468
  colorimetric method using sodium naphthoqui-
    none-4-sulfonate 8:467
  commercial specifications 8:466
  crystalline form 8:463
  derivatives 8:466
  gas chromatography 8:429
  in gelatin 12:552
  hydrazine hydrolysis 8:467
  ion exchange chromatography 8:415, 421
  manufacture 8:466
  occurrence 8:466
  optical rotation 8:444
  paper chromatography 8:423
  properties 8:463
  solubility 8:465
  in yeast 19:529
Cystosine
  in gelatin 12:552
  ion exchange chromatography 12:552
  ultraviolet spectroscopy 12:553
Cytisine 4:590
  analysis 4:609
Cyzone IP 6:74

D

Dacron. (See also Polyester fibers.)
mixtures, determination in 4:353
solubility 4:353
DADA. See Dianisidinediamine.
DAIH
colorimetric method for mercury 16:22
Danburite 7:316
D & C Blue No. 3. See D & C Violet No. 2.
D &C Blue No. 4. (See also Certified colors.)
classification 10:452
commercial specifications 10:465
impurities, determination of 10:500, 519
liquid chromatography 10:500
manufacture 10:452
structure 10:452
titanous chloride titration 10:485
visible spectroscopy 10:486
D & C Blue No. 6. (See also Certified colors.)
classification 10:452
commercial specifications 10:465
liquid chromatography 10:500
structure 10:452
titanous chloride titration 10:485
visible spectroscopy 10:486
D & C Blue No. 9. (See also Certified colors.)
classification 10:452
commercial specifications 10:465
impurities, determination of 10:500
liquid chromatography 10:500
manufacture 10:452
structure 10:452
D & C colors. See Certified colors; and individual
compounds.
D & C Green No. 5. (See also Certified colors.)
classification 10:452
commercial specifications 10:465
impurities, determination of 10:500, 508, 519
liquid chromatography 10:500
manufacture 10:452
paper chromatography 10:475
structure 10:452
titanous chloride titration 10:485
visible spectroscopy 10:486
D & C Green No. 6. (See also Certified colors.)
classification 10:452
commercial specifications 10:465
impurities, determination of 10:500, 508
liquid chromatography 10:500
manufacture 10:452
paper chromatography 10:475
structure 10:452

titanous chloride titration 10:485
visible spectroscopy 10:486
D & C Green No. 8. (See also Certified colors.)
classification 10:52
commercial specifications 10:465
impurities, determination of 10:500, 508
liquid chromatography 10:500
manufacture 10:452
structure 10:452
visible spectroscopy 10:486
D & C Lakes. (See also Certified colors.)
commercial specifications 10:470
D & C Orange No. 4. (See also Certified colors.)
classification 10:454
commercial specifications 10:465
impurities, determination of 10:500, 519
in lipsticks 11:69
liquid chromatography 10:500
manufacture 10:454
structure 10:454
thin-layer chromatography 11:69
titanous chloride titration 10:485
visible spectroscopy 10:486; 11:70
D & C Orange No. 5. (See also Certified colors.)
classification 10:454
commercial specifications 10:466
impurities, determination of 10:501, 520
in lipsticks 11:69
liquid chromatography 10:501
manufacture 10:454
structure 10:454
thin-layer chromatography 11:69
visible spectroscopy 10:486; 11:70
D & C Orange No. 10. (See also Certified colors.)
classification 10:454
commercial specifications 10:466
impurities, determination of 10:501
liquid chromatography 10:500
manufacture 10:454
structure 10:454
visible spectroscopy 10:486
D & C Orange No. 11. (See also Certified colors.)
classification 10:454
commercial specifications 10:466
impurities, determination of 10:501
liquid chromatography 10:501
manufacture 10:454
structure 10:454
D & C Orange No. 17. (See also Certified colors.)
classification 10:454
commercial specifications 10:466

thin-layer chromatography 11:69
visible spectroscopy 10:486; 11:70
D & C Red No. 22. (See also Certified colors.)
  classification 10:456
  commercial specifications 10:468
  gravimetric determination 10:487
  impurities, determination of 10:501, 523
  liquid chromatography 10:501
  manufacture 10:456
  structure 10:456
  visible spectroscopy 10:486
D & C Red No. 27. (See also Certified colors.)
  classification 10:456
  commercial specifications 10:468
  in lipsticks 11:69
  liquid chromatography 10:501
  manufacture 10:456
  structure 10:456
  thin-layer chromatography 11:69
  visible spectroscopy 10:486; 11:70
D & C Red No. 28. (See also Certified colors.)
  classification 10:458
  commercial specifications 10:468
  gravimetric determination 10:487
  liquid chromatography 10:501
  manufacture 10:458
  structure 10:458
  visible spectroscopy 10:486
D & C Red No. 30. (See also Certified colors.)
  classification 10:458
  commercial specifications 10:469
  in lipsticks 11:69
  liquid chromatography 10:501
  manufacture 10:458
  structure 10:458
  thin-layer chromatography 11:69
  titanous chloride titration 10:485
  visible spectroscopy 10:486; 11:70
D & C Red No. 31. (See also Certified colors.)
  classification 10:458
  commercial specifications 10:469
  impurities, determination of 10:501, 522
  liquid chromatography 10:501
  manufacture 10:458
  structure 10:458
  titanous chloride titration 10:485
  visible spectroscopy 10:486
D & C Red No. 33. (See also Certified colors.)
  classification 10:458
  commercial specifications 10:469
  impurities, determination of 10:502, 523
  liquid chromatography 10:502
  manufacture 10:458
  structure 10:458
  titanous chloride titration 10:485
  visible spectroscopy 10:486
D & C Red No. 34. (See also Certified colors.)

classification 10:458
commercial specifications 10:469
impurities, determination of 10:502, 504
in lipsticks 11:69
liquid chromatography 10:502
manufacture 10:458
structure 10:458
thin-layer chromatography 11:69
titanous chloride titration 10:485
visible spectroscopy 10:486; 11:70
D & C Red No. 36. (See also Certified colors.)
  classification 10:458
  commercial specifications 10:469
  impurities, determination of 10:502
  in lipsticks 11:69
  liquid chromatography 10:502
  manufacture 10:458
  structure 10:458
  thin-layer chromatography 11:69
  titanous chloride titration 10:485
  visible spectroscopy 10:486; 11:70
D & C Red No. 37. (See also Certified colors.)
  classification 10:458
  commercial specifications 10:469
  impurities, determination of 10:502, 510
  liquid chromatography 10:502
  manufacture 10:458
  structure 10:458
  titanous chloride titration 10:485
  visible spectroscopy 10:486
D & C Violet No. 2. (See also Certified colors.)
  classification 10:458
  commercial specifications 10:469
  impurities, determination of 10:502, 508, 510
  liquid chromatography 10:502
  manufacture 10:458
  paper chromatography 10:475
  structure 10:458
  visible spectroscopy 10:486
D & C Yellow No. 7. (See also Certified colors.)
  classification 10:460
  commercial specifications 10:470
  impurities, determination of 10:502
  liquid chromatography 10:502
  manufacture 10:460
  structure 10:460
  titanous chloride titration 10:485
  visible spectroscopy 10:486
D & C Yellow No. 8. (See also Certified colors.)
  classification 10:460
  commercial specifications 10:470
  in D & C Red No. 22 10:523
  impurities, determination of 10:502
  liquid chromatography 10:502, 523
  manufacture 10:460
  structure 10:460
  titanous chloride titration 10:485

visible spectroscopy 10:486, 523
D & C Yellow No. 10. (See also Certified colors.)
  classification 10:460
  commercial specifications 10:470
  impurities, determination of 10:502, 524
  liquid chromatography 10:502
  manufacture 10:460
  structure 10:460
  visible spectroscopy 10:486
D & C Yellow No. 11. (See also Certified colors.)
  classification 10:460
  commercial specifications 10:470
  impurities, determination of 10:502
  liquid chromatography 10:502
  manufacture 10:460
  structure 10:460
  visible spectroscopy 10:486
Daphnetin. See 7,8-Dihydroxycoumarin.
DAPTA. See DPTA.
Darvan. (See also *Acrylic and modacrylic fibers.*)
    4:220, 247
  infrared spectrum 4:242
  microscopy 4:264
  moisture regain 4:333
  staining tests 4:252
Dates
  diethylene glycol in 12:332
  ethylene chlorohydrin in 12:332
  ethylene glycol in 12:332
  ethylene oxide in 12:332; 13:192
Datolite 7:316; 8:75
Davidite
  composition 19:225, 321
  occurrence 19:225, 321
DBX. (See also High-explosive mixtures.)
  analysis 12:448
  composition 12:448
  uses 12:448
DCPD. See *N,N'*-Dicyclohexyl-*p*-phenylene-
    diamine.
DCYTA. (See also *EDTA and related com-
    pounds.*)
  cobaltous nitrate–ferric chloride colorimetric
    method 12:28
  structure 12:24
  thorium nitrate–hexamethylenetetraamine colori-
    metric method 12:28
  volumetric method for titanium 19:107
DDD. See TDE.
DDE
  gas chromatography 19:389
  in raw milk 16:129
  in water 19:389
D-D soil fumigant 13:181
  composition 13:181
  gas chromatography 13:183
  manufacture 13:181

properties 13:181
solubility 13:181
toxicity 13:181
uses 13:181
DDT. (See also *Pesticides and insecticides.*)
  commercial grades and specifications 19:593
  composition 19:588, 596
  dosage-mortality of houseflies 7:161
  gas chromatography 6:210; 14:376; 19:389
  manufacture 19:596
  properties 19:588
  in raw milk 16:129
  solubility 19:587
  structure 19:580
  toxicity 19:591, 592
  in water 19:389
DDVP. See Dichlorvos.
Deacetylcephaloglycin. (See also Cephalosporins.)
  assay 16:579
Deacetylcephalosporin C. (See also Cephalosporins.)
  assay 16:579
Dean and Starks method
  for water in furfural 13:239
Decaborane
  acid–base titration, nonaqueous 7:346
  in air 7:349
  analysis 7:345
  colorimetric methods 7:347
  infrared spectra 7:354
  iodometric titration 7:346
  monitors for detection in air 7:349
  near-infrared spectrum 7:355
Decaborane(14) 7:342, 354
  properties 7:343
Decaborane(16) 7:342
Decaboranes
  gas chromatography 7:353
Decachlorobiphenyls. (See also Chlorinated
    biphenyls.)
  properties 9:492
trans, trans-2,4-Decadienal
  in black tea 18:511
  gas chromatography 18:511
Decaffeinated coffee 10:376
  analysis 10:440
  caffeine content 10:409
  gas chromatography 10:440
  impurities, determination of 10:440
  manufacture 10:376
Decafluorobutane. (See also *Fluorocarbons.*)
  properties 13:60
Decahydronaphthalene. (See also Hydrogenated
    naphthalenes.)
  gas chromatography 16:220
  manufacture 16:219
  properties 16:219
  uses 16:219

Decalin. See Decahydronaphthalene.
Decamethylenediamine
  gas chromatography 17:290
Decanal
  2,4-dinitrophenylhydrazine colorimetric method
      8:363
  gas chromatography 8:355
n-Decane. (See also Alkanes.)
  commercial specifications 14:305
  gas chromatography 7:8; 13:352; 14:333, 336
  in gasoline 13:352, 354
  occurrence 14:303
  properties 13:352
1-Decanethiol. See Decyl mercaptan.
Decanol. (See also Alcohols, higher; Fatty alcohols.)
  commercial grades 4:555, 563
  properties 4:552, 561
  uses 4:556
1-Decanol
  gas chromatography 7:96; 8:355
n-Decanonitrile. (See also Nitriles.)
  derivatives for identification 16:402
  properties 16:396
1-Decene. (See also Alpha olefins.)
  commercial specifications 14:314
Decolorizing carbon. See Carbon, activated.
Decolorizing efficiency
  of activated carbon 4:448
  of fine mesh clays 4:447
Decortin. See Desoxycorticosterone.
Decyl alcohol. See Decanol.
Decyl aldehyde
  in grapefruit oil 12:282
  hydroxylamine hydrochloride volumetric method
      12:282
  in mandarin oil 12:282
  in orange oil 12:282
n-Decylbenzene
  liquid chromatography 7:13, 14, 15, 16
Decyl 3,5-dinitrobenzoate
  thin-layer chromatography 16:94
Decyl mercaptan. (See also Mercaptans.)
  properties 15:552
n-Decyl methacrylate
  thin-layer chromatography 16:90
Deenax 6:74
  colorimetric method using diazotized p-nitro-
      aniline 6:106
Deep maroon. See D & C Red No. 34.
Deformation tests
  of dental elastomers 11:285
  of dental impression materials, agar-based
      11:280
  of dental impression materials, alginate-based
      11:282
DEGN. (See also High explosives.)
  infrared spectrum 12:456

manufacture 12:445
melting point 12:461
properties 12:432
solubility 12:445
uses 12:445
Degreasers, engine 6:355
  analysis 6:367
  performance tests 6:378
Degrees Baumé 18:341
  of starch hydrolysates 11:42
Degrees Brix 18:341
Degrees of freedom 7:171
De Haens salt 6:49
Dehydroacetic acid
  in beer 7:596
Dehydroallohyodeoxycholic acid
  properties 7:143
  structure 7:142
Dehydrocholic acid
  acid–base titration 7:149
  color tests 7:147
  commercial specifications 7:145
  impurities, determination of 7:150
  properties 7:143
  structure 7:142
11-Dehydrocorticosterone. (See also Steroids.)
  structure 14:184
Dehydrodeoxycholic acid
  color tests 7:147
  properties 7:143
  structure 7:142
Dehydroepiandrosterone
  gas chromatography 14:216
11-Dehydro-17-hydroxycorticosterone. (See also
      Steroids.)
  structure 14:184
α-Dehydrohyodeoxycholic acid
  color tests 7:147
  properties 7:143
  structure 7:143
Dehydroindigo. (See also Indigoid dyes.)
  manufacture 14:500
  structure 14:500
De-icers 6:356
  analysis 6:370
  performance tests 6:378
Delphinidin
  structure 18:461; 19:461
  in wines 19:461
Delsan. See Thiram.
Deltacor. See Prednisolone.
Deltasone. See Prednisone.
Deltastab. See Prednisolone.
Delusterant
  in acrylic and modacrylic fibers 4:329
Dely-type gas buret 5:303
Demerol. See Meperidine; Meperidine hydrochloride.

Demosan. See Chloroneb.
Demulsibility tests
  for emulsified asphalts 6:327
Denatured alcohol
  denaturants 4:511
  uses 4:507
Denier uniformity
  of polyamide fibers 17:323
Densite. (See also Dental gypsum products.)
  uses 11:263
Density. (See also Bulk density; Skeletal density;
    and Specific gravity.)
  of abrasive grains 10:267
  of adsorbents 4:443
  of ammonium hydroxide 5:317
  of benzene–acetone–water mixtures 7:38
  of benzene–carbon tetrachloride–acetic acid
    mixtures 7:37
  of carbon black 8:230
  of cellulose acetate polymers 4:75
  of chlorine 9:307
  of coal and coke 10:226
  of copper metal 10:627
  of corn products 11:15
  of detergents 11:384
  of diethylenetriamine 11:409
  of dipropylene glycol 13:584
  of enamels 12:178
  of ethanol 19:432
  of ethylene and propylene polymers 12:381
  of ethylenediamine 11:407
  of glass and glass-ceramics 13:473
  of glycols 13:584
  of grinding wheels 7:296
  of hexylene glycol 13:584
  of hydrocarbon resins 14:267, 294
  linear, of fibers 4:89
  of lubricants 15:342
  of pentaethylenehexamine 11:409
  of pesticides and insecticides 19:617
  of pigments 17:207
  of polyamide fibers 17:314
  of polytetrafluoroethylene 13:80, 85
  of polyurethane foams 19:281
  of propylene glycol 13:584
  of pulp handsheets 7:256
  of semicrystalline polyacrylonitrile 4:318
  of sugar syrup 8:177
  of tetraethylenepentamine 11:409
  of triethylene glycol 13:584
  of triethylenetetraamine 11:409
Density gradient method
  specific gravity of acrylic and modacrylic fibers
    4:259
Density gradient ultracentrifugation
  measurement of microgel in polymers 4:310
Density, temperature of maximum

  of beer 7:502
Dental alloys. (See also Cobalt alloys; Dental
    amalgams; Dental cast gold alloys; Dental
    chromium–cobalt alloys; and Dental wrought
    gold alloys.) 10:290; 11:287–295
  composition 10:290
  gold content 14:36
Dental amalgams 11:288
  amalgamation time 11:290
  analysis 11:288
  assay 11:288
  commercial grades and specifications 11:257,
    288
  composition 11:288
  copper content 11:289
  flow tests 11:290
  manufacture 11:287
  non-volatile materials in 11:291
  setting expansion 11:291
  silver content 11:289
  specification tests 11:290
  tin content 11:288
  uses 11:287
  zinc content 11:288
Dental cast gold alloys 11:292
  analysis 11:292
  commercial grades and specifications 11:257,
    292
  elongation 11:293
  fusion temperature 11:293
  hardness 11:293
  tensile strength 11:293
  types 11:288
  uses 11:288
Dental casting golds. See Dental cast gold alloys.
Dental ceramics. (See also Dental gypsum pro-
    ducts; Dental investments; Dental porcelains;
    Dental silicate cements; and Dental zinc phos-
    phate cements.) 11:257–278
Dental chromium–cobalt alloys 11:291
  analysis 11:291
  commercial grades and specifications 11:257,
    291
  composition 11:291
  elongation 11:292
  fusion temperature 11:292
  hardness 11:292
  tensile strength 11:292
  uses 11:288
  yield strength 11:291
Dental colloids. (See also Dental impression ma-
    terials, agar-based; Dental impression materials;
    alginate-based.) 11:278–283
Dental cream. (See also *Detergents.*) 11:325
  composition 11:326
Dental elastomers. (See also Dental impression
    materials, polysulfide-based; Dental impres-

sion materials, silcone-based.) 11:283–287
analysis 11:284
commercial specifications 11:257
compressive strength 11:286
consistency 11:285
deformation tests 11:285
flow tests 11:286
gypsum compatibility 11:286
metalizing bath compatibility 11:287
stability, dimensional 11:286
standard mix 11:285
working time 11:285
Dental gypsum products 11:263
analysis 11:263
commercial specifications 11:263
compressive strength 11:266
cone penetration test 11:265
consistency 11:264
detail reproduction 11:267
fineness 11:266
properties 11:264
sample preparation 11:264
setting expansion 11:266
slump test 11:264
time of setting 11:266
Dental impression compounds 11:317
analysis 11:317
commercial grades and specifications 11:257, 317
composition 11:316
flow tests 11:317
impression test 11:317
Dental impression materials, agar-based 11:278
analysis 11:278
commercial specifications 11:257, 278
composition 11:278
compressive strength 11:280
consistency 11:279
deformation tests 11:280
gelation temperature 11:279
gypsum compatibility 11:279
Dental impression materials, alginate-based 11:280
analysis 11:280
commercial specifications 11:257
composition 11:280
compressive strength 11:283
deformation tests 11:282
deterioration 11:283
fineness 11:281
gypsum compatibility 11:282
setting time 11:281
working time 11:281
Dental impression materials, polysulfide-based. (See also Dental elastomers.) 11:283
analysis 11:284
commercial grades and specifications 11:284
composition 11:283

uses 11:283
Dental impression materials, silicone-based. (See also Dental elastomers.) 11:284
analysis 11:284
commercial grades and specifications 11:284
composition 11:284
uses 11:283
Dental investments 11:267
analysis 11:267
commercial grades and specifications 11:268
composition 11:267
compressive strength 11:270
consistency 11:268
fineness 11:268
MOD die test 11:271
setting expansion 11:269
thermal expansion 11:269
time of setting 11:269
uses 11:267
Dental materials. (See also Dental alloys; Dental ceramics; Dental colloids; Dental elastomers; Dental plastics; Dental resins; Dental waxes; and Metal-organic dental products.) 11:256–321
classification 11:257
commercial specifications 11:257
Dental metals. See Dental alloys.
Dental plastics. (See also Dental restorative materials, composite; Methyl methacrylate dental materials.) 11:299–316
Dental porcelains 11:271
abrasion 11:276
analysis 11:271
anchorage strength 11:273
composition 11:271
flexural strength 11:272
fluorescence 11:278
impact resistance 11:275
porosity 11:276
properties 11:273
solubility 11:278
splitting strength 11:274
thermal expansion 11:276
thermal shock resistance 11:275
Dental resins. (See also Dental impression compounds.) 11:316–318
Dental restorative materials, composite 11:308
abrasion 11:315
analysis 11:308
ash content 11:310
color stability 11:313
composition 11:309
composition, determination of 11:310
crushing strength 11:313
identification 11:310
impurities, determination of 11:315
infrared spectrum 11:311

MEHQ content 11:312
oxirane oxygen content 11:312
properties 11:310
separation 11:310
shrinkage 11:314
solubility 11:314
specification tests 11:313
thermal expansion 11:314
water sorption test 11:314
working time 11:313
WPE number 11:312
Dental silicate cements 11:257
analysis 11:257
arsenic in 11:261
commercial specifications 11:257, 258
composition 11:257
compressive strength 11:259
consistency 11:259
disintegration test 11:260
opacity 11:260
sample preparation 11:258
time of setting 11:259
uses 11:258
Dental waxes 11:318–321
analysis 11:319
commercial grades and specifications 11:257, 319
composition 11:318
non-volatile materials in 11:320
properties 11:320
thermal expansion 11:320
types 11:319
uses 11:318
Dental wrought gold alloys 11:294
analysis 11:294
commercial grades and specifications 11:257, 295
composition 11:294
elongation 11:295
fusion temperature 11:295
properties 11:294
tensile strength 11:295
types 11:294
uses 11:288
yield strength 11:295
Dental zinc phosphate cements 11:261
analysis 11:261
arsenic in 11:263
commercial specifications 11:257, 261, 262
composition 11:261
compressive strength 11:263
consistency 11:262
disintegration test 11:263
film thickness 11:263
sample preparation 11:262
time of setting 11:263
uses 11:261

Denture base polymers. See Methyl methacrylate dental materials.
Denture base temporary relining material. (See also Methyl methacrylate dental materials.)
specification tests 11:305
Denture cold curing repair resins. (See also Methyl methacrylate dental materials.)
specification tests 11:305
Deodorants. (See also *Cosmetics.*) 11:86
aluminum hydroxychloride content 11:88
analysis 11:87
composition 11:86
dichlorophene content 11:88
in embalming fluids 12:167
hexachlorophene content 11:87
nonvolatile matter in 11:87
zinc phenolsulfonate content 11:87
Deoxycholic acid
acid–base titration 7:149
colorimetric method using salicylaldehyde 7:152
color tests 7:147
commercial specifications 7:145
fluorometry 7:152
preparation 7:144
properties 7:143
structure 7:142
uses 7:144
visible spectroscopy 7:151
Depilatory powders
sulfide content 18:395
Deposit modifiers
in gasoline 13:359
Derivatives for identification
of acetals 4:67
of acetone 4:110, 114, 116
of acetonitrile 16:401, 402
of acetophenone 4:123
of acid halides 8:572
of acrylonitrile 16:402, 407
of adipic acid 4:412
of alcohols 4:67, 516, 558
of aldehydes 4:67, 111, 116
of allyl halides 5:101
of amides 8:601
of amines 5:264
of 4-aminodiphenylamine 11:438
of aniline and its derivatives 5:429
of anthracene 16:208
of benzidine 11:438
of benzoic acid 7:62
of benzoic acid esters 7:80
of benzoic anhydride 7:73
of benzonitrile 16:402
of benzyl alcohol 7:93, 95
of *p*-bromobenzonitrile 16:402
of 1-bromo-2-butanol 14:160
of 2-bromo-1-butanol 14:160

of 1-bromo-2-propanol 14:160
of 2-bromo-1-propanol 14:160
of 1,3-butanediol 13:568
of 1,4-butanediol 13:568
of 2,3-butanediol 13:568
of n-butyl mercaptan 15:559
of butylphenols 17:21
of n-butyronitrile 16:401, 402
of carboxylic acids 7:62; 8:374
of 1-chloro-2-butanol 14:160
of 2-chloro-1-butanol 14:160
of 3-chloro-1,2-propanediol 14:160
of 1-chloro-2-propanol 14:160
of 2-chloro-1-propanol 14:160
of cinnamaldehyde 10:45
of cinnamic acid 10:58
of cinnamyl alcohol 10:78
of cresols 17:21, 28
of crotonaldehyde 11:177
of cyanoacetamide 15:442
of cyanoacetic acid 15:439
of cyclobutanol 11:189
of cyclodecanol 11:189
of cyclododecanol 11:189
of cycloeicosanol 11:189
of cycloheptadecanol 11:189
of cyclohexadecanol 11:189
of cyclohexanol 11:187, 189
of cyclohexanone 11:204
of cyclohexylamine 11:211
of cyclohexyl mercaptan 15:559
of cyclononadecanol 11:189
of cyclononanol 11:189
of cyclooctadecanol 11:189
of cyclooctanol 11:189
of cyclopentadecanol 11:189
of cyclopentanol 11:189
of cyclopropanol 11:189
of cyclotetradecanol 11:189
of cyclotridecanol 11:189
of cycloundecanol 11:189
of n-decanonitrile 16:402
of diamines and higher amines, aromatic 11:437
of diaminobiphenyls 11:438
of p,p'-diaminodiphenylmethane 11:438
of diaminonaphthalenes 11:438
of dibenzylamine 11:438
of 1,3-dibromo-2-propanol 14:160
of 2,3-dibromo-1-propanol 14:160
of 1,3-dichloro-2-propanol 14:160
of 2,3-dichloro-1-propanol 14:160
of diethylene glycol 13:568
of 2,3-diiodo-1-propanol 14:160
of dimethisterone 10:580
of unsym-dimethylhydrazine 14:235
of dimethylphenols 17:21
of dimethyl terephthalate 17:165

of diphenylamine 11:438
of N,N'-diphenyl-1,4-diaminobenzene 11:438
of ditolylamines 11:438
of esters 8:582
of ethinyl estradiol 10:570
of ethylene bromohydrin 14:160
of ethylene chlorohydrin 14:160
of ethylene fluorohydrin 14:160
of ethylene glycol 13:568
of ethyl mercaptan 15:559
of ethylphenols 17:21
of ethylpyridines 17:426
of fatty acids 12:491
of 1-fluoro-2-propanol 14:160
of formaldehyde 13:107
of formals 4:68
of furfural 13:234
of glutaraldehyde 12:585
of glycerol 13:568
of glycols 13:567
of halohydrins 14:157
of n-heptanonitrile 16:402
of heptyl mercaptan 15:559
of hexahydrobenzene 17:21
of n-hexanonitrile 16:401, 402
of hexyl mercaptan 15:559
of hydrazine 14:235
of hydroquinone 17:21
of hydroxyhydroquinone 17:21
of 1-iodo-2-butanol 14:160
of 2-iodo-1-butanol 14:160
of 2-iodo-1-ethanol 14:160
of 3-iodo-2,4-propanediol 14:160
of 1-iodo-2-propanol 14:160
of 2-iodo-1-propanol 14:160
of isoamyl mercaptan 15:559
of isobutyl mercaptan 15:559
of isohexanonitrile 16:401, 402
of isopentanonitrile 16:402
of isophthalic acid 17:167
of isopropyl mercaptan 15:559
of 4-isopropylphenol 17:21
of ketals 4:111, 116
of ketones 4:67, 111, 116
of lutidines 17:426
of malachite green 19:168
of malonic acid 15:427
of mercaptans 15:559
of 4-methyl-3-ethylpyridine 17:426
of methylhydrazine 14:235
of methylisopropylphenols 17:21
of methyl mercaptan 15:559
of methylnaphthalenes 16:208
of 4-methylphenyl mercaptan 15:559
of naphthalene 16:208
of naphthalenediols 16:230
of naphthalenedisulfonic acids 16:223

of naphthalenemonosulfonic acid 16:223
of naphthalenesulfonic acids 16:222
of 1,3,5,7-naphthalenetetrasulfonic acid 16:223
of naphthalenetrisulfonic acid 16:223
of naphthols 16:230
of α-naphthylacetonitrile 16:402
of naphthylamines 16:226
of nicitinonitrile 16:402
of nitriles 16:399, 401
of nitrogen 16:453
of n-nonanonitrile 16:402
of nonyl mercaptan 15:559
of n-octanonitrile 16:402
of octyl mercaptan 15:559
of pentaerythritol 13:568
of 1,5-pentanediol 13:568
of n-pentanonitrile 16:401, 402
of n-pentyl mercaptan 15:559
of phenanthrene 16:208
of phenindione 10:181
of phenol 17:21
of phenols 17:20
of phenothiazine 11:438
of phenylacetonitrile 16:402
of phenylenediamines 11:438
of β-phenylethyl alcohol 7:93, 95
of phenyl mercaptan 15:559
of N-phenylnaphthylamines 11:438
of phloroglucinol 17:21
of phthalic acid 17:153
of phthalic anhydride 17:153
of picolines 17:426
of polycarbonates 12:589
of polyesters 12:589
of polyhydric alcohols 13:567
of polystyrene 12:589
of propionitrile 16:401, 402
of propylene glycol 13:568
of n-propyl mercaptan 15:559
of propylphenols 17:21
of pyridine 17:426
of pyridine and its derivatives 17:425
of pyrocatechol 17:21
of pyrogallol 17:21
of resorcinol 17:21
of terephthalic acid 17:160
of n-tetradecanonitrile 16:402
of 2,3,4,6-tetramethylphenol 17:21
of toluenediamines 11:438, 453
of o-toluidine 11:438
of m-tolunitrile 16:402
of o-tolunitrile 16:399
of p-tolunitrile 16:402
of p-tolyl-2-naphthylamine 11:438
of triethylene glycol 13:568
of trimellitic anhydride 17:171
of trimethylene glycol 13:568

of trimethylphenols 17:21
of 2,4,6-trimethylpyridine 17:426
Descloizite
composition 19:321
occurrence 19:321
Desmethylmorphine. See Normorphine.
Desolvated viscosity
of silicone fluids 18:113
Desoxycorticosterone. (See also Steroids.)
manufacture 14:193
structure 14:184, 185
trade name 14:192
Desoxycorticosterone acetate. (See also Steroids.)
manufacture 14:194
properties 14:194
solubility 14:194
uses 14:194
Desoxyephedrine
gas chromatography 6:209
Destriol. See Estriol.
DETA. See Diethylenetriamine.
Detail reproduction
of dental gypsum products 11:267
Detergent products. (See also Detergents.) 11:324
Detergents. (See also Amphoteric detergents;
    Anionic detergents; Cationic detergents;
    Nonionic detergents; Surfactants; and individual
    compounds.) 11:322–406
abrasive content 11:349
alkalinity 11:385
ammonia content 11:350
analysis 11:326
available chlorine content 11:355
bicarbonate content 11:353
carbonate content 11:351
carboxymethylcellulose content 11:353
car washes 6:354
classification 11:322
composition, determination of 11:328
cooling system cleaners 6:352
corrosion tests 11:398
density 11:384
in drycleaning solvents, used 11:609
EDTA and related compounds in 12:43
in engine tune-up products 6:356
enzyme content 11:356
ethanol content 11:359
ethylenediaminetetraacetate content 11:363
fatty acid content 11:363
foam tests 11:390
gas chromatography 11:329
in gasoline 13:359
glycerine content 11:365
history 11:322
hydrocarbon oil content 11:381
hydrotrope content 11:366
identification 11:327

impurities, determination of 11:367
infrared spectroscopy 11:328, 330
inorganic salt content 11:380
interfacial tension 11:386
isopropyl alcohol content 11:359
in oil conditioners 6:357
particle size 11:385
perborate content 11:368
performance tests 11:387
pH 11:385
phosphate content 11:369
phosphorus content 11:369
properties, determination of 11:383
sampling 11:326
separation 11:327
silicate content 11:379
silicon content 18:30
soap content 11:363
in soaps 18:182, 204
sodium chloride content 11:336, 355
sodium toluene sulfonate content 11:366
sulfate content 11:381
surface tension 11:386
surfactant content 11:335
upholstery cleaners 6:354
viscosity 11:386
water content 11:382
wetting tests 11:394
whitewall tire cleaners 6:354
x-ray diffraction 11:333
Detergent toilet bars. (See also *Detergents.*)
    11:326
  composition 11:326
Deuterium
  gas chromatography 14:386
  occurrence 14:377
  properties 14:378
Devarda reduction method
  nitrogen content of ammonium nitrate 5:341
Devarda's alloy
  in nitrate reduction 9:291
  volumetric method for nitrate 16:467, 502
Developed dyes 12:3
Developers
  in hair dyes 11:107
  in photographic films 12:586
Deviation, standard 7:161
Dew point
  of water in natural gas 13:327
  of water in noble gases 13:303
Dew point analyzers 16:532
Dew point method
  apparatus 5:308; 8:257
  for water in ammonia 5:306
  for water in carbon dioxide 8:257
  for water in hydrocarbons 14:345
Dexamethasone. (See also Steroids.)

manufacture 14:195
properties 14:195
solubility 14:195
structure 14:194
uses 14:195
Dexchlorpheniramine maleate. (See also *Anti-
    histamines.*) 6:32
  properties 6:32
Dexon 13:220
  analysis 13:220
  colorimetric determination 13:221
  in crops 13:220
  manufacture 13:220
  properties 13:220
  resorcinol colorimetric method 13:221
  in seeds 13:220
  solubility 13:220
  solvent extraction 13:221
  toxicity 13:220
  uses 13:220
Dextran–charcoal adsorption
  of insulin 14:578
Dextrin
  adhesives 4:404
  in beer 7:555, 564
  in cooling system sealers 6:362
  in finishing agents 11:126, 132
  hydrolysis 7:564
  identification 6:362
  iodine colorimetric method 7:556; 11:127, 132
  paper chromatography 7:569
  in roasted coffee 10:434
  solubility 11:126; 13:539, 552
Dextromoramide. (See also *Narcotics.*)
  structure 16:245
Dextrose. (See also Reducing sugars; Starch
    hydrolysates; and *Sugar.*) 18:337
  in beer 7:558, 564
  in cocoa beans 9:561
  in cocoa products 9:578
  copper acetate volumetric method 9:579
  enzyme methods 11:39
  iodometric titration 9:578
  Munson-Walker values 9:583
  in starch hydrolysates 11:38
Dextrose equivalent
  of starch hydrolysates 11:40
DHEG. (See also *EDTA and related compounds.*)
  cupric nitrate titration 12:35
  structure 12:24
Diacetatozirconic acid. See Zirconyl acetate.
Diacetone
  in enamels 12:175
  properties 12:175
Diacetone alcohol
  acidity 8:358
  azeotropes 8:336

in brake fluids 7:423, 424
commercial specifications 8:341
Karl Fischer titration 8:360
water in 8:360
Diacetyl
in beer 7:582, 608
in black tea 18:510
in cocoa bean vapor 9:570
determination as dimethylglyoxime 7:583
gas chromatography 12:204; 18:510; 19:457
mass spectroscopy 9:570
α-naphthol–creatine colorimetric method
19:457
structure 19:456
in wines 19:452, 456
Diacetyldichlorofluorescin
fluorometric method for hydrogen peroxide
14:431, 434
Diacetylenic ketone
derivative of amines 5:88
Diacetylmonoxime–nickel
colorimetric method for hydroxylamine 11:142;
14:447
Diacetylmorphine. (See also Heroin.)
gas chromatography 6:210, 211
Dial 5:91
Dialdehydes
in embalming fluids 12:165
Dialkanolamines. (See also Alkanolamines.)
manufacture 5:3
mixtures, determination in 5:19
Diallyl. See Biallyl.
Diallylamine
acid–base titration 5:89
azeotropes 5:261
gas chromatography 5:88
paper chromatography 5:89
polarography 5:89
properties 5:87, 260
Diallylbarbituric acid (Allobarbital) 6:468
analysis 5:92
bromate–bromide titration 6:497
bromine monochloride method 5:92
N-bromosuccinimide titration 5:92
derivatives for identification 6:477
extraction 6:491
gas chromatography 6:483, 484
melting point 5:91
paper chromatography 6:479
properties 6:470
solubility 6:476
thin-layer chromatography 6:480, 481, 482
Diallyl disulfide
boiling point 5:104
gas chromatography 5:105
Diallyl ether
in allyl alcohol 5:79

boiling point 5:98
gas chromatography 5:79
mercurimetric methods 5:100
Diallyl ether dioxide. (See also Epoxy compounds.)
infrared spectroscopy 12:200
Diallylisocyanurate 5:93
Diallyl maleate 5:94
N,N-Diallylmelamine 5:87
Diallylphenol sulfide 6:76
Diallyl phthalate 5:94
N,N'-Diallylpiperazine 5:87
Dialysis methods
for cocoa products 9:593
Diamagnetic susceptibility
of polyethylene 12:385
Diamines
gas chromatography 6:404
identification 6:410
in phenylenediamines 11:447
m-Diamines
colorimetric method using p-toluidine 6:410,
411
o-Diamines
quinoxaline derivatives 6:411
p-Diamines
indophenol test 6:410, 411
Diamines and higher amines, aliphatic. (See also
Ethyleneamines; and individual compounds.)
11:407–431
analysis 11:420
impurities, determination of 11:428
manufacture 11:417
properties 11:409
Diamines and higher amines, aromatic. (See also
individual compounds.) 11:432–462
analysis 11:436
assay 11:439
colorimetric determination 11:438
derivatives for identification 11:437
freezing point 11:439
gas chromatography 11:439, 440, 443
identification 11:437
impurities, determination of 11:440
infrared spectroscopy 11:442
melting point 11:440
polarography 11:443
properties 11:432
sampling 11:436
separation 11:437
toxicity 11:436
ultraviolet spectroscopy 11:439, 442
uses 11:432
visible spectroscopy 11:439, 442
volumetric determination 11:440, 442
2,4'-Diamino-4-(4''-aminobenzyl)diphenylmethane
thin-layer chromatography 19:306
2,4-Diaminoanisole

in hair dyes 11:103
thin-layer chromatography 11:103
3,3'-Diaminobenzidine
colorimetric method for selenium 10:625;
17:599
structure 17:599
Diaminobenzidine hydrochloride
colorimetric method for vanadium 19:334
Diaminobiphenyls. (See also *Diamines and
higher amines, aromatic.*)
derivatives for identification 11:438
properties 11:434
solubility 11:435
Diaminochrysazin
colorimetric method for boron 7:410
1,2-Diaminocyclohexanetetraacetic acid. (See also
DCYTA.)
in beer 7:585
4,4'-Diamino-1,1'-dianthraquinoylamine
colorimetric method for boron 7:389, 410
p-Diaminodiphenyl. See Benzidine.
2,4'-Diaminodiphenylamine. (See also *Diamines
and higher amines, aromatic.*) 11:433
paper chromatography 6:85
4,4'-Diaminodiphenylamine. (See also *Diamines
and higher amines, aromatic.*)
properties 11:434
solubility 11:435
2,7-Diaminodiphenylene oxide
colorimetric method for chromium 11:140, 144
p,p'-Diaminodiphenylmethane. (See also *Diamines
and higher amines, aromatic.*) 6:76
derivatives for identification 11:438
nuclear magnetic resonance spectroscopy 19:302
paper chromatography 6:85
in polyurethanes 19:302
structure 19:304
thin-layer chromatography 19:303, 306
3,3'-Diaminodipropylamine. See Iminobispropylamine.
(Diaminoethyl ether)tetraacetic acid. See Chel ME.
Diaminonaphthalenes. (See also *Diamines and
higher amines, aromatic.*)
colorimetric method for selenium 15:413
derivatives for identification 11:438
2,4-Diaminophenol
in hair dyes 11:103
thin-layer chromatography 11:103
1,3-Diaminopropane. (See also *Diamines and
higher amines, aliphatic.*)
properties 11:410
(1,3-Diaminopropan-2-ol)tetraacetic acid. See
DPTA.
2,4-Diaminotoluene 6:75
Diaminotoluenes. (See also Toluenediamines.)
in hair dyes 11:103
thin-layer chromatography 11:103

3,3'-Diammoniumbenzidine hydrochloride
colorimetric method for selenium 10:302
Diammonium citrate. (See also Citric acid and
salts.)
commercial grades and specifications 10:91
impurities, determination of 10:93
uses 10:86
Diammonium ethylenebisdithiocarbamate. (See
also Dithiocarbamic acid salts.)
uses 11:577
Diammonium hydrogen phosphate
gravimetric method for magnesium 10:122, 640;
12:536; 15:362
gravimetric method for magnesium oxide 9:116;
13:421; 14:95
gravimetric method for titanium 15:83
gravimetric method for zinc 19:559
gravimetric method for zirconium 9:666; 10:649;
19:102
Diammonium phosphate. (See also Ammonium
phosphate, dibasic.)
analysis 15:535
commercial grades and specifications 15:520
in matches 15:520
Diammonium phosphate–hydrogen peroxide
gravimetric method for zirconium 16:341
Diamond 4:31
Diamond abrasives 7:294
Diamond grit, metal-clad 11:468
composition 11:468
Diamonds, industrial 11:462–484
analysis 11:470
ash content 11:481
bulk density 11:477
edge characteristics 11:465
friability 11:465, 478
history 11:462
in metal-bonded products 11:465
microscopic methods 11:471
occurrence 11:462, 464
particle size distribution 11:474
properties 11:463
radiography 11:480
reclaiming of 11:481
refractive index 11:471
in resinoid-bonded products 11:468
sampling 11:470
shape 11:465, 473
surface characteristics 11:465
thermal stability 11:481
types 11:465
uses 11:462, 465
in vitrified-bonded products 11:468
x-ray diffraction 11:471
Diampromid. (See also *Narcotics.*)
structure 16:245
Diamylamines

commercial grades and specifications 5:264
properties 5:260
uses 5:263
2,5-Di-*tert*-amylhydroquinone
paper chromatography 6:85
Diamyl ketone
properties 8:335
Dianilinogossyptol
colorimetric method for gossyptol 10:543
o-Dianisidine
colorimetric method for aldehydes 17:298
colorimetric method for phosphate 11:142
gravimetric method for phosphomolybdic acid
17:61
Dianisidinediamine
in polyurethanes 19:304
structure 19:304
Dianisidine diisocyanate. (See also Isocyanates;
Isocyanic acid esters.)
properties 15:95
structure 19:271
trade names 15:103; 19:271
1,1'-Dianthrimide
colorimetric method for boron 7:389, 407, 410
Diantipyrylmethane
colorimetric method for titanium 19:107, 118
Diantipyrylmethane–thiocyanate
colorimetric method for niobium 16:370
Diarylamines. (See also *Diamines and higher amines,
aromatic*; and individual compounds.)
properties 11:433
uses 11:432
Diastase method
for carbohydrate determination 4:474
for starch in chocolate liquor 9:586
Diastatic activity
of flour 12:609
Diatomaceous earth. (See also Diatomites.) 11:
484–500
analysis 11:489; 15:529
in automobile polishes 6:353
color 15:529
commercial grades and specifications 15:512
composition, determination of 11:489
emission spectroscopy 11:492
formation 11:484
ignition loss 11:489
impurities, determination of 11:490, 494
in matches 15:512
moisture content 11:489
occurrence 11:485
particle size 15:529
properties 11:486
silica content 11:489
uses 11:486
vanadium content 19:340
water absorption 15:529

Diatomaceous silica. (See also *Diatomaceous
earth.*)
commercial specifications 17:192
impurities, determination of 17:192
Diatomites. (See also *Diatomaceous earth.*)
composition 11:488
manufacture 11:486
properties 11:486
solubility tests 11:494
Diazinon. (See also *Pesticides and insecticides.*)
commercial grades and specifications 19:594
composition 19:589, 597
manufacture 19:597
properties 19:589
solubility 19:587
structure 19:582
toxicity 19:592
O-Diazoacetyl-L-serine. See Azaserine.
Diazoamines
colorimetric method for estrogens 14:207
p-Diazobenzenesulfonic acid
colorimetric method for nitroethane 16:433
colorimetric method for nitromethane 16:433
colorimetric method for nitroparaffins 16:443
colorimetric method for 1-nitropropane 16:433
Diazodinitrophenol. (See also High explosives.)
12:416
analysis 12:416
commercial specifications 12:416
crystal size 12:416
in explosive priming mixtures 12:422
granulation 12:416
impurities, determination of 12:416
properties 12:406, 416
specification tests 12:416
uses 12:416
x-ray diffraction 12:418
Diazomethane
esterification method for lactic acid esters 15:
135
methylation reagent for carboxylic acids 8:385
Diazonium
colorimetric method for coumarin 11:166
Diazonium salt of H-acid. See H-acid diazonium
salt.
Diazotization
of acetaminophen 6:195
of acetanilide 6:194
of *p*-aminophenol 6:195
of aniline and its derivatives 5:429, 445
of ethyl-*p*-aminobenzoate 5:389
manufacture of azo dyes 6:402
of phenacetin 6:194
of *p*-phenetidine 6:194
procaine hydrochloride 5:406, 413
propoxycaine hydrochloride 5:410
tetracaine 5:416

Diazotized dianisidine
  colorimetric method for bishydroxycoumarin
    10:161
Diazotized dinitroaniline
  colorimetric method for aldrin 19:611
Diazotized p-nitroaniline
  colorimetric method for ethyl biscoumacetate
    10:163
  colorimetric method for phenprocoumon 10:166
  titration of 2-naphthol 16:232
Diazotized sulfanilic acid
  colorimetric method for cyanoacetic acid 15:439
Dibasic acids 5:22
  in alkyd resins 5:34, 47
  chemical methods of analysis 5:34
  gas chromatography 5:25
  infrared spectroscopy 5:46
  Kappelmeier separation 5:33
  paper chromatography 5:49
Dibenzalacetone
  for identification of acetone 4:111
Dibenzopyridine. See Acridine.
Di-2-benzothiazyl disulfide
  in elastomers, synthetic 12:143
  ultraviolet spectrum 12:143
4,5,4′,5′-Dibenzothioindigo. See CI Vat Brown 5.
Dibenzoylmethane
  colorimetric method for uranium 19:247, 249
Dibenzylamine. (See also Diamines and higher
    amines, aromatic.)
  derivatives for identification 11:438
Dibenzylamine benzylpenicillin. (See also Benzyl-
    penicillin.)
  structure 16:567
Dibenzylamine penicillin G. See Dibenzylamine
    benzylpenicillin.
N,N′-Dibenzylethylenediamine penicillin G. See
    Benzathine benzylpenicillin.
Dibenzyl ketone
  hydroxylamine hydrochloride method 8:357
Diborane 7:340, 344
  in air 7:349
  analysis 7:345
  gas chromatography 7:353
  infrared spectrum 7:354
  manufacture 7:343
  monitors for detection in air 7:349
  properties 7:343
Diboron tetraboride 7:337
Diboron tetrachloride 7:337
Diboron tetrafluoride 7:337
Diboron tetraiodide 7:337
Dibromochloropropane 13:182
  in cherries 13:185
  in food crops 13:185
  gas chromatography 13:183, 185
  properties 13:182

solubility 13:182
toxicity 13:182
uses 13:182
1,2-Dibromo-3-chloropropane. See Dibromochloro-
    propane.
1,2-Dibromo-2,2-dichloroethyl dimethyl phosphate.
    See Naled.
Dibromodifluoromethane
  gas chromatography 13:64
4,5-Dibromo-3,6-fluorandiol. See D & C Orange
    No. 5.
Dibromofluoresceins. (See also D & C Orange
    No. 5.)
  in D & C Red No. 21 10:523
  paper chromatography 10:520, 523
  visible spectroscopy 10:523
5,5′-Dibromoindigo. (See also Indigoid dyes.)
  properties 14:496
1,3-Dibromo-2-propanol. (See also Glycerol
    halohydrins.)
  derivatives for identification 14:160
2,3-Dibromo-1-propanol. (See also Glycerol
    halohydrins.)
  derivatives for identification 14:160
5,7-Dibromo-8-quinolinol
  colorimetric method for indium 14:534
2,6-Dibromoquinone chloroimide
  colorimetric method for phenol 8:146
2,6-Dibromoquinone chloroimine
  colorimetric method for phenols 17:29
3,5-Dibromosalicylanilide. (See also Salicyl-
    anilides.)
  ultraviolet spectroscopy 11:552
4′,5-Dibromosalicylanilide. (See also Salicyl-
    anilides.)
  nuclear magnetic resonance spectroscopy 11:551
  properties 11:551
  structure 11:551
  ultraviolet spectroscopy 11:552
Dibromosilane
  properties 18:40
Dibromosulfogalein. See Bromopyrogallol red.
1,2-Dibromo-1,1,2,2-tetrafluoroethane
  gas chromatography 11:98
5,5′-Dibromothioindigo. (See also Thioindigoid
    dyes.)
  properties 14:496
Dibucaine
  analysis 5:386
  gas chromatography 6:210
  properties 5:386
Dibucaine hydrochloride 5:375
  analysis 5:386
  properties 5:386
Dibutylamine
  properties 5:260
  uses 5:263

volumetric method for isocyanates 19:272
volumetric method for isocyanic acid esters
15:112
volumetric method for polymethylene polyphenyl
diisocyanate 15:113
volumetric method for tolylene diisocyanates
15:112
Di-*n*-butylamine
azeotropes 5:261
manufacture 5:262
Di-*n*-butylamine–acetic anhydride
volumetric method for epoxy compounds 12:
216
volumetric method for 1,2-epoxy-2,4,4-trimethyl-
pentane 12:215
2,6-(Di-*tert*-butyl)-4-aminophenol
in photographic films 12:592
Di-*tert*-butylchlorobenzene
mass spectroscopy 9:407
2,6-Di-*tert*-butyl-*p*-cresol 6:72
in impact polystyrene 18:333
in photographic films 12:592
2,6-Di-*tert*-butyl-α-dimethylamino-*p*-cresol. (See
also *Phenols.*) 6:74
structure 17:2
Dibutylethanolamine
azeotropes 5:3
N,N-Dibutylethanolamine
properties 5:2
solubility 5:1
1,1-Di-*tert*-butylethylene oxide. (See also *Epoxy
compounds.*)
infrared spectroscopy 12:200
2,5-Di-*tert*-butylhydroquinone 6:76
paper chromatography 6:85
3,5-Di-*tert*-butyl-4-hydroxytoluene. See BHT.
N,N-Dibutylisopropanolamine
properties 5:2
solubility 5:1
Dibutyl ketone
properties 8:335
2,6-Di-*tert*-butyl-4-methyl phenol
in elastomers, synthetic 12:142
gas chromatography 16:95
in methacrylate polymers 16:95
paper chromatography 6:85
thin-layer chromatography 16:95
ultraviolet spectra 12:142
2,6-Di-*tert*-butylphenol. (See also *Phenols.*)
4-aminoantipyrine colorimetric method 17:18
commercial grades and specifications 17:15
Millon's reagent color test 17:18
Dibutyl phthalate
absorption by carbon black 8:191, 225
commercial specifications 8:582
in enamels 12:176
gas chromatography 16:97

infrared spectrum 8:585
properties 12:176
saponification method 8:586
solubility 13:539, 552
uses 15:302
Dibutyl sebacate
gas chromatography 16:97
1,3-Dibutylthiourea 6:75
Dibutyl tin laurate
in enamels 12:177
properties 12:177
Dibutyl tin maleate
in enamels 12:177
properties 12:177
Dicalcium phosphate
manufacture 17:121
occurrence 17:121
uses 17:121
Dicalcium silicate
in hydraulic cements 9:107
light microscopy 9:103
in portland cements 9:103
Dicarboxylic acids 8:369
gas chromatography 19:294
identification 8:375; 19:293
ion exchange methods 19:292
in polyurethanes 19:291
solvent extraction 19:292
thin-layer chromatography 19:293
Dichlone 13:201
analysis 13:201
dimethylamine colorimetric method 13:201
manufacture 13:201
properties 13:201
solubility 13:201
toxicity 13:201
uses 13:201
Dichloricide. See *p*-Dichlorobenzene.
Dichloroacetic acid
Fujiwara colorimetric method 9:386
properties 8:396
Dichloroacetylene. (See also C₂ Chlorinated hydrocarbons.)
mass spectroscopy 9:401
properties 9:466
2,4-Dichloroaniline
derivatives for identification 5:429
reaction with benzoic anhydride 7:74
2,5-Dichloroaniline
azo dyes for identification 5:430
derivatives for identification 5:429
properties 5:423
3,4-Dichloroaniline
properties 5:423
Dichloroanilines
gas chromatography 5:441
manufacture 5:426

2,4-Dichlorobenzaldehyde
  commercial specifications 6:609
  infrared spectroscopy 6:617
2,6-Dichlorobenzaldehyde
  commercial specifications 6:607
  infrared spectroscopy 6:616
3,4-Dichlorobenzaldehyde
  commercial specifications 6:608
Dichlorobenzene
  in metal cleaners 6:355
1,2-Dichlorobenzene. See o-Dichlorobenzene.
1,3-Dichlorobenzene. See m-Dichlorobenzene.
1,4-Dichlorobenzene. See p-Dichlorobenzene.
m-Dichlorobenzene. (See also Dichlorobenzenes.)
  9:478
  gas chromatography 9:378
  infrared spectroscopy 9:394, 481
  mass spectroscopy 9:405
  properties 9:479
o-Dichlorobenzene. (See also Dichlorobenzenes.)
  9:478; 13:182
  azeotropes 9:479
  Beilstein detection 9:382
  gas chromatography 9:378; 13:183
  infrared spectroscopy 9:394, 481
  mass spectroscopy 9:405
  properties 9:479; 13:182
  solubility 13:182, 539, 552
  toxicity 9:480; 13:182
  uses 9:478; 13:182
p-Dichlorobenzene. (See also Dichlorobenzenes.)
  9:478; 13:182
  azeotropes 9:479
  Beilstein detection 9:382
  commercial grades and specifications 9:480
  crystallizing point 9:481
  gas chromatography 9:378; 13:183
  infrared spectroscopy 9:394, 481
  mass spectroscopy 9:405
  properties 9:479; 13:182
  solubility 13:182
  specification tests 9:481
  toxicity 9:480; 13:182
  uses 9:478; 13:182
Dichlorobenzenes. (See also individual compounds.)
  9:478
  analysis 9:480
  composition 9:411
  infrared spectroscopy 9:480
  manufacture 9:480
  solubility 9:478
2,4-Dichlorobenzoic acid
  properties 8:402
3,4-Dichlorobenzoic acid
  paper chromatography 8:379
4,4'-Dichlorobiphenyl
  mass spectroscopy 9:408

Dichlorobiphenyls. (See also Chlorinated bi-
    phenyls.)
  gas chromatography 9:495
  properties 9:492
1,1-Dichloro-2,2-bis(p-chlorophenyl)ethane. See
    TDE.
1,1-Dichloro-2,2-bis(p-chlorophenyl)ethylene. See
    DDE.
1,1-Dichloro-2,2-bis(p-ethylphenyl)ethane. See
    Perthane.
1,1-Dichlorobutane. (See also C$_4$ Chlorinated
    hydrocarbons.)
  gas chromatography 9:474
  infrared spectroscopy 9:390
  mass spectroscopy 9:397
1,2-Dichlorobutane. (See also C$_4$ Chlorinated
    hydrocarbons.)
  gas chromatography 9:474
  infrared spectroscopy 9:390
  mass spectroscopy 9:396
1,3-Dichlorobutane. (See also C$_4$ Chlorinated
    hydrocarbons.)
  gas chromatography 9:474
  infrared spectroscopy 9:390
1,4-Dichlorobutane. (See also C$_4$ Chlorinated
    hydrocarbons.)
  gas chromatography 9:474
  infrared spectroscopy 9:390
  mass spectroscopy 9:397
2,2-Dichlorobutane. (See also C$_4$ Chlorinated
    hydrocarbons.)
  gas chromatography 9:474
  infrared spectroscopy 9:390
2,3-Dichlorobutane. (See also C$_4$ Chlorinated
    hydrocarbons.)
  gas chromatography 9:474
  infrared spectroscopy 9:390
  mass spectroscopy 9:398
1,3-Dichloro-2-butene. (See also C$_4$ Chlorinated
    hydrocarbons.)
  infrared spectroscopy 9:390
1,4-Dichloro-2-butene. (See also C$_4$ Chlorinated
    hydrocarbons.)
  infrared spectroscopy 9:390
  mass spectroscopy 9:399
3,4-Dichloro-1-butene. (See also C$_4$ Chlorinated
    hydrocarbons.)
  infrared spectroscopy 9:390
  mass spectroscopy 9:399
1,4-Dichlorobutyne
  mass spectroscopy 9:400
Dichlorochinoxaline dyes
  identification 12:20
2,4-Dichloro-6-(o-chloranilino)-s-triazine. See
    Dyrene.
1,2-Dichloro-4,1-chloroethylbenzene
  mass spectroscopy 9:406

1,4-Dichloro-2,1-chloroethylbenzene
mass spectroscopy 9:406
1,4-Dichloro-2,2-chloroethylbenzene
mass spectroscopy 9:406
3,6-Dichlorochromotropic acid
colorimetric method for titanium 19:98
2,5-Dichlorocumene
mass spectroscopy 9:407
3,4-Dichlorocumene
mass spectroscopy 9:406
4,4'-Dichloro-5',5'-dibromoindigo. (See also
Indigoid dyes.)
properties 14:496
1,2,1,1-Dichlorodifluoroethane. (See also Chlorofluorocarbons.)
properties 13:61
Dichlorodifluoromethane. (See also Aerosols;
Chlorofluorocarbons.)
in aerosols 11:95
commercial specifications 13:63
gas chromatography 11:89, 98; 12:324; 13:64
in hair sprays 11:89
properties 11:95; 13:61
1,4-Dichloro-2,5-dimethoxybenzene. See Chloroneb.
Dichlorodimethoxybenzenes
gas chromatography 9:519
Dichlorodimethylhydantoin
available chlorine 7:224, 228
iodometric titration 7:228
separation 7:225
1,1-Dichloro-2,2-dimethylpropane
mass spectroscopy 9:401
5,5'-Dichloro-7,7'-dimethylthioindigo. See CI Vat
Violet 2.
6,6'-Dichloro-4,4'-dimethylthioindigo. See CI Vat
Red 1; D & C Red No. 30.
Dichlorodiphenyltrichloroethane. See DDT.
1,1-Dichloroethane. (See also C$_2$ Chlorinated
hydrocarbons.)
Beilstein detection 9:382
in chloroform 9:433
gas chromatography 9:374, 375, 376, 377, 442
in hydrochloric acid 9:392
infrared spectroscopy 9:390, 392
mass spectroscopy 9:398
properties 9:466
in 1,1,1-trichloroethane 9:442
1,2-Dichloroethane. (See also C$_2$ Chlorinated
hydrocarbons; Ethylene dichloride.) 9:437
analysis 9:439
appearance 9:439
azeotropes 9:438
Beilstein detection 9:382
in carbon tetrachloride 8:325
in chloroform 9:433
color 9:439
commercial grades and specifications 9:439

composition 9:411
distillation range 9:439
gas chromatography 9:375, 376, 377, 380, 381,
442; 14:375
in hydrochloric acid 9:392
impurities, determination of 9:439
infrared spectroscopy 8:325; 9:390, 392
manufacture 9:438
mass spectroscopy 9:398
properties 9:437
solubility 9:437
specification tests 9:439
toxicity 9:438
in 1,1,1-trichloroethane 9:442
uses 9:437
sym-Dichloroethane. See 1,2-Dichloroethane.
1,2-Dichloroethylbenzene
mass spectroscopy 9:404
1,2-Dichloro-3-ethylbenzene
mass spectroscopy 9:406
1,2-Dichloro-4-ethylbenzene
mass spectroscopy 9:406
1,3-Dichloro-2-ethylbenzene
mass spectroscopy 9:406
1,4-Dichloro-2-ethylbenzene
mass spectroscopy 9:406
1,1-Dichloroethylene. (See also C$_2$ Chlorinated
hydrocarbons.)
Beilstein detection 9:382
gas chromatography 9:374, 375, 376, 377, 380,
381, 442
infrared spectroscopy 9:390
mass spectroscopy 9:397
properties 9:466
in 1,1,1-trichloroethane 9:442
1,2-Dichloroethylene. (See also C$_2$ Chlorinated
hydrocarbons.)
composition 9:411
gas chromatography 9:374, 375, 376, 377, 380,
381, 442
infrared spectroscopy 9:390, 394
mass spectroscopy 9:398
in 1,1,1-trichloroethane 9:442
Dichloroethyl ether 13:181
gas chromatography 13:183
infrared spectroscopy 7:30
manufacture 13:181
properties 13:181
solubility 13:181, 539, 552
toxicity 13:181
uses 13:181
sym-Dichloroethyl ether. See Dichloroethyl ether.
1,2-Dichloroethyltoluene
mass spectroscopy 9:405
Dichlorofluoromethane. (See also Chlorofluorocarbons.)
gas chromatography 13:64

properties 13:61
5,7-Dichloro-8-hydroxyquinoline
  gravimetric method for niobium 16:367
3,3'-Dichloroindanthrene. See D & C Blue No. 9.
2,6-Dichloroindophenol
  colorimetric method for reducing substances in
    beer 7:618
  indicator time test 7:618
Dichloroisocyanuric acid
  available chlorine 7:224, 229
  iodometric titration 7:229
Dichloromethane. (See also Methylene chloride.)
  in ampicillin 16:577
  in sodium ampicillin 16:566
Dichloromethoxyphenols. (See also Chlorophenols.)
  gas chromatography 9:519
  infrared spectroscopy 9:515
  ion exchange chromatography 9:523
  in 2,4,5-trichlorophenol 9:523
1,2-Dichloro-2-methylpropane
  mass spectroscopy 9:396
Dichloromonofluoromethane
  gas chromatography 11:98
Dichloronaphthalenes. (See also Chlorinated
    naphthalenes.)
  properties 9:489
2,3-Dichloro-1,4-naphthoquinone. See Dichlone.
2,6-Dichloro-4-nitroaniline. (See also Dichloran.)
  infrared spectroscopy 5:454
  manufacture 5:426
  properties 5:423
1,1-Dichloro-1-nitroethane 13:189
  analysis 13:189
  in corn 13:189
  gas chromatography 13:190; 16:427, 432
  manufacture 13:189
  properties 13:189
  solubility 13:189
  sulfuric acid–resorcinol colorimetric method
    13:189
  toxicity 13:189
  uses 13:189
1,1-Dichloro-1-nitropropane
  gas chromatography 16:432
1,5-Dichloropentane
  mass spectroscopy 9:397
Dichlorophene. (See also Bisphenols.)
  acid–base titration 11:547
  in deodorants 11:88
  gas chromatography 11:88
  melting range 11:547
  specification tests 11:547
  ultraviolet spectroscopy 11:549
  uses 11:546
2,3-Dichlorophenol. (See also *Phenols.*)
  gas chromatography 17:36
2,4-Dichlorophenol. (See also *Phenols.*)

gas chromatography 17:36
2,5-Dichlorophenol. (See also Dichlorophenols;
    Phenols.)
  gas chromatography 17:36
  ion exchange chromatography 9:523
  in 2,4,5-trichlorophenol 9:523
2,6-Dichlorophenol. (See also *Phenols.*)
  gas chromatography 17:36
2,6-Dichlorophenol indophenol
  colorimetric method for phytonadione 10:191
  titration of ascorbic acid 19:424
  titration of phytonadione 10:190
Dichlorophenols. (See also *Chlorophenols.*)
  commercial grades and specifications 9:513
  gas chromatography 9:518; 19:386
  infrared spectroscopy 9:515
  ion exchange chromatography 9:521
  manufacture 9:512
  properties 9:511
  ultraviolet spectroscopy 9:521
  uses 9:511
2,4-Dichlorophenoxyacetic acid (2,4-D) 8:406
  analysis 8:406
  properties 8:402
1,6-Di-(4-chlorophenyldiguanido)hexane. See
    Chlorohexidine.
2,5-Dichlorophenyl isocyanate. (See also *Isocyanic
    acid esters.*)
  infrared spectroscopy 15:106
3,4-Dichlorophenyl isocyanate. (See also *Isocyanic
    acid esters.*)
  commercial sources 15:103
  properties 15:95
Dichloro-*o*-phenylphenol
  in embalming fluids 12:165
Dichlorophenyltrichlorosilane. (See also Organo-
    chlorosilanes.)
  properties 18:54
1,1-Dichloropropane
  mass spectroscopy 9:398
1,2-Dichloropropane. (See also C$_3$ Chlorinated
    hydrocarbons; Propylene dichloride.) 9:465
  azeotropes 9:467
  Beilstein detection 9:382
  composition 9:411
  gas chromatography 9:375, 376, 381, 442, 472
  infrared spectroscopy 9:390
  manufacture 9:468
  mass spectroscopy 9:398
  properties 9:467
  toxicity 9:468
  in 1,1,1-trichloroethane 9:442
  uses 9:465
1,3-Dichloropropane. (See also C$_3$ Chlorinated
    hydrocarbons.)
  infrared spectroscopy 9:392
  mass spectroscopy 9:399

properties 9:471
2,2-Dichloropropane. (See also C$_3$ Chlorinated
   hydrocarbons.)
  infrared spectroscopy 9:390
2,3-Dichloropropane
  in epichlorohydrin 14:175
  gas chromatography 14:175
1,3-Dichloro-2-propanol. (See also Glycerol
   1,3-dichlorohydrin.)
  acetic anhydride volumetric method 14:161
  derivatives for identification 14:160
  structure 14:167
2,3-Dichloro-1-propanol. (See also Glycerol
   1,2-dichlorohydrin.)
  derivatives for identification 14:160
  structure 14:167
1,1-Dichloropropene. (See also C$_3$ Chlorinated
   hydrocarbons.)
  infrared spectroscopy 9:390
  properties 9:471
1,2-Dichloropropene. (See also C$_3$ Chlorinated
   hydrocarbons.)
  infrared spectroscopy 9:390
  properties 9:471
1,3-Dichloropropene. (See also C$_3$ Chlorinated
   hydrocarbons.) 13:180
  gas chromatography 9:472
  infrared spectroscopy 9:390
  manufacture 13:180
  properties 9:471; 13:180
  solubility 13:181
  toxicity 13:181
  uses 13:180
2,3-Dichloropropene. (See also C$_3$ Chlorinated
   hydrocarbons.)
  gas chromatography 9:376
  infrared spectroscopy 9:390
  mass spectroscopy 9:399
  properties 9:471
3,3-Dichloropropene. (See also C$_3$ Chlorinated
   hydrocarbons.)
  properties 9:471
3,3-Dichloro-1-propene
  in allyl chloride 5:102
  gas chromatography 5:102
α,γ-Dichloropropylene. See 1,3-Dichloropropene.
Dichloropyridazone dyes
  identification 12:20
Dichloroquinonechloroimide
  colorimetric method for BHA 9:243
2,6-Dichloroquinone-4-chloroimine
  colorimetric method for phenol 12:237
Dichlorosilane. (See also Chlorosilanes.)
  commercial grades and specifications 18:49
  gas chromatography 18:51
  infrared spectroscopy 18:50
  properties 18:40, 48

safety 18:48
2,5-Dichlorostyrene
  mass spectroscopy 9:406
3,4-Dichloro-α-styrene
  mass spectroscopy 9:407
Dichlorotetrafluoroethane
  in aerosols 11:95
  gas chromatography 11:98; 12:324
  properties 11:95
1,1-Dichlorotetrafluoroethane. (See also Chlorofluorocarbons.)
  properties 13:61
1,2-Dichlorotetrafluoroethane. (See also Chlorofluorocarbons.)
  commercial specifications 13:63
  properties 13:61
α,α-Dichlorotoluene. (See also Benzal chloride.)
  mass spectroscopy 9:404
2,4-Dichlorotoluene
  mass spectroscopy 9:404
2,6-Dichlorotoluene
  mass spectroscopy 9:404
3,4-Dichlorotoluene
  mass spectroscopy 9:404
o,α-Dichlorotoluene
  mass spectroscopy 9:404
Dichlorotoluenes
  gas chromatography 9:488
4,4'-Dichloro-α-(trichloromethyl)benzhydrol. See
  Dicofol.
1,4-Dichloro-2,3,5-triethylbenzene
  mass spectroscopy 9:408
1,2,1,1,2-Dichlorotrifluoroethane. (See also
  Chlorofluorocarbons.)
  properties 13:61
4,4'-Dichloro-3(trifluoromethyl) carbanilide
  in soaps 18:200
  ultraviolet spectroscopy 18:200
2,2-Dichlorovinyl dimethyl phosphate. See Dichlorvos.
Dichlorvos. (See also *Pesticides and insecticides.*)
  commercial grades and specifications 19:594
  composition 19:597
  manufacture 19:597
  properties 19:588
  solubility 19:587
  structure 19:582
  toxicity 19:592
  uses 19:586, 600
α,α-Dichloro-m-xylene
  mass spectroscopy 9:405
α,α-Dichloro-p-xylene
  mass spectroscopy 9:405
Dicholine citrate. (See also *Choline and its
  derivatives.*)
  cadmium chloride gravimetric method 9:623
Dichromate

in cesium dichromate 9:288
oxidation method for ethanol 4:550
Dickite. (See also *Clays and clay minerals.*)
  composition 10:103
Dicloran 13:216
  analysis 13:217
  colorimetric determination 13:217
  gas chromatography 13:217
  liquid chromatography 13:217
  manufacture 13:217
  properties 13:217
  solubility 13:217
  toxicity 13:217
  uses 13:216
Dicloxacillin. (See also Isoazolylpenicillins;
  Sodium dicloxacillin.)
  analysis 16:577
  assay 16:577
  bioassay 16:577
  manufacture 16:577
  microbiological methods 16:577
  properties 16:577
  specification tests 16:577
  structure 16:555
  ultraviolet spectroscopy 16:565
  uses 16:577
Dicofol. (See also *Pesticides and insecticides.*)
  commercial grades and specifications 19:593
  composition 19:596
  manufacture 19:596
  properties 19:588
  solubility 19:587
  structure 19:581
  toxicity 19:592
Dicoumarol. See Bishydroxycoumarin.
2,6-Di-*tert-p*-cresol
  gas chromatography 18:333
Dicresylolpropane
  in photographic films 12:592
Dicyandiamide 5:276
Dicyclohexylamine 11:238–241
  acid–base titration 11:217
  acid–base titration, nonaqueous 11:240
  analysis 11:239
  assay 11:239
  commercial grades and specifications 11:239
  in cyclohexylamine 11:217
  gas chromatography 11:212, 213, 240, 245
  identification 11:239
  infrared spectroscopy 11:240
  properties 5:260; 11:238
  solubility 11:238
  toxicity 11:239
  uses 5:264; 11:238
N,N'-Dicyclohexyl-*p*-phenylenediamine (DCPD)
  gas chromatography 6:81
Dicyclohexylthallic carbonate

gravimetric method for nitrates 16:501
Dicyclopentadiene 11:511
  analysis 11:512
  azeotropes 11:511
  manufacture 11:511
  properties 11:511
Dicyclopentadienyl iron. (See also Iron compounds.)
  uses 15:86, 88
Didymium. (See also *Rare earths.*) 17:470, 499
  commercial specifications 17:471
Didymium carbonate. (See also Rare earth carbonates.)
  commercial specifications 17:499
Didymium chloride. (See also Rare earth chlorides.)
  commercial specifications 17:499
Didymium fluoride. (See also Rare earth fluorides.)
  commercial specifications 17:499
Didymium hydrate. (See also Rare earth hydrates.)
  commercial specifications 17:499
Didymium nitrate. (See also Rare earth nitrates.)
  commercial specifications 17:499
Didymium oxide. (See also Rare earth oxides.)
  commercial specifications 17:499
  uses 17:509
Dieldrin
  bioassay 7:183
  gas chromatography 19:389
  in raw milk 16:128
  in water 19:389
Dielectric breakdown machines 18:125
Dielectric ceramics. (See also *Ceramics.*)
  properties 9:209
  uses 9:209
Dielectric constants
  of lubricants 15:343
  of polyethylene 12:384
  of polypropylene 12:384
  for silicone fluids 18:116
  of tolylene diisocyanates 15:117
Dielectric materials
  cerium content of 9:260
Dielectric strength
  of silicone fluids 18:115
Dienes. (See also individual compounds.) 11:500–
  515; 12:86
  manufacture 11:500
  properties 11:500
  separation 11:501
  uses 11:501
1,2,3,4-Diepoxybutene. (See also *Epoxy compounds.*)
  properties 12:195
1,2,4,5-Diepoxycyclohexane. (See also *Epoxy
  compounds.*)
  infrared spectroscopy 12:200

1,2,5,6-Diepoxypentane. (See also *Epoxy compounds.*)
  infrared spectroscopy 12:200
9,10, 12, 13-Diepoxystearic acid
  thin-layer chromatography 12:205
5,6,5',6'-Diepoxyzeaxanthin. See Violaxanthin.
Dierythritol
  gas chromatography 13:578
Diesel engine oil. (See also *Lubricants.*)
  commercial grades and specifications 15:312
Dietary yeast tablets. (See also *Yeast.*) 19:518
  composition 19:518
  species 19:516
  uses 19:516, 518
Diethanolamine. (See also Ethanolamines.)
  assay 12:292
  infrared spectrum 5:7
  manufacture 5:3
  properties 5:2; 12:291
  solubility 13:539, 552
  specifications 5:4
  volumetric determination 12:292
Diethanol glycine
  in beer 7:585
Diethers
  Zeisel method 4:70
3,4-Diethoxybenzaldehyde. See Protocatech-
  aldehyde.
3,3'-Diethoxy-4,4'-diisocyanatodiphenyl. (See also
  Isocyanic acid esters.)
  infrared spectroscopy 15:106
6,6'-Diethoxythioindigo. See CI Vat Orange 5.
Diethyl acetal. (See also *Acetals and ketals.*) 4:62
  properties 4:63
  specifications 4:65
Diethylallylcyclohexenyl malonate
  properties 15:423
Diethylallylisobutyl malonate
  properties 15:423
Diethylallylisopropyl malonate
  properties 15:423
Diethylallyl-1-methylbutyl malonate
  properties 15:423
Diethylamine
  azeotropes 5:261
  commercial grades and specifications 5:264
  gas chromatography 11:245
  manufacture 5:262
  in methanol 16:114
  properties 5:260
  thin-layer chromatography 19:140
  in tobacco 19:140
p-Diethylaminobenzaldehyde 6:596
  commercial specifications 6:607
  derivatives for identification 6:612
  manufacture 6:600
  properties 6:598

Diethylaminodiethyldithiocarbamate. (See also
  Dithiocarbamic acid salts.)
  ultraviolet spectroscopy 11:583
2-(Diethylamino)ethanol
  infrared spectrum 5:8
  properties 5:2
m-Diethylaminophenol
  in D & C Red No. 19 10:498, 509
  in D & C Red No. 37 10:498, 510
  ultraviolet spectroscopy 10:510
Diethylaminopropylamine. (See also *Diamines
  and higher amines, aliphatic.*)
  properties 11:410
Diethylammonium diethyldithiocarbamate
  colorimetric method for copper 14:543
  extraction method for lead 15:178
Diethyl-*n*-amyl malonate
  properties 15:423
N,N-Diethylaniline 5:421
  acid–base titration 5:447
  azeotropes 5:424
  chronopotentiometry 5:454
  colorimetric methods 5:452
  color reactions for identification 5:431
  derivatives for identification 5:431
  gas chromatography 5:438
  infrared spectrum 5:436
  manufacture 5:425
  properties 5:422
  ultraviolet spectroscopy 5:432
5,5-Diethylbarbituric acid 5:91
Diethylbenzene
  in ethylbenzene 18:309
  gas chromatography 7:8; 18:309
Diethylbenzenes
  gas chromatography 13:352
  in gasoline 13:352, 355
  properties 13:352
Diethyl-*n*-butyl malonate
  properties 15:423
2,4-Diethylchlorobenzene
  mass spectroscopy 9:405
N,N-Diethylcyclohexylamine
  gas chromatography 11:245
Diethyldecaborane
  colorimetric method using β-naphthoquinoline
  7:347
Diethyldiallyl malonate
  properties 15:423
Diethyldiethoxymethyl malonate
  in diethylethoxymethylene malonate 15:437
  gas chromatography 15:437
Diethyldiethyl malonate
  properties 15:423
2,2'-Diethyldiphenylamine. (See also *Diamines
  and higher amines, aromatic.*)
  properties 11:434

solubility 11:435
Diethyldithiocarbamate
  colorimetric method for bismuth 7:215
  colorimetric method for tellurium 17:606
  extraction of aluminum 5:205
1,4-Diethylene dioxide. See *Dioxane.*
Diethylene ether. See *Dioxane.*
Diethylene glycol. (See also Polyglycols.)
  azeotropes 13:554
  commercial grades and specifications 13:558,
    560
  in dates 12:332
  derivatives for identification 13:568
  distillation range 13:602
  in ethylene glycol 13:600
  freezing point 6:7; 13:535
  gas chromatography 5:32; 6:5; 7:427; 13:503,
    576, 589; 19:298, 300, 301
  liquid chromatography 13:580
  manufacture 13:553
  paper chromatography 13:581
  phthalic anhydride volumetric method 13:586
  in polyurethanes 19:297
  potassium dichromate volumetric method 13:
    599
  properties 6:2; 13:549
  Rosen color test 13:573
  solubility 13:552
  uses 13:555
Diethylene glycol dinitrate. (See also DEGN.)
  gas chromatography 13:577
Diethylene glycol ethers
  in brake fluids 7:425
Diethylene glycol monobutyl ether. (See also
    Ethers.)
  commercial specifications 12:299
  in enamels 12:175
  properties 12:175
Diethylene glycol monoethyl ether. (See also
    Carbitol; *Ethers.*)
  commercial specifications 12:299
  in enamels 12:175
  properties 12:175
Diethylene glycol monomethyl ether. (See also
    Ethers.)
  commercial specifications 12:299
  in enamels 12:175
  properties 12:175
Diethylene glycol phenyl ether. (See also Phenyl
    ethers.)
  gas chromatography 12:310
  phenol in 12:310
Diethylenetriamine. (See also Ethyleneamines.)
  acid–base titration 11:424, 425
  amine value 11:425
  commercial specifications 11:418
  density 11:409

gas chromatography 11:421
infrared spectroscopy 11:419
latent heat 11:417
nitrogen content 11:427
properties 11:408
specific heat 11:415
thermal conductivity 11:416
vapor pressure 11:410
viscosity 11:412, 413
N,N,N',N'',N''-Diethylenetriaminepentaacetic acid.
  See DTPA.
Diethylethanolamine
  azeotropes 5:3
  specifications 5:4
N,N-Diethylethanolamine. See (Diethylamino)-
  ethanol.
Diethyl ether. See Ethyl ether.
Diethylethoxymethylene malonate
  analysis 15:437
  commercial specifications 15:425
  gas chromatography 15:437
  impurities, determination of 15:437, 438
  infrared spectroscopy 15:437
  manufacture 15:424
  properties 15:421
  saponification methods 15:438
  solubility 15:422
  ultraviolet spectroscopy 15:437
Diethylethyl-*n*-amyl malonate
  properties 15:423
Diethylethylcyclohexenyl malonate
  properties 15:423
Diethylethyl-*n*-hexyl malonate
  properties 15:423
Diethylethylisoamyl malonate
  properties 15:423
Diethylethylisopropyl malonate
  properties 15:423
Diethylethyl malonate
  properties 15:423
Diethylethyl-1-methylbutenyl malonate
  properties 15:423
Diethylethylmethylbutyl malonate
  properties 15:423
Diethylethyl-1-methylpropyl malonate
  properties 15:423
Diethylethylphenyl malonate
  gas chromatography 15:435
  properties 15:423
  saponification methods 15:435
O,O-Diethyl *S*-2-(ethylthio)ethyl phosphorothioate.
  See Disulfoton.
O,O-Diethyl *S*-(ethylthio)methyl phosphorodithio-
  ate. See Phorate.
Diethyl formal. See Ethylal.
N,N-Di(2-ethylhexyl)ethanolamine
  properties 5:2

solubility 5:1
Di(2-ethylhexyl) phthalate
in enamels 12:176
properties 12:176
Di(2-ethylhexyl) sebacate
gas chromatography 16:97
properties 8:577
uses 15:302
N,N-Diethylhydroxylamine. (See also N-Alkylhydroxylamines.)
commercial grades and specifications 14:446
toxicity 14:444
uses 14:443
Diethylisopropyl malonate·
properties 15:423
O,O-Diethyl O-2-isopropyl-4-methyl-6-pyrimidyl
phosphorothioate. See Diazinon.
Diethyl ketone
in cigarette smoke 10:31
derivatives for identification 8:343
gas chromatography 10:31
properties 8:335
N,N-Diethyl-d-lysergamide. (See also LSD.)
bioassay 7:182
Diethyl maleate
commercial specifications 8:581
in diethyl succinate 8:587
gas chromatography 8:587
properties 8:577
Diethyl malonate
analysis 15:431
assay 15:433
colorimetric determination 15:431
colorimetric method for menadione 10:197
commercial specifications 15:425
in diethylethoxymethylene malonate 15:437
in dimethyl malonate 15:435
m-dinitrobenzene colorimetric method 15:431
3,5-dinitrotoluene–sodium nitroprusside colorimetric method 15:431
in ethyl cyanoacetate 15:441
gas chromatography 15:432, 437, 441
identification 15:431
impurities, determination of 15:433, 434
infrared spectroscopy 15:431
manufacture 15:424
mass spectrum 15:432
properties 15:421
resorcinol colorimetric method 15:431
saponification methods 15:433
thin-layer chromatography 15:433
Diethylmethylbutyl malonate
properties 15:423
Diethylmethylcyclohexenyl malonate
properties 15:423
Diethyl-1-methylpropyl malonate
properties 15:423

O,O-Diethyl O-p-(methylsulfinyl)phenyl phosphorothioate. See Fensulfothion.
N,N-Diethylnicotinamide
colorimetric method for decaborane 7:347
O,O-Diethyl O-p-nitrophenyl phosphorothioate.
See Parathion.
Diethylphenyl malonate
gas chromatography 15:435
properties 15:423
Diethyl phthalate
in enamels 12:176
in essential oils 12:276
gas chromatography 16:97
molecular weight 12:276
properties 8:577; 12:176
N,N-Diethyl-1,3-propanediamine. See Diethylaminopropylamine.
2,2-Diethyl-1,3-propanediol. (See also Glycols.)
solubility 13:538
Diethyl succinate
commercial specifications 8:581
gas chromatography 8:587
properties 8:577
Diethyl sulfate
in diethyl malonate 15:433
gas chromatography 15:433
Diethyl sulfide
gas chromatography 15:558
Differential infrared spectroscopy. See Infrared
spectroscopy, differential.
Differential reaction rate analysis hydrolysis
of amides 4:167
Differential scanning calorimetry
of polypropylene 12:383
of polyurethanes 19:289
Differential thermal analysis
of barium 6:578
of carbon fibers 4:21
of ethylene–propylene copolymers 12:384
of polyamide fibers 17:320
of polycarbonates 17:335
of polyethylene 12:382
of polymers 4:261
of polypropylene 12:382
Diffusion coefficient
of polycarbonates 17:344
Diffusion constant technique molecular weight
determination of acrylic polymers 4:313
Diffusion methods
for hydrogen 18:232
Diffusion theory of adhesion 4:391
Difluoroethane
in aerosols 11:95
gas chromatography 11:98
properties 11:95
1,1-Difluoroethane. (See also Fluorocarbons.)
properties 13:60

Difluoromethane. (See also *Fluorocarbons.*)
    properties 13:60
Difluorophosphoric acid. (See also Fluorophos-
        phoric acids.) 13:37, 42
    commercial grades and specifications 13:44
    composition, determination of 13:46
    in hexafluorophosphoric acid 13:47
    manufacture 13:45
    in monofluorophosphoric acid 13:45, 46
    nitron gravimetric method 13:46
    properties 13:44
    uses 13:43
Difluorosilane
    properties 18:40
Difolatan 13:209
    analysis 13:211
    gas chromatography 13:212
    manufacture 13:210
    properties 13:210
    solubility 13:210
    thin-layer chromatography 13:211
    toxicity 13:210
    uses 13:209
Difolatan epoxide
    gas chromatography 13:212
    thin-layer chromatography 13:211
Di-FP acid. See Difluorophosphoric acid.
Digalloylbisepigallocatechin
    in black tea 18:493
    structure 18:493
    thin-layer chromatography 18:495, 496
Digitonin
    gravimetric method for sterols 19:149
Diglucosides
    liquid chromatography 19:466
    separation 19:465
    thin-layer chromatography 19:465
    in wines 19:464
Diglycerol
    gas chromatography 13:578
    paper chromatography 13:581
Diglycerol tetraacetate
    gas chromatography 13:505
Diglycidal ether of bisphenol A and methacrylic
        acid adducts. See Dental restorative materials,
        composite.
Diglycolic acid
    gas chromatography 5:25
Diheptyldiphenylamine
    paper chromatography 6:85
Diheptyl ketone
    properties 8:335
Dihexyl ketone
    properties 8:335
Dihydrocaffeic acid
    in green coffee 10:396
Dihydrocodeine. (See also *Narcotics.*)

    ammonium molybdate colorimetric method
        16:246
    ammonium vanadate colorimetric method 16:246
    color tests 16:246
    formaldehyde–sulfuric acid colorimetric method
        16:246
    microchemical tests 16:249
    potassium–bismuth iodide microchemical method
        16:249
    structure 16:244
    thin-layer chromatography 16:256
    ultraviolet spectroscopy 16:251
Dihydrocodeinone. See Hydrocodone.
Dihydrocortisone
    liquid chromatography 14:210
2,3-Dihydro-2,2-dimethyl-7-benzofuranyl methyl-
        carbamate. See Carbofuran.
Dihydroergotamine
    thin-layer chromatography 15:295
Dihydrohydroxycodeinone. See Oxycodone.
Dihydrohydroxymorphinone. See Oxymorphone.
9,10-Dihydro-9-ketoanthracene. See Anthrone.
Dihydromethylmorphinone. See Metapon.
5,6-Dihydro-2-methyl-1,4-oxathiin-3-carboxamilide.
        See Carboxin.
5,6-Dihydro-2-methyl-1,4-oxathiin-3-carboxamilide-
        4,4-dioxide. See Oxycarboxin.
Dihydromorphine. (See also *Narcotics.*)
    ammonium molybdate colorimetric method
        16:246
    ammonium vanadate colorimetric method 16:246
    color tests 16:246
    formaldehyde–sulfuric acid colorimetric method
        16:246
    gas chromatography 16:255
    microchemical tests 16:249
    platinic iodide microchemical method 16:249
    potassium ferricyanide fluorometric method
        16:263
    potassium–mercuric iodide microchemical method
        16:249
    structure 16:244
    ultraviolet spectroscopy 16:251
Dihydromorphinone. See Hydromorphone.
Dihydronicotyrine
    in tobacco 19:136
9,10-Dihydro-9-oxoanthracene. See Anthrone.
Dihydrostreptomycin 5:473, 579
    analysis 5:582, 589, 595
    colorimetric method 5:595
    infrared spectrum 5:583
    microbiological assay 5:589
    properties 5:579
    solubility 5:580
    structure 5:473
    ultraviolet spectrum 5:583
    uses 5:579

Dihydrotachysterol. (See also Parathyroid hormones.)
  manufacture 14:190
  properties 14:190
  solubility 14:191
  structure 14:191
  uses 14:191
6a, 7-Dihydro-3,4,6a,10-tetrahydroxybenz[b]-indeno[1,2-d]-pyran-9(6H)-one. See Hematein.
1,2-Dihydro-2,2,4-trimethyl-6-dodecylquinoline 6:76
1,2-Dihydro-2,2,4-trimethyl-6-ethoxyquinoline (DTEQ) 6:76, 111
1,2-Dihydro-2,2,4-trimethyl-6-phenylquinoline 6:72
Dihydroxyandrostanes
  liquid chromatography 14:209
Dihydroxyandrostanones
  liquid chromatography 14:209
Dihydroxyandrostenes
  liquid chromatography 14:209
1,4-Dihydroxyanthraquinone
  in D & C Green No. 5; and No. 6 10:498, 508
  in D & C Violet No. 2 10:498, 508
2,2'-Dihydroxy-1,1'-azonaphthalene-4-sulfonic acid, sodium salt. See Pontachrome Blue Black R.
2,2'-Dihydroxy-1,1'-azonaphthalene-5-sulfonic acid, sodium salt. See Pontachrome Violet SW.
2,4-Dihydroxybenzaldehyde 6:596
  derivatives for identification 6:612
  fluorometry 6:626
  infrared spectroscopy 6:617
  properties 6:598
  ultraviolet spectroscopy 6:632
2,5-Dihydroxybenzaldehyde 6:596
  fluorometry 6:625, 627
  properties 6:598
  ultraviolet spectroscopy 6:632
3,4-Dihydroxybenzaldehyde (Protocatechualdehyde) 6:596
  derivatives for identification 6:612
  fluorometry 6:627
  properties 6:598
  ultraviolet spectroscopy 6:632
1,2-Dihydroxybenzene. See Pyrocatechol.
1,3-Dihydroxybenzene. See Resorcinol.
1,4-Dihydroxybenzene. See Hydroquinone.
1,2-Dihydroxybenzene-3,5-disulfonate. See Tiron.
Dihydroxybenzenes. (See also individual compounds.)
  commercial grades and specifications 17:15
  solubility 17:5
4,5-Dihydroxy-m-benzenesulfonic acid disodium salt. See Tiron.
2,4-Dihydroxybenzoic acid (β-Resorcyclic acid) 8:400

properties 8:400, 402
3,4-Dihydroxybenzoic acid. See Protocatechuic acid.
2,4-Dihydroxybenzophenone
  thin-layer chromatography 16:96
  ultraviolet spectroscopy 16:95
4,4'-Dihydroxybiphenyl
  paper chromatography 6:85
3,3'-Dihydroxy-α-carotene. See Lutein.
3,3'-Dihyroxy-β-carotene. See Zeaxanthin.
3,4-Dihydroxycinnamic acid. See Caffeic acid.
3-(3,4-Dihydroxycinnamoyl)quinic acid. See Chlorogenic acid.
Dihydroxycoumarins. (See also Coumarins.)
  gas chromatography 11:168, 170
2,2'-Dihydroxy-4,4'-dimethoxybenzophenone
  ultraviolet spectroscopy 16:95
(N,N'-Dihydroxyethyl)ethylenediaminediacetic acid. See HEEDA.
Di(2-hydroxyethyl)glycine. See DHEG.
Dihydroxyketosteroids
  gas chromatography 14:218
3,4-Dihydroxymandelic acid
  gas chromatography 8:389
2,2'-Dihydroxy-4-methoxybenzophenone
  thin-layer chromatography 16:96
  ultraviolet spectroscopy 16:95
2,6-Dihydroxy-4-methylbenzaldehyde 6:596
  properties 6:598
7,8-Dihydroxy-4-methylcoumarin
  gravimetric method for niobium 16:366
  gravimetric method for tantalum 16:366
1,2-Dihydroxynaphthalene
  gravimetric method for niobium 16:366
2,7-Dihydroxynaphthalene
  colorimetric method for carbon tetrachloride 9:387
  colorimetric method for carboxymethylcellulose 11:354
  colorimetric method for chloroform 9:387
  colorimetric method for glycollic acid 9:92
  colorimetric method for methylene chloride 9:387
  color test for formaldehyde 13:106
1,8-Dihydroxynaphthalene derivatives
  color reagents for boron 7:387
1,8-Dihydroxynaphthalene-3,6-disulfonic acid. See Chromotropic acid.
4,5-Dihydroxy-2,7-naphthalenedisulfonic acid. See Chromotropic acid.
1,8-Dihydroxynaphthalene-3,6-disulfonic acid-2,7-bis(1-azo-2) phenylarsonic acid. See Arsenazo III.
3α,12α-Dihydroxy-7-oxo-5β-cholanic acid
  color tests 7:147
  properties 7:143
2,5-Dihydroxyphenylacetic acid.

See Homogentisic acid.
3,4-Dihydroxyphenylacetic acid. See Homoprotocatechuic acid.
3,4-Dihydroxyphenyl-4-azobenzene–thiocyanate
colorimetric method for niobium 16:370, 375
3-(3,4-Dihydroxyphenyl)propenoic acid. See
Caffeic acid.
3-(3,4-Dihydroxyphenyl)propionic acid. See Dihydrocaffeic acid.
Dihydroxypregnanes
liquid chromatography 14:209
9,10-Dihydroxystearic acid
thin-layer chromatography 12:205
Dihydroxysuccinic acid. See Tartaric acid.
Dihydroxytartaric acid osazone
identification reagent for calcium 8:99
4,5-Diiodo-3,6-fluorandiol. See D & C Orange
No. 10.
Diiodofluoresceins. (See also D & C Orange No.
10.)
in F D & C Red No. 3 10:513
liquid chromatography 10:514
paper chromatography 10:514
visible spectroscopy 10:514
2,3-Diiodo-1-propanol. (See also Glycerol halohydrins.)
derivatives for identification 14:160
Diiodosilane
properties 18:40
Diiodothyronine. (See also Thyroid hormones.)
14:183
structure 14:183
Diisobutylamine
properties 5:260
uses 5:263
Diisobutyl carbinol. (See also Alcohols, higher.)
phthalic anhydride method 4:559
properties 4:552
specifications 4:557
uses 4:556
1,2-Diisobutylene. (See also Epoxy compounds.)
near-infrared spectroscopy 12:217
1,2-Diisobutylene oxide. (See also Epoxy compounds.)
zinc bromide–2,4-dinitrophenylhydrazine gravimetric method 12:214
Diisobutyl ketone
acidity 8:358
alcohol in 8:360
azeotropes 8:336
commercial specifications 8:341
Diisobutyl phthalate
in enamels 12:176
properties 12:176
Diisocyanates
gas chromatography 19:313

infrared spectroscopy 15:96
manufacture 15:96
in polyurethanes 19:313
1,4-Diisocyanatobenzol. (See also Isocyanic acid
esters.)
infrared spectroscopy 15:106
4,4'-Diisocyanatodicyclohexylmethane. (See also
Isocyanic acid esters.)
infrared spectroscopy 15:106
Diisodecyl phthalate
in enamels 12:176
properties 12:176
Diisopropanolamine
acetylation 5:20
properties 5:2
specifications 5:3
Diisopropylamine
azeotropes 5:261
properties 5:260
uses 5:263
Diisopropylchlorobenzene
mass spectroscopy 9:407
Diisopropylethanolamine
azeotropes 5:3
N,N-Diisopropylethanolamine
properties 5:2
solubility 5:1
2,6-Diisopropylphenol. (See also Phenols.)
4-aminoantipyrine colorimetric method 17:18
commercial grades and specifications 17:15
Millon's reagent color test 17:18
1,3-Diisopropylphenylene-4,6-diisocyanate. (See
also Isocyanic acid esters.)
near-infrared spectroscopy 15:105
Dikar 13:219
Dilatometer method
for melting range of margarine 15:502
Dilaudid. See Hydromorphone.
Dill oil. (See also Essential oils.)
carvone content 12:282
Diluents
in brake fluids 7:424
for epoxy resins 12:236
Dilution test
for disinfectants and antiseptics 11:539
Dimedon
gravimetric method for formaldehyde 13:107
Dimedone derivative
of acetals for identification 4:67
2,3-Dimercapto-1-propanol. (See also Mercaptans.)
p-chloromercuribenzoic acid colorimetric method
15:561
uses 15:555
Dimeryl diisocyanate. (See also Isocyanates.)
structure 19:271
trade names 19:271

Dimesoperiodic acid. (See also Periodates.) 14: 609
 manufacture 14:609
Dimethallyl
 halogenation methods 5:104
 properties 5:103
Dimethallylamine 5:87
 acid–base titration 5:89
N,N'-Dimethallylethylenediamine 5:87
Dimethindene maleate. (See also *Antihistamines.*) 6:27
Dimethisoquin hydrochloride 5:375
 analysis 5:387
 properties 5:387
Dimethisterone. (See also Steroids.) 10:579
 analysis 10:580, 587
 assay 10:580
 derivatives for identification 10:580
 identification 10:580, 587
 infrared spectroscopy 10:588
 isoniazid spectroscopic method 10:589
 isonicotinyl hydrazide colorimetric method 14: 207
 properties 10:579
 solubility 10:579
 ultraviolet spectroscopy 10:580
Dimethoate. (See also *Pesticides and insecticides.*)
 commercial grades and specifications 19:594
 composition 19:597
 manufacture 19:597
 properties 19:589
 solubility 19:587
 structure 19:582
 toxicity 19:592
2,3-Dimethoxybenzaldehyde
 fluorometry 6:625, 626
 ultraviolet spectroscopy 6:633
2,4-Dimethoxybenzaldehyde
 fluorometry 6:626
 ultraviolet spectroscopy 6:633
2,5-Dimethoxybenzaldehyde
 fluorometry 6:625, 626, 628
 ultraviolet spectroscopy 6:633
3,4-Dimethoxybenzaldehyde (Veratraldehyde) 6: 596
 column chromatography 6:618
 derivatives for identification 6:612
 fluorometry 6:626
 properties 6:598
 ultraviolet spectroscopy 6:633
3,5-Dimethoxybenzaldehyde
 fluorometry 6:625, 626, 628
 ultraviolet spectroscopy 6:633
3,4-Dimethoxybenzoic acid
 properties 8:402
Dimethoxycoumarins. (See also *Coumarins.*)
 gas chromatography 11:168, 170

3,3'-Dimethoxy-4,4'-diisocyanatodiphenyl. (See also *Isocyanic acid esters.*)
 infrared spectroscopy 15:106
p,p'-Dimethoxydiphenylamine
 paper chromatography 6:85
Dimethoxyethane
 in black tea 18:510
 gas chromatography 16:109; 18:510
 in methanol 16:109
3,5-Dimethoxy-4-hydroxybenzaldehyde. See Syringaldehyde.
3,5-Dimethoxy-4-hydroxycinnamic acid. See Sinapic acid.
5,7-Dimethoxy-6-hydroxycoumarin. (See also *Coumarins.*)
 gas chromatography 11:170
N-(Dimethoxymethylsilylisobutyl)ethylenediamine. (See also Organofunctional silanes.)
 properties 18:80
3,4-Dimethoxyphenylacetic acid. See Homoveratric acid.
1-(2,5-Dimethoxyphenylazo)-2-naphthol. See Citrus Red No. 2.
2,5-Dimethoxyphenyl isocyanate. (See also *Isocyanic acid esters.*)
 infrared spectroscopy 15:106
2,6-Dimethoxyphenylpenicillin. See Methicillin.
2,2-Dimethoxypropane
 gas chromatography 16:109
 in methanol 16:109
Dimethyl acetal. (See also *Acetals and ketals.*) 4:62
 melting point of derivatives 4:68
 properties 4:63
Dimethylacetamide
 commercial specifications 8:599
 infrared spectrum 8:603
 properties 8:598
2,4-Dimethylacetophenone
 absorption spectroscopy 9:564
Dimethyl acetylene. See 2-Butyne.
9,9-Dimethylacridanes. See Acetone–diphenylamine reaction products.
9,9-Dimethylacridine
 in acetone–diphenylamine reaction products 6:110
 gas chromatography 6:111
Dimethylaesculetin. See 6,7-Dimethoxycoumarin.
N,N-Dimethylallylamine 5:87
Dimethyl allyl cyanurate 5:93
Dimethylamine
 colorimetric method for dichlone 13:201
 commercial grades and specifications 5:264
 gas chromatography 5:268
 manufacture 5:262
 in methanol 16:114
 properties 5:260

thin-layer chromatography 19:140
in tobacco 19:140
uses 5:263
p-Dimethylaminoaniline
   colorimetric method for sulfur in acrylic poly-
      mers 4:285
p-Dimethylaminobenzaldehyde 6:596
   absorption spectroscopy 9:564
   colorimetric method 15:291, 298
   colorimetric method for alcohols 4:551
   colorimetric method for anileridine 16:246
   colorimetric method for antipyrine 6:214
   colorimetric method for apomorphine 16:277
   colorimetric method for butyl acrylate 4:216
   colorimetric method for carboxin 13:213
   colorimetric method for codeine 16:246, 270
   colorimetric method for ergot alkaloids 4:600
   colorimetric method for fusel oil 4:485
   colorimetric method for gramicidin 5:268
   colorimetric method for isocyanate 19:282
   colorimetric method for maleic hydrazine 19:
      155
   colorimetric method for morphine 16:247
   colorimetric method for oxycarboxin 13:213
   colorimetric method for *p*-phenylenediamine 11:
      448
   colorimetric method for polyurethanes 19:282
   colorimetric method for pyrroles 17:298
   colorimetric method for staphylomycin 5:563
   colorimetric method for thiophene 18:541
   colorimetric method for tryptophan 8:534
   colorimetric method for tyrothricin 5:549
   commercial specifications 6:607
   derivatives for identification 6:612
   in F D & C Violet No. 1 10:498, 507
   infrared spectroscopy 6:616
   liquid chromatography 10:507
   manufacture 6:600
   properties 6:598
   ultraviolet spectroscopy 6:633; 10:507
   uses 6:606
p-Dimethylaminobenzaldehyde–chloramine-T
   colorimetric method for hydroxyproline 13:372
p-Dimethylaminobenzaldehyde–hydrogen peroxide
   colorimetric method for hydroxyproline 13:365
p-Dimethylaminobenzaldehyde isonicotinoyl hydra-
      zone. See DAIH.
p-Dimethylaminobenzaldehyde–sulfuric acid
   colorimetric method for fusel oil 19:439
p-Dimethylaminobenzalrhodanine
   colorimetric method for silver 18:164, 168
p-Dimethylaminobenzenediazo sodium sulfonate.
   See Dexon.
p-Dimethylaminobenzoic acid
   in F D & C Violet No. 1 10:498, 507
   liquid chromatography 10:507
   ultraviolet spectroscopy 10:507

5-(Dimethylaminobenzylidine)rhodamine
   colorimetric method for gold 14:25
p-Dimethylaminobenzylidinerhodanine
   colorimetric method for mercury 11:140; 16:18,
      20, 42
   colorimetric method for silver 10:303
p-Dimethylaminocinnamaldehyde
   colorimetric method for hetacillin 16:577
Dimethylaminoethanol 5:2
   gas chromatography 5:5
Dimethylaminoethyl methacrylate. (See also
      Acrylic and methacrylic acids and esters.)
   commercial grades and specifications 4:187
   nuclear magnetic resonance spectroscopy 16:86
   polymer in 4:200
   properties 4:182
   saponification method 4:196
   solubility 4:181, 184
1-Dimethylamino-2-methyl-2-propanol
   gas chromatography 5:5
2-Dimethylamino-2-methyl-1-propanol
   gas chromatography 5:5
β-Dimethylaminopropionitrile
   catalyst for production of AM-9 chemical grout
      4:160, 167
Dimethylaminopropylamine. (See also *Diamines
      and higher amines, aliphatic.*)
   properties 11:410
N,N-Dimethylaniline 5:421
   acid–base titration 5:448
   azeotropes 5:424
   chronopotentiometry 5:454
   colorimetric method for chlorinated naphthalenes
      9:491
   colorimetric method for nitrite 16:494
   colorimetric methods 5:452
   color reactions for identification 5:431
   gas chromatography 5:268, 438
   infrared spectroscopy 5:434, 453
   manufacture 5:425
   properties 5:422
   separation 5:435
   ultraviolet spectroscopy 5:432, 450
   use in manufacture of azo dyes 6:403
   uses 5:427
   x-ray diffraction 5:435
Dimethylanilines
   derivatives for identification 5:429
   gas chromatography 5:268
2,2-Dimethylaziridine. (See also *Imines.*)
   properties 14:468
   structure 14:468
3,4-Dimethylbenzoic acid. See Veratric
      acid.
2,2-Dimethylbutane. (See also Hexanes.)
   commercial specifications 14:304
   gas chromatography 14:330

properties 14:303
2,3-Dimethylbutane. (See also Hexanes.)
  commercial specifications 14:304
  gas chromatography 13:319; 14:330, 335
  properties 14:303
Dimethylbutanes
  gas chromatography 13:350
  in gasoline 13:350, 353
  properties 13:350
2,3-Dimethyl-2-butene
  liquid chromatography 7:13, 14, 15, 16
Dimethylbutenes
  gas chromatography 13:350
  in gasoline 13:350, 353
  properties 13:350
1,3-Dimethylbutylamine
  azeotropes 5:261
  properties 5:260
Dimethyl-*t*-butylbenzene
  gas chromatography 13:352
  in gasoline 13:352, 355
  properties 13:352
N-(1,3-Dimethylbutyl)-*N*-phenyl-*p*-phenylene-
    diamine 6:74, 113
2,3-Dimethylchlorobenzene
  mass spectroscopy 9:402
2,4-Dimethylchlorobenzene
  mass spectroscopy 9:402
2,5-Dimethylchlorobenzene
  mass spectroscopy 9:402
2,6-Dimethylchlorobenzene
  mass spectroscopy 9:402
Dimethylchlorosilane. (See also Organo-chloro-
    silanes.)
  gas chromatography 18:61
  mass spectrum 18:59
1,2-Dimethylcyclohexane. (See also *Hydrocarbons.*)
  commercial grades and specifications 14:308
  uses 14:307
Dimethylcyclohexanedione
  turbidimetric method for aldehydes 12:309
Dimethylcyclohexanes
  gas chromatography 13:351
  in gasoline 13:351, 354
  properties 13:351
N,N-Dimethylcyclohexylamine 11:250
  gas chromatography 11:251
  infrared spectroscopy 11:250
  manufacture 11:250
  properties 11:242
  solubility 11:251
  toxicity 11:251
  uses 11:251
Dimethylcyclopentanes
  in benzene 7:11
  gas chromatography 7:11; 13:350; 14:330, 332
  in gasoline 13:350, 353

  in *n*-heptane 14:332
  properties 13:350
3,3-Dimethylcyclopentanone
  derivatives for identification 8:343
3,3'-Dimethyl-4,4'-diaminodiphenylinethane
  in polyurethanes 19:304
  structure 19:304
  thin-layer chromatography 19:306
O,O-Dimethyl *S*-(1,2-dicarbethoxyethyl)phos-
    phorodithioate. See Malathion.
Dimethyldichlorosilane. (See also Organo-chloro-
    silanes.)
  commercial specifications 18:56
  gas chromatography 18:51, 60, 64
  impurities, determination of 18:64
  infrared spectrum 18:56
  manufacture 18:55
  mass spectrum 18:59
  in methyltrichlorosilane 18:64
  properties 18:54
  in tetrachlorosilane 18:51
Dimethyldiethoxysilane. (See also Organosilane
    esters.)
  commercial specifications 18:71
  infrared spectrum 18:72
  properties 18:70
Dimethyldiethyllead
  atomic absorption spectroscopy 13:358
  gas chromatography 13:358
Dimethyldimethoxymethyl malonate
  in dimethylmethoxymethylene malonate 15:437
  gas chromatography 15:437
Dimethyldimethoxysilane. (See also Organosilane
    esters.)
  properties 18:70
N,N'-Dimethyl-*N*-*N'*-di(1-methylpropyl)-*p*-phenyl-
    enediamine 6:74
  gas chromatography 6:82
3,6-Dimethyl-*p*-dioxane-2,5-dione. See Lactide.
4,4'-Dimethyldiphenylamine. (See also *Diamines
    and higher amines, aromatic.*)
  properties 11:434
  solubility 11:435
2,9-Dimethyl-4,7-diphenyl-1,10-phenanthroline.
    See Bathocuproine.
Dimethyl disulfide
  in cocoa bean vapor 9:570
  mass spectroscopy 9:570
Dimethylene glycol
  gas chromatography 13:579
sym-Dimethylene glycol. See 2,3-Butanediol.
N,N-Dimethylethanolamine. See 2-(Dimethyl-
    amino)ethanol.
Dimethyl ether
  in methanol 16:114
1,1-Dimethylethylamine. See *tert*-Butylamine.
Dimethylethylbenzenes

gas chromatography 13:352
in gasoline 13:352, 355
properties 13:352
N,N-Dimethylethylenediamine. (See also *Imines.*)
  manufacture 14:473
  properties 14:470
  structure 14:470
Dimethylethylpentanes
  gas chromatography 13:351
  in gasoline 13:351, 354
  properties 13:351
2,6-Dimethyl-4-ethynl-heptan-4-ol. (See also
  Acetylenic alcohols.)
  silver acetylide method 4:578
Dimethylfluorone
  colorimetric method for tantalum 16:368; 18:
  448
Dimethylformamide
  colorimetric method for dinocap 13:220
  color test for reactive dyes 12:19
  commercial specifications 8:599, 600
  in enamels 12:175
  properties 8:558, 597; 12:175
  residual in acrylic fibers 4:343
Dimethyl fumarate 5:26
Dimethylglyoxime
  colorimetric method for nickel 5:139; 6:44,
    227; 7:116, 661; 10:301, 624; 11:141; 14:
    132, 542; 15:388; 16:299, 316
  colorimetric method for nickel oxide 13:441
  color test for palladium 17:249
  commercial grades and specifications 16:304
  extraction of nickel 10:129
  gravimetric method for nickel 5:138; 9:672;
    10:317, 330, 642; 12:533; 15:387; 16:296,
    299, 302, 304, 305, 306, 307
  gravimetric method for palladium 17:249, 253,
    257
Dimethylglyoxime–nickel biuret
  color test for malonic acid 15:424
Dimethylheptanes
  gas chromatography 13:351
  in gasoline 13:351, 354
  properties 13:351
2,6-Dimethyl-4-heptanol. See Diisobutyl carbinol.
N,N'-Di-3-(5-methylheptyl)-*p*-phenylenediamine
  6:114
  acid–base titration 6:115
  analysis 6:115
  gas chromatography 6:81
  identification 6:115
  manufacture 6:115
  properties 6:114
  separation 6:115
2,5-Dimethyl-1,5-hexadiene. See Dimethallyl.
2,2-Dimethylhexane
  gas chromatography 14:330, 332

in *n*-heptane 14:332
in isooctane 14:332
2,4-Dimethylhexane
  gas chromatography 14:332, 335
  in isooctane 14:332
2,5-Dimethylhexane
  gas chromatography 14:332
  in isooctane 14:332
3,3-Dimethylhexane
  gas chromatography 14:330
Dimethylhexanes
  gas chromatography 13:350
  in gasoline 13:350, 353
  properties 13:350
2,3-Dimethyl-1-hexyne-3,4-diol
  thin-layer chromatography 13:582
unsym-Dimethylhydrazine
  analysis 14:234
  assay 14:235
  colorimetric determination 14:235
  commercial grades and specifications 14:234
  coulometric method 14:238
  derivatives for identification 14:235
  gas chromatography 14:237, 239
  identification 14:235
  impurities, determination of 14:239
  infrared spectroscopy 14:235
  manufacture 14:233
  polarography 14:238
  potassium iodate titration 14:236
  properties 14:232
  proton magnetic resonance spectroscopy 14:237
  solubility 14:232
  toxicity 14:233
  uses 14:232
  volumetric determination 14:237
N'-(1,1-Dimethyl-2-hydroxyethyl)-2-methyl-1,2-
  propanediamine 5:5
N,N-Dimethylhydroxylamine. (See also *N*-Alkyl-
  hydroxylamines.)
  properties 14:443
O,N-Dimethylhydroxylamine. (See also *Hydroxylamines and their salts.*)
  properties 14:443
Dimethyl isobutyral
  gas chromatography 16:109
  in methanol 16:109
2,2-Dimethyl-3-isocyanatobutane. (See also
  Isocyanic acid esters.)
  infrared spectroscopy 15:106
Dimethyl ketone. (See also *Acetone.*)
  properties 8:335
Dimethyl maleate 5:26
Dimethyl malonate
  analysis 15:431
  assay 15:433
  colorimetric determination 15:431

commercial specifications 15:425
in diethyl malonate 15:433, 435
in dimethylmethoxymethylene malonate 15:437
gas chromatography 15:432, 437, 441
identification 15:431
impurities, determination of 15:433, 434
infrared spectroscopy 15:431
manufacture 15:424
mass spectrum 15:432
in methyl cyanoacetate 15:441
potassium methoxide titration 15:435
properties 15:421
saponification methods 15:433
solubility 15:422
thin-layer chromatography 15:433
Dimethylmethoxymethylene malonate
analysis 15:437
commercial specifications 15:425
gas chromatography 15:437
impurities, determination of 15:437, 438
infrared spectroscopy 15:437
manufacture 15:424
properties 15:421
saponification methods 15:438
O,O-Dimethyl *S*-(*N*-methylcarbamoyl)methyl phosphorodithioate. See Dimethoate.
O,O-Dimethyl *O*-(4-methylthio-*m*-tolyl)phosphoro-
thioate. See Fenthion.
3,3'-Dimethylnaphidine
colorimetric method for vanadium 19:331, 333
Dimethylnaphthalenes. (See also Alkylnaphthalenes.)
boiling point 16:209
gas chromatography 16:217
liquid chromatography 7:13, 14, 15, 16
properties 16:217
3,3'-Dimethylnaphthidine
colorimetric method for oxygen 16:531
colorimetric method for vanadium 14:68
O,O-Dimethyl *O-p*-nitrophenyl phosphorothioate.
See Methyl parathion.
2,2-Dimethyl-3,4-octadienal
ultraviolet spectroscopy 6:632
3,7-Dimethyl-2,6-octadienal. See Citral.
Dimethyloctanes
gas chromatography 13:351
in gasoline 13:351, 354
properties 13:351
2,2-Dimethylol butane. See 2-Methyl-2-ethyl-1,3-
propanediol.
Dimethylol ethyl carbamate
in cotton 11:138
infrared spectroscopy 11:139
Dimethylol ethyleneurea 5:276
in cotton 11:138
infrared spectroscopy 11:139
Dimethylol ethyl triazone

in cotton 11:138
infrared spectroscopy 11:139
Dimethylol perhydropyrimidinone
in cotton 11:138
infrared spectroscopy 11:139
Dimethylol propyleneurea 5:726
Dimethylol urea 5:276
in cotton 11:138
infrared spectroscopy 11:139
O,O-Dimethyl *S*-4-oxo-1,2,3-benzotriazin-3-(4*H*)-
ylmethyl phosphorodithioate. See Azinphosmethyl.
β-(2-(3,5-Dimethyl-2-oxocyclohexyl)-2-hydroxy-
ethylglutarimide. See Cycloheximide.
6-(2,2-Dimethyl-5-oxo-4-phenyl-1-imidazolidinyl)
penicillanic acid. See Hetacillin.
2,2-Dimethylpentane
gas chromatography 14:330
2,3-Dimethylpentane
gas chromatography 14:330, 332
in *n*-heptane 14:332
2,4-Dimethylpentane
gas chromatography 14:330, 335
3,3-Dimethylpentane
gas chromatography 14:330
Dimethylpentanes
in benzene 7:11
gas chromatography 7:11; 13:350
in gasoline 13:350, 353
properties 13:350
2,6-Dimethyl-4-pentanol
phthalic anhydride method 4:536
Dimethylpentenes
gas chromatography 13:350
in gasoline 13:350, 353
properties 13:350
1,4-Dimethylpentylamine 5:260
3,4-Dimethyl-1-pentyn-3-ol. (See also Acetylenic
alcohols.)
silver acetylide method 4:578
2,9-Dimethyl-1,10-phenanthroline. (See also
Neocuproine.)
colorimetric method for copper 10:614
2,3-Dimethylphenol. (See also 2,3-Xylenol.)
derivatives for identification 17:21
properties 17:4
thin-layer chromatography 17:24
2,4-Dimethylphenol. (See also 2,4-Xylenol.)
4-aminoantipyrine colorimetric method
17:18
derivatives for identification 17:21
Millon's reagent color test 17:18
properties 17:4
thin-layer chromatography 17:24
2,5-Dimethylphenol. (See also 2,5-Xylenol.)
4-aminoantipyrine colorimetric method
17:18

derivatives for identification 17:21
Millon's reagent color test 17:18
properties 17:4
thin-layer chromatography 17:24
2,6-Dimethylphenol. (See also 2,6-Xylenol.)
    4-aminoantipyrine colorimetric method 17:18
    derivatives for identification 17:21
    Millon's reagent color test 17:18
    properties 17:4
    thin-layer chromatography 17:24
3,4-Dimethylphenol. (See also 3,4-Xylenol.)
    4-aminoantipyrine colorimetric method 17:18
    derivatives for identification 17:21
    Millon's reagent color test 17:18
    properties 17:4
    thin-layer chromatography 17:24
3,5-Dimethylphenol. (See also 3,5-Xylenol.)
    4-aminoantipyrine colorimetric method 17:18
    derivatives for identification 17:21
    Millon's reagent color test 17:18
    properties 17:4
    thin-layer chromatography 17:24
Dimethylphenols. See Xylenols; and individual
    compounds.
Dimethyl-p-phenylenediamine
    colorimetric method based on formation of
        Wurster salts 6:101, 103
    colorimetric method for cystine and cysteine
        8:468
    colorimetric method for mercaptans 15:570
Dimethylphenyl polysiloxane
    uses 15:302
Dimethyl phthalate
    in enamels 12:176
    in essential oils 12:276
    gas chromatography 16:97
    molecular weight 12:276
    properties 8:577; 12:176
2,2-Dimethylpropane. See Neopentane.
N,N-Dimethyl-1,3-propanediamine. See Dimethyl-
    aminopropylamine.
2,2-Dimethyl-1,3-propanediol. (See also Glycols.)
    commercial grades and specifications 13:547
    gas chromatography 13:577
    manufacture 13:545
    properties 13:537
    structure 13:534
    uses 13:545
Di-(1-methylpropyl)-p-phenylenediamine 6:77
    gas chromatography 6:82
2,3-Dimethylpyridine. (See also Pyridine and
    pyridine derivatives.)
    commercial specifications 17:423
    derivatives for identification 17:415
    gas chromatography 17:437
    ionization constants 17:416
    properties 17:415

2,4-Dimethylpyridine. (See also Pyridine and
    pyridine derivatives.)
    commercial specifications 17:423
    derivatives for identification 17:415
    ionization constants 17:416
    properties 17:415
2,5-Dimethylpyridine. (See also Pyridine and
    pyridine derivatives.)
    derivatives for identification 17:415
    gas chromatography 17:437
    ionization constants 17:416
    properties 17:415
2,6-Dimethylpyridine. (See also Pyridine and
    pyridine derivatives.)
    commercial specifications 17:423
    derivatives for identification 17:415
    gas chromatography 17:437
    ionization constants 17:416
    properties 17:415
3,4-Dimethylpyridine. (See also Pyridine and
    pyridine derivatives.)
    derivatives for identification 17:415
    ionization constants 17:416
    properties 17:415
3,5-Dimethylpyridine. (See also Pyridine and
    pyridine derivatives.)
    derivatives for identification 17:415
    ionization constants 17:416
    properties 17:415
Dimethylpyridines. See Lutidines; and individual
    compounds.
Dimethylsilicone fluids. (See also Silicone fluids.)
    commercial grades and specifications 18:102,
        104
    gas chromatography 18:106
    gel permeation chromatography 18:111
    infrared spectroscopy 18:105
    manufacture 18:101
    properties 18:97
    refractive index 18:97
Dimethyl sulfate
    gravimetric method for barium 6:575
Dimethyl sulfide
    in black tea 18:510
    in cocoa bean vapor 9:570
    gas chromatography 15:558; 18:510
    mass spectroscopy 9:570
Dimethyl sulfoxide
    in methanol 16:115
Dimethyl terephthalate 17:163–166
    acid number 17:165
    acid number stability 17:165
    analysis 17:165
    assay 17:165
    color tests 17:166
    commercial specifications 17:165
    derivatives for identification 17:165

freezing point 17:165
gas chromatography 17:166
identification 17:165
impurities, determination of 17:165
manufacture 17:163, 164
nuclear magnetic resonance spectroscopy 17:165
properties 17:164
saponification equivalent 17:165
sodium methylate titration 17:165
solubility 17:164
uses 17:163
3,5-Dimethyltetrahydro-1,3,5(2*H*)thiadiazine-2-
thione. See Mylone.
O,O-Dimethyl 2,2,2-trichloro-1-hydroxyethyl phos-
phonate. See Trichlorfon.
Dimethyl uron
in cotton 11:138
infrared spectroscopy 11:139
1,7-Dimethylxanthine. See Theophylline.
3,7-Dimethylxanthine. See Theobromine.
unsym-Dimethylydrazine hydrochloric acid
titration of cinnamaldehyde 10:51
N,N'-Di-2-naphthyl-4-aminoaniline
in photographic films 12:592
N,N'-Di-2-naphthyl-*p*-phenylenediamine. (See also
Diamines and higher amines, aromatic.)
column chromatography 6:89
paper chromatography 6:85
properties 11:434
solubility 11:435
ultraviolet spectroscopy 6:97
Di-β-naphthylthiocarbazone
colorimetric method for mercury 16:42
Dinicotinic acid. (See also *Pyridine and pyridine
derivatives.*) 17:413
Dinitraniline orange. (See also *Pigments.*)
composition 12:173
in enamels 12:173, 184
identification 12:184
structure 17:180
Dinitraniline reds
in enamels 12:182
identification 12:182
2,4-Dinitroaniline. (See also High explosives.)
12:435
in D & C Orange No. 17 10:498, 508
derivatives for identification 5:429
manufacture 5:426; 12:435
properties 5:423
solubility 12:435
ultraviolet spectroscopy 5:433, 450; 10:509
uses 12:435
Dinitroanilines
separation 5:435
2,4-Dinitroanisole. (See also High explosive.) 12:
435
manufacture 12:435

melting point 12:461
properties 12:433
solubility 12:436
uses 12:435
2,4-Dinitrobenzaldehyde. (See also High explosives.)
melting point 12:461
properties 12:433
ultraviolet spectroscopy 6:634, 635
Dinitrobenzene. (See also High explosives.) 12:
436
identification reagent for threonine 8:528
manufacture 12:436
melting point 12:461
properties 12:433
solubility 12:436
uses 12:436
1,2-Dinitrobenzene
colorimetric method for *N*-alkylhydroxylamines
14:453
1,3-Dinitrobenzene
colorimetric method for diethyl malonate 15:431
colorimetric method for 3-oxocholanic acid derivatives 7:148
3,5-Dinitrobenzoate
reagent for formation of derivatives of esters
8:584
3,5-Dinitrobenzoate derivatives
of acetals 4:67
acid–base titration, nonaqueous 7:93
of alcohols 7:93
of alcohols, higher 4:558
of alcohols, lower 4:517
of benzoic acid esters 7:83
gas chromatography 7:93, 96
paper chromatography 7:96
Dinitrobenzoic acids. (See also High explosives.)
melting point 12:461
properties 12:433
solubility 12:436
uses 12:436
3,5-Dinitrobenzoyl chloride
gravimetric method for cinnamyl alcohol 10:78
method for alcohols 4:547
reaction of amides with 8:608
4,6-Dinitro-*o*-cresol. (See also High explosives.)
melting point 12:461
properties 12:434
solubility 12:437
uses 12:437
1,1-Dinitrocyclohexane. (See also Nitroparaffins.)
properties 16:414
1,1-Dinitroethane. (See also Nitroparaffins.)
properties 16:414
1,2-Dinitroethane. (See also Nitroparaffins.)
properties 16:414
2,4-Dinitroethylaniline. (See also High explosives.)

melting point 12:461
properties 12:432
2,4-Dinitroethylbenzene. (See also High explosives.)
properties 12:433
2,4-Dinitrofluorobenzene
colorimetric method for lysine 8:499
ultraviolet spectroscopy of aniline 5:449
Dinitrogen pentoxide. (See also Nitrogen oxides.)
manufacture 16:476
properties 16:474
Dinitrogen tetroxide. (See also Nitrogen oxides.) 16:475
analysis 16:478
assay 16:479
colorimetric determination 16:479
manufacture 16:476
properties 16:474
thermal equilibrium 16:475
toxicity 16:476
uses 16:476
volumetric determination 16:479
Dinitrogen trioxide. (See also Nitrogen oxides.) 16:475
manufacture 16:475
properties 16:474
Dinitroglycerin. (See also High explosives.) 12:444
properties 12:444
solubility 12:444
uses 12:444
Dinitromethane. (See also Nitroparaffins.)
properties 16:414
2,4-Dinitromethylaniline. (See also High explosives.)
melting point 12:461
properties 12:432
Dinitronaphthalenes. (See also High explosives; Nitronaphthalenes.) 12:437
infrared spectra 12:458
manufacture 12:437
melting point 12:461; 16:224
properties 12:433
solubility 12:437
uses 12:437
1,5-Dinitropentanediol
gas chromatography 13:577
2,4-Dinitrophenol. (See also High explosives.)
manufacture 12:438
melting point 12:461
properties 12:434
solubility 12:439
toxicity 12:438
uses 12:438
1-(2,4-Dinitrophenylazo)-2-naphthol. See D & C Orange No. 17.
2,4-Dinitrophenyl dimethyldithiocarbamate.

See Safex.
2,4-Dinitrophenylhydrazine. (See also High explosives.)
colorimetric method for aldehydes 17:298
colorimetric method for ascorbic acid 19:425
colorimetric method for aspartic acid 8:461
colorimetric method for carbonyl compounds 8:364; 11:521
colorimetric method for hydroquinone and/or quinone 4:206
colorimetric method for menadione 10:196
colorimetric method for phytonadione 10:193
colorimetric method for polyethylene glycols 12:578
gravimetric method for cinnamaldehyde 10:45
identification reagent for carbonyl compounds 8:342
melting point 12:461
properties 12:434
reaction of carbonyl compounds with 8:333
2,4-Dinitrophenylhydrazine hydrochloride
gravimetric method for formaldehyde 13:107
2,4-Dinitrophenylhydrazine–potassium hydroxide
colorimetric method for glycols 13:574
colorimetric method for polyhydric alcohols 13:574
2,4-Dinitrophenylhydrazone
colorimetric method for acetone 4:116
colorimetric method for secondary alcohols in the presence of primary alcohols 4:548
gravimetric method for acetone 4:116
gravimetric method for acetophenone 4:125
2,4-Dinitrophenylhydrazone derivative
of acetals 4:67, 70
of acetone 4:110, 114
of acetophenone 4:123, 125
of alcohols 4:516
of aldehydes and ketones 4:111, 116
2,4-Dinitrophenylhydrazones
of benzaldehydes 6:611, 612
column chromatography 6:617
thin-layer chromatography 6:620
3,6-Dinitrophthalic acid
colorimetric method for sugar 18:354
2,2-Dinitropropane. (See also Nitroparaffins.)
properties 16:414
1,2-Dinitropropanediol
gas chromatography 13:577
2,4-Dinitroresorcinol. (See also High explosives.)
melting point 12:461
properties 12:434
2,4-Dinitrotoluene. (See also Dinitrotoluenes.)
infrared spectrum 12:459
Dinitrotoluenes. (See also High explosives.)

commercial specifications 12:440
melting point 12:461
polarography 11:454
properties 12:434
solubility 12:439
in toluenediamines 11:454
uses 12:439
3,5-Dinitrotoluene–sodium nitroprusside
colorimetric method for diethyl malonate 15:431
Dinocap 13:219
analysis 13:220
N,N-dimethylformamide colorimetric method
13:220
in fruit 13:220
liquid chromatography 13:220
manufacture 13:219
properties 13:219
solubility 13:219
toxicity 13:219
uses 13:219
4,4'-Dioctyldiphenylamine. (See also *Diamines and
higher amines, aromatic.*) 6:72
properties 11:434
solubility 11:435
N,N'-Di-2-octyl-*p*-phenylenediamine (DOPD)
gas chromatography 6:81
Dioctyl sodium sulfosuccinate titration
of codeine 16:279
of meperidine 16:279
of methadone 16:279
Diolefins. (See also Dienes.) 14:314
Diols. See Glycols.
Dionin. See Ethylmorphine.
Diopside 11:464
Diopside–hedenbergite 8:75
Dioxane 11:516–524
analysis 11:519
azeotropes 11:516
colorimetric determination 11:520
commercial grades and specifications 11:519
gas chromatography 11:519
impurities 11:518
impurities, determination of 11:520
infrared spectroscopy 11:519
manufacture 11:517
nuclear magnetic resonance spectroscopy 11:520
properties 11:516
separation 11:517
solubility 11:521
specification tests 11:520
structure 11:516
toxicity 11:517
ultraviolet spectroscopy 11:520
uses 11:516
S,S'-p-Dioxane-2,3-diyl *O,O*-diethyl phosphorodithioate. See Dioxathion.
Dioxathion. (See also *Pesticides and insecticides.*)

commercial grades and specifications 19:594
composition 19:590, 597
manufacture 19:597
properties 19:590
solubility 19:587
structure 19:582
toxicity 19:592
Dioxazine dyes. (See also *Azine and related dyes.*)
6:382, 383
identification 6:389
manufacture 6:383, 385
solubility 6:383
Dioxazine violet. (See also *Pigments.*)
structure 17:184
Dioxyethylene ether. See *Dioxane.*
Dioxygen difluoride
manufacture 13:10
properties 13:10
Dipaxin. See Diphenadione.
Dipentaerythritol. (See also Polyhydric alcohols.)
5:22
commercial grades and specifications 13:566
gas chromatography 13:578, 594
manufacture 13:565
properties 13:562
solubility 13:561
structure 13:561
uses 13:565
Dipentaerythritol hemiacetal
gas chromatography 13:594
Dipentaerythritol hexanitrate. (See also High
explosives.) 12:445
manufacture 12:445
properties 12:445
uses 12:445
Dipentene. (See also Terpenes.)
commercial grades and specifications 14:265, 316
gas chromatography 12:104
occurrence 14:316
uses 14:266
Diphenadione 10:177
analysis 10:178
ferric chloride–dioxane colorimetric method
10:179
identification 10:178
infrared spectrum 10:178
properties 10:177
solubility 10:177
ultraviolet spectroscopy 10:178, 179
Diphenhydramine
gas chromatography 6:210, 211
Diphenhydramine hydrochloride. (See also
Antihistamines.) 6:27
properties 6:32
thin-layer chromatography 6:35
Diphenic acid
gravimetric method for thorium 19:9

Diphenols. (See also *Phenols*; and individual compounds.)
    analysis 17:46
    manufacture 17:10
Diphenoxylate. (See also Diphenoxylate hydrochloride; *Narcotics*.)
    iodine–potassium iodide microchemical method 16:249
    microchemical tests 16:249
    potassium iodide microchemical method 16:249
    structure 16:245
Diphenoxylate hydrochloride. (See also Diphenoxylate.)
    ultraviolet spectroscopy 16:251
Diphenyl 11:524–532; 13:224
    analysis 11:526; 13:224
    assay 11:527
    ceric sulfate–nitric acid colorimetric method 11:526
    in citrus fruit 11:527
    colorimetric determination 13:225
    commercial grades and specifications 11:526
    distillation range 11:527
    eutectics 11:525
    formaldehyde–sulfuric acid colorimetric method 11:526
    freezing point 11:527
    in fruit 13:224
    gas chromatography 9:495; 11:528; 16:217
    identification 11:526
    impurities, determination of 11:528
    infrared spectroscopy 11:527
    liquid chromatography 7:16; 11:529
    manufacture 11:524; 13:224
    phosphorimetric method 11:528
    properties 11:524; 13:224
    solubility 11:525; 13:224
    solvent extraction 13:224
    tetracyanoethylene colorimetric method 11:526
    thin-layer chromatography 11:526, 530; 13:225
    toxicity 11:525; 13:224
    ultraviolet spectroscopy 11:527
    uses 11:524; 13:224
2-(Diphenylacetyl)indane-1,3-dione. See Diphenadione.
Diphenylamine. (See also *Diamines and higher amines, aromatic*.) 6:72; 11:455–460
    acid–base titration 5:444
    analysis 11:457
    assay 11:458
    azeotropes 11:436
    in brake fluids 7:428
    bromination methods 5:447
    colorimetric method for cellulose nitrate 11:127, 132
    colorimetric method for nitrate 6:12; 15:401
    colorimetric method for sucrose 19:411
    colorimetric methods 5:450
    color reactions for identification 5:431
    commercial grades and specifications 11:457
    derivatives for identification 5:429; 11:438
    in Ext D & C Yellow No. 1 10:498, 510
    freezing point 11:458
    gas chromatography 6:111; 11:458
    identification 5:432; 11:457
    impurities, determination of 11:458
    infrared spectrum 5:436
    manufacture 5:425; 11:456
    p-nitrophenol colorimetric method 11:457
    oxalic acid color test 11:457
    polarography 5:454
    properties 5:422; 11:434, 456
    solubility 11:435
    toxicity 11:456
    ultraviolet spectroscopy 5:432, 450; 10:510
    use in manufacture of BLE 6:110
    uses 5:426; 11:455
    vapor in air 5:455
Diphenylamine–acetone
    in elastomers, synthetic 12:141
    ultraviolet spectrum 12:141
Diphenylamine acetone resin 6:76
Diphenylamine–sulfuric acid
    color test for cellulose nitrate 12:588
    color test for chlorate 12:428
Diphenyl boric acid ethanolamine complex. See Flavognost.
Diphenylcarbazide
    colorimetric method for chromic oxide 13:440
    colorimetric method for chromium 5:126; 7:115, 657; 9:636, 642, 643, 644, 645, 646; 10:195, 297, 621; 11:140; 15:381
    colorimetric method for molybdenum 9:642
sym-Diphenylcarbazide. (See also Diphenylcarbazide.)
    colorimetric method for chromium 16:326
1,5-Diphenylcarbazide
    colorimetric method for chromium 14:129
Diphenylcarbazone
    colorimetric method for mercury 11:140; 16:18, 20, 22, 42
N,N'-Diphenyl-1,4-diaminobenzene. (See also *Diamines and higher amines, aromatic*.)
    derivatives for identification 11:438
Diphenyldichlorosilane. (See also Organo-chlorosilanes.)
    commercial specifications 18:56
    gas chromatography 18:60
    properties 18:54
Diphenyldiethoxysilane. (See also Organosilane esters.)
    properties 18:70
Diphenyldimethoxysilane. (See also Organosilane esters.)

properties 18:70
Diphenyldithiocarbazone. See Dithizone.
1,4-Diphenyl-3,5-endanilo-4,5-dihydro-1,2,4-triazole. See Nitron.
1,4-Diphenyl-1,3,5-endanilotriazoline. See Nitron.
Diphenylene oxide
gas chromatography 16:218
Diphenyl ether. (See also *Ethers.*)
properties 12:297
N,N'-Diphenylethylenediamine 6:77
paper chromatography 6:85
Diphenylguanidine
in elastomers, synthetic 12:143
ultraviolet spectrum 12:143
Diphenylhydantoin
gas chromatography 6:210
Diphenyline
colorimetric method for tungsten 11:141
Diphenyline chloride
gravimetric method for tungsten 19:202
Diphenyl mercury. (See also Organomercury compounds.)
commercial grades 16:54
Diphenylmercury adipate 4:429
Diphenylmethane diisocyanate. (See also *Isocyanates; Isocyanic acid esters.*)
composition, determination of 19:275
gas chromatography 19:275, 315
impurities, determination of 19:275
isomer content 19:275
near-infrared spectroscopy 15:105
nuclear magnetic resonance spectroscopy 19:302
in polyurethanes 19:315
properties 15:95
structure 19:270
trade names 15:103; 19:270
Diphenylmethane dyes 12:7
Diphenylmethylmonochlorosilane. (See also Organo-chlorosilanes.)
properties 18:54
Diphenylol propane. See Bisphenol A.
Diphenyl oxide
gas chromatography 11:529
4,7-Diphenyl-1,10-phenanthroline. See Bathophenanthroline.
N,N'-Diphenyl-*p*-phenylenediamine. (See also *Diamines and higher amines, aromatic*; DPPD.)
benzoyl peroxide colorimetric method 4:210
colorimetric method for chloranil 13:200
in elastomers, synthetic 12:140
properties 11:434
solubility 11:435
ultraviolet spectrum 12:141
1,4-Diphenyl-3,5-(phenylimino)-Δ^2-1,2,4-triazoline. See Nitron.
Diphenylpiazselenol
structure 17:599

1,3-Diphenyl-2-propanone
absorption spectroscopy 9:564
N,N'-Diphenylpropylenediamine 6:77
Diphenylpyraline hydrochloride. (See also *Antihistamines.*) 6:27
Diphenyls and terphenyls. (See also Diphenyls; Methyldiphenyls; and Terphenyls.) 11:524–537
Diphenylthiocarbazone. (See also Dithizone.) 4:476
1,4-Diphenylthiosemicarbazide
colorimetric method for ruthenium 17:248
2,4-Diphenylthiosemicarbazide
colorimetric method for rhenium 17:541
1,3-Diphenylurea. See Carbanilide.
Diphosphorus tetraiodide. (See also Phosphorus halides.)
properties 17:108
Dipicolinic acid. (See also *Pyridine and pyridine derivatives.*) 17:413
Dipicrylamine
colorimetric method for acetylcholine chloride 9:617
colorimetric method for amines 5:89
colorimetric method for bethanechol chloride 9:617
colorimetric method for carbamylcholine chloride 9:617
colorimetric method for choline chloride 9:617
colorimetric method for methacholine chloride 9:617
colorimetric method for β-methylcholine chloride 9:617
colorimetric method for potassium 17:383
colorimetric method for succinyldicholine chloride 9:617
colorimetric method for succinylmonocholine iodide 9:617
gravimetric method for potassium 17:368
structure 17:368
Dipotassium orthophosphate
properties 17:120
solubility 17:120
Dipropylamine 5:260
Di-*n*-propylamine 5:263
Dipropylbarbituric acid
infrared spectrum 6:489
Dipropyl disulfide 5:105
Dipropylene glycol. (See also Polyglycols.)
azeotropes 13:555
in brake fluids 7:424
commercial grades and specifications 13:558, 560
density 13:584
distillation range 13:602
gas chromatography 5:22; 13:576, 589; 19:298
isomers 13:556
manufacture 13:556

phthalic anhydride volumetric method 13:586
in polyurethanes 19:297
properties 13:549
refractive index 13:584
Rosen color test 13:573
solubility 13:552
uses 13:556
Dipropylene glycol monomethyl ether. (See also *Ethers.*)
in brake fluids 7:425
commercial specifications 12:299
gas chromatography 7:427
Dipropylene glycol salicylate
in sunscreens 11:92
ultraviolet spectroscopy 11:92
Dipropyl ketone
properties 8:335
2,2'-Dipyridyl. (See also *Pyridine and pyridine derivatives.*)
colorimetric method for iron 6:262, 459; 7:660; 9:242; 11:140, 498
colorimetric method for vanadium 19:331, 332
structure 17:414
uses 17:414
2,3'-Dipyridyl
in tobacco 19:136
4,4'-Dipyridyl. (See also *Pyridine and pyridine derivatives.*)
structure 17:414
α,β-Dipyridyl–pyrocatechol
colorimetric method for niobium 16:370, 375
Diquat
structure 17:414
uses 17:414
2,2'-Diquinolyl. (See also Cuproine.)
colorimetric method for copper 11:140
Direct dyes 6:402; 12:3
Directional triangle tests 17:627
Dirt
in casein 9:12
in cyclamates 11:230
in fatty alcohols 4:573
in greases 15:326
N,N'-Disalicylideneethylenediamine 6:74
N,N'-Disalicylidenepropylenediamine 6:74
Disc centrifuges 9:173, 174, 175
characteristics 9:160
settling curves 9:161
Dishwashing detergents. (See also *Detergents.*) 11:325
composition 11:325
performance tests 11:388
types 11:325
Disinfectants. (See also *Disinfectants and antiseptics.*)
definition 11:537
Disinfectants and antiseptics. (See also

Aldehydes; Alcohols; Bisphenols; Epoxides; Halogen donors; Halogens; Guanidine derivatives; Mercurials; *Phenols*; Quaternary ammonium compounds; and individual compounds.) 11:537–572
bacteriological methods 11:538
chlorine germicidal equivalence test 11:539
classifications 11:537
dilution test 11:539
evaluation methods 11:538
phenol coefficient test 11:539
Disintegration test
for dental silicate cements 11:260
for dental zinc phosphate cements 11:263
Disintegrator for preparation of pulp handsheets 7:241, 242
Disodium 2,2'-[(diazoamino)-di-*p*-phenylene]-bis[6-methyl-7-benzothiazolsulfonate]. See Titan yellow.
Disodium 2',7'-dibromo-4'-(hydroxymercuri)fluorescin. See Merbromin.
Disodium dihydrogen pyrophosphate. (See also Sodium acid pyrophosphate.)
properties 17:123
uses 17:123
Disodium ethylenebisdithiocarbamate. (See also Dithiocarbamic acid salts; Nabam.)
uses 11:577
xanthate method 11:580
Disodium ethylenediaminetetraacetate titration of zinc 18:205
Disodium methylarsenate 6:229
Disodium orthophosphate
manufacture 17:125
Disodium phosphate
spectroscopic method for anionic surfactants 12:575
uses 17:120
Disodium salt of 8-amino-2-phenylazo-1-naphthol-3,6-disulfonic acid. See D & C Red No. 33.
Disodium salt of 1,4-bis(*o*-sulfo-*p*-toluino) anthraquinone. See D & C Green No. 5.
Disodium salt of 3-carboxy-5-hydroxy-1-phenyl-4-*p*-sulfo-phenylazopyrazole
in F D & C Yellow No. 5 10:517
liquid chromatography 10:517
paper chromatography 10:518
visible spectroscopy 10:517
Disodium salt of 3-carboxy-5-hydroxy-1-*p*-sulfo-phenyl-4-phenylazopyrazole
in F D & C Yellow No. 5 10:517
liquid chromatography 10:517
paper chromatography 10:518
visible spectroscopy 10:517
Disodium salt of 9-*o*-carboxyphenyl-6-hydroxy-4,5-diiodo-3-isoxanthone. See D & C Orange No. 11.

Disodium salt of 9-*o*-carboxyphenyl-6-hydroxy-3-
isoxanthone. See D & C Yellow No. 8.
Disodium salt of 9-*o*-carboxyphenyl-6-hydroxy-
2,3,5,7-tetraiodo-3-isoxanthone. See F D & C
Red No. 3.
Disodium salt of 2,4-dinitro-1-naphthol-7-sulfonic
acid. See Ext D & C Yellow No. 7.
Disodium salt of disulfonic acid of 2-(2-quinolyl)-
1,3-indandione. See D & C Yellow No. 10.
Disodium salt of 4-{[4-(*N*-ethyl-*m*-sulfobenzyl-
amino)phenyl] (4-hydroxy-2-sulfoniumphenyl)-
methylene} -[1-(*N*-ethyl-*N*-*m*-sulfobenzyl)-$\Delta^{2,5}$-
cyclohexadienimine]. See F D & C Green No.
3.
Disodium salt of 4-{[4-(*N*-ethyl-*m*-sulfobenzyl-
amino)phenyl] (2-sulfoniumphenyl)methylene}
[1-(*N*-ethyl-*N*-*m*-sulfobenzyl)-$\Delta^{2,5}$ -cyclo-
hexadienimine]. See F D & C Blue No. 1.
Disodium salt of 5,5′-indigo-tindisulfonic acid. See
F D & C Blue No. 2.
Disodium salt of 1-(phenylazo)-2-naphthol-6-
sulfonic acid
in F D & C Yellow No. 6 10:518
visible spectroscopy 10:518
Disodium salt of 1-4-sulfophenylazo-2-naphthol-
6-sulfonic acid. (See also F D & C No. 6.)
in F D & C Yellow No. 6 10:518
visible spectroscopy 10:518
Disodium salt of 1-(4-sulfophenyl)-3-ethylcarboxy-
4-(4-sulfonaphthylazo)-5-hydroxypyrazole.
See Orange B.
Disodium salt of 4-(*o*-sulfo-*p*-tolylazo)-3-hydroxy-
2-naphthoic acid. See D & C Red No. 6.
Disodium salt of 2-(5-sulfo-2,4-xylylazo)-1-naphthol-
4-sulfonic acid. See F D & C Red No. 4.
Disodium salt of 2,4,5,7-tetrabromo-9-*o*-carboxy-
phenyl-6-hydroxy-3-isoxanthone. See D & C
Red No. 22.
Disodium salt of 2,4,5,7-tetrabromo-9-(3,4,5,6-
tetrachloro-*o*-carboxyphenyl)-6-hydroxy-3-
isoxanthone. See D & C Red No. 28.
Dispersed dyes 6:402; 12:11
identification 12:21
Dispersibility
of pigments 17:207
Displacement balance method
for specific gravity of natural gas 13:330
Dissipation factors
for silicone fluids 18:116
Dissociation constants
of amino acids 8:445
of ion exchange resins 15:14
Distearin
thin-layer chromatography 12:205
Distillation. (See also Azeotropic distillation.)
of ammonium bicarbonate 5:329
of ammonium chloride 5:335
of ammonium nitrate 5:340
of ammonium sulfate 5:351
of antifreezes 6:369
of arsenic 6:237
of benzene 7:32, 46
of bituminous materials 6:313, 317
of boron 16:325
of carbonyl compounds 8:358
of cut-back asphalts 6:313
of ethanol 11:362
of ethanol in wines 19:429
of hydrocarbons in phenols 17:37
of isopropyl alcohol 11:362
of methyl borate 7:391
of organic acids 9:75
of recovery of asphalt from bituminous paving
mixtures 6:333
of residue of emulsified asphalts 6:325
of rhenium 17:522
of tars 6:315
of technetium 17:522
of volatile matter in bituminous paving mixtures
6:333
of water 18:191
of water in bituminous materials 6:316, 317
of water in emulsified asphalts 6:324
of water in lubricants 15:330
Distillation curves
of gasoline 13:340
Distillation range
of acetone 4:113
of acrylonitrile 4:376
of alcohols 4:542
of alkanolamines 5:14
of carbon disulfide 8:274
of carbon tetrachloride 8:326
of chlorophenols 9:526
of 1,2-dichloroethane 9:439
of diethylene glycol 13:602
of diphenyl 11:527
of dipropylene glycol 13:602
of epichlorohydrin 12:227
of ethylene glycol 13:602
of 2-ethyl-1,3-hexanediol 13:602
of glycols 13:601
of hydrocarbons 14:341
of mercaptans 15:571
of methanol 16:113
of methylene chloride 9:429
of 2-methyl-2,4-pentanediol 13:602
of morpholine 16:199
of 1,5-pentanediol 13:602
of polyhydric alcohols 13:601
of propylene glycol 13:602
of propylene oxide 12:225
of Stoddard solvent 11:606
of tetraethylene glycol 13:602

of triethylene glycol 13:602
of tripropylene glycol 13:602
Distillation test
for drycleaning detergents 11:611
Distilled spirits
methanol content 16:119
Distiller's beer 4:462
analysis 4:480
quality control 4:468
Distiller's dried feeds. (See also *Yeast.*) 19:518
manufacture 19:518
species 19:516
uses 19:516
Distiller's feeds
analysis 4:469
quality control 4:467
Distribution, normal 7:161, 163
Distribution, standard 7:161
Disulfatozirconic acid. See Zirconium sulfate.
Disulfides
in elastomers, synthetic 12:128
in phenols 17:38
5,7'-Disulfonated indigo
in F D & C Blue No. 2 10:512
liquid chromatography 10:512
Disulfoton. (See also *Pesticides and insecticides.*)
commercial grades and specifications 19:594
composition 19:590, 597
manufacture 19:597
properties 19:590
solubility 19:587
structure 19:582
toxicity 19:592
Disulfur decafluoride
properties 13:11
toxicity 13:13
Disulfur difluoride
properties 13:11
Disulfuric acid
manufacture 18:413
Dithane D-14. See Nabam.
Dithane M-22. See Maneb.
Dithane Z-78. See Zineb.
Dithiocarbamate esters 11:586-591
acid hydrolysis method 11:589
analysis 11:589
manufacture 11:586
properties 11:587, 588, 589
ultraviolet spectroscopy 11:590
uses 11:586
Dithiocarbamates. (See also Dithiocarbamate
esters; Dithiocarbamic acid salts; Thiuram disulfides; and individual compounds.) 11:573-
602; 13:202-205
analysis 13:205
uses 11:573
Dithiocarbamate salts. See Dithiocarbamic

acid salts; and individual compounds.
Dithiocarbamic acid salts. (See also individual
compounds.) 11:573-586
acid hydrolysis method 11:574
amine determination 11:582
analysis 11:574
gas chromatography 11:582
heavy metal titration 11:585
infrared spectroscopy 11:583
liquid chromatography 11:584
manufacture 11:573
oxidation methods 11:585
paper chromatography 11:584
polarography 11:586
properties 11:575
thin-layer chromatography 11:584
ultraviolet spectroscopy 11:580, 583
uses 11:573
xanthate method 11:578
Dithiocarbonic anhydride. See *Carbon disulfide.*
Dithiol
colorimetric method for molybdenum 19:213
colorimetric method for tin 19:37, 55
colorimetric method for tungsten 19:208, 213
Dithiooxamide. See Rubeanic acid.
Dithiourethanes. See Dithiocarbamate esters.
Dithizone
colorimetric method for bismuth 7:215; 15:177
colorimetric method for cadmium 8:53
colorimetric method for copper 7:116; 10:613;
15:177; 16:344
colorimetric method for EDTA-type compounds
7:586
colorimetric method for gold 14:25
colorimetric method for lead 6:43, 226, 231;
8:440; 10:299, 623, 668; 15:190, 193, 225,
385; 16:331
colorimetric method for mercury 10:496; 16:18,
20, 36, 38
colorimetric method for phenylmercuric acetate
11:564
colorimetric method for silver 18:164, 170
colorimetric method for thimerosal 11:564
colorimetric method for zinc 5:149; 7:664;
10:125, 305, 331; 11:141, 144; 13:268;
15:491; 16:340; 19:556, 563
colorimetric test using carbon tetrachloride
8:327
extraction method for lead 15:176
extraction of heavy metals 10:94
volumetric method for gold 14:23
volumetric method for organomercury compounds
16:64
volumetric method for silver 6:44, 227
Dithizone–chloroform extraction
of lead 10:494
Dithizone–oxine

colorimetric method for indium 13:396
Dithol
  colorimetric method for tin 11:141
Ditolylamines. (See also *Diamines and higher
    amines, aromatic.*)
  derivatives for identification 11:438
N,N'-Di-*o*-tolylethylenediamine 6:77
  paper chromatography 6:85
1,3-Di-*o*-tolylguanidine 4:273
Di-*o*-tolylguanidine salt of dicatechol borate
    6:75
  color test 6:96
Di-*p*-tolyl mercury. (See also Organomercury com-
    pounds.)
  commercial grades 16:54
N,N'-Di-*o*-tolyl-*p*-phenylenediamine. (See also
    Diamines and higher amines, aromatic.)
  properties 11:434
  solubility 11:435
N,N'-Di-*p*-tolyl-*p*-phenylenediamine. (See also
    Diamines and higher amines, aromatic.)
  properties 11:434
  solubility 11:435
Ditungsten carbide. (See also Tungsten compounds.)
  manufacture 19:221
  properties 19:221
  solubility 19:221
  uses 19:221
Divinyl acetylene
  in acrylonitrile 4:376
Divinylbenzene. (See also *Hydrocarbons.*)
  gas chromatography 18:299, 308
  in styrene 18:299, 308
  uses 14:326
Divinylmethane
  properties 5:103
Dixylymercury adipate 4:429
DMDEL. See Dimethyldiethyllead.
DMF. See Dimethylformamide.
Dobner's violet. (See also *Triphenylmethane and
    related dyes.*)
  paper chromatography 19:172
"Doctors"
  in confectionery products 10:552
Doctor test
  for mercaptans 15:560
Dodecafluoropentane. (See also *Fluorocarbons.*)
  properties 13:61
Dodecamethylenediamine
  gas chromatography 17:290
n-Dodecane. (See also Alkanes.)
  commercial specifications 14:305
  gas chromatography 13:352
  in gasoline 13:352, 355
  occurrence 14:303
  properties 13:352
Dodecanedioic acid 4:416

1-Dodecanethiol. See Dodecyl mercaptan.
Dodecanoic acid
  gas chromatography 17:290
Dodecanol. (See also Alcohols, higher.)
  commercial grades 4:555
1-Dodecanol. See Lauryl alcohol.
n-Dodecanonitrile. (See also *Nitriles.*)
  properties 16:396
1-Dodecene. (See also Alpha olefins.)
  commercial specifications 14:314
Dodecyl aziridinylformate. (See also *Imines.*)
  manufacture 14:472
n-Dodecylguanidine acetate. See Dodine.
n-Dodecyl isocyanate. (See also *Isocyanic acid
    esters.*)
  near-infrared spectroscopy 15:105
Dodecyl mercaptan. (See also *Mercaptans.*)
  properties 15:552
  volumetric method for acrylonitrile 16:408
Dodecyl mercaptan addition. (See also Mercaptan
    addition.)
  for allyl esters 5:97
Dodecyl methacrylate
  thin-layer chromatography 16:91
p-Dodecylphenyl isocyanate. (See also *Isocyanic
    acid esters.*)
  near-infrared spectroscopy 15:105
Dodine 13:205
  analysis 13:205
  bromocresol purple colorimetric method 13:205
  in fruit 13:205
  manufacture 13:205
  properties 13:205
  solubility 13:205
  uses 13:205
Dolan. (See also *Acrylic and modacrylic fibers.*)
    4:220, 249
Dolantin. See Meperidine hydrochloride.
Dolerophane. See Basic copper sulfate.
Dolomite 8:75, 244; 15:357, 358, 400
  composition 9:203
  properties 9:204
Dolomitic limestone 15:203
Domiphen. See β-(Phenoxyethyl)dimethyl-
    dodecylammonium bromide.
DOPD. See *N,N'*-Di-2-octyl-*p*-phenylenediamine.
Doré beads
  gold content 14:32
Doré metal 18:160
Dorlone 13:181
Dosage
  in bioassay 7:156, 157
  relation to effect 7:159
Dose, effective
  in bioassay 7:158
Double polarization method
  for sucrose in chocolate products 9:578

Dough. (See also *Flour.*)
  rheological properties 12:612
Doughnuts 6:444, 447
Dough raising ability
  of yeast 19:536
Dowex resins. (See also *Ion exchange resins.*) 15:2
Dowfume 13:181
Doxylamine succinate. (See also *Antihistamines.*)
  6:27
  column chromatography 6:150
  properties 6:27
  ultraviolet spectroscopy 6:150
DPA. See Diphenylamine.
DPPD (*N,N'*-Diphenyl-*p*-phenylenediamine) 6:72
  colorimetric method based on formation of
    Wurster salts 6:101, 103
  color test 6:93, 94
  column chromatography 6:89
  gas chromatography 6:81, 108
  paper chromatography 6:85
DPTA. (See also *EDTA and related compounds.*)
  structure 12:25
Draft resistance
  of cigarette filters 10:39
  of cigarettes 10:4
Dragendorff's reagent
  colorimetric test for acetylcholine chloride 9:
    617
  colorimetric test for bethanechol chloride 9:617
  colorimetric test for carbamylcholine chloride 9:
    617
  colorimetric test for choline chloride 9:617
  colorimetric test for methacholine chloride 9:
    617
  colorimetric test for β-methylcholine chloride 9:
    617
  colorimetric test for succinyldicholine
    chloride 9:617
  colorimetric test for succinylmonocholine iodide
    9:617
Drag test
  of brake linings 7:469
Dralon. (See also *Acrylic and modacrylic fibers*;
    Acrylic fibers.) 4:220
  cross section 4:225, 263
  end groups by infrared spectroscopy 4:296
  microscopy 4:264
  precipitation fractionation methods 4:299
  properties 4:222
  solubility 4:249
  staining tests 4:255
Dresinate
  ultraviolet spectroscopy 6:97
Dried eggs. See *Eggs and egg products.*
Dried food
  β-carotene content 10:545
Drierite 8:97

Driers
  for alkyd resins 5:22
  in enamels 12:176
Dropping point
  of greases 15:349
Dross
  antimony in 6:58
  sample preparation 6:59, 236
Drugs
  colors in 10:527
  cyclohexylamine content 11:217
Dry ashing
  of beer 7:553
  of plant material 7:400, 416
Drycleaning agents. (See also Drycleaning deter-
    gents; Drycleaning solvents, used; Perchloro-
    ethylene; and Stoddard solvent.) 11:603–617
Drycleaning detergents 11:610–616
  acid number 11:612
  analysis 11:611
  classification 11:611
  composition 11:611
  distillation test 11:611
  flash point 11:611
  light transmittance 11:613
  methylene blue titration 11:611
  performance tests 11:613
  pH 11:611
  redeposition control 11:613
  soil removal tests 11:613, 615
  solubility 11:611
  specification tests 11:611
  specific gravity 11:611
  water-holding capacity 11:612
Drycleaning solvents, used 11:609
  acid number 11:609
  color 11:609
  impurities, determination of 11:609
  odor 11:609
Dry combustion method
  for sulfur compounds, organic 18:380
Dry density
  of ion exchange resins 15:11
Dry ice 8:244, 247
Drying agents
  for carbon dioxide 8:255
  efficiency 8:256
Drying alkyds 5:22
Drying apparatus 7:531
Drying, loss on. See Loss on drying.
Drying methods
  for corn products 11:2
Drying oils
  analysis 7:444
  in brake linings 7:441, 444
  gas chromatography 7:444
Drying plates

for preparation of pulp handsheets 7:247
Drying rings
  for preparation of pulp handsheets 7:247
Drying time
  of enamels 12:189
  of silicone resins 18:124
Dry milk 16:144
  analysis 16:145
  manufacture 16:144
  milkfat content 16:145
  moisture content 16:145
  sampling 16:145
  solids content 16:145
  solubility index 16:146
  whey protein content 16:146
Dry mold type brake linings 7:441
DTEQ. See 1,2-Dihydro-2,2,4-trimethyl-6-
  ethoxyquinoline.
DTMC. See Dicofol.
DTPA. (See also EDTA and related compounds.)
  in beer 7:585
  in calcium chelates 12:50
  europium oxide titration 12:402
  ferric chloride titration 12:53
  iodometric method 12:28
  in iron chelates 12:52
  lanthanum oxide titration 15:152
  lanthanum titration 15:159
  in magnesium chelates 12:50
  magnesium chloride titration 12:52
  in manganese chelates 12:50
  rare earths titration 17:478
  scandium nitrate colorimetric method 12:28
  strontium titration 18:275
  structure 12:24
  thorium nitrate titration 12:36
  thorium titration 15:159
  zinc chloride titration 12:33
  zinc nitrate titration 12:36
DTPA, free
  in iron chelates 12:53
Ducan multiple range test 17:642
Ductile iron. (See also Iron metal.)
  composition 15:27
  uses 15:27
Ductility
  of bituminous materials 6:302
  testing machine 6:302
Duff formylation
  of aniline derivatives 5:432
Dumas method 4:278
  for alkanolamines 5:8
  for nitrogen 8:281, 292; 16:468
  for nitrogen in indigoid dyes 14:514
  for nitrogen in polyethylenimine 14:483
Dumortierite 7:316
1-Duodecanol. See Eicosyl alcohol.

Duolite resins. (See also *Ion exchange resins.*)
  15:2
Duo-trio testing 17:627
Durability
  of bituminous materials 6:306
  of brake linings 7:452
  of yeast 19:540
Duradene
  gel permeation chromatography 12:120
  intrinsic viscosity 12:119
  thermal analysis, differential 12:133
Duragen 12:86
Duranickel alloys. (See also Nickel alloys.)
  composition 16:290
Duranit 12:86
Durene. (See also *Hydrocarbons.*)
  commercial specifications 14:323
  uses 14:324
Dusting tests
  for pesticides and insecticides 19:617
Dutch liquid. See Ethylene dichloride.
Dutch process
  for chocolate and cocoa products 9:545
Dutral 12:86
Dydrogesterone. (See also Steroids.)
  manufacture 14:201
  properties 14:201
  solubility 14:201
  structure 14:201
  uses 14:201
Dye capacity, basic
  of acrylonitrile copolymer fibers 4:328
Dyeing test
  for triphenylmethane and related dyes 19:164
Dye modifiers
  ion exchange chromatography 17:299
  in polyamides 17:299
Dyes. (See also *Azine and related dyes*; Azo dyes;
    and individual compounds.) 12:1–22
  analysis 12:11
  classification 12:2
  identification 12:13
  in lipsticks 11:68
  manufactured from benzaldehyde 6:606
  in photographic films 12:572
  polyfiber fabric test 12:13
  separation 12:11
  solubility 12:11
Dyes, antihalation
  paper chromatography 12:587
  in photographic films 12:587
Dye solubilization test
  for surfactants 12:573
Dyfonate. (See also *Pesticides and insecticides.*)
  commercial grades and specifications 19:594
  composition 19:590, 597
  manufacture 19:597

  properties 19:590
  solubility 19:587
  structure 19:583
  toxicity 19:592
Dynamite. (See also Nitroglycerin.) 12:444
Dynamometer testing
  advantages 7:452
  bench 7:453, 457
  of brake linings 7:452
  full-scale 7:453
Dynel. (See also *Acrylic and modacrylic fibers*;
    Acrylonitrile polymers; Acrylonitrile–vinyl
    chloride polymers; and Modacrylic fibers.)
    4:220
  birefringence 4:258
  composition 4:223
  cross section 4:226, 263
  infrared spectra 4:242, 243, 385
  melting point 4:262
  microscopy 4:262
  mixtures, determination in 4:353
  moisture regain 4:333
  properties 4:222
  solubility 4:247, 251, 353
  specific gravity 4:258
  staining tests 4:252
Dyrene 13:221
  analysis 13:222
  in food crops 13:222
  manufacture 13:222
  N-1-naphthylenediamine colorimetric method
    13:222
  properties 13:222
  solubility 13:222
  solvent extraction 13:222
  toxicity 13:222
  uses 13:221

Dysprosium. (See also Dyprosium metal; *Rare
    earths.*)
  absorption spectroscopy 17:483
  in aluminum metal 5:129
  atomic absorption spectroscopy 17:494
  bead test 6:587
  emission spectroscopy 15:153, 155; 17:501
  in erbium oxide 17:503
  in europium oxide 17:502
  flame photometry 17:492
  in gadolinium oxide 17:502
  in holmium oxide 17:503
  in lanthanum oxide 15:153, 155; 17:501
  in terbium oxide 17:502
  x-ray fluorescent spectroscopy 17:490
  in yttrium oxide 17:501
Dysprosium fluorides
  carrier gas distillation fusion methods 8:636
  oxygen in 8:635, 636
  vacuum distillation fusion methods 8:635
Dysprosium metal. (See also Rare earth metals.)
  carrier gas fusion methods 8:627, 682
  hot extraction methods 8:684
  hydrogen in 8:623, 680, 684
  nitrogen in 8:623, 627
  oxygen in 8:623, 627, 629, 680, 682
  properties 17:471
  vacuum fusion methods 8:623, 680
  vacuum hot extraction methods 8:638
Dysprosium oxide. (See also Rare earth oxides.)
  emission spectroscopy 17:502
  impurities, determination of 17:502
  properties 17:497
  uses 17:509
Dysprosium perchlorate. (See also Rare earth
    compounds.)
  absorption spectrum 17:484

E

Earth acids. (See also Niobic acid; Tantalic acid.) 16:354

Earthenware. (See also *Ceramics.*) 9:191
  composition 9:194

Eastozone 6:34, 74

Eastozone 30 (31 and 32) 6:74
  colorimetric method based on formation of Wurster salts 6:101, 103

Eau de Javelle 7:221

Eau de Labarraque 7:221

Ebulliometers 14:487; 19:428

Ecgonine. (See also *Narcotics.*)
  microchemical tests 16:249
  platinic iodide microchemical method 16:249
  potassium-bismuth iodide microchemical method 16:249
  structure 16:244
  ultraviolet spectroscopy 16:251

EDA. See Ethylenediamine.

EDDHA. See Chel 138.

Edge curl of fabrics 4:25

EDHPA. (See also Chel 138.)
  in iron chelates 12:52

EDNA. (See also High explosives.) 12:442
  infrared spectrum 12:453
  manufacture 12:442
  melting point 12:461
  properties 12:432
  solubility 12:443
  uses 12:442

EDTA. (See also *EDTA and related compounds.*)
  aluminum titration 5:164, 218, 221; 6:432; 12:154; 13:445, 446; 14:141, 257
  aminopyrine titration 6:216
  barium titration 6:577; 12:154; 13:445
  bismuth titration 7:213; 13:445
  bismuth oxychloride titration 11:72
  bismuth-pyrocatechol violet colorimetric method 12:40
  cadmium titration 8:50; 10:635; 13:445
  calcium carbonate volumetric method 12:32
  calcium chloride-ferric chloride volumetric method 12:51
  calcium chloride titration 12:51
  calcium in beer titration 7:656
  calcium lactate titration 15:134
  calcium oxide titration 9:128; 10:119, 121
  calcium titration 6:264; 8:82, 87, 90, 93, 95, 104; 9:343; 10:96; 11:44, 225, 490, 495; 12:153, 154; 13:445, 447; 19:363, 492
  chromium titration 9:641; 13:445

cobalt(II) nitrate colorimetric method 12:41
cobaltous nitrate-ferric chloride colorimetric method 12:28
cobalt titration 9:639; 10:365; 13:445
colorimetric method for chromium 9:648
copper-8-hydroxyquinoline spectroscopic method 12:41
copper in copper(II) acetate titration 10:653
copper sulfate spectroscopic method 12:41
copper sulfate titration 12:43
copper titration 10:611; 13:445
cupric acetate titration 12:38
cupric nitrate titration 12:35
gallium titration 13:258, 272, 273, 445
gravimetric method for aluminum 10:634
hafnium titration 13:445
indium titration 13:445; 14:530, 548, 561
iodometric method 12:28
iron(III)-ammonium thiocyanate colorimetric method 12:40
iron(III)-salicylic acid colorimetric method 12:40
iron titration 9:639; 12:53, 154; 13:445; 15:40, 45
iron(II)-TPTZ colorimetric method 12:39
lanthanum nitrate-sodium hydroxide colorimetric method 12:28
lanthanum titration 13:445
lead titration 13:42, 445; 15:181
magnesium chloride titration 12:49
magnesium in beer titration 7:656
magnesium oxide titration 9:129; 10:120, 121; 15:219
magnesium titration 5:135; 8:87; 9:343; 10:96; 11:490; 12:153, 154; 13:445, 447; 15:364; 19:363
manganese titration 13:445; 15:493
mercury acetate titration 12:38, 53
mercury titration 13:445; 16:31
metals in glass titration 13:444
nickel sulfate electrophoretic method 12:42
nickel sulfate-potassium dithiooxalate colorimetric method 12:46, 48
nickel titration 9:638; 12:533; 13:445; 16:313
phosphorus titration 17:65
rare earths titration 13:445; 17:478, 512
scandium titration 13:445; 17:571
strontium titration 13:445; 18:275
structure 12:24
sulfate titration 18:391
sulfur content of acrylic polymers titration 4:286
tellurium titration 13:445

thallium titration 18:528

thorium nitrate–hexamethylenetetraamine colorimetric method 12:28, 30

thorium nitrate titration 12:36

thorium titration 13:445; 19:13

titanium titration 9:639

uranium titration 19:247

vanadium titration 13:445

volumetric method for aluminum 13:269

volumetric method for borate 15:160

volumetric method for gallium 13:269, 274

yttrium titration 13:445

zinc chloride titration 12:33

zinc nitrate titration 12:36, 38

zinc oxide titration 15:530

of zinc phenolsulfonate titration 11:87

zinc titration 12:153, 532; 13:445, 448; 15:395; 19:562

zirconium titration 13:445; 14:115, 143; 15:396

zirconium–xylenol orange complex colorimetric method 12:38, 45

EDTA and related compounds. (See also EDTA; and individual compounds.) 12:23–56

analysis 12:27

assay 12:31

in beer 7:584; 12:47

in biological fluids 12:47

calcium carbonate titration 12:32

in calcium chelates 12:50

calcium chloride titration 12:31

in cobalt chelates 12:51

colorimetric determination 12:28

commercial grades and specifications 12:27

in copper chelates 12:49

in detergents 12:43

dithizone color test 7:586

in embalming fluids 12:167

in fish products 12:46

in food products 12:44

identification 12:28

in iron chelates 12:51

in magnesium chelates 12:50

in manganese chelates 12:50

manufacture 12:26

in metal chelates 12:49

nomenclature 12:26

paper electrophoresis 12:31

in plasma 12:48

separation 12:28

in soaps 12:43; 18:204

structure 12:24

thin-layer chromatography 12:30

in urine 12:48

uses 12:23

in water 12:47

in zinc chelates 12:49

zinc–dithiozonate colorimetric method 12:47

EDTA-8-hydroxyquinoline colorimetric method for titanium 19:97

Effectiveness tests of brake linings 7:454

EGDN. (See also High explosives.)

manufacture 12:445

melting point 12:461

properties 12:432

uses 12:445

Egg albumen. (See also Eggs; *Eggs and egg products;* and Egg white.)

composition 12:58

in confectionery products 10:551

fat content 12:69

foaming test 12:72

foam stability 12:72

insoluble material in 12:69

liquid chromatography 12:60

whipping test 12:72

Eggs. (See also *Eggs and egg products;* Egg white; and Egg yolk.)

analysis 12:59

appearance 12:77

in cakes 6:446

composition 12:57

composition, determination of 12:59

Haugh unit values 12:76

mercury content 16:44

shell quality 12:77

uses 12:57

yolk index 12:75

Eggs and egg products. (See also Egg albumen; Eggs; Egg white; and Egg yolk.) 12:57–81

analysis 12:59

ash content 12:64

color 12:67

composition 12:57

composition, determination of 12:59

emulsifying capacity 12:72

fat content 12:63

flavor deterioration 12:70

impurities, determination of 12:66

lactic acid content 15:141

metal content 12:65

palatability 12:80

pH 12:68

protein content 12:62

quality control tests 12:67

sodium chloride content 12:66

solid content 12:59

solubility 12:68

sugar content 12:63

ultraviolet spectroscopy 12:70

uses 12:57

viscosity 12:74
vitamin content 12:66
Egg shell. (See also Eggs.)
  porosity 12:78
  specific gravity 12:77
  strength 12:78, 79
  thickness 12:78
Egg white. (See also Egg albumen; Eggs; and *Eggs and egg products*.)
  composition 12:58
  egg yolk in 12:69
  protein content 12:58, 60
  sodium lauryl sulfate in 12:65
  surface tension 12:75
  viscosity 12:75
Egg yolk. (See also Eggs; *Eggs and egg products*.)
  cholesterol content 12:63
  composition 12:58
  in egg albumen 12:69
  liquid chromatography 12:61
  protein content 12:58, 61
  viscosity 12:75
EGTA. (See also *EDTA and related compounds*.)
  cadmium nitrate volumetric method 12:47
  calcium oxide titration 15:216
  calcium titration 8:105
  iodometric method 12:28
  magnesium chloride colorimetric method 12:28
  structure 12:25
  thorium nitrate–hexamethylenetetraamine colorimetric method 12:28
Eichwaldite 7:315
Eicosanol. (See also Alcohols, higher; Fatty alcohols.)
  commercial grades 4:555
  3,5-dinitrobenzoyl chloride method for mixed isomers 4:547
  properties 4:561
  solubility 4:562
1-Eicosene. (See also Alpha olefins.)
  commercial specifications 14:314
Eicosenoic acid. (See also Fatty acids.)
  properties 12:473
Eicosonyl alcohol. (See also Fatty alcohols.)
  properties 4:561
Eicosyl alcohol. See Eicosanol.
Einsteinium. (See also Transuranium elements.)
  manufacture 19:259
Elaprim 12:86
Elasticity
  of chromium 9:635
  of fibers 4:89
Elastomers
  analysis 7:444
  in brake linings 7:441, 444
  gas chromatography 7:444
Elastomers, dental. See Dental elastomers.

Elastomers, synthetic. (See also individual polymers.) 12:81–161
  additives, determination of 12:140
  analysis 12:88, 109
  bromine content 12:111
  carbon content 12:109
  chlorine content 12:111
  color tests 12:91
  combustion methods 12:109
  commercial 12:86
  composition, determination of 12:109
  crosslink density 12:124
  density 12:91
  differential thermal analysis 12:132
  gas chromatography 12:101
  gel content 12:131
  gel permeation chromatography 12:119
  glass transition temperature 12:132
  hydrogenation 12:124
  hydrogen content 12:109
  hydroperoxides in 12:114
  identification 12:90
  impurities, determination of 12:156
  intrinsic viscosity 12:118
  iodine monochloride volumetric method 12:122
  molecular weight distribution 12:118
  nitrogen content 12:112
  nuclear magnetic resonance spectroscopy 12:103
  oxygen content 12:113
  ozonization 12:124
  perbenzoic acid volumetric method 12:121
  phosphorus content 12:116
  polyisoprene content 12:138
  polymer content 12:135
  pyridinium bromide perbromide volumetric method 12:123
  pyrolytic methods 12:135
  Raman spectroscopy 12:109
  sample preparation 12:89
  solvent extraction 12:89
  strain modulus 12:126
  sulfur content 12:117, 128
  sulfur crosslinkages 12:128
  sulfur, elemental, in 12:146
  swelling volume 12:125
  torsional modulus 12:134
  types 12:82, 86
  unsaturation, determination of 12:120
  uses 12:82
  vanadium content 19:339
Elastothane 12:86
Electrical conductivity
  of copper metal 10:627
  of water 19:359
Electrical properties
  of polyamide fibers 17:309

Electrical resistance
  of carbon and graphite fabrics 4:25
Electrical resistance alloys. (See also Cobalt
     alloys.) 10:290
  composition 10:290
  uses 10:290
Electrical resistivity
  of graphite 14:82
Electrical tests
  for silicone resins 18:124
  for silicone rubbers 18:139
Electric conductivity
  of polyethylene 12:384
Electric motor oil. (See also *Lubricants.*)
  commercial grades and specifications 15:312
Electrobalance 8:282
Electrochemical methods
  for hydrogen sulfide 19:504
  for mercury 16:36
  for nitrate 16:506
  for oxygen 19:502
  for triphenylmethane and related dyes 19:169
  for zinc 19:566
Electrodeposition
  of anionic materials in finishing agents 11:131
  of bismuth 7:213
  of cadmium 8:39, 40, 50, 51
  of cationic materials in finishing agents 11:131
  of cobalt 10:353, 358
  of copper 10:603, 606, 654, 663, 670; 15:65;
     17:195
  of indium 14:530
  of lead 10:495; 15:179
  of mercury 16:17, 24
  of silver 18:163, 165, 171
  of strontium 18:270
  of thallium 18:526
Electrolytic iron. (See also Iron metal.)
  manufacture 15:27
  uses 15:28
Electrolytic methods
  for chloroplatinic acid 17:273
  for cobalt 16:303
  for copper 10:311; 12:534; 16:301, 328
  for gold 14:40
  for lithium 10:327
  for magnesium 15:361
  for nickel 10:317; 16:297, 301, 304, 305,
     307
  for potassium 10:327
  for sodium 10:327
  for titanium 10:321
  for uranium 19:255
  for vanadium 10:324; 19:331
  for water in fluorocarbons 13:68
  for water in liquefied petroleum gas 15:252
  for water in oxygen 16:532

  for zinc 19:555
  for zirconium 10:325
Electrolytic process
  for chromium 9:634
Electrolytic separation
  of aluminum 5:199
  of lead 5:134
Electrometric titration
  of alcohols 4:538
Electron diffraction
  crystallinity of polymers 4:316
  electrostatic adhesion 4:392
Electronics
  ceramics for 9:209, 212
Electronics, alloys for 14:546
Electron micrograph
  of amosite asbestos 6:258
  of chrysotile asbestos 6:255
  of crocidolite asbestos 6:258
Electron microprobe methods
  for glass and glass-ceramics 13:474
Electron microscopy
  of carbon black 8:180, 202
  of cements 9:146
  of ceramics 9:223
  of coal and coke 10:226
  of concrete 9:146
  of glass ceramics 9:228
  of glass and glass-ceramics 13:474
  of glaze ceramics 9:226
  of mortars 9:146
Electron paramagnetic resonance spectroscopy
  of coal and coke 10:214
Electrophoresis
  of certified colors 10:482
  of F D & C Blue No. 1 10:482
  of F D & C Blue No. 2 10:482, 483
  of F D & C Green No. 1 10:482
  of F D & C Green No. 2 10:482
  of F D & C Green No. 3 10:482
  of F D & C Orange No. 1 10:482
  of F D & C Red No. 1 10:483
  of F D & C Red No. 2 10:482, 483
  of F D & C Red No. 3 10:483
  of F D & C Red No. 4 10:483
  of F D & C Violet No. 1 10:482
  of F D & C Yellow No. 1 10:482
  of F D & C Yellow No. 5 10:482
  of F D & C Yellow No. 6 10:482
  of gold 14:18
  of sodium heparin 10:142
Electrophoresis, high-voltage
  of amino acids 8:427
Electrophoresis, paper
  of amino acids 8:427
Electroplating solutions
  zinc content 19:559, 562, 566

Electrostatic cling
  of polyamide fibers 17:326
α-Eleostearic acid. (See also Fatty acids.)
    properties 12:473
Elgiloy. (See also Cobalt alloys.) 10:290
    composition 10:290
Elmendorf tearing strength tester 7:283
Elongation
  of fibers 4:89
  of silicone rubbers 18:131, 138
  of vinylidene fluoride–hexafluoropropylene
      copolymer 13:99
Elongation test
  for dental cast gold alloys 11:293
  for dental chromium–cobalt alloys 11:292
  for dental wrought gold alloys 11:295
Elutography
  of aniline and its derivatives 5:435
Embafume. See Methyl bromide.
Embalming chemicals. (See also Embalming
    fluids.) 12:162–169
  anticoagulants 12:166
  coloring agents 12:167
  commercial specifications 12:162
  composition 12:163
  modifying agents 12:165
  preservatives 12:164
  reodorants 12:167
  solvents 12:168
  surfactants 12:167
  types 12:162
Embalming fluids. (See also Embalming chemicals.)
  aldehyde content 12:165, 168
  amides in 12:166
  assay 12:168
  borates in 12:166
  carbonates in 12:166
  coal tar dyes in 12:167
  deodorants in 12:167
  dialdehydes in 12:165
  EDTA content 12:167
  formaldehyde donors in 12:165
  formalin content 12:164, 168
  gum content 12:166
  humectants in 12:166
  isopropyl alcohol content 12:165
  paraformaldehyde content 12:164
  pH 12:166
  phenols in 12:165
  quaternary ammonium compounds in 12:165
  solvents in 12:168
  surfactants in 12:167
  trioxane content 12:164
Emerich method
  for sugar 18:354
Emery 4:31
Emetine hydrochloride 4:592

analysis 4:614
Emission spectroscopy
  acrylic and modacrylic polymers and fibers,
      metals in 4:345
  adipic acid, metals in 4:418
  of alkali barium silicate glass 9:218
  of aluminum 5:211, 242; 9:346; 11:152, 493;
      13:277; 14:34, 48, 74, 122, 540; 15:155,
      375; 16:160, 387
  of aluminum oxide 9:218
  aluminum oxide abrasives, impurities in 4:51
  of antimony 6:57; 14:34; 15:197; 16:387
  of antimony trioxide 9:218
  of arsenic 6:237; 14:34; 15:144, 197; 16:387
  of arsenic oxide 9:218
  of barium 6:579; 15:156, 187; 16:160
  of barium oxide 9:218
  of beryllium 7:138; 13:277; 15:375
  of beryllium oxide 9:218
  of bismuth 7:209, 216; 11:152; 13:277; 14:48,
      540; 15:187, 197; 16:387
  of boron 7:411; 14:34, 48, 122; 15:155, 187;
      16:387
  of boron trioxide 9:218
  of cadmium 8:54; 11:152; 13:277; 14:48, 122,
      538; 15:187, 197; 16:387; 19:570
  of calcium 8:112; 9:346; 11:493; 14:48, 74,
      122; 15:144, 155, 375; 16:160, 387
  of calcium oxide 9:218
  of ceramics 9:217
  of ceric oxide 9:258
  of cerium 15:153, 155, 375; 17:488, 501
  of cerium compounds 9:256
  of cerium metal 9:249, 275
  of cerium oxide 17:501
  of cesium 9:281, 301
  of chromium 9:346, 635, 653; 11:152; 13:277;
      14:48, 74, 122; 15:155, 187; 16:160, 387
  of cobalt 9:346; 10:348, 371; 11:152; 13:277;
      14:48, 122; 15:156, 187; 16:160, 387;
      18:451
  of copper 9:346; 10:602, 616; 11:153; 14:48,
      122, 536; 15:144, 155, 187, 197, 375;
      16:160, 387; 19:570
  of copper oxide 9:218
  of cotton 11:145, 147
  of diatomaceous earth 11:492
  of dysprosium 15:153, 155; 17:501
  of dysprosium oxide 17:502
  of erbium 15:153, 155; 17:501
  of erbium oxide 17:503
  of europium 15:153, 155; 17:501
  of europium oxide 17:502
  of explosive priming mixtures 12:417
  of ferric oxide 9:218
  of fluoride 9:218
  of gadolinium 14:123; 15:153, 155; 17:488, 501

of gadolinium oxide 17:502
of gallium 13:256, 260, 265, 271; 14:34, 538; 15:187
of gallium arsenide 13:275
of germanium 13:385; 14:34, 538
of glass and glass-ceramics 13:453
of gold 14:2, 5, 27, 29, 30, 32, 43, 45; 17:270
of hafnium 14:107, 119, 121; 16:387
of holmium 15:153, 155; 17:501
of holmium oxide 17:503
of impurities in aluminum silicate 5:190
of impurities in asbestos 6:267
of impurities in boron 7:326
of impurities in calcium metal 8:76
of impurities in chrysotile 6:267
of impurities in silica ablators 4:3
of impurities in silicon carbide abrasives 4:56
of indium 13:277; 14:34, 48, 526, 534, 536
of inorganic fluorides 13:14
of iridium 14:48; 17:246
of iron 9:346; 11:153, 493; 14:48, 74, 122, 540; 15:33, 144, 155, 197, 375; 16:160, 387; 19:570
of lanthanum 15:151, 155, 158, 187, 375; 17:488, 501
of lanthanum oxide 17:501
of lead 9:346; 11:153; 13:277; 14:48, 74, 122, 536; 15:144, 155, 173, 187, 375; 16:160, 387; 19:570
of lead oxide 9:218
of lithium 15:266, 271
of lithium oxide 9:218
of lutecium 15:155
of lutetium 17:501
of lutetium oxide 17:503
of magnesium 9:346; 11:493; 13:277; 14:48, 74, 122, 540; 15:144, 155, 367, 372, 374; 16:160, 387; 19:570
of magnesium oxide 9:218
of manganese 9:346; 11:153; 13:277; 14:48, 74, 122; 15:155, 187, 375, 456, 477; 16:160, 387
of manganese dioxide 9:218
of mercury 16:37
of metals in beer 7:650
of molybdenum 9:346; 13:277; 14:74, 122; 15:155, 187; 16:159, 161, 387
of molybdenum trioxide 16:160
of neodymium 15:153, 155, 375; 17:488, 501
of neodymium oxide 17:501
of nickel 9:346; 11:153; 13:277; 14:48, 74, 122, 538; 15:155, 187, 197, 375; 16:160, 387
of nickel oxide 9:218
of niobium 14:122; 16:362; 18:451
of nitrogen 16:454
of noble gases 13:292

of osmium 17:246
of oxygen 14:134
of palladium 14:48; 17:246, 270; 18:451
of phosphorus 17:67
of platinum 14:48; 17:246, 270
of potassium 15:155, 226; 16:160; 17:367, 371
of potassium oxide 9:218
of praseodymium 15:153, 155, 375; 17:488, 501
of praseodymium oxide 17:501
of radium 17:452
of rare earth compounds 17:500
of rare earths 14:123; 15:153, 155; 17:479, 487
of rhenium 17:530
of rhodium 14:48; 17:246, 270
of rubidium 17:550
of rubidium oxide 9:218
of ruthenium 14:48; 17:246
of samarium 14:123; 15:153, 155; 17:488, 501
of samarium oxide 17:502
of scandium 17:568, 577
of silica 9:218
of silicon 9:346; 13:277; 14:48, 74, 122; 15:155, 375; 16:160, 387; 18:16
of silver 14:48, 538; 15:155, 187, 197; 17:270; 18:163, 172
of sodium 11:493; 15:226; 16:160; 18:217, 218
of sodium oxide 9:218
of stannic oxide 9:218
of strontium 9:346; 11:493; 15:156, 187; 16:160; 18:270, 277, 281
of strontium oxide 9:218
of tantalum 14:122; 16:387; 18:445, 450
of tellurium 14:48; 16:387; 17:592
of terbium 15:153, 155; 17:501
of terbium oxide 17:502
of thallium 14:48, 538; 18:525
of thorium 9:257; 14:123; 15:375; 17:488, 502; 19:10
of thulium 15:153, 155; 17:501
of thulium oxide 17:503
of tin 14:34, 48, 122, 538; 15:144, 156, 187, 197, 375; 16:160, 387; 19:38, 64, 570
of titanium 9:346; 11:493; 13:277; 14:48, 74, 122; 15:155, 187; 16:387; 18:445; 19:99, 119
of titanium oxide 9:218
of tungsten 14:122; 15:156; 16:160, 387
of uranium 19:241
of vanadium 9:346; 14:74, 122; 15:155, 187; 16:160, 387; 19:334, 342
of ytterbium 15:153, 155; 17:503
of ytterbium oxide 17:503
of yttrium 15:155; 17:488, 501

of yttrium oxide 17:501
of zinc 11:153; 13:277; 14:48, 122, 538;
    15:144, 187, 197, 375; 16:387; 19:558, 568,
    569
of zirconium 13:277; 14:107, 119, 121; 15:155,
    375; 16:387
of zirconium oxide 9:218
Emmi. (See also Mercury-containing compounds.)
    13:199
    manufacture 13:199
    properties 13:199
    solubility 13:199
    toxicity 13:199
    uses 13:199
Emulsified asphalts. See Asphalts, emulsified.
Emulsifiers
    in automobile polishes 6:363
    bromphenol blue color test 16:98
    in confectionery products 10:552
    in creams and lotions 11:93
    in methacrylate polymers 16:97
    methylene blue color test 16:98
Emulsifying capacity
    of eggs and egg products 12:72
Emulsion characteristics
    of lubricants 15:337
Emulsion performance
    of gelatin 12:555
Emulsion stability
    of pesticides and insecticides 19:616
    of silicone fluids 18:116
Enamels. (See also Ceramics.) 9:193; 12:170–192
    acid dissolution 9:215
    adhesion 12:180
    analysis 12:174
    anti-skinning agents 12:177
    application tests 12:189
    color 12:179
    composition 9:195; 12:170
    density 12:178
    drier additives 12:176
    drying time 12:189
    filler content 12:174
    flash point 12:181
    flexibility 12:180
    fluxes for 9:216
    gloss 12:190
    hardness 12:179
    hiding power 12:178
    infrared spectroscopy 12:188
    metal content 12:183
    mildew resistance 12:181
    nonvolatile content 12:175, 188
    oil content 12:170
    pigment content 12:173, 181
    pigment dispersion 12:179
    plasticizers in 12:176

    properties 12:171
    resin content 12:170
    sample preparation 12:181
    solvent content 12:170, 175, 188
    stabilizers in 12:177
    thickeners in 12:176
    thixotropy 12:178
    toughness 12:180
    types 12:170
    viscosity 12:178
    volatile content 12:175
    weathering 12:181
Enanthic acid
    properties 8:558
Enanthic anhydride
    properties 8:558
n-Enanthonitrile. See n-Heptanonitrile.
Enargite
    copper content 10:600
    occurrence 13:376
End count of raw roving 4:25
End groups
    in acrylic and modacrylic fibers 4:227, 288
    Gibson method 4:295
    by infrared spectroscopy 4:296
    Kirby method 4:293
    Lunney method 4:291
    Philipp methods 4:290
Endosulfan. (See also Pesticides and insecticides.)
    commercial grades and specifications 19:593
    composition 19:588, 596
    gas chromatography 19:389
    manufacture 19:596
    properties 19:588
    solubility 19:587
    structure 19:581
    toxicity 19:592
    in water 19:389
Endrin (See also Pesticides and insecticides.)
    gas chromatography 19:389
    in water 19:389
Engine degreasers 6:355
    analysis 6:367
    performance tests 6:378
Engine starting fluids 6:356
    analysis 6:370
Engine tune-up products 6:356
    analysis 6:370
Engler specific viscosity
    of bituminous materials 6:296
Engler viscometer 6:296
Enjay butyl 12:86
Enjay elastomers 12:86
Enocianin
    ammonium hydroxide colorimetric method
        10:538
    in meat 10:538

Enocianina. See Grape skin extract.
Enovid. See Norethynodrel.
Enzymatic decarboxylation
    of arginine 8:455
    of aspartic acid 8:461
    of glutamic acid 8:474
    of lysine 8:498
    of tyrosine 8:539
Enzymatic methods
    for alanine 8:450
    for carbon dioxide 19:501
    for ethanol 19:436
    for fructose 19:410
    for glucose 19:409
    for glycerol 13:510; 19:445
    for malic acid 19:421
    for protein 8:411
    for sorbitol 19:451
    for sucrose 19:410
Enzyme activity
    in beer 7:586
Enzyme detergents. (See also *Detergents.*)
    11:325
    composition 11:325
Enzyme methods
    for β-carotene 10:545
    for dextrose 11:39
    for sodium heparin 10:149
Enzymes
    biochemical method 11:356
    in detergents 11:356
    in green coffee 10:414
    in green tea leaf 18:482
    in yeast 19:533
Eosin. See D & C Red No. 22.
Eosin A. (See also *Triphenylmethane and related dyes.*)
    titanous chloride titration 19:177
Eosine G. See D & C Red No. 22.
Eosine lakes
    in enamels 12:183
    identification 12:183
Ephedra alkaloids 4:590
    analysis 4:606
Ephedrine 4:590
    analysis 4:606
    codeine content 16:270
    gas chromatography 6:209
    ion exchange chromatography 6:500
Ephedrine hydrochloride
    colorimetric method for osmium 17:248
Ephedrone
    analysis 4:607
1-Ephenamine benzylpenicillin. (See also Benzyl-penicillin.)
    solubility 16:557
    structure 16:567

1-Ephenamine penicillin G. See 1-Ephenamine benzylpenicillin.
Epiandrosterone
    gas chromatography 14:216
Epibromohydrin. (See also Epichlorohydrin; *Epoxy compounds.*)
    near-infrared spectroscopy 12:217; 14:176
    Raman spectroscopy 14:176
Epicatechin
    in black tea 18:495
    gas chromatography 18:471
    in green tea leaf 18:457, 465
    liquid chromatography 18:465
    paper chromatography 18:468
    structure 18:461
    thin-layer chromatography 18:463, 470, 495, 496
    ultraviolet spectroscopy 18:465
Epicatechin gallate
    in black tea 18:495
    gas chromatography 18:472
    in green tea leaf 18:457, 465
    liquid chromatography 18:465
    paper chromatography 18:468
    structure 18:461
    thin-layer chromatography 18:463, 470, 495, 496
    ultraviolet spectroscopy 18:465
Epichlorohydrin. (See also *Epoxy compounds;* Halohydrins.) 12:226; 14:172
    in air 14:176
    ammonium acetate–sodium arsenite colorimetric method 14:176
    analysis 12:227; 14:175
    assay 12:227
    azeotropes 14:174
    chlorine content 12:227
    commercial grades and specifications 12:227
    commercial specifications 14:175
    distillation range 12:227
    gas chromatography 14:171, 175
    handling precautions 14:174
    hydrochlorination methods 12:208
    impurities, determination of 12:227; 14:175
    infrared spectroscopy 12:200
    manufacture 14:174
    near-infrared spectroscopy 12:217, 218; 14:176
    nuclear magnetic resonance spectrum 14:177
    properties 12:226; 14:173
    Raman spectroscopy 14:176
    solubility 12:226; 14:173
    specification tests 12:227
    toxicity 14:174
    uses 14:172
    Volhard titration 12:227
Epichlorohydrin elastomers. See Polyepichloro-hydrins.

Epidote 8:75
Epietiocholanolone
  gas chromatography 14:216
Epigallocatechin
  in black tea 18:495
  gas chromatography 18:471
  in green tea leaf 18:457, 465
  liquid chromatography 18:465
  paper chromatography 18:468
  structure 18:461
  thin-layer chromatography 18:463, 470, 495, 496
  ultraviolet spectroscopy 18:465
Epigallocatechin gallate
  in black tea 18:495
  gas chromatography 18:472
  in green tea leaf 18:457, 465
  liquid chromatography 18:465
  paper chromatography 18:468
  structure 18:461
  thin-layer chromatography 18:463, 470, 495, 496
  ultraviolet spectroscopy 18:465
Epihalohydrins. (See also Halohydrins.) 14:153
Epiiodohydrin. (See also Epichlorohydrin.)
  Raman spectroscopy 14:176
Epinephrine
  analysis 14:220
  fluorometric methods 14:220
  structure 14:185
  uses 14:185
Epitestosterone
  gas chromatography 14:224
Epoxide equivalent
  of epoxy resins 12:241
Epoxides. (See also *Ethylene oxide;* Propylene oxide; and individual compounds.) 13:191
  acid–base titration 11:568
  analysis 11:568
  assay 11:568
  in disinfectants and antiseptics 11:567
  modifier for alkyd resins 5:24
1,2-Epoxides. See *Epoxy compounds.*
α-Epoxides. See *Epoxy compounds.*
Epoxidized soybean oil. See Epoxy plasticizers.
Epoxy alkanes
  use in manufacture of alkanolamines 5:1
1,2-Epoxy-*n*-butane. (See also *Epoxy compounds.*)
  gas chromatography 12:204
2,3-Epoxybutanes. (See also *Epoxy compounds.*)
  gas chromatography 12:204
  infrared spectroscopy 12:200
1,2-Epoxy-3-butanone. (See also *Epoxy compounds.*)
  infrared spectroscopy 12:200
1,2-Epoxy-3-butene. (See also *Epoxy compounds.*)

  infrared spectroscopy 12:200
Epoxy compounds. (See also individual compounds.) 12:192–232
  analysis 12:196
  argentimetric method 12:212
  assay 12:205
  colorimetric determination 12:217
  di-*n*-butylamine–acetic anhydride volumetric method 12:216
  gas chromatography 12:201
  hydrochloric acid–acetic acid titration 12:211
  hydrochloric acid–dioxane volumetric method 12:209
  hydrochlorination methods 12:207
  hydrogen bromide–acetic acid titration 12:206, 211
  hydrohalogenation 12:205
  identification 12:196
  infrared spectroscopy 12:198
  manufacture 12:193
  near-infrared spectroscopy 12:217
  oxirane oxygen content 12:205
  periodic-silver nitrate gravimetric method 12:197
  picric acid colorimetric method 12:217
  properties 12:193, 194
  pyridinium chloride–chloroform volumetric method 12:210
  pyridinium chloride–pyridine titration 12:209
  separation 12:201
  sodium thiosulfate–acetic acid volumetric method 12:216
  thin-layer chromatography 12:204
  types 12:193
  ultraviolet spectroscopy 12:201, 202
  zinc bromide–2,4-dinitrophenylhydrazine gravimetric method 12:214
Epoxy content
  of epoxy resins 12:242
  of resins 4:24, 401
2,3-Epoxycyclohexanone. (See also *Epoxy compounds.*)
  infrared spectroscopy 12:200
β-(3,4-Epoxycyclohexyl)ethyltrimethoxysilane. (See also Organofunctional silanes.)
  commercial specifications 18:82
  gas chromatography 18:85
  properties 18:80
1,2-Epoxyethane. See *Ethylene oxide.*
1,2-Epoxyethylbenzene. See Styrene oxide.
1,2-Epoxy-3-hexyne. (See also *Epoxy compounds.*)
  infrared spectroscopy 12:200
1,2-Epoxy-3-isopropoxypropane. (See also *Epoxy compounds.*)
  near-infrared spectroscopy 12:217
2,3-Epoxy-1-methoxybutane. (See also

Epoxy compounds.)
  infrared spectroscopy 12:200
2,3-Epoxy-2-methylbutane. (See also *Epoxy compounds.*)
  gas chromatography 12:204
1,2-Epoxy-2-methyl-3-butanone. (See also *Epoxy compounds.*)
  infrared spectroscopy 12:200
2,3-Epoxy-2-methylheptane. (See also *Epoxy compounds.*)
  infrared spectroscopy 12:200
2,3-Epoxy-2-methyl-4-pentanone. (See also *Epoxy compounds.*)
  infrared spectroscopy 12:200
Epoxy-2-methyl propane. (See also *Epoxy compounds.*)
  gas chromatography 12:204
Epoxy novolac resins. (See also *Epoxy resins.*) 12:234
  infrared spectra 12:239
  manufacture 12:234
  properties 12:235
1,2-Epoxy-3-phenoxypropane. (See also *Epoxy compounds;* Phenyl glycidyl ether.)
  near-infrared spectroscopy 12:217
Epoxy plasticizers. (See also *Epoxy compounds.*) 12:230
  commercial specifications 12:230
  uses 12:230
1,2-Epoxypropane. See Propylene oxide.
Epoxy resins. 12:232-259
  ablators 4:16, 22
  adhesives 4:400
  analysis 12:236
  chlorine content 12:250
  color 12:253
  colorimetric determination 12:238
  conversion, degree of 12:254
  crosslinking, degree of 12:257
  cure evaluation tests 12:253
  cure temperatures 12:255
  curing agents 12:255
  diluents 12:236
  epoxide equivalent 12:241
  epoxy content of 4:24, 401; 12:242
  esterification value 12:249
  evaluation tests 12:253
  ferric sulfate gravimetric method 12:237
  formaldehyde colorimetric method 12:237
  free phenol in 4:24
  α-glycol content 12:249
  heat-distortion temperature 12:253
  hydroxyl content 12:244
  identification 12:236
  infrared spectroscopy 12:240, 254, 256
  iodine number 12:251
  lepidine–diethylene glycol monoethyl ether

  colorimetric method 12:237
  manufacture 12:232
  mercuric sulfate gravimetric method 12:239
  molecular weight distribution 12:253
  near-infrared spectroscopy 12:243
  p-phenylenediamine colorimetric method 12:236
  properties, determination of 12:241
  pyrolytic methods 12:237
  quality control tests 4:22
  separation 12:240
  softening point 12:252, 257
  solid content 12:251
  solubility 12:232
  specific gravity 12:253
  swelling value 12:257
  tetraethylammonium bromide–perchloric acid volumetric method 12:243
  thin-layer chromatography 12:240
  types 12:233
  unsaturation, degree of 12:251
  use of aniline in 5:427
  viscosity 12:251
9,10-Epoxystearic acid
  thin-layer chromatography 12:205
cis-9,10-Epoxystearyl alcohol
  thin-layer chromatography 12:205
Epoxy-2,4,4-trimethylpentanes. (See also *Epoxy compounds.*)
  di-*n*-butylamine–acetic anhydride volumetric method 12:215
  infrared spectroscopy 12:200
E.P. potential. See Extreme pressure potential.
Epsilon acid. See 1-Naphthol-3,8-disulfonic acid.
Epsomite 15:357
Epsom salts. See Magnesium sulfate.
Epsyn 12:86
Equilenin. (See also Steroids.)
  occurrence 14:186
  structure 14:186
Equilin. (See also Steroids.)
  occurrence 14:186
  structure 14:186
Erbium. (See also Erbium metal; *Rare earths.*)
  absorption spectroscopy 17:483
  atomic absorption spectroscopy 17:494
  in dysprosium oxide 17:502
  emission spectroscopy 15:153, 155; 17:501
  flame photometry 17:492
  in holmium oxide 17:503
  in lanthanum oxide 15:153, 155; 17:501
  in lutetium oxide 17:503
  in thulium oxide 17:503
  x-ray fluorescent spectroscopy 17:490
  in ytterbium oxide 17:503
  in ytterium oxide 17:501
Erbium fluorides
  carrier gas distillation fusion methods 8:636

oxygen in 8:635, 636
vacuum distillation fusion methods 8:635
Erbium metal. (See also Rare earth metals.)
  carrier gas fusion methods 8:626, 627, 682
  hot extraction methods 8:684
  hydrogen in 8:623, 626, 638, 680, 684
  nitrogen in 8:623, 627
  oxygen in 8:623, 627, 680, 682
  properties 17:471
  vacuum fusion methods 8:623, 680
  vacuum hot extraction methods 8:638
Erbium oxide. (See also Rare earth oxides.)
  emission spectroscopy 17:503
  impurities, determination of 17:503
  properties 17:497
  uses 17:509
Erbium perchlorate. (See also Rare earth com-
    pounds.)
  absorption spectrum 17:485
Ergocistine
  liquid chromatography 15:292
Ergometrine. See Ergonovine.
Ergometrinine
  liquid chromatography 15:292
Ergonovine 4:588
  analysis 4:600
  liquid chromatography 15:293
  thin-layer chromatography 15:295
Ergot alkaloids 4:588
  analysis 4:600
Ergotamine 4:588
  analysis 4:600
  liquid chromatography 15:293
  thin-layer chromatography 15:295
Ergotaminine
  liquid chromatography 15:292
Ergotoxine 4:588
  liquid chromatography 15:293
Eriochrome black 100
  colorimetric method for tantalum 16:368
Eriochrome Black T
  colorimetric method for calcium 15:369
  colorimetric method for magnesium 15:368
Eriochromecyanin
  colorimetric method for tantalum 16:368
Eriochrome Cyanine R. (See also *Triphenyl-
    methane and related dyes.*)
  reagent for identification of aluminum 5:213,
    226
  visible spectroscopy 19:167
Erioglaucine. See D & C Blue No. 4.
Erioglaucine A. (See also *Triphenylmethane and
    related dyes.*)
  visible spectroscopy 19:167
Erioglaucine O. (See also *Triphenylmethane and
    related dyes.*)
  ceric sulfate titration 19:175

Eriogreen B. (See also *Triphenylmethane and
    related dyes.*)
  ceric sulfate titration 19:175
Erlenmeyer senior's isocinnamic acid 10:56
Erucic acid. (See also Fatty acids.)
  properties 12:473
Erythrite 10:284
Erythritol. (See also Polyhydric alcohols.)
  catechol colorimetric method 13:572
  gas chromatography 13:578; 19:300
  liquid chromatography 13:503, 580
  paper chromatography 13:501, 581
Erythrodextrin
  iodine color test 7:556
Erythromycin 5:464, 517
  acid–base titration 5:531
  analysis 5:517, 526, 531
  bioautography 5:522
  colorimetric methods 5:531
  identification 5:521
  infrared spectrum 5:522
  microbiological assay 5:526
  paper chromatography 5:521
  properties 5:518
  separation 5:521
  solubility 5:519
  structure 5:464
  ultraviolet spectroscopy 5:523, 532
  uses 5:517
Erythrosine. (See also F D & C Red No. 3; *Tri-
    phenylmethane and related dyes.*)
  paper chromatography 19:172
Erythrosine yellowish. See D & C Orange No. 11.
Escalol 506. See Amyl-*p-N,N*-dimethylamino-
    benzoate.
Eschka method
  for chlorine in coal and coke 10:254
  for sulfur in carbon 14:72
Eschka mixture 8:236; 18:380
Essential oils. (See also individual oils.) 12:259–
    290
  acid content 12:274
  acidity of 7:84
  acid number 12:274
  alcohol content 12:277
  aldehyde content 12:280
  analysis 12:263
  cineole content 12:285
  commercial specifications 12:260
  composition, determination of 12:274
  definition 12:259
  diluent content 12:284
  ester content 12:275
  ester number 12:275
  flavor 12:264
  gas chromatography 12:265
  identification 12:264

impurities, determination of 12:288
ketone content 12:280
mass spectroscopy 12:267
odor 12:264
optical rotation 12:270
organoleptic tests 12:264
phenol content 12:283
properties, determination of 12:269
refractive index 12:271
residue on evaporation 12:273
saffrole content 12:286
sampling 12:263
separation 12:264
solubility 12:272
specific gravity 12:269
sugar syrup test 12:264
water flavor test 12:264
Esso elastomer 12:86
Estatite
  in chromite 9:700
Esterification
  of carboxylic acids 8:394
  preparation of alkyd resins 5:23
Esterification methods
  for glycols 13:586
Esterification process
  manufacture of acrylates and methacrylates
    4:186
Esterification value
  of epoxy resins 12:249
  pyridinium chloride–acetic anhydride volumetric
    method 12:250
Ester plasticizers
  infrared spectroscopy 18:327
  in styrene polymers 18:327
Esters 8:576
  in alcoholic beverages 4:484, 490
  analysis 8:582
  commercial grades and specifications 8:581
  derivatives for identification 8:582
  derivatives for identification of carboxylic acids
    8:374, 376
  in essential oils 12:275
  in ethanol 4:544
  ferric hydroxamate colorimetric method 8:561,
    590
  formation from carboxylic acids 8:372
  gas chromatography 8:585, 586
  identification 8:582
  impurities, determination of 8:587
  infrared spectroscopy 12:576
  manufacture 8:580
  properties 8:558, 576
  saponification methods 8:585; 12:275
  separation 8:585
  in shampoos 11:76
  in styrene polymers 18:321

in sunscreens 11:91
  thin-layer chromatography 8:585
  use in preparation of carboxylic acids 8:372
Estinyl. See Ethinyl estradiol.
Estradiol. (See also Steroids.)
  analysis 14:220
  gas chromatography 14:219
  infrared spectrum 14:199
  liquid chromatography 14:221
  manufacture 14:199
  occurrence 14:199
  paper chromatography 14:220
  properties 14:199
  solubility 14:199
  structure 14:186
  trade name 14:192
  uses 14:199
Estradiol methyl ether. (See also Steroids.)
  liquid chromatography 14:221
1,3,5(10)-Estratriene-3,17β-diol. See Estradiol.
1,3,5(10)-Estratriene-3-ol-17-one. See
    Estrone.
1,3,5(10)-Estratriene-3,17β,16α-triol. See
    Estriol.
Estriol. (See also Steroids.)
  analysis 14:220
  liquid chromatography 14:221
  paper chromatography 14:220
  structure 14:186
  trade names 14:192
Estriol methyl ether. (See also Steroids.)
  liquid chromatography 14:221
Estrogenic hormones. See Estrogens.
Estrogens. (See also Ethinyl estradiol; Mestranol;
    Steroids; and individual compounds.) 14:186
  analysis 10:585; 14:220
  colorimetric determination 14:222
  in contraceptives 10:569, 585
  diazoamine colorimetric method 14:207
  fluorometric methods 14:222
  gas chromatography 14:215
  paper chromatography 14:220
  properties 14:199
  sources 14:199
  thin-layer chromatography 14:213, 222
  uses 14:186, 199
Estrone. (See also Steroids.)
  analysis 14:220
  infrared spectrum 14:200
  liquid chromatography 14:221
  manufacture 14:200
  occurrence 14:199
  paper chromatography 14:220
  properties 14:200
  solubility 14:200
  structure 14:186
  trade names 14:192

uses 14:200
Estrone methyl ether. (See also Steroids.)
  liquid chromatography 14:221
Etamycin 5:469, 549
  analysis 5:555, 559, 562
  bioautography 5:555
  countercurrent distribution 5:555
  identification 5:555
  microbiological assay 5:559
  paper chromatography 5:555
  properties 5:552
  separation 5:555
  solubility 5:551
  structure 5:469
  uses 5:549
Etchants
  for glass and glass-ceramics 13:472
Etchants, chemical
  for ceramics 9:224
Etching test
  for inorganic fluorides 13:14
Ethanamide. See Acetamide.
Ethane. (See also Alkanes.) 14:299
  in air 16:485
  in ammonia 5:313
  analysis 14:328
  calorific value 13:331
  in carbon monoxide 8:311
  commercial specifications 14:300
  in ethylene oxide 12:338
  gas chromatography 4:212; 5:313; 8:262, 311;
    12:338; 13:317, 318; 14:328, 329, 384;
    15:241; 16:456, 485
  in hydrogen 14:380, 384
  in natural gas 13:320
  in propane 15:241
  properties 13:311; 14:299
  solubility 14:299
  specific gravity 13:329; 15:245
  thermal conductivity 14:380
  uses 14:299
  vapor pressure 15:249
1,2-Ethanediol. See Ethylene glycol.
Ethanethiol. See Ethyl mercaptan.
Ethanoic acid. See *Acetic acid.*
Ethanoic anhydride. See *Acetic anhydride.*
Ethanol. (See also *Alcoholic beverages, distilled;*
    Alcohols, lower.) 4:496
  acetylation with acetic anhydride 4:530
  in acrolein 4:154
  in acrylic monomers 4:212
  antifreeze 6:3
  azeotropes 4:499, 502
  in beer 7:541
  in black tea 18:510
  boiling point 5:1
  boiling point determination 19:427

  by boiling point difference 7:549
  in brake fluids 7:423, 424
  in cocoa bean vapor 9:570
  commercial grades and specifications 4:513
  denatured 4:507
  density 19:432
  derivatives for identification 4:518
  in detergents 11:359
  dichromate oxidation method 4:550
  in diethylethoxymethylene malonate 15:437
  in diethyl malonate 15:433
  3,5-dinitrobenzoyl chloride method 4:547
  distillation method 11:362; 19:429
  in enamels 12:175, 189
  enzymatic method 19:436
  in essential oils 12:285
  in ethyl chloride 5:391
  in ethyl cyanoacetate 15:441
  in ethyl ether 12:310
  freezing point 6:7
  gas chromatography 4:154, 212; 5:5; 6:363;
    7:96; 8:351; 11:73, 74, 89, 362; 12:204, 300,
    302; 15:433, 437, 441; 16:107, 110, 425;
    18:510; 19:436, 458
  in hair sprays 11:89
  infrared spectroscopy 7:30
  iodoform test 6:4
  manufacture 4:504; 19:426
  mass spectroscopy 9:570
  in methanol 16:107, 110
  in nail enamels 11:73
  nitrochromic acid colorimetric method 12:310
  oxidation methods 19:432
  phase separation method 19:436
  phthalic anhydride method 4:535
  potassium dichromate–ferrous ammonium sulfate
    volumetric method 19:435
  properties 4:496; 6:2; 12:175
  pyridine content 17:425
  refractive index 19:432
  in shampoos 11:74
  in soaps 18:196
  solvent extraction 19:436
  by specific gravity 4:539; 7:544; 12:272
  in sunscreens 11:90
  uses 4:507
  in wines 19:425
Ethanolamine
  azeotropes 5:3
  infrared spectrum 5:7
  ion exchange chromatography 9:618
  manufacture 5:3
  properties 5:2
  specifications 5:4
  titration of 5:11
Ethanolamine–butyl Cellosolve
  colorimetric method for heptachlor 19:611

Ethanolamines. (See also Diethanolamine; Mono-
    ethanolamine; and Triethanolamine.) 12:291–
    295
  acid–base titration 12:295
  analysis 12:291
  assay 12:292
  equivalent weight, determination of 12:295
  gas chromatography 12:292
  identification 12:291
  impurities, determination of 12:295
  manufacture 12:291
  properties 12:291
  separation 12:291
  solubility 5:1
  titration of 5:11
Ethanol-insoluble matter
  in car washes 6:364
Ethanol solubles
  in benzaldehydes 6:625
Ethanol–water mixtures
  specific gravity 7:544; 11:360
Ethchlorvynol
  gas chromatography 6:209
Ethene. See Ethylene.
Ethers. (See also individual compounds.) 12:296–
    316
  acetic acid–hydroxylamine chloride colorimetric
    method 12:299
  analysis 12:299
  assay 12:302
  commercial grades and specifications 12:299
  ethoxyl content 12:304
  ferric hexathiocyanatoferriate colorimetric
    method 12:300
  formaldehyde–sulfuric acid colorimetric method
    12:300
  gas chromatography 12:300
  identification 12:299
  impurities, determination of 12:308
  infrared spectroscopy 12:300
  iodometric methods 12:304
  manufacture 12:298
  methoxyl content 12:304
  oxyalkalene content 12:305
  properties 12:296
  properties, determination of 12:308
  pyridine–hydriodic acid volumetric method
    12:302
  separation 12:299
  specification tests 12:309
  toxicity 12:296
  uses 12:296
Ethide. See 1,1-Dichloro-1-nitroethane.
17α-Ethinyl-4-androsten-17β-ol-3-one. See Ethyl-
    testosterone.
Ethinyl estradiol. (See also Steroids.) 10:569
  analysis 10:569, 585, 587, 596

  assay 10:571, 588, 590
  derivatives for identification 10:570
  fluorometry 10:573, 592
  gas chromatography 10:574, 588
  identification 10:569, 587, 590
  infrared spectroscopy 10:569, 570, 574
  manufacture 14:200
  paper chromatography 10:570
  properties 10:569; 14:200
  separation 10:569, 590
  sodium sulfate colorimetric method 10:588
  solubility 10:569; 14:200
  structure 14:200
  sulfuric acid colorimetric method 10:569, 571,
    585, 590
  thin-layer chromatography 10:590
  trade names 14:192
  ultraviolet spectroscopy 10:569, 571, 573
  Urbanyi and Rehm colorimetric method 10:573
  uses 14:200
17α-Ethinylestradiol-3-methyl ether. See Mes-
    tranol.
17-Ethinyl-1,3,5(10)-estratrene-3,17β-diol. See
    Ethinyl estradiol.
17α-Ethinyl-4-estren-17-ol-3-one. See 17-Ethinyl-
    19-nortestosterone.
17α-Ethinyl-5(10)-estren-17-ol-3-one. See Nor-
    ethynodrel.
17-Ethinyl-19-nortestosterone. (See also Steroids.)
  trade names 14:192
Ethisterone. (See also Steroids.)
  manufacture 14:201
  properties 14:201
  structure 14:201
  uses 14:201
3-Ethochloride of 9-*o*-carboxyphenyl-6-diethyl-
    amino-3-ethylimino-3-isoxanthene. See D &
    C Red No. 19.
Ethohexadiol. See 2-Ethyl-1,3-hexanediol.
3-Ethostearate of 9-*o*-carboxyphenyl-6-diethyl-
    amino-3-ethylimino-3-isoxanthene. See D &
    C Red No. 37.
Ethoxide content
  of triethylaluminum 5:171
Ethoxyacetylene
  infrared spectroscopy 4:142
  properties 4:137
2-Ethoxybenzaldehyde 6:596
  derivatives for identification 6:612
  properties 6:598
3-Ethoxybenzaldehyde 6:596
  properties 6:598
4-Ethoxybenzaldehyde 6:596
  derivatives for identification 6:612
  properties 6:598
Ethoxy content
  of phenacetin 6:190

Zeisel method 6:190
6-Ethoxy-1,2-dihydro-2,2,4-trimethylquinoline
  in elastomers, synthetic 12:141
  ultraviolet spectrum 12:141
β-Ethoxyethyl lactate. (See also Lactic acid
    esters.)
  properties 15:131
β-Ethoxyethyl lactate acetate. (See also Lactic
    acid esters.)
  properties 15:131
3-Ethoxy-4-hydroxybenzaldehyde. See Ethyl
    vanillin.
Ethoxylated alkyl phenols
  infrared spectroscopy 11:331
Ethoxylated fatty acids
  infrared spectroscopy 11:330
Ethoxylated fatty alcohols
  infrared spectroscopy 11:330
Ethoxyl content
  of ethers 12:304
  of ethylcellulose 9:77
α-(β-Ethoxyl)naphthylpenicillin. See Nafcillin.
4-Ethoxy-3-methoxybenzaldehyde
  fluorometry 6:626
  ultraviolet spectroscopy 6:633
2-Ethoxy-1-naphthaldehyde
  infrared spectroscopy 6:616
p-Ethoxyphenyl isocyanate. (See also Isocyanic
    acid esters.)
  infrared spectroscopy 15:106
  near-infrared spectroscopy 15:105
Ethoxy triglycol
  gas chromatography 7:427
Ethylac. (See also Dithiocarbamate esters.)
  properties 11:589
  structure 11:589
  uses 11:589
Ethyl acetate
  in black tea 18:510
  commercial specifications 8:578, 580
  in decaffeinated coffee 10:440
  in enamels 12:175, 189
  in essential oils 12:276
  gas chromatography 4:212; 8:351, 352; 11:73;
    12:300, 302; 16:108, 109; 18:510
  in methanol 16:108, 109, 114
  molecular weight 12:276
  in nail enamels 11:73
  properties 8:558, 577; 12:175
  saponification method 8:586
  solubility 13:540
  in wines 19:413
Ethyl acetate–orthophosphoric acid
  colorimetric method for pyrethrins 19:611
Ethyl acetoacetate
  commercial specifications 8:578
  2-hydroxy-1-naphthaldehyde colorimetric meth-
    od 11:204

  properties 8:577
Ethyl acetylene. See 1-Butyne.
Ethyl acid malate
  paper chromatography 19:418
Ethyl acid tartrate
  paper chromatography 19:418
Ethyl acrylate. (See also Acrylic and methacrylic
    acids and esters.)
  azeotropes 4:184
  bromate–bromide method 4:191
  commercial grades and specifications 4:187
  gas chromatography 4:212
  mercaptan-addition method 4:192
  mercuric acetate method 4:193
  permanganate reduction 4:216
  polarography 4:341
  polymer in 4:200
  properties 4:182
  residual monomer in polymers 4:214, 341
  saponification method 4:196, 341
  solubility 4:184
Ethyl acrylate polymers
  gas chromatography 4:286
Ethylal. (See also Acetals and ketals.) 4:62
  melting points of derivatives 4:68
  properties 4:63
Ethyl alcohol. See Ethanol.
Ethyl α-allylacetoacetate 5:106
Ethylamine
  commercial grades and specifications 5:264
  manufacture 5:262
  in methanol 16:114
  properties 5:260
  titration of 5:11
  uses 5:263
Ethyl-p-aminobenzoate 5:375
  acid–base titration 5:388
  analysis 5:388
  anesthetic 5:388
  Bratton-Marshall procedure 5:389
  diazotization method 5:389
  gas chromatography 6:187
  potentiometric titration 5:389
  properties 5:388
Ethyl n-amyl ether. (See also Ethers.)
  properties 12:297
2-Ethylaniline
  gas chromatography 5:268
4-Ethylaniline
  gas chromatography 5:268
N-Ethylaniline
  azeotropes 5:424
  chronopotentiometry 5:454
  colorimetric methods 5:452
  derivatives for identification 5:420
  gas chromatography 5:438
  infrared spectroscopy 5:436, 453

manufacture 5:425
near-infrared spectroscopy 5:443, 453
polarography 5:454
properties 5:422
ultraviolet spectroscopy 5:432
uses 5:427
x-ray diffraction 5:435
2-Ethylaziridine. (See also *Imines*.)
  properties 14:468
  structure 14:468
N-Ethylaziridine. (See also *Imines*.)
  properties 14:468
  structure 14:468
  toxicity 14:467
Ethylbenzene. (See also *Hydrocarbons*.)
  azeotropes 14:321
  commercial specifications 14:321
  gas chromatography 7:7, 8, 9, 10; 13:351;
    14:335, 336, 337; 18:298
  in gasoline 13:351, 354
  infrared spectroscopy 7:30
  liquid chromatography 7:13, 15
  manufacture 14:319
  in paraffins 14:336
  properties 13:351; 14:319
  in styrene 18:298
  toxicity 14:320
  ultraviolet spectroscopy 7:40
  uses 14:321
Ethyl benzoate. (See also Benzoic acid esters.)
  commercial grades and specifications 7:78
  3,5-dinitrobenzoate derivative 7:83
  in essential oils 12:276
  molecular weight 12:276
  properties 7:77; 8:577
  saponification method 7:84
Ethylbenzylanilinesulfonic acid
  in F D & C Violet No. 1 10:498, 507
  liquid chromatography 10:507
  ultraviolet spectroscopy 10:507
3-(α-Ethylbenzyl)-4-hydroxycoumarin. See Phen-
    procoumon.
Ethyl biscoumacetate. (See also *Coumarins*.)
    10:161; 11:160
  acid–base titration 10:163
  analysis 10:162
  assay 10:163
  colorimetric determination 10:162
  diazotized *p*-nitroaniline colorimetric method
    10:163
  ferric chloride colorimetric method 10:162
  identification 10:151, 162
  infrared spectroscopy 10:163
  iodoform test 10:162
  melting point 10:162
  paper chromatography 10:152
  properties 10:161

solubility 10:161
thin-layer chromatography 10:151
2,2,4-trimethylpentane extraction method
    10:163
ultraviolet spectroscopy 10:151, 162
uses 11:159
Ethyl bromide
  commercial specifications 8:12
  properties 8:12
2-Ethylbutanal
  infrared spectroscopy 6:616
2-Ethylbutanol. (See also Alcohols,
    higher.)
  azeotropes 4:554
  phthalic anhydride method 4:536, 559
  properties 4:552
  specifications 4:556
  uses 4:554
2-Ethyl-1-butene
  gas chromatography 13:350
  in gasoline 13:350, 353
  properties 13:350
Ethyl *n*-butyl ether. (See also *Ethers*.)
  properties 12:297
Ethyl butyl ketone
  azeotropes 8:336
  commercial specifications 8:341
  Karl Fischer titration 8:360
  water in 8:360
2-Ethyl-2-butyl-5-methyl-3,4-hexadienal
  ultraviolet spectroscopy 6:632
2-Ethylbutyraldehyde
  commercial specifications 8:340
Ethyl butyrate
  gas chromatography 18:510
  properties 8:558, 577
Ethyl caprate
  in essential oils 12:276
  molecular weight 12:276
Ethyl caproate
  in essential oils 12:276
  molecular weight 12:276
Ethyl caprylate
  in essential oils 12:276
  molecular weight 12:276
Ethyl carbamate. See Urethane.
Ethyl Cellosolve
  gas chromatography 11:73
  in nail enamels 11:73
Ethyl cellulose. (See also *Cellulose derivatives*.)
    9:76
  analysis 9:77
  colorimetric determination 11:126, 130
  ethoxyl content 9:77
  in finishing agents 11:126, 130
  impurities, determination of 9:77
  manufacture 9:77
  properties 9:76

solubility 11:126, 130
uses 9:76
viscosity 9:62
Ethyl chloride. (See also C$_2$ Chlorinated hydro-
    carbons.) 5:375; 9:435
    analysis 5:390
    azeotropes 9:436
    Beilstein detection 9:382
    composition 9:411
    gas chromatography 9:377, 380, 381
    manufacture 9:436
    properties 5:390; 9:435
    solubility 9:436
    toxicity 9:436
    uses 9:435
2-Ethylchlorobenzene
    mass spectroscopy 9:404
3-Ethylchlorobenzene
    mass spectroscopy 9:404
4-Ethylchlorobenzene
    mass spectroscopy 9:404
Ethyl chloroformate (Ethyl chlorocarbonate)
    properties 8:570
Ethylchlorosilanes. (See also individual com-
    pounds.)
    manufacture 18:55
Ethyl cinnamate. (See also *Cinnamic acid and
    its derivatives.*)
    commercial specifications 10:66
    in essential oils 12:276
    molecular weight 12:276
    properties 10:65
    saponification methods 10:67
    ultraviolet spectroscopy 10:69
Ethyl cyanoacetate
    analysis 15:440
    assay 15:441
    colorimetric determination 15:440
    colorimetric method for menadione 10:197
    colorimetric method for menadione sodium bi-
        sulfite 10:201, 202
    commercial specifications 15:425
    in diethyl malonate 15:433
    gas chromatography 15:433, 441
    identification 15:440
    impurities, determination of 15:441, 442
    infrared spectroscopy 15:440
    manufacture 15:424
    mass spectrum 15:441
    properties 15:421
    solubility 15:422
    trinitrobenzene–dimethylformamide colorimetric
        method 15:440
Ethylcyclohexane
    gas chromatography 13:351
    in gasoline 13:351, 354
    properties 13:351

4-Ethyl-1-cyclohexene
    liquid chromatography 7:13,
        15
N-Ethylcyclohexylamine 11:245
    gas chromatography 11:245
    infrared spectroscopy 11:245
    manufacture 11:245
    properties 11:242
    solubility 11:246
    toxicity 11:246
    uses 11:245
Ethylcyclopentane
    gas chromatography 13:350
    in gasoline 13:350, 353
    properties 13:350
Ethyldecaborane
    colorimetric method using β-naphthoquinoline
        7:347
Ethyl decanoate
    gas chromatography 18:511
1-Ethyl-4-dichloromethylbenzene
    mass spectroscopy 9:405
N-Ethyldicyclohexylamine
    gas chromatography 11:245
N-Ethyldiethanolamine 5:2
Ethyl 3,5-dinitrobenzoate
    thin-layer chromatography 16:94
Ethylene. (See also *Hydrocarbons.*) 14:309
    in air 16:485
    in ammonia 5:313
    analysis 5:362; 14:327
    anesthetic 5:362
    assay 5:362
    in carbon monoxide 8:310
    commercial grades and specifications 14:310
    in ethylene oxide 12:338
    flammability 14:310
    gas chromatography 5:313; 8:262, 310; 12:101,
        338; 14:327, 328, 329, 384; 15:241; 16:456,
        485
    in hydrogen 14:384
    impurities, determination of 5:362
    manufacture 14:310
    in propane 15:241
    properties 5:357, 362; 14:309
    specific rotation 12:382
    uses 14:309
Ethylene–acetylene copolymers. (See also *Ethyl-
    ene and propylene polymers.*) 12:344
    infrared spectrum 12:366
    properties 12:344
Ethylene acrylate ionomers. (See also *Ethylene
    and propylene polymers.*) 12:344
Ethylene–acrylic acid ionomers. (See also *Ethyl-
    ene and propylene polymers.*) 12:344
    infrared spectrum 12:367
    properties 12:344

Ethylene–acrylonitrile copolymers. (See also *Ethylene and propylene polymers.*) 12:344
manufacture 12:344
Ethylene–amide copolymers. (See also *Ethylene and propylene polymers.*) 12:344
properties 12:344
Ethyleneamines. (See also *Diamines and higher amines, aliphatic;* and individual compounds.)
acid–base titration 11:422, 425
acid–base titration, nonaqueous 11:425
AHEW 11:430
amine value 11:425
analysis 11:420
application tests 11:430
assay 11:421
color 11:430
gas chromatography 11:421
identification 11:420
impurities, determination of 11:428
manufacture 11:417
nitrogen content 11:426
properties 11:409
solubility 11:407
specific gravity 11:429
viscosity 11:429
Ethylene and propylene polymers. (See also Polyethylene; Polypropylene; and individual copolymers.) 12:341–395
analysis 12:346
chemical methods 12:381
color 12:347
crosslink density 12:380
crosslinking reactions 12:346
crystallinity, degree of 12:381
density 12:381
elastic performance 12:385
flame tests 12:347
fractionation methods 12:375
gel permeation chromatography 12:379
intrinsic viscosity 12:370
liquid chromatography 12:376
mechanical testing 12:385
melt index 12:374
melting point 12:347
melt viscosity 12:373
microscopy 12:348
molecular weight, determination of 12:370
nuclear magnetic resonance spectroscopy 12:369
pH 12:347
pyrolysis–gas chromatography 12:365
pyrolysis–mass spectroscopy 12:369
refractive index 12:382
separation 12:346
solubility 12:346, 347
solubility temperatures 12:348
specific refraction 12:382

specific viscosity 12:370
surface reactions 12:345
thermal methods 12:382
turbidity temperatures 12:348
viscosity-average molecular weight 12:372
N,N′-Ethylenebis[2-(*o*-hydroxyphenyl)] glycine. See Chel 138.
Ethylene-1,2-bis(thiocarbamoyldimethylthiocarbamoyl)disulfide. (See also Thiuram disulfides.)
structure 11:592
uses 11:592
Ethylene bromide 8:6
analysis 8:13
commercial specifications 8:12
gas chromatography 13:183, 184
iodometric method 8:13
properties 8:12
in soil 13:184
uses 8:1, 4
Ethylene bromohydrin. (See also Halohydrins.)
derivatives for identification 14:160
infrared spectrum 14:158
Ethylene–butadiene copolymers. (See also *Ethylene and propylene polymers.*) 12:344
manufacture 12:344
Ethylene-1-butene copolymers. (See also *Ethylene and propylene polymers.*) 12:344
infrared spectrum 12:366
properties 12:344
Ethylene-2-butene copolymers. (See also *Ethylene and propylene polymers.*) 12:344
glass transition temperature 12:384
infrared spectra 12:366
manufacture 12:344
Ethylene–carbon monoxide copolymers. (See also *Ethylene and propylene polymers.*) 12:344
infrared spectrum 12:366
manufacture 12:344
Ethylene–carboxylate copolymers. (See also *Ethylene and propylene polymers.*) 12:344
Ethylene chloride. See 1,2-Dichloroethane; Ethylene dichloride.
Ethylene chlorobromide 13:180
properties 13:180
solubility 13:180
toxicity 13:180
uses 13:180
Ethylene chlorohydrin. (See also Halohydrins.) 14:162
acetic anhydride volumetric method 14:161
in air 14:155
analysis 14:164
azeotropes 14:163
commercial grades and specifications 14:164
in dates 12:332

derivatives for identification 14:160
gas chromatography 14:165
infrared spectrum 14:158
manufacture 14:164
nuclear magnetic resonance spectrum 14:162
paper chromatography 14:157
properties 14:162
silver nitrate volumetric method 14:164
sodium hydroxide colorimetric method 14:155
sodium hydroxide volumetric method 14:165
solubility 14:162
structure 14:153
toxicity 14:162
uses 14:162
volumetric determination 14:164
Ethylene copolymers. (See also *Ethylene and propylene polymers.*) 12:344
manufacture 12:344
properties 12:344
types 12:344
Ethylene cyanohydrin
acrylic acid and acrylates, manufacture 4:185
acrylonitrile, manufacture 4:370
Ethylene–cyclooctene copolymers. (See also *Ethylene and propylene polymers.*) 12:344
manufacture 12:344
Ethylene–cyclopentene copolymer. (See also *Ethylene and propylene polymers.*) 12:344
properties 12:344
Ethylenediamine. (See also Ethyleneamines.)
acid–base titration 11:423, 425
amine value 11:425
azeotropes 11:412
boiling point 11:414
commercial grades and specifications 11:418
density 11:407
in ethylenimine 14:478
freezing point 11:413
gas chromatography 11:421; 14:478; 16:198
heat capacity 11:415
infrared spectroscopy 11:419
ion exchange chromatography 6:499
latent heat 11:416
in methanol 16:114
nitrogen content 11:427
properties 11:408
specific gravity 11:414
specific heat 11:415
thermal conductivity 11:416
vapor pressure 11:410, 411
viscosity 11:411, 413
Ethylenediamine-di[(*o*-hydroxyphenyl acetic acid)]. See Chel 138.
Ethylenediamine dinitrate. (See also High explosives.)
melting point 12:461
properties 12:434

Ethylenediaminetetraacetate
copper sulfate titration 11:363
in detergents 11:363
N,N,N′,N′-Ethylenediaminetetraacetic acid. See EDTA.
Ethylene dibromide. (See also Ethylene bromide.)
in flour 13:184
gas chromatography 13:184, 185
Ethylene dichloride. (See also 1,2-Dichloroethane.) 13:180
in flour 13:184
gas chromatography 13:183, 184, 185; 15:444
manufacture 13:180
properties 13:180
solubility 13:180
toxicity 13:180
uses 13:180
Ethylene–diethyl fumarate copolymers. (See also *Ethylene and propylene polymers.*)
infrared spectrum 12:367
Ethylene–dimethyl fumarate copolymers. (See also *Ethylene and propylene polymers.*)
infrared spectrum 12:367
4,4′-Ethylenedimorpholine
gas chromatography 16:200
in morpholine 16:200
Ethylenedinitramine. See EDNA.
(Ethylenedinitrilo)tetraacetic acid. See EDTA.
2,2′-Ethylene{dioxybis[ethyliminodi(acetic acid)]}. See EGTA.
Ethylene–ethyl acrylate copolymers. (See also *Ethylene and propylene polymers.*) 12:344
infrared spectrum 12:367
properties 12:344
Ethylene–ethylenephosphoric diethyl ester copolymers. (See also *Ethylene and propylene polymers.*) 12:344
manufacture 12:344
Ethylene fluorohydrin. (See also Halohydrins.)
derivatives for identification 14:160
infrared spectrum 14:158
Ethylene–formaldehyde copolymers. (See also *Ethylene and propylene polymers.*) 12:344
manufacture 12:344
Ethylene glycol. (See also Glycols.) 5:38
in alkyd resins 5:38
antifreeze 6:1
azeotropes 13:541
azeotropic distillation 13:507
in brake fluids 7:425
in choline and its derivatives 9:627
codeine color test 13:570
colorimetric determination 13:570
commercial grades and specifications 13:546
in dates 12:332
derivatives for identification 13:568
in diethylene glycol 13:600

distillation range 13:602
in enamels 12:175
freezing point 6:7; 13:535
gas chromatography 5:32; 6:5; 13:504, 507, 576, 589; 19:298, 300, 301
in glycerol 13:507
gravimetric determination 13:570
humectant values 13:542
identification 6:4
iodometric titration 9:627
liquid chromatography 13:503, 580
manufacture 13:538
β-naphthol color test 13:570
oxalic acid colorimetric method 13:572
paper chromatography 13:501, 581
periodic acid volumetric method 13:597
phenylhydrazine acetate gravimetric method 13:570
in polyurethanes 19:297
properties 6:2; 12:175; 13:536
refractive index 6:8
in resins 13:596
resorcinol color test 13:570
Rosen color test 13:573
sodium periodate colorimetric method 13:571
solubility 13:535, 539
structure 13:534
thin-layer chromatography 13:502, 582
thymol color test 13:570
uses 13:541
Ethylene glycol bis(β-aminoethyl ether)tetraacetic acid. See EGTA.
Ethylene glycol dimethacrylate
nuclear magnetic resonance spectroscopy 16:86
thin-layer chromatography 16:90
Ethylene glycol dinitrate. See EGDN.
Ethylene glycol ethers
in brake fluids 7:425
Ethylene glycol monobutyl ether. (See also Ethers.)
commercial specifications 12:299
in enamels 12:175
gas chromatography 8:351
properties 12:175
Ethylene glycol monoethyl ether. (See also Cellosolve; Ethers.)
commercial specifications 12:299
in enamels 12:175
gas chromatography 8:351
properties 12:175
Ethylene glycol monomethyl ether. (See also Methyl Cellosolve; Ethers.)
commercial specifications 12:299
in enamels 12:175
properties 12:175
Ethylene glycol nitrates. See also DEGN; EGDN;

and TEGN.) 12:445
Ethylene glycol phenyl ether. (See also Phenyl ethers.)
gas chromatography 12:310
phenol in 12:310
Ethylene glycol stearate
solubility 13:539, 552
Ethylene iodide. (See also Iodine compounds, organic.)
properties 14:612
Ethylene–maleic anhydride copolymers. (See also Ethylene and propylene polymers.) 12:344
infrared spectrum 12:367
manufacture 12:344
Ethylene–methacrylate ionomers. (See also Ethylene and propylene polymers.) 12:344
Ethylene–methacrylic acid ionomers. (See also Ethylene and propylene polymers.) 12:344
infrared spectrum 12:367
Ethylene–methyl acrylate copolymers. (See also Ethylene and propylene polymers.)
infrared spectrum 12:367
Ethylene–methyl methacrylate copolymers. (See also Ethylene and propylene polymers.) 12:344
manufacture 12:344
Ethylene oxide. (See also Epoxides; Epoxy compounds.) 12:219, 317–340; 13:191
acid–base titration 11:568
in air 12:223; 13:191
analysis 12:221, 319; 13:191
assay 11:568; 12:221, 321
azeotropes 12:317
barium chloride–hydrochloric gravimetric method 12:574
in coconut oil 12:330
colorimetric determination 12:319; 13:191
commercial grades and specifications 12:220, 319
in copra 12:330
Critchfield and Johnson colorimetric methods 12:328
in dates 12:332; 13:192
in disinfectants and antiseptics 11:567
in flour 13:184
in foodstuffs 12:327
gas chromatography 11:568; 12:324, 332, 338; 13:184
gasometric method 12:325
in gum rubber 12:331
hydrochlorination methods 12:208
hydrogen bromide titration 12:321, 331; 13:191
identification 12:319
impurities, determination of 12:221, 333
infrared spectroscopy 12:199, 200, 320, 577
lepidine colorimetric method 12:198
Lubatti method 12:322

magnesium bromide–sulfuric acid volumetric method 12:330
manufacture 12:220, 317; 13:191
mass spectrum 12:320
in medical supplies 12:327
morpholine volumetric method 12:322
periodic acid volumetric method 12:221
in photographic films 12:574
in polyethylene 12:331
potassium thiocyanate–phenolphthalein color test 12:223, 321
properties 12:194, 220, 317; 13:191
pyridine–hydrogen chloride volumetric method 12:322
sampling 12:319
sodium chloride colorimetric method 12:320
sodium sulfite volumetric method 12:323
solubility 12:220, 317; 13:191
solvent extraction 12:332
specification tests 12:221
toxicity 12:317; 13:191
in Tygon 12:331
ultraviolet spectroscopy 12:335, 575
use in manufacture of acrylic acid and acrylates 4:185
use in manufacture of alkanolamines 5:1
uses 12:219, 317; 13:191
Ethylene oxide derivatives
in brake fluids 7:425
Ethylene oxide polymers
gravimetric method for silica 18:28
Ethylene–polystyrene copolymers. (See also *Ethylene and propylene polymers.*) 12:345
manufacture 12:345
properties 12:345
Ethylene–propylene copolymers. (See also *Ethylene and propylene polymers.*) 12:343
glass transition temperature 12:384
infrared spectroscopy 12:362, 363, 368
properties 12:344
Young's modulus, dynamic 12:386, 388
Ethylene–propylene–cyclooctadiene terpolymers. (See also *Ethylene and propylene polymers.*)
infrared spectrum 12:368
Ethylene–propylene–dicyclopentadiene terpolymers. (See also *Ethylene and propylene polymers.*)
infrared spectrum 12:368
Ethylene–propylene–diene rubber
properties 18:314
Ethylene–propylene–1,4-hexadiene terpolymers. (See also *Ethylene and propylene polymers.*) 12:344
infrared spectrum 12:368
uses 12:344
Ethylene–propylene–natural rubber blends. (See also *Ethylene and propylene polymers.*) 12:345

properties 12:345
Ethylene–propylene–norbornene terpolymers. (See also *Ethylene and propylene polymers.*) 12:344
uses 12:344
Ethylene–propylene rubber. (See also *Elastomers, synthetic;* Ethylene–propylene copolymers.)
commercial elastomers 12:86
density 12:91
glass transition temperature 12:133
infrared spectroscopy 12:93, 100
pyrolysis–gas chromatography 12:107
Ethylene–propylene–4-vinylcyclohexene terpolymers. (See also *Ethylene and propylene polymers.*) 12:344
uses 12:344
Ethylene–styrene copolymers. (See also *Ethylene and propylene polymers.*) 12:344
infrared spectrum 12:366
manufacture 12:344
Ethylene–tetrafluoroethylene copolymers. (See also *Ethylene and propylene polymers.*) 12:344
manufacture 12:344
Ethylene–trichlorofluoroethylene copolymer. (See also *Ethylene and propylene polymers.*)
infrared spectrum 12:368
Ethylene–triethyl tin methacrylate copolymer. (See also *Ethylene and propylene polymers.*) 12:344
properties 12:344
Ethylene–trifluoropropylene copolymers. (See also *Ethylene and propylene polymers.*)
infrared spectrum 12:368
Ethylene–vinyl acetate copolymers. (See also *Elastomers, synthetic; Ethylene and propylene polymers.*) 12:344
commercial elastomers 12:87
glass transition temperatures 12:384
infrared spectroscopy 12:93, 366
properties 12:344
specific rotation 12:382
uses 12:344
Ethylene–vinyl alkyl ether copolymers. (See also *Ethylene and propylene polymers.*) 12:344
manufacture 12:344
Ethylene–N-vinyl carbazole copolymer. See Carbathene.
Ethylene–vinyl carbonate copolymers
properties 12:344
Ethylene–vinyl chloride copolymers. (See also *Ethylene and propylene polymers.*) 12:344
infrared spectrum 12:368
manufacture 12:344
Ethylene–vinylene carbonate copolymers. (See also *Ethylene and propylene polymers.*) 12:344

Ethylene–vinylidene chloride copolymers. (See also *Ethylene and propylene polymers.*) 12:344
   manufacture 12:344
Ethylene–vinylpyrrolidone copolymers. (See also *Ethylene and propylene polymers.*) 12:344
   manufacture 12:344
Ethylene–vinyl sulfonic acid copolymers. (See also *Ethylene and propylene polymers.*)
   infrared spectrum 12:368
Ethylene–vinyl triethoxysilane copolymers. (See also *Ethylene and propylene polymers.*) 12:344
   manufacture 12:344
Ethylenic alcohols 4:575
   analysis 4:577
   manufacture 4:575
   properties 4:576
Ethylenimine. (See also *Imines.*)
   acid–base titration 14:476
   in air 14:478
   analysis 14:473
   assay 14:474
   colorimetric determination 14:477
   gas chromatography 14:478, 480
   handling 14:465, 473
   hydrogen bromide titration 14:475
   identification 14:473
   impurities, determination of 14:480
   infrared spectroscopy 14:473
   manufacture 14:472
   melting point 14:474
   4-(*p*-nitrobenzyl)pyridine colorimetric method 14:477
   properties 14:461
   separation 14:474
   specific heat 14:461
   structure 14:460
   toxicity 14:465
   uses 14:460
   viscosity 14:463
   volumetric determination 14:475
   in water 14:478
Ethylenimine derivatives. (See also *Imines.*)
   acid–base titration 14:476
   analysis 14:492
   gas chromatography 14:493
   infrared spectroscopy 14:493
   manufacture 14:472
Ethylenimine dimer. (See also *Imines.*)
   gas chromatography 14:493
Ethylenimine trimer. (See also *Imines.*)
   gas chromatography 14:493
Ethyl esters of carboxylic acids
   properties 8:558
17-α-Ethyl-4-estren-17-ol-3-one. See 17-Ethyl-19-nortestosterone.

Ethyl ether. (See also *Ethers.*)
   analysis 5:361; 6:370
   commercial grades and specifications 12:298
   engine starting fluid 6:356
   in ethanol 4:544
   gas chromatography 12:204, 300; 16:108
   impurities, determination of 12:308
   iodometric method 12:306
   in methanol 16:108, 114
   properties 5:357, 361; 12:297
   solubility 13:539, 552
   specification tests 12:309
Ethylethoxymethylene malonate. See Diethylethoxymethylene malonate.
Ethyl α-ethoxypropionate. (See also Lactic acid esters.)
   properties 15:131
Ethyl formate. (See also Formic acid esters.)
   in black tea 18:510
   commercial specifications 8:578; 13:132
   in essential oils 12:276
   gas chromatography 16:108, 109, 114; 18:510
   in methanol 16:108, 109
   molecular weight 12:276
   properties 8:558, 577; 13:131
   saponification method 13:132
   solubility 13:131, 134, 540
   uses 13:120
2-Ethylfuran
   in black tea 18:510
   gas chromatography 18:510
γ-*N*-Ethyl glutamine. See Theanine.
Ethyl groups
   infrared spectroscopy 12:356
   in polyethylene 12:356
Ethylheptanes
   gas chromatography 13:351
   in gasoline 13:351, 354
   properties 13:351
Ethyl heptanoate
   gas chromatography 18:510
5-Ethyl-2-heptanol
   acetylation with acetic anhydride 4:530
   3,5-dinitrobenzoyl chloride method 4:547
Ethyl hexaldehyde
   in acetylenic chemicals 4:145
2-Ethylhexaldehyde
   commercial specifications 8:340
2-Ethylhexanal
   infrared spectroscopy 6:616
Ethylhexane
   gas chromatography 13:351
   in gasoline 13:351, 354
   properties 13:351
2-Ethyl-1,3-hexanediol. (See also Glycols.)
   azeotropes 13:541
   commercial grades and specifications 13:547

distillation range 13:602
manufacture 13:548
phthalic anhydride volumetric method 13:586
properties 13:537
solubility 13:538
structure 13:534
uses 13:548
Ethyl hexanoate
gas chromatography 18:510
2-Ethylhexanoic acid 5:22
2-Ethylhexanol. (See also Alcohols, higher.)
acetylation with acetic anhydride 4:530
azeotropes 4:554
3,5-dinitrobenzoyl chloride method 4:547
gas chromatography 16:98
phthalic anhydride method 4:535
properties 4:552
specifications 4:557
uses 4:554
2-Ethylhexene
gas chromatography 14:334
2-Ethyl-1-hexene
liquid chromatography 7:13, 14, 15, 16
2-Ethylhexyl acetate
commercial specifications 8:578
2-Ethylhexyl acrylate. (See also *Acrylic and methacrylic acids and esters.*)
gas chromatography 4:213, 286, 342
commercial grades and specifications 4:187
polymer in 4:200
polymers 4:286
properties 4:182
residual monomer 4:342
saponification method 4:196
solubility 4:184
2-Ethylhexylamine
azeotropes 5:261
properties 5:260
Ethyl-*n*-hexylbarbituric acid
paper chromatography 6:479
thin-layer chromatography 6:480
Ethylhydroxyethylcellulose. See Hydroxyethylcellulose and its derivatives.
Ethylidene chloride. (See also 1,1-Dichloroethane.)
gas chromatography 16:91
Ethyl iodide. (See also Iodine compounds, organic.)
gas chromatography 16:91, 92
properties 14:612
Ethyl isobutyrate
properties 8:577
Ethyl isocyanate. (See also *Isocyanic acid esters.*)
commercial sources 15:102
infrared spectroscopy 15:109
near-infrared spectroscopy 15:105
properties 15:95
Ethyl isonicotinate. (See also *Pyridine and*

pyridine derivatives.)
trisodium pentacyanoammino ferrate colorimetric method 17:433
Ethyl isopropyl ketone
properties 8:335
Ethyl lactate. (See also Lactic acid esters.)
gas chromatography 19:446, 449
properties 15:131
solubility 15:130
Ethyl lactate acetate. (See also Lactic acid esters.)
properties 15:131
Ethyl laureate
gas chromatography 18:511
N-Ethylmaleimide
colorimetric method for glutathione 15:561
colorimetric method for mercaptans 15:561
colorimetric method for mercaptoethanol 15:561
colorimetric method for thioglycolic acid 15:561
Ethyl mercaptan. (See also *Mercaptans.*)
in carbon disulfide 8:274
derivatives for identification 15:559
gas chromatography 8:274; 15:557
properties 15:552
uses 15:555
N-Ethylmercuri-1,2,3,6-tetrahydro-3,6-endo-methano-3,4,5,6,7,7-hexachlorophthalimide. See Emmi.
Ethyl methacrylate. (See also *Acrylic and methacrylic acids and esters.*)
bromate–bromide method 4:191
commercial grades and specifications 4:187
gas chromatography 4:212
mercuric acetate method 4:193
nuclear magnetic resonance spectroscopy 16:86
polymer in 4:200
properties 4:182
saponification method 4:196
solubility 4:184
Ethyl α-methoxypropionate. (See also Lactic acid esters.)
properties 15:131
5-Ethyl-2-methylpyridine. (See also *Pyridine and pyridine derivatives.*)
derivatives for identification 17:415
gas chromatography 17:437
properties 17:415
Ethylmorphine. (See also *Narcotics.*)
ammonium molybdate colorimetric method 16:246
color tests 16:246
formaldehyde–sulfuric acid colorimetric method 16:246
gas chromatography 6:210; 16:254
iodine–potassium iodide microchemical method 16:249

mercuric chloride microchemical method 16:249
microchemical tests 16:249
polarography 16:268
structure 16:243
thin-layer chromatography 16:256
ultraviolet spectroscopy 16:251
Ethyl myristate
  gas chromatography 18:511
Ethylnaphthalenes. (See also Alkylnaphthalenes.)
  gas chromatography 16:217
  properties 16:217
Ethyl nicotinate. (See also *Pyridine and pyridine derivatives.*)
  trisodium pentacyanoammino ferrate colorimetric method 17:433
Ethyl nonanoate
  gas chromatography 18:511
17-Ethyl-19-nortestosterone. (See also Steroids.)
  trade names 14:192
Ethyl octanoate
  gas chromatography 18:511
3-Ethylpentane
  in benzene 7:11
  gas chromatography 7:11; 14:330, 332
  in *n*-heptane 14:332
Ethylpentanes
  gas chromatography 13:350
  in gasoline 13:350, 353
  properties 13:350
Ethylpentenes
  gas chromatography 13:350
  in gasoline 13:350, 353
  properties 13:350
3-Ethyl-1-pentyn-3-ol. (See also Acetylenic alcohols.)
  silver acetylide method 4:578
m-Ethylphenol. (See also *Phenols.*)
  derivatives for identification 17:21
  ferric chloride colorimetric method 17:16
  gas chromatography 17:33
  infrared spectroscopy 17:29
  properties 17:4
o-Ethylphenol. (See also *Phenols.*)
  derivatives for identification 17:21
  gas chromatography 17:33
  infrared spectroscopy 17:29
  paper chromatography 17:24
  properties 17:4
  in 2,6-xylenol 17:33
p-Ethylphenol. (See also *Phenols.*)
  derivatives for identification 17:21
  gas chromatography 17:33
  infrared spectroscopy 17:29
  paper chromatography 17:24
  properties 17:4
Ethyl phenylacrylate. See Ethyl cinnamate.

N-Ethyl-*N*-phenylbenzylamine
  manufacture 5:425
  properties 5:422
O-Ethyl *S*-phenylethyl phosphonothionate. See Dyfonate.
N-Ethylpiperidine
  gravimetric method for benzylpenicillin 16:570
  titration of phthalic acid 17:156
N-Ethylpiperidine benzylpenicillin. (See also Benzylpenicillin.)
  structure 16:567
N-Ethylpiperidine penicillin G. See *N*-Ethylpiperidine benzylpenicillin.
Ethyl propionate
  gas chromatography 18:510
  properties 8:558, 577
Ethyl propyl ketone
  properties 8:335
2-Ethylpyridine. (See also *Pyridine and pyridine derivatives.*)
  derivatives for identification 17:415, 426
  gas chromatography 17:437
  ionization constants 17:416
  properties 17:415
3-Ethylpyridine. (See also *Pyridine and pyridine derivatives.*)
  derivatives for identification 17:415, 426
  ionization constants 17:416
  properties 17:415
4-Ethylpyridine. (See also *Pyridine and pyridine derivatives.*)
  derivatives for identification 17:415, 426
  gas chromatography 17:437
  ionization constants 17:416
  properties 17:415
  trisodium pentacyanoammino ferrate colorimetric method 17:433
Ethyl silicate. (See also Silicon compounds.)
  toxicity 18:8
Ethyl stearate
  uses 15:302
N-Ethyl-*N*-(3-sulfobenzyl)sulfanilic acid.
  in F D & C Blue No. 1 10:498, 506
  liquid chromatography 10:506
  ultraviolet spectroscopy 10:507
17-Ethyltestosterone. (See also Steroids.)
  trade names 14:192
Ethyl thiocyanate. (See also *Isocyanic acid esters.*)
  gas chromatography 15:110
Ethyltoluenes
  gas chromatography 7:7, 8; 14:337; 18:298
  in styrene 18:298
N-Ethyl *p*-toluene sulfonamide
  in polyamides 17:300
  solvent extraction 17:300
Ethyltrichlorosilane. (See also Organo-chlorosilanes.)

commercial specifications 18:56
gas chromatography 18:62, 66
manufacture 18:55
properties 18:54
in vinyltrichlorosilane 18:66
Ethyltriethoxysilane. (See also Organosilane esters.)
commercial specifications 18:71
properties 18:70
Ethyl undecanoate
gas chromatography 18:511
Ethyl valerate
in black tea 18:510
gas chromatography 18:510
Ethyl vanillin (3-Ethoxy-4-hydroxybenzaldehyde) 6:597
column chromatography 6:618
commercial grades and specifications 6:609
derivatives for identification 6:612
gas chromatography 6:620; 11:167
manufacture 6:603
properties 6:598
solubility 13:540
uses 6:607
p-Ethylvinylbenzene
gas chromatography 18:299
in styrene 18:299
Ethyl violet. (See also *Triphenylmethane and related dyes.*)
ceric sulfate titration 19:175
manufacture 19:163
thin-layer chromatography 19:173
visible spectroscopy 19:167
Ethyne. See *Acetylene.*
Ethynodiol diacetate 10:581
acetic anhydride colorimetric method 10:587
acid–base titration 10:581
analysis 10:581, 587
identification 10:581
properties 10:581
solubility 10:581
ultraviolet spectroscopy 10:581
Ethynyl butyl ether. (See also *Ethers.*)
properties 12:297
1-Ethynylcyclohexanol. (See also Acetylenic alcohols.)
properties 4:576
silver acetylide method 4:578
Ethynylcyclohexyl carbamate
gas chromatography 6:209
1-Ethynylcyclopentanol. (See also Acetylenic alcohols.)
silver acetylide method 4:578
Ethynyl ethyl ether. (See also *Ethers.*)
properties 12:297
Eticylol. See Ethinyl estradiol.
Etiocholanolone

gas chromatography 14:216
Etryptamine
thin-layer chromatography 15:295
Eucalyptol (See also Cineole.)
gas chromatography 8:353; 12:266
in peppermint oil 12:266
solubility 13:540
Eucalyptus oil. (See also *Essential oils.*)
commercial specifications 12:262
refractive index 12:271
Eugenol
Fourier transform spectra 13:155, 156, 158
gas chromatography 8:355; 11:167
uses in manufacture of vanillin 6:604
Euroacril. (See also *Acrylic and modacrylic fibers.*) 4:220
Europium. (See also Europium metal; *Rare earths.*) 12:395–404
absorption spectroscopy 17:483
atomic absorption spectroscopy 12:399; 17:494
bead test 6:587
emission spectroscopy 12:403; 15:153, 155; 17:501
in europium-activated phosphors 12:399
in europium oxide–gadolinium oxide phosphors 12:399
in europium oxide–yttrium oxide phosphors 12:399
in europium oxysulfide–yttrium oxysulfide phosphors 12:399
in europium yttrium vanadate 12:399
ferric chloride volumetric method 12:397
flame photometry 17:492
in gadolinium oxide 17:502
iodometric titration 12:397
in lanthanum oxide 15:153, 155; 17:501
in neodymium oxide 17:501
occurrence 12:395
polarography 12:397; 17:493
in rare earths 12:398
in samarium oxide 17:502
in terbium oxide 17:502
volumetric determination 12:397
x-ray fluorescent spectroscopy 12:399; 17:490
Europium-activated phosphors
europium content 12:399
Europium metal. (See also *Europium;* Rare earth metals.)
analysis 12:396
hydrogen in 8:617, 681
oxygen in 8:681
production 12:395
properties 12:396; 17:471
sample preparation 8:641
uses 12:396
vacuum fusion methods 8:617, 681
Europium oxide. (See also Rare earth

oxides.) 12:400
analysis 12:402
assay 12:402
atomic absorption spectrum 12:401
commercial specifications 12:400
DTPA titration 12:402
emission spectroscopy 17:502
identification 12:402
impurities, determination of 12:402; 17:502
properties 12:400; 17:497
solubility 12:400
uses 12:400; 17:509
Europium oxide–gadolinium oxide phosphors
europium content 12:399
Europium oxide–yttrium oxide phosphors
europium content 12:399
x-ray fluorescence spectroscopy 12:399
Europium oxysulfide–yttrium oxysulfide
phosphors
europium content 12:399
Europium perchlorate
atomic absorption spectrum 12:401
Europium yttrium vanadate
europium content 12:399
Europrene 12:86
Euxenite 17:470, 560
composition 16:349
decomposition 19:235
Evaporated milk. (See also Condensed milk.)
commercial specifications 16:144
sampling 16:145
Evaporation loss
of lubricants 15:345
Evaporation method
for platinum group metals 17:244
for residue of emulsified asphalts 6:326
Evaporation residue. See Residue on evaporation.
Evaporation tube for ammonia 5:302
Exhaust gases
hydrogen chloride content 14:395
Exlan. (See also *Acrylic and modacrylic fibers;*
Acrylic fibers.) 4:220, 249
properties 4:222
Expansive cements. (See also Hydraulic cements.)
9:95
composition 9:96
identification 9:108
Explosion point
of tetracene 12:416
Explosive D. See Ammonium picrate.
Explosive priming mixtures 12:417–431
analysis 12:417, 421
composition, determination of 12:417, 422,
424
emission spectroscopy 12:417
identification 12:417
microscopy 12:417

sampling 12:421
solvent extraction 12:422
typical mixtures 12:424
x-ray methods 12:417
Explosives. (See also Explosive priming mixtures;
Explosives, primary; High-explosive mixtures;
High explosives; and individual compounds.)
12:405–471
definition 12:405
physical testing 12:408
safety precautions 12:405
Explosives, primary. (See also Diazodinitrophenol;
Lead azide; Lead picrate; Lead styphnate;
Mercury fulminate; and Tetracene.) 12:408–
431
properties 12:406
Ext D & C colors. See Certified colors; and
individual compounds.
Ext D & C Green No. 1. (See also Certified
colors.)
classification 10:460
commercial specifications 10:470
impurities, determination of 10:502
liquid chromatography 10:502
manufacture 10:460
structure 10:460
titanous chloride titration 10:485
visible spectroscopy 10:486
Ext D & C Violet No. 2. (See also Certified
colors.)
paper chromatography 10:475
Ext D & C Yellow No. 1. (See also Certified
colors.)
classification 10:460
commercial specifications 10:470
impurities, determination of 10:502, 510
liquid chromatography 10:502
manufacture 10:460
structure 10:460
titanous chloride titration 10:485
visible spectroscopy 10:486
Ext D & C Yellow No. 7. (See also Certified
colors.)
classification 10:460
commercial specification 10:471
liquid chromatography 10:502
manufacture 10:460
structure 10:460
titanous chloride titration 10:485
visible spectroscopy 10:486
Extenders
in chrome greens 17:200
Extinction coefficient
of coal and coke 10:224
Extractable matter
in certified colors 10:489
Extractable resin content

of brake linings 7:448
Extract content
  of wines 19:405
Extraction
  of acetylsalicylic acid 6:133, 158,
    166
  of aminopyrine 6:213
  of antidegradants from rubber 6:86, 108
  of arsenic 6:240
  of asphalt from bituminous paving mixtures
    6:334
  of barbiturates 6:475, 491, 497
  of bituminous paving mixtures 6:330
  of butylated hydroxytoluene 6:116
  of caffeine 6:128, 133

of calcium and strontium from barium 6:573
of certified colors 10:490
of crude fat in grains 4:473
of phenacetin 6:128, 133
of phenobarbital 6:491, 497, 521
of Polygard 6:119
of stearic acid 6:169
of textile finishes and coatings 4:329
Extracts of beer 7:495
Extreme pressure potential
  of lubricants 15:328
Extruded-type brake linings 7:442
Extrusion rheometer test
  for caulking compounds 9:19
Ezcurrite 7:314

F

Fabric finishes. See Finishing agents.

Fabrics
  quality control tests 4:25

Fabrics, bleached
  blue reflectance 7:233
  whiteness 7:233

Face powder
  silicon content 18:30
  titanium content 19:106

Fade 7:471
  of brake linings 7:455, 468

Fajans titration
  of chloride 9:342

Falling ball method
  for cuen viscosity of pulp 7:270
  of cellulose derivatives 9:66

Falling number test
  for viscosity of flour paste 12:611

False set
  of cements 9:133

Fanchon maroon. See D & C Red No. 34.

Fansteel alloys
  composition 16:351

Farinogram 12:614

Farinograph 12:612

Fast acid fuchsine B. See D & C Red No. 33.

Fast blue B
  colorimetric method for nitroparaffins 16:422
  colorimetric method for *m*-phenylenediamine
    11:449

F. A. S. T. dynamometer. See Friction assessment
  screening test dynamometer.

Fast Green. See F D & C Green No. 3.

Fast green FCF. (See also *Triphenylmethane and
    related dyes.*)
  liquid chromatography 19:173

Fastness
  of pigments 17:208

Fast red E. (See also *Triphenylmethane and
    related dyes.*)
  paper chromatography 19:172

Fast Scarlet R
  colorimetric method for 2,4-toluenediamine
    11:455

Fat content, crude
  of grains 4:492

Fat extractors 19:532

Fats. (See also *Fats and fatty oils.*)
  in aluminum powder 17:203
  in bakery products 6:449
  in bread 6:440, 449

in cakes 6:446
in casein 9:9
in cereals 9:237
in chocolate 9:540, 550
in chocolate and cocoa products 9:538, 575,
  590
in cocoa beans 9:548, 549, 555
in confectionery products 10:550
in corn products 11:7
in corn starch 11:17
in egg albumen 12:69
in eggs and egg products 12:63
in glues of animal origin 13:493
glycerol content 13:507, 512
Goldfisch method 19:531
n-hexane extraction 9:562
hydrolysis 11:17
Kleinert refractometric method 9:576
Knorr extraction tube method 9:555, 575
in margarine 15:502
monomolecular film test 12:69
Nadj and Weeden method 9:576
in roasted coffee 10:435
in soaps 18:201
solvent extraction 11:7; 12:63; 19:531
in yeast 19:531

Fats and fatty oils. (See also individual oils.)
    12:472–493
  acid–base titration 12:477
  acid–base titration, nonaqueous 12:478
  acid value 12:477
  analysis 12:475, 484
  color 12:477
  commercial grades and specifications 12:474
  composition 12:472
  composition, determination of 12:484
  cyclopropene group content 12:484
  definition 12:472
  esterification 12:484
  extraction 12:474
  fatty acid content 12:489
  gas chromatography 12:484
  hydroxyl value 12:478
  infrared spectroscopy 12:490
  iodine value 12:481
  iodometric methods 12:481
  oxidized fatty acid content 12:482
  oxirane oxygen content 12:482
  peroxide value 12:482
  polyunsaturated acid content 12:489
  processing 12:474

properties 12:427, 473
properties, determination of 12:475
refining 12:474
refining loss 12:492
refractive index 12:475
sampling 12:475
saponification value 12:479
setting point 12:476
solidification temperature 12:476
specific gravity 12:475
titer test 12:476
ultraviolet spectroscopy 12:489
unsaponifiable matter in 12:479
uses 12:472
Fatty acid esters
 in glycerol 13:527
Fatty acid methyl esters
 analyses 12:487
 gas chromatography 12:487
 manufacture 12:484
Fatty acids 8:369
 acid–base titration 15:498; 18:189, 205
 acid–base titration, nonaqueous 11:365
 acid value 18:184
 in alkyd resins 5:37
 boron trifluoride–methanol esterification
  11:364
 classification 12:474
 in creams and lotions 11:95
 derivatives for identification 12:491
 in detergents 11:363
 esterification 12:484
 in fats and fatty oils 12:489
 gas chromatography 5:25, 28; 8:383; 11:364;
  12:488; 18:187
 infrared spectroscopy 5:46, 12:490
 iodine value 18:185
 Kappelmeier separation 5:33
 low-temperature crystallization 12:490
 in metal cleaners 6:368
 in metallic soaps 18:204
 paper chromatography 5:49
 properties 12:474
 rosin content 18:186
 saponification value 18:184
 separation 18:184
 in shampoos 11:81
 in soaps 18:184
 solvent extraction 11:81, 363; 13:596; 18:188
 titer test 18:186
Fatty acids, free
 acid–base titration 11:57
 in corn oil 11:57
 in margarine 15:498
Fatty alcohols. (See also individual compounds.)
  4:560
 analysis 4:563

azeotropes 4:560
cloud point 4:571
commercial grades 4:563
gas chromatography 4:566; 11:346
hydroxyl number 4:568
impurities, determination of 4:572
iodine number 4:569
liquid–solid transition 4:570
manufacture 4:560, 562
melting point (slip point) 4:571
occurrence 4:560, 562
properties 4:560
sampling 4:565
setting point 4:570
in soaps 18:201
specifications 4:564
thin-layer chromatography 12:204
uses 4:562
Fatty alcohols, free
 in detergents 11:368
Fatty alcohol sulfates. (See also Anionic detergents.) 11:323
 infrared spectroscopy 11:330
 manufacture 11:323
 molecular weight distribution 11:346
 solvent extraction 11:345
 uses 11:323
Fatty alkyl sulfonates
 in car washes 6:364
Fatty amides
 in styrene polymers 18:328
Fatty oils. (See also *Fats and fatty oils.*)
 in essential oils 12:285
F D & C Blue No. 1. (See also Certified colors;
  Triphenylmethane and related dyes.)
 classification 10:448
 commercial specifications 10:463
 electrophoresis 10:482
 impurities, determination of 10:500, 506, 511
 infrared spectrum 10:472
 in lipsticks 11:69
 liquid chromatography 10:479, 500
 manufacture 10:448
 paper chromatography 10:475
 sodium carbonate–potassium carbonate fusion
  10:494
 structure 10:448
 thin-layer chromatography 11:69; 19:173
 titanous chloride titration 10:485
 visible spectroscopy 10:486; 11:70
 visible spectrum 10:472
F D & C Blue No. 2. (See also Certified colors.)
 classification 10:448
 commercial specifications 10:463
 electrophoresis 10:482, 483
 impurities, determination of 10:500, 511
 infrared spectrum 10:473

liquid chromatography 10:500
manufacture 10:448
paper chromatography 10:475
structure 10:448
titanous chloride titration 10:485
visible spectroscopy 10:486
visible spectrum 10:473
F D & C colors. See Certified colors; and indi-
      vidual compounds.
F D & C Green No. 1. (See also Certified colors.)
   electrophoresis 10:482
   in F D & C Blue No. 1 10:511
   liquid chromatography 10:479
   paper chromatography 10:475
F D & C Green No. 2. (See also Certified colors.)
   electrophoresis 10:482
   liquid chromatography 10:479
   paper chromatography 10:475
F D & C Green No. 3. (See also Certified colors.)
   classification 10:448
   commercial specifications 10:463
   electrophoresis 10:482
   impurities, determination of 10:500, 513
   infrared spectrum 10:474
   liquid chromatography 10:479, 500
   manufacture 10:448
   paper chromatography 10:475
   structure 10:448
   titanous chloride titration 10:485
   visible spectroscopy 10:486
   visible spectrum 10:474
F D & C Greens. (See also *Triphenylmethane and
      related dyes.*)
   thin-layer chromatography 19:173
F D & C lakes. (See also Certified colors.)
   commercial specifications 10:464
   manufacture 10:450
F D & C Orange No. 1. (See also Certified colors.)
   electrophoresis 10:482
   paper chromatography 10:475
F D & C Red No. 1. (See also Certified colors.)
   electrophoresis 10:483
   paper chromatography 10:475
F D & C Red No. 2. (See also Certified colors.)
   classification 10:448
   commercial specifications 10:463
   electrophoresis 10:482, 483
   impurities, determination of 10:500, 513
   infrared spectrum 10:475
   liquid chromatography 10:479, 500
   manufacture 10:448
   paper chromatography 10:475
   structure 10:448
   titanous chloride titration 10:485
   visible spectroscopy 10:486
   visible spectrum 10:476
F D & C Red No. 3. (See also Certified colors;

Triphenylmethane and related dyes.)
   classification 10:448
   commercial specifications 10:463
   electrophoresis 10:483
   gravimetric determination 10:487
   impurities, determination of 10:500
   infrared spectrum 10:476
   liquid chromatography 10:500
   manufacture 10:448
   paper chromatography 10:475
   structure 10:448
   thin-layer chromatography 19:173
   visible spectroscopy 10:486; 11:70
   visible spectrum 10:477
F D & C Red No. 4. (See also Certified colors.)
   classification 10:448
   commercial specifications 10:463
   electrophoresis 10:483
   impurities, determination of 10:500, 515
   infrared spectrum 10:477
   liquid chromatography 10:500
   manufacture 10:448
   paper chromatography 10:475
   structure 10:448
   titanous chloride titration 10:485
   visible spectroscopy 10:486
   visible spectrum 10:478
F D & C Violet No. 1. (See also Certified colors;
      Triphenylmethane and related dyes.)
   classification 10:450
   commercial specifications 10:464
   electrophoresis 10:482
   impurities, determination of 10:500, 517
   infrared spectrum 10:478
   liquid chromatography 10:479, 500
   manufacture 10:450
   paper chromatography 10:475
   structure 10:450
   thin-layer chromatography 19:173
   titanous chloride titration 10:485
   visible spectroscopy 10:486
   visible spectrum 10:479
F D & C Yellow No. 1. (See also Certified colors.)
   electrophoresis 10:482
   paper chromatography 10:475
F D & C Yellow No. 5. (See also Certified colors.)
   classification 10:450
   commercial specifications 10:464
   electrophoresis 10:482
   impurities, determination of 10:500, 517
   infrared spectrum 10:479
   in lipsticks 11:69
   liquid chromatography 10:479, 500
   manufacture 10:450
   paper chromatography 10:475
   structure 10:450
   thin-layer chromatography 11:69

titanous chloride titration 10:485
visible spectroscopy 10:486; 11:70
visible spectrum 10:480
F D & C Yellow No. 6. (See also Certified colors.)
  classification 10:450
  commercial specifications 10:464
  electrophoresis 10:482
  impurities, determination of 10:500, 518
  infrared spectrum 10:480
  liquid chromatography 10:500
  manufacture 10:450
  paper chromatography 10:475
  structure 10:450
  titanous chloride titration 10:485
  visible spectroscopy 10:486
  visible spectrum 10:481
F D & C Yellows. (See also *Triphenylmethane and
    related dyes.*)
  titanous chloride titration 19:177
Fecal coliform count of water 19:392
Feces
  methadone content 16:279
Feed products. (See also *Corn products.*)
  analysis 11:48
  carotene content 11:52
  crude fiber content 11:49
  lactic acid content 11:51
  silver content 18:172
  starch content 11:48
  xanthophyll content 11:52
Feed yeast. (See also *Yeast.*) 19:517
  manufacture 19:517
  species 19:516
  uses 19:516
Fehling's reagent
  color test for aldehydes 8:344
  color test for cane sugar 11:127, 132
  color test for glucose 11:127, 132
  gravimetric method for sugar 15:147
Feigl color test
  for barbiturates 6:477
Feldspar 4:31, 33; 15:261; 17:545
  in chrome cleaners 6:353
  in clay 9:191
Felt 12:494–514
  acid content 12:509
  air permeability 12:512
  analysis 12:494
  ash content 12:509
  breaking strength 12:510
  bursting strength 12:511
  carbon tetrachloride soluble content 12:494
  commercial grades and specifications 12:496
  definition 12:494
  fire retardance 12:513
  manufacture 12:496
  oil retention 12:512

pH 12:509
physical tests 12:510
properties 12:496
shrinkage 12:511
splitting resistance 12:510
stiffness 12:513
tensile strength 12:510
thickness 12:510
types 12:494, 496
uses 12:494, 496
water absorption 12:511
water soluble content 12:495
water thickness swell 12:511
weight 12:510
wool content 12:495
Felt apparel. See *Felt.*
Fenazaflor. (See also *Pesticides and insecticides.*)
  19:586
Fennel oil. (See also *Essential oils.*)
  refractive index 12:271
Fensulfothion. (See also *Pesticides and insec-
    ticides.*)
  commercial grades and specifications 19:594
  composition 19:598
  manufacture 19:597
  properties 19:589
  solubility 19:587
  structure 19:583
  toxicity 19:592
Fentanyl. (See also *Narcotics.*)
  ammonium molybdate colorimetric method
    16:246
  color tests 16:246
  formaldehyde–sulfuric acid colorimetric method
    16:246
  iodine–potassium iodide microchemical method
    16:249
  microchemical tests 16:249
  picric acid microchemical method 16:249
  structure 16:245
  ultraviolet spectroscopy 16:251
Fenthion. (See also *Pesticides and insecticides.*)
  commercial grades and specifications 19:595
  composition 19:598
  manufacture 19:598
  properties 19:589
  solubility 19:587
  structure 19:583
  toxicity 19:592
Ferason. See Thiram.
Ferbam. (See also Ferric dimethyl dithiocarba-
    mate.) 13:202
  manufacture 13:202
  properties 13:202
  solubility 13:202
  toxicity 13:202
  uses 13:202

Ferbeck. See Ferbam.
Ferberite 19:190
  decomposition 19:197
Fergusonite
  composition 16:349
Fermate. See Ferbam.
Fermentation
  of bread 6:442
Fermentation ability
  of yeast 19:534, 540
Fermentation, degree of wort 7:496, 501
Fermented milk products. (See also individual
    products.) 16:134
  acidity 16:135
  analysis 16:135
  manufacture 16:134
  pH 16:135
  solids content 16:136
Fermenting liquids. (See also *Yeast*.) 19:519
  analysis 19:519
  uses 19:516
  species 19:516
Fermentometer flasks 19:535
Fermium. (See also Transuranium elements.)
  manufacture 19:259
Fernichrome. (See also Cobalt alloys.) 10:290
  composition 10:290
Fernico. (See also Cobalt alloys.) 10:290
  composition 10:290
Ferradow. See Ferbam.
Ferric acetate, basic. (See also Iron compounds.)
  uses 15:85
Ferric acetylacetonate. (See also Iron compounds.)
  uses 15:86
Ferric alum. See Ferric ammonium sulfate.
Ferric ammonium citrate. (See also Citric acid
    and salts; Iron compounds.)
  uses 10:86; 15:86
Ferric ammonium oxalate. (See also Iron com-
    pounds.)
  uses 15:86
Ferric ammonium sulfate. (See also Iron com-
    pounds.)
  colorimetric method for salicylic acid 10:166
  color test for thiocyanate 12:428
  gravimetric method for phosphorus 16:334
  titration of titanium 19:104, 106
  titration of titanous chloride 19:180
  uses 15:85
  volumetric method for titanium 9:261; 10:321,
    647
Ferric ammonium sulfate–potassium permanganate
  volumetric method for hydroxylamine and its
    salts 14:447
Ferric benzoyl acetonate. (See also Iron com-
    pounds.)
  uses 15:86

Ferric cacodylate. (See also Iron compounds.)
  uses 15:86
Ferric chloride. (See also Iron compounds.)
  colorimetric method for aminopyrine 6:217
  colorimetric method for benzoic acid 19:485
  colorimetric method for casein 11:127, 132
  colorimetric method for casein–formaldehyde
    resins 11:127, 133
  colorimetric method for catechol 11:507
  colorimetric method for cresols 11:126, 131;
    17:16
  colorimetric method for ethyl biscoumacetate
    10:162
  colorimetric method for 3-ethylphenol 17:16
  colorimetric method for ferrocyanide 15:90
  colorimetric method for hexahydrobenzene
    17:16
  colorimetric method for hydroquinone 17:16
  colorimetric method for hydroxyhydroquinone
    17:16
  colorimetric method for iodide 12:543
  colorimetric method for lactic acid 15:138, 140
  colorimetric method for 1-methyl-4,6-di-*tert*-
    butylphenol 17:16
  colorimetric method for 3-methyl-6-*tert*-butyl-
    phenol 17:16
  colorimetric method for 3-methyl-6-isopropyl-
    phenol 17:16
  colorimetric method for naphthols 16:231
  colorimetric method for phenol 17:16
  colorimetric method for phenols 11:126, 131;
    17:16
  colorimetric method for phloroglucinol 17:16
  colorimetric method for 4-propylphenol 17:16
  colorimetric method for pyrocatechol 17:16
  colorimetric method for pyrogallol 17:16
  colorimetric method for resorcinol 17:16
  colorimetric method for R salt 16:235
  colorimetric method for salicylates 16:95
  colorimetric method for salicylic acid 19:484
  colorimetric method for sodium heparin 10:143
  colorimetric method for tannins 7:647
  color test for acetate 14:581
  color test for natural color materials 10:537
  commercial grades 15:88
  extraction of iodide 9:347
  gravimetric method for carrageenin 12:582
  gravimetric method for naphthols 16:231
  gravimetric method for tin 16:336
  gravimetric method for tragacanth 12:582
  reduction method for copper(I) oxide 10:671
  titration of calcium chelates 12:50
  titration of DTPA 12:53
  titration of magnesium chelates 12:50
  titration of manganese chelates 12:50
  uses 15:85, 88
  volumetric method for europium 12:397

volumetric method for titanium 8:136
volumetric method for zinc metal 17:204
Ferric chloride–ammonium hydroxide
colorimetric method for synthetic resins 10:282
Ferric chloride-2,2'-bipyridine
colorimetric method for BHA and BHT 9:243
Ferric chloride–ceric ammonium nitrate
volumetric method for cuprous oxide 17:195
volumetric method for hydroxylamine and its
salts 14:448
Ferric chloride–dioxane
colorimetric method for diphenadione 10:179
Ferric chloride, hexahydrate
properties 9:334
solubility 9:336
Ferric chloride–sulfuric acid
colorimetric method for codeine 16:246
Ferric choline citrate. (See also *Choline and its
derivatives.*)
uses 9:612
Ferric dimethyl dithiocarbamate. (See also Iron
compounds.)
uses 15:86
Ferric ferrocyanide. (See also Ferrocyanides.)
uses 15:85
Ferric fluoride. (See also Iron compounds.)
etchant for ceramics 9:225
uses 15:85
Ferric gluconate. (See also Iron compounds.)
uses 15:86
Ferric hexathiocyanatoferriate
colorimetric method for ethers 12:300
Ferric hydroxamate colorimetric method
for acid anhydrides 8:569, 590
for amides 8:601
for carboxylic acid derivatives 8:561, 590
for cyclic alcohols 11:195
for esters 8:590
Ferric hydroxide. (See also Iron compounds.)
gravimetric method for arsenic 10:294
gravimetric method for bismuth 10:295
uses 15:85
Ferric hydroxide–sulfuric acid
colorimetric method for alginates 9:589
Ferric hypophosphite. (See also Iron compounds.)
uses 15:85
Ferric linoleate. (See also Iron compounds.)
uses 15:86
Ferric nitrate. (See also Iron compounds.)
colorimetric method for acetylsalicylic acid
6:148
colorimetric method for thiocyanate 18:435,
436
uses 15:85
Ferric nitrate, basic
in uranium 19:238

Ferric oleate. (See also Iron compounds.)
uses 15:86
Ferric oxalate. (See also Iron compounds.)
uses 15:86
Ferric oxide. (See also Iron compounds.)
in alkali barium silicate glass 9:218
ammonium hydroxide gravimetric method
9:112
ammonium thiocyanate colorimetric method
17:192
in barium sulfate 17:192
bathophenanthrolinedisulfonic acid colorimetric
method 15:209
in cements 9:112, 114, 126
in ceramics 9:194
commercial grades and specifications 15:89
cupferron colorimetric method 9:126
emission spectroscopy 9:218
in glass 9:200
in glazes 9:196
in hydrated lime 9:191
in hydraulic cements 9:96, 100
hydrogen sulfide gravimetric method 15:210
in lime and limestone 9:191; 15:208
potassium dichromate titration 9:114
potassium permanganate titration 15:210
in quicklime 9:191
in refractories 9:206
uses 15:85
Ferric oxide–sulfuric acid
colorimetric method for alginate 12:582
Ferric perchlorate. (See also Iron compounds.)
uses 15:85
Ferric periodate
gravimetric method for lithium 15:265
Ferric phenanthroline
colorimetric method for zinc metal 19:575
Ferric phosphate. (See also Iron compounds.)
uses 15:85
Ferric phosphate casse
in wines 19:498
Ferric phosphate–potassium permanganate
volumetric method for niobium 16:381
Ferric pyrophosphate. (See also Iron compounds.)
uses 15:85
Ferric salt of 6-sodiumsulfo-1-isonitroso-1,2-naph-
thoquinone. See Ext D & C Green No. 1.
Ferric sodium oxalate. (See also Iron compounds.)
uses 15:86
Ferric stearate. (See also Iron compounds.)
uses 15:86
Ferric sulfate. (See also Iron compounds.)
gravimetric method for epoxy resins 12:237
titration of titanium 19:105
uses 15:85
Ferric thiocyanate
volumetric method for mercury 16:34

Ferricyanides. (See also individual compounds.)
  analysis 15:91
  colorimetric method for hydrogen peroxide
    14:434
  colorimetric method for zinc 19:556
  gravimetric method for silver 18:164
  iodometric methods 15:91
  titration of cobalt 16:344
  titration of manganese 16:344
  volumetric determination 15:91
  volumetric method for reducing value of corn
    starch 11:27
Ferrimycin 5:477, 599
  analysis 5:604, 610, 614
  countercurrent distribution 5:604
  identification 5:604
  ion exchange separation 5:610
  microbiological assay 5:610
  paper chromatography 5:605
  properties 5:600
  separation 5:604
  solubility 5:601, 602
  uses 5:599
Ferrites 12:514–538
  acid dissolution 9:215
  analysis 12:520
  commercial grades and specifications 12:518
  composition 12:520
  composition, determination of 12:522
  copper content 12:534
  fluxes for 9:216
  in hydraulic cements 9:107
  iron content 12:522
  light microscopy 9:103
  magnesium content 12:536
  magnetic properties 12:515
  manganese content 12:531
  manufacture 12:518
  nickel content 12:533
  in portland cement 9:103
  properties 12:514
  raw materials 12:519
  sample dissolution 12:521
  sampling 12:521
  uses 12:517
  zinc content 12:532
Ferroactinolite 8:75
Ferro alloys. (See also Iron alloys.) 5:57, 59
  hydrogen in 8:638
  oxygen in 8:621
  vacuum fusion method 8:621
  vacuum hot extraction methods 8:638
Ferroboron. (See also Iron alloys.) 5:59
  analysis 15:83
  boron content 15:83
  commercial specifications 15:73
  composition 5:61; 15:73

composition, determination of 15:83
Ferroboron alloys 7:323
  composition 7:322
  manufacture 7:323
  sample decomposition 7:324
Ferrocene. See Dicyclopentadienyl iron.
Ferrochrome. (See also Ferrochromium.)
  nitrogen in 8:616, 617, 618, 619, 620, 629,
    632, 633
  oxygen in 8:616, 617, 618, 619, 620, 629,
    632, 633
  vacuum fusion methods 8:616, 617, 618, 619,
    620
Ferrochrome–silicon. (See also Silicon alloys.)
  composition 18:2
Ferrochromium. (See also Iron alloys; Steel.)
    5:61; 9:660; 18:262
  analysis 15:79
  carbon content 15:80
  chromium content 15:79
  commercial grades and specifications 9:661;
    15:72
  composition 5:62; 9:661; 15:72
  composition, determination of 15:79
  manufacture 9:660
  silicon content 15:80
  uses 9:662
Ferrochromsilicon. (See also Iron alloys.)
  commercial specifications 15:73
  composition 15:73
Ferrocolumbium. (See also Iron alloys.)
  commercial specifications 15:73
  composition 15:73
Ferrocyanides. (See also individual compounds.)
  ammonium molybdate gravimetric method
    15:89
  analysis 15:89
  N-chlorosuccinimide spot test 15:89
  copper chloride–silver nitrate volumetric
    method 15:90
  ferric chloride colorimetric method 15:90
  ferrous sulfate–sodium hydroxide colorimetric
    method 19:483
  gravimetric determination 15:90
  gravimetric method for silver 18:164
  volumetric determination 15:90
  in wines 19:482
  zinc chloride titration 15:90
  zinc oxide titration 17:188
Ferrodipyridyl
  gravimetric method for niobium 16:367
  gravimetric method for tantalum 16:367
Ferroin. See Ferrophenanthroline.
Ferroludwegite 7:315
Ferromagnetic alloys 14:547
  composition 14:547
Ferromanganese. (See also Iron alloys; Manganese

alloys; and *Steel*.) 5:61; 18:262
analysis 15:76
carbon content 15:78
commercial grades and specifications 15:72, 453
composition 5:64; 15:72, 451
composition, determination of 15:76
manganese content 15:76, 460
phosphorus content 15:78; 17:86
silicon content 15:78
sulfur content 15:78; 18:382
Ferromolybdenum. (See also Iron alloys; Molybdenum alloys.) 5:61
commercial specifications 15:73
composition 5:64; 16:171
manufacture 16:171
molybdenum content 16:171
phosphorus content 17:88
sample decomposition 18:24
silicon content 18:24
sulfur content 18:382
uses 16:171
Ferronickel. (See also Nickel alloys.)
decomposition 16:295
solubility 16:295
uses 16:288
Ferroniobium. (See also Niobium alloys.)
analysis 16:389
commercial grades and specifications 16:360
composition, determination of 16:389
impurities, determination of 16:388, 389
ion exchange methods 16:390
manufacture 16:359
sample preparation 16:362
uses 16:350
Ferrophenanthroline
gravimetric method for tantalum 16:367
Ferrophosphorus
phosphorus content 17:84
Ferropyridyl
gravimetric method for tantalum 18:445
Ferrosilicon. (See also Iron alloys; Silicon alloys; and *Steel*.) 5:61; 18:262
analysis 15:73
carbon content 15:74
commercial grades and specifications 18:2
commercial specifications 15:72
composition 5:66; 15:72; 18:2
composition, determination of 15:74
phosphorus content 15:75; 17:86
sample decomposition 18:13, 26
silicon content 15:74; 18:26
sodium peroxide fusion 18:13
sulfur content 15:76; 18:382
uses 18:2
Ferrotitanium. (See also Iron alloys; Titanium alloys.) 5:61

analysis 15:82
commercial specifications 15:73
composition 5:67; 15:73; 19:101
composition, determination of 15:82
sample decomposition 18:24
silicon content 18:24
titanium content 15:82; 19:101, 105
Ferrotungsten (See also Iron alloys; Tungsten alloys.) 5:61
analysis 15:81
commercial specifications 15:73
composition 5:67; 15:73
composition, determination of 15:81
phosphorus content 17:88
sample decomposition 18:24
silicon content 18:24, 30
sulfur content 18:382, 385
tin content 19:48
tungsten content 15:81; 19:206
Ferrous acetate. (See also Iron compounds.)
uses 15:86
Ferrous alloys. (See also Pig iron; *Steel*.) 5:57
niobium content 16:380
Ferrous ammonium sulfate. (See also Iron compounds.)
potassium dichromate titration 19:127
titration of potassium dichromate 19:376
titration of vanadium 19:102, 337
uses 15:85
volumetric method for cerium 9:271
volumetric method for chromate 6:358
volumetric method for chromium 9:655; 15:79
volumetric method for manganese 15:76
volumetric method for titanium 15:82; 19:127
volumetric method for vanadium 15:81
Ferrous ammonium sulfate–ceric sulfate method for chlorate 9:360
Ferrous ammonium sulfate–potassium dichromate titration
of chromium 9:640
Ferrous ammonium sulfate–potassium permanganate
volumetric method for chromium 10:636
volumetric method for vanadium 16:339
Ferrous ammonium sulfate–titanous chloride
volumetric method for nitrogen 12:465
Ferrous carbonate. (See also Iron compounds.)
uses 15:85
Ferrous chloride. (See also Iron compounds.)
properties 9:334
solubility 9:336
uses 15:85
Ferrous ethylenediammonium sulfate. (See also Iron compounds.)
uses 15:86
Ferrous 2-ethylhexoate. (See also Iron compounds.)

uses 15:86
Ferrous fluoroborate. (See also Iron compounds.)
  uses 15:85, 88
Ferrous fumarate. (See also Iron compounds.)
  uses 15:86
Ferrous gluconate. (See also Iron compounds.)
  assay 10:540
  ceric sulfate titration 10:540
  commercial specifications 10:531
  impurities, determination of 10:543
  structure 10:539
  uses 15:86
Ferrous iodide. (See also Iron compounds.)
  uses 15:85
Ferrous ion content
  of iron 15:43
Ferrous iron
  hydrofluoric acid–boric acid volumetric deter-
    mination 15:43
Ferrous lactate. (See also Iron compounds.)
  uses 15:86
Ferrous materials
  carrier gas fusion methods 8:682
  hot extraction methods 8:684
  hydrogen in 8:680, 684
  oxygen in 8:680, 682
  vacuum fusion methods 8:680
Ferrous metals
  chromium content 9:650
  impurities, determination of 17:78
  isobutyl acetate extraction 9:650
Ferrous nitrate–sulfur
  colorimetric method for cyanoacetic acid
    15:439
Ferrous oxalate. (See also Iron compounds.)
  uses 15:86
Ferrous oxide
  in glass 9:200
  in lime and limestone 15:220
  potassium permanganate titration 15:220
Ferrous perchlorate. (See also Iron compounds.)
  uses 15:85
Ferrous phenanthroline
  colorimetric method for cyclamates 11:234
Ferrous phosphate. (See also Iron compounds.)
  uses 15:85
Ferrous sulfate. (See also Iron compounds.)
  colorimetric method for anthocyanogens
    19:468
  gravimetric method for potassium chlorate
    12:420
  gravimetric method for silver 18:165
  manufacture 15:84, 88
  uses 15:85, 88
  volumetric method for ammonium dichromate
    9:685
  volumetric method for chromic acid 9:685
  volumetric method for chromic oxide 9:703

volumetric method for nitrate 7:125
volumetric method for potassium chlorate
  15:527
volumetric method for potassium dichromate
  9:685
volumetric method for sodium chromate 9:685
volumetric method for sodium dichromate
  9:685
volumetric method for vanadium 8:137
Ferrous sulfate–permanganate
  volumetric method for nitrates 16:502
Ferrous sulfate–sodium hydroxide
  colorimetric method for cyanide 19:483
  colorimetric method for ferrocyanide 19:483
Ferrous sulfate–sulfuric acid
  colorimetric method for nitrate 11:142
Ferrous sulfide. (See also Iron compounds.)
  uses 15:85
Ferrovanadium (See also Iron alloys; Vanadium
    alloys.) 5:61
  analysis 15:80; 19:326
  commercial specifications 15:72
  composition 5:67; 15:72
  composition, determination of 15:80
  dissolution 19:327
  manufacture 19:325
  phosphorus content 17:87
  sample decomposition 18:24
  sampling 19:326
  silicon content 18:24
  sulfur content 19:382
  uses 19:322
  vanadium content 15:80; 19:338
Ferrox. See Ferric hexathiocyanatoferriate.
Ferrozine
  colorimetric method for sulfur dioxide 18:408
Ferruccite 7:315
Fertilizers. (See also Phosphate fertilizers.)
  analysis 17:363
  boron in 7:407, 410
  iron content 15:56
  potassium content 17:363, 405
  sample preparation 17:363
  sulfur content 18:373, 384
Ferulic acid. (See also *Cinnamic acid and its
    derivatives*.) 10:63
Fiber classification
  of cellulose 9:52
Fiber color value 4:328
Fiber, crude. (See also Crude fiber.)
  in bakery products 6:455
Fiber finish
  in polyamide fibers 17:322
Fibers. (See also Yarns.)
  bleaching 7:221
  in brake linings 7:441
  content of grains 4:472

density, linear 4:89
elasticity 4:89
heat, effect of 4:260
infrared spectroscopy 4:230
length and length distribution 4:89
microscopy 4:263
multiple internal reflection 4:246
quality control tests 4:25
in roasted coffee 10:375, 425
shrinkage 4:89
solubility 4:247
specific gravity 4:258
strength and elongation 4:89
Fibril formation
  of insulin 14:574
Ficin
  in beer 7:588
Ficusin. See Psoralen.
Fiehe's solution
  colorimetric method for hydroxymethylfurfural
    19:459
Field tests
  for cobalt 10:371
Filipin 5:472, 564
  analysis 5:569, 573, 578
  bioautography 5:569
  identification 5:569
  microbiological assay 5:573
  paper chromatography 5:569
  properties 5:565
  separation 5:569
  solubility 5:566, 567
  structure 5:472
  ultraviolet spectroscopy 5:578
Fillers
  in brake linings 7:441
  in cloth backing for coated abrasives 10:278,
    283
  in enamels 12:174
  for grinding wheels 7:294
Fillers, inorganic
  in elastomers, synthetic 12:151
  infrared spectroscopy 12:152
Film bases, photographic 12:559
  analysis 12:588
  flame test 12:588
  heating test 12:588
  identification 12:588
  infrared spectroscopy 12:589
  solubility 12:589
Films, photographic 12:539–599
  additives 12:558, 570
  analysis 12:561, 568, 570, 588
  antihalation dyes in 12:587
  antioxidants in 12:592
  antistatic agents in 12:587
  ash content 12:592
classification 12:539
coating defects 12:561
composition 12:539
developing agents in 12:586
film bases in 12:559, 588
gelatin 12:546
gelatin coating weight 12:568
gelatin substitutes 12:558
halide ratio 12:565
H and D curve 12:595
hardeners in 12:584
humectants in 12:587
impurities, determination of 12:594
melting point 12:592
metallic nuclei in 12:586
paper 12:559
photographic performance 12:595
plasticizers in 12:590
polysaccharides in 12:579
in polyurethanes 19:312
preservatives in 12:587
properties, determination of 12:592
raw materials 12:540
sensitizers in 12:570
silver coating weight 12:561
silver halide content 12:561
stabilizers in 12:579, 592
storage quality 12:594
surfactants in 12:572
swelling 12:593
water absorption 12:593
Film technique
  polymer samples for infrared spectroscopy
    4:237
Film test
  for cellulose nitrate 9:68
Film thickness
  of dental zinc phosphate cements 11:263
Filterability
  of activated carbon 8:151
Filterable residue
  of water 19:381
Filter membrane technique
  for coliform bacteria in water 19:391
Filtration rate
  of adsorbents 4:450
Fineness
  of dental gypsum products 11:266
  of dental impression materials, alginate based
    11:281
  of dental investments 11:269
  of gold 14:2
Finish
  spin 4:91
  textile 4:329
Finishing agents. (See also *Cotton.*)
  for cotton 11:126, 129

FMIR 11:137
    identification 11:126, 129
    infrared spectroscopy 11:136
    infrared spectroscopy, differential 11:136
    Krammes and Maresh identification 11:126, 130
    separation 11:126, 129
    Skinkle technique 11:126, 130
    solvent extraction 11:126, 130
    ultraviolet spectroscopy 11:133
    visible spectroscopy 11:133
Fire assay
    of gold 10:621
    of gold ores 14:12
    of platinum 10:644
    of platinum group metals 17:250
    of silver 15:196
Firebrick. (See also Refractories.)
    composition 9:194, 208
    properties 9:208
Fireclay. (See also Refractories.)
    composition 9:194, 202, 208
    properties 9:204, 208
Fireclay brick. (See also Refractories.)
    composition 9:206
Fire-holding capacity
    of tobacco 19:158
Fire point
    of hydrocarbon resins 14:294
    of lubricants 15:345
    of paraffin wax 15:533
Fire red
    in enamels 12:182
    identification 12:182
Fire retardance
    of felt 12:513
Fire retention
    of cigarettes 10:4
Fish
    mercury content 16:44, 51
    sodium content 18:219, 220
Fisher-Johns method
    for melting point of polytetrafluoroethylene
        13:85
Fisher-Serfass hydrogen analyzer 8:667
Fish glues, evaluation 4:401
Fish-kill method
    for antimycin 5:613
Fish products
    EDTA and related compounds in 12:46
Fiske cryoscope 16:124
Fissium
    technetium content 17:541
Fivonite. (See also High explosives.)
    melting point 12:461
    properties 12:432
Fixed residue
    of water 19:381

Flame emission spectroscopy
    of ceramics 9:217
    of cesium 13:452; 17:552
    of gallium 13:260
    of glass and glass-ceramics 13:449
    of lithium 13:452
    of potassium 13:452; 18:245
    of rubidium 13:452; 17:551
    of silver 18:172
    of sodium 13:452
    of strontium 18:277
    of titanium 19:120
    of tungsten 19:218
Flame ionization detection
    of hydrocarbons in inert gases 13:303
Flame ionization methods
    for methane 16:532
Flame photometry
    of alkali 6:266
    of alkali and alkaline earth metals in beer, 7:651
    of alkali metals 5:186
    of aluminum 5:212, 244; 15:287
    of barium 6:579
    of bismuth 7:218
    of borate 11:75
    of boron 7:390, 411
    of boron compounds, organic 7:351
    of boron hydrides 7:411
    of calcium 7:381, 382; 9:666; 11:494, 498;
        15:280, 399
    of calcium(III) 8:595
    of cesium 9:283, 294, 300; 17:555
    of cesium alum 9:293
    of chlorinated compounds 10:84
    of chlorocarbons 9:382
    of chromium 9:652
    of copper 10:602, 617
    of dysprosium 17:492
    of erbium 17:492
    of europium 17:492
    of gadolinium 17:492
    of gallium 13:270
    of holmium 17:492
    of indium 14:525, 534
    of lanthanum 15:158; 17:492
    of lithium 5:134; 15:266, 270; 17:555
    of lutetium 17:493
    of magnesium 9:666; 15:280, 359, 372
    of manganese 15:476
    of neodymium 17:492
    of palladium 17:247
    of phosphorus 17:67
    of platinum 17:247
    of potassium 6:266; 8:595; 9:242, 349; 11:75;
        15:280, 399; 17:369, 381, 387, 388, 407,
        555; 19:492
    of potassium oxide 9:121; 10:123

of praseodymium 17:492
of rare earths 17:490
of rhodium 17:247
of rubidium 17:548, 555
of ruthenium 17:247
of samarium 17:492
of scandium 17:493
of sodium 6:266; 8:595; 9:242, 349, 362, 699;
    10:303; 11:75, 226; 14:606; 15:272, 280,
    399, 528; 17:381, 388, 400, 555; 18:218;
    19:492
of sodium oxide 9:121; 10:123
of strontium 8:595; 15:399
of terbium 17:492
of thulium 17:493
of ytterbium 17:493
of yttrium 17:493
Flame tests 6:588
for borate 18:184
for boron 7:389
for calcium 18:270
for casein 12:588
for cellulose 12:588
for cellulose acetate 12:588
for cellulose acetobutyrate 12:588
for cellulose nitrate 12:588
for copper phthalocyanine blue 17:205
for ethylene and propylene polymers 12:347
identification of acrylic and modacrylic fibers
    4:260
for photographic film bases 12:588
for polyacrylates 12:588
for polyamides 12:588
for polystyrene 12:588
for potassium 17:367
for strontium 18:270
for thallium 18:525
for vinyl polymers 12:588
for warfarin potassium 10:175
for warfarin sodium 10:167
Flaming red. See D & C Red No. 36.
Flammability
of antifreezes 6:6
Flash point
of bituminous materials 6:320
of drycleaning detergents 11:611
of hydrocarbon resins 14:294
of hydrocarbons 14:343
of lubricants 15:345
of mercaptans 15:571
of paraffin wax 15:533
of Stoddard solvent 11:605
of terpenes 14:284
Flash tendency
of pesticides and insecticides 19:618
Flask combustion method
for sulfur compounds, organic 18:378

Flavanols
gas chromatography 18:471, 503
in green tea leaf 18:465
liquid chromatography 18:465
paper chromatography 18:467
solvent extraction 18:465, 468, 473
thin-layer chromatography 18:470
ultraviolet spectroscopy 18:465
Flavanones
in wines 19:461
Flavognost
colorimetric method for theaflavin 18:500
Flavone derivatives
color reagents for boric acid 7:387
Flavonoid phenols
colorimetric determination 19:467
formaldehyde gravimetric method 19:466
in wines 19:466
Flavonol glycoside content
of green tea leaf 18:457, 474
Flavonol glycosides
bathochromic shifts 18:474
liquid chromatography 18:474
thin-layer chromatography 18:474
Flavonols
in wines 19:461
Flavonthrone yellow. (See also *Pigments*.)
structure 17:186
Flavor
of cocoa beans 9:567
of cocoa products 9:592
of essential oils 12:264
Flavoring agents
in confectionery products 10:552
Flavor profiling 17:643
Flavors
coumarin content 11:166
Flavor testing. See *Sensory testing methods*.
Flavylogens 18:475
in green tea leaf 18:475
leucoanthocyanin colorimetric method 18:475
liquid chromatography 18:475
solvent extraction 18:475
Flax
microscopy 4:265
Flaxseed
cooling system sealer 6:352
Flectol B 6:74
color test 6:94
Flectol H 6:74
colorimetric method using diazotized *p*-nitro-
    aniline 6:98
color test 6:95
column chromatography 6:89
Flexamine 6:74
colorimetric method based on formation of
    Wurster salts 6:101, 103

colorimetric method using diazotized *p*-nitro-
   aniline 6:98
Flexamine G 6:74
Flex-crack inhibitors 6:78
Flexibility
   of enamels 12:180
Flexural strength
   of dental porcelains 11:272
   of graphite 14:79
Flexzone 6:71, 74
Flexzone 3C (and 6H) 6:74
   colorimetric method based on formation of
      Wurster salts 6:101, 103
Flint clays. (See also *Clay*.)
   composition 9:202
Float test
   apparatus 6:299
   for bituminous materials 6:299
Flock point
   of lubricants 15:332, 333
Floor cleaners. (See also *Detergents*.) 11:325
   composition 11:325
Flotation method
   specific gravity of acrylic and modacrylic fibers
      4:258
Flour. (See also Dough; Wheat.) 12:599–618
   α-amylase content 12:607
   analysis 12:602
   ash content 12:602
   baking tests 12:615
   in bread 6:439
   bromate or iodate in 8:8
   in cakes 6:445
   carbohydrate content 12:606
   classification 12:599
   commercial grades and specifications 12:601
   composition 12:601
   composition, determination of 12:602
   crude fiber content 12:606
   diastatic activity 12:608
   ethylene dibromide content 13:184
   ethylene dichloride content 13:184
   ethylene oxide content 13:184
   filth tests 6:461
   hot-paste viscosity 12:609
   iron content 15:32
   lipid content 12:604
   methyl bromide content 13:184
   milling 12:600
   moisture content 12:602
   pH 12:612
   pigment content 12:612
   protein content 12:602
   uses 12:600
   water absorption 12:613
   wet gluten method 12:603
Flour–water suspensions

viscosity 12:604
Flow cup tests
   for brake lining formulas 7:448
Flow curves
   for polyethylene 12:374
Flowers of sulfur. (See also Sulfur, elemental.)
   18:365
Flow test, inclined plate
   of resins 7:444
Flow tests
   for dental amalgams 11:290
   for dental elastomers 11:286
   for dental impression compounds 11:317
   for pesticides and insecticides 19:617
Fludrocortisone acetate. (See also Steroids.)
   manufacture 14:195
   properties 14:195
   solubility 14:195
   structure 14:195
Fludrocortisone hemisuccinate. (See also
   Steroids.)
   solubility 14:195
   uses 14:196
Flue dust
   antimony in 6:58
   sample preparation 6:59, 236
   thallium content 18:524
Fluidity. (See also Viscosity.)
   of brake fluids 7:434
   of cellulose dispersions 7:235
Fluoberyllates 7:122
   analysis 7:123
Fluoborates 7:337
   analysis 7:339
   electrolytic reduction to elemental boron
      7:321
   occurrence 7:313
Fluoboric acid. (See also Fluoroboric acid.)
   7:336
   in antimony fluoborate 6:55
   manufacture 7:336
   nitron volumetric method 13:41
   potentiometric methods 13:42
   properties 13:38
   sulfuric acid content 13:40
   uses 13:37
Fluoborite 7:315
Fluocinoline acetonide. (See also Steroids.)
   manufacture 14:196
   properties 14:196
   solubility 14:196
   structure 14:196
   uses 14:196
Fluon. See Polytetrafluoroethylene.
3,6-Fluorandiol. See D & C Yellow No. 7.
Fluorapatite 13:2; 17:143
Fluorel. (See also Vinylidene fluoride–hexa-

fluoropropylene copolymer.) 12:86
Fluorene
 in diphenyl 11:528
 gas chromatography 11:528; 16:218
 thin-layer chromatography 11:532
Fluorescein. (See also D & C Yellow No. 7.)
 colorimetric method for bromide 8:15
 colorimetric method for halogens 11:141
 in D & C Red No. 21 10:523
 in F D & C Red No. 3 10:513
 liquid chromatography 10:514
 paper chromatography 10:514, 523
 structure 17:7
 visible spectroscopy 10:514, 523
Fluorescein methylene iminodiacetic acid. See
 Calcein.
Fluorescence
 of dental procelains 11:278
Fluorescence methods
 for boron 7:385, 410
 for optical bleaches 18:183
 for rare earth compounds 17:504
 for rare earths 15:158
 for thallium 18:526
Fluorescence spectroscopy
 of coumarins 11:162
 of menthadione 10:195
 of warfarin sodium 10:168
Fluorides. (See also *Fluorine compounds,*
 inorganic; and individual compounds.)
 acid–base titration 19:75
 in air 13:26, 29
 alizarin fluorine blue colorimetric method
  13:28
 in alkali barium silicate glass 9:218
 in barite 6:558, 562
 boric oxide colorimetric method 13:414
 calcium chloride gravimetric method 13:42
 carrier gas fusion methods 8:683
 in ceric hydroxide 9:257
 in cesium fluoride 9:289
 colorimetric determination 13:28
 emission spectroscopy 9:218; 13:14
 etching test 13:14
 in fluoroboric acid 13:42
 of fluorosulfuric acid 13:57
 fusion methods 8:634
 in glass 9:200
 in glass and glass-ceramics 13:413, 414, 415,
  433
 gravimetric determination 13:23
 in gypsum 14:98
 in hafnium 14:130
 hanging drop test 13:14
 hexafluorosilicic acid distillation method 13:16
 hydrogen in 8:681
 in inorganic fluorine compounds 13:14, 20

ion exchange chromatography 13:15, 25
ion exchange methods 19:75
lanthanum nitrate titration 13:23
lead chloronitrate gravimetric method 13:23
in minerals 13:29
neutron activation analysis 14:155
oxygen in 8:681, 683
in phosphate rocks 13:27
potentiometric methods 13:20, 24
pyrohydrolytic methods 13:18; 14:130
8-quinolinol fluorimetric method 13:15
in rocks 13:29
in silicate glasses 9:198
in silicon 18:22
silicon content 18:32
silicon tetrafluoride gasometric method 13:15,
 16
in stannous fluoride 19:75
steam distillation 13:433
thorium nitrate titration 9:289; 13:45, 434
toxicity 13:13
in tungsten 13:27
in uranium 19:244
uranium oxide pyrohydrolytic method 13:435
in urine 13:13, 26
vacuum fusion methods 8:681
in water 13:24, 30
in zircaloy 14:130
in zirconium 14:130
zirconium–eriochrome cyanine R colorimetric
 method 13:32
zirconyl chloride–alizarin red S colorimetric
 method 14:98
Fluorinated propellants
 properties 4:453
Fluorine. (See also Fluorine, elemental.) 13:1–5
 alizarin red S colorimetric method 15:227
 alizarin–zirconium oxide colorimetric method
  14:155
 in alumina 5:116
 in aluminates 5:160
 in aluminum fluoride 5:178
 in aluminum oxide 5:187
 in antimony fluoborate 6:54
 in antimony trifluoride 6:50
 in cements 9:131
 in chlorine 9:321
 in diatomaceous earth 11:493
 in fluorocarbons 13:66
 in fluorophosphoric acids 13:45
 fluosilicic acid distillation 17:507
 gas chromatography 13:64
 gravimetric determination as lead chlorofluoride
  5:178
 in halohydrins 13:154, 157
 in lime and limestone 15:226
 Megregian-Maier colorimetric method 11:493

in monofluorophosphoric acid 13:45
in rare earth fluorides 17:507
silicon tetrafluoride distillation method 15:226
in sodium fluoaluminate 5:178
SPADNS colorimetric method 11:493, 494
3-(4-sulfophenylazo) chromotropic acid colori-
    metric method 14:157
thorium nitrate volumetric method 9:322;
    17:507
tungstic oxide pyrolysis 11:494
Fluorine compounds, inorganic. (See also Fluor-
    ides; and individual compounds.) 13:6–36
analysis 13:14
emission spectroscopy 13:14
etching test 13:14
fluoride content 13:14, 20
fluorometric method 13:15
hanging drop test 13:15
identification 13:14
laboratory precautions 13:14
safety precautions 13:12
separation 13:15
specification tests 13:33
toxicity 13:12
Fluorine-containing polymers. (See also *Fluoro-*
    carbons, polymers.) 13:93–100
Fluorine, elemental
in air 13:4
analysis 13:3
commercial specifications 13:3
gas chromatography 13:4
history 13:1
impurities, determination of 13:3
manufacture 13:3
occurrence 13:2
properties 13:1
radioactive method 13:4
toxicity 13:1
Fluorine monoxide. See Oxygen difluoride.
Fluorite 8:75; 13:2
Fluoro acids, inorganic. (See also individual com-
    pounds.) 13:37–58
Fluoroapatites 17:51
in silicate rocks 17:71
Fluoroborates
toxicity 13:13
Fluoroboric acid 13:37
acid–base titration 13:39
acidimetric methods 13:39
analysis 13:39
assay 13:39
boric acid content 13:39
commercial grades and specifications 13:38
fluorosilicic acid content 13:40
gravimetric determination 13:41
hydrofluoric acid content 13:39
impurities, determination of 13:42

manufacture 13:37
Fluorocarbon elastomers. See *Fluorocarbons,*
    polymers.
Fluorocarbons 13:58–73
analysis 13:62
boiling point 13:66
boiling range 13:66
combustion methods 13:66
fluorine content 13:66
fusion methods 13:66
gas chromatography 13:62
impurities, determination of 13:67
infrared spectroscopy 13:66
manufacture 13:62
nomenclature 13:59
properties 13:59
specification tests 13:66
toxicity 13:59
uses 13:58
Fluorocarbons, polymers. (See also individual
    polymers.) 13:73–100
ablators 4:4
analysis 13:95
Mooney viscosity 13:96
physical tests 13:97
specification tests 13:95
viscosity 13:96
volatile content 13:96
Fluorocinnamic acid. (See also *Cinnamic acid and*
    its derivatives.)
properties 10:63
Fluoroelastomers. (See also *Elastomers, synthetic.*)
commercial elastomers 12:86
density 12:91
identification 12:90
1-Fluoro-2,3-epoxypropane. See Propylene
    epifluorohydrin.
Fluoroethane. (See also *Fluorocarbons.*)
properties 13:60
Fluorohydrocarbon polymers. (See also *Elasto-*
    mers, synthetic.)
identification 12:90
9-Fluoro-16-hydroxy prednisolone. (See also
    Steroids.)
trade names 14:192
Fluorometholone. (See also Steroids.)
manufacture 14:196
properties 14:196
solubility 14:196
structure 14:196
uses 14:196
Fluorometric methods
for adrenocorticoids 14:219
for aluminum 5:236
for antimycin 5:613
for beryllium 5:118
for chromomycin A_3 5:507

for dibucaine 5:387
for emetine 4:615
for epinephrine 14:220
for estrogens 14:222
for gallium 5:129
for inorganic fluorides 13:15
for LSD 15:300
for malvidin diglucoside 19:465
for morphine and codeine 4:600
for noscopine 16:272
for papaverine 16:272
for pentamycin 5:579
for pesticides and insecticides 19:611
for progesterone 14:222
for quinoline alkaloids 4:605
for reserpine and rescinnamine 4:604
for riboflavin 6:457
for rifamycin B 5:508
for steroids 14:207
for streptomycin 5:598
for thiamine hydrochloride 6:456
for thorium 19:17
for uranium 19:251
Fluorometry
  of benzaldehydes 6:625
  of benzo[a]pyrene 10:20
  of bile acids 7:152
  of codeine 16:265
  of ethinyl estradiol 10:573, 592
  of magnesium 15:370
  of mestranol 10:576
  of morphine 16:263
  of norethindrone acetate 10:592
  of tyrosine 8:539
  of uranium 14:138
Fluorophosphoric acids. (See also Difluorophos-
    phoric acid; Hexafluorophosphoric acid; and
    Monofluorophosphoric acid.) 13:42
  analysis 13:45
  fluorine content 13:45
  handling 13:43
  phosphorus content 13:45
  storage 13:43
Fluoroplast. See Polytetrafluoroethylene.
9α-Fluoro-1,4-pregnadiene-11β,16α,17α,21-tetrol-
    3,20-dione. See 9-Fluoro-16-hydroxypredni-
    solone.
1-Fluoro-2-propanol. (See also Halohydrins.)
  derivatives for identification 14:160
Fluoropyridines. (See also *Pyridine and pyridine
    derivatives.*)
  ionization constants 17:416
Fluorosilane
  properties 18:40
Fluorosilicate
  acid–base titration 13:33
  silicon content 18:32

in sodium fluoride 13:33
toxicity 13:13
Fluorosilicic acid. (See also Fluosilicic acid.)
    13:37, 47
  acid–base titration 13:40, 48
  analysis 13:48
  assay 13:48
  commercial grades and specifications 13:48
  in fluoroboric acid 13:40
  in fluorosulfuric acid 13:56
  handling 13:47
  in hydrofluoric acid 13:50
  hydrofluoric acid content 13:49
  impurities, determination of 13:49, 50
  manufacture 13:47
  properties 13:48
  silicic acid content 13:49
  specific gravity 13:48
  toxicity 13:47
Fluorosilicones. (See also *Elastomers, synthetic.*)
  identification 12:90
Fluorosulfonic acid. See Fluorosulfuric acid.
Fluorosulfuric acid 13:37, 50
  analysis 13:52
  assay 13:52
  commercial grades and specifications 13:52
  conductometric titration 13:52
  fluoride content 13:57
  handling 13:51
  impurities, determination of 13:55, 57
  manufacture 13:50
  properties 13:51
  solubility 13:51
  specific conductivity 13:54
  sulfate content 13:56
  uses 13:50
Fluorotitanates. (See also Titanium compounds.)
  manufacture 19:87
Fluoroxene
  analysis 5:363
  properties 5:357, 363
Fluorspar 13:2, 12
  toxicity 13:13
Fluosilicic acid. (See also Fluorosilicic acid;
    Silicon compounds.)
  acid–base titration 18:7
  distillation method for fluorine 17:507
  ion exchange chromatography 18:15
  manufacture 18:5
  properties 18:7
Fluosulfonic acid. See Fluorosulfuric acid.
Fluothane. See Halothane.
Fluoxymestrone. (See also Steroids.)
  manufacture 14:203
  properties 14:203
  solubility 14:203
  structure 14:204

uses 14:203
Flushes
alkaline-type 6:352, 359, 373
analysis 6:358
for cooling systems 6:351, 358, 373
performance tests 6:373
solvent-type 6:352, 361, 373
Fluxes
for ceramics 9:215
Flux fusion
of platinum group metals 17:238
FMIR
of cotton 11:124
of finishing agents 11:137
spectra of Plexiglas 16:78
Foam content
of adhesives 10:275
of glues of animal origin 13:492
Foam evaluation tests
for organosilicon copolymers 18:149
Foaming characteristics
of lubricants 15:338
Foaming tendency
of antifreezes 6:6
of automotive chemicals 6:357
of beer 7:535, 538
foam flashing method 7:537
sigma value method 7:536
Foam ratio
of glues of animal origin 15:525
Foam stability
of adhesives 15:535
of egg albumen 12:72
of glues of animal origin 15:524, 536
Foam stability loss
of glue 15:532
Foam tests
for detergents 11:390
for egg albumen 12:72
Fobbing
of beer 7:538
Folding endurance
Massachusetts Institute of Technology tester
7:289
of paper 7:287
of pulp handsheets 7:255
Schopper tester 7:287
Folic acid
in cocoa beans 9:540
in eggs and egg products 12:66
Folin-Ciocalteu
colorimetric method for bishydroxycoumarin
10:160
colorimetric method for phenols 19:462
colorimetric method for warfarin sodium
10:173
Folin Denis reagent

colorimetric method for phenols 19:463
colorimetric method for tannin 8:174
Folin's reagent
colorimetric method for chlorogenic acids
19:147
colorimetric method for phenols 19:147
Follicle-stimulating hormone. (See also Gonadotropic hormones.) 14:180
uses 14:180
Folosan. See Terraclor.
Folpet 13:210
analysis 13:210
gas chromatography 13:212
liquid chromatography 13:211
manufacture 13:210
properties 13:210
resorcinol colorimetric method 13:210
solubility 13:210
thin-layer chromatography 13:211
toxicity 13:210
uses 13:210
Fondants. (See also *Confectionery products.*)
10:548
manufacture 10:553
Food
bromide residue in 8:25
decomposition 19:328
lactic acid content 15:133, 137
lead content 15:191
phosphorus content 17:99
selenium content 17:598
strontium content 18:269
tin content 19:39, 53
vanadium content 19:328
Food additives
sulfite content 18:405
sulfurous acid content 18:403, 405, 406
Food crops
chloranil content 13:200
1-chloro-2-nitropropane content 13:186
dexon content 13:220
dibromochloropropane content 13:185
dyrene content 13:222
terraclor content 13:215
Food dyes
in cereals 9:244
Food energy content. (See also Caloric content.)
of chocolate and cocoa products 9:538
Food products
benzoic acid in 7:68
cyclohexylamine content 11:217
EDTA and related compounds in 12:44
iron content 15:48
Foodstuffs
benzylpenicillin content 16:568
ethylene oxide in 12:327
phenols in 11:541

Food yeast. (See also *Yeast*.) 19:517
  composition 19:517
  manufacture 19:517
  species 19:516
  uses 19:516, 517
Foreign particles
  in adipic acid 4:416
Formaldehyde. (See also Aldehydes.) 13:100–
    117, 194
  absorption spectroscopy 9:564
  in acetone 4:114
  acetylacetone–ammonium acetate colorimetric
    method 11:566; 13:106, 107
  acetylacetone colorimetric reagent 11:73
  acid–base titration 11:566
  acidity 8:359
  in amino resins 5:282, 287
  ammonium chloride volumetric method 13:111
  analysis 5:280; 13:105
  assay 11:566; 13:108
  chromotropic acid colorimetric method 11:134;
    12:585; 13:106, 107
  colorimetric determination 13:106, 112
  colorimetric method for epoxy resins 12:237
  commercial grades and specifications 13:105
  commercial specifications 8:340
  in cotton 11:133
  derivatives for identification 8:343; 13:107
  2,7-dihydroxynaphthalene color test 13:106
  dimedon gravimetric method 13:107
  2,4-dinitrophenylhydrazine colorimetric
    method 8:363
  2,4-dinitrophenylhydrazine hydrochloride
    gravimetric method 13:107
  in disinfectants and antiseptics 11:565
  distillation 13:102
  in embalming fluids 12:168
  gas chromatography 8:354; 11:567; 13:108,
    113; 16:108, 119, 425
  gravimetric method for flavonoid phenols
    19:466
  gravimetric method for gelatin 13:364
  gravimetric method for nonflavonoid phenols
    19:466
  gravimetric method for potassium chlorate
    12:420
  gravimetric method for silver 18:165
  Hantzsch colorimetric method 13:106, 107
  history 13:100
  hydrogen peroxide volumetric method 12:168;
    13:109
  2-hydroxycarbazole–sulfuric acid color test
    13:106
  hydroxylamine volumetric methods 13:111
  hydroxylammonium chloride volumetric
    method 16:446

  identification 13:106
  identification reagent for carbon disulfide
    8:273
  identification test using chromotropic acid
    8:344
  impurities, determination of 13:114
  infrared spectroscopy 13:108
  iodometric method 13:109
  liquid chromatography 13:503
  manufacture 8:339; 13:104, 194
  in maple syrup 11:567
  mass spectroscopy 13:108
  mercurimetric method 13:112
  in methanol 16:108, 114
  methone gravimetric method 13:107
  morphine hydrochloride–sulfuric acid color test
    13:106
  in nail enamels 11:72
  β-naphthol gravimetric method 13:107
  p-nitrophenylhydrazine hydrochloride gravi-
    metric method 13:107
  nuclear magnetic resonance spectroscopy
    13:108
  phenylhydrazine color test 13:106
  in photographic films 12:585
  polarography 13:112
  in polyamides 17:301
  polymerization 13:102
  potassium cyanide volumetric method 13:110
  potassium ferricyanide–phenylhydrazine hydro-
    chloride color test 13:106
  in production of acrolein 4:150
  properties 8:334; 13:100, 194
  Schiff's fuchsin–bisulfite colorimetric method
    13:106
  separation 13:105
  sodium sulfite volumetric method 12:168;
    13:108; 17:301
  solubility 13:101, 194
  in styrene 18:306
  Tollen's reagent color test 13:106
  toxicity 13:104, 194
  uses 8:332; 13:100, 194
  in vaccines 11:566
  in water 16:119
Formaldehyde–aniline resins. See Aniline–for-
    maldehyde resin.
Formaldehyde donor products
  in embalming fluids 12:165
Formaldehyde donors
  in photographic films 12:585
Formaldehyde, free
  in amino resin adhesives 4:399
  in amino resins 5:280
  in nitro alcohols 16:446
  in phenolic resin adhesives 4:402

Formaldehyde–melamine resin. See Melamine-
  formaldehyde resin.
Formaldehyde method
  for ammonium chloride 5:336
  for ammonium salts 5:336
  for ammonium sulfate 5:352
Formaldehyde–phenol resin. See Phenol–formaldehyde resin.
Formaldehyde polymers 13:102
Formaldehyde–sulfuric acid colorimetric method
  for alphaprodine 16:246
  for apomorphine 16:246
  for benzylmorphine 16:246
  for codeine 16:246
  for dihydrocodeine 16:246
  for dihydromorphine 16:246
  for diphenyl 11:526
  for ethers 12:300
  for ethylmorphine 16:246
  for fentanyl 16:246
  for heroin 16:246
  for hydrocodone 16:246
  for hydromorphone 16:246
  for meperidine 16:246
  for metazocine 16:246
  for metopon 16:247
  for morphine 16:247
  for nalorphine 16:247
  for narceine 16:247
  for *l*-narcotine 16:247
  for norlevophanol 16:247
  for oxycodone 16:247
  for oxymorphone 16:247
  for papaverine 16:247
  for phenazocine 16:247
  for pholcodine 16:247
  for piminodine 16:247
  for terphenyls 11:526
  for thebaine 16:247
Formaldehyde–triazine resin. See Triazine-
  formaldehyde resin.
Formaldehyde–urea resin. See Urea-formaldehyde
  resin.
Formaldoxime
  colorimetric method for manganese 15:457,
  475
Formalin. (See also *Formaldehyde.*)
  commercial specifications 12:164
  in embalming fluids 12:164
Formals. (See also *Acetals and ketals.*)
  chromotropic acid colorimetric method 4:71
  hydrolysis 4:68, 70
  identification tests 4:68
Formamide
  acid hydrolysis 8:606
  commercial specifications 8:599, 600

derivatives for identification 8:601
  properties 8:558, 597, 598
Formanilide
  infrared spectrum 5:438
  manufacture 5:426
  properties 5:422
  ultraviolet spectroscopy 5:433
  uses 5:428
Formate
  in ammonium formate 13:137
  iodometric titration 13:137
Formate esters. See Formic acid esters.
Formate salts. See Formic acid salts.
Formazin turbidity units 7:523
Formazone
  colorimetric method for niobium 16:370, 375
Formez 12:86
Formic acid 8:395; 13:117–138
  in acetic acid 4:99
  acid–base titration 13:125
  analysis 13:124
  assay 13:125
  azeotropes 13:119
  chromotropic acid color test 13:125
  commercial grades and specifications 13:122
  copper acetate-benzidine acetate color test 13:125
  density 13:119
  derivatives for identification 8:376
  gas chromatography 8:384
  handling 13:122
  history 13:117
  identification 13:124
  impurities, determination of 13:127
  infrared spectroscopy 13:119
  manufacture 13:121
  mercury(II) chloride color test 13:124
  in methanol 16:114
  N-methylquinaldinium *p*-toluenesulfonate color
    test 13:125
  molybdenum blue color test 13:125
  occurrence 13:117
  potassium permanganate volumetric method
    13:126
  properties 8:371, 396, 558; 13:118
  sampling 13:124
  shipping data 13:123
  solubility 13:118
  storage 13:122
  thin-layer chromatography 8:381; 13:125
  ultraviolet spectroscopy 13:119
  uses 8:395; 13:120
Formic acid esters 13:131–134
  analysis 13:132
  assay 13:132
  impurities, determination of 13:133
  saponification method 13:132

solubility in ethanol 13:133
in tobacco 19:148
uses 13:120
Formic acid salts. (See also individual compounds.)
analysis 13:135
uses 13:120
Formic aldehyde. See *Formaldehyde.*
Formol nitrogen
in beer 7:615
titration method 7:615
Formol titration
of amino acids 19:474
4-Formylbenzaldehyde
infrared spectroscopy 6:616
Formylpyridines. See Pyridinealdehydes.
Forsterite
composition 9:208
properties 9:204, 208, 211
in whitewares 9:191
Fortisan 4:76
Fossil flour. See *Diatomaceous earth.*
Fossil silica. See *Diatomaceous earth.*
Foundry slags
phosphorus content 17:85
Fourier spectroscopy. See *Fourier transform
spectroscopy.*
Fourier transform computers. (See also Fourier
transform spectrometers.) 13:156
Fourier transform spectra
of eugenol 13:155, 156, 158
of hydrogen chloride 13:167
of longifolene 13:167
of methyl ethyl ketone 13:154, 156
of nitric oxide 13:168
of nitrous oxide 13:174
of polystyrene 13:164, 165, 166, 168
of water 13:172, 173
Fourier transform spectrometers. (See also
Interferometers.)
advantages 13:159
Beckman 13:149, 160, 161
Block Engineering 13:168
CODERG 13:170, 171, 172
commercial instruments 13:161
components 13:147
Digilab 13:161, 163
Grubb Parsons 13:168, 169
Idealab 13:169
Fourier transform spectroscopy. (See also
Fourier transform spectrometers.) 13:139–
176
beam splitter effects 13:148
computers 13:156
history 13:139
instrumentation 13:147
mirror displacement 13:146, 150
multiple scanning 13:158

resolution 13:144
theory 13:140
Fractionation methods
for cellulose 9:40
of ethylene and propylene polymers 12:376
for plasticizers in photographic films 12:591
of polyethylene 12:376
of polypropylene 12:376
by ultracentrifugation 9:180
Fractionation, molecular weight
distribution by of acrylamide polymers 4:174
distribution by of acrylic polymers 4:298
distribution by of acrylonitrile polymers 4:389
distribution by chromatographic methods
4:301
distribution by data treatment 4:301
distribution by precipitation methods 4:298
Fradiomycin 5:475
Framycetin 5:475
Franchimont color test
for nitrate 12:420
Francolite 17:143
Franklandite 7:314
Franklinite 14:519
composition 19:545
occurrence 19:545
Fraxetin. See 6-Methoxy-7,8-dihydroxycoumarin.
Fraxinol. See 5,7-Dimethoxy-6-hydroxycoumarin.
Free-swelling index
of coal and coke 10:245
Freeze-out method
for air in aerosols 4:461
Freezing point
of beer 7:504
of benzene 7:46
of chlorophenols 9:527
of diamines and higher amines, aromatic 11:439
of dimethyl terephthalate 17:165
of diphenyl 11:527
of diphenylamine 11:458
of ethylenediamine 11:413
of hydrocarbons 14:341
of isoprene 11:509
of naphthalene 16:210
of phenylenediamines 11:446
of phthalic anhydride 17:155
of pyridine 17:439
of raw milk 16:124
of styrene 18:299
of trimellitic anhydride 17:173
Freezing point determination
of acetic acid 4:97
of acrylic acid 4:199
of adipic acid 4:412
of alkanolamines 5:14
of antifreezes 6:7
apparatus 6:8

of methacrylic acid 4:199
Freons. (See also Chlorofluorocarbons; *Fluoro-carbons.*)
  commercial specifications 13:63
  gas chromatography 13:65
Friability
  of aluminum oxide abrasive grain 7:308
  of diamonds, industrial 11:465, 478
Friction assessment screening test (F.A.S.T.)
  dynamometer 7:462
  schematic diagram 7:463
Friction coefficient
  of brake linings 7:452
  of polyamide fibers 17:325
Friction materials. (See also Brake linings.)
    7:440
Friction materials test dynamometer 7:457
  schematic diagram 7:458
Friction modifiers
  in brake linings 7:441, 447
FRN 12:86
Frolovite 7:314
Froodite
  composition 17:215
Frozen desserts. (See also Ice cream.) 16:142
  analysis 16:143
  homogenization efficiency 16:143
  milkfat content 16:143
  solids content 16:143
FR-S 12:86
Fructose. (See also Reducing sugar; *Sugar.*)
  in beer 7:566
  in cocoa beans 9:558
  enzymatic method 19:410
  in green tea leaf 18:478
  paper chromatography 7:569; 18:478; 19:142
  refractive index 18:343
  thin-layer chromatography 13:502
  in tobacco 19:142
  in wines 19:410
Fruit
  arsenic in 6:248
  cyclamate content 11:232
  cycloheximide content 13:223
  dinocap content 13:220
  diphenyl content 13:224
  dodine content 13:205
  glyodin content 13:208
  lactic acid content 15:140
  potassium content 17:363, 386
  sodium content 18:219, 220
  sulfite content 18:405
  tin content 19:40, 61
Fruit juices
  caramel content 10:538
  composition 10:531
  tin content 19:65

Frustrated multiple internal reflectance. See
    FMIR.
FSH. See Follicle-stimulating hormone.
Fuchsin
  colorimetric method for tantalum 16:370
  identification reagent for aldehydes 8:344
Fuchsine. (See also *Triphenylmethane and related dyes.*)
  color test for sulfite 18:402
  paper chromatography 19:172
  structure 19:161
  thin-layer chromatography 19:173
Fuchsine base. (See also *Triphenylmethane and related dyes.*)
  thin-layer chromatography 19:173
Fuchsine NJ. (See also *Triphenylmethane and related dyes.*)
  visible spectroscopy 19:167
Fudge. (See also *Confectionery products.*)
    10:548
  manufacture 10:554
Fuel
  potassium content 17:390
Fuel gases
  gas chromatography 8:311
Fuel oil
  sodium content 18:219
  vanadium content 19:339, 342
Fugamillin 5:472, 564
  analysis 5:569, 573, 578
  countercurrent distribution 5:569
  identification 5:569
  infrared spectrum 5:569
  microbiological assay 5:573
  properties 5:565
  separation 5:569
  solubility 5:566, 567
  structure 5:472
  ultraviolet spectroscopy 5:569, 578
  uses 5:564
Fujisaki method
  light scattering of acrylic polymers 4:312
Fujiwara colorimetric method
  for benzal chloride 9:386
  for benzotrichloride 9:386
  for bromoform 9:386
  for carbon tetrabromide 9:386
  for carbon tetrachloride 9:385, 386
  for chloral hydrate 9:386
  for chlorocarbons 9:384
  for chloroform 9:385, 386, 433
  for dichloroacetic acid 9:386
  for hexachloroethane 9:386
  for iodoform 9:386
  for methyl chloride 9:386
  for pentachloroethane 9:386
  for 1,1,2,2-tetrachloroethane 9:385, 386

for tetrachloroethylene 9:386
for tribromoethylene 9:386
for trichloroacetic acid 9:386
for 1,1,1-trichloroethane 9:386
for trichloroethylene 9:385, 386
Fuller's earth. (See also *Clays and clay minerals.*)
    10:106
  bulk density 4:445
  chemical analysis 4:439
  efficiency for percolation processes 4:449
  hardness 4:446
  pH 4:437
  uses 4:434; 10:106
  water soluble materials 4:438
Fulminate content
  of mercury fulminate 12:409
Fulvene
  colorimetric method for cyclopentadiene
    11:512
Fumaric acid 5:22; 8:399
  in alkyd resins 5:37
  derivatives for identification 8:376
  gas chromatography 5:25; 15:132; 19:294
  liquid chromatography 15:133
  paper chromatography 15:133
  preparation 8:373
  properties 8:396, 399
  thin-layer chromatography 8:381; 15:133
  uses 8:399
Fumaronitrile. (See also *Nitriles.*)
  polarography 16:398
Fumazone. See Dibromochloropropane.
Fumigants. (See also individual compounds.)
    13:176–197
  bromine content of 8:13
  detection 13:176
  residues 8:14, 25, 26, 27
  toxicity 13:176
  uses 13:176
Fuming nitric acid 16:495
Functioning properties
  of matches 15:543, 544
  of match strikers 15:548
Fungicides 13:198–227
  mercury content 16:44
  in wines 19:487
Fural. See Furfural.
2-Furaldehyde. (See also Furfural.)
  absorption spectroscopy 9:564
Furan 13:228–230
  aldehyde content of cocoa beans 9:563
  aldehydes by TBA colorimetric method
    9:563
  analysis 13:229
  assay 13:229
  in cocoa bean vapor 9:570
  commercial grades and specifications 13:229

gas chromatography 13:229
impurities, determination of 13:229, 230
manufacture 13:229
mass spectroscopy 9:570
properties 13:228
safety precautions 13:228
solubility 13:228
specific gravity 13:230
structure 13:227
toxicity 13:228
2-Furanacrylaldehyde
  absorption spectroscopy 9:564
Furan and furan derivatives. (See also Furan; and
    individual compounds.) 13:227–250
  structure 13:227
2-Furancarboxaldehyde. See Furfural.
2-Furancarboxylic acid
  properties 8:396
2-Furanmethanol. See Furfuryl alcohol.
Furan resin adhesives
  evaluation 4:401
Furfural 13:232
  acidity 13:239
  analysis 13:234
  assay 13:235
  in black tea 18:511
  bromometric titration 13:236
  colorimetric determination 13:235
  commercial grades and specifications 13:234
  derivatives for identification 13:234
  in distilled alcoholic beverages 4:488, 490
  in embalming fluids 12:165
  in furan 13:229
  in furfuryl alcohol 13:242
  gas chromatography 8:353; 13:229, 238;
    18:511
  handling 13:234
  Hughes-Acree volumetric method 13:236
  hydroxylamine hydrochloride method 8:357
  identification 13:234
  impurities, determination of 13:235, 238
  infrared spectroscopy 13:235
  manufacture 13:234
  nomograph for sodium bisulfite method
    13:237
  phloroglucinol gravimetric method 13:235
  properties 13:232
  sodium bisulfite volumetric method 13:237,
    242
  solubility 13:233
  structure 13:228
  ultraviolet spectroscopy 6:632; 13:238
  uses 13:232
Furfuraldehyde. See Furfural.
Furfuran. See Furan.
Furfuryl acetate
  gas chromatography 8:353

Furfuryl alcohol 13:240
  analysis 13:241
  assay 13:241
  cloud point 13:242
  commercial grades and specifications 13:240
  gas chromatography 8:353; 13:241, 244
  impurities, determination of 13:242
  manufacture 13:240
  properties 13:240
  solubility 13:240
  structure 13:228
  in tetrahydrofurfuryl alcohol 13:244
  ultraviolet spectroscopy 13:241
  uses 13:240
α-Furildioxime
  colorimetric method for rhenium 17:540
Furnace carbon 8:179
  commercial grades and specifications 8:196
  composition 8:191
  manufacture 8:194
  particle size distribution 8:185
  properties 8:195
  types 8:195
Furnaces
  for carrier gas and vacuum fusion systems 8:642
  degassing 8:669
Furoic acid 8:407
N-Furoylphenylhydroxylamine-thiocyanate
  colorimetric method for niobium 16:370, 375
β-2-Furylacrylophenone
  absorption spectroscopy 9:564
4-(2-Furyl)-3-buten-2-one
  absorption spectroscopy 9:564
Fused alumina. (See also Refractories.)
  composition 9:203
Fusel oil 4:462
  in alcoholic beverages 4:485, 492
  amyl alcohols preparation 4:507

analysis 4:480
in brandies 19:439
p-dimethylaminobenzaldehyde–sulfuric acid
  colorimetric method 19:439
gas chromatography 19:441
pyridine content 17:425
quality control 4:469
in wines 19:439
Fusibility
  of coal ash 10:243
Fusidic acid 5:463, 508
  analysis 5:510, 513, 517
  colorimetric methods 5:517
  identification 5:510
  Liebermann-Burchard test 5:517
  microbiological assay 5:513
  paper chromatography 5:510
  properties 5:509
  separation 5:510
  solubility 5:509, 510
  structure 5:463
  thin-layer chromatography 5:511
  uses 5:508
Fusion cast refractories. (See also Refractories.)
  composition 9:208
  microscopy 9:227
  properties 9:208
Fusion methods
  for ceramics 9:214
  for fluorocarbons 13:66
  for gold 14:40
Fusion techniques. See Carrier gas fusion; Vacuum
  fusion.
Fusion temperature
  of dental cast gold alloys 11:293
  of dental chromium–cobalt alloys 11:292
  of dental wrought gold alloys 11:295
Fyrite gas analyzer 8:261

G

Gadolinium. (See also Gadolinium metal; *Rare earths.*)
  absorption spectroscopy 17:483
  in aluminum metal 5:152
  atomic absorption spectroscopy 17:494
  bead test 6:587
  emission spectroscopy 14:123; 15:153, 155; 17:488, 501
  in europium oxide 17:502
  flame photometry 17:492
  in hafnium 14:123
  in lanthanum oxide 15:153, 155; 17:501
  in samarium oxide 17:502
  in terbium oxide 17:502
  x-ray fluorescent spectroscopy 17:490
  in yttrium oxide 17:501
  in zirconium 14:123
Gadolinium alloys
  carrier gas fusion methods 8:627, 682
  hydrogen in 8:680
  nitrogen in 8:623, 627
  oxygen in 8:623, 627, 680, 682
  vacuum fusion methods 8:623, 680
Gadolinium fluorides
  carrier gas distillation fusion methods 8:636
  oxygen in 8:635, 636
  vacuum distillation fusion methods 8:635
Gadolinium metal. (See also Rare earth metals.)
  carrier gas fusion methods 8:626, 627, 682
  hot extraction methods 8:684
  hydrogen in 8:623, 626, 638, 680, 684
  nitrogen in 8:623, 627
  oxygen in 8:623, 627, 629, 680, 682
  properties 17:471
  vacuum fusion methods 8:623, 680
  vacuum hot extraction methods 8:638
Gadolinium oxide. (See also Rare earth oxides.)
  emission spectroscopy 17:502
  impurities, determination of 17:502
  properties 17:497
  uses 17:509
Gadolinium perchlorate. (See also Rare earth compounds.)
  absorption spectrum 17:483
Galactitol
  liquid chromatography 13:503
Galactose
  anthrone colorimetric method 12:550
  gas chromatography 13:507; 19:300
  in gelatin 12:550
D-Galacturonic acid
  thin-layer chromatography 8:381

Galena. (See also Lead sulfide.) 15:166; 18:360
Gallates 13:281
Gallic acid (3,4,5-Trihydroxybenzoic acid) 8:401
  analysis 8:401
  gravimetric method for niobium 16:366
  gravimetric method for tantalum 16:366
  in green tea leaf 18:458, 475
  paper chromatography 8:379
  potassium permanganate titration 18:459
  potassium titanium oxalate colorimetric method 18:458
  properties 8:401, 402
  structure 17:2
  thin-layer chromatography 18:475
  use in preparation of gallocyanine 6:385
Gallic acid–ammonium persulfate
  colorimetric method for vanadium 19:341
Gallite 13:251
Gallium. (See also Gallium alloys; Gallium metal.) 13:251–285
  in aluminum 13:262
  in aluminum and aluminum alloys 5:129
  ammonium hydroxide gravimetric method 13:253, 257
  in antimony–aluminum–gallium alloys 13:269
  atomic absorption spectroscopy 13:260
  in bauxite 5:115
  camphoric acid gravimetric method 13:253, 258
  colorimetric determination 13:256, 262
  in copper–gallium alloys 13:270
  cupferron gravimetric method 13:253, 258
  EDTA titration 13:258, 272, 273, 445
  EDTA volumetric method 13:269, 274
  emission spectroscopy 13:256, 260, 265, 271; 14:34, 538, 15:187
  flame emission spectroscopy 13:260
  flame spectroscopy 13:270
  flame tests 6:592
  fluorometric method using sulfonaphthol–azo-resorcinol 5:129
  in gallium arsenide 13:275
  gallium ferrocyanide gravimetric method 13:253
  in gallium phosphide 13:281
  gas chromatography 13:264
  in germanium 13:395
  in glass and glass-ceramics 13:445
  in gold alloys 14:34
  in gold–gallium alloys 13:270
  gravimetric determination 13:253, 257
  in hydrochloric acid 14:374
  identification 13:256
  impurities, determination of 13:265
  in indium 13:262; 14:538

264

in indium alloys 14:555
in indium–gallium alloys 13:271
ion exchange chromatography 13:254
in iron 13:261
in lead 13:262
lumogallion fluorometric method 13:264
in magnesium alloys 15:384
in magnesium–gallium alloys 13:272
mass spectroscopy 13:265
neutron activation analysis 13:262, 265
in niobium–gallium alloys 13:272
in plutonium 13:264
in plutonium–gallium alloys 13:273
polarography 13:260
Pontachrome Blue Black R colorimetric method 13:257
potassium ferrocyanide titration 13:259
1-(2-pyridylazo)-2-naphthol colorimetric method 13:270
8-quinolinol colorimetric method 13:257, 263, 395
8-quinolinol volumetric method 13:259
Rhodamine B colorimetric method 13:256, 262
in rocks 15:187
separation from aluminum 5:210
in silicon 13:262
sodium camphorate gravimetric method 13:253, 258
in steel 13:261
tannin gravimetric method 13:253, 258
in thallium 13:262
titanous chloride–monochlorobenzene colorimetric method 14:555
in uranium–gallium alloys 13:274
volumetric determination 13:258
x-ray fluorescence spectroscopy 13:264
x-ray methods 14:556
Gallium alloys. (see also Gallium metal; and individual alloys.) 13:269–275
carrier gas fusion methods 8:682
hydrogen in 8:616, 681
oxygen in 8:616, 681, 682
vacuum fusion methods 8:616, 681
Gallium antimonide 13:280
properties 13:280
Gallium antimonide semiconductors. (See also *Ceramics.*)
properties 9:213
uses 9:213
Gallium arsenide 13:275
analysis 13:275
assay 13:275
emission spectroscopy 13:275
gallium content 13:275
impurities, determination of 13:265, 275
neutron activation analysis 13:265, 279
properties 13:275
Gallium arsenide semiconductors. (See also

Ceramics.)
properties 9:213
uses 9:213
Gallium bromide
properties 13:280
Gallium chloride
impurities, determination of 13:280
properties 13:280
Gallium complexes 13:282
Gallium compounds. (See also individual compounds.) 13:275–283
Gallium compounds, organic 13:282
properties 13:283
Gallium ferrocyanide
gravimetric method for gallium 13:253
Gallium fluoride
properties 13:280
Gallium halides 13:280
Gallium hydroxide
manufacture 13:281
solubility 13:281
Gallium iodide
properties 13:280
Gallium metal. (See also Gallium alloys.) 13:251–269
analysis 13:253
assay 13:257
commercial grades and specifications 13:253
extraction methods 13:254
history 13:251
hydrogen in 8:616, 681
manufacture 13:253
occurrence 13:251
oxygen in 8:616, 681
properties 13:252
separation 13:253
solubility in gold 14:9
uses 13:251
vacuum fusion methods 8:616, 681
Gallium nitrate 13:280
manufacture 13:280
solubility 13:280
Gallium nitride 13:280
properties 13:280
Gallium oxides 13:281
properties 13:281
solubility 13:281
Gallium perchlorate 13:280
Gallium phosphate 13:281
Gallium phosphide 13:281
gallium content 13:281
impurities, determination of 13:281
manufacture 13:281
properties 13:281
Gallium phosphide semiconductors. (See also *Ceramics.*)
properties 9:213

uses 9:213
Gallium selenates 13:282
Gallium selenides 13:282
Gallium sulfates 13:282
Gallium sulfides 13:282
    properties 13:282
Gallium tellurides 13:282
    properties 13:282
Gallocatechin
    in black tea 18:495
    in green tea leaf 18:457, 465
    liquid chromatography 18:465
    paper chromatography 18:468
    structure 18:461
    thin-layer chromatography 18:463, 470, 495, 496
    ultraviolet spectroscopy 18:465
Gallocatechin gallate
    in green tea leaf 18:457
    structure 18:461
    thin-layer chromatography 18:463
Gallocyanine 6:382, 383, 385
    manufacture 6:385
    solubility 6:383
    visible spectroscopy 6:395
    volumetric method using titanous chloride 6:396
Galloylbisepigallocatechin
    in black tea 18:493
    structure 18:493
    thin-layer chromatography 18:495, 496
3-Galloylquinic acid. See Theogallin.
Gamma-ray spectroscopy
    of radium 17:453, 456
Ganister
    composition 9:202
Gardner color scale 14:293
Gardner-Holdt method
    bubble tube 4:399
    for viscosity of hydrocarbon resins 14:295
Garnet 4:31, 33; 11:464
Garnierite
    composition 16:286
Gas absorption method
    apparatus 8:249
    for carbon dioxide 8:249, 260
Gas acetylene 4:130
Gas adsorption
    by carbon black 8:186, 214
Gas chromatography
    of acenaphthene 16:217
    of acetal 16:109
    of acetaldehyde 10:28, 31; 11:181; 12:204, 300, 339; 16:108, 109, 425; 18:510; 19:455
    of acetal polymers 4:66
    of acetals 4:66
    of acetanilide 6:187

of acetic acid 11:181; 13:127; 18:511; 19:449
of acetoin 19:449, 457
of acetone 4:113, 118; 10:28, 31; 11:73, 362; 12:204, 300, 302, 332; 16:109, 425, 432; 18:510; 19:458
of acetone–diphenylamine reaction products 6:111
of acetonitrile 10:31, 16:404, 425
of acetophenone 4:124
of acetovanillone 11:167
of acetylene 7:53; 12:338; 14:327; 16:458, 485
of acetylenic alcohols 4:578
of acetylenic chemicals 4:143
of acetylpropionyl 18:510
of acetylsalicylic acid 6:135, 138, 156, 158
of acrolein 4:153; 10:28, 31; 16:108
of acrylamide 4:163
of acrylic monomers 4:211
of acrylonitrile 4:371; 12:105; 16:404; 18:335
of acrylonitrile polymers 4:387
of active amyl alcohol 19:442
of adipic acid 17:290; 19:294
of adipic acid esters 4:426
of adiponitrile 16:404
of adrenocorticoids 14:216
of air 8:263; 12:324, 339; 14:328, 329, 393, 415; 16:425, 485
of β-alanine 17:290
of alcoholic beverages 4:490
of alcohols 4:523, 558, 566; 18:75, 77
of alcohols, lower 6:369
of aldrin 19:389
of alkaloids 4:594
of alkanolamines 5:4
of alkyd resins 5:25
of alkylbenzenes 14:333
of alkyl iodides 16:92
of alkylnaphthalenes 16:216
of allotetrahydrocortisol 14:218
of allyl alcohol 5:79
of allylamines 5:88
of allyl ethers 5:100
of allyl olefins 5:103
of allyl sulfur compounds 5:105
of alphaprodine 16:255
of aluminum 13:264
of amides 8:605
of amines 5:267; 19:305
of amino acids 8:428; 19:473
of γ-amino-n-butyric acid 17:290
of ϵ-amino-n-caproic acid 17:290
of η-amino-n-caprylic acid 17:290
of N-β-(aminoethyl)-γ-aminopropyltrimethoxy-silane 18:85
of 4-(2-aminoethyl)morpholine 16:200
of γ-amino-n-lauric acid 17:290
of N-(3-aminopropyl)cyclohexylamine 11:249

of κ-amino-*n*-undecanoic acid 17:290
of δ-amino-*n*-valeric acid 17:290
of ammonia 14:239; 16:485
of amyl acetate 11:73
of amyl alcohol 11:73
of *n*-amyl alcohol 16:107; 19:442, 446, 449
of 2-amyl furan 18:510
of anabasine 19:137
of anatabine 19:137
of 5 α-androstan-17-one 14:224
of Δ⁴-androstenedione 14:224
of androsterone 14:216
of angelicin 11:168, 170
of aniline 11:212, 213, 241, 247, 459
of aniline and its derivatives 5:436, 445
of antidegradants in elastomers 12:142
of antifreezes 6:5
of antioxidants and antiozonants 6:81, 108
of apomorphine 16:254
of arabinose 13:507; 19:300
of argon 13:297, 307; 16:485, 523
of aromatic hydrocarbons 14:331, 335, 336,
    337
of automobile exhaust gases 7:7
of azelaic acid 17:290; 19:294
of barbiturates 6:482
of benzal chloride 9:488
of benzaldehyde 18:511
of benzaldehydes 6:619
of benzene 7:7; 9:378; 10:31; 11:73, 98;
    13:350; 14:330, 333, 335, 336, 337, 375;
    16:220; 18:63, 299
of benzocaine 6:87
of benzoic acid 7:72; 17:154, 162
of benzo[a]pyrene 10:20
of benzyl alcohol 7:96, 99; 16:98; 18:511
of benzyl chloride 9:488
of bergapten 11:168, 170
of BHC 19:389
of bile acids 7:148
of biphenyl 9:495; 16:217
of bis(dipentaerythritol) monoformal 13:578,
    594
of *N*,*N*-(bis-2-hydroxyethyl)cyclohexylamine
    11:247
of bis(isocyanatocyclohexyl)methane 19:315
of bis(pentaerythritol) monoformal 13:578
of bispentaerythritol monoformal hexaacetate
    13:578
of bisphenol A 17:338
of bisphenols 11:550
of bistrichlorosilylethane 18:62
of bistrichlorosilylethylene 18:62
of black tea volatiles 18:508
of boron compounds, organic 7:353
of brake fluids 7:427
of brake linings 7:449

of bromotrifluoromethane 13:64
of butadiene 12:103, 104, 105
of 1,3-butadiene 11:505; 14:329
of butadiene dimer 11:506
of butanal 18:510
of butane 15:240
of *n*-butane 11:98; 13:317, 318, 350; 14:329
of 1,2-butanediol 13:504
of 1,3-butanediol 13:504, 576
of 1,4-butanediol 13:504, 507, 577
of 2,3-butanediol 13:504, 576, 577; 19:446,
    447
of 1,4-butanediol dipropylene glycol 13:579
of butanediols 19:298, 300, 301
of butanol 9:475; 11:73; 16:98, 107
of 2-butanol 18:510
of *n*-butanol 15:565; 19:442
of *sec*-butanol 9:475
of 2-butanone 18:510
of 1-butene 11:506; 14:329; 15:240
of 1,3-butenediol 13:576
of 2-butene-1,4-diol 13:576
of butenes 13:350
of 2-butenes 11:506; 14:329; 15:240
of butyl acetate 11:73
of *n*-butyl acetate 16:108
of *sec*-butyl acetate 16:108
of *n*-butyl alcohol 12:204
of *sec*-butyl alcohol 12:204; 16:107
of *tert*-butyl alcohol 12:204
of butylated hydroxytoluene 6:116
of butylbenzenes 13:352
of butyl benzyl phthalate 16:97
of butyl Cellosolve 11:73
of *n*-butyl formate 16:108
of *n*-butyl iodide 16:92
of *n*-butyl mercaptan 15:558
of *sec*-butyl mercaptan 15:558, 565
of *tert*-butyl mercaptan 15:557, 565
of *n*-butyl thiocyanate 15:110
of l-butyne 11:510
of 2-butyne 11:510
of *n*-butyraldehyde 12:204; 16:108, 109
of butyric acid 18:511
of γ-butyrolactone 18:511
of butyronitrile 16:404
of caffeic acid 19:146
of caffeine 6:135, 156; 8:59; 18:491
of campesterol 19:150
of caprolactam 17:297
of γ-caprolactone 18:511
of captan 13:212
of captan epoxide 13:212
of Carbitol 11:73
of carbohydrates 7:566
of carbon dioxide 12:324, 338; 13:297, 317,
    318; 14:384, 392, 416; 15:557; 16:456, 485;

19:502
of carbon dioxide–hydrocarbon mixture 8:262
of carbon disulfide 8:274
in carbon, hydrogen, and nitrogen microanalysis
   8:283
of carbon monoxide 8:310, 318, 650; 10:26;
   13:71; 14:384, 392, 415; 16:454, 485
of carbon tetrachloride 8:324; 9:374, 375, 376,
   377, 380, 381, 442
of carbon tetrafluoride 13:64
of carbonyl compounds 8:351, 366
of carbonyl sulfide 14:392
of carboxylic acid anhydrides 8:562
of carboxylic acids 5:25; 8:383
of carburetor cleaners 6:370
of caryophylene 12:266, 268
of catechin 18:471
of C_3 chlorinated hydrocarbons 9:470
of C_4 chlorinated hydrocarbons 9:474
of cetane 14:334
of *n*-cetane 14:330, 334
of 2-chlorallyl alcohol 14:175
of chlordane 19:389
of chlorinated biphenyls 9:494
of chlorinated solvents 6:367
of chlorinated toluenes 9:487
of chlorine 9:320; 13:64; 14:392, 416
of chlorine trifluoride 13:64
of chlormadinone acetate 10:577
of *p*-chloroacetanilide 6:187
of chlorobenzene 9:378; 14:375
of chlorobenzylchlorides 9:378
of chlorobiphenyls 9:495
of 2-chloro-1,3-butadiene 9:375
of 1-chlorobutane 9:376, 474, 475
of 2-chlorobutane 9:474, 475
of 1-chloro-1-butene 9:375
of 1-chloro-2-butene 9:375
of 2-chloro-2-butene 9:375
of 3-chloro-1-butene 9:375
of 1-chloro-1,1-difluoroethane 11:98
of chlorodifluoromethane 13:64
of chlorofluorocarbons 13:62
of chloroform 9:375, 376, 377, 380, 381, 433,
   442; 13:64; 14:375
of chlorogenic acids 10:401; 19:146
of 1-chloro-3-methylbutane 9:376
of 2-chloro-2-methylbutane 9:376
of 1-chloro-3-methylpropane 9:376
of 2-chloro-2-methylpropane 9:475
of 3-chloro-2-methylpropene 9:376
of chloroneb 13:218
of 1-chloro-1-nitroethane 16:432
of 1-chloro-1-nitropropane 16:432
of 1-chloro-2-nitropropane 13:186
of 2-chloro-1-nitropropane 16:432
of 2-chloro-2-nitropropane 16:427, 432

of 3-chloro-1-nitropropane 16:432
of 1-chloropentane 9:376
of *p*-chlorophenol 11:544
of chlorophenols 9:518; 17:36; 19:386
of 4-chloro-2-phenylphenol 11:544
of 6-chloro-2-phenylphenol 11:544
of chloropicrin 8:13; 13:183; 16:427
of chloroprene 9:375
of 1-chloropropane 9:376
of 2-chloropropane 9:376
of 1-chloropropene 9:376
of 2-chloropropene 9:376
of 3-chloropropene 9:376
of chlorosilanes 18:50
of chlorothene 11:73
of chlorotoluenes 9:378
of chlorotrifluoroethylene 13:64
of 1-chloro-1,2,2-trifluoroethylene 11:98
of chlorotrifluoromethane 13:64
of cholesterol 19:150
of cigarette smoke 10:30, 31
of cinnamaldehyde 10:49
of cinnamic acid 10:60
of cinnamic acid and its derivatives 10:69
of citraconic acid 17:154
of citric acid 19:149
of cocaine 16:254
of cocoa bean headspace 9:567, 569, 570
of codeine 16:254, 266
of codeine phosphate 16:269
of Compound S 14:218
of corticosteroids 14:215
of cortisol 14:216, 219
of cortisone 14:216, 219
of cortolones 14:218
of cortols 14:218
of coumarin 11:163, 166, 168
of coumarins 11:163, 167, 170
of *m*-cresol 14:580; 19:386
of cresols 6:367; 17:33
of crotonaldehyde 11:181; 16:108
of crotonitrile 16:404
of crotyl acetate 11:181
of *N*-(2-cyanoethyl)cyclohexylamine 11:249
of cyanogen 14:416
of cyanuric chloride 14:416
of cyclamates 11:235
of cyclic alcohols 11:193
of cycloheptane 13:351
of cyclohexane 11:73, 192, 212, 213; 13:319,
   350; 14:330
of cyclohexanediol 13:579
of cyclohexanol 11:192, 212, 213, 245, 249
of cyclohexanone 11:192, 206, 241
of cyclohexene 11:212, 213
of cyclohexyl acetate 11:192
of cyclohexylamine 11:212, 213, 214, 215, 245

of *N*-cyclohexylidene aniline 11:212, 241
of *N*-cyclohexylidenecyclohexylamine 11:212, 241
of cycloparaffins 14:330
of cyclopentadiene 11:510
of cyclopentane 13:350; 14:330
of cyclopentanol 11:192
of cyclopentanone 11:192
of cyclopentene 13:350
of *p*-cymene 12:266, 268
of DDE 19:389
of D-D soil fumigant 13:183
of DDT 14:376; 19:389
of *trans, trans*-2,4-decadienal 18:511
of decaffeinated coffee 10:440
of decahydronaphthalene 16:220
of decamethylenediamine 17:290
of *n*-decane 13:352; 14:333, 336
of dehydroepiandrosterone 14:216
of detergents 11:329
of deuterium 14:386
of diacetyl 12:204; 18:510; 19:457
of diamines 6:404
of diamines and higher amines, aromatic 11:439, 440, 443
of dibasic acids 4:416; 5:25
of dibromochloropropane 13:183, 185
of dibromodifluoromethane 13:64
of 1,2-dibromo-1,1,2,2-tetrafluoroethane 11:98
of 2,6-di-*tert*-butyl-*p*-cresol 18:333
of dibutyl phthalate 16:97
of dibutyl sebacate 16:97
of dicarboxylic acids 19:294
of *o*-dichlorobenzene 13:183
of *p*-dichlorobenzene 13:183
of dichlorobenzenes 9:378
of dichlorobiphenyls 9:495
of dichlorobutanes 9:474, 475
of dichlorodifluoromethane 11:89, 98; 12:324; 13:64
of dichlorodimethoxybenzenes 9:519
of 1,1-dichloroethane 9:374, 375, 376, 377, 442
of 1,2-dichloroethane 9:375, 376, 377, 380, 381, 442; 14:375
of 1,1-dichloroethylene 9:374, 375, 376, 377, 380
of 1,2-dichloroethylene 9:374, 375, 376, 377, 380, 381, 442
of dichloroethyl ether 13:183
of dichlorofluoromethane 13:64
of dichloromethoxyphenols 9:519
of dichloromonofluoromethane 11:98
of 1,1-dichloro-1-nitroethane 13:190; 16:427, 432
of 1,1-dichloro-1-nitropropane 16:432
of dichlorophene 11:88

of dichlorophenols 9:518; 17:36; 19:386
of 1,2-dichloropropane 9:375, 376, 381, 442, 472
of 2,3-dichloropropane 14:175
of 1,3-dichloropropene 9:472
of 2,3-dichloropropene 9:376
of dichlorosilane 18:51
of dichlorotetrafluoroethane 12:324
of 1,2-dichloro-1,1,2,2-tetrafluoroethane 11:98
of dichlorotoluenes 9:488
of dicloran 13:217
of dicyclohexylamine 11:212, 213, 240, 245
of dieldrin 19:389
of dierythritol 13:578
of diethylamine 11:245
of diethylbenzene 18:309
of diethylbenzenes 13:352
of *N,N*-diethylcyclohexylamine 11:245
of diethyldiethoxymethyl malonate 15:437
of diethylene glycol 12:332; 13:503, 576, 589; 19:298, 300, 301
of diethylene glycol dinitrate 13:577
of diethylenetriamine 11:421
of diethylethoxymethylene malonate 15:437
of diethylethylphenyl malonate 15:435
of di(2-ethylhexyl) sebacate 16:97
of diethyl ketone 10:31
of diethyl malonate 15:432, 437, 441
of diethylphenyl malonate 15:435
of diethyl phthalate 16:97
of diethyl succinate 8:587
of diethyl sulfate 15:433
of diethyl sulfide 15:558
of 1,2-difluoroethane 11:98
of difolatan 13:212
of difolatan epoxide 13:212
of diglycerol 13:578
of diglycerol tetraacetate 13:505
of dihydromorphine 16:255
of dihydroxycoumarins 11:168, 170
of dihydroxyketosteroids 14:218
of diisocyanates 19:313
of diisopropyl ether 12:300, 302
of dimethoxycoumarins 11:168, 170
of dimethoxyethane 16:109
of 1,1-dimethoxyethane 18:510
of 5,7-dimethoxy-6-hydroxycoumarin 11:170
of 2,2-dimethoxypropane 16:109
of 2,2-dimethylbutane 14:330
of 2,3-dimethylbutane 14:330, 335
of dimethylbutanes 13:319, 350
of dimethylbutenes 13:350
of dimethyl-*t*-butylbenzene 13:352
of dimethylchlorosilane 18:61
of dimethylcyclohexanes 13:351
of *N,N*-dimethylcyclohexylamine 11:251
of dimethylcyclopentanes 13:350; 14:330, 332

of dimethyldichlorosilane 18:51, 60, 64
of dimethyldiethyllead 13:358
of dimethyldimethoxymethyl malonate 15:437
of dimethylene glycol 13:579
of dimethylethylbenzenes 13:352
of dimethylethylpentanes 13:351
of dimethylheptanes 13:351
of 2,2-dimethylhexane 14:330, 332
of 2,4-dimethylhexane 14:332, 335
of 2,5-dimethylhexane 14:332
of 3,3-dimethylhexane 14:330
of dimethylhexanes 13:350
of *unsym*-dimethylhydrazine 14:237, 239
of dimethyl isobutyral 16:109
of dimethyl malonate 15:432, 437, 441
of dimethylmethoxymethylene malonate 15:437
of dimethylnaphthalenes 16:217
of dimethyloctanes 13:351
of 2,2-dimethylpentane 14:330
of 2,3-dimethylpentane 14:330, 332
of 2,4-dimethylpentane 14:330, 335
of 3,3-dimethylpentane 14:330
of dimethylpentanes 13:350
of dimethylpentenes 13:350
of dimethyl phthalate 16:97
of 2,2-dimethyl-1,3-propanediol 13:577
of dimethylpyridines 17:437
of dimethylsilicone fluids 18:106
of dimethyl sulfide 15:558; 18:510
of dimethyl terephthalate 17:166
of 3,5-dinitrobenzoate derivatives of alcohols 7:93, 96
of 1,5-dinitropentanediol 13:577
of 1,2-dinitropropanediol 13:577
of dioxane 11:519
of dipentaerythritol 13:578, 594
of dipentaerythritol hemiacetal 13:594
of dipentene 12:104
of diphenylamine 11:458
of diphenyldichlorosilane 18:60
of diphenylene oxide 16:218
of diphenylmethane-4,4′-diisocyanate 19:275, 315
of diphenyl oxide 11:529
of dipropylene glycol 13:576, 589; 19:298
of dithiocarbamic acid salts 11:582
of divinylbenzene 18:299, 308
of dodecamethylenediamine 17:290
of *n*-dodecane 13:352
of dodecanoic acid 17:290
of elastomers, synthetic 12:101
of endosulfan 19:389
of endrin 19:389
of engine degreasers 6:367
of engine starting fluids 6:370
of epiandrosterone 14:216

of epicatechin 18:471
of epicatechin gallate 18:472
of epichlorohydrin 14:171, 175
of epietiocholanolone 14:216
of epigallocatechin 18:471
of epigallocatechin gallate 18:472
of epitestosterone 14:224
of 1,2-epoxy-*n*-butane 12:204
of *trans*-2,3-epoxy-*n*-butane 12:204
of epoxy compounds 12:201
of β-(3,4-epoxycyclohexyl)ethyltrimethoxysilane 18:85
of 2,3-epoxy-2-methylbutane 12:204
of erythritol 13:578; 19:300
of essential oils 12:265
of esters 8:585, 586
of estradiol 14:219
of estrogens 14:215
of ethane 12:338; 13:317, 318; 14:328, 329, 384; 15:241; 16:456, 485
of ethanol 6:363; 11:73, 74, 89, 362; 12:204, 300, 302; 15:433, 437, 441; 16:107, 110, 425; 18:510; 19:436, 458
of ethanolamines 12:292
of ethers 12:300
of ethinyl estradiol 10:574, 588
of ethyl acetate 11:73; 12:300, 302; 16:108, 109; 18:510
of ethylbenzene 13:351; 14:335, 336, 337; 18:298
of 2-ethyl-1-butene 13:350
of ethyl butyrate 18:510
of ethyl Cellosolve 11:73
of ethyl chloride 9:377, 381
of ethyl cyanoacetate 15:433, 441
of ethylcyclohexane 13:351
of *N*-ethylcyclohexylamine 11:245
of ethylcyclopentanes 13:350
of ethyl decanoate 18:511
of *N*-ethyldicyclohexylamine 11:245
of ethylene 12:107, 338; 14:327, 328, 329, 384; 15:241; 16:456, 485
of ethyleneamines 11:421
of ethylene bromide 13:183, 184
of ethylene chlorohydrin 12:332; 14:165
of ethylenediamine 11:421; 14:478; 16:198
of ethylene dibromide 13:184, 185
of ethylene dichloride 13:183, 184, 185; 15:444
of 4,4′-ethylenedimorpholine 16:200
of ethylene glycol 12:332; 13:504, 507, 576, 589; 19:298, 300, 301
of ethylene oxide 11:568; 12:324, 332, 338; 13:184
of ethylenic alcohols 4:578
of ethylenimine 14:478, 480
of ethylenimine derivatives 14:493

of ethylenimine dimer 14:493
of ethylenimine trimer 14:493
of ethyl ether 12:204, 300; 16:108
of ethyl formate 16:108, 109; 18:510
of 2-ethylfuran 18:510
of ethylheptanes 13:351
of ethyl heptanoate 18:510
of ethylhexane 13:351
of ethyl hexanoate 18:510
of 2-ethyl hexanol 16:98
of 2-ethylhexene 14:334
of ethylidene chloride 16:91
of ethyl iodide 16:91, 92
of ethyl lactate 19:446, 449
of ethyl laurate 18:511
of ethyl mercaptan 15:557
of 3-ethyl-4-methylpyridine 17:437
of ethylmorphine 16:254
of ethyl myristate 18:511
of ethylnaphthalenes 16:217
of ethyl nonanoate 18:511
of ethyl octanoate 18:511
of 3-ethylpentane 14:330, 332
of ethylpentanes 13:350
of ethylpentenes 13:350
of ethylphenols 17:33
of ethylpropionate 18:510
of ethylpyridines 17:437
of ethyl thiocyanate 15:110
of ethyltoluenes 14:337; 18:298
of ethyltrichlorosilane 18:62, 66
of ethyl undecanoate 18:511
of ethyl valerate 18:510
of ethylvanillin 11:167
of p-ethylvinylbenzene 18:299
of etiocholanolone 14:216
of eucalyptol 12:266
of eugenol 11:167
of fats and fatty oils 12:484
of fatty acid methyl esters 11:364; 12:487
of fatty acids 18:187
of fatty alcohols 4:566; 11:346
of flavanols 18:471, 503
of fluorene 11:528; 16:218
of fluorine, elemental 13:4, 64
of fluorocarbons 13:62
of fluroxene 5:363
of folpet 13:212
of formaldehyde 11:567; 13:108, 113; 16:108,
   425
of freons 13:65
of fuel gases 8:311
of fumaric acid 15:132; 19:294
of furan 13:229
of furfural 13:229, 239; 18:511
of furfuryl alcohol 13:241, 244
of fusel oil 19:441

of galactose 13:507; 19:300
of gallium 13:264
of gasoline 13:320, 342, 349
of geraniol 18:511
of geranyl formate 18:511
of glucose 19:300
of glutaric acid 17:290
of glycerin 14:171
of glycerol 10:29; 13:503, 576; 19:300, 301,
   445
of glycerol dichlorohydrins 14:171, 175
of glycerol 1-monochlorohydrin 14:171
of glycerol triacetate 13:505
of glycerol trimethylsilyl ether 13:505
of γ-glycidoxypropyltrimethoxysilane 18:85
of glycine 17:290
of glycol ethers 13:577
of glycols 6:369; 13:575, 589; 19:298, 300,
   301
of glyoxal 12:204
of guaiacol 11:167
of halogenated hydrocarbons 13:182
of Halothane 5:364
of helium 13:297, 306
of heptachlor 19:389
of heptachlor epoxide 19:389
of trans,trans-2,4-heptadienal 18:511
of heptamethylenediamine 17:290
of heptanal 18:510
of n-heptane 11:73; 13:319, 350; 14:330,
   332; 16:91
of heptanes 13:317; 18:64
of 1-heptene 14:334
of 3-heptene 14:334
of heptenes 13:350
of n-heptyl p-hydroxybenzoate 7:600
of heroin 16:254, 271
in Hewlett-Packard model 185 analyzer 8:286
of hexachlorobenzene 9:378
of hexachloroethane 9:377, 380
of hexachlorophene 11:88, 550
of 1,2-hexadecanediol 13:577
of trans,trans-2,4-hexadienal 18:510
of hexamethyldisilazane 18:94
of hexamethyldisiloxane 18:94
of hexamethylenediamine 17:290
of hexanal 18:510
of n-hexane 11:73; 13:319, 350; 14:330,
   335
of 1,6-hexanediol 13:504, 576, 577; 19:298
of 2,5-hexanediol 13:576
of hexanes 13:317; 18:61, 64
of hexanoic acid 18:511
of hexanol 16:98; 18:510; 19:442, 446, 449
of trans-2-hexenal 18:510
of 1-hexene 14:334
of hexenes 13:350

of 2-hexenes 14:334
of *trans*-2-hexenoic acid 18:511
of *cis*-3-hexenoic acid 18:511
of *trans*-2-hexen-1-ol 18:510
of 3-hexen-1-ol 18:510
of *cis*-3-hexenyl acetate 18:510
of hydrazine 14:237, 239
of hydrocarbons 7:7; 14:326; 16:456
of hydrocarbons, aromatic 7:7
of hydrocarbons in ammonia 5:313
of hydrocarbons, nonaromatic 7:47
of hydrocodone 16:254
of hydrogen 8:650; 13:297; 14:383; 18:67
of hydrogenated naphthalenes 16:220
of hydrogen chloride 14:392
of hydrogen cyanide 14:415; 16:425
of hydrogen deuteride 14:386
of hydrogen fluoride 13:64
of hydrogen isomers 14:386
of hydrogen sulfide 14:393; 15:557
of hydromorphone 16:254
of 3β-hydroxy-5α-androstan-16-one 14:224
of *p*-hydroxybenzaldehyde 11:167
of *p*-hydroxybenzoates 11:545
of 6β-hydroxycortisol 14:219
of 6β-hydroxycortisone 14:219
of hydroxycoumarins 11:168, 170
of 11-hydroxyctiocholanolone 14:219
of *N*-(2-hydroxyethyl)cyclohexylamine 11:246
of hydroxymethoxycoumarins 11:168, 170
of 2(2′-hydroxy-5′-methylphenyl)benzotriazole
  16:96
of imidazole 18:94
of impurities in carbon dioxide 8:251
of impurities in carbon disulfide 8:274
of impurities in carbon monoxide 8:310
of impurities in carbon tetrachloride 8:324
of indane 13:352; 18:299
of indene 18:299
of indium 13:264
of inhibitors in 1,1,1-trichloroethane 9:442
of α-ionone 18:511
of β-ionone 18:511
of isoamyl alcohol 16:107; 18:510; 19:442
of isoamyl thiocyanate 15:110
of isobergapten 11:168, 170
of isobutane 11:98; 13:317; 318, 350; 14:329;
  15:240
of isobutene 15:240
of isobutyl acetate 16:108
of isobutyl alcohol 9:475; 12:204; 16:107;
  19:442
of isobutylbenzene 13:352
of isobutyl chloride 9:475
of isobutylene 11:506; 12:105; 13:350; 14:329
of isobutyl iodide 16:92
of isobutyl mercaptan 15:558, 565

of isobutyraldehyde 12:204; 16:109
of isobutyric acid 18:511
of isocaproic acid 18:511
of isocrotonitrile 16:404
of isocyanates 19:272, 276, 313
of isocyanic acid esters 15:110
of isoeugenol 11:167
of isomenthone 12:266
of isooctane 11:73; 14:330, 332
of isopentane 13:317, 318, 350; 14:329, 335;
  15:240
of isopentylbenzene 13:352
of isophthalic acid 17:158
of isopimpinellin 11:168, 170
of isoprene 11:509; 12:104
of isopropenylacetylene 11:510
of *p*-isopropenyltoluene 18:299
of isopropyl acetate 16:108
of isopropyl alcohol 6:363; 11:73; 12:204;
  16:107, 425
of isopropylbenzene 13:351; 14:333, 337;
  18:299
of isopropylcyclopentane 13:351
of isopropyl ether 16:108
of isopropyl iodide 16:92
of isopropyl mercaptan 15:557
of *o*-isopropyl phenol 17:33
of 4-isopropylpyridine 17:437
of isopropyl thiocyanate 15:110
of isovaleraldehyde 18:510
of isovaleric acid 18:511
of isovaleronitrile 18:510
of isovanillin 11:167
of *cis*-jasmone 18:511
of Kelthane 19:389
of ketals 4:66
of 11-ketoctiocholanolone 14:219
of krypton 13:297, 307
of lactic acid 15:132
of lactic acid esters 15:136
of lead additives 13:357
of lemon oil 12:267
of levorphanol 16:255
of limonene 12:266, 268; 18:510
of linalool 18:511
of linalool oxide 18:511
of lindane 19:389
of liquefied petroleum gas 15:239
of LSD 15:298
of majoram, oil of 12:268
of maleic anhydride 17:154
of malic acid 19:149
of malonates, substituted 15:435
of malonic acid 15:132
of malonic acid esters 15:433
of malononitrile 15:443
of mannitol 13:594; 19:451

of mannose 19:300
of menadione 10:200
of menthofurane 12:266
of menthol 10:21; 12:266
of menthone 12:266
of meperidine 16:254
of mercaptans 15:257, 557, 565
of γ-mercaptopropyltrimethoxysilane 18:85
of mesitylene 16:220
of mestranol 10:576, 577
of methacrylate polymers 16:87
of methacrylonitrile 16:404
of γ-methacryloxypropyltrimethoxysilane
    18:85
of γ-methacryloxypropyl-tris-(2-methoxyethoxy)-
    silane 18:85
of methadone 16:254
of methane 12:339; 13:307, 316, 317, 318;
    14:327, 328, 384, 416; 15:241; 16:425, 454
    457, 485
of methanol 10:31; 11:181; 12:204; 13:576;
    15:433, 437, 441; 16:107, 108; 17:40;
    18:510; 19:439
of methoxychlor 19:389
of 7-methoxycoumarin 11:168, 170
of 6-methoxy-7,8-dihydroxycoumarin 11:170
of methoxyflurane 5:364
of 6-methoxy-7-hydroxycoumarin 11:168, 170
of methyl acetate 11:181; 12:300; 16:108, 109
of methylacetylene 11:510
of methylal 16:109
of methyl bromide 8:13; 13:183, 184
of 2-methyl-1,3-butadiene 13:350
of 2-methyl butanal 18:510
of 2-methylbutane 14:330
of 3-methyl-1-butene 14:329
of methylbutenes 13:350
of 2-methyl-3-buten-2-ol 18:510
of methyl-t-butylbenzenes 13:352
of 2-methylbutyric acid 18:511
of α-methylbutyronitrile 18:510
of methyl caprate 12:488
of methyl caprylate 12:488
of methyl Cellosolve 11:73
of methyl chloride 9:381, 424; 18:61, 64
of methylchlorosilanes 18:51, 60
of methyl cyanoacetate 15:433, 441
of methylcyclohexane 13:350; 14:330, 332
of methylcyclopentane 13:319, 350; 14:330
of methylcyclopentenes 13:350
of methyldichlorosilane 18:60, 64
of 2-methyl-1,3-dioxolane 11:522
of methyldiphenyls 11:528
of methylene chloride 9:374, 375, 376, 377,
    381, 428, 442; 11:98; 16:91
of methylethylbenzenes 13:351
of methylethylcyclopentanes 13:351

of methylethylhexanes 13:351
of methyl ethyl ketone 10:31; 11:73; 12:204;
    16:432
of methylethylpentanes 13:319, 351
of 2-methyl-5-ethylpyridine 17:437
of methyl ethyl sulfide 15:558
of methyl formate 11:181; 16:108
of methyl furan 13:229
of methylfurfural 13:239
of 5-methylfurfural 18:511
of methylheptanes 13:351
of 2-methyl-2-hepten-6-one 18:510
of 2-methylhexane 14:330, 332
of 3-methylhexane 14:330, 332
of methylhexanes 13:319, 350
of methylhexenes 13:350
of methylhydrazine 14:237, 239
of methylindanes 13:352
of methyl iodide 16:91, 92
of methyl isobutyl ketone 11:73; 16:432
of methyl isobutyrate 16:109
of methylisopropylbenzenes 13:352
of methyl isopropyl ketone 12:204
of methyl laurate 12:488
of methyl linoleate 12:488
of methyl linolenate 12:488
of methyl mercaptan 15:557
of methyl myristate 12:488; 15:136
of methylnaphthalenes 11:528; 16:217
of methylnonanes 13:351
of methyloctanes 13:351
of methyl oleate 12:488
of methyl palmitate 12:488, 15:136
of methyl palmitoleate 12:488
of methylparaben 14:581
of 3-methylpentadecane 14:334
of 5-methylpentadecane 14:334
of 7-methylpentadecane 14:334
of 2-methylpentane 14:330
of 3-methylpentane 14:330
of methylpentanes 13:350
of methylpentenes 13:350
of 4-methyl-3-penten-2-one 18:510
of 4-methyl-4-penten-2-one 18:510
of 2-methylpropanal 18:510
of 2-methyl-1-propanol 18:510
of methyl propionate 16:108, 109, 425
of methyl-n-propylbenzenes 13:352
of methyl salicylate 18:511
of methyl stearate 12:488; 15:136
of α-methylstyrene 13:65; 18:298
of β-methylstyrene 18:299
of methyl thiocyanate 15:110
of methyltrichlorosilane 18:60, 64
of methyltriethoxysilane 18:71
of methyltrimethoxysilane 18:75
of metopon 16:255

of monochlorodifluoromethane 11:98
of monochlorophenols 9:518
of monochlorosilane 18:51
of monochlorotoluenes 9:488
of monoethylene glycol dinitrate 13:577
of morphine 16:254, 266
of morpholine 16:197, 200
of 4-morpholinecarboxaldehyde 16:200
of 4-morpholineethanol 16:200
of 2-(2-morpholinoethylamino)ethanol 16:200
of myosmine 19:137
of myrcene 12:266, 268
of nalorphine 16:254
of naphthalene 11:528; 13:352; 16:209, 217, 220; 17:154
of 1,4-naphthoquinone 17:154
of narcotics 16:253
of natural gas 13:315
of neohexane 13:317, 319
of neon 13:297, 307
of neopentane 13:350
of neopentyl glycol 13:579; 19:298, 300, 301
of nerol 18:511
of nerolidol 18:511
of nicotine 19:137
of nitric oxide 14:416; 16:485
of nitriles 16:403
of nitro alcohols 16:441
of 1-nitrobutane 16:425
of 2-nitrobutane 16:425
of 2-nitro-1-butanol 16:444
of nitroethane 16:425
of 2-nitro-1-ethanol 16:444
of 2-nitro-2-ethyl-1,3-propanediol 16:444
of nitrogen 8:650; 13:71, 297, 307, 316, 318; 14:384, 392, 416; 16:454, 485, 523
of nitrogen dioxide 16:485
of nitrogen oxides 16:485
of nitrogen trifluoride 13:64
of nitromethane 13:65; 16:425
of 1-nitro-2-methylpropane 16:425
of 2-nitro-2-methylpropane 16:425
of 2-nitro-2-methyl-1,3-propanediol 16:444
of 2-nitro-2-methyl-1-propanol 16:444
of nitroparaffins 16:422, 425, 431
of 1-nitropropane 16:425
of 2-nitropropane 16:425, 432
of 1-nitro-2-propanol 16:444
of 2-nitro-1-propanol 16:444
of nitrous oxide 13:65; 16:485
of noble gases 13:292, 294, 306; 14:385
of nonamethylenediamine 17:290
of *n*-nonane 13:351; 14:336
of 1-nonanol 18:511
of nonflavonoid phenols 19:468
of norethindrone 10:577
of normorphine 16:255

of nornicotine 19:137
of 19-nortestosterone 14:224
of ocimene 18:510
of *trans,trans*-3,5-octadien-2-one 18:511
of octafluorocyclobutane 11:98
of octamethylenediamine 17:290
of octanal 18:510
of *n*-octane 13:351
of octanes 13:319
of 1-octanol 18:511
of *trans*-2-octenal 18:511
of 1-octene 14:334
of 2-octenes 14:334
of 1-octen-3-ol 18:511
of octylene glycol 13:576
of olefins 14:329, 331, 334
of organic impurities in hydrochloric acid 9:351
of organochlorine pesticides 19:157, 389
of organo-chlorosilanes 18:59
of organofunctional silanes 18:83
of organophosphorus pesticides 19:157
of organosilane esters 18:73
of organosilicon copolymers 18:147
of osthol 11:168, 170
of oxalic acid 15:132; 19:149
of oxosteroids 14:215
of oxycodone 16:254
of oxygen 8:650; 13:71, 297, 307, 316; 14:384, 392, 416; 16:454, 470, 485, 523
of oxymorphone 16:255
of palmitoyl dilactoyl lactylate 15:136
of palmitoyl lactoyl lactylate 15:136
of palmitoyl lactylate 15:136
of palmitoyl tetralactoyl lactylate 15:136
of palmitoyl trilactoyl lactylate 15:136
of papaverine 16:254
of paraffins 14:329, 330, 331, 335, 336
of paraldehyde 11:181
of penicillins 16:559
of pentachlorobenzene 9:378
of pentachloroethane 9:377, 380
of pentachlorophenol 9:520
of pentadiene 10:31
of pentaerythritol 13:578, 594; 19:300, 301
of pentaerythritol cyclic monoformal 13:578, 594
of pentaerythritol cyclic monoformal diacetate 13:578
of pentaerythritol dicyclic diformal 13:578
of pentaerythritol–dipentaerythritol monoformal 13:578
of pentaethylenehexamine 11:421
of pentamethylenediamine 17:290
of pentamethylene glycol 13:576
of *n*-pentane 13:183, 317, 318, 350; 14:329, 330, 335; 15:240
of 1,2-pentanediol 13:244

of β-sitosterol 16:139; 19:150
of sodium borohydride 14:253
of solvents 6:361, 363, 364, 365
of sorbitol 19:451
of sphondin 11:168, 170
of stearic acid 18:325
of stearoyl dilactoyl lactylate 15:136
of stearoyl lactoyl lactylate 15:136
of stearoyl lactylate 15:136
of stearoyl trilactoyl lactylate 15:136
of steroids 7:148; 14:213
of stigmasterol 19:150
of Strobane 19:389
of styrene 11:506; 12:104; 18:297, 311, 327, 335
of styrene dimers 18:309
of styrene trimers 18:309
of suberic acid 17:290
of succinic acid 15:132; 17:290; 19:296
of succinonitrile 16:404
of sulfur dioxide 14:392
of sulfur hexafluoride 13:64
of sulfuryl fluoride 13:196
of Sulphenone 19:389
of syringaldehyde 11:167
of TDE 19:389
of terephthalic acid 17:158
of terpenes 14:333
of terphenyls 11:528, 535
of α-terpineol 18:511
of testosterone 14:219, 224
of tetrachlorobenzenes 9:378
of tetrachlorobisphenol A 17:338
of 1,1,1,2-tetrachloroethane 9:377, 380
of 1,1,2,2-tetrachloroethane 9:375, 377, 381
of tetrachloroethylene 9:375, 376, 377, 380, 381, 442
of tetrachlorophenols 9:519
of tetrachlorosilane 18:51, 60, 64
of 1,2-tetradecanediol 13:577
of tetraerythritol 13:578
of tetraethoxysilane 18:75
of tetraethylene glycol 13:577, 589
of tetraethylenepentamine 11:421
of tetraethyllead 13:358
of tetrahydrocortisone 14:218
of tetrahydrofuran 13:229, 231; 16:109
of tetrahydrofurfuryl alcohol 13:243
of 1,2,3,4-tetrahydronaphthalene 16:220
of tetrahydrophthalimide 13:212
of tetralin 13:352
of tetramethylbenzenes 13:352
of 2,2,3,3-tetramethylbutane 14:330
of tetramethylcyclopentane 13:351
of tetramethyldichlorodisilane 18:61
of tetramethylenediamine 17:290

of tetramethylene glycol 13:576
of tetramethylhexanes 13:351
of tetramethyllead 13:358
of tetrapentaerythritol 13:594
of theaflavin 18:503
of theaflavin-3,3'-digallate 18:503
of theaflavin gallate 18:503
of thebaine 16:254
of thiophene 18:538
of thiophene derivatives 18:538
of thorium 19:17
of threitol 13:507; 19:300
of tobacco humectants 13:503
of toluene 10:31; 11:73; 13:128, 351; 14:332, 335, 336, 337; 16:432; 18:75, 299
of toluenediamines 11:451
of toluenediisocyanate 19:313
of o-toluic acid 17:154
of p-toluic acid 17:162
of tolylene diisocyanates 15:110
of tolylisocyanates 15:110
of toxaphene 19:389
of 2,4,5-trichloroanisole 9:519
of trichlorobenzenes 9:378
of 1,1,1-trichloroethane 9:374, 375, 376, 377, 381, 442
of 1,1,2-trichloroethane 9:375, 376, 377, 380, 381; 16:92
of trichloroethylene 9:374, 375, 376, 377, 380, 381
of trichlorofluoromethane 11:89, 98; 12:324; 13:64
of trichlorophenols 9:519
of 1,2,3-trichloropropane 9:381; 14:175
of trichlorosilane 18:51, 60
of 1,1,2-trichloro-1,2,2-trifluoromethane 11:98
of trierythritol 13:578
of triethylamine 11:245
of triethylene glycol 13:576, 579, 589; 19:300
of triethylene glycol dinitrate 13:577
of triethylenetetraamine 11:421
of triethylmethyllead 13:358
of trifluoromethane 13:64
of 4,5,7-trihydroxycoumarin 11:170
of 2,4,6-triisocyanato toluene 15:110
of trimethylbenzenes 13:351; 14:333, 337
of 2,2,3-trimethylbutane 13:350; 14:330
of 2,3,3-trimethyl-1-butene 13:350
of trimethylchlorosilane 18:51, 60, 64
of trimethylcyclohexane 13:351
of 2,2,6-trimethylcyclohexanone 18:511
of 1,1,3-trimethylcyclopentane 14:332
of trimethylcyclopentanes 13:350
of trimethylenediamine 17:290
of trimethylene glycols 13:576
of trimethylethyllead 13:358

of trimethylheptanes 13:351
of trimethyl hexanes 13:351
of 3,5,5-trimethyl hexanol 16:98
of trimethylnaphthalene 16:218
of trimethylolethane 13:579; 19:300, 301
of trimethylolpropane 13:579; 19:300, 301
of 2,2,3-trimethylpentane 14:332
of 2,3,3-trimethylpentane 14:332
of 2,3,4-trimethylpentane 14:332
of trimethylpentanes 13:350
of trimethylpentenes 14:334
of trimethylphenols 17:33, 35
of 2,4,6-trimethylpyridine 17:437
of N-trimethylsilylacetamide 18:94
of N-trimethylsilylimidazole 18:94
of tripentaerythritol 13:578, 594
of triphenylmethane 9:520
of tripropylene glycol 13:577, 589; 19:298
of tris(hydroxymethyl)nitromethane 16:444
of tris(pentaerythritol) diformal 13:578, 594
of n-undecane 13:352
of valeraldehyde 18:510
of valeric acid 18:511
of vanillin 11:67; 19:349
of vegetable fat 16:139
of veratraldehyde 11:167
of vinyl acetate 8:587; 11:181
of vinyl chloride 9:374, 375, 377, 380, 381;
   11:98; 13:65
of vinyl cyclohexane 12:103, 105
of vinylidene chloride 9:442
of vinyl propionate 11:181
of vinyltoluenes 18:299
of vinyltriacetoxysilane 18:85
of vinyltrichlorosilane 18:62, 66
of vinyltriethoxysilane 18:85
of vinyltrimethoxysilane 18:85
of vinyl-tris(2-methoxyethoxy)silane 18:85
of warfarin sodium 10:169
of water 10:14; 11:215, 441; 12:339; 13:68,
   129, 576; 14:171, 175, 237, 239, 385, 393,
   415, 480; 15:252; 16:108, 425, 485; 17:40;
   19:458
of windshield de-icers 6:370
of windshield washer solvents 6:366
of xanthotoxin 11:168, 170
of xanthyletin 11:170
of xenon 13:297, 308
of xylene 11:73; 16:92, 220, 432; 17:154
of o-xylene 13:183
of xylenes 13:351; 14:333, 335, 336, 337;
   18:298
of xylenols 17:33
of xylose 19:300
Gases
   in glass and glass-ceramics 13:478

Gases, acidic
   in hydrogen 14:381
Gases, noble. (See also individual gases.)
   13:286–309
   analysis 13:291
   assay 13:293
   emission spectroscopy 13:292
   gas chromatography 13:292, 294, 306;
      14:385
   history 13:286
   in hydrogen 14:385
   identification 13:291
   impurities, determination of 13:293
   infrared spectroscopy 13:294
   manufacture 13:290
   mass spectroscopy 13:292, 301, 304
   in nitrogen 16:470
   occurrence 13:286, 290
   properties 13:287
   solubility 13:288
   toxicity 13:290
   uses 13:286
Gas, natural 13:310–339
   analysis 13:312
   calorific gravity 13:331
   composition 13:310
   composition, determination of 13:314
   ethane content 13:320
   gas chromatography 13:315
   impurities, determination of 13:320
   infrared spectroscopy 13:315
   mass spectroscopy 13:315
   measurement 13:312
   methane content 13:320
   occurrence 13:311
   production 13:311
   propane content 13:320
   properties 13:311
   properties, determination of 13:329
   sampling 13:312
   specific gravity 13:329
   toxicity 13:311
   transmission 13:311
   utilization 13:312
   volumetric–chemical determination 13:314
Gasoline 13:339–361
   additives 13:359
   API gravity 13:360
   aromatic hydrocarbon content 14:335
   charcoal adsorption 13:320
   commercial grades and specifications 13:340,
      353
   distillation curve 13:340
   gas chromatography 13:320, 342, 349
   hydrocarbon composition 13:345
   impurities, determination of 13:359

isopropanol content 16:119
lead additives 13:352
lead content 13:355; 15:184, 189, 194
liquid chromatography 13:347
manufacture 13:339
mercaptan content 15:560
methanol content 16:19
in natural gas 13:320
octane number 13:344
properties 13:340
properties, determination of 13:340
Reed vapor pressure test 13:341
sulfur content 18:374
test car method 13:320
tetraethyllead content 15:184, 185, 194
tetramethyllead content 15:185
vapor pressure 13:340
Gasoline additive AK 33X
analysis 15:493
composition 15:490
manganese content 15:493
Gasometric methods
for carbon dioxide 12:325; 15:220; 19:501
for ethylene oxide 12:325
for hydrogen in manganese 15:482
for inert gas 12:325
for nitrate 16:506
Gas sampling apparatus 5:299
Gas sampling tubes 5:300, 301
Gastric juices
hydrochloric acid content 14:367
Gas turbine fuel
sodium content 18:222
vanadium content 19:339
Gauging plaster content
of white coat plaster 14:101
Gauze
iodoform content 14:614
Gay-Lussac method
volumetric method for silver 18:167
Gehaltemeter
for carbon dioxide 7:571
Gehlenite 8:75
Gelatin 12:546; 13:361–375
analysis 12:547, 568; 13:363
assay 13:365
Bloom ratings 13:368
commercial grades and specifications 13:362
in confectionery products 10:551
in cottage cheese 13:364
emulsion performance 12:555
formaldehyde gravimetric method 13:364
gel strength 13:368
gravimetric determination 12:581; 13:363
gravimetric method for silica 18:28
hydroxyproline content 13:365, 372
in ice cream 13:363

identification 13:363
impurities, determination of 12:547; 13:369
infrared spectroscopy 12:583
jelly strength 13:368
manufacture 12:546; 13:361, 362
in meat extracts 13:364
in milk 13:363
Millon reagent gravimetric test 10:435
moisture content 12:547
nitrogen content 12:569; 13:367
occurrence 13:362
in paper 13:364
potassium dichromate gravimetric method
13:363
properties 12:546; 13:362
properties, determination of 13:368
in roasted coffee 10:435
separation 13:364
silver halide sol test 12:555
sodium tungstate–sulfuric acid gravimetric
method 13:364
solubility 13:362
tannic acid gravimetric method 13:364
tannic acid turbidimetric method 13:363
trinitrophenol gravimetric method 13:363
turbidity diagrams 12:556
ultraviolet spectroscopy 12:553
uses 12:546; 13:361, 365
Gelatin coating weight
of photographic films 12:568
Gelatins, chemically modified
analysis 12:570
Gelatin substitutes
in photographic films 12:558
Gelation temperature
of dental impression materials, agar-based
11:279
Gel characteristics
of adhesives 10:273
Gel content
of elastomers, synthetic 12:131
of impact polystyrene 18:329
of polyamide fibers 17:320
of SBR 12:132
Gel formers
in confectionery products 10:551
Gelometers 10:272
Gel permeation chromatography 4:298, 389
of cellulose 9:41
of dimethylsilicone fluids 18:111
of elastomers, synthetic 12:119
of ethylene and propylene polymers 12:379
of polybutadienes 12:119
of polycarbonates 17:346
of polyisoprenes 12:119
of polyurethanes 19:290
of rubber, natural 12:119

of SBR 12:119
of silicone fluids 18:110
of styrene polymers 18:323
Gel strength
  of corn starch 11:25
  of gelatin 13:368
  of glues of animal origin 13:492
  Hjermstad method 11:26
  Saare and Martens method 11:27
Gel time
  of adhesives 10:274
  of phenols 17:45
Generator oil. (See also *Lubricants*.)
  commercial grades and specifications 15:312
Gentamycin 5:477, 599
  analysis 5:605, 610, 614
  identification 5:605
  microbiological assay 5:610
  paper chromatography 5:605
  properties 5:600
  separation 5:605
  solubility 5:601, 602
  uses 5:599
Gentamycin sulfate
  infrared spectrum 5:604
Genthane 12:86
Gentisinaldehyde. See 2,5-Dihydroxybenzalde-
  hyde.
Gentro 12:86
Geological samples
  gold content 14:26
Geothite
  composition 10:104
Geraniol
  in black tea 18:507, 511
  gas chromatography 18:511
Geranium oil. (See also *Essential oils*.)
  menthone content 12:282
  refractive index 12:271
Geranyl acetate
  in essential oils 12:276
  molecular weight 12:276
Geranyl benzoate. (See also Benzoic acid
  esters.)
  commercial grades and specifications 7:79
  3,5-dinitrobenzoate derivative 7:83
  properties 7:77
  saponification method 7:84
Geranyl formate. (See also Formic acid esters.)
  in black tea 18:511
  commercial specifications 13:132
  in essential oils 12:276
  gas chromatography 18:511
  molecular weight 12:276
  properties 13:131
  solubility 13:131, 134
Geranyl valerate

  in essential oils 12:276
  molecular weight 12:276
Gerber test
  for milkfat in raw milk 16:127
Germanite 13:251
  occurrence 13:376
Germanium. (See also Germanium alloys; Germanium, elemental.) 13:375–402
  ammonium molybdate–acetone colorimetric
    method 13:401
  colorimetric determination 13:386
  distillation methods 13:381, 383
  emission spectrography 13:385; 14:34, 538
  in gallium phosphide 13:282
  in germanium alloys 13:398
  in germanium dioxide 13:400
  in germanium–gold alloys 13:399
  in gold alloys 14:34
  gravimetric determination 13:382, 398
  hematein colorimetric method 13:379
  hydrogen sulfide gravimetric method 13:382
  identification 13:379
  impurities, determination of 13:389
  in indium 14:538
  in indium alloys 14:557
  iodometric titration 13:384
  ion exchange chromatography 13:381
  iron(III) hydroxide gravimetric method 13:381
  mannitol–germanic acid volumetric method
    13:399
  manufacture 13:376
  neutron activation analysis 13:265
  nitric–hydrofluoric–sulfuric acid gravimetric
    method 13:398
  occurrence 13:375
  in organogermanium compounds 13:400
  oxidized hematoxylin colorimetric method
    13:388
  phenylfluorone colorimetric method 13:386
  solvent extraction 13:380
  separation 13:380
  uses 13:376
  volumetric determination 13:383, 399
  x-ray methods 14:557
Germanium alloys 13:398
  carrier gas fusion methods 8:682
  germanium content 13:398
  oxygen in 8:682
  uses 13:377
Germanium compounds. (See also individual
  compounds.) 13:399
  uses 13:377
Germanium dioxide 13:399
  analysis 13:399
  impurities, determination of 13:389, 399
  properties 13:399
  specification tests 13:399

uses 13:399
Germanium, elemental. (See also Germanium
    alloys.)
    analysis 13:378
    commercial grades and specifications 13:378
    history 13:375
    impurities, determination of 13:378
    isotopes 13:377
    properties 13:377
Germanium–gold alloys
    germanium content 13:399
Germanium metal
    hydrogen in 8:680, 681
    oxygen in 8:616, 680, 681
    vacuum fusion methods 8:616, 680, 681
Germanium semiconductors. (See also *Ceramics*.)
    properties 9:213
    uses 9:213
Germanium tetrachloride 13:400
    properties 13:400; 19:96
    uses 13:400
    vapor pressure 13:400
German silver 5:68
Germicides. See Disinfectants.
Gersdorffite
    composition 16:286
Gibberellins
    bioassay 7:182
Gibbsite
    composition 10:104
Gibbs reagent
    colorimetric method for phenols 17:29
    color test for milk 16:133
Gibson method
    determination of end groups in acrylic polymers
        4:295
Gillmore time-of-set test
    for cements 9:136
Gilsonite 6:274
Gin 4:465
    analysis 4:480
    quality control 4:470
Ginger root
    cooling system sealer 6:352
Ginorite 7:314
Glacial acetic acid. See *Acetic acid.*
Glass. (See also *Ceramics; Glass and glass-
    ceramics*.) 9:197
    acid dissolution 9:214
    boron in 7:394, 396, 400, 411, 413, 417
    carrier gas fusion methods 8:626, 683
    cerium content 9:260
    color 9:200
    commercial grades and specifications 9:198
    composition 9:198
    in explosive priming mixtures 12:417, 421, 424,
        425, 427

fluxes for 9:216
    impurities in 9:200
    microscopy 9:224; 12:417, 424
    oxygen in 8:626, 683
    refractive index 12:417
    silicon content 18:30
    sodium carbonate fusion 7:400
    solubility 12:421
    uses 9:198
    x-ray diffraction 9:221
Glass and glass-ceramics 13:403–487
    aluminum oxide content 13:413, 415, 418, 445,
        446
    analysis 13:412
    antimony trioxide content 13:413, 415, 428
    arsenic trioxide content 13:413, 415, 428
    atomic absorption spectroscopy 13:450
    barium oxide content 13:413, 415, 418
    boric oxide content 13:413, 415, 417
    bromide content 13:413, 415, 435
    calcium oxide content 13:413, 415, 418, 445,
        447
    ceric oxide content 13:415, 442
    chelometry 13:444
    chloride content 13:413, 415, 435
    chromic oxide content 13:413, 415, 440
    cobalt oxide content 13:413, 415, 441
    color 13:408
    composition 13:407, 409
    composition, determination of 13:413, 414
    constitution 13:405
    coulometry 13:458
    decomposition 13:415
    definition 13:403
    density 13:473
    electron microprobe methods 13:474
    electron microscopy 13:474
    emission spectroscopy 13:453
    etchants 13:472
    flame emission spectroscopy 13:449
    fluoride content 13:413, 414, 415, 433
    history 13:403
    impurities, determination of 13:439, 478
    infrared spectroscopy 13:459
    iodide content 13:413, 415, 435
    iron oxide content 13:413, 415, 439
    lead oxide content 13:413, 415, 418
    lithium oxide content 13:413, 415, 423
    magnesium oxide content 13:413, 415, 418,
        445, 447
    manganese oxide content 13:415, 442
    manufacture 13:406
    mass spectroscopy 13:456
    metal content 13:444
    microscopy 13:471
    nickel oxide content 13:413, 415, 440
    occurrence 13:403

phosphorous pentoxide content 13:413, 415, 432
polarography 13:457
potassium oxide content 13:413, 415, 423
properties 13:403, 467
properties, determination of 13:459
redox state determination 13:443
refractive index 13:473, 475
sample preparation 13:413
sampling 13:413
selenium content 13:413, 415, 439
silica content 13:413, 415, 416
sodium oxide content 13:413, 415, 423
spark source spectroscopy 13:456
spectroscopy 13:459
sulfur content 13:415, 436
thermal expansion 13:404
titanium dioxide content 13:413, 415, 425
uses 13:467
viscosity 13:404
visible spectroscopy 13:461
x-ray diffraction 13:465
x-ray emission spectroscopy 13:454
zinc oxide content 13:413, 415, 418, 445, 448
zirconium dioxide content 13:413, 415, 427
Glass-ceramics. (See also *Ceramics; Glass and glass-ceramics*.) 9:197, 200
acid dissolution 9:215
commercial grades and specifications 9:201
composition 9:199; 13:407, 409, 412
electron microscopy 9:228
fluxes for 9:216
manufacture 9:200; 13:411
microscopy 9:224
properties 9:201
uses 9:199, 201
x-ray diffraction 9:221
Glass, high-silica
quality control tests 4:25
silica content of 4:26
Glass sands
bismuth in 7:216
Glass-to-metal seals. (See also Cobalt alloys.) 10:290
composition 10:290
Glass transition temperature
of elastomers, synthetic 12:132
of ethylene–butene copolymers 12:384
of ethylene–propylene copolymers 12:133, 384
of ethylene–vinyl acetate copolymers 12:384
of NBR 12:133
of polyamides 17:283
of polybutadienes 12:133
of polychloroprene 12:133
of polydimethylsiloxane 12:133
of polyisobutene 12:133
of polyisoprene 12:133
of polymers 4:261
of polypropylene 12:383
of SBR 12:133
Glassy carbon. (See also *Graphite and related carbons*.) 14:59
manufacture 14:59
properties 14:59
Glatwasser 7:476
Glauber's salt. (See also Sodium sulfate.) 18:243
Glauconite. (See also *Clays and clay minerals*.)
composition 10:102
Glaze ceramics. (See also *Ceramics*.)
electron microscopy 9:226
microscopy 9:226
Glazes. (See also *Ceramics*.) 9:197
acid dissolution 9:214
composition 9:196
in confectionery products 10:552
fluxes for 9:216
Globulin
in egg white 12:58, 61
liquid chromatography 12:60
properties 12:59
Gloss
of enamels 12:190
Glucagon. (See also Pancreatic hormones.) 14:182
analysis 14:206
bioassay 14:206
composition 14:182
occurrence 14:189
properties 14:189
structure 14:182
uses 14:182
Glucitol
liquid chromatography 13:503
Gluconic acid (Glyconic acid) 8:398
in beer 7:585
properties 8:396, 398
Glucose. (See also Reducing sugars; *Sugar.*)
in beer 7:566
in cocoa beans 9:558
colorimetric method using anthrone 7:557
in egg yolk 12:58
enzymatic method 19:409
Fehling's reagent color test 11:127, 132
in finishing agents 11:126, 132
gas chromatography 19:300
in glycerol 13:527
in green tea leaf 18:459, 478
hydroxylamine hydrochloride method 8:357
invertase activity 7:587
paper chromatography 7:569; 18:478; 19:142
potassium permanganate titration 18:459
potassium titanium oxalate colorimetric method 18:459
refractive index 18:343

in roasted coffee 10:435
solubility 11:126
thin-layer chromatography 13:502
in tobacco 19:142
in wines 19:409
d-Glucose
  Quisumbing and Thomas determination 10:7
  in tobacco for cigarettes 10:7
Glucuronic acids
  paper chromatography 12:582
  thin-layer chromatography 8:381
Glue. (See also Animal glues; Blood glues; *Casein*;
    Fish glues; *Glues of animal origin*; and
    Soybean glues.)
  in elastomers, synthetic 12:147
  solvent extraction 12:147
Glues of animal origin 13:487–494
  analysis 13:491, 15:524, 536
  ash content 13:493
  commercial grades and specifications 13:488;
    15:507, 520
  composition 13:488
  copper sulfate color test 13:491
  fat content 13:493
  foam content 13:492
  foam ratio 15:525
  foam stability 15:524, 536
  gel strength 13:492
  identification 13:491
  isoelectric point 13:492
  jelly strength 15:536
  manufacture 13:489
  in matches 15:506, 524
  in match strikers 15:520, 536
  melting point 15:525
  moisture content 13:493
  odor 13:493
  pH 13:492
  properties 13:488
  properties, determination of 13:491
  sampling 13:491
  solubility 13:488, 539, 552
  uses 13:487
  viscosity 13:491
Glutamate–pyruvate transaminase 8:451
Glutamic acid. (See also *Carboxylic acids, amino*.)
    8:412, 468
  acid–base titration 8:473
  analysis 8:473
  in black tea 18:507
  in cocoa beans 9:556
  commercial specifications 8:473
  composition 8:412
  crystalline form 8:470
  enzymatic decarboxylation 8:474
  gas chromatography 8:429
  in green tea leaf 18:459

identification 8:473
infrared spectrum 8:471
ion exchange chromatography 8:415, 421
manufacture 8:471
metal complexes 8:459
ninhydrin test 8:473
nuclear magnetic resonance spectrum 8:472
optical rotation 8:444
paper chromatography 8:422
potassium titanium oxalate colorimetric method
    18:459
properties 8:469
in protein 8:414
solubility 8:471, 473
in yeast 19:529
Glutamic acid salts 8:468
Glutaraldehyde. (See also Aldehydes.)
  assay 11:566
  commercial specifications 8:340
  derivatives for identification 12:585
  in disinfectants and antiseptics 11:565
  in embalming fluids 12:165
  hydroxylamine hydrochloride volumetric
    method 11:566
  in photographic films 12:585
  ultraviolet spectroscopy 12:585
Glutaric acid
  in adipic acid 4:415
  gas chromatography 17:290
  properties 8:396
Glutaric anhydride
  infrared spectroscopy 8:562
Glutaronitrile. (See also *Nitriles.*)
  properties 16:396
Glutathione. (See also *Mercaptans.*)
  p-chloromercuribenzoic acid colorimetric
    method 15:561
  N-ethylmaleimide colorimetric method 15:561
Gluten. (See also *Flour.*) 12:599
Gluten feed. See Feed products.
Glutethimide
  gas chromatography 6:210, 211
Glyceraldehyde
  in gelatin 12:550
  3-methyl-2-benzothiazolone hydrazone–ferric
    chloride colorimetric method 12:550
Glycerin. (See also *Glycerol.*) 13:494
  in detergents 11:365
  in embalming fluids 12:166
  in finishing agents 11:126, 131
  gas chromatography 14:171
  in insulin 14:580
  potassium hydrogen sulfate colorimetric
    methods 11:126, 131
  sodium periodate volumetric method 11:365;
    14:580
  solubility 11:126

in sunscreens 11:91
Glycerine nitrates. (See also Dinitroglycerin;
    Nitroglycerin; and Tetranitrodiglycerin.)
    12:443
Glycerol. (See also Glycerin.) 5:22; 13:494–532
    acetin volumetric method 13:508
    acrolein-2,7-dihydroxynaphthalene colorimetric
        method 13:501
    in alkyd resins 5:38
    ammonium perchloratocerate volumetric method
        13:512
    analysis 13:501
    anthrone colorimetric method 12:587
    ash content 13:521
    azeotropes 13:499
    azeotropic distillation 13:507
    benzanthrone fluorometric method 13:508
    in blood 13:520
    in brake fluids 7:423
    catechol colorimetric method 13:572
    in cigarette smoke 10:29
    color 13:530
    colorimetric determination 19:443
    commercial grades and specifications 13:500
    copper(II) chloride colorimetric method
        13:509
    copper(II) chloride volumetric method 13:509
    in creams and lotions 11:95
    derivatives for identification 13:568
    enzymatic method 13:511; 19:445
    in fats and oils 13:507, 512
    gas chromatography 5:32; 10:29; 13:503, 576;
        19:300, 301, 445
    history 13:494
    identification 13:501
    impurities, determination of 13:508, 521
    iodometric titrations 18:201
    liquid chromatography 13:502, 580
    manufacture 13:498, 561
    miscibility 13:495
    near-infrared spectroscopy 13:527
    occurrence 13:498, 561
    oxidation methods 13:512; 19:443
    paper chromatography 13:501, 581
    perchloratocerate titration 13:510
    periodic acid volumetric method 13:514, 597
    in pharmaceuticals 13:520
    phloroglucinol colorimetric method 19:443
    in photographic films 12:587
    in polyurethanes 19:299
    potassium dichromate–sulfuric acid colorimetric
        method 13:508
    potassium iodate colorimetric method 11:82
    properties 6:2; 13:495
    quinoline fluorometric method 13:509
    refractive index 13:530
    residue on ignition 13:521

    in resins 13:596
    separation 13:501; 19:444
    in shampoos 11:82
    in soaps 18:182, 201
    sodium periodate colorimetric method 13:571
    sodium periodate volumetric method 13:518
    solubility 13:496, 560
    solvent extraction 13:507
    specific gravity 13:528, 530
    structure 13:561
    thin-layer chromatography 12:587; 13:501, 582
    uses 13:495
    vanadium volumetric method 13:520
    in wines 19:425, 443
Glycerol alkyds
    infrared spectroscopy 5:46
Glycerol 1,2-dichlorohydrin. (See also 2,3-Di-
    chloro-1-propanol.)
    azeotropes 14:170
    properties 14:169
Glycerol 1,3-dichlorohydrin. (See also 1,3-Di-
    chloro-2-propanol.)
    azeotropes 14:170
    nuclear magnetic resonance spectrum 14:168
    properties 14:169
Glycerol dichlorohydrine
    colorimetric method for Vitamin A 12:66
Glycerol dichlorohydrins. (See also individual
    compounds.) 14:168
    analysis 14:171
    calcium hydroxide volumetric method 14:172
    commercial specifications 14:169
    in epichlorohydrin 14:175
    gas chromatography 14:171, 175
    manufacture 14:169
    paper chromatography 14:157
    properties 14:168
    solubility 14:168
    toxicity 14:169
Glycerol halohydrins. (See also Glycerol dichloro-
    hydrins; Glycerol monochlorohydrins; and
    individual compounds.) 14:167
    uses 14:167
Glycerol 2-monobromohydrin. See 2-Bromo-1,3-
    propanediol.
Glycerol 1-monochlorohydrin. (See also 3-Chloro-
    1,2-propanediol.) 14:167
    analysis 14:171
    commercial specifications 14:169
    gas chromatography 14:171
    manufacture 14:169
    periodic acid–sodium thiosulfate volumetric
        method 14:171
Glycerol 2-monochlorohydrin. (See also 2-Chloro-
    1,3-propanediol.) 14:167
Glycerol monochlorohydrins. (See also individual
    compounds.) 14:167

analysis 14:171
calcium hydroxide volumetric method 14:172
manufacture 14:169
toxicity 14:169
Glycerol monosalicylate 6:76
Glycerols. See Polyhydric alcohols.
Glycerol triacetate. (See also Triacetin.)
gas chromatography 13:505
Glycerol trimethylsilyl ether
gas chromatography 13:505
Glyceryl dinitrate. See Dinitroglycerin.
Glyceryl monostearates
infrared spectroscopy 11:331
Glyceryl monostearate–soap emulsions identification 11:94
Glyceryl trinitrate. See Nitroglycerin.
Glycidaldehyde. (See also Epoxy compounds.)
infrared spectroscopy 12:200
Glycidol. (See also Epoxy compounds.)
infrared spectroscopy 12:200
Glycidonitrile. (See also Epoxy compounds.)
infrared spectroscopy 12:200
γ-Glycidoxypropyltrimethoxysilane. (See also Organofunctional silanes.)
commercial specifications 18:82
gas chromatography 18:85
infrared spectroscopy 18:83
properties 18:80
Glycidyl allyl ether. (See also Epoxy compounds.)
hydrochlorination methods 12:208
Glycidyl benzoate. (See also Epoxy compounds.)
hydrochlorination methods 12:208
Glycidyl tert-butyl ether. (See also Epoxy compounds.)
hydrochlorination methods 12:208
Glycidyl p-chlorophenyl ether. (See also Epoxy compounds.)
hydrochlorination methods 12:208
Glycidyl o-cresyl ether. (See also Epoxy compounds.)
hydrochlorination methods 12:208
Glycidyl dichlorophenyl ether. (See also Epoxy compounds.)
hydrochlorination methods 12:208
Glycidyl diethylcarbinyl ether. (See also Epoxy compounds.)
hydrochlorination methods 12:208
Glycidyl isopropyl ether. (See also Epoxy compounds.)
hydrochlorination methods 12:208
Glycidyl methacrylate. (See also Acrylic and methacrylic acids and esters.)
saponification method 4:196
Glycidyl phenyl ether. (See also Epoxy compounds.)
hydrochlorination methods 12:208
Glycine. (See also Carboxylic acids, amino.)

7:141; 8:412, 476
acid–base titration, nonaqueous 8:482
analysis 8:481
calcium nitrate volumetric method 12:37
in cocoa beans 9:556
colorimetric method using o-phthaldialdehyde 8:482
commercial specifications 8:481
composition 8:412
copper nitrate volumetric method 12:37
crystalline form 8:478
gas chromatography 8:429; 17:290
in green tea leaf 18:459
identification 8:481
infrared spectrum 8:478
ion exchange chromatography 8:415, 421, 450, 482; 9:618
lead nitrate volumetric method 12:37
manufacture 8:480
metal complexes 8:459
nuclear magnetic resonance spectrum 8:479
paper chromatography 8:422
potassium permanganate titration 18:459
potassium titanium oxalate colorimetric method 18:459
properties 8:477
in protein 8:414
solubility 8:478
uses 8:477
in yeast 19:529
Glycocholic acid 7:141
liquid chromatography 7:148
Glycodeoxycholic acid
liquid chromatography 7:149
Glycol ethers
antifreezes 6:1
in brake fluids 7:423, 424, 425
in carburetor cleaners 6:356
gas chromatography 13:577
in photographic films 12:573
pyrolytic method 12:573
α-Glycol groups
benzyltrimethylammonium periodate volumetric method 12:249
in epoxy resins 12:249
Glycolithocholic acid
liquid chromatography 7:149
Glycols. (See also Glycols and polyhydric alcohols; and individual compounds.) 13:533–549
acid number 13:587
alkaline number 13:587
analysis 6:5
antifreezes 6:1
assay 13:584
in brake fluids 7:424, 425
bromination method 4:541
colorimetric determination 13:598

commercial grades and specifications
13:549
in de-icers 6:356
density 13:584
derivatives for identification 13:567
distillation range 13:601
in embalming fluids 12:166
esterification methods 13:586
in finishing agents 11:126, 131
gas chromatography 13:575, 589; 19:298, 300,
301
hydroxyl number 13:587
identification 6:4; 19:296
infrared spectroscopy 11:83; 13:583, 590
ion exchange methods 19:292
oxidation 6:4
paper chromatography 13:581
periodate spectroscopic method 13:599
periodic acid volumetric method 13:592
in photographic films 12:587
phthalic anhydride volumetric method 13:586
in polyurethanes 19:291, 296
potassium triiodide gravimetric method 13:571
properties 13:534
pyromellitic dianhydride volumetric method
13:587
refractive index 13:584
silver nitrate–periodic acid gravimetric method
13:572
sodium metaperiodate volumetric method 13:
591
sodium periodate–chromotropic acid colorimetric method 13:598
sodium periodate colorimetric method 13:571
solubility 11:126; 13:535, 538
solvent extraction 19:292
thin-layer chromatography 12:587; 13:582
toxicity 13:534
uses 13:535
in windshield washer antifreezes 6:356, 369
Glycols, acetylenic. See Acetylenic glycols.
Glycols and polyhydric alcohols. (See also individual compounds.) 13:533–607
assay 13:584
colorimetric determination 13:570
derivatives for identification 13:567
2,4-dinitrophenylhydrazine–potassium hydroxide
colorimetric method 13:574
distillation range 13:601
identification 13:567
impurities, determination of 13:600
infrared spectroscopy 13:583
liquid chromatography 13:581
nuclear magnetic resonance spectroscopy 13:
583
paper chromatography 13:583
polarography 13:596

properties, determination of 13:601
in resins 13:596
separation 13:567
specific gravity 13:602
thin-layer chromatography 13:583
volumetric determination 13:595
Glyconic acid. See Gluconic acid.
Glyodin 13:208
analysis 13:208
bromocresol purple colorimetric method
13:208
in fruit 13:208
manufacture 13:208
properties 13:208
solubility 13:208
toxicity 13:208
uses 13:208
Glyoxal
anthrone colorimetric method 12:585
in embalming fluids 12:165
gas chromatography 12:204
in methanol 16:114
in photographic films 12:585
Glyoxal–bis(2-hydroxyanil)
colorimetric method for calcium 11:140, 145,
491
Glyoxalidine. See Glyodin.
Glyoxide. See Glyodin.
Glyoxylic acid. (See also Hydroxyacetic acid.)
properties 8:397
thin-layer chromatography 8:381
Glyoxylic acid 2,4-dinitrophenylhydrazone
thin-layer chromatography 8:381
Goethite 15:21; 16:286
composition 15:22
Goetz centrifuge tube 7:531
Gogan hardness tester 7:449
Gold. (See also Gold alloys; Gold metal.) 14:1–51
ascorbic acid titration 14:22
in aluminum metal 5:152
atomic absorption spectroscopy 14:26, 30, 41,
43, 47
bead test 6:588
in beryllium 14:42
Chiddy gravimetric method 14:37
Chow and Beamish sampling method 14:10
colorimetric determination 14:11, 23
in copper 14:42
in copper alloys 10:638
copper(I) chloride gravimetric method 14:38
in copper metal 10:621
copper(I) sulfate gravimetric method 14:39
in cupellation beads 14:30
in cyanide solutions 14:37
in dental alloys 14:36
5-(dimethylaminobenzylidine) rhodamine
colorimetric method 14:25

dithizone colorimetric method 14:25
dithizone titration 14:23
in doré beads 14:32
electrolytic method 14:40
electrophoresis 14:18
emission spectroscopy 14:2, 5, 27, 29, 30, 32, 43; 17:270
fire assay 10:621
flame tests 6:589
fusion method 14:40
in geological samples 14:26
in glass 9:200
in glass and glass-ceramics 13:413
in gold alloys 14:32
in gold ores 14:10, 12, 19, 28
in gold–silver alloys 14:14
gravimetric determination 14:11, 18, 35
hydrobromic acid colorimetric method 14:23, 31
in hydrochloric acid 14:374
hydroquinone gravimetric method 14:16, 19
hydroquinone titration 14:20, 44
identification 14:11
iodometric titration 14:22
ion exchange chromatography 12:570; 14:17
isotopes 14:5
in jewelry alloys 14:36
in lead 14:42
manganese(II) sulfate titration 14:23
mercuric nitrate color test 14:11
in meteorites 14:43
methyl violet colorimetric method 14:24
Mohr's salt titration 14:23, 37
neutron activation analysis 13:265; 14:28, 42
occurrence 14:1
in ores 17:266
oxalic acid gravimetric method 10:638; 14:19
oxidation potentials 18:158
paper chromatography 14:18, 43
in photographic films 12:570
in platinum 14:42
in platinum group metals 17:229, 239
polarography 14:44
potassium iodide titration 14:21
α-pyridyl ketoxine colorimetric method 14:25
Reinecke's salt gravimetric method 14:19
Rhodamine B colorimetric method 12:570
Rhodamine B spectroscopic reagent 14:11
separation 14:12
in silver beads 14:30
sodium nitrite gravimetric method 14:19, 36
solvent extraction 14:14
Strasheim and Van Wamelen sampling method 14:10
sulfur dioxide gravimetric method 10:638; 14:15, 19
sulfuric acid gravimetric method 14:39

tetraphenylarsonium chloride colorimetric method 14:25
thioglycollic acid gravimetric method 14:19
thiophenol gravimetric method 14:19
o-toluidine colorimetric method 14:24
in urine 14:19, 21, 44
volumetric determination 14:20
in water 14:43
x-ray emission spectroscopy 14:5
x-ray fluorescence spectroscopy 14:31
Gold alloys. (See also Gold metal.) 5:65; 14:8
analysis 14:9
composition 5:68
gold content 14:32
sampling 14:9
uses 14:1, 8
Gold alloys, dental. See Dental cast gold alloys; Dental wrought gold alloys.
Gold bromide
microchemical method for oxymorphone 16:250
Gold chloride
colorimetric method for arsenic 11:140
microchemical method for apomorphine 16:249
microchemical method for cocaine 16:249
Gold compounds 14:8
Gold–copper alloys
uses 14:8
Goldfisch method
for crude fat in grains 4:473
for fat in yeast 19:531
Gold–gallium alloys 13:270
analysis 13:270
gallium content 13:270
Gold metal. (See also Gold alloys.)
analysis 14:9
assay 14:18
carrier gas fusion methods 8:682
commercial grades and specifications 14:7
cupellation 14:13
dissolution 14:10
fineness 14:2
history 14:1
hydrogen in 8:680
impurities, determination of 14:44
manufacture 14:6
oxygen in 8:680, 682
properties 14:2
reaction medium for fusion methods 8:633
sampling 14:9
silver content 18:172
solubility 14:6, 9
specific heat 14:4
uses 14:1
vacuum fusion methods 8:680
Gold–nickel alloys
composition 14:8

uses 14:8
Gold ores
  fire assay 14:12
  gold content 14:10, 12, 19, 28
  iron-nail fire assay 14:14
  sampling 14:10
Gold–palladium alloys
  properties 14:9
  uses 14:9
Gold stamping
  copper sulfate test 11:115
Gonadotropic hormones. (See also Pituitary
    hormones.) 14:180
  analysis 14:205
  bioassay 14:205
  uses 14:180
Goslarite 14:519
  composition 19:545, 573
Gossypol
  colorimetric method for niobium 16:368, 375
Gossyptol
  in cottonseed flour 10:542
  dianilinogossyptol colorimetric method 10:543
Gowerite 7:314
Graduate density
  of ion exchange resins 15:11
Graft copolymers 12:345
  manufacture 12:345
  properties 12:345
  types 12:345
Graham's salt 17:124
Grain inhibitors
  in confectionery products 10:552
Grain neutral spirits 4:463
  analysis 4:480
  quality control 4:469
Grains
  bromide residue in 8:26
  mercury content 16:36
Grains, for distilled alcoholic beverages
  analysis 4:469
  quality control 4:467
Grain size, of abrasives 4:34
Grain structure
  of copper metal 10:627
Gramicidin 5:466, 533
  analysis 5:540, 544, 548
  colorimetric method 5:548
  identification 5:540
  microbiological assay 5:544
  properties 5:534
  separation 5:540
  solubility 5:535, 536
  ultraviolet spectroscopy 5:548
  uses 5:533
Gramicidin D
  infrared spectrum 5:539

ultraviolet spectrum 5:539
Gramicidin J
  assay 5:548
  microbiological assay 5:545
  structure 5:467
Gramicidin S
  structure 5:467
Grandidierite 7:316
Granite 19:225
Granulation
  of diazodinitrophenol 12:416
  of mercury fulminate 12:410
  of tetracene 12:416
Grapefruit oil. (See also *Essential oils.*)
  decyl aldehyde content 12:282
  optical rotation 12:270
  residue on evaporation 12:273
  ultraviolet spectroscopy 12:288
Grape juice. See Musts.
Grape pigments
  solvent extraction 19:463
  thin-layer chromatography 19:463
  in wines 19:463
Grapes
  sample preparation 19:400
  sampling 19:399
  sugar content 19:405
Grape skin extract
  commercial specifications 10:531
  composition 10:531
Graphite. (See also *Carbon; Graphite and related
    carbons.*) 11:464
  as ablator 4:7
  in brake linings 7:441
  commercial grades and specifications 14:52
  crystalline structure 14:53
  crystallite structure 8:140, 141
  in enamels 12:173
  in explosive priming mixtures 12:421, 424
  identification 12:421
  impurities in 4:14
  in iron 15:70
  Langmuir vaporization of 4:12
  manufacture 14:54
  mass spectroscopy 4:10, 12
  in metal borides 7:363
  occurrence 14:54
  oxidation of 4:7
  physical tests 4:13
  properties 14:52
  reaction medium for fusion methods 8:631,
    633
  silicon content 18:30, 37
  solubility 12:421
  thermogravimetry 4:8
  uses 14:54
  vanadium content 19:342

vapor pressure 4:11
Graphite and Related Carbons. (See also *Carbon;*
    Graphite.) 14:52–87
    analysis 14:60
    crystallite size 14:76
    electrical resistivity 14:82
    flexural strength 14:79
    impurities, determination of 14:61, 63
    interlayer spacing 14:76
    manufacture 14:54
    modulus of elasticity 14:77, 79, 80
    porosity 14:82
    properties 14:52
    properties, determination of 14:76
    sampling 14:73
    shear strength 14:80
    spectrographic methods 14:73
    tensile strength 14:77
    thermal expansion 14:84
    x-ray diffraction 14:76
Graphite composites, fiber reinforced. (See also
    Graphite and related carbons.) 14:59
    uses 14:60
Graphite–refractory composites. (See also
    Graphite and related carbons.) 14:59
    composition 14:59
    properties 14:59
Gravel
    composition 9:202
Gravimetry
    of ceramics 9:217
Gray iron. (See also Iron metal.)
    composition 15:27
    properties 15:27
Gray-King test
    for coal and coke 10:246
Gray tin. (See also Tin metal.) 19:26
Greases. (See also *Lubricants.*) 15:302
    in adhesives 10:275
    classification 15:302
    composition 15:302
    copper corrosion test 15:335
    dirt content 15:326
    dropping point 15:349
    lead content 15:171, 186
    manufacture 15:308
    oxidation characteristics 15:341
    penetration value 15:347
    properties 15:303
Greenalite
    composition 15:22
Green coffee 10:375
    analysis 10:377
    ash content 10:391
    average weight 10:382
    caffeine content 10:404
    carbohydrate content 10:391

chlorogenic acid content 10:398, 413
choline content 10:411
composition 10:374
enzyme content 10:414
lipid content 10:394
macroscopic examination 10:380
nitrogen compound content 10:410
olfactory examination 10:380
oligosaccharide content 10:392
organic acid content 10:395
phenolic acid content 10:396
polysaccharide content 10:393
protein content 10:403
sampling 10:377, 382
size of beans 10:380
sucrose content 10:392
trigonelline content 10:410, 411
visual examination 10:379
water content 10:383
Green fade 7:471
Greenockite 8:47
Green organic pigments
    in enamels 12:187
    identification 12:187
Green pigments. (See also *Pigments;* and indi-
    vidual compounds.) 17:200
Green reflectance 7:233
Green S. (See also *Triphenylmethane and related
    dyes.*)
    paper chromatography 19:172
Green's method
    identification of dyes 6:389
Green tea. (See also *Tea.*)
    manufacture 18:455
Green tea leaf 18:455
    amino acid content 18:457, 476
    analysis 18:456
    carbohydrate content 18:457, 477
    carotenoid content 18:479
    chlorophyll content 18:478
    composition, determination of 18:457
    enzyme content 18:482
    flavanol content 18:465
    flavonol glycoside content 18:457, 474
    flavylogen content 18:475
    mineral content 18:481
    paper chromatography 18:462
    phenolic acid content 18:457, 475
    polyphenol content 18:457
    protein content 18:476
    sampling 18:456
    solvent extraction 18:462, 465
    thin-layer chromatography 18:462
Griess color test
    for nitrite 12:420
Griess-Ilosvay method
    color test for nitrite 16:492

Griess-Saltzman reaction
  colorimetric method for nitrogen dioxide
    16:479, 484
Grignard reaction
  of nitriles 16:400, 406
Grignard reagent
  apparatus for preparation 7:81, 82
  preparation 8:583
  reaction of carbon black with 8:191
  use in preparation of carboxylic acids 8:372
Grinding wheels 7:293
  analysis 7:295
  classification 7:293
  composition 7:294
  density 7:296
  grinding efficiency 7:301
  grinding operation 7:293
  metal-bonded 7:299
  performance testing 7:299
  resinoid 7:298
  safety checks 7:300
  vitrified 7:297
Griseofulvin 5:461, 494
  analysis 5:498, 502, 508
  column chromatography 5:498
  identification 5:498
  infrared spectrum 5:498
  microbiological assay 5:502
  paper chromatography 5:498
  properties 5:493
  separation 5:498
  solubility 5:494
  structure 5:462
  ultraviolet spectroscopy 5:508
  uses 5:494
Grit
  in carbon black 8:237
Gross alpha radioactivity
  in water 19:394
Gross beta radioactivity
  in water 19:394
Grossularite 8:75
Ground glass
  commercial grades and specifications 15:513,
    522
  in matches 15:513
  in match strikers 15:522
  particle size 15:529
Groundwood pulp
  tearing strength 7:255
Growth hormone. (See also Pituitary hormones.)
    14:179
  analysis 14:205
  bioassay 14:205
  uses 14:179
Gruenerite
  composition 15:22

G salt
  analysis 16:293
  bromometric titration 16:237, 239
  commercial grades and specifications 16:234
  impurities, determination of 16:239
  manufacture 16:233
  naphtholsulfonic acid content 16:239
  in R salt 16:237
Guaiacol
  color test for propylene glycol 13:570
  gas chromatography 11:167
  in green tea leaf 18:459
  potassium titanium oxalate colorimetric method
    18:459
  use in manufacture of vanillin 6:604
Guanamines 5:276
Guanidine 5:276
Guanidine derivatives. (See also Chlorhexidine.)
  analysis 11:569
  in disinfectants and antiseptics 11:569
Guanidine nitrate. (See also High explosives.)
  melting point 12:461
  properties 12:434
Guanidine orthophosphate
  uses 17:122
Guanidines. (See also individual compounds.)
    13:205-209
Guanidine thiocyanate. (See also Thiocyanates.)
  manufacture 18:434
Guanine
  in gelatin 12:552
  ion exchange chromatography 12:552
  ultraviolet spectroscopy 12:553
Guano-urea orthophosphate
  uses 17:122
4-Guanyl-1-(nitrosoaminoguanyl)-1-tetrazene.
    See Tetracene.
Guar gum
  infrared spectroscopy 12:583
Guignet's green. See Hydrated chromium oxide.
Guinea Green B. (See also F D & C Green No. 1;
    Triphenylmethane and related dyes.)
  liquid chromatography 19:173
Gum arabic 6:362
  composition 12:583
  in confectionery products 10:551
  gravimetric determination 12:581
  infrared spectroscopy 12:583
  Millon's reagent gravimetric method 12:580
  in photographic films 12:580
Gum chicle
  in confectionery products 10:552
Gum damar
  solubility 13:539, 552
Gum ghatti
  gravimetric spectroscopy 12:583
  infrared spectroscopy 12:583

Gum rubber
  ethylene oxide in 12:331
Gums. (See also *Confectionery products.*)
    10:548
  analysis 15:527
  colorimetric determination 12:584
  in embalming fluids 12:166
  in explosive priming mixtures 12:425, 426
  in finishing agents 11:126, 132
  in gasoline 13:359
  infrared spectroscopy 12:583
  lead acetate gravimetric method 11:127, 132
  in matches 15:508
  paper chromatography 12:582
  in shampoos 11:75
  solubility 11:126
  in sunscreens 11:91
  viscosity 15:527
Gums, natural
  in cooling system sealer 6:353, 362
Gum tragacanth
  in confectionery products 10:552
Gushing
  of beer 7:538, 571
  measurement 7:539
Gustafsson reduction method
  for sulfur determination of acrylic polymers
    4:283
Gutzeit method
  ACS method 6:249
  AOAC method 6:246
  for arsenic 6:246
Guymon method
  for proofing 4:483
Gypsite 14:88
Gypsum. (See also Ceramics; *Gypsum and
    gypsum products.*) 8:75; 9:190; 18:360
  in cements 9:141
  commercial grades and specifications 14:89
  fluxes for 9:216
  history 14:87
  manufacture 14:89
  occurrence 14:88, 89
  properties 14:88
  solubility 14:88
  uses 14:87
Gypsum and gypsum products. (See also

  Gypsum.) 14:87–102
  aluminum oxide content 14:94, 96
  analysis 14:90
  assay 14:91
  calcium oxide content 14:95
  calcium sulfate content 14:91
  carbon dioxide content 14:93
  chloride content 14:96
  commercial grades and specifications 14:89
  composition, determination of 14:93, 97
  impurities, determination of 14:98
  iron oxide content 14:94, 96
  loss on ignition 14:97
  magnesium oxide content 14:95
  manufacture 14:89
  occurrence 14:89
  properties 14:88
  sampling 14:90
  silicon dioxide content 14:94, 95
  sulfur trioxide content 14:96
  water content 14:93
Gypsum cement. (See also Hydraulic cements.)
    9:95
  commercial specifications 9:99, 100, 101
  identification 9:107
Gypsum compatibility
  of dental elastomers 11:286
  of dental impression materials, agar-based
    11:279
  of dental impression materials, alginate-based
    11:282
  of zinc oxide–eugenol dental products
    11:297
Gypsum plasters. (See also Hydraulic cements.)
    9:95
  commercial grades and specifications 9:99, 100,
    101
  composition 9:96
  hardening 9:97
Gypsum products. (See also *Gypsum and gypsum
    products.*)
  analysis 14:99
  gauging plaster content 14:101
  sand content 14:100
  wood fiber content 14:100
Gypsum products, dental. See Dental gypsum
    products.

H

Haber-Bosch process
for ammonia 5:296
H acid. See 4-Amino-5-hydroxy-2,7-naphthalene-
sulfonic acid.
H-acid diazonium salt
colorimetric method for cyclohexanone 11:194,
205
Haematoxylin, azo dye derivatives
color reagents for boron 7:389
Hafnia. See Hafnium dioxide.
Hafnium. (See also Hafnium alloys; Hafnium
metal.)
atomic absorption spectroscopy 14:108, 120,
124
colorimetric determination 14:110, 117
EDTA titration 13:445
emission spectroscopy 14:107, 119, 121;
16:387
in ferroniobium 16:387
in glass and glass-ceramics 13:445
gravimetric determination 14:108, 110, 114
history 14:103
identification 14:107
ion exchange chromatography 14:113, 114
ion exchange methods 16:390
mandelic acid gravimetric method 14:115, 140
in niobium 14:116; 16:387
occurrence 14:104
separation 14:110
solvent extraction 14:111, 114
in steel 14:116
in uranium 14:116
x-ray fluorescence spectroscopy 14:108, 116
xylenol orange colorimetric method 14:117
in zircon 14:139
in zirconium 14:114
Halfnium alloys. (See also Hafnium metal.)
14:106
Hafnium and zirconium. (See also Hafnium;
Zirconium.) 14:103–152
Hafnium compounds. (See also individual compounds.) 14:138
Hafnium dioxide 14:145
analysis 14:145
manufacture 14:145
properties 14:145
uses 14:145
Hafnium metal. (See also Hafnium alloys.)
carrier gas fusion methods 8:622, 626, 682
commercial grades and specifications 14:106
hydrogen in 8:622, 680

impurities, determination of 14:121
manufacture 14:105
nitrogen in 8:622, 623, 627
oxygen in 8:622, 623, 626, 627, 680, 682
properties 14:104
vacuum fusion methods 8:623, 680
Hafnium tetrachloride 14:146
analysis 14:147
manufacture 14:146
properties 14:146
sample preparation 14:147
uses 14:146
Hafnyl chloride 14:146
manufacture 14:146
solubility 14:146
Hair dyes. (See also *Cosmetics.*) 11:98
ammonia content 11:99
analysis 11:99
composition 11:98
dye developers 11:107
dye intermediates in 11:103
hair coloring evaluation 11:108
hydrogen peroxide content 11:107
stability 11:108
thin-layer chromatography 11:102
Hair sprays. (See also *Cosmetics.*) 11:89
analysis 11:89
composition 11:89
dichlorodifluoromethane content 11:89
ethanol content 11:89
lanolin in 11:90
moisture content 11:90
nonvolatile matter in 11:90
plasticizers in 11:90
proteins in 11:90
quaternary ammonium compounds in 11:90
resins in 11:90
silicone compounds in 11:90
sunscreens in 11:90
trichlorofluoromethane content 11:89
Haleite. See EDNA.
Half-life
of silver 18:155
Halide lamps 9:382
Halide detector 8:13
Halide meters 9:382
Halide ratio
of photographic films 12:565
Halides
in fluorocarbons 13:69
in iodides 14:597

in lime and limestone 15:228
mercuric nitrate titration 15:228
in potassium metaperiodate 14:610
in silver halides 12:566
silver nitrate gravimetric method 13:69; 14:597
silver nitrate turbidimetric method 14:610
in sodium metaperiodate 14:610
in uranium 19:238
in vinbarbital 6:529
volumetric determination 12:566
Halite. (See also Rock salt; Sodium chloride.)
9:304, 337
Halloysite. (See also *Clays and clay minerals*.)
composition 10:103
Halogenated hydrocarbons. (See also individual
compounds.) 13:179–186
in air 13:182
analysis 13:182
gas chromatography 13:182
in soil 13:183
Halogenated propellants 4:453
Halogenation methods. (See also Bromination
method; Iodine monochloride method; and
Wijs method.)
for allyl esters 5:96
for allyl ethers 5:98
for allyl halides 5:102
for allyl olefins 5:104
for allyl sulfides 5:105
Halogen donors. (See also individual compounds.)
analysis 11:553
available halogen content 11:553
in disinfectants and antiseptics 11:552
Halogen fluorides. (See also individual compounds.) 13:7
manufacture 13:9
properties 13:8
uses 13:9
Halogen, free
4,4'-bis-dimethylaminothiobenzophenone
colorimetric method 11:142
in bromohydrocarbons 8:13
in cotton 11:141
fluorescein colorimetric method 11:141
in methylene chloride 9:430
Halogenopyridines. (See also *Pyridine and pyridine derivatives.*)
manufacture 17:422
Halogens. (See also individual elements.)
argentometric determination 14:156
Beilstein flame test 12:289
in cocoa products 9:595
in cyclopropane 5:360
in disinfectants and antiseptics 11:552
in essential oils 12:288
in halohydrins 14:154
in indigoid dyes 14:513

in nitrous oxide 5:366
silver nitrate titration 14:513
volumetric determination 11:553
Halogen volatilization methods
for tin 19:34, 71, 75
Halohydrins. (See also Epichlorohydrin; Ethylene
chlorohydrin; Glycerol halohydrins; Propylene
chlorohydrins; and individual compounds.)
14:153–178
analysis 14:154
ceric nitrate colorimetric method 14:157
chlorine content 14:157
chlorohydrin content 14:155
derivatives for identification 14:157
fluorine content 14:157
halogen content 14:154
hydroxyl content 14:157
infrared spectroscopy 14:157
manufacture 14:154
neutron activation analysis 14:155
paper chromatography 14:157
properties 14:153
solubility 14:153
toxicity 14:154
Halon. See Polytetrafluoroethylene.
Halothane
analysis 5:363
anesthetic 5:363
gas chromatography 5:364
infrared spectroscopy 5:364
properties 5:357, 363
solubility 5:363
Halphen test
for cyclopropene groups 12:484
Hambergite 7:315
H and D curve
for photographic film 12:595
Handsheets
apparatus for preparation of 7:242
conditioning 7:250
for optical testing of pulp 7:241, 251
for physical testing of pulp 7:241, 248, 253
shrinkage 7:251
Hanging drop test
for inorganic fluorides 13:15
Hansa orange. (See also *Pigments.*)
structure 17:180
Hansa yellow
in enamels 12:173, 184
identification 12:184
Hansa yellow G. (See also *Pigments.*)
structure 17:181
Hansa yellow 10G. (See also *Pigments.*)
structure 17:181
Hantzsch test
colorimetric method for formaldehyde 13:106,
107

Hard candy. (See also *Confectionery products.*) 10:548
  manufacture 10:553
Hardeners
  in photographic films 12:584
Hard facing alloys. (See also Cobalt alloys.) 10:289
  composition 10:289
Hardness
  of abrasive grains 7:309
  of acrylic resin teeth 11:307
  of adsorbents 4:445
  of brake linings 7:449
  of dental cast gold alloys 11:293
  of dental chromium–cobalt alloys 11:292
  of enamels 12:179
  of fibers 4:25
  of match heads 15:544
  of match strikers 15:547
  of methyl methacrylate dental materials 11:306
  of pigments 17:208
  of silicone rubbers 18:131, 139
  of vinylidene fluoride–hexafluoropropylene
    copolymer 13:98
  of water 19:362
  of zinc oxide–eugenol dental products 11:297
Hardness number
  of activated carbon 8:153
Hardness scales 4:31
Hardware bronze 5:68
Hardy sectioning technique of fibers for microscopy 4:265
Hassid method
  for reducing sugars in cocoa beans 9:559
Hastelloy alloys. (See also Nickel alloys.)
  composition 16:290, 351
Hastelloy X. (See also Cobalt alloys.)
  composition 10:286
Hatchett's blue. See Copper ferrocyanide.
Haugh unit value
  of eggs 12:76
Hausmannite. (See also Mangano–manganic
  oxide.) 15:448
Hayesine 7:314
Haynes alloys. (See also Cobalt alloys.)
  composition 10:286, 290
Haze
  in beer 7:519, 532
  measurement 7:521
Haze number
  of organosilicon copolymers 18:148
  of silicone resins 18:124
HBX. (See also High-explosive mixtures.)
  analysis 12:448
  composition 12:448
  uses 12:448
Heads (in alcoholic beverages) 4:462

analysis 4:480
quality control 4:469
Headspace analysis
  of black tea 18:508
  of cocoa beans 9:567
Heat capacity
  of ethylenediamine 11:415
Heat-distortion temperature
  of epoxy resins 12:253
Heat, effect of
  identification of acrylic and modacrylic fibers
    4:260
Heat fade 7:471
Heating sources
  for carrier gas and vacuum fusion systems
    8:641
Heating tests
  for acrylates 12:588
  for cellulose derivatives 12:588
  for photographic film bases 12:588
  for polyacrylonitrile 12:588
  for polyamides 12:588
  for polystyrene 12:588
Heat of fusion
  of polyethylene 12:383
  of polypropylene 12:383
Heat of hydration
  of cements 9:141
Heat of reaction
  apparatus 7:24
  of benzene with nitrating acid 7:25
Heat of sublimation
  of carbon 4:11
Heat stability
  of adipic acid 4:417
  of brake fluids 7:435
  of cellulose acetate 9:72
  of cellulose nitrate 9:64
  of hot-melt adhesives 4:402
  of lactic acid 15:147
  of organic friction modifiers 7:447
  of pesticides and insecticides 19:616, 618
  of polymers and fibers 4:263
Heavy metals
  in acetic acid 4:99
  in acid anhydrides 8:568
  in acid halides 8:575
  in acid salts 8:596
  in adsorbents 4:439
  in aluminum acetate 5:163
  in aluminum chloride 5:176
  in aluminum hydroxide 5:188
  in aluminum sulfate 5:193
  in alums 5:193
  in amino acids 8:438
  in ammonium acetate 5:326
  in ammonium carbonate 5:333

in ammonium chloride 5:337
in ammonium nitrate 5:343
in ammonium oxalate 5:345
in ammonium persulfate 5:347
in ammonium phosphate, dibasic 5:350
in ammonium sulfate 5:352
in ammonium thiocyanate 5:354
in antibiotics 5:492
in arsenic(V) oxide 6:233
in barium acetate 6:537
in barium carbonate 6:544
in barium chloride 6:546
in barium hydroxide 6:550, 551
in barium nitrate 6:552
in barium peroxide 6:555
in barium sulfate 6:567
in borax 7:383
in boric acid 7:382
in bromine 8:6
in cadmium bromide 8:38
in calcium acetate 8:79
in calcium carbonate 8:84
in carbonyl compounds 8:363
in certified colors 10:494
in choline and its derivatives 9:628
in chromium compounds 9:699
in cinnamic acid and its derivatives 10:70
in citrate esters 10:98
in citric acid and salts 10:94
in clays and clay minerals 10:130
in copper(II) acetate 10:655
in copper(II) nitrate 10:668
dithizone colorimetric method 19:222
dithizone extraction 10:94
in essential oils 12:289
in esters 8:590
in fluorosilicic acid 13:50
in formaldehyde 13:114
in formic acid 13:130
in formic acid esters 13:134
in gelatin 13:369
in glycerol 13:526
in gypsum 14:98
in hydrochloric acid 9:340, 351; 14:373
in hydrofluoric acid 13:34
in hydrogen peroxide 14:437
hydrogen sulfide colorimetric method 9:349,
    628; 10:99; 12:289; 13:33, 34, 130, 369;
    14:373, 437, 454; 15:143, 281, 399, 402,
    411, 490; 16:61, 537; 17:111, 403; 18:234,
    283; 19:19
hydrogen sulfide gravimetric method 12:415
in hydroxylamine salts 14:454
in lactic acid and its derivatives 15:143
in lithium carbonate 15:281
in magnesium acetate 15:399
in magnesium carbonate 15:402

in magnesium chloride 15:405
in magnesium nitrate 15:407
in magnesium oxide 15:411
in magnesium perchlorate 9:364, 366
in magnesium sulfate 15:412
in manganese sulfate 15:490
in methylcellulose 9:81
in mercuric iodide 16:61
in nicotinamide 17:439
in nicotinic acid 17:439
in nitric acid 16:498
in penicillins 16:566
in perchloric acid 9:364, 365
in phosphorus pentoxide 17:111
in potassium chlorate 9:359, 362
in potassium chloride 9:339
in potassium hydrogen phthalate 17:403
in sodium arsenate 6:234
in sodium benzoate 7:87
in sodium chloride 9:337, 349
in sodium fluoride 13:33
in sodium metal 18:234
in sodium peroxide 16:537
in sodium tungstate 19:222
solvent extraction 19:222
in strontium chloride 18:283
in strontium nitrate 18:284
in tetracene 12:415
in thorium nitrate 19:19
titration of dithiocarbamic acid salts 11:585
in vinbarbital 6:529
Heavy metal soaps. (See also *Soaps*.)
    manufacture 18:179
    uses 18:179
Heavy spar 18:360
Heazlewoodite
    composition 16:286
Hecla alloys. (See also Cobalt alloys.)
    composition 10:287
Hectorite. (See also *Clays and clay minerals*.)
    composition 10:102
HEDTA. See HOEDTA.
Hedulin. See Phenindione.
HEEDA. (See also *EDTA and related com-
    pounds*.)
    structure 12:25
HEEDTA. See HOEDTA.
HEIDA. See HIMDA.
Heintzite 7:314
Helindone pink. See D & C Red No. 30.
Helio Bordeaux B. L.
    in enamels 12:182, 186
    identification 12:182, 186
Helio fast red R.M.T.A.
    in enamels 12:183
    identification 12:183
Helio fast rubin

in enamels 12:186
identification 12:186
Helio fast rubine 4BL. (See also *Pigments*.)
structure 17:182
Heliotropin (3,4-Methylenedioxybenzaldehyde;
    Piperonal) 6:597
commercial grades and specifications 6:609
derivatives for identification 6:612
fluorometry 6:626
hydroxylamine hydrochloride method 6:620
manufacture 6:601
near-infrared spectroscopy 6:629
properties 6:598
ultraviolet spectroscopy 6:633
uses 6:606
Helium. (See also *Gases, noble*.)
in argon 13:299
commercial specifications 13:287
gas chromatography 13:297, 306
in hydrogen 14:380
impurities, determination of 13:298
infrared spectroscopy 13:295
isotopic abundance 13:289
in krypton 13:300
manufacture 13:290
mass spectroscopy 13:301
in neon 13:298
occurrence 13:290
properties 13:288
purification of carrier gas 8:672
solubility 13:289
thermal conductivity 14:380
in xenon 13:300
Helium group. See *Gases, noble*.
Hematein
colorimetric method for germanium 13:379
colorimetric method for niobium 16:369
colorimetric method for tantalum 16:369
Hematite 15:21; 19:81
commercial grades and specifications 17:194
composition 10:104; 15:22
Hematoxylin
colorimetric method for niobium 16:369, 375
colorimetric method for tantalum 16:369;
    18:449
reagent for identification of aluminum 5:213
Hemicellulose content
of green coffee 10:374
of roasted coffee 10:375
Hemimellitic acid chromatography
paper chromatography 8:379
Hemimorphite
composition 19:545
Hemoglobin
reaction with carbon monoxide 8:304
Hemolysis inhibition method
for insulin 14:575

Hempel apparatus for gas absorption 4:128
Henna
commercial specifications 10:535
composition 10:535
1-Hentriacontanol. See Melissyl alcohol.
Heparin. See Sodium heparin.
Heptabarbital 6:468
properties 6:470
solubility 6:476
thin-layer chromatography 6:481
Heptachlor. (See also *Pesticides and insecticides*.)
commercial grades and specifications 19:593
composition 19:588, 596
ethanolamine–butyl Cellosolve colorimetric
    method 19:611
gas chromatography 19:389
manufacture 19:596
properties 19:588
in raw milk 16:129
solubility 19:587
structure 19:581
toxicity 19:592
in water 19:389
Heptachlor epoxide
gas chromatography 19:389
in raw milk 16:129
in water 19:389
Heptachlorobiphenyls. (See also Chlorinated bi-
    phenyls.)
properties 9:492
1,4,5,6,7,8,8-Heptachloro-2,3-epoxy-3a,4,7,7a-
    tetrahydro-4,7-methanoindan. See Heptachlor
    epoxide.
Heptachloronaphthalene. (See also Chlorinated
    naphthalenes.)
properties 9:490
Heptachloropropanes. (See also C_3 Chlorinated
    hydrocarbons.)
properties 9:471
1,4,5,6,7,8,8-Heptachloro-3a,4,7,7a-tetrahydro-
    4,7-methanindan. See Heptachlor.
1-Heptadecanol. See Margaryl alcohol.
2-Heptadecanone
absorption spectroscopy 9:564
2-hydroxy-1-naphthaldehyde colorimetric
    method 11:204
n-Heptadecanonitrile. (See also *Nitriles*.)
properties 16:396
2-Heptadecyl-2-imidazoline acetate. See Glyodin.
trans,trans-2,4-Heptadienal
in black tea 18:511
gas chromatography 18:511
Heptafluoroniobic acid
manufacture 16:356
Heptaldehyde
properties 8:334
Heptamethylenediamine

gas chromatography 17:290
2,2,4,4,6,8,8-Heptamethylnonane. (See also
    Alkanes.) 14:306
    commercial specifications 14:
        306
    uses 14:306
Heptanal
    in black tea 18:510
    2,4-dinitrophenylhydrazine colorimetric
        method 8:363
    gas chromatography 8:355; 18:
        510
    infrared spectroscopy 6:616
n-Heptane. (See also Heptanes.)
    analysis 14:332
    in benzene 7:11
    commercial grades and specifications 14:305,
        306
    gas chromatography 7:11; 11:73; 13:319, 350;
        14:330, 332; 16:91
    in gasoline 13:350, 353
    infrared spectroscopy 7:30
    in isooctane 14:332
    in nail enamels 11:73
    properties 13:350
    toxicity 14:306
    uses 14:303
Heptanes. (See also Alkanes; and individual com-
        pounds.) 14:303
    in dimethyldichlorosilane 18:64
    flammability 14:306
    gas chromatography 13:317; 18:64
    in natural gas 13:320
    solubility 13:539, 552
1-Heptanethiol. See Heptyl mercaptan.
Heptanoic acid
    properties 8:397
1-Heptanol. (See also Fatty alcohols.)
    gas chromatography 7:96
    properties 4:561
    pyromellitic dianhydride method 4:537
    solubility 4:562
2-Heptanol
    gas chromatography 7:96
    phthalic anhydride method 4:536
3-Heptanol
    2,4-dinitrophenylhydrazone colorimetric meth-
        od 4:549
4-Heptanol
    2,4-dinitrophenylhydrazone colorimetric meth-
        od 4:549
2-Heptanone
    gas chromatography 8:355
4-Heptanone
    absorption spectroscopy 9:564
Heptanones
    2-hydroxy-1-naphthaldehyde colorimetric

method 11:204
n-Heptanonitrile. (See also Nitriles.)
    derivatives for identification 16:402
    properties 16:396
Heptenes. (See also Hydrocarbons.)
    commercial grades and specifications 14:
        313
    gas chromatography 13:350; 14:334
    in gasoline 13:350, 353
    infrared spectroscopy 7:30
    properties 13:350
Heptylamines
    manufacture 5:262
Heptyldiphenylamine
    paper chromatography 6:85
n-Heptyl ether (See also Ethers.)
    properties 12:297
n-Heptyl p-hydroxybenzoate
    in beer 7:594
    colorimetric method using aminoantipyrine
        7:602
    gas chromatography 7:600
    hydrolysis 7:601
    surface tension method 7:595
    ultraviolet spectroscopy 7:602
Heptyl mercaptan. (See also Mercaptans.)
    derivatives for identification 15:559
    properties 15:552
n-Heptylpenicillin. (See also Penicillins.)
    in benzylpenicillin 16:566, 570
    solvent extraction 16:571
    structure 16:554
Heraeus-Feichtinger vacuum fusion analyzer
        8:657, 658
Herich-Toth-Osztrovszky method for carbohydrate
        determination 4:474
Herniarin. See 7-Methoxycoumarin.
Heroin. (See also Narcotics.) 16:271
    acid–base titration 16:271
    ammonium molybdate colorimetric method
        16:246
    ammonium vanadate colorimetric method
        16:246
    analysis 16:271
    assay 16:271
    color tests 16:246
    formaldehyde–sulfuric acid colorimetric meth-
        od 16:246
    gas chromatography 16:254, 271
    impurities 16:271
    ion-exchange methods 16:271
    manufacture 16:271
    mercuric chloride microchemical method
        16:249
    microchemical tests 16:249
    paper chromatography 16:255
    platinic chloride microchemical method 16:249

polarography 16:268
properties 16:271
solvent extraction 16:271
structure 16:244
thin-layer chromatography 16:256
ultraviolet spectroscopy 16:251, 271
Herring oil. (See also *Fats and fatty oils*.)
  properties 12:473
Hetacillin. (See also Ampicillin.)
  analysis 16:575
  assay 16:576
  p-dimethylaminocinnamaldehyde colorimetric
    method 16:577
  microbiological methods 16:576
  properties 16:556, 575
  specification tests 16:576
  structure 16:554
Heterocyclic antibiotics 5:462, 509, 515
  properties 5:507, 510
  solubility 5:507, 508
Heterogenite 10:284
Heteropoly blue colorimetric method
  for silica 6:260
Heteropoly molybdates. See Molybdates; and
    individual compounds.
Heulandite 8:75
Heusler alloys. See Ferromagnetic alloys.
Hewlett-Packard model 185 C,H, and N analyzer
    8:286
  operational details 8:287, 291
  schematic diagram 8:287, 291
Hexaborane(10) 7:341
  properties 7:343
4,5,7,4',5',7'-Hexabromoindigo. (See also *Indi-
    goid dyes*.)
  properties 14:496
Hexachlorobenzene. (See also Chlorinated ben-
    zenes.) 9:483
  analysis 9:483
  in chlorine 9:321
  commercial grades and specifications 9:483
  composition 9:411
  gas chromatography 9:378
  impurities, determination of 9:483
  mass spectroscopy 9:408
  melting point 9:483
  properties 9:483
  solubility 9:483
  specification tests 9:483
  toxicity 9:483
  uses 9:483
Hexachlorobiphenyls. (See also Chlorinated bi-
    phenyls.)
  properties 9:492
Hexachloro-1,3-butadiene. (See also C_4 Chlo-
    rinated hydrocarbons.)
  properties 9:473

1,2,3,4,5,6-Hexachlorocyclohexane. See BHC;
    Lindane.
1,2,3,4,10,10-Hexachloro-6,7-epoxy-1,4,4a,5,6,7,
    8,8a-octahydro-1,4-*endo-exo*-5,8-dimethano-
    naphthalene. See Dieldrin; Endrin.
Hexachloroethane. (See also C_2 Chlorinated
    hydrocarbons.) 9:449
  analysis 9:451
  azeotropes 9:450
  in chlorine 9:321
  commercial grades and specifications 9:450
  composition 9:411
  Fujiwara colorimetric method 9:386
  gas chromatography 9:377, 380
  impurities, determination of 9:451
  manufacture 9:450
  mass spectroscopy 9:403
  melting point 9:451
  properties 9:449
  solubility 9:449
  specification tests 9:451
  toxicity 9:450
  uses 9:449
1,2,3,4,10,10-Hexachloro-1,4,4a,5,8,8a-hexahydro-
    exo-1,4-*endo*-5,8-dimethanonaphthalene. See
    Aldrin.
6,7,8,9,10,10-Hexachloro-1,5,5a,6,9,9a-hexahydro-
    6,9-methano-2,4,3-benzodioxathiepin-3-oxide.
    See Endosulfan.
Hexachloronaphthalene. (See also Chlorinated
    naphthalenes.)
  toxicity 9:490
1,4,5,6,7,7-Hexachloro-5-norbornene-2,3-dicar-
    boxylic acid. See Chlorendic acid.
Hexachlorophene. (See also Bisphenols.)
  acid-base titration 11:547
  analysis 11:546
  assay 11:547
  in blood 11:550
  in deodorants 11:87
  gas chromatography 11:88, 550
  identification 11:546
  impurities, determination of 11:546
  loss on drying 11:546
  melting range 11:547
  in pharmaceuticals 11:548
  properties 11:546
  residue on ignition 11:547
  in soap 11:550; 18:199
  solubility 11:546
  specification tests 11:546
  structure 17:3
  ultraviolet spectroscopy 11:87, 548, 549;
    18:199
  in urine 11:550
  uses 11:546; 17:3
Hexachloroplatinic acid. See Chloroplatinic acid.

Hexachloropropanes. (See also C₃ Chlorinated hydrocarbons.)
  properties 9:471
Hexachloropropene. (See also C₃ Chlorinated hydrocarbons.)
  properties 9:471
1-Hexacosanol. See Ceryl alcohol.
Hexadecafluoroheptane. (See also *Fluorocarbons.*)
  properties 13:61
n-Hexadecane. See *n*-Cetane.
1,2-Hexadecanediol
  gas chromatography 13:577
1-Hexadecanol. (See also Alcohols, higher; Cetyl alcohol.)
  commercial grades 4:555
n-Hexadecanonitrile. (See also *Nitriles.*)
  derivatives for identification 16:402
  properties 16:396
1-Hexadecene. (See also Alpha olefins.)
  commercial specifications 14:314
  liquid chromatography 7:13, 14, 15, 16
1-Hexadecylpyridinium chloride. See Cetyl-pyridinium chloride.
Hexadecyltrimethylammonium chloride
  gravimetric method for boron 7:402
8-Hexadecynl-1-ol. See Palmitoleyl alcohol.
2,4-Hexadienal
  ultraviolet spectroscopy 6:632
trans,trans-2,4-Hexadienal
  in black tea 18:510
  gas chromatography 18:510
1,5-Hexadiene. (See also Biallyl.)
  liquid chromatography 7:13, 14, 15, 16
Hexa(2-ethylbutoxy)disiloxane
  uses 15:302
Hexafluoroethane. (See also *Fluorocarbons.*)
  properties 13:60
Hexafluorophosphoric acid. (See also Fluoro-phosphoric acids.) 13:37, 42
  commercial grades and specifications 13:44
  composition, determination of 13:46
  gravimetric determination 13:46
  impurities, determination of 13:47
  manufacture 13:45
  in monofluorophosphoric acid 13:46
  nitron gravimetric method 13:46
  properties 13:44
  tetraphenylarsonium chloride gravimetric method 13:46
  uses 13:43
Hexafluorosilicic acid. (See also Fluorosilicic acid.)
  distillation method for fluorides 13:16
Hexaglycerol. See Trimethylolpropane.
Hexahydroaniline. See Cyclohexylamine.
Hexahydrobenzene
  ferric chloride colorimetric method 17:16

Hexahydro-1,3,4,5-tetrahydroxybenzoic acid. See Quinic acid.
Hexahydroxybenzene. (See also *Phenols.*)
  derivatives for identification 17:21
  properties 17:5
  solubility 17:5
Hexametaphosphate 17:124
Hexamethyldisilazane. (See also Silylation reagents.)
  commercial grades and specifications 18:92
  gas chromatography 18:94
  impurities, determination of 18:94
  infrared spectroscopy 18:92
  manufacture 18:91
  properties 18:90
  uses 18:89
Hexamethyldisiloxane
  in *N,O*-bis(trimethylsilyl)acetamide 18:94
  gas chromatography 18:94
  in hexamethyldisilazane 18:94
  in silylation reagents 18:93
Hexamethylenediamine. (See also *Diamines and higher amines, aliphatic.*) 4:423, 428
  gas chromatography 17:290
  in polyurethanes 19:304
  properties 11:410
  structure 19:304
  thin-layer chromatography 19:305
Hexamethylene diisocyanate. (See also Isocyanates; *Isocyanic acid esters.*)
  distillation methods 19:316
  in polyurethanes 19:313
  properties 15:95
  thin-layer chromatography 19:315
  trade names 15:102; 19:271
Hexamethylene glycol. See 1,6-Hexanediol.
Hexamethylenetetramine. (See also Hexamine; Urotropine.)
  in beer 7:585
  titration method for allyl halides 5:102
Hexamine
  gravimetric method for thorium 17:495
Hexanal
  absorption spectroscopy 9:564
  in black tea 18:510
  2,4-dinitrophenylhydrazine colorimetric method 8:363
  gas chromatography 8:355; 18:510
n-Hexane. (See also Hexanes.)
  azeotropes 14:303
  calorific value 13:331
  commercial specifications 14:304
  in engine starting fluids 6:356
  extraction of fats from cocoa beans 9:562

gas chromatography 11:73; 13:319, 350;
    14:330, 335
in gasoline 13:350, 353
in nail enamels 11:73
properties 13:350; 14:303
uses 14:303
1,6-Hexanediamine. See Hexamethylenediamine.
Hexanedioic acid. See *Adipic acid.*
1,6-Hexanediol. (See also Glycols.)
commercial grades and specifications 13:547
gas chromatography 13:504, 576, 577; 19:298
manufacture 13:545
in polyurethanes 19:297
properties 13:537
solubility 13:538
structure 13:534
uses 13:545
2,5-Hexanediol. (See also Glycols.)
commercial grades and specifications 13:547
gas chromatography 13:576
manufacture 13:545
properties 13:536
solubility 13:538
structure 13:534
uses 13:545
Hexanediols. (See also Glycols.)
paper chromatography 13:581
Hexanes. (See also Alkanes; and individual com-
    pounds.) 14:303
commercial specifications 14:303
flammability 14:303
gas chromatography 13:317; 18:61, 64
in methyldichlorosilane 18:64
in methyltrichlorosilane 18:64
in natural gas 13:320
occurrence 14:303
specific gravity 13:329
toxicity 14:303
in trimethylchlorosilane 18:64
uses 14:303
1-Hexanethiol. See Hexyl mercaptan.
1,2,6-Hexanetriol. (See also Polyhydric alcohols.)
commercial grades and specifications 13:566
manufacture 13:564
in polyurethanes 19:299
properties 13:562
solubility 13:560
structure 13:561
uses 13:564
Hexanitrodiphenylamine. (See also High explosives.)
melting point 12:461
properties 12:434
2,2',4,4',6,6'-Hexanitrodiphenylamine. See Di-
    picrylamine.
Hexanoic acid
in black tea 18:511
gas chromatography 18:511

1-Hexanol. (See also Alcohols, higher; Fatty alco-
    hols.)
acetylation with acetic anhydride 4:530
azeotropes 4:554, 560
in black tea 18:510
in brake fluids 7:424
in brandies 19:440
commercial grades 4:555
3,5-dinitrobenzoyl chloride method 4:547
gas chromatography 7:96; 16:98; 18:510;
    19:442, 446, 449
infrared spectroscopy 7:30
in methanol 16:114
phthalic anhydride method 4:559
properties 4:552, 561
solubility 4:552, 562
specifications 4:556
uses 4:554
in wines 19:425, 440
2-Hexanol
2,4-dinitrophenylhydrazone colorimetric meth-
    od 4:549
gas chromatography 7:96
Hexanols. (See also Alcohols, higher.)
azeotropes 4:554
properties 4:552
specifications 4:556
uses 4:554
n-Hexanonitrile. (See also *Nitriles.*)
derivatives for identification 16:401, 402
properties 16:396
trans-2-Hexenal
in black tea 18:510
gas chromatography 18:510
1-Hexene. (See also Alpha olefins.)
commercial specifications 14:314
gas chromatography 14:334
liquid chromatography 7:13, 14, 15, 16
Hexenes
gas chromatography 13:350, 353
in gasoline 13:350, 353
properties 13:350
2-Hexenes
gas chromatography 14:334
trans-2-Hexenoic acid
in black tea 18:511
gas chromatography 18:511
cis-3-Hexenoic acid
in black tea 18:511
gas chromatography 18:511
trans-2-Hexen-1-ol
in black tea 18:510
gas chromatography 18:510
3-Hexen-1-ol
in black tea 18:510
gas chromatography 18:510
cis-3-Hexen-1-ol

gas chromatography 7:96
5-Hexen-2-one
  2-hydroxy-1-naphthaldehyde colorimetric method 11:204
cis-3-Hexenyl acetate
  in black tea 18:510
  gas chromatography 18:510
Hexethal
  gas chromatography 6:483, 484
  thin-layer chromatography 6:481
Hexethal sodium 6:489
  infrared spectrum 6:489
Hexobarbital 6:468, 517
  derivatives for identification 6:477
  extraction 6:491
  gas chromatography 6:483, 484
  paper chromatography 6:479
  properties 6:470
  solubility 6:476
  thin-layer chromatography 6:480, 481
Hexobarbital sodium 6:517
  analysis 6:517
  commercial grades and specifications 6:516
  impurities, determination of 6:517
Hexogen. See RDX.
Hexylamines
  manufacture, methods of 5:262
Hexylcaine hydrochloride 5:375
  analysis 5:391
  colorimetric method using methyl orange 5:392
  properties 5:391
  ultraviolet spectroscopy 5:391
α-*n*-Hexyl cinnamaldehyde
  commercial specifications 10:44
Hexyl 3,5-dinitrobenzoate
  thin-layer chromatography 16:94
Hexylene glycol. (See also 2-Methyl-2,4-pentanediol.)
  density 13:584
  refractive index 13:584
2,5-Hexylene glycol. See 2,5-Hexanediol.
n-Hexyl ether. (See also *Ethers.*)
  properties 12:297
Hexyl mercaptan. (See also *Mercaptans.*)
  derivatives for identification 15:559
  properties 15:552
n-Hexyl methacrylate
  nuclear magnetic resonance spectroscopy 16:86
  thin-layer chromatography 16:90
Hexyl nicotinate. (See also *Pyridine and pyridine derivatives.*)
  trisodium pentacyanoammino ferrate colorimetric method 17:433
4-Hexylresorcinol
  colorimetric method for crotonaldehyde 11:180

1-Hexyne
  liquid chromatography 7:13, 15
HHDN. See Aldrin.
Hiding power
  of enamels 12:178
High-alumina brick. (See also Refractories.)
  composition 9:206
High-alumina cement. (See also Hydraulic cements.) 9:95
  commercial specifications 9:99, 100, 101
  composition 9:96
High-alumina chemical ware. (See also *Ceramics.*)
  composition 9:194
High-alumina refractories. (See also Refractories.)
  composition 9:208
  fluxes for 9:216
  properties 9:208
High-explosive mixtures. (See also High explosives.) 12:452
  analysis 12:446, 460
  ash content 12:467
  assay 12:463
  bulk density 12:467
  commercial specifications 12:446
  composition 12:446
  moisture content 12:467
  quality control 12:467
  separation 12:461
  solidification point 12:468
  solvent extraction 12:462
  specific gravity 12:467
  trade names 12:446
  uses 12:446
High explosives. (See also High-explosive mixtures; and individual compounds.) 12:431–470
  additives 12:435
  analysis 12:452
  colorimetric determination 12:421
  color tests 12:458
  in explosive priming mixtures 12:421, 422
  identification 12:421, 452
  infrared spectroscopy 12:452
  melting point 12:455
  nitrogen content 12:460
  solubility 12:421, 462
High-frequency discharge method
  specimen preparation for electron microscopy 8:203
High-lead glass. (See also Silicate glasses.) 9:200
  composition 9:198
  uses 9:198
High-silica ceramics. (See also *Ceramics.*)
  acid dissolution 9:215
High-speed steel (See also Cobalt alloys; Tungsten alloys.) 10:289
  composition 10:289; 19:192
  uses 19:192

High-temperature alloys. (See also Cobalt alloys.)
10:286, 288
High-temperature combustion method
for sulfur compounds, organic 18:379
High-temperature stability
of brake fluids 7:435
High wines
analysis 4:480
definition 4:462
quality control 4:469
High-zirconia refractories. (See also Refractories.)
fluxes for 9:216
Hilgardite 7:315
HIMDA. (See also *EDTA and related compounds.*)
structure 12:25
Hippuric acid 7:60
gas chromatography 8:389
in manufacture of benzoic acid 7:60
Histamine 6:24
Histidine. (See also *Carboxylic acids, amino.*)
8:412, 482
analysis 8:486
in cocoa beans 9:556
colorimetric methods 8:487
commercial specifications 8:486
composition 8:412
crystalline form 8:484
gas chromatography 8:429
ion exchange chromatography 8:415, 421
manufacture 8:485
occurrence 8:485
paper chromatography 8:422
properties 8:483
in protein 8:414
solubility 8:485
structure 8:483
in yeast 19:529
Hjermstad method
for gel strength of corn starch 11:26
HMF. See Hydroxymethylfurfural.
HMS
in roasted cocoa beans 9:542
HMX. (See also High explosives.) 12:440
chromotropic acid color test 12:440
commercial specifications 12:441
impurities, determination of 12:441
infrared spectrum 12:454
manufacture 12:441
melting point 12:461
properties 12:432, 440
solubility 12:441
uses 12:440
HOEDTA. (See also *EDTA and related compounds.*)
structure 12:24
zinc chloride titration 12:33
Hoeppler viscometer method 7:515

Hoesch reaction
for nitriles 16:400
Hoffman and Lundell method
for silicon 18:22
Hog bile
in preparation of hyodeoxycholic acid 7:144
Holmium. (See also Holmium metal; *Rare earths.*)
absorption spectroscopy 17:483
atomic absorption spectroscopy 17:494
bead test 6:587
in dysprosium oxide 17:502
emission spectroscopy 15:153, 155; 17:501
in erbium oxide 17:503
flame photometry 17:492
in gadolinium oxide 17:502
in lanthanum oxide 15:153, 155; 17:501
in thulium oxide 17:503
x-ray fluorescent spectroscopy 17:490
in yttrium oxide 17:501
Holmium fluorides
carrier gas distillation fusion methods 8:636
oxygen in 8:635, 636
vacuum distillation fusion methods 8:635
Holmium metal. (See also Rare earth metals.)
carrier gas fusion methods 8:627, 682
hot extraction methods 8:684
hydrogen in 8:623, 680, 684
nitrogen in 8:623, 627
oxygen in 8:623, 627, 680, 682
properties 17:471
vacuum fusion methods 8:623, 680
vacuum hot extraction methods 8:684
Holmium oxide. (See also Rare earth oxides.)
emission spectroscopy 17:503
impurities, determination of 17:503
properties 17:497
Holmium perchlorate. (See also Rare earth compounds.)
absorption spectrum 17:485
Holocellulose content
of green coffee 10:374
Homatropine 4:587
analysis 4:594
Homilite 7:316
Homogenization efficiency
of frozen desserts 16:143
of ice cream 16:143
of milk 16:133
Homogenized milk. See Milk.
Homogentisic acid
gas chromatography 8:389
Homomenthyl cinnamate. (See also *Cinnamic acid and its derivatives.*)
uses 10:69
Homopolystyrene. See Polystyrene.
Homoprotocatechuic acid
gas chromatography 8:389

Homovanillic acid
  gas chromatography 8:389
Homoveratic acid
  gas chromatography 8:389
Honey
  as antifreeze 6:1
Hoolamite detector
  for carbon monoxide 8:305
Hopeite
  composition 19:545
Hops
  in beer 7:603
Hormestral. See Estrone.
Hormones. (See also Nonsteroid hormones;
    Steroids.) 14:179–231
  analysis 14:205
  classification 14:179
  definition 14:179
  properties 14:187
  sources 14:187
  uses 14:187
Hornblende 8:75
Horn silver. See Cerargyrite.
Hortvet cryoscope 16:124
Hostaflon. See Polytetrafluoroethylene.
Hot extraction
  gasometric method for hydrogen in hydrides
    14:249
Hot-melt adhesives, evaluation 4:401
Hot-paste viscosity
  amylograph method 12:611
  falling number test 12:611
  of flour 12:609
Hot-point hardness tester 12:256
Household bleach. See Hypochlorite bleaches.
Howlite 7:316
HTA. (See also High-explosive mixtures.)
  analysis 12:449
  composition 12:449
  uses 12:449
Hubach apparatus 19:483
Hubnerite 14:519; 19:190
  decomposition 19:197
Huggins equation
  for polycarbonates 17:343
Hughes-Acree
  volumetric method for furfural 13:236
Hulsite 7:315
Human tissues
  organomercury content 16:64
Humectants
  in cigarettes 10:29
  in confectionery products 10:552
  in embalming fluids 12:166
Humulone
  in beer 7:604
Huntite 8:75

Hutchinsonite 18:519
Huttonite
  composition 19:2
  thorium content 19:2
Hveem tests
  for cohesion and stability of bituminous paving
    mixtures 6:341
Hyalotekite 7:316
Hyamine 1622 titration
  of anionic surfactants 11:78
Hycar 12:86
Hydantoin formation 8:337
Hydantoins, chlorinated
  available chlorine 7:228
  bleaching agents 7:222
Hydrabamine benzylpenicillin. (See also Benzyl-
    penicillin.)
  properties 16:556
  solubility 16:557
  structure 16:567
Hydrabamine penicillin G. See Hydrabamine
    benzylpenicillin.
Hydrabamine penicillin V. See Hydrabamine
    phenoxymethylpenicillin.
Hydrabamine phenoxymethylpenicillin. (See
    also Phenoxymethylpenicillin.)
  iodometric methods 16:562
  solubility 16:557
Hydrated ceric oxide. See Ceric hydroxide.
Hydrated chromium oxide
  analysis 17:201
  properties 17:201
  uses 17:201
Hydrated lime. (See also Lime.)
  commercial grades and specifications 9:191
Hydraulic cements. (See also Cements.) 9:94
  alumina content 9:96, 100
  calcium aluminate content 9:107
  calcium aluminum silicate content 9:107
  calcium oxide content 9:96, 100, 107
  commercial grades and specifications 9:99
  composition 9:96
  composition, determination of 9:102
  dicalcium silicate content 9:107
  ferric oxide content 9:96, 100
  ferrite content 9:107
  hardening 9:97
  magnesia content 9:96, 100
  pozzolan content 9:107
  setting 9:97
  silica content 9:96, 100
  slag content 9:107
  tricalcium silicate content 9:107
Hydraulic fluids. See Brake fluids
Hydraulic fluids, nonpetroleum 7:423
Hydraulic oil. (See also *Lubricants.*)
  commercial grades and specifications 15:312

Hydrazine 14:232-241; 16:471
  analysis 14:234
  assay 14:236
  colorimetric determination 14:235, 238
  commercial grades and specifications 14:234
  coulometric method 14:238
  derivatives for identification 14:235
  gas chromatography 14:237, 239
  gravimetric method for selenium 17:591, 592
  hydrolysis reagent for cystine and cysteine
    8:467
  identification 14:235
  impurities, determination of 14:239
  infrared spectroscopy 14:235
  manufacture 14:233; 16:471
  polarography 14:238
  potassium iodate titration 14:236
  properties 14:232; 16:471
  proton magnetic resonance spectroscopy
    14:237
  solubility 14:232
  toxicity 14:233
  uses 14:232; 16:471
  volumetric determination 14:237
p-Hydrazinebenzenesulfonate
  colorimetric method for selenium 13:398
Hydrazine hydrate
  in *o*-toluidine 5:446
Hydrazines
  reaction of carbonyl compounds with 8:333
Hydrazine sulfate-iodine
  volumetric method for nitrite 16:492
  volumetric method for potassium nitrite
    16:492
  volumetric method for sodium nitrite 16:492
Hydrazoic acid 16:472
  acid-base titration 16:472
  ammonium hexanitratocerate(IV) volumetric
    method 16:472
  analysis 16:472
  manufacture 16:472
  properties 16:472
  sodium nitrite titration 16:472
Hydrides. (See also individual compounds.)
    14:242-260
  in aluminum alkyls 5:171
  analysis 14:247
  assay 14:247
  classification 14:242
  combustion method 14:249
  hydrogen content of 8:614
  hydrogen evolution method 14:247
  in metals 8:613, 639
  properties 14:243
  thermal decomposition 14:249
  uses 14:242
Hydrin 12:86

Hydriodic acid. (See also Iodides.)
  assay 14:598
  impurities, determination of 14:603
  Volhard titration 14:598
  volumetric method for alkoxyl groups 16:92
  volumetric method for oxyethylene-oxypro-
    pylene copolymers 19:308
  volumetric method for polyethers 19:308
Hydriodic acid–hypophosphoric acid reduction
  of nitrogen, organic 10:488
Hydroabietyl alcohol
  modifier for alkyd resins 5:24
Hydroboracite 7:314
Hydrobromic acid
  acid-base titration 8:10
  analysis 8:10
  colorimetric method for copper 10:615;
    15:382
  colorimetric method for gold 14:23, 31
  commercial grades and specifications 8:8, 10
  distillation of antimony trioxide 13:430
  formation from bromine 8:3
  impurities, determination of 8:11
  properties 8:9
  uses 8:4
Hydrocal. (See also Dental gypsum products.)
  uses 11:263
Hydrocarbon oils. (See also *Lubricants.*) 15:301
  composition 12:145; 15:301
  in detergents 11:381
  in elastomers, synthetic 12:145
  infrared spectra 12:144
  uses 15:301
Hydrocarbon resin emulsions. (See also *Hydrocarbon resins.*)
  commercial grades and specifications 14:281
Hydrocarbon resins 14:261-297
  acid number 14:284
  analysis 14:281
  aniline point 14:285
  ash content 14:284
  bromine number 14:286
  m-chloroperbenzoic acid volumetric method
    14:288
  cloud point 14:292
  color 14:266, 281, 293
  commercial grades and specifications 14:274
  density 14:267, 294
  fire point 14:294
  flash point 14:294
  history 14:261
  impurities, determination of 14:284
  infrared spectroscopy 14:268
  iodine number 14:286, 288
  manufacture 14:261, 271
  nonvolatile content 14:284
  oxygen absorption method 14:289

polymer distribution 14:269
polymer structure 14:267
properties 14:261, 266
properties, determination of 14:281, 284, 290
raw materials 14:263, 282
refractive index 14:294
saponification number 14:284
softening point 14:266, 281, 290
solubility 14:270, 282
specification tests 14:282
types 14:261, 262
unsaturation, degree of 14:285, 288
viscosity 14:282, 294
Hydrocarbon resin solutions. (See also *Hydrocarbon resins.*)
commercial grades and specifications 14:280
Hydrocarbons. (See also individual compounds.)
14:298–354
acid absorption 14:340
in ammonia 5:313
analysis 14:326
aniline point 14:348
bromine index 14:339
bromine number 14:336
bromometric titration 14:336
in carbon dioxide 8:252, 259
in carbon monoxide 8:310
in cigarette smoke 10:13
color 14:343
composition, determination of 14:326
copper corrosion test 14:347
in creams and lotions 11:94
degreasers 6:355
distillation methods 17:37
distillation range 14:341
in enamels 12:189
in fatty alcohols 4:573
flash point 14:342
freezing point 14:341
by gas chromatography 5:313; 7:7; 8:262; 14:326; 16:456
in gasoline 13:345
hydrogen flame ionization detector method 8:259
impurities, determination of 14:344
infrared spectroscopy 8:252
kauri-butanol value 14:348
liquid chromatography 7:14; 14:335
mass spectroscopy 7:31
in methanol 16:113
naphthene content 14:341
in noble gases 13:303
occurrence 14:298
in oxygen 16:532
in phenols 17:37
properties 14:298
in shampoos 11:76

in soaps 18:195, 201
solvent extraction 14:349
specification tests 14:341
specific gravity 14:341
steam distillation 18:195
in sunscreens 11:91
toxicity 14:298
unsaturation, determination of 14:336
uses 14:298
vapor pressure 14:342
water in 7:53; 14:349
Hydrocarbons, aliphatic. (See also individual compounds.)
cooling system flushes 6:352
tar removers 6:354
Hydrocarbons. aromatic. (See also individual compounds.)
acid-wash color 7:42; 14:345
in carburetor cleaners 6:356
copper corrosion test 14:347
degreasers 6:355
gas chromatography 7:7; 14:331, 335, 336, 337
in gasoline 14:335
impurities, determination of 14:345
liquid chromatography 14:336
manufacture 14:317
tar removers 6:354, 366
ultraviolet spectroscopy 7:39, 40
Hydrocarbons, nonaromatic. (See also individual compounds.)
in benzene 7:47
gas chromatography 7:47
Hydrocarbon solvents. (See also *Hydrocarbons.*)
14:326
Hydrocarbons, unsaturated. (See also individual compounds.)
in cyclopropane 5:361
in Stoddard solvent 11:606
sulfuric acid absorption 11:606
Hydrocarbon vapors
in empty tanks 14:349
Hydrochloric acid 9:336; 14:354–376
acid–base titration 14:361, 366, 368
in acid mixtures 14:367
in acids 14:367
analysis 14:361
assay 9:349; 14:361
azeotropes 14:376
barium sulfate decomposition 14:368
in chlorides 14:366
in chlorinated paraffins 9:498
in chlorine solutions 14:366
color 9:340, 350
colorimetric method for iron 18:53, 79
colorimetric method for niobium 16:372
colorimetric method for rhodium 17:248
colorimetric method for ruthenium 17:247

colorimetric method for whey protein nitrogen 16:147
color test for natural coloring materials 10:537
commercial grades and specifications 9:340, 349; 14:360
conductometric method 14:365
coulometric method 14:365
effervescence test for bicarbonate 18:183
effervescence test for carbonate 18:183
etchant for ceramics 9:224
in gastric juices 14:367
gravimetric method for antimony sulfide 12:420, 424
gravimetric method for calcium silicide 12:421, 424
gravimetric method for lead 15:172
gravimetric method for mercury 16:6
gravimetric method for radium 17:463
gravimetric method for silicon 18:19
gravimetric method for silver 10:645; 18:166
gravimetric method for titanium 19:93
gravimetric method for tungsten 16:172
history 14:354
hydrogen chloride content 9:340, 349
hydrolysis of nitrogen trichloride 9:323
impurities, determination of 9:349, 392; 14:369
ion exchange methods 14:367
manufacture 14:354, 359
mercurimetric method 14:367
in methanol 16:115
microchemical method for apomorphine 16:249
microchemical method for benzylmorphine 16:249
properties 9:336; 14:355
refractive index 14:361
Segal color test 14:367
solubility 14:356
specification tests 14:369
specific gravity 14:361
toxicity 14:358
turbidimetric method for silver 18:171
in uranium 14:366
Hydrochloric acid–acetic acid titration of epoxy compounds 12:211
Hydrochloric acid–dioxane volumetric method for epoxy compounds 12:209
Hydrochloric acid extractable content of activated carbon 8:158
Hydrochloric acid insolubles
in acid salts 8:593
in antimony halides 6:51
in antimony oxides 6:53
in arsenic(III) chloride 6:234
in arsenic(III) oxide 6:231
in barium carbonate 6:540, 543
in barium chromate 6:547
in barium hydroxide 6:550, 551
in barium peroxide 6:554
in barium sulfate 6:567, 568
in bismuth chloride 7:200
in calcium acetate 8:77
in calcium carbonate 8:83
in calcium oxide 8:93
in calcium phosphate 8:96
Hydrochloric acid–nitric acid gravimetric method for tungsten 16:332, 338
Hydrochloric acid, substances not precipitated by
in silver nitrate 12:540; 18:175
in silver sulfate 18:177
Hydrochlorination methods
for butadiene monoxide 12:208
for epichlorohydrin 12:208
for epoxy compounds 12:207
for ethylene oxide 12:208
for glycidyl allyl ether 12:208
for glycidyl benzoate 12:208
for glycidyl-t-butyl ether 12:208
for glycidyl p-chlorophenyl ether 12:208
for glycidyl o-cresyl ether 12:208
for glycidyl dichlorophenyl ether 12:208
for glycidyl diethylcarbinyl ether 12:208
for glycidyl isopropyl ether 12:208
for glycidyl phenyl ether 12:208
for isobutylene oxide 12:208
for propylene oxide 12:208
for styrene oxide 12:208
7-Hydrocholanic acid derivatives
color tests 7:146, 147
Hydrocinnamaldehyde
absorption spectroscopy 9:564
commercial specifications 10:44
Hydrocinnamic acid. (See also *Cinnamic acid and its derivatives.*)
paper chromatography 8:379
properties 8:402
Hydrocinnamyl acetate
commercial specifications 10:78
saponification methods 10:83
Hydrocinnamyl alcohol
commercial specifications 10:78
Hydrocinnamyl cinnamate. (See also *Cinnamic acid and its derivatives.*)
commercial specifications 10:66, 78
properties 10:65
saponification methods 10:68, 83
Hydrocinnamyl formate
commercial specifications 10:78
saponification methods 10:83
Hydrocodone. (See also Hydrocodone bitartrate; *Narcotics.*) 16:272
ammonium molybdate colorimetric method 16:246

analysis 16:272
in antihistamines 16:272
assay 16:272
color tests 16:246
formaldehyde–sulfuric acid colorimetric method
16:246
gas chromatography 16:254
liquid chromatography 16:272
mercuric chloride microchemical method
16:249
microchemical tests 16:249
potassium–bismuth iodide microchemical method 16:249
selenious acid–sulfuric acid colorimetric method 16:246
structure 16:244
thin-layer chromatography 16:256
ultraviolet spectroscopy 16:252
uses 16:272
Hydrocodone bitartrate. (See also Hydrocodone.)
ultraviolet spectroscopy 16:252
Hydrocortamate chloride. (See also Steroids.)
uses 14:195
Hydrocortisone. (See also Steroids.)
manufacture 14:195
properties 14:195
structure 14:195
trade names 14:192
Hydrocortisone acetate. (See also Steroids.)
infrared spectrum 14:194
uses 14:195
Hydrocortisone sodium succinate. (See also
Steroids.)
uses 14:195
Hydrocortone. See Hydrocortisone.
Hydrocyanic acid. (See also *Hydrogen cyanide.*)
in acrylonitrile 4:376
in benzaldehyde 6:624
Hydrocyanite. See Copper sulfate.
Hydrofluoric acid 13:6
in antimony fluoborate 6:56
commercial grades and specifications 13:7
composition 13:7
etchant for ceramics 9:224
in fluoroboric acid 13:39
in fluorosilicic acid 13:49
fluorosilicic acid content 13:50
in fluorosulfuric acid 13:55
gasometric method for sulfide 13:438
gravimetric method for rare earth metals 9:268,
269
gravimetric method for rare earths 17:476
gravimetric method for silica 9:111
gravimetric method for silicon 9:674, 704;
10:320, 645
gravimetric method for thorium 19:9
in hexafluorophosphoric acid 13:47

impurities, determination of 13:34
in nitric acid 13:27
silica content 18:30
specification tests 13:34
titration of chloride 14:127
toxicity 13:13
turbidimetric method for silicate 18:16
use in gravimetric determination of silicon
5:142
Hydrofluoric acid–boric acid
volumetric method for ferrous ion 15:43
Hydrofluoric acid–sulfuric acid
gravimetric method for silica 13:416;
17:507
gravimetric method for silicate 18:193
gravimetric method for silicates 11:379
gravimetric method for silicon 12:186; 15:63,
74; 16:184; 18:7, 17, 45, 87
gravimetric method for silicon dioxide 14:95,
139; 15:207
Hydrofluorosilicic acid. See Fluorosilicic acid.
acid.)
acid–base titration 13:34
in hydrofluoric acid 13:34
Hydrofluosilicic acid. See Fluorosilicic acid.
Hydrogen. (See also Carbon and hydrogen; *Carbon,
hydrogen, and nitrogen microanalysis;* and
Hydrogen, elemental.) 14:377–389
in aluminum metal 5:130
in argon 13:299
in boron compounds, organic 7:352
in calcium hydride 14:247
in carbon dioxide 8:251
in carbon monoxide 8:310, 313
carrier gas fusion methods 8:624
in chlorine 9:320
in chromium 9:635
in cigarette smoke 10:13
in coal and coke 10:237
in cobalt alloys 10:313
in cobalt metal 10:293, 298
in copper alloys 10:643
diffusion method 18:232
in elastomers, synthetic 12:109
evolution method for hydrides 14:247
evolution method for silanic hydrogen 18:67
evolution method for titanium 19:128
in fluorine, elemental 13:3
in gallium arsenide 13:279
in gallium metal 13:266
gas chromatography 8:251, 262, 310, 313, 650;
13:247; 18:67
gasometric method 15:482
gasometric method for lithium hydride 15:285
in hafnium 14:131
in helium 13:298
history 14:377

hot extraction methods 8:638, 684; 14:249
in hydrides 8:614
isotopes 14:378
in krypton 13:300
in lithium aluminum hydride 14:247
in lithium borohydride 14:247
in lithium hydride 14:247
in magnesium alloys 15:384
in magnesium hydride 14:247
in manganese 15:482
mass spectroscopy 13:302
measurement 8:649
mercury content 16:44, 45, 46
in metals 8:613, 639
in neon 13:298
in nickel 16:333
nitrogen content 16:454
in noble gases 13:294, 297, 302
occurrence 14:377
in organic materials 8:280
in potassium borohydride 14:247
in potassium hydride 14:247
in rubidium 17:555
in sodium aluminum hydride 14:247
in sodium borohydride 14:247
in sodium hydride 14:247
in sodium metal 18:232
in titanium hydride 14:249
vacuum fusion methods 8:616, 680; 15:384
in xenon 13:300
in zirconium 14:131
in zirconium hydride 14:144, 249
Hydrogenated naphthalenes. (See also Decahydronaphthalene; 1,2,3,4-Tetrahydronaphthalene.)
 16:219
analysis 16:220
gas chromatography 16:220
manufacture 16:219
Hydrogenation
apparatus for electrolytically generated hydrogen 7:26
of benzene 7:25
of elastomers, synthetic 12:124
of sulfur 15:253
Hydrogen bromide
commercial specifications 8:9
impurities, determination of 8:11
properties 8:9
titration of ethylene oxide 12:321, 331; 13:191
titration of ethylenimine 14:475
titration of propylene oxide 13:191
Hydrogen bromide–acetic acid titration
of epoxy compounds 12:206, 211
Hydrogen chloride 9:336; 14:389–403; 18:63
acid–base titration 9:330; 18:63
acid–base titration, nonaqueous 7:76
in acid chlorides 7:76; 8:574

in air 14:395
analysis 14:392
assay 14:392
in chlorine 9:329
in chlorine dioxide absorption tower solution
 7:227
in chlorosulfonic acid 9:533
colorimetric determination 14:395
commercial grades and specifications 14:392
in exhaust gases 14:395
Fourier transform spectra 13:167
gas chromatography 14:392
gravimetric method for sodium 15:267; 18:228
in hydrochloric acid 9:340
in hydrogen bromide 8:11
hydrogen chloride content 14:396
identification 14:392
impurities, determination of 14:395
infrared spectrum 14:391
nephelometric method 14:395
occurrence 14:389
properties 9:334; 14:389
sodium carbonate volumetric method 14:396
solubility 9:336; 14:391
toxicity 14:391
volumetric determination 14:395
x-ray fluorescence method 14:395
Hydrogen cyanide. (See also Cyanides.) 13:176;
 14:403–426.
in air 14:411, 412
Aldridge colorimetric method 14:418
analysis 14:409
assay 14:416
benzidine–copper acetate color test 14:411
bromine–benzidine colorimetric method 14:418
in cigarette smoke 10:13
colorimetric determination 14:410, 418
commercial grades and specifications 14:409
diffusion methods 14:415
distillation methods 14:413
electrometric determination 14:421
gas chromatography 14:415; 16:425
identification 14:410
impurities, determination of 14:416, 421
manufacture 13:176; 14:408
methyl orange–mercuric chloride color test
 14:411
p-nitrobenzaldehyde–dinitrobenzene colorimetric method 14:420
phenolphthalein–copper sulfate colorimetric
 method 14:418
p-phenylenediamine–pyridine colorimetric
 method 14:419
properties 13:176; 14:403, 404
Prussian blue color test 14:410
pyridine–pyrazolone colorimetric method
 14:420

safety precautions 14:405
sampling 14:409
separation 14:412
silver nitrate titration 14:417
sodium hydroxide absorption 14:412
solubility 13:176; 14:405
solvent extraction 14:412
specific gravity 14:406, 416
toxicity 13:176; 14:405
use in manufacture of acrylic and methacrylic
    acids and esters 4:185
uses 13:176; 14:403; 16:473
volumetric determination 14:417
in water 14:416, 417
Hydrogen deuteride
gas chromatography 14:386
Hydrogen, elemental
analysis 14:380
gas chromatography 14:383
impurities, determination of 14:380
isomers 14:377
manufacture 14:379
Orsat technique 14:380
properties 14:377
solubility 14:378
thermal conductivity 14:380
Hydrogen fluoride 13:6
commercial specifications 13:7
in fluorine, elemental 13:4
gas chromatography 13:64
manufacture 13:7
properties 13:6
solubility 13:6
toxicity 13:13
uses 13:7
Hydrogen iodide. (See also Iodides.)
    14:594
commercial grades and specifications
    14:594
properties 14:594
solubility 14:594
Hydrogen isomers
gas chromatography 14:386
Hydrogen 3-nitrophthalate
of higher alcohols 4:558
of lower alcohols 4:518, 520
Hydrogen peroxide 14:427–439
active oxygen 7:224
analysis 14:431
bleaching agents 7:222
ceric ammonium sulfate titration 11:107
ceric titration 14:433
colorimetric determination 14:431, 434
colorimetric method for chromium 19:108
colorimetric method for iron 19:98
colorimetric method for molybdenum 19:108
colorimetric method for niobium 10:319;

16:365, 371, 372
colorimetric method for tantalum 18:449
colorimetric method for titanium 4:40; 5:146;
    10:321, 647; 11:141; 12:186; 14:67; 16:176,
    382, 391; 19:20, 97, 107
colorimetric method for titanium dioxide
    9:123; 10:118; 15:211
colorimetric method for uranium 19:248
colorimetric method for vanadium 10:324;
    19:98, 108, 331, 340
commercial grades and specifications 14:430
diacetyldichlorofluorescin fluorometric method
    14:431, 434
electrometric determination 14:435
etchant for ceramics 9:225
ferricyanide colorimetric method 14:434
gasometric determination 14:433
gasometric method for catalase 16:124
in gelatin 12:552
in hair dyes 11:107
identification 14:431
impurities, determination of 14:436
iodometric method 14:433
luminol chemiluminescent method 12:552
manufacture 14:429; 16:535
physical determination 14:435
polarography 14:435
potassium permangante titration 14:432
properties 14:427; 16:535
safety precautions 14:429
sodium chromate colorimetric method 14:431
specification tests 14:436
titanium(IV) sulfate colorimetric method
    14:434
toxicity 14:429
ultraviolet spectroscopy 14:435
uses 14:427
volumetric determination 14:432
volumetric method for formaldehyde 12:168;
    13:109
volumetric method for sulfur dioxide 18:408
volumetric method for tantalum 18:446
Hydrogen peroxide–diammonium phosphate
gravimetric method for zirconium 10:325
Hydrogen peroxide–nitric acid
gravimetric method for thorium 19:9
Hydrogen peroxide–potassium hydroxide
volumetric method for nitriles 16:405
Hydrogen peroxide–sodium hydroxide
volumetric method for molybdenum chlorides
    16:182
volumetric method for molybdenum oxychlo-
    rides 16:182
Hydrogen peroxide–sulfuric acid
colorimetric method for titanium 16:337
colorimetric method for titanium dioxide
    17:302; 18:194

colorimetric method for vanadium 16:339
Hydrogen phosphide. See Phosphine.
Hydrogen phthalate
  of higher alcohols 4:559
  of lower alcohols 4:519
Hydrogen selenide. (See also Selenium compounds.)
  manufacture 17:584
  properties 17:584
  toxicity 17:584, 586
Hydrogen sulfide. (See also Sulfides.) 18:392–401
  absorption methods 18:394
  in acetylene 4:134
  in air 18:394
  p-aminodimethylaniline–ferric chloride colorimetric method 18:394, 398
  analysis 18:393
  in beer 7:640
  in benzene 7:47
  in 1,3-butadiene 11:504
  cadmium sulfate gravimetric method 15:568
  in carbon disulfide 8:274, 275, 278
  colorimetric determination 18:398
  colorimetric method for heavy metals 5:163;
    9:349, 628; 10:99; 12:289; 13:33, 34, 130,
    369; 14:373; 15:143, 281, 399, 402, 411,
    490; 16:61, 537; 17:111, 403; 18:234, 283;
    19:19, 74
  colorimetric method for hydroxylamine salts
    14:454
  colorimetric method for lead 12:186; 15:281
  colorimetric method for mercury 16:42
  colorimetric method for nickel 10:655, 668;
    15:491
  colorimetric method for silver 18:176
  colorimetric method using p-aminodimethyl-
    aniline sulfate 7:641
  detection 18:393
  determination of impurities in antimony halides
    6:51
  electrochemical methods 19:504
  in ethylene 5:362
  evolution method for sulfur 18:384
  evolution method for sulfurous acid 18:403
  evolution method for thiophene 18:544
  gas chromatography 8:274; 14:393; 15:557
  gravimetric method 18:395
  gravimetric method for antimony 12:151, 420
  gravimetric method for cadmium 12:151
  gravimetric method for cobalt 19:554
  gravimetric method for copper 10:603, 608;
    15:65; 16:302
  gravimetric method for ferric oxide 15:210
  gravimetric method for germanium 13:382
  gravimetric method for heavy metals 12:415
  gravimetric method for indium 14:529
  gravimetric method for iridium 17:251, 260

gravimetric method for lead 12:151, 420;
  15:172
gravimetric method for mercuric chloride
  11:563; 16:59
gravimetric method for mercury 16:6, 7, 16
gravimetric method for molybdenum 10:315;
  16:153, 332; 19:200
gravimetric method for nickel 16:292, 295
gravimetric method for osmium 17:251, 256
gravimetric method for palladium 17:251
gravimetric method for platinum 10:320;
  17:251, 258
gravimetric method for platinum group metals
  17:250
gravimetric method for rhenium 17:524, 526,
  530, 533
gravimetric method for rhodium 17:251
gravimetric method for ruthenium 17:251
gravimetric method for selenious acid 17:591
gravimetric method for silver 18:164
gravimetric method for technetium 17:525, 526
gravimetric method for tellurous acid 17:591
gravimetric method for thallium 18:523
gravimetric method for tin 15:67; 19:32, 39
gravimetric method for tungsten 19:199, 206
gravimetric method for vanadium 19:329
gravimetric method for zinc 12:151; 16:298;
  19:554, 558
gravimetric method for zinc oxide 13:419
in hydrocarbons 14:346
iodine–azide gasometric method 18:393
iodometric methods 15:569; 18:395
iodometric titration 13:322
lead acetate color test 13:321; 14:346; 15:255;
  18:393
in liquefied petroleum gas 15:255
manufacture 18:365, 392
mercury(II) acetate titration 18:544
methylene blue colorimetric method 13:323;
  15:255
in natural gas 13:321; 15:569
occurrence 18:369
in petroleum products 15:567
precipitation 8:100
properties 18:392
reduction method for iron 5:132; 15:41
from reduction of mercaptans 7:642
separation 18:394
silver cyanide color test 13:321; 15:255; 18:394
silver nitrate titration 19:504
toxicity 18:367; 19:504
ultraviolet spectroscopy 8:278; 18:398
uses 18:392
volumetric determination 18:395, 544
volumetric method for sulfide 15:480
volumetric method for sulfur 16:213
in water 18:399

in wines 19:500, 504
Hydrogen sulfide, substances not precipitated by
  in bismuth chloride 7:200
  in copper(II) acetate 10:655
  in copper(II) bromide 10:657
  in copper(II) carbonate, basic 10:659
  in copper(I) chloride 10:661
  in copper(II) chloride 10:662
  in copper(I) cyanide 10:664
  in copper(II) nitrate 10:668
  in copper(II) oxide 10:675
  in stannous chloride 19:74
Hydrogen sulfide–sulfuric acid
  gravimetric method for antimony trioxide
    13:430
  gravimetric method for lead oxide 13:422
Hydrogen telluride. (See also Tellurium com-
    pounds.)
  manufacture 17:584
  properties 17:584
  toxicity 17:584, 586
Hydrohalogenation
  of epoxy compounds 12:205
Hydrolysis
  of acetals and ketals 4:67, 69
  of acetylsalicylic acid 6:136
  of alkyl sulfates 11:83
  of aluminum alkyls 5:168
  of amides 8:606
  of amides, differential reaction rate 4:167
  of azo dyes 6:407
  of cellulose 9:54
  of cyanoacetamide 15:443
  of fat in corn starch 11:17
  of formals 4:68, 70
  of pesticides and insecticides 19:608
  of polyamides 17:295
  of polycarbonates 17:337
  of polyurethanes 19:291
  of protein 8:411
  of starch 19:142
  for vinyl acetate content of copolymers 4:274
Hydrolysis, acid
  of acetal and ketals 4:68
  of acrylonitrile copolymers 4:270
Hydrolysis, basic
  of acrylonitrile copolymers 4:271
Hydrolytic stability
  of lubricants 15:339
Hydrometer method
  for proofing 4:481
  for specific gravity of ammonium hydroxide
    5:319
  for specific gravity of antifreezes 6:5, 17
  for specific gravity of beer 7:486
  for specific gravity of bituminous materials
    6:319

for specific gravity of liquefied petroleum gas
    15:246
Hydromorphone. (See also Narcotics.) 16:272
  ammonium molybdate colorimetric method
    16:246
  analysis 16:272
  assay 16:272
  color tests 16:246
  formaldehyde–sulfuric acid colorimetric method
    16:246
  gas chromatography 16:254
  microchemical tests 16:250
  sodium nitroprusside dihydrate microchemical
    method 16:250
  structure 16:244
  thin-layer chromatography 16:256
  ultraviolet spectroscopy 16:252
  uses 16:272
Hydronitrous acid
  properties 16:490
Hydroperoxides
  in elastomers, synthetic 12:114
  iodometric method 17:298
  in polyamides 17:298
  sodium iodide colorimetric method 12:115
Hydroquinone. (See also Dihydroxybenzenes.)
  in acrolein 4:158
  analysis 17:47
  assay 17:47
  n-butylamine method 4:204
  in carbon black 8:192
  caustic color method 4:202
  ceric sulfate titration 8:589; 17:47
  colorimetric method for niobium 16:368, 373,
    383
  colorimetric method for o-phthalic acid 5:47
  colorimetric method for tantalum 16:368
  commercial grades and specifications 17:15
  derivatives for identification 17:21
  2,4-dinitrophenylhydrazine method 4:206
  ferric chloride colorimetric method 17:16
  gravimetric method for gold 14:16, 19
  in green tea leaf 18:459
  in hair dyes 11:103
  inhibitor in acrylic monomers 4:188, 201
  iodine method 4:203
  manufacture 17:1, 10
  in methanol 16:115
  Millon's reagent colorimetric method 4:158
  morpholine method 4:204
  occurrence 17:1
  paper chromatography 17:24
  phthalic anhydride colorimetric method 17:19
  polarographic method 4:207
  potassium titanium oxalate colorimetric method
    18:459
  properties 17:5

sodium cyanide colorimetric method 17:20
solubility 17:5
thin-layer chromatography 11:103; 17:24
titration of gold 14:20, 44
ultraviolet spectroscopy 4:205
uses 17:2
  in vinyl acetate 8:589
Hydroquinone, monobenzyl ether
  paper chromatography 6:85
Hydroquinone, monomethyl ether
  inhibitor in acrylic monomers 4:188, 201
  sodium nitrite, reaction with 4:208
  voltammetry 4:209
Hydrotrope content
  of detergents 11:366
Hydroxamic acid
  colorimetric method for acetanilide 6:193
  method for acrylates and methacrylates 4:216
  method for esters in alcoholic beverages 4:484
  method for vinyl acetate content of copolymers
    4:274
Hydroxides 16:535
  in lithium hydride 14:251
  in water 19:362
p-Hydroxyacetanilide. See Acetaminophen.
Hydroxyacetic acid (Glycolic acid) 8:395
  properties 8:395
  uses 8:395
Hydroxy acids 8:370
Hydroxyalkoxyl content
  of starch derivatives 11:31
3β-Hydroxy-5α-androstan-16-one
  gas chromatography 14:224
Hydroxyandrostanones
  liquid chromatography 14:208, 209
Hydroxyandrostenenones
  liquid chromatography 14:209
Hydroxyanisole, butylated 6:77
Hydroxyanthraquinones
  color reagents for boric acid 7:386
Hydroxybenzaldehyde
  hydroxylamine hydrochloride method 8:357
  properties 8:334
2-Hydroxybenzaldehyde. (See also Salicylalde-
    hyde.)
  absorption spectroscopy 9:564
3-Hydroxybenzaldehyde 6:597
  column chromatography 6:618
  derivatives for identification 6:612
  fluorometry 6:626
  properties 6:598
  ultraviolet spectroscopy 6:632
4-Hydroxybenzaldehyde 6:597
  column chromatography 6:618
  derivatives for identification 6:612
  fluorometry 6:626
  gas chromatography 11:167

infrared spectroscopy 6:616
paper chromatography 6:620
properties 6:598
ultraviolet spectroscopy 6:632
4-Hydroxybenzaldehyde-3-sulfonic acid
  colorimetric method for fusel oil 4:486
p-Hydroxybenzoates. (See also *Phenols*.)
  in creams and lotions 11:93
  gas chromatography 11:545
  in pharmaceuticals 11:545, 548
m-Hydroxybenzoic acid 8:400
  gas chromatography 8:388
  ion exchange chromatography 6:144
  paper chromatography 8:379, 400
  properties 8:403
  in salicylic acid 6:143
o-Hydroxybenzoic acid. See Salicyclic acid.
p-Hydroxybenzoic acid 8:400
  gas chromatography 8:388
  ion exchange chromatography 6:144
  paper chromatography 8:379, 400
  properties 8:402
  in salicyclic acid 6:143
  ultraviolet spectroscopy 6:145
p-Hydroxybenzylpenicillin. (See also Penicil-
    lins.)
  structure 16:554
p-Hydroxybiphenyl
  colorimetric method for lactic acid 15:137
3-Hydroxy-2-butanone. See Acetoin.
Hydroxybutyric acid
  properties 8:397
  thin-layer chromatography 8:381
2-Hydroxycarbazole–sulfuric acid
  color test for formaldehyde 13:106
3-Hydroxy-β-carotene. See Cryptoxanthin.
Hydroxycinnamic acid. See Coumaric acid.
17-Hydroxycorticosterone. (See also Steroids.)
  structure 14:185
17-Hydroxycorticosterone acetate. (See also
    Steroids.)
  structure 14:184
6β-Hydroxycortisol
  gas chromatography 14:219
6β-Hydroxycortisone
  gas chromatography 14:219
Hydroxycoumarins. (See also *Coumarins*.)
  gas chromatography 11:168, 170
11-Hydroxyctiocholanolone
  gas chromatography 14:219
o-(2-Hydroxy-4-diethylaminobenzoyl)benzoic acid
  in D & C Red No. 19 10:498, 510
  ultraviolet spectroscopy 10:510
4-Hydroxy-3,5-dimethoxybenzaldehyde
  fluorometry 6:627
dl-2'-Hydroxy-5,9-dimethyl-2-phenethyl-6,7-
    benzomorphan. See Phenazocine.

p-Hydroxydiphenyl
  colorimetric method for lactic acid 19:422
p-Hydroxydiphenylamine
  benzoyl peroxide colorimetric method 4:210
  nitric acid colorimetric method 4:210
o-(2-Hydroxy-3,6-disulfo-1-naphthylazo)-benzene-
    arsonic acid. See Thoron.
o-[2-(Hydroxy-3,6-disulfo-1-naphthyl)azo] benzene
    arsonic acid disodium salt. See Thorin.
4-Hydroxy-3-ethoxybenzaldehyde
  fluorometry 6:627
  ultraviolet spectroscopy 6:633
N-(2-Hydroxyethyl)aziridine. (See also *Imines.*)
  properties 14:469
  structure 14:469
Hydroxyethylcellulose and its derivatives. (See
    also *Cellulose derivatives.*) 9:82
  analysis 9:83
  composition, determination of 9:84
  impurities, determination of 9:83
  manufacture 9:83
  properties 9:83
  uses 9:82
  viscosity 9:62, 84
N-(2-Hydroxyethyl)cyclohexylamine 11:246
  gas chromatography 11:246
  infrared spectroscopy 11:247
  manufacture 11:246
  properties 11:244
  solubility 11:246
  toxicity 11:246
  uses 11:246
(*N*-Hydroxyethyl)ethylenediamine-*N*,*N'*,*N'*-tri-
    acetic acid. See HOEDTA.
2-Hydroxyethyl methacrylate. (See also *Acrylic
    and methacrylic acids and esters.*)
  commercial grades and specifications 4:187
  properties 4:182
  saponification method 4:196
  solubility 4:184
β-Hydroxyethyltrimethyl ammonium hydroxide.
    See Choline.
Hydroxyflavones
  color reagents for boric acid 7:387
Hydroxyhydroquinone. (See also *Trihydroxy-
    benzenes.*)
  derivatives for identification 17:21
  ferric chloride colorimetric method 17:16
  manufacture 17:10
  properties 17:5
  solubility 17:5
Hydroxyisobutyric acid
  thin-layer chromatography 8:381
Hydroxyisocaproic acid
  thin-layer chromatography 8:381
p-Hydroxyisophthalic acid
  in salicylic acid 6:145

  thin-layer chromatography 6:145
Hydroxyisovaleric acid
  thin-layer chromatography 8:381
Hydroxyketone dyes 12:9
Hydroxylamine. (See also *Hydroxylamines and
    their salts.*) 14:440; 16:471
  analysis 14:446
  colorimetric determination 14:451
  colorimetric method for acetanilide 5:452
  in cotton 11:142
  in determination of diacetyl 7:583
  diacetylmonoxime–nickel colorimetric method
    11:142; 14:447
  gravimetric method for cinnamaldehyde
    10:45
  identification 14:446
  manufacture 14:444; 16:471
  α-naphthylamine colorimetric method 14:446,
    451
  properties 14:440; 16:471
  reaction of aldehyde and ketones with 8:333
  sodium bromide titration 14:452
  solubility 14:441
  toxicity 14:443
  uses 14:440
  volumetric methods for formaldehyde 13:111
Hydroxylamine–ferric ammonium sulfate
  colorimetric method for cephalosporins 16:563
  colorimetric method for penicillins 16:563
Hydroxylamine–ferric chloride
  colorimetric method for cellulose esters 12:588
  colorimetric method for nitriles 16:400
  color test for polyurethanes 19:282
Hydroxylamine hydrochloride. See Hydroxylam-
    monium chloride.
Hydroxylamine hydrochloride method
  for acetals 4:70
  for acetone 4:112
  for acrolein 4:152, 157
  for aldehydes 10:83; 12:281; 18:306
  for anisaldehyde 6:620
  for benzaldehyde 6:620; 12:282
  for benzoic acid 19:485
  for carbonyl compounds 4:124, 145, 375; 8:356
  for carvone 12:282
  for cinnamic aldehyde 12:282
  for citral 12:282
  for citronellal 12:282
  for cuminal 6:620
  for cuminic aldehyde 12:282
  for decyl aldehyde 12:282
  for formaldehyde 16:446
  for free formaldehyde in phenolic resins 4:402
  for glutaraldehyde 11:566
  for heliotropin 6:620
  for ketones 12:281
  for magnesium 10:300

for menthone 12:282
for 2-methyl-1,3-dioxolane 11:522
for methyl nonyl ketone 12:282
for pulegone 12:282
for selenium 17:591, 592, 596, 597
for tellurium 17:591
for thujone 12:282
for p-tolualdehyde 6:620
Hydroxylamine hydrochloride–bathophenanthro-
    line
    colorimetric method for iron 9:366
Hydroxylamine hydrochloride–α,α'-dipyridyl
    colorimetric method for iron 10:150
Hydroxylamine hydrochloride–ferric perchlorate
    colorimetric method for cycloheximidine 13:
    224
Hydroxylamine hydrochloride oximation
    of crotonaldehyde 11:178
Hydroxylamine hydrochloride–o-phenanthroline
    colorimetric method for iron 12:157, 554;
    13:114
Hydroxylamine hydrochloride–potassium hydroxide
    titration
    of cinnamaldehyde 10:50
Hydroxylamine hydrochloride–sodium hydroxide
    titration
    of cinnamaldehyde 10:50
Hydroxylamines and their salts. (See also Hydroxyl-
    amine; and individual compounds.) 14:440–
    460
    analysis 14:446
    assay 14:447
    commercial grades and specifications 14:445
    ferric ammonium sulfate–potassium permanga-
        nate volumetric method 14:447
    ferric chloride–ceric ammonium nitrate volu-
        metric method 14:448
    impurities, determination of 14:454
    manufacture 14:444
    properties 14:440
    Raschig volumetric method 14:447
    toxicity 14:443
    uses 14:440
Hydroxylammonium acid sulfate. (See
    also Hydroxylamines and their
    salts.)
    commercial specifications 14:445
    impurities, determination of 14:455
    infrared spectrum 14:442
    properties 14:441
    toxicity 14:444
Hydroxylammonium arsenate. (See also Hydroxyl-
    amines and their salts.)
    properties 14:441
Hydroxylammonium chloride. (See also Hydroxyl-
    amines and their salts.)
    commercial grades and specifications 14:445

impurities, determination of 14:455
infrared spectrum 14:442
manufacture 14:444
properties 14:442
toxicity 14:444
uses 14:440
Hydroxylammonium oxalate. (See also Hydroxyl-
    amines and their salts.)
    properties 14:441
Hydroxylammonium perchlorate. (See also Hy-
    droxylamines and their salts.)
    properties 14:441
Hydroxylammonium phosphate. (See also Hy-
    droxylamines and their salts.)
    properties 14:441
Hydroxylammonium sulfate. (See also Hydroxyl-
    amines and their salts.)
    commercial specifications 14:445
    impurities, determination of 14:455
    infrared spectrum 14:442
    manufacture 14:444
    properties 14:441
    toxicity 14:444
    uses 14:440
Hydroxyl groups
    acetic anhydride acylation 10:212
    acetyl chloride–sodium hydroxide volumetric
        method 12:248
    in carbon black 8:192
    in cellulose acetate 9:72
    in coal and coke 10:211
    in epoxy resins 12:244
    in halohydrins 14:157
    infrared spectroscopy 12:249
    lithium aluminum hydride method 12:245
    near-infrared spectroscopy 12:248
    in pentaerythritol 13:593
    trimethylsilylation 10:212
Hydroxyl number
    of alkyd resins 5:51
    of castor oil 12:478
    of fats and fatty oils 12:478
    of fatty alcohols 4:568
    of glycols 13:587
    of organosilicon copolymers 18:147
    of polyols 19:278
p-Hydroxymandelic acid
    gas chromatography 8:389
2-Hydroxy-3-methoxybenzaldehyde 6:597
    column chromatography 6:618
    infrared spectroscopy 6:616
    properties 6:598
3-Hydroxy-4-methoxybenzaldehyde 6:597
    properties 6:598
4-Hydroxy-3-methoxybenzaldehyde. See Vanillin.
2-Hydroxy-4-methoxybenzophenone
    thin-layer chromatography 16:96

ultraviolet spectroscopy 16:95

Hydroxymethoxycoumarins. (See also *Coumarins.*)
gas chromatography 11:168, 170

2-Hydroxy-4-methoxy-4'-methylbenzophenone
thin-layer chromatography 16:96

4-Hydroxymethyl-2,6-di-*t*-butylphenol. (See also
Phenols.)
structure 17:2

Hydroxymethylcellulose
viscosity 9:62

3-Hydroxy-*N*-methyl-*cis*-crotonamide dimethyl
phosphate. See Monocrotophos.

2-Hydroxymethyl-2-ethyl-1,3-propanediol. (See
also Trimethylolpropane.) 13:561

5-Hydroxymethyl-2-furaldehyde. See HMF

Hydroxymethylfurfural
barbituric acid–*p*-toluidine colorimetric method
19:459
colorimetric determination 19:459
Fiehe's solution colorimetric method 19:459
solvent extraction 19:459
structure 19:459
ultraviolet spectroscopy 6:632
in wines 19:452, 459

2-Hydroxymethyl-2-methyl-1,3-propanediol. See
Trimethylolethane.

l-(1-3-Hydroxy-*N*-methylmorphinan.) See Levorphanol.

4-Hydroxy-4-methyl-2-pentanone
absorption spectroscopy 9:564

2(2'-Hydroxy-5'-methylphenyl)benzotriazole
gas chromatography 16:96
in methacrylate polymers 16:96
ultraviolet spectroscopy 16:95

17-Hydroxy-6α-methylpregn-4-ene-3,20-dione
acetate. See Medroxyprogesterone acetate.

17β-Hydroxy-6α-methyl-17-(1-propynyl)androst-
4-en-3-one. See Dimethisterone.

2-Hydroxy-1-naphthaldehyde colorimetric method
for acetone 11:204
for 2-butanone 11:204
for cyclobutanone 11:204
for cycloheptanone 11:204
for cyclohexanone 11:204
for cyclononanone 11:204
for cyclooctanone 11:204
for cyclopentadecanone 11:204
for cyclopentanone 11:204
for ethyl acetoacetate 11:204
for 2-heptadecanone 11:204
for 2-heptanone 11:204
for 3-heptanone 11:204
for 5-hexen-2-one 11:204
for 2-nonadecanone 11:204
for 2-nonanone 11:204
for 2-octanone 11:204
for 2,4-pentanedione 11:204

for 2-pentanone 11:204
for 3-pentanone 11:204
for 2-tridecanone 11:204
for 2-undecanone 11:204

3-Hydroxy-2,7-naphthalenedisulfonic acid
as colorimetric reagent 5:450, 452
colorimetric method for arylamines 6:408

β-Hydroxynaphthoic acid
properties 8:402, 406

3-Hydroxy-2-naphthoic anilide
use in manufacture of azo dyes 6:403

2-(2-Hydroxy-1-naphthylazo)-6-and-8-naphthal-
enesulfonic acids
in D & C Red Nos. 10–13 10:521
visible spectroscopy 10:522

1-(2-Hydroxy-1-naphthylazo)-2-naphthol-4-sul-
fonic acid. See Pontachrome Blue Black R.

4-Hydroxy-3-nitrophenylarsonic acid
colorimetric method for tin 10:626

4-Hydroxy-3-nitrophenyl phosphoric acid
volumetric method for aluminum 5:218

17-Hydroxy-19-nor-17α-pregn-4-en-20-yn-3-one.
See Norethindrone.

17-Hydroxy-19-nor-17α-pregn-5(10)-en-20-yn-3-
one. See Norethynodrel.

17β-Hydroxy-19-nor-17α-pregn-4-en-20-yn-3-one
acetate. See Norethindrone acetate.

2-Hydroxy-4-octoxybenzophenone
thin-layer chromatography 16:95

3α-Hydroxy-12-oxo-5β-cholanic acid
color tests 7:147
properties 7:143

Hydroxyphenamate
gas chromatography 6:210, 211

o-Hydroxyphenols
ammonium molybdate colorimetric reagent
17:20

Hydroxyphenylacetic acid
gas chromatography 8:388, 389

p-Hydroxyphenylarsonic acid
gravimetric method for titanium 19:96

2-(*o*-Hydroxyphenyl)benzoxazole
gravimetric method for cadmium 8:51

4-Hydroxyphenylfluorone
colorimetric method for niobium 16:368, 375

p-Hydroxyphenyllactic acid
gas chromatography 8:389

N-*p*-Hydroxyphenylmorpholine 6:76

p-Hydroxyphenylpropionic acid. See Phloretic
acid.

p-Hydroxyphenylpyruvic acid
gas chromatography 8:389

3-Hydroxy-5-pregnan-20-ones
liquid chromatography 14:208

Hydroxyprogesterones
liquid chromatography 14:209

Hydroxyproline. (See also *Carboxylic acids.*

amino.) 8:412, 513
analysis 8:518
colorimetric methods 8:518; 13:365
commercial specifications 8:518
composition 8:412
crystalline form 8:515
p-dimethylaminobenzaldehyde–chloramine-T
  colorimetric method 13:372
p-dimethylaminobenzaldehyde–hydrogen
  peroxide colorimetric method 13:365
gas chromatography 8:429
in gelatin 13:365, 372
ion exchange chromatography 8:421
manufacture 8:517
naphthoquinone sulfate colorimetric method
  8:518
paper chromatography 8:422
properties 8:513
solubility 8:514
structure 8:514
in yeast 19:529
2-Hydroxy-1,2,3-propanetricarboxylic acid. See
  Citric acid.
2-Hydroxypropionic acid. See Lactic acid.
Hydroxypropyl content
of methylhydroxypropylcellulose 9:81
Hydroxypropyl methacrylate. (See also *Acrylic
  and methacrylic acids and esters.*)
commercial grades and specifications 4:187
properties 4:182
saponification method 4:196
solubility 4:184
3-Hydroxypyridine. (See also *Pyridine and pyridine derivatives.*)
trisodium pentacyanoammino ferrate colorimetric method 17:433
Hydroxypyridines. (See also *Pyridine and pyridine derivatives;* and individual compounds.)
ionization constants 17:415
manufacture 17:422
8-Hydroxyquinaldine
reagent for separation of aluminum 5:204
8-Hydroxyquinoline. (See also Oxine; 8-Quinoli-
  nol.)
colorimetric method for aluminum 5:228;
  10:293; 15:377
colorimetric method for cerium 9:275
colorimetric method for manganese 15:457
colorimetric method for vanadium 11:491;
  19:331, 33, 340
colorimetric method for zirconium 14:118
extraction method for aluminum 5:204
extraction method for cerium 9:270
extraction method for thorium 17:477
extraction method for vanadium 9:644; 10:130
gravimetric method for aluminum 5:178, 217;
  9:664, 704; 14:256; 16:324

gravimetric method for indium 14:529
gravimetric method for magnesium 9:704;
  15:360, 362, 369
gravimetric method for magnesium oxide
  10:120
gravimetric method for niobium 16:367, 374
gravimetric method for scandium 17:570
gravimetric method for titanium 9:664; 19:95
gravimetric method for zirconium 9:664
method for chromium 9:639
volumetric method for aluminum 5:222
8-Hydroxyquinoline oxide
colorimetric method for tungsten 11:141
8-Hydroxyquinolino–silver nitrate
gravimetric method for chloride 11:141
12-Hydroxystearic acid
thin-layer chromatography 12:205
3β-Hydroxysterols. See Phytosterols.
Hydroxysuccinic acid. See Malic acid.
3-Hydroxy-5,7,3',4'-tetramethoxyquercetin
colorimetric method for tantalum 16:368
Hydroxytoluene, butylated 6:73
inhibitor in acrylic monomers 4:188
in methyl methacrylate 4:210
Hydroxy-*p*-toluic acid
gas chromatography 8:385
1-Hydroxy-4-*p*-toluinoanthraquinone
in D & C Violet No. 2 10:498, 510
ultraviolet spectroscopy 10:510
β-Hydroxytriarballylic acid. See Citric acid.
Hydrozincite
composition 19:545
Hygromycin B 5:473, 579
analysis 5:583, 589, 595
identification 5:583
infrared spectrum 5:584
ion exchange separation 5:583
microbiological assay 5:589
properties 5:579
separation 5:583
solubility 5:580, 581
uses 5:579
Hyocholic acid
properties 7:143
structure 7:142
α-Hyodeoxycholic acid
color tests 7:147
preparation 7:144
properties 7:143
structure 7:142
Hyoscyamine 4:587, 595
Hypalon. (See also Chlorosulfonated polyethylene.) 12:86
infrared spectroscopy 12:93
Hyperglycemic factor. See Glucagon.
Hypobromites
formation from bromine 8:3

Hypochlorite bleaches 9:352
  alkalinity 9:354
  analysis 9:353
  available chlorine 9:353
  chlorine content 9:354
  impurities, determination of 9:354
  sodium chlorate content 9:353
  sodium chloride content 9:354
Hypochlorites. (See also *Chlorine compounds, inorganic;* Hypochlorite bleaches; and individual compounds.) 9:351–355
  analysis 9:353
  in cotton 11:142
  hypochlorite content 9:353
  manufacture 9:352
  zinc chloride colorimetric method 11:142
Hypochlorous acid. (See also Hypochlorites.) 9:351
  bleaching agent 7:223
  manufacture 9:351
Hyponitrous acid and salts 16:490
  manufacture 16:490
  properties 16:490
Hypophosphite
  ceric ammonium nitrate volumetric method 17:113

colorimetric determination 17:61
copper sulfate gravimetric method 17:113
detection 17:61, 113
iodometric method 17:114
mercuric chloride gravimetric method 17:113
molybdic acid colorimetric method 17:116
sodium hypochlorite volumetric method 17:115
Hypophosphites. (See also individual compounds.) 17:111
  manufacture 17:112
  uses 17:112
Hypophosphorous acid
  analysis 17:113
  commercial specifications 17:112
  gravimetric method for selenium 17:591
  gravimetric method for tellurium 10:646; 17:591
  manufacture 17:112
  properties 17:111
Hypovanadates. (See also Vanadium compounds.)
  manufacture 19:344
Hypovanadic oxide. See Vanadium tetroxide.
Hypoxanthine
  in black tea 18:488

I

Ice cream. (See also Frozen desserts.) 16:142
  analysis 16:143
  commercial specifications 16:142
  gelatin content 13:363
  homogenization efficiency 16:143
  manufacture 16:143
  milkfat content 16:143
  solids content 16:143
Iceland spar 15:203
Icnonel. (See also Cobalt alloys.)
  composition 10:287
IDA. (See also *EDTA and related compounds.*)
  calcium nitrate volumetric method 12:37
  copper nitrate volumetric method 12:37
  lead nitrate volumetric method 12:37
  structure 12:25
Identical pH method
  for boron 7:405
Idocrase 8:75
Ignition loss
  of diatomaceous earth 11:489
Ignition, loss on. See Loss on ignition.
Ignition, residue on. See Residue on ignition.
Ignition temperature
  of matches 15:549
Illite. (See also *Clays and clay minerals.*) 9:191
  composition 10:102
Illium alloys. (See also Cobalt alloys; Nickel alloys.)
  composition 10:287; 16:290
Ilmenite. (See also Ferrous titanate.) 11:464; 19:81
  carrier gas fusion methods 8:624
  composition 10:104; 15:22
  nitrogen in 8:624
  occurrence 19:81
  oxygen in 8:624
  titanium content 19:104
Ilosvay's reagent
  color test for acetylene 16:532
IMDA. See IDA.
Imidazole
  gas chromatography 18:94
  in *N*-trimethylsilylimidazole 18:94
Imidazole alkaloids 4:590
  analysis 4:608
Imidazoles. (See also individual compounds.) 13:205–209
β-(4-Imidazolyl)ethylamine. See Histamine.
Imides 8:599
Imine derivatives. (See also *Imines.*)
  handling 14:470

toxicity 14:470
Imines. (See also Ethylenimine; Ethylenimine derivatives; Polyethylenimine; and individual compounds.) 14:460–495
  handling 14:464
  manufacture 14:472
  properties 14:461
  structure 14:460
  toxicity 14:464
Iminobispropylamine. (See also *Diamines and higher amines, aliphatic.*)
  properties 11:410
Iminodiacetic acid. See HIMDA.
2,2'-Iminodiethanol. See Diethanolamine.
Immunological methods
  for insulin 14:575
Impact grading test
  for vitrified grinding wheels 7:302
Impact polystyrene. (See also *Styrene polymers.*)
  analysis 18:328
  butadiene copolymer content 18:330
  composition, determination of 18:329
  2,6-di-*tert*-butyl-*p*-cresol content 18:333
  gel content 18:329
  infrared spectroscopy 18:318
  manufacture 18:314
  polybutadiene content 18:332
  properties 18:314
  solution viscosity 18:329
  styrene–butadiene rubber content 18:332
  styrene graft copolymer content 18:330
  swelling index 18:329
  tris(nonylated phenyl) phosphate content 18:332
  unsaturation, determination of 18:329
Impact resistance
  of dental porcelains 11:275
Impingement carbon 8:179
  commercial grades and specifications 8:196
  manufacture 8:193
Impression compounds, dental. See Dental impression compounds.
Impression test
  of dental impression compounds 11:317
Inclined plate flow test
  of resins 7:444
Inco 717. (See also Cobalt alloys.)
  composition 10:287
Incoloy alloys. (See also Nickel alloys.)
  composition 16:290
Inconel alloys. (See also Nickel alloys.)
  composition 16:290, 351

Incoro. (See also Indium alloys.)
  composition 14:547
  properties 14:547
  uses 14:547
Incosil. (See also Indium alloys.)
  composition 14:547
  properties 14:547
  uses 14:547
Indalloy. (See also Indium alloys.)
  composition 14:547
  properties 14:547
  uses 14:547
Indamine dyes 12:8
Indane
  gas chromatography 13:352; 18:299
  in gasoline 13:352, 355
  liquid chromatography 7:13
  properties 13:352
  in styrene 18:299
Indane-1,2,3-trione hydrate. See Ninhydrin.
Indanthrene blue. (See also D & C Blue No. 9.)
  in enamels 12:173
Indanthrene maroon
  in enamels 12:186
  identification 12:186
Indanthrone blue. (See also *Pigments.*)
  structure 17:185
Indene
  boiling point 16:209
  gas chromatography 18:299
  liquid chromatography 7:13, 14, 15, 16
  in styrene 18:299
Indene–phenolic resins. (See also *Hydrocarbon resins.*)
  commercial grades and specifications 14:274, 279
Inderborite 7:314
Inderite 7:313, 314
Indian red
  composition 12:173
  in enamels 12:173
Indian tea. (See also Black tea.) 18:455
Indigo. (See also D & C Blue No. 6; *Indigoid dyes.*)
  assay 14:511
  history 14:498
  infrared spectrum 14:504
  manufacture 14:499
  occurrence 14:498
  potassium permanganate titration 14:511
  properties 14:496
  structure 14:495
  sublimation test 14:504
  titanous chloride titration 14:512
  uses 14:498
  visible spectroscopy 14:508
Indigo carmine. (See also *Indigoid dyes.*)
  colorimetric method for nitrate 10:659

  colorimetric method for oxygen 16:529
  colorimetric method for oxygen in beer 7:624
  potassium permanganate titration 14:511
  properties 14:496
  titanous chloride titration 14:512
  titration of auramine 19:175
  titration of magenta 19:175
  visible spectroscopy 14:511
Indigoid dyes. (See also individual compounds.)
  12:10; 14:495–518
  analysis 14:503
  application methods 14:515
  assay 14:509
  classification 14:503
  commercial grades and specifications 14:502
  composition, determination of 14:513
  Dumas method 14:514
  dyeing methods 14:515
  halogen content 14:513
  identification 14:503
  impurities, determination of 14:515
  infrared spectroscopy 14:505
  liquid chromatography 14:508
  manufacture 14:499
  nitrogen content 14:514
  paper chromatography 14:509
  Parr peroxide bomb method 14:513
  printing methods 14:515
  properties 14:496
  sodium hydrosulfite color test 14:503
  solubility 14:503
  solvent extraction 14:510
  structure 14:495
  sulfur content 14:514
  uses 14:498
  vatting test 14:509
  visible spectroscopy 14:507, 510
Indigo test
  for acetone 4:112
Indigotin. See D & C Blue No. 6.
Indigotine. See F D & C Blue No. 2.
Indigo white. (See also *Indigoid dyes.*)
  manufacture 14:500
  structure 14:500
Indirubin. (See also *Indigoid dyes.*)
  structure 14:495
Indium. (See also Indium alloys; Indium metal.)
  14:518–565
  alizarin colorimetric method 14:526
  in aluminum alloys 5:131
  ammonium hydroxide gravimetric method 14:528
  in antimony–indium alloys 14:548
  atomic absorption spectroscopy 14:535
  5,7-dibromo-8-quinolinol colorimetric method 14:534
  dithizone–oxime colorimetric method 13:396

EDTA titration 13:445; 14:530, 548, 561
electrodeposition 14:530
emission spectroscopy 13:277; 14:34, 48, 526,
    534, 536
flame photometry 14:525, 534
flame tests 6:589, 592
in gallium arsenide 13:279
gallium content 13:262
in gallium metal 13:266
gas chromatography 13:264
in germanium 13:396
in germanium dioxide 13:397
in glass and glass-ceramics 13:445
in gold 14:45
in gold alloys 14:34
gravimetric determination 14:526, 527, 548
in hydrochloric acid 14:374
hydrogen sulfide gravimetric method 14:529
8-hydroxyquinoline gravimetric method 14:529
identification 14:525
in indium alloys 14:548
in indium compounds 14:561
in indium metal 14:526
ion exchange chromatography 13:254
in lead alloys 14:548
in magnesium alloys 15:384
morin fluorometric method 14:526
neutron activation analysis 13:265; 14:535
occurrence 14:519
polarography 13:219; 14:531, 549
potassium cyanate gravimetric method 14:528
potassium ferrocyanide titration 14:530
pyridine gravimetric method 14:529
quinalizarin colorimetric method 14:526
separation 14:526
separation from aluminum 5:210
solvent extraction 14:526, 533
urotropine gravimetric method 14:526, 529
uses 14:518
Indium alloys. (See also Indium metal.) 14:545
    aluminum content 14:553
    analysis 14:548
    antimony content 14:551
    arsenic content 14:550
    bismuth content 14:551
    cadmium content 14:551
    carrier gas fusion methods 8:682
    composition 14:547, 548
    copper content 14:551
    gallium content 14:555
    germanium content 14:557
    hydrogen in 8:616, 681
    indium content 14:548
    lead content 14:551
    manufacture 14:545
    oxygen in 8:616, 681, 682
    properties 14:547, 548

    silver content 14:555
    tin content 14:553
    trade names 14:547
    types 14:545
    uses 14:520, 545, 547
    vacuum fusion methods 8:616, 681
    zinc content 14:552
Indium antimonide. (See also Indium alloys.)
    manufacture 14:546
    properties 14:546
    uses 14:546
Indium antimonide semiconductors. (See also
    Ceramics.)
    properties 9:213
    uses 9:213
Indium arsenide. (See also Indium alloys.)
    manufacture 14:546
    properties 14:546
    uses 14:546
Indium arsenide semiconductors. (See also
    Ceramics.)
    properties 9:213
    uses 9:213
Indium bromides. (See also Indium compounds.)
    14:559
    manufacture 14:559
    properties 14:558
    solubility 14:559
Indium chlorides. (See also Indium compounds.)
    14:559
    manufacture 14:559
    properties 14:558
    solubility 14:559
Indium compounds. (See also individual com-
    pounds.) 14:558
    analysis 14:561
    indium content 14:561
    properties 14:558
    toxicity 14:524
Indium fluoride. (See also Indium compounds.)
    14:560
    properties 14:558
    solubility 14:560
Indium–gallium alloys 13:271
    analysis 13:271
    gallium content 13:271
Indium 8-hydroxyquinolate
    potassium bromate titration 14:531
Indium iodides. (See also Indium compounds.)
    14:560
    manufacture 14:560
    properties 14:558
    solubility 14:560
Indium metal. (See also Indium alloys.)
    analysis 14:525
    commercial grades and specifications 14:525
    hydrogen in 8:616, 681

impurities, determination of 14:536
indium content 14:526
manufacture 14:524
oxygen in 8:616, 681
properties 14:521
solubility 14:523
toxicity 14:524
uses 14:520
vacuum fusion methods 8:616, 681
Indium nitride. (See also Indium compounds.)
    14:560
manufacture 14:560
properties 14:558
solubility 14:560
Indium oxides. (See also Indium compounds.)
    14:559
manufacture 14:559
properties 14:558
solubility 14:559
Indium phosphide. (See also Indium alloys.)
manufacture 14:546
properties 14:546
uses 14:546
Indium phosphide semiconductors. (See also
    Ceramics.)
properties 9:213
uses 9:213
Indium sulfates. (See also Indium compounds.)
    14:560
manufacture 14:560
solubility 14:560
Indium sulfides. (See also Indium compounds.)
    14:560
manufacture 14:560
properties 14:558
solubility 14:560
Indole
thin-layer chromatography 15:295
in tryptophan 8:533
2-Indole-1'-acenapthene
structure 14:498
2-Indole-1'-acenaphthene indigo. (See also Indigoid
    dyes.)
Indole alkaloids 4:587
analysis 4:600
Indole–thionaphthene dyes. (See also Indigoid
    dyes; and individual compounds.) 14:495
properties 14:497
2'-Indole-2-thionaphthene indigo. (See also Indigoid
    dyes.)
structure 14:495
3'-Indole-2-thionaphthene indigo. (See also Indigoid
    dyes.)
structure 14:495
Indophenol blue
colorimetric method for nitrogen 12:113
Indophenol dyes 12:8

Indophenol test
for p-diamines 6:410, 411
Induction time
of styrene 18:311
Indulines. (See also Azine and related dyes.) 6:
    381, 382
identification 6:389
manufacture 6:383, 384
solubility 6:383
Industrial detergents. (See also Detergents.) 11:
    326
Inert gases. (See also Gases, noble.)
in fluorine, elemental 13:4
gasometric method 12:325
Inert gas fusion
of carbides 8:135
of ceramics 9:217
Infrared detector methods
for sulfur 15:480
Infrared milk analyzer 16:128
Infrared radiation, polarized
structure of crystalline polymers and fibers
    4:316
Infrared spectra
of acenocoumarol 10:153
of acetaldehyde 8:345
of acetamide 8:602
of acetaminophen 6:183
of acetanilide 5:438; 6:182
of acetic anhydride 8:563
of acetone 8:346
of acetophenone 8:347
of acetylsalicylic acid 6:128
of acrolein 4:153
of acrylamide 4:163
of acrylic and modacrylic polymers and fibers
    4:240
of acrylic fibers 4:241, 245, 385
of acrylonitrile 4:372
of acrylonitrile copolymers 4:245
of actinomycin D 5:533
of adipic acid 4:414
of alanine 8:448
of alkyd resins 5:41, 43, 44, 45, 47
2-amino-1-butanol 5:9
2-(2-aminoethylamino)ethanol 5:8
of N-(2-aminoethyl)aziridine 14:492
2-amino-2-ethyl-1,3-propanediol 5:9
2-amino-2-methyl-1,3-propanediol 5:10
2-amino-2-methyl-1-propanol 5:10
1-amino-2-propanol 5:9
of N-(3-aminopropyl)cyclohexylamine 11:248
of aminopyrine 6:213
of ammonia, anhydrous 5:304
of ammonium acetate 5:326
of ammonium bicarbonate 5:328
of ammonium carbonate 5:332

of ammonium chloride 5:334
of ammonium nitrate 5:339; 12:459
of ammonium oxalate 5:344
of ammonium persulfate 5:346
of ammonium phosphate, dibasic 5:348
of ammonium picrate 12:460
of ammonium sulfate 5:351
of ammonium thiocyanate 5:353
of amobarbital sodium 6:486
of amphomycin 5:554
of amphotericins 5:568
of amyl acetate 8:584
of aniline 5:434
of *m-(o-,p-)*anisidine 5:439
of anisindione 10:176
of anisomycin 5:603
of antipyrine 6:208
of arginine 8:454
of bacitracin 5:537
of benzaldehyde 6:614; 8:346
of benzenes 7:28, 29
of benzyl alcohol 7:97
of bis(3,4-epoxy-6-methylcyclohexylmethyl)
    adipate 12:240
of bisphenol A epoxy resins 12:238
of bisphenol A polycarbonate 17:332
of bisphenol Z polycarbonate 17:332
of biurets 15:96
of borax 7:380
of boric acid 7:380
of 1-bromo-2-propanol 14:159
of BTTN 12:455
of butallylonal 6:487
of butethal 6:487
of butylated urea–formaldehyde resin 5:283
of caffeine 6:129
of calcium amphomycin 5:554
of carbomycin 5:521
of carbon dioxide 8:253, 264
of carbon disulfide 8:272
of carbon monoxide 8:314, 317
of casein 9:7
of cellulose acetate 4:79
of cellulose nitrate 12:455
of cellulose triacetate 4:79
of china clay 12:152
of chloramphenicol 5:497
of chloramphenicol palmitate 5:497
of *m-(o-,p)*chloraniline 5:440
of chlorinated polyethylene 12:367
of chlorinated polypropylene 12:367
of *p*-chlorobenzoyl chloride 8:573
of 3-chloro-1,2-propanediol 14:160
of 3-chloro-1-propanol 14:159
of 1-chloro-2-propanol 14:159
of chlorosulfonated polyethylene 12:368
of choline chloride 9:612

of cinchophen 6:200
of cinnamic acid 10:60
of CI Pigment Red 88 14:507
of CI Vat Black 1 14:507
of CI Vat Blue 5 14:505
of CI Vat Brown 5 14:505
of CI Vat Orange 5 14:507
of CI Vat Red 1 14:506
of CI Vat Violet 2 14:506
of colistin sulfate 5:538
of copper(II) hydroxide 10:666
of cortisone acetate 14:193
of crotonaldehyde 8:345
of cuminal 6:614
of *N*-(2-cyanoethyl)cyclohexylamine 11:248
of cyclamic acid 11:237
of cyclobarbital 6:488
of cyclohexanol 11:189, 196
of 2-cyclohexylamino-1-phenylethanol 11:250
of cyclopal sodium 6:488
of cycloserine 5:511
of decaborane 7:354
of DEGN 12:456
of dental restorative materials, composite 11:311
of diborane 7:354
of dibutyl phthalate 8:585
of dicyclohexylamine 11:240
of diethanolamine 5:7
of 2-(diethylamino)ethanol 5:8
of *N,N*-diethylaniline 5:436
of diethylenetriamine 11:419
of dihydrostreptomycin sulfate 5:583
of dimethylacetamide 8:603
of *N,N*-dimethylaniline 5:434
of *N,N*-dimethylcyclohexylamine 11:250
of dimethyldichlorosilane 18:56
of dimethyldiethoxysilane 18:72
of dinitronaphthalenes 12:458
of 2,4-dinitrotoluene 12:459
of diphenadione 10:178
of diphenylamine 5:436
of dipropylbarbituric acid 6:489
of Dynel acrylic fiber 4:385
of EDNA 12:453
of epoxy novolac resins 12:239
of erythromycin 5:522
of estradiol 14:199
of estrone 14:200
of *N*-ethylaniline 5:436
of *N*-ethylcyclohexylamine 11:245
of ethylene–acetylene copolymers 12:366
of ethylene–acrylic acid ionomers 12:367
of ethylene bromohydrin 14:158
of ethylene–butene copolymers 12:366
of ethylene–carbon monoxide copolymers 12:
    366
of ethylene chlorohydrin 14:158

of ethylene–chlorotrifluoroethylene copolymer 12:368
of ethylenediamine 11:419
of ethylene–diethyl fumarate copolymers 12:367
of ethylene–dimethyl fumarate copolymers 12:367
of ethylene–ethyl acrylate copolymers 12:367
of ethylene fluorohydrin 14:158
of ethylene–maleic anhydride copolymers 12:367
of ethylene–methacrylic acid ionomers 12:367
of ethylene–methyl acrylate copolymers 12:367
of ethylene oxide 12:320
of ethylene–propylene–cyclooctadiene terpolymer 12:368
of ethylene–propylene–dicyclopentadiene terpolymer 12:368
of ethylene–propylene–1,4-hexadiene terpolymer 12:368
of ethylene–styrene copolymers 12:366
of ethylene trifluoropropylene copolymer 12:368
of ethylene–vinyl acetate copolymers 12:366
of ethylene–vinyl chloride copolymer 12:368
of ethylene–vinyl sulfonic acid copolymer 12:368
of F D & C Blue No. 1 10:472
of F D & C Blue No. 2 10:473
of F D & C Green No. 3 10:474
of F D & C Red No. 2 10:475
of F D & C Red No. 3 10:476
of F D & C Red No. 4 10:477
of F D & C Violet No. 1 10:478
of F D & C Yellow No. 5 10:479
of F D & C Yellow No. 6 10:480
of formanilide 5:438
of fumagillin 5:569
of gentamycin sulfate 5:604
of glutamic acid 8:471
of glycerol alkyds 5:47
of glycerol and pentaerythritol alkyds 5:47
of glycine 8:478
of gramicidin D 5:539
of griseofulvin 5:498
of hexethal sodium 6:489
of HMX 12:454
of hydrocortisone acetate 14:194
of hydrogen chloride 14:391
of N-(2-hydroxyethyl)cyclohexylamine 11:247
of hydroxylammonium acid sulfate 14:442
of hydroxylammonium chloride 14:442
of hydroxylammonium sulfate 14:442
of hygromycin B 5:584
of indigo 14:504
of isoleucine 8:492
of isophthalic alkyd 5:41

of N-isopropylhydroxylammonium oxalate 14:443
of kanamycin B sulfate 5:584
of lactic acid 8:377
of leucine 8:492
of lincomycin hydrochloride monohydrate 5:585
of lycine 8:497
of mannitol hexanitrate 12:456
of MEDINA 12:453
of melamine–formaldehyde resin 5:284, 285
of mepivacaine hydrochloride 5:397
of metanilic acid 5:441
of methane in carbon dioxide 8:253
of methionine 8:502
of 2-(methylamino)ethanol 5:8
of N-methylaniline 5:434
of methyldichlorosilane 18:57
of 4,4'-methylenedianiline 5:441
of methyl hydrogen silicone fluids 18:107
of methyl methacrylate–n-butyl acrylate copolymers 16:75
of methyl methacrylate–n-butyl methacrylate copolymers 16:75, 76
of methyl methacrylate–ethyl acrylate copolymers 16:74
of methylphenyldiethoxysilane 18:72
of methyltriethoxysilane 18:73
of metriol trinitrate 12:456
of monoethanolamine 5:7
of neocinchophen 6:201
of neomycin palmitate 5:586
of neomycin sulfate 5:586
of neomycin undecylenate 5:587
of m-(o-,p-)nitroaniline 5:442
of 1-nitrobutane 16:421
of 2-nitrobutane 16:421
of 2-nitro-1-butanol 16:442
of nitroethane 16:420, 436
of 2-nitro-2-ethyl-1,3-propanediol 16:443
of nitroglycerin 12:457
of nitromethane 16:417, 436
of 1-nitro-2-methylpropane 16:421
of 2-nitro-2-methylpropane 16:422
of 2-nitro-2-methyl-1,3-propanediol 16:442
of 2-nitro-2-methyl-1-propanol 16:442
of 1-nitropropane 16:420, 436
of 2-nitropropane 16:420, 436
of 2-nitro-1-propanol 16:441
of novobiocin 5:499
of nystatin 5:570
of oleandomycin chloroform adduct 5:523
of oxalic acid 8:377
of oxyethylene–oxypropylene copolymers 19:307
of paromomycin sulfate 5:588
of pentaerythritol alkyds 5:47

of pentaethylenehexamine 11:420
of PETN 12:457
of phenacaine hydrochloride 5:402
of phenacetin 6:129
of phenylalanine 8:508
of N-phenylcyclohexylamine 11:251
of β-phenylethyl alcohol 7:98
of phenylethylmethylsilicone fluids 18:108
of phenylmethyldichlorosilane 18:58
of phenylmethylsilicone fluids 18:107
of phenyltrichlorosilane 18:58
of phenyltriethoxysilane 18:71
of o-phthalic alkyds 5:41, 43, 45, 47
of phthalic anhydride 8:563
of phytonadione 10:186
of polyacrylamide 4:172
of polyacrylonitrile 4:231, 384
of poly(n-butyl methacrylate) 16:73
of poly(n-dodecyl methacrylate) 16:73
of polyethylene–polyisobutene blends 12:366
of poly(ethyl methacrylate) 16:72
of poly(isobutyl methacrylate) 16:73
of poly(methacrylic acid) 16:74
of poly(methyl methacrylate) 16:72, 77, 78
of polymyxin B sulfate 5:540
of polypropylene–polyisobutene blends 12:366
of poly(n-propyl methacrylate) 16:72
of prednisolone acetate 14:196
of prednisone 14:197
of probarbital calcium 6:490
of procaine hydrochloride 5:405
of proline 8:516
of prophenal 6:490
of propionic acid 8:376
of propionyl chloride 8:572
of propoxycaine hydrochloride 5:411
of propylene–styrene copolymers 12:366
of pyruvic acid 8:377
of RDX 12:453
of salicylaldehyde 6:615
of salicylamide 6:129
of serine 8:521
of silica fillers 12:152
of sodium barbital 6:486
of sodium colistimethate 5:538
of sodium hexobarbital 5:368
of sodium salicylate 6:128
of streptomycin sulfate 5:588
of sulfanilic acid 5:441
of TEGN 12:457
of testosterone 14:203
of tetracaine hydrochloride 5:416
of tetrachlorobisphenol A polycarbonate 17:332
of tetraethylenepentamine 11:420
of tetrafluoroethylene–hexafluoro-propylene co-
     polymer 13:91
of 1,3,6,8-tetranitrocarbazole 12:460

of tetryl 12:455
of thioindigo 14:505
of threonine 8:526
of thyroxine 14:190
of TNT 12:459
of m-tolualdehyde 6:615
of m-(o-,p-)toluidine 5:437
of triacetyloleandomycin 5:524
of triamcinolone 14:198
of triamcinolone acetonide 14:199
of triazine–formaldehyde resin 5:284
of tribromoethanol 5:373
of trichloroacetic acid 8:378
of triethanolamine 5:7
of triethylenetetraamine 11:419
of trinitronaphthalene 12:458
of tris(hydroxymethyl)aminoethane 5:10
of tris(hydroxymethyl)nitromethane
     16:443
of tryptophan 8:530
of tylosin 5:524
of tyrocidine hydrochloride 5:540
of tyrosine 8:535
of tyrothricin 5:541
of urea–formaldehyde resin 5:282, 283
of urea–melamine resin 5:284
of valine 8:541
of vancomycin sulfate 5:606
of vinyltrichlorosilane 18:57
of viomycin sulfate 5:557
of warfarin sodium 10:167
of whiting 12:152
of 2,4-xylidine 5:438
Infrared spectroscopy
of acetals 4:66, 71
of acetate and triacetate fibers 4:78
of acetylene 13:294
of acetyl isocyanate 15:106
of acetylsalicylic acid 6:162
of acid anhydrides 8:562
of acid halides 8:573
of acid salts 8:592
of acrylate rubbers 12:94
of acrylic and modacrylic polymers and fibers
     4:229
of acrylic monomers 4:215
of acrylonitrile 4:378; 16:411
of acrylonitrile–butadiene–styrene polymer
     18:319
of acrylonitrile polymers 4:384
of agar 12:583
of alcohol polyether sulfates 11:331
of alcohols 4:523, 525
of aliphatic quaternary ammonium chlorides
     11:331
of alkanolamines 5:7
of alkyd resins 5:40

of alkylaryl sulfonates 12:576
of alkylbenzene sulfonates 11:330
of alkylolamines 12:576
of alkyl sulfates 12:576
of alkyl sulfonates 12:576
of allophanates 15:94
of allyl chloride 5:102; 9:394
of amides 8:602, 609; 12:576
of γ-aminopropyltriethoxysilane 18:83
of ammonia 14:458
of n-amyl isocyanate 15:106
of aniline and its derivatives 5:433, 453
of antidegradants in elastomers 12:141
of antihistamines 6:25
of antioxidants and antiozonants 6:97
of aureothricin 5:515
of azo dyes 6:404
of barbiturates 6:484, 492, 496
of benzal chloride 9:488
of benzaldehydes 6:615
of benzenes 7:28; 9:394
of benzoic acid 7:65
of benzyl alcohol 7:97, 100
of benzyl isocyanate 15:106
of biscyclohexylsulfamide 11:226
of bishydroxycoumarin 10:157, 159; 11:162
of 1,3-bis(isocyanatomethyl)benzol 15:106
of 1,1-bis(4-isocyanatophenyl)cyclohexane
    15:106
of bis(4-isocyanatophenyl)methane 15:106
of 2,2-bis(4-isocyanatophenyl)propane 15:106
of bis(trimethylsilyl)acetamide 18:92
of boric oxide 13:459
of boron compounds, organic 7:353, 355
of butadiene dioxide 12:200
of butadiene–methacrylic acid copolymer 12:96
of butadiene–styrene–vinyl-2-pyridine terpolymer
    12:97
of n-butyl glycidyl ether 12:200
of butyl isocyanates 15:106
of butyl rubber 12:93, 97
of caffeine 6:162
of caprolactam 8:122; 17:296, 322
of carbodiimides 15:98
of carbon dioxide 8:252, 265; 13:294
of carbon disulfide 18:543
of carbon monoxide 8:313, 318, 653; 13:294
of carbon tetrachloride 8:328; 9:390, 392, 394,
    433
of carbon tetrafluoride 13:294
of carbonyl compounds 8:345
of carboxylic acids 4:416; 7:65; 8:378; 12:576
of carrageenin 12:583
of cellulose acetate 12:590
of cellulose acetobutyrate 12:590
of cellulose butyrate 12:590
of cellulose derivatives 12:590

of cellulose diacetate 12:590
of cellulose esters 12:590
of cellulose nitrate 12:590
of cellulose propionate 12:590
of cellulose triacetate 12:590
of cephalosporins 16:559
of ceramics 9:217
of chlorinated toluenes 9:487
of chlorine 9:321
of 1-chlor-6-isocyanato-n-hexane 15:106
of chlorobenzene 9:394
of chlorobutadienes 9:390
of chlorobutanes 9:390
of chlorobutenes 9:390
of chlorocarbons 9:389
of chloroform 9:390, 392, 394
of chlorophenyl isocyanates 15:106
of chloropicrin 8:13
of chloropropanes 9:390
of chloropropenes 9:390, 393, 394
of chlorosilanes 18:49
of chlorosulfonated polyethylene 12:99
of cinnamaldehyde 10:46
of coal and coke 10:212
of cocoa products 9:593
of codeine phosphate 6:163
of cotton 11:118
of cresols 17:29
of 1-cyan-5-isocyanato-n-pentane 15:106
of 1-cyan-3-isocyanatopropane 15:106
of cyanoacetamide 15:442
of cyanoacetic acid 15:439
of 1-cyanocyclohexene oxide 12:200
of cyclamates 11:236
of cyclic alcohols 11:190
of cyclodecene oxides 12:200
of cyclohexanone 11:196
of cyclohexylamine 11:211
of cyclohexylethylene oxide 12:200
of cyclohexyl isocyanate 15:106
of cyclooctene oxides 12:200
of detergents 11:328, 330
of diallyl ether dioxide 12:200
of diamines and higher amines, aromatic 11:442
of 1,1-di-t-butylethylene oxide 12:200
of dichlorobenzenes 9:394, 480, 481
of dichlorobutanes 9:390
of dichlorobutenes 9:390
of dichloroethanes 9:390, 392
of dichloroethylenes 9:390, 394
of dichloromethoxyphenols 9:515
of dichlorophenols 9:515
of 2,5-dichlorophenyl isocyanates 15:106
of dichloropropanes 9:390, 393
of dichloropropenes 9:390
of dichlorosilane 18:50
of 1,2,4,5-diepoxycyclohexane 12:200

of 1,2,5,6-diepoxypentane 12:200
of 3,3'-diethoxy-4,4'-diisocyanatodiphenyl
    15:106
of diethylethoxymethylene malonate 15:437
of diethyl malonate 15:431
of diisocyanates 15:96
of 1,4-diisocyanatobenzol 15:106
of 4,4'-diisocyanatodicyclohexylmethane 15:
    106
of dimethisterone 10:588
of 3,3'-dimethoxy-4,4'-diisocyanatodiphenyl
    15:106
of 2,5-dimethoxyphenyl isocyanate 15:106
of *unsym*-dimethylhydrazine 14:235
of 2,2-dimethyl-3-isocyanatobutane 15:106
of dimethyl malonate 15:431
of dimethylmethoxymethylene malonate 15:437
of dimethylol ethyl carbamate 11:139
of dimethylol ethyleneurea 11:139
of dimethylol ethyl triazone 11:139
of dimethylol perhydropyrimidinone 11:139
of dimethylol urea 11:138
of dimethylsilicone fluids 18:105
of dimethyl uron 11:139
of dioxane 11:519
of diphenyl 11:527
of dithiocarbamic acid salts 11:583
of elastomers, synthetic 12:92
of enamels 12:188
for end groups in acrylic polymers 4:296
of epichlorohydrin 12:200
of 2,3-epoxybutanes 12:200
of 1,2-epoxy-3-butanone 12:200
of 1,2-epoxy-3-butene 12:200
of epoxy compounds 12:198
of 2,3-epoxycyclohexanone 12:200
of 1,2-epoxy-3-hexyne 12:200
of 2,3-epoxy-1-methoxybutane 12:200
of 1,2-epoxy-2-methyl-3-butanone 12:200
of 2,3-epoxy-2-methylheptane 12:200
of 2,3-epoxy-2-methyl-4-pentanone 12:200
of epoxy resins 12:240, 254, 256
of epoxy-2,4,4-trimethylpentanes 12:200
of ester plasticizers 18:327
of esters 8:584; 12:576
of ethers 12:300
of ethinyl estradiol 10:569, 570, 574
of ethoxyacetylene 4:142
of ethoxylated alkyl phenols 11:331
of ethoxylated fatty acids 11:330
of ethoxylated fatty alcohols 11:330
of 3-ethoxyphenyl isocyanate 15:106
of ethyl biscoumacetate 10:163
of ethyl cyanoacetate 15:440
of ethylene oxide 12:199, 200, 577
of ethylene–propylene copolymers 12:362, 363
of ethylene–propylene rubber 12:93, 100

of ethylene–vinyl acetate copolymers 12:93
of ethylenimine 14:473
of ethylenimine derivatives 14:493
of ethyl groups 12:356
of ethyl isocyanate 15:109
of ethylphenols 17:29
of fats and fatty oils 12:490
of fatty acids 12:490
of fatty alcohol sulfates 11:330
of finishing agents 11:136
of fluorocarbons 13:66
of formaldehyde 13:108
of formic acid 13:119
of furfural 13:235
of gelatins 12:583
of glass and glass-ceramics 13:459
of glyceryl monostearates 11:331
of glycidaldehyde 12:200
of glycidol 12:200
of glycidonitrile 12:200
of γ-glycidoxypropyltrimethoxysilane 18:83
of glycols 11:83; 13:583, 590
of guar gum 12:583
of gum arabic 12:583
of gum ghatti 12:583
of gums 12:583
of halohydrins 14:157
of Halothane 5:364
of helium 13:295
of hexamethyldisilazane 18:92
of 1,6-hexamethylene diisocyanate 19:316
of high explosives 12:452
of hydrazine 14:235
of hydrocarbon resins 14:268
of hydrocarbons in carbon dioxide 8:252
of hydroxyl groups 12:249
of hypalon 12:93
of impact polystyrene 18:318
of impurities in carbon monoxide 8:313
of impurities in carbon tetrachloride 8:325
of indigoid dyes 14:505
of inorganic fillers in elastomers 12:152
of iron oxide 13:464
of isobutylene oxide 12:200
of isobutyl isocyanate 15:106
of 2-isocyanatochryson 15:106
of 1-isocyanato-2-ethoxy-5,6,7,8-tetrahydro-
    naphthalin 15:106
of 3-isocyanato-*n*-pentane 15:106
of 3-isocyanatotetramethylensulfone 15:106
of isocyanic acid esters 15:107
of isocyanurates 15:97
of isoprene–acrylonitrile copolymer 12:99
of isopropyl chloride 9:394
of isopropyl isocyanate 15:109
of karaya 12:583
of lactose 16:128

of limonene dioxide 12:200
of locust bean gum 12:583
of LSD 15:295
of malonic acid 15:426
of malononitrile 15:443
of menthene oxide 12:200
of mestranol 10:577
of methacrylate polymers 16:71
of methane 13:294
of methane in carbon dioxide 8:252
of methanol 16:119
of 3-methoxyphenyl isocyanate 15:106
of methylcellulose 12:583
of methyl cyanoacetate 15:440
of 1-methyl-3,5-diethyl-2,4-diisocyanatobenzol
    15:106
of 1-methyldiisocyanatobenzols 15:106
of 1-methyl-2,4-diisocyanatocyclohexane 15:106
of methyldiphenyls 11:527
of methylene chloride 9:390, 392, 429, 430; 13:
    294
of methylenecyclohexane oxide 12:200
of methyl ethyl ketone 7:100, 101
of methyl groups 12:350
of methylhydrazine 14:235
of methyl isocyanate 15:108, 109
of 2-methyl-3-isocyanatobutane 15:106
of methyl isothiocyanate 13:193
of methyl methacrylate–lauryl methacrylate copoly-
    mers 16:77
of methyl methacrylate–methyl acrylate copolymers
    16:75
of methylol melamine 11:139
of methylphenylsilicone resin 18:120
of α-methylstyrene oxide 12:200
of milkfat 16:128
of milk protein 16:128
of mono-alkylol amides 11:331
of monochlorophenol 9:515
of monochlorotoluenes 9:488
of morpholine 16:195
of 2-naphthylamine 16:229
of 1-naphthyl isocyanate 15:106
of natural gas 13:315
of natural rubber 12:92, 98
of NBR 12:93, 95
of nitramines 12:452
of nitrates 12:452
of nitric oxide 16:478, 485
of nitriles 8:609; 16:397, 403
of nitro aromatics 12:452
of 3-nitro-4-chlorophenyl isocyanate 15:106
of 1-nitro-2-methyl-4,6-diethyl-3,5-diisocyanato-
    benzol 15:106
of 1-nitro-3-methyl-4,6-diisocyanatobenzol 15:106
of nitroparaffins 16:417, 436
of 4-nitrophenyl isocyanate 15:106

of nitrous oxide 13:294; 16:485
of noble gases 13:294
nondispersive 8:266
of norethindrone acetate 10:590
of n-octyl alcohol 7:100, 101
of oil 8:275
of organic materials in hydrogen chloride 14:401
of organo-chlorosilanes 18:56
of organofunctional silanes 18:83
of organosilane esters 18:71, 75
of pectin 12:583
of penicillin O 16:571
of penicillins 16:559
of pentachlorobenzene 9:394
of pentachloroethane 9:390
of pentachlorophenol 9:515
of pentamethylene oxide 12:199
of pesticides and insecticides 19:610
of phenacaine hydrochloride 5:400
of phenacetin 6:162
of phenethicillin 16:573
of phenol 17:29
of phenolic resins 7:444
of phenols 17:22, 28
of β-phenylethyl alcohol 7:97
of phenyl isocyanate 15:106
of 1-phenylpropene oxides 12:200
of photographic film bases 12:589
of α-pinene oxide 12:200
of β-pinene resins 14:268
of polyacrylamide 4:171
of polyacrylonitrile 18:319
of polyamide fibers 17:317
of polyamides 17:284
of polybutadiene 12:93, 94; 18:318, 332
of polycarbonates 12:590; 17:331
of polychloroprene 12:93
of polyepichlorohydrin 12:99
of polyesters 12:590
of polyethers 17:300
of polyethylene 12:350, 590
of polyethylene glycols 13:590
of polyethylene–polypropylene blends 12:362,
    363, 368
of polyethylenimine 14:481
of polyglycols 13:590
of polyhydric alcohols 13:583
of polyisoprene 12:98, 101
of poly(lauryl methacrylate) 16:77
of polyoxyethylene compounds 11:83
of polypropylene 12:361
of polypropylene glycols 13:590
of polysaccharides 12:583
of polystyrene 12:590; 18:316
of polysulfide rubber 12:100
of polytetrafluoroethylene 13:80
of polyurethanes 19:282

of poly(vinyl chloride) 12:93, 99
preparation of resin samples 5:41
of n-propyl chloride 9:394
of propylene oxide 12:200
of pyridine and its derivatives 17:426, 436
of pyrrocaine hydrochloride 5:414
of pyrrole 17:436
of quaternary ammonium compounds 12:577
of resins 7:444
of rhenium 17:540
sample preparation for polymers and fibers 4:232
of saponins 12:577
of SBR 12:93, 95
of silicone fluids 18:104
of silicone resins 18:119
of silylation reagents 18:92
of soap 11:331
of sodium heparin 10:143
of sodium malonate 15:426
of sodium warfarin 11:162
standards for polymer analysis 4:231
of stearyl isocyanate 15:106
of steroids 14:207
of styrene 18:294
of styrene–acrylonitrile copolymer 18:319
of styrene–butadiene copolymer 18:332
of styrene–methyl methacrylate copolymers 16:74
of styrene oxide 12:200
of styrene polymers 18:316
of sulfate 18:423
of sulfated monoglycerides 11:330
of sulfated phenoxy ethers 11:330
of sulfonamides 12:576
of sulfonated amides 11:330
of sulfonated esters 11:330
of sulfosuccinates 12:576
of sulfur hexafluoride 13:294
of surfactants 12:575
of technetium 17:540
of terpene resins 14:268
of terphenyls 11:534
of tetrachlorobenzenes 9:394
of 1,1,2,2-tetrachloroethane 9:394
of tetrachloroethylene 9:390, 394
of tetrachlorophenols 9:515
of tetramethylene oxide 12:199
of textile finishes and coatings 4:329
of thiophene 18:542
of thiuram disulfides 11:596
of toluene diisocyanate 19:273
of tolylene diisocyanates 15:113
of tolyl isocyanates 15:106
of tragacanth 12:583
of trichlorobenzenes 9:394
of trichloroethanes 9:390, 394

of trichloroethylene 9:390, 394
of trichlorophenols 9:515
of trichloropropanes 9:390
of trichloropropenes 9:390
of trichlorosilane 18:50
of 4,4′,4″-triisocyanatotriphenylmethane 15:106
of trimethylene oxide 12:199
of N-trimethylsilylimidazole 18:92
of triphenylmethane and related dyes 19:168
of trisaziridinyl phosphine oxide 11:139
of 1,3,5-trivinylbenzene trioxide 12:200
of ureas 15:96
of urethanes 15:94
of vinylcyclohexane oxide 12:200
of vinylene groups 12:357
of vinyl groups 12:357
of vinylidene groups 12:357, 361
of vulcanization accelerators 12:143
of water 13:68, 459, 462; 14:239
of xylenols 17:29
Infrared spectroscopy, differential
of finishing agents 11:136
Ingot iron. (See also Iron metal.)
commercial grades and specifications 15:28
manufacture 15:27
uses 15:27
Inherent viscosity 4:303
Inhibitors
in 1,3-butadiene 11:503, 504, 507
gas chromatography 9:442
in isoprene 11:510
in 1,1,1-trichloroethane 9:442
visible spectroscopy 11:511
Inhibitors, corrosion
in brake fluids 7:424, 425
Inhibitors, flex-crack 6:78
Ink
carbon black in 8:241
in polyurethanes 19:312
solvent extraction 18:320
in styrene polymers 18:320
Inorganic alkaline detergents. (See also Detergents.) 11:324
Inorganic material
sulfur content 18:380
Inorganic oxides
carrier gas fusion methods 8:625
oxygen in 8:625
Inorganic pigments
in matches 15:516
Inorganic salts
in certified colors 10:491
in detergents 11:380
Inositol
in green leaf tea 18:478
paper chromatography 18:478
in wines 19:425, 451

Insecticides. (See also *Pesticides and insecticides.*)
  bioassay 7:171, 181, 183
  in tobacco 19:155
Insoluble ash content
  of black tea 18:487
Insoluble matter
  in aluminum potassium sulfate 5:194
  in ammonium acetate 5:326
  in ammonium carbonate 5:333
  in ammonium chloride 5:338
  in ammonium iodide 14:604
  in ammonium nitrate 5:343
  in ammonium oxalate 5:345
  in ammonium persulfate 5:347
  in ammonium phosphate, dibasic 5:350
  in ammonium thiocyanate 5:354
  in certified colors 10:489
  in chromium compounds 9:686, 693
  in citric acid 10:93
  in cobalt(II) acetate 10:329
  in cobalt carbonate 10:333
  in cobalt chloride 10:336
  in cobalt nitrate 10:340
  in cobalt(II) sulfate 10:343, 344
  in copper(II) acetate 10:654
  in copper(II) bromide 10:656
  in copper(II) carbonate, basic 10:658
  in copper(I) chloride 10:661
  in copper(II) chloride 10:662
  in copper(II) nitrate 10:668
  in copper(I) oxide 10:672
  in copper(II) oxide 10:674
  in copper(II) sulfate 10:678
  in egg albumen 12:69
  in G salt 16:239
  in hexachlorobenzene 9:484
  in magnesium acetate 15:398
  in magnesium carbonate 15:401
  in magnesium chloride 15:404
  in magnesium nitrate 15:407
  in magnesium oxide 15:409
  in magnesium perchlorate 9:364, 366
  in magnesium sulfate 15:412
  in manganese sulfate 15:490
  in mercuric chloride 16:60
  in mercuric iodide 16:60
  in mercuric oxide 16:61
  in mercury fulminate 12:409
  in molybdenum disulfide 16:179
  in 2-naphthol 16:232
  in nickel sulfate 16:344
  in pesticides and insecticides 19:614
  in phenyl mercuric acetate 16:63
  in phosphorus pentoxide 17:111
  in portland cement 9:109
  in potassium chlorate 9:359, 361
  in potassium chloride 9:339

  in potassium hydrogen phthalate 17:403
  in potassium iodate 14:608
  in potassium iodide 14:605
  in potassium permanganate 15:492
  in red lead 17:196
  in R salt 16:238
  in Schaeffer's salt 16:236
  in silver nitrate 18:175
  in silver sulfate 18:176
  in sodium chlorate 9:360, 361
  in sodium chloride 9:338, 347
  in sodium fluoride 13:33
  in sodium iodide 14:605
  in thorium nitrate 19:18
  in zinc chromate 9:707
Instant coffee 10:376
  analysis 10:439
  caffein content 10:409
  water content 10:440
Instant tea
  analysis 18:512
  light-scattering 18:513
  manufacture 18:456
  solubility 18:513
  turbidity 18:513
Insulating firebrick. (See also Refractories.)
  composition 9:207
Insulin. (See also Pancreatic hormones.) 14:182,
    565–583
  amino acid composition 14:566
  analysis 14:206, 571
  assay 14:206
  bioassay 14:206, 573, 575
  commercial grades and specifications 14:188, 570
  composition 14:182
  dextran–charcoal adsorption 14:578
  fibril formation 14:574
  gravimetric determination 14:577
  hemolysis inhibition method 14:575
  history 14:565
  identification 14:579
  immunological determination 14:575
  impurities, determination of 14:579
  manufacture 14:569
  nitrogen content 14:580
  occurrence 14:182, 569
  paper chromatography 14:573
  in plasma 14:575
  properties 14:566
  radioimmunological methods 14:576
  separation 14:572
  in serum 14:572
  solubility 14:567
  solvent extraction 14:572
  specification tests 14:579
  structure 14:182, 188, 567
  in tissue 14:572, 575

in tumors 14:575
ultraviolet spectroscopy 14:568
uses 14:182, 189
Intene 12:86
gel permeation chromatography 12:120
intrinsic viscosity 12:119
Interfacial tension
of detergents 11:386
of lubricants 15:346
Interference spectroscopy. See *Fourier transform spectroscopy.*
Interferograms. (See also Fourier transform spectra.)
Interferometers. (See also Fourier transform spectrometers.) 13:141, 147, 151, 152, 153, 169
Interferometry. See *Fourier transform spectroscopy.*
Interlayer spacing
of graphite 14:76
Intermediate compounds
in certified colors 10:499
Intermediates for distilled alcoholic beverages 4: 468, 480
Intermedin. (See also Pituitary hormones.) 14:180
composition 14:180
uses 14:180
Internal bond
of paper matchboard 15:534
Internal combustion engine lubricants. (See also *Lubricants.*)
commercial grades and specifications 15:311
Internal tearing resistance
of paper 7:282
Intex 12:86
Intol 12:86
gel permeation chromatography 12:120
intrinsic viscosity 12:119
thermal analysis, differential 12:133
Intrinsic viscosity 4:303
of acrylamide polymers 4:174
of acrylic polymers 4:302
of cellulose 9:34, 37
of cellulose acetate 9:62, 71
of cellulose acetate polymers 4:82
of elastomers, synthetic 12:118
of ethylene and propylene polymers 12:370
Huggins equation 17:343
polyacrylonitrile, constants for 4:305
of polybutadienes 12:119
of poly(butyl methacrylate) 16:71
of polycarbonates 17:343
of poly(dodecyl methacrylate) 16:71
of polyethylene 12:371
of poly(ethyl methacrylate) 16:71
of polyisoprenes 12:119
of polymers 4:80

of poly(methyl methacrylate) 16:71
of poly(nitrile methacrylate) 16:71
of poly(octadecyl methacrylate) 16:71
of poly(octyl methacrylate) 16:71
of polypropylene 12:371
of regenerated cellulose 4:82
of rubber, natural 12:119
of SBR 12:119
Schultz-Blaschke equation 17:343
of silicone fluids 18:114
Invertase
activity in beer 7:586
hydrolysis of sugar 18:347
optical rotation method 19:534
sugar reduction method 19:534
in yeast 19:534
Invert soaps. See Cationic detergents.
Invert sugar. (See also *Sugar.*) 18:337
refractive index 18:343
Investments, dental. See Dental investments.
Inyoite 7:314
Iodates. (See also individual compounds.) 14:607–609
analysis 14:607
in cesium iodate 9:290
in cotton 11:142
gravimetric method for cerium 9:272
gravimetric method for thorium 19:9
potassium cyanide–sodium thiosulfate volumetric method 14:599
in potassium iodide 12:545; 14:605
in potassium metaperiodate 14:610
potassium thiocyanate colorimetric method 11:142
pyrogallol colorimetric method 11:142
in sodium iodide 14:605
in sodium metaperiodate 14:610
starch–sulfuric acid colorimetric method 12:545; 14:605
sulfuric acid–sodium thiosulfate volumetric method 14:607
Iodeosin
colorimetric method for morpholine 16:200
Iodic acid. (See also Iodates.) 14:607
properties 14:607
Iodide chromium. See Chromium metal.
Iodides. (See also individual compounds.) 14:594–607
in ammonium bromide 12:543
analysis 14:596
in animal feed 14:601
in brine 14:600
bromine–sodium thiosulfate volumetric method 14:600
in cadmium iodide 8:43
in calcium iodide 8:92
ceric ammonium sulfate–arsenious acid

colorimetric method 14:602
in chlorides 9:340
chlorine–chloroform color test 14:596
chlorine–starch color test 14:596
citric acid colorimetric method 14:608
colorimetric method for bismuth 7:215
ferric chloride colorimetric method 12:543
ferric chloride extraction 9:347
in glass and glass-ceramics 9:200; 13:413, 415, 435
gravimetric determination 14:596
gravimetric method for silver 18:164
halide content 14:597
identification 14:596
impurities, determination of 14:597, 603
iodide content 14:596
in iodides 14:597
in iodine 14:592
method for copper 5:128
neutron activation method 12:568
in nitrates 16:505
palladous chloride gravimetric method 14:596
polarography 12:543, 545, 566
in potassium chloride 9:339; 12:545
potassium cyanide–sodium thiosulfate volumetric method 14:599
in potassium iodate 14:608
potassium iodate titration 14:592
potassium persulfate redox reagent 9:340
in salt 14:601
in silver iodide 12:566
silver nitrate gravimetric method 14:596
silver nitrate titration 13:436
in sodium chloride 9:337, 347
starch–sulfuric acid color test 14:609
in water 14:600, 602
Iodinated substances
in F D & C Red No. 3 10:513
liquid chromatography 10:514
paper chromatography 10:513
Iodine. (See also Iodine, elemental.) 14:584–616
adsorption by activated carbon 8:149
adsorption by carbon black 8:187, 224
in bromine 8:5
Carius method 14:613
in chlorine 9:321
colorimetric method for acetylcholine chloride 9:617
colorimetric method for agar 12:582
colorimetric method for bethanechol chloride 9:617
colorimetric method for carbamylcholine chloride 9:617
colorimetric method for choline chloride 9:617
colorimetric method for coproporphyrin 7:638
colorimetric method for dextrin 11:127, 132
colorimetric method for locust bean gum 12:580

colorimetric method for methacholine chloride 9:617
colorimetric method for β-methylcholine chloride 9:617
colorimetric method for modified starches 11:132
colorimetric method for starch 7:555; 11:127, 132
colorimetric method for succinyldicholine chloride 9:617
colorimetric method for succinylmonocholine iodide 9:617
coulometrically generated 7:53
gravimetric method for starch 19:142
history 14:584
method for hydroquinone in acrylic monomers 4:203
neutron activation analysis 14:155
occurrence 14:584
in organic iodine compounds 14:613
oxidation in presence of bromine 8:23
in sodium chloride 9:344
sodium thiosulfate titration 9:322, 344
test for caffeine 8:58
titration of antimony 6:49, 53, 55, 65
titration of arsenic(III) 6:241
titration of boron hydrides 7:346
titration of caffeine 6:166
titration of cinchophen 6:204
titration of phenacetin 6:190
titration of thiosulfate 12:586
visible spectroscopy 8:21
volumetric method for caffeine 8:65
volumetric method for carbon disulfide 8:273
Iodine–azide
gasometric method for hydrogen sulfide 18:393
gasometric method for thiosulfate 12:549
Iodine–barium chloride
gravimetric method for sulfur dioxide 13:371
Iodine compounds, inorganic. (See also Iodates; Iodides; Iodine halides; Iodine oxides; and Periodates.) 14:593–613
Iodine compounds, organic. (See also individual compounds.) 14:613–615
analysis 14:613
iodine content 14:613
properties 14:613
uses 14:613
Iodine-consuming substances
in ammonium thiocyanate 5:354
in carbon tetrachloride 8:327
Iodine, elemental
analysis 14:591
commercial grades and specifications 14:590
identification 14:591
impurities, determination of 14:592
iodide content 14:592

manufacture 14:589
potassium arsenite titration 14:592
properties 14:585
sodium thiosulfate titration 14:591
solubility 14:586
starch color test 14:591
toxicity 14:589
uses 14:588, 589, 590
Iodine halides. (See also individual compounds.)
    14:593–607
    toxicity 14:593
Iodine heptafluoride. (See also Iodine halides.)
    14:593
    manufacture 13:9
    properties 13:8; 14:594
Iodine–lime water
    gravimetric method for acetone 12:188
Iodine monobromide. (See also Iodine halides.)
    14:593
    properties 14:594
Iodine monochloride. (See also Iodine halides; Wijs
    reagent.) 14:593
    color method for antioxidants 6:92
    method for allyl alcohol 5:84
    method for allyl esters 5:97
    method for unsaturated compounds 5:84
    properties 9:334; 14:593
    solubility 9:336
    uses 14:593
    volumetric method for elastomers, synthetic
        12:122
    volumetric method for polybutadiene 12:123
Iodine number 14:589
    of activated carbon 8:149
    of carbon fabrics 4:26
    of cocoa beans 9:548
    of epoxy resins 12:251
    of fatty alcohols 4:569
    of hydrocarbon resins 14:286
    of margarine 15:499
    Wijs volumetric method 14:286
Iodine oxides. (See also Iodine pentoxide.) 14:
    611–613
Iodine pentafluoride. (See also Iodine halides.)
    14:593
    manufacture 13:9
    properties 13:8; 14:594
Iodine pentoxide 14:611
    analysis 14:611
    commercial grades and specifications 14:611
    manufacture 14:611
    potassium iodide–sodium thiosulfate volumetric
        method 14:611
    properties 14:611
    solubility 14:611
    uses 14:611
    volumetric method for carbon monoxide 8:316

Iodine–potassium iodide
    gravimetric method for morpholine 16:195
    microchemical method for codeine 16:249
    microchemical method for diphenoxylate 16:249
    microchemical method for ethylmorphine 16:249
    microchemical method for fentanyl 16:249
    microchemical method for morphine 16:250
    microchemical method for narceine 16:250
    microchemical method for oxycodone 16:250
Iodine–pyrocatechol
    colorimetric method for phloroglucinol 17:48
    colorimetric method for resorcinol 17:47
Iodine–resorcinol
    colorimetric method for pyrocatechol 17:46
Iodine staining method
    for corn starch 11:16
Iodine–tetramethyldiaminodiphenylmethane
    color test for sulfur dioxide 18:403
Iodine tribromide. (See also Iodine halides.) 14:
    593
    properties 14:594
Iodine trichloride. (See also Iodine halides.) 14:
    593
    properties 14:594
    uses 14:593
Iodine value
    of corn oil 11:55
    of fats and fatty oils 12:481
    of fatty acids 18:185
    Wijs method 12:481
    Wijs reagent volumetric method 11:55
Iodized methylene blue
    colorimetric reagent for mercury 16:18, 20
Iodoacetic acid
    properties 8:397
o-Iodoanisol. (See also Iodine compounds, organic.)
    properties 14:612
Iodoantimonite
    colorimetric method for antimony 6:67
3-Iodobenzaldehyde
    infrared spectroscopy 6:617
Iodobenzene. (See also Iodine compounds, organic.)
    properties 14:612
Iodobromite 14:585
1-Iodo-2-butanol. (See also *Halohydrins.*)
    derivatives for identification 14:160
2-Iodo-1-butanol. (See also *Halohydrins.*)
    derivatives for identification 14:160
2-Iodo-1-ethanol. (See also *Halohydrins.*)
    derivatives for identification 14:160
Iodofluoresceins
    in F D & C Red No. 3 10:513
    liquid chromatography 10:514
    paper chromatography 10:514
    visible spectroscopy 10:514
Iodoform. (See also Iodine compounds, organic.)
    analysis 14:614

assay 14:614
in chloroform 9:434
commercial specifications 14:613
Fujiwara colorimetric method 9:386, 434
in gauze 14:614
gravimetric method for propylene glycol 11:74
identification test for carbonyl compounds 8:343
method for acetals 4:68
method for acetone 4:111
in ointments 14:614
properties 14:612
reaction 8:338
silver nitrate gravimetric method 14:614
specification tests 14:614
test for ethanol 6:4
test for ethyl biscoumacetate 10:162
test for warfarin potassium 10:175
test for warfarin sodium 10:167
Volhard titration 14:614
Iodometric methods
for acetaldehyde 19:453
for acrylonitrile 16:408
for aldehydes 12:222, 226, 335
for alkoxy groups in organosilane esters 18:75
for alkyl iodides 16:93
for aluminum benzylpenicillin 16:562
for ammonium dichromate 9:684
for ampicillin 16:562
for arsenic 14:550; 16:188
for ascorbic acid 19:425
for available chlorine in detergents 11:355
for benzathine benzylpenicillin 16:562
for benzathine phenoxymethylpenicillin 16:562
for benzothiazyl disulfide 15:571
for benzylpenicillin 16:571
for beryllium 7:134
for bisulfite 19:416
for boron hydrides 7:346
for bromate 9:361
for carboxylic acids 8:394
for cephaloridine 16:562
for cephalosporins 16:562
for cesium 17:555
for chlorine 9:330; 14:370, 395
for chlorine-containing bleaches 7:225
for chlorogenic acid 10:413
for chloroisocyanuric acids and salts 7:229
for chloroprocaine penicillin O 16:562
for chromic acid 9:684
for chromium 9:641
for cinchophen 6:203
for cobalt 10:332, 366
for copper 10:312, 608; 11:289, 315
for copper(I) chloride 10:660
for copper in copper(II) carbonate, basic 10:658
for copper naphthenate 10:679

for copper oleate 10:679
for dextrose 9:578
for DTPA 12:28
for EDTA 12:28
for EGTA 12:28
for ethers 12:304
for ethylene glycol 9:627
for ethyl ether 12:306
for europium 12:397
for fats and fatty oils 12:481
for formaldehyde 13:109
for formate 13:137
for germanium 13:384
for glycerol 18:201
for gold 14:22
for hydrabamine phenoxymethylpenicillin 16:562
for hydrogen peroxide 14:433
for hydrogen sulfide 13:322; 15:569; 18:395
for hydroperoxide 17:298
for hypophosphites 17:114
for isopropyl alcohol 10:174
for lead chromate 13:357
for lithium hypochlorite 15:287
for magnesium 15:366
for malonic acid 15:430
for margarine 15:499
for menadione 10:198, 199
for menadione sodium bisulfite 10:202
for mercaptans 15:564, 570
for mercuric cyanide 16:23
for mercurous mercury 16:61
for mercury 16:33
for methanol 16:118
for methionine 8:505, 506
for methylbenzethonium chloride 11:558
for 2-naphthol 16:236
for NTA 12:28
for oxidizing agents 18:183
for oxidizing agents in detergents 11:334
for oxygen 16:524, 530, 533
for oxygen in water 19:372
for ozone 16:546, 548
for PDTA 12:28
for penicillin O 16:571
for penicillins 16:561
for penicilloic acid 16:561, 566
for 1,2-pentanediol 13:244
for peroxide 11:522; 13:232; 18:307
for peroxide in corn oil 11:56
for phenoxymethylpenicillanic acid 16:562
for potassium benzylpenicillin 16:562
for potassium dichromate 9:684
for potassium monopersulfate 7:230
for potassium phenethicillin 16:562
for potassium phenoxymethylpenicillin 16:562
for procaine benzylpenicillin 16:562

for red lead 17:196
for reducing sugars 9:560; 10:428
for rubidium 17:555
for selenium 17:598, 600
for sodium 18:229
for sodium ampicillin 16:562
for sodium benzylpenicillin 16:562
for sodium borohydride 14:253
for sodium cephalothin 16:562
for sodium chlorite 9:356
for sodium chromate 9:684
for sodium cloxacillin 16:562
for sodium dichromate 9:684
for sodium dicloxacillin 16:562
for sodium dioxide bleaches 9:356
for sodium formate 13:135
for sodium hypochlorite 9:356
for sodium metabisulfite 18:395
for sodium metaborate 14:255
for sodium methicillin 16:562
for sodium nafcillin 16:562
for sodium oxacillin 16:562
for sodium penicillin O 16:562
for sodium sulfide 18:395
for sodium sulfite 18:395
for sodium thiosulfate 18:432
for sugar 18:353, 356
for sulfide 9:119; 15:480; 18:390, 395; 19:365
for sulfite 9:350; 13:33, 34; 14:371; 18:403,
    404; 19:366
for sulfur 13:437, 438; 15:480; 18:387, 395
for sulfur dioxide 10:303; 14:136; 18:387, 395,
    408, 418; 19:479, 481
for sulfurous acid 18:395, 407; 19:415
for thallium 18:529, 532
for thiocyanate 18:436
for thiosulfate 18:432
for tin 10:646; 14:136, 553; 19:38, 42
for trigonelline 10:413
for tryptophan 8:506
for TTHA 12:28
for tyrosine 8:506
for vanillin 19:351
for vinyl ethers 12:311
for water in ethylene oxide 12:221
Iodophenols. (See also Iodine compounds, organic.)
  properties 14:612
Iodophors. (See also Halogen donors.)
  in disinfectants and antiseptics 11:553
Iodoplatinate
  colorimetric method for acetylcholine chloride
    9:617
  colorimetric method for bethanechol chloride
    9:617
  colorimetric method for carbamylcholine chlo-
    ride 9:617
  colorimetric method for choline chloride 9:617

colorimetric method for methacholine chloride
  9:617
colorimetric method for β-methylcholine chloride
  9:617
colorimetric method for succinyldicholine chloride 9:617
colorimetric method for succinylmonocholine
  iodide 9:617
3-Iodo-1,2-propanediol. (See also Glycerol halo-
    hydrins.)
  derivatives for identification 14:160
1-Iodo-2-propanol. (See also Halohydrins.)
  derivatives for identification 14:160
2-Iodo-1-propanol. (See also Halohydrins.)
  derivatives for identification 14:160
Iodopyridines. (See also Pyridine and pyridine
    derivatives.)
  ionization constants 17:416
Iodosilane
  properties 18:40
Iodyrite 14:585
Ionac resins. (See also Ion exchange resins.) 15:2
Ion exchange chromatography
  of acetaminophen 6:182
  of acetyl codeine 16:270
  of adenine 12:552
  of aldehydes and ketones 4:116
  of alkaline earths 18:268
  of alkalis 18:268
  of alkyd resins 5:48
  of alkyl sulfates 11:85
  of aluminum 5:208, 230; 13:254; 14:113, 140
  of amides 8:610
  of amines 19:292
  of amino acids 8:414; 12:552; 19:474
  of amino acids in cocoa beans 9:557
  of p-aminophenol 6:182
  of aminophylline 6:499
  of ammonia 19:473
  of ammonium chloride 5:336
  of ammonium nitrate 5:340
  of ammonium sulfate 5:352; 11:84
  of ammonium thiocyanate 5:353
  of antihistamines 6:34
  of antimony 13:254
  of antipyrine 6:218
  automated analyzer 8:415
  of barbiturates 6:499
  of barium 6:573; 18:268
  of beryllium 18:268
  of betaine 9:618
  of bismuth 13:255
  of boron 7:396; 15:84
  of brominated salicylanilides 11:552
  of butyrobetaine 9:618
  of cadmium 13:255; 17:364
  of caffeine 6:182; 18:489, 490

of calcium 8:101; 13:255; 18:268; 19:492
of carboxylic acids 8:382
of carnitine 9:618
of cerium 15:157
of cesium 9:300; 17:550
of chloramphenicol 5:496
of chloride 19:497
of chlorine 12:251
of chlorophenols 9:521
of choline 9:617, 618
of chromium 9:637; 13:255
of cinnamic acid 10:60
of cobalt 10:356; 13:255; 17:526
of cocaine hydrochloride 5:383
of codeine 16:270
of copper 13:254; 17:526
of coumarins 11:166
of cyclamates 11:234
of cystosine 12:552
of dicarboxylic acids 19:292
of dichloromethoxyphenols 9:523
of dichlorophenols 9:521, 523
of dye modifiers in polyamides 17:299
of ethanolamine 9:618
of ethylenediamine 6:499
of ferrimycin 5:610
of ferroniobium 16:390
of fluorides 13:15, 25
of fluosilicic acid 18:15
of food dyes 9:244
of gallium 13:254
of germanium 13:381
of glycine 8:482; 9:618
of glycols 19:292
of gold 12:570; 14:17, 43
of guanine 12:552
of hafnium 14:113; 16:390
of heroin 16:271
of hygromycin B 5:583
of indium 13:254
of iron 12:523; 13:254; 14:113, 140; 17:526
of kanamycins 5:583
of lactic acid 19:420, 422
of lanthanum 13:255
of lead 13:254
of leucomycin 5:523
of lithium 15:266; 17:364
of magnesium 13:255; 15:361; 17:364; 18:268
of malic acid 19:421
of manganese 13:255; 17:526
of mannitol 19:450
of mepivacaine hydrochloride 5:398
of mercury 10:495; 16:17, 36
of *N*-methylnicotinamide 9:618
of molybdenum 14:113; 16:153, 162, 171, 185,
    390; 17:526
of monochlorophenols 9:521

of morphine 16:267
of neodymium 15:157
of neomycin 5:596
of niacin 9:618
of nickel 13:255; 16:298; 17:526
of nicotinamide 9:618
of niobium 10:318; 14:113; 16:153, 333, 364,
    377, 385, 390; 17:527
of niobium alloys 16:390
of nitrite 12:551
of nonionic detergents 11:348
of opium alkaloids 4:598
of orthophosphates 17:131
of pentachlorophenol 9:521
of phenobarbital 6:499
of phenol 9:521
of phosphates 11:372; 17:131
of phosphorus 16:185
of platinum group metals 17:252
of polyamides 17:292
of polymyxin 5:548
of potassium 17:364, 407, 550
of praseodymium 15:157
of procaine hydrochloride 5:409
of proline 9:618
of protocatechuic acid 8:401
of pyrophosphates 17:131
of radium 17:455, 467, 468; 18:269
of rare earths 17:472, 564; 18:268
of rhenium 17:526, 533, 537
of rubidium 17:550
of salicylamide 6:182
of salicylanilides 11:552
of salicylic acid 6:144
of scandium 17:564, 575
of selenium 14:17
of shampoos 11:85, 86
of silica 18:15
of silicates 17:364
of silicic acid 18:15
of silicon 13:281; 18:15
of silver 18:163
of sodium 16:188; 17:364, 550; 18:217
of sodium heparin 10:140
of sorbitol 19:450
of stachydrine 9:618
of streptomycin 5:588
of strontium 13:255; 18:268
of strychnine and brucine 4:603
of succinic acid 19:423
of sugar 19:409
of sulfate 18:431
of sulfide 18:431
of sulfite 18:431
of surfactants 6:365; 11:80, 81; 12:574
of tartaric acid 19:420
of tantalum 10:318; 14:113; 16:153,

333, 364, 377, 385, 390; 17:526; 18:445
of technetium 17:526
of tellurium 14:17
of tetrachlorophenols 9:521
of tetrametaphosphates 17:131
of tetrapolyphosphates 17:131
of thallium 13:254, 255
of thiosulfate 18:431
of thorium 17:495, 575; 19:10
of tin 19:37
of titanium 13:255; 14:113, 140; 16:153, 385,
    390; 17:526; 19:97
of 2,4,5-trichloroanisole 9:523
of trichlorophenols 9:521, 523, 526
of trigonelline 9:618
of trimetaphosphates 17:131
of tripolyphosphates 17:131
of tungsten 14:113; 16:153, 162, 390; 17:526
of uranium 13:255; 17:575; 19:231, 236, 237,
    238, 250
of vanadium 14:113; 16:185, 390; 19:331
of ytterbium 13:255
of zinc 12:532; 13:254; 19:555, 564
of zirconium 13:255; 14:113; 16:153, 390; 17:
    526
Ion exchange materials 4:434
Ion exchange resins 15:1–20
  acid–base titration 15:14
  analysis 15:1
  anion exchange capacity 15:19
  cation exchange capacity 15:16
  classification 15:1
  column density 15:9
  commercial products 15:2
  definition 15:1
  dissociation constants 15:14
  dry density 15:11
  functional groups, determination of 15:13
  graduate density 15:11
  particle size distribution 15:7
  properties, determination of 15:7, 12
  sample preparation 15:4, 12
  solids content 15:12
  trademarks 15:4
  types 15:2
  volumetric determination 15:13
  wet density 15:10
Ionization constants
  of nitroethane 16:415
  of nitromethane 16:415
  of nitropropanes 16:415
  of sulfuric acid 18:412
Ionol. (See also 2,6-Di-*tert*-butyl-4-methylphenol.)
    6:74
  colorimetric method using diazotized *p*-nitro-
    aniline 6:106
α-Ionone

in black tea 18:508, 511
  gas chromatography 18:511
β-Ionone
  in black tea 18:508, 511
  gas chromatography 18:511
Ionox 330 6:74
  gas chromatography 6:84
Ion selective electrodes
  for calcium 8:106
  for sodium 18:222
IPDA. See Isophorone diamine.
Ipecac alkaloids 4:592
  analysis 4:614
IPPD. See *N*-Isopropyl-*N'*-phenyl-*p*-phenylene-
    diamine.
IPS. See Impact polystyrene.
Iridium. (See also Iridium compounds; Iridium
    metal; and *Platinum group metals*.)
  bead test 6:588
  colorimetric determination 17:248, 261, 266
  copper reduction 17:253
  detection 17:248
  emission spectroscopy 14:48; 17:246
  flame tests 6:589
  in gold 14:45
  gravimetric determination 17:237, 260
  history 17:215
  hydrogen sulfide gravimetric method 17:251,
    260
  identification 17:248
  iridium chloride gravimetric method 17:260
  isotopes 17:215
  neutron activation analysis 13:265
  nitric acid colorimetric method 17:248
  p-nitrosodimethylaniline colorimetric method
    17:249
  occurrence 17:214
  oxidation numbers 17:220
  oxidation potentials 17:221
  separation 17:253, 260, 266
  sodium chloride colorimetric method 17:261
  tin chloride colorimetric method 17:266
  uses 17:217
Iridium alloys. See Iridium metal.
Iridium chloride
  gravimetric method for iridium 17:260
  manufacture 17:226
  properties 17:226
Iridium compounds. (See also individual com-
    pounds.)
  manufacture 17:226
Iridium hydroxide
  manufacture 17:226
  properties 17:226
Iridium metal
  carrier gas fusion methods 8:682
  hydrogen in 8:680

oxygen in 8:680, 682
properties 17:219, 220, 226
solubility in gold 14:9
uses 17:220
vacuum fusion methods 8:680
Iridium oxide
manufacture 17:226
properties 17:226
Iridium sulfate
manufacture 17:227
Iridium sulfide
manufacture 17:227
properties 17:227
Iridosmine
composition 17:215, 216
sample decomposition 17:231
Irish whisky 4:465
analysis 4:480
Iron. (See also Iron alloys; Iron metal.) 15:21–93
in acetic acid 4:100
in acetone 4:114
in acid anhydrides 8:568
in acid halides 8:575
in acid salts 8:595
in acrylamide 4:166
in acrylonitrile 4:376
in adipic acid 4:418
in alcoholic beverages 4:481, 489
in alumina 5:116
in aluminates 5:160
in aluminum and aluminum alloys 5:131
in aluminum chloride 5:176
in aluminum oxide 5:186
in aluminum oxide abrasives 4:39, 49
in aluminum sulfate 5:194
in alums 5:194
in amino acids 8:440
in ammonium acetate 5:326
in ammonium bicarbonate 5:330
in ammonium carbonate 5:333
in ammonium chloride 5:337, 338
ammonium chloride–hydrogen peroxide gravimetric method 16:238
in ammonium chromate 9:699
in ammonium dichromate 9:699
in ammonium hydroxide 5:320
ammonium hydroxide gravimetric method 9:
670; 10:313, 638; 12:151; 16:164, 180
in ammonium nitrate 5:343
in ammonium oxalate 5:345
in ammonium persulfate 5:347
ammonium persulfate–ammonium thiocyanate
colorimetric method 19:19, 74
ammonium persulfate–potassium thiocyanate
colorimetric method 9:349, 362, 629; 10:95,
330; 13:34; 14:373, 438; 15:399, 402, 405,
408, 410, 491; 16:334, 537; 17:403; 18:176,

177, 234, 283, 284
in ammonium phosphate, dibasic 5:350
in ammonium sulfate 5:352
ammonium sulfide–tartrate gravimetric method
15:33
in ammonium thiocyanate 5:354
ammonium thiocyanate colorimetric method
4:114; 7:196, 200, 205, 382, 383; 8:440;
9:533; 10:655; 13:130
ammonium thioglycolate colorimetric reagent
15:51
in amosite 6:267
in aniline 5:446
in antimony metal 6:43
in antimony pentachloride 6:51
in arsenic(III) chloride 6:234
in arsenic(III) oxide 6:231
in arsenic(V) oxide 6:233
in asbestos 6:262, 267
in asbestos–cement products 6:270, 272
atomic absorption spectroscopy 7:200; 9:346;
10:131, 655, 661; 11:157; 14:49, 125; 15:56,
195, 196; 16:117
in bakery products 6:455, 458
in barite 6:557, 559
in barium acetate 6:537
in barium carbonate 6:542, 544
in barium chloride 6:546
in barium hydroxide 6:550
in barium nitrate 6:553
in barium peroxide 6:555
in barium sulfate 6:567
bathophenanthroline colorimetric method 13:
267; 15:54
in bauxite 5:116
bead tests 12:14
in beer 7:650, 660
in beryllium metal 7:114
in biological materials 15:53
2,2′-bipyridine colorimetric method 4:376
bismuth in 7:215
in bismuth chloride 7:200
in bismuth metal 7:194
in bismuth nitrate 7:203
in bismuth oxide 7:205
in bismuth subcarbonate 7:196
in bismuth subnitrate 7:203
in borax 7:383
borax bead test 6:584
in boric acid 7:382
in bread 15:32
butyl acetate extraction 9:645, 650
in cadmium bromide 8:38
in cadmium chloride 8:42
in cadmium nitrate 8:45
cadmium sulfide gravimetric method 15:48
in calcium acetate 8:78

in calcium carbonate 8:83
in calcium chloride 8:88
in calcium fluoride 8:90
in calcium metal 8:76
in calcium oxide 8:94
in caprolactam 8:124
in carbon 14:65, 74
in carbonated beverages 8:174
in carbonyl compounds 8:362
in cellulose 9:50
in cellulose acetate polymers 4:89
in cement 15:56
in cereals 9:233, 242
in ceric ammonium nitrate 9:252
ceric sulfate titration 12:523, 526; 15:38
in cerous chloride 9:260
in chlorinated paraffins 9:499
in chlorocarbons 9:409
in chlorosilanes 18:53
in chlorosulfonic acid 9:533
in chocolate and cocoa products 9:538, 595,
    598
in choline and its derivatives 9:629
in chrome ores 9:701, 703
in chrome steels 9:662
in chrome tanning compounds 9:691
in chromic oxide 9:692
in chromium 9:635, 684, 693
in chromium alloys 9:670
in chromium compounds 9:699
chromium content 9:645, 647, 650
in chrysotile 6:262, 267
in citric acid 10:95
in clays and clay minerals 10:117, 130, 136
in cobalt(II) acetate 10:330
in cobalt alloys 10:313, 355
in cobalt carbonate 10:333
in cobalt chloride 10:337
in cobalt compounds 10:328
in cobalt metal 10:293, 298
in cobalt nitrate 10:340
in cobalt ores 10:355
in cobalt(II) orthophosphate 10:343
in cobalt oxides 10:342
in cobalt(II) sulfate 10:343, 344
colorimetric determination 15:33, 45; 19:494
colorimetric method using bathophenanthroline
    8:174
of constant modulus alloys 10:290
in copper(II) acetate 10:655
in copper alloys 10:638
in copper(II) bromide 10:657
in copper(II) carbonate, basic 10:659
in copper(I) chloride 10:661
in copper(II) chloride 10:663
in copper(I) cyanide 10:664
in copper(II) hydroxide 10:666

in copper metal 10:622
in copper(II) nitrate 10:668
in cotton 11:140, 153, 154, 156
in crocidolite 6:267
cupferron–chloroform extraction 9:645
cupferron gravimetric method 15:46
in cyanoacetic acid 15:439
in dental alloys 10:290
in dental restorative materials, composite 11:316
in detergents 11:367
in diamines and higher amines, aromatic 11:441
in diatomaceous earth 11:495, 496, 497, 498
dipyridyl colorimetric method 6:262, 459; 7:
    660; 9:242; 11:140, 498
dithizone colorimetric method 19:222
EDTA titration 9:639; 12:154; 13:445; 15:40,
    45
in eggs and egg products 12:66
in elastomers, synthetic 12:151
in electrical resistance alloys 10:290
emission spectroscopy 4:3, 56, 345; 7:650;
    8:76; 9:346; 11:153, 493; 14:48, 74, 122, 540;
    15:33, 144, 155, 197, 375; 16:160, 387; 19:
    570
in enamels 9:195
ether extraction 10:355
in ethyleneamines 11:429
in ferrites 12:522
in ferroniobium 16:387
ferrous ion content 15:43
ferrous iron determination 6:263
in fertilizers 15:56
flame tests 6:589
in flour 15:32
in fluorosilicic acid 13:50
in food products 15:48
in formaldehyde 13:114
in formic acid 13:130
gallium content 13:261
in gallium metal 13:266, 267
in gallium phosphide 13:282
in gelatin 12:554; 13:371
in glass and glass-ceramics 13:445
in gold 14:45, 49
in graphite 14:65, 74
gravimetric determination 15:32, 34
in hafnium 14:122, 124, 131
in hard facing alloys 10:289
in hexachloroethane 9:452
in high-temperature alloys 10:286
history 15:21
in hydrochloric acid 9:340, 351; 14:372, 375
hydrochloric acid colorimetric method 18:53,
    79
in hydrofluoric acid 13:34
in hydrogen peroxide 14:438
hydrogen peroxide colorimetric method 19:98

hydrogen sulfide–potassium permanganate method 5:132
hydrogen sulfide reduction method 15:41
hydroxylamine hydrochloride–bathophenanthroline colorimetric method 9:366
hydroxylamine hydrochloride-α,α'-dipyridyl colorimetric method 10:150
in hydroxylammonium acid sulfate 14:455
in hydroxylammonium sulfate 14:455
in hypochlorite bleaches 9:355
identification 15:33
impurities, determination of 17:78
in indium 14:540
in iodides 14:600
ion exchange chromatography 12:523; 13:254; 14:113, 140; 17:526
in iron ores 15:40
in iron oxide blacks 17:203
isotopes 15:22
in lactic acid and its derivatives 15:143, 144
in lanthanum oxide 15:153, 155
in lead 15:195, 196
in lead alloys 15:195
in lithium 15:274
in lithium carbonate 15:280
in lithium hydroxide 15:286
in low expansion alloys 10:290
in lubricating oils 15:57
in magnesium 15:375, 384
in magnesium acetate 15:399
in magnesium alloys 15:375, 384
in magnesium carbonate 15:402
in magnesium chloride 15:405
in magnesium nitrate 15:408
in magnesium oxide 15:410
in magnesium perchlorate 9:364, 366
in magnesium sulfate 15:412
in manganese 15:478
in manganese sulfate 15:491
in mercuric chloride 16:60
in mercuric oxide 16:62
in metal complex dyes 12:14
in methanol 16:117
in methyl cellulose 9:80
microcosmic salt bead test 6:587
in minerals 15:31
in molybdenum 16:161, 164
in molybdenum disulfide 16:180
in molybdenum trioxide 16:161
neutron activation analysis 13:265
in nickel 16:297, 302, 309, 329
in nickel alloys 16:329
in nickel–copper alloys 15:49
in nickel sulfate 16:344
in niobium 16:387, 389
in nitric acid 16:498
occurrence 15:21

in organic materials 15:32
in organosilane esters 18:79
oxidation potentials 15:24
in perchloric acid 9:364, 366
permanganate titration method 6:262
in petroleum 15:57
phenanthroline colorimetric method 4:39, 89, 166, 348, 418; 5:132, 331, 350; 6:43; 7:114, 660; 9:50, 242, 252, 344, 499; 10:298, 622; 11:46, 367; 12:157, 554; 13:114; 14:65, 131, 141, 372, 455; 15:52, 143, 154, 274, 384, 478; 16:329; 18:248, 418; 19:370
in plants 15:57; 18:221
polarographic method 12:554
in polymers 4:345
in potassium chlorate 9:359, 362
in potassium chloride 9:339
potassium dichromate–ceric sulfate volumetric method 15:44
potassium dichromate–ferrous sulfate titration 10:117
potassium dichromate titration 9:670, 703; 10:313, 639; 15:38, 41; 16:164, 180
potassium ferrocyanide colorimetric method 10:544; 11:140
in potassium hydrogen phthalate 17:403
potassium permanganate titration 14:600; 15:37
potassium thiocyanate colorimetric method 7:194; 9:361, 452; 11:140, 316; 15:45, 47; 19:494
in rare earth chlorides 17:506
in rare earth fluorides 17:508
in reducing water 4:480
rhenium content 17:523
in R salt 16:238
in sand 15:53
separation 15:32, 46
in serum 15:32, 54
in silica ablators 4:3
in silicate minerals 15:56
in silicate rocks 9:644
in silicates 19:113
in silicon carbide abrasives 4:56
in silver 18:173
in silver nitrate 18:176
silver reduction methods 15:36
in silver sulfate 18:177
in soaps 18:202
in sodium arsenate 6:235
in sodium arsenite 6:235
in sodium carbonate 18:248
in sodium chlorate 9:360, 361
in sodium chloride 9:337, 338, 344, 346, 349
in sodium dichromate 9:684, 699
in sodium fluoride 13:34
in sodium heparin 10:150

in sodium metal 18:234
in sodium peroxide 16:537
sodium thiocyanate colorimetric method 15:49
in sodium tungstate 19:222
in soft magnetic alloys 10:288
solvent extraction 12:554; 15:46; 16:297, 302,
    309; 19:222, 494
in spring steels 10:290
in stannous chloride 19:74
stannous chloride–potassium dichromate
    volumetric method 16:330
stannous chloride reduction method 15:39, 42
in stannous fluoride 19:75
in starch hydrolysates 11:46
in strontium chloride 18:283
in strontium nitrate 18:284
in styrene 18:309
sulfosalicylic acid colorimetric method 15:439
in sulfuric acid 18:418
in surgical alloys 10:290
in talc 10:544
by thiocyanate colorimetric method 5:321, 326,
    330, 333, 345, 347, 352, 354; 6:231, 233, 234,
    235, 542, 543
thioglycolic acid–ammonium hydroxide colori-
    metric method 11:429
in thorium nitrate 19:19
in tin metal 19:66, 67
tiron colorimetric method 19:113
in titanium metal 19:121
in titanium–zirconium alloys 15:57
in tobacco 19:152
TPTZ colorimetric method 11:497
in urine 15:32, 54, 57
uses 15:21
vanadium content 9:645
volumetric determination 15:33, 34
in water 15:54; 19:370
in wine 15:32, 54; 19:491, 494
x-ray diffraction 15:34
by x-ray emission spectroscopy 4:49
x-ray fluorescence spectroscopy 10:136; 11:156
x-ray spectrography 9:409; 15:33
in yeast 19:531
in zinc metal 19:570
in zinc powder 17:204
zinc reduction methods 15:35
in zircaloy 14:131
in zircon 14:139
in zirconium 14:122, 124, 131
Iron alloys. (See also Ferro alloys; Ferrochrome;
    Ferrous materials; Iron metal; and individual
    compounds.) 15:72–84
aluminum in 5:228
analysis 15:73
carrier gas fusion methods 8:624
carrier gas hot extraction methods 8:638

commercial grades and specifications 15:72
hydrogen in 8:638
oxygen in 8:624
Iron–ammonium thiocyanate
    colorimetric method for EDTA 12:40
Iron–base alloys. (See also Cobalt alloys.)
    composition 10:289
Iron blue. (See also Iron compounds.)
    analysis 17:201
    assay 17:201
    in chrome greens 17:201
    commercial specifications 12:173; 17:201
    composition 17:201
    in enamels 12:173
    impurities, determination of 17:201
    uses 15:85
Iron, cast
    carrier gas fusion methods 8:624
    hydrogen in 8:616, 638
    nitrogen in 8:616
    oxygen in 8:616, 624
    vacuum fusion methods 8:616
    vacuum hot extraction methods 8:638
Iron chelates. (See also Iron compounds.)
    colorimetric determination 12:52
    DTPA content 12:52
    DTPA, free content 12:53
    EDHPA content 12:52
    EDTA content 12:51
    iron, free content 12:53
    NTA content 12:53
    uses 15:86
    visible spectroscopy 12:52
Iron–chromium–nickel stainless steel. See Chrome
    steels.
Iron–chromium stainless steel. See Chrome steels.
Iron citrate complex
    stability 10:89
Iron–cobalt alloys. (See also Cobalt alloys.)
    composition 10:288
Iron–cobalt–molybdenum alloys. (See also Cobalt
    alloys.) 10:285
    composition 10:285
Iron–cobalt powders. (See also Cobalt alloys.)
    10:288
    composition 10:288
Iron–cobalt–vanadium alloys. (See also Cobalt
    alloys.) 10:285
    composition 10:285
Iron compounds. (See also individual compounds.)
    15:84–91
    analysis 15:89
    assay 15:89
    impurities, determination of 15:89
    nomenclature 15:26
    properties 15:25
    solubility 15:25

Iron dextran injection. (See also Iron compounds.)
  uses 15:86
Iron dibutyldithiocarbamate. (See also Dithio-
    carbamic acid salts.)
  properties 11:575
Iron diethyldithiocarbamate. (See also Dithio-
    carbamic acid salts.)
  properties 11:575
Iron dimethyldithiocarbamate. (See also Dithio-
    carbamic acid salts.)
  properties 11:575
  uses 11:576
  xanthate method 11:580
Iron–dimethylglyoxime
  colorimetric method for vanadium 19:331, 332
Iron ethylenebisdithiocarbamate. (See also Dithio-
    carbamic acid salts.)
  uses 11:577
Iron ferrite
  Curie temperature 9:212
  magnetic constants 9:212
Iron filings
  cooling system sealer 6:353
Iron fluoride
  properties 13:11
Iron, free
  in iron chelates 12:53
Iron glycerophosphates. (See also Iron compounds.)
  uses 15:86
Iron hydroxide
  gravimetric method for germanium 13:381
Iron lactate. (See also Lactic acid salts.)
  uses 15:130
Iron magnesium silicate. See Amosite.
Iron–manganese alloys 5:61
  composition 5:64
Iron–manganese–zinc ferrite. (See also *Ferrites.*)
  composition 12:520
Iron metal. (See also Iron alloys.)
  analysis 15:30
  assay 15:44
  carrier gas fusion methods 8:626
  commercial grades and specifications 15:28
  dissolution 19:326
  hydrogen in 8:619
  impurities, determination of 15:58
  manufacture 15:26
  nitrogen in 8:619
  oxygen in 4:619, 626
  properties 15:22
  reaction medium for fusion methods 8:629, 631,
    632
  reducing power 15:44
  sample decomposition 18:26, 27
  sample preparation 15:31
  sampling 15:30
  silicon content 18:26, 27, 38

  solubility in gold 14:9
  sulfur content 18:382, 385
  titanium content 19:111, 120
  tungsten content 19:198, 206
  uses 15:27
  vacuum fusion methods 8:619
  zinc content 19:559
Iron-nail fire assay
  of gold 14:14
Iron naphthenate. (See also Iron compounds.)
  in enamels 12:176
  uses 15:87
Iron nitrides
  carrier gas fusion methods 8:625
  oxygen in 8:625
Iron ores
  aluminum in 5:228
  chromium content 9:644
  impurities, determination of 17:73
  iron content 15:40
  manganese content 15:465
  isobutyl acetate extraction 9:650
  sample preparation 15:31
  sodium peroxide fusion 9:644
Iron oxide. (See also Iron compounds.)
  acid dissolution 9:215
  ammonium hydroxide gravimetric method 14:94
  beta-ratio method 13:464
  in calcium oxide 8:94
  carrier gas fusion methods 8:626
  in clays and clay minerals 10:116
  commercial grades and specifications 10:531,
    534; 12:519; 17:196
  composition 10:531; 12:519
  in enamels 12:173, 186
  in glass and glass-ceramics 13:413, 415, 439,
    464
  in gypsum and gypsum products 14:94, 96
  hydrogen in 8:616
  infrared spectroscopy 13:464
  in iron oxide pigments 17:194
  nitrogen in 8:616
  oxidation state 13:464
  oxygen in 8:616, 626
  phenanthroline colorimetric method 10:116;
    13:439
  potassium dichromate titration 10:117
  in styrene polymers 18:320
  ultraviolet spectroscopy 13:464
  uses 15:85, 88
  vacuum fusion methods 8:616
  Zimmerman-Reinhardt volumetric method 17:
    194
Iron oxide blacks
  analysis 17:203
  commercial specifications 17:203
  iron content 17:203

uses 17:203
Iron oxide pigments
  analysis 17:194
  commercial grades and specifications 17:193,
    194
  impurities, determination of 17:194
  iron oxide content 17:194
  types 17:192
Iron pentacarbonyl. (See also Iron compounds.)
  uses 15:85
Iron-1,10-phenanthroline
  colorimetric method for vanadium 19:331, 332
Iron phosphate-potassium permanganate
  volumetric method for molybdenum 16:157
Iron, pig
  hydrogen in 8:616
  nitrogen in 8:616
  oxygen in 8:616
  vacuum fusion methods 8:616
Iron pyrites
  in concrete 9:142
Iron resinate. (See also Iron compounds.)
  uses 15:87
Iron-salicylic acid
  colorimetric method for EDTA 12:40
Iron silicate. See Amosite; Crocidolite.
Iron sorbitex injection. (See also Iron compounds.)
  uses 15:87
Iron sulfate-EDTA
  gravimetric method for silver 18:164
Iron sulfide
  in glass 9:200
  in match strikers 15:522, 536
  particle size 15:536
Iron tallate. (See also Iron compounds.)
  uses 15:87
Iron-TPTZ complex
  colorimetric method for EDTA 12:39
Iron trichloride
  properties 19:96
Isakemithal
  extraction 6:491
Isatin
  colorimetric method for proline and hydroxy-
    proline 8:518
  colorimetric method for thiophene 7:48; 8:276;
    18:537, 540
Isatinsulfonic acid
  in F D & C Blue No. 2 10:512
  liquid chromatography 10:512
  ultraviolet spectroscopy 10:512
Isoamyl alcohol
  in black tea 18:510
  in brandies 19:440
  gas chromatography 16:107; 18:510; 19:442
  in methanol 16:107
  in wines 19:425, 439, 440, 442

Isoamyl benzoate. (See also Benzoic acid esters.)
  commercial specifications 7:79
  3,5-dinitrobenzoate derivative 7:83
  properties 7:77
  saponification method 7:84
Isoamyl ether. (See also Ethers.)
  properties 12:297
Isoamyl formate
  solubility 13:540
Isoamyl mercaptan. (See also Mercaptans.)
  derivatives for identification 15:559
Isoamyl thiocyanate. (See also Isocyanic acid
    esters.)
  gas chromatography 15:110
Isobergapten. (See also Coumarins.) 11:160
  gas chromatography 11:168, 170
Isobucaine hydrochloride 5:375
  analysis 5:394
  properties 5:394
Isobutanal
  in black tea 18:507
Isobutane. (See also Butanes.)
  in aerosols 11:95
  in butane 15:240
  calorific value 13:331
  commercial specifications 14:301
  gas chromatography 11:98; 13:317, 318, 350;
    14:329; 15:240; 18:52
  in gasoline 13:350, 353
  in natural gas 13:320
  properties 11:95; 13:350; 14:301
  specific gravity 15:245
  in trichlorosilane 18:52
  vapor pressure 15:249
Isobutanol. See Isobutyl alcohol.
Isobutene
  in butane 15:240
  gas chromatography 15:240
Isobutene polymers. See Butyl rubber; Polyiso-
    butenes.
Isobutyl acetate
  commercial specifications 8:578
  extraction of iron 9:650
  gas chromatography 16:108
  in methanol 16:108
  properties 8:577
Isobutyl alcohol
  in brandies 19:440
  gas chromatography 9:475; 12:204; 16:107;
    19:442
  in methanol 16:107, 114
  in wines 19:425, 440, 442
Isobutyl amine
  manufacture, methods of 5:262
  in methanol 16:114
  properties 5:260
  uses 5:263

Isobutylbenzene. (See also Butylbenzenes.)
commercial grades and specifications 14:322
gas chromatography 13:352
in gasoline 13:352, 354
properties 13:352
Isobutyl benzoate. (See also Benzoic acid esters.)
commercial specifications 7:79
3,5-dinitrobenzoate derivative 7:83
properties 7:77
saponification method 7:84
Isobutyl chloride. (See also C_4 Chlorinated
hydrocarbons.)
gas chromatography 9:475
properties 9:473
Isobutyl cinnamate. (See also *Cinnamic acid and
its derivatives.*)
commercial specifications 10:66
properties 10:65
saponification methods 10:67
Isobutylene. (See also *Hydrocarbons.*)
in 1,3-butadiene 11:504, 506
commercial grades and specifications 14:313
gas chromatography 11:506; 12:105; 13:350;
14:329
in gasoline 13:350, 353
in manufacture of methacrylic acid and methacrylates 4:186
properties 13:350; 14:312
Isobutylene oxide. (See also *Epoxy compounds.*)
hydrochlorination methods 12:208
infrared spectroscopy 12:200
Isobutyl ether. (See also *Ethers.*)
properties 12:297
Isobutyl heptyl ketone
acidity 8:358
alcohol in 8:360
azeotropes 8:336
commercial specifications 8:341
Karl Fischer titration 8:360
specific gravity 8:358
water in 8:360
Isobutyl iodide
gas chromatography 16:92
Isobutyl isocyanate. (See also *Isocyanic acid esters.*)
infrared spectroscopy 15:106
Isobutyl lactate. (See also Lactic acid esters.)
properties 15:131
Isobutyl lactate acetate. (See also Lactic acid
esters.)
properties 15:131
Isobutyl mercaptan. (See also *Mercaptans.*)
in *n*-butyl mercaptan 15:565
derivatives for identification 15:559
gas chromatography 15:558, 565
properties 15:552
Isobutyl methacrylate
nuclear magnetic resonance spectroscopy 16:86

Isobutyl salicylate
in sunscreens 11:92
ultraviolet spectroscopy 11:92
Isobutyraldehyde
absorption spectroscopy 9:564
in cocoa vapor 9:570
commercial specifications 8:340
derivatives for identification 8:343
gas chromatography 12:204; 16:109
hydroxylamine hydrochloride method 8:357
mass spectroscopy 9:570
in methanol 16:109
properties 8:334
Isobutyramide
properties 8:597
Isobutyric acid
in black tea 18:511
derivatives for identification 8:376
gas chromatography 18:511
properties 8:396
thin-layer chromatography 8:381
Isobutyric aldehyde
infrared spectroscopy 6:616
Isobutyronitrile. (See also *Nitriles.*)
properties 16:396
Isocapric acid
properties 8:396
thin-layer chromatography 8:381
Isocaproic acid
in black tea 18:511
gas chromatography 18:511
Isocinchomeronic acid. (See also *Pyridine and
pyridine derivatives.*)
uses 17:413
Isocitric acid
thin-layer chromatography 8:381
Isoclasite 17:121
Isocothan. See Dinocap.
Isocrotonitrile. (See also *Nitriles.*)
gas chromatography 16:404
Isocrotylbenzene
gas chromatography 7:8
Isocyanate requirement
of polyols 19:278
Isocyanates. (See also *Isocyanic acid esters.*)
19:268
amine equivalent 19:269, 271, 272
assay 19:269
composition, determination of 19:273
dibutylamine volumetric method 19:272
p-dimethylaminobenzaldehyde colorimetric
method 19:282
gas chromatography 19:272, 276, 313
impurities, determination of 19:272, 276
isocyanate content 19:269
in isocyanic acid esters 15:111, 112
modifier for alkyd resins 5:24

in polyurethanes 19:312
properties 19:269
in tolylene diisocyanates 15:112
trade names 19:270
2-Isocyanatobicyclohexane. (See also *Isocyanic acid esters.*)
near-infrared spectroscopy 15:105
2-Isocyanatochryson. (See also *Isocyanic acid esters.*)
infrared spectroscopy 15:106
1-Isocyanato-2-ethoxy-5,6,7,8-tetrahydronaph-thalin. (See also *Isocyanic acid esters.*)
infrared spectroscopy 15:106
3-Isocyanato-*n*-pentane. (See also *Isocyanic acid esters.*)
infrared spectroscopy 15:106
3-Isocyanatotetramethylensulfone. (See also *Isocyanic acid esters.*)
infrared spectroscopy 15:106
Isocyanic acid esters. (See also individual compounds.) 15:94–124
amine equivalent 15:112
analysis 15:101
aniline gravimetric method 15:111
assay 15:111
commercial grades and specifications 15:101
dibutylamine volumetric method 15:112
2-*N*-diisobutylamine volumetric method 15:111
gas chromatography 15:110
identification 15:104
impurities, determination of 15:118
infrared spectroscopy 15:107
isocyanate content 15:111
isocyanate number 15:112
manufacture 15:98
near-infrared spectroscopy 15:104
Parr oxygen bomb method 15:120
piperidine volumetric method 15:111
properties 15:94
separation 15:104
ultraviolet spectroscopy 15:104
volumetric determination 15:111
Isocyanurates
infrared spectroscopy 15:97
manufacture 15:97
Isocyanuric acids, chlorinated
available chlorine 7:228
bleaching agents 7:222
water in 7:232
Isodecanoic acid 5:22
Isodecanol (mixed isomers). (See also Alcohols, higher.)
properties 4:552
specifications 4:557
uses 4:556
Isodibenzanthrone violet. (See also *Pigments.*)
structure 17:186

Isoelectric point
of glues of animal origin 13:492
Isoeugenol
gas chromatography 11:167
Isoheptane. (See also Heptanes.)
commercial specifications 14:305
Isoheptyl alcohol
gas chromatography 7:96
Isohexanonitrile. (See also *Nitriles.*)
derivatives for identification 16:401, 402
properties 16:396
Isohumulones
in beer 7:603
ultraviolet spectroscopy 7:604
Isoleucine. (See also *Carboxylic acids, amino.*) 413, 488
analysis 8:494
in black tea 18:507
in cocoa beans 9:556
commercial specifications 8:494
composition 8:488
crystalline form 8:491
gas chromatography 8:429
infrared spectra 8:492
ion exchange chromatography 8:415, 421
manufacture 8:493
paper chromatography 8:422
properties 8:489
in protein 8:414
solubility 8:490
structure 8:489
toxicity 8:493
in yeast 19:529
Isolysergic acid
structure 15:290
Isomenthone
gas chromatography 8:353
Isomeric colors
in certified colors 10:511
Isomethadone. (See also *Narcotics.*)
ammonium vanadate colorimetric method 16:246
color tests 16:246
microchemical tests 16:250
picric acid microchemical method 16:250
platinic chloride microchemical method 16:250
ultraviolet spectroscopy 16:252
Isoniazid
spectroscopic method for dimethisterone 10:589
Isonicotinamide. (See also *Pyridine and pyridine derivatives.*)
ionization constants 17:417
Isonicotine acid hydrazide. (See also *Pyridine and pyridine derivatives.*)
structure 17:413
uses 17:413
Isonicotinic acid. (See also *Pyridine*

and pyridine derivatives.) 8:407
structure 17:412
in tobacco 19:136
trisodium pentacyanoammino ferrate colorimetric
method 17:433
Isonicotinic acid amide. (See also *Pyridine and
pyridine derivatives.*)
trisodium pentacyanoammino ferrate colorimetric
method 17:433
Isonicotinic hydrazide. (See also *Pyridine and
pyridine derivatives.*)
colorimetric method for dimethisterone 14:207
colorimetric method for medroxyprogesterone
acetate 14:207
colorimetric method for norethindrone 14:207
colorimetric method for norethindrone acetate
14:207
colorimetric method for norethynodrel 14:207
trisodium pentacyanoammino ferrate colorimetric
method 17:433
Isonipecaine. See Meperidine.
Isonipecotic acid. (See also *Pyridine and pyridine
derivatives.*) 17:413
Isooctane. (See also Octanes.)
analysis 14:332
commercial grades and specifications 14:305,
306
gas chromatography 11:73; 14:330, 332
in *n*-heptane 14:332
in nail enamels 11:73
uses 14:303
Isooctanic acid 5:22
Isooctene. (See also *Hydrocarbons.*)
commercial grades and specifications 14:313
Isooctyl alcohol (mixed isomers). (See also
Alcohols, higher.)
phthalic anhydride method 4:559
properties 4:552
specifications 4:557
uses 4:556
Isopentane. (See also Pentanes.)
in butane 15:240
calorific value 13:331
commercial grades and specifications 14:302
gas chromatography 13:317, 318, 350; 14:329,
335; 15:240; 18:52
in gasoline 13:350, 353
in natural gas 13:320
properties 13:350; 14:302
specific gravity 13:329
structure 14:301
in trichlorosilane 18:52
uses 14:302
Isopentanonitrile. (See also *Nitriles.*)
derivatives for identification 16:402
properties 16:396
Isopentylbenzene

gas chromatography 13:352
in gasoline 13:352, 355
properties 13:352
Isophorone
acidity 8:358
azeotropes 8:336
commercial specifications 8:341
in enamels 12:175
Karl Fischer titration 8:360
properties 12:175
specific gravity 8:358
water in 8:360
Isophorone diamine
in polyurethanes 19:304
structure 19:304
Isophorone diisocyanate. (See also Isocyanates.)
structure 19:271
trade names 19:271
Isophthalic acid 5:22; 17:166–170
acid–base titration 17:158
acid number 17:168
in alkyd resins 5:37
analysis 17:158, 167
assay 17:167
color tests 17:168
commercial grades and specifications 17:167
derivatives for identification 17:167
gas chromatography 5:25; 17:158
identification 17:167
impurities, determination of 17:168
manufacture 17:166, 167
properties 17:166
solubility 17:166
thin-layer chromatography 19:293
uses 17:166
Isophthalic alkyd
infrared spectroscopy 5:41
Isophthalonitrile. (See also *Nitriles.*)
polarography 16:398
Isopimpinellin. (See also *Coumarins.*) 11:161
gas chromatography 11:168, 170
Isopolymolybdates. (See also Molybdates.)
manufacture 16:183
Isoprene 11:507–511
analysis 11:509
assay 11:509
azeotropes 11:508
in 1,3-butadiene 11:504
in cigarette smoke 10:31
commercial grades and specifications 11:509
freezing point 11:509
gas chromatography 10:31; 11:509; 12:104
impurities, determination of 11:510
properties 11:508
uses 11:507
Isoprene-acrylonitrile copolymer. (See also
Elastomers, synthetic.)

infrared spectroscopy 12:99
uses 12:85
Isoprene–isobutene copolymer. (See also
    Elastomers, synthetic.)
commercial elastomers 12:86
density 12:91
Isoprene polymers. See Polyisoprenes; Rubber,
    natural.
Isopropanol. See Isopropyl alcohol.
Isopropanolamine
    properties 5:2
    solubility 5:1
    specifications 5:4
Isopropenylactetylene. (See also 2-Methyl-1-
    butene-3-yne.)
    gas chromatography 11:510
    in isoprene 11:510
p-Isopropenyltoluene
    gas chromatography 18:299
    in styrene 18:299
4-Isopropoxydiphenylamine 6:72
p-Isopropoxydiphenylamine
    paper chromatography 6:85
Isopropyl acetate
    commercial specifications 8:578
    gas chromatography 16:108
    in methanol 16:108
    properties 8:577
    saponification method 8:586
Isopropyl alcohol. (See also Alcohols, lower;
    Propyl alcohols.)
    acetylation with acetic anhydride 4:530
    in acrolein 4:154
    in alcoholic beverages 4:489
    in allyl alcohol 5:79
    in black tea 18:510
    in brake fluids 7:424
    in detergents 11:359
    3,5-dinitrobenzoyl chloride method 4:547
    2,4-dinitrophenylhydrazone colorimetric
        method 4:549
    distillation method 11:362
    in embalming fluids 12:165
    in enamels 12:175
    gas chromatography 4:154; 5:5, 79; 6:363; 7:96;
        11:73; 12:204; 16:107, 425; 18:510
    gas line antifreeze 6:355
    in gasoline 16:119
    iodometric titration 10:174
    manufacture 4:505
    in methanol 16:107, 114
    in nail enamels 11:73
    phthalic anhydride method 4:535
    properties 6:2; 12:175
    pyromellitic dianhydride method 4:537
    in soaps 18:196
    specific gravity 4:540

in warfarin sodium 10:174
Isopropyl alcohol–water solutions
    specific gravity 11:361
Isopropylamine
    manufacture 5:262
    properties 5:260
    uses 5:263
Isopropylbenzene. (See also Cumene.)
    gas chromatography 7:8; 13:351; 14:333, 337;
        18:299
    in gasoline 13:351, 354
    properties 13:351
    in styrene 18:299
Isopropyl chloride. (See also 2-Chloropropane.)
    in allyl chloride 5:102
    composition 9:411
    gas chromatography 5:102
    infrared spectroscopy 9:394
2-Isopropylchlorobenzene
    mass spectroscopy 9:405
4-Isopropylchlorobenzene
    mass spectroscopy 9:405
Isopropyl cinnamate. (See also *Cinnamic acid
    and its derivatives.*)
    ultraviolet spectroscopy 10:69
N-Isopropylcyclohexylamine
    properties 11:242
Isopropylcyclopentane
    gas chromatography 13:351
    in gasoline 13:351, 354
    properties 13:351
Isopropyl ether. (See also *Ethers.*)
    gas chromatography 12:300, 302; 16:108
    in methanol 16:108, 114
    properties 12:297
N-Isopropylhydroxylammonium oxalate. (See
    also *N*-Alkylhydroxylamines.)
    infrared spectrum 14:443
4,4'-Isopropylidenediphenol. See Bisphenol A.
4,4'-Isopropylidenediphenol, polybutylated 6:75,
    77
Isopropyl iodide
    gas chromatography 16:92
Isopropyl isocyanate. (See also *Isocyanic acid
    esters.*)
    commercial sources 15:102
    infrared spectroscopy 15:109
    near-infrared spectroscopy 15:105
    properties 15:95
Isopropyl lactate. (See also Lactic acid esters.)
    properties 15:131
Isopropyl mercaptan. (See also *Mercaptans.*)
    derivatives for identification 15:559
    gas chromatography 15:557
    properties 15:552
Isopropyl methacrylate
    nuclear magnetic resonance spectroscopy 16:86

Isopropylmethylchlorobenzene
  mass spectroscopy 9:405
N'-Isopropyl-2-methyl-1,2-propanediamine 5:5
m-Isopropylphenol. (See also *Phenols.*)
  properties 17:4
  thin-layer chromatography 17:24
o-Isopropylphenol. (See also *Phenols.*)
  commercial grades and specifications 17:15
  gas chromatography 17:33
  properties 17:4
  thin-layer chromatography 17:24
p-Isopropylphenol. (See also *Phenols.*)
  derivatives for identification 17:21
  properties 17:4
  thin-layer chromatography 17:24
N-Isopropyl-*N'*-phenyl-*p*-phenylenediamine. (See
    also *Diamines and higher amines, aromatic.*) 6:74
  properties 11:434
  solubility 11:435
Isopropylpyridines. (See also *Pyridine and pyridine
    derivatives.*)
  gas chromatography 17:437
  ionization constants 17:416
Isopropyl thiocyanate. (See also *Isocyanic acid
    esters.*)
  gas chromatography 15:110
Isoquinoline
  volumetric method for aluminum alkyls 5:170
Isosafrole
  gas chromatography 8:355
Isotactic content
  of polypropylene 12:361
Isotheaflavin
  liquid chromatography 18:501
  manufacture 18:497
Isotope abundance methods
  for boron 7:326
Isotope dilution methods
  for amino acids 8:433
  for strontium 18:281
Isotope exchange

  in mercury-containing compounds 13:199
Isotopes
  of cesium 9:280
  of chlorine 9:304
Isovaleraldehyde
  in black tea 18:507, 510
  in cocoa bean vapor 9:570
  gas chromatography 18:510
  mass spectroscopy 9:570
  properties 8:334
Isovaleric acid
  in black tea 18:511
  gas chromatography 18:511
  properties 8:397
  thin-layer chromatography 8:381
Isovaleric aldehyde
  in black tea 18:507
  infrared spectroscopy 6:616
Isovaleronitrile
  in black tea 18:510
  gas chromatography 18:510
Isovaleryl halides
  derivatives for identification 8:572
Isovanillin
  gas chromatography 11:167
Isoxazolylpenicillins. (See also *Penicillins and
    related compounds*; and individual compounds.)
    16:577
  analysis 16:577
  assay 16:577
  bioassay 16:577
  microbiological methods 16:577
  properties 16:577
  specification tests 16:577
  uses 16:577
Itaconic acid 5:25
  preparation from citric acid 8:373
  properties 8:397, 400
Itobarbital
  column chromatography 6:152
  gas chromatography 6:484

J

Jaborandi alkaloids 4:590
  analysis 4:608
Jamesonite 6:39
Jasmine tea. (See also *Tea*.)
  manufacture 18:456
cis-Jasmone
  in black tea 18:511
  gas chromatography 18:511
Jellies. (See also *Confectionery products*.)
  10:548
  manufacture 10:554
  potassium content 17:387
Jelly strength
  of adhesives 10:273
  of gelatin 13:368
  of glues of animal origin 4:399; 15:536
Jeremejewite 7:315

Jetalloy. (See also Cobalt alloys.)
  composition 10:287
Jewelry alloys
  composition 14:2
  gold content 14:36
Jewelry enamel. (See also *Enamels*.)
  composition 9:195
Johachidolite 7:315
Jones reductor 15:35
  volumetric method for molybdenum 17:197
  volumetric method for titanium 17:187
Josephinite 15:21; 16:285
JSR elastomer 12:86
Juerst ebulliometer 7:548
Juices
  β-carotene content 10:545
  carotenoid content 10:545

K

Kaempferol
  structure 18:461
Kaempferol-3-glucoside
  in green tea leaf 18:461
  thin-layer chromatography 18:463
Kaempferol-3-rhamnoglucoside
  in green tea leaf 18:461
  thin-layer chromatography 18:463
Kainite 15:357; 17:353
  potassium content 17:405
Kaliborite 7:314
Kanamycin 5:473, 579
  analysis 5:583, 590, 596
  colorimetric methods 5:595
  column chromatography 5:583
  identification 5:583
  ion exchange separation 5:583
  microbiological assay 5:590
  paper chromatography 5:584
  properties 5:579
  separation 5:583
  solubility 5:580, 581
  structures 5:473
  uses 5:579
Kanamycin B sulfate
  infrared spectrum 5:584
Kanekalon. (See also *Acrylic and modacrylic
    fibers*; *Acrylonitrile polymers*; Acrylonitrile–
    vinyl acetate polymers; and Modacrylic fibers.)
    4:220
  properties 4:222
Kanthal. (See also Cobalt alloys.) 10:290
  composition 10:290
Kaolin 4:33, 434
  composition 9:202
Kaolinite. (See also *Clays and clay minerals.*)
    9:191
  composition 10:103
Kaolins. (See also *Clays and clay minerals.*)
    10:106
  uses 10:106
Kapeller-Adler reaction
  for histidine 8:487
Kappa number
  of pulp 9:52
Kappelmeier separation
  of alkyd resins 5:33
Karathane. See Dinocap.
Karaya
  colorimetric determination 12:582
  composition 12:583

  in embalming fluids 12:166
  gravimetric determination 12:581
  infrared spectroscopy 12:583
  in matches 15:508
  in photographic films 12:582
  potassium hydroxide gravimetric reagent 12:582
  properties 15:508
Karaya gum
  identification 6:362
Karbam black. See Ferbam.
Karl Fischer titration 4:115, 460; 5:337; 6:17,
    358; 7:53; 8:359, 588
  of corn products 11:3
  of crystal violet 19:187
  of cyclamates 11:230
  of cyclohexylamine 11:214
  of ethyl ether 12:310
  of fluorocarbons 13:67
  of formic acid 13:128
  of glycerol 13:526
  of hydrogen cyanide 14:422
  of hydroxylammonium acid sulfate 14:455
  of liquefied petroleum gas 15:252
  of lubricants 15:330
  of nitroparaffins 16:437
  of phenols 17:37
  of polyamides 17:301
  of polyethylenimine 14:490
  of polyols 19:279
  of polytetramethylene ether glycols 13:249
  of pyridine 17:438
  of terephthalic acid 17:161
Kauffmann and Hartweg method
  for allyl compounds 5:82
  for allyl esters 5:96
  for methallyl chloride 5:103
  for olefins 5:82
Kauri-butanol value
  of hydrocarbons 14:348
Kauri gum
  solubility 13:539, 552
Keatite. (See also Silica.) 18:5
Keene's cement. (See also Hydraulic cements.)
    9:95
  commercial specifications 9:99, 100, 101
  identification 9:108
Kel F elastomer 12:86
Keltan 12:86
Kelthane
  gas chromatography 19:389
  in water 19:389

Kemate. See Dyrene.
Kemithal 6:468
  properties 6:470
Kermesite 6:39
Kernite 7:313, 377
  occurrence 7:313, 314, 376
Kerosene
  in crankcase cleaners 6:357
  degreasers 6:355
Ketals. See Acetals and ketals.
Keto acids 8:370
Ketocholanic acids 7:142
11-Ketoctiocholanolone
  gas chromatography 14:219
Keto–enol equilibrium 8:338
α-Ketoglutaric acid
  thin-layer chromatography 8:381
  in wines 19:423
α-Ketoglutaric acid 2,4-dinitrophenylhydrazone
  thin-layer chromatography 8:381
Ketones. (See also Carbonyl compounds; Methyl
  ketones; and individual compounds.) 8:331
  in acetylenic chemicals 4:145
  in alcoholic beverages 4:489, 490
  analysis 8:339
  azeotropes 8:336
  in carburetor cleaners 6:356
  in chloroform 5:359
  derivatives for identification 4:67, 111
  2,4-dinitrophenylhydrazone colorimetric method
    4:116
  in enamels 12:189
  in essential oils 12:280
  hydroxyamine hydrochloride volumetric method
    4:124, 145; 12:281
  ion exchange chromatography 4:116
  manufacture 8:339
  nomenclature 8:331
  properties 8:333
  sodium sulfite volumetric method 12:281
  Stillman-Reed volumetric method 12:283
  uses 8:332
Kieselguhr. See Diatomaceous earth.
Kieserite 15:357; 18:360
Kimberlite 11:462
Kinematic method
  for viscosity of epoxy resins 12:252
Kinematic viscosity 4:307
  of asphalt cements 6:300
  of bituminous materials 6:300
  of silicone fluids 18:113
  temperature susceptibility 6:305
Kinetic energy method
  for specific gravity of natural gas 13:330
Kirby method
  determination of end groups in acrylic polymers
    4:293

Kjeldahl method 16:159
  for acrylonitrile content of copolymers 18:334
  for acrylonitrile polymers 4:270, 278, 384
  for alkanolamines 5:9
  for amino resins 5:286
  for ammonia 19:382
  for azo compounds 16:460
  for caffeine in black tea 18:490
  for nitrate 16:460
  for nitrogen in coal and coke 10:239
  for nitrogen in cocoa beans 9:554
  for nitrogen in elastomers, synthetic 12:112
  for nitrogen in ethyleneamines 11:428
  for nitrogen in gelatin 12:569; 13:367
  for nitrogen in hafnium and zirconium 14:33
  for nitrogen in manganese 15:484
  for nitrogen in yeast 19:528
  for nitrogeneous compounds in tobacco 19:137
  for organic nitrogen 19:382
  for protein in casein 9:9
  for protein in cereals 9:236
  for protein in corn products 11:5
  for protein in eggs and egg products 12:62
  for protein in worts and beers 7:609
Kjeldahl-Wilfarth-Gunning method
  for protein in flour 12:603
Kleinert refractometric method
  for fat in cocoa products 9:576
Knight and Allen method
  for sugar 18:355
Knoop reaction
  for histidine 8:487
Knoop scale of hardness 4:32
Knorr extraction tube method
  for fat in cocoa beans 9:555
  for fat in cocoa products 9:575
Koch's acid. See 1-Naphthylamine-3,6,8-trisulfonic
  acid.
Kohman procedure
  for butter 16:137
König reaction
  for antihistamines 6:35
  color test for pyridine and its derivatives 17:424,
    432
  nicotine determination 4:610
Kornerupine 7:316
Kotoite 7:313, 315
Kovar. (See also Cobalt alloys.) 10:290
  composition 10:290
Kraft pulp
  bleaching 7:222
Kramerite 7:314
Krammes and Maresh
  identification of finishing agents 11:126, 130
Kraton 12:86
  thermal analysis, differential 12:133
  torsional modulus 12:136
Krenerite 14:1

Krigbaum method
  light scattering of acrylic polymers 4:313
  osmotic pressure of acrylic polymers 4:311
Krylene 12:86
Krynac 12:87
Krynol 12:86
Krypton. (See also *Gases, noble.*)
  commercial specifications 13:287
  gas chromatography 13:297, 307
  in helium 13:298
  impurities, determination of 13:300
  isotopic abundance 13:289
  manufacture 13:290
  mass spectroscopy 13:301, 305
  in neon 13:299
  occurrence 13:290
  in oxygen 16:531
  properties 13:288
  solubility 13:289
  in xenon 13:301
Kurnakovite 7:314
K value
  of sunscreens 11:91
Kyanite
  composition 9:203
Kynar. See Poly(vinylidene fluoride.)

L

Labdanol. See Isobutyl cinnamate.

Label tests 11:114

Laboratory Equipment Corp.
  carrier gas fusion apparatus 8:661
  direct reading nitrogen determinator 8:666
  hot extraction hydrogen analyzer 8:667
  induction furnace 8:661
  vacuum fusion analyzer 8:656

L-acid. See 1-Naphthol-5-sulfonic acid.

Lacquers
  in chrome protectors 6:354
  in explosive priming mixtures 12:431
  solvent extraction 12:431

Lactase
  sugar reduction method 19:534
  in yeast 19:534

Lactic acid. (See also *Lactic acid and its
  derivatives.*) 8:398
  acid–base titration 15:133, 136
  assay 15:133
  in beer 7:579
  in biological fluids 15:137
  in canned food 15:140
  color 15:147
  commercial grades and specifications 15:127,
   129
  composition 8:281; 15:125
  derivatives for identification 8:376
  in eggs and egg products 15:141
  in feed products 11:51
  ferric chloride colorimetric method 15:138, 140
  in food 15:133, 137
  in fruit 15:140
  gas chromatography 15:132
  heat stability 15:147
  p-hydroxydiphenyl colorimetric method 15:137;
   19:422
  infrared spectrum 8:377
  ion exchange methods 19:420, 422
  in lactic acid esters 15:136
  liquid chromatography 15:133
  manufacture 15:128
  in meat 15:133
  in milk and milk products 15:139
  miscibility tests 15:148
  in musts 19:412
  oxidation–distillation method 11:51
  oxidation methods 19:422
  paper chromatography 15:133; 19:418
  properties 8:398; 15:125, 126
  sodium nitroprusside colorimetric method 19:422

solvent extraction 15:139, 140, 141
specific gravity 15:147
structure 15:125
thin-layer chromatography 8:381; 15:133;
  19:419
toxicity 15:128
uses 15:125
in wines 15:141; 19:412, 414, 418, 422

Lactic acid and its derivatives. (See also Lactic
  acid; Lactic acid esters; Lactic acid salts; and
  individual compounds.) 15:125–149
analysis 15:130
assay 15:133
impurities, determination of 15:141

Lactic acid esters. (See also *Lactic acid and its
  derivatives*; and individual compounds.)
assay 15:135
diazomethane esterification method 15:135
gas chromatography 15:136
lactic acid content 15:136
manufacture 15:130
solubility 15:130
uses 15:130

Lactic acid salts. (See also *Lactic acid and its
  derivatives*; and individual compounds.) 15:
  129
assay 15:134

Lactide. (See also *Lactic acid and its derivatives.*)
manufacture 15:128
structure 15:128

β-Lactones
use in preparation of carboxylic acids 8:373

Lactose
in bakery products 6:452
in casein 9:10
in chocolate 9:550
in cocoa products 9:581
cuprous oxide gravimetric method 9:581
infrared spectroscopy 16:128
Munson-Walker values 9:583
in raw milk 16:128

Lactose hydrate. (See also *Sugar.*)
refractive index 18:343

Lactyllactic acid. (See also *Lactic acid and its
  derivatives.*)
manufacture 15:127
structure 15:125

Lake Bordeau B. See D & C Red No. 34.

Lake Red C. (See also D & C Red No. 8; and No.
  9.)
in enamels 12:183, 184, 185

identification 12:183, 184, 185
Lake Red C amine
  in D & C Red No. 8 10:498, 509
  in D & C Red No. 9 10:498, 509
  ultraviolet spectroscopy 10:509
Lake Red D
  in enamels 12:185
  identification 12:185
Lamination technique
  polymer samples for infrared spectroscopy 4:
    236
Lampblack 8:179
  commercial grades and specifications 8:196
  commercial specifications 12:173
  in enamels 12:173
  manufacture 8:193
Lamp method
  for sulfur compounds, organic 18:378
  for sulfur in hydrocarbons 14:346
  for sulfur in naphthalene 16:213
Lane and Eynon method
  for reducing sugars 10:562; 11:40
  for sugar 18:348
  for sugar in cereals 9:239
  for sugar in wines 19:408
Langbeinite 17:398, 405
  potassium content 17:405
Langmuir vaporization of graphite 4:12
Lanolin
  acetic anhydride colorimetric method 11:82
  in creams and lotions 11:94
  in hair sprays 11:90
  in shampoos 11:76, 82
  in sunscreens 11:91
Lanstan. See 1-Chloro-2-nitro-propane.
Lanthanide fluorides
  carrier gas distillation fusion methods 8:636
  carrier gas fusion methods 8:625
  nitrogen in 8:635
  oxygen in 8:625, 630, 636, 637
Lanthanide metals
  carrier gas fusion methods 8:625
  impurities in 8:613
  manufacture 8:613
  nitrogen in 8:630, 631
  oxygen in 8:613, 625, 630, 631
Lanthanides. (See also *Rare earths.*) 17:470, 560
Lanthanide sulfides
  carrier gas fusion methods 8:624
  oxygen in 8:624
Lanthanons 17:560
Lanthanum. (See also Lanthanum metal; *Rare earths.*) 15:149–161
  in aluminum metal 5:133
  atomic absorption spectroscopy 15:159; 17:494
  in cerium oxide 17:501
  DTPA titration 15:159

EDTA titration 13:445
emission spectroscopy 15:151, 155, 158, 187,
  375; 17:488, 501
in europium oxide 17:502
flame photometry 15:158; 17:492
in glass and glass-ceramics 13:445
ion exchange chromatography 13:255
in lanthanum compounds 15:151
in lanthanum glass 15:159
in magnesium 15:375
in magnesium alloys 15:375, 385
occurrence 15:149
in praseodymium oxide 17:501
in rare earths 15:158
in rocks 15:187
x-ray fluorescence spectroscopy 15:158; 17:490
in yttrium oxide 17:501
zirconium content 14:120
Lanthanum acetate. (See also Lanthanum compounds.)
  uses 15:150
Lanthanum alloys
  carrier gas fusion methods 8:627, 682
  hydrogen in 8:680
  nitrogen in 8:623, 627
  oxygen in 8:623, 627, 680, 682
  vacuum fusion methods 8:623, 680
Lanthanum compounds. (See also individual
  compounds.)
  analysis 15:151
  lanthanum content 15:151
  properties 15:150
  solubility 15:150
Lanthanum fluorides
  carrier gas distillation fusion methods 8:636
  oxygen in 8:635, 636
  vacuum distillation fusion methods 8:635
Lanthanum glass
  analysis 15:159
  borate content 15:160
  lanthanum content 15:159
  properties 15:159
  sample preparation 15:159
  thorium content 15:159
Lanthanum hydroxide. (See also Rare earth
  hydroxides.)
  manufacture 17:510
Lanthanum metal. (See also *Lanthanum*; Rare
  earth metals.)
  analysis 15:151
  carrier gas fusion methods 8:627, 682
  hydrogen in 8:618, 619, 629, 680
  nitrogen in 8:618, 623, 627, 629
  oxygen in 8:618, 623, 627, 629, 680, 682
  properties 15:150; 17:471
  sample preparation 8:641
  uses 15:149; 17:470

vacuum fusion methods 8:618, 619, 623, 682
Lanthanum nitrate
  titration of fluoride 13:23
Lanthanum nitrate–iodine
  colorimetric method for cellulose esters 12:589
  colorimetric method for cellulose nitrate 11:132
Lanthanum nitrate–sodium hydroxide
  colorimetric method for EDTA 12:28
  colorimetric method for NTA 12:28
  colorimetric method for PDTA 12:28
Lanthanum oxide. (See also Lanthanum com-
    pounds; Rare earth oxides.)
  analysis 15:151
  assay 15:151
  carrier gas fusion methods 8:626
  commercial specifications 15:151
  DTPA titration 15:152
  emission spectroscopy 17:501
  impurities, determination of 15:152; 17:501
  oxalic acid gravimetric method 15:152
  oxygen in 8:626
  properties 15:151; 17:497
  uses 15:150
Lanthanum series. (See also *Rare earths.*) 17:470
Lapsang Soochong tea. (See also *Tea.*)
  manufacture 18:456
Larderellite 7:314
Lard oil
  solubility 13:539, 552
Larnite 8:75
Larue method
  color test for pyridine and its derivatives 17:424,
    433
Larvacide. See Chloropicrin.
LAS. (See also Linear alkyl aryl sulfonates.) 19:
  385
Lasché test for malt 4:478
Latent heat
  of diethylenetriamine 11:417
  of ethylenediamine 11:416
  of tetraethylenepentamine 11:417
  of triethylenetetraamine 11:417
Laterite 16:286
Lateritic ores 16:285
Laubenheimer reaction
  for thiophene 18:537
Laundry bleaches. (See also *Detergents*; Hypo-
    chlorite bleaches.) 11:325
  composition 11:325
Laundry detergents. (See also *Detergents.*) 11:324
  composition 11:324
  performance tests 11:389
  types 11:324
Laundry soda. (See also Sodium carbonate.)
  commercial grades and specifications 18:237
Laurent's acid. See 1-Naphthylamine-5-sulfonic
  acid.

Lauric acid. (See also Fatty acids.) 5:25
  gas chromatography 8:382, 388
  properties 8:558; 12:473
Lauric anhydride
  properties 8:558
Lauronitrile. See *n*-Dodecanonitrile.
N-Lauroyl-*p*-aminophenol 6:77
Lauryl alcohol. (See also Alcohols, higher.)
  commercial grades 4:563
  properties 4:552, 561
  solubility 4:562
  specifications 4:564
  uses 4:562
Lauryl cetyl dimethylamine oxide. (See also
    Cationic detergents.)
  acid–base titration, nonaqueous 11:348
Lauryl dimethylamine oxide. (See also Cationic
    detergents.)
  acid–base titration, nonaqueous 11:348
Laurylguanidine acetate. See Dodine.
Lauryl methacrylate. (See also *Acrylic and meth-
    acrylic acids and esters.*)
  commercial grades and specifications 4:187
  polymer in 4:200
  properties 4:182
  saponification method 4:196
  solubility 4:184
Lauryl sulfate
  titration of cationic detergents 11:347
Lautarite 14:584
Lavender oil. (See also *Essential oils.*)
  commercial specifications 12:262
  refractive index 12:271
Lawrencium. (See also Transuranium elements.)
  manufacture 19:259
Lead. (See also Heavy metals; Lead alloys; and
    Lead metal.) 15:161–201
  in air 15:192
  in aluminum alloys 5:133
  in aluminum metal 5:133
  in aluminum sulfate 5:194
  in alums 5:194
  in amino acids 8:440
  ammonium molybdate gravimetric method
    15:179
  ammonium molybdate titration 15:180
  amperometry 15:189
  in antimony metal 6:43, 46
  in arsenic metal 6:226
  in arsenic(III) oxide 6:231
  in arsenic(V) oxide 6:233
  atomic absorption spectroscopy 6:46; 7:200;
    9:346; 10:668; 11:157; 12:572; 14:49; 15:185,
    196; 16:167
  in barite 6:558, 559
  bead test 15:172
  in beer 7:651

benzidine colorimetric method 15:172
in biological materials 15:191
bismuth in 7:216
in bismuth chloride 7:200
in bismuth metal 7:194
in bismuth nitrate 7:203
in bismuth oxide 7:205
in bismuth subcarbonate 7:196
in bismuth subnitrate 7:204
in blood 15:186, 189
in blue lead 17:202
borax bead test 6:584
in brake linings 7:441
in cadmium chloride 8:42
in carbon 14:74
in cements 9:131
in certified colors 10:494
in chrome yellow 9:705
in chromic acid 9:684
in chromium metal 9:635
in chromium pigments 9:707
in cinnamic acid and its derivatives 10:70
in citric acid and salts 10:95
in cobalt(II) acetate 10:330
in cobalt alloys 10:314
in cobalt carbonate 10:333
in cobalt compounds 10:328
in cobalt metal 10:293, 298
in cobalt nitrate 10:340
in cobalt(II) orthophosphate 10:343
in cobalt oxides 10:342
in cobalt(II) sulfate 10:344
in cocoa products 9:595, 598
colorimetric determination 15:172, 190
in copper alloys 10:639
in copper metal 10:622
in copper(II) nitrate 10:668
in cotton 11:153, 154, 156
in dental restorative materials, composite 11:
    315
diethylammonium diethyldithiocarbamate
    extraction method 15:178
displacement of copper 10:604
dithizone–chloroform extraction 10:494
dithizone colorimetric method 6:43, 226, 231;
    8:440; 10:299, 623, 668; 15:176, 190, 193,
    225, 385; 16:331
EDTA titration 13:42, 445; 15:181
in elastomers, synthetic 12:151
electrolytic deposition 5:134; 10:495; 15:179
emission spectroscopy 7:651; 9:346; 11:153;
    13:277; 14:48, 74, 122, 536; 15:144, 155,
    173, 187, 375; 16:160, 387; 19:570
in enamels 12:186
in ferrochromium 9:661
in ferroniobium 16:387
flame tests 6:589, 592

in fluoroboric acid 13:42
in food 15:191
in formic acid esters 13:134
gallium content 13:262
in gallium metal 13:266
in gallium phosphide 13:282
in gasoline 13:355; 15:184, 189, 194
in gelatin 12:554; 13:371
in glass 9:200
in glass and glass-ceramics 13:445
in gold 14:45, 49
gold content 14:42
in graphite 14:74
gravimetric determination 15:172, 174, 178;
    16:20
in grease 15:171, 186
in hafnium 14:122
history 15:161
in hydrochloric acid 14:374
hydrochloric acid gravimetric method 15:172
hydrogen sulfide colorimetric method 12:186;
    15:281; 18:176
hydrogen sulfide gravimetric method 12:151,
    420; 15:172
identification 15:172
in indium 14:536
in indium alloys 14:551
iodometric titration 13:357
ion exchange chromatography 13:254
in iron 15:59
in lactic acid and its derivatives 15:144
in lanthanum oxide 15:155
in lead alloys 15:195
in lead azide 12:419
in lead chromate 17:197
in leaded zinc oxide 17:189
in lead nitrate 12:420
in lead picrate 12:419
in lead silico–chromate 9:705
in lead styphnate 12:414, 419
in lead thiocyanate 12:421
in lime and limestone 15:225
in lithium carbonate 15:281
in lubricating oils 15:171, 186
in magnesium 15:375, 385
in magnesium alloys 15:375, 385
in manganese 15:479
in mercury 16:20
microcosmic salt bead test 6:587
in molybdenum 16:161, 167
in molybdenum disulfide 16:167, 178
in molybdenum trioxide 16:161, 167
in nickel 16:331
in nickel alloys 16:331
in niobium 16:387
occurrence 15:161, 166
in organic materials 15:170

in photographic films 12:571, 592
polarography 10:299; 12:541, 542, 555, 571;
    15:188
potassium chromate gravimetric method 15:172
potassium chromate turbidimetric method 10:
    330, 340
potassium dichromate gravimetric method 12:
    186, 419; 13:357; 15:175, 179; 17:188
potassium dichromate titration 15:190
potassium iodide colorimetric method 12:186
potassium iodide color test 15:173
potassium iodide extraction method 15:178
potassium iodide gravimetric method 15:172
potassium nitrate colorimetric method 15:173
properties 15:161
in red lead 17:196
reduction method for tin 15:68
in rocks 15:187
sample preparation 15:169
separation 15:174
in silicate white lead 17:188
in silver 18:173
in silver nitrate 12:542; 18:176
in sodium arsenite 6:235
in sodium chloride 9:346
in sodium dichromate 9:684
sodium diethyldithiocarbamate extraction
    method 15:177
sodium rhodizonate colorimetric method 15:173
sodium sulfide colorimetric method 15:192
sodium sulfide gravimetric method 10:95
in soil 15:186
solvent extraction 15:175; 16:167
standard potentials 15:163
in stannous fluoride 19:75
in sulfate white lead 17:188
sulfuric acid gravimetric method 10:639; 15:172,
    174, 178, 530; 17:189, 196, 197
tetrahydroxy benzoquinone color test 15:193
tetrahydroxyquinone color test 15:193
in tin metal 19:67
toxicity 15:165
ultraviolet spectroscopy 11:315
in urine 15:185, 186, 189
uses 15:161
volumetric determination 15:180
in water 15:173, 186, 189
x-ray emission spectroscopy 15:183
x-ray fluorescence spectroscopy 11:156
in zinc metal 19:570
in zinc oxide 15:530
in zinc powder 17:204
in zirconium 14:122
Lead-210
in tobacco 19:153
Lead acetate. (See also Lead compounds.)
clarifying reagent for sugar 18:345

color test for hydrogen sulfide 13:321; 14:346;
    15:255; 18:393
color test for sulfide 15:71
gravimetric method for albumen 11:127, 132
gravimetric method for chlorogenic acid 10:398
gravimetric method for gums 11:127, 132
gravimetric method for molybdenum 16:156
gravimetric method for mosses 11:127, 132
gravimetric method for pyrocatechol 17:46
properties 15:164
solubility 15:165
Lead acetate, basic. See Lead subacetate.
Lead acetate trihydrate. (See also Lead compounds.)
colorimetric method for sulfur 12:545
properties 15:164
solubility 15:165
Lead additives
atomic absorption spectroscopy 13:358
gas chromatography 13:357
in gasoline 13:352
Lead alloys. (See also Lead metal.) 5:65
aluminum in 5:226
antimony in 6:58, 60
bismuth in 7:214
commercial grades and specifications 15:169
composition 5:69
impurities, determination of 15:194
indium content 14:548
lead content 15:195
sample preparation 6:60; 15:170
uses 15:161
Lead arsenate. (See also *Pesticides and insecticides.*)
    6:229
arsenic content 19:596
commercial grades and specifications 19:593
composition 19:580, 586
manufacture 19:595
solubility 19:587
toxicity 19:592
Lead azide. (See also Lead compounds.) 12:410
analysis 12:411
azide content 12:411
ceric ammonium nitrate volumetric method
    12:411, 419
commercial grades and specifications 12:410
in explosive priming mixtures 12:419, 423, 426,
    429
lead content 12:419
manufacture 12:410
properties 12:406, 410; 15:164
silver nitrate gravimetric method 12:419, 423
sodium nitrate titration 16:473
solubility 12:419; 15:165
solvent extraction 12:426
x-ray diffraction 12:418
Lead–barium glass. (See also *Glass and glass-
    ceramics.*)

composition 13:407
Lead bromate. (See also Lead compounds.)
  properties 15:164
  solubility 15:165
Lead bromide. (See also Lead compounds.)
  properties 15:164
  solubility 15:165
Lead bullion
  bismuth in 7:211
  sample preparation 7:210
Lead carbonate. (See also Lead compounds.)
  in carbonate white lead 17:188
  properties 15:164
  solubility 15:165
Lead chlorate. (See also Lead compounds.)
  properties 15:164
  solubility 15:165
Lead chloride. (See also Lead compounds.)
  properties 9:335; 15:164
  solubility 9:336; 15:165
Lead chlorofluoride precipitation method
  for fluorine 6:50
Lead chloronitrate
  gravimetric method for fluoride 13:23
Lead chromate. (See also Lead compounds.)
  analysis 9:693; 17:197
  assay 9:693; 17:197
  in chrome greens 17:200
  commercial grades and specifications 9:695
  impurities, determination of 17:197
  lead content 17:197
  in ochres 17:194
  properties 15:164
  solubility 15:165
  uses 17:197
Lead citrate
  colorimetric method for sulfate 18:427
Lead citrate complex
  stability 10:89
Lead compounds. (See also individual compounds.)
  manufacture 15:167
  properties 15:164
  sample preparation 15:170
  solubility 15:165
  toxicity 15:165
  uses 15:161
Lead dibutyldithiocarbamate. (See also Dithio-
    carbamic acid salts.)
  properties 11:575
Lead diethyldithiocarbamate. (See also Dithio-
    carbamic acid salts.)
  properties 11:575
Lead dimethyldithiocarbamate. (See also Dithio-
    carbamic acid salts.)
  properties 11:575
  uses 11:576
Lead 2,4-dinitroresorcinol

  in explosive priming mixtures 12:418
  x-ray diffraction 12:418
Lead dioxide. (See also Lead compounds.)
  oxidation of cobalt 10:370
  properties 15:164
  solubility 15:165
Leaded zinc oxide. (See also Zinc compounds.)
  analysis 17:189
  commercial specifications 17:189
  impurities, determination of 17:189
  lead content 17:189
  manufacture 19:571
  sulfur content 17:189
  zinc content 17:189
Lead fluoride. (See also Lead compounds.)
  properties 13:11; 15:164
  solubility 15:165
Lead fluorochloride. (See also Lead compounds.)
  solubility 15:165
Lead fusion
  of platinum group metals 17:233
Lead glass. (See also *Glass and glass-ceramics.*)
  composition 13:407
Lead hydroxide. (See also Lead compounds.)
  properties 15:164
  solubility 15:165
Lead iodate. (See also Lead compounds.)
  properties 15:164
  solubility 15:165
Lead iodide. (See also Lead compounds.)
  microchemical method for cocaine 16:249
  microchemical method for levomethorphan
    16:250
  properties 15:164
  solubility 15:165
Lead metal. (See also Lead alloys.)
  aluminum in 5:225
  commercial grades and specifications 15:168
  copper content 10:617
  impurities, determination of 15:194
  manufacture 15:167
  properties 15:162
  sample preparation 15:169
  silver content 18:171, 172
  tin content 19:64
  uses 15:161
  zinc content 19:568
Lead molybdate. (See also Lead compounds;
    Molybdates.)
  properties 15:164
  solubility 15:165
  uses 16:184
Lead mononitroresorcinol
  in explosive priming mixtures 12:418
  x-ray diffraction 12:418
Lead monoxide. (See also Lead compounds.)
  manufacture 15:167

properties 15:164
in red lead 17:196
solubility 15:165
Lead naphthenate
in enamels 12:176
Lead nitrate. (See also Lead compounds.)
in explosive priming mixtures 12:417, 420, 423
lead content 12:420
microscopy 12:417
nitrate content 12:420
properties 15:164
refractive index 12:417
solubility 12:420; 15:165
solvent extraction 12:423
titration of sulfate 18:425
titration of sulfide 17:512
volumetric method for glycine 12:37
volumetric method for IDA 12:37
volumetric method for NTA 12:37
volumetric method for sulfur 11:37
x-ray diffraction 12:418
Lead ores
recovery of bismuth from 7:189
Lead orthoarsenate. (See also Lead compounds.)
solubility 15:165
Lead orthophosphate. (See also Lead compounds.)
solubility 15:165
properties 15:164
Lead oxide
in alkali barium silicate glass 9:218
emission spectroscopy 9:218
in enamels 9:195
in glass and glass-ceramics 13:413, 415, 418
in glazes 9:196
hydrogen sulfide–sulfuric acid gravimetric method
13:422
in silicate glasses 9:198
sulfuric acid gravimetric method 13:419
Lead perchlorate. (See also Lead compounds.)
solubility 15:165
titration of sulfur dioxide 18:391
Lead phosphate, dibasic
in enamels 12:177
properties 12:177
Lead phosphosilicate
silicon content 18:30
Lead picrate
colorimetric determination 12:419
in explosive priming mixtures 12:419
lead content 12:419
solubility 12:419
uses 12:439
Lead, pig
bismuth in 7:215
Lead powder
analysis 17:203
commercial grades and specifications 17:203

uses 17:203
Lead, refined
bismuth in 7:211
sample preparation 7:210
Lead salicylate
in enamels 12:177
properties 12:177
Lead selenide semiconductors. (See also Ceramics.)
properties 9:213
uses 9:213
Lead sesquioxide. (See also Lead compounds.)
properties 15:164
solubility 15:165
Lead silico-chromate. (See also Chromium pigments.)
analysis 9:705; 17:197
chromium content 9:705
commercial grades and specifications 9:706;
17:198
lead content 9:705
silica content 9:707; 17:199
silicon content 18:30
uses 9:705; 17:197
Lead stearate, basic
in enamels 12:177
properties 12:177
Lead styphnate 12:413
ammonium acetate colorimetric method 12:429
analysis 12:414
assay 12:414
commercial grades and specifications 12:414
in explosive priming mixtures 12:417, 419, 429,
430
impurities, determination of 12:415
lead content 12:414, 419
manufacture 12:414
microscopy 12:417
nitrogen content 12:415
properties 12:406, 413
solubility 12:413, 419
solvent extraction 12:429
styphnate content 12:415
uses 12:414
x-ray diffraction 12:418
Lead subacetate. (See also Lead compounds.)
properties 15:164
solubility 15:165
Lead suboxide. (See also Lead compounds.)
properties 15:164
Lead sulfate. (See also Lead compounds.)
properties 15:164
solubility 15:165
Lead sulfide. (See also Lead compounds.)
in glass 9:200
properties 15:164
solubility 15:165
sulfur content 18:383

Lead sulfide semiconductors. (See also *Ceramics.*)
  properties 9:213
  uses 9:213
Lead telluride semiconductors. (See also *Ceramics.*)
  properties 9:213
  uses 9:213
Lead tetraacetate
  colorimetric method for citric acid 19:423
Lead tetrachloride. (See also Lead compounds.)
  properties 15:164
Lead tetroxide. (See also Lead compounds.)
  properties 15:164
  solubility 15:165
Lead thiocyanate
  colorimetric determination 12:420
  in explosive priming mixtures 12:418, 420, 423,
    426, 427, 429, 430
  gravimetric determination 12:420, 423
  lead content 12:421
  potassium dichromate gravimetric method 12:
    426
  silver nitrate gravimetric method 12:427, 430
  solubility 12:420
  solvent extraction 12:423, 430
  thiocyanate content 12:421, 423
  x-ray diffraction 12:418
Lead titanate zirconate
  in ceramics 9:225
  microscopy 9:225
Lead 2,4,6-trinitroresorcinate. See Lead styphnate.
Lead zirconate titanate
  properties 9:212
Leakage tester 6:374
Leak detection lamps 9:381
Least squares method
  mathematical analysis of bioassay data 7:172
Leather
  application of dyes to 6:398
Leavening agents 6:425
Lecithin
  in cocoa products 9:587
  in egg yolk 12:58
  molybdenum blue colorimetric method 9:587
Leco automatic sulfur determinator 15:479
Leco method
  for sulfur in carbon 14:71
Lee spectroscopic method
  for caffeine 10:433
Lehigh method
  for basicity of basic chromium salts 9:691
  for basicity of chrome tanning compounds 9:
    691
Leithner's blue. See Cobalt aluminate.
Lemonade
  caramel content 10:538
Lemongrass oil. (See also *Essential oils.*)
  citral content 12:282

refractive index 12:271
Lemon oil. (See also *Essential oils.*)
  citral content 12:282
  commercial specifications 12:261
  composition 12:267
  gas chromatography 12:267
  optical rotation 12:270
  refractive index 12:271
  residue on evaporation 12:273
  solubility 13:540
  ultraviolet spectroscopy 12:288
Length and length distribution of fibers 4:89
Lepidine
  colorimetric method for ethylene oxide 12:198
Lepidine–diethylene glycol monoethyl ether
  colorimetric method for epoxy resins 12:237
Lepidolite 9:278; 13:2; 15:261; 17:547
  in cocoa beans 9:556
  composition 17:544
  occurrence 17:544, 547
Lesserite 7:314
Lethane. (See also *Pesticides and insecticides.*)
  19:586
Leucine. (See also *Carboxylic acids, amino.*) 8:
  413, 488
  acid–base titration, nonaqueous 8:494
  analysis 8:494
  in black tea 18:507
  commercial specifications 8:494
  composition 8:413
  crystalline form 8:491
  gas chromatography 8:429
  infrared spectrum 8:492
  ion exchange chromatography 8:415, 421
  manufacture 8:493
  methionine in 8:495
  nuclear magnetic resonance spectrum 8:493
  optical rotation 8:444
  paper chromatography 8:422
  properties 8:489
  in protein 8:414
  solubility 8:490
  structure 8:489
  tyrosine in 8:495
  in yeast 19:529
Leucite 9:278; 17:353
  composition 17:544
  occurrence 17:544
Leucoanthocyanidins. See Anthocyanogens.
Leucoanthocyanin
  colorimetric method for flavylogens 18:475
  in green tea leaf 18:457
  thin-layer chromatography 18:463
Leuco base
  in F D & C Blue No. 1 10:498, 510
  in F D & C Violet No. 1 10:498, 510
  ultraviolet spectroscopy 10:511

Leuco content
  of triphenylmethane colors 10:510
Leucocyanin
  structure 18:461
Leucodelphinin
  structure 18:461
Leuco ester vat dyes 12:4
  colorimetric determination 12:18
  identification 12:18
  sodium nitrate color test 12:18
Leucomycin 5:464, 519
  analysis 5:523, 527, 532
  bioautography 5:524
  colorimetric methods 5:532
  column chromatography 5:523
  countercurrent distribution 5:523
  identification 5:523
  ion exchange chromatography 5:523
  microbiological assay 5:527
  paper chromatography 5:523
  properties 5:518
  separation 5:523
  solubility 5:519, 520
  uses 5:519
Levapren 12:87
Levomethorphan. (See also *Narcotics.*)
  ammonium molybdate colorimetric method
    16:246
  color tests 16:246
  lead iodide microchemical method 16:250
  microchemical tests 16:250
  platinic chloride microchemical method 16:250
  structure 16:245
  ultraviolet spectroscopy 16:252
Levorphanol. (See also Levorphanol tartrate;
    Narcotics.) 16:280
  ammonium molybdate colorimetric method
    16:246
  analysis 16:280
  color tests 16:246
  gas chromatography 16:255
  microchemical tests 16:250
  platinic bromide microchemical method 16:250
  platinic chloride microchemical method 16:250
  structure 16:245
  thin-layer chromatography 16:256
  ultraviolet spectroscopy 16:252
  uses 16:280
Levorphanol tartrate. (See also Levorphanol.)
  acid–base titration, nonaqueous 16:280
  solvent extraction 16:280
  ultraviolet spectroscopy 16:252
Levulinic acid
  thin-layer chromatography 8:381
Levulinic acid 2,4-dinitrophenylhydrazone
  thin-layer chromatography 8:381
Levulose. (See also *Sugar.*) 18:337

  color reagent for bile acids 7:146
LH. See Luteinizing hormone.
Liberman-Storch test
  for rosin acids 5:38
Librium
  gas chromatography 6:210, 211
Lidocaine
  analysis 5:395
  gas chromatography 6:210, 211
  properties 5:394
Lidocaine hydrochloride 5:375
  analysis 5:395
  properties 5:395
Liebermann-Burchard test
  for fusidic acid 5:517
Liebermann reaction
  color test for phenols 17:19
  for thiophene 18:537
Liebermann's allocinnamic acid 10:56
Liebermann's reagent
  color test for antioxidants 6:92
Light green SF yellowish. (See also *Triphenyl-
    methane and related dyes.*)
  liquid chromatography 19:173
Light microscopy
  of calcium aluminates 9:103
  of calcium oxide 9:103
  of dicalcium silicate 9:103
  of ferrites 9:103
  of hydraulic cements 9:102
  of magnesia 9:103
  of portland cement 9:103, 104, 105, 106, 107
  of tricalcium silicate 9:103
Light scattering 4:312
  of polycarbonates 17:341
Light-scattering methods
  for instant tea 18:513
Light scattering photometry
  of styrene polymers 18:322
Light transmittance
  of drycleaning detergents 11:613
Lignins
  in green coffee 10:374
  solubility 19:145
  solvent extraction 19:147
  sulfuric acid gravimetric method 19:147
  in tobacco 19:143, 145, 147
Lignite 19:225
  in concrete 9:142
Lilial
  commercial specifications 10:44
Lime. (See also Calcium oxide; *Ceramics; Lime
    and limestone.*) 8:72, 92; 9:190
  acid–base titration 9:109
  acid dissolution 9:215
  in cements 9:141
  identification 15:205

manufacture 15:203
in portland cement 9:108
slaking rate 15:230
standard consistency 15:230
Lime–aluminoborosilicate glass. (See also
  Silicate glasses.)
composition 9:199
uses 9:199
Lime and limestone. (See also Lime; Limestone.)
  15:202–233
acid–base titration 15:229
acid-insoluble content 15:206
ammonium hydroxide group content 15:208
analysis 15:204
available lime content 15:229
bulk density 15:202
calcium carbonate equivalent 15:230
calcium oxide content 15:215
carbon dioxide content 15:220
color 15:202
composition, determination of 15:205
crystal structure 15:202
ferric oxide content 15:209
handling precautions 15:204
hardness 15:202
heat of formation 15:202
history 15:202
identification 15:205
impurities, determination of 15:211
loss on ignition 15:206
moisture content 15:228
neutralization coefficient 15:229
porosity 15:202
properties 15:202
refractive index 15:202
sampling 15:204
silicon dioxide content 15:207
sodium carbonate fusion 15:206
solubility 15:202
texture 15:202
uses 15:203, 204
Lime, available
in lime and limestone 15:229
Lime cement. (See also Hydraulic cements.)
  9:96
commercial specifications 9:99, 100, 101
composition 9:96
identification 9:108
Lime oil. (See also *Essential oils.*)
citral content 12:282
residue on evaporation 12:273
ultraviolet spectroscopy 12:288
Limestone 8:244; 9:190
color 15:231
commercial grades and specifications 9:191
identification 15:205
occurrence 15:203

oil absorption 15:232
particle size distribution 15:231
Limit-dextrin
colorimetric method for α-amylase 12:607
Limonene
commercial grades and specifications 14:265
in black tea 18:510
gas chromatography 8:353; 12:266, 268; 18:510
in majoram, oil of 12:268
in peppermint oil 12:266
structure 14:266
uses 14:266
Limonene dioxide. (See also *Epoxy compounds.*)
infrared spectroscopy 12:200
Limonite 15:21
composition 15:22
Linaloe oil. (See also *Essential oils.*)
refractive index 12:271
Linalool
in black tea 18:511
gas chromatography 12:268; 18:511
in majoram, oil of 12:268
Linalool oxide
in black tea 18:511
gas chromatography 18:511
Linalyl acetate
in essential oils 12:276
molecular weight 12:276
Linalyl benzoate. (See also Benzoic acid esters.)
commercial grades and specifications 7:79
properties 7:77
saponification method 7:84
Lincomycin 5:474, 582
analysis 5:584, 591, 595
bioautography 5:585
identification 5:584
microbiological assay 5:591
paper chromatography 5:584
properties 5:579
separation 5:584
solubility 5:580, 581
structure 5:474
uses 5:582
Lincomycin hydrochloride
countercurrent distribution 5:584
Lincomycin hydrochloride monohydrate
infrared spectrum 5:585
Lindane
gas chromatography 19:389
in raw milk 16:129
in water 19:389
Lindane residues
bioassay 7:183
Lindo-Gladding
gravimetric method for potassium 17:405
Linear alkyl aryl sulfonates
methylene blue colorimetric method 19:385

in water 19:385
Linear density
  of polyamide fibers 17:323
Linen
  bleaching 7:221
Linen fiber
  mixtures, determination in 4:353
  solubility 4:353
Linnaeite 10:284
Linoleic acid. (See also Fatty acids.) 5:25
  gas chromatography 8:388
  properties 12:473
Linolenic acid. (See also Fatty acids.) 5:25
  gas chromatography 8:388
  properties 12:473
Linseed oil. (See also *Fats and fatty oils.*) 5:23
  in brake linings 7:441
  properties 12:473
  solubility 13:539, 552
Linseed oil fatty acids
  gas chromatography 5:25
Lipids
  in beer 7:604
  classification 19:149
  in egg yolk 12:58
  in flour 12:604
  in green coffee 10:394
  in roasted coffee 10:430
  solubility 19:149
  solvent extraction 12:604
  in tobacco 19:149
Lipoproteins
  in egg yolk 12:58
Lipowitz's alloy 5:65
Lipsticks. (See also *Cosmetics.*) 11:65
  analysis 11:65
  colors in 10:529
  composition 11:65
  dyes in 11:68
  identification 11:65
  liquid chromatography 11:66
  oils in 11:67
  quality control 11:70
  separation 11:65
  thin-layer chromatography 11:68
  waxes in 11:67
Liquefied petroleum gas 15:234–260
  analysis 15:234
  commercial grades and specifications 15:234
  composition 15:234
  composition, determination of 15:239
  copper corrosion test 15:257
  gas chromatography 15:239
  hydrocarbon content 15:239
  impurities, determination of 15:251
  manufacture 15:234
  mercury freeze method 15:241

oil ring method 15:244
  properties, determination of 15:245
  sampling 15:234
  specific gravity 15:245
  uses 15:234
  vapor pressure 15:248
  weathering test 15:242
Liquemar. See Phenprocoumon.
Liquers 4:466
  analysis 4:480
Liquid absorption
  of polyamide fibers 17:308
Liquid bleach. See Hypochlorite bleaches.
Liquid carrier distillation
  of water in hydrocarbons 14:345
Liquid chlorine. See Chlorine, elemental.
Liquid chromatography. (See also Column chro-
    matography; and individual compounds to be
    analyzed.)
  of 21-acetoxy-3-hydroxy-5-pregnane-11,20-diones
    14:208
  of acid colors 10:527
  of androsta-3,5-dien-17-one 14:209
  of 5-androstane-3,17-diones 14:208
  of 5-androst-2-ene-17-one 14:209
  of androsterone 14:209
  of arabinitol 13:503
  of aromatic hydrocarbons 14:336
  of basic colors 10:527
  of benzyl violet 4B 19:173
  of bile acids 7:148
  of bishydroxycoumarin 10:159
  of brilliant blue FCF 19:173
  of 1,2-butanediol 13:580
  of 2,3-butanediol 13:580
  of butter 10:545; 16:138
  of caffeine 8:59; 10:407, 408
  of calcium 10:37
  of captan 13:211
  of caramel 10:544
  of carotene 11:53
  of α-carotene 10:542
  of β-carotene 10:542, 545
  of carotenoids 10:545
  of catechin 18:465
  of certified colors 10:476, 524
  of 3-chloro-androst-5-ene-17-one 14:209
  of chlorogenic acid 10:400
  of chlorogenic acids 18:476
  of cholesterol 14:209
  of Chromotrope 2R 10:523
  of cinnamaldehyde 10:49
  of citric acid 15:133; 19:149
  of coal tar dyes 10:525
  of conalbumin 12:60
  of cortexone 14:210
  of corticosteroids 14:209, 210
  of corticosterone 14:209, 210
  of cortisol 14:209, 210

of cortisone 14:210
of coumarins 11:164
of *p*-coumarylquinic acid 18:476
of cyanoacetic acid 15:439
of D & C Yellow No. 8 10:523
of dicloran 13:217
of diethylene glycol 13:580
of diglucosides 19:466
of dihydrocortisone 14:210
of dihydroxyandrostanes 14:209
of dihydroxyandrostanones 14:209
of dihydroxyandrostenes 14:209
of dihydroxypregnanes 14:209
of diiodofluoresceins 10:514
of *p*-dimethylaminobenzaldehyde 10:507
of *p*-dimethylaminobenzoic acid 10:507
of dinocap 13:220
of diphenyl 11:529
of disodium salt of 3-carboxy-5-hydroxy-1-
    phenyl-4-*p*-sulfophenylazopyrazole 10:517
of disodium salt of 3-carboxy-5-hydroxy-1-*p*-
    sulfophenyl-4-phenylazopyrazole 10:517
of dithiocarbamic acid salts 11:584
of egg albumen 12:60
of egg yolk 12:61
of epicatechin 18:465
of epicatechin gallate 18:465
of epigallocatechin 18:465
of epigallocatechin gallate 18:465
of ergocistine 15:292
of ergometrinine 15:292
of ergonovine 15:293
of ergotamine 15:293
of ergotaminine 15:292
of ergotoxine 15:293
of erythritol 13:503, 580
of estradiol 14:221
of estradiol methyl ether 14:221
of estriol 14:221
of estriol methyl ether 14:221
of estrone 14:221
of estrone methyl ether 14:221
of ethylbenzylaniline sulfonic acid 10:507
of ethylene and propylene polymers 12:376
of ethylene glycol 13:503, 580
of *N*-ethyl-*N*-(3-sulfobenzyl)-sulfanilic acid
    10:506
of fast green FCF 19:173
of F D & C Blue No. 1 10:479
of F D & C Green No. 1 10:479
of F D & C Green No. 3 10:479
of F D & C Red No. 2 10:479
of F D & C Violet No. 1 10:479
for F D & C Yellow No. 5 10:479
of flavanols 18:465
of flavonol glycosides 18:474
of flavylogens 18:475

of fluorescein 10:514
of folpet 13:210
of formaldehyde 13:503
of fumaric acid 15:133
of galactitol 13:503
of gallocatechin 18:465
of gasoline 13:347
of globulin 12:60
of glucitol 13:503
of glycerol 13:502, 580
of glycols 13:581
of guinea green B 19:173
of hydrocarbons 14:335
of hydrocodone 16:272
of hydroxyandrostanones 14:208, 209
of hydroxyandrosteneones 14:209
of 3-hydroxy-5-pregnan-20-ones 14:208
of hydroxyprogesterones 14:209
of indigoid dyes 14:508
of iodinated substances 10:514
of iodofluoresceins 10:514
of isatinsulfonic acid 10:512
of isotheaflavin 18:501
of lactic acid 15:133
of light green SF yellowish 19:173
of lipsticks 11:66
of LSD 15:292
of lysergic acid 15:292
of lysozyme 12:60
of malic acid 15:133; 19:149
of malonic acid 15:429
of mannitol 13:503
of mercury 16:17
of methyl acrylate 16:99
of methylergonovine 15:293
of methyl methacrylate 16:99
of methysergide 15:293
of morphine 16:265
of oils 11:67
of oil-soluble colors 10:525
of olefins 14:336
of organochlorine pesticides 19:156
of organophosphorus pesticides 19:156
of ovalbumin 12:60
of ovomucoid 12:60
of oxalic acid 19:149
of 17-oxo-androsenes 14:209
of 17-oxo-androstan-3,5-diene 14:209
of γ-oxoglutaric acid 15:133
of γ-oxopyruvic acid 15:133
of γ-oxovaleric acid 15:133
of penicillins 16:559
of pesticide residues 16:130
of pesticide residues in eggs 12:67
of pesticides 19:156
of phenolic acids 18:476
of phenolic steroids 14:210

of phenols 17:25
of phytonadione 10:187, 188
of pigments in tobacco 19:152
of polyethylene 12:378
of polyglycols 13:580
of polyhydric alcohols 13:581
of polyoxyethylene glycol 13:580
of polypropylene 12:378
of potassium 10:37
of 5-pregnane-3,20-diones 14:208
of pregnanetriol 14:209, 210
of pregnenolone 14:209
of progesterone 14:209, 210
of propylene glycol 13:580
of pyruvic acid 15:133
of quaternary ammonium compounds 11:556
of 2-(2-quinolyl-6,8-disulfonic acid)-1,3-
    indandione 10:524
of 2-(2-quinolyl-6-sulfonic acid)-1,3-indandione
    10:524
of 2-(2-quinolyl-8-sulfonic acid)-1,3-indandione
    10:524
of 2-(2-quinolyl-6-sulfonic acid)-1,3-indandione-
    5-sulfonic acid 10:524
of 2-(2-quinolyl-8-sulfonic acid)-1,3-indandione-
    5-sulfonic acid 10:524
of ribitol 13:503
of sodium 10:37
of sodium iodide 10:493
of steroids 14:209
of succinic acid 15:133
of m-sulfobenzaldehyde 10:507
of sulfobenzaldehydes 10:506
of 2-(4-sulfo-1-naphthylazo)-1-naphthol-4-
    sulfonic acid 10:516
of sulfonated compounds 10:512
of 2-(3-sulfo-2,6-xylylazo)-1-naphthol-4-sulfonic
    acid 10:516
of 2-(6-sulfo-2,4-xylylazo)-1-naphthol-4-sulfonic
    acid 10:516
of terphenyls 11:529, 536
of terraclor 13:215
of testosterone 14:210
of tetrahydrocortisone 14:210
of 2,4,5,7-tetraiodofluorescein 10:514
of theaflavin 18:497, 501
of theaflavin-3,3'-digallate 18:502
of theaflavin gallate 18:502
of thearubigens 18:501
of theogallin 18:476
of thiophosphates 19:156
of triethyl rhodamine 10:522
of trihydroxypregnanes 14:209
of trihydroxypregnanones 14:209, 210
of triiodofluoresceins 10:514
of triphenylmethane and related dyes 19:173
of triphenylmethane colors 10:477

of trisodium salts of 1-(4-sulfophenylazo)-2-
    naphtholdisulfonic acids 10:518
of vegetable fat 16:138
of water-soluble colors 10:525
of waxes 11:67
of xanthophylls 11:53
of xylitol 13:503
Liquid soaps. (See also Soaps.)
    composition 18:180
    uses 18:180
Liquid-solid transition
    of fatty alcohols 4:570
Liquid sugar. (See also Sugar.) 18:337
Litchfield-Wilcoxon method
    mathematical analysis of bioassay data 7:173,
        174
Litharge. (See also Lead monoxide.)
    in brake linings 7:441
Lithiophylite 15:261
Lithium. (See also Lithium metal.) 15:260–289
    absorption method for nitrogen 13:293
    absorption method for oxygen 13:293
    absorption method for water vapor 13:293
    in alkali salts 15:266
    in aluminum and aluminum alloys 5:134
    in aluminum oxide 5:186
    ammonium stearate gravimetric method 15:265
    atomic absorption spectroscopy 15:266, 386
    in beer 7:652
    in cements 9:131
    in cesium metal 9:282, 284
    in cobalt compounds 10:327
    colorimetric determination 15:268
    electrolytic method 10:327
    emission spectroscopy 15:266, 271
    ferric periodate gravimetric method 15:265
    flame emission spectroscopy 13:452
    flame photometry 5:134; 7:652; 9:284; 15:266,
        270; 17:555
    flame tests 6:592
    in glass and glass-ceramics 13:452
    gravimetric determination 14:257; 15:265, 267
    history 15:260
    identification 15:265
    ion exchange chromatography 15:266; 17:364
    in lithium aluminum hydride 14:257
    lithium chloride gravimetric method 15:266,
        267
    in magnesium 15:386
    in magnesium alloys 15:386
    occurrence 15:261
    perchloric acid gravimetric reagent 15:267
    in plants 18:221
    potassium ferric periodate colorimetric method
        15:269
    potassium metaperiodate gravimetric method
        15:265

potassium metaperiodate volumetric method
15:268
in rubidium 17:555
separation 15:266
in spodumene 15:270
thoron colorimetric method 15:269
in water 18:221
x-ray emission spectroscopy 15:272
Lithium acetylide
titration method 4:142
xylene dispersions, analysis of 4:146
Lithium alloys
hydrogen in 8:617, 681
oxygen in 8:681
uses 15:261
vacuum fusion methods 8:617, 681
Lithium aluminate. (See also Lithium compounds.)
uses 15:279
Lithium aluminum hydride. (See also *Hydrides*;
Lithium compounds.)
aluminum content 14:256
analysis 14:247, 256
assay 14:247, 256
commercial grades and specifications 14:247;
15:278
composition, determination of 14:247, 256
hydrogen content 14:247
lithium aluminum hydride content 14:256
manufacture 14:246
properties 14:243, 245
reducing agent for amides 8:607
uses 14:243; 15:278
volumetric method for sulfur 12:129
Lithium aluminum hydride method
for hydroxyl groups 12:246
Lithium aluminum hydroxide
volumetric method for nitriles 16:405
Lithium amide. (See also Lithium compounds.)
ammonia volumetric method 15:285
analysis 15:285
assay 15:285
commercial grades and specifications 15:278
uses 15:278
Lithium borohydride. (See also *Hydrides*.) 7:343,
344
analysis 14:247
assay 14:247, 252
commercial grades and specifications 14:246
composition, determination of 14:247
hydrogen content 14:247
manufacture 14:245
properties 14:243, 244
Lithium bromide. (See also Lithium compounds.)
analysis 15:282
assay 15:282
commercial grades and specifications 15:278
commercial specifications 8:10

impurities, determination of 15:282
Mohr titration 15:282
properties 8:9
solubility 15:277
uses 15:277, 278
Lithium carbonate. (See also Lithium compounds.)
acid–base titration 15:279
analysis 15:279
assay 15:279
commercial grades and specifications 15:278
impurities, determination of 15:279
properties 15:277
uses 15:277, 278
Lithium chlorate. (See also Chlorates.)
properties 9:358
solubility 9:359
Lithium chloride. (See also Lithium compounds.)
9:339
commercial grades and specifications 15:278
commercial specifications 9:339
electrical conductivity method for moisture in
natural gas 13:326
gravimetric method for lithium 15:266, 267
impurities, determination of 9:345
properties 9:335, 339
solubility 9:336; 15:277
uses 15:277, 278
Lithium chromate
commercial specifications 9:697
solubility 9:697
Lithium citrate tetrahydrate. (See also Citric acid
and salts.)
uses 10:86
Lithium cobaltite. (See also Lithium compounds.)
uses 15:279
Lithium compounds. (See also individual compounds.) 15:277
analysis 15:279
properties 15:277
solubility 15:277
uses 15:277
Lithium cyanide. (See also Cyanides.) 14:424
properties 14:424
uses 14:424
Lithium cyanoborohydride 7:344
Lithium ferrite. (See also *Ferrites*.)
magnetic properties 12:515
Lithium fluoberyllate 7:123
Lithium fluoride. (See also Lithium compounds.)
analysis 15:283
assay 15:283
carrier gas distillation fusion methods 8:636
in ceramics 9:225
commercial grades and specifications 15:278
impurities, determination of 15:283
microscopy 9:225
properties 13:11

solubility 15:277
sulfuric acid gravimetric method 15:283
uses 15:278
Lithium gallium hydride 13:280
  uses 13:280
Lithium greases. (See also Lithium compounds.)
  uses 15:277
Lithium hydride. (See also *Hydrides*; Lithium com-
    pounds.)
  analysis 14:247, 250; 15:285
  assay 14:247; 15:285
  commercial grades and specifications 14:246;
    15:278
  composition, determination of 14:247, 251
  hydrogen content 14:247
  hydrogen gasometric method 15:285
  impurities, determination of 14:251
  manufacture 14:245
  properties 14:243, 244; 15:277
  uses 14:242; 15:277, 278
Lithium hydroxide. (See also Lithium com-
    pounds.)
  acid–base titration 15:285
  analysis 15:285
  assay 15:285
  commercial grades and specifications 15:278
  impurities, determination of 15:285
  properties 15:277
  uses 15:279
Lithium hypochlorite. (See also Hypochlorites;
    Lithium compounds.)
  analysis 15:287
  assay 15:288
  available chlorine 7:224; 15:287
  iodometric method 15:287
  properties 9:352
  uses 9:352; 15:277
Lithium imide. (See also Lithium compounds.)
  manufacture 15:277
Lithium iodide. (See also Lithium compounds.)
  commercial grades and specifications 15:279
  solubility 15:277
  uses 15:279
Lithium manganite. (See also Lithium compounds.)
  uses 15:279
Lithium metaborate. (See also Lithium com-
    pounds.)
  fusion of refractory minerals 15:32
  fusion of silicates 15:32
  uses 15:279
Lithium metal 15:260
  analysis 15:265
  commercial grades and specifications 15:264
  hydrogen in 8:617, 681
  impurities, determination of 15:272
  manufacture 15:263
  oxygen in 8:681

properties 15:261
  uses 15:261
  vacuum fusion methods 8:617, 681
Lithium metasilicate. (See also Lithium com-
    pounds.)
  uses 15:279
Lithium methoxide
  titration of ampicillin 16:577
  volumetric method for bishydroxycoumarin
    10:159
Lithium molybdate. (See also Lithium compounds.)
  uses 15:279
Lithium–nickel ferrite. (See also *Ferrites.*)
  commercial specifications 12:520
  composition 12:520
Lithium nitrate. (See also Lithium compounds.)
  commercial grades and specifications 15:279
  uses 15:279
Lithium nitride. (See also Lithium compounds.)
  in lithium 15:275
  Nessler reagent color test 15:275
Lithium oxide
  in alkali barium silicate glass 9:218
  emission spectroscopy 9:218
  in enamels 9:195
  in glass and glass-ceramics 13:413, 415, 423
  in silicate glasses 9:198
  zinc uranyl acetate gravimetric method 13:424
Lithium perchlorate. (See also Lithium compounds;
    Perchlorates.)
  commercial grades and specifications 15:279
  properties 9:363
  solubility 9:364; 15:277
  uses 15:279
Lithium silicates. (See also Silicates.) 18:42
  properties 18:42
Lithium soaps. (See also Lithium compounds.)
  uses 15:277
Lithium sulfate. (See also Lithium compounds.)
  commercial grades and specifications 15:279
  properties 15:277
  uses 15:279
Lithium tantalate. (See also Tantalum compounds.)
  uses 18:453
Lithium tetraborate. (See also Lithium compounds.)
  uses 15:279
Lithium tetra-*p*-tolylboron
  gravimetric method for sodium 18:229
Lithium titanate. (See also Lithium compounds.)
  uses 15:279
Lithium zirconate. (See also Lithium compounds.)
  uses 15:279
Lithocholic acid
  liquid chromatography 7:149
  preparation 7:145
  properties 7:143
  structure 7:142

Lithol maroon
  in enamels 12:186
  identification 12:186
Lithol Red. See D & C Red No. 10; No. 11; No.
  12; and No. 13.
Lithol red 2G. (See also *Pigments.*)
  structure 17:178
Lithol red R. (See also *Pigments.*)
  structure 17:178
Lithol reds
  in enamels 12:182, 185
  identification 12:182, 185
Lithol Rubin B. See D & C Red No. 6; and No. 7.
Lithol Rubine. (See also *Pigments.*)
  in enamels 12:173
  structure 17:179
Lithol rubin lake
  in enamels 12:182, 185
  identification 12:182, 185
Lithopone
  analysis 17:190
  commercial grades and specifications 17:190;
    19:572
  manufacture 19:572
Litmus 6:383
Live cell content
  of yeast 19:524
Livetin
  in egg yolk 12:58
Livingstonite 6:39
Lizardite
  composition 10:103
Lobeline 4:591
  analysis 4:611
Locust bean gum
  composition 12:583
  gravimetric determination 12:581
  infrared spectroscopy 12:583
  iodine colorimetric method 12:580
  in photographic films 12:580
  sodium tetraborate gel test 12:580
Lodestone 12:514
Loellingite 6:223
  metallurgy 6:224
Logit 7:165
Logwood
  identification 10:537
Logwood black. (See also *Pigments.*)
  structure 17:186
Longifolene
  Fourier transform spectra 13:167
Lorite
  composition 12:174
  in enamels 12:174
Loss on drying
  of amino acids 8:442
  of amobarbital 6:504

  of antibiotics 5:492
  of aprobarbital 6:512
  of bithionol 11:547
  of bituminous materials 6:307
  of calcium cyclobarbital 6:516
  of chloroisocyanuric acids and salts 7:232
  of hexachlorophene 11:546
  of iodoform 14:615
  of mephobarbital 6:518
  of phenobarbital 6:523
  of secobarbital 6:525
  of sodium benzoate 7:87
  of sodium butabarbital 6:515
  of sodium fluoride 13:33
  of sodium hexobarbital 6:517
  of sodium pentobarbital 6:520
  of sodium phenobarbital 6:523
  of sodium secobarbital 6:525
  of sodium thiamylal 6:526
  of sodium thiopental 6:528
  of vinbarbital 6:529
Loss on heating
  of sodium carbonate 18:246
Loss on ignition
  of adsorbents 4:435
  of aluminum hydroxide 5:189
  of aluminum oxide 5:189
  of asbestos 6:259
  of asbestos–cement products 6:270
  of barite 6:556, 558
  of barium carbonate 6:540
  of barium sulfate 6:567
  of cadmium carbonate 8:39
  of calcium oxide 8:93
  of chrysotile 6:259
  of clays and clay minerals 10:111
  of crocidolite 6:267
  of fibers 4:25
  of gypsum and gypsum products 14:97
  of lime and limestone 15:206
  of magnesium hydroxide 15:406
  of magnesium oxide 15:409
  of magnesium sulfate 15:412
  of manganese sulfate 15:490
Lotions. See Creams and lotions.
Lotusate 5:91
Lovibond method
  for color of distilled alcoholic beverages and
    intermediates 4:481, 487
  for color of margarine 15:502
Low consistency hydrocal. See Densite.
Low-expansion alloys. (See also Cobalt alloys.)
    10:290
  composition 10:290
Low-temperature crystallization
  of fatty acids 12:490
Low-temperature properties

of brake fluids 7:434
Lozenges. (See also *Confectionery products.*)
   10:548
LSD 15:290–300
   analysis 15:291
   assay 15:298
   p-dimethylaminobenzaldehyde colorimetric
      method 15:291, 298
   fluorometric methods 15:300
   gas chromatography 15:298
   identification 15:291
   infrared spectroscopy 15:295
   liquid chromatography 15:292
   mass spectroscopy 15:297
   properties 15:290
   separation 15:291
   solvent extraction 15:291
   structure 15:290
   thin-layer chromatography 15:294
   ultraviolet spectroscopy 15:299
   uses 15:290
LSD citrate. (See also *LSD.*)
   solubility 15:290
LSD maleate. (See also *LSD.*)
   solubility 15:290
LSD tartrate. (See also *LSD.*)
   mass spectroscopy 15:298
   properties 15:290
   solubility 15:290
LtH. See Luteotropic hormone.
Lubatti method
   for ethylene oxide 12:322
Lubricant additives 15:304
   types 15:303
   uses 15:303
Lubricants 15:301–355
   acid–base titration 15:314, 322
   analysis 7:426; 15:313
   aniline point 15:319
   in antirusts 6:351
   application tests 15:331
   blending 15:308
   in brake fluids 7:424, 426, 431
   brass corrosion test 15:337
   carbon residue 15:323
   chlorine content 15:327
   cloud point 15:331, 332
   commercial grades and specifications 15:308,
      310
   composition analysis 15:313
   copper corrosion test 15:336
   corrosion tests 15:333
   density 15:342
   dielectric constants 15:343
   emulsion characteristics 15:337
   in engine starting fluids 6:356
   in engine tune-up products 6:356

evaporation loss 15:345
extreme pressure potential 15:328
fire point 15:345
flash point 15:345
flock point 15:332, 333
foaming characteristics 15:338
hydrolytic stability 15:339
interfacial tension 15:346
manufacture 15:305
neutralization number 15:313
oxidation characteristics 15:340
physical tests 15:342
potassium content 17:390
pour point 15:331, 332
precipitation number 15:321
properties 15:308
saponification number 15:321
sodium content 17:390
specific gravity 15:342
steam emulsion number 15:337
steel corrosion test 15:336
sulfur content 15:328
types 15:301
uses 15:301, 304, 308
viscosity 15:350
viscosity index 15:351
water content 15:329
Lubricants, synthetic. (See also *Lubricants.*)
   15:302
   classification 15:302
   manufacture 15:308
Lubricating oils
   iron content 15:57
   lead content 15:171, 186
   phosphorus content 17:93
   sodium content 18:220, 226
   zinc content 19:559, 566
Lucas test
   for alcohols 4:522
Ludwigite 7:315
Lumiflavin
   in riboflavin 10:544
Luminal 5:91
Luminol
   chemiluminescent method for hydrogen peroxide
      12:552
Luminous reflectance 7:233
Lumogallion
   colorimetric method for niobium 16:369, 374
   fluorometric method for gallium 13:264
Lundin protein fractions
   of beer 7:612
Luneburgite 7:315
Lunney method
   determination of end groups in acrylic polymers
      4:291
Lupanine 4:590

analysis 4:609
Lupetidines. (See also *Pyridine and pyridine derivatives.*) 17:413
Lupin alkaloids 4:590
analysis 4:609
Lupinine 4:590
analysis 4:609
Luster
of acetate and triacetate fibers 4:76, 87
Lutecium
emission spectroscopy 15:155
in lanthanum oxide 15:155
Lutein
in green tea leaf 18:480
thin-layer chromatography 18:480
Luteinizing hormone. (See also Gonadotropic hormones.) 14:180
uses 14:180
Luteophosphomolybdic acid. See 9-Phosphomolybdic acid.
Luteotropic hormone. (See also Gonadotropic hormones.) 14:180
uses 14:180
Luter water
analysis 4:480
quality control 4:468
Lutetium. (See also Lutetium metal; *Rare earths.*)
atomic absorption spectroscopy 17:494
emission spectroscopy 17:501
flame photometry 17:493
in lanthanum oxide 17:501
in thulium oxide 17:503
x-ray fluorescent spectroscopy 17:490
in ytterbium oxide 17:503
Lutetium fluorides
carrier gas distillation fusion methods 8:636
oxygen in 8:635, 636
vacuum distillation fusion methods 8:635
Lutetium hydroxide. (See also Rare earth hydroxides.)
manufacture 17:510
Lutetium metal. (See also Rare earth metals.)
carrier gas fusion methods 8:627, 682
hot extraction methods 8:684
hydrogen in 8:623, 638, 680, 684
nitrogen in 8:623, 627
oxygen in 8:623, 627, 629, 680, 682
properties 17:471

vacuum fusion methods 8:623, 680
vacuum hot extraction methods 8:638
Lutetium oxide. (See also Rare earth oxides.)
emission spectroscopy 17:503
impurities, determination of 17:503
properties 17:497
Lutidines. (See also Dimethylpyridines; *Pyridine and pyridine derivatives.*)
derivatives for identification 17:426
manufacture 17:420
uses 17:412
Lutidinic acid. (See also *Pyridine and pyridine derivatives.*) 17:413
Lututrin. (See also Steroids.)
occurrence 14:186
Lycopene
in green tea leaf 18:480
thin-layer chromatography 18:480
Lysergic acid 4:588
analysis 4:600
liquid chromatography 15:292
structure 15:290
Lysergic acid diethylamide. See *LSD.*
Lysine. (See also *Carboxylic acids, amino.*) 8:413, 495
analysis 8:498
in bakery products 6:460
in cocoa beans 9:556
colorimetric methods 8:499
commercial specifications 8:498
composition 8:413
crystalline form 8:496
derivatives 8:496
2,4-dinitrofluorobenzene colorimetric method 8:499
enzymatic decarboxylation 8:498
gas chromatography 8:429
infrared spectrum 8:497
ion exchange chromatography 8:415, 421
manufacture 8:497
nuclear magnetic resonance spectrum 8:497
paper chromatography 8:422
properties 8:495
in protein 8:414
uses 8:495
in yeast 19:529
Lysozyme
in egg white 12:58, 61
liquid chromatography 12:60
properties 12:58

M

Macaroni
  colors in 10:527
Mace oil. (See also *Essential oils.*)
  refractive index 12:271
Macerals
  in coal and coke 10:222
Mack's cement. (See also Hydraulic cements.)
  9:95
Macrolide antibiotics 5:463, 517
  properties 5:518
  solubility 5:519, 520
Madder lake R
  in enamels 12:183, 185
  identification 12:183, 185
Magenta. (See also *Triphenylmethane and related dyes.*)
  ceric sulfate titration 19:175
  indigo carmine titration 19:175
  titanous chloride titration 19:177
Magenta P. (See also *Triphenylmethane and related dyes.*)
  visible spectroscopy 19:167
Magnesia. (See also Magnesium oxide.) 4:434
  gravimetric method for phosphorus 11:371; 16:185
Magnesia-ammonium hydroxide
  gravimetric method for phosphorus pentoxide 13:432
Magnesia cement. (See also Hydraulic cements.)
  9:96
Magnesia chrome ore refractories. (See also Refractories.)
  microscopy 9:227
Magnesite 15:357, 400
  composition 9:203, 208; 10:104
  properties 9:208
Magnesium. (See also Magnesium alloys; Magnesium metal.) 15:356-420
  in acid salts 8:594, 595
  in aluminates 5:160
  in aluminum alloys 5:135; 15:371
  in aluminum metal 5:135
  in aluminum oxide 5:187
  in aluminum oxide abrasives 4:41, 51
  in ammonium chloride 5:337
  ammonium hydroxide gravimetric method 12:186
  ammonium oxalate gravimetric method 14:606
  ammonium phosphate gravimetric method 12:153; 15:360
  in amosite 6:267
  in asbestos 6:263, 267

in asbestos–cement products 6:270, 271
atomic absorption spectroscopy 9:346, 667; 10:133; 11:157; 14:126; 15:157, 359, 370
in bauxite 5:116
in beer 7:656
in blood serum 15:371; 18:222
borax bead test 6:584
brilliant yellow colorimetric method 15:368
in calcium acetate 8:79
in calcium carbonate 8:84
in calcium chloride 8:87
in calcium metal 8:76
in calcium oxide 8:94
in carbon 14:74
in cast iron 15:371
in cement 15:371
in cereals 9:233
in chrome ores 9:701, 704
in chromic acid 9:684
in chromium alloys 9:666
in chromium metal 9:635
in chrysotile 6:263, 267
in citric acid 10:96
in clays and clay minerals 10:133, 136
in cobalt alloys 10:314
in cobalt compounds 10:327
in cobalt metal 10:293, 300
in cocoa products 9:595
colorimetric determination 15:359, 367
in copper alloys 10:640
in copper metal 10:620
in cotton 11:140, 157
diammonium hydrogen phosphate gravimetric method 10:640; 12:536; 15:362
in diatomaceous earth 11:490
EDTA titration 7:656; 8:87; 9:343; 10:96; 11:490; 12:153, 154; 15:364; 19:363
electrolytic methods 15:361
in elastomers, synthetic 12:153
emission spectroscopy 4:3, 51, 56; 8:76; 9:346; 11:493; 13:277; 14:48, 74, 122, 540; 15:144, 155, 367, 372, 374; 16:160, 387; 19:570
in enamels 12:186
Eriochrome Black T colorimetric method 15:368
in explosive priming mixtures 12:423
in ferrites 12:536
in ferroniobium 16:387
flame photometry 9:666; 15:280, 359, 372
flame tests 6:589
fluorometry 15:370
in gallium arsenide 13:276

369

in gold 14:45
in graphite 14:74
gravimetric determination 15:360, 361
gravimetric method for tellurium 17:591
in hafnium 14:122
history 15:356
in hydrochloric acid 14:375
hydroxylamine hydrochloride colorimetric
  method 10:300
8-hydroxyquinoline colorimetric method
  15:369
8-hydroxyquinoline gravimetric method 9:704;
  15:360, 362
identification 15:359
in indium 14:540
iodometric method 15:366
ion exchange chromatography 13:255; 15:361;
  17:364; 18:268
in iron 15:59
in lactic acid and its derivatives 15:144
in lanthanum oxide 15:155
in lithium carbonate 15:280
in magnesium acetate 15:398
in magnesium alloys 15:374
magnesium ammonium phosphate volumetric
  methods 15:363
in magnesium carbonate 15:401
in magnesium chloride 15:404
in magnesium hydroxide 15:406
in magnesium nitrate 15:407
in magnesium oxide 12:537; 15:409
microcosmic salt bead test 6:587
in minerals 15:358
in molybdenum 16:161
in molybdenum trioxide 16:161
neutron activation analysis 15:373
in nickel 16:331
in nickel alloys 16:331
in niobium 16:387
p-nitrobenzeneazo-α-naphthol colorimetric
  method 11:140
p-nitrobenzeneazoresorcinol colorimetric method
  15:359
occurrence 15:356
in organic materials 15:358
in plants 15:371; 18:221
polarography 15:374
in potassium chlorate 9:359, 361
in potassium chloride 9:339
in potassium iodide 14:606
quinalizarin colorimetric method 11:140
in rare earth chlorides 17:506
rare earth content 17:473
in rare earth fluorides 17:508
separation 15:360
in silica ablators 4:3, 4
in silicon carbide abrasives 4:56, 59

in sodium chloride 9:337, 338, 343, 346, 349
sodium hydroxide gravimetric method 15:360
in sodium iodide 14:606
in soil 15:371; 18:220
solvent extraction 12:423; 15:360
in strontium chloride 18:283
in strontium nitrate 18:284
in titanium 19:121
titan yellow colorimetric method 15:359, 367
in tobacco 19:152
in urine 15:371; 18:222
uses 15:356
volumetric determination 15:363
in water 18:221; 19:362
in wines 19:491, 493
x-ray fluorescent spectroscopy 10:136; 15:367
in zinc metal 19:570
in zirconium 14:122
Magnesium acetate 15:398
  analysis 15:398
  commercial grades and specifications 15:398
  impurities, determination of 15:398
  magnesium content 15:398
  manufacture 15:398
  properties 15:398
  solubility 15:398
  specification tests 15:398
  uses 15:398
Magnesium alloys. (See also Magnesium metal.)
  5:65
  aluminum in 5:245; 15:375
  analysis 15:374
  beryllium content 15:375, 378
  calcium content 15:375, 380
  calcium in 8:111
  casting 5:70
  cerium content 15:375, 380
  composition 5:69, 70, 71
  hydrogen in 8:617, 633, 681
  impurities, determination of 15:374
  iron content 15:375, 384
  lanthanum content 15:375, 385
  lithium content 15:386
  magnesium content 15:374
  manganese content 15:375, 386
  nickel content 15:375, 387
  nitrogen in 8:617, 633
  oxygen in 8:616, 617, 633, 681
  physical properties 5:70, 71
  rare earth content 15:375, 389
  silver content 15:392
  temper designations 5:69
  thorium content 15:375, 393; 19:12
  tin content 15:375, 394; 19:53
  uses 15:356
  vacuum fusion methods 8:616, 617,
    681

wrought alloys 7:71
  yttrium content 15:394
  zinc content 15:375, 394
  zirconium content 15:375, 396
Magnesium aluminosilicate glass. (See also
  Silicate glasses.)
  composition 9:199
  uses 9:199
Magnesium ammonium phosphate
  gravimetric method for phosphorus 17:60, 62,
  101
Magnesium borates 7:313
Magnesium bromide–sulfuric acid
  volumetric method for ethylene oxide 12:330
Magnesium calcium silicate
  commercial specifications 17:191
Magnesium carbonate 15:399
  acid–base titration 15:401
  analysis 15:401
  assay 15:401
  commercial grades and specifications 15:400
  impurities, determination of 15:401
  magnesium content 15:401
  manufacture 15:400
  occurrence 15:400
  properties 15:400
  specification tests 15:401
  uses 15:399; 17:192
Magnesium chelates
  DTPA content 12:50
  EDTA content 12:50
  ferric chloride titration 12:50
Magnesium chlorate. (See also Chlorates.)
  properties 9:358
  solubility 9:359
Magnesium chloride 15:402
  analysis 15:404
  antifreeze 6:1
  colorimetric method for EGTA 12:28
  colorimetric method for NTA 12:28
  commercial grades and specifications 15:403
  magnesium content 15:404
  manufacture 15:403
  occurrence 15:403
  properties 9:335; 15:403
  solubility 9:336
  specification tests 15:404
  titration of DTPA 12:52
  titration of EDTA 12:49
  uses 15:403
  volumetric determination 15:363
Magnesium chloride–ammonium chloride
  gravimetric method for orthophosphate 17:61
  gravimetric method for phosphorus 17:63, 101
Magnesium chromate
  commercial specifications 9:697
  solubility 9:697

Magnesium citrate complex
  stability 10:89
Magnesium complexes 8:98
Magnesium compounds. (See also individual
  compounds.) 15:397–413
  uses 15:397
Magnesium dichromate
  commercial specifications 9:697
  solubility 9:697
Magnesium DTPA. See DTPA.
Magnesium ferrite. (See also Ferrites.)
  in ceramics 9:225
  Curie temperature 9:212
  magnetic constants 9:212
  magnetic properties 12:515
  microscopy 9:225
Magnesium ferrosilicon. (See also Silicon alloys.)
  composition 18:2
Magnesium fluoride
  properties 13:11
Magnesium–gallium alloys 13:272
  gallium content 13:272
Magnesium hydride. (See also Hydrides.)
  analysis 14:247, 252
  assay 14:247, 252
  composition, determination of 14:247, 252
  hydrogen content 14:247
  magnesium content 14:252
  manufacture 14:245
  properties 14:243, 244
Magnesium hydroxide 15:405
  acid–base titration 15:363
  analysis 15:406
  commercial grades and specifications 15:406
  impurities, determination of 15:406
  magnesium content 15:406
  manufacture 15:405
  properties 15:405
  uses 15:406
Magnesium iodate. (See also Iodates.) 14:607
  solubility 14:607
Magnesium metal. (See also Magnesium alloys.)
  analysis 15:374
  carrier gas distillation fusion methods 8:636
  carrier gas fusion method 8:682
  commercial grades and specifications 15:358
  gravimetric method for thallium 18:523
  hydrogen in 8:617, 681
  impurities, determination of 15:374
  manufacture 15:356
  oxygen in 8:636, 681, 682
  properties 15:356
  silicon content 18:34
  solubility in gold 14:9
  tin content 19:53
  vacuum fusion methods 8:617, 681
Magnesium-1-naphthylamine-8-sulfonate octahydrate

gravimetric method for sodium 18:229
volumetric method for sodium 18:229
Magnesium nitrate 15:406
  analysis 15:407
  commercial grades and specifications 15:407
  fusion of sulfur compounds, organic 18:379
  impurities, determination of 15:407
  magnesium content 15:407
  manufacture 15:406
  pH 15:407
  properties 15:406
  specification tests 15:407
  uses 15:406
Magnesium oxide 15:408
  acid–base titration 15:363
  in alkali barium silicate glass 9:218
  analysis 15:409
  carrier gas fusion methods 8:624, 626
  in cements 9:116, 128
  in ceramics 9:194, 225
  in clays and clay minerals 10:119
  in cobalt(II) orthophosphate 10:343
  in cobalt oxides 10:342
  in cobalt(II) sulfate 10:344
  commercial grades and specifications 15:409
  diammonium hydrogen phosphate gravimetric
    method 9:116; 10:122; 13:421; 14:95
  EDTA titration 9:128; 10:120, 121; 13:445,
    447; 15:219
  emission spectroscopy 9:218
  in glass and glass-ceramics 13:413, 415, 418,
    445, 447
  in glazes 9:196
  in gypsum and gypsum products 14:95
  in hydrated lime 9:191
  in hydraulic cements 9:96, 100
  hydrogen in 8:616
  8-hydroxyquinoline gravimetric method 10:120
  impurities, determination of 15:409
  light microscopy 9:103
  in lime and limestone 9:191; 15:218
  magnesium content 12:537; 15:409
  manufacture 15:408
  microscopy 9:225
  nitrogen in 8:616
  oxygen in 8:616, 624, 626
  in portland cement 9:103
  properties 9:204; 15:408
  in quicklime 9:191
  in refractories 9:206
  in silicate glasses 9:198
  sodium ammonium hydrogen phosphate
    gravimetric method 15:218
  specification tests 15:409
  uses 15:409
  vacuum fusion methods 8:616
Magnesium oxide reflectance standards 7:260, 262

Magnesium oxychloride cement. See Sorel cement.
Magnesium perchlorate. (See also Perchlorates.)
  commercial grades and specifications 9:364
  gravimetric method for water 14:397
  solubility 9:364
  uses 9:362
  water absorption 9:326
Magnesium pyrophosphate
  gravimetric method for phosphorus 10:424
Magnesium salts
  in calcium lactate 15:144
Magnesium silicate. (See also Chrysotile; Silicates.)
  analysis 17:191
  in chromite 9:700
  commercial specifications 12:174; 17:191
  in enamels 12:174
  uses 18:44
  in zinc sulfide pigments 17:190
Magnesium silicate pigment
  silicon content 18:29
Magnesium stearate
  in enamels 12:176
Magnesium sulfate 15:411
  analysis 15:412
  assay 15:412
  commercial grades and specifications 15:411
  impurities, determination of 15:412
  manufacture 15:411
  pH 15:412
  specification tests 15:412
  uses 15:412
Magnesium sulfate–sodium thiosulfate
  volumetric method for styrene oxide 12:229
Magnesium uranyl acetate
  colorimetric method for sodium 18:227
  gravimetric method for sodium 18:218, 223, 227
Magnetic ceramics. (See also *Ceramics.*)
  properties 9:209
  uses 9:209
Magnetic constants
  of cobalt ferrite 9:212
  of copper ferrite 9:212
  of iron ferrite 9:212
  of magnesium ferrite 9:212
  of manganese ferrite 9:212
  of nickel ferrite 9:212
Magnetic content
  of abrasive grain 7:311
Magnetite 11:464; 12:514; 19:81
  in chrysotile 6:266
  composition 10:104; 15:22
  uses 17:203
Magnolia 5:69
Majoram, oil of. (See also *Essential oils.*)
  composition 12:268
  gas chromatography 12:268
  mass spectroscopy 12:268

Malachite 10:657
  copper content 10:600
Malachite green. (See also *Pigments*; *Triphenyl-
    methane and related dyes.*)
  colorimetric method for end groups in acrylic
    fibers 4:289
  colorimetric method for sulfur dioxide 12:551
  colorimetric method for tantalum 16:370
  colorimetric method for tungsten 19:202
  color test for sulfite 18:402
  derivatives for identification 19:168
  in enamels 12:187
  identification 12:187
  orange II titration 19:174
  paper chromatography 19:172
  structure 17:182
  thin-layer chromatography 19:173
  titanous chloride titration 19:177
  visible spectroscopy 19:166, 167, 168
Malacon 14:104
Malanocerite 7:316
Malathion. (See also *Pesticides and insecticides.*)
  colorimetric determination 19:611
  commercial grades and specifications 19:595
  composition 19:589, 598
  manufacture 19:598
  properties 19:589
  solubility 19:587
  structure 19:583
  toxicity 19:592
Maleic acid 8:399
  in alkyd resins 5:37
  derivatives for identification 8:376
  gas chromatography 5:25
  paper chromatography 15:133
  in phthalic anhydride 17:156
  properties 8:397, 399
  thin-layer chromatography 8:381
  uses 8:399
Maleic anhydride 5:22
  absorption method for 1,3-butadiene 11:505
  gas chromatography 5:28; 17:154
  infrared spectroscopy 8:562
  maleic acid in 8:567
  manufacture 8:561
  in phthalic anhydride 17:154, 156
  potassium permanganate titration 17:156
  preparation 8:373
  properties 8:559
Maleic hydrazine
  p-dimethylaminobenzaldehyde colorimetric
    reagent 19:155
  distillation 19:155
  in tobacco 19:155
Malic acid (Hydroxysuccinic acid) 8:399
  acid–base titration 19:149
  in confectionery products 10:552

enzymatic methods 19:421
gas chromatography 19:149
in grapes 19:420
in green coffee 10:374
ion exchange methods 19:421
liquid chromatography 15:133; 19:149
in musts 19:412, 420
2,7-naphthalene disulfonic acid colorimetric
    method 19:421
oxidation methods 19:421
paper chromatography 15:133; 19:418
properties 8:397, 399
in roasted coffee 10:375
solvent extraction 19:149
thin-layer chromatography 8:373, 397; 15:133;
    19:419
in tobacco 19:148
in wines 19:412, 418, 420
DL-Malic acid
  in green tea leaf 18:459
  potassium permanganate titration 18:459
  potassium titanium oxalate colorimetric method
    18:459
Malleable iron. (See also Iron metal.)
  manufacture 15:27
  properties 15:27
  uses 15:27
Malonamide
  commercial specifications 15:425
  manufacture 15:424
  properties 15:421
  solubility 15:422
Malonates, substituted. (See also individual com-
    pounds.)
  analysis 15:435
  assay 15:435
  gas chromatography 15:435
  saponification methods 15:435
Malonic acid
  acetic anhydride colorimetric method 15:424
  acid–base titration 15:429
  analysis 15:424
  anthrazinedione color test 15:424
  colorimetric determination 15:424
  commercial specifications 15:425
  derivatives for identification 15:427
  in diethyl malonate 15:434
  dimethylglyoxime–nickel biuret color test 15:424
  in dimethyl malonate 15:434
  gas chromatography 15:132
  identification 15:424
  impurities, determination of 15:431
  infrared spectroscopy 15:426
  iodometric methods 15:430
  liquid chromatography 15:429
  manufacture 15:424
  near-infrared spectroscopy 15:426

paper chromatography 15:427
preparation 8:373, 397
properties 15:421, 422
pyridyl pyridinium dichloride color test 15:426
separation 15:429
solubility 15:422
thin-layer chromatography 8:381; 15:428
volumetric determination 15:429
Malonic acid and derivatives. (See also individual
   compounds.) 15:420–447
commercial grades and specifications 15:424
manufacture 15:423
properties 15:422
uses 15:420
Malonic acid esters. (See also individual compounds.)
gas chromatography 15:433
use in preparation of barbiturates 6:470
Malononitrile. (See also *Nitriles.*)
analysis 15:443
assay 15:444
colorimetric determination 15:443
commercial specifications 15:425
gas chromatography 15:443
identification 15:443
impurities, determination of 15:444
infrared spectroscopy 15:443
manufacture 15:424
nuclear magnetic resonance spectroscopy 15:443
properties 15:421, 423; 16:396
solubility 15:422
tetrabutylammonium hydroxide titration 15:
   444
thin-layer chromatography 15:444
toxicity 15:423
ultraviolet spectroscopy 15:443
visible spectroscopy 15:443
Malt
analysis 4:469
bioassay 7:159
quality control 4:467
Malt beverage 7:476
Maltol assay
for streptomycin 5:596
Maltose
in beer 7:555, 563, 566
in green tea leaf 18:478
Munson-Walker method 7:563
paper chromatography 7:569; 18:478
sodium thiosulfate titration 12:607, 609
Maltose hydrate. (See also *Sugar.*)
refractive index 18:343
Malt whisky 4:464
analysis 4:480
Malvidin
structure 19:461
in wines 19:461

Malvidin diglucoside
fluorometric method 19:465
thin-layer chromatography 19:465
in wines 19:464
Manandonite 7:316
Mandarin oil. (See also *Essential oils.*)
decyl aldehyde content 12:282
residue on evaporation 12:273
ultraviolet spectroscopy 12:288
Mandelic acid 8:405
analysis 8:405
gas chromatography 8:388
gravimetric method for hafnium 14:115, 140
gravimetric method for zirconium 7:362; 14:109,
   115, 140
gravimetric method for zirconium dioxide 13:
   427
paper chromatography 8:379
properties 8:402
dl-Mandelic acid
gravimetric method for zirconium oxide 9:262
Mandelonitrile
preparation from benzaldehyde 6:595
Maneb 13:203
manufacture 13:203
properties 13:203
solubility 13:203
toxicity 13:203
uses 13:203
Manganese. (See also Manganese alloys; Manganese
   metal.) 15:447–496
in acrylic fibers 4:349
in alumina 5:116
in aluminum 15:456, 476
in aluminum alloys 5:136; 15:462, 477
in aluminum metal 5:136
in aluminum oxide 5:187
ammonium chloride–diammonium phosphate
   gravimetric method 15:457, 460
in ammonium persulfate 5:347
ammonium persulfate colorimetric method
   9:49; 15:473
in amosite 6:267
in antimony 15:456
in arsenic 15:456
in asbestos 6:263, 267
atomic absorption spectroscopy 9:346, 671;
   11:157; 15:386, 478
in barium 15:456
benzidine colorimetric method 11:141; 15:457
in beryllium 15:456, 476
in beryllium metal 7:115
in biological materials 15:472, 476, 477
in bismuth 15:456
borax bead test 6:584
in boron 15:456
in brass 15:462, 466

bromine–ammonium hydroxide gravimetric
   method 16:297
in bronze 15:462, 466
in cadmium 15:456
in calcium 15:456
in calcium metal 8:76
in carbon 14:74; 15:456
in cast iron 15:467, 476
in caustic soda 15:476
in cellulose 9:48
in ceramics 15:476
in chrome steels 9:662
in chromic acid 9:684
in chromium 15:456
in chromium alloys 9:670
in chromium metal 9:635
in chromium ores 9:701, 704
in chrysotile 6:263, 267
in cobalt alloys 10:314
in cobalt compounds 10:328
in cobaltic oxide 15:476
in cobalt metal 10:293, 300
in cobalt(II) orthophosphate 10:343
in cobalt oxides 10:342
in cobalt(II) sulfate 10:343, 344
in cocoa products 9:595, 598
colorimetric determination 15:457, 458, 469
colorimetric method based on oxidation to
   permanganate 7:115
in copper 15:456, 476
in copper alloys 10:640; 15:466
in copper metal 10:623
in cotton 11:141, 153, 154, 157
in cupro–nickel alloys 15:476
in cutting alloys 10:289
in dental alloys 10:290
EDTA titration 15:493
in elastomers, synthetic 12:157
emission spectroscopy 4:4; 8:76; 9:346; 11:153;
   13:277; 14:48, 74, 122, 542; 15:155, 187, 375,
   456, 477; 16:160, 387
ferricyanide titration 16:344
in ferrites 12:531
in ferrochromium 9:661
in ferromanganese 15:76, 460
in ferroniobium 16:387
ferrous ammonium sulfate volumetric method
   15:76
flame photometry 15:478
flame tests 6:592
formaldoxime colorimetric method 15:457, 475
in gallium arsenide 13:276
in gallium metal 13:266
in gallium phosphide 13:282
in gasoline additive AK 33X 15:493
in glass and glass-ceramics 9:200; 13:445
in gold 14:45; 15:456

in graphite 14:74
gravimetric determination 15:454, 457, 459
gravimetric separation 6:267
in hafnium 14:122
in hard facing alloys 10:289
in high-temperature alloys 10:286
history 15:447
in hydrochloric acid 14:375
8-hydroxyquinoline colorimetric method 15:457
identification 15:456
in indium 14:542
ion exchange chromatography 13:255; 17:526
in iron 15:59, 71, 456, 476, 478
in iron ores 15:465
in lanthanum oxide 15:155
in lead 15:456
in lithium 15:456
in magnesium 15:375, 386, 456, 476
in magnesium acetate 15:399
in magnesium alloys 15:375, 386
in magnesium chloride 15:405
in magnesium nitrate 15:408
in magnesium oxide 15:410
in magnesium sulfate 15:412
in manganese alloys 15:454, 459, 463
in manganese brown 17:194
in manganese metal 15:454
in manganese ores 15:454, 459, 464, 470, 476
in manganese(II) oxalate dihydrate 15:457
in mercury 15:456
microcosmic salt bead test 6:587
in minerals 15:476
in molybdenum 15:456; 16:161
in molybdenum trioxide 16:161
in natural rubber 15:472
neutron activation analysis 4:349; 13:265
in nickel 15:476, 478; 16:297, 331
in nickel alloys 15:466, 476; 16:331
in nickel–chromium–iron alloys 15:461
in nickel sulfate 16:344
in niobium 16:387
nitric acid–potassium chlorate gravimetric
   method 15:455
nitrogen content 15:484
occurrence 15:447, 448
in organic materials 15:454, 476
oxidation potentials 15:450
perchloric acid gravimetric method 15:77
periodate–tetrabase colorimetric method 11:141
permanganate colorimetric method 19:371
phosphoric acid colorimetric method 14:611
in plant materials 15:475, 476, 478
in plants 18:221
in platinum 15:456
in plutonium 15:476
in potassium 15:456
potassium bromate–ferrous ammonium

sulfate volumetric method 15:467
potassium chlorate–nitric acid gravimetric
    method 16:297
in potassium metaperiodate 14:611
potassium periodate colorimetric method 9:671;
    10:300, 623; 12:157; 15:59, 71, 386, 399, 405,
    408, 412, 469; 16:331
potassium permanganate–sodium pyrophosphate
    volumetric method 12:531
in rocks 15:187, 476
sample decomposition 18:27
separation 15:454
in siennas 17:194
in silica 15:476, 478
in silica ablators 4:4
in silicomanganese 15:460
in silicon 15:456
silicon content 18:27
in silver 15:456; 18:173
in slag 15:468, 473, 476
in soda ash 15:476
in sodium 15:456
sodium arsenite volumetric method 15:465
sodium bismuthate colorimetric method 7:207;
    10:151; 15:218, 457
sodium bismuthate–ferrous ammonium sulfate
    volumetric method 17:194
sodium bismuthate volumetric method 10:314,
    641; 15:460
sodium chlorate–nitric acid gravimetric method
    16:297
in sodium chloride 9:346
in sodium dichromate 9:684
in sodium heparin 10:151
in sodium metaperiodate 14:611
sodium nitrite volumetric method 15:467
sodium pyrophosphate volumetric method 15:
    458, 462
solvent extraction 15:157, 455
in Spiegeleisen 15:460
in spring steels 10:290
in steel 15:476, 478
in strontium 15:456
in surgical alloys 10:290
in tantalum 15:456
in textiles 15:472
in tin 15:456
in titanium 15:456, 476
in titantium alloys 15:466
in titanium metal 19:121
in tobacco 19:152
toxicity 15:447
in tungsten 15:456
in umbers 17:194
in uranium 15:476
in uranium oxide 15:476
uses 15:447

volumetric determination 15:457, 460
in water 15:471, 474; 19:371
in wear-resistant alloys 10:289
x-ray fluorescence spectroscopy 11:154, 157;
    15:456, 477
in zinc 15:456, 476
in zirconium 14:122; 15:456, 470
Manganese alloys. (See also Manganese metal.)
    5:61, 64
analysis 15:454
carrier gas fusion methods 8:624, 682
hydrogen in 8:680
manganese content 15:454, 459, 463
manufacture 15:453
oxygen in 8:624, 631, 635, 636, 680, 682
sampling 15:454
vacuum distillation fusion methods 8:635
vacuum fusion methods 8:680
Manganese–boron alloy
composition 7:322
uses 7:323
Manganese bronze. (See also Manganese alloys.)
    5:68
aluminum in 5:248
composition 15:452
Manganese brown
commercial specifications 17:194
manganese content 17:194
Manganese–calcium–silicon alloys 5:61
composition 5:64
Manganese carbonate. (See also Manganese com-
    pounds.) 15:485
commercial grades and specifications 12:519
composition 12:519
manufacture 15:485
uses 15:485
Manganese chelates
DTPA content 12:50
EDTA content 12:50
ferric chloride titration 12:50
Manganese chloride. (See also Manganese com-
    pounds.) 15:485
manufacture 15:485
properties 15:485
uses 15:485
Manganese compounds. (See also individual
    compounds.) 15:484–493
analysis 15:490
uses 15:485
Manganese–copper ferrite. (See also *Ferrites.*)
commercial specifications 12:520
composition 12:520
Manganese dioxide 15:487
in alkali barium silicate glass 9:218
analysis 15:492, 531
assay 15:531
commercial grades and specifications 15:487, 515

coprecipitation of antimony 6:63
emission spectroscopy 9:218
in enamels 9:195
gravimetric method for antimony 10:294
manufacture 15:487
in matches 15:515
in match strikers 15:522
oxygen content 15:492
particle size 15:531
properties 15:450
sensitizing ability 15:531
uses 15:487
Manganese dioxide–manganese(III) oxide fusion
  of nitrogen 16:454
Manganese ethylenebisdithiocarbamate. (See also
  Dithiocarbamic acid salts; Maneb; and
  Manganese compounds.) 15:485
manufacture 15:485
uses 11:577; 15:485
xanthate method 11:580
Manganese ethylenediaminedithiocarbamate. (See
  also Dithiocarbamic acid salts.)
properties 11:575
Manganese ferrite. (See also Ferrites.)
Curie temperature 9:212
magnetic constants 9:212
magnetic properties 12:515
Manganese fluoride. (See also Manganese compounds.) 15:486
manufacture 15:486
properties 13:11
uses 15:486
Manganese hydroxide. (See also Manganese compounds.)
properties 15:450
Manganese–iron alloys 5:61
composition 5:64
Manganese–magnesium ferrite. (See also Ferrites.)
commercial specifications 12:520
composition 12:520
Manganese metal. (See also Manganese alloys.)
  15:447–484
analysis 15:454
carrier gas fusion methods 8:625, 682
commercial grades and specifications 15:453
hydrogen in 8:618, 681
impurities, determination of 15:478
manganese content 15:454
manufacture 15:452
nitrogen in 8:618
oxygen in 8:617, 618, 625, 630, 681, 682
properties 15:449
sampling 15:454
solubility in gold 14:9
thallium content 18:524
uses 15:451
vacuum fusion methods 8:617, 618, 681

Manganese metal, alloys, and compounds 15:447-496
Manganese naphthenate
in enamels 12:176
Manganese ores
classification 15:448
composition 15:448
manganese content 15:454, 459, 464, 470
potassium content 17:379
sampling 15:454
Manganese oxalate dihydrate. (See also Manganese
  compounds.)
manganese content 15:457
Manganese oxide
commercial grades and specifications 12:519
composition 12:519
phosphoric acid–potassium periodate colorimetric
  method 13:442
Manganese(II)(III) oxide. (See Mangano–manganic
  oxide.)
Manganese(III) oxide. (See also Manganese compounds.)
in glass and glass-ceramics 13:415, 442, 461
manufacture 15:451
properties 15:451
Manganese oxide pigments
analysis 17:194
commercial grades and specifications 17:194
impurities, determination of 17:194
types 17:192
Manganese sesquioxide. (See also Manganese
  compounds.) 15:486
manufacture 15:486
uses 15:487
Manganese–silicon. (See also Silicon alloys.)
phosphorus content 17:86
sample decomposition 18:24
silicon content 18:24
Manganese soaps. (See also Manganese compounds.)
  15:486
uses 15:486
Manganese steel. (See also Manganese alloys.)
carrier gas fusion methods 8:625
composition 15:452
oxygen in 8:630, 652
Manganese sulfate 15:486
analysis 15:490
commercial grades and specifications 15:486
impurities, determination of 15:490
manufacture 15:486
properties 15:450
specification tests 15:490
titration of gold 14:23
uses 15:486
Manganese triacetate. (See also Manganese
  compounds.)
uses 15:487
Manganese trioxide

in glass 9:200
Manganese–zinc ferrite. (See also *Ferrites.*)
   commercial grades and specifications 12:520
   composition 12:517, 520
   properties 12:517
Manganic oxide
   in cements 9:119
   potassium permanganate gravimetric method
      9:119
Manganin. (See also Manganese alloys.) 5:68
   composition 15:452
Manganite 15:448
Mangano–manganic oxide. (See also Manganese
      compounds.) 15:486
   manufacture 15:486
   uses 15:486
Manganosite. See Manganous oxide.
Manganotantalite
   composition 18:441
Manganous ethylene bisdithiocarbamate. See
      Maneb.
Manganous oxide. (See also Manganese com-
      pounds.) 15:485
   commercial grades 15:485
   in lime and limestone 15:212
   manufacture 15:485
   potassium periodate colorimetric method 15:
      212
   sodium bismuthate–ferrous ammonium sulfate
      volumetric method 15:212
   uses 15:485
Manganous phosphate. (See also Manganese
      compounds.) 15:485
   manufacture 15:485
   uses 15:485
Manganous sulfate–diphenylcarbazide
   colorimetric method for chlorate 11:141
Manganous sulfate–sulfuric acid
   colorimetric method for bromate 11:141
Manganous sulfide. (See also Manganese com-
      pounds.)
   solubility 15:450
Mannan
   in cellulose 9:46
Mannitol. (See also Polyhydric alcohols.) 5:31
   catechol colorimetric method 13:572
   gas chromatography 13:594; 19:451
   ion exchange methods 19:450
   liquid chromatography 13:503
   polarography 19:449
   separation 19:444
   structure 19:449
   thin-layer chromatography 19:450
   titration of boric acid in presence of 7:361, 402,
      404, 405
   use in polarimetric determination of boron 7:
      412

volumetric method for borax 18:192
volumetric method for boron 14:144; 16:325
in wines 19:425, 444, 449
Mannitol–germanic acid
   volumetric method for germanium 13:399
Mannitol hexanitrate. (See also High explosives.)
      12:451
   infrared spectrum 12:456
   manufacture 12:451
   melting point 12:461
   properties 12:432
   solubility 12:451
   uses 12:451
Mannose
   gas chromatography 19:300
   in green tea leaf 18:459
   potassium permanganate titration 18:459
   potassium titanium oxalate colorimetric method
      18:459
Manzate D. See Maneb.
Maple syrup
   caramel content 10:544
   formaldehyde in 11:567
Maraging-type steels. (See also Cobalt alloys.)
      10:290
   composition 10:290
Marcasite 15:166
   composition 15:22
Margarite 8:75
   composition 10:103
Margarine. (See also *Fats and fatty oils.*) 15:496–
      504
   analysis 15:497
   color 15:502
   commercial grades and specifications 15:496
   composition 15:496
   composition, determination of 15:501
   fat content 15:502
   free fatty acid content 15:496
   iodine number 15:499
   iodometric method 15:499
   manufacture 15:497
   melting range 15:502
   moisture content 15:502
   oxidative rancidity, resistance to 15:501
   peroxide value 12:482; 15:500
   potassium iodide volumetric method 15:500
   salt content 15:502
   sampling 15:497
   solvent extraction 12:482
   2-thiobarbituric acid colorimetric method 15:501
   thiocyanogen volumetric method 15:499
   unsaturation, degree of 15:498
Margaronitrile. See *n*-Heptadecanonitrile.
Margaryl alcohol. (See also Fatty alcohols.)
   properties 4:561
Markol test

for packages 11:115
Marl 15:203
Marmatite
  composition 19:544, 545
Maroon organic pigments
  in enamels 12:186
  identification 12:186
Maroon pigments. (See also *Pigments*; and
    individual compounds.) 17:192
Marshall test
  for bituminous paving mixtures 6:338
Marshite 14:585
Marshmallows. (See also *Confectionery products.*)
    10:548
  manufacture 10:554
Martin's cement. (See also Hydraulic cements.)
    9:95
Masonry cements. (See also Hydraulic cements.)
    9:96
  commercial specifications 9:99, 100, 101
  composition 9:96
  identification 9:108
Massachusetts Institute of Technology folding
    endurance tester 7:289
Massicot. See Lead monoxide.
Mass pyrograms. See Pyrolysis–mass spectroscopy.
Mass spectra. (See also mass spectroscopy.)
  of *sec*-amylchlorobenzene 9:405
  of benzyl alcohol 7:96
  of benzyl chloride 9:400
  of bis-*p*-chlorophenylmethane 9:407
  of *n*-butane 16:418
  of *t*-butylchlorobenzene 9:405
  of carbon tetrachloride 9:403
  of chloroacetylene 9:397
  of 3-chloroallylbenzene 9:403
  of chlorobenzene 9:403
  of 2-chlorobiphenyl 9:407
  of 4-chlorobiphenyl 9:407
  of 2-chloro-1,3-butadiene 9:396
  of 1-chlorobutane 9:397
  of 2-chlorobutane 9:396
  of 2-chloro-2-butene 9:397
  of *β*-chlorocumene 9:402
  of chlorocyclohexane 9:398
  of chlorocyclopentane 9:398
  of 1-chlorodecane 9:401
  of *p*-chloro-*β*,*β*-dichlorocumene 9:404
  of 1-chlorododecane 9:401
  of chloroethane 9:398
  of 1-chloroethylbenzene 9:402
  of 2-chloroethylbenzene 9:401
  of *α*-chloroethylbenzene 9:402
  of *β*-chloroethylbenzene 9:400
  of *p*-chloroethylbenzene 9:404
  of *α*-chloroethyldimethylbenzene 9:404
  of 1-chloro-2-ethylhexane 9:397

  of chloroform 9:399
  of 1-chloroheptane 9:400
  of 2-chloroheptane 9:397
  of 1-chlorohexadecane 9:397
  of 1-chlorohexane 9:400
  of 1-chloroindane 9:403
  of *β*-chloroisopropylbenzene 9:402
  of chloromethylbenzene 9:400
  of 1-chloro-3-methylbutane 9:396
  of 2-chloro-2-methylbutane 9:399
  of *α*-chloromethylethylbenzene 9:403
  of chloromethyl-2-methylbenzene 9:402
  of chloromethyl-3-methylbenzene 9:402
  of chloromethyl-4-methylbenzene 9:402
  of chloromethylnaphthalene 9:407
  of 1-chloromethylnaphthalene 9:405
  of 3-chloro-3-methylpentane 9:396
  of 1-chloro-2-methylpropane 9:396
  of 2-chloro-2-methylpropane 9:397
  of 3-chloro-2-methylpropene 9:396
  of *α*-chloromethyl-2,4,6-triisopropylbenzene
      9:408
  of 1-chloronaphthalene 9:406
  of 2-chloronaphthalene 9:406
  of 1-chlorononane 9:401
  of 1-chlorooctadecane 9:397
  of 1-chlorooctane 9:401
  of 3-chlorooctane 9:397
  of 1-chloropentane 9:396, 399
  of 2-chloropentane 9:399
  of 3-chloropentane 9:399
  of *p*-chlorophenylacetylene 9:404
  of chloroprene dimer 9:400
  of 1-chloropropane 9:396
  of 2-chloropropane 9:396
  of 1-chloro-1-propene 9:396
  of 3-chloropropene 9:396
  of 3-chloropropenylbenzene 9:403
  of *β*-chloro-*n*-propylbenzene 9:401
  of 2-chloro-*n*-propylbenzene 9:401
  of 3-chloropropylbenzene 9:401
  of 3-chloropropyne 9:396
  of *p*-chlorostyrene 9:405
  of *β*-chlorostyrene 9:401
  of 1-chlorotetradecane 9:397
  of 2-chloro-*m*-xylene 9:402
  of 4-chloro-*m*-xylene 9:402
  of *α*-chloro-*m*-xylene 9:402
  of *α*-chloro-*o*-xylene 9:402
  of 2-chloro-*p*-xylene 9:402
  of *α*-chloro-*p*-xylene 9:402
  of cyclobutanol 11:190
  of cycloheptanol 11:190
  of cyclohexanol 11:190
  of cyclohexanone 11:202
  of 2-(1-cyclohexenyl)cyclohexanone 11:202
  of cyclooctanol 11:190

of cyclopentanol 11:190
of di-*t*-butylchlorobenzene 9:407
of dichloroacetylene 9:401
of *m*-dichlorobenzene 9:405
of *o*-dichlorobenzene 9:405
of *p*-dichlorobenzene 9:405
of 4,4′-dichlorobiphenyl 9:408
of 1,1-dichlorobutane 9:397
of 1,2-dichlorobutane 9:396
of 1,4-dichlorobutane 9:397
of 2,3-dichlorobutane 9:398
of 3,4-dichloro-1-butene 9:399
of 1,4-dichloro-2-butene 9:399
of 1,4-dichlorobutyne 9:400
of 1,2-dichloro-4,1-chloroethylbenzene 9:406
of 1,4-dichloro-2,1-chloroethylbenzene 9:406
of 1,4-dichloro-2,2-chloroethylbenzene 9:406
of 2,5-dichlorocumene 9:407
of 3,4-dichlorocumene 9:406
of 1,1-dichloro-2,2-dimethylpropane 9:401
of 1,1-dichloroethane 9:398
of 1,2-dichloroethane 9:398
of 1,2-dichloroethylbenzene 9:404
of 1,2-dichloro-3-ethylbenzene 9:406
of 1,2-dichloro-4-ethylbenzene 9:406
of 1,3-dichloro-2-ethylbenzene 9:406
of 1,4-dichloro-2-ethylbenzene 9:406
of 1,1-dichloroethylene 9:397
of 1,2-dichloroethylene 9:398
of 1,2-dichloroethyltoluene 9:405
of 1,2-dichloro-2-methylpropane 9:396
of 1,5-dichloropentane 9:397
of 1,1-dichloropropane 9:398
of 1,2-dichloropropane 9:398
of 1,3-dichloropropane 9:399
of 2,3-dichloropropene 9:399
of 2,5-dichlorostyrene 9:406
of 3,4-dichloro-α-styrene 9:407
of 2,4-dichlorotoluene 9:404
of 2,6-dichlorotoluene 9:404
of 3,4-dichlorotoluene 9:404
of *o*,α-dichlorotoluene 9:404
of α,α-dichlorotoluene 9:404
of 1,4-dichloro-2,3,5-triethylbenzene 9:408
of α,α-dichloro-*m*-xylene 9:405
of α,α-dichloro-*p*-xylene 9:405
of 2,4-diethylchlorobenzene 9:405
of diethyl malonate 15:432
of diisopropylchlorobenzene 9:407
of 2,3-dimethylchlorobenzene 9:402
of 2,4-dimethylchlorobenzene 9:402
of 2,5-dimethylchlorobenzene 9:402
of 2,6-dimethylchlorobenzene 9:402
of dimethylchlorosilane 18:59
of dimethyldichlorosilane 18:59
of dimethyl malonate 15:432
of 2-ethylchlorobenzene 9:404

of 3-ethylchlorobenzene 9:404
of 4-ethylchlorobenzene 9:404
of ethyl cyanoacetate 15:441
of 1-ethyl-4-dichloromethylbenzene 9:405
of ethylene oxide 12:320
of hexachlorobenzene 9:408
of hexachloroethane 9:403
of 2-isopropylchlorobenzene 9:405
of 4-isopropylchlorobenzene 9:405
of isopropylmethylchlorobenzene 9:405
of methyl chloride 9:396
of 2-methylchlorobenzene 9:400
of 3-methylchlorobenzene 9:400
of 4-methylchlorobenzene 9:400
of methyl cyanoacetate 15:441
of methyldichlorosilane 18:59
of methylene chloride 9:396
of methyltrichlorosilane 18:59
of morpholine 16:195
of 1-nitrobutane 16:418
of 2-nitrobutane 16:418
of nitroethane 16:418
of nitromethane 16:418
of 1-nitro-2-methylpropane 16:418
of 2-nitro-2-methylpropane 16:418
of 1-nitropropane 16:418
of 2-nitropropane 16:418
of pentachlorobutadiene 9:407
of 1,1,2,3,3-pentachloropropane 9:401
of pentachloropropene 9:407
of phenyl-1,2-dichloroethane 9:404
of β-phenylethyl alcohol 7:97
of picolines 17:428
of 2-*n*-propylchlorobenzene 9:404
of 4-*n*-propylchlorobenzene 9:404
of pyridine 17:428
of styrene 18:296
of 1,2,3,4-tetrachlorobenzene 9:408
of 1,2,3,5-tetrachlorobenzene 9:408
of 1,2,4,5-tetrachlorobenzene 9:399
of 1,2,2,4-tetrachlorobutane 9:403
of 1,2,3,4-tetrachlorocyclo-1,3-pentadiene
   9:406
of 1,1,1,2-tetrachloroethane 9:404
of 1,1,2,2-tetrachloroethane 9:399
of tetrachloroethylene 9:406
of 1,1,3,3-tetrachloro-2-methylpropane 9:403
of 1,1,1,5-tetrachloropentane 9:401
of 1,1,1,3-tetrachloropropane 9:403
of 1,1,2,3-tetrachloropropane 9:403
of 1,2,3,3-tetrachloropropene 9:405
of *o*,α,α,α-tetrachlorotoluene 9:407
of tetramethylsilane 18:59
of tri-*sec*-butylchlorobenzene 9:407
of 1,2,4-trichlorobenzene 9:407
of 1,3,5-trichlorobenzene 9:407
of trichlorobutane 9:398

of 2,4,6-trichlorocumene 9:407
of 1,1,1-trichloroethane 9:401
of 1,1,2-trichloroethane 9:400
of trichloroethylene 9:401
of 1,1,3-trichloro-2-methylpropene 9:403
of 3,3,3-trichloro-2-methylpropene 9:403
of 1,1,5-trichloro-1-pentene 9:398
of 1,1,2-trichloropropane 9:398
of 1,2,3-trichloropropane 9:399
of 1,1,2-trichloropropene 9:403
of 3,3,3-trichloropropene 9:402
of β,2,4-trichlorostyrene 9:407
of β,2,5-trichlorostyrene 9:407
of trichlorotoluene 9:406
of α,2,4-trichlorotoluene 9:406
of α,3,4-trichlorotoluene 9:406
of α,α,α-trichloro-m-xylene 9:406
of α,α,α-trichloro-p-xylene 9:406
of triethylchlorobenzene 9:407
of triisopropylchlorobenzene 9:408
of 2,4,6-trimethylchlorobenzene 9:403
of trimethylchlorosilane 18:59
of vinyl chloride 9:398
of vinylchloropropene 9:399
Mass spectrometer, Nier-type 7:338, 339
Mass spectroscopy. (See also mass spectra.)
of acetals 4:66, 72
of acetonitrile 14:422
of alcohols 4:526
of argon 13:292, 301, 304
of benzene 7:31
of benzyl alcohol 7:95
of boron compounds, organic 7:355
of boron hydrides 7:346
of boron trifluoride 7:338
of carbon dioxide 8:255, 266; 13:302, 306
of carbon disulfide 8:272
of carbon monoxide 8:315, 650; 13:302
chlorine isotope peaks 9:395
of chlorocarbons 9:395, 408
of chlorosilanes 18:50
of cigarette smoke 10:30
of coal and coke 10:213
of cocoa bean headspace 9:567, 570
of codeine 16:270
of cyclic alcohols 11:190
of essential oils 12:267
of formaldehyde 13:108
of gallium 13:265
of glass and glass-ceramics 13:456
of graphite 4:10
of helium 13:301
of hydrocarbons 7:31
of hydrogen 13:302
of impurities in carbon dioxide 8:255
of impurities in carbon monoxide 8:315
of krypton 13:301, 305

of LSD 15:297
of LSD tartrate 15:298
of majoram, oil of 12:268
of methacrylate polymers 16:79
of methane 13:302
of methanol 16:119
of methylchlorosilanes 18:59
of natural gas 13:315
of neon 13:292, 301, 304
of nitriles 16:398
of nitrogen 8:650; 13:302
of nitromethane 16:417
of nitroparaffins 16:417, 423
in nitrous oxide 13:302
of noble gases 13:292, 301, 304
of organo-chlorosilanes 18:59
of oxygen 8:650; 13:302
of phenols 17:22
of β-phenylethyl alcohol 7:95
of pyridine and its derivatives 17:427
of thiuram disulfides 11:597
of thorium 19:17
of triphenylmethane and related dyes 19:168
of xenon 13:301, 305
Masstone
of carbon black 8:181, 211
Mastitis tests
for raw milk 16:124
Match book covers. (See also *Matches*.)
analysis 15:537
commercial grades and specifications 15:522
composition 15:522
curl test 15:537
Matches 15:504–551
analysis 15:523
appearance 15:543
ash forming content 15:512
binders 15:506
burning rate 15:547
coloring agent content 15:516
combustible content 15:510
comeback test 15:545
commercial specifications 15:542
composition 15:505
flame transfer agents 15:517
functioning properties 15:543, 544
glow inhibitors in 15:519
heads 15:538, 540
history 15:504
humidity resistance 15:548
ignition temperature 15:549
neutralizers in 15:513
oxidizers 15:509
particle size 15:523
properties 15:540
raw materials 15:506
sampling 15:523

sensitizers in 15:514
specification tests 15:543, 548
splints 15:518, 534, 540
splint strength 15:544
spontaneous ignition 15:549
strikers 15:520, 535, 541
thickeners 15:506
Match glue. (See also *Glues of animal origin.*)
  commercial grades and specifications 15:507
  properties 15:506
Match heads. (See also *Matches.*)
  analysis 15:539
  composition 15:537
  functioning test 15:540
  hardness 15:544
  manufacture 15:537
  properties 15:540
  specific gravity 15:539
  viscosity 15:539
Match splints. (See also *Matches.*)
  analysis 15:534
  properties 15:540
  raw materials 15:518
  types 15:518
Match strikers. (See also *Matches.*)
  abrasive content 15:522
  adhesion 15:548
  analysis 15:523, 535, 540
  appearance 15:547
  binders in 15:520
  composition 15:538
  extenders in 15:522
  functioning properties 15:548
  functioning test 15:540
  hardness 15:547
  manufacture 15:520, 539
  moisture resistance 15:548
  particle size 15:523
  properties 15:541
  raw materials 15:520
  red phosphorus content 15:521
  sampling 15:523
  sensitizers in 15:522
  thickness 15:547
  types 15:541
  viscosity 15:540
Matte
  antimony in 6:58
  sample preparation 6:59, 236
Matte glaze. (See also *Glazes.*)
  composition 9:196
Maucherite
  composition 16:286
Mauveine 6:381
Maximum likelihood method 7:172
  mathematical analysis of bioassay data 7:172
Mayonnaise. See *Eggs and egg products.*

MDA. See Diaminodiphenylmethane.
MDI. See Diphenylmethane-4,4'-diisocyanate.
Meat
  lactic acid content 15:133
Meat extracts
  gelatin content 13:364
Mechanical adhesion 4:391
Mechanical felt. See *Felt.*
Mechanical stability
  of rubber-based adhesives 4:403
Meclizine
  gas chromatography 6:210
Medical supplies
  ethylene oxide in 12:327
MEDINA. (See also High explosives.)
  infrared spectrum 12:453
  melting point 12:461
  properties 12:432
Medroxyprogesterone acetate. (See also Steroids.)
    10:581
  acetic acid–sulfuric acid colorimetric method
    10:582
  analysis 10:582, 596
  assay 10:582
  identification 10:582
  isonicotinyl hydrazide colorimetric method
    14:207
  manufacture 14:201
  properties 10:581; 14:201
  solubility 10:581; 14:201
  specific rotation 10:582
  structure 14:202
  ultraviolet spectroscopy 10:582
  uses 14:201
MEDTA. See PDTA.
Megregian-Maier
  colorimetric method for fluorine 11:493
MEHQ
  in dental restorative materials, composite
    11:312
  potassium nitrite colorimetric method 11:312
Melamine 5:276
  in alkyd resins 5:44
  analysis 5:280
Melamine–formaldehyde resin
  in cotton 11:134
  curing range 5:279
  in finishing agents 11:126, 133
  infrared spectrum 5:284, 285
  properties 5:277, 281
  solubility 11:126
  structure 5:279
  sulfuric acid–carbazole colorimetric method
    11:133
  ultraviolet spectroscopy 11:134
  uses 5:280
Melamine orthophosphate

uses 17:122
Melamine resin adhesives, evaluation 4:399
Melamine resins
  in alkyd resins 5:24
Melamine–urea resin. See Urea–melamine resin.
Melanocyte-stimulating hormone. See Intermedin.
Melaven mercury-cathode cell 5:200
Meldola's Blue 6:381, 382, 383, 385
  application method 6:398
  manufacture 6:385
  solubility 6:383
  volumetric method using titanous chloride 6:395
Melilite 8:75
Melissyl alcohol. (See also Fatty alcohols.) 4:561
Mellitic acid 8:405
  paper chromatography 8:377
  properties 8:403
Melprex. See Dodine.
Melt flow
  of tetrafluoroethylene–hexafluoropropylene
    copolymer 13:92
Melt index
  of ethylene and propylene polymers 12:374
Melting point
  of anisindione 10:177
  of cocoa beans 9:548
  of diamines and higher amines, aromatic 11:440
  of ethylbiscoumacetate 10:162
  of ethylene and propylene polymers 12:347
  of ethyleneimine 14:474
  Fisher-Johns method 13:85
  of glues of animal origin 15:525
  of menadiol sodium diphosphate 10:203
  of menadione sodium bisulfite 10:201
  of norethindrone 10:583
  of paraffin wax 15:533
  of phosphorus sesquisulfide 15:531
  of photographic films 12:592
  of picric acid 12:419
  of polyethylene 12:349, 383
  of polypropylene 12:349
  of polytetrafluoroethylene 13:84
  of styphnic acid 12:419
  of tetracene 12:416
  of warfarin potassium 10:175
  of warfarin sodium 10:167
Melting point determination
  of acrylic acid 4:199
  of acrylic and modacrylic fibers 4:260
  of alkanolamines 5:14
  of amobarbital 6:505
  of cellulose acetate polymers 4:76
  of methacrylic acid 4:199
Melting range
  of bithionol 11:547
  capillary tube method 15:503
  of dichlorophene 11:547

dilatometer method 15:502
  of hexachlorophene 11:547
  of margarine 15:502
Melting range determination
  of amobarbital 6:505
  of aprobarbital 6:512
  of mephobarbital 6:518
  of phenobarbital 6:523
  of vinbarbital 6:529
Melting transition
  of polypropylene 12:383
Melt viscosity
  of ethylene and propylene polymers 12:373
  of polycarbonates 17:347
Menadiol sodium diphosphate 10:203
  ammonium molybdate gravimetric method
    10:203
  analysis 10:203
  assay 10:203
  ceric sulfate titration 10:204
  identification 10:203
  melting point 10:203
  properties 10:203
  solubility 10:203
  ultraviolet spectroscopy 10:204
Menadione 10:194
  analysis 10:195
  ceric sulfate titration 10:198
  colorimetric determination 10:195
  cysteine colorimetric method 10:196
  diethyl malonate colorimetric method 10:197
  2,4-dinitrophenylhydrazine colorimetric method
    10:196
  ethyl cyanoacetate colorimetric method 10:197
  fluorescence spectroscopy 10:195
  gas chromatography 10:200
  identification 10:195
  impurities, determination of 10:195
  infrared spectrum 10:194, 195
  iodometric titration 10:198, 199
  paper chromatography 10:199
  polarography 10:197
  properties 10:194
  solubility 10:194
  thin-layer chromatography 10:199
  titanous chloride titration 10:198
  ultraviolet spectrum 10:194, 195
  zinc chloride–hydrochloric acid colorimetric
    method 10:196
Menadione sodium bisulfite 10:200
  analysis 10:200
  assay 10:201
  ceric sulfate titration 10:201
  ethyl cyanoacetate colorimetric method 10:201,
    202
  identification 10:200
  impurities, determination of 10:201

iodometric titration 10:202
melting point 10:201
properties 10:200
solubility 10:200
sulfur dioxide test 10:201
titanous chloride titration 10:202
Menaphthone. See Menadione.
Mendelevium. (See also Transuranium elements.)
  manufacture 19:259
1-Menthene oxide. (See also *Epoxy compounds.*)
  infrared spectroscopy 12:200
Menthofurane
  gas chromatography 8:353; 12:266
  in peppermint oil 12:266
Menthol
  in cigarette smoke 10:21
  gas chromatography 8:353; 10:21; 12:266
  in peppermint oil 12:266
Menthone
  gas chromatography 8:353; 12:266
  in geranium oil 12:282
  hydroxylamine hydrochloride volumetric method
    12:282
  in peppermint oil 12:266, 282
Menthyl acetate
  in essential oils 12:276
  gas chromatography 8:353
  molecular weight 12:276
Meperidine. (See also Meperidine hydrochloride;
    Narcotics.) 16:274
  analysis 16:274
  in biological materials 16:275
  color tests 16:246
  dioctyl sodium sulfosuccinate titration 16:279
  formaldehyde–sulfuric acid colorimetric method
    16:246
  gas chromatography 6:209; 16:254
  manufacture 16:274
  methyl orange colorimetric method 16:275
  picric acid microchemical method 16:250
  picrolonic acid microchemical method 16:250
  solvent extraction 16:275
  structure 16:245
  thin-layer chromatography 16:256
  ultraviolet spectroscopy 16:252
  in urine 16:276
  uses 16:274
  volumetric determination 16:276
Meperidine hydrochloride. (See also Meperidine;
    Pyridine and pyridine derivatives.)
  acid–base titration 16:274
  properties 16:274
  solubility 16:274
  solvent extraction 16:275
  structure 17:413
  uses 17:413
Mephobarbital 6:468, 517

acid–base titration, nonaqueous 6:517
analysis 6:517
commercial grades and specifications 6:517
derivatives for identification 6:477
extraction 6:491
gas chromatography 6:483, 484
impurities, determination of 6:518
properties 6:470
thin-layer chromatography 6:481
Mepivacaine hydrochloride 5:376
  acid–base titration 5:396
  analysis 5:396
  colorimetric method using reineckate 5:397
  gravimetric method 5:398
  infrared spectroscopy 5:396
  infrared spectrum 5:397
  ion exchange separation 5:398
  properties 5:396
  solubility 5:396
  ultraviolet spectroscopy 5:396, 398
  ultraviolet spectrum 5:397
Meprobamate
  gas chromatography 6:210
Meprylcaine hydrochloride 5:376
  acid–base titration 5:398
  analysis 5:398
  properties 5:398
  ultraviolet absorption spectroscopy 5:398
Merbromin. (See also Mercurials; Organomercury
    compounds.)
  ammonium thiocyanate volumetric method
    11:564
  assay 11:564
  commercial specifications 11:562
  structure 11:562; 16:54, 57
  zinc reduction 11:564
Mercaptan-addition method
  for acrylic monomers 4:192
  for acrylonitrile 4:373, 380
  for acrylonitrile in air (colorimetric) 4:378
  for acrylonitrile monomer in polymers 4:335
Mercaptans. (See also individual compounds.)
    15:551–574
  in air 15:570
  analysis 15:555
  in antifreezes 6:16; 15:571
  antirusts 6:351
  assay 15:561
  in beer 7:642
  in 1,3-butadiene 11:504
  in carbon disulfide 8:274, 276
  p-chloromercuribenzoic acid colorimetric
    method 15:461
  colorimetric determination 15:560
  coulometric methods 15:569, 571
  derivatives for identification 15:559
  N,N-dimethyl-*p*-phenylenediamine

colorimetric method 15:570
distillation range 15:571
doctor test 15:560
N-ethylmaleimide colorimetric method 15:561
flash point 15:571
gas chromatography 8:276; 15:257, 557, 565
in gasoline 15:560
gravimetric determination 15:571
identification 15:556
iodometric methods 15:570
iodometric titration 15:564
in liquefied petroleum gas 15:256
manufacture 15:554
mercuric perchlorate titration 15:568
in methacrylate polymers 16:99
in natural gas 15:557, 569
occurrence 15:551
palladium chloride color test 15:256
in petroleum products 15:566
phospho-18-tungstic acid colorimetric method
    15:461
properties 15:553
properties, determination of 15:571
reduction to hydrogen sulfide 7:642
refractive index 15:571
sampling 15:555
separation 15:556
silver dithizonate colorimetric method 15:461
silver nitrate titration 8:276; 15:561, 562, 566;
    16:99
in solvents 15:560
specific gravity 15:571
in Stoddard solvent 11:606
ultraviolet spectroscopy 15:559
uses 15:555
volumetric determination 15:561
in water 15:570
Mercaptoacetic acid. (See also Mercaptans.)
    uses 15:555
2-Mercaptoacetic acid. (See also Thioglycolic acid.)
    14:19
2-Mercaptobenzamidazole
    colorimetric method for phosphine 17:61
Mercaptobenzimidazole
    in elastomers, synthetic 12:141
    ultraviolet spectrum 12:141
Mercaptobenzothiazole. (See also Mercaptans.)
    in antifreezes 6:16
    antirusts 6:351
    in elastomers, synthetic 12:143
    gravimetric determination 15:571
    ultraviolet spectrum 12:143
    uses 15:555
2-Mercapto-4,5-dimethylthiazole
    colorimetric method for rhodium 17:265
Mercaptoethanol. (See also Mercaptans.)
    N-ethylmaleimide colorimetric method 15:561

Mercaptoethylamine hydrochloride. (See also
    Imines.)
    properties 14:471
    structure 14:471
Mercaptoglycolic acid. See Thioglycolic acid.
α-Mercapto-N,2-naphthylacetamide. See
    Thionalide.
γ-Mercaptopropyltrimethoxysilane. (See also
    Organofunctional silanes.)
    commercial specifications 18:82
    gas chromatography 18:85
    properties 18:80
Mercurials. (See also individual compounds.)
    analysis 11:563
    assay 11:563
    in disinfectants and antiseptics 11:561
    paper chromatography 11:565
    polarographic methods 11:565
    volumetric determination 11:564
Mercuric acetate. (See also Organomercury
    compounds.)
    commercial grades and specifications 16:54,
        58
    gravimetric method for thiophene 18:539
    manufacture 16:57
    method for acrylate esters 4:193
    solubility 16:57
    titration of chloride 13:69
    titration of EDTA 12:38, 53
    titration of hydrogen sulfide 18:544
    titration of NTA 12:53
    uses 16:58
Mercuric acetate–methanol
    volumetric method for polytetramethylene
        ether glycols 13:247
Mercuric amidochloride. See Ammoniated
    mercury.
Mercuric bromide. (See also Mercury compounds.)
    16:55
    colorimetric method for arsenic 13:526
    commercial grades and specifications 16:54,
        55
Mercuric chloride. (See also Mercurials; Mercury
    compounds.) 16:55
    acidimetric method for thiosulfate 18:431
    analysis 16:59
    assay 11:563; 16:59
    commercial grades and specifications 16:54, 55
    commercial specifications 11:562
    gravimetric method for ammonium sulfamate
        11:127
    gravimetric method for hypophosphites 17:113
    gravimetric method for tin 19:37, 41
    hydrogen sulfide gravimetric method 11:563;
        16:59
    impurities, determination of 16:60
    manufacture 16:55

microchemical method for cotarnine 16:249
microchemical method for ethylmorphine
   16:249
microchemical method for heroin 16:249
microchemical method for hydrocodone
   16:249
microchemical method for methadone 16:250
microchemical method for oxymorphone 16:250
properties 9:335; 16:6
residue on ignition 16:60
solubility 9:336; 16:60
uses 16:55
Mercuric cyanide. (See also Mercury compounds.)
   commercial grades 16:54
   copper(II) acetate–benzidine acetate colorimetric
      method 16:23
   identification 16:23
   iodometric method 16:23
   silver argentocyanide gravimetric method 16:23
Mercuric cyclohexane butyrate. (See also Organo-
      mercury compounds.)
   commercial grades 16:54
Mercuric fluoride. (See also Mercury compounds.)
   commercial grades 16:54
Mercuric iodide. (See also Mercury compounds.)
   16:55
   analysis 16:60
   assay 16:60
   commercial grades and specifications 16:54, 56
   impurities, determination of 16:60
   manufacture 16:55
   potassium iodate volumetric method 16:60
   solubility 16:60
   uses 16:55
Mercuric nitrate. (See also Mercury compounds.)
   color test for gold 14:11
   commercial grades 16:54
   titration of bromine 12:111
   titration of chlorine 12:111
   titration of halides 15:228
   titration of papaverine hydrochloride 16:272
   titration of sodium chloride 9:360
   volumetric method for chloride 7:581
   volumetric method for chlorine 14:157
Mercuric oxide. (See also Mercurials; Mercury
      compounds.) 16:55
   analysis 16:61; 17:196
   ammonium thiocyanate titration 11:563
   assay 11:563
   commercial grades and specifications 16:54, 56
   commercial specifications 11:562; 17:196
   impurities, determination of 16:61; 17:196
   manufacture 16:56
   mercury content 17:196
   uses 16:56; 17:196
Mercuric perchlorate. (See also Mercury com-
      pounds.)

commercial grades 16:54
titration of chlorine 12:251
titration of mercaptans 15:568
Mercuric sulfate. (See also Mercury compounds.)
   16:57
   commercial grades and specifications 16:54,
      57
   gravimetric method for epoxy resins 12:239
   manufacture 16:57
   uses 16:57
Mercuric sulfide. (See also Mercury compounds.)
   commercial grades 16:54
Mercuric thiocyanate. (See also Mercury com-
      pounds.)
   commercial grades 16:54
   spectroscopic method for sodium sulfite 18:408
Mercurimetric methods
   for allyl alcohol 5:85
   for allyl aromatic compounds 5:91
   for allyl esters 5:85, 96
   for allyl ethers 5:85, 100
   for formaldehyde 13:112
   for hydrochloric acid 14:367
Mercurimetric titration
   of chloride 9:342
   of thiophene 18:540
   of vinyl ethers 12:312
Mercurin. (See also Organomercury compounds.)
   structure 16:54, 57
Mercurochrome. See Merbromin.
Mercurophen. (See also Organomercury com-
      pounds.)
   structure 16:54, 57
Mercurous chloride. (See also Mercury compounds.)
   16:55
   commercial grades and specifications 16:54,
      55
   manufacture 16:55
   properties 9:335
   solubility 9:336
   uses 16:55
Mercurous nitrate
   colorimetric method for resorcinol 17:20
Mercurous perchlorate. (See also Mercury com-
      pounds.)
   commercial grades 16:54
Mercurous sulfate. (See also Mercury compounds.)
   16:57
   uses 16:57
Mercury. (See also Amalgams; Mercury metal.)
   16:1–67
   in acetic acid 4:100
   in air 16:22, 42, 44, 45, 46
   in aluminum metal 5:152
   ammonia–iodide colorimetric method 16:42
   in ammoniated mercury ointment 16:28
   ammonium hydroxide gravimetric method 16:6, 7

ammonium sulfide gravimetric method 16:27
ammonium thiocyanate titration 11:564
aniline–sulfanilamide gravimetric method 16:22
in animal tissue 16:15
anodic-stripping voltammetry 16:53
atomic absorption spectroscopy 16:22, 43
in biological materials 16:39
4,4′-bis(dimethylamino)diphenylamine colori-
    metric method 16:42
in brine 16:44, 45, 46
cadmium sulfide adsorption method 16:16
in caustic 16:44
in certified colors 10:495
in coal 16:41
in cobalt 16:44
in cocoa products 9:595, 598
colorimetric determination 16:18, 20, 38
copper(I) iodide colorimetric method 16:18,
    21, 42
copper reduction method 16:7
copper(II) sulfite–iodide colorimetric method
    16:42
in cotton 11:140, 154, 156
DAIH colorimetric method 16:22
p-dimethylaminobenzylidenerhodanine colori-
    metric method 11:140; 16:18, 20, 42
di-β-naphthylthiocarbazone colorimetric method
    16:42
diphenylcarbazone colorimetric method 11:140;
    16:18, 20, 22, 42
dithizone colorimetric method 10:496; 16:18,
    20, 36, 38
EDTA titration 13:445; 16:31
in eggs 16:44
electrochemical methods 16:36
electrodeposition 16:17, 24
emission spectroscopy 14:542; 16:20, 37
ferric–thiocyanate volumetric method 16:34
in fish 16:44, 51
flame tests 6:592
in fungicides 16:44
in glass and glass-ceramics 13:445
in grain 16:36
gravimetric determination 16:6, 10, 12, 16, 24
in hydrochloric acid 14:374
hydrochloric acid gravimetric reagent 16:6
in hydrogen 16:44, 45, 46
hydrogen sulfide colorimetric method 16:42
hydrogen sulfide gravimetric method 16:6, 7, 16
identification 16:20
in indium 14:542
iodized methylene blue colorimetric method
    16:18, 20
iodometric titration 16:33
ion exchange chromatography 10:495; 16:17,
    36
isotopes 16:2

liquid chromatography 16:17
in mercuric oxide 17:196
in mercury-containing compounds 13:199
methylene blue–mercury(II) iodide colorimetric
    method 16:42
methylxylenol blue colorimetric method 16:42
micrometric method 16:37
mining 16:8
in mud 16:44
neutron activation analysis 13:265; 16:53
in nickel 16:44
in nitrates 16:505
occurrence 16:1
in organic materials 16:23
oxidation potentials 16:5
in paint 16:25
palladium chloride colorimetric method 16:23
in paperboard 16:44
phosphomolybdic acid colorimetric method
    16:42
phosphorous acid reduction method 16:16
photometric mercury vapor method 10:496
polarography 16:52
potassium chromate colorimetric method 16:7
potassium ferrocyanide-α,α′-dipyridyl colori-
    metric method 16:19, 21
potassium iodide gravimetric method 16:6, 7
in pulp 16:44
in rice 16:40
in rocks 16:44
Rupp iodometric method 16:33
safety precautions 16:8
sample decomposition 16:12, 29
in sediment 16:45, 46
in seed grains 4:476
selenium sulfide colorimetric method 16:23
separation 16:16, 36
in sludges 16:44, 45, 46
sodium carbonate gravimetric method 16:6, 7
sodium chloride gravimetric method 16:29
in sodium hydroxide 16:45, 46
sodium hydroxide gravimetric method 16:6, 7
sodium iodide titration 16:37
sodium nitrite gravimetric method 16:7
sodium nitrite–silver nitrite colorimetric method
    16:19, 21
in soil 16:41
solvent extraction 16:16
thiocyanate titration 17:196
thiosalicylamide colorimetric method 16:42
tin(II) chloride–aniline colorimetric method
    16:19, 21
tin(IV) chloride–cacotheline colorimetric method
    16:19, 21
tin(II) chloride gravimetric method 16:6, 7
toxicity 16:8
turbidimetric method for sulfur 15:256

uses 16:1
volatilization methods 16:16, 20
Volhard titration 16:29
volumetric determination 16:12, 29
in water 16:22, 36, 44, 45, 46, 49
x-ray emission spectroscopy 10:495
x-ray fluorescence spectroscopy 11:156
Mercury alloys. (See also Amalgams.) 5:57
Mercury-cathode cells
dual unitized 5:201, 202
for electrolytic separation of aluminum 5:199
Mercury chloride
color test for formic acid 13:124
Mercury compounds. (See also Organomercury
compounds; and individual compounds.)
16:53–64
analysis 16:59
assay 16:59
properties 16:5, 53
safety precautions 16:59
solubility 16:6, 53
toxicity 16:53
uses 16:9, 53
Mercury-containing compounds. (See also
individual compounds.) 13:198–200
analysis 13:199
isotope exchange 13:199
mercury content 13:199
Mercury dibutyldithiocarbamate. (See also Dithiocarbamic acid salts.)
properties 11:575
Mercury diethyldithiocarbamate. (See also Dithiocarbamic acid salts.)
properties 11:575
Mercury dimethyldithiocarbamate. (See also
Dithiocarbamic acid salts.)
properties 11:575
Mercury displacement method
for specific gravity of explosives 12:467
Mercury extraction
of sodium metal 18:231
Mercury, free
in mercuric oxide 16:62
in mercury fulminate 12:409
Mercury freeze method
for liquefied petroleum gas 15:241
Mercury fulminate
fulminate content 12:409
granulation 12:410
identification 12:419
impurities, determination of 12:409
manufacture 12:409
microscopy 12:417
properties 12:406, 408
sodium thiosulfate-hydrochloric acid volumetric
method 12:409
solubility 12:419

specification tests 12:409
stannous chloride gravimetric method
12:419
thiosulfate colorimetric method 12:420
uses 12:408
x-ray diffraction 12:418
Mercury, mercurous
iodometric titration 16:61
in mercuric iodide 16:61
Mercury metal. (See also Amalgams.)
analysis 16:10
commercial grades and specifications 16:10
impurities, determination of 16:24
manufacture 16:8
in mercuric oxide 17:196
properties 16:1, 3
solubility in gold 14:9
turbidimetric method for sulfur 18:374
uses 16:3, 9
Wichers test 16:24
Mercury ores
sample decomposition 16:13
Mercury porosimetry 8:141
Mercury salts
derivatives for identification of amides 8:601
in mercuric iodide 16:61
in phenyl mercuric acetate 16:63
Mercury selenide semiconductors. (See also
Ceramics.)
properties 9:213
uses 9:213
Mercury telluride semiconductors. (See also
Ceramics.)
properties 9:213
uses 9:213
Mercury thiocyanate
colorimetric method for cobalt 19:557
gravimetric method for zinc 19:557
Merkel method for fiber identification
by solubility 4:247
by staining 4:252
Mersalyl. (See also Organomercury compounds.)
structure 16:54, 57
Merthiolate. (See also Organomercury compounds.)
structure 16:54, 57
Mesaconic acid
properties 8:397
Mescalin
bioassay 7:181
Mesitol. See 2,4,6-Trimethylphenol.
Mesitylene. (See also *Hydrocarbons.*)
commercial grades and specifications 14:323,
324
gas chromatography 16:220
uses 14:324

Mesityl oxide
  absorption spectroscopy 9:564
  acidity 8:358
  azeotropes 8:336
  commercial specifications 8:341
  Karl Fischer titration 8:360
  properties 8:335
Mesotartaric acid
  properties 8:397
Mestranol. (See also Steroids.) 10:575
  acid–base titration 10:577
  analysis 10:575, 585, 594, 596
  assay 10:575, 594
  fluorometry 10:576
  gas chromatography 10:576, 577
  infrared spectroscopy 10:577
  manufacture 14:201
  properties 10:575
  sample preparation 10:575
  solubility 10:575; 14:201
  structure 14:200
  sulfuric acid colorimetric method 10:576, 585,
    594
  thin-layer chromatography 10:577
  uses 14:201
Metabisulfites. See Pyrosulfites; and individual
    compounds.
Metaboric acid 7:370
Metabutethamine
  colorimetric method using sodium 1,2-naphtho-
    quinone-4-sulfonate 5:405, 407
Metabutethamine hydrochloride 5:376
  analysis 5:399
  gravimetric method 5:399
  properties 5:399
  ultraviolet absorption spectroscopy 5:399
Metabutoxycaine
  colorimetric method using sodium 1,2-naphtho-
    quinone-4-sulfonate 5:405, 407
Metabutoxycaine hydrochloride 5:376
  acid–base titration 5:400
  analysis 5:399
  properties 5:399
  ultraviolet absorption spectroscopy 5:399
Metacinnabar 16:1
Metahalloysite. (See also Clays and clay minerals.)
  composition 10:103
Metal-bonded grinding wheels
  analysis 7:299
Metal-bonded products
  diamond reclaiming from 11:482
  diamonds, industrial in 11:465
Metal borates. (See also individual compounds.)
  7:313
Metal borides. (See also individual compounds.)
  7:331, 358
  analysis 7:360

  boron content of 7:361
  boron (free) in 7:360
  carbon in 7:363
  manufacture 7:358
  nitrogen in 7:364
  oxygen in 7:365
  properties 7:359, 360
  uses 7:359
Metal chelates
  EDTA content 12:49
  metal content 12:54
Metal complex dyes 12:3
  bead tests 12:14
  identification 12:14
Metal complexes
  of aspartic acid 8:459
  of glutamic acid 8:459
  of glycine 8:459
Metal content
  of glass and glass-ceramics 13:444
  of metal chelates 12:54
Metal deactivators
  in gasoline 13:359
Metal haze
  in beer 7:520
Metalizing bath compatibility
  of dental elastomers 11:287
Metallic abrasives 4:31, 34
Metallic browns
  commercial specifications 17:194
Metallic nuclei
  in photographic films 12:587
Metallic powders 17:203
Metallic soaps. (See also Soaps.)
  analysis 18:204
  fatty acid content 18:204
Metallurgical microscopy. See Light microscopy.
Metal-organic complexes, dental. See Metal-organic
    dental products.
Metal-organic dental products. (See also Calcium
    hydroxide–salicylate dental products; Zinc
    oxide–eugenol dental products; and Zinc oxide–
    methacrylate dental products.) 11:295–299
Metal oxides
  in brake linings 7:441, 447
Metal oxides, abrasives 4:31, 34
Metal parts cleaners 6:355
  analysis 6:367
Metals. (See also Heavy metals; and individual
    elements.)
  in acetic acid 4:100
  in adipic acid 4:418
  in beer 7:651
  in brake linings 7:441, 447
  in carbonated beverages 8:174
  in carbon black 8:235
  carrier gas fusion methods 8:624

in detergents 11:367
EDTA titration 13:444
in explosive priming mixtures 12:423
in fatty alcohols 4:575
hydrogen in 8:613
hydrogen sulfide colorimetric reagent 19:74
nitrogen in 8:613, 628
oxygen in 4:15; 8:613; 16:524
in photographic films 12:570
in polyamides 17:303
in polymers and fibers 4:343
in resins 4:24
sample preparation 6:236; 15:31
selenium content 17:589
solvent extraction 12:423
in stannous chloride 19:74
sulfur content 18:380
tellurium content 17:589
in terephthalic acid 17:162
tin content 19:62
vacuum fusion methods 8:616
Metal salts of carboxylic acids. See Acid salts.
Metals, dental. See Dental alloys.
Metanicotine
in tobacco 19:136
Metanilic acid 5:421
azo dye colorimetric method 5:445, 451
azo dyes for identification 5:430
bromination method 5:444
color reactions for identification 5:431
identification 5:432
infrared spectrum 5:441
manufacture 5:426
properties 5:423
sulfanilic acid in 5:448
Metanil yellow. See Ext D & C Yellow No. 1.
Metaperiodic acid. (See also Periodates.) 14:609
manufacture 14:609
Metaphosphates. (See also individual compounds.)
17:117
in sodium tripolyphosphate 17:137
Metaphosphol
gravimetric method for niobium 16:367
Metaphosphoric acid. (See also Phosphoric acid.)
acid–base titration 17:131
analysis 17:131
assay 17:131
commercial specifications 17:126
Metapyrilene
ninhydrin method 6:35
Metastannic acid. (See also Stannic oxide,
hydrated.)
gravimetric method for tin 19:33
Metatitanates. (See also Titanates.) 19:124
Metavanadates. (See also Vanadates.) 19:344
uses 19:345
Metazocine. (See also *Narcotics.*)

ammonium molybdate colorimetric method
16:246
ammonium vanadate colorimetric method 16:
246
color tests 16:246
formaldehyde–sulfuric acid colorimetric method
16:246
microchemical tests 16:250
platinic chloride microchemical method 16:250
structure 16:245
Meteorites
gold content 14:43
titanium content 19:96
Methacholine chloride. (See also *Choline and its
derivatives.*)
dipicrylamine colorimetric method 9:617
Dragendorff's colorimetric reagent 9:617
iodine colorimetric method 9:617
iodoplatinate colorimetric method 9:617
properties 9:613
thin-layer chromatography 9:616
Methacrylamide. (See also *Acrylic and methacrylic acids and esters.*) 4:160
bromate–bromide method 4:191
nuclear magnetic resonance spectroscopy 16:86
Methacrylamide–polyethylene copolymers. (See
also *Ethylene and propylene polymers.*)
12:345
manufacture 12:345
Methacrylate polymers. (See also individual
compounds.) 16:67–102
additives, determination of 16:94
alkoxyl group content 16:92
analysis 16:70
antioxidant content 16:94
burning characteristics 16:70
carboxyl group content 16:91
commercial sources 16:69
emulsifier content 16:97
functional groups, determination of 16:91
gas chromatography 16:87
history 16:67
identification 16:70
impurities, determination of 16:94
infrared spectroscopy 16:71
manufacture 16:68
mass spectroscopy 16:78
nuclear magnetic resonance spectroscopy 16:79
oil conditioner 6:357
opalescent content 16:97
plasticizer content 16:96
potassium hydroxide saponification method
16:90
properties 16:68
pyrolysis–gas chromatography 16:87
sample preparation 16:71
solvent content 16:98

solvent extraction 16:71
structure 16:67
thin-layer chromatography 16:90
trade names 16:69
ultraviolet radiation absorber content 16:96
uses 16:67
zinc–nitric acid color test 16:70
Methacrylates. See *Acrylic and methacrylic acids and esters.*
Methacrylic acid. (See also *Acrylic and methacrylic acids and esters.*) 8:398
acid–base titration 4:197
bromine–bromide method 4:192
commercial grades and specifications 4:187
freezing point 4:199
infrared spectroscopy 4:215
manufacture 4:185
melting point 4:199
nuclear magnetic resonance spectroscopy 16:86
polymer in 4:200
properties 4:182; 8:397
solubility 4:181
thin-layer chromatography 16:90
Methacrylic acid and diglycidal ether of bisphenol A adducts. See Dental restorative materials, composite.
Methacrylic acid chloride
nuclear magnetic resonance spectroscopy 16:86
Methacrylic acid–polypropylene copolymers. (See also *Ethylene and propylene polymers.*) 12:345
manufacture 12:345
Methacrylonitrile. (See also *Nitriles.*)
gas chromatography 16:404
morpholine volumetric method 16:410
nuclear magnetic resonance spectroscopy 16:86
polarography 4:355; 16:398
residual monomer in polymers 4:335
γ-Methacryloxypropyltrimethoxysilane. (See also Organofunctional silanes.)
commercial specifications 18:82
gas chromatography 18:85
properties 18:80
solubility 18:89
specification tests 18:89
γ-Methacryloxypropyl-tris(2-methoxyethoxy) silane. (See also Organofunctional silanes.)
commercial specifications 18:82
gas chromatography 18:85
properties 18:80
Methadone. (See also Methadone hydrochloride; *Narcotics.*) 16:277
ammonium vanadate colorimetric method 16:246
analysis 16:277
in biological materials 16:278
colorimetric determination 16:278

color tests 16:246
dioctyl sodium sulfosuccinate titration 16:279
in feces 16:279
gas chromatography 6:210; 16:254
gravimetric determination 16:279
mercuric chloride microchemical method 16:250
microchemical tests 16:250
platinic bromide microchemical method 16:250
potassium nitrate–potassium hydroxide colorimetric method 16:278
solvent extraction 16:278
structure 16:245
thin-layer chromatography 16:256
ultraviolet spectroscopy 16:252
in urine 16:279
uses 16:277
volumetric determination 16:279
Methadone hydrochloride. (See also Methadone.)
acid–base titration, nonaqueous 16:277
solvent extraction 16:277
Methallatal
gas chromatography 6:483
Methallyl alcohol
analysis 5:78
bromination method 5:83
manufacture, methods of 5:76
properties 5:77
solubility 5:76
toxicity 5:78
Methallylamine
acid–base titration 5:89
properties 5:87
Methallylamines. (See also Allylamines.)
acid–base titration 5:88
Methallyl bromide 5:101
Methallyl chloride
halogenation methods 5:102
properties 5:101
Methallyl compounds. See *Allyl compounds;* and individual compounds.
Methallyl cyanide
analysis 5:92
commercial specifications 5:92
properties 5:92
Methallyl iodide 5:101
Methallylphenylamine 5:87
Methamphetamine. See Desoxyephedrine.
Metham sodium. See Sodium methyl dithiocarbamate.
Methanal. See *Formaldehyde.*
Methanamide. See Formamide.
Methandriol. (See also Steroids.)
trade names 14:192
Methandrostenolone. (See also Steroids.)
manufacture 14:204
properties 14:204
solubility 14:204

structure 14:204
uses 14:204
Methane. (See also Alkanes.) 14:299
  in air 16:485
  analysis 14:328
  calorific value 13:331
  in carbon dioxide 8:251
  in carbon monoxide 8:310, 312, 313
  in cigarette smoke 10:13
  commercial specifications 14:299
  flame ionization methods 16:532
  gas chromatography 4:212; 8:251, 262, 310,
    312, 313; 12:339; 13:307, 316, 317, 318;
    14:327, 328, 384, 416; 15:241; 16:415, 454,
    457, 485
  in hydrogen 14:380, 384
  infrared spectroscopy 8:252, 253, 313; 13:294
  mass spectroscopy 13:302
  in natural gas 13:320
  in noble gases 13:294, 302, 307
  in oxygen 16:532
  in propane 15:241
  properties 13:311; 14:299
  solubility 14:299
  specific gravity 13:329
  thermal conductivity 14:380
  uses 14:299
Methanedicarboxylic acid. See Malonic acid.
Methanethiol. See Methyl mercaptan.
Methanoic acid. See *Formic acid.*
Methanol. (See also Alcohols, lower.) 16:102–122
  acetylation with acetic anhydride 4:530
  acidity 16:112
  in acrolein 4:154
  in alcoholic beverages 4:488, 492
  alkalinity 16:112
  analysis 16:106
  antifreeze 6:1, 355
  appearance 16:113
  assay 16:107
  azeotropes 4:499, 502; 16:103
  in black tea 18:510
  chromotropic acid colorimetric method 4:550;
    15:146; 19:437
  in cigarette smoke 10:31
  in cocoa bean vapor 9:570
  color 16:113
  commercial grades and specifications 4:512;
    16:106
  in *o*-cresol 17:39
  density 16:104
  derivatives for identification 4:518
  in dimethyl malonate 15:433
  in dimethylmethoxymethylene malonate 15:437
  3,5-dinitrobenzoyl chloride method 4:547
  distillation range 16:113
  in distilled spirits 16:119

  in enamels 12:188
  extraction of choline 9:620
  in formaldehyde 13:115
  freezing point 6:7; 16:105
  gas chromatography 4:154, 212; 5:5; 7:96;
    10:31; 11:181; 12:204; 13:576; 15:433, 437,
    441; 16:107, 108, 118; 17:40; 18:510; 19:439
  gas line antifreeze 6:355
  in gasoline 16:119
  gravimetric method for boron 18:13
  history 16:102
  identification 6:4; 16:107
  impurities, determination of 16:108, 112
  infrared spectroscopy 16:119
  in lactic acid 15:146
  manufacture 4:501, 504; 16:105
  mass spectroscopy 9:570; 16:119
  in methyl chloride 9:425
  in methyl cyanoacetate 15:441
  nuclear magnetic resonance spectroscopy 16:119
  odor 16:115
  oxidation methods 19:437
  permanganate time 16:115
  in phenols 17:39
  phthalic anhydride method 4:535
  properties 4:496; 6:2; 16:103
  refractive index 16:104
  sampling 16:106
  separation 16:107
  solubility 13:539, 552; 16:103
  specification tests 16:112
  specific gravity 4:538; 16:104, 116
  surface tension 16:104
  toxicity 16:105
  turbidimetric method for styrene polymers
    18:304
  ultraviolet spectroscopy 16:107
  uses 4:507; 16:102
  viscosity 16:104
  visible spectroscopy 16:119
  in water 16:119
  in wines 19:425, 437
Methanol insolubles
  in benzoic acid 7:66
  in boric acid 7:382
Methanol soluble material
  in styrene polymers 18:325
Methapyrilene
  gas chromatography 6:210
Methapyrilene hydrochloride. (See also *Anti-
    histamines.*) 6:28
  column chromatography 6:150
  ninhydrin method 6:35
  properties 6:32
  thin-layer chromatography 6:35
  ultraviolet spectroscopy 6:150
Metharbital 6:468

derivatives for identification 6:477
gas chromatography 6:483, 484
properties 6:470
solubility 6:476
thin-layer chromatography 6:481
Methdilazine hydrochloride. (See also *Antihist-*
   amines.) 6:28
Methicillin. (See also Penicillins; Sodium methicillin.) 16:573
analysis 16:573
assay 16:573
bioassay 16:573
manufacture 16:573
microbiological methods 16:573
properties 16:573
specification tests 16:574
structure 16:554
ultraviolet spectroscopy 16:560, 565, 574
uses 16:573
Methine dyes 12:8
Methionine. (See also *Carboxylic acids, amino.*)
   8:413, 500
analysis 8:505
in cocoa beans 9:556
commercial specifications 8:504
composition 8:413
crystalline forms 8:501
cupric sulfate color test 8:505
gas chromatography 8:429
identification 8:505
infrared spectrum 8:502
iodometric titration 8:505, 506
ion exchange chromatography 8:415, 421
in leucine 8:495
nuclear magnetic resonance spectrum 8:502
occurrence 8:504
optical rotation 8:444
paper chromatography 8:422
peroxide oxidation method 8:505
properties 8:500
sodium nitroprusside color test 8:505, 506
solubility 8:503
uses 8:500
in yeast 19:529
Methitural 6:468
properties 6:470
solubility 6:476
Methomyl. (See also *Pesticides and insecticides.*)
commercial grades and specifications 19:594
composition 19:590, 599
manufacture 19:599
properties 19:590
solubility 19:587
structure 19:585
toxicity 19:592
Methone
gravimetric method for formaldehyde 13:107

Methostem. See Methandriol.
Methoxalen. See Xanthotoxin.
2-Methoxybenzaldehyde
column chromatography 6:597
derivatives for identification 6:618
fluorometry 6:612
infrared spectroscopy 6:626
near-infrared spectroscopy 6:629
properties 6:598
ultraviolet spectroscopy 6:633
3-Methoxybenzaldehyde 6:597
column chromatography 6:618
derivatives for identification 6:612
fluorometry 6:625, 626, 628
properties 6:598
ultraviolet spectroscopy 6:633
4-Methoxybenzaldehyde. (See also Anisaldehyde.)
absorption spectroscopy 9:564
p-Methoxybenzenethiohydroxamic acid
gravimetric method for niobium 16:367
p-Methoxybenzoic acid (Anisic acid) 8:405
properties 8:403, 405
paper chromatography 8:379
Methoxycarbonylpyridines. (See also *Pyridine*
   and pyridine derivatives.)
ionization constants 17:417
Methoxychlor
gas chromatography 19:389
in raw milk 16:129
in water 19:389
Methoxy cinnamic acid. (See also *Cinnamic acid*
   and its derivatives.)
properties 10:63
7-Methoxycoumarin. (See also *Coumarins.*)
gas chromatography 11:168, 170
6-Methoxy-7,8-dihydroxycoumarin. (See also
   Coumarins.)
gas chromatography 11:170
2-Methoxyethyl mercuric chloride. (See also
   Mercury-containing compounds.) 13:198
manufacture 13:198
properties 13:198
solubility 13:198
toxicity 13:198
uses 13:198
Methoxyflurane
analysis 5:364
gas chromatography 5:364
properties 5:357, 363
3-Methoxy-4-hydrocinnamic acid. See Ferulic acid.
3-Methoxy-4-hydroxybenzaldehyde. See *Vanillin.*
3-Methoxy-4-hydroxybenzoic acid. See Vanillic
   acid.
6-Methoxy-7-hydroxycoumarin. (See also
   Coumarins.)
gas chromatography 11:168, 170
3-Methoxy-4-hydroxymandelic acid. See

Vanilmandelic acid.
3-Methoxy-4-hydroxyphenylacetic acid. See
    Homovanillic acid.
Methoxyl content
  of ethers 12:304
  of methylcellulose 9:81
  of methylhydroxypropylcellulose 9:81
Methoxymercuration
  of allyl chloride 9:470
  of C_3 chlorinated hydrocarbons 9:470
3-Methoxy-2-nitrosophenol
  colorimetric method for cobalt 7:658
3-Methoxy-19-nor-17α-pregna-1,3,5(10)-trien-20-
    yn-17-ol. See Mestranol.
DL-Methoxyphenylacetic acid
  gravimetric method for sodium 18:218, 228
4-p-Methoxyphenylethynylpyridine. (See also
    Pyridine and pyridine derivatives.)
  ionization constants 17:416
2-(p-Methoxyphenyl)indane-1,3-dione. See
    Anisindine.
3-Methoxyphenyl isocyanate. (See also Isocyanic
    acid esters.)
  infrared spectroscopy 15:106
Methoxyphenyl isocyanates. (See also Isocyanic
    acid esters.)
  near-infrared spectroscopy 15:105
Methoxypyridines. (See also Pyridine and pyridine
    derivatives.)
  ionization constants 17:417
4-p-Methoxystyrylpyridine. (See also Pyridine
    and pyridine derivatives.)
  ionization constants 17:416
Methoxy triglycol
  gas chromatography 7:427
Methyl acetate
  in cocoa bean vapor 9:570
  in enamels 12:188
  gas chromatography 11:181; 12:300; 16:108,
    109
  mass spectroscopy 9:570
  in methanol 16:108, 109
  potassium hydroxide volumetric method 12:188
  properties 8:558, 577
Methyl acetoacetate
  commercial specifications 8:578
  properties 8:577
p-Methylacetophenone
  absorption spectroscopy 9:564
Methyl p-acetoxycinnamate. (See also Cinnamic
    acid and its derivatives.)
  ultraviolet spectroscopy 10:69
Methyl acetylene. See Propyne.
  gas chromatography 11:510
  in isoprene 11:510
β-Methylacrolein. See Crotonaldehyde.
Methyl acrylate. (See also Acrylic and

methacrylic acids and esters.)
  azeotropes 4:184
  bromate–bromide method 4:191
  commercial grades and specifications 4:187
  gas chromatography 4:212
  liquid chromatography 16:99
  mercaptan-addition method 4:192
  mercuric acetate method 4:193
  in methacrylic polymers 16:99
  polymer in 4:200
  properties 4:182
  saponification method 4:196
  sodium sulfite addition 4:341
  solubility 4:184
Methyl acrylate–acrylonitrile polymers. See
    Acrylonitrile–methyl acrylate polymers.
7-Methylaesculetin. See 6-Hydroxy-7-methoxy-
    coumarin.
Methylal. (See also Acetals and ketals; Dimethyl
    acetal.) 4:62
  gas chromatography 16:109
  melting point of derivatives 4:68
  in methanol 16:109, 114
  properties 4:63
  specifications 4:65
Methyl alcohol. See Methanol.
Methyl aldehyde. See Formaldehyde.
Methylamine
  gas chromatography 5:268
  manufacture, methods of 5:262
  properties 5:260
  uses 5:263
Methylamines
  commercial grades and specifications 5:264
  gas chromatography 5:268
  manufacture, methods of 5:262
2-(Methylamino)ethanol
  infrared spectrum 5:8
  properties 5:2
p-Methylaminophenol
  in hair dyes 11:103
  thin-layer chromatography 11:103
p-Methylaminophenol sulfate
  colorimetric method for phosphate 7:382, 383
4-Methylamino-1-(3-pyridyl)-1-butanol
  in tobacco 19:136
Methyl amyl acetate
  commercial specifications 8:578
  gas chromatography 8:351
Methyl amyl alcohol. (See also Alcohols, higher.)
  in brake fluids 7:424
  phthalic anhydride method 4:559
  properties 4:552
  specifications 4:556
  use in manufacture of fatty alcohols 4:560
  uses 4:554
Methyl amyl ketone

azeotropes 8:336
commercial specifications 8:341
gas chromatography 8:351, 353
properties 8:335
N-Methyl anabasine
in tobacco 19:136
17α-Methyl-5-androstene-3β,17β-diol. See Methandriol.
17α-Methyl-4-androsten-17β-ol-3-one. See Methyltestosterone.
2-(3-; 4-)Methylaniline
gas chromatography 5:268
N-Methylaniline
acid–base titration 5:448
azeotropes 5:424
chronopotentiometry 5:454
colorimetric methods 5:452
color reactions for identification 5:431
derivatives for identification 5:429
gas chromatography 5:268, 438
infrared spectroscopy 5:434, 453
manufacture 5:425
polarography 5:454
properties 5:422
separation 5:435
ultraviolet spectroscopy 5:432
uses 5:427
x-ray diffraction 5:435
Methyl anthranilate
in essential oils 12:276
molecular weight 12:276
solubility 13:540
Methylated urea 5:276
2-Methylaziridine. (See also *Imines.*)
properties 14:468
structure 14:468
N-Methylaziridine. (See also *Imines.*)
properties 14:468
structure 14:468
2-Methylbenzaldehyde. See *o*-Tolualdehyde.
3-Methylbenzaldehyde. See *m*-Tolualdehyde.
4-Methylbenzaldehyde. See *p*-Tolualdehyde.
Methylbenzene. See Toluene.
Methylbenzethonium chloride. (See also Quaternary ammonium compounds.)
assay 11:557
commercial specifications 11:555
in disinfectants and antiseptics 11:555
iodometric titration 11:558
properties 11:555
Methyl benzoate. (See also Benzoic acid esters.) 7:78
commercial grades and specifications 7:80
3,5-dinitrobenzoate derivative 7:83
in essential oils 12:276
gas chromatography 7:72; 8:564
molecular weight 12:276

in phthalic anhydride 8:564
properties 7:77; 8:577
saponification method 7:84
3-Methyl-2-benzothiazolone hydrazone–ferric chloride
colorimetric method for glyceraldehyde 12:550
3-Methylbenzothiol-2-one hydrazone
colorimetric method for acetaldehyde 19:454
Methyl borate
distillation 7:391
flame test 7:390
Methyl bromide 13:179
in air 13:183
analysis 8:13, 25
commercial specifications 8:12
in flour 13:184
gas chromatography 8:13, 14; 13:183, 184
impurities, determination of 8:13
manufacture 13:179
properties 8:12; 13:179
in soil 13:184
solubility 13:179
toxicity 13:179
uses 13:179
van der Meulen method 8:25
Methyl *m*-bromocinnamate. (See also *Cinnamic acid and its derivatives.*)
ultraviolet spectroscopy 10:69
2-Methyl-1,3-butadiene. (See also Isoprene.)
gas chromatography 13:350
in gasoline 13:350
properties 13:350
2-Methylbutanal
in black tea 18:507, 510
gas chromatography 18:510
2-Methylbutane. (See also Isopentane.)
gas chromatography 14:330
2-Methyl-1-butanol. (See also Amyl alcohol, active.)
gas chromatography 7:96
3-Methyl-1-butanol. (See also Isoamyl alcohol.)
gas chromatography 7:96
3-Methyl-2-butanol
gas chromatography 7:96
3-Methyl-2-butanone
absorption spectroscopy 9:564
3-Methyl-1-butene. (See also *Hydrocarbons.*)
gas chromatography 14:329
2-Methyl-2-butene. (See also *Hydrocarbons.*)
commercial grades and specifications 14:312, 313
3-Methyl-3-butenenitrile. See Methallyl cyanide.
Methylbutenes. (See also *Hydrocarbons.*)
gas chromatography 13:350
in gasoline 13:350, 353
properties 13:350

2-Methyl-1-butene-3-yne
  properties 4:137
  silver acetylide method 4:139
  solubility 4:138
Methyl butenol. (See also Ethylenic alcohols.)
  properties 4:576
  uses 4:576
2-Methyl-3-buten-2-ol
  in black tea 18:510
  gas chromatography 18:510
Methyl α-butoxypropionate. (See also Lactic
    acid esters.)
  properties 15:131
Methyl-t-butylbenzenes
  gas chromatography 13:352
  in gasoline 13:352, 355
  properties 13:352
Methyl 1-(butylcarbamoyl)-2-benzimidazole-
    carbamate. See Benomyl.
Methyl n-butyl ether. (See also Ethers.)
  properties 12:297
Methyl tert-butyl ether. (See also Ethers.)
  properties 12:297
Methyl butyl ketone
  properties 8:335
2-Methyl-6-tert-butylphenol. (See also Phenols.)
  commercial grades and specifications 17:15
3-Methyl-6-tert-butylphenol. (See also Phenols.)
  ferric chloride colorimetric method 17:16
Methyl butynol. (See also Acetylenic alcohols.)
  azeotropes 4:139
  commercial grades and specifications 4:139
  gas chromatography 4:144
  properties 4:136, 576
  silver acetylide method 4:578
  solubility 4:138
  uses 4:576
3-Methylbutyraldehyde
  derivatives for identification 8:343
Methyl butyrate
  properties 8:558, 577
2-Methylbutyric acid
  in black tea 18:511
  gas chromatography 18:511
α-Methylbutyronitrile
  in black tea 18:510
  gas chromatography 18:510
Methyl caprate
  gas chromatography 8:355; 12:488
Methyl caprylate
  gas chromatography 12:488
Methyl Carbitol
  antifreeze 6:1
  gas chromatography 7:427
Methyl Cellosolve
  antifreeze 6:1
  gas chromatography 7:427; 8:352; 11:73

  in nail enamels 11:73
Methylcellulose
  composition 12:583
  gravimetric determination 12:581
  infrared spectroscopy 12:583
  in matches 15:509
Methylcellulose and its derivatives. (See also
    Cellulose derivatives.) 9:76
  analysis 9:79
  hydroxylpropyl content 9:81
  impurities, determination of 9:79
  manufacture 9:79
  methoxyl content 9:81
  properties 9:78
  solubility 9:79
  uses 9:78
  viscosity 9:62
Methyl chloride. (See also Chlorocarbons.) 9:422
  analysis 9:423
  Beilstein detection 9:382
  boiling range 9:425
  in carbon tetrachloride 8:325
  color 9:425
  commercial grades and specifications 9:423
  composition 9:411
  Fujiwara colorimetric method 9:386
  gas chromatography 8:325; 9:381, 424; 18:61,
    64
  impurities, determination of 9:424
  manufacture 8:323; 9:423
  mass spectroscopy 9:396
  in methyldichlorosilane 18:64
  odor 9:425
  properties 9:422
  specification tests 9:424
  toxicity 9:423
  uses 9:422
2-Methylchlorobenzene
  mass spectroscopy 9:400
3-Methylchlorobenzene
  mass spectroscopy 9:400
4-Methylchlorobenzene
  mass spectroscopy 9:400
Methyl p-chlorocinnamate. (See also Cinnamic
    acid and its derivatives.)
  ultraviolet spectroscopy 10:69
Methylchloroform. See 1,1,1-Trichloroethane.
4-(5-Methyl-3,2'-chlorophenyl)isoxazolylpenicillin.
    See Cloxacillin.
Methylchlorosilanes. (See also individual com-
    pounds.)
  gas chromatography 18:51, 60
  manufacture 18:55
  mass spectroscopy 18:59
  in tetrachlorosilane 18:51
  uses 18:55
β-Methylcholine chloride. (See also

Choline and its derivatives.)
dipicrylamine colorimetric method 9:617
Dragendorff's colorimetric method 9:617
iodine colorimetric method 9:617
iodoplatinate colorimetric method 9:617
thin-layer chromatography 9:616
α-Methyl cinnamaldehyde
commercial specifications 10:44
Methyl cinnamate. (See also *Cinnamic acid and its derivatives.*)
commercial specifications 10:66
nuclear magnetic resonance spectra 10:67
properties 10:65
saponification methods 10:67
in sunscreens 11:92
ultraviolet spectroscopy 10:68; 11:92
Methylcinnamic acid. (See also *Cinnamic acid and its derivatives.*)
properties 10:63
Methyl cinnamylate. See Methyl cinnamate.
Methylcoumarones
boiling point 16:209
Methyl cyanoacetate
analysis 15:440
assay 15:441
colorimetric determination 15:440
commercial specifications 15:425
in dimethyl malonate 15:433
gas chromatography 15:433, 441
identification 15:440
impurities, determination of 15:441, 442
infrared spectroscopy 15:440
manufacture 15:424
mass spectrum 15:441
properties 15:421
solubility 15:422
trinitrobenzene–dimethylformamide colorimetric method 15:440
Methylcyclohexane. (See also *Hydrocarbons.*)
in benzene 7:11
commercial grades and specifications 14:307, 308
gas chromatography 7:11; 13:350; 14:330, 332
in gasoline 13:350, 353
in *n*-heptane 14:332
infrared spectroscopy 7:30
in isooctane 14:332
properties 13:350
uses 14:307
1-Methylcyclohexene. (See also *Hydrocarbons.*)
commercial grades and specifications 14:314, 315
uses 14:314
N-Methylcyclohexylamine
properties 11:242
Methylcyclopentadienylmanganese tricarbonyl. (See also Gasoline additive AK 33X.) 15:490

properties 15:489
uses 15:490
Methylcyclopentane. (See also *Hydrocarbons.*)
in benzene 7:11
commercial grades and specifications 14:307, 308
gas chromatography 7:11; 13:319, 350; 14:330
in gasoline 13:350, 353
properties 13:350
toxicity 14:307
uses 14:307
Methylcyclopentenes
gas chromatography 13:350, 353
in gasoline 13:350
properties 13:350
N-Methyldiallylamine 5:87
Methyl diallyl cyanurate 5:93
1-Methyl-4,6-di-*tert*-butylphenol. (See also *Phenols.*)
ferric chloride colorimetric method 17:16
4-Methyl-2,6-di-*tert*-butylphenol. (See also *Phenols.*)
4-aminoantipyrine colorimetric method 17:18
Millon's reagent color test 17:18
4-(5-Methyl-3,2',6'-dichlorophenyl)isoxazolyl-penicillin. See Dicloxacillin.
Methyldichlorosilane. (See also Organo-chloro-silanes.)
commercial specifications 18:56
gas chromatography 18:52, 60, 64
impurities, determination of 18:64
infrared spectrum 18:59
mass spectrum 18:59
properties 18:54
in trichlorosilane 18:52
in trimethylchlorosilane 18:64
Methyldiethanolamine
properties 5:2
specifications 5:4
1-Methyl-3,5-diethyl-2,4-diisocyanatobenzol. (See also *Isocyanic acid esters.*)
infrared spectroscopy 15:106
1-Methyldiisocyanatobenzols. (See also *Isocyanic acid esters.*)
infrared spectroscopy 15:106
1-Methyl-2,4-diisocyanatocyclohexane. (See also *Isocyanic acid esters.*)
infrared spectroscopy 15:106
4-Methyl-1,2-dimercaptobenzene. See Dithol.
Methyl 3,5-dinitrobenzoate
thin-layer chromatography 16:94
2-Methyl-1,3-dioxolane. (See also *Acetals and ketals.*) 4:62
in dioxane 11:522
gas chromatography 11:522
hydroxylamine hydrochloride volumetric method 11:522

properties 4:63
Methyldiphenylamines. (See also *Diamines and higher amines, aromatic.*)
  properties 11:434
  solubility 11:435
Methyldiphenyls
  in diphenyl 11:528
  gas chromatography 11:528
  infrared spectroscopy 11:527
  tetracyanoethylene colorimetric method 11:527
  ultraviolet spectroscopy 11:527
N-Methyldithiocarbamic acid sodium salt. See Sodium methyl dithiocarbamate.
N,N'-Methylenebisacrylamide
  use in AM-9 chemical grout 4:160, 167
2,2'-Methylenebis(6-*tert*-butyl-*p*-cresol) (2,2'-Methylenebis(6-*tert*-butyl-4-methylphenol)
  paper chromatography 6:85
  thin-layer chromatography 6:89
  ultraviolet spectroscopy 6:116
4,4'-Methylenebis(6-*tert*-butyl-*o*-cresol) 6:74
4,4'-Methylene-bis(*o*-chloroaniline)
  in polyurethanes 19:304
  structure 19:304
2,2'-Methylenebis(4-chlorophenol). See Dichlorophene.
4,4'-Methylenebis(2,6-di-*tert*-butylphenol). (See also *Phenols.*) 6:73
  structure 17:2
2,2'-Methylenebis(4-ethyl-6-*tert*-butylphenol) 6:73, 76
3,3'-Methylenebis(4-hydroxycoumarin). See Bishydroxycoumarin.
2,2'-Methylenebis(4-methyl-6-*tert*-butylphenol) 6:73
2,2'-Methylenebis(6-(1-methylcyclohexyl)-*p*-cresol)
  thin-layer chromatography 6:89
2,2'-Methylenebis(4-methyl-6-nonylphenol) 6:74
1,1'-Methylenebis-2-naphthol 6:73
2,2'-Methylenebis(3,4,6-trichlorophenol). See Hexachlorophene.
Methylene blue. (See also *Triphenylmethane and related dyes.*) 6:381, 382, 383, 386
  adsorption by activated carbon 8:150
  application method 6:398
  colorimetric method for alkyl sulfates 11:93
  colorimetric method for anionic detergents 11:328, 339
  colorimetric method for hydrogen sulfide 13:323; 15:255
  colorimetric method for linear alkyl aryl sulfonates 19:385
  colorimetric method for polythionates 12:594
  colorimetric method for selenium 17:592
  colorimetric method for silver sulfide 12:586
  colorimetric method for sulfur 12:560
  colorimetric method for sulfur content of acrylic polymers 4:285
  colorimetric method for thiosulfate 12:549, 594; 13:371
  color test for anionic detergents 18:183
  color test for emulsifiers 16:98
  commercial specifications 6:387
  manufacture 6:386
  solubility 6:383
  thin-layer chromatography 19:173
  titration of anionic detergents 11:337
  titration of cationic detergents 11:347
  titration of drycleaning detergents 11:611
  visible spectroscopy 6:395
  volumetric method using titanous chloride 6:395
Methylene blue–mercury iodide
  colorimetric method 16:42
Methylene blue–pyrocatechol violet
  colorimetric method for surfactants 12:573
Methylene bromide
  commercial specifications 8:12
  properties 8:12
Methylene chloride. (See also *Chlorocarbons.*) 9:425
  in aerosols 11:95
  analysis 9:428
  azeotropes 9:427
  Beilstein detection 9:382
  in carbon tetrachloride 8:325
  in chlorine 9:321
  in chloroform 9:433
  commercial grades and specifications 9:427
  composition 9:411
  cyclopentanol–sodium hydroxide colorimetric method 9:387
  2,7-dihydroxynaphthalene colorimetric method 9:387
  distillation range 9:429
  gas chromatography 8:325; 9:374, 375, 376, 377, 381, 428, 442; 11:98; 16:91
  in hydrochloric acid 9:392
  impurities, determination of 9:429
  infrared spectroscopy 13:294
  liquid chromatography 7:15
  manufacture 8:323; 9:427
  mass spectroscopy 9:396
  in metal cleaners 6:355
  α-naphthol colorimetric method 9:387
  in noble gases 13:294
  odor 9:429
  phenolphthalein colorimetric method 9:387
  properties 9:426; 11:95
  solubility 9:426
  specification tests 9:429
  toxicity 9:426
  in 1,1,1-trichloroethane 9:442

ultraviolet spectroscopy 9:430
uses 9:425
Methylenecyclohexane oxide. (See also *Epoxy compounds.*)
infrared spectroscopy 12:200
4,4'-Methylenedianiline. (See also *Diamines and higher amines, aromatic.*) 5:421; 11:433
acid–base titration 5:443
derivatives for identification 5:429
gas chromatography 5:440
infrared spectroscopy 5:441, 454
manufacture 5:425
properties 5:423; 11:434
solubility 11:435
ultraviolet spectroscopy 5:432
uses 5:426
Methylene dichloride. See Methylene chloride.
Methylenedinitramine. See MEDINA.
3,4-Methylenedioxybenzaldehyde. See Heliotropin.
Methylene green. (See also *Triphenylmethane and related dyes.*)
paper chromatography 19:172
Methylene iodide. (See also Iodine compounds, organic.)
properties 14:612
Methylene oxide. See *Formaldehyde.*
Methylergonovine
thin-layer chromatography 15:295
Methyl esters of carboxylic acids
gas chromatography 8:388
preparation 8:385
properties 8:558
N-Methylethanolamine. See 2-(Methylamino)-ethanol.
Methyl ether. (See also *Ethers.*)
properties 12:297
Methyl ether of hydroquinone. See MEHQ.
Methylethylamine
thin-layer chromatography 19:140
in tobacco 19:140
4-(1-Methylethyl)benzaldehyde. See Cuminal.
Methylethylbenzenes
gas chromatography 7:9; 13:351
in gasoline 13:351, 354
properties 13:351
Methylethylcyclopentanes
gas chromatography 13:351
in gasoline 13:351, 354
properties 13:351
Methylethylhexanes
gas chromatography 13:351
in gasoline 13:351, 354
properties 13:351
Methyl ethyl ketone
in acetylenic chemicals 4:145
acidity 8:358

azeotropes 8:336
in cigarette smoke 10:31
colorimetric method for benzene 7:17
commercial specifications 8:341
derivatives for identification 8:343
in enamels 12:175
Fourier transform spectra 13:154, 156
gas chromatography 10:31; 11:73; 12:204; 16:432
infrared spectroscopy 7:30, 100, 101
Karl Fischer titration 8:360
in methanol 16:114
in nail enamels 11:73
properties 8:335; 12:175
specific gravity 8:358
water in 8:360
Methyl ethyl ketoxime
in enamels 12:177
Methylethylpentanes
gas chromatography 13:351
in gasoline 13:351, 354
properties 13:351
2-Methyl-2-ethyl-1,3-propanediol. (See also Glycols.)
properties 13:537
solubility 13:538
structure 13:534
2-Methyl-5-ethylpyridine. (See also *Pyridine and pyridine derivatives.*)
manufacture 17:421
uses 17:412
4-Methyl-3-ethylpyridine. (See also *Pyridine and pyridine derivatives.*)
derivatives for identification 17:426
gas chromatography 17:437
Methyl ethyl sulfide
gas chromatography 15:558
Methyl fluoride. (See also *Fluorocarbons.*)
properties 13:60
Methyl formate. (See also Formic acid esters.)
commercial specifications 13:132
gas chromatography 4:212; 11:181; 16:108
manufacture 13:121
in methanol 16:108, 114
properties 8:558, 577; 13:131
saponification method 13:132
solubility 13:131
uses 13:120
5-Methyl-2-furaldehyde
absorption spectroscopy 9:564
Methyl furan
in cocoa bean vapor 9:570
in furan 13:229
gas chromatography 13:229
mass spectroscopy 9:570
Methylfurfural
in black tea 18:511

in furfural 13:239
gas chromatography 13:239; 18:511
ultraviolet spectroscopy 6:632
Methyl groups
infrared spectroscopy 12:350
in polyethylene 12:350
Methylheptanes
gas chromatography 13:351
in gasoline 13:351, 354
properties 13:351
2-Methyl-2-hepten-6-one
in black tea 18:510
gas chromatography 18:510
1-Methylheptylamine
spectroscopic method for anionic detergents
11:340
2-(1-Methylheptyl)-4,6-dinitrophenyl crotonate.
See Dinocap.
1-Methylhexahydroindane-3-carboxylic acid
structure 15:302
2-Methylhexane. (See also Isoheptane.)
gas chromatography 14:330, 332
in n-heptane 14:332
3-Methylhexane
gas chromatography 14:330, 332
in n-heptane 14:332
Methylhexanes
in benzene 7:11
gas chromatography 7:11; 13:319, 350
in gasoline 13:350, 353
properties 13:350
Methylhexenes
gas chromatography 13:350
in gasoline 13:350, 353
properties 13:350
3-Methyl-1-hexyn-3-ol. (See also Acetylenic
alcohols.)
silver acetylide method 4:578
Methylhydrazine
analysis 14:234
assay 14:236
colorimetric determination 14:235
commercial grades and specifications 14:234
coulometric method 14:238
derivatives for identification 14:235
gas chromatography 14:237, 239
identification 14:235
impurities, determination of 14:239
infrared spectroscopy 14:235
manufacture 14:233
polarography 14:238
potassium iodate titration 14:236
properties 14:232
proton magnetic resonance spectroscopy
14:237
solubility 14:232
toxicity 14:233

uses 14:232
volumetric determination 14:237
Methyl hydrogen silicone fluids. (See also
Silicone fluids.)
infrared spectrum 18:107
Methyl-p-hydroxybenzoate. (See also Phenols.)
in beer 7:597
ultraviolet spectroscopy 11:548
Methyl hydroxycinnamate. (See also Cinnamic
acid and its derivatives.)
ultraviolet spectroscopy 10:69
Methyl 3-hydroxy-α-crotonate dimethyl phosphate.
See Mevinphos.
N-Methylhydroxylamine. (See also N-Alkylhydroxylamines.)
properties 14:443
O-Methylhydroxylamine. (See also Hydroxylamines and their salts.)
manufacture 14:445
properties 14:443
N-Methylhydroxylammonium sulfate. (See also
N-Alkylhydroxylamines.)
manufacture 14:444
Methylindanes
gas chromatography 13:352
in gasoline 13:352, 355
properties 13:352
Methyl iodide. (See also Iodine compounds,
organic.)
gas chromatography 16:91, 92
properties 14:612
Methyl isoamyl ketone
commercial specifications 8:341
in enamels 12:175
Karl Fischer titration 8:360
properties 12:175
water in 8:360
Methyl isobutyl ketone
in acetylenic chemicals 4:145
acidity 8:358, 359
azeotropes 8:337
commercial specifications 8:341
in enamels 12:175
extraction of chromium 9:652
gas chromatography 8:352; 11:73; 16:432
hydroxylamine hydrochloride method 8:357
Karl Fischer titration 8:360
in nail enamels 11:73
properties 8:335; 12:175
uses 8:332
Methyl isobutyrate
gas chromatography 16:109
in methanol 16:109
properties 8:577
Methyl isocyanate. (See also Isocyanic acid
esters.)
commercial sources 15:102

infrared spectroscopy 15:108, 109
properties 15:95
2-Methyl-3-isocyanato butane. (See also
   Isocyanic acid esters.)
infrared spectroscopy 15:106
1-Methyl-4-isopropylbenzene. See *p*-Cymene.
Methylisopropylbenzenes
   gas chromatography 13:352
   in gasoline 13:352, 355
   properties 13:352
Methyl isopropyl ketone
   gas chromatography 12:204
   properties 8:335
1-Methyl-4-isopropylphenol. (See also *Phenols.*)
   paper chromatography 17:24
2-Methyl-5-isopropylphenol. (See also *Phenols.*)
   4-aminoantipyrine colorimetric method 17:18
   derivatives for identification 17:21
   Millon's reagent color test 17:18
   properties 17:4
2-Methyl-6-isopropylphenol. (See also *Phenols.*)
   thin-layer chromatography 17:24
3-Methyl-6-isopropylphenol. (See also *Phenols*;
   Thymol.)
   4-aminoantipyrine colorimetric method
      17:18
   ferric chloride colorimetric method 17:16
   Millon's reagent color test 17:18
5-Methyl-2-isopropylphenol. (See also *Phenols.*)
   derivatives for identification 17:21
Methyl isothiocyanate
   analysis 13:192
   infrared spectroscopy 13:193
   pentacyanoamine ferrate colorimetric method
      13:192
   in soil 13:192
   thin-layer chromatography 13:193
Methyl ketones
   indigo test 4:112
   iodoform test 4:111
   oxidation 8:338
   sodium nitroprusside test 4:111
Methyl lactate. (See also Lactic acid esters.)
   in lactic acid 15:146
   properties 15:131
   solubility 15:130
Methyl lactate acetate. (See also Lactic acid
   esters.)
   properties 15:131
Methyl lactate propionate. (See also Lactic acid
   esters.)
   properties 15:131
Methyl laurate
   gas chromatography 12:488
Methyl linoleate
   gas chromatography 12:488
Methyl linolenate

gas chromatography 12:488
Methylmalonic acid
   properties 8:397
Methyl mercaptan. (See also *Mercaptans.*)
   in carbon disulfide 8:274
   derivatives for identification 15:559
   gas chromatography 8:274; 15:557
   manufacture 15:554
   properties 15:552
   sampling 15:555
   uses 15:555
Methylmercury dicyandiamide. See Panogen.
Methyl methacrylate. (See also *Acrylic and methacrylic acids and esters.*)
   in air 4:216
   azeotropes 4:184
   bromate–bromide method 4:191
   butylated hydroxytoluene in 4:210
   commercial grades and specifications 4:187
   gas chromatography 4:212
   hydroquinone in 4:205
   liquid chromatography 16:99
   mercaptan-addition method 4:192
   mercuric acetate method 4:193
   in methacrylate polymers 16:99
   modifier for alkyd resins 5:24
   near-infrared spectroscopy 16:99
   nuclear magnetic resonance spectroscopy 16:86
   permanganate reduction 4:216
   polarography 4:216, 336
   polymer in 4:200
   in poly(methyl methacrylate) 16:99
   properties 4:182
   pyridine sulfate dibromide method 4:190, 336
   residual monomer in polymer 4:190, 215, 336
   saponification method 4:196
   solubility 4:184
   thin-layer chromatography 16:90
Methyl methacrylate–acrylonitrile polymers. See
   Acrylonitrile–methyl methacrylate polymers.
Methyl methacrylate-*n*-butyl acrylate copolymers.
   (See also *Methacrylate polymers.*)
   infrared spectrum 16:75
Methyl methacrylate-*n*-butyl methacrylate copolymers. (See also *Methacrylate polymers.*)
   infrared spectra 16:75, 76
   nuclear magnetic resonance spectra 16:83, 84
Methyl methacrylate dental materials. (See also
   Acrylic resin teeth.) 11:299
   analysis 11:299
   color stability 11:304
   commercial grades and specifications 11:300
   composition 11:299
   consistency 11:305
   hardness 11:306
   packing test 11:301
   sample preparation 11:302

solubility 11:303
specification tests 11:300, 305, 306
temperature rise test 11:305
transverse deflection 11:303
uses 11:299
water sorption 11:302, 306
Methyl methacrylate dimer
thin-layer chromatography 16:90
Methyl methacrylate–ethyl acrylate copolymers.
(See also *Methacrylate polymers.*)
infrared spectrum 16:74
pyrolysis–gas chromatography 16:88
Methyl methacrylate–lauryl methacrylate copolymers. (See also *Methacrylate polymers.*)
infrared spectroscopy 16:77
Methyl methacrylate–methacrylic acid copolymers.
(See also *Methacrylate polymers.*)
nuclear magnetic resonance spectroscopy 16:79
Methyl methacrylate–methyl acrylate copolymers.
(See also *Methacrylate polymers.*)
infrared spectroscopy 16:75
Methyl methacrylate polymers. (See also
Acrylonitrile–methyl methacrylate polymers.)
gas chromatography 4:286
Methyl methacrylate–polypropylene copolymers.
(See also *Ethylene and propylene polymers.*)
12:345
manufacture 12:345
properties 12:345
5-Methyl-2-methoxyaniline. See Cresidine.
Methyl methoxycinnamate. (See also *Cinnamic
acid and its derivatives.*)
ultraviolet spectroscopy 10:69
Methylmethoxymethylene malonate. See Dimethylmethoxymethylene malonate.
Methyl α-methoxypropionate. (See also Lactic
acid esters.)
properties 15:131
S-Methyl-N-(methylcarbamoyl)oxy thioacetimi-
date. See Methomyl.
Methyl p-methylcinnamate. (See also *Cinnamic
acid and its derivatives.*)
ultraviolet spectroscopy 10:69
Methyl methylergonovine
thin-layer chromatography 15:295
2-Methyl-2-(methylthio)propionaldehyde *O*-
(methylcarbamoyl)oxime. See Aldicarb.
Methylmorphine. See Codeine.
N-Methyl myosmine
in tobacco 19:136
Methyl myristate
gas chromatography 12:488; 15:136
2-Methyl-1,4-naphthahydroquinone-3-sulfonate
in menadione sodium bisulfite 10:201
phenanthroline indicator gravimetric method
10:201
1-Methylnaphthalene. (See also Alkylnaphthalenes.)

boiling point 16:209
derivatives for identification 16:208
gas chromatography 16:217
properties 16:217
2-Methylnaphthalene. (See also Alkylnaphthalenes.)
boiling point 16:209
derivatives for identification 16:208
gas chromatography 16:217
occurrence 16:216
properties 16:217
uses 16:216
Methylnaphthalenes. (See also Alkylnaphthalenes;
and individual compounds.)
in diphenyl 11:528
gas chromatography 11:528
liquid chromatography 7:13, 14, 15, 16
thin-layer chromatography 11:532
2-Methyl-1,4-naphthoquinone. See Menadione.
2-Methyl-1,4-naphthoquinone sodium bisulfite.
See Menadione sodium bisulfite.
N-Methylnicotinamide
ion exchange chromatography 9:618
Methylnonanes
gas chromatography 13:351
in gasoline 13:351, 354
properties 13:351
Methyl nonyl ketone
hydroxylamine hydrochloride volumetric
method 12:282
in rue oil 12:282
3-Methyl-1-nonyn-3-ol. (See also Acetylenic
alcohols.)
silver acetylide method 4:578
Methyloctadecyldichlorosilane. (See also
Organo-chlorosilanes.)
properties 18:54
Methyloctanes
gas chromatography 13:351
in gasoline 13:351, 354
properties 13:351
Methyl oleate
gas chromatography 12:488
Methylol melamine
in cotton 11:138
infrared spectroscopy 11:139
Methylolpolyacrylamide 4:170
Methyl orange
colorimetric method for alkanolamines 5:392
colorimetric method for amines 5:273
colorimetric method for cocaine 5:384
colorimetric method for cyclohexylamine
11:217, 228
colorimetric method for meperidine 16:275
colorimetric method for morphine 16:263
colorimetric method for oleandomycin 5:532
colorimetric method for tetracaine 5:414
solubility 13:539, 552

Methyl orange–mercuric chloride
  color test for hydrogen cyanide 14:411
Methyl oxalate
  gravimetric method for rare earth metals 9:266
Methyl oxirane. See Propylene oxide.
Methyl palmitate
  gas chromatography 12:488; 15:136
Methyl palmitoleate
  gas chromatography 12:488
Methylparaben
  gas chromatography 14:581
  in insulin 14:581
Methyl parathion. (See also *Pesticides and insecticides.*)
  commercial grades and specifications 19:595
  composition 19:589, 598
  manufacture 19:598
  properties 19:589
  safety 19:605
  solubility 19:587
  structure 19:583
  toxicity 19:592
Methylpentadecanes
  in cetane 14:334
  gas chromatography 14:334
2-Methyl-1,3-pentadiene
  liquid chromatography 7:13, 15
2-Methylpentaldehyde
  commercial specifications 8:340
2-Methyl-1,3-pentanediol. (See also Glycols.)
  solubility 13:538
2-Methyl-2,4-pentanediol. (See also Glycols.)
  commercial grades and specifications 13:547
  distillation range 13:602
  manufacture 13:545
  properties 13:537
  solubility 13:538
  structure 13:534
  thin-layer chromatography 13:582
  uses 13:548
Methylpentanes. (See also Hexanes.)
  commercial specifications 14:304
  gas chromatography 13:319, 350; 14:330
  in gasoline 13:350, 353
  properties 13:350; 14:303
2-Methyl-1,2,3-pentanetriol
  thin-layer chromatography 13:582
2-Methyl-1-pentanol. (See also Alcohols, higher.)
  phthalic anhydride method 4:559
  properties 4:552
  specifications 4:556
4-Methyl-2-pentanol. See Methyl amyl alcohol.
2-Methyl pentan-3-one
  hydroxylamine hydrochloride method 8:357
2-Methyl-1-pentene
  liquid chromatography 7:13, 15
4-Methyl-1-pentene. (See also *Hydrocarbons.*)

commercial grades and specifications 14:312, 313
4-Methyl-2-pentene
  liquid chromatography 7:14, 16
Methylpentenes
  gas chromatography 13:350
  in gasoline 13:350, 353
  properties 13:350
4-Methyl-3-penten-2-one
  in black tea 18:510
  gas chromatography 18:510
4-Methyl-4-penten-2-one
  in black tea 18:510
  gas chromatography 18:510
Methyl pentynol
  azeotropes 4:139
  properties 4:136
  solubility 4:138
3-Methyl-1-pentyn-3-ol. (See also *Acetylenic alcohols.*)
  properties 4:576
  silver acetylide method 4:578
Methylphenobarbital
  paper chromatography 6:479
  thin-layer chromatography 6:480
Methylphenyldiethoxysilane. (See also Organo-silane esters.)
  infrared spectrum 18:72
  properties 18:70
Methylphenyldimethoxysilane. (See also Organo-silane esters.)
  properties 18:70
4-*p*-Methylphenylethynylpyridine. (See also *Pyridine and pyridine derivatives.*)
  ionization constants 17:416
4-(5-Methyl-3-phenyl)isoxazolylpenicillin. See Oxacillin.
Methyl phenyl ketone. See Acetophenone.
4-Methylphenyl mercaptan. (See also *Mercaptans.*)
  derivatives for identification 15:559
Methylphenyl polysiloxane
  uses 15:302
2-Methyl-1-phenylpropane. See Isobutylbenzene.
2-Methyl-2-phenylpropane. See *tert*-Butylbenzene.
3-Methyl-1-phenyl-5-pyrazolone
  use in manufacture of azo dyes 6:403
Methylphenyl silicone. (See also Silicone fluids.)
  commercial specifications 18:104
Methylphenylsilicone resin. (See also Silicone resins.)
  infrared spectroscopy 18:120
2-Methyl-3-phytyl-1,4-naphthoquinone. See Phytonadione.
Methyl picolinate. See Methoxycarbonyl-pyridines.
2-Methylpropanal
  in black tea 18:510

gas chromatography 18:510
2-Methylpropane. See Isobutane.
2-Methyl-1-propanethiol. See Isobutyl mercaptan.
2-Methyl-2-propanethiol. See *tert*-Butyl
  mercaptan.
2-Methyl-1-propanol. (See also Alcohols, lower;
  Butyl alcohols; and Isobutanol.)
  in black tea 18:510
  gas chromatography 7:96; 18:510
  phthalic anhydride method 4:535
2-Methylpropene. See Isobutylene.
Methyl propionate
  gas chromatography 16:108, 109, 425
  in methanol 16:108, 109
  properties 8:558, 577
1-Methylpropylamine. See *sec*-Butylamine.
2-Methylpropylamine. See Isobutylamine.
N-Methyl-*n*-propylamine
  thin-layer chromatography 19:140
  in tobacco 19:140
Methyl-*n*-propylbenzenes
  gas chromatography 13:352
  in gasoline 13:352, 355
  properties 13:352
Methylpropyldichlorosilane. (See also Organo-
  chlorosilanes.)
  properties 18:54
Methyl *n*-propyl ether. (See also *Ethers*.)
  properties 12:297
Methyl propyl ketone 8:331
  azeotropes 8:337
  commercial specifications 8:341
  derivatives for identification 8:343
  hydroxylamine hydrochloride method 8:357
  properties 8:335
Methylprylon
  gas chromatography 6:209, 211
Methylpyridines. See Picolines.
1-Methyl-2(3-pyridyl)pyrrolidine. See Nicotine.
2-Methyl-6-(3-pyridyl)tetrahydro-1,2-oxazine
  hydrochloride
  in tobacco 19:136
N-Methylquinaldinium *p*-toluenesulfonate
  color test for formic acid 13:125
4-Methylquinoline. See Lepidine.
6-Methyl-2,3-quinoxalinedithiol cyclic *S,S*-dithio-
  carbamate. See Chinomethionate.
Methyl salicylate
  in black tea 18:511
  in essential oils 12:276
  gas chromatography 18:511
  ion exchange chromatography 6:144
  molecular weight 12:276
  phenol in 12:289
  solubility 13:539
  ultraviolet spectroscopy 16:95
Methylsalicylic acid. See *o*-Cresotic acid.

Methylsilicones. (See also Silicone fluids.)
  properties 18:97
  viscosity 18:97
Methyl stearate
  gas chromatography 12:488; 15:136
α-Methylstyrene. (See also *Hydrocarbons.*)
  gas chromatography 13:65; 18:298
  properties 14:326
  in styrene 18:298
β-Methylstyrene
  gas chromatography 18:299
  in styrene 18:299
α-Methylstyrene oxide. (See also *Epoxy compounds.*)
  infrared spectroscopy 12:200
  zinc bromide–2,4-dinitrophenylhydrazine
    gravimetric method 12:214
4-*p*-Methylstyrylpyridine. (See also *Pyridine and
    pyridine derivatives.*)
  ionization constants 17:416
Methylsuccinic acid
  properties 8:397
Methyl sulfate
  gravimetric method for strontium 18:268
Methyl sulfide
  in methanol 16:115
Methyltestosterone. (See also Steroids.)
  manufacture 14:204
  properties 14:204
  solubility 14:204
  structure 14:204
  trade names 14:192
  uses 14:187, 204
Methyl thiocyanate. (See also *Isocyanic acid
    esters.*)
  gas chromatography 15:110
Methylthiopyridines. (See also *Pyridine and
    pyridine derivatives.*)
  ionization constants 17:417
Methylthymol blue
  colorimetric method for tantalum 16:369
Methyl toluate
  gas chromatography 7:71
Methyltrichlorosilane. (See also Organo-chloro-
    silanes.)
  commercial specifications 18:56
  in dimethyldichlorosilane 18:64
  gas chromatography 18:60, 64
  impurities, determination of 18:64
  manufacture 18:55
  mass spectrum 18:59
  properties 18:54
  in trimethylchlorosilane 18:64
Methyltriethoxysilane. (See also Organosilane
    esters.)
  commercial specifications 18:71

gas chromatography 18:74
infrared spectrum 18:73
properties 18:70
9-Methyl-2,3,7-trihydroxy-6-fluorone
colorimetric method for antimony 11:141
Methyltrimethoxysilane. (See also Organosilane
esters.)
commercial specifications 18:71
gas chromatography 18:75
impurities, determination of 18:75
properties 18:70
α-Methyltrimethylene glycol. See 1,3-Butanediol.
N-Methyl-N-trimethylsilylacetamide. (See also
Silylation reagents.)
properties 18:90
Methylvinyldichlorosilane. (See also Organo-
chlorosilanes; Organofunctional silanes.)
commercial specifications 18:56, 82
properties 18:54, 80
Methyl vinyl ketone
in acrylonitrile 4:377
in methanol 16:114
Methylvinylpyridine
acid–base titration, nonaqueous 4:341
2-Methyl-5-vinyl pyridine. (See also *Pyridine
and pyridine derivatives*.)
manufacture 17:421
uses 17:412
Methylvinylpyridine–acrylonitrile polymers. See
Acrylonitrile–methylvinyl pyridine polymers.
Methylvinylpyridine–acrylonitrile–vinyl acetate
polymers
precipitation fractionation methods 4:299
Methyl violet. (See also *Triphenylmethane and
related dyes*.)
colorimetric method for gold 14:24
colorimetric method for tantalum 16:370;
18:449
in enamels 12:187
identification 12:187
naphthol yellow S titration 19:175
thin-layer chromatography 19:173
visible spectroscopy 19:167
Methylxylenol blue
colorimetric method for mercury 16:42
Methysergide
liquid chromatography 15:293
Meticortin. See Prednisone.
Metopon. (See also *Narcotics*.)
ammonium molybdate colorimetric method
16:247
ammonium vanadate colorimetric method
16:247
color tests 16:247
formaldehyde–sulfuric acid colorimetric method
16:247
gas chromatography 16:255

microchemical tests 16:250
picrolonic acid microchemical method 16:250
potassium–bismuth iodide microchemical method
16:250
structure 16:244
ultraviolet spectroscopy 16:252
Metriol trinitrate. (See also High explosives.)
12:451
infrared spectrum 12:456
manufacture 12:451
melting point 12:461
properties 12:432
solubility 12:451
uses 12:451
Mevinphos. (See also *Pesticides and insecticides*.)
commercial grades and specifications 19:595
composition 19:598
manufacture 19:598
properties 19:588
solubility 19:587
structure 19:583
toxicity 19:592
uses 19:586
Meyerhofferite 7:314
Mica 15:358
commercial specifications 12:174; 17:191
in enamels 12:174
for lamination of infrared samples 4:236
Michigan solvent immersion method
for specific gravity of bituminous paving mixtures
6:337
Michigan specific gravity bottle 6:338
Michinerite
composition 17:215
Microanalysis
automated instruments 8:281
definition 8:280
history 8:280
sources of error 8:280
Microbiological assay
of actinomycin 5:558
of amphomycin 5:558
of amphotericins 5:571
of anisomycin 5:607
of antibiotics 5:479
of antimycin 5:607
of aterrimin 5:608
of aureothricin 5:511
of azaserine 5:608
of bacitracin 5:543
of bakery products 6:455, 460
of candicidins 5:572
of capreomycin 5:559
of carbomycin 5:525
of carzinophilin 5:609
of chloramphenicol 5:501
of chromomycin A$_3$ 5:501

of colistin 5:543
of cycloheximide 5:512
of cycloserine 5:513
of dihydrostreptomycin 5:589
of erythromycin 5:526
of etamycin 5:559
of ferrimycin 5:610
of filipin 5:573
of fumagillin 5:573
of fusidic acid 5:513
of gentamycin 5:610
of gramicidin 5:544
of griseofulvin 5:502
of hygromycin B 5:589
of kanamycins 5:590
of leucomycin 5:527
of lincomycin 5:591
of mikamycin 5:559
of mitomycin C 5:503
of neomycin 5:592
of nisin 5:546
of novobiocin 5:504
of nystatin 5:575
of oleandomycin 5:528
of ostreogrycins 5:560
of paromomycin 5:593
of pentamycin 5:576
of pimaricin 5:576
of polymyxin 5:547
of pyostacine 5:560
of rifamycin 5:505
of ristocetin 5:610
of sarkomycin 5:514
of spiramycins 5:529
of staphylomycin 5:560
of stendomycin 5:611
of streptomycin 5:593
of thiazolidomycin 5:515
of thiostrepton 5:561
of trichomycin 5:576
of tylosin 5:529
of tyrocidine 5:547
of tyrothricin 5:547
of vancomycin 5:612
of variotin 5:577
of viomycin 5:561
of xanthocillin 5:505
Microbiological methods
for ampicillin 16:575
for benzylpenicillin 16:568
for carbenicillin 16:574
for cephalosporins 16:578
for chocolate and cocoa products 9:599
for choline and its derivatives 9:624
for cloxacillin 16:577
for dicloxacillin 16:577
for hetacillin 16:576

for isoxazolylpenicillins 16:577
for methicillin 16:573
for nafcillin 16:578
for niacin 10:434
for oxacillin 16:577
for penicillins 16:560, 561
for phenethicillin 16:573
for phenoxymethylpenicillin 16:572
Microchemical balance 8:282
Microchemical tests
for narcotics 16:248
Microcline 17:353
Microcosmic salt bead test 6:587
Microdiffusion method
for threonine 8:528
Microgel
measurement of 4:320
Microlite
composition 18:441
Micrometric method
for mercury 16:37
Microorganisms
in carbonated beverages 8:167
Microprojection method
for particle size distribution in chocolate
9:571
Microscopy
of acrylic and modacrylic fibers 4:263
of bakery chemicals 6:430
of barium nitrate 12:417
for calcium silicate cement 9:227
of carborundum 12:417
for ceramics 9:222
of chemical etchants 9:224
for concrete 9:145, 146
for diamonds, industrial 11:417
of ethylene and propylene polymers 12:348
of explosive priming mixtures 12:417
for fused-cast refractories 9:227
of glass and glass-ceramics 12:417; 13:471
for glaze ceramics 9:226
of lead nitrate 12:417
of lead styphnate 12:417
for magnesia–chrome ore refractories 9:227
of mercury fulminate 12:417
of milk 16:133
for mortars 9:145, 146
particle size of aerosols 4:458
of polycarbonates 17:334
of polyethylene 12:348
of polypropylene 12:348
for silicon 9:227
of tetracene 12:417
of yeast 19:523
Microstructure
of carbon black 8:187
Microtomes 4:265

Miersite 14:585
Mikamycin 5:469, 549
  analysis 5:555, 559, 563
  bioautography 5:555
  column chromatography 5:555
  identification 5:555
  microbiological assay 5:559
  paper chromatography 5:555
  properties 5:550
  separation 5:555
  solubility 5:551, 552
  ultraviolet spectroscopy 5:563
  uses 5:549
Mikamycin B
  structure 5:470
Mildex. See Dinocap.
Milk. (See also Condensed milk; Cream; Dry
    milk; and Raw milk.) 16:131
  analysis 16:131
  in bakery products 6:452
  in bread 6:439
  in cakes 6:446
  commercial specifications 16:123
  gelatin content 13:363
  Gibbs phenol reagent color test 16:133
  homogenization efficiency 16:133
  lactic acid content 15:139
  manufacture 16:131
  microscopy 16:133
  milkfat content 16:131
  pasteurization 16:131
  pasteurization efficiency 16:133
  phosphatase content 16:133
  potassium content 17:363
  sodium content 18:220
Milk and milk products. (See also Butter; Cheese;
    Condensed milk; Cream; Dry milk; Fermented
    milk products; Frozen desserts; Ice cream;
    Milk; and Raw milk.) 16:123–149
Milk chocolate. See Chocolate.
Milkfat
  Babcock test 16:127, 132, 141, 143
  in butter 16:138
  in cheese 16:141
  in chocolate 9:580
  in chocolate milk 16:132
  in condensed milk 16:145
  in cream 16:131
  in dry milk 16:145
  in frozen desserts 16:143
  Gerber test 16:127
  in ice cream 16:143
  infrared spectroscopy 16:128
  in milk 16:131
  in raw milk 16:127
  Reichert-Meissel number 9:580
  Roese-Gottlieb test 16:132

  in skim milk 16:132
  solvent extraction 16:132, 138, 141, 143, 145
  ultrasonic determination 16:128
Milk of sulfur. (See also Sulfur, elemental.)
    18:365
Milko-tester 16:128
Milk powder. See Dry milk.
Milk products
  in confectionery products 10:551
Milk protein
  in chocolate 9:540, 580
  sodium oxalate extraction 9:580
Milk solids content
  of chocolate 9:540
  of cocoa products 9:580
Millboard-type brake linings 7:440, 443
Millerite
  composition 16:286
Millon's reagent
  colorimetric determination of hydroquinone
    4:158
  color test for antioxidants 6:92
  color test for 4-*tert*-butylphenol 17:18
  color test for cresols 17:18
  color test for 2,6-di-*tert*-butylphenol 17:18
  color test for 2,6-diisopropylphenol 17:18
  color test for dimethylphenols 17:18
  color test for 4-methyl-2,6-di-*tert*-butylphenol
    17:18
  color test for methylisopropylphenols 17:18
  color test for phenol 11:542; 17:18
  color test for phenols 17:17
  color test for pyrocatechol 17:18
  color test for resorcinol 17:18
  color test for tyrosine 17:17
  gravimetric test for albumin 10:435
  gravimetric test for gelatin 10:435
  gravimetric test for gum arabic 12:580
Milo, for distilled alcoholic beverages
  analysis 4:469
  quality control 4:467
Milori blue. See Iron blues.
Mimosa
  color method for boric acid 7:389
Mineral oil
  in antifreezes 6:15
  in antirusts 6:351, 358
  in automobile polishes 6:353
  in essential oils 12:285
  refractive index 12:285
  sulfuric acid volumetric method 12:285
Mineral residue
  in bakery products 6:454
Minerals
  aluminum in 5:239, 246
  boron in 7:416
  in cereals 9:233, 241

fluoride content 13:29
in green tea leaf 18:481
iron content 15:31
magnesium content 15:358
niobium content 16:365
radium content 17:453
sample preparation 15:31
thallium content 18:522, 524
titanium content 19:96
Mineral spirits. (See also *Hydrocarbons.*)
commercial grades and specifications 14:325
degreasers 6:355
in enamels 12:175
properties 12:175
Minium. See Lead tetroxide.
Minnesotaite
composition 15:22
Minol. (See also High-explosive mixtures.)
analysis 12:449
composition 12:449
uses 12:449
Mirabilite. (See also Sodium sulfate.) 18:208, 244
Miradon. See Anisindione.
Misch metal 17:470
cerium content of 9:247
commercial grades and specifications 17:471
composition 9:246
properties 9:247
rare earth content of 9:248
uses 9:247
Mispickel 6:223
metallurgy 6:224
Mitiram 13:204
properties 13:204
solubility 13:204
toxicity 13:204
uses 13:204
Mitomycin C 5:461, 494
analysis 5:499, 503, 507
countercurrent distribution 5:499
identification 5:499
microbiological assay 5:503
properties 5:493
separation 5:499
solubility 5:494
structure 5:462
uses 5:494
Mitscherlich flash test
for white phosphorus 15:536
Mixed acids
acid–base titration 16:498
analysis 16:498
composition 16:495
nitric acid content 16:500
nitrosylsulfuric acid content 16:500
sulfuric acid content 16:499

total acidity 16:498
water content 16:500
Mix proportions test
for concrete 9:149
for mortar 9:149
Mix water
for concrete 9:143
MOCA. See 4,4'-Methylene-bis(*o*-chloroaniline.)
Modacrylic fibers. (See also *Acrylic and modacrylic fibers*; *Acrylonitrile polymers.*) 4:219, 383
heat and flame test, identification by 4:261
microscopy 4:263
properties 4:223
MOD die test
for dental investments 11:271
Modified cotton. See *Cotton.*
Modified soda. (See also Sodium sesquicarbonate.) 18:237, 239
Modified starches
in finishing agents 11:132
iodine colorimetric method 11:132
solubility 11:126
Modifiers
for alkyd resins 5:22
Modulus of elasticity
of graphite 14:77, 79, 80
Mohavite 7:314
Mohler color test
for cinnamic acid 10:60
Mohr salt titration
of gold 14:23, 37
Mohr titration
of chloride 9:342
of lithium bromide 15:282
Mohs' scale of hardness 4:31
Moissan boron
chemical reactions 7:320
manufacture 7:319
Moisture. See Water.
Moisture absorption
of polyamide fibers 17:306
Moisture conditioning
of cigarettes 10:10
Moisture meters 18:487
Moisture resistance
of match strikers 15:548
Moisture sorption
in cellulose 9:54
Molasses
adsorption by activated carbon 8:148
assay 18:341
sucrose content 18:341
Molasses value
activated carbon 8:148
Molding compounds
quality control tests 4:23

Molding index 4:23
Mold inhibitors
   in bakery products 6:454
Molds
   in casein 9:13
   in cocoa products 9:600
   in penicillins 16:566
Molecular sieves 4:434
Molecular weight
   of polyamide fibers 17:320
   of poly(butyl methacrylate) 16:71
   of poly(dodecyl methacrylate) 16:71
   of poly(ethyl methacrylate) 16:71
   of poly(methyl methacrylate) 16:71
   of poly(nitrile methacrylate) 16:71
   of poly(octadecyl methacrylate) 16:71
   of poly(octyl methacrylate) 16:71
Molecular weight, apparent
   of beer-dissolved solids 7:505
Molecular weight, average
   of polyamides 17:293
Molecular weight determination
   of acrylamide polymers 4:173
   of acrylic polymers 4:302
   of acrylonitrile polymers 4:388
   boiling point elevation 14:486
   of polyethylenimine 14:486
   of polymers 4:80
   of polytetramethylene ether glycols 13:245
Molecular weight distribution
   of acrylic polymers 4:298
   of cellulose 9:34, 40
   of elastomers, synthetic 12:118
   of epoxy resins 12:253
   of ethylene and propylene polymers 12:370
   of fatty alcohol sulfates 11:346
   of polybutadienes 12:119
   of polyisoprenes 12:119
   of polymers 4:80
   of rubber, natural 12:119
   of SBR 12:119
   of silicone fluids 18:110
   of styrene polymers 18:321
Molybdate chrome orange. See Molybdenum
   orange.
Molybdate orange
   commercial specifications 12:173; 17:197,
      198
   composition 12:173
   in enamels 12:173
   uses 17:197
Molybdates. (See also individual compounds.)
      16:183
   ammonia content 16:189
   analysis 16:184
   in antifreezes 6:12
   colorimetric method for silica 4:38

colorimetric method for silicon 4:404
gravimetric method for phosphorus 11:490
impurities, determination of 16:188
manufacture 16:183
properties 16:183
sodium content 16:188
solubility 16:183
uses 16:184
Molybdenite 16:151
   rhenium content 17:523, 529, 535
Molybdenum. (See also Molybdenum alloys;
      Molybdenum metal.) 16:150-191
   in aluminum alloys 5:138
   ammonium sulfide gravimetric method 16:153
   atomic absorption spectroscopy 9:346; 14:126;
      16:160
   in bauxite 5:116
   in beer 7:651
   benzoinoxime gravimetric method 10:316;
      16:155
   in biological material 19:217
   borax bead test 6:584
   in carbon 14:74
   in cements 9:131
   in chrome steels 9:662
   in chromium 9:639
   in chromium alloys 9:672
   in cobalt alloys 10:315
   in cocoa products 9:595, 598
   colorimetric determination 16:153, 158
   in copper alloys 10:641
   in dental alloys 10:290
   diphenylcarbazide colorimetric method 9:642
   dithiol colorimetric method 19:213
   emission spectroscopy 7:651; 9:346; 13:277;
      14:74, 122; 15:155, 187; 16:159, 387
   in ferrochromium 9:661
   in ferromolybdenum 16:171
   in ferroniobium 16:389
   flame tests 6:589, 592
   in graphite 14:74
   gravimetric determination 16:153, 154
   in hafnium 14:122, 126
   in hard facing alloys 10:289
   in high-speed steels 10:289
   in high-temperature alloys 10:286
   history 16:150
   in hydrochloric acid 14:375
   hydrogen peroxide colorimetric method 19:108
   hydrogen sulfide gravimetric method 10:315;
      16:153, 332; 19:200
   identification 16:153
   ion exchange chromatography 14:113; 16:153,
      162, 171, 185, 390; 17:526
   in iron 15:59
   iron(III) phosphate–potassium permanganate
      volumetric method 16:157

Jones reductor volumetric method 17:197
in lanthanum oxide 15:155
lead acetate gravimetric method 16:156
in lead chromate 17:197
in manganese 15:479
microcosmic salt bead test 6:587
in molybdenum alloys 16:171
in molybdenum chlorides 16:182
in molybdenum disulfide 16:178
in molybdenum oxychlorides 16:182
in molybdovanadophosphoric acid and salts
    16:185
in nickel 16:332
in nickel alloys 16:332
in niobium 16:387, 389
occurrence 16:151
in organomolybdenum compounds 16:191
potassium thiocyanate colorimetric method
    10:641
potassium xanthate–sulfuric acid colorimetric
    method 19:222
rhenium content 17:529
in rocks 15:187
separation 16:153; 19:213
in silicate rocks 19:214
in sodium chloride 9:346
in sodium dichromate 9:684
in sodium molybdate 16:184
sodium pyrophosphate fusion 16:172
sodium thiocyanate colorimetric method 16:158
sodium thiocyanate–stannous chloride colori-
    metric method 16:382
in sodium tungstate 19:222
solvent extraction 19:213, 214
in spring steels 10:290
in steel 16:382
sulfur content 18:383
in surgical alloys 10:290
in terephthalic acid 17:162
thiocyanate–stannous chloride colorimetric
    method 10:315
in titanium metal 19:121
in tungsten carbide 19:219
in uranium 19:239
x-ray fluorescent method 17:162
in zirconium 14:122, 126
of zirconium–copper–molybdenum alloys
    14:126
Molybdenum acetylacetonate. (See also Organo-
    molybdenum compounds.)
uses 16:190
Molybdenum alloys. (See also Molybdenum metal;
    and individual alloys.) 5:61, 64; 16:170–176
aluminum in 5:248
analysis 16:171
commercial grades and specifications 16:170
hydrogen in 8:638, 681

impurities, determination of 16:171
manufacture 16:170
molybdenum content 16:171
nitrogen in 8:638
oxygen in 8:620, 638, 681
titanium content 16:174
tungsten content 16:172
uses 16:150, 170
vacuum fusion methods 8:620, 681
vacuum hot extraction methods 8:638
zirconium content 16:173
Molybdenum-α-benzoin oxime
gravimetric method for tungsten 10:322
Molybdenum blue
colorimetric method for arsenic 6:244, 245;
    10:294, 619; 13:391
colorimetric method for inorganic phosphate
    6:13
colorimetric method for lecithin 9:587
colorimetric method for organic phosphate
    6:17
colorimetric method for phosphate 17:102
colorimetric method for phosphorus 7:77;
    13:392; 17:65, 72, 82, 98, 100; 19:529
colorimetric method for silica 10:114
colorimetric method for silicon 18:34
color test for formic acid 13:125
color test for vanadium 19:334
manufacture 16:177
properties 16:177
Molybdenum bromides
properties 16:182
Molybdenum chlorides. (See also individual
    compounds.)
analysis 16:182
chlorine content 16:182
hydrogen peroxide–sodium hydroxide volumetric
    method 16:182
molybdenum content 16:182
Molybdenum-cobalt steel. (See also Cobalt
    alloys.)
composition 10:289
Molybdenum compounds. (See also individual
    compounds.) 16:177–191
Molybdenum dichloride. (See also Molybdenum
    chlorides.)
solubility 16:182
Molybdenum dioxide
manufacture 16:177
properties 16:177
solubility 16:177
Molybdenum disilicide. (See also Silicon com-
    pounds.)
sodium peroxide fusion 18:13
Molybdenum disulfide
acid number 16:181
analysis 16:178

impurities, determination of 16:167, 178
manufacture 16:178
molybdenum content 16:178
properties 16:178
solubility 16:178
sulfur content 16:178
uses 16:178
Molybdenum dithiocarbamate. (See also Organo-
    molybdenum compounds.)
  uses 16:190
Molybdenum halides. (See also individual com-
    pounds.) 16:182
Molybdenum heteropoly blue
  colorimetric method for silicon 10:302
Molybdenum hexafluoride
  uses 16:182
Molybdenum hydroxides
  manufacture 16:177
  properties 16:178
  solubility 16:177
Molybdenum metal. (See also Molybdenum
    alloys.) 16:150-170
  analysis 16:161
  carrier gas fusion methods 8:625, 626, 627,
    682
  commercial grades and specifications 16:152
  hot extraction methods 8:684
  hydrogen in 8:618, 619, 620, 621, 622, 623,
    629, 630, 632, 638, 680, 684
  impurities, determination of 16:161
  manufacture 16:151
  nitrogen in 8:618, 619, 620, 621, 622, 623,
    629, 630, 638, 639
  oxygen in 8:613, 617, 618, 619, 620, 621,
    622, 623, 625, 626, 627, 629, 630, 631,
    632, 633, 638, 639, 680, 682
  properties 16:150
  solubility 16:153
  uses 16:150
  vacuum fusion methods 8:617, 618, 619, 620,
    621, 622, 623, 680
  vacuum hot extraction methods 8:638
Molybdenum orange. (See also Chromium pig-
    ments.)
  commercial grades and specifications 9:706
  molybdenum content 9:707
  uses 9:705
Molybdenum oxalate. (See also Organo-
    molybdenum compounds.)
  uses 16:190
Molybdenum oxides. (See also individual com-
    pounds.) 16:177
Molybdenum oxychlorides
  analysis 16:182
  chlorine content 16:182
  hydrogen peroxide–sodium hydroxide volu-
    metric method 16:182

molybdenum content 16:182
  oxygen content 16:182
Molybdenum pentachloride. (See also Molyb-
    denum chlorides.)
  manufacture 16:182
  properties 16:182
  solubility 16:182
Molybdenum sulfides. (See also individual
    compounds.) 16:178
Molybdenum tetrachloride. (See also Molyb-
    denum chlorides.)
  properties 16:182
Molybdenum–titanium alloys. (See also Molyb-
    denum alloys.)
  commercial specifications 16:170
  uses 16:170
Molybdenum–titanium–zirconium alloys. (See
    also Molybdenum alloys.)
  commercial specifications 16:170
  composition 16:170
  titanium content 16:176
  uses 16:170
Molybdenum trichloride. (See also Molybdenum
    chlorides.)
  manufacture 16:182
  properties 16:182
Molybdenum trioxide
  benzoin anti-oxime gravimetric method 16:181
  emission spectroscopy 16:160
  impurities, determination of 16:161
  manufacture 16:177
  in molybdenum disulfide 16:181
  properties 16:177
  solubility 16:177
Molybdenum trisulfide
  properties 16:178
Molybdenum–tungsten alloys. (See also Molyb-
    denum alloys.)
  commercial specifications 16:170
  properties 16:170
  uses 16:170
Molybdenum yellow
  colorimetric method for silicon 5:141; 15:391
Molybdic acid
  colorimetric method for hypophosphite 17:116
  colorimetric method for silica 9:126
Molybdic acid–ammonium hydroxide
  gravimetric method for phosphorus 17:101
Molybdic acid–ammonium nitrate
  gravimetric method for orthophosphate 17:61
Molybdic acid–hydrofluoric acid
  colorimetric method for silicon 14:135
Molybdic acid–sulfuric acid
  colorimetric reagent for nalorphine 16:247
Molybdisilicic acid
  colorimetric method for silicon 10:625
Molybdivanadate

colorimetric method for phosphorus 10:124
Molydivanadophosphoric acid
  colorimetric method for phosphorus 10:624,
  644
Molybdoarsenic acid. (See also Molybdates.)
  analysis 16:187
  arsenic content 16:187
Molybdosilicic acid. (See also Molybdates.)
  analysis 16:184
  colorimetric method for cyclohexanone 11:204
  silicon content 16:184
Molybdovanadate
  colorimetric method for phosphate 10:666
Molybdovanadophosphate
  colorimetric method for phosphate 17:104
Molybdovanadophosphoric acid and salts. (See
    also Molybdates.)
  analysis 16:185
  molybdenum content 16:185
  phosphorus content 16:185
  vanadium content 16:185
Monazide 17:470
Monazite 9:246; 12:395; 15:149; 17:71, 499;
    19:2
  decomposition 19:8, 12, 235
  occurrence 19:3
  rare earth content 19:12
  thorium content 19:6, 8, 12
  uses 19:3, 6
Monel 5:72
  bismuth in 7:216
Monel alloys. (See also Nickel alloys.)
  composition 16:290
Monel metal. (See also Copper alloys.)
  composition 10:631
  properties 10:631
Monier-Williams method
  for sulfur dioxide in corn products 11:9
  for sulfurous acid 18:406
6-Monoacetylmorphine. (See also Morphine.)
  potassium ferricyanide fluorometric method
    16:263
Monoalkanolamines. (See also *Alkanolamines.*)
  manufacture 5:3
  mixtures, determination in 5:15
Monoalkylol amides
  infrared spectroscopy 11:331
Monoammonium phosphate
  analysis 15:535
  commercial grades and specifications 15:520
  in matches 15:520
Monobenzyl ether hydroquinone 6:72
Monobromacetic acid
  in wines 19:486
4-Monobromofluorescein
  in D & C Red No. 21 10:523
  paper chromatography 10:523

visible spectroscopy 10:523
Monobromomethane. See Methyl bromide.
Monocalcium phosphate
  manufacture 17:121, 125
  uses 17:121
Monocarbonyl content
  of cocoa 9:543
Monochloracetic acid 8:395
  properties 8:395
  solvent extraction 19:486
  uses 8:395
  Volhard titration 19:486
  in wines 19:486
Monochlorobenzene. See Chlorobenzene.
Monochlorodifluoroethane
  in aerosols 11:95
  properties 11:95
Monochlorodifluoromethane
  gas chromatography 11:98
Monochlorohydrins. See Glycerol monochloro-
    hydrins.
Monochloromethane. See Methyl chloride.
Monochlorophenols. (See also *Chlorophenols.*)
  commercial grades and specifications 9:513
  gas chromatography 9:518
  infrared spectroscopy 9:515
  ion exchange chromatography 9:521
  manufacture 9:512
  properties 9:511
  ultraviolet spectroscopy 9:521
  uses 9:511
Monochlorosilane. (See also Chlorosilanes.)
  gas chromatography 18:51
  properties 18:48
Monochlorotoluenes. (See also Chlorinated
    toluenes.) 9:486
  composition 9:411
  gas chromatography 9:378, 488
  infrared spectroscopy 9:488
  manufacture 9:486
  properties 9:487
  uses 9:487
Monoclinic sulfur. (See also Sulfur, elemental.)
    18:364
Monocrotophos. (See also *Pesticides and
    insecticides.*)
  commercial grades and specifications 19:595
  composition 19:598
  manufacture 19:598
  properties 19:588
  solubility 19:587
  structure 19:583
  toxicity 19:592
Monoethanolamine. (See also *Ethanolamines.*)
  assay 12:292
  colorimetric determination 12:292
  properties 12:291

solubility 13:539, 552
Monoethylene glycol dinitrate
  gas chromatography 13:577
Monoethylthionine
  colorimetric method for boron 7:410
Monofluorophosphoric acid. (See also Fluoro-
    phosphoric acids.) 13:37, 42
  commercial grades and specifications 13:44
  composition, determination of 13:45
  difluorophosphoric acid content 13:45
  fluorine content 13:45
  hexafluorophosphoric acid content 13:46
  impurities, determination of 13:46
  manufacture 13:44
  monofluorophosphoric acid content 13:45
  phosphorus content 13:46
  properties 13:43
  solubility 13:44
  uses 13:43
Mono-FP acid. See Monofluorophosphoric acid.
Monoglycerides
  periodic acid volumetric method 13:515
α-Monoglycerides
  in bakery products 6:450
Monohydroxybenzene. See Phenol.
Monomers
  in methacrylate polymers 16:99
  in polyamides 17:295
2-Monomethylamino-2-methyl-1-propanol 5:5
Monomethylhydrazine. See Methylhydrazine.
Monomethylol urea 5:276
Monomethylphenols. See Cresols.
Monomolecular film test
  for fat in albumen 12:69
Mononitroanilines. (See also High explosives.)
  melting point 12:461
  properties 12:432
Mononitroanisoles. (See also High explosives.)
  melting point 12:461
  properties 12:433
  solubility 12:436
Mononitrobenzene. See Nitrobenzene.
Mononitrobenzoic acids. (See also High explo-
    sives.)
  melting point 12:461
  properties 12:433
  solubility 12:436
Mononitrophenol. See Nitrophenol.
Mononitrotoluenes. (See also High explosives.)
  melting point 12:461
  properties 12:434
  solubility 12:439
  uses 12:439
Monophenols. (See also Phenols; and individual
    compounds.)
  analysis 17:25
Monopotassium orthophosphate

properties 17:120
solubility 17:120
Monosodium acetylide. (See also Sodium com-
    pounds.)
  manufacture 18:213
Monosodium citrate. (See also Sodium citrate.)
    10:86
Monosodium glutamate
  analysis 8:473
  commercial specifications 8:473
  crystalline form 8:470
  properties 8:469
  solubility 8:471
Monosodium orthophosphate
  manufacture 17:125
  uses 17:120
Monosodium salt of 1-(4-chloro-o-sulfo-5-
    tolylazo)-2-naphthol. See D & C Red No. 8.
Monosodium salt of 4-{[4-(N-ethyl-m-sulfobenzyl-
    amino) phenyl]-[4-(N-ethyl-m-sulfonium-
    benzylamino) phenyl]-methylene}-(N,N-di-
    methyl-Δ^{2,5}-cyclohexadienimine. See F D &
    C Violet No. 1.
Monosodium salt of 2-(2-hydroxy-1-1-naphthylazo)-
    1-naphthalenesulfonic acid. See D & C Red No.
    10.
Monosodium salt of 4-m-sulfophenylazodiphenyl-
    amine. See Ext D & C Yellow No. 1.
Monosodium salt of 1-p-sulfophenylazo-2-naphthol.
    See D & C Orange No. 4.
Monostearin
  thin-layer chromatography 12:205
Monticellite 8:75
Montmorillonite. (See also Clays and clay
    minerals.) 9:191
  composition 10:102
Mooney viscosity
  of fluorocarbons, polymers 13:96
Mordant dyes 6:402
  identification test 6:390, 391
Morin
  colorimetric method for aluminum 11:140
  colorimetric method for boric acid 7:387, 410
  colorimetric method for tantalum 18:449
  colorimetric method for titanium 19:98
  colorimetric method for zirconium 11:141
  fluorometric method for aluminum 5:237,
    240
  fluorometric method for beryllium 5:118;
    7:138; 15:378
  fluorometric method for indium 14:526
  gravimetric method for niobium 16:366
  gravimetric method for tantalum 16:366, 368
Morphine. (See also Morphine hydrochloride;
    Morphine sulfate; and Narcotics.) 4:588;
    7:156; 16:258–268
  acid–base titration 16:259, 261, 267

acid–base titration, nonaqueous 16:260
4-amino-6-chloro-*m*-benzenedisulfonamide
  colorimetric method 16:262
ammonium molybdate colorimetric method
  16:247
ammonium vanadate colorimetric method
  16:247
analysis 4:594, 597; 16:259
in atropine sulfate tablets 16:261
1-chloro-2,4-dinitrobenzene gravimetric method
  16:267
colorimetric determination 16:261
color tests 16:247
commercial grades and specifications 16:258
p-dimethylaminobenzaldehyde colorimetric
  method 16:247
fluorometry 16:263
formaldehyde–sulfuric acid colorimetric method
  16:247
gas chromatography 6:210, 211; 16:254, 266
history 16:258
iodine–potassium iodide microchemical method
  16:250
ion exchange chromatography 16:267
liquid chromatography 16:265
methyl orange colorimetric method 16:263
microchemical tests 16:250
in morphine injection 16:261, 262
in morphine sulfate 16:261
in opium 16:259, 265, 266, 267
in paregoric 16:262, 265
phosphorimetry 16:265
polarography 16:267
potassium–cadmium iodide microchemical
  method 16:250
potassium ferricyanide fluorometric method
  16:263
properties 16:258
sodium nitrite colorimetric method 16:262
solubility 16:258
solvent extraction 16:259, 261
specification tests 16:261
structure 16:243
thin-layer chromatography 16:256
ultraviolet spectroscopy 16:252
in urine 16:263, 266
Morphine hydrochloride. (See also Morphine.)
  acid–base titration 16:260, 261
  solvent extraction 16:260
Morphine hydrochloride–sulfuric acid
  color test for formaldehyde 13:106
Morphine injection. (See also Morphine.)
  morphine content 16:261, 262
Morphine sulfate. (See also Morphine.)
  acid–base titration 16:260
  assay 16:261
  morphine content 16:261

phosphorimetry 16:264
solvent extraction 16:260
ultraviolet spectroscopy 16:252
Morpholine 16:192–203
  acid–base titration 16:198, 201
  addition method for acrylamide 4:164
  addition method for acrylic monomers 4:194,
    341
  addition method for acrylonitrile 4:374
  analysis 16:194
  assay 16:198
  azeotropes 16:193
  bismuth potassium iodide gravimetric method
    16:195
  bromometric titration 16:200
  carbon disulfide colorimetric method 16:200
  in citrus fruit 16:202
  color 16:199
  colorimetric method for hydroquinone 4:204
  commercial grades and specifications 16:194
  copper dimorpholinodithiocarbamate colori-
    metric method 16:202
  in corrosion inhibitors 16:200
  in cosmetics 16:201
  derivatives for identification 16:194, 197
  distillation range 16:199
  gas chromatography 16:197, 200
  gravimetric determination 16:195, 201
  history 16:192
  identification 16:194
  impurities, determination of 16:200
  infrared spectroscopy 16:195
  iodeosin colorimetric method 16:200
  iodine–potassium iodide gravimetric method
    16:195
  manufacture 16:192
  mass spectrum 16:195
  nuclear magnetic resonance spectroscopy 16:195
  properties 16:192
  properties, determination of 16:199
  reagent for amide formation with acid anhydrides
    8:564, 565
  separation 16:197
  sodium 1,2-naphthaquinone-4-sulfonate colori-
    metric method 16:201
  solubility 16:193
  specification tests 16:199
  specific gravity 16:199
  structure 16:192
  thin-layer chromatography 16:198
  toxicity 16:194
  uses 16:193
  volumetric method for acrylonitrile 16:409
  volumetric method for 3-butenenitrile 16:410
  volumetric method for ethylene oxide 12:322
  volumetric method for methacrylonitrile 16:410
  in water 16:200

x-ray powder diffraction 16:197
4-Morpholinecarboxaldehyde
  gas chromatography 16:200
  in morpholine 16:200
4-Morpholineethanol
  gas chromatography 16:200
  in morpholine 16:200
Morpholine oleates
  in automobile polishes 6:353
Morpholine-reaction method
  for acetic, propionic, or butyric anhydrides
    4:103
Morpholine stearates
  in automobile polishes 6:353
2-(2-Morpholinoethylamino)ethanol
  gas chromatography 16:200
  in morpholine 16:200
Mortars. (See also Cements; Concrete.) 9:94
  acid dissolution 9:215
  analysis 9:144, 147, 149
  compressive strength 9:136
  electron microscopy 9:146
  microscopic methods 9:145, 146
  mix proportions test 9:149
  tensile strength 9:136
  Willis-Hime test 9:148
Mosaic gold. See Stannic sulfide.
Mosses
  in finishing agents 11:126, 132
  lead acetate gravimetric method 11:127,
    132
  solubility 11:126
Motor oil
  thiophene content 18:542
MOX. (See also High-explosive mixtures.)
  analysis 12:449
  composition 12:449
  uses 12:449
MSH. See Intermedin.
Mucochloric acid
  in photographic films 12:585
  ultraviolet spectroscopy 12:585
Mud
  mercury content 16:44
Mulling technique
  specimen preparation for electron microscopy
    8:205
Mullite
  in abrasive grains 7:295
  in ceramics 9:225
  composition 9:203, 208
  microscopy 9:225
  properties 9:204, 208
  in whitewares 9:191
Mull method
  polymer samples for infrared spectroscopy
    4:233

Multiple Internal Reflection
  polymers and fibers 4:246
Multiplex spectroscopy. See Fourier transform
    spectroscopy.
Munson-Walker method
  for dextrose 4:474
  for reducing sugars 6:452; 7:563
  for starch 18:197
  for sugar 18:196, 351
Munson-Walker values
  for reducing sugars 9:583
Muntz metal 5:68
Murexide
  colorimetric method for strontium 18:276
  test for caffeine 8:58
Muriate of potash. See Potassium chloride.
Muriatic acid. See Hydrochloric acid.
Muscovite 15:261; 17:353
  composition 10:102; 17:544
  occurrence 17:544
Mustard oil. (See also Essential oils.)
  refractive index 12:271
Musts. (See also Grapes; Wines.)
  acid–base titration 19:411
  acid content 19:411
  acidity 19:411
  ash content 19:488
  definition 19:399
  nitrogen compound content 19:469
  pH 19:412
  refractive index 19:403
  sample preparation 19:400
  sampling 19:399
  soluble solid content 19:401
  specific gravity 19:402
  sucrose content 19:410
  sugar content 19:402, 405
Mylone. (See also Dithiocarbamate esters.)
  analysis 11:589
  properties 11:589
  structure 11:589
  uses 11:589
Myosmine
  in tobacco 19:136
Myrcene
  gas chromatography 12:266, 268
  in majoram, oil of 12:268
  in peppermint oil 12:266
Myricetin
  structure 18:461
Myricetin-3-glucoside
  in green tea leaf 18:461
  thin-layer chromatography 18:463
Myricetin glycosides
  colorimetric determination 18:474
Myricetin-3-rhamnoglucoside
  in green tea leaf 18:461

thin-layer chromatography 18:463
Myristic acid 5:25
  gas chromatography 8:382, 388
Myristic aldehyde
  thin-layer chromatography 12:205
Myristica oil. See Nutmeg oil.

Myristonitrile. See *n*-Tetradecanonitrile.
Myristyl alcohol. (See also Fatty alcohols.)
  commercial grades 4:563
  properties 4:561
  solubility 4:562
  specifications 4:564

N

Nabam 13:202
  manufacture 13:202
  properties 13:202
  toxicity 13:202
  uses 13:202
Nacreous sulfur. (See also Sulfur, elemental.)
  18:364
Nacrite. (See also *Clays and clay minerals.*)
  composition 10:103
Nadj and Wieden method
  for fat in cocoa products 9:576
Nafcillin. (See also Penicillins; Sodium nafcillin.)
  16:578
  assay 16:578
  bioassay 16:578
  commercial grades 16:578
  microbiological methods 16:578
  properties 16:578
  specification tests 16:578
  structure 16:555
  ultraviolet spectroscopy 16:578
Naigite 14:104
Nail enamels. (See also *Cosmetics.*) 11:70
  analysis 11:71
  bismuth oxychloride content 11:72
  composition 11:70
  formaldehyde content 11:72
  nonvolatile matter in 11:71
  solvent content 11:73
Naled. (See also *Pesticides and insecticides.*)
  commercial grades and specifications 19:595
  composition 19:598
  manufacture 19:598
  properties 19:588
  set point determination 19:609
  solubility 19:587
  structure 19:584
  toxicity 19:592
Nalorphine. (See also Nalorphine hydrobromide;
    Nalorphine hydrochloride; and *Narcotics.*)
  16:268
  ammonium molybdate colorimetric method
    16:247
  analysis 16:268
  in blood plasma 16:268
  color tests 16:247
  formaldehyde–sulfuric acid colorimetric method
    16:247
  gas chromatography 16:254
  microchemical tests 16:250
  molybdic acid–sulfuric acid colorimetric

  method 16:247
  potassium chromate microchemical method
    16:250
  potassium–mercuric iodide microchemical
    method 16:250
  radiochemical methods 16:268
  structure 16:244
  thin-layer chromatography 16:256
  ultraviolet spectroscopy 16:252
  uses 16:268
Nalorphine hydrobromide. (See also Nalorphine.)
  acid–base titration 16:268
  assay 16:268
  solvent extraction 16:268
Nalorphine hydrochloride. (See also Nalorphine.)
  assay 16:268
  ultraviolet spectroscopy 16:252, 268
Nandrolone phenpropionate. (See also Steroids.)
  manufacture 14:204
  properties 14:204
  solubility 14:204
  structure 14:204
Nantokite 10:659
Naphtha. (See also *Hydrocarbons.*)
  benzene in 7:31
  commercial grades and specifications 14:325
  composition 14:326
Naphthaldehyde
  properties 8:334
1-Naphthaldehyde
  near-infrared spectroscopy 6:629
2-Naphthaldehyde
  fluorometry 6:625, 627
  infrared spectroscopy 6:617
  ultraviolet spectroscopy 6:632
Naphthalene. (See also *Hydrocarbons.*) 13:194;
    14:325; 16:204–215
  in ammonium hydroxide 5:322
  analysis 16:208
  azeotropes 16:206
  benzylbenzoic acid–sulfuric acid colorimetric
    method 16:208
  boiling point 16:209
  color 16:212
  commercial grades and specifications 16:207
  derivatives for identification 16:208
  in diphenyl 11:528
  freezing point 16:210
  gas chromatography 7:72; 8:564; 11:528;
    13:352; 16:209; 17:154
  in gasoline 13:352, 355

history 16:204
identification 16:208
impurities, determination of 16:210, 212
liquid chromatography 7:13, 14, 15, 16
manufacture 13:194; 16:207
occurrence 14:325; 16:207
in phthalic anhydride 8:564; 17:154
picric acid volumetric method 16:209
properties 13:194, 352; 16:204
properties, determination of 16:210
separation 16:208
solidification 16:211
solubility 13:194; 16:205, 211
specification tests 16:210
structure 16:204
tetracyanoethylene colorimetric method 11:527
thin-layer chromatography 11:532
toxicity 13:194; 16:206
uses 13:194; 14:325
vapor pressure 16:205
Naphthaleneacetic acid 8:406
analysis 8:406
properties 8:403, 406
Naphthalene and naphthalene derivatives. (See
    also Naphthalene; and individual compounds.)
    16:204–242
manufacture 16:206
Naphthalene derivatives 16:216–240
1,5-Naphthalenediamine
in polyurethanes 19:304
structure 19:304
1,8-Naphthalenediamine
colorimetric method for nitrite 11:142
1,5-Naphthalene diisocyanate. (See also Iso-
    cyanates.)
trade names 19:271
1,5-Naphthalenediol
in hair dyes 11:103
thin-layer chromatography 11:103
Naphthalenediols. (See also Naphthols.)
derivatives for identification 16:230
properties 16:230
2,7-Naphthalene disulfonic acid
colorimetric method for malic acid 19:421
Naphthalenedisulfonic acids. (See also Naph-
    thalenesulfonic acids.)
derivatives for identification 16:223
Naphthalene red B. See D & C Red No. 33.
1-Naphthalenesulfonic acid. (See also Naph-
    thalenesulfonic acids.)
derivatives for identification 16:223
manufacture 16:222
properties 16:222
2-Naphthalenesulfonic acid. (See also
    Naphthalenesulfonic acids.)
derivatives for identification 16:223
manufacture 16:222

properties 16:222
volumetric method for rosin 18:186
Naphthalenesulfonic acids. (See also individual
    compounds.) 16:222
analysis 16:222
derivatives for identification 16:222
uses 16:222
Naphthalene-sulfonyl chloride
properties 8:570
1,3,5,7-Naphthalenetetrasulfonic acid. (See
    also Naphthalenesulfonic acids.)
derivatives for identification 16:223
Naphthalenetrisulfonic acids. (See also
    Naphthalenesulfonic acids.)
derivatives for identification 16:223
Naphtharsen. See Thorin.
Naphthenes. (See also Cycloalkanes.)
in hydrocarbons 14:341
Naphthenic lubricants. (See also *Lubricants.*)
commercial grades and specifications 15:307
manufacture 15:306
properties 15:307
Naphthenic olefins. (See also *Hydrocarbons*;
    and individual compounds.) 14:314
Naphthoic acid 8:369, 406
analysis 8:406
properties 8:403, 406
Naphthol
in enamels 12:184
identification 12:184
α-Naphthol. (See also Naphthols; *Phenols.*)
colorimetric determination 17:30
colorimetric method for acetanilide 6:194
colorimetric method for *p*-aminophenol 6:156
colorimetric method for antipyrine 6:214
colorimetric reagent for arginine 8:455
colorimetric method for carbon tetrachloride
    9:387
colorimetric method for chloroform 9:388
colorimetric method for cinchophen 6:205
colorimetric method for methylene chloride
    9:387
colorimetric method for trichloroethylene
    9:387
colorimetric method using diazotized *p*-nitro-
    aniline 6:106
ferric chloride colorimetric method 16:231
ferric chloride gravimetric method 16:231
in hair dyes 11:103
manufacture 16:229
in β-naphthol 16:231
properties 16:229
solubility 16:229
thin-layer chromatography 11:103
use in manufacture of azo dyes 6:403
β-Naphthol. (See also Naphthols.) 6:73
analysis 16:231

assay 16:232
in certified colors 10:503
colorimetric method for arylamines 6:409
colorimetric method for chloramphenicol 5:506
colorimetric method for *p*-phenetidine 6:194
colorimetric method using diazotized *p*-nitro-
    aniline 6:106
color test for ethylene glycol 13:570
commercial grades and specifications 16:229
in D & C Orange No. 4 10:498
in D & C Orange No. 17 10:498
in D & C Red No. 8 10:498
in D & C Red No. 9 10:498
in D & C Red No. 17 10:498
in D & C Red No. 36 10:498, 504
diazotized *p*-nitroaniline titration 16:232
ferric chloride colorimetric method 16:231
ferric chloride gravimetric method 16:231
gravimetric method for formaldehyde 13:107
impurities, determination of 16:231
iodometric titration 16:236
manufacture 16:230
p-nitrobenzenediazonium chloride chlorimetric
    method 10:504
properties 16:229
in Schaeffer's salt 16:236
solubility 16:229
specification tests 16:231
titanous chloride titration 10:503
toxicity 16:230
use in manufacture of azo dyes 6:403
use in preparation of Meldola's Blue 6:385
1-Naphthol. See α-Naphthol.
2-Naphthol. See β-Naphthol.
1-Naphthol-8-amino-3,6-disulfonic acid diazonium
    salt. See H-acid diazonium salt.
α-Naphthol–creatine
  colorimetric method for acetoin 19:457
  colorimetric method for diacetyl 19:457
2-Naphthol-3,6-disodium acid disodium salt. See
    R salt.
1-Naphthol-3,8-disulfonic acid. (See also Naph-
    tholsulfonic acids.)
  commercial grades and specifications 16:233
  solubility 16:233
2-Naphthol-3,6-disulfonic acid. (See also
    Naphtholsulfonic acids.)
  manufacture 16:233
2-Naphthol-6,8-disulfonic acid. (See also Naph-
    tholsulfonic acids.) 16:234
  manufacture 16:233
2-Naphthol-6,8-disulfonic acid dipotassium salt.
    See G salt.
Naphthol dyes 12:5
Naphthol green B. See Ext D & C Green No. 1.
Naphthols. (See also individual compounds.)
    16:229–232

analysis 16:230
derivatives for identification 16:230
ferric chloride colorimetric method 16:231
ferric chloride gravimetric method 16:231
identification 16:230
uses 16:229
1-Naphthol-4-sulfonic acid. (See also Naphthol-
    sulfonic acids.)
  commercial grades and specifications 16:233
  properties 16:233
  solubility 16:233
1-Naphthol-5-sulfonic acid. (See also Naphthol-
    sulfonic acids.)
  properties 16:233
1-Naphthol-7-sulfonic acid. (See also Naphthol-
    sulfonic acids.)
  properties 16:233
  solubility 16:233
2-Naphthol-1-sulfonic acid. (See also Naphthol-
    sulfonic acids.)
  manufacture 16:233
  properties 16:233
2-Naphthol-6-sulfonic acid. (See also Naphthol-
    sulfonic acids.)
  properties 16:233
  solubility 16:233
Naphtholsulfonic acids. (See also individual
    compounds.) 16:232–240
  analysis 16:234
  assay 16:234
  in G salt 16:239
  manufacture 16:232
  in R salt 16:237
  uses 16:232
2-Naphthol-6-sulfonic acid sodium salt. See
    Schaeffer's salt.
1-Naphthol–sulfuric acid
  color test for cellulose derivatives 12:588
Naphthol yellow S. (See also Ext D & C Yellow
    No. 7.)
  titration of methyl violet 19:175
β-Naphthoquinoline
  colorimetric method for decaboranes 7:347
  gravimetric method for niobium 16:367
  gravimetric method for tantalum 16:367
1,4-Naphthoquinone
  gas chromatography 7:72; 8:564; 17:154
  in phthalic anhydride 8:564; 17:154, 156
  stannous chloride colorimetric method 17:157
  ultraviolet spectroscopy 7:70
Naphthoquinone sulfonate
  colorimetric method for proline and hydroxy-
    proline 8:518
1,2-Naphthoquinone-4-sulfonic acid
  colorimetric method for polyethylenimine
    14:490
1-Naphthylacetic acid

paper chromatography 8:379
α-Naphthylacetonitrile. (See also *Nitriles.*)
  derivatives for identification 16:402
Naphthylamine
  use in manufacture of azo dyes 6:403
1-Naphthylamine. (See also Naphthalamines.)
  colorimetric method for hydroxylamine
    14:446, 451
  colorimetric method for nitrite 12:551
  colorimetric method for terraclor 13:215
  derivatives for identification 16:226
  manufacture 16:226
  properties 16:226
  solubility 16:226
  toxicity 16:226
2-Naphthylamine. (See also Naphthylamines.)
  in aminonaphthalenesulfonic acids 16:228
  derivatives for identification 16:226
  infrared spectroscopy 16:229
  manufacture 16:226
  properties 16:226
  solubility 16:226
  solvent extraction 16:228
  toxicity 16:226
2-Naphthylamine-4,8-disulfonic acid. (See also
    Aminonaphthalenesulfonic acids.)
  commercial grades and specifications 16:228
α-Naphthylamine maroon toner
  in enamels 12:182
  identification 12:182
Naphthylamines. (See also individual compounds.)
    16:226
  analysis 16:226
  colorimetric determination 16:227
  colorimetric method using chloroform 5:274
  identification 16:226
  manufacture 16:226
  sodium nitrite titration 16:227
  uses 16:226
α-Naphthylamine–sulfanilic acid
  colorimetric method for nitrite 19:367
1-Naphthylamine-3-sulfonic acid. (See also
    Aminonaphthalenesulfonic acids.)
  commercial specifications 16:227
  properties 16:227
1-Naphthylamine-4-sulfonic acid. (See also
    Aminonaphthalenesulfonic acids.)
  uses 16:227
1-Naphthylamine-5-sulfonic acid. (See also
    Aminonaphthalenesulfonic acids.)
  commercial grades and specifications 16:227
1-Naphthylamine-6-sulfonic acid. (See also
    Aminonaphthalenesulfonic acids.)
  commercial grades and specifications 16:227
1-Naphthylamine-7-sulfonic acid. (See also
    Aminonaphthalenesulfonic acids.)
  commercial grades and specifications 16:227

1-Naphthylamine-8-sulfonic acid. (See also
    Aminonaphthalenesulfonic acids.)
  commercial grades and specifications 16:228
2-Naphthylamine-1-sulfonic acid. (See also
    Aminonaphthalenesulfonic acids.)
  commercial grades and specifications 16:228
  dissolution 16:228
  properties 16:228
2-Naphthylamine-6-sulfonic acid. (See also
    Aminonaphthalenesulfonic acids.)
  commercial grades and specifications 16:228
1-Naphthylamine-3,6,8-trisulfonic acid. (See
    also Aminonaphthalenesulfonic acids.)
  commercial grades and specifications 16:228
1-Naphthylamine-4,6,8-trisulfonic acid
  colorimetric reagent for osmium 17:248
1-(Naphthylazo)-2-naphthol
  in D & C Red Nos. 10-13 10:521
N-(1-Naphthyl)-N-diethylpropylenediamine
  colorimetric method for acetaminophen 6:195
N-1-Naphthylenediamine
  colorimetric method for dyrene 13:222
N-(1-Naphthyl)ethylenediamine. (See Bratton-
    Marshall reaction.)
1-Naphthylethylenediamine dihydrochloride
  colorimetric method for p-aminobenzoic acid
    esters 5:389
  colorimetric method for nitrate 16:506
  colorimetric method for nitrogen dioxide 16:
    480, 484
  colorimetric method for nitrogen oxides 10:27
N-Naphthylethylenediamine hydrochloric acid
  colorimetric method for propoxycaine hydro-
    chloride 5:412
N-(1-Naphthyl)ethylenediamine hydrochloride
  colorimetric method for chloropicrin 13:187
1-Naphthyl isocyanate. (See also *Isocyanic acid
    esters.*)
  gravimetric method for cinnamyl alcohol 10:79
  infrared spectroscopy 15:106
1-Naphthyl methylcarbamate. See Carbaryl.
α-Naphthylurethane
  of higher alcohols 4:559
  of lower alcohols 4:519
Napoli extractor 6:167
Narceine. (See also *Narcotics.*) 4:588
  ammonium molybdate colorimetric method
    16:247
  ammonium vanadate colorimetric method 16:
    247
  analysis 4:597
  color tests 16:247
  formaldehyde–sulfuric acid colorimetric method
    16:247
  iodine–potassium iodide microchemical method
    16:250
  microchemical tests 16:250

paper chromatography 16:255
phosphorimetry 16:264
platinic chloride microchemical method 16:250
structure 16:244
thin-layer chromatography 16:256
ultraviolet spectroscopy 16:252
zinc–potassium iodide microchemical methods 16:250
Narcotics. (See also individual compounds.) 16:243–284
acid–base titration 16:255
analysis 16:246
color tests 16:246
definition 16:243
gas chromatography 16:253
microchemical tests 16:248
paper chromatography 16:255
thin-layer chromatography 16:255
ultraviolet spectroscopy 16:253
uses 16:243
Narcotine. (See also Narcotics; Noscapine.) 4:588
ammonium hydroxide microchemical method 16:250
ammonium molybdate colorimetric method 16:247
analysis 4:597
color tests 16:247
formaldehyde–sulfuric acid colorimetric method 16:247
microchemical tests 16:250
paper chromatography 16:255
phosphorimetry 16:264
polarography 16:268
potassium hydroxide microchemical method 16:250
sulfuric acid colorimetric method 16:247
thin-layer chromatography 16:256
ultraviolet spectroscopy 16:252
Natsyn 12:87
gel permeation chromatography 12:121
intrinsic viscosity 12:119
thermal permeation, differential 12:133
Natural black 3. See Logwood black.
Natural cement. (See also Hydraulic cements.) 9:96
commercial specifications 9:99, 100, 101
composition 9:96
Natural coloring materials
color tests 10:537
Natural gas. (See also Gas, natural.)
hydrogen sulfide content 15:569
mercaptan content 15:557, 569
Natural pigments
in soaps 18:201
Natural rubber. See Rubber, natural.
Naugatex 12:87
Naugawhite 6:74

colorimetric method using diazotized p-nitro-aniline 6:106
Naugawhite Powder 6:74
NBC 6:74
NBR. (See also Elastomers, synthetic; Acrylo-nitrile–butadiene copolymers.)
colorimetric determination 12:91
commercial grades and specifications 12:83
glass transition temperature 12:133
identification 12:90
infrared spectroscopy 12:93, 95
NBS sodium oxalate
volumetric method for oxygen in manganese dioxide 15:492
Near-infrared spectra
of aniline 5:443
of benzaldehyde 6:629
of benzyl alcohol 7:98
of decaborane 7:355
of N-ethylaniline 5:443
Near-infrared spectroscopy
of acetic acid in acetic anhydride 4:104
of n-amyl isocyanate 15:105
of aniline and its derivatives 5:434, 453
of benzaldehydes 6:628
of o-biphenyl isocyanate 15:105
of p-bisphenyl isocyanate 15:105
of p-bromophenyl isocyanate 15:105
of butadiene monoxide 12:217, 218
of n-butyl isocyanate 15:105
of chlorophenyl isocyanates 15:105
of cyclic alcohols 11:196
of 1,2-diisobutylene 12:217
of 1,3-diisopropylphenylene-4,6-diisocyanate 15:105
of 4,4′-diphenylmethane diisocyanate 15:105
of n-dodecyl isocyanate 15:105
of p-dodecylphenyl isocyanate 15:105
of epibromohydrin 12:217; 14:176
of epichlorohydrin 12:217, 218; 14:176
of epoxy compounds 12:217
of 1,2-epoxy-3-isopropoxypropane 12:217
of 1,2-epoxy-3-phenoxypropane 12:217
of epoxy resins 12:243
of p-ethoxyphenyl isocyanate 15:105
of ethyl isocyanate 15:105
of glycerol 13:527
of hydroxyl groups 12:248
of 2-isocyanatobicyclohexane 15:105
of isocyanic acid esters 15:104
of isopropylisocyanate 15:105
of malonic acid 15:426
of methoxyphenyl isocyanates 15:105
of methyl methacrylate 16:99
of nitrobenzaldehydes 6:629
of nitrophenyl isocyanates 15:105
of n-octyl isocyanate 15:105

of *m*-phenylene diisocyanate 15:105
of phenyl isocyante 15:105
of sodium malonate 15:426
of styrene oxide 12:217
of *n*-tetradecylisocyanate 15:105
of tolylene diisocyanates 15:105
of tolylisocyanates 15:105
of 4,4',4''-triisocyanatotriphenylmethane
    15:105
of water 13:527
of *m*-xylylene diisocyanate 15:105
Near-ultraviolet spectroscopy
of 1,3-butadiene 11:505
Needle felt. See *Felt.*
Nelson's chromogenic reagent 7:587
Nemafume. See Dibromochloropropane.
Nemagon. See Dibromochloropropane.
Neochlorogenic acid
    in green tea leaf 18:475
    structure 18:462
    thin-layer chromatography 18:475
Neocinchophen 6:198
    analysis 6:201
    commercial grades and specifications 6:199
    gravimetric method 6:202
    impurities, determination of 6:205
    infrared spectrum 6:201
    manufacture 6:199
    properties 6:198
    saponification method 6:204
    ultraviolet spectroscopy 6:202, 204
Neocuproine. (See also 2,9 Dimethyl-1,10-phenan-
    throline.)
    colorimetric method for copper 6:42, 226;
        9:669; 13:395; 14:129; 15:383
Neodymium. (See also Neodymium metal;
    Rare earths.)
    absorption spectroscopy 17:483
    in aluminum and aluminum alloys 5:152
    atomic absorption spectroscopy 17:494
    borax bead test 6:584
    in cerium oxide 17:501
    emission spectroscopy 15:153, 155, 375; 17:488,
        501
    in europium oxide 17:502
    flame photometry 17:492
    in gadolinium oxide 17:502
    ion exchange chromatography 15:157
    in lanthanum oxide 15:153, 155; 17:501
    in magnesium 15:375
    in magnesium alloys 15:375
    in praseodymium oxide 17:501
    in samarium oxide 17:502
    visible spectroscopy 15:154
    x-ray fluorescent spectroscopy 17:490
    in yttrium oxide 17:501
Neodymium fluorides

carrier gas distillation fusion methods 8:636
    oxygen in 8:635, 636
    vacuum distillation fusion methods 8:635
Neodymium metal. (See also Rare earth metals.)
    carrier gas fusion methods 8:627, 682
    hydrogen in 8:680
    nitrogen in 8:623, 627
    oxygen in 8:623, 627, 680, 682
    properties 17:471
    sample preparation 8:641
    vacuum fusion methods 8:623, 680
Neodymium oxide. (See also Rare earth oxides.)
    emission spectroscopy 17:501
    impurities, determination of 17:501
    properties 17:497
    uses 17:509
Neodymium perchlorate. (See also Rare earth
    compounds.)
    absorption spectrum 17:482
Neohexane
    gas chromatography 13:317, 319
Neoisothujyl alcohol
    gas chromatography 12:268
    in majoram, oil of 12:268
Neomycin 5:474, 582
    analysis 5:585, 592, 595
    bioautography 5:586
    colorimetric methods 5:595
    column chromatography 5:585
    countercurrent distribution 5:585
    identification 5:585
    ion exchange separation 5:596
    microbiological assay 5:592
    paper chromatography 5:585
    properties 5:579
    separation 5:585
    solubility 5:580, 581
    structure 5:474
    thin-layer chromatography 5:586
    uses 5:582
Neomycin palmitate
    infrared spectrum 5:586
Neomycin sulfate
    infrared spectrum 5:586
    ultraviolet spectrum 5:587
Neomycin undecylenate
    infrared spectrum 5:587
Neon. (See also *Gases, noble.*)
    in argon 13:299
    commercial specifications 13:287
    gas chromatography 13:297, 307
    in helium 13:298
    impurities, determination of 13:298
    isotopic abundance 13:289
    in krypton 13:300
    manufacture 13:290
    mass spectroscopy 13:292, 301, 304

occurrence 13:290
properties 13:288
solubility 13:289
in xenon 13:300
Neopentane. (See also Pentanes.)
   calorific value 13:331
   commercial grades and specifications 14:302
   gas chromatography 13:350
   in gasoline 13:350, 353
   properties 13:350; 14:302
   structure 14:301
Neopentyl glycol. (See also 2,2-Dimethyl-1,3-
      propanediol.) 5:22
   gas chromatography 5:31; 13:579; 19:298,
      300, 301
   paper chromatography 13:581
   in polyurethanes 19:297
Neoprene. (See also Polychloroprenes.) 12:87
Neoxanthene
   in green tea leaf 18:480
   thin-layer chromatography 18:480
Neozone A 6:75
   colorimetric method using diazotized p-nitro-
      aniline 6:98
Neozone B (and Special) 6:75
Neozone C 6:75
   colorimetric method using diazotized p-nitro-
      aniline 6:98
Neozone D 6:75
   colorimetric method using diazotized p-nitro-
      aniline 6:99
Neozone HF
   column chromatography 6:89
Nepheline syenite
   in clay 9:191
Nephelometry
   of cyclamates 11:238
   of haze in beer 7:525
   of hydrogen chloride 14:395
   of lupin alkaloids 4:609
   of polyvinylpyrrolidone in beer 7:637
Neptunium. (See also Transuranium elements.)
   half-life 19:259
   manufacture 19:259
   separation 19:259
Neptunium oxide
   carrier gas fusion methods 8:626
   oxygen in 8:626
Nerol
   in black tea 18:511
   gas chromatography 12:268; 18:511
   in majoram, oil of 12:268
Nerolidol
   in black tea 18:511
   gas chromatography 18:511
Neryl acetate
   in essential oils 12:276

molecular weight 12:276
Nessler's reagent
   color test for acetone 16:112
   color test for aldehydes 16:112
   color test for ammonia 5:305; 10:336; 19:383
   color test for ammonium group 14:372, 437;
      15:404, 407, 412; 17:111
   color test for lithium nitride 15:275
   color test for nitriles 16:399
   color test for nitrogen 15:71; 16:461; 18:233
   color test for nitrogen compounds 9:348, 351,
      365; 10:674; 13:130; 14:605; 15:398, 492;
      16:62, 344, 536; 19:222
   for polyacrylamide in water 4:178
Nessler's reagent–sodium chloride
   turbidimetric method for aldehydes 12:309
Net weight delivery
   of aerosols 11:97
Neutrality. (See also pH.)
   of carbon tetrachloride 8:326
   of dioxane 11:521
   of magnesium chloride 15:405
   of magnesium sulfate 15:412
Neutralization equivalent
   of aminoalkyl esters 4:198
   of baking acids 6:425
   of lime and limestone 15:229
Neutralization number
   of lubricants 15:313
Neutral oil
   in detergents 11:368
   in phenols 17:37
Neutron absorption
   apparatus 7:413, 415
   of boron 7:413
   of boron compounds, organic 7:351
Neutron activation analysis
   of antimony 13:265
   of barium 6:580
   of bromide residues 8:27
   of bromine 14:155
   of cadmium 8:54; 13:265
   of cesium 9:300
   of chloride 12:568
   of chlorine 14:155
   of chromium 13:265
   of cobalt 13:265
   of copper 13:265; 15:382
   of fluoride 14:155
   of gallium 13:262, 265
   of gallium arsenide 13:265, 278
   of germanium 13:265
   of gold 13:265; 14:28, 42
   of halohydrins 14:155
   of indium 13:265; 14:535
   of iodide 12:568
   of iodine 14:155

of iridium 13:265
of iron 13:265
of magnesium 15:373
of manganese 13:265
of mercury 13:265; 16:53
of metals in polymers 4:349
of nickel 13:265
of oxygen 14:134
of pesticides and insecticides 19:613
of scandium 13:265; 17:577
of selenium 13:265
of silver 13:265
of silver halides 12:568
of strontium 18:271, 281
of tantalum 13:265; 18:452
of tellurium 13:265
of thorium 19:17
of tin 13:265
of tungsten 13:265
of zinc 13:265
Nevastain A (and B) 6:75
  colorimetric method using diazotized *p*-
    nitroaniline 6:106
Nevile and Winther's acid. See 1-Naphthol-4-
    sulfonic acid 16:233
Niacin
  in bakery products 6:455, 458
  in cereals 9:233, 245
  in chocolate and cocoa products 9:538
  colorimetric method using cyanogen bromide
    6:458
  in eggs and egg products 12:66
  ion exchange chromatography 9:618
  microbiological method 10:434
  in roasted coffee 10:434
Niacinamide. See Nicotinamide.
Niccolite
  composition 16:286
Nichrome 5:72; 9:663
  composition 9:663
  uses 9:663
Nickel. (See also Nickel alloys; Nickel metal.)
    16:284–348
  in aluminum 5:138; 16:300
  in aluminum alloys 5:138; 16:310, 319
  ammonia gravimetric method 16:296
  ammonium hydroxide gravimetric method
    16:292, 296
  ammonium sulfide gravimetric method 16:292
  in antimony 16:300
  in antimony metal 6:44, 46
  in arsenic metal 6:227
  in asbestos 6:266, 267
  atomic absorption spectroscopy 6:46; 9:346;
    10:132; 14:125; 15:196; 16:321
  in bauxite 5:116
  bead tests 12:14; 16:300

  in beer 7:651, 661
  in beryllium metal 7:116
  in biological materials 16:320
  borax bead test 6:584
  in brass 16:300
  in bromine 8:6
  bromine–dimethylglyoxime colorimetric
    method 10:330
  in bronze 16:300
  in calcium metal 8:76
  in carbon 14:74
  in cements 9:131
  in chlorocarbons 9:409
  in chrome steels 9:662
  in chromic acid 9:684
  in chromite 9:700
  in chromium 9:638
  in chromium alloys 9:672
  in chrysotile 6:266, 267
  in citric acid 10:96
  in clays and clay minerals 10:128, 130
  in cobalt(II) acetate 10:330
  in cobalt alloys 10:316
  in cobalt carbonate 10:333
  in cobalt chloride 10:337
  in cobalt compounds 10:328
  in cobalt metal 10:293, 300
  in cobalt nitrate 10:341
  in cobalt(II) orthophosphate 10:343
  in cobalt oxides 10:342
  in cobalt samples 10:360
  in cobalt(II) sulfate 10:343, 344
  colorimetric determination 16:299, 316
  colorimetric method using dimethylglyoxime
    5:139; 7:116, 661
  in constant modulus alloys 10:290
  in copper 16:300
  in copper(II) acetate 10:655
  in copper alloys 10:642
  in copper–beryllium alloys 16:300
  in copper metal 10:624
  in copper–nickel–zinc alloys 16:300
  in copper(II) nitrate 10:668
  in cotton 11:141, 153
  in dental alloys 10:290
  dimethylglyoxime colorimetric method 6:44, 227;
    7:116, 661; 10:301, 624; 11:141; 14:132, 542;
    15:388; 16:299, 316
  dimethylglyoxime extraction 10:128
  dimethylglyoxime gravimetric method 5:138;
    9:672; 10:317, 330, 642; 12:533; 15:387;
    16:296, 299, 302, 304, 306, 307
  EDTA titration 9:638; 12:533; 13:445; 16:313
  in electrical resistance alloys 10:290
  electrolysis 10:317; 16:297, 301, 304, 305,307
  emission spectroscopy 7:651; 8:76; 9:346;
    11:153; 13:277; 14:48, 74, 122, 538;

15:155, 187, 197, 375; 16:160, 387
in ferrites 12:533
in ferrochromium 9:661
in ferroniobium 16:387
in ferrous alloys 16:306, 320
in ferrous metals 16:300
flame tests 6:589
in gallium arsenide 13:276
in gallium metal 13:266
in gallium phosphide 13:282
in glass and glass-ceramics 13:445
in gold 14:45
in graphite 14:74
gravimetric determination 6:266; 16:291, 296,
   300
in hafnium 14:122, 124, 132
in hard facing alloys 10:289
in high-temperature alloys 10:286
history 16:284
in hydrochloric acid 14:374
hydrogen sulfide colorimetric method 10:655,
   668; 15:491
hydrogen sulfide gravimetric method 16:292, 295
in hypochlorite bleaches 9:355
identification 16:299
in indium 14:538, 542
ion exchange chromatography 13:255; 16:298;
   17:526
in iron 15:59; 16:306, 318
isotopes 16:285
in lanthanum oxide 15:155
in lead 15:196; 16:300
in lead alloys 15:195
in low-expansion alloys 10:290
in magnesium 15:375, 387; 16:300
in magnesium alloys 15:375, 387
in manganese 15:479
in manganese sulfate 15:491
in maraging-type steels 10:290
mercury content 16:44
in metal complex dyes 12:14
microcosmic salt bead test 6:587
in molybdenum 16:161, 300
in molybdenum trioxide 16:161
neutron activation analysis 13:265
in nichrome 9:663
in nickel alloys 16:300, 322
in nickel–chromium alloys 16:300, 304
in nickel–chromium–iron alloys 16:300, 304
in nickel–copper alloys 16:300, 305
in nickel metal 16:322
in nickel-plating solutions 16:307, 313, 315
in nickel steel 16:312
in niobium 16:387
nioxime colorimetric method 10:96
in nonferrous alloys 16:320
occurrence 16:285

in organic materials 16:295
oxidation potentials 16:291
paper chromatography 16:299
polarography 16:321
potassium cyanide titration 16:310
in rocks 15:187
separation 16:295
in silicate rocks 16:320
in silver 18:173
sodium carbonate fusion 16:300
sodium carbonate gravimetric method 16:292
in sodium chloride 9:345, 346
in sodium dichromate 9:684
sodium diethyldithiocarbamate colorimetric
   method 10:128
sodium hydroxide gravimetric method 16:297
sodium peroxide gravimetric method 16:297
sodium potassium tartrate colorimetric method
   9:345
in soft magnetic alloys 10:288
in soil 16:299
solvent extraction 15:157; 16:297, 298, 312,
   313
in spring steels 10:290
in stannous fluoride 19:75
in steel 16:306, 314
in stream sediment 16:299
in surgical alloys 10:290
in tin 16:300; 19:67
toxicity 16:285
in vegetation 16:299
visible spectroscopy 16:300, 321
volumetric determination 16:310
in water 16:299
x-ray spectroscopy 9:409; 16:300, 321
zinc oxide gravimetric method 16:297
in zirconium 14:122, 124, 132; 16:300
Nickel acetate. (See also Nickel compounds.)
   manufacture 16:343
   uses 16:343
Nickel alloys. (See also Nickel metal; and individual alloys.) 5:72; 16:284–341
   aluminum in 5:228
   analysis 16:294, 322
   composition 5:72; 16:290
   hydrogen in 8:617, 681
   impurities, determination of 16:323
   manganese content 15:466
   manufacture 16:293
   nickel content 16:300, 322
   nitrogen in 8:617
   oxygen in 8:616, 617, 681
   properties 16:289
   sampling 16:294
   solubility 16:294
   titanium content 19:120
   uses 16:287

vacuum fusion methods 8:616, 617, 681
Nickel ammonium sulfate. (See also Nickel
    compounds.)
  uses 16:342
Nickel azo yellow. (See also *Pigments.*)
  structure 17:181
Nickel–boron alloys
  composition 7:322
  uses 7:323
Nickel brass. (See also Nickel alloys.)
  uses 16:289
Nickel bromide. (See also Nickel compounds.)
  uses 16:342
Nickel bronze. (See also Nickel alloys.)
  uses 16:289
Nickel carbonate. (See also Nickel compounds.)
  commercial grades and specifications 12:519
  composition 12:519; 16:342
  uses 16:341, 342
Nickel carbonyl. (See also Nickel compounds.)
  manufacture 16:341
  toxicity 16:285
  uses 16:288, 341
Nickel chloride. (See also Nickel compounds.)
  commercial specifications 16:343
  manufacture 16:342
  properties 9:335
  solubility 9:336
  uses 16:342
Nickel–chromium alloys. (See also Nickel alloys.)
  composition 16:290
  nickel content 16:300, 304
  titanium content 19:112
  uses 16:289
Nickel–chromium–iron alloys. (See also Nickel
    alloys.)
  nickel content 16:300, 304
  manganese content 15:461
  titanium content 19:112
Nickel citrate complex
  stability 10:89
Nickel coinage 5:68
Nickel compounds. (See also individual com-
    pounds.) 16:341–345
  analysis 16:344
  uses 16:341
Nickel–copper alloys. (See also Nickel
    alloys.)
  copper content 16:305
  iron content 15:49
  nickel content 16:300, 305
  sample decomposition 18:25
  silicon content 18:25
  uses 16:289
Nickel cyanide (ammoniacal)
  gravimetric method for benzene 7:25
Nickel dibutyldithiocarbamate. (See also

Dithiocarbamic acid salts.) 6:74
  uses 11:576
Nickel–dimethylglyoxime
  colorimetric method for silver 18:164
Nickel ferrite. (See also *Ferrites.*)
  Curie temperature 9:212
  magnetic constants 9:212
  magnetic properties 12:515
Nickel fluoroborate. (See also Nickel compounds.)
  commercial grades 16:342
  manufacture 16:342
  uses 16:342
Nickel formate. (See also Nickel compounds.)
  manufacture 16:343
  uses 13:120, 134; 16:343
Nickel hydroxide. (See also Nickel compounds.)
  manufacture 16:342
  solubility 16:342
  uses 16:342
Nickeliferous limonite
  composition 16:286
Nickeliferous pyrrhotite 16:285
Nickel–iron alloys. (See also Nickel alloys.)
  uses 16:289
Nickel–iron–cobalt alloys. (See also Cobalt
    alloys.)
  composition 10:288
Nickel–magnesium alloys. (See also Nickel alloys.)
  uses 16:288
Nickel–magnesium–silicon alloys. (See also Nickel
    alloys.)
  uses 16:288
Nickel–manganese alloys. (See also Nickel alloys.)
  uses 16:289
Nickel metal. (See also Nickel alloys.) 16:284–
    341
  analysis 16:294, 322
  carrier gas fusion methods 8:624, 682
  commercial grades and specifications 16:293
  hot extraction methods 8:684
  hydrogen in 8:618, 638, 680, 681, 684
  impurities, determination of 16:322
  manufacture 16:292
  nickel content 16:322
  in nickel oxide 16:322
  nitrogen in 8:618
  oxygen in 8:616, 618, 624, 680, 681, 682
  properties 16:288
  reaction medium for fusion methods 8:629,
    632
  sampling 16:294
  silicon content 18:37
  solubility 16:294, 322
  solubility in gold 14:9
  specification tests 16:322
  sulfur content 18:382
  titanium content 19:113, 120

tungsten content 19:207
uses 16:286
vacuum fusion methods 8:616, 618, 680, 681
vacuum hot extraction methods 8:638
Nickel–molybdenum–chromium–iron alloys. (See also Nickel alloys.)
uses 16:289
Nickel–molybdenum–iron alloys. (See also Nickel alloys.)
uses 16:289
Nickel nitrate. (See also Nickel compounds.)
manufacture 16:342
uses 16:342
Nickel ores
decomposition 16:294
sampling 16:294
Nickel oxide. (See also Nickel compounds.)
in alkali barium silicate glass 9:218
commercial grades and specifications 12:519
commercial specifications 16:343
composition 12:519
dimethylglyoxime colorimetric method 13:441
emission spectroscopy 9:218
in glass and glass-ceramics 9:200; 13:413, 415, 440, 461
manufacture 16:342
in nickel metal 16:322
in nickel sulfide 16:322
properties 16:342
solvent extraction 13:440
uses 16:288, 342
Nickel-plating solutions
nickel content 16:307, 313, 315
Nickel–silicon alloys. (See also Nickel alloys.)
uses 16:289
Nickel silver. (See also Nickel alloys.)
uses 16:289
Nickel stearate. (See also Nickel compounds.)
uses 16:343
Nickel steel. (See also Nickel alloys.)
nickel content 16:312
Nickel sulfamate. (See also Nickel compounds.)
manufacture 16:342
properties 16:342
uses 16:342
Nickel sulfate
analysis 16:344
commercial grades and specifications 16:343
electrophoretic method for EDTA 12:42
impurities, determination of 16:344
specification tests 16:344
uses 16:342
Nickel sulfate–potassium dithiooxalate colorimetric method for EDTA 12:46, 48
Nickel titanate
properties 17:200
uses 17:200

Nickel–zinc ferrite. (See also Ferrites.)
commercial specifications 12:520
Nickel–zinc–manganese ferrite. (See also Ferrites.)
commercial specifications 12:520
composition 12:520
Nicotinamide. (See also Pyridine and pyridine derivatives.)
acid–base titration 17:435
assay 17:435
color 17:439
commercial specifications 17:424
impurities, determination of 17:439
ion exchange chromatography 9:618
ionization constants 17:417
properties 8:597
structure 17:413
in tobacco 19:136
uses 17:413
Nicotine. (See also Pyridine and pyridine derivatives.) 4:591
analysis 4:594, 610
in cigarette smoke 10:16
gas chromatography 6:209; 19:137
paper chromatography 19:136
structure 17:414
in tobacco 19:135
trisodium pentacyanoammino ferrate colorimetric method 17:433
ultraviolet spectroscopy 10:16
uses 17:414
Nicotinic acid. (See also Niacin; Pyridine and pyridine derivatives.)
acid–base titration 17:435
assay 17:435
in cocoa beans 9:540
commercial grades and specifications 17:423
cyanogen bromide colorimetric method 17:432
impurities, determination of 17:439
manufacture 17:422
specification tests 17:439
structure 17:412
in tobacco 19:136
trisodium pentacyanoammino ferrate colorimetric method 17:433
ultraviolet spectroscopy 17:439
uses 17:413
Nicotinic acid amide. (See also Pyridine and pyridine derivatives.)
trisodium pentacyanoammino ferrate colorimetric method 17:433
Nicotinic acid diethylamide. (See also Pyridine and pyridine derivatives.)
structure 17:413
uses 17:413
Nicotinonitrile. (See also Nitriles.)
derivatives for identification 16:402
Nicotyrine

in tobacco 19:136
Nicrotung. (See also Cobalt alloys.)
  composition 10:286
Night Blue. (See also *Triphenylmethane and related dyes.*)
  thin-layer chromatography 19:173
Nigrometer 8:213
Nigrosines. (See also *Azine and related dyes.*) 6:381, 382
  identification 6:389
  manufacture 6:383, 384
  solubility 6:383
Nile Blue 6:383
  volumetric method using titanous chloride 6:395
Nilevar. See 17-Ethyl-19-nortestosterone.
Nimocast alloys. (See also Cobalt alloys.)
  composition 10:286
Nimonic alloys. (See also Cobalt alloys.)
  composition 10:286
Ninhydrin
  colorimetric method for alanine 8:450
  colorimetric method for amino acids 8:416, 432; 19:474
  colorimetric method for arginine 8:455
  colorimetric method for glutamic acid 8:473
  colorimetric method for kanamycin 5:595
  colorimetric method for metapyrilene 6:35
  colorimetric method for polymyxin 5:548
  colorimetric method for proline 8:518; 19:474
  colorimetric method for proteins 11:75
  colorimetric method for valine 8:544
  color test for propylene glycol 10:30
  spectroscopic method for thiophene 18:542
Niobates
  acid dissolution 9:215
  fluxes for 9:216
  manufacture 16:354
Niobic acid
  manufacture 16:354
  properties 16:354
Niobium. (See also Niobium alloys; Niobium metal.) 16:348–395
  acid chrome violet K colorimetric method 16:369, 374
  in aluminum and aluminum alloys 5:139, 152
  alizarin red S colorimetric method 16:368, 375
  ammonium benzeneseleninate gravimetric method 16:367
  ammonium naphthaleneseleninate gravimetric method 16:367
  ammonium pyrrolidinedithiocarbamate gravimetric method 16:366
  ammonium salicylate gravimetric method 16:367

  ammonium sulfosalicylate colorimetric method 16:368
  ammonium sulfosalicylate gravimetric method 16:367
  ammonium thiocyanate colorimetric method 16:378
  anthracene chrome violet B colorimetric method 16:369, 374
  arsenazo I colorimetric method 16:369, 374
  ascorbic acid colorimetric method 16:368, 375
  benzohydroxamic acid gravimetric method 16:367
  N-benzoylphenylhydroxylamine gravimetric method 16:366
  N-benzoylphenylhydroxylamine–thiocyanate colorimetric method 16:370, 375
  bromopyrogallol red colorimetric method 16:368, 373, 384
  in chrome steels 9:662, 673
  in chromium 9:639
  in chromium alloys 9:673
  chromotropic acid colorimetric method 16:369, 374
  N-cinnamoylphenylhydroxylamine gravimetric method 16:366
  cinnamylhydroxamic acid gravimetric method 16:367
  in cobalt alloys 10:318
  colorimetric determination 16:365, 368
  cupferron gravimetric method 10:318; 16:333, 366, 378, 386, 392
  diantipyrylmethane–thiocyanate colorimetric method 16:370
  5,7-dichloro-8-hydroxyquinoline gravimetric method 16:367
  7,8-dihydroxy-4-methylcoumarin gravimetric method 16:366
  1,2-dihydroxynaphthalene gravimetric method 16:366
  3,4-dihydroxyphenyl-4-azobenzene–thiocyanate colorimetric method 16:370, 375
  α,β-dipyridyl–pyrocatechol colorimetric method 16:370, 375
  emission spectroscopy 14:122; 16:362; 18:451
  ferric phosphate–potassium permanganate volumetric method 16:381
  ferrodipyridyl gravimetric method 16:367
  in ferroniobium 16:390
  in ferrous alloys 16:380
  formazone colorimetric method 16:370, 375
  N-furoylphenylhydroxylamine–thiocyanate colorimetric method 16:370, 375
  gallic acid gravimetric method 16:366
  gossypol colorimetric method 16:368, 375
  gravimetric determination 16:364, 366
  in hafnium 14:122

hafnium content 14:116
hematein colorimetric method 16:369
hematoxylin colorimetric method 16:369, 375
in high-temperature alloys 10:286
history 16:348
hydrochloric acid colorimetric method 16:372
hydrogen peroxide colorimetric method 10:319; 16:365, 371, 372
hydroquinone colorimetric method 16:368, 373, 383
4-hydroxyphenylfluorone colorimetric method 16:368, 375
8-hydroxyquinoline colorimetric method 16:374
8-hydroxyquinoline gravimetric method 16:367
ion exchange chromatography 10:318; 14:113; 16:153, 333, 364, 377, 385, 390; 17:527
isotopes 16:349
lumogallion colorimetric method 16:369, 374
metaphosphol gravimetric method 16:367
p-methoxybenzenethiohydroxamic acid gravimetric method 16:367
in minerals 16:365
morin gravimetric method 16:366
β-naphthoquinoline gravimetric method 16:367
in nickel 16:332
in nickel alloys 16:332
in niobium carbide 8:133, 135
in niobium–zirconium alloys 14:132
4-nitrobenzene-(-azo-1)3′,4′-dihydroxybenzene colorimetric method 16:369
nitrogen content 16:464
occurrence 18:440
oxidation potentials 16:352
phenylarsonic acid gravimetric method 16:363, 366
phenylfluorone colorimetric method 16:368, 374
phosphomolybdic heteropolyacid colorimetric method 16:372
picramin R colorimetric method 16:369, 374
potassium bisulfate fusion 16:333
potassium pyrogallosulfonate colorimetric method 16:368, 374
potassium pyrosulfate fusion 16:371
potassium thiocyanate colorimetric method 16:364, 372
4-(2-pyridylazo)resorcinol colorimetric method 16:369, 371, 373, 376, 383
pyrocatechol colorimetric method 16:368, 373
pyrocatechol violet colorimetric method 16:368
pyrogallol colorimetric method 16:368, 373
pyrogallol gravimetric method 16:367
quercetinsulfonic acid colorimetric method 16: 368, 374
quinaldinehydroxamic acid gravimetric method 16:367

in rocks 16:365, 376, 378
salicylhydroxamic acid gravimetric method 16:367
selenious acid gravimetric method 18:446
sodium diethyldithiocarbamate gravimetric method 16:366
sodium piperidinedithiocarbamate gravimetric method 16:366
stannous chloride–hydroquinone colorimetric method 16:333
in steel 16:362, 364, 376, 380, 383
2-sulfo-4-chlorophenol S colorimetric method 16:369, 374
2,4-sulfophenol R colorimetric method 16:369
sulfosalicylic acid colorimetric method 16:375
sulfurous acid gravimetric method 16:380
tannin gravimetric method 16:363, 366, 370
2-thenolyltrifluoroacetone gravimetric method 16:366
tiron colorimetric method 16:368, 373
in titanium 16:385
in titanium alloys 16:377
in titanium metal 19:121
toxicity 16:349
tribromopyrogallol colorimetric method 16:368, 373
in tungsten carbide 19:219
in uranium 16:385
urotropin–pyrogallol gravimetric method 16:366
volumetric determination 16:365
x-ray emission spectroscopy 16:362
x-ray fluorescence spectroscopy 14:132; 18:452
xylenol orange colorimetric method 16:369, 373
in zirconium 14:122; 16:385
zirconium content 14:116
Niobium alloys. (See also Ferroniobium; Niobium metal; and individual alloys.)
analysis 16:360, 388
carrier gas fusion methods 8:682
composition 16:351
ion exchange chromatography 16:390
oxygen in 8:682
properties 16:350
sample preparation 16:362
sampling 16:360
uses 16:350
Niobium carbide. (See also Carbides; Niobium alloys.)
carbon content 8:134
carrier gas fusion methods 8:624
formation 8:130
free carbon in 8:134
hydrogen in 8:624
niobium content 8:133, 135
nitrogen in 8:624
oxygen in 8:624

properties 8:131, 132; 16:350
uses 8:132; 16:350
Niobium compounds. (See also individual compounds.) 16:393
Niobium dichloride
  manufacture 16:356
Niobium–gallium alloys 13:272
  analysis 13:272
  gallium content 13:272
Niobium metal
  analysis 16:360, 388
  carrier gas fusion methods 8:624, 625, 626, 627, 682
  carrier gas hot extraction methods 8:638
  commercial grades and specifications 16:358
  hot extraction methods 8:684
  hydrogen in 8:620, 622, 624, 638, 680, 684
  impurities, determination of 16:388
  manufacture 16:349, 356
  nitrogen in 8:620, 622, 623, 624, 626, 627
  occurrence 16:349
  oxygen in 8:617, 618, 619, 620, 621, 622, 623, 624, 625, 626, 627, 632, 633, 638, 680, 682
  potassium pyrosulfate fusion 16:363
  properties 16:348, 350, 352
  sample preparation 16:362
  sampling 16:360
  solubility 16:352
  sources 16:349
  vacuum fusion methods 8:617, 618, 619, 620, 621, 622, 623, 680
  vacuum hot extraction methods 8:638
  uses 16:350; 18:442
Niobium metal, alloys, and compounds 16:348–395
Niobium ores
  analysis 16:360
  potassium pyrosulfate fusion 16:361
  sample decomposition 16:361
  sampling 16:360
  sodium pyrosulfate fusion 16:361
Niobium oxide
  carrier gas fusion methods 8:626
  oxygen in 8:626
Niobium oxychloride
  properties 16:355
Niobium oxyfluoride
  manufacture 16:356
Niobium pentabromide
  properties 16:355
Niobium pentachloride
  manufacture 16:355
  properties 16:355; 19:96
Niobium pentafluoride
  manufacture 16:356
  properties 16:355
Niobium pentaiodide

properties 16:355
Niobium pentoxide 16:393
  commercial grades and specifications 16:393
  manufacture 16:354
  properties 16:354
  solubility 16:354
  uses 16:350, 393
Niobium pentoxide hydrate. See Niobic acid.
Niobium stannide
  properties 16:350
  uses 16:350
Niobium–tantalum–tungsten alloys. (See also Niobium alloys.)
  composition 16:351
Niobium tetrachloride
  manufacture 16:356
  properties 16:355
Niobium–tin alloy. (See also Tin alloys.)
  properties 19:25
  uses 19:25
Niobium tribromide
  properties 16:355
Niobium trichloride
  manufacture 16:356
  properties 16:355
Niobium–vanadium alloys. (See also Niobium alloys.)
  composition 16:351
Niobium–zirconium alloys
  niobium content 14:132
Ni-o-nel alloys. (See also Nickel alloys.)
  composition 16:290
Nioro. (See also Indium alloys.)
  composition 14:547
  properties 14:547
  uses 14:547
Nioxime
  colorimetric method for nickel 10:96
Nipol 12:87
Nisin 5:467, 533
  analysis 5:540, 546, 548
  countercurrent distribution 5:540
  identification 5:540
  microbiological assay 5:546
  properties 5:534
  separation 5:540
  solubility 5:535, 536
  uses 5:535
Niter cake. (See also Sodium bisulfate.) 18:422
Nitramines. (See also EDNA; HMX; Nitroguanidine; RDX; and Tetryl.) 12:440
  color tests 12:459
  infrared spectroscopy 12:452
o-Nitraniline orange
  in enamels 12:184
  identification 12:184
Nitraniline reds

in enamels 12:182
identification 12:182
Nitrated benzenes. (See also individual com-
    pounds.) 13:214–222
Nitrate D P viscosity
of cellulose 9:34, 39
Nitrate esters, organic. (See also BTTN; Cellulose
    nitrate; Ethylene glycol nitrates; Glycerin
    nitrates; Mannitol hexanitrate; Mitriol tri-
    nitrate; and Pentaerythritol nitrates.) 12:443
color tests 12:459
infrared spectroscopy 12:452
nitrogen content 12:463
Nitrates. (See also individual compounds.) 16:497
in acetone 4:113
in acid salts 8:594
in aluminum nitrate 5:181
1-aminopyrine colorimetric method 16:504
in ammonium acetate 5:326
in ammonium phosphate, dibasic 5:350
in ammonium sulfate 5:352
analysis 16:500
in antifreezes 6:12
antirusts 6:351
in arsenic(V) oxide 6:233
in barium nitrate 12:420
biochemical methods 19:138
in bismuth chloride 7:200
in bismuth oxide 7:205
in bismuth subcarbonate 7:196
brown ring test 15:501
brucine colorimetric method 11:142; 16:503
brucine–sulfanilic acid colorimetric method 16:
    504
brucine sulfate colorimetric method 9:348;
    10:329; 15:404, 410
in cadmium acetate 8:37
cadmium acetate–α-naphthylamine colorimetric
    method 19:476
in cadmium chloride 8:42
in cadmium sulfate 8:47
in calcium acetate 8:78
in calcium carbonate 8:84
in cesium nitrate 9:291
in chromic acid 9:699
in cobalt(II) acetate 10:329
in cobalt carbonate 10:333
in cobalt chloride 10:336
in cobalt(II) sulfate 10:343
colorimetric determination 16:502
in copper(II) carbonate, basic 10:658
in copper(II) chloride 10:662
in cotton 11:142
detection 16:500
Devarda's alloy reduction 9:291
Devarda's alloy volumetric method 16:467, 502

dicyclohexylthallic carbonate gravimetric method
    16:501
diphenylamine colorimetric method 15:401
electrochemical methods 16:506
ferrous sulfate–permanganate volumetric method
    16:502
ferrous sulfate–sulfuric acid colorimetric method
    11:142
Franchimont color test 12:420
gasometric methods 16:506
gravimetric determination 7:125; 16:501
in hydrogen peroxide 14:437
indigo carmine colorimetric method 10:659
Kjeldahl method 16:458
in lead nitrate 12:420
in magnesium carbonate 15:401
in magnesium chloride 15:404
in magnesium oxide 15:410
in magnesium sulfate 15:412
in musts 19:470, 476
1-naphthylethylenediamine dihydrochloride
    colorimetric method 16:506
in nitrites 16:505
nitron gravimetric method 16:501
nitroxylenol colorimetric method 16:504
phenoldisulfonic acid colorimetric method
    12:420; 14:437; 16:502; 18:176
in potassium chloride 9:339
in potassium pyroantimonate 6:56
properties 16:497
reduction methods 19:138
in sea water 16:505
in silver sulfate 18:176
in sodium arsenate 6:235
in sodium chloride 9:337, 348
in sodium nitrate 12:587
solubility 16:497
sulfamic acid–sodium nitrite volumetric method
    16:502
in tobacco 19:138
toluene spectrophotometric method 16:504
volumetric determination 7:125; 16:502
in water 19:366
in wines 19:470, 476
3,4-xylenol colorimetric method 12:587
zinc–α-naphthylamine colorimetric method
    19:366
in zirconyl nitrate 14:148
Nitration method
for phenylalanine 8:511
Nitric acid 16:494
acid–base titration 14:367; 16:497
in acid mixtures 14:367
in adipic acid by column chromatography 4:415
analysis 16:497
assay 16:497

colorimetric method for apomorphine 16:246
colorimetric method for codeine 16:246
colorimetric method for p-hydroxydiphenyl-
    amine 4:210
colorimetric method for iridium 17:248
colorimetric method for warfarin sodium
    10:173
color test for vat dyes 12:18
commercial grades and specifications 16:495
etchant for ceramics 9:224
gravimetric method for cellulose 19:142
gravimetric method for phosphorus pentoxide
    15:213
gravimetric method for strontium 18:274
gravimetric method for tin 11:289; 19:33, 41
hydrofluoric acid content 13:27
impurities, determination of 16:497
manufacture 16:494
in mixed acids 16:500
properties 16:490, 496
reaction with azo dyes 6:407
specification tests 16:498
toxicity 16:497
uses 16:494
Nitric acid insolubles
in bismuth nitrate 7:203
in bismuth subcarbonate 7:196
in bismuth subnitrate 7:204
Nitric acid–potassium chlorate
gravimetric method for manganese 15:455
Nitric acid–potassium chlorate–bromine oxidation
of sulfur 18:383
Nitric anhydride. See Dinitrogen pentoxide.
Nitric-hydrochloric acid–potassium chlorate
    oxidation
of sulfur 18:383
Nitric-hydrofluoric-sulfuric acid
gravimetric method for germanium 13:398
Nitric oxide. (See also Nitrogen oxides.) 16:474
in air 16:484, 485
analysis 16:477
brown ring test 16:477
colorimetric determination 16:478
Fourier transform spectra 13:168
gas chromatography 14:416; 16:485
gasometric determination 16:477
infrared spectroscopy 16:478, 485
manufacture 16:474
properties 16:474
ultraviolet spectroscopy 16:478
volumetric determination 16:478
volumetric method for nitrogen 12:463
Nitric phosphates
manufacture 17:143
Nitrides. (See also individual compounds.) 16:470
analysis 16:471
manufacture 16:470

in metals 8:613
nitrogen content of 8:614, 628
sample decomposition 16:470
in steel 16:461
uses 16:470
Nitrile rubber. (See also Butadiene–styrene
    polymers; *Elastomers, synthetic*.) 4:383
pyrolysis–gas chromatography 12:105
Nitriles. (See also individual compounds.)
    16:396–411
in amides 8:609
analysis 16:397
assay 16:403
copper ethylacetoacetate–4,4'-tetramethyl-
    diaminodiphenylmethane colorimetric method
    16:399
derivatives for identification 16:399, 401
distillation 16:403
gas chromatography 16:403
Grignard reaction 16:400, 406
history 16:397
Hoesch reaction 16:400
hydrogen peroxide–potassium hydroxide volu-
    metric method 16:405
hydrolytic methods 16:399
hydroxylamine–ferric chloride colorimetric
    method 16:400
identification 16:397
infrared spectroscopy 8:609; 16:397, 403
ion exchange methods 16:399
lithium aluminum hydroxide volumetric method
    16:405
manufacture 16:397
mass spectroscopy 16:398
Nessler's reagent color test 16:399
nitrogen content 16:407
occurrence 16:397
polarography 16:398, 407
properties 16:396
separation 16:403
sodium reduction 16:399
solubility 16:397
spot tests 16:398
toxicity 16:397
uses 16:396
water–boron trifluoride volumetric method
    16:406
2,2',2''-Nitriloethanol. See Triethanolamine.
Nitrilotriacetic acid. (See also NTA.)
in beer 7:585
Nitrilotrispropionamide
in acrylamide 4:166
Nitrites. (See also individual compounds.) 16:490
in ammonium nitrate 5:343; 12:451
analysis 16:491
in antifreezes 6:12
antirusts 6:351

colorimetric determination 16:492
in cotton 11:142
dimethylaniline colorimetric method 16:494
in gelatin 12:550
Griess-Ilosvay colorimetric method 12:420;
    16:492
hydrazine sulfate–iodine volumetric method
    16:492
ion exchange chromatography 12:551
manufacture 16:490
1,8-naphthalenediamine colorimetric method
    11:142
α-naphthylamine-sulfanilic acid colorimetric
    method 12:551; 16:492; 19:367
in nitrates 16:505
phenylazoaniline–dimethylformamide colori-
    metric method 16:494
potassium chlorate–silver nitrate volumetric
    method 16:491
potassium iodide–starch colorimetric method
    11:142
potassium permanganate titration 16:491;
    19:368
in sewage 16:492
sulfanilamide–N-(1-naphthyl)-ethylenediamine
    dihydrochloride colorimetric method 16:494
thioglycolic acid colorimetric method 16:494
volumetric determination 16:491
in water 16:491, 492; 19:367
2,6-xylenol colorimetric method 16:494
Nitrition methods
for platinum group metals 17:251
Nitro alcohols. (See also individual compounds.)
    16:439–447
in amino alcohols 16:445
analysis 16:441
chromotropic acid colorimetric method 16:443
color 16:447
colorimetric determination 16:443
commercial grades and specifications 16:440
gas chromatography 16:441
identification 16:441
impurities, determination of 16:446
manufacture 16:440
polarography 16:445
properties 16:439
safety 16:440
solubility 16:439
toxicity 16:440
uses 16:412
Nitro alkanes. (See also Nitroparaffins.)
manufacture of primary alkanolamines 5:3
m-Nitroaniline
acid–base titration 5:447
azo dye colorimetric method 5:445
azo dyes for identification 5:430
bromination method 5:445

derivatives for identification 5:429
identification 5:432
infrared spectrum 5:442
manufacture 5:426
near-infrared spectroscopy 5:433
properties 5:423
ultraviolet spectroscopy 5:433
uses 5:428
o-Nitroaniline
azo dye colorimetric method 5:445
azo dyes for identification 5:430
bromination method 5:445
derivatives for identification 5:429
identification 5:432
infrared spectroscopy 5:442, 454
manufacture 5:426
near-infrared spectroscopy 5:435
properties 5:423
ultraviolet spectroscopy 5:433
p-Nitroaniline 5:421
acid–base titration 5:444
azo dye colorimetric method 5:445
azo dyes for identification 5:430
in beer 7:592
colorimetric method for amine antioxidants
    6:98
colorimetric method for aminopyrine 6:217
colorimetric method for PBNA 6:109
colorimetric method for phenolic antioxidants
    6:103
color reactions for identification 5:431
derivatives for identification 5:429
identification 5:432
infrared spectroscopy 5:442, 454
manufacture 5:426
near-infrared spectroscopy 5:435
properties 5:423
ultraviolet spectroscopy 5:433
vapor in air 5:455
visible spectroscopy 7:592
Nitroanilines. (See also Dinitroanilines; Mono-
    nitroanilines; 2,3,4,6-Tetranitroaniline; and
    2,4,6-Trinitroaniline.) 12:435
acid–base titration 5:445, 447
identification 5:432
separation 5:435
ultraviolet spectroscopy 5:450
Nitroanisoles. (See also 2,4-Dinitroanisole; Mono-
    nitroanisoles; Tetranitroanisoles; and 2,4,6-
    Trinitroanisole.) 12:435
Nitro aromatics. (See also Nitroanilines; Nitro-
    anisoles; Nitrobenzenes; Nitrobenzoic acids;
    Nitrocarbazoles; Nitrocresols; Nitronaph-
    thalenes; Nitrophenols; Nitrotoluenes; and
    Picrates.) 12:435
color tests 12:458
infrared spectroscopy 12:452

nitrogen content 12:466
o-Nitrobenzalacetophenone oxides. (See also
    Epoxy compounds.)
  ultraviolet spectroscopy 12:203
Nitrobenzaldehyde. (See also High explosives.)
  hydroxylamine hydrochloride method 8:357
  melting point 12:461
  properties 8:355; 12:433
m-Nitrobenzaldehyde 6:597
  commercial specifications 6:610
  derivatives for identification 6:612
  infrared spectroscopy 6:616
  near-infrared spectroscopy 6:629
  oxidation to m-nitrobenzoic acid 6:612
  properties 6:598
  ultraviolet spectroscopy 6:634, 635
o-Nitrobenzaldehyde 6:597
  colorimetric method for polyamides 17:282
  commercial grades and specifications 6:610
  derivatives for identification 6:612
  infrared spectroscopy 6:616
  manufacture 6:601
  in manufacture of o-aminobenzaldehyde 6:600
  near-infrared spectroscopy 6:629
  properties 6:598
  ultraviolet spectroscopy 6:634
p-Nitrobenzaldehyde 6:597
  commercial specifications 6:610
  derivatives for identification 6:612
  infrared spectroscopy 6:616
  manufacture 6:602
  in manufacture of p-aminobenzaldehyde 6:600
  near-infrared spectroscopy 6:629
  properties 6:598
  ultraviolet spectroscopy 6:634
p-Nitrobenzaldehyde–dinitrobenzene
  colorimetric method for hydrogen cyanide 14:420
Nitrobenzaldehydes. (See also individual com-
    pounds.)
  manufacture 6:601
  near-infrared spectroscopy 6:629
  ultraviolet spectroscopy 6:634
  uses 6:606
Nitrobenzene. (See also Dinitrobenzene; High
    explosives; Tetranitrobenzene; and Trinitro-
    benzene.) 12:436
  in aniline 5:446, 449
  in benzaldehyde 6:624
  manufacture 12:436
  manufacture of aniline and its derivatives
    5:424
  melting point 12:461
  polarography 7:34
  properties 12:433, 436
  solubility 12:436
  ultraviolet spectroscopy 5:449
  uses 12:436; 16:412

visible spectroscopy 7:17
p-Nitrobenzeneazochromotropic acid. (See also
    Chromotrope-2B.)
  colorimetric method for borate 11:141
4-Nitrobenzene-(1-azo-1)3′,4′-dihydroxybenzene
  colorimetric method for niobium 16:369
p-Nitrobenzeneazo-α-naphthol
  colorimetric method for magnesium 11:140
p-Nitrobenzeneazoorcinol
  colorimetric method for beryllium 10:309
p-Nitrobenzeneazoresorcinol
  colorimetric method for magnesium 15:359
p-Nitrobenzenediazonium
  colorimetric method for β-naphthol 10:504
p-Nitrobenzenediazonium chloride
  color method for antioxidants 6:92, 94
p-Nitrobenzenediazonium fluoroborate
  colorimetric method for Polygard 6:119
  color test for polyurethanes 19:282
p-Nitrobenzoate
  of higher alcohols 4:558
  of lower alcohols 4:518
Nitrobenzoic acid 8:404
  paper chromatography 8:379, 404
  properties 8:402, 403, 404
  uses 8:404
m-Nitrobenzoic acid
  preparation from m-nitrobenzaldehyde 6:614
Nitrobenzoic acids. (See also Dinitrobenzoic
    acids; Mononitrobenzoic acids; and 2,4,6-
    Trinitrobenzoic acid.) 12:436
p-Nitrobenzoyl chloride
  gravimetric method for cinnamyl alcohol 10:78
  method for alcohols 4:548
p-Nitrobenzyl chloride
  gravimetric method for cinnamic acid 10:59
p-Nitrobenzyl derivatives
  of barbiturates 6:478
  of benzoic acid 7:64
p-Nitrobenzyl ester
  of adipic acid 4:412
  derivatives for identification of carboxylic
    acids 8:376
4-(p-Nitrobenzyl)pyridine
  colorimetric method for ethylenimine 14:477
1-Nitrobutane. (See also Nitroparaffins.)
  gas chromatography 16:425
  infrared spectrum 16:421
  mass spectrum 16:418
  properties 16:414
  sodium hypochlorite volumetric method 16:431
2-Nitrobutane. (See also Nitroparaffins.)
  gas chromatography 16:425
  infrared spectrum 16:421
  mass spectrum 16:418
  properties 16:414
  sodium hypochlorite volumetric method 16:431

2-Nitro-1-butanol. (See also Nitro alcohols.)
  chromotropic acid colorimetric method
    16:444
  commercial specifications 16:441
  gas chromatography 16:444
  infrared spectrum 16:442
  properties 16:439
  safety 16:440
  toxicity 16:440
Nitrocarbazoles. (See also Tetranitrocarbazole.)
  12:436
Nitrocellulose. (See also Cellulose nitrate.)
  in alkyd resins 5:24
  solubility 13:539, 552
Nitrocellulose lacquers
  in chrome protectors 6:354
Nitrochloroform. See Chloropicrin.
3-Nitro-4-chlorophenyl isocyanate. (See also
    Isocyanic acid esters.)
  infrared spectroscopy 15:106
Nitrochromic acid
  colorimetric method for ethanol 12:310
Nitrocinnamic acid. (See also Cinnamic acid
    and its derivatives.)
  properties 10:63
Nitro compounds
  in phenylenediamines 11:448
  polarography 11:454
  in toluenediamines 11:453, 454
Nitro compounds, organic. (See also Nitro
    alcohols; Nitrobenzene; Nitroparaffins;
    Nitrophenols; Nitrotoluenes; and individual
    compounds.) 16:412–448
Nitrocresols. (See also Cresylite; 4,6-Dinitro-o-
    cresol; and 2,4,5,6-Tetranitro-m-cresol.)
  12:437
Nitrocyclohexane. (See also Nitroparaffins.)
  properties 16:414
m-Nitrodimethylaniline. (See also High explosives.)
  melting point 12:461
  properties 12:432
Nitro dyes 12:6
Nitroethane. (See also Nitroparaffins.)
  commercial grades and specifications 16:417
  p-diazobenzenesulfonic acid colorimetric method
    16:433
  gas chromatography 5:5; 16:425
  infrared spectrum 16:420, 436
  ionization constants 16:415
  manufacture of alkanolamines 5:3
  mass spectrum 16:418
  in nitromethane 16:436
  properties 16:413
  sodium hydroxide–ferric chloride colorimetric
    method 16:434
  sodium hypochlorite volumetric method 16:431
  toxicity 16:416

  ultraviolet spectrum 16:423
2-Nitro-1-ethanol. (See also Nitro alcohols.)
  gas chromatography 16:444
Nitroethylbenzene. (See also High explosives.)
  melting point 12:461
  properties 12:433
2-Nitro-2-ethyl-1,3-propanediol. (See also Nitro
    alcohols.)
  chromotropic acid colorimetric method 16:444
  commercial specifications 16:441
  gas chromatography 16:444
  infrared spectrum 16:443
  properties 16:439
  toxicity 16:440
Nitroferricyanide. See Sodium nitroprusside.
2-Nitro-5-fluorobenzaldehyde
  infrared spectroscopy 6:617
Nitrogen. (See also Nitrogen, elemental.) 16:
    449–470
  acetic anhydride–perchloric acid volumetric
    method 11:427
  in acetylene 4:130
  in acrylonitrile polymers 4:270, 278, 384
  in alkanolamines 5:8
  in aluminum metal 5:140
  in aluminum nitrate 5:181
  Amerine volumetric method 19:471
  in amides 8:607
  in amine curing agents 4:24
  in amino acids 8:434
  ammonia gasometric methods 16:454
  ammonia volumetric method 14:580
  in ammonium nitrate 5:340
  in argon 13:299
  in beer 7:608
  in beryllium metal 7:117
  in boron compounds, organic 7:352
  in carbide 8:135
  in carbon dioxide 8:351
  in carbon fibers 4:21
  in carbon monoxide 8:310, 312
  carrier gas and vacuum fusion methods 16:469
  carrier gas fusion methods 8:624
  in cellulose nitrate 9:66
  in cesium 9:283
  in chlorine 9:309, 320
  in chrome steels 9:662
  in chromium 9:635
  chromous reduction method 5:341
  in cigarette smoke 10:13
  in coal and coke 10:239
  in cobalt alloys 10:319
  in cobalt metal 10:301
  in cocoa beans 9:548, 549, 554
  in cocoa products 9:575
  in copper alloys 10:643
  Devarda reduction method 5:341

in diethylenetriamine 11:427
Dumas method 8:281, 292; 14:483; 16:468
in elastomers, synthetic 12:112
in ethyleneamines 11:426
in ethylenediamine 11:427
in ferrochromium 9:661
ferrous ammonium sulfate–titanous chloride
    volumetric method 12:465
in fluorocarbons 13:70
gas chromatography 8:251, 310, 312, 650;
    13:71, 297, 307, 316, 318; 14:384, 392,
    416; 16:454, 485, 523
in gelatin 12:569; 13:367
in hafnium 14:133
in helium 13:298
in high explosives 12:460
history 16:449
hot extraction methods 8:638
in hydrogen 14:380, 381, 384
impurities, determination of 16:470
in indigoid dyes 14:514
indophenol blue colorimetric method 12:113
in insulin 14:580
in iron 15:71
isotopes 16:450
Kjeldahl method 4:270, 278, 384; 9:554;
    10:239; 11:428; 12:112, 569; 13:367;
    14:133; 15:484; 16:458; 19:528
in krypton 13:300
in lead styphnate 12:415
in lithium 15:275
lithium absorption method 13:293
in lithium hydride 14:251
in magnesium 15:389
in manganese 15:484
manganese dioxide-manganese oxide fusion 16:454
mass spectroscopy 8:650; 13:302
measurement 8:648
in metal borides 7:364
in metals 8:613; 16:461
microanalysis 8:280–290
in musts 19:471
in natural gas 13:320
in neon 13:299
Nessler's reagent color test 15:71; 16:461; 18:233
in nickel 16:333
in niobium 16:389, 464
in nitrate esters, organic 12:463
nitric oxide volumetric method 12:463
in nitrides 8:614, 628
in nitriles 16:407
in nitro aromatics 12:466
in nitroguanidine 12:442
in nitronaphthalenes 16:224
in noble gases 13:293, 294, 297, 302, 307
occurrence 16:449
in organic materials 8:280

in organofunctional silanes 18:85
Orsat technique 14:381
in oxygen 16:531
in pentaethylenehexamine 11:427
2,4-pentanedione–sodium methylate volumetric
    method 11:426
perchloric acid–acetic anhydride volumetric
    method 14:485
in polyethylenimine 14:483
potassium fusion 16:454
pyrolitic hydrogenation 16:468
in pyrrocaine hydrochloride 5:414
in RDX 12:440
in refractory carbides 16:467
in refractory materials 4:59
in roasted coffee 10:417
in silicon carbide abrasives 4:59
in silylation reagents 18:94
sodium fusion 16:454
in sodium heparin 10:150
in sodium metal 18:233
specific gravity 13:329
in steel 8:665; 16:461
in tetracene 12:415
in tetraethylenepentamine 11:427
in tetryl 12:442
thermal conductivity 14:380
in titanium 16:465; 19:121
titanium absorption method 13:293
in titanium boride 7:367
titanous chloride–ferric ammonium sulfate
    volumetric method 12:467
in triethylenetetraamine 11:427
vacuum fusion methods 8:616, 628; 10:301
volumetric determination 19:471
in wines 19:471
in xenon 13:300
in yeast 19:528
in zirconium 14:133; 16:464
in zirconium nitride 14:144
Nitrogen adsorption
    by carbon black 8:186, 214
Nitrogen analyzers 16:484
Nitrogen and nitrogen compounds, inorganic
    16:449–512
Nitrogen compounds. (See also Nitrogen com-
    pounds, inorganic; Nitrogen compounds,
    organic; and individual compounds.)
aluminum–sodium hydroxide color test 14:605
in barium chromate 9:699
in barium peroxide 6:554
in copper(II) oxide 10:674
in diethyl malonate 15:434
in dimethyl malonate 15:434
in formic acid 13:130
in green coffee 10:410
Kjeldahl method 19:137

in magnesium acetate 15:398
in mercuric oxide 16:62
in musts 19:469
Nessler's reagent color test 9:348, 365; 10:674;
    13:130; 14:605; 15:398, 492; 16:62, 344,
    536; 19:222
in nickel sulfate 16:344
in perchloric acid 9:364, 365
in potassium chlorate 9:359, 361
in potassium chloride 9:339
in potassium iodate 14:609
in potassium iodide 14:605
in potassium permanganate 15:492
in roasted coffee 10:417
in sodium chloride 9:337, 348
in sodium iodide 14:605
in sodium peroxide 16:536
in sodium tungstate 19:222
in tobacco 19:137
toxicity 16:452
volumetric determination 15:434
in water 19:382
in wines 19:469
Nitrogen compounds, inorganic. (See also
    individual compounds.) 16:470–509
Nitrogen dioxide. (See also Nitrogen oxides.)
    16:475
in air 16:479, 484, 485
ammonium hexanitratocerate(IV) titration
    16:479
analysis 16:478
colorimetric determination 16:479
gas chromatography 16:485
Griess-Saltzman colorimetric method 16:479,
    484
manufacture 16:476
N-(1-naphthyl)-ethylenediamine dihydrochloride
    colorimetric method 16:480, 484
properties 16:474
thermal equilibrium 16:475
toxicity 16:476
uses 16:476
visible spectroscopy 16:485
Nitrogen, elemental
absorption methods 16:454
analysis 16:453
commercial grades and specifications 16:453
derivatives for identification 16:453
detection 16:453
emission spectroscopy 16:454
in hydrogen 16:454
manufacture 16:452
occurrence 16:449
in oxygen 16:454
properties 16:450
toxicity 16:452
uses 16:450

Nitrogen Engineering Corp.
    process for ammonia synthesis 5:297
Nitrogen fertilizers 16:506–509
    ammonia content 16:509
    analysis 16:508
    commercial grades and specifications 16:507
    composition 16:508
    specification tests 16:509
Nitrogen fluorides. (See also individual compounds.)
    13:9
Nitrogen-free extract content
    of yeast 19:533
Nitrogen, organic. (See also Nitrogen.)
    in certified colors 10:488
    Kjeldahl method 19:382
    hydriodic acid–hypophosphoric acid reduction
        10:488
    sulfuric acid digestion 10:489
    in water 19:382
Nitrogeneous base
    acid–dye method 4:592
Nitrogen oxides. (See also individual compounds.)
    16:473
    analysis 16:477, 483
    in automobile exhaust 16:481
    catalytic reduction 16:483
    in cigarette smoke 10:26
    gas chromatography 16:485
    gasometric determination 16:483
    N-(1-naphthyl)-ethylenediamine dihydro-
        chloride colorimetric method 10:27
    in nitric acid 16:498
    nitrogen oxide analyzers 16:484
    phenoldisulfonic acid colorimetric method
        16:481
    potassium permanganate titration 16:498
    separation 16:483
    volumetric determination 16:483
Nitrogen peroxides. (See also Nitrogen dioxide.)
    16:476
    properties 16:474
Nitrogen sesquioxide. (See Dinitrogen trioxide.)
Nitrogen trichloride
    in chlorine 9:309, 323
    hydrochloric acid hydrolysis 9:323
    properties 9:323
Nitrogen trifluoride
    gas chromatography 13:64
    manufacture 13:9
    properties 13:9
    solubility 13:9
    uses 13:9
Nitroglycerin. (See also High explosives.) 12:443
    commercial specifications 12:444
    infrared spectrum 12:457
    manufacture 12:443
    melting point 12:461

properties 12:432
solubility 12:443
Nitroguanidine. (See also High explosives.) 12:441
   commercial specifications 12:442
   manufacture 12:441
   nitrogen content 12:442
   properties 12:441
   solubility 12:441
   uses 12:441
4-Nitro-3-hydroxymercuri-*o*-cresol anhydride.
   See Nitromersol.
Nitromannite. See Mannitol hexanitrate.
Nitromersol. (See also Mercurials.)
   assay 11:564
   commercial specifications 11:562
   structure 11:562
Nitrometer 9:65; 12:463
Nitromethane. (See also Nitroparaffins.)
   commercial grades and specifications 16:417
   p-diazobenzene sulfonic acid colorimetric
     method 16:433
   gas chromatography 13:65; 16:425
   infrared spectrum 16:417, 436
   ionization constants 16:415
   mass spectroscopy 16:417
   properties 16:413
   sodium hypochlorite volumetric method 16:431
   sodium 1,2-naphthoquinone-4-sulfonate colori-
     metric method 16:435
   structure 16:414
   toxicity 16:416
   ultraviolet spectrum 16:423
   uses 16:412
m-Nitromethylaniline. (See also High explosives.)
   melting point 12:461
1-Nitro-2-methyl-4,6-diethyl-3,5-diisocyanato-
   benzol. (See also *Isocyanic acid esters.*)
   infrared spectroscopy 15:106
1-Nitro-3-methyl-4,6-diisocyanatobenzol. (See
   also *Isocyanic acid esters.*)
   infrared spectroscopy 15:106
1-Nitro-2-methylpropane. (See also Nitroparaffins.)
   gas chromatography 16:425
   infrared spectrum 16:421
   mass spectrum 16:418
   properties 16:414
   sodium hypochlorite volumetric method 16:431
2-Nitro-2-methylpropane. (See also Nitroparaffins.)
   gas chromatography 16:425
   infrared spectrum 16:422
   mass spectrum 16:418
   properties 16:414
2-Nitro-2-methyl-1,3-propanediol. (See also Nitro
   alcohols.)
   chromotropic acid colorimetric method 16:444
   commercial specifications 16:441

gas chromatography 16:444
infrared spectrum 16:442
properties 16:439
toxicity 16:440
2-Nitro-2-methyl-1-propanol. (See also Nitro
   alcohols.)
   acid–base titration 16:445
   chromotropic acid colorimetric method 16:444
   commercial specifications 16:441
   gas chromatography 16:444
   infrared spectrum 16:442
   properties 16:439
   toxicity 16:440
Nitron
   gravimetric method for boron 7:401
   gravimetric method for difluorophosphoric
     acid 13:46
   gravimetric method for hexafluorophosphoric
     acid 13:46
   gravimetric method for nitrates 7:125; 16:501
   gravimetric method for rhenium 17:532
   gravimetric method for technetium 17:531
   volumetric method for fluoroboric acid 13:41
α-Nitronaphthalene. (See also High explosives.)
   manufacture 12:437
   melting point 12:461
   properties 12:433
   solubility 12:437
   uses 12:437
1-Nitronaphthalene. (See also Nitronaphthalenes.)
   manufacture 16:224
   melting point 16:224
   properties 16:224
   solubility 16:224
   titanium dichloride–ferric ammonium sulfate
     volumetric method 16:225
2-Nitronaphthalene. (See also Nitronaphthalenes.)
   manufacture 16:224
   melting point 16:224
Nitronaphthalenes. (See also Dinitronaphthalene;
   α-Nitronaphthalene; Tetranitronaphthalene;
   Trinitronaphthalene; and individual compounds.) 12:437; 16:224
   analysis 16:224
   manufacture 16:224
   melting points 16:224
   nitrogen content 16:224
   tin–hydrochloric acid gravimetric method 16:225
Nitronaphthalenesulfonic acids 16:225
   manufacture 16:225
   uses 16:225
6-Nitro-2-naphthylamine-8-sulfonic acid ammonium
   salt
   fluorimetric method for tin 19:38
Nitroparaffins. (See also individual compounds.)
   16:413

analysis 16:417
classification 16:412
color 16:438
colorimetric determination 16:433
color tests 16:422
commercial grades and specifications 16:417
p-diazobenzenesulfonic acid colorimetric
method 16:433
fast blue B colorimetric method 16:422
gas chromatography 16:422, 425, 431
identification 16:417
impurities, determination of 16:437
infrared spectroscopy 16:417, 436
manufacture 16:416
mass spectroscopy 16:417, 423
properties 16:413
resorcinol–sulfuric acid colorimetric method
16:435
sodium hydroxide–ferric chloride colorimetric
method 16:434
sodium hypochlorite volumetric method 16:428
solubility 16:414
toxicity 16:416
ultraviolet spectroscopy 16:417
uses 16:412
volumetric determination 16:428
Nitrophenol. (See also 2,4-Dinitrophenol; High
explosives; and Picric acid.) 12:438
manufacture 12:438
melting point 12:461
properties 12:434
solubility 12:439
uses 12:438; 16:412
p-Nitrophenylazoorcinol
colorimetric method for beryllium 7:136;
10:620
Nitrophenylenediamines
in hair dyes 11:103
thin-layer chromatography 11:103
4-p-Nitrophenylethynylpyridine. (See also
Pyridine and pyridine derivatives.)
ionization constants 17:416
p-Nitrophenylhydrazine
gravimetric method for cinnamaldehyde 10:45
p-Nitrophenylhydrazine hydrochloride
gravimetric method for formaldehyde 13:107
p-Nitrophenylhydrazone
of acetone 4:111
of benzaldehydes 6:612, 613
p-Nitrophenylhydrazone of diacetylmonoxime
colorimetric method for cobalt 10:348
Nitrophenyl isocyanates. (See also Isocyanic
acid esters.)
infrared spectroscopy 15:106
near-infrared spectroscopy 15:105
Nitrophenylpyridines. (See also Pyridine and
pyridine derivatives.)

ionization constants 17:416
p-Nitrophenylurethane
of higher alcohols 4:559
of lower alcohols 4:519
4-Nitrophthalimide derivative of allyl halides
5:101
1-Nitropropane. (See also Nitroparaffins.)
commercial grades and specifications 16:417
p-diazobenzenesulfonic acid colorimetric method
16:433
gas chromatography 5:5; 16:425
infrared spectrum 16:420, 436
ionization constants 16:415
mass spectrum 16:418
properties 16:413
sodium hydroxide–ferric chloride colorimetric
method 16:434
sodium hypochlorite volumetric method 16:431
toxicity 16:416
2-Nitropropane. (See also Nitroparaffins.)
commercial grades and specifications 16:417
gas chromatography 16:425, 432
infrared spectra 16:420, 436
ionization constants 16:415
mass spectrum 16:418
in nitromethane 16:436
properties 16:413
resorcinol–sulfuric acid colorimetric method
16:436
sodium hypochlorite volumetric method 16:431
toxicity 16:416
ultraviolet spectrum 16:423
2-Nitropropane monoethanolamine
gas chromatography 5:5
1-Nitro-2-propanol. (See also Nitro alcohols.)
gas chromatography 16:444
properties 16:439
2-Nitro-1-propanol. (See also Nitro alcohols.)
gas chromatography 16:444
infrared spectrum 16:441
Nitropyridines. (See also Pyridine and pyridine
derivatives.)
ionization constants 17:417
manufacture 17:422
5-Nitro-8-quinolinol
color method for antioxidants 6:92
Nitrosation method
monomethyl ether of hydroquinone 4:208
p-Nitrosodimethylaniline
colorimetric method for iridium 17:249
colorimetric method for palladium 17:265
colorimetric method for platinum 17:265
use in preparation of gallocyanine 6:385
use in preparation of Meldola's Blue 6:385
Nitrosodiphenylamines
sulfuric acid color test 11:458
Nitroso dyes 12:6

α-Nitroso-β-naphthol
  colorimetric method for palladium 17:249,
    271
  colorimetric method for thiophene 18:542
  color method for antioxidants 6:92
  gravimetric method for cobalt 10:351, 637;
    16:297, 327
β-Nitroso-α-naphthol
  colorimetric method for cobalt 10:369
  colorimetric method for zirconium 11:141
1-Nitroso-2-naphthol-3,6-disulfonic acid disodium
    salt. See Nitroso-R-salt.
2-Nitroso-1-naphthol-4-sulfonic acid
  colorimetric method for cobalt 10:334, 370
p-Nitrosophenol
  colorimetric method for diphenylamine 11:457
Nitroso-R-salt
  colorimetric method for antioxidants 6:92
  colorimetric method for cobalt 5:127; 10:348,
    367, 621; 15:64, 381; 16:327
  colorimetric method for cobalt oxides 13:441
4-p-Nitrostyrylpyridine. (See also Pyridine and
    pyridine derivatives.)
  ionization constants 17:416
Nitrosyl chloride
  in chlorine 9:329
Nitrosyl peroxide. (See also Nitrogen oxides.)
  atomic absorption spectroscopy 16:476
Nitrosylsulfuric acid
  in mixed acids 16:500
  potassium permanganate titration 16:500
o-Nitrotoluene
  in manufacture of o-nitrobenzaldehyde 6:601
Nitrotoluenes. (See also Dinitrotoluenes; Mono-
    nitrotoluenes; TNT; and Trinitrotoluenes.)
    12:439
  uses 16:412
  visible spectroscopy 7:17
Nitrous acid 16:490
  manufacture 16:490
  properties 16:490
  titration of cyclamates 11:231
Nitrous oxide. (See also Nitrogen oxides.)
    16:473
  in air 16:477, 485
  analysis 5:364; 16:477
  anesthetic 5:364
  assay 5:364; 16:485
  in blood 16:477
  catalytic decomposition 16:477
  commercial grades and specifications 16:474
  Fourier transform spectra 13:174
  gas chromatography 13:65; 16:485
  gasometric determination 16:477
  impurities, determination of 5:364
  infrared spectroscopy 13:294; 16:485
  manufacture 16:473

  mass spectroscopy 13:302
  in noble gases 13:294, 302
  in oxygen 16:531, 533
  properties 5:357, 364; 16:474
  solubility 16:474
  uses 16:474
Nitrous vitriol 18:414
Nitroxylenol
  colorimetric method for nitrates 16:504
Nivco. (See also Cobalt alloys.)
  composition 10:287
Nobelium. (See also Transuranium elements.)
  manufacture 19:259
Noble gases. See Gases, noble.
Nobleite 7:314
Noble metals
  in uranium 19:239
Noise tests
  of brake linings 7:455, 457
Nomersan. See Thiram.
Nomex
  quality control tests 4:25
Nomograph
  for sodium bisulfite assay of furfural 13:237
Nonaborane(15) 7:341
1-Nonadecanol. (See also Fatty alcohols.) 4:561
2-Nonadecanone
  absorption spectroscopy 9:564
  2-hydroxy-1-naphthaldehyde colorimetric
    method 11:204
Nonamethylenediamine
  gas chromatography 17:290
Nonanal
  2,4-dinitrophenylhydrazine colorimetric method
    8:363
  gas chromatography 8:355
n-Nonane. (See also Alkanes.)
  in benzene 7:47
  commercial specifications 14:305
  gas chromatography 7:10; 13:351; 14:336
  in gasoline 13:351, 354
  properties 13:351
Nonanoic acid
  properties 8:397
1-Nonanol. (See also Fatty alcohols.) 4:561
  in black tea 18:511
  gas chromatography 7:96; 18:511
2-Nonanone
  absorption spectroscopy 9:564
  2-hydroxy-1-naphthaldehyde colorimetric
    method 11:204
n-Nonanonitrile. (See also Nitriles.)
  derivatives for identification 16:402
  properties 16:396
Nonbasic inerts
  in ammonia 5:310
  liquid phase absorption method 5:311

vapor phase absorption method 5:310
Noncarbonate hardness
  of water 19:362
Noncharring ablators 4:4
Noncondensable matter
  in methyl chloride 9:425
Nondrying alkyds 5:22
Nonfat dry milk. (See also Dry milk.)
  commercial specifications 16:144
  manufacture 16:145
Nonferrous metals
  thallium content 18:524
Nonfilterable residue
  of water 19:381
Nonflavonoid phenols
  formaldehyde gravimetric method 19:466
  gas chromatography 19:468
  separation 19:469
  in wines 19:466, 468
Nonionic detergents. (See also *Detergents.*)
    11:323
  cloud point 11:383
  cobalt nitrate–ammonium thiocyanate colori-
    metric method 11:328; 12:578; 18:183
  ion exchange chromatography 11:348
  in photographic films 12:578
  in soaps 18:183, 204
  uses 11:324
Nonionic surfactants. (See also Surfactants.)
  ion exchange chromatography 11:81
  solubility 11:79
Nonox AN (B, BL, CC, Dl, D, DPPD, EX, EXN,
    NS, S, WSL, WSP, and ZA) 6:75
Nonox ZA 6:75
  colorimetric method based on formation of
    Wurster salts 6:101, 103
Nonsteroid hormones. (See also Pancreatic
    hormones; Parathyroid hormones; Pituitary
    hormones; and Thyroid hormones.) 14:179
  analysis 14:205
  properties 14:187
  sources 14:187
  uses 14:187
Nontronite. (See also *Clays and clay minerals.*)
  composition 10:102
Nonvolatile content
  of alkyd resins 5:51
  of amino resin adhesives 4:399
  of caulking compounds 9:24
  of phenolic resin adhesives 4:402
  of vinyl resin adhesives 4:404
Nonvolatile matter
  in acetone 4:115
  in acrylonitrile 4:377
  in alcohols 4:543
  in aluminum chloride 5:177
  in ammonia 5:315

in ammonium bicarbonate 5:330
in ammonium carbonate 5:333
in ammonium chloride 5:337
in brake fluids 7:426, 431
in 1,3-butadiene 11:504, 507
in carbon tetrachloride 8:326
in chlorine 9:309, 325
in chloroform 5:358; 9:434
in creams and lotions 11:92
in dental amalgams 11:291
in dental waxes 11:320
in deodorants 11:87
in dioxane 11:521
in drycleaning solvents, used 11:609
in enamels 12:180
in ethyl chloride 5:390
in ethylene oxide 12:223, 333
in ethylenimine 14:480
in ethyl ether 5:361
in fluorosulfuric acid 13:55, 57
in furfural 13:238
in glycerol 13:521
in hair sprays 11:90
in hydrocarbon resins 14:284
in hydrocarbons 14:344
in hydrochloric acid 14:369
in hydrogen peroxide 14:436
in metal cleaners 6:368
in methanol 16:115
in methyl chloride 9:425
in methylene chloride 9:429
in molybdenum 16:169
in nail enamels 11:71
in naphthalene 16:211
in perchloroethylene 11:608
in propylene oxide 12:225
in shampoos 11:74
in Stoddard solvent 11:606
in sulfuric acid 18:421
in tetrachloroethylene 9:464
in tetrahydrofurfuryl alcohol 13:244
in 1,1,1-trichloroethane 9:443
in trichloroethylene 5:367; 9:458
in tungsten metal 19:218
in vinyl ether 5:367
in volatile esters 8:588
in waxes and polishes 6:363
in windshield washer antifreezes 6:369
Nonvolatiles with methanol
  in boric acid 7:382
Nonyl aldehyde
  ultraviolet spectroscopy 6:632
Nonyl mercaptan. (See also *Mercaptans.*)
  derivatives for identification 15:559
Nonylphenol
  colorimetric method using diazotized *p*-
    nitroaniline 6:106

Nordel 12:87
Nordenskioldine 7:315
Norepinephrine. (See also Steroids.)
  analysis 14:220
Norethandrolone. (See also Steroids.)
  manufacture 14:204
  properties 14:204
  solubility 14:204
  structure 14:204
Norethindrone. (See also Steroids.)
  analysis 10:583, 596
  assay 10:583
  blue tetrazolium colorimetric method 14:207
  gas chromatography 10:577
  identification 10:583
  isonicotinyl hydrazide colorimetric method
      14:207
  manufacture 14:202
  melting point 10:583
  in mestranol 10:577
  properties 10:583; 14:202
  separation 10:583
  solubility 10:583; 14:202
  structure 14:202
  thin-layer chromatography 10:583
  ultraviolet spectroscopy 10:584
  uses 14:202
Norethindrone acetate. (See also Steroids.)
      10:584
  analysis 10:584
  assay 10:584, 591
  blue tetrazolium colorimetric method 14:207
  fluorometry 10:592
  identification 10:584, 590
  infrared spectroscopy 10:590
  isonicotinyl hydrazide colorimetric method
      14:207
  properties 10:584
  separation 10:590
  solubility 10:584
  thin-layer chromatography 10:584, 590
  ultraviolet spectroscopy 10:584, 591
Norethynodrel. (See also Steroids.) 10:585
  analysis 10:596
  isonicotinyl hydrazide colorimetric method
      14:207
  manufacture 14:202
  properties 10:585; 14:202
  solubility 10:585; 14:202
  structure 14:202
  trade names 14:192
  uses 14:202
  visible spectroscopy 10:596
Norgestrel. (See also Steroids.)
  blue tetrazolium colorimetric method 14:207
Norleucine
  ion exchange chromatography 8:415, 421

paper chromatography 8:422
Norlevorphanol. (See also *Narcotics.*)
  ammonium molybdate colorimetric method
      16:247
  color tests 16:247
  formaldehyde–sulfuric acid colorimetric
      method 16:247
  microchemical tests 16:250
  platinic chloride microchemical method 16:250
Norlutin. See 17-Ethinyl-19-nortestosterone.
Normal distribution 7:161, 163
Normorphine. (See also *Narcotics.*)
  potassium ferricyanide fluorometric method
      16:263
  structure 16:244
  thin-layer chromatography 16:256
Nornicotine
  analysis 4:610
  gas chromatography 19:137
  paper chromatography 19:136
  in tobacco 19:135
Nornicotyrine
  in tobacco 19:136
Norpethidine. (See also *Narcotics.*)
  ultraviolet spectroscopy 16:252
19-Nor-17α-pregna-1,3,5 (10)-trien-20-yne-3,17-
      diol. See Ethinyl estradiol.
19-Nor-17α-pregn-4-en-20-yne-3β-17-diol diacetate.
      See Ethynodiol diacetate.
19-Nortestosterone
  gas chromatography 14:224
Noscopine. (See also *Narcotics;* Narcotine; and
      Noscopine hydrochloride.) 16:271
  acid–base titration 16:272
  analysis 16:271
  fluorometric method 16:272
  occurrence 16:271
  structure 16:244
  uses 16:271
Noscopine hydrochloride. (See also Noscopine.)
  acid–base titration 16:271
  assay 16:271
Nougat. (See also *Confectionery products.*)
      10:548
  manufacture 10:554
Novobiocin 5:461, 494
  analysis 5:500, 504, 507
  bioautography 5:500
  colorimetric method 5:507
  identification 5:500
  infrared spectrum 5:499
  microbiological assay 5:504
  paper chromatography 5:500
  properties 5:493
  separation 5:500
  solubility 5:494
  structure 5:462

ultraviolet spectroscopy 5:499, 507
uses 5:495
NRC vacuum fusion apparatus 8:654
NTA. (See also *EDTA and related compounds.*)
  calcium chloride volumetric method 12:33
  calcium nitrate volumetric method 12:37
  copper nitrate volumetric method 12:37
  cupric acetate titration 12:38
  cupric chloride electrophoretic method 12:44
  cupric chloride titration 12:34
  cupric nitrate titration 12:35
  iodometric method 12:28
  in iron chelates 12:53
  lanthanum nitrate–sodium hydroxide colori-
    metric method 12:28
  lead nitrate volumetric method 12:37
  magnesium chloride colorimetric method 12:28
  mercury acetate titration 12:53
  polarographic method 12:42
  structure 12:24
  thorium nitrate–hexamethylenetetraamine
    colorimetric method 12:28
  zinc nitrate titration 12:38
  zinc–Zincon colorimetric method 12:48
  zirconium–xylenol orange complex colorimetric
    method 12:38
Nuclear control rod alloy. (See also Indium
  alloys.)
  composition 14:547
  properties 14:547
  uses 14:547
Nuclear fuel
  uranium content 19:255
Nuclear magnetic resonance spectra
  of acetylsalicylic acid 6:172
  of benzyl alcohol 7:98
  of caffeine 6:173
  of cinnamyl alcohol 10:81
  of cyclohexanone 11:201
  of epichlorohydrin 14:177
  of ethylene chlorohydrin 14:162
  of glutamic acid 8:472
  of glycerol 1,3-dichlorohydrin 14:168
  of glycine 8:479
  of leucine 8:493
  of lysine 8:497
  of methionine 8:502
  of methyl cinnamate 10:67
  of methyl methacrylate–*n*-butyl methacrylate
    copolymers 16:83, 84
  of phenacetin 6:172
  of phenylalanine 8:509
  of β-phenylethyl alcohol 7:99
  of poly(*n*-butyl methacrylate) 16:82
  of poly(ethyl methacrylate) 16:81
  of poly(isobutyl methacrylate) 16:83
  of poly(*n*-propyl methacrylate) 16:82

of serine 8:521
of tyrosine 8:536
of valine 8:542
Nuclear magnetic resonance spectroscopy
  of acetals 4:66, 71
  of acetylsalicylic acid 6:174
  of alcohols 19:297
  of alkyd resins 5:50
  of bisphenol A polycarbonate 17:333
  of 1,4-butanediol dimethacrylate 16:86
  of *n*-butyl methacrylate 16:86
  of *sec*-butyl methacrylate 16:86
  of *tert*-butyl methacrylate 16:86
  of caffeine 6:174
  of carbanilides 11:551
  of carbonyl compounds 8:348
  of cephalosporins 16:560
  of cinnamaldehyde 10:48
  of coal and coke 10:214
  of diaminodiphenylmethane 19:302
  of 4′,5-dibromosalicylanilide 11:551
  of dimethylaminoethyl methacrylate 16:86
  of dimethyl terephthalate 17:165
  of dioxane 11:520
  of diphenylmethane-4,4′-diisocyanate 19:302
  of elastomers, synthetic 12:103
  of ethylene and propylene polymers 12:369
  of ethyl methacrylate 16:86
  of formaldehyde 13:108
  of glycol dimethacrylate 16:86
  of glycols 13:583
  of *n*-hexyl methacrylate 16:86
  of isobutyl methacrylate 16:86
  of isopropyl methacrylate 16:86
  of malononitrile 15:443
  of methacryl amide 16:86
  of methacrylate polymers 16:79
  of methacrylic acid 16:86
  of methacrylic acid chloride 16:86
  of methacrylonitrile 16:86
  of methanol 16:119
  of methyl methacrylate 16:86
  of methyl methacrylate–methacrylic acid
    copolymers 16:79
  of morpholine 16:195
  of oxyethylene–oxypropylene copolymers
    19:307
  of penicillins 16:560
  of phenacetin 6:174
  of phenols 17:22
  of polyamides 17:287
  of polybutadiene 12:106
  of polycarbonates 17:333
  of polyethers 17:300
  of polyethylene 12:369
  of polyhydric alcohols 13:583
  of polyisoprene 12:106

of poly(methyl methacrylate) 16:80, 84
of polypropylene 12:370
of polyurethanes 19:285
of *n*-propyl methacrylate 16:86
of pyridine and its derivatives 17:430
of salicylanilides 11:551
of silicone resins 18:121
of styrene 18:294
of styrene–butadiene copolymers 12:104
of *n*-tetradecyl methacrylate 16:86
of toluene diisocyanate 19:302
of 3,5,4'-tribromosalicylanilide 11:551
of 3,4,4'-trichlorocarbanilide 11:551
of 3-trifluoromethyl-4,4'-dichlorocarbanilide
  11:551
of triphenylmethane and related dyes 19:168
Nuclear reactor materials
  rare earth content 17:473, 477
Nutmeg oil. (See also *Essential oils.*)
  commercial specifications 12:262
  solubility 13:540
NW acid. See 1-Naphthol-4-sulfonic acid.
Nylon. (See also *Polyamides.*)
  as ablator 4:15
  microscopy 4:264
  mixtures, determination in 4:353

quality control tests 4:25
  solubility 4:248, 250, 353
Nylon fibers. See Polyamide fibers.
Nylon-6 8:117
Nystatin 5:472, 564
  acid–base titration 5:578
  analysis 5:569, 575, 578
  bioautography 5:570
  colorimetric methods 5:578
  countercurrent distribution 5:569
  identification 5:569
  infrared spectrum 5:570
  microbiological assay 5:575
  paper chromatography 5:569
  properties 5:565
  separation 5:569
  solubility 5:566, 567
  thin-layer chromatography 5:570
  ultraviolet spectroscopy 5:570, 578
  uses 5:564
Nytox ADP 6:75
Nytril. (See also *Acrylic and modacrylic fibers.*)
  heat and flame test, identification by
    4:261
  microscopy 4:264
  solubility 4:248, 250

O

Oats
  composition 9:233
  mineral content 9:233
  vitamin content 9:233
Obsidian 13:403
Ocher
  commercial specifications 17:197
  lead chromate content 17:194
Ocimene
  in black tea 18:510
  gas chromatography 18:510
Octachlorobiphenyls. (See also Chlorinated
    biphenyls.)
  properties 9:492
1,2,4,5,6,7,8,8-Octachloro-2,3,3a,4,7,7a-hexahydro-
    4,7-methanoindene. See Chlordane.
Octachloronaphthalene. (See also Chlorinated
    naphthalenes.)
  properties 9:490
  toxicity 9:490
Octachloropropane. (See also C_3 Chlorinated
    hydrocarbons.)
  properties 9:471
1,2,4,5,6,7,8,8-Octachloro-3a,4,7,7a-tetrahydro-
    4,8-methanoindane. See Chlordane.
1-Octadecanol. See Alcohols, higher; Stearyl
    alcohol.
1-Octadecene. (See also Alpha olefins.)
  commercial specifications 14:314
9-Octadecenyl-1-ol. See Oleyl alcohol.
Octadecyl isocyanate. (See also *Isocyanic acid
    esters.*)
  properties 15:95
  trade names 15:102
Octadecyltrichlorosilane. (See also Organo-
    chlorosilanes.)
  properties 18:54
1,7-Octadiene 11:512
  assay 11:512
  commercial specifications 11:513
  manufacture 11:512
  properties 11:512
  toxicity 11:512
trans,trans-3,5-Octadien-2-one
  in black tea 18:511
  gas chromatography 18:511
Octafluorocyclobutane. (See also *Fluorocarbons.*)
  in aerosols 11:95
  gas chromatography 11:98
  properties 11:95; 13:60
Octafluoropropane. (See also *Fluorocarbons.*)

properties 13:60
Octahydrophenazine
  in caprolactam 8:127
Octamethylenediamine
  gas chromatography 17:290
Octamine 6:75
  colorimetric method using diazotized *p*-
    nitroaniline 6:98
Octanal
  in black tea 18:510
  2,4-dinitrophenylhydrazine colorimetric
    method 8:363
  gas chromatography 18:510
n-Octane. (See also Octanes.)
  commercial specifications 14:305
  gas chromatography 7:10; 13:351
  in gasoline 13:351, 354
  infrared spectroscopy 7:30
  occurrence 14:303
  properties 13:351
  toxicity 14:306
Octane number 14:303
  of gasoline 13:344
Octanes. (See also Alkanes; and individual
    compounds.)
  in benzene 7:11
  flammability 14:306
  gas chromatography 7:11; 13:319
1-Octanethiol. See Octyl mercaptan.
Octanoic acid
  properties 8:397
1-Octanol. (See also Alcohols, higher; Fatty
    alcohols; and Octyl alcohols.)
  in black tea 18:511
  gas chromatography 7:96; 8:355; 18:511
  infrared spectroscopy 7:100, 101
  pyromellitic dianhydride method 4:537
  solubility 4:562
2-Octanol. (See also Alcohols, higher; Fatty
    alcohols; Octyl alcohols.)
  2,4-dinitrophenylhydrazone colorimetric
    method 4:549
  phthalic anhydride method 4:535
  properties 4:561
  pyromellitic dianhydride method 4:537
  solubility 4:562
4-Octanol
  2,4-dinitrophenylhydrazone colorimetric
    method 4:549
2-Octanone
  gas chromatography 8:355

445

2-hydroxy-1-naphthaldehyde colorimetric
  method 11:204
n-Octanonitrile. (See also *Nitriles.*)
  derivatives for identification 16:402
  properties 16:396
trans-2-Octenal
  in black tea 18:507, 511
  gas chromatography 18:511
1-Octene. (See also Alpha olefins.)
  commercial specifications 14:314
  gas chromatography 14:334
  infrared spectroscopy 7:30
2-Octene. (See also *Hydrocarbons.*)
  commercial grades and specifications 14:313
  gas chromatography 14:334
  infrared spectroscopy 7:30
4-Octene
  liquid chromatography 7:13, 15
1-Octen-3-ol
  in black tea 18:511
  gas chromatography 18:511
Octol. (See also High-explosive mixtures.)
  analysis 12:449
  composition 12:449
  uses 12:449
n-Octyl alcohol. See 1-Octanol.
Octyl alcohols. (See also Alcohols, higher;
  Fatty alcohols.)
  azeotropes 4:554
  in brake fluids 7:425
  commercial grades 4:555, 564
  properties 4:552
  specifications 4:557, 564
  uses 4:554
Octylamines
  manufacture, methods of 5:262
Octyl 3,5-dinitrobenzoate
  thin-layer chromatography 16:94
Octylene glycol
  gas chromatography 13:576
n-Octyl ether. (See also *Ethers.*)
  properties 12:297
Octyl formate. (See also Formic acid esters.)
  commercial specifications 13:132
  properties 13:131
  solubility 13:131, 134
n-Octyl isocyanate. (See also *Isocyanic acid
  esters.*)
  near-infrared spectroscopy 15:105
Octyl mercaptan. (See also *Mercaptans.*)
  derivatives for identification 15:559
  properties 15:552
Odor
  of benzene 7:47
  in casein 9:14
  of chloroform 9:434
  of drycleaning solvents, used 11:609

  of essential oils 12:264
  of ethylene oxide 12:223
  of ethyl ether 12:309
  of glues of animal origin 13:493
  of methanol 16:115
  of methyl chloride 9:425
  of methylene chloride 9:429
  of oxygen 16:532
  of perchloroethylene 11:608
  of pesticides and insecticides 19:618
  of phenol in methyl salicylate 12:290
  of Stoddard solvent 11:605
  of tetrachloroethylene 9:463
  of trichloroethylene 9:458
  of water 19:357
Odor-causing substances
  adsorption by activated carbon 8:148
Odor determination
  of acetone 4:115
  of acetophenone 4:125
  of alcohols 4:543
Odor evaluation. See *Sensory testing methods.*
Odor, foreign
  in chloroform 5:359
  in ethyl chloride 5:390
  in ethyl ether 5:361
  in vinyl ether 5:367
Odor permeation
  of packages 11:114
Ofner method
  for sugar 18:353
OHMI. See Chlordane.
Oil absorption
  of copper hydroxide 10:666
  of limestone 15:232
  of pigments 17:207
Oil conditioners 6:357
  analysis 6:371
Oil-gas tars. (See also Bituminous materials.)
  6:277
  properties 6:276
Oil immersion tests
  for silicone rubbers 18:140
Oil of vitriol. (See also Sulfuric acid.) 18:411
Oil retention
  of adsorbents 4:450
  of felt 12:512
Oil ring method
  for liquefied petroleum gas 15:244
Oils. (See also *Fats and fatty oils.*)
  adsorption by carbon black 8:226
  in aluminum powder 17:203
  in ammonia 5:313
  in antifreezes 6:14
  in carbon dioxide 8:258
  in carbon disulfide 8:275
  carbon disulfide in 8:278

in casein 9:10
colors in 10:527
decomposition 19:328
in enamels 12:170
glycerol content 13:507, 512
in green coffee 10:374
infrared spectroscopy 8:275
in lipsticks 11:67
liquid chromatography 11:67
in molybdenum disulfide 16:181
in paraffin wax 15:533
pigment content 17:177
in polyamide fibers 17:322
in roasted coffee 10:375
in seeds 12:491
vanadium content 19:328
in zinc powder 17:204
Oils, essential. See *Essential oils.*
Oil-soluble colors. (See also Certified colors;
    and individual compounds.)
  liquid chromatography 10:525
Oil tars. (See also Bituminous materials.)
    6:276
Oil well cements. (See also Hydraulic cements.)
    9:96
  commercial specifications 9:99, 100, 101
  composition 9:96
  identification 9:108
Ointments
  iodoform content 14:614
Oiticica oil
  in brake linings 7:441
Oleandomycin 5:464, 519
  analysis 5:524, 528, 532
  bioautography 5:524
  colorimetric methods 5:532
  identification 5:524
  microbiological assay 5:528
  paper chromatography 5:524
  properties 5:518
  separation 5:524
  solubility 5:519, 520
  structure 5:464
  uses 5:519
Oleandomycin chloroform adduct
  infrared spectrum 5:523
Olefins. (See also *Hydrocarbons*; and individual
    compounds.) 14:309
  in aromatic hydrocarbons 14:345
  gas chromatography 14:329, 331, 334
  in hydrogen 14:381
  in liquefied petroleum gas 15:244
  liquid chromatography 14:336
  manufacture 14:309
  properties 14:309
  silver–mercuric nitrate gasometric method
    15:244

uses 14:309, 312
1-Olefins. See Alpha olefins.
Oleic acid. (See also Fatty acids.) 5:25
  gas chromatography 8:388
  properties 12:473
Oleum. (See also Sulfuric acid.) 18:411
  boiling point 18:413
  density 18:412
  freezing point 18:413
  vapor pressure 18:412
Oleyl alcohols. (See also Fatty alcohols.)
  commercial grades 4:563
  properties 4:561
  specifications 4:564
  thin-layer chromatography 12:204
Oliensis spot test
  for bituminous materials 6:309
Oligomers
  in polyamides 17:295
Oligosaccharide content
  of green coffee 10:392
Olive oil. (See also *Fats and fatty oils.*)
  in alkyd resins 5:23
  properties 12:473
  solubility 13:539, 552
Olivine 11:464; 15:357, 358
Omphacite 8:75
Onyon method
  osmotic pressure of acrylic polymers 4:312
Oolong tea. (See also *Tea.*)
  manufacture 18:456
Opacimeter 10:32
Opacity
  of cigarette paper 10:31
  of dental silicate cements 11:260
  of pigments 17:206
Opalescents
  in methacrylate polymers 16:97
Opal glass. (See also *Glass and glass-ceramics.*)
  composition 13:407, 410
  uses 13:410
Opaline
  composition 10:104
Opaque glazes. (See also Glazes.)
  composition 9:196
Opium. (See also Morphine; *Narcotics.*) 7:155;
    16:258–268
  analysis 16:259
  codeine content 16:266
  composition 16:258
  manufacture 16:258
  morphine content 16:259, 265, 266, 267
Opium alkaloids 4:587
  analysis 4:597
Oppanol 12:87
Optical barium crown glass. (See also *Glass
    and glass-ceramics.*)

composition 13:409
Optical bleaches
  fluorescence test 18:183
  in soaps 18:183
Optical brighteners
  in detergents 11:334
Optical emission spectroscopy
  of ceramics 9:217
Optical glass. (See also *Glass and glass-ceramics.*)
  composition 13:408, 410
  uses 13:411
Optical rotation
  of amino acids 8:443
  of essential oils 12:270
  of grapefruit oil 12:270
  of invertase 19:534
  of lemon oil 12:270
  of orange oil 12:270
Orange B. (See also Certified colors.)
  classification 10:450
  commercial specifications 10:464
  impurities, determination of 10:500
  liquid chromatography 10:500
  structure 10:450
  titanous chloride titration 10:485
  visible spectroscopy 10:486
Orange oil. (See also *Essential oils.*)
  commercial specifications 12:261
  decyl aldehyde content 12:282
  optical rotation 12:270
  refractive index 12:271
  residue on evaporation 12:273
  solubility 13:540
  ultraviolet spectroscopy 12:288
Orange organic pigments
  in enamels 12:184
  identification 12:184
Orange pekoe tea. (See also *Tea.*) 18:485
Orange pigments. (See also *Pigments*; and
    individual compounds.) 17:196
Oranges
  colors in 10:527
Orange toner
  in enamels 12:182
  identification 12:182
Orange II. (See also D & C Orange No. 4.)
  colorimetric method for cationic surfactants
    11:135
  titration of brilliant green 19:175
  titration of malachite green 19:174
Orcein 6:383
Orcinol
  colorimetric method for alginate 12:584
Ores
  antimony in 6:58
  boron in 7:413, 416
  gold content 17:266

palladium content 17:266
platinum content 17:266
radium content 17:453
rhodium content 17:266
sample preparation 6:59; 7:209; 17:268
selenium content 17:589
silver content 17:266; 18:172
sulfur content 18:372, 380
tantalum content 18:451
tellurium content 17:589
thallium content 18:524
thorium content 19:7, 16
Organic acids
  acid–base titration 12:147; 19:148
  in cocoa beans 9:565
  in elastomers, synthetic 12:147
  in glycerol 13:524
  in green coffee 10:395
  paper chromatography 9:567
  in roasted coffee 10:430
  Rohan and Stewart method 9:566
  steam distillation 19:148
  in tobacco 19:148
Organic additives
  solvent extraction 18:321
  in styrene polymers 18:321
Organic carbon, total
  in water 19:379
Organic chlorides
  in methanol 16:118
Organic chlorosilanes. See Organo-chlorosilanes.
Organic colors
  in chromium pigments 9:707
  in pigments 17:177
Organic compounds
  in bromine 8:6
  bromine in 8:18, 22
  carrier gas fusion methods 8:627
  decomposition with sodium diphenyl 8:18
  oxygen content 16:533
  oxygen in 8:627
  peroxide bomb method 8:18
  pyrolytic methods 16:533
Organic dyes
  analysis 15:532
  foam-stability loss of glue 15:532
  in matches 15:517, 532
  potassium dichromate compatibility 15:532
  solubility 15:532
  tinctorial strength 15:532
Organic esters
  in glycerol 13:524
Organic lakes
  in chromium pigments 9:707
Organic materials
  adsorption methods 19:379
  in ammonium hydroxide 5:324

analysis 17:363
antimony in 6:58, 62
in asbestos 6:266
in barium sulfate 6:567
carbon, hydrogen, and nitrogen content 8:280
in chrysotile 6:266
cobalt in 10:349
combustion 8:282
condensation methods 14:399
decomposition 10:349, 602
in diatomaceous earth 11:497
in diphenylamine 11:459
in furfural 13:239
in furfuryl alcohol 13:242
in glycerol 13:521, 527
in hydrochloric acid 14:375
in hydrogen chloride 14:399
infrared spectroscopy 14:401
iron content 15:32
lead content 15:170
magnesium content 15:359
manganese content 15:454
mercury content 16:23
nickel content 16:295
oxidation methods 19:379
in platinum group metals 17:245
in portland cement 9:110
potassium content 17:363
potassium dichromate oxidation 19:375
sample decomposition 18:378
sample preparation 6:62, 237; 17:363
Schöninger method 8:22
selenium content 17:589
separation 19:379
in shampoos 11:75
solvent extraction 19:379
sulfur content 18:378
sulfuric acid–persulfate digestion 19:369
tellurium content 17:589
in tetrahydrofuran 13:231
thallium content 18:523
titanium content 19:113
in water 19:369, 371, 378, 382, 384
Organic materials, extractable
gas chromatography 9:350
in hydrochloric acid 9:340, 350
Organic mercury compounds. See Organomercury
compounds.
Organic nitro compounds. See *Nitro compounds,
organic.*
Organic pigments
analysis 15:532
commercial grades and specifications 15:517
in matches 15:516
particle size 15:532
tinctorial strength 15:532
Organic reducing substances

in carbon dioxide 8:260
Organic silane esters. See Organosilane esters.
Organoaluminum compounds 5:165
Organoaminosilanes. (See also Silylation reagents;
and individual compounds.) 18:89–94
properties 18:89
uses 18:89
Organobismuth compounds. See Bismuth
compounds, organic.
Organoboron compounds. See Boron compounds,
organic.
Organochlorine pesticides. (See also Chlorinated
hydrocarbon pesticides.)
gas chromatography 19:157, 389
liquid chromatography 19:156
solvent extraction 19:156, 388
in tobacco 19:155
in water 19:388
Organo-chlorosilanes. (See also individual
compounds.) 18:53–69
analysis 18:56
chloride content 18:63
commercial grades and specifications 18:55
composition, determination of 18:56
gas chromatography 18:59
hydrogen evolution method 18:67
identification 18:56
impurities, determination of 18:64
infrared spectroscopy 18:56
manufacture 18:55
mass spectroscopy 18:59
properties 18:53
properties, determination of 18:63
refractive index 18:63
silanic hydrogen content 18:67
specific gravity 18:63
uses 18:55
Organofunctional silanes. (See also individual
compounds.) 18:79–89
alkoxy content 18:85
analysis 18:83
carbon content 18:87
commercial grades and specifications 18:83
composition, determination of 18:83
gas chromatography 18:83
identification 18:83
impurities, determination of 18:83, 87
infrared spectroscopy 18:83
manufacture 18:81
nitrogen content 18:85
pH 18:88
potassium hydroxide fusion 18:85
properties 18:79
silicon content 18:87
solubility 18:89
specification tests 18:88
toxicity 18:79

uses 18:79
Organogermanium compounds 13:400
  germanium content 13:400
Organolithium compounds. (See also Lithium
    compounds; and individual compounds.)
    15:288
  properties 15:277, 288
  solubility 15:277
  uses 15:288
Organomanganese compounds. (See also Man-
    ganese compounds; and individual compounds.)
    15:489
Organomercury compounds. (See also Mercury
    compounds.) 16:57
  in animal tissues 16:64
  dithizone volumetric method 16:64
  in human tissues 16:64
  sample decomposition 16:14
  solvent extraction 16:63
  toxicity 16:53, 59
  uses 16:57
Organomolybdenum compounds. (See also
    individual compounds.) 16:190
  analysis 16:191
  molybdenum content 16:191
Organonickel compounds. (See also Nickel
    compounds; and individual compounds.)
  uses 16:343
Organoniobium compounds 16:393
Organophosphorus pesticides. (See also
    individual compounds.) 19:581
  colorimetric determination 19:611
  commercial grades and specifications 19:594
  gas chromatography 19:157
  liquid chromatography 19:156
  manufacture 19:597
  properties 19:588
  solubility 19:587
  solvent extraction 19:156
  in tobacco 19:155
  toxicity 19:579, 592
  uses 19:579
Organosilamines. (See also Silylation reagents;
    and individual compounds.) 18:89–94
  properties 18:89
  uses 18:89
Organosilane esters. (See also individual com-
    pounds.) 18:69–79
  alkoxy content 18:75
  analysis 18:71
  commercial grades and specifications 18:71
  composition, determination of 18:71
  gas chromatography 18:73
  identification 18:71
  impurities, determination of 18:75, 78
  infrared spectroscopy 18:71, 75
  manufacture 18:69

  properties 18:69
  sample decomposition 18:78
  silicon content 18:77
  sodium peroxide fusion 18:78
  transesterification 18:77
  uses 18:69
Organosilicon copolymers 18:141–151
  acid–base titration 18:147
  acid number 18:147
  analysis 18:147
  cloud point 18:148
  composition 18:141
  composition, determination of 18:147
  foam evaluation tests 18:149
  functionality 18:148
  gas chromatography 18:147
  haze number 18:148
  hydroxyl number 18:147
  impurities, determination of 18:147
  manufacture 18:146
  pour point 18:148
  properties 18:141
  properties, determination of 18:148
  silane hydrogen content 18:148
  solubility 18:141
  structure 18:146
  toxicity 18:141
  types 18:141, 146
  in urethane foam 18:141
  uses 18:141
  viscosity 18:148
Organotin compounds. (See also Tin compounds;
    and individual compounds.) 19:76
  analysis 19:77
  manufacture 19:77
  in polyurethanes 19:312
  properties 19:76
  sample decomposition 19:32, 54, 56, 77
  solvent extraction 19:312
  thin-layer chromatography 19:312
  tin content 19:54, 56, 77
  uses 19:77
Organotitanium compounds 19:129
  classification 19:129
  decomposition 19:92
  manufacture 19:130
  properties 19:129
  sampling 19:90
  toxicity 19:87
  uses 19:130
Orientation
  of polymer and fibers 4:316
Origanum oil. (See also *Essential oils.*)
  refractive index 12:271
Orla-lutin. See 17-Ethyltestosterone.
Orlon. (See also *Acrylic and modacrylic fibers*;
    Acrylic fibers.) 4:221

birefringence 4:258
cross section 4:225, 263
end groups 4:289, 296
infrared spectroscopy 4:240, 246, 296
melting point 4:262
microscopy 4:264
mixtures, determination in 4:353
moisture regain 4:333
precipitation fractionation methods 4:299
properties 4:222
solubility 4:247, 353
staining tests 4:252
Ornithine
ion exchange chromatography 8:421
paper chromatography 8:422
Orpiment 6:223
Orsat technique
for acetylene 14:381
for carbon dioxide 14:381
for carbon monoxide 14:381
for ethylene 14:381
for hydrocarbons 14:381
for hydrogen assay 14:380
for nitrogen 14:381
for olefins 14:381
for oxygen 14:381
for paraffins 14:381
for propylene 14:381
Orthanilic acid
azo dyes for identification 5:430
color reactions for identification 5:431
manufacture 5:426
properties 5:423
Orthoarsenous acid 6:228
Orthoboric acid. See Boric acid.
Orthocide. See Captan.
Orthoclase 17:353
composition 10:104
Orthophosphate. (See also Phosphates.) 17:117
acid–base titration 17:139
ammonium molybdate colorimetric method
17:61
ammonium vanadate–ammonium molybdate
colorimetric method 17:61
colorimetric determination 17:61
detection 17:61
gravimetric determination 17:61
ion exchange chromatography 17:131
magnesium–ammonium chlorides gravimetric
method 17:61
manufacture 17:125
molybdic acid–ammonium nitrate gravimetric
method 17:61
paper chromatography 17:135
in sodium tripolyphosphate 17:137
in water 19:368
Orthophosphoric acid. (See also

Phosphoric acid.) 17:119
acid–base titration 17:120, 128
analysis 17:128
assay 17:128
commercial specifications 17:126
properties 17:119
specific gravity 17:129
uses 17:119
Orthotitanates. (See also Titanates.) 19:124
Orthotolidine dihydrochloride
colorimetric method for chlorine 19:377
Orthotolidine–sodium arsenite
colorimetric method for chlorine 19:377
Orthovanadates. (See also Vanadates.) 19:344
properties 19:345
Oscillometry
of benzene–methyl ethyl ketone–water
mixtures 7:33
Osmiridium
composition 17:215
sample decomposition 17:231
Osmium. (See also Osmium compounds; Osmium
metal; and *Platinum group metals.*)
bead test 6:588
in catalysts 17:272
colorimetric determination 17:248, 256,
263
detection 17:248
emission spectroscopy 17:246
ephedrine hydrochloride colorimetric method
17:248
gravimetric determination 17:256, 263
history 17:215
hydrogen sulfide gravimetric method 17:251,
256
identification 17:248, 256
isotopes 17:215
1-naphthylamine-4,6,8-trisulfonic acid colorimetric method 17:248
occurrence 17:214
osmium tetroxide volatilization 17:252, 254,
256, 262
oxidation numbers 17:220
oxidation potentials 17:221
2-phenylbenzothiazole gravimetric method
17:263
separation 17:252, 254, 262
thiourea colorimetric method 17:248, 256,
263
uses 17:217
Osmium alloys. See Osmium metal.
Osmium chloride
manufacture 17:226
properties 17:226
Osmium compounds. (See also individual
compounds.)
manufacture 17:225

Osmium hexafluoride
  manufacture 17:226
  properties 17:226
Osmium metal
  carrier gas fusion methods 8:682
  hydrogen in 8:680
  oxygen in 8:680, 682
  properties 17:219, 220, 225
  solubility 17:244
  uses 17:218
  vacuum fusion methods 8:680
Osmium oxide
  manufacture 17:225
  properties 17:225
Osmium tetroxide
  colorimetric method for pyrogallol 17:20, 47
  manufacture 17:225
  properties 17:225
  solubility 17:226
  toxicity 17:225, 228
  uses 17:225, 273
  volatilization method for osmium 17:252, 254,
    256, 262
Osmotic pressure
  measurements 4:310
  of polycarbonates 17:341
Osthol. (See also *Coumarins.*) 11:160
  gas chromatography 11:168, 170
Ostreogrycins 5:470, 549
  analysis 5:556, 560, 563
  column chromatography 5:556
  countercurrent distribution 5:556
  identification 5:556
  microbiological assay 5:560
  properties 5:550
  separation 5:556
  solubility 5:551
  structures 5:470
  uses 5:549
Ostwald-Cannon-Fenske viscometer 4:308;
    7:235, 236, 267
  calibration 7:269
Ostwald viscometer method 7:516
Ovalbumin
  in egg white 12:58, 61
  liquid chromatography 12:60
  properties 12:58
Ovarian hormones. See Estrogens; Progestational
    hormones; and individual compounds.
Ovesterin. See Estriol.
Ovocylin. See Estradiol.
Ovomucin
  in egg white 12:58
  properties 12:58
Ovomucoid
  in egg white 12:58, 61
  liquid chromatography 12:60

  properties 12:58
Oxacillin. (See also Isoxazolylpenicillins;
    Sodium oxacillin.)
  analysis 16:577
  assay 16:577
  bioassay 16:577
  manufacture 16:577
  microbiological methods 16:577
  properties 16:577
  specification tests 16:577
  structure 16:554
  ultraviolet spectroscopy 16:565
  uses 16:577
Oxalates
  acid–base titration method 7:616
  in ammonium oxalate 5:344
  in beer 7:616
  calcium oxalate turbidimetric method 10:93
  in citric acid 10:93
  derivatives of amides 8:601
  in diammonium citrate 10:93
  gravimetric method for barium 6:572
  gravimetric method for calcium 12:151
  gravimetric method for uranium 19:245
  haze in beer 7:520
  potassium permanganate titration 17:511
  precipitation of calcium 7:361; 8:82, 102
  in rare earth oxalates 17:511
  redox titration method 7:617
Oxalic acid 8:398
  acid–base titration 19:149
  in adipic acid 4:415
  calcium acetate turbidimetric method 10:543
  calcium hydroxide turbidimetric method
    15:145
  in chocolate 9:541
  colorimetric method for ethylene glycol
    13:572
  color test for diphenylamine 11:457
  in cooling system cleaners 6:351, 359
  in ferrous gluconate 10:543
  gas chromatography 15:132; 19:149
  gravimetric method for calcium oxide 10:119
  gravimetric method for gold 10:638; 14:19
  gravimetric method for lanthanum oxide
    15:152
  gravimetric method for rare earths 9:265;
    17:473, 507
  gravimetric method for thorium 16:335; 19:9
  in green coffee 19:374
  infrared spectrum 8:377
  in lactic acid and its derivatives 15:145
  liquid chromaotgraphy 19:149
  paper chromatography 15:133
  preparation 8:373, 397
  properties 8:399
  in roasted coffee 10:375

solvent extraction 19:149
thin-layer chromatography 8:381
in tobacco 19:148
uses 8:399
volumetric method using potassium permanganate 6:359
Oxalic acid–ammonium hydroxide
gravimetric method for calcium oxide 14:95
Oxathiins. (See also individual compounds.)
   13:212–214
analysis 13:213
Oxazine 6:381
Oxazine dyes. (See also *Azine and related
   dyes.*) 6:381, 382; 12:8
analysis 6:388
identification 6:388
manufacture 6:383, 385
solubility 6:383
Ox bile 7:144
oxidation 7:145
Oxidation, accelerated
of trichloroethylene 9:458
Oxidation characteristics
of greases 15:341
of lubricants 15:340
of turbine oil 15:340
Oxidation–distillation method
for lactic acid 11:52
Oxidation dyes 12:5
structure 12:10
Oxidation haze
in beer 7:520
Oxidation methods
for allyl alcohol 5:86
for ascorbic acid 19:425
for boron compounds, organic 7:349, 399
for boron, elemental 7:324
for boron, free 7:324
for 2,3-butanediol 19:447
for carbon 19:379
for carbonate 19:379
for dithiocarbamic acid salts 11:585
for ethanol 19:432
for glycerol 19:444
identification test for dyes 6:390
for lactic acid 19:422
for malic acid 19:421
for methanol 19:437
for organic matter 19:379
for phenindione 10:182
for thiophene 18:543
Oxidation potentials
of copper 10:651; 18:158
of gold 18:158
of iridium 17:221
of iron 15:24
of manganese 15:450

of mercury 16:5
of nickel 16:291
of niobium 16:352
of osmium 17:221
of oxygen 16:518, 542
of ozone 16:542
of palladium 17:221
of platinum 17:221
of radium 17:448
of rhodium 17:221
of ruthenium 17:221
of silver 18:158, 165
of uranium 19:229
Oxidation–reduction methods
for oxygen 19:502
Oxidation–reduction titration. See Redox
   titration.
Oxidation resistance of brake fluids 7:432
Oxidative rancidity, resistance to
of margarine 15:501
Oxides
carrier gas fusion methods 8:683
hydrogen in 8:681
in metals 8:613
oxygen content of 8:614, 616, 681, 683
sample preparation 15:31
in sodium metal 18:231
in sodium–potassium alloy 18:231
vacuum fusion methods 8:616, 681
in vitrified grinding wheels 7:298
Oxidizable substances
in beer 7:618
in benzoic acid 7:66
in ketones 8:361
Oxidized fatty acid
in fats and fatty oils 12:482
solvent extraction 12:483
Oxidized hematoxylin
colorimetric method for germanium 13:388
Oxidizing agents
in detergents 11:334
iodometric method 11:334; 18:183
in soaps 18:183
Oxidizing capacity
of ammonium persulfate 5:347
Oxidizing substances
in barium acetate 6:536
in barium carbonate 6:544
in barium chloride 6:546
in cadmium bromide 8:38
in calcium carbonate 8:84
in caprolactam 8:123
in nitrous oxide 5:366
Oximation
of carbonyl compounds 8:356
Oxine. (See also 8-Hydroxyquinoline;
   8-Quinolinal.)

gravimetric method for tungsten 19:201, 203
gravimetric method for uranium 19:242, 245
gravimetric method for zinc 19:560
Oxirane. See *Ethylene oxide.*
Oxirane compounds. See *Epoxy compounds.*
Oxirane oxygen
  in dental restorative materials, composite
    11:312
  in epoxy compounds 12:205
  in fats and fatty oils 12:482
  pyridinium chloride volumetric method 11:312
17-Oxo-androsenes
  liquid chromatography 14:209
17-Oxo-androstan-3,5-diene
  liquid chromatography 14:209
3-Oxocholanic acid derivatives color tests
    7:147, 148
γ-Oxoglutaric acid
  liquid chromatography 15:133
Oxomethane. See *Formaldehyde.*
Oxo process
  manufacture of amyl alcohols 4:506
γ-Oxopyruvic acid
  liquid chromatography 15:133
Oxosteroids
  gas chromatography 14:215
  thin-layer chromatography 14:213
γ-Oxovaleric acid
  liquid chromatography 15:133
Oxyalkalene content
  of ethers 12:305
Oxycarboxin 13:213
  analysis 13:213
  in cottonseed meal 13:213
  p-dimethylaminobenzaldehyde colorimetric
    method 13:213
  manufacture 13:213
  in plants 13:213
  in potatoes 13:213
  properties 13:213
  solubility 13:213
  in sugar beets 13:213
  toxicity 13:213
  uses 13:213
Oxychloride method
  for bismuth 5:121
Oxycodone. (See also *Narcotics.*)
  ammonium molybdate colorimetric method
    16:247
  color tests 16:247
  formaldehyde–sulfuric acid colorimetric method
    16:247
  gas chromatography 16:254
  iodine–potassium iodide microchemical reagent
    16:250
  microchemical tests 16:250
  sodium carbonate microchemical method 16:250

structure 16:244
ultraviolet spectroscopy 16:252
Oxydipropanediols
  paper chromatography 13:581
Oxyethylene–oxypropylene copolymers
  cleavage methods 19:308
  composition, determination of 19:308
  hydriodic acid volumetric method 19:308
  infrared spectrum 19:307
  manufacture 19:307
  nuclear magnetic resonance spectroscopy
    19:307
  phosphoric acid colorimetric method 19:308
  in polyurethanes 19:307
Oxyfluoroniobic acid
  manufacture 16:356
Oxygen. (See also Oxygen, elemental.) 16:514–
    538
  in acetylene 4:130
  in alumina 5:116
  in aluminum and aluminum alloys 5:140
  in aluminum hydroxide 5:185
  in aluminum oxide 5:185
  in aluminum powder 5:153
  in beer 7:620
  in beer package headspace 7:629
  in boron 7:327
  in 1,3-butadiene 11:503, 504, 507
  in carbides 8:135
  in carbonaceous materials 4:15
  in carbon black 8:191
  in carbon dioxide 8:251, 307, 310, 312, 315
  carbon dioxide gasometric methods 14:133
  in carbon fibers 4:21
  carrier gas distillation fusion methods 8:636
  carrier gas fusion methods 8:624, 682;
    16:169
  in cerium metal 9:249
  in cesium 9:283
  in chlorine 9:309, 320
  in chromium 9:635
  in chromium alloys 9:673
  in cigarette smoke 10:13
  in coal and coke 10:241
  in cobalt alloys 10:319
  in cobalt metal 10:293, 301
  conductometric methods 8:650
  in copper alloys 10:643
  in copper metal 10:624
  coulometric methods 8:649
  in elastomers, synthetic 12:113
  electrochemical methods 19:502
  emission spectroscopy 14:134
  etchant for ceramics 9:224
  in fluorine, elemental 13:3, 4
  in fluorocarbons 13:70
  in gallium arsenide 13:279

in gallium metal 13:266
gas chromatography 8:251, 310, 312, 650;
    13:71, 297, 307, 316; 14:384, 392, 416;
    16:454, 470, 485, 523
in gas mixtures 16:522
in hafnium 14:133
in helium 13:298
history 16:515
hot extraction methods 8:638
in hydrogen 14:380, 381, 384
indigo carmine colorimetric method 7:622
infrared spectroscopy 8:653
iodometric methods 16:533; 19:372
isotopes 16:514
in krypton 13:300
lithium absorption method 13:293
in lithium hydride 14:251
in magnesium alloys 15:389
in manganese 15:482
in manganese dioxide 15:492
mass spectroscopy 8:315, 650; 13:302
measurement 8:648
in metal borides 7:365
in metals 4:15; 8:613; 16:524
in molybdenum 16:169
in molybdenum oxychlorides 16:182
NBS sodium oxalate volumetric method 15:492
in neon 13:299
neutron activation analysis 14:134
in nickel 16:333
in niobium 16:389
in nitrogen 16:470
nitrogen content 16:454
in noble gases 13:293, 294, 297, 302, 307
occurrence 16:514
in organic compounds 16:533
Orsat technique 14:381
oxidation potentials 16:518
oxidation–reduction methods 19:502
in oxides 8:614, 616, 681, 683
in oxyhalides 8:614
polarography 7:626; 18:305
pyrogallic acid absorption method 8:307
in refractory compounds 4:15
in rubidium 17:555
Schütze-Unterzacher volumetric method 16:533
in steel 8:665
in styrene 18:305
thermal conductivity 14:380
in titanium 19:121
in titanium p-boride 7:367
in tungsten metal 19:218
vacuum fusion distillation methods 8:635
vacuum fusion methods 8:616, 680; 10:301
in volatile materials 8:634
in water 19:372
in wines 19:500, 502

in xenon 13:300
in zirconium 14:133
Oxygen absorption
    for hydrocarbon resins 14:289
    of polymers 6:79
Oxygen, available
    in sodium perborate 7:381
Oxygen bomb method 4:377
    for boron compounds, organic 7:400
    for sulfur compounds, organic 18:378
    for sulfur in carbon 14:71
Oxygen combustion flasks 9:416
Oxygen compounds. (See also individual
        compounds.) 16:534–537
    classification 16:534
    properties 16:534
Oxygen-containing bleaches
    active oxygen 7:224, 229
    analysis 7:229
    identification 7:225
    permanganate titration 7:229
Oxygen demand, theoretical
    of water 19:375
Oxygen difluoride
    in air 13:4
    manufacture 13:10
    properties 13:10
    toxicity 13:13
    uses 13:10
Oxygen, dissolved
    in water 19:372
Oxygen electrodes 19:503
Oxygen, elemental
    absorption methods 16:522
    Alsterberg titration 16:525
    analysis 16:522
    commercial grades and specifications 16:520
    detection 16:522
    3,3'-dimethylnaphthidine colorimetric
        method 16:531
    impurities, determination of 16:531
    indigo carmine colorimetric method 16:529
    iodometric methods 16:524, 530
    manufacture 16:519
    odor 16:532
    polarography 16:531
    Pomeroy-Kirshman-Alsterberg titration 16:528
    potassium permanganate volumetric method
        16:526
    properties 16:517
    Rideal-Stewart titration 16:526
    safety 16:519
    in sewage 16:524
    sodium azide–iodide volumetric method
        16:525, 528
    sodium hypochlorite volumetric method 16:527
    toxicity 16:519

uses 16:515
in water 16:524, 529
Winkler titration 16:524
Oxygen flask method
  for boron compounds, organic 7:399
  chlorine content of vinyl chloride and vinylidene
    chloride polymers 4:280
  of chlorocarbons 9:416
Oxygen fluorides. (See also individual compounds.)
    13:10
Oxyhalides
  oxygen content of 8:614
Oxyhydrogen combustion
  apparatus 7:350
  of boron compounds, organic 7:349
  of sulfur 15:253
  of sulfur compounds, organic 18:379
Oxymetholone. (See also Steroids.)
  manufacture 14:205
  properties 14:205
  solubility 14:205
  structure 14:204
  uses 14:205
Oxymethylene. See *Formaldehyde*.
Oxymorphone. (See also *Narcotics*.)
  ammonium molybdate colorimetric method
    16:247
  color tests 16:247
  formaldehyde–sulfuric acid colorimetric method
    16:247
  gas chromatography 16:255
  gold bromide microchemical method 16:250
  mercuric chloride microchemical method
    16:250
  microchemical tests 16:250

structure 16:244
thin-layer chromatography 16:256
ultraviolet spectroscopy 16:253
Oxynicotine
  in tobacco 19:136
Oxytocin. (See also Pituitary hormones.) 14:181
  analysis 14:206
  assay 14:206
  commercial specifications 14:188
  manufacture 14:188
  occurrence 14:188
  structure 14:181
  uses 14:181, 188
Ozone 16:538–551
  in air 16:484, 546
  analysis 16:546
  history 16:538
  iodometric methods 16:546, 548
  manufacture 16:544
  occurrence 16:538
  oxidation potentials 16:542
  properties 16:538, 540
  safety 16:543
  sampling 16:547
  solubility 16:540
  o-tolidine-arsenite colorimetric method 16:549
  o-tolidine-manganese colorimetric method
    16:549
  toxicity 16:543
  ultraviolet spectroscopy 16:538, 546, 548
  uses 16:539
  in water 16:548
Ozonization
  of elastomers, synthetic 12:124
Ozono 6:75

P

Packages 11:109
  cap tests 11:114
  collapsible tube tests 11:114
  compatibility with product 11:111
  evaluation for cosmetics 11:109
  gold stamping test 11:115
  label tests 11:114
  leakage tests 11:112
  Markol test 11:115
  odor permeation 11:113
  perspiration tests 11:115
  physical tests 11:112
  plastic 11:113
  seal tests 11:111
Package tests
  for carbonated beverages 8:175
Packing test
  for methyl methacrylate dental materials
    11:301
Paigeite 7:315
Paint
  carbon black in 8:241
  mercury content 16:25
Paintability
  of caulking compounds 9:26
Paint solvent
  gas chromatography 8:351
Paired comparison testing 17:629
Paired difference testing 17:630
Paired intensity testing 17:630
Paired preference test 17:633
Palladium. (See also Palladium alloys; Palladium
    compounds; Palladium metal; and *Platinum
    group metals.*)
  bead test 6:588
  in catalysts 17:272
  colorimetric determination 17:249, 265
  detection 17:249
  dimethylglyoxime color test 17:249
  dimethylglyoxime gravimetric method 17:249,
    253, 257
  emission spectroscopy 14:48; 17:246, 270;
    18:451
  flame photometry 17:247
  flame tests 6:589
  in gold 14:45
  gravimetric determination 17:249, 253, 257
  history 17:215
  hydrogen sulfide gravimetric method 17:251
  identification 17:249
  isotopes 17:215

p-nitrosodimethylaniline colorimetric method
  17:265
α-nitroso-β-naphthol colorimetric method
  17:249, 271
  occurrence 17:214
  in ores 17:266
  oxidation numbers 17:220
  oxidation potentials 17:221
  paper chromatography 14:18
  phenyl-α-pyridyl ketoxime colorimetric method
    17:249
  potassium iodide gravimetric method 17:237,
    249, 257
  separation 17:253, 264
  in silver 18:173
  sodium formate gravimetric method 17:258
  sodium metabisulfite gravimetric method
    17:264
  in titanium alloys 17:271
  in titanium metal 19:121
  uses 17:217
Palladium alloys. (See also Palladium metal.)
  solubility 17:244
Palladium azide
  sodium nitrite titration 16:473
Palladium chloride
  colorimetric method for mercury 16:23
  color test for mercaptans 15:256
  manufacture 17:225
Palladium compounds. (See also individual
    compounds.)
  manufacture 17:225
Palladium diiodide
  manufacture 17:225
  solubility 17:225
Palladium gold 5:68
Palladium metal
  carrier gas fusion methods 8:682
  hydrogen in 8:680
  oxygen in 8:616, 621, 623, 680, 682
  properties 17:219, 220, 225
  solubility 17:244
  solubility in gold 14:9
  uses 17:218
  vacuum fusion methods 8:616, 621, 623, 680
Palladium monoxide
  manufacture 17:225
Palladium sulfide
  manufacture 17:225
Palladous chloride
  gravimetric method for iodides 14:596

identification reagent for carbon monoxide
8:304
Palmitamide
derivatives for identification 8:601
Palmitic acid. (See also Fatty acids.) 5:25
composition 8:281
gas chromatography 8:382, 388
properties 12:473
Palmitoleyl alcohol. (See also Fatty alcohols.)
properties 4:561
Palmitonitrile. See *n*-Hexadecanonitrile.
Palmitoyl dilactoyl lactylate. (See also Lactic
acid esters.)
gas chromatography 15:136
Palmitoyl lactoyl lactylate. (See also Lactic
acid esters.)
gas chromatography 15:136
Palmitoyl lactylate. (See also Lactic acid esters.)
gas chromatography 15:136
Palmitoyl tetralactoyl lactylate. (See also Lactic
acid esters.)
gas chromatography 15:136
Palmitoyl trilactoyl lactylate. (See also Lactic
acid esters.)
gas chromatography 15:136
Palmityl alcohol. See Cetyl alcohol.
Palm oil. (See also *Fats and fatty oils.*)
properties 12:473
PAN
colorimetric method for cobalt 10:370
colorimetric method for gallium 13:270
PANA 6:75
Pancreatic hormones. (See also individual
compounds.) 14:182
analysis 14:206
properties 14:188
sources 14:188
uses 14:188
Pandermite 7:314
Panogen. (See also Mercury-containing compounds.) 13:199
manufacture 13:199
properties 13:199
solubility 13:199
toxicity 13:199
uses 13:199
Pantothenic acid
in cocoa beans 9:540
in eggs and egg products 12:66
Papain
in beer 7:588
Papaverine. (See also *Narcotics*; and Papaverine
hydrochloride.) 4:588; 16:271
ammonium molybdate colorimetric method
16:247
ammonium vanadate colorimetric method
16:247

analysis 4:597; 16:271
cobalt nitrate–ammonium thiocyanate colorimetric method 16:272
colorimetric determination 16:272
color tests 16:247
coulometric methods 16:272
fluorometric method 16:272
formaldehyde–sulfuric acid colorimetric method
16:247
gas chromatography 16:254
gravimetric determination 16:272
microchemical tests 16:250
occurrence 16:271
paper chromatography 16:255
potassium–cadmium iodide microchemical
method 16:250
potassium chromium thiocyanate gravimetric
method 16:272
structure 16:244
thin-layer chromatography 16:256
ultraviolet spectroscopy 16:253
uses 16:271
volumetric determination 16:272
zinc chloride microchemical reagent 16:250
Papaverine hydrochloride. (See also Papaverine.)
acid–base titration 16:271
assay 16:271
mercuric nitrate titration 16:272
phosphorimetry 16:264
Paper
application of dyes to 6:398
bleaching 7:222
brightness reversion 7:263
carbon black in 8:241
gelatin content 13:364
physical testing 7:273
sample conditioning 7:276
sampling 7:273
Paper backing
for coated abrasives 10:276
Paper, bleached
properties, determination of 7:241
Paper blue R. (See also *Triphenylmethane and
related dyes.*)
visible spectroscopy 19:184
Paperboard. (See also Paper.)
mercury content 16:44
physical testing 7:273
Paper chromatography
of acenocoumarol 10:152
of acetals and ketals 4:65
of acetic acid 15:133
of acetylsalicylic acid 6:183
of acrylonitrile 4:372, 380
of actinomycins 5:553
of adipic acid 15:133
of alcohols 4:527

of alkyd resins 5:49
of alkylaryl sulfates 12:575
of N-alkyl pyridinium halides 12:575
of alkyl sulfates 12:575
of alkyltrimethyl pyridinium halides 12:575
of allylamines 5:89
of amines 5:87
of amino acids 8:422, 435
of amino acids in cocoa beans 9:557
of aminobenzoic acids 8:404
of amobarbital 6:183
of amphomycin 5:555
of amphotericins 5:568
of anabasine 19:136
of aniline and its derivatives 5:435
of anionic surfactants 12:575
of anisomycin 5:602
of annatto 10:541
of anthraquinone dyes 10:475
of antihalation dyes 12:587
of antihistamines 6:35
of antimycin 5:602
of antioxidants and antiozonants 6:84
of apomorphine 16:255
of arabinose 18:478
of aterrimin 5:603
of auramine G 19:172
of auramine O 19:172
of aureothricin 5:510
of bacitracin 5:538
of barbiturates 6:478
of barium 6:574; 18:270
of benzylpenicillin 16:558, 575
of bishydroxycoumarin 10:152, 156, 158, 161
of blue VRS 19:172
of boron 7:397
of brominated compounds in D & C Red No. 21
    10:523
of bromofluoresceins 10:520, 523
of 1,3-butanediol 13:581
of 1,4-butanediol 13:581
of 2,3-butanediol 13:581
of 1,2,4-butanetriol 13:581
of butyric acid 15:133
of caffeine 6:183; 8:59; 10:412
of candicidin 5:568
of capreomycin 5:555
of carbenicillin 16:575
of carbohydrates 7:567
of carbomycin 5:521
of carboxylic acids 8:380
of carboxymethylcellulose 12:582
of catechin 18:468
of cationic surfactants 12:575
of cellulose derivatives 12:582
of cephalosporin C 16:559
of cephalosporins 16:558

of cephalothin 16:559
of certified colors 10:475
of chloramphenicol 5:496
of chlorobenzoic acids 8:400
of chlorogenic acids 19:146
of choline 9:615
of chromomycin 5:497
of cinnamic acid 10:59
of cinnamic acid and its derivatives 10:64
of citric acid 15:133; 19:418
of codeine 6:183; 16:255
of colistin 5:538
of coumarins 10:152; 11:164
of cresols 17:24
of crotonaldehyde 11:180
of crystal violet 19:172
of cyanoacetic acid 15:439
of cyclamates 11:234
of cycloheximide 5:510
of cyclohexylamine 11:211
of D & C Green No. 5 10:475
of D & C Green No. 6 10:475
of D & C Violet No. 2 10:475
of dibromofluoresceins 10:520, 523
of diethylene glycol 13:581
of diglycerols 13:581
of diiodofluoresceins 10:514
of 2,4-dinitrophenylhydrazones of aldehydes
    and ketones 4:116
of disodium salt of 3-carboxy-5-hydroxy-1-
    phenyl-4-p-sulfophenylazopyrazole 10:518
of disodium salt of 3-carboxy-5-hydroxy-1-p-
    sulfophenyl-4-phenylazopyrazole 10:518
of dithiocarbamic acid salts 11:584
of Dobner's violet 19:172
of epicatechin 18:468
of epicatechin gallate 18:468
of epigallocatechin 18:468
of epigallocatechin gallate 18:468
of erythritol 13:501, 581
of erythromycin 5:522
of erythrosine 19:172
of estradiol 14:220
of estriol 14:220
of estrone 14:220
of etamycin 5:555
of ethinyl estradiol 10:570
of ethyl acid malate 19:418
of ethyl acid tartrate 19:418
of ethyl biscoumacetate 10:152
of ethylene chlorohydrin 14:157
of ethylene glycol 13:501, 581
of ethylphenols 17:24
of Ext D & C Violet No. 2 10:475
of fast red E 19:172
of F D & C Blue No. 1 10:475
for F D & C Blue No. 2 10:475

of F D & C Green No. 1 10:475
of F D & C Green No. 2 10:475
of F D & C Green No. 3 10:475
of F D & C Orange No. 1 10:475
of F D & C Red No. 1 10:475
of F D & C Red No. 2 10:475
of F D & C Red No. 3 10:475
of F D & C Red No. 4 10:475
of F D & C Violet No. 1 10:475
of F D & C Yellow No. 1 10:475
of F D & C Yellow No. 5 10:475
of F D & C Yellow No. 6 10:475
of ferrimycin 5:604
of filipin 5:569
of flavanols 18:467
of fluorescein 10:514, 523
of fructose 18:478; 19:142
of fuchsine 19:172
of fumaric acid 15:133
of fusidic acid 5:510
of gallocatechin 18:468
of gentamycin 5:605
of glucose 18:478; 19:142
of glucuronic acids 12:582
of glycerol 13:501, 581
of glycerol dichlorohydrin 14:157
of glycols 13:581
of gold 14:18
of green S 19:172
of griseofulvin 5:498
of gums 12:582
of halohydrins 14:157
of heroin 16:255
of hexanediols 13:581
of hydroquinone 17:24
of *p*-hydroxybenzaldehyde 6:620
of hydroxybenzoic acids 8:400
of indigoid dyes 14:509
of inositol 18:478
of insulin 14:573
of iodinated substances 10:513
of iodofluoresceins 10:514
of kanamycin 5:584
of lactic acid 15:133; 19:418
of leucomycin 5:523
of lincomycin 5:584
of malachite green 19:172
of maleic acid 15:133
of malic acid 15:133; 19:418
of malonic acid 15:427
of maltose 18:478
of menadione 10:199
of mercurials 11:565
of methylene green 19:172
of 1-methyl-4-isopropylphenol 17:24
of mikamycin 5:555
of narceine 16:255

of narcotics 16:255
of narcotine 16:255
of neomycins 5:585
of neopentyl glycol 13:581
of nickel 16:299
of nicotine 19:136
of nitrobenzoic acids 8:404
of nornicotine 19:136
of novobiocin 5:500
of nystatin 5:569
of organic acids 9:567
of orthophosphates 17:135
of oxalic acid 15:133
of oxydipropanediols 13:581
of palladium 14:18
of papaverine 16:255
of parafuchsine 19:172
of paromomycin 5:586
of penicillins 16:558
of pentaerythritol 13:581
of pesticides and insecticides 19:612
of phenacetin 6:183
of phenindione 10:181
of phenol 17:24
of phenols 17:23
of phenoxymethylpenicillin 16:558
of phenprocoumon 10:152
of β-phenylethyl alcohol 7:95
of phloroglucinol 17:24
of phosphates 17:135
of phytonadione 10:186
of pimaricin 5:571
of plasticizers 12:592
of platinum 14:18
of polyamides 17:289
of polyethylene glycols 12:575; 13:581
of polyhydric alcohols 13:581
of polymyxin 5:541
of polyoxyethylene glycol 13:581
of polyphenols 19:146
of polyphenols in tea 18:462
of polyphosphates 17:135
of polysaccharides 12:582
of polysaccharides in cellulose 9:46
of polyvinylpyrrolidone 7:636
of ponceau MX 19:172
of ponceau 3R 19:172
of ponceau 4R 19:172
of ponceau SX 19:172
of pregnanetriol 14:211
of pregnenediol 14:211
of progesterone 14:211
of propionic acid 15:133
of 1,2-propylene chlorohydrin 14:157
of propylene glycol 13:501, 581
of 4-*n*-propylphenol 17:24
of pyrocatechol 17:24

of pyrophosphates 17:135
of quaternary ammonium compounds 11:556
of quinine 6:183
of raffinose 18:478
of red 6B 19:172
of red 10B 19:172
of red FB 19:172
of red 2G 19:172
of resorcinol 17:24
of rhamnose 18:478
of rhodium 14:18
of ribose 18:478
of rifamycins 5:500
of ruthenium 14:18
of rutin 19:146
of scandium 17:568
of scopoletin 19:146
of scopolin 19:146
of selenium 14:18
of semisynthetic penicillins 16:558
of sodium alkyl sulfonates 12:575
of sodium heparin 10:141
of sodium tripolyphosphate 17:137
of sorbitol 13:501
of spiramycins 5:525
of staphylomycin 5:556
of steroids 14:210
of streptomycin 5:587
of strontium 18:270
of succinic acid 15:133; 19:418
of sucrose 18:478; 19:142
of sugar 18:477
of sugar derivatives 12:582
of surfactants 12:575
of syringaldehyde 6:620
of tartaric acid 15:133; 19:418
of tartrazine 19:172
of tellurium 14:18
of 2,4,5,7-tetrabromofluorescein 10:520,
   523
of tetracaine hydrochloride 5:418
of 2,4,5,7-tetraiodofluorescein 10:514
of tetrametaphosphates 17:135
of thebaine 16:255
of thiazolidomycin 5:511
of thiostrepton 5:556
of tribromofluoresceins 10:520, 523
of trichomycins 5:571
of triethylene glycol 13:581
of trigonelline 10:412
of triiodofluoresceins 10:514
of trimetaphosphates 17:135
of trimethylene glycol 13:581
of trimethylolethane 13:581
of trimethylolpropane 13:581
of triphenylmethane and related dyes 12:587;
   19:171

of uronic acids 12:582
of vancomycin 5:605
of vanillic acid 8:379, 406
of vanillin 6:620
of victoria blue B 19:172
of violet BNP 19:172
of viomycin 5:557
of warfarin sodium 10:167
of xanthocillins 5:500
of xylose 18:478
of yellow 2G 19:172
of yellow RFS 19:172
of yellow RY 19:172
Paper electrophoresis
  of amino acids 8:427
  of EDTA and related compounds 12:31
  of rifamycins 5:500
Paper matchboard 15:518
  analysis 15:534
  basis weight 15:534
  brightness 15:534
  caliper 15:534
  commercial grades and specifications 15:519
  cutting quality 15:534
  internal bond 15:534
  moisture content 15:534
  properties 15:518
  splint strength 15:534
  water drop test 15:535
Paper, photographic 12:559
  composition 12:559
  impurities, determination of 12:560
  properties 12:559
Paper pulp
  sodium sulfite content 18:408
Paprika
  composition 10:532
Paprika oleoresin
  composition 10:532
PAR. (See also 4-(2-Pyridylazo)resorcinol.)
  colorimetric method for tantalum 18:449
Paracide. See p-Dichlorobenzene.
Paraconic acid
  preparation from citric acid 8:373
Paracril 12:87
Paraffinic hydrocarbons. See Alkanes.
Paraffinic lubricants. (See also *Lubricants.*)
  commercial grades and specifications 15:307
  manufacture 15:306
  properties 15:307
Paraffin oil
  solubility 13:539, 552
Paraffins. (See also Alkanes.)
  analysis 14:336
  in aromatic hydrocarbons 14:345
  gas chromatography 14:329, 330, 331, 335,
    336

in hydrogen 14:381
in naphthalene 16:212
Paraffin wax
  analysis 15:533
  in automobile waxes 6:353
  commercial grades and specifications 15:517
  fire point 15:533
  flash point 15:533
  manufacture 15:517
  in matches 15:517
  melting point 15:533
  oil content 15:533
Paraform
  in brake linings 7:441
Paraformaldehyde 13:103
  in embalming fluids 12:164
  manufacture 13:103
  properties 13:103
  toxicity 13:104
  uses 13:194
Parafuchsine. (See also *Triphenylmethane and
  related dyes.*)
  paper chromatography 19:172
Paragonite
  composition 10:102
Parahilgardite 7:315
Paraldehyde
  commercial specifications 8:340
  gas chromatography 11:181
  gravimetric method for caramel 10:536
Paramoth. See *p*-Dichlorobenzene.
Paraperiodic acid. (See also Periodates.) 14:609
  properties 14:609
  solubility 14:609
Paraquat
  structure 17:414
  uses 17:414
Para red. (See also *Pigments.*)
  commercial specifications 12:173; 17:204
  composition 12:173
  in enamels 12:173, 185
  identification 12:185; 17:205
  potassium hydroxide color test 17:205
  solubility 17:205
  specification tests 17:205
  structure 17:178
Pararosaniline. See *p*-Rosaniline.
Parathion. (See also *Pesticides and insecticides.*)
  commercial grades and specifications 19:595
  composition 19:589, 598
  manufacture 19:598
  potassium hydroxide colorimetric method
    19:611
  properties 19:589
  safety 19:605
  solubility 19:587
  structure 19:584

toxicity 19:592
zinc-*N*-1-naphthylethylenediamine colorimetric
  method 19:611
Parathion residues
  bioassay 7:183
Parathyroid hormones. (See also individual
  compounds.) 14:183
  analysis 14:207
  assay 14:207
  commercial specifications 14:190
  composition 14:183
  properties 14:190
  sources 14:190
  uses 14:183, 190
Parazone. See *p*-Phenylphenol.
Paregoric
  assay 16:262
  morphine content 16:262, 265
Parian cement. (See also Hydraulic cements.)
  9:95
Paris green. See Copper acetoarsenite.
Paromomycin 5:475, 582
  analysis 5:586, 593, 596
  column chromatography 5:586
  identification 5:586
  microbiological assay 5:593
  paper chromatography 5:586
  properties 5:579
  separation 5:586
  solubility 5:580
  structure 5:475
  thin-layer chromatography 5:587
  uses 5:582
Paromomycin sulfate
  infrared spectrum 5:588
Parr bomb method
  of amino resins 5:287
  apparatus 5:288
  for indigoid dyes 14:513
  for isocyanic acid esters 15:120
  for sulfur in naphthalene 16:213
Parri colorimetric method
  for barbiturates 6:477, 491
Particle charge test
  for cationic asphalt emulsions 6:330
Particle density
  of adsorbents 4:443
Particle size
  of abrasive powders 15:536
  of activated carbon 8:151
  of adsorbents 4:439
  of aerosols 4:458
  of antimony sulfide 15:536
  of black iron oxide 15:531
  of calcium carbonate 15:530
  of carbon black 8:180, 202
  of cements 9:131

of chocolate 10:564
of corn products 11:14
of copper(I) oxide 10:672
of detergents 11:385
of diatomaceous earth 15:529
of ground glass 15:529
of iron sulfide 15:536
of limestone 15:231
of manganese dioxide 15:531
of match raw materials 15:523
of organic pigments 15:532
of pesticides and insecticides 19:617
of potassium dichromate 15:530
of pulverized flint 15:529
of pulverized silica 15:529
of red phosphorus 15:536
of rosin 15:529
of sand 15:529
of sulfur 15:529
of titanium dioxide 15:531
of vinsol 15:529
of zinc oxide 15:529
Particle size distribution
of abrasive grains 7:303; 10:264
of abrasive powders 10:266
of activated carbon 8:151
of carbon black 8:184; 12:150
by centrifugal separation 9:169, 178
of chocolate 9:571
of cocoa products 9:568
of diamonds, industrial 11:474
of ground coffee 10:420
microprojection method 9:571
of pigments 17:206
of polytetrafluoroethylene 13:87
Particulate matter
in cigarette smoke 10:13, 14
Parzate. See Nabam.
Parzate C and D. See Zineb.
Passenger car brake system road test 7:455,
465
Pasteurization efficiency
of beer 7:586
Gibbs phenol color test 16:133
of milk 16:133
Scharer method 16:133
Pastilles. (See also Confectionery products.)
10:548
Pastry 6:447
Patchouly oil. (See also Essential
oils.)
refractive index 12:271
Paternoite 7:314
Patronite
composition 19:321
occurrence 19:321
Pauly reaction

for histidine 8:487
PBNA. See N-Phenyl-2-naphthylamine.
PBX. (See also High-explosive mixtures.)
analysis 12:450
composition 12:450
uses 12:450
PCNB. See Terraclor.
PDA-8 (10 and 14) 6:75
PDTA. (See also EDTA and related compounds.)
cobaltous nitrate–ferric chloride colorimetric
method 12:28
iodometric method 12:28
lanthanum nitrate–sodium hydroxide colorimetric method 12:28
structure 12:25
thorium nitrate–hexamethylenetetraamine
colorimetric method 12:28
Peacock blues
in enamels 12:187
identification 12:187
Peanut oil. (See also Fats and fatty oils.)
in alkyd resins 5:23
properties 12:473
Pearl ash. See Potassium carbonate.
Pectic acid
in tobacco 19:145
Pectin
in confectionery products 10:552
gravimetric determination 12:581
in green coffee 10:374
infrared spectroscopy 12:583
in photographic films 12:582
solvent extraction 19:143
thorium nitrate gravimetric method 12:582
in tobacco 19:143
Pectin extractor 19:143, 144
Peebles method
light scattering of acrylic polymers 4:313
osmotic pressure of acrylic polymers 4:312
Pegmatite 17:470; 19:225
PEHA. See Pentaethylenehexamine.
Pelargonamide
derivatives for identification 8:601
Pelargonic acid 5:22
gas chromatography 5:25
properties 8:558
n-Pelargonitrile. See n-Nonanonitrile.
Pelletiernine 4:591
Penetration test
for asphalt residue 6:393
for bituminous materials 6:291
for custards 12:72
temperature susceptibility 6:304
Penetration value
of greases 15:347
Penetrometer 6:291; 15:347
Penicillenic acid

in penicillins 16:561, 566
ultraviolet spectroscopy 16:561, 566
Penicillin
    in benzylpenicillin 16:570
    bioassay 7:182
    in carbenicillin 16:575
    in penicillin O 16:571
    in phenoxymethylpenicillin 16:572
    synthesis 8:542
Penicillin F. See 2-Pentenylpenicillin.
Penicillin G. See Benzylpenicillin.
Penicillin K. See *n*-Heptylpenicillin.
Penicillin O. (See also Chloroprocaine penicillin O;
        Penicillins; and Sodium penicillin O.) 16:571
    analysis 16:571
    assay 16:571
    bioassay 16:571
    bromometric method 16:571
    impurities, determination of 16:566, 571
    infrared spectroscopy 16:571
    iodometric method 16:571
    manufacture 16:553
    penicillin content 16:571
    properties 16:571
    solvent extraction 16:571
    structure 16:554
    ultraviolet spectroscopy 16:565
    uses 16:571
Penicillins and related compounds. (See also
        Cephalosporins; Isoxazolylpencillins; and
        individual compounds.) 5:464; 16:552–583
    acid–base titration 16:563
    analysis 16:558
    assay 16:560
    bioautography 16:559
    biological tests 16:566
    chromotropic acid colorimetric method 16:560
    degradation products 16:558
    gas chromatography 16:559
    history 16:553
    hydroxylamine–ferric ammonium sulfate colori-
        metric method 16:563
    identification 16:558
    impurities, determination of 16:561, 565
    infrared spectroscopy 16:559
    iodometric methods 16:561
    liquid chromatography 16:559
    manufacture 16:552, 553
    microbiological methods 16:560, 561
    nuclear magnetic resonance spectroscopy
        16:560
    paper chromatography 16:558
    properties 16:552, 553
    safety test 16:566
    separation 16:558
    solubility 16:553
    structure 16:552, 554

thin-layer chromatography 16:559
ultraviolet spectroscopy 16:560, 564
Penicillin V. See Phenoxymethylpenicillin.
Penicillin V acid. See Phenoxymethylpenicillanic
        acid.
Penicillin X. See *p*-Hydroxybenzylpenicillin.
Penicilloic acid
    iodometric titration 16:561, 566
    in penicillins 16:561, 566
Penillic acid
    in penicillins 16:561, 566
Pennox A (B, C, and D) 6:75
Pennyroyal oil. (See also *Essential oils.*)
    pulegone content 12:282
Pennzone B (and E) 6:75
Pensky-Martens closed cup method
    flash point of bituminous materials 6:323
    flash tester used for 6:322
Pentaborane 7:340, 341, 344
    in air 7:349
    colorimetric method using pyridine 7:348
    gas chromatography 7:353
    monitors for detection in air 7:349
    properties 7:343
Pentabromoacetone
    gravimetric method for citric acid 10:92
Pentachlorobenzene. (See also Chlorinated
        benzenes.)
    gas chromatography 9:378
    infrared spectroscopy 9:394
Pentachlorobiphenyls. (See also Chlorinated
        biphenyls.)
    properties 9:492
Pentachlorobutadiene
    mass spectroscopy 9:407
Pentachloroethane. (See also C_2 Chlorinated
        hydrocarbons.)
    Fujiwara colorimetric method 9:386
    gas chromatography 9:377, 380
    infrared spectroscopy 9:390
    properties 9:466
1,1,1,2,2-Pentachlorofluoroethane. (See also
        Chlorofluorocarbons.)
    properties 13:61
Pentachloronaphthalenes. (See also Chlorinated
        naphthalenes.)
    properties 9:490
    toxicity 9:490
Pentachloronitrobenzene. See Terraclor.
Pentachlorophenol. (See also *Chlorophenols.*)
    acid–base titration 9:514
    commercial grades and specifications 9:513
    gas chromatography 9:520
    ion exchange chromatography 9:521
    impurities, determination of 9:520
    infrared spectroscopy 9:517
    manufacture 9:512

in photographic films 12:587
properties 9:511
ultraviolet spectroscopy 9:521
1,1,2,3,3-Pentachloropropane
  mass spectroscopy 9:401
Pentachloropropanes. (See also C₃ Chlorinated
    hydrocarbons.)
  properties 9:471
Pentachloropropene. (See also C₃ Chlorinated
    hydrocarbons.)
  mass spectroscopy 9:407
  properties 9:471
Pentacyanoamine ferrate
  colorimetric method for methyl isothiocyanate
    13:192
Pentacyano-*p*-nitrosodimethylanilinoferrate
  colorimetric method for zinc 19:557
1-Pentadecanol. (See also Fatty alcohols.)
  properties 4:561
Pentadiene
  in cigarette smoke 10:31
  gas chromatography 10:31
1,3-Pentadiene. See Piperylene.
1,4-Pentadiene. See Divinylmethane.
Pentadigalloylglucose. See Tannin.
Pentaerythritol. (See also Polyhydric alcohols.)
    5:22; 13:561
  acetic anhydride–pyridine volumetric method
    13:593
  in alkyd resins 5:40
  assay 13:592
  benzaldehyde gravimetric method 13:592
  catechol colorimetric method 13:572
  color 13:601
  commercial grades and specifications 13:566
  derivatives for identification 13:568
  gas chromatography 5:31; 13:578, 594; 19:300,
    301
  hydroxyl content 13:593
  manufacture 13:564
  paper chromatography 13:581
  phthalic anhydride volumetric method 13:586
  platinum–cobalt colorimetric method 13:601
  in polyurethanes 19:299
  properties 13:562
  in resins 13:596
  solubility 13:561, 563
  structure 13:561
  thin-layer chromatography 13:582
  uses 13:564
Pentaerythritol alkyds
  infrared spectroscopy 5:46
Pentaerythritol, allyl ethers
  gas chromatography 5:100
Pentaerythritol cyclic monoformal
  gas chromatography 13:578, 594
Pentaerythritol cyclic monoformal diacetate

gas chromatography 13:578
Pentaerythritol dicyclic diformal
  gas chromatography 13:578
Pentaerythritol dinitrate. (See also High
    explosives.) 12:445
Pentaerythritol–dipentaerythritol monoformal
  gas chromatography 13:578
Pentaerythritol nitrates. (See also Dipentaery-
    thritol hexanitrate; Pentaerythritol dinitrate;
    Pentaerythritol trinitrate; PETN; and Tri-
    pentaerythritol octanitrate.) 12:445
Pentaerythritol tetranitrate. See PETN.
Pentaerythritol trinitrate. (See also High
    explosives.)
  manufacture 12:445
  properties 12:445
  uses 12:445
Pentaethylenehexamine. (See also Ethylene-
    amines.)
  amine value 11:425
  commercial specifications 11:418
  density 11:409
  gas chromatography 11:421
  infrared spectroscopy 11:420
  nitrogen content 11:427
  properties 11:408
  vapor pressure 11:410
  viscosity 11:412
Pentafluoroethane. (See also *Fluorocarbons.*)
  properties 13:60
Pentafluoropropionamides
  gas chromatography 8:604
Pentaglycerol. See Trimethylolethane.
Pentahydroxybenzene
  solubility 17:5
2',3,4',5,7-Pentahydroxyflavone. See Morin.
3,5,7,3',4'-Pentahydroxyflavone. See Quercetin.
Pentalin. See Pentachloroethane.
Pentamethylene diamine
  gas chromatography 17:290
Pentamethylene glycol. See 1,5-Pentanediol.
Pentamethylene oxide. (See also *Epoxy com-
    pounds.*)
  infrared spectroscopy 12:199
Pentamycin 5:472, 564
  analysis 5:571, 576, 579
  composition 5:472
  fluorometry 5:579
  identification 5:571
  microbiological assay 5:576
  properties 5:565
  separation 5:571
  solubility 5:566
  uses 5:564
n-Pentane. (See also Pentanes.)
  azeotropes 14:302
  in butane 15:240

calorific value 13:331
commercial grades and specifications 14:302
in engine starting fluids 6:356
gas chromatography 13:183, 317, 318, 350;
  14:329, 330, 335; 15:240; 18:52
in gasoline 13:350, 353
in natural gas 13:320
properties 13:350; 14:302
specific gravity 13:329
structure 14:301
in trichlorosilane 18:52
1,2-Pentanediol
  gas chromatography 13:244
  iodometric titration 13:244
  in tetrahydrofurfuryl alcohol 13:244
1,5-Pentanediol. (See also Glycols.)
  commercial grades and specifications 13:547
  derivatives for identification 13:568
  distillation range 13:602
  gas chromatography 13:504, 576; 19:300
  manufacture 13:545
  phthalic anhydride volumetric method 13:586
  properties 13:536
  solubility 13:538
  structure 13:534
  thin-layer chromatography 13:582
  uses 13:545
2,4-Pentanediol. (See also Glycols.)
  gas chromatography 13:576
  solubility 13:538
2,4-Pentanedione. See Acetylacetone.
2,4-Pentanedione–sodium methylate
  volumetric method for nitrogen 11:426
Pentanes. (See also Alkanes; and individual
    compounds.) 14:301
  occurrence 14:302
  properties 14:301
  uses 14:302
1-Pentanethiol. See Amyl mercaptan.
Pentanoic acid
  derivatives for identification 8:376
1-Pentanol
  in black tea 18:510
  gas chromatography 7:96; 18:510
2-Pentanol
  gas chromatography 7:96; 16:107
  in methanol 16:107
3-Pentanol. (See also Alcohols, lower; Amyl
    alcohols.)
  2,4-dinitrophenylhydrazone colorimetric
    method 4:549
  pyromellitic dianhydride method 4:537
Pentanols. (See also Alcohols, lower; Amyl
    alcohols.)
  gas chromatography 16:98
2-Pentanone 8:331
3-Pentanone

absorption spectroscopy 9:564
  reaction rate analysis 8:364
Pentanones
  2-hydroxy-1-naphthaldehyde colorimetric
    method 11:204
n-Pentanonitrile. (See also *Nitriles.*)
  derivatives for identification 16:401, 402
  properties 16:396
Pentasodium tripolyphosphate
  properties 17:123
  uses 17:124
Pentenes
  gas chromatography 13:350; 14:334
  in gasoline 13:350, 353
  properties 13:350
1-Penten-3-ol
  in black tea 18:510
  gas chromatography 18:510
cis-2-Penten-1-ol
  in black tea 18:510
  gas chromatography 18:510
trans-3-Penten-2-one
  in black tea 18:510
  gas chromatography 18:510
2-Pentenylpenicillin. (See also Penicillins and
    related compounds.)
  structure 16:554
Pentlandite 16:285; 18:360
  composition 16:286
Pentobarbital
  derivatives for identification 6:477
  gas chromatography 6:210, 211, 483, 484
  paper chromatography 6:478, 479
  properties 6:470
  in sodium pentobarbital 6:520
  thin-layer chromatography 6:480, 481
  ultraviolet spectroscopy 6:501
Pentolite. (See also High-explosive mixtures.)
  analysis 12:450
  composition 12:450
  uses 12:450
Pentosans
  in cellulose 9:46
  in furfural 13:235
  in green coffee 10:374
  phloroglucinol gravimetric method 13:235
Pentoses
  in furfural 13:235
  phloroglucinol gravimetric method 13:235
n-Pentylbenzene
  gas chromatography 13:352
  in gasoline 13:352, 355
  properties 13:352
Pentyl formate. (See also Formic acid esters.)
  commercial specifications 13:132
  properties 13:131
  saponification method 13:132

solubility 13:131, 134
uses 13:120
1-Pentyne
  gas chromatography 11:510
  in isoprene 11:510
Peonidin
  structure 19:461
  in wines 19:461
Peppermint oil. (See also *Essential oils.*)
  commercial specifications 12:261
  composition 12:266
  gas chromatography 8:353; 12:266
  menthone content 12:282
  refractive index 12:271
Pepsin
  in beer 7:588
Peptide antibiotics 5:466, 533, 549
Peptides
  in beer 7:613
Peracetic acid
  preparation 8:373
Perbenzoic acid
  volumetric method for elastomers, synthetic
    12:121
  volumetric method for polybutadienes 12:121
Perborates
  in detergents 11:368
  potassium permanganate volumetric method
    7:229; 11:369
Perbunan 12:87
Percarbonates
  permanganate titration 7:229
Perchlorates. (See also individual compounds.)
    9:362–366
  analysis 9:362
  gravimetric method for cesium 9:298
  identification 9:362
  perchloric acid content 9:362
  uses 9:362
Perchloratocerate
  titration of glycerol 13:510
Perchloric acid. (See also Perchlorates.) 9:362
  analysis 9:365
  assay 9:365
  catalyst for acetylation of alcohols 4:529, 533
  commercial grades and specifications 9:364
  gravimetric method for bismuth 15:199
  gravimetric method for carbonate 18:255
  gravimetric method for cesium 15:267
  gravimetric method for chromium 9:654
  gravimetric method for lithium 15:267
  gravimetric method for manganese 15:77
  gravimetric method for potassium 15:267;
    17:368, 376, 386
  gravimetric method for rubidium 15:267
  gravimetric method for silica 13:417
  gravimetric method for silicon 18:21

impurities, determination of 9:365
  in perchlorates 9:362
  properties 9:363
  solubility 9:364
  titration of cobalt acetate 10:329
  uses 9:362
Perchloric acid–acetic anhydride
  volumetric method for nitrogen in poly-
    ethylenimine 14:485
Perchloric acid digestion
  ashing of polymers and fibers 4:283, 344
Perchloric acid–hydrochloric acid
  gravimetric method for chromium 16:298
Perchloric acid–hydrofluoric acid
  gravimetric method for silicon 15:391
Perchloric acid–hydrogen peroxide
  colorimetric method for titanium dioxide
    13:426
Perchloroethane. See Hexachloroethane.
Perchloroethylene. (See also Tetrachloro-
    ethylene.) 11:607
  absorption method for air in aerosols 4:460
  in carbon tetrachloride 8:325
  commercial grades and specifications 11:607
  gas chromatography 8:325
  impurities, determination of 11:608
  infrared spectroscopy 8:325
  odor 11:608
  properties 11:607
  specification tests 11:608
  stability 11:608
  toxicity 11:607
Perchloromethane. See Carbon tetrachloride.
Perchloryl fluoride
  properties 13:9
  solubility 13:9
  uses 13:9
Percodan. See Oxycodone.
Percolation processes
  efficiency of granular adsorbents for 4:449
PerFlectol 6:75
  colorimetric method based on formation of
    Wurster salts 6:101, 103
Perfumes
  coumarins in 11:167
Peri acid. See 1-Naphthylamine-8-sulfonic acid.
Periclase
  in cements 9:141
Perinone orange. (See also *Pigments.*)
  structure 17:184
Periodate oxidation
  of serine 8:523, 524
  of threonine 8:523, 524
Periodates. (See also individual compounds.)
    14:609–611
  analysis 14:609
  colorimetric method for manganese 5:136

gravimetric method for cesium 9:299
spectroscopic method for glycols 13:599
Periodate–tetrabase
colorimetric method for manganese 11:141
Periodic acid
gravimetric method for cesium 9:299
oxidation reagent for glycols 6:4
reaction alkanolamines, determination 5:19
spectroscopic method for propylene glycol
13:599
volumetric method for ethylene glycol 13:597
volumetric method for ethylene oxide 12:221
volumetric method for glycerol 13:514, 597
volumetric method for glycols 13:592
volumetric method for monoglycerides 13:515
volumetric method for polyglycerol 13:514
volumetric method for trimethylene glycol
13:514
Periodic acid–silver nitrate
gravimetric method for epoxy compounds
12:197
Periodic acid–sodium thiosulfate
volumetric method for glycerol 1-monochloro-
hydrin 14:171
Perkin-Elmer model 240 C, H, and N analyzer
8:292
Permalux 6:75
color test 6:96
Permanent hardness
of water 19:362
Permanent magnet alloys. (See also Cobalt
alloys.) 10:285
Permanganate number
of caprolactam 8:123
of pulp 9:52
Permanganate reducing substances
in acid anhydrides 8:568
in acid salts 8:596
Permanganate, substances reducing
in manganese sulfate 15:491
Permanganate time test
for acetic acid 4:100
for acetic anhydride 4:106
for acetone 4:115
for alcoholic beverages 4:487
for alcohols 4:543
for benzoic acid 7:66
for methanol 16:115
Permatone orange. See D & C Orange No. 17.
Permeability of activated carbon 8:151
Permendur. (See also Cobalt alloys.) 10:288
Perminvar. (See also Cobalt alloys.) 10:288
Pernitrous acid
properties 16:490
Perovskite. (See also Calcium titanate.) 8:75;
19:81
Peroxidase

ascorbic acid–benzidine colorimetric method
18:485
in green tea leaf 18:482
solvent extraction 18:485
Peroxide bomb method
for bromide 8:18
Peroxide decomposition of beer 7:554, 650
Peroxide fusion
of chromic oxide 9:690
Peroxides. (See also individual compounds.)
16:535
in acrylonitrile 4:377
in alcohols 4:543
ammonium sulfate–formic acid colorimetric
method 17:298
ammonium vanadate colorimetric method
11:522
bleaching agent 7:222, 223
in 1,3-butadiene 11:503, 504, 507
colorimetric determination 17:298
in dioxane 11:518, 522
in ethers 12:308
in ethyl ether 5:362; 12:308
iodometric titration 11:522; 13:232; 18:307
oxidation of methionine 8:505
permanganate titration 7:229
polarography 18:307
in polyamides 17:298
potassium iodide colorimetric method 12:308
properties 16:535
sodium thiosulfate titration 14:283
in styrene 18:291, 307
in terpenes 14:283
in tetrahydrofuran 13:231
titanium tetrachloride colorimetric method
12:308
Peroxide value
of corn oil 11:56
of fats and fatty oils 12:482
iodometric titration 11:56
of margarine 12:482; 15:500
of terpenes 14:283
Peroxyniobic acid
manufacture 16:355
properties 16:355
Peroxynitric acid
properties 16:490
Peroxytantalic acid
manufacture 16:355
properties 16:355
Peroxyuranates. (See also Uranium oxides.)
19:256
manufacture 19:256
Perrhenate
properties 17:519
Persian orange lake
in enamels 12:183, 184

identification 12:183, 184
Perspiration tests
  for packages 11:115
Persulfate fusion
  of boron samples 7:325
Persulfate method
  for chromium 9:689
Persulfate potentiometric titration
  of chromium 5:125
Pertechnetate
  properties 17:517
  uses 17:517
Perthane
  gas chromatography 19:389
  in water 19:389
Pertitanic acid
  colorimetric method for titanium 12:155
Pervanadic acid. (See also Vanadium compounds.)
    19:345
  colorimetric method for vanadium 9:687
Perylene maroon. (See also *Pigments.*)
  structure 17:185
Perylene scarlet. (See also *Pigments.*)
  structure 17:185
Pesticide residues
  bioassay 7:181
  in butter 16:139
  in cheese 16:142
  in eggs and egg products 12:66
  gas chromatography 12:67; 16:131
  liquid chromatography 12:67; 16:130
  in raw milk 16:128
  solvent extraction 16:129
Pesticides and insecticides. (See also individual
    compounds.) 19:578–619
  analysis 19:603
  bioassay 7:183; 19:614
  chlorine content 19:608
  classification 19:578
  in cocoa products 9:595
  colorimetric determination 19:611
  commercial grades and specifications 19:593
  composition 19:578
  cryoscopic methods 19:609
  definition 19:578
  density 19:617
  dusting tests 19:617
  emulsion stability 19:616
  flash tendency 19:618
  flow tests 19:617
  fluorometric methods 19:611
  formulations 19:593, 599
  gas chromatography 19:157, 487, 612
  in grapes 19:487
  heat stability 19:616, 618
  history 19:579
  hydrolytic methods 19:608

impurities, determination of 19:614
infrared spectroscopy 19:610
liquid chromatography 19:156
manufacture 19:591
neutron activation analysis 19:613
odor 19:618
paper chromatography 19:612
particle size 19:617
phosphorus content 19:609
polarography 19:613
properties 19:586
properties, determination of 19:615
radiochemical methods 19:613
safety 19:603
sample preparation 19:603
sampling 19:603
separation 19:607
solvent extraction 19:156, 487
staining characteristics 19:618
sulfur content 19:609
thin-layer chromatography 19:612
in tobacco 19:155
toxicity 19:591
ultraviolet spectroscopy 19:610
uses 19:578, 600
volumetric determination 19:609
wetting characteristics 19:616
in wines 19:487
Petalite 9:278; 15:261; 17:545
Pethidine. See Meperidine.
Pethidine derivatives. (See also individual
    compounds.)
  structure 16:244
Petitgrain oil. (See also *Essential oils.*)
  refractive index 12:271
PETN. (See also High explosives.) 12:445
  colorimetric determination 12:421
  commercial specifications 12:444
  in explosive priming mixtures 12:418, 421,
    422
  infrared spectrum 12:457
  manufacture 12:445
  melting point 12:461
  properties 12:432
  solubility 12:421, 445
  uses 12:445
  x-ray diffraction 12:418
Petrographic microscopy. See Light microscopy.
Petroleum 14:263
  chromium content 9:655
  composition 14:263
  cracking 14:263
  iron content 15:57
  water content of 6:317
Petroleum asphalts. (See also Bituminous
    materials.) 6:275
  manufacture 6:277

Petroleum ether. (See also *Hydrocarbons.*)
  commercial grades and specifications 14:325
  composition 14:326
Petroleum gas. (See also *Liquefied petroleum gas.*)
  copper corrosion test 14:347
Petroleum lubricants. (See also *Lubricants.*)
  manufacture 15:305
  properties 15:307
Petroleum oil
  in oil conditioners 6:357
Petroleum products
  analysis 14:282
  antifreezes 6:1
  hydrogen sulfide content 15:567
  manufacture 14:263, 271
  mercaptan content 15:566
  sulfur content 18:374, 378
  uses 14:261
Petroleum resins. See *Hydrocarbon resins.*
Petroleum resins, nonaromatic. (See also
    Hydrocarbon resins.) 14:262
  commercial grades and specifications 14:274,
    279
  density 14:267
  manufacture 14:262, 271
  polymer distribution 14:269
  structure 14:267
Petroleum resins, reactive. (See also *Hydrocarbon
    resins.*) 14:262
  commercial grades and specifications 14:273,
    279
  density 14:267
  manufacture 14:262, 271
  structure 14:267
Petroleum solvents
  copper corrosion test 14:347
Petroleum solvents, aliphatic
  degreasers 6:355
Petroleum sulfonate
  antirusts 6:351
Petroleum tars. (See also Bituminous materials.)
    6:275
Petroleum waxes 6:353
Pettenkofer color reaction
  for cholic acid 7:146
Petunidin
  structure 19:461
  in wines 19:461
Petzite 14:1
Pewter. (See also Copper alloys; Tin alloys.)
  composition 10:631; 19:24
  uses 19:25
pH. (See also Acidity; Basicity; and Neutrality.)
  of activated carbon 8:160
  of adhesives 10:274
  of adsorbents 4:437
  of alcoholic beverages 4:480, 487

of ammonium acetate solution 5:327
of ammonium chloride solution 5:338
of ammonium nitrate solution 5:343
of ammonium phosphate, dibasic solution
    5:350
of ammonium sulfate solution 5:352
of ammonium thiocyanate solution 5:354
of antibiotics 5:492
of antifreezes 6:8
of automotive chemical specialities 6:357
of barium chloride solution 6:546
of barium nitrate solution 6:552
of beer 7:507
of brake fluids 7:435
of carbon black 8:192, 229
of casein 9:14
of cereals 9:236
of cigarette smoke 10:21, 24, 25
of citric acid 10:88
of cocoa products 9:573
of corn products 11:8
of detergents 11:385
of drycleaning detergents 11:611
of eggs and egg products 12:68
of embalming fluids 12:166
of ethylene and propylene polymers 12:347
of fabrics 4:26
of felt 12:509
of fermented milk products 16:135
of flour 12:612
of glues of animal origin 13:492
of magnesium nitrate 15:407
of magnesium perchlorate 9:364, 366
of magnesium sulfate 15:412
methods for boric acid based on change in
    7:390
of musts 19:412
of organofunctional silanes 18:88
of pigments 17:177
of polyethylenimine 14:492
of potassium chloride 9:339
of potassium hydrogen phthalate 17:403
of reducing water 4:479
of resins 4:23
of soaps 18:200
of sodium chloride 9:337, 347
of sulfur dioxide 19:478
of tobacco 19:158
of water 19:360
of wines 19:412
Phaeophytin
  colorimetric method for chlorophyll 18:478
Phalthan. See Folpet.
Pharmaceuticals
  bithionol in 11:548
  caffeine in 8:63
  cresol in 11:548

glycerol content 13:520
hexachlorophene in 11:548
p-hydroxybenzoates in 11:545, 548
phenolic preservatives in 11:544
phenol in 11:548
resorcinol in 11:548
Phase separation methods
  for ethanol 19:436
Phelam. See Phenyl mercuric dimethyldithiocarbamate.
α-Phellandrene
  gas chromatography 12:268
  in majoram, oil of 12:268
β-Phellandrene
  gas chromatography 12:268
  in majoram, oil of 12:268
  structure 14:266
Phellopterin. (See also *Coumarins.*) 11:161
  gas chromatography 11:170
Phenacaine hydrochloride 5:376
  analysis 5:400
  anesthetic 5:400
  gravimetric method 5:401
  infrared spectroscopy 5:402
  infrared spectrum 5:402
  properties 5:400
  solubility 5:400
  ultraviolet spectra 5:400, 401
Phenacetin 6:177
  acid–base titration, nonaqueous 6:166, 170
  analysis 6:181
  codeine content 16:270
  codeine phosphate content 16:269
  colorimetric method using chromic acid 6:195
  colorimetric method using β-naphthol 6:194
  column chromatography 6:36, 128, 134, 150,
    152
  commercial grades and specifications 6:180
  diazotization method 6:194
  ethoxy content 6:190
  extraction 6:128, 133
  gas chromatography 6:135, 156, 187, 209, 211;
    8:603
  gravimetric method 6:189
  identification 6:182
  impurities, determination of 6:187, 196
  infrared spectrum 6:129
  ion exchange chromatography 8:62
  liquid chromatography 8:59, 60
  manufacture 6:178
  nuclear magnetic resonance spectroscopy
    6:172, 174
  paper chromatography 6:183
  properties 6:178; 8:597
  separation 6:128, 182
  solvent extraction 8:61
  thin-layer chromatography 8:59

ultraviolet spectroscopy 6:132, 133, 147, 149,
  150, 154, 159, 162, 182; 8:67
volumetric method using iodine 6:190
volumetric method using sodium nitrite 6:190
Phenacite 7:118
Phenanthraquinone
  colorimetric method for thiophene 18:537
Phenanthrene
  derivatives for identification 16:208
  liquid chromatography 7:16
Phenanthrene-9-aldehyde
  fluorometry 6:625, 627
  ultraviolet spectroscopy 6:632
o-Phenanthroline
  colorimetric method for iron 4:39, 89, 166,
    348, 418; 5:132, 176, 331, 350; 6:43; 7:114,
    660; 9:50, 242, 252, 344, 499; 10:116, 298,
    622; 11:46, 367; 14:65, 131, 141, 372, 455;
    15: 52, 143, 154, 274, 384, 478; 16:329;
    18:248, 418; 19:370
  colorimetric method for sulfate 18:427
  colorimetric method for sulfur dioxide 18:408
o-Phenanthroline–ferrous sulfate
  colorimetric method for cerium 9:274
1, 10-Phenanthroline hydrochloride
  colorimetric method for iron oxide 13:439
Phenanthroline indicator solution
  gravimetric method for 2-methyl-1,4-naphtha-
    hydroquinone-3-sulfonate 10:201
Phenazocine. (See also *Narcotics.*)
  ammonium molybdate colorimetric method
    16:247
  color tests 16:247
  formaldehyde–sulfuric acid colorimetric method
    16:247
  gas chromatography 16:255
  microchemical tests 16:250
  potassium iodide microchemical method 16:250
  sodium carbonate microchemical method 16:250
  structure 16:245
  thin-layer chromatography 16:256
  ultraviolet spectroscopy 16:253
Phenazocine hydrobromide. (See also *Narcotics.*)
  ultraviolet spectroscopy 16:253
Phenethicillin. (See also *Penicillins and related
  compounds*; Potassium phenethicillin.)
    16:573
  analysis 16:573
  assay 16:573
  bioassay 16:573
  commercial grades and specifications
    16:573
  manufacture 16:573
  microbiological methods 16:573
  properties 16:573
  specification tests 16:573
  structure 16:554

ultraviolet spectroscopy 16:565, 573
uses 16:573
β-Phenethyl alcohol. See 2-Phenyl ethanol.
Phenethylamine
  gas chromatography 6:209
Phenethylazocine. See Phenazocine.
Phenethyl cinnamate. (See also *Cinnamic acid and its derivatives.*)
  commercial specifications 10:66
  properties 10:65
  saponification methods 10:68
m-Phenetidine
  derivatives for identification 5:429
  near-infrared spectroscopy 5:435
  properties 5:422
o-Phenetidine
  azeotropes 5:424
  azo dyes for identification 5:430
  derivatives for identification 5:429
  infrared spectroscopy 5:454
  near-infrared spectroscopy 5:435
  properties 5:422
p-Phenetidine
  azeotropes 5:424
  azo dyes for identification 5:430
  colorimetric methods 5:452
  colorimetric method using chromic acid 6:195
  colorimetric method using β-naphthol 6:194
  derivatives for identification 5:429
  diazotization method 6:194
  gas chromatography 6:187
  near-infrared spectroscopy 5:435
  in phenacetin 6:196
  properties 5:422
Phenetidines 5:421
  manufacture 5:425
Phenindamine
  column chromatography 6:36
Phenindamine tartrate. (See also *Antihistamines.*)
  6:28
  acid–base titration, nonaqueous 6:170
  column chromatography 6:150
  properties 6:32
  ultraviolet spectroscopy 6:150
Phenindione 10:180
  acid–base titration 10:182
  analysis 10:180
  assay 10:181
  bromination method 10:181
  derivatives for identification 10:181
  identification 10:180
  infrared spectrum 10:180
  oxidation method 10:182
  paper chromatography 10:181
  in plasma 10:183
  polarography 10:182
  properties 10:180

separation 10:180
  in serum 10:183
  silver nitrate titration 10:182
  sodium hydroxide colorimetric method
    10:182, 183
  solubility 10:180
  sulfuric acid colorimetric method 10:181
  ultraviolet spectroscopy 10:180, 182
Pheniramine
  gas chromatography 6:210, 211
Pheniramine maleate. (See also *Antihistamines.*)
  6:28
  properties 6:32
Phenobarbital 6:468, 520
  acid–base titration 6:171, 521
  analysis 6:521
  column chromatography 6:150, 495, 498
  commercial grades and specifications 6:520
  derivatives for identification 6:477
  extraction 6:491, 497, 521
  gas chromatography 6:210, 211, 483, 484
  impurities, determination of 6:523
  infrared spectroscopy 6:492
  ion exchange chromatography 6:499
  manufacture 6:474
  paper chromatography 6:478, 479
  polarography 6:494
  properties 6:470
  in sodium phenobarbital 6:524
  solubility 6:476
  thin-layer chromatography 6:480, 481, 482
  ultraviolet spectroscopy 6:150, 501, 521
Phenol. (See also *Phenols.*)
  adsorption by activated carbon 8:145
  4-aminoantipyrine colorimetric method
    12:310; 17:18
  in aniline 5:446
  in carbon black 8:192
  colorimetric method for alkyd resin 11:127
  color test for phthalate plasticizers 16:96
  commercial grades and specifications 17:10
  commercial specifications 11:540
  in cresols 11:542
  derivatives for identification 17:21
  2,6-dichloroquinone-4-chloroimine colorimetric
    method 12:237
  in essential oils 12:283
  ferric chloride colorimetric method 17:16
  gas chromatography 9:519; 12:310; 14:580;
    17:33; 19:386
  history 17:1
  identification reagent for amides 8:600
  infrared spectroscopy 17:29
  in insulin 14:580
  ion exchange chromatography 6:144; 9:521
  manufacture 17:8
  in manufacture of azo dyes 6:403

in manufacture of salicylic acid 6:124
in methanol 16:115
in methyl salicylate 12:289
Millon's reagent color test 11:542; 17:18
occurrence 17:1, 9
odor 12:290
paper chromatography 17:24
in pharmaceuticals 11:548
in phenyl ethers 12:310
phthalic anhydride colorimetric method 17:19
properties 17:4
in salicylic acid 6:143
solidification point 17:40
solubility 13:539, 552; 17:3, 5
specific gravity 17:44
thin-layer chromatography 17:24
ultraviolet spectroscopy 9:521; 11:548
uses 17:2
in wines 19:460
Phenol absorption
  of carbon fibers 4:22
Phenol acetate
  in essential oils 12:284
Phenol cake 18:244
Phenol coefficient test
  for disinfectants and antiseptics 11:539
Phenoldisulfonic acid
  colorimetric method for nitrate 12:420;
    14:437; 16:502; 18:76
  colorimetric method for nitrogen oxides 16:481
Phenol–formaldehyde resin
  as ablator 4:19
  in finishing agents 11:126, 132
  properties 5:277, 281
  pyrolysis–gas chromatography 7:446
  solubility 11:126
  sulfuric acid–carbazole colorimetric method
    11:127, 132
  uses 5:280
Phenol, free
  in epoxy hardeners 4:24
Phenolic acids
  in green coffee 10:396
  in green tea leaf 18:475
  liquid chromatography 18:476
  thin-layer chromatography 18:475
Phenolic antioxidants. (See also Antioxidants
    and antiozonants.)
  in brake fluids 7:424, 425, 428
  colorimetric method using diazotized p-
    nitroaniline 6:103
  gas chromatography 6:84
  paper chromatography 6:85
  ultraviolet spectroscopy 6:116
Phenolic compounds
  in aniline 5:446
  in water 19:386

Phenolic content
  of alkyd resins by infrared spectroscopy
    5:46
  of tobacco 19:145
Phenolic glycosides
  in cocoa products 9:593
Phenolic hydroxyl groups
  in polycarbonates 17:339
  titanium chloride colorimetric method 17:339
Phenolic material
  in engine degreasers 6:355
Phenolic modifiers
  in alkyd resins 5:24
Phenolic plasticizers
  in polyamides 17:300
  ultraviolet spectroscopy 17:300
Phenolic preservatives
  in pharmaceuticals 11:544
Phenolic resins
  as ablator 4:15, 23
  adhesives, evaluation of 4:402
  analysis 7:444
  in brake linings 7:441, 444
  carbon fiber reinforced 4:22
  gas chromatography 7:445
  infrared spectroscopy 7:444
  manufacture 17:7
  nylon reinforced 4:15
  quality control tests 4:22
  silica reinforced 4:16
  structure 17:7
  thermogravimetry 4:19
Phenolic steroids
  gas chromatography 14:215
  liquid chromatography 14:210
  thin-layer chromatography 14:213
Phenolphthalein. (See also Triphenylmethane
    and related dyes.)
  colorimetric method for carbon tetrachloride
    9:387
  colorimetric method for hydrocyanic acid
    4:376
  colorimetric method for methylene chloride
    9:387
  colorimetric method for trichloroethylene
    9:387
  manufacture 17:6
  structure 17:7; 19:161
  titanous chloride titration 19:177
Phenolphthalein–copper sulfate
  colorimetric method for hydrogen cyanide
    14:418
Phenolphthalein–sodium carbonate
  colorimetric method for carbonate 11:141
Phenol red
  colorimetric method for bromine 8:23
Phenols. (See also individual compounds.) 17:1–50

acid–base titration, nonaqueous 17:27
4-aminoantipyrine colorimetric method 11:544;
    17:17, 30; 19:386
analysis 11:540; 17:15, 25
bromine gravimetric method 16:212
bromometric titration 17:26
in cigarette smoke 10:23
cloud point 17:42
colorimetric determination 17:28
color standards 17:44
color tests 17:16
commercial grades and specifications 17:10
commercial specifications 11:540
derivatives for identification 17:20
2,6-dibromoquinone chlorimine colorimetric
    method 17:29
in disinfectants and antiseptics 11:539
distillation analysis 17:25
in embalming fluids 12:165
ferric chloride colorimetric method 11:126,
    131; 17:16
in finishing agents 11:126, 131
Folin-Ciocalteu colorimetric method 19:462
Folin-Denis colorimetric method 19:462
Folin's colorimetric method 19:147
in food 11:541
gas chromatography 10:24; 11:541, 543; 17:32;
    19:386
gel time 17:45
Gibbs colorimetric method 17:29
identification 11:540; 17:15
impurities, determination of 17:36
infrared spectroscopy 17:22, 28
Liebermann color test 17:19
liquid chromatography 17:25
manufacture 17:8
mass spectroscopy 17:22
Millon's reagent color test 17:17
in naphthalene 16:212
nuclear magnetic resonance spectroscopy 17:22
paper chromatography 17:23
in photographic films 12:587
phthalic anhydride colorimetric method 17:18
potassium hydroxide volumetric method 12:283
properties 17:3, 43
in qualitative test for alcohols 4:522
separation 11:540; 17:23
solidification point 17:25, 40
solubility 11:126; 17:3
specific gravity 17:43
thin-layer chromatography 11:540; 17:25
in tobacco 19:147
p-toluene diazonium chloride colorimetric
    method 17:29
toxicity 17:7
ultraviolet spectroscopy 17:22, 28
uses 17:2

in water 17:28; 19:386
Phenol value
of activated carbon 8:147
Phenothiazine. (See also *Diamines and higher
    amines, aromatic.*)
derivatives for identification 11:438
inhibitor in acrylic monomers 4:188, 211
properties 11:434
solubility 11:435
Phenothiazine derivatives
column chromatography 6:152
Phenoxyacetic acid
properties 8:403
β-(Phenoxyethyl)dimethyldodecylammonium
    bromide. (See also Quaternary ammonium
    compounds.)
ultraviolet spectroscopy 11:556
Phenoxyethylpenicillin. See Phenethicillin.
Phenoxymethylpenicillanic acid. (See also
    Phenoxymethylpenicillin.)
iodometric methods 16:562
solubility 16:557
Phenoxymethylpenicillin. (See also Benzathine
    phenoxymethylpenicillin; Hydrabamine
    phenoxymethylpenicillin; *Penicillins and
    related compounds*; Phenoxymethylpeni-
    cillanic acid; and Potassium phenoxymethyl-
    penicillin.) 16:572
analysis 16:572
assay 16:572
bioassay 16:572
manufacture 16:553, 572
microbiological methods 16:572
paper chromatography 16:558
penicillin content 16:572
properties 16:572
specification tests 16:572
structure 16:554
thin-layer chromatography 16:559, 572
ultraviolet spectroscopy 16:560, 565,
    572
uses 16:572
Phenprocoumon. (See also *Coumarins.*) 10:164;
    11:160
acid–base titration 10:165
analysis 10:164
diazotized p-nitroaniline colorimetric method
    10:166
identification 10:151, 164
impurities, determination of 10:166
paper chromatography 10:152
properties 10:164
pyridine–ferric chloride colorimetric method
    10:164
solubility 10:164
specification tests 10:166
thin-layer chromatography 10:151

ultraviolet spectroscopy 10:151, 164
uses 11:159
Phentanyl. See Fentanyl.
Phenylacetaldehyde
    absorption spectroscopy 9:564
    in black tea 18:507, 511
    gas chromatography 18:511
    properties 8:334
N-Phenylacetamide. See Acetanilide.
Phenylacetic acid 8:405
    analysis 8:405
    gas chromatography 8:388
    paper chromatography 8:379
    properties 8:403, 405
Phenylacetonitrile. (See also *Nitriles.*)
    derivatives for identification 16:402
    properties 16:396
Phenylacetylene
    gas chromatography 18:308
    silver nitrate–ammonium thiocyanate volumetric
        method 18:308
    in styrene 18:308
Phenylacetylhydroxamic acid
    gravimetric method for tantalum 16:367
Phenylacrolein. See *Cinnamaldehyde.*
3-Phenylacrylic acid. See Cinnamic acid.
Phenylalanine. (See also *Carboxylic acids, amino.*)
        8:413, 507
    acid–base titration, nonaqueous 8:511
    analysis 8:511
    in black tea 18:507
    in cocoa beans 9:556
    colorimetric methods 8:511
    commercial specifications 8:511
    composition 8:413
    crystalline form 8:508
    gas chromatography 8:429
    infrared spectrum 8:508
    ion exchange chromatography 8:415, 421
    manufacture 8:510
    nitration 8:511
    nuclear magnetic resonance spectrum 8:509
    occurrence 8:510
    paper chromatography 8:422
    properties 8:508
    in protein 8:414
    solubility 8:508
    structure 8:508
    in yeast 19:529
Phenylalkylamine alkaloids 4:590
    analysis 4:606
γ-Phenylallyl alcohol. See *Cinnamyl alcohol.*
Phenylarsonic acid
    gravimetric method for niobium 16:363, 366
    gravimetric method for tantalum 16:363
        366
Phenylazoaniline–dimethylformamide

colorimetric method for nitrite 16:494
1-(Phenylazo)-2-naphthol
    in D & C Orange No. 4 10:519
    in D & C Red No. 17 10:522
    in D & C Red No. 31 10:522
    thin-layer chromatography 10:522
Phenylbarbituric acid
    in phenobarbital 6:523
N-Phenylbenzamide. See Benzanilide.
Phenylbenzene. See Diphenyl.
2-Phenylbenzothiazole
    gravimetric method for osmium 17:263
Phenylbutane. See Butylbenzenes.
1-Phenyl-2-butene
    gas chromatography 7:8
1-Phenyl-3-butene
    gas chromatography 7:8
4-Phenyl-3-buten-2-one
    absorption spectroscopy 9:564
Phenyl chloride. See Chlorobenzene.
Phenylchlorosilanes. (See also individual
        compounds.)
    manufacture 18:55
Phenyl cinnamate. (See also *Cinnamic acid and
        its derivatives.*)
    ultraviolet spectroscopy 10:69
α-Phenylcinnamonitrile. (See also *Nitriles.*)
    polarography 16:398
N-Phenylcyclohexylamine 11:251
    gas chromatography 11:212, 241, 251
    infrared spectroscopy 11:251
    manufacture 11:251
    properties 11:244
    solubility 11:251
    toxicity 11:251
    uses 11:251
N-Phenyl-*N*′-cyclohexyl-*p*-phenylenediamine
    column chromatography 6:89
    paper chromatography 6:85
Phenyl-1,2-dichloroethane
    mass spectroscopy 9:404
Phenyldichlorosilane. (See also Organo-
        chlorosilanes.)
    gas chromatography 18:60, 66
    in phenyltrichlorosilane 18:66
Phenyldiethoxysilane. (See also Organosilane
        esters.)
    gas chromatography 18:75
    in phenyltriethoxysilane 18:75
m-Phenylenediamine. (See also Phenylene-
        diamines.)
    in air 11:449
    Fast Blue B colorimetric method 11:449
    manufacture 11:445
    sodium nitrite colorimetric method 11:446
    uses 11:444
o-Phenylenediamine. (See also Phenylenediamines.)

bismuth nitrate colorimetric method 11:446
in green tea leaf 18:459
manufacture 11:445
potassium titanium oxalate colorimetric method 18:459
uses 11:443
p-Phenylenediamine. (See also Phenylene-diamines.)
in benzidine 11:448
colorimetric method based on formation of Wurster salts 6:101, 103
colorimetric method for amine antioxidants 6:94
colorimetric method for epoxy resins 12:236
p-dimethylaminobenzaldehyde colorimetric method 11:448
manufacture 11:445
in polyurethanes 19:304
potassium permanganate colorimetric method 11:446
structure 19:304
toxicity 11:444
uses 11:444
p-Phenylenediamine derivatives
antiozonants 6:78
colorimetric method based on formation of Wurster salts 6:100
m-Phenylenediamine dihydrochloride
colorimetric method for crotonaldehyde 11:179
p-Phenylenediamine–pyridine
colorimetric method for hydrogen cyanide 14:419
Phenylenediamines. (See also *Diamines and higher amines, aromatic*; and individual compounds.)
11:443–449
acid–base titration 11:447
analysis 11:446
commercial grades and specifications 11:445
derivatives for identification 11:438
diamine content 11:447
freezing point 11:446
gas chromatography 11:447
in hair dyes 11:103
impurities, determination of 11:447
polarography 11:448
properties 11:434, 444
solubility 11:435
thin-layer chromatography 11:103; 19:306
toxicity 11:444
uses 11:443
visible spectroscopy 11:448
m-Phenylene isocyanate. (See also *Isocyanic acid esters*.)
properties 15:95
Phenylephrine
codeine content 16:270

Phenyl ethers. (See also *Ethers*; and individual compounds.)
gas chromatography 12:310
phenol in 12:310
Phenylethyl acetate
in essential oils 12:276
molecular weight 12:276
Phenylethyl alcohol
solubility 13:540
β-Phenylethyl alcohol 7:89
acetylation method 7:99
analysis 7:91
azeotropes 7:93
in black tea 18:507, 511
in brandies 19:440
commercial grades and specifications 7:91, 95
derivatives for identification 7:93, 95
gas chromatography 7:93, 96, 99; 18:511; 19:442
identification 7:93
impurities, determination of 7:99
infrared spectroscopy 7:97, 98
manufacture 7:91
mass spectroscopy 7:95, 97
nuclear magnetic resonance spectrum 7:99
paper chromatography 7:95
properties 7:90
properties, determination of 7:99
separation 7:93
uses 7:89
in wines 19:425, 440
2-Phenylethyl alcohol. See β-Phenylethyl alcohol.
N-Phenylethylaziridine. (See also *Imines*.)
properties 14:469
structure 14:469
5-Phenyl-5-ethylbarbituric acid 5:91
Phenylethyl benzoate. (See also Benzoic acid esters.)
commercial specifications 7:80
3,5-dinitrobenzoate derivative 7:83
properties 7:77
saponification method 7:84
Phenylethylene. See *Styrene*.
Phenyl ethyl ether. (See also *Ethers*.)
properties 12:297
Phenylethylmethylsilicone fluids. (See also Silicone fluids.)
infrared spectrum 18:108
4-Phenylethynylpyridine. (See also *Pyridine and pyridine derivatives*.)
ionization constants 17:416
Phenylfluorone
colorimetric method for antimony 6:67
colorimetric method for germanium 13:386
colorimetric method for niobium 16:368, 374

colorimetric method for tantalum 16:368; 18:
    449
colorimetric method for tin 6:46, 228; 19:62
Phenyl glycidyl ether
    volumetric method for AHEW 11:430
Phenyl groups
    in polycarbonates 17:339
    ultraviolet spectroscopy 17:339
Phenylhydrazine
    color test for formaldehyde 13:106
    gravimetric method for cinnamaldehyde 10:45
Phenylhydrazine acetate
    gravimetric method for ethylene glycol 13:570
    gravimetric method for propylene glycol 13:570
Phenylhydrazine hydrochloride–sodium acetate
    gravimetric method for caramel 10:538
Phenylhydrazone
    of acetophenone 4:124
    of benzaldehydes 6:612, 613
2-(Phenyl)indane-1,3-dione. See Phenindione.
Phenyl isocyanate. (See also *Isocyanic acid
    esters.*)
    commercial sources 15:103
    gravimetric method for cinnamyl alcohol
        10:79
    infrared spectroscopy 15:106
    near-infrared spectroscopy 15:105
    properties 15:95
    volumetric method for polytetramethylene
        ether glycols 13:246
Phenyllactic acid
    gas chromatography 8:388
Phenyl mercaptan. (See also *Mercaptans.*)
    derivatives for identification 15:559
    manufacture 15:555
    properties 15:552
Phenylmercuric acetate. (See also Mercurials;
    Organomercury compounds.)
    ammonium thiocyanate volumetric method
        11:564; 16:63
    analysis 16:63
    assay 11:564; 16:63
    commercial grades and specifications 16:54, 58
    dithizone colorimetric method 11:564
    impurities, determination of 16:63
    manufacture 16:58
    solubility 16:58, 63
    ultraviolet spectroscopy 11:565
    uses 16:58
    zinc reduction 11:564
Phenylmercuric benzoate. (See also Organo-
    mercury compounds.)
    commercial grades 16:54
Phenylmercuric borate. (See also Organomercury
    compounds.)
    commercial grades 16:54
Phenylmercuric chloride. (See also Organo-

mercury compounds.)
    commercial grades 16:54
Phenylmercuric dimethyldithiocarbamate. (See
    also Organomercury compounds.)
    commercial grades 16:54
Phenylmercuric lactate. (See also Organomercury
    compounds.)
    commercial grades 16:54
Phenylmercuric nitrate. (See also Mercurials;
    Organomercury compounds.)
    ammonium thiocyanate volumetric method
        11:564
    assay 11:564
    commercial grades 16:54
    commercial specifications 11:562
    zinc reduction 11:564
Phenylmercuric oleate. (See also Organo-
    mercury compounds.)
    commercial grades 16:54
Phenylmercuric triethanol ammonium lactate.
    (See also Organomercury compounds.)
    commercial grades 16:54
Phenylmethyldichlorosilane. (See also Organo-
    chlorosilanes.)
    gas chromatography 18:67
    impurities, determination of 18:67
    infrared spectrum 18:58
    manufacture 18:55
    properties 18:54
Phenylmethyldiethoxysilane. (See also
    Organosilane esters.)
    gas chromatography 18:75
    in phenyltriethoxysilane 18:75
Phenyl methyl ether. (See also *Ethers.*)
    properties 12:297
Phenylmethylsilicone fluids. (See also Silicone
    fluids.)
    commercial specifications 18:103
    infrared spectrum 18:107
    properties 18:97
    refractive index 18:97
    viscosity 18:97
N-Phenyl-1-naphthylamine. (See also *Diamines
    and higher amines, aromatic.*) 6:72, 74, 75
    analysis 7:428
    in brake fluids 7:425, 428
    color test 6:93
    column chromatography 6:89
    paper chromatography 6:85
    ultraviolet spectroscopy 11:442
Phenyl-β-naphthylamine. (See also *Diamines
    and higher amines, aromatic.*) 6:72, 73,
    75, 77, 107; 11:436
    analysis 6:108
    colorimetric method using diazotized *p*-
        nitroaniline 6:99, 109
    color tests 6:90

column chromatography 6:88
commercial grades and specifications 6:108
gas chromatography 6:81, 108
identification 6:108
manufacture 6:108
mixtures, determination in 6:108
paper chromatography 6:85
properties 6:107
separation 6:108
ultraviolet spectroscopy 6:97
N-Phenylnaphthylamines. (See also Diamines
    and higher amines, aromatic.)
    derivatives for identification 11:438
    in elastomers, synthetic 12:140
    properties 11:434
    solubility 11:435
    thin-layer chromatography 12:140
    ultraviolet spectra 12:141
Phenyloxirane. See Styrene oxide.
2-Phenylpentane. See sec-Amylbenzene.
p-Phenylphenacyl derivative
    of adipic acid 4:412
    of benzoic acid 7:64
p-Phenylphenol
    colorimetric method using diazotized p-
        nitroaniline 6:106
N-Phenyl-p-phenylenediamine
    in hair dyes 11:103
    thin-layer chromatography 11:103
1-Phenylpropene oxides. (See also Epoxy
    compounds.)
    infrared spectroscopy 12:200
3-Phenylpropenoic acid. See Cinnamic acid.
3-Phenyl-2-propenol-1. See Cinnamyl alcohol.
Phenylpropionic acid
    gas chromatography 8:388
Phenylpropyl cinnamate. See Hydrocinnamyl
    cinnamate.
1-Phenyl-2-propynol-1. (See also Acetylenic
    alcohols.)
    silver acetylide method 4:578
Phenylpyridines. (See also Pyridine and pyridine
    derivatives.)
    ionization constants 17:416
    manufacture 17:422
Phenyl-α-pyridyl ketoxime
    colorimetric method for palladium 17:249
Phenylpyruvic acid
    gas chromatography 8:388
Phenyl salicylate
    ion exchange chromatography 6:144
    in photographic films 12:592
    in salicylic acid 6:143
    ultraviolet spectroscopy 16:95
Phenylthiourea
    derivatives of amines 5:266
    derivatives of aniline and its derivatives 5:429

N-Phenyl-N'-p-tolyl-p-phenylenediamine. (See
    also Diamines and higher amines, aromatic.)
    properties 11:434
    solubility 11:435
Phenyltrichlorosilane. (See also Organo-
    chlorosilanes.)
    commercial specifications 18:56
    gas chromatography 18:60, 66, 67
    impurities, determination of 18:66
    infrared spectrum 18:58
    manufacture 18:55
    in phenylmethyldichlorosilane 18:67
    properties 18:54
Phenyltriethoxysilane. (See also Organosilane
    esters.)
    commercial specifications 18:71
    gas chromatography 18:75
    impurities, determination of 18:75
    infrared spectrum 18:71
    properties 18:70
Phenyltrimethoxysilane. (See also Organosilane
    esters.)
    properties 18:70
Phenylurethane
    of higher alcohols 4:558
    of lower alcohols 4:519
Philipp methods
    for end groups in acrylonitrile polymers
        4:290
Phillipsite 8:75
Philprene 12:87
Phlogopite 11:464; 14:519
    composition 10:102
Phloretic acid
    gas chromatography 8:389
Phloroglucinol. (See also Trihydroxybenzenes.)
    analysis 17:48
    colorimetric method for allyl chloride 5:102
    colorimetric method for caramel 10:538
    colorimetric method for glycerol 19:443
    commercial grades and specifications 17:15
    derivative of acetals 4:67
    derivatives for identification 17:21
    ferric chloride colorimetric method 17:16
    gravimetric method for furfural 13:235
    gravimetric method for pentosans 13:235
    gravimetric method for pentoses 13:235
    in green tea leaf 18:459
    iodine–pyrocatechol colorimetric method
        17:48
    manufacture 17:10
    occurrence 17:2
    paper chromatography 17:24
    potassium permanganate titration 18:459
    potassium titanium oxalate colorimetric
        method 18:459
    properties 17:5

solubility 17:5
thin-layer chromatography 17:24
Phlorol. See o-Ethylphenol.
Phloxine
  in enamels 12:185
  identification 12:185
Phloxine B. See D & C Red No. 28.
Phobophine
  in cacao 9:536
Pholcodine. (See also *Narcotics*.)
  ammonium molybdate colorimetric method
    16:247
  color tests 16:247
  formaldehyde–sulfuric acid colorimetric method
    16:247
  microchemical tests 16:250
  platinic chloride microchemical method 16:250
Phorate. (See also *Pesticides and insecticides*.)
  commercial grades and specifications 19:595
  composition 19:589, 598
  manufacture 19:598
  properties 19:589
  solubility 19:587
  structure 19:584
  toxicity 19:592
Phorone
  derivatives for identification 8:343
  properties 8:335
Phosgene (Carbonyl chloride)
  in chlorine 9:320, 321
  in chloroform 4:435; 5:359
  gas chromatography 14:392
  properties 8:570
Phosphatase
  in milk 16:133
Phosphate fertilizers 17:141–149
  analysis 17:144
  chemically available phosphorus content 17:149
  citrate–insoluble phosphorus content 17:148
  citrate–soluble phosphorus content 17:149
  composition 17:141
  manufacture 17:143
  phosphorus content 17:144
  phosphorus content, soluble 17:148
  sample preparation 17:144
  water-soluble phosphorus content 17:148
Phosphate opal glass. (See also *Glass and glass-
    ceramics*.)
  composition 13:409
Phosphate rock
  commercial grades and specifications 17:143
  fluoride content 13:27
  impurities, determination of 17:71
  occurrence 17:51, 143
  phosphorus content 17:62, 70
  uses 17:53, 141
Phosphates. (See also Orthophosphates;

and individual compounds.) 17:117
  in acid anhydrides 8:567
  acid–base titration 17:139
  in acid halides 8:575
  in acid salts 8:593
  in alkaline flushes 6:359
  in ammonium chloride 5:338
  in ammonium iodide 14:604
  ammonium molybdate–aminonaphthol
    sulfonic acid colorimetric method 17:102;
    19:498
  ammonium molybdate–ammonium vanadate
    colorimetric method 10:93, 666; 11:377;
    17:104; 19:369
  ammonium molybdate–benzidine colorimetric
    method 11:142
  ammonium molybdate gravimetric method
    11:75, 132; 18:183
  ammonium molybdate-p-methylaminophenol
    sulfate colorimetric method 7:382, 383;
    9:348; 13:34; 14:437; 15:401, 404; 16:536;
    18:233
  ammonium molybdate–sodium bisulfite
    colorimetric method 14:604
  ammonium molybdate–stannous chloride
    colorimetric method 9:251; 14:604; 17:105
  ammonium molybdate volumetric method
    17:507
  in ammonium nitrate 5:343
  in ammonium sulfate 5:352
  analysis 17:128, 131
  antirusts 6:351
  in bakery chemicals 6:433
  in barium sulfate 6:567, 568
  in borax 7:383
  in boric acid 7:382
  in calcium acetate 8:79
  in calcium carbonate 8:84
  in calcium phosphate 8:96
  in ceric ammonium nitrate 9:251
  in ceric hydroxide 9:257
  in cerous chloride 9:260
  in citric acid 10:93
  in copper(II) hydroxide 10:666
  in cotton 11:142
  in creams and lotions 11:92
  in detergents 11:369
  in diammonium citrate 10:93
  o-dianisidine molybdate–hydrazine colorimetric
    method 11:142
  in ethylene 5:362
  in finishing agents 11:132
  gravimetric determination as ammonium
    phosphomolybdate 6:360
  gravimetric method for silver 18:164
  in hydrofluoric acid 13:34
  in hydriodic acid 14:604

in hydrogen peroxide 14:437
ion exchange chromatography 11:372; 17:131
ion exchange methods 17:58, 60
in magnesium carbonate 15:401
in magnesium chloride 15:404
in magnesium nitrate 15:407
in magnesium sulfate 15:412
in methacrylate polymers 16:96
in methanol 16:117
molybdenum blue colorimetric method
　17:102
paper chromatography 17:135
in perchloric acid 9:364, 365
in potassium chloride 9:339
in potassium iodide 14:606
in rare earth chlorides 17:506
in rare earth fluorides 17:507
in shampoos 11:75
in soaps 18:182, 183, 193
in sodium chloride 9:337, 348
in sodium iodide 14:606
in sodium metal 18:233
sodium molybdate–chlorostannous acid colori-
　metric method 16:117
in sodium peroxide 16:536
solubility 11:126; 17:58
in uranium 19:239
volumetric determination 19:498
in water 17:102; 19:368
in wines 19:498
zinc sulfate volumetric method 18:194
Phosphates, amorphous condensed 17:124
Phosphates, condensed 17:118
classification 17:118
manufacture 17:125
Phosphates, inorganic
ammonium molybdate gravimetric method
　11:335
in antifreezes 6:13
in detergents 11:335
Phosphides
acid dissolution 9:215
fluxes for 9:216
Phosphine 13:194
of acetylene 4:132
in carbon dioxide 8:260
detection 17:61
flammability 13:194
2-mercaptobenzamidazole colorimetric method
　17:61
properties 13:194
silver nitrate color test 17:61
toxicity 13:194; 17:57
uses 13:194
Phosphites. (See also individual compounds.)
　17:111
ceric ammonium nitrate volumetric

method 17:113
colorimetric determination 17:61
detection 17:61
sodium hypochlorite volumetric method
　17:115
uses 17:113
Phospholipids
in egg yolk 12:58
Phosphomolybdic acid
colorimetric method for caffeine 8:66, 167
colorimetric method for codeine 16:270
colorimetric method for mercury 16:42
colorimetric method for tin 19:37
colorimetric method for vanadium 19:331
o-dianisidine gravimetric method 17:61
gravimetric method for cesium 17:553
gravimetric method for protein 19:475
gravimetric method for rubidium 17:553
gravimetric method for thallium 18:526
Phosphomolybdic heteropolyacid
colorimetric method for niobium 16:372
Phosphomolybdotungstic acid
colorimetric method for thallium 18:533
Phosphoproteins 9:4
Phosphoric acid. (See also Metaphosphoric
　acid; Orthophosphoric acid; and Pyro-
　phosphoric acid.) 17:117
in acid mixtures 14:367
analysis 17:128
calcium hydroxide turbidimetric method
　15:145
cerous nitrate volumetric method 14:367
colorimetric method for manganese 14:611
colorimetric method for oxyethylene–oxy-
　propylene copolymers 19:308
colorimetric method for polyethers 19:308
commercial grades and specifications 17:125,
　126
in cyclamates 11:231
desulfonation method for anionic detergents
　11:329
etchant for ceramics 9:224
in lactic acid and its derivatives 15:145
manufacture 17:58, 124
in methanol 16:115
in monofluorophosphoric acid 13:46
silver orthophosphate volumetric method
　13:46
thin-layer chromatography 8:381
toxicity 17:56
Phosphoric acid–hydrochloric acid
colorimetric method for sodium sulfate
　9:360
Phosphoric acid–potassium periodate
colorimetric method for manganese oxide
　13:442
Phosphoric acids, condensed 17:122

manufacture 17:122
properties 17:122
Phosphoric acid–sodium tungstate
  colorimetric method for vanadium 19:339
Phosphoric anhydride
  gravimetric method for water in aerosols
    4:460
Phosphorimetry
  of apomorphine hydrochloride 16:264
  of codeine 16:264
  for diphenyl 11:528
  of morphine 16:265
  of morphine sulfate 16:264
  in narceine 16:264
  of narcotine 16:264
  of papaverine hydrochloride 16:264
  of thebaine 16:264
Phosphorous acid
  analysis 17:113
  commercial specifications 17:112
  gravimetric method for selenium 17:591
  manufacture 17:112
  properties 17:112
  reduction method for mercury 16:16
  uses 17:113
Phosphorus. (See also Phosphorus compounds;
    Phosphorus, elemental.) 17:51–150
  acid–base titration 17:64, 65, 74, 80
  in alumina 5:116
  in aluminum oxide abrasives 4:47
  ammonium molybdate–ammonium vanadate
    colorimetric method 10:124, 624, 644;
    11:36; 17:66, 70, 71, 76, 91, 94, 146
  ammonium molybdate colorimetric method
    9:242, 296
  ammonium molybdate–hydrazine sulfate
    colorimetric method 11:369; 15:60; 17:83
  ammonium molybdate–hydroquinone colori-
    metric method 17:98
  ammonium molybdate–magnesia gravimetric
    method 17:78, 95, 101
  ammonium molybdate volumetric method
    9:295; 15:75; 17:74, 80, 86, 96, 97, 145
  ammonium nitromolybdate volumetric method
    17:85
  ammonium phosphomolybdate gravimetric
    method 17:59, 62, 101
  ammonium phosphomolybdate volumetric
    method 17:64, 189
  analysis 17:58
  in antifreezes 6:17
  in asbestos 6:261
  ascorbic acid–ammonium molybdate colorimetric
    method 17:72
  in baking chemicals 6:433
  in bauxite 5:115
  in beer 7:633

in benzoyl chloride 7:77
in brass 17:89
in brazing alloys 17:93
in bronze 17:89
in carbonate rocks 17:72
in cement 17:97
in cereals 9:233, 242
chlorostannous acid colorimetric method
  10:301
in chocolate and cocoa products 9:538, 595
in chrome steels 9:662
in chromium 9:635
in chrysotile 6:261
in clays and clay minerals 10:123
in coal 17:96
in cobalt alloys 10:308, 319
in cobalt metal 10:301
in coke 17:96
colorimetric determination 17:65
in copper 17:91
in copper alloys 10:644; 17:91
in copper metal 10:624
in cotton 11:154
detection 17:60
in detergents 11:369
in diatomaceous earth 11:490
EDTA titration 17:65
in elastomers, synthetic 12:116
emission spectroscopy 17:67
ferric ammonium sulfate gravimetric method
  16:334
in ferroalloys 17:85
in ferrochromium 9:661
in ferromanganese 15:78; 17:86
in ferromolybdenum 17:88
in ferrophosphorus 17:84
in ferrosilicon 15:75; 17:86
in ferrotungsten 17:88
in ferrous metals 17:78
in ferrovanadium 17:87
in fertilizers 17:148, 149
flame photometry 17:67
flame tests 6:592
in fluorophosphoric acids 13:45
in food 17:99
in foundry slags 17:85
in gallium arsenide 13:279
in gallium metal 13:266
in germanium 13:392
in germanium dioxide 13:393
gravimetric determination 17:59, 62
in hafnium 14:134
ion exchange methods 16:185
in iron 15:59, 60; 17:59, 78
in iron ore 17:73
in lithium fluoride 15:284
in lubricating oils 17:93

magnesia gravimetric method 11:371;
16:185
in magnesium alloys 15:389
magnesium ammonium phosphate gravimetric
method 17:60, 62, 101
magnesium chloride–ammonium chloride
gravimetric method 17:63, 101
magnesium pyrophosphate gravimetric method
10:424
in manganese 15:482
in manganese–silicon 17:86
molybdate gravimetric method 11:490
molybdenum blue colorimetric method 13:392;
17:65, 72, 82, 98, 100; 19:529
molybdic acid–ammonium hydroxide gravimetric
method 17:101
in molybdovanadophosphoric acid and salts
16:185
in monofluorophosphoric acid 13:46
in nickel 16:333
in nickel alloys 16:333
occurrence 17:51
in organic compounds 17:68
oxidation numbers 17:55
in pesticides and insecticides 19:609
in phosphate fertilizers 17:144
in phosphate rocks 17:62, 70
in plants 17:97
in pollucite 9:295
in Polygard 6:118
in polyphosphates 17:139
quimociac gravimetric method 6:433;
17:147
radioactivation methods 17:68
in roasted coffee 10:424
sample decomposition 17:58
separation 17:59
in silicate rocks 17:71
in silicate white lead 17:189
in silicomanganese 17:86
sodium molybdate–citric acid gravimetric method
17:98
sodium molybdate–hydrazine sulfate colorimetric
method 15:284
sodium molybdate–quinoline gravimetric method
17:68
sodium molybdate–quinoline hydrochloride
volumetric method 12:116
sodium molybdate–stannous chloride colorimetric
method 14:134; 16:334
in starch derivatives 11:35
in steel 17:78
in tobacco 19:152
volumetric determination 17:64
in water 17:60, 101
in yeast 19:529
in zircaloy 14:135

in zirconium 14:134
Phosphorus acids, lower. (See also Hypophosphorous acid; Phosphorous acid.)
17:111–116
analysis 17:113
Phosphorus alloys
carrier gas fusion methods 8:682
oxygen in 8:682
Phosphorus, black. (See also Phosphorus,
elemental.)
manufacture 17:54
properties 17:54
Phosphorus compounds. (See also Phosphorus
compounds, inorganic; Phosphorus compounds, organic.)
toxicity 17:54
uses 17:52
Phosphorus compounds, inorganic. (See also
individual compounds.) 17:107–150
toxicity 17:56
Phosphorus compounds, organic. (See also
individual compounds.)
phosphorus content 17:68
sample decomposition 17:59, 68
sodium peroxide fusion 17:68
Phosphorus content of phosphate fertilizers
chemically available 17:149
citrate insoluble 17:148
citrate soluble 17:149
water soluble 17:148
Phosphorus, elemental. (See also Phosphorus,
black; Phosphorus, red; and Phosphorus,
white.)
allotropes 17:54
analysis 17:58
commercial grades and specifications
17:58
detection 17:60
isotopes 17:53
manufacture 17:57
properties 17:53
toxicity 17:56
uses 17:52
Phosphorus halides. (See also individual
compounds.) 17:107–109
manufacture 17:108
properties 17:107
toxicity 17:107
uses 17:109
Phosphorus heptasulfide
commercial specifications 17:117
properties 17:117
Phosphorus–nitrogen compounds. (See also
individual compounds.) 17:109
uses 17:109
Phosphorus oxides. (See also

individual compounds.) 17:109
manufacture 17:109
Phosphorus oxychloride. (See also Phosphorus
    halides.)
    commercial specifications 17:109
    manufacture 17:108
    properties 17:107
    uses 17:109
Phosphorus oxyfluoride. (See also Phosphorus
    halides.)
    properties 17:108
Phosphorus pentabromide. (See also Phosphorus
    halides.)
    properties 17:107
Phosphorus pentachloride. (See also Phosphorus
    halides.)
    commercial specifications 17:109
    manufacture 17:108
    properties 9:335; 17:107
    solubility 9:336
    uses 17:109
Phosphorus pentafluoride. (See also Phosphorus
    halides.)
    properties 17:108
Phosphorus pentasulfide
    commercial specifications 17:117
    properties 17:117
    uses 17:117
Phosphorus pentoxide
    absorption method for water in fluorocarbons
        13:67
    acid–base titration 17:111
    ammonium molybdate colorimetric method
        9:130
    ammonium molybdate tetrahydrate gravimetric
        method 13:432
    ammonium molybdate volumetric method
        15:214
    ammonium vanadate–ammonium molybdate
        colorimetric method 15:213
    analysis 17:110
    assay 17:111
    in cements 9:130
    in ceramics 9:194
    commercial grades and specifications 17:110
    in enamels 9:195
    in glass and glass-ceramics 13:413, 415, 432
    impurities, determination of 17:111
    in lime and limestone 9:191; 15:213
    magnesia–ammonium hydroxide gravimetric
        method 13:432
    manufacture 17:110
    nitric acid gravimetric method 15:213
    properties 17:110
    specification tests 17:110
    uses 17:110
    vanadic acid–sodium molybdate colorimetric

method 13:433
Phosphorus, red. (See also Phosphorus,
    elemental.)
    analysis 15:536
    commercial grades and specifications 15:521;
        17:58
    impurities, determination of 15:536
    manufacture 15:521; 17:54, 58
    in match strikers 15:521, 536
    particle size 15:536
    properties 17:54
    safety precautions 15:521
    uses 17:52
Phosphorus sesquisulfide
    analysis 15:531
    commercial grades and specifications 15:516
    commercial specifications 17:117
    impurities, determination of 15:531
    in matches 15:515
    melting point 15:531
    particle size 15:531
    properties 17:116
    safety precautions 15:515
    toxicity 17:117
    uses 15:516; 17:117
Phosphorus sulfides. (See also individual
    compounds.) 17:116
    manufacture 17:116
Phosphorus sulfochloride. (See also Phosphorus
    halides.)
    manufacture 17:109
    properties 17:108
Phosphorus sulfofluoride. (See also Phosphorus
    halides.)
    properties 17:107
Phosphorus tribromide. (See also Phosphorus
    halides.)
    manufacture 17:109
    properties 17:108
    uses 17:109
Phosphorus trichloride. (See also Phosphorus
    halides.)
    commercial specifications 17:109
    manufacture 17:108
    properties 9:335; 17:108
    solubility 9:336
    uses 17:109
Phosphorus trifluoride. (See also Phosphorus
    halides.)
    properties 17:108
Phosphorus triiodide. (See also Phosphorus
    halides.)
    properties 17:108
Phosphorus trioxide
    in phosphorus pentoxide 17:111
Phosphorus, white. (See also Phosphorus,
    elemental.)

commercial grades and specifications 17:58
handling 17:57
manufacture 17:57
properties 17:54
solubility 17:54
uses 17:52, 58
Phosphorus, yellow. See Phosphorus, white.
Phosphoryl chloride. See Phosphorus oxychloride.
Phosphotungstic acid
colorimetric method for ascorbic acid 19:424
colorimetric method for cysteine 15:561
colorimetric method for mercaptans 15:561
colorimetric method for thioglycolic acid
15:561
colorimetric method for vanadium 10:130
gravimetric method for bacitracin 5:547
gravimetric method for polymyxin B 5:548
gravimetric method for trigonelline 10:413
gravimetric method for viomycin 5:564
Phosphotungstovanadic acid
colorimetric method for vanadium 5:147;
9:677
Phostoxin
composition 13:194
Phosvitin
in egg yolk 12:58
Photochromic glasses. (See also *Glass and
glass-ceramics.*)
composition 13:411
uses 13:411
Photographic film bases. See Film bases, photographic.
Photographic films. See *Films, photographic.*
Photographic paper. See Paper, photographic.
Photographic performance
of photographic films 12:595
Photographic sensitizers 12:570
Photographic solutions
silver content 18:172
Photometric method
for mercury vapor 10:496
Photomicrograph
of pyrographite 14:58
Photosensitive glass. (See also *Glass and
glass-ceramics.*)
manufacture 13:411
uses 13:411
Phthalaldehyde
infrared spectroscopy 6:617
Phthalate plasticizers
gas chromatography 12:591; 16:96, 98
in methacrylate polymers 16:96
phenol color test 16:96
in photographic films 12:591
polarography 16:97
pyrolysis–gas chromatography 16:96
resorcinol color test 16:96

saponification methods 16:97
o-Phthaldialdehyde
colorimetric method for glycine 8:482
Phthalic acid 5:22; 8:404; 17:150–158
acid–base titration 17:158
in alkyl resins 5:34; 7:67
analysis 17:153, 158
colorimetric method 5:47
derivatives for identification 8:376; 17:153
N-ethylpiperidine titration 17:156
gas chromatography 5:25; 17:154, 158
gravimetric methods 5:34
identification 17:153
manufacture 17:150
in manufacture of benzoic acid 7:60
nuclear magnetic resonance spectroscopy
5:50
in phthalic anhydride 17:154, 156
properties 8:403; 17:151
solubility 17:152
structure 17:150
thin-layer chromatography 19:293
ultraviolet spectroscopy 5:35; 7:70
uses 17:150
Phthalic acids and derivatives. (See also
individual compounds.) 17:150–175
in certified colors 10:505
in D & C Orange No. 5 10:498
in D & C Orange No. 10 10:498
in D & C Orange No. 11 10:498
in D & C Orange No. 17 10:498
in D & C Red No. 19 10:498
in D & C Red No. 21 10:498
in D & C Red No. 22 10:498
in D & C Yellow No. 7 10:498
in D & C Yellow No. 8 10:498
in D & C Yellow No. 10 10:498
in D & C Yellow No. 11 10:498
in F D & C Red No. 3 10:498
ultraviolet spectroscopy 10:505
Phthalic anhydride 4:535, 559; 5:22; 17:150–158
acid–base titration 17:154
in alkyd resins 5:42
analysis 17:153
assay 17:153
benzoic acid in 7:70
colorimetric method 5:47
colorimetric method for *o*-cresol 17:19
colorimetric method for hydroquinone 17:19
colorimetric method for phenol 17:19
colorimetric method for phenols 17:18
colorimetric method for pyrogallol 17:19
colorimetric method for resorcinol 17:19
colorimetric method for thymol 17:19
color tests 17:155
commercial grades and specifications 8:560;
17:153

derivatives for identification 17:153
freezing point 17:155
gas chromatography 7:72; 8:563; 17:154
identification 17:153
impurities in 8:564; 17:154
infrared spectroscopy 5:42; 8:562
infrared spectrum 8:563
manufacture 8:561; 17:152
in manufacture of benzoic acid 7:61
phthalic acid in 8:567
properties 8:559; 17:151
solubility 17:152
structure 17:150
uses 17:150
volumetric method for diethylene glycol
   13:586
volumetric method for dipropylene glycol
   13:586
volumetric method for 2-ethyl-1,3-hexanediol
   13:586
volumetric method for glycols 13:586
volumetric method for pentaerythritol 13:586
volumetric method for 1,5-pentanediol 13:586
Phthalide
   gas chromatography 7:71; 17:154
   in phthalic anhydride 17:154
Phthalimide
   gas chromatography 13:212
   thin-layer chromatography 13:211
Phthalocyanine blue
   in chrome greens 17:201
   commercial specifications 12:173
   composition 12:173
   in enamels 12:173, 187
   identification 12:187
Phthalocyanine dyes 12:10
Phthalocyanine green
   composition 12:173
   in enamels 12:173, 187
   identification 12:187
Phthalonitrile. (See also *Nitriles.*)
   polarography 16:398
Phylloquinone. See Phytonadione.
Physiological methods
   for choline and its derivatives 9:626
Phytoene
   in green tea leaf 18:480
   thin-layer chromatography 18:480
Phytofluene
   in green tea leaf 18:480
   thin-layer chromatography 18:480
Phytomenadione. See Phytonadione.
Phytonadione 10:185
   analysis 10:186
   assay 10:187
   2,6-dichlorophenol indophenol colorimetric
     method 10:191

2,6-dichlorophenol indophenol titration
   10:190
2,4-dinitrophenylhydrazine colorimetric
   method 10:193
gas chromatography 10:193
identification 10:186
infrared spectrum 10:186
liquid chromatography 10:187, 188
paper chromatography 10:186
properties 10:185
separation 10:186
solubility 10:185
ultraviolet spectroscopy 10:186, 187
xanthane hydride colorimetric method 10:193
Phytosterol
   in butter 16:138
   gas chromatography 16:139
   in tobacco 19:149
Picfum. See Chloropicrin.
Picocurie 17:445
Picolinamide. (See also *Pyridine and pyridine
   derivatives.*)
   ionization constants 17:417
β-Picoline. (See also *Pyridine and pyridine
   derivatives.*)
   manufacture 17:420
   in tobacco 19:136
2-Picoline. (See also *Pyridine and pyridine
   derivatives.*)
   commercial specifications 17:423
   derivatives for identification 17:415
   gas chromatography 17:437
   impurities, determination of 17:436
   ionization constants 17:416
   manufacture 17:420
   mass spectrum 17:428
   properties 17:415
3-Picoline. (See also *Pyridine and pyridine
   derivatives.*)
   commercial specifications 17:423
   derivatives for identification 17:415
   gas chromatography 17:437
   ionization constants 17:416
   manufacture 17:420
   mass spectrum 17:429
   properties 17:415
   trisodium pentacyanoammino ferrate
     colorimetric method 17:433
4-Picoline. (See also *Pyridine and pyridine
   derivatives.*)
   commercial specifications 17:423
   derivatives for identification 17:415
   gas chromatography 17:437
   ionization constants 17:416
   manufacture 17:420
   mass spectrum 17:429
   properties 17:415

trisodium pentacyanoammino ferrate colorimetric method 17:433

Picoline borane 7:344

Picoline-*N*-oxides. (See also *Pyridine and pyridine derivatives.*)
commercial grades and specifications 17:423

Picolines. (See also *Pyridine and pyridine derivatives.*)
derivatives for identification 17:426
uses 17:412

Picolinic acid. (See also *Pyridine and pyridine derivatives; and individual compounds.*)
structure 17:412

Picolylamine. (See also 2-Aminomethylpyridine.)
ionization constants 17:417

Picramide. See 2,4,6-Trinitroaniline.

Picramin R
colorimetric method for niobium 16:369, 374

Picrates. (See also individual compounds.) 12:438
derivative of amines 5:267
derivative of aniline and its derivatives 5:429

Picratol. (See also High-explosive mixtures.)
analysis 12:450
composition 12:450
uses 12:450

Picric acid. (See also High explosives.) 12:438
acid–base titration 12:419
ammonium sulfide colorimetric method 12:419
colorimetric method for amines 16:117
colorimetric method for epoxy compounds 12:217
colorimetric method for pyridine derivatives 11:127. 132
commercial specifications 12:438
etchant for ceramics 9:224
gravimetric method for pyridine and its derivatives 17:425
manufacture 12:439
melting point 12:419, 461
microchemical method for alphaprodine 16:249
microchemical method for fentanyl 16:249
microchemical method for isomethadone 16:250
microchemical method for meperidine 16:250
potassium cyanide colorimetric method 12:419, 439
properties 12:434, 438
solubility 12:439
solvent extraction 12:419
uses 12:438
volumetric method for naphthalene 16:209

Picric acid–sodium hydroxide
colorimetric method for chlormadinone acetate 10:595

Picrite. See Nitroguanidine.

Picrolam. See Tordan.

Picrolonic acid

microchemical method for meperidine 16:250
microchemical method for methadone 16:250

Picromerite 17:353

Picryl chloride
reagent for determination of aniline 5:447

Piezoelectric analyzers 16:532

Piezoelectric ceramics. (See also *Ceramics*; and individual compounds.)
properties 9:209, 212
uses 9:209

Pig iron. (See also Iron metal.) 5:57
commercial grades and specifications 15:28
composition 5:58
manufacture 15:26

Pigment blue 15. See Copper phthalocyanine blue.

Pigment blue 19. See Alkali blue.

Pigment dispersion
of enamels 12:179

Pigment green 3. (See also *Pigments.*)
structure 17:186

Pigment green 4. See Malachite green.

Pigment green 7. See Copper phthalocyanine green.

Pigment green 8. See Pigment green 3.

Pigment green 10. See Nickel azo yellow.

Pigment, inorganic
in yarn 4:91

Pigment orange 1. See Hansa orange.

Pigment orange 5. See Dinitraniline orange.

Pigment orange 13. See Benzidine orange.

Pigment red 1. See Para red.

Pigment red 3. See Toluidine red.

Pigment red 4. See Chlorinated para red.

Pigment red 14. See Arylide maroon.

Pigment red 38. See Pyrazolone red.

Pigment red 49. See Lithol red R.

Pigment red 52. See Lithol red 2G.

Pigment red 57. See Lithol rubine.

Pigment red 63. See BON maroon.

Pigment red 69. See Lithol red 2G.

Pigment red 81. See Rhodamine 6G.

Pigment red 83. See Alizarine red B.

Pigments. (See also individual compounds.)
12:5; 17:176–213
acidity 17:177
alkalinity 17:177
analysis 17:177
ash content 17:177
classification 17:176
color 17:205
commercial specifications 17:176
definition 17:176
density 17:207
dispersibility 17:207
in enamels 12:173, 181
extenders 17:190

fastness 17:208
in flour 12:612
hardness 17:208
identification 12:21, 182
liquid chromatography 19:152
moisture content 17:177
oil absorption 17:207
in oils 17:177
opacity 17:206
organic color content 17:177
particle size distribution 17:206
pH 17:177
in polyurethanes 19:312
properties, determination of 17:205
rheology 17:207
separation 12:21
solubility 12:21; 17:177, 207
solvent extraction 12:181, 612; 19:152
stability 17:208
texture 17:207
tinting strength 17:206
in tobacco 19:152
uses 17:176
visible spectroscopy 19:152
in wines 19:460
Pigment scarlet lake
in enamels 12:182, 185
identification 12:182, 185
Pigments, organic. (See also *Pigments*; and
    individual compounds.)
analysis 17:204
identification 17:205
structures 17:178
types 17:178
Pigment violet 5. See Helio fast rubine 4BL.
Pigment violet 23. See Dioxazine violet.
Pigment yellow 1. See Hansa yellow G.
Pigment yellow 3. See Hansa yellow 10G.
Pigment yellow 13. See Benzidine yellow GR.
Pig tin. (See also Tin metal.) 19:29
commercial grades and specifications 19:30
Pilbarite
composition 19:2
thorium content 19:2
Pilocarpidine 4:590
analysis 4:608
Pilocarpine 4:590
analysis 4:608
Pilosine 4:590
analysis 4:608
Pimaricin 5:473, 564
analysis 5:571, 576, 579
bioautography 5:571
identification 5:571
microbiological assay 5:576
paper chromatography 5:571
properties 5:565

separation 5:571
solubility 5:566, 567
structure 5:473
ultraviolet spectroscopy 5:579
uses 5:564
Pimelic acid
gas chromatography 17:290
properties 8:397
Pimenta oil. (See also *Essential oils.*)
refractive index 12:271
Piminodine. (See also *Narcotics.*)
color tests 16:247
formaldehyde–sulfuric acid colorimetric
    method 16:247
gas chromatography 16:255
microchemical tests 16:250
potassium chromate microchemical method
    16:250
potassium iodide microchemical method
    16:250
structure 16:245
ultraviolet spectroscopy 16:253
Piminodine ethanesulfonate. (See also *Narcotics.*)
ultraviolet spectroscopy 16:253
Pimpinellin. (See also *Coumarins.*) 11:160
gas chromatography 11:168, 170
Pinacol. (See also Glycols.)
azeotropes 13:541
Pinakiolite 7:315
α-Pinene. (See also Terpenes.)
commercial grades and specifications 14:265,
    316
gas chromatography 8:353; 12:266, 268
in majoram, oil of 12:268
in peppermint oil 12:266
structure 14:266
uses 14:266
β-Pinene. (See also Terpenes.)
commercial grades and specifications 14:265,
    316
gas chromatography 8:353; 12:266
in peppermint oil 12:266
structure 14:266
uses 14:266
α-Pinene oxide. (See also *Epoxy compounds.*)
infrared spectroscopy 12:200
zinc bromide–2,4-dinitrophenylhydrazine
    gravimetric method 12:214
α-Pinene resins. (See also *Hydrocarbon resins.*)
commercial grades and specifications 14:275
β-Pinene resins. (See also *Hydrocarbon resins.*)
commercial grades and specifications 14:275
infrared spectroscopy 14:268
polymer distribution 14:269
Pine oil
solubility 13:539, 552
Pinnoite 7:314

Pipe
asbestos–cement 6:255, 270
Pipecolines. (See also *Pyridine and pyridine derivatives.*) 17:413
Piperazine
gas chromatography 16:198
Piperidine. (See also *Pyridine and pyridine derivatives.*)
colorimetric method for 2,3-butanediol 19:447
gas chromatography 16:197
reineckate gravimetric method 9:620
structure 17:413
uses 17:413
volumetric method for isocyanic acid esters 15:111
Piperidine salt of *N,N*-pentamethylene dithiocarbamic acid. (See also Dithiocarbamic acid salts.)
uses 11:577
Piperidine–sodium nitroprusside
colorimetric method for acetaldehyde 19:455
Piperine. (See also *Pyridine and pyridine derivatives.*) 4:591
structure 17:413
uses 17:413
Piperonal. See Heliotropin.
Piperylene. (See also 1,3-Pentadiene.) 11:512
assay 11:512
azeotropes 11:512
commercial specifications 11:513
toxicity 11:512
Piria's acid. See 1-Naphthylamine-4-sulfonic acid.
Pirimicarb. (See also *Pesticides and insecticides.*) 19:586
Pitchblend 17:445; 19:224, 225
composition 17:450; 19:225, 255
decomposition 19:235
occurrence 19:225
radium content 17:463
Pitches. (See also Bituminous materials.) 6:274, 277
commercial specifications 6:285
composition 6:275
manufacture 6:279
properties 6:275
specific gravity 6:319
Pitches, hardwood
in brake linings 7:441
Pituitary hormones. (See also individual compounds.) 14:179
analysis 14:205
properties 14:187
sources 14:187
uses 14:187
Plagioclase 8:75
Plant material
boron in 7:407, 409, 410, 416, 417

calcium in 8:110
dry ashing 7:400, 416
manganese content 15:475, 476
Plants
calcium content 18:221
carboxin content 13:213
copper content 18:221
coumarins in 11:167
decomposition 19:327
iron content 15:57; 18:221
lithium content 18:221
magnesium content 15:371; 18:221
manganese content 18:221
oxycarboxin content 13:213
phosphorus content 17:97
potassium content 17:384; 18:221
sample decomposition 18:28
selenium content 17:596
silicon content 18:28
sodium content 17:384; 18:221, 227
strontium content 18:280
sulfur content 18:379
zinc content 18:221
Plant sprays
sulfur content 18:377
Plant tissues
benomyl content 13:206
chloroneb content 13:218
Plantvax. See Oxycarboxin.
Plasma
EDTA and related compounds in 12:48
insulin content 14:575
phenindione content 10:183
Plastanox 6:76, 425
Plaster. See *Gypsum and gypsum products.*
Plaster of Paris. (See also Dental gypsum products.)
uses 11:263
Plastic clays
composition 9:202
Plasticity
of coal and coke 10:249
Plasticizers
acid number 8:393
column chromatography 12:592
in enamels 12:176
fractionation methods 12:591
gas chromatography 12:591
in hair sprays 11:90
identification 12:591
in methacrylate polymers 16:96
paper chromatography 12:592
in photographic films 12:590
in polyamides 17:300
separation 12:590
solvent extraction 12:590
in styrene polymers 18:321

thin-layer chromatography 12:592
vacuum distillation 12:591
Plastic packages
composition, determination of 11:113
Plastics
carbon black in 8:238
pyrolysis 8:238
Plastic sulfur. (See also Sulfur, elemental.)
  18:365
Plating solutions
silver content 18:172
tin content 19:64
Platinic bromide microchemical method
for levorphanol 16:250
for methadone 16:250
Platinic chloride microchemical method. (See
  also Chloroplatinic acid.)
for cocaine 16:249
for cotarnine 16:249
for heroin 16:249
for isomethadone 16:250
for levomethorphan 16:250
for levorphanol 16:250
for metazocine 16:250
for narceine 16:250
for norlevorphanol 16:250
for pholcodine 16:250
for thebaine 16:250
Platinic iodide microchemical method
for dihydromorphine 16:249
for ecgonine 16:249
Platinum. (See also Platinum alloys; Platinum
  compounds; Platinum group metals; and
  Platinum metal.)
ammonium chloride gravimetric method
  17:258
ammonium hexachloroplatinate gravimetric
  method 17:254
bead test 6:588
in catalysts 17:272
in cobalt–platinum alloys 10:319
colorimetric determination 17:249, 259
in copper alloys 10:644
detection 17:249
emission spectroscopy 14:48; 17:246, 270
fire assay 10:644
flame photometry 17:247
flame tests 6:589
in gallium metal 13:266
in gold 14:45
gold content 14:42
gravimetric determination 17:249, 254,
  258
history 17:214
hydrogen sulfide gravimetric method 10:319;
  17:251, 258
identification 17:249

isotopes 17:215
p-nitrosodimethylaniline colorimetric method
  17:265
occurrence 17:214
in ores 17:266
oxidation numbers 17:220
oxidation potentials 17:221
paper chromatography 14:18
recovery from ammonium chloroplatinate
  17:406
recovery from potassium chloroplatinate
  17:407
separation 17:254, 258, 264
sodium formate gravimetric method 17:260
sodium metabisulfite gravimetric method
  17:264
thallium nitrate–tin chloride color test 17:249
tin chloride colorimetric method 17:249, 259
uses 17:217
Platinum alloys. (See also Platinum metal.)
solubility 17:244
uses 17:218
Platinum chloride
manufacture 17:227
properties 17:227
Platinum–cobalt
color test for pentaerythritol 13:601
Platinum–cobalt alloys. (See also Cobalt alloys.)
  10:285
Platinum–cobalt color standards
comparison test for hydrocarbons 14:344
comparison test for hydrogen cyanide 14:423
Platinum complexes 17:228
Platinum compounds. (See also individual
  compounds.)
manufacture 17:227
Platinum gold 5:68
Platinum group metals. (See also Iridium;
  Osmium; Palladium; Platinum; Rhodium;
  and Ruthenium.) 17:214–275
alkali fusion 17:231
analysis 17:229
assay 17:229
chlorination methods 17:245
commercial grades and specifications 17:229
copper chloride–chlorine fusion 17:240
copper chloride–chlorine–potassium chloride
  fusion 17:242
detection 17:245
emission spectroscopy 17:246
evaporation methods 17:244
fire assay 17:250
flame photometry 17:247
flux fusion 17:238
gold content 17:229, 239
hydrogen sulfide gravimetric method 17:250
identification 17:254

impurities, determination of 17:272
ion exchange methods 17:252
lead fusion 17:233
manufacture 17:228
nitrition methods 17:251
occurrence 17:214
organic material content 17:245
properties 17:214, 220
pyrosulfate fusion 17:232
sample decomposition 17:231
sampling 17:229
separation 17:250, 252, 262
silver content 17:229
sodium chloride–chlorine fusion 17:238
sodium formate reduction 17:250
solubility 17:244
solvent extraction 17:252
toxicity 17:228
uses 17:217, 218
volatilization methods 17:245
zinc–chlorine fusion 17:244
zinc metal fusion 17:231
zinc reduction 17:250
Platinum group metals, compounds of. (See also
    individual compounds.) 17:272–274
uses 17:273
Platinum hydroxide
manufacture 17:227
properties 17:227
Platinum metal
carrier gas fusion methods 8:624, 682
commercial grades and specifications 17:229
hydrogen in 8:681
oxygen in 8:616, 624, 681, 682
properties 17:219, 220, 227
reaction medium for fusion methods 8:629,
    630, 632
solubility 17:244
solubility in gold 14:9
uses 17:218
vacuum fusion methods 8:616, 681
Platinum oxides
carrier gas fusion methods 8:626
oxygen in 8:626
Platinum–rhodium alloy
uses 17:218
Platinum sulfides
manufacture 17:227
Plato tables of sugar solutions 7:487
Plattnerite. See Lead dioxide.
Plexiglas. (See also Poly(methyl methacrylate).)
FMIR spectrum 16:78
Plictan. (See also *Pesticides and insecticides.*)
    19:586
Plioflex 12:87
Plumber's solder 5:69
identification reagent for carbon disulfide 8:273

Plutonium. (See also Transuranium elements.)
gallium content 13:264
manufacture 19:259
occurrence 19:258, 259
uses 19:259
Plutonium alloys
aluminum in 5:230
carrier gas fusion methods 8:682
oxygen in 8:682
Plutonium carbide
hydrogen in 8:681
nitrogen in 8:619
oxygen in 8:619, 681
vacuum fusion methods 8:619, 681
Plutonium–gallium alloys 13:273
analysis 13:273
gallium content 13:273
Plutonium metal
carrier gas fusion methods 8:626, 682
hydrogen in 8:619, 622, 680
nitrogen in 8:619, 622, 626
oxygen in 8:619, 622, 626, 680, 682
vacuum fusion methods 8:619, 622, 680
Plutonium nitride
carrier gas fusion methods 8:626, 682
hydrogen in 8:626, 681
nitrogen in 8:619, 626
oxygen in 8:619, 626, 681, 682
vacuum fusion methods 8:619, 681
Plutonium oxide
carrier gas fusion methods 8:626
oxygen in 8:626
PLX. (See also High-explosive mixtures.)
analysis 12:450
composition 12:450
uses 12:450
PMA. See Phenyl mercuric acetate.
PMDA. See Pyromellitic dianhydride.
PMMA. (See also Poly(methyl methacrylate).)
uses 16:67
Polarimetry
of boric acid 7:412
of boron 7:412
of starch 11:48
of sucrose 18:344
of sugar 18:344
Polarography 8:512
of acenocoumarol 10:154
of acetaldehyde 4:71
of acetone 4:119
of acrolein 4:153
of acrylamide 4:164
of acrylamide monomer in polymers 4:175,
    337
of acrylates and methacrylates in polymers
    4:216
of acrylonitrile 4:374, 379; 16:398

of acrylonitrile monomer in polymers 4:333
of alcohols 4:546
of alkyd resins 5:49
of allyl alcohol 5:82
of allyl esters 5:97
of amines, secondary 5:89
of aminocaproic acid 17:299
of aniline and its derivatives 5:455
of antimony 6:58
of antipyrine 6:215
of apomorphine 16:268
of aspartic acid 8:461
of barium 6:578
of benzene 7:34
of benzoyl peroxide 16:99
of bismuth 7:217; 13:279
of boron 7:412
of bromate 8:27; 12:543
of bromide 12:545
of cadmium 5:123; 8:52; 12:571; 13:279
of carbon disulfide 8:273
of carbonyl compounds 8:365
of 4-carboxybenzaldehyde 17:162
of carboxylic acids 8:394
of ceramics 9:217
of chromium 9:653
of cinnamaldehyde 10:53
of cinnamonitrile 16:398
of cobalt 10:370
of cocaine 16:274
of codeine 16:268, 270
of copper 10:297, 616; 12:541, 542, 571;
  13:279, 370
of coumarins 11:163
of crotonaldehyde 11:182
of crystal violet 19:182
of cyclamates 11:237
of cyclohexanone 11:207
of 2-(1-cyclohexenyl)cyclohexanone 11:207
of 2-cyclohexylidenecyclohexanone 11:207
of diamines and higher amines, aromatic 11:443
of dibucaine hydrochloride 5:387
of unsym-dimethylhydrazine 14:238
of m-dinitrobenzene 7:34
of dinitrotoluene 11:454
of dithiocarbamic acid salts 11:586
of ethylmorphine 16:268
of europium 12:397; 17:493
of formaldehyde 13:112
of fumaronitrile 16:398
of gallium 13:260
of glass and glass-ceramics 13:457
of glycols 13:596
of gold 14:44
of heroin 16:268
of hydrazine 14:238
of hydrogen peroxide 14:435

of hydroquinone and/or quinone in acrylic
  monomers 4:207
of indium 13:279; 14:531, 549
of iodide 12:543, 545, 566
for iron 12:554
of isophthalonitrile 16:398
of lead 10:299; 12:541, 542, 571; 15:188
of magnesium 15:374
of mannitol 19:449
of menadione 10:197
of mercurials 11:565
of mercury 16:52
of methacrylonitrile 16:398
of methacrylonitrile monomer in polymers
  4:335
of methylhydrazine 14:238
of methyl methacrylate monomer in polymers
  4:336
of methyl vinyl ketone in acrylonitrile 4:377
of morphine 4:600; 16:267
of narcotine 16:268
of nickel 16:321
of nitriles 16:398, 407
of nitro alcohols 16:445
of nitro compounds in toluenediamines
  11:454
of NTA 12:43
of oxygen 16:531; 18:305
of oxygen in beer 7:626
of peroxide 18:307
of pesticides and insecticides 19:613
of phenindione 10:182
of phenobarbital 6:494
of phenylalanine 8:512
of α-phenylcinnamonitrile 16:398
of phenylenediamines 11:448
of phthalate plasticizers 16:97
of phthalonitrile 16:398
of polyhydric alcohols 13:596
of polymerization initiators 16:99
of polyphenol oxidase 18:482
of rare earths 17:493
of rhodamine WT 19:182
of samarium 17:493
of silver 12:567
for silver halides 12:563, 566
of sodium heparin 10:142
of sodium thiopental 5:372
of sorbitol 19:449
of steroids 14:208
of sulfur 12:146, 560
of sulfur dioxide 12:551; 13:57
of terephthalic acid 17:167
of terephthalonitrile 16:398
of terraclor 13:216
of tetracaine hydrochloride 5:415, 418
of tetraethylthiuram disulfide 11:598

of thallium 12:571; 18:534
of thimerosal 11:565
of thiophene 18:542
of thiosulfate 12:548, 586
of thiuram disulfides 11:597
of tin 7:662; 19:66
of titanium 19:119
of triphenylmethane and related dyes 19:170, 182
of tryptophan 8:512
of tyrosine 8:512
of uranium 19:234, 255
of warfarin sodium 10:168
of ytterbium 17:493
of zinc 5:150; 12:555; 19:185, 566
Polished surface analysis
of coal and coke 10:219
Polishes
analysis 6:362
automobile 6:353, 362
in confectionery products 10:552
performance tests 6:375
Pollucite. (See also *Cesium*; Cesium metal.)
9:278, 294; 17:546
alumina content 9:295
analysis 9:294
cesium content 9:294
composition 9:293; 17:544
impurities, determination of 9:295
occurrence 17:544
sample preparation 9:294
silica content 9:294
solubility 9:294
Polonium-210
in tobacco 19:153
Polyacrylamide. (See also *Acrylamide polymers*.)
4:168
analysis 4:171
carboxyl content 4:172
distillation–nesslerization method 4:178
fractionation of 4:174
infrared spectroscopy 4:171
modifications 4:170
molecular weight determination 4:173
preparation 4:168
properties 4:169
residual monomer in 4:175, 338
turbidimetric method 4:178
ultraviolet spectroscopy 4:171
uses 4:169
viscosity 4:173
in water 4:177
Polyacrylates
flame test 12:588
in photographic films 12:588
residual monomer in 4:214
Polyacrylonitrile. (See also *Acrylic and*

modacrylic fibers; *Acrylonitrile polymers*.)
acrylonitrile content 4:271
chromatographic fractionation methods
4:301
crystallinity 4:315
density 4:318
end groups 4:288
fiber in mixtures 4:360
glass transition temperatures 4:262
heating test 12:588
heat stability 4:263
infrared spectra 4:231, 241, 244, 384
infrared spectroscopy 4:234, 238, 296, 384;
18:318
microgel 4:320
nitrogen content 4:278
osmotic pressure measurements 4:310
in photographic films 12:588
precipitation fractionation methods 4:299
preparation 4:382
properties 4:222, 382
solubility 18:316
structure 4:316, 382
uses 4:383
viscosity 4:302, 389
Polyallomers. (See also *Ethylene and propylene polymers*.) 12:344
properties 12:344
Polyamide fibers. (See also *Polyamides*.) 17:305–329
additives, determination of 17:322
analysis 17:314
birefringence 17:314
composition 17:305
composition, determination of 17:320
crystallinity 17:314
denier uniformity 17:323
density 17:314
differential thermal analysis 17:320
electrical properties 17:309
electrostatic cling 17:326
fiber characterization 17:314
friction coefficient 17:325
gel content 17:320
history 17:305
identification 17:314
impurities, determination of 17:322
infrared spectroscopy 17:317
linear density 17:323
liquid absorption 17:308
manufacture 17:312
moisture absorption 17:306
molecular weight 17:320
properties 17:305, 323
quality control 17:326
refractive index 17:314
shrinkage tests 17:325

sonic velocity 17:315
staining tests 17:316
tensile properties 17:309
tensile testing 17:323
twist level 17:326
uses 17:305
x-ray diffraction 17:319
Polyamide resin adhesives, evaluation 4:402
Polyamides. (See also Polyamide fibers.)
8:596; 17:275–329
acid–base titration 17:293
acid–dye response 17:299
additives, determination of 17:295
amine end group content 17:293
analysis 17:280
ash content 17:303
carboxyl end group content 17:293
color tests 17:282
commercial grades and specifications 17:280
degradation products, determination of 17:298
flame test 12:588
gas chromatography 17:289, 296
glass transition temperature 17:283
heating test 12:588
hydrolysis methods 17:295
identification 17:280
impurities, determination of 17:295
infrared spectroscopy 17:284
ion exchange chromatography 17:292
manufacture 17:278
melting point 17:282
molecular weight, average 17:293
o-nitrobenzaldehyde colorimetric method
17:282
nomenclature 17:275
nuclear magnetic resonance spectroscopy
17:287
paper chromatography 17:289
in photographic films 12:588
potassium iodobismuthate colorimetric method
17:282
properties 17:276
pyrolysis–gas chromatography 17:291
separation 17:280
solubility 17:277, 280
specific gravity 17:282
thermal methods 17:282
thin-layer chromatography 17:291
types 17:275
ultraviolet spectroscopy 17:298
uses 17:276
viscosity 17:294
Polyaromatic amines. (See also *Diamines and
higher amines, aromatic*; and individual
compounds.)
properties 11:433
Polyaspartic acid 8:459

Polybasic acids. See Dibasic acids.
Polybutadienes. (See also *Elastomers, synthetic.*)
12:82
commercial elastomers 12:86
density 12:91
gel permeation chromatography 12:119
glass transition temperature 12:133
in impact polystyrene 18:332
infrared spectroscopy 12:93, 94; 18:318, 332
intrinsic viscosity 12:119
iodine monochloride volumetric method
12:123
molecular weight distribution 12:119
nuclear magnetic resonance spectroscopy
12:106
perbenzoic acid volumetric method 12:121
properties 12:82
Raman spectroscopy 12:109
solubility 18:316
solvent extraction 18:332
unsaturation, determination of 12:121
uses 12:82
Polybutenes 14:314
properties 14:314
uses 14:314
Poly(*n*-butyl methacrylate). (See also *Methacrylate polymers.*)
composition 16:70
infrared spectrum 16:73
intrinsic viscosity 16:71
molecular weight 16:71
nuclear magnetic resonance spectrum 16:82
properties 16:68
pyrolysis–gas chromatography 16:87
Polycarbonates. (See also individual compounds.)
17:329–352
analysis 17:331
4-*tert*-butylphenyl terminal group content
17:339
carbonate content 17:340
chloroformate terminal group content 17:339
composition, determination of 17:338
crystallinity index 17:333
derivatives for identification 12:589
differential thermal analysis 17:335
diffusion coefficient 17:344
gas chromatography 17:335, 349
gel permeation chromatography 17:346
hydrolysis 17:337
identification 17:330
impurities, determination of 17:348
infrared spectroscopy 12:590; 17:331
intrinsic viscosity 17:343
light scattering 17:341
manufacture 17:330
melt viscosity 17:347
microscopy 17:334

molecular weight, determination of 17:340
nuclear magnetic resonance spectroscopy
17:333
osmotic pressure 17:341
phenolic hydroxyl terminal group content
17:339
phenyl terminal group content 17:339
in photographic films 12:589
precipitation fractionation 17:344
properties 17:330
sedimentation velocity 17:344
solubility 12:589; 17:330
solution fractionation 17:346
structure 17:329
terminal group analysis 17:341
thermal methods 17:335
thin-layer chromatography 17:338
trade names 17:329
uses 17:329
vapor pressure 17:341
viscosity 17:342
x-ray diffraction 17:334
Polycarboxylic acids 8:369
Polychloronaphthalenes. See Chlorinated
naphthalenes.
Polychloroprenes. (See also *Elastomers, synthetic.*)
12:84
colorimetric determination 12:91
commercial elastomers 12:87
composition 12:84
density 12:91
glass transition temperature 12:133
identification 12:90
infrared spectroscopy 12:93
pyrolysis–gas chromatography 12:106
uses 12:84
Polychlorotrifluoroethylene 13:93
analysis 13:95
commercial grades and specifications 13:94
manufacture 13:93
properties 13:93
specification tests 13:95
uses 13:93
zero strength time 13:95
Polycrase
composition 16:349
Poly(cyanoethyl aziridine). (See also *Imines.*)
properties 14:470
structure 14:470
Poly(*n*-decyl methacrylate). (See also *Methacrylate polymers.*)
composition 16:70
properties 16:68
Polydimethylsiloxane
glass transition temperature 12:133
Poly(dodecyl methacrylate). (See also *Methacrylate polymers.*)

infrared spectrum 16:73
intrinsic viscosity 16:71
molecular weight 16:71
Polydymite
composition 16:286
Polyene antibiotics 5:472, 564
analysis 5:568, 571, 577
identification 5:568
microbiological assay 5:571
properties 5:565
separation 5:568
solubility 5:566
structures 5:472
uses 5:564
Polyepichlorohydrins. (See also *Elastomers,
synthetic.*) 12:85
commercial elastomers 12:86
density 12:91
infrared spectroscopy 12:99
properties 12:85
types 12:85
uses 12:85
Polyester color test 4:413
Polyester fibers
microscopy 4:264
solubility 4:248, 250
Polyester foams. (See also Polyurethane
foams.)
composition 19:264
manufacture 19:265
properties 19:281
uses 19:265
Polyester resin adhesives 4:403
Polyesters. (See also Polyols.) 19:277
analysis 19:277
derivatives for identification 12:589
infrared spectroscopy 12:590
in photographic films 12:589
Polyester, styrenated
as ablator 4:19
Polyether foams. (See also Polyurethane foams.)
composition 19:264
manufacture 19:264
properties 19:265, 281
uses 19:265
Polyethers. (See also Polyols.) 19:277
ammonium thiocyanate–cobalt chloride
colorimetric method 17:300
analysis 19:277
cleavage methods 19:308
gas chromatography 19:309
hydriodic acid volumetric method 19:308
identification 19:307
infrared spectroscopy 17:300
manufacture 19:307
nuclear magnetic resonance spectroscopy
17:300

phosphoric acid colorimetric method 19:308
in polyamides 17:300
in polyurethanes 19:307
pyrolysis gas chromatography 19:309
Polyethylene. (See also *Ethylene and propylene polymers*.) 12:341
as ablator 4:4
antioxidants 12:592
crystallinity 12:361
diamagnetic susceptibility 12:385
dielectric constant 12:384
differential thermal analysis 12:382
electric conductivity 12:384
ethylene oxide in 12:331
ethyl group content 12:356
flow curves 12:374
fractionation methods 12:376
heat of fusion 12:383
for impregnation of infrared samples 4:236
infrared spectroscopy 12:350, 590
intrinsic viscosity 12:371
liquid chromatography 12:378
manufacture 12:341, 342
melting point 12:349, 383
methyl group content 12:350
microscopy 12:348
nuclear magnetic resonance spectroscopy 12:369
PBNA in 6:109
in photographic films 12:589
properties 12:341, 342
pyrolysis–gas chromatography 12:365
solubility 12:589
specific viscosity 12:373
specific volume 12:381
types 12:341, 342
unsaturated group content 12:357, 360
x-ray diffraction 12:350
Young's modulus, dynamic 12:387
Polyethylene–butyl rubber blends. (See also *Ethylene and propylene polymers*.) 12:345
properties 12:345
Polyethylene fibers
microscopy 4:264
Polyethylene glycols. (See also *Polyglycols*.)
in brake fluids 7:424, 425
commercial grades and specifications 13:559, 560
2,4-dinitrophenylhydrazine colorimetric method 12:578
gas chromatography 13:575
infrared spectroscopy 13:590
manufacture 13:557
paper chromatography 12:575; 13:581
in photographic films 12:578
properties 13:550
Rosen color test 13:573

solubility 13:553
thin-layer chromatography 13:582
uses 13:557
Polyethylene oxide
in beer 7:634
colorimetric method using polyacrylic acid 7:635
Polyethylene–polyisobutylene blends. (See also *Ethylene and propylene polymers*.)
infrared spectrum 12:366
Polyethylene–polyorganosiloxane copolymers. (See also *Ethylene and propylene polymers*.) 12:345
properties 12:345
Polyethylene–polypropylene blends. (See also *Ethylene and propylene polymers*.)
infrared spectroscopy 12:362, 363
Polyethylene–polypropylene glycol polymers. (See also *Polyglycols*.)
properties 13:551
uses 13:560
Polyethylene–polypropylene mixed glycols. (See also *Polyglycols*.)
manufacture 13:558
solubility 13:551, 553
uses 13:558
Polyethylene–poly(vinyl chloride) blends. (See also *Ethylene and propylene polymers*.) 12:345
properties 12:345
Polyethylene terephthalate
in photographic films 12:589
solubility 12:589
Polyethylenimine. (See also *Imines*.)
analysis 14:481
composition 14:461
handling 14:467
impurities, determination of 14:490
infrared spectroscopy 14:481
manufacture 14:473
molecular weight, determination of 14:486
1,2-naphthoquinone-4-sulfonic acid colorimetric method 14:490
nitrogen content 14:483
pH 14:492
properties 14:461, 464, 466, 491
solubility 14:466, 492
specific gravity 14:491
toxicity 14:467
uses 14:461
viscosity 14:491
in water 14:489
Poly(ethyl methacrylate). (See also *Methacrylic polymers*.)
infrared spectrum 16:72
intrinsic viscosity 16:71
molecular weight 16:71

nuclear magnetic resonance spectrum 16:81
properties 16:68
Polyfiber fabric test
for dyes 12:13
Polyflex 6:76, 111
Polyflon. See Polytetrafluoroethylene.
Polygard 6:76, 117
analysis 6:118
colorimetric method using diazotized *p*-
nitroaniline 6:106
colorimetric method using *p*-nitrobenzene-
diazonium fluoborate 6:119
color tests 6:90
commercial specifications 6:118
identification 6:118
manufacture 6:118
properties 6:117
separation 6:118
uses 6:118
Polyglycerol
periodic acid volumetric method 13:515
Polyglycols. (See also *Glycols and polyhydric
alcohols*; and individual compounds.)
13:549–560
analysis 7:427
in brake fluids 7:423, 424, 427
colorimetric determination 13:599
commercial grades and specifications 13:560
gas chromatography 13:579
gravimetric determination 13:600
identification 13:572
infrared spectroscopy 13:590
liquid chromatography 13:580
properties 13:549
Rosen color test 13:573
solubility 13:551
Polyhalite 17:353
Poly(*n*-hexyl methacrylate). (See also *Methacrylate polymers*.)
properties 16:68
Polyhydric alcohols. (See also *Glycols and polyhydric alcohols*; Polyols; and individual
compounds.) 13:560–567
assay 13:592
commercial grades and specifications 13:565
derivatives for identification 13:567
in embalming fluids 12:166
gas chromatography 13:575; 19:300, 301
identification 19:296, 299
paper chromatography 13:581
in polyurethanes 19:291, 296, 299
properties 13:560
solubility 13:560
thin-layer chromatography 13:582
Polyhydroxy antibiotics 5:473, 579
analysis 5:579, 589, 595
identification 5:582

microbiological assay 5:589
properties 5:579
separation 5:582
solubility 5:580
structures 5:473
uses 5:579
Polyhydroxy compounds
in creams and lotions 11:94
in shampoos 11:82
Poly(*N*-(2-hydroxyethyl)aziridine). (See also
Imines.)
properties 14:470
structure 14:470
Polyhydroxyphenols. (See also individual
compounds.)
solubility 17:5
Polyisobutenes. (See also Butyl rubber; *Elastomers,
synthetic*; and Polybutenes.) 12:84
commercial elastomers 12:86
density 12:91
glass transition temperature 12:133
gravimetric determination 12:92
uses 12:84
Poly(isobutyl methacrylate). (See also *Methacrylate polymers*.)
infrared spectrum 16:73
nuclear magnetic resonance spectrum 16:83
properties 16:68
Polyisocyanate. (See also *Isocyanic acid esters*.)
commercial sources 15:102
Polyisoprenes. (See also *Elastomers, synthetic*;
Rubber, natural.) 12:85
chromic acid volumetric method 12:138
colorimetric determination 12:91, 92
commercial elastomers 12:86
density 12:91
in elastomers, synthetic 12:138
gel permeation chromatography 12:119
glass transition temperature 12:133
infrared spectroscopy 12:98, 101
intrinsic viscosity 12:119
molecular weight distribution 12:119
nuclear magnetic resonance spectroscopy
12:106
Raman spectroscopy 12:108
Poly(isopropyl methacrylate). (See also
Methacrylate polymers.)
properties 16:68
Polylactic acids. (See also *Lactic acid and its
derivatives*.)
manufacture 15:127
structure 15:127
Poly(lauryl methacrylate). (See also *Methacrylate
polymers*.)
infrared spectroscopy 16:77
Polylite 6:76
colorimetric method using diazotized

p-nitroaniline 6:98
Polylithonite 15:261
Polymer blends 12:345
  manufacture 12:345
  properties 12:345
Polymer distribution
  of aromatic petroleum resins 14:269
  of coumarone–indene resins 14:270
  of hydrocarbon resins 14:269
  of nonaromatic petroleum resins 14:269
  of *β*-pinene resins 14:269
Polymerization initiators
  in methacrylate polymers 16:99
  polarography 16:99
Polymers
  in acrylic monomers 4:200
  crystallinity 4:86
  in elastomers, synthetic 12:135
  in furfuryl alcohol 13:242
  gel permeation chromatography 4:298, 389
  glass-transition temperature 4:261
  heat stability 4:263
  infrared spectroscopy 4:230
  intrinsic viscosity 4:80, 82
  molecular weight characterization 4:80, 297
  multiple internal reflectance 4:246
  quality control tests 4:23
  refractive index 4:177
Polymers, acidic
  acid–base titration of 4:387
Polymers, basic
  acid–base titration 4:273, 386
Polymers, synthetic. See *Elastomers, synthetic.*
Polymer stabilizers 6:78
Polymer structure
  of hydrocarbon resins 14:267
Poly(methacryl amide). (See also *Methacrylate polymers.*)
  composition 16:70
Polymethacrylates. See *Methacrylate polymers.*
Poly(methacrylic acid). (See also *Methacrylate polymers.*)
  acid–base titration, nonaqueous 16:91
  composition 16:70
  cupric nitrate–sodium diethyl dithiocarbamate colorimetric method 16:92
  infrared spectrum 16:74
Poly(methacrylonitrile). (See also *Methacrylate polymers.*)
  intrinsic viscosity 16:71
  molecular weight 16:71
  properties 16:68
Polymethine dyes 12:8
Polymethylene polyphenyl diisocyanate. (See also *Isocyanic acid esters.*)
  dibutylamine volumetric method 15:113
  trade names 15:103

Polymethylene polyphenyl isocyanates. (See also Isocyanates.)
  structure 19:270
  trade names 19:270
Poly(methyl methacrylate). (See also *Methacrylate polymers.*)
  composition 16:70
  infrared spectra 16:72, 77, 78
  intrinsic viscosity 16:71
  manufacture 16:68
  molecular weight 16:71
  nuclear magnetic resonance spectroscopy 16:80, 84
  properties 16:68
  pyrolysis–gas chromatogram 16:87
  residual monomer 4:214, 216
Poly(methyl vinyl ether–maleic anhydride)
  benzene in 7:12
Polymyxin 5:467, 535
  analysis 5:541, 547, 548
  bioautography 5:541
  colorimetric method using ninhydrin 5:548
  identification 5:541
  ion exchange separation 5:548
  microbiological assay 5:547
  paper chromatography 5:541
  properties 5:534
  separation 5:541
  solubility 5:535, 536
  uses 5:535
Polymyxin B
  gravimetric method using phosphotungstic acid 5:548
Polymyxin B_1
  structure 5:467
Polymyxin B sulfate
  infrared spectrum 5:541
Polymyxin E
  structure 5:466
Poly(octadecyl methacrylate). (See also *Methacrylate polymers.*)
  composition 16:70
  intrinsic viscosity 16:71
  molecular weight 16:71
Poly(octyl methacrylate). (See also *Methacrylate polymers.*)
  intrinsic viscosity 16:71
  molecular weight 16:71
Polyolefins. See individual compounds.
Polyolefin tow filters for cigarettes 10:2
Polyols 5:22; 19:277
  acid–base titration 19:278
  acid number 19:278
  alkalinity number 19:278
  in alkyd resins 5:38
  analysis 19:277
  gas chromatography 5:31

hydroxyl number 19:278
infrared spectroscopy 5:46
isocyanate requirement 19:278
Kappelmeier separation 5:33
paper chromatography 5:49
thin-layer chromatography 5:49
water content 19:279
Polyorganosilazanes. (See also Silylation reagents; and individual compounds.) 18:89-94
properties 18:89
uses 18:89
Polyoxyethylene
in creams and lotions 11:95
silicotungstic acid gravimetric method 11:95
Polyoxyethylene compounds
infrared spectroscopy 11:83
in shampoos 11:82, 83
silicotungstic acid gravimetric method 11:82
Polyoxyethylene ethers
in automobile polishes 6:353
Polyoxyethylene glycol
liquid chromatography 13:580
paper chromatography 13:581
Polyoxymethylenes 13:103
properties 13:103
solubility 13:103
Polyoxypropylenes
pyrolysis–gas chromatography 19:309
Polypeptides
in beer 7:613
Polyphenol oxidase
in green tea leaf 18:482
polarography 18:482
purpurogallin colorimetric method 18:484
solvent extraction 18:483
Polyphenols. (See also individual compounds.)
Arnow's colorimetric method 19:146
in beer 7:552, 604, 646
in black tea 18:492
gas chromatography 19:146
in green tea leaf 18:457
identification 18:461
manufacture 17:10
paper chromatography 18:462; 19:146
potassium permanganate titration 18:458
potassium titanium oxalate colorimetric method 18:458
separation 18:461, 496
solvent extraction 18:462, 492, 494; 19:146
thin-layer chromatography 18:462, 495, 496
in tobacco 19:145
Polyphenylene
as ablator 4:19
Poly(phenyl ether)
uses 15:302
Polyphosphates. (See also individual compounds.) 17:117

acid–base titration 17:139
chain length 17:139
hydrolysis 19:368
paper chromatography 17:135
phosphorus content 17:139
properties 17:122
uses 17:123
in water 19:368
Polypropylene. (See also *Ethylene and propylene polymers.*) 12:341
atactic content 12:361
dielectric constant 12:384
differential scanning calorimetry 12:383
differential thermal analysis 12:382
fractionation methods 12:376
heat of fusion 12:383
infrared spectroscopy 12:359, 360, 361
intrinsic viscosity 12:371
isotactic content 12:361
liquid chromatography 12:378
manufacture 12:341, 343
melting point 12:349
melting transition 12:383
microscopy 12:348
nuclear magnetic resonance spectroscopy 12:370
properties 12:343
solubility 12:343
specific heat 12:384
specific viscosity 12:373
specific volume 12:381
titanium content 19:120
types 12:343
unsaturated group content 12:361
Young's modulus, dynamic 12:385
Polypropylene–fluoroolefin copolymers. (See also *Ethylene and propylene polymers.*) 12:345
manufacture 12:345
Poly(propylene glycol). (See also Polyglycols.)
in brake fluids 7:424
commercial grades and specifications 13:560
infrared spectroscopy 13:590
manufacture 13:557
properties 13:551; 18:145
solubility 13:553
surfactants in 18:145
uses 13:558; 15:302
Polypropylene–1-phthalimido-1,3-butadiene copolymers. (See also *Ethylene and propylene polymers.*) 12:345
properties 12:345
Polypropylene–polyisobutylene blends. (See also *Ethylene and propylene polymers.*) 12:345
infrared spectrum 12:366
properties 12:345

Polypropylene–polyorganosiloxane copolymers.
(See also *Ethylene and propylene polymers.*)
12:345
properties 12:345
Polypropylene–poly(vinyl chloride) blends. (See
also *Ethylene and propylene polymers.*)
12:345
properties 12:345
Poly(*n*-propyl methacrylate). (See also *Methacrylate polymers.*)
infrared spectrum 16:72
nuclear magnetic resonance spectrum 16:82
properties 16:68
Polyram. See Mitiram.
Polyram combi. See Mitiram.
Polysaccharides
colorimetric determination 12:584
gravimetric determination 12:581
in green coffee 10:393
identification 12:580
infrared spectroscopy 12:583
paper chromatography 12:582
in photographic films 12:579
separation 12:580
solvent extraction 12:580
Polysaccharides, noncellulosic
in cellulose 9:46
gas chromatography 9:46
paper chromatography 9:46
Polysar 12:87
Polysilanes
properties 18:41
Polysiloxanes. See Silicones.
Polyspiroacetal resins
manufacture 13:565
uses 13:565
Polystyrene. (See also *Styrene polymers.*)
additives, determination of 18:325
analysis 18:323
derivatives for identification 12:589
in finishing agents 11:126, 130
flame test 12:588
Fourier transform spectra 13:164, 165, 166, 168
heating test 12:588
impurities, determination of 18:325
infrared spectroscopy 12:590; 18:316
manufacture 18:314
in methacrylate polymers 16:97
in photographic films 12:589
properties 18:314
pyrolysis–gas chromatography 16:97
solubility 11:126, 130; 12:589; 18:316
solution viscosity 18:323
ultracentrifugation techniques 18:323
uses 18:285, 314
Polysulfide-based impression materials. See
Dental impression materials, polysulfide-based.

Polysulfide elastomers. See Polysulfide rubber.
Polysulfide rubber. (See also *Elastomers,
synthetic.*) 12:85
commercial elastomers 12:87
density 12:91
identification 12:90
infrared spectroscopy 12:100
uses 12:85
Polysulfides
in elastomers, synthetic 12:128
Polytetrafluoroethylene 13:73–88
as ablator 4:4
analysis 13:80
assay 13:81
color 13:88
commercial grades and specifications 13:78, 80
crystallinity 13:80
density 13:80, 85
fillers 13:77
infrared spectroscopy 13:80
manufacture 13:76
melting point 13:84
particle size distribution 13:87
properties 13:73
pyrolysis 13:81
sample preparation 13:84
solids content 13:82
solubility 13:73, 76
specification tests 13:84
specific gravity 13:83, 85
specific surface area 13:88
thermal analysis, differential 13:84
thermal instability 13:88
thermal properties 4:5
toxicity 13:76
trademarks 13:75
uses 13:75
water content 13:88
wetting agent content 13:82
Polytetramethylene ether glycols 13:244–249
acetylation method 13:246
acid number 13:247
analysis 13:245
commercial grades and specifications 13:245
flash point 13:245
impurities, determination of 13:248
manufacture 13:245
mercuric acetate–methanol volumetric method
13:247
molecular weight 13:245
phenyl isocyanate volumetric method 13:246
properties 13:244
solubility 13:245
specific gravity 13:244
unsaturation 13:247
uses 13:244
viscosity 13:245

Polythionate
  methylene blue colorimetric method 12:594
  in photographic films 12:594
Polytrifluoroethylene
  as ablator 4:4
Polyunsaturated acids
  in fats and fatty oils 12:489
  ultraviolet spectroscopy 12:489
Polyurethane coatings. (See also Polyurethanes.)
  composition 19:266
  manufacture 19:267
  properties 19:267
  types 19:267
  uses 19:267
Polyurethane elastomers. (See also Polyurethanes.)
  analysis 19:282
  characterization 19:282
  color 19:282
  hardness 19:282
  manufacture 19:263
  properties 19:263
  uses 19:263, 267
Polyurethane foams. (See also Polyurethanes.)
  analysis 19:281
  cell structure 19:280
  characterization 19:281
  color 19:281
  composition 19:264
  density 19:281
  manufacture 19:263
  preliminary tests 19:281
  properties 19:265
  sodium hydroxide spot test 19:282
  texture 19:281
  types 19:265
  uses 19:265
Polyurethanes. (See also *Elastomers, synthetic*;
    Polyurethane coatings; Polyurethane elastomers;
    Polyurethane foams; and *Urethane polymers.*)
    19:279
  additive content 19:311
  analysis 19:279
  commercial elastomers 12:86
  composition, determination of 19:281, 290
  density 12:91
  differential scanning calorimetry 19:289
  p-dimethylaminobenzaldehyde colorimetric
    method 19:282
  gel permeation chromatography 19:290
  identification 12:90
  history 19:262
  hydrolytic methods 19:291
  hydroxylamine hydrochloride–ferric chloride
    color test 19:282
  impurities, determination of 19:312
  infrared spectroscopy 19:282
  manufacture 19:262, 263

p-nitrobenzene diazonium fluoroborate color
  test 19:282
nuclear magnetic resonance spectroscopy
  19:285
properties 19:263
pyrolysis–gas chromatography 19:290
raw materials 19:268
solvent content 19:310
spot tests 19:281
thermal methods 19:288
types 19:263
uses 19:263
Poly(vinyl acetate). (See also Vinyl acetate polymers.)
  colorimetric determination 11:126, 130
  in finishing agents 11:126, 130
  solubility 11:126, 130
Poly(vinyl chloride). (See also *Elastomers,
    synthetic.*) 12:85
  chlorine determination 4:280
  colorimetric determination 12:92
  identification 12:90
  infrared spectroscopy 4:239, 242; 12:93, 99
  solubility 4:330
  torsional modulus 12:136
  uses 12:85
Poly(vinyl fluoride) 13:94
  as ablator 4:4
  properties 13:94
Poly(vinylidene chloride)
  chlorine determination 4:280
  infrared spectroscopy 4:239, 243
Poly(vinylidene fluoride) 13:95
  as ablator 4:4
  properties 13:95
  uses 13:95
Polyvinylpyrrolidone
  in beer 7:636
  in embalming fluids 12:166
  nephelometric method 7:637
  paper chromatography 7:636
Pomasol. See Thiram.
Pomeroy-Kirshman-Alsterberg titration
  of oxygen 16:528
Pomorsol. See Thiram.
Ponceau. See F D & C Red No. 4.
Ponceau MX. (See also *Triphenylmethane and
    related dyes.*)
  paper chromatography 19:172
Ponceau 3R. (See also *Triphenylmethane and
    related dyes.*)
  paper chromatography 19:172
Ponceau 4R. (See also *Triphenylmethane and
    related dyes.*)
  paper chromatography 19:172
Ponceau SX. (See also *Triphenylmethane and
    related dyes.*)
  paper chromatography 19:172

Pontachrome Blue Black R
  colorimetric method for gallium 13:257
  fluorometric method for aluminum 5:236, 238;
    15:376
  for identification of aluminum 5:214
Pontachrome Violet SW
  fluorometric method for aluminum 5:237, 240
  use in polarography of aluminum 5:250
  use in voltammetry of aluminum 5:251, 254
Porcelain. (See also *Ceramics.*) 9:191
  acid dissolution 9:215
  composition 9:194
  fluxes for 9:216
  microscopy 9:225
Porcelains, dental. See Dental porcelains.
Pore volume
  of activated carbon 8:140
Pore volume distribution
  of activated carbon 8:141
Porosimeters 14:83
Porosity
  of cigarette paper 10:32
  of dental porcelains 11:276
  of egg shell 12:78
  of graphite 14:82
Porosity densometer 10:33
Porphyrinogen
  in beer 7:638
  iodine colorimetric method 7:638
Portland blast-furnace slag cement. (See also
    Slag cements.)
  commercial grades and specifications 9:99, 100,
    101
  impurities, determination of 9:110
Portland cement. (See also Hydraulic cements.)
    9:96
  analysis 9:108
  in asbestos–cement products 6:270, 271
  commercial grades and specifications 9:99, 100,
    101
  composition 9:96
  hardening 9:97
  impurities, determination of 9:109
  light microscopy 9:103
  lime content 9:108
  manufacture 9:190
  organic materials in 9:110
  x-ray diffraction 9:103
Potash. (See also Potassium carbonate; Potassium
    chloride; and Potassium oxide.) 9:338; 17:404
Potash alum
  aluminum content 12:584
  in photographic films 12:584
Potash feldspar 9:278
Potash fertilizers. See Potassium fertilizers.
Potash–lead glass. (See also Silicate glasses.)
  composition 9:199

uses 9:199
Potash lye. See Potassium hydroxide.
Potash soaps. See Potassium soaps.
Potash–soda–barium glass. (See also Silicate
    glasses.)
  composition 9:199
  uses 9:199
Potash–soda–lead glass. (See also Silicate glasses.)
  composition 9:198
  uses 9:198
Potassium. (See also Potassium alloys; Potassium
    compounds; and Potassium metal.)
  in acid salts 8:595
  in alcoholic beverages 17:363, 390
  in alkali salts 15:267
  in alloys 17:363
  in aluminum and aluminum alloys 5:152
  in aluminum oxide 4:44, 46; 5:186
  ammonium sulfate gravimetric method 17:554
  in asbestos 6:266
  atomic absorption spectroscopy 10:37, 131;
    11:157; 15:157; 16:165; 17:372, 390; 19:492
  in baking powder 17:388
  in beer 7:652
  in biological materials 17:383
  in blood serum 18:222
  in butter 17:363
  in calcium carbonate 8:84
  in cereals 9:233, 242
  in cesium 9:282, 284
  in cheese 17:363
  chloroplatinic acid gravimetric method 17:273,
    367, 372, 384, 386, 388, 405; 18:190
  in chocolate and cocoa products 9:538, 595
  in chrysotile 6:266
  in cigarette paper 10:36
  in clays and clay minerals 10:130, 136
  in cobalt compounds 10:327
  column chromatography 10:6
  in cotton 11:154, 157
  in cotton hull ash 17:406
  in creams and lotions 11:92
  detection 17:367
  in diatomaceous earth 11:492
  dipicrylamine colorimetric method 17:383
  dipicrylamine gravimetric method 17:368
  electrolytic method 10:327
  emission spectroscopy 4:3; 15:155; 16:160;
    17:367, 371
  in fertilizers 17:363, 405
  flame emission spectroscopy 13:452; 18:245
  flame photometry 6:266; 7:652; 8:84, 595;
    9:242, 284, 349; 11:75; 15:280, 399; 17:369,
    381, 387, 407, 555; 19:492
  flame tests 6:583, 592; 17:367
  in fruit 17:363, 386
  in fuels 17:390

in glass and glass-ceramics 13:452
gravimetric determination 17:367, 372; 19:492
history 17:353
ion exchange chromatography 17:364, 407, 550
isotopes 17:354
in jellies 17:387
in kainite 17:405
in langbeinite 17:405
in lanthanum oxide 15:155
Lindo-Gladding gravimetric method 17:405
liquid chromatography 10:37
in lithium 15:272
in lithium carbonate 15:280
in lithium hydride 14:251
in lithium hydroxide 15:286
in lubricants 17:390
in magnesium acetate 15:399
in magnesium chloride 15:405
in magnesium nitrate 15:408
in magnesium oxide 15:411
in magnesium sulfate 15:413
in manganese ores 17:379
in milk 17:363
in molybdenum 16:161, 165
in molybdenum trioxide 16:161, 165
occurrence 17:353
in organic materials 17:363
oxygen in 8:635, 636
perchloric acid gravimetric method 15:267; 17:368, 376, 386
in plants 17:384; 18:221
in potassium chloride 17:405
in potassium nitrate 17:406
in potassium–sodium nitrate 17:406
in potassium sulfate 17:405
radiochemical methods 17:372
in rubidium 17:555
separation 17:363
in shampoos 11:75
in silica ablators 4:3
in silicate rocks 17:381
in soaps 18:190
sodium bitartrate gravimetric method 14:606
in sodium chloride 9:337, 349
sodium cobaltinitrite colorimetric method 17:377
sodium cobaltinitrite gravimetric method 17:368, 376
sodium cobaltinitrite volumetric method 17:377
in sodium compounds 18:245
in sodium dichromate 9:684
in sodium hydroxide 18:245
in sodium iodide 14:606
in sodium oxalate 18:245
sodium tetraphenylboron colorimetric

method 17:381
sodium tetraphenylboron gravimetric method 16:166; 17:367, 378; 19:492
sodium tetraphenylboron volumetric method 17:379, 409
in soil 17:363, 384; 18:220
sulfuric acid gravimetric method 10:175
in syrups 17:387
in terephthalic acid 17:163
in tobacco 10:6; 17:390; 19:152
ultraviolet spectroscopy 17:375
in urine 18:222
uses 17:353
vacuum distillation fusion methods 8:635
in warfarin potassium 10:175
in water 17:388; 18:221
in wines 18:221; 19:490
in wood ash 17:406
x-ray emission spectroscopy 17:371
x-ray fluorescence spectroscopy 10:136; 11:154, 157
Potassium acetate. (See also Potassium compounds.) 17:392
chloride in 8:593
commercial specifications 8:591; 17:399
manufacture 17:392
properties 8:591; 17:392
uses 17:392
Potassium acid fluoride fusion
of silicon 18:14
Potassium acid saccharate
in beer 7:585
Potassium acid tartrate
bakery chemical 6:425
Potassium adipate 4:429
Potassium alloys. (See also Potassium metal; Potassium–sodium alloys.)
potassium content 17:363
Potassium alum. See Aluminum potassium sulfate.
Potassium aluminate. (See also Aluminates; Potassium compounds.) 5:158; 17:393
manufacture 17:393
properties 5:158
uses 17:393
Potassium aluminum fluoride
etchant for ceramics 9:224
Potassium amalgam. (See also Amalgams.)
melting point 16:4
Potassium and potassium compounds. (See also Potassium; Potassium compounds.) 17:353–411
Potassium arsenite
titration of iodine 14:592
Potassium benzoate
properties 8:591
Potassium benzylpencillin. (See also Benzylpencillin.)

iodometric methods 16:562
properties 16:556
structure 16:567
Potassium bicarbonate. (See also Potassium
  compounds.) 17:393
  commercial grades and specifications 17:393,
    399
  impurities, determination of 17:400
  manufacture 17:393
  uses 17:393
Potassium bifluoride. (See also Potassium
  compounds.) 17:393
  uses 17:393
Potassium–bismuth iodide
  microchemical method for dihydrocodeine
    16:249
  microchemical method for ecgonine 16:249
  microchemical method for hydrocodone
    16:249
  microchemical method for metopon 16:250
Potassium bisulfate. (See also Potassium compounds.) 17:394
  etchant for ceramics 9:224
  flux for ceramics 9:216
  fusion of niobium 16:333
  manufacture 17:394
  properties 17:394
  uses 17:394
Potassium borates 7:380
Potassium borohydride. (See also *Hydrides.*)
  7:344
  analysis 14:247
  assay 14:247, 252
  commercial grades and specifications 14:246
  composition, determination of 14:247
  hydrogen content 14:247
  manufacture 14:246
  properties 14:243, 244
Potassium boron tetrafluoride
  identification 7:391
Potassium bromate. (See also Potassium compounds.)
  analysis 8:8
  commercial grades and specifications 8:7
  commercial specifications 17:399
  impurities, determination of 17:400
  properties 8:7
  titration of antimony 6:64; 10:633; 14:551
  titration of antimony trioxide 13:430
  titration of arsenic 6:241; 10:634
  titration of arsenic trioxide 13:430
  titration of indium 8-hydroxyquinolate 14:531
  titration of thallium 18:528
Potassium bromate–ferrous ammonium sulfate
  volumetric method for manganese 15:467
Potassium bromide. (See also Potassium compounds.) 12:544; 17:393

commercial grades and specifications 8:8, 9,
  10; 12:544
commercial specifications 17:399
impurities, determination of 12:545
manufacture 17:393
pellet technique 4:234
properties 8:9
uses 17:393
Potassium bromide–methylamino-4-phenyl sulfate
  colorimetric method for silver 18:164
  microchemical method for codeine 16:249
  microchemical method for morphine 16:250
  microchemical method for papaverine 16:250
Potassium carbonate. (See also Potassium compounds.) 17:394
  commercial grades and specifications 17:393,
    394, 399
  fusion of silicon 18:12
  fusion of tungsten 19:200
  manufacture 17:394
  safety 17:357
  sampling 17:359
  solubility 17:394
  uses 17:394
Potassium chlorate. (See also Potassium compounds.) 9:359
  analysis 9:361; 15:527
  assay 15:527
  chlorate content 12:420, 423
  commercial grades and specifications 9:359;
    15:510
  commercial specifications 17:399
  in explosive priming mixtures 12:418, 420, 423,
    425, 427, 431
  ferrous sulfate gravimetric method 12:420
  ferrous sulfate volumetric method 15:527
  formaldehyde gravimetric method 12:420
  gravimetric determination 12:420
  impurities, determination of 9:361; 15:528;
    17:400
  in matches 15:509
  potassium thiocyanate gravimetric method 12:427
  properties 9:358
  silver nitrate gravimetric method 12:425, 427
  sodium cobaltic nitrate gravimetric method 12:420
  solubility 9:359; 12:420
  solvent extraction 12:423, 430
  uses 9:359
  x-ray diffraction 12:418
Potassium chlorate–nitric acid
  gravimetric method for manganese 16:297
  volumetric method for potassium nitrite 16:492
  volumetric method for sodium nitrite 16:492
Potassium chloride. (See also Potassium compounds.) 9:338; 12:545; 17:394
  carrier gas fusion methods 8:624
  colorimetric method for rhenium 17:531

commercial grades and specifications 9:338,
   349; 12:544; 17:395, 399, 404
gravimetric method for caprolactam oligomers
   17:322
gravimetric method for tartaric acid 19:419
impurities, determination of 9:349; 12:545
manufacture 17:394
oxygen in 8:624
pellet technique 4:235
potassium content 17:405
properties 9:335
solubility 9:336
specific conductivity 13:56
uses 9:338; 17:394, 404
Potassium chloroplatinate
platinum recovery from 17:407
solubility 17:374
Potassium chromate. (See also Potassium
   compounds.)
assay 9:693
colorimetric method for mercury 16:7
colorimetric method for silver 18:164
commercial grades and specifications 9:695
commercial specifications 17:399
gravimetric method for lead 15:172
gravimetric method for thallium 18:523, 527
microchemical method for nalorphine 16:250
microchemical method for piminodine 16:250
turbidimetric method for barium 15:402
turbidimetric method for lead 10:330, 340
uses 9:693
Potassium chromium thiocyanate
colorimetric method for bismuth 11:140
gravimetric method for papaverine 16:272
Potassium citrate monohydrate. (See also Citric
   acid and salts.)
commercial grades and specifications 10:91
uses 10:86
Potassium cobalticyanide. See Cobalt complexes.
Potassium cobaltocyanide. See Cobalt complexes.
Potassium compounds. (See also individual
   compounds.) 17:391, 410
analysis 17:398
impurities, determination of 17:399
safety 17:357
solubility 17:356
toxicity 17:357
uses 17:391
Potassium cyanate
gravimetric method for indium 14:528
Potassium cyanide. (See also Cyanides; Potassium
   compounds.) 14:424
colorimetric method for picric acid 12:419, 439
commercial specifications 14:424; 17:399
fusion of tungsten 19:200
impurities, determination of 14:425; 17:400
properties 14:424

solubility 14:424
titration of cobalt 10:363
titration of copper 10:611
titration of formaldehyde 13:110
titration of nickel 16:310
titration of silver 18:168
titration of silver halides 12:562, 566
uses 14:424
Potassium cyanide–ferric chloride
colorimetric method for sulfur 18:371
Potassium cyanide–silver nitrate
titration of thiuram disulfides 11:595
Potassium cyanide–sodium thiosulfate
volumetric method for iodate 14:599
volumetric method for iodide 14:599
Potassium dibutyldithiocarbamate. (See also
   Dithiocarbamic acid salts.)
uses 11:576
Potassium dichloroisocyanurate
available chlorine 7:224, 229
identification 7:225
iodometric titration 7:229
Potassium dichromate. (See also Chromium
   compounds; Potassium compounds.) 9:681
analysis 9:683; 15:530
assay 9:684, 693
chromium content 9:684
colorimetric method for pyrocatechol 17:20
commercial grades and specifications 9:683,
   695; 15:514
commercial specifications 17:399
ferrous ammonium sulfate titration 19:376
ferrous sulfate titration 9:685
gravimetric method for barium 12:186
gravimetric method for gelatin 13:363
gravimetric method for lead 12:186, 419;
   13:357; 15:175, 179; 17:188
gravimetric method for lead thiocyanate 12:427
impurities, determination of 17:400
iodometric titration 9:684
in matches 15:514
organic dye compatibility 15:532
oxidation of organic matter in water 19:375
particle size 15:530
properties 9:682
sensitizing ability 15:530
turbidimetric method for barium 14:607;
   15:399, 404, 410
uses 9:692
volumetric method for caprolactam 17:322
volumetric method for caprolactam oligomers
   17:322
volumetric method for diethylene glycol 13:599
volumetric method for ethanol 19:435
volumetric method for iron 9:670, 703; 10:117,
   313, 639; 15:38, 41, 44; 16:164, 180
volumetric method for iron oxide 9:114; 10:117

volumetric method for lead 15:190
volumetric method for strontium nitrate 18:275
volumetric method for tellurium 17:602
volumetric method for triethylene glycol 13:599
volumetric method for uranium 19:246
volumetric method for volatile matter in roasted coffee 10:437
Potassium dichromate–sulfuric acid
  colorimetric method for glycerol 13:508
Potassium dihydrogen phosphate. (See also Potassium compounds.)
  commercial specifications 17:399
Potassium dimethyldithiocarbamate. (See also Dithiocarbamic acid salts.)
  uses 11:576
Potassium dinitrobenzofuroxane
  in explosive priming mixtures 12:418
  x-ray diffraction 12:418
Potassium dipicrylaminate
  structure 17:368
Potassium ferric periodate
  colorimetric method for lithium 15:269
Potassium ferricyanide. (See also Ferricyanides.)
  colorimetric method for sugar 19:141
  fluorometric method for N-allylmorphine 16:263
  fluorometric method for dihydromorphine 16:263
  fluorometric method for 6-monoacetylmorphine 16:263
  fluorometric method for morphine 16:263
  fluorometric method for normorphine 16:263
  titration of cobalt 10:361
  uses 15:85
Potassium ferricyanide–α,α'-dipyridyl
  colorimetric method for silver 18:164
Potassium ferricyanide–phenylhydrazine hydrochloride
  color test for formaldehyde 13:106
Potassium ferricyanide–potassium hydroxide
  gravimetric method for thallium 18:528
Potassium ferrocyanide
  colorimetric method for copper 7:194, 196, 199; 10:330
  colorimetric method for iron 10:544; 11:140
  colorimetric method for technetium 17:531
  colorimetric method for uranium 19:242
  gravimetric method for zinc 12:186
  microchemical method for cotarnine 16:249
  titration of gallium 13:259
  titration of indium 14:530
  titration of zinc 10:649; 14:552; 19:186, 561
  titration of zinc oxide 17:190
Potassium ferrocyanide–diethylaniline
  colorimetric method for zinc 19:556

Potassium ferrocyanide–3,3'-dimethyl-naphthidine chloride
  colorimetric method for zinc 19:557
Potassium ferrocyanide–α,α'-dipyridyl
  colorimetric method for mercury 16:19, 21
Potassium ferrocyanide–sodium nitroprusside
  colorimetric method for sulfite 11:142
Potassium fertilizers. (See also individual compounds.) 17:404–410
  analysis 17:405
  manufacture 17:404
  occurrence 17:404
  potassium content 17:405
  uses 17:404
Potassium fluoberyllate 7:123
Potassium fluoborate 7:337
Potassium fluoride
  carrier gas distillation fusion methods 8:636
  commercial grades and specifications 13:12
  properties 13:11
  volumetric method for aluminum 5:217
Potassium fluoroaluminate. (See also Potassium compounds.) 17:393
  manufacture 17:393
  uses 17:393
Potassium fluoroborate
  properties 13:11
Potassium fluorotantalate. (See also Tantalum compounds.) 18:452
  gravimetric method for tantalum 18:445
  properties 18:452
  solubility 18:452
  uses 18:452
Potassium formate
  uses 13:121
Potassium fusion
  of nitrogen 16:454
Potassium hexachloroplatinate
  uses 17:227
Potassium hexacyanocobaltate. See Cobalt complexes.
Potassium hexafluorosmate
  manufacture 17:226
  properties 17:226
Potassium hexafluosilicate
  volumetric method for silicon 18:30
Potassium hexaniobate
  properties 16:355
  solubility 16:355
Potassium hexatantalate
  properties 16:355
  solubility 16:355
Potassium hydride. (See also *Hydrides.*)
  analysis 14:247, 250
  assay 14:247
  commercial grades and specifications 14:246
  composition, determination of 14:247

hydrogen content 14:247
manufacture 14:245
properties 14:243, 244
Potassium hydrogen phthalate. (See also Potassium
  compounds.) 17:396
  acid–base titration 17:401
  analysis 17:401
  assay 17:401
  commercial grades and specifications 17:396,
    399
  impurities, determination of 17:403
  pH 17:403
  specification tests 17:401
  uses 17:396
Potassium hydrogen sulfate
  colorimetric method for glycerine 11:131
  etchant for ceramics 9:224
  fusion of tantalum 16:393
Potassium hydroxide. (See also Caustic potash;
  Potassium compounds.) 17:395
  absorbing agent for carbon dioxide 8:306
  acid–base titration 17:401; 18:250
  alkalinity 18:250
  analysis 17:401; 18:248
  assay 17:401; 18:250
  colorimetric method for aldehydes 12:309
  colorimetric method for chlorophyll 10:538
  colorimetric method for parathion 19:611
  color test for para red 17:205
  color test for toluidine red 17:205
  commercial grades and specifications 17:396,
    399
  etchant for ceramics 9:224
  flux for ceramics 9:216
  fusion of organofunctional silanes 18:85
  fusion of refractory ores 19:236
  fusion of silicon 18:12
  gravimetric method for karaya 12:582
  gravimetric method for thorium 19:14
  handling 18:249
  impurities, determination of 17:401; 18:252
  manufacture 17:395
  microchemical method for narcotine 16:250
  in potassium antimonate 6:56
  safety 17:357; 18:249
  sampling 18:249
  saponification method for methacrylate polymers
    16:90
  specification tests 17:401; 18:250
  uses 17:396
  volumetric method for amyl acetate 12:189
  volumetric method for carbonate 17:340
  volumetric method for methyl acetate 12:189
  volumetric method for phenols 12:283
Potassium hydroxide–N-1-naphthylethylenediamine
  colorimetric method for azinphosmethyl 19:611
Potassium hydroxide–silver nitrate

volumetric method for chlorohydrin 14:156
Potassium hypochlorite. (See also Hypochlorites.)
  properties 9:352
Potassium iodate. (See also Iodates; Potassium
  compounds.) 14:607
  analysis 14:607
  assay 14:607
  colorimetric method for glycerol 11:82
  commercial grades and specifications 14:607
  commercial specifications 17:399
  gravimetric method for thorium 19:14
  impurities, determination of 14:608; 17:400
  potassium iodide–sodium thiosulfate volumetric
    method 14:608
  properties 14:607
  solubility 14:607
  specification tests 14:608
  titration of aryl hydrazines 14:236
  titration of benzalkonium chloride 11:557
  titration of calcium iodide 8:92
  titration of copper thiocyanate 10:610
  titration of unsym-dimethylhydrazine 14:236
  titration of hydrazine 14:236
  titration of mercuric iodide 16:60
  titration of methylhydrazine 14:236
  titration of potassium iodide 14:592, 599
  titration of sodium iodide 14:592, 599
  titration of stannic oxide 9:263
  titration of sulfur 9:675
  titration of tin 5:145; 15:68; 16:336
Potassium iodide. (See also Iodides; Potassium
  compounds.) 12:545; 14:595; 17:397
  assay 14:599
  colorimetric method for bismuth 7:215; 10:295,
    620; 15:198
  colorimetric method for lead 12:186; 15:173
  colorimetric method for peroxides 12:308
  colorimetric method for rhenium 17:531
  colorimetric method for tellurium 17:606
  commercial grades and specifications 12:544;
    14:595
  commercial specifications 17:399
  extraction method for lead 15:178
  gravimetric method for lead 15:172
  gravimetric method for mercury 16:6, 7
  gravimetric method for palladium 17:237,
    249, 257
  gravimetric method for selenium 17:591
  gravimetric method for thallium 18:526, 527,
    535
  impurities, determination of 12:545; 14:605
  manufacture 17:397
  microchemical method for apomorphine 16:249
  microchemical method for benzylmorphine
    16:249
  microchemical method for diphenoxylate 16:249
  microchemical method for phenazocine 16:250

microchemical method for piminodine 16:250
potassium iodate titration 14:599
properties 14:595
titration of gold 14:21
uses 14:595; 17:397
volumetric method for margarine 15:501
Potassium iodide–sodium thiosulfate
volumetric method for iodine pentoxide 14:611
volumetric method for potassium iodate 14:608
volumetric method for potassium metaperiodate
14:610
volumetric method for sodium metaperiodate
14:610
Potassium iodide–starch
colorimetric method for bromate 12:543
colorimetric method for nitrite 11:142
Potassium iodobismuthate
colorimetric method for polyamides 17:282
Potassium mercuric iodide
gravimetric method for pyridine and its derivatives 17:425
microchemical method for dihydromorphine
16:249
microchemical method for nalorphine 16:250
Potassium metal. (See also Potassium alloys.)
analysis 17:359
commercial grades and specifications 17:359
impurities, determination of 17:390
manufacture 17:357
properties 17:355
safety 17:357
sample preparation 17:359
sampling 17:359
uses 17:355
Potassium metaniobate
solubility 16:355
Potassium metaperiodate. (See also Periodates.)
14:609
assay 14:610
commercial grades and specifications 14:609
gravimetric method for lithium 15:265
impurities, determination of 14:610
potassium iodide–sodium thiosulfate volumetric
method 14:610
properties 14:609
solubility 14:609
uses 14:609
volumetric method for lithium 15:268
Potassium metatantalate
solubility 16:355
Potassium methoxide
titration of dimethyl malonate 15:435
Potassium methyldithiocarbamate. (See also
Dithiocarbamic acid salts.)
uses 11:576
Potassium monopersulfate
active oxygen 7:224, 230

iodometric titration 7:230
Potassium monoxide
in potassium chloride 9:339
Potassium muriate. See Potassium chloride.
Potassium nitrate. (See also Potassium compounds.) 17:397
colorimetric method for lead 15:173
commercial grades and specifications 16:497
commercial specifications 17:399
manufacture 17:397, 405
potassium content 17:406
uses 17:391, 397, 405
Potassium nitrate–potassium hydroxide
colorimetric method for methadone 16:278
Potassium nitrite. (See also Potassium compounds.)
17:397
analysis 16:491
assay 16:491
colorimetric method for MEHQ 11:312
colorimetric method for thiophene 18:537
commercial grades and specifications 16:490
commercial specifications 17:399
gravimetric method for cobalt 10:352; 16:297
hydrazine sulfate–iodine volumetric method
16:492
manufacture 17:397
potassium chlorate–silver nitrate volumetric
method 16:492
potassium permanganate titration 16:491
Potassium orthophosphates. (See also individual
compounds.) 17:120
properties 17:120
solubility 17:120
Potassium oxalate. (See also Potassium compounds.) 17:397
commercial specifications 17:399
impurities, determination of 17:400
manufacture 17:397
properties 8:591
Potassium oxide
in alkali barium silicate glass 9:218
in cements 9:120
in ceramics 9:194
chloroplatinic acid gravimetric method 13:424
in clays and clay minerals 10:123
emission spectroscopy 9:218; 15:226
in enamels 9:195
flame photometry 9:121; 10:123
in glass and glass-ceramics 13:413, 415, 423
in glazes 9:196
in hydraulic cements 9:100
in lime and limestone 9:191; 15:226
in refractories 9:206
in silicate glasses 9:198
in zinc chromate 9:707
Potassium penicillin G. See Potassium benzylpenicillin.

Potassium penicillin V. See Potassium phenoxymethylpenicillin.
Potassium perchlorate. (See also Perchlorates.)
   properties 9:363
   solubility 9:364; 17:376
Potassium periodate
   colorimetric method for cerium 15:152
   colorimetric method for manganese 5:137;
      9:671; 10:300, 623; 12:157; 15:59, 71,
      386, 399, 405, 408, 412, 469; 16:331
   colorimetric method for manganous oxide
      15:212
Potassium permanganate 15:488
   analysis 15:492
   colorimetric method for cocaine hydrochloride
      16:246
   colorimetric method for manganese 19:371
   colorimetric method for p-phenylenediamine
      11:446
   color test for sulfate 18:423
   color test for sulfite 18:402
   commercial grades and specifications 15:489
   gravimetric method for manganic oxide 9:119
   identification method for alanine 8:450
   identification method for serine and threonine
      8:522
   impurities, determination of 15:492
   manufacture 15:489
   oxidation of cobalt 10:370
   oxidation of glycols 6:4
   properties 15:450
   qualitative test for alcohols 4:523
   reduction method for acrylic monomers 4:216
   specification tests 15:492
   titration of antimony 6:49, 66
   titration of antimony oxide 17:190
   titration of arsenic 6:242
   titration of ascorbic acid 18:459
   titration of calcium 10:296
   titration of calcium oxalate 8:106
   titration of calcium oxide 10:121; 15:216
   titration of ceric oxide 9:261
   titration of chromium 10:310
   titration of cresols 18:459
   titration of ferric oxide 15:210
   titration of ferrous oxide 15:220
   titration of formic acid 13:126
   titration of gallic acid 18:459
   titration of glucose 18:459
   titration of glycine 18:459
   titration of hydrogen peroxide 14:432
   titration of indigo 14:511
   titration of indigo carmine 14:511
   titration of iron 5:132; 6:262, 263; 14:600;
      15:37
   titration of maleic anhydride 17:156
   titration of DL-malic acid 18:459

titration of mannose 18:459
titration of nitrite 16:491; 19:368
titration of nitrogen oxides 16:498
titration of nitrosylsulfuric acid 16:500
titration of oxygen 16:526
titration of oxygen-containing bleaches
   7:230, 229
titration of perborates 11:369
titration of phloroglucinol 18:459
titration of polyphenols in tea 18:458
titration of potassium nitrite 16:491
titration of protocatechuic acid 18:459
titration of pyrite in coal and coke 10:253
titration of pyrocatechol 18:459
titration of pyrogallol 18:459
titration of rare earth oxalates 17:511
titration of selenium 17:597
titration of sodium nitrite 16:491
titration of sodium peroxide 16:536; 18:258
titration of strontium oxalate 18:275
titration of tellurium 17:603
titration of titanium 19:103
titration of vanadium 10:324; 16:186; 19:335
titration of volatile matter in roasted coffee
   10:436
uses 15:488
Potassium permanganate–ammonium thiocyanate
   colorimetric method for iron 18:283, 284
Potassium permanganate–barium chloride
   colorimetric method for sulfate 11:142
Potassium permanganate–sodium pyrophosphate
   volumetric method for manganese 12:531
Potassium permanganate, substance reducing. See
   Permanganate time test.
Potassium perrhenate
   toxicity 17:519
Potassium persulfate
   redox reagent for iodide 9:340
Potassium persulfate–silver sulfate
   colorimetric method for ceric oxide 13:442
Potassium persulfate–sulfuric acid
   redox reagent for bromide 9:340
Potassium phenethicillin. (See also Phenethicillin.)
   iodometric methods 16:562
   properties 16:556
   solubility 16:557
Potassium phenoxyethylpenicillin. (See
   Potassium phenethicillin.)
Potassium phenoxymethylpenicillin. (See also
   Phenoxymethylpenicillin.)
   iodometric methods 16:562
   properties 16:556
   solubility 16:557
Potassium phosphate dibasic
   commercial specifications 17:127
Potassium phosphate monobasic
   commercial specifications 17:127

Potassium phosphates. (See also Potassium compounds; and individual compounds.) 17:397
  in bread 6:454
Potassium pyroantimonate 6:52, 56
  analysis 6:56
  antimony content 6:56
  gravimetric method for sodium 18:229
  impurities, determination of 6:56
  sample preparation 6:62
Potassium pyrogallosulfonate
  colorimetric method for niobium 16:368, 374
  colorimetric method for tantalum 16:368;
    18:448
Potassium pyrosulfate. (See also Potassium compounds.)
  commercial specifications 17:399
  flux for ceramics 9:216
  fusion of ammonium hydroxide group 15:208
  fusion of niobium 16:363, 371
  fusion of niobium ores 16:361
  fusion of tantalum ores 16:361
  fusion of titanium 16:176; 19:102, 111
  fusion of titanium minerals 19:92
  fusion of zirconium 16:340
Potassium rhodizonate
  gravimetric method for strontium 18:268
  structure 17:456
Potassium silicates. (See also Silicates.) 18:42
  properties 18:42
Potassium soaps. (See also *Soaps.*)
  manufacture 18:179
  uses 18:179
Potassium–sodium alloys. (See also Potassium metal.)
  analysis 17:359
  commercial grades and specifications 17:359
  composition 17:355
  impurities, determination of 17:390
  properties 17:355
  sample preparation 17:359
  uses 17:355
Potassium–sodium chlorides
  sodium content 18:229
Potassium sodium ferrocyanide. (See also Ferrocyanides.)
  uses 15:85
Potassium–sodium nitrate
  potassium content 17:406
Potassium–sodium perchlorates
  sodium content 18:229
Potassium stannate. (See also Stannates.)
  alkali content 19:72
  analysis 19:72
  commercial grades and specifications 19:72
  impurities, determination of 19:72
  manufacture 19:72
  tin content 19:72

uses 19:72
Potassium sulfate. (See also Potassium compounds; Sulfates.) 17:398
  in ammonium dichromate 9:687
  commercial grades and specifications 17:398, 399; 18:422
  manufacture 17:398, 405
  potassium content 17:405
  turbidimetric method for barium 14:605
  uses 17:398, 405; 18:422
Potassium superoxide. (See also Potassium compounds.) 17:398
  handling 17:398
  manufacture 17:398
  properties 17:398
  uses 17:398
Potassium tantalum fluoride. See Potassium fluorotantalate.
Potassium tantalum heptafluoride
  properties 16:355
Potassium thiocarbonate
  colorimetric method for cobalt 10:358
Potassium thiocyanate. (See also Potassium compounds; Thiocyanates.)
  colorimetric method for ethylene oxide 12:321
  colorimetric method for iodate 11:142
  colorimetric method for iron 9:361, 452;
    11:140, 316; 15:45, 47; 19:494
  colorimetric method for molybdenum 10:641
  colorimetric method for niobium 16:364, 372
  colorimetric method for ruthenium 17:247
  colorimetric method for technetium 17:531, 538
  commercial grades and specifications 18:434
  commercial specifications 17:399
  gravimetric method for potassium chlorate 12:427
  manufacture 18:433
  properties 18:434
  uses 18:433
Potassium thiocyanate–phenolphthalein
  color test for ethylene oxide 12:223
Potassium titanate. (See also Titanates.)
  composition 19:124
Potassium titanium oxalate
  colorimetric method for ascorbic acid 18:458
  colorimetric method for cresols 18:459
  colorimetric method for gallic acid 18:458
  colorimetric method for glucose 18:459
  colorimetric method for L-glutamic acid 18:459
  colorimetric method for glycine 18:459
  colorimetric method for guaiacol 18:459
  colorimetric method for hydroquinone 18:459
  colorimetric method for DL-malic acid 18:459
  colorimetric method for mannose 18:459
  colorimetric method for *o*-phenylene diamine 18:459

colorimetric method for phloroglucinol 18:459
colorimetric method for polyphenols 18:458
colorimetric method for protocatechuic acid 18:458
colorimetric method for pyrocatechol 18:458
colorimetric method for pyrogallol 18:458
colorimetric method for resorcinol 18:459
colorimetric method for L-tyrosine 18:459
Potassium triiodide
gravimetric method for glycols 13:571
Potassium xanthate
colorimetric method for technetium 17:531
Potassium xanthate–sulfuric acid
colorimetric method for molybdenum 19:222
Potatoes
carboxin content 13:213
oxycarboxin content 13:213
Potency, relative
in bioassay 7:175, 176
Potency tests
for protamine sulfate 10:184
for sodium heparin 10:143
Potentiometric titration. (See also Acid–base titration.)
of alkaloids 4:593
of allyl alcohol 5:82
of allylamine 5:89
of allylbarbituric acids 5:92
of chromium 5:125
of ethyl-*p*-aminobenzoate 5:389
of fluorides 13:20, 24
of fluoroboric acid 13:42
Pour point
of lubricants 15:331, 332
of organosilicon copolymers 18:148
of silicone fluids 18:115
Powdered cleansers. (See also *Detergents.*) 11:325
composition 11:325
Power and Chestnut method
for caffeine 10:431
Power factors
for silicone fluids 18:116
Pozzolanic cements. (See also Hydraulic cements.) 9:97
commercial grades and specifications 9:99, 100, 101
composition 9:96
hardening 9:98
in hydraulic cements 9:107
Pramoxine hydrochloride 5:376
acid–base titration 5:402
analysis 5:402
anesthetic 5:401
properties 5:401
ultraviolet absorption spectroscopy 5:402, 403
Praseodymium. (See also Praseodymium metal; *Rare earths.*)
absorption spectroscopy 17:483
in aluminum and aluminum alloys 5:152
atomic absorption spectroscopy 17:494
borax bead test 6:584
in cerium oxide 17:501
emission spectroscopy 15:153, 155, 375; 17:488, 501
in europium oxide 17:502
flame photometry 17:492
ion exchange chromatography 15:157
in lanthanum oxide 15:153, 155; 17:501
in magnesium 15:375
in magnesium alloys 15:375
in neodymium oxide 17:501
visible spectroscopy 15:154
x-ray fluorescent spectroscopy 17:490
in yttrium oxide 17:501
Praseodymium fluoride
carrier gas distillation fusion method 8:636
oxygen in 8:636
Praseodymium metal. (See also Rare earth metals.)
carrier gas fusion methods 8:627, 682
hydrogen in 8:680
nitrogen in 8:623, 627
oxygen in 8:623, 627, 637, 680, 682
properties 17:471
sample preparation 8:641
vacuum fusion methods 8:623, 680
Praseodymium oxides. (See also Rare earth oxides.)
emission spectroscopy 17:501
impurities, determination of 17:501
properties 17:497, 509
uses 17:509
Praseodymium perchlorate. (See also Rare earth compounds.)
absorption spectrum 17:481
Precipitation fractionation
data treatment 4:301
molecular weight distribution of acrylic polymers 4:298
of polycarbonates 17:344
Precipitation number
of lubricants 15:321
Precision
of bioassay 7:179
Prednisolone. (See also Steroids.)
gas chromatography 14:219
manufacture 14:196
properties 14:196
solubility 14:197
trade names 14:192
Prednisolone acetate. (See also Steroids.)
infrared spectrum 14:196
solubility 14:197
uses 14:197

Prednisolone butylacetate. (See also Steroids.)
  solubility 14:197
  uses 14:197
Prednisolone 21-(3,3-dimethyl)-butyrate. See
    Prednisolone butylacetate.
Prednisolone 21-disodium phosphate. See
    Prednisolone sodium phosphate.
Prednisolone sodium phosphate. (See also
    Steroids.)
  solubility 14:197
  uses 14:197
Prednisone. (See also Steroids.)
  gas chromatography 14:219
  infrared spectrum 14:197
  manufacture 14:197
  properties 14:197
  solubility 14:197
  structure 14:197
  trade names 14:192
  uses 14:197
Pregl method
  for carbon and hydrogen 8:281, 292
1,4-Pregnadien-17α,21-diol-3,11,20-trione. See
    Prednisone.
1,4-Pregnanediene-11β,17α,21-triol-3,20-dione.
    See Prednisolone.
Pregnanediol
  gas chromatography 14:216, 223
5-Pregnane-3,20-diones
  liquid chromatography 14:208
Pregnane-3γ,11β,17γ,21-tetrol-20-one. See THF.
Pregnanetriol
  gas chromatography 14:223
  liquid chromatography 14:209, 210
  paper chromatography 14:211
Pregnane-3α,17α,21-triol-11,20-dione. See THE.
Pregnanolones
  gas chromatography 14:223
Pregnenediol
  paper chromatography 14:211
4-Pregnene-17α, 21-diol-3, 20-dione. See Com-
    pound S.
4-Pregnene-17α,21-diol-3,11,20-trione. See
    Cortisone.
4-Pregnene-3,20-dione. See Progesterone.
4-Pregnene-11β,17α,21-triol-3,20-dione. See
    Hydrocortisone.
4-Pregnen-21-ol-3,20-dione. See Desoxycortico-
    sterone.
Pregnenolone
  liquid chromatography 14:209
Prehnite 8:75
Prehnitic acid
  paper chromatography 8:379
Preobrazhenskite 7:314
Preservatives
  in confectionery products 10:552

in photographic films 12:587
Press
  for preparation of pulp handsheets 7:245
Pressed pellet technique 4:234
Press template
  for preparation of pulp handsheets 7:247
Pressure balance method
  for specific gravity of natural gas 13:330
Pressure of aerosols 4:547
Priceite 7:314
Primary explosives. See Explosives, primary.
Primene JM-T
  manufacture, methods of 5:262
  properties 5:260
Primene 81-R
  manufacture, methods of 5:262
  properties 5:260
Priming mixtures, explosive. See Explosive priming
    mixtures.
Printing ink resins. (See also *Hydrocarbon resins.*)
  commercial grades and specifications 14:274,
    279
Probarbital 6:468
  derivatives for identification 6:477
  gas chromatography 6:209, 211, 483, 484
  properties 6:470
Probarbital calcium 6:490
  infrared spectrum 6:490
Probertite 7:314
Probits 7:164
Procaine
  gas chromatography 6:210, 211
  mixtures, determination in 5:385
Procaine benzylpenicillin. (See also Benzyl-
    penicillin.)
  iodometric methods 16:562
  properties 16:556
  solubility 16:557
  structure 16:567
Procaine hydrochloride 5:377
  acid–base titration 5:408
  analysis 5:404
  anesthetic 5:404
  assay 5:405
  colorimetric method using sodium 1,2-naphtho-
    quinone-4-sulfonate 5:407
  diazotization method 5:406, 413
  ion exchange chromatography 5:409
  properties 5:404
  solubility 5:404
  tetraphenylboron titration 5:407
  thin-layer chromatography 5:409
  ultraviolet spectra 5:404, 405
  ultraviolet spectroscopy 5:404, 408, 419
Procaine penicillin G. See Procaine benzylpenicillin.
Producer gas
  carbon disulfide in 8:278

composition 8:302
manufacture 8:302
Progestational hormones. (See also Steroids;
    and individual compounds.) 14:186
  properties 14:201
  sources 14:201
  uses 14:201
Progesterone. (See also Steroids.)
  analysis 14:222
  fluorometric methods 14:222
  gas chromatography 14:219, 222
  liquid chromatography 14:209, 210
  manufacture 14:202
  occurrence 14:186, 202
  paper chromatography 14:211
  properties 14:202
  solubility 14:202
  structure 14:185, 186, 202
  thin-layer chromatography 14:213
  trade names 14:192
  ultraviolet spectroscopy 14:222
  uses 14:186, 202
Progestin. See Progesterone.
Progestogens. (See also Chlormadione acetate;
    Dimethisterone; Ethynodiol diacetate;
    Medroxyprogesterone acetate; Norethindrone;
    Norethindrone acetate; and Norethynodrel.)
  analysis 10:585
  in contraceptives 10:578
Progestral. See 17-Ethyltestosterone.
Progynon. See Estradiol.
Proline. (See also *Carboxylic acids, amino.*) 8:413,
    512
  analysis 8:518
  in cocoa beans 9:556
  colorimetric methods 8:518
  commercial specifications 8:518
  composition 8:413
  crystalline form 8:515
  gas chromatography 8:429
  infrared spectrum 8:516
  ion exchange chromatography 8:415, 421; 9:
    618
  manufacture 8:517
  in musts 19:469, 474
  naphthoquinone sulfate colorimetric method
    8:518
  ninhydrin colorimetric method 19:474
  paper chromatography 8:422
  properties 8:513
  in protein 8:414
  solubility 8:514
  structure 8:514
  in wines 19:469, 474
  in yeast 19:529
Prolutin. See Progesterone.
Promazine

gas chromatography 6:210, 211
Promethazine hydrochloride. (See also *Anti-
    histamines.*) 6:28
  properties 6:32
  thin-layer chromatography 6:35
Promethium. (See also Promethium metal;
    Rare earths.) 17:470
  half-life 17:470
Promethium metal. (See also Rare earth metals.)
  properties 17:471
Promethium oxide. (See also Rare earth oxides.)
  properties 17:497
Prominal
  gas chromatography 6:210, 211
Proof 4:462
  determination of 4:480, 481
  of wines 19:434
Proof gallon 4:463
Propallylonal 6:468
  derivatives for identification 6:477
  gas chromatography 6:484
  properties 6:470
  solubility 6:476
Propane. (See also Alkanes; *Liquefied petroleum
    gas.*) 14:300
  in aerosols 11:95
  in ammonia 5:313
  in butane 15:242
  butane content 15:241
  calorific value 13:331
  commercial grades and specifications 14:300;
    15:235
  gas chromatography 5:313; 11:98; 13:317, 318,
    350; 14:329; 15:241; 16:425; 18:52; 19:298,
    300
  in gasoline 13:350, 353
  moisture content 14:345
  in natural gas 13:320
  occurrence 14:300
  properties 11:95; 13:350; 14:300
  specific gravity 13:329; 15:245
  toxicity 14:300
  in trichlorosilane 18:52
  uses 14:300
  vapor pressure 15:249
Propane–butane mixtures. (See also *Liquefied
    petroleum gas.*)
  commercial grades and specifications
    15:235
1,2-Propanediamine. See Propylenedi-
    amine.
Propanedioic acid. See Malonic acid.
1,2-Propanediol. See Propylene glycol.
1,3-Propanediol. See Trimethylene glycol.
1-Propanethiol. (See also *n*-Propyl mercaptan.)
    15:552
2-Propanethiol. See Isopropyl mercaptan.

Propane-2-thiol–piperidine
  extraction of sulfur from elastomers 12:130
1,2,3-Propanetriol. See *Glycerol.*
1-Propanol. See *n*-Propyl alcohol.
2-Propanol. See Isopropyl alcohol.
2-Propanolamines. See Isopropanolamines.
2-Propanone. See *Acetone.*
Propargyl alcohol. (See also Acetylenic alcohols.)
  azeotropes 4:139
  commercial grades 4:139
  gas chromatography 4:578
  manufacture 4:139, 575
  properties 4:136, 576
  silver acetylide method 4:578
  solubility 4:138
  specifications 4:139, 577
  uses 4:576
Propargyl aldehyde
  properties 8:334
Propellants 4:453
  composition, determination of 11:98
  in cosmetic aerosol preparations 11:95, 96, 98
  gas chromatography 11:98
  halogenated 4:453
  nonhalogenated 4:454
  mixtures 4:456
  specific gravity 11:97
2-Propenal. See *Acrolein.*
Propene. See Propylene.
Propene nitrile. See *Acrylonitrile.*
Propene oxide. See Propylene oxide.
2-Propenol-1. See Allyl alcohol.
Prophenal 6:490
  infrared spectrum 6:490
β-Propiolactone
  use in manufacture of acrylic acid and acrylates
    4:185
  use in manufacture of carboxylic acids 8:373
Propionaldehyde
  absorption spectroscopy 9:564
  in acrolein 4:154
  in black tea 18:510
  in cigarette smoke 10:31
  in cocoa bean vapor 9:570
  commercial specifications 8:340
  derivatives for identification 8:343
  in embalming fluids 12:165
  gas chromatography 10:31; 12:204, 300, 302;
    16:92, 108, 109; 18:510
  infrared spectroscopy 6:616
  mass spectroscopy 9:570
  in methanol 16:108, 109, 114
  in oxyethylene–oxypropylene copolymers
    19:308
  properties 8:334
  sodium nitroprusside–diethanolamine colorimetric
    method 19:308

ultraviolet spectroscopy 6:632
Propionamide
  derivatives for identification 8:601
  properties 8:558, 597
Propionic acid 8:398
  acid–base titration 14:457
  in black tea 18:507, 511
  derivatives for identification 8:376
  distillation 14:456
  gas chromatography 8:383, 384; 18:511
  in hydroxylammonium sulfate 14:456
  infrared spectrum 8:376
  in methanol 16:114
  paper chromatography 15:133
  properties 8:397, 398, 558
  thin-layer chromatography 8:381
  uses 8:398
Propionic anhydride
  morpholine-reaction method 4:103, 104
  properties 8:558, 559
  propionic acid in 4:104, 105
Propionitrile. (See also *Nitriles.*)
  derivatives for identification 16:401, 402
  gas chromatography 16:404, 425
  properties 16:396
Propionoin
  in black tea 18:511
  gas chromatography 18:511
Propionyl chloride
  infrared spectrum 8:573
  properties 8:558, 570
Propionyl halides
  derivatives for identification 8:572
Propoxycaine hydrochloride 5:377
  analysis 5:409
  anesthetic 5:409
  assay 5:410
  colorimetric method using Bratton-Marshall
    reagent 5:412
  diazotization method 5:410
  infrared spectrum 5:411
  mixtures, determination in 5:412
  properties 5:409
  solubility 5:409
  ultraviolet spectra 5:410, 411
  ultraviolet spectroscopy 5:409, 411, 413
Propoxyphene
  gas chromatography 6:210, 211
n-Propyl acetate
  commercial specifications 8:578
  gas chromatography 16:108
  in methanol 16:108
  properties 8:577
Propyl acrylate. (See also *Acrylic and methacrylic
    acids and esters.*)
  gas chromatography 4:212
n-Propyl alcohol. (See also

Alcohols, lower; Propyl alcohols.)
acetylation with acetic anhydride 4:530
in acrolein 4:154
in allyl alcohol 5:79
in brandies 19:440
gas chromatography 5:5, 79; 7:96; 8:355;
   12:204; 16:107, 425; 19:442
manufacture 4:505
in methanol 16:107, 114
phthalic anhydride method 4:535
in wines 19:425, 440
Propyl alcohols. (See also Alcohols, lower.)
   4:496
azeotropes 4:500, 503
commercial grades and specifications 4:513, 515
derivatives for identification 4:518
gas chromatography 4:212
manufacture 4:505
properties 4:496
uses 4:512
Propylals. See *Acetals and ketals.*
Propylamine
in methanol 16:114
properties 5:260
uses 5:263
Propylamines
manufacture, methods of 5:262
uses 5:263
n-Propylarsonic acid
gravimetric method for tantalum 16:366
n-Propylbenzene
gas chromatography 7:7, 8, 9; 13:351; 14:337;
   18:298
in gasoline 13:351, 354
properties 13:351
in styrene 18:298
n-Propyl chloride. (See also 1-Chloropropane.)
in allyl chloride 5:102
gas chromatography 5:102
2-*n*-Propylchlorobenzene
mass spectroscopy 9:404
4-*n*-Propylchlorobenzene
mass spectroscopy 9:404
n-Propylcyclopentane
gas chromatography 13:351
in gasoline 13:351, 354
properties 13:351
Propylene. (See also *Hydrocarbons.*) 14:310
in ammonia 5:313
commercial grades and specifications 14:311
in cyclopropane 5:361
in ethylene 14:327
flammability 14:310
gas chromatography 5:313; 12:107; 14:327,
   329; 15:241
in hydrogen 14:381
manufacture 14:310

in production of acrolein 4:149
in propane 15:241, 244
properties 14:310
silver–mercuric nitrate gasometric method
   15:244
specific gravity 15:245
uses 14:310
vapor pressure 15:249
Propylene–1-butene copolymers. (See also
   Ethylene and propylene polymers.) 12:345
Propylene chloride. See 1,2-Dichloropropane.
1,2-Propylene chlorohydrin. (See also Halohydrins.)
paper chromatography 14:161
Propylene chlorohydrins. (See also Halohydrins.)
   14:165
analysis 14:166
manufacture 14:165
solubility 14:166
uses 14:165
Propylene copolymers. (See also *Ethylene and
   propylene polymers.*) 12:345
manufacture 12:345
properties 12:345
types 12:345
Propylenediamine. (See also *Diamines and
   higher amines, aliphatic.*)
properties 11:410
1,2-Propylenediaminetetraacetic acid. See PDTA.
Propylene dichloride 13:181
gas chromatography 13:183
manufacture 13:181
properties 13:181
solubility 13:181
toxicity 13:181
uses 13:181
Propylenedinitrilotetraacetic acid. See PDTA.
Propylene epifluorohydrin. (See also Halohydrins.)
structure 14:153
Propylene glycol. (See also Glycols.) 5:22
antifreeze 6:1; 13:599
azeotropes 13:541
in brake fluids 7:424
catechol colorimetric method 11:74
in cigarette smoke 10:29
codeine color test 13:570
colorimetric determination 13:570
commercial grades and specifications 13:546,
   548; 14:312
in creams and lotions 11:92
density 13:584
derivatives for identification 13:568
distillation range 13:602
in enamels 12:175
freezing point 6:7; 13:535
gas chromatography 5:32; 6:5; 11:75; 13:503,
   504, 576, 589

gravimetric determination 13:570
guaiacol color test 13:570
humectant values 13:542
identification 6:4
iodoform gravimetric method 11:74
liquid chromatography 13:580
manufacture 13:542
ninhydrin reagent color test 10:30
paper chromatography 13:501, 502, 581
periodic acid spectroscopic method 13:599
phenylhydrazine acetate gravimetric method
    13:570
in polyurethanes 19:297
properties 6:2; 12:175; 13:536
refractive index 13:584
resorcinol color test 13:570
Rosen color test 13:573
in shampoos 11:74
sodium periodate colorimetric method 13:571
solubility 13:535, 539, 540
structure 13:534
in sunscreens 11:90
thin-layer chromatography 13:582
thymol color test 13:570
uses 13:542
violet ninhydrin colorimetric method 13:599
Propylene glycol alginate
in beer 7:639
Propylene glycol isobutyl ether. (See also *Ethers.*)
commercial specifications 12:299
Propylene glycol monomethyl ether. (See also
    Ethers.)
commercial specifications 12:299
gas chromatography 7:427
Propylene–maleic anhydride copolymer. (See also
    Ethylene and propylene polymers.) 12:345
manufacture 12:345
Propylene oxide. (See also Epoxides; *Epoxy
    compounds.*) 12:224; 13:191
acid–base titration 11:568
in acrolein 4:154
analysis 12:225; 13:191
commercial grades and specifications 12:225
in disinfectants and antiseptics 11:567
distillation range 12:225
gas chromatography 11:569; 12:204
hydrochlorination methods 12:208
hydrogen bromide titration 13:191
impurities, determination of 12:225
infrared spectroscopy 12:200
manufacture 12:224; 13:191
properties 12:194, 224; 13:191
solubility 12:224; 13:191
specification tests 12:225
toxicity 13:191
uses 5:1; 12:224; 13:191
Propylene oxide derivatives

in brake fluids 7:424
Propylene oxide–ethylene oxide polymers. (See
    also Polyglycols.)
Rosen color test 13:573
Propylene–styrene copolymers. (See also
    Ethylene and propylene polymers.) 12:345
infrared spectrum 12:366
n-Propyl ether. (See also *Ethers.*)
properties 12:297
n-Propyl formate
gas chromatography 16:108
in methanol 16:108
uses 13:120
Propyl gallate 6:77
n-Propyl *p*-hydroxybenzoate
in beer 7:597
n-Propyl iodide. (See also Iodine compounds,
    organic.)
gas chromatography 16:92
properties 14:612
n-Propyl isocyanate. (See also *Isocyanic acid
    esters.*)
commercial sources 15:102
properties 15:95
Propyl lactate. (See also Lactic acid esters.)
properties 15:131
solubility 15:130
n-Propyl lactate acetate. (See also Lactic acid
    esters.)
properties 15:131
Propyl methacrylate. (See also *Acrylic and
    methacrylic acids and esters.*)
gas chromatography 4:212
nuclear magnetic resonance spectroscopy
    16:86
n-Propyl mercaptan. (See also *Mercaptans.*)
derivatives for identification 15:559
gas chromatography 15:557
properties 15:552
2-*n*-Propylphenol. (See also *Phenols.*)
derivatives for identification 17:21
properties 17:4
3-*n*-Propylphenol. (See also *Phenols.*)
properties 17:4
4-*n*-Propylphenol. (See also *Phenols.*)
derivatives for identification 17:21
ferric chloride colorimetric method 17:16
paper chromatography 17:24
properties 17:4
2-*n*-Propylpyridine. (See also *Pyridine and
    pyridine derivatives.*)
derivatives for identification 17:415
ionization constants 17:416
properties 17:415
4-*n*-Propylpyridine. (See also *Pyridine and
    pyridine derivatives.*)
gas chromatography 17:437

n-Propyltrichlorosilane. (See also Organo-
   chlorosilanes.)
   manufacture 18:55
   properties 18:54
Propyne
   gas chromatography 4:144
   properties 4:137
2-Propynol-1. See Propargyl alcohol.
Protamine sulfate 10:184
   analysis 10:184
   commercial specifications 10:184
   manufacture 10:184
   potency tests 10:184
   turbidimetric method for sodium heparin
     10:149
Protease activity
   ammonium sulfate test 7:594
   axocoll color test 7:589
   in beer 7:588
   n-benzoyl-*dl*-arginine-*p*-nitroanilide hydrolysis
     method 7:591
   casein digestion method 7:590
Protein
   automated analysis 7:665
   in bakery products 6:453, 460
   in beer 7:608, 611, 612, 665
   biuret test 11:75; 19:476
   in casein 9:9
   in cereals 9:236
   in chocolate 9:540
   in chocolate and cocoa products 9:538
   in confectionery products 10:564
   in corn products 11:5
   in eggs and egg products 12:62
   in egg white 12:58, 60
   in egg yolk 12:58, 61
   in flour 12:602
   fractionation 7:612
   in gelatin 12:552
   in grains 4:472
   gravimetric determination 12:552
   in green coffee 10:374, 403
   in green tea leaf 18:476
   in hair sprays 11:90
   hydrolysis 8:411
   Kjeldahl volumetric method 11:5; 12:62
   Kjeldahl-Wilfarth-Gunning method 12:603
   infrared spectroscopy 16:128
   in musts 19:469, 475
   ninhydrin colorimetric method 11:75
   phosphomolybdic acid gravimetric method
     19:475
   in raw milk 16:128
   in roasted coffee 10:375
   in shampoos 11:75
   in sodium heparin 10:150
   sodium hydroxide–cupric sulfate colorimetric

     method 19:476
   solvent extraction 19:138
   in tobacco 19:138
   trichloroacetic acid gravimetric method 19:475
   trichloroacetic acid turbidimetric method
     10:150
   wet gluten method 12:603
   in wines 19:469, 475
   in yeast 19:528
Protein haze
   in beer 7:520
Protein hydrolysates
   in tobacco 19:140
Protocatechualdehyde (3,4-Diethoxybenzalde-
   hyde) 6:596
   derivatives for identification 6:612
   properties 6:598
Protocatechuic acid (3,4-Dehydroxybenzoic acid)
   8:401
   analysis 8:401
   gas chromatography 8:389
   in green tea leaf 18:458
   paper chromatography 8:379
   potassium permanganate titration 18:459
   potassium titanium oxalate colorimetric method
     18:458
   properties 8:401, 403
Proton magnetic resonance spectroscopy
   of cyclic alcohols 11:191
   of *unsym*-dimethylhydrazine 14:237
   of hydrazine 14:237
   of methylhydrazine 14:237
Protopectin
   in tobacco 19:145
Proustite 18:155
Prussian blue. (See also Ferric ferrocyanide;
   Iron salts.)
   color test for hydrogen cyanide 14:410
   in wines 19:483
Prussic acid. See *Hydrogen cyanide.*
P-salts 17:273
Pseudocumene. (See also *Hydrocarbons.*)
   commercial specifications 14:323
   uses 14:323
Pseudocumenol. (See also 2,4,5-Trimethylphenol;
   Phenols.)
   properties 17:4
Psilomelane 15:448
Psoralen. (See also *Coumarins.*) 11:161
   gas chromatography 11:168, 170
Psychometric chart 7:276
PTMEG. See Polytetramethylene ether glycols.
PTX. (See also High-explosive mixtures.)
   analysis 12:450
   composition 12:450
   uses 12:450
Pulegone

gas chromatography 8:353
hydroxylamine hydrochloride volumetric
    method 12:282
in pennyroyal oil 12:282
Pulp
  bleaching 7:222
  bleaching evaluation 7:233
  brightness 7:257, 262
  handsheets for optical testing 7:241, 251
  handsheets for physical testing 7:241, 248
  mercury content 16:44
  physical testing 7:253
  sample dissolving tube 7:266
  viscosity, disperse 7:264
Pulp, bleached
  properties, determination of 7:241
Pulp, unbleached
  viscosity 7:269
Pulverized flint
  in matches 15:513
  in match strikers 15:522
  particle size 15:529
Pulverized silica
  commercial grades and specifications 15:512
  in matches 15:512
  in match strikers 15:522
  particle size 15:529
Pumice 4:31, 33
  in chrome cleaners 6:353
  commercial specifications 17:191
  uses 17:191
Pumicite 4:31, 33
Pump, hand-operated
  for preparation of pulp handsheets 7:246
Puralin. See Thiram.
Purines 8:55
  bioassay 7:182
Purity
  of yeast 19:541
Purple lake M
  in enamels 12:183
  identification 12:183
Purple organic pigments
  in enamels 12:187
  identification 12:187
Purple pigments. (See also *Pigments*; and
    individual compounds.) 17:201
Purpurin
  color method for boric acid 7:387
Purpurogallin
  colorimetric method for polyphenol oxidase
    18:484
Pycnometer method
  for proofing 4:483
  for specific gravity of abrasive grains 7:296
  for specific gravity of beer 7:486
  for specific gravity of bituminous materials 6:318

for specific gravity of liquefied petroleum
    gas 15:247
Pyostacine 5:471, 551
  analysis 5:556, 560, 563
  microbiological assay 5:560
  properties 5:553
  solubility 5:551, 552
  structure 5:471
  uses 5:551
Pyranine concentrated. See D & C Green No. 8.
Pyrathiazine hydrochloride. (See also *Anti-
    histamines*.) 6:29
Pyrazolone
  composition 12:174
  in enamels 12:174
  structure 17:179
Pyrazolone red. See *Pigments*.
Pyrene
  in D & C Green No. 8 10:498, 508
Pyrethrins. (See also *Pesticides and insecticides*.)
  commercial grades and specifications 19:593
  composition 19:586
  ethyl acetate–orthophophoric acid colorimetric
    method 19:611
  gravimetric determination 19:609
  manufacture 19:596
  occurrence 19:596
  properties 19:586
  solubility 19:587
  structure 19:580
  toxicity 19:592
Pyridine. (See also *Pyridine and pyridine
    derivatives*.)
  in ammonium hydroxide 5:323
  analysis 17:424
  chloroauric acid gravimetric method 17:434
  1-chloro-2,4-dinitrobenzene colorimetric
    method 17:424
  colorimetric determination 17:424
  colorimetric method for pentaborane 7:348
  colorimetric method for tetrachloroethylene
    9:462
  commercial grades and specifications 17:423
  cyanogen bromide method 6:35
  derivatives for identification 17:426
  in ethanol 4:545
  freezing point 17:439
  gas chromatography 17:437
  gravimetric method for indium 14:529
  identification 17:424
  impurities, determination of 17:438
  infrared spectroscopy 17:436
  manufacture 17:420
  mass spectrum 17:428
  in methanol 16:115
  in 2-methylpyridine 17:436
  properties 17:414

reineckate gravimetric method 9:620
residue after evaporation 17:438
solubility 17:415, 438
specification tests 17:438
structure 17:412
trisodium pentacyanoammino ferrate colorimetric method 17:433
Pyridinealdehydes. (See also *Pyridine and pyridine derivatives.*)
ionization constants 17:417
Pyridine alkaloids 4:591
analysis 4:610
Pyridine and pyridine derivatives. (See also individual compounds.) 17:412–442
acid–base titration 17:434
analysis 17:424
assay 17:434
bromine–potassium cyanide colorimetric method 17:424
colorimetric determination 17:424, 432
commercial grades and specifications 17:423
cyanogen bromide colorimetric method 17:424, 432
derivatives for identification 17:425
in ethanol 17:425
in fusel oil 17:425
gas chromatography 17:426, 436
gravimetric determination 17:434
identification 17:424
infrared spectroscopy 17:426, 436
ionization constants 17:416
König reaction color test 17:424, 432
Larue colorimetric method 17:424, 433
manufacture 17:420
mass spectroscopy 17:427
nuclear magnetic resonance spectroscopy 17:430
occurrence 17:412
picric acid gravimetric method 17:425
potassium mercuric iodide gravimetric method 17:425
properties 17:414
specification tests 17:438
tetracyanoethylene colorimetric method 17:425
trisodium pentacyanoammino ferrate colorimetric method 17:425, 433
uses 17:412
volumetric determination 17:434
Pyridine bases
acid–base titration 17:38
in phenols 17:38
Pyridine–benzidine
colorimetric method for cyanides 13:179
Pyridine carboxaldehydes
ultraviolet spectroscopy 6:634
Pyridine carboxylic acids. (See also *Pyridine and pyridine derivatives.*)

manufacture 17:422
Pyridine derivatives
in finishing agents 11:126, 132
picric acid colorimetric method 11:127, 133
solubility 11:126
2,3-Pyridinedicarboxylic acid. See Quinolinic acid.
2,4-Pyridinedicarboxylic acid. See Lutidinic acid.
2,5-Pyridinedicarboxylic acid. See Isocinchomeronic acid.
2,6-Pyridinedicarboxylic acid. See Dipicolinic acid.
3,4-Pyridinedicarboxylic acid. See Cinchomeronic acid.
3,5-Pyridinedicarboxylic acid. See Dinicotinic acid.
Pyridine–ferric chloride
colorimetric method for phenprocoumon 10:164
Pyridine–hydroidic acid
volumetric method for ethers 12:302
Pyridine–hydrogen chloride
volumetric method for ethylene oxide 12:322
Pyridine-*N*-oxide. (See also *Pyridine and pyridine derivatives.*)
commercial specifications 17:423
manufacture 17:422
structure 17:413
uses 17:413
Pyridine–pyrazolone
colorimetric method for cyanides 13:178
colorimetric method for hydrogen cyanide 14:420
colorimetric method for thiocyanate 14:420
Pyridine-3-sodium sulfonate. (See also *Pyridine and pyridine derivatives.*)
trisodium pentacyanoammino ferrate colorimetric method 17:433
Pyridine sulfate dibromide
volumetric method for acrylic monomers 4:190, 336
volumetric method for allyl ethers 5:99
Pyridinium bromide perbromide
volumetric method for elastomers, synthetic 12:123
Pyridinium chloride
volumetric method for oxirane oxygen 11:312
Pyridinium chloride–acetic anhydride
volumetric method for esterification value of epoxy resins 12:250
Pyridinium chloride–chloroform
volumetric method for epoxy compounds 12:210
Pyridinium chloride–pyridine
titration of epoxy resins 12:209
Pyridoxine. (See also *Pyridine and pyridine derivatives.*)

in cocoa beans 9:540
in eggs and egg products 12:66
structure 17:412
uses 17:413
Pyridoxine hydrochloride. (See also *Pyridine and pyridine derivatives.*)
    acid–base titration 17:435
    assay 17:435
1-(2-Pyridylazo)-2-naphthol. See PAN.
4-(2-Pyridylazo)resorcinol
    colorimetric method for niobium 16:369, 371, 373, 376, 383
    colorimetric method for tantalum 16:369
3-(2-Pyridyl)-5,6-bis-4-phenylsulfonic acid disodium salt. See Ferrozine.
α-Pyridyl ketoxime
    colorimetric method for gold 14:25
Pyridyl methyl ketones. (See also *Pyridine and pyridine derivatives.*)
    trisodium pentacyanoammino ferrate colorimetric method 17:433
Pyridylpyridines. (See also *Pyridine and pyridine derivatives.*)
    ionization constants 17:416
Pyridyl pyridinum dichloride
    color test for malonic acid 15:426
Pyrilamine maleate. (See also *Antihistamines.*) 6:29
    column chromatography 6:150
    properties 6:32
    thin-layer chromatography 6:35
    ultraviolet spectroscopy 6:150
Pyrimidines
    bioassay 7:182
Pyrite. (See also Iron compounds.) 15:21, 166; 18:360
    in coal and coke 10:252
    commercial specifications 17:194
    composition 15:22
    potassium permanganate titration 10:253
    sulfur content 18:369
    thallium content 18:522
    uses 15:85
Pyrocarbon. (See also *Graphite and related carbons.*) 14:57
    manufacture 14:57
Pyrocatechin
    gas chromatography 18:472
Pyrocatechol. (See also *Dihydroxybenzenes.*)
    4-aminoantipyrine colorimetric method 17:18
    analysis 17:46
    assay 17:46
    colorimetric method for niobium 16:368, 373
    colorimetric method for tantalum 16:368; 18:448
    commercial grades and specifications 17:15
    derivatives for identification 17:21

ferric chloride colorimetric method 17:16
in green tea leaf 18:458
iodine–resorcinol colorimetric method 17:46
lead acetate gravimetric method 17:46
manufacture 17:1, 10
Millon's reagent color test 17:18
occurrence 17:1
paper chromatography 17:24
potassium dichromate colorimetric method 17:20
potassium permanganate titration 18:459
potassium titanium oxalate colorimetric method 18:458
properties 17:5
separation 17:23
solubility 17:5
thin-layer chromatography 17:24
Pyrocatechol, azo dye derivatives
    color reagents for boron 7:389
Pyrocatechol-3,5-disulfonic acid. See Tiron.
Pyrocatecholphthalein
    structure 17:7
Pyrocatecholsulfonphthalein
    colorimetric method for aluminum 7:652
Pyrocatechol violet
    colorimetric method for niobium 16:368
    colorimetric method for tantalum 16:368; 18:448
    colorimetric method for tin 19:55, 61
    colorimetric method for zirconium 14:110, 119
    colorimetric method for zirconium dioxide 13:428
    titration of thorium 19:13
Pyrochlore
    composition 16:349
Pyrogallic acid
    absorbing agent for oxygen 8:307
Pyrogallol. (See also Trihydroxybenzenes.)
    analysis 17:47
    colorimetric method for iodate 11:142
    colorimetric method for niobium 16:368, 373
    colorimetric method for tantalum 10:319; 16:368, 392; 18:447
    commercial grades and specifications 17:15
    derivatives for identification 17:21
    ferric chloride colorimetric method 17:16
    gravimetric method for niobium 16:367
    gravimetric method for tantalum 16:367
    in green tea leaf 18:458
    in hair dyes 11:103
    manufacture 17:1, 10
    occurrence 17:1
    osmium tetroxide colorimetric method 17:20, 47
    phthalic anhydride colorimetric method 17:19
    potassium permanganate titration 18:459

potassium titanium oxalate colorimetric
method 18:458
properties 17:5
separation 17:23
solubility 17:5
thin-layer chromatography 11:103; 17:24
uses 17:2
Pyrogallol red
colorimetric method for silver 18:164
Pyrogens
in penicillins 16:566
Pyrogen test 7:151
for impurities in antibiotics 5:491
Pyrograms. See Pyrolysis–gas chromatography.
Pyrographite. (See also *Graphite and related
carbons.*) 14:57
manufacture 14:57
photomicrograph 14:58
properties 14:58
Pyrohydrolysis
apparatus 7:326
of boron 7:394
of boron, elemental 7:325
of ceramics 9:217
Pyrohydrolytic method
for fluoride 13:18; 14:130
Pyrolusite. (See also Manganese dioxide.)
15:448, 515
rhenium content 17:523, 538
Pyrolysis
of polytetrafluoroethylene 13:81
Pyrolysis–gas chromatograms
of acrylic automotive lacquer 16:89
of acrylic coatings 16:88
of methyl methacrylate–ethyl acrylate copolymers 16:88
of poly(*n*-butyl methacrylate) 16:87
of poly(methyl methacrylate) 16:87
Pyrolysis–gas chromatography
of brake linings 7:449
of butyl rubber 12:102, 105
of carbon in trichlorosilane 18:53
of chlorosulfonated polyethylene 12:107
of ethylene and propylene polymers 12:365
of ethylene–propylene rubber 12:107
of methacrylate polymers 16:87
of natural rubber 12:102, 104
of nitrile rubber 12:105
of phthalate plasticizers 16:96
of polyamides 17:291
of polychloroprene 12:106
of polyethers 19:309
of polyethylene 12:365
of polymers 7:444
of polyoxypropylenes 19:309
of polystyrene 16:97
of polyurethanes 19:290

of styrene–butadiene copolymers 12:103, 104
Pyrolysis–mass spectroscopy
of ethylene and propylene polymers 12:369
Pyrolytic graphite. See Pyrographite.
Pyrolytic graphite electrode
for voltammetry of aluminum 5:254
Pyrolytic hydrogenation
of nitrogen 16:468
Pyrolytic methods
for elastomers, synthetic 12:135
for epoxy resins 12:237
for glycol ethers 12:573
for organic compounds 16:533
for zinc oxide 12:156
Pyromellitic acid 8:404
analysis 8:405
paper chromatography 8:379
properties 8:403, 404
Pyromellitic dianhydride 4:535, 537
volumetric method for glycols 13:587
Pyromorphite 15:166
Pyronine G. (See also *Triphenylmethane and
related dyes.*)
thin-layer chromatography 19:173
Pyrophosphate. (See also individual compounds.)
17:118
in bakery chemicals 6:433
gravimetric method for uranium 19:245
ion exchange chromatography 17:131
paper chromatography 17:135
in sodium tripolyphosphate 17:137
Pyrophosphoric acid. (See also Phosphoric
acid.)
manufacture 17:122
properties 17:122
Pyrophylite
commercial specifications 10:534
composition 10:103, 534
Pyrosulfate fusion
of platinum group metals 17:232
of silicon 18:14
of tungsten minerals 19:197
Pyrosulfites. (See also Sulfites.)
manufacture 18:401
properties 18:402
Pyrosulfuric acid. See Disulfuric acid.
Pyrosulfuryl chloride
in chlorosulfonic acid 9:533
Pyrovanadates. (See also Vanadates.) 19:344
Pyroxenes 15:358
Pyrozolone derivatives. (See also individual
compounds.) 6:205
Pyrrhotite 14:519
composition 15:22
Pyrrobutamine phosphate. (See also *Antihistamines.*) 6:29
properties 6:32

Pyrrocaine hydrochloride 5:377
  acid–base titration 5:414
  analysis 5:413
  anesthetic 5:413
  assay 5:413
  infrared spectroscopy 5:413
  properties 5:413
  solubility 5:413
  ultraviolet spectroscopy 5:413
Pyrrole
  colorimetric determination 17:298
  p-dimethylaminobenzaldehyde colorimetric
    method 17:298
  infrared spectroscopy 17:436
  in 2-methylpyridine 17:436
  in polyamides 17:298

Pyrrole aldehydes
  in cocoa beans 9:563
  TBA colorimetric method 9:563
Pyruvic acid
  infrared spectrum 8:377
  liquid chromatography 15:133
  structure 19:446
Pyruvic acid 2,4-dinitrophenylhydra-
    zone
  thin-layer chromatography 8:381
Pyruvic aldehyde
  in embalming fluids 12:165
Pyruvic isonicotinyl hydrazone. (See also
    Pyridine and pyridine derivatives.)
  trisodium pentacyanoammino ferrate
    colorimetric method 17:433

Q

Quartz. (See also Silica.) 18:1, 5
  as ablator 4:2
  as abrasive 4:31, 33
  in asbestos–cement products 6:270
  in clay 9:191
  composition 10:104
  decomposition 19:235
  silicon content 18:27
α-Quartz
  in explosive priming mixtures 12:418
  x-ray diffraction 12:418
Quartzite 4:31, 33
Quaternary ammonium compounds. (See also
  Benzalkonium chloride; Benzethonium
  chloride; Cetylpyridinium chloride; Methylbenzethonium chloride; and β-(Phenoxyethyl)-
  dimethyldodecylammonium bromide.) 11:
  554–561
  analysis 11:556
  assay 11:557
  bromphenol blue colorimetric method 11:82,
    93
  bromphenol blue volumetric method
    11:560
  colormetric determination 11:559
  in creams and lotions 11:93
  in disinfectants and antiseptics 11:554
  in embalming fluids 12:165
  gas chromatography 11:556
  gravimetric determination 11:560
  in hair sprays 11:90
  identification 11:556
  infrared spectroscopy 12:577
  liquid chromatography 11:556
  paper chromatography 11:556
  separation 11:556
  in shampoos 11:82
  in sunscreens 11:91
  ultraviolet spectroscopy 11:556
  volumetric determination 11:560
Quaternary ammonium silicates. (See also
  Silicates.) 18:42
  properties 18:42
Quaternary methyl iodide
  derivatives of aniline and its derivatives
    5:429
Quaterphenyls
  gas chromatography 11:535
  in terphenyls 11:535
  thin-layer chromatography 11:532
Quercetin

  colorimetric method for tantalum 16:368;
    18:449
  gas chromatography 18:472
  gravimetric method for tantalum 16:366
  structure 18:461
Quercetin-3-glucoside
  in green tea leaf 18:461
  thin-layer chromatography 18:463
Quercetin, pentamethyl ester
  colorimetric method for boric acid 7:387,
    410
Quercetin-3-rhamnoglucoside
  in green tea leaf 18:461
  thin-layer chromatography 18:463
Quercetinsulfonic acid
  colorimetric method for niobium 16:368, 374
Quicklime. (See also Lime.)
  commercial grades and specifications 9:191
Quimociac
  gravimetric method for phosphate 5:349
  gravimetric method for phosphorus 6:433;
    17:147
Quinacridone magenta. (See also *Pigments.*)
  structure 17:183
Quinacridone maroon. (See also *Pigments.*)
  structure 17:184
Quinacridone red
  in enamels 12:174
Quinacridone scarlet. (See also *Pigments.*)
  structure 17:184
Quinaldinehydroxamic acid
  gravimetric method for niobium 16:367
  gravimetric method for tantalum 16:367
Quinalizarin
  colorimetric method for aluminum 11:140
  colorimetric method for boric acid 7:386, 407
  colorimetric method for indium 14:526
  colorimetric method for magnesium 11:140
  colorimetric method for tantalum 16:368
Quinic acid
  in roasted coffee 10:375, 396
  structure 17:2
Quinidine 4:589
  analysis 4:605
  gas chromatography 6:210
  thin-layer chromatography 15:295
Quinine 4:589
  analysis 4:605
  gas chromatography 6:210
  paper chromatography 6:183
  thin-layer chromatography 15:295

ultraviolet spectroscopy 16:271
Quinizarin
    structure 17:7
2-Quinizarin sulfonic acid
    colorimetric method for aluminum 14:66
Quinoline
    colorimetric method for decaborane 7:347
    in enamels 12:184
    fluorometric method for glycerol 13:509
    gas chromatography 11:529
    identification 12:184
Quinoline alkaloids 4:589
    analysis 4:605
Quinoline derivatives. (See also individual com-
    pounds.) 6:198
    antiozonants 6:78
Quinoline dyes 12:7
Quinoline silico molybdate
    gravimetric method for silica 12:155
Quinoline yellow. See D & C Yellow No. 10;
    and No. 11.
Quinolinic acid. (See also *Pyridine and pyridine
    derivatives.*) 17:413
Quinolinium molybdate method
    phosphate content of ammonium phosphate,
        dibasic 5:349
8-Quinolinol. (See also 8-Hydroxyquinoline.)
    colorimetric method for aluminum 13:267
    colorimetric method for gallium 13:257, 263,
        395
    colorimetric method for tin 19:64
    fluorimetric method for fluorides 13:15
    gravimetric method for aluminum 9:292

volumetric method for gallium 13:259
2-(2-Quinolyl-6,8-disulfonic acid)-1,3-indandione
    in D & C Yellow No. 10 10:524
    liquid chromatography 10:524
2-(2-Quinolyl)-1,3-indandione. See D & C
    Yellow No. 11.
2-(2-Quinolyl-6-sulfonic acid)-1,3-indandione
    in D & C Yellow No. 10 10:524
    liquid chromatography 10:524
2-(2-Quinolyl-8-sulfonic acid)-1,3-indandione
    in D & C Yellow No. 10 10:524
    liquid chromatography 10:524
2-(2-Quinolyl-6-sulfonic acid)-1,3-indandione-
    5-sulfonic acid
    in D & C Yellow No. 10 10:524
    liquid chromatography 10:524
2-(2-Quinolyl-8-sulfonic acid)-1,3-indandione-
    5-sulfonic acid
    in D & C Yellow No. 10 10:524
    liquid chromatography 10:524
Quinones. (See also individual compounds.)
    13:200–202
    in carbon black 8:192
    2,4-dinitrophenylhydrazine method 4:206
    polarographic method 4:207
Quinoxaline derivatives
    of *o*-diamines 6:411
Quinoxalinedithiol
    colorimetric method for cobalt 7:658
Quinozarin green. See D & C Green No.
    6.
Quisumbing and Thomas
    determination of *d*-glucose 10:7

R

Racemethorphan. (See also *Narcotics.*)
  ammonium molybdate colorimetric method
    16:247
  color tests 16:247
  microchemical tests 16:250
  structure 16:245
  trinitrobenzoic acid microchemical method
    16:250
  ultraviolet spectroscopy 16:253
Racemorphan. (See also *Narcotics.*)
  ammonium molybdate colorimetric method
    16:247
  color tests 16:247
  microchemical tests 16:250
  sodium carbonate microchemical method
    16:250
  structure 16:245
  ultraviolet spectroscopy 16:253
R acid. See 2-Naphthol-3,6-disulfonic acid.
Radioactive decay
  of radium 17:443
  of thorium 17:446
  of uranium 17:444
Radioactive methods
  for adipic acid 4:419
  for fluorine, elemental 13:4
  for phosphorus 17:68
Radioactivity
  of technetium 17:531
  of thorium 19:17
  of tobacco 19:153
  of water 19:393
Radiochemical methods
  for aldosterone 14:219
  for beryllium 7:139
  for bismuth 7:218
  for carbon dioxide 19:502
  for cobalt 10:371
  for copper 10:618
  for cyclamates 11:238
  for nalorphine 16:268
  for pesticides and insecticides 19:613
  for potassium 17:372
  for thallium 18:534
Radiography
  of diamonds, industrial 11:480
Radioimmunological methods
  for insulin 14:576
Radiometer haze meter 7:525
Radium. (See also Radium compounds; Radium
    metal.) 17:443–469

alpha-particle spectroscopy 17:453, 456,
    459, 462
in biological samples 17:454, 465
in bone 17:465
coprecipitation 17:454, 457
detection 17:452
emission spectroscopy 17:452
gamma-ray spectroscopy 17:453, 456
history 17:443
hydrochloric acid gravimetric method 17:463
identification 17:453
ion exchange chromatography 17:455, 467,
    468; 18:269
isotopes 17:443
in minerals 17:453
occurrence 17:445
in ores 17:453
oxidation potentials 17:448
in pitchblend 17:463
radioactive decay 17:443
radon de-emanation method 17:460, 464,
    465
safety 17:449
sample preparation 17:453
separation 17:454
in sludges 17:464
in soil 17:453
toxicity 17:449
in urine 17:456, 468
uses 17:447
in water 17:456, 457; 19:394, 395
Radium-226
  in tobacco 19:153
Radium amalgam. (See also Radium metal.)
  manufacture 17:448
Radium bromide
  solubility 17:448
Radium carbonate
  solubility 17:448
Radium chloride
  solubility 17:448
Radium chromate
  solubility 17:448
Radium compounds. (See also individual
    compounds.)
  solubility 17:448
Radium fluoride
  solubility 17:448
Radium fluoroberyllate
  solubility 17:448
Radium hydroxide

solubility 17:448
Radium iodate
  properties 17:447
  solubility 17:448
Radium metal
  analysis 17:452
  manufacture 17:448, 450
  properties 17:447
Radium nitrate
  solubility 17:448
Radium oxalate
  solubility 17:448
Radium phosphate
  solubility 17:448
Radium selenate
  solubility 17:448
Radium sulfate
  solubility 17:448
  uses 17:447
Radium tellurate
  solubility 17:448
Radon. (See also Gases, noble.)
  commercial specifications 13:287
  de-emanation method for radium 17:460,
    464, 465
  half-life 17:450
  isotopic abundance 13:289
  manufacture 13:290
  occurrence 13:290
  properties 13:288
Raffinose
  in green coffee 10:393
  in green tea leaf 18:478
  paper chromatography 18:478
Ramachandran extractions
  of cocoa products 9:593
Raman spectroscopy
  of acetals 4:66, 71
  of acrylonitrile 4:372
  of benzene 7:36
  of carbon disulfide 8:272
  of elastomers, synthetic 12:109
  of epibromohydrin 14:176
  of epichlorohydrin 14:176
  of epiiodohydrin 14:176
  of polybutadiene 12:109
  of polyisoprene 12:108
  of styrene 18:294
Rammelsbergite
  composition 16:286
Ramsbottom method
  for carbon residue of lubricants 15:323
Raney nickel. (See also Nickel alloys.)
  manufacture 16:290
Ranking test 17:633
  analysis 17:634
  design 17:633

Rape oil. (See also *Fats and fatty oils.*)
  properties 12:473
Rape seed displacement
  of cakes 12:71
Rapid dyeing test
  for triacetate in acetate fibers 4:79
Rare earth alloys. (See also Rare earth metals.)
    17:470
Rare earth carbonates. (See also Rare earth
    compounds.) 17:508
  analysis 17:508
  assay 17:508
  carbonate content 17:508
  commercial specifications 17:498
  manufacture 17:508
  properties 17:508
  solubility 17:508
  uses 17:508
Rare earth chlorides. (See also Rare earth
    compounds.) 17:505
  analysis 17:506
  assay 17:506
  chloride content 17:506
  commercial specifications 17:498
  impurities, determination of 17:506
  manufacture 17:505
  properties 17:505
  rare earth content 17:506
  solubility 17:505
  uses 17:506
Rare earth compounds. (See also individual
    compounds.) 17:497–513
  analysis 17:499
  commercial grades and specifications 17:498
  emission spectroscopy 17:500
  fluorescence methods 17:504
  impurities, determination of 17:500
  properties 17:497
  solubility 17:498
  x-ray fluorescent spectroscopy 17:504
Rare earth fluorides. (See also Rare earth
    compounds.) 17:506
  analysis 17:506
  assay 17:507
  commercial specifications 17:498
  fluorine content 17:507
  impurities, determination of 17:507
  manufacture 17:506
  rare earth content 17:506
  solubility 17:506
  uses 17:506
Rare earth hydrates. (See also Rare earth
    compounds.)
  commercial specifications 17:498
Rare earth hydroxides. (See also Rare earth
    compounds; and individual compounds.)
    17:510

analysis 17:510
manufacture 17:510
properties 17:510
solubility 17:510
uses 17:510
Rare earth nitrates. (See also Rare earth compounds.) 17:511
analysis 17:511
manufacture 17:511
solubility 17:511
uses 17:511
Rare earth oxalates. (See also Rare earth compounds.) 17:510
analysis 17:511
commercial specifications 17:498
manufacture 17:510
oxalate content 17:511
solubility 17:511
uses 17:511
Rare earth oxides. (See also Rare earth compounds; and individual compounds.) 17:509
analysis 17:510
assay 17:510
commercial specifications 17:498
impurities, determination of 17:510
manufacture 17:509
properties 17:509
sample preparation 17:510
solubility 17:509
uses 17:509
Rare earth perchlorates
atomic absorption spectrum 12:401
Rare earths. (See also Rare earth alloys; Rare earth compounds; and individual elements.) 17:470–515
absorption spectroscopy 17:480
alizarin red S colorimetric method 19:19
ammonium fluoride gravimetric method 17:477
ammonium hydroxide gravimetric method 17:475
ammonium sebacate–oxalic acid gravimetric method 15:389
analysis 17:480
arsenazo colorimetric method 17:479
atomic absorption spectroscopy 17:491
in beryllium 17:473
carrier distillation methods 17:479
in cerium compounds 9:255
in cerium metal 9:248
colorimetric determination 17:479
DTPA titration 9:248; 17:478
EDTA titration 13:445; 17:478, 512
emission spectroscopy 9:256; 12:403; 14:123; 15:153, 155; 17:479, 487
europium content 12:398
in europium oxide 12:402

flame photometry 17:490
flame tests 6:589
fluorescence methods 15:158
fluoride precipitation 9:268
in gallium arsenide 13:278
in glass 9:200
in glass and glass-ceramics 13:445
gravimetric determination 17:472
in hafnium 14:123
hydrofluoric acid gravimetric method 17:476
hydroxide precipitation 9:266
ion exchange chromatography 17:472, 564; 18:268
lanthanum content 15:158
in lanthanum oxide 15:153, 155
in magnesium 17:473
in magnesium alloys 15:389
in Misch metal 9:248
in monazite 19:12
in nuclear reactor materials 17:473, 477
occurrence 17:470
oxalate precipitation 9:264
oxalic acid gravimetric method 17:472, 507
polarography 17:493
properties 17:471
in rare earth chlorides 17:506
in rare earth fluorides 17:506
in rare earth selenides 17:512
in rare earth sulfates 17:512
in rare earth sulfides 17:512
in rare earth tellurides 17:512
separation 9:264; 17:472
sodium hydroxide gravimetric method 17:475
solvent extraction 17:567; 19:19
in steel 17:477
thorium content 17:477, 495
in thorium nitrate 19:19
in titanium 17:477
in uranium 17:477
uses 17:470
volumetric determination 17:478
x-ray fluorescent spectroscopy 17:489
in zirconium 14:123; 17:477
Rare earth selenides. (See also Rare earth compounds.)
analysis 17:512
rare earth content 17:512
sample preparation 17:512
selenide content 17:512
Rare earths, heavy. See Yttrium earth group.
Rare earths, light. See Cerium earth group.
Rare earth sulfates. (See also Rare earth compounds.) 17:511
analysis 17:511
properties 17:511
rare earth content 17:512

sample preparation 17:511
solubility 17:511
sulfate content 17:512
uses 17:511
Rare earth sulfides. (See also Rare earth
  compounds.)
analysis 17:512
rare earth content 17:512
sample preparation 17:512
sulfide content 17:512
Rare earth tellurides. (See also Rare earth
  compounds.)
analysis 17:512
rare earth content 17:512
sample preparation 17:512
telluride content 17:512
Rare gases. See *Gases, noble.*
Raschig method
  gravimetric method for *m*-cresol 17:28
  for hydroxylamine and its salts 14:447
  for salicylaldehyde 6:603
Rasorite 7:314
Rating tests 17:638
Rauwolfia alkaloids 4:589
  analysis 4:603
Raw milk. (See also Milk.) 16:123–131
  analysis 16:123
  California mastitis test 16:125
  catalase content 16:124
  freezing point 16:124
  impurities, determination of 16:128
  lactose content 16:129
  mastitis tests 16:124
  milkfat content 16:127
  protein content 16:128
  sampling 16:123
  sediment content 16:127
  solids-not-fat content 16:128
  water content 16:124
  Whiteside method 16:125
  Wisconsin mastitis test 16:125
Raw sienna
  commercial specifications 17:193, 197
  impurities, determination of 17:194
Raw umber
  commercial specifications 17:193
  impurities, determination of 17:194
Rayon cake 18:244
Rayon fibers
  microscopy 4:264
  mixtures, determination in 4:353
  solubility 4:250, 353
RDX. (See also High explosives.) 12:440
  chromotropic acid colorimetric method 12:421,
    460
  colorimetric determination 12:421
  commercial specifications 12:441

in explosive priming mixtures 12:418, 421,
    422
  in HMX 12:441
  infrared spectrum 12:453
  manufacture 12:440
  melting point 12:461
  nitrogen content 12:440
  properties 12:432, 440
  solubility 12:421, 440
  uses 12:440
  x-ray diffraction 12:418
Reaction, heat of
  apparatus 7:24
  of benzene with nitrating acid 7:25
Reaction rate analysis
  of carbonyl compounds 8:364
Reactive dyes 6:402; 12:4
  acid cross staining test 12:20
  colorimetric determination 12:20
  dimethylformamide color test 12:19
  identification 12:19
  sodium hydroxide color test 12:19
  solvent test 12:19
Readily carbonizable substances. See Carbonizable
  substances.
Realgar 6:223
Recovery tests
  of brake linings 7:455, 468
Red 6B. (See also *Triphenylmethane and
  related dyes.*)
  paper chromatography 19:172
Red 10B. (See also *Triphenylmethane and
  related dyes.*)
  paper chromatography 19:172
Red brass 5:68
Red cake 19:325
  manufacture 19:325
Red FB. (See also *Triphenylmethane and
  related dyes.*)
  paper chromatography 19:172
Red 2G. (See also *Triphenylmethane and
  related dyes.*)
  paper chromatography 19:172
Red lead. (See also Lead tetroxide.) 17:196
  analysis 17:196
  assay 17:196
  commercial specifications 17:196
  impurities, determination of 17:196
  iodometric method 17:196
  lead content 17:196
  silicon content 18:29
  solubility 17:196
  uses 17:196
Red manganese oxide. See Mangano–manganic
  oxide.
Redon. (See also *Acrylic and modacrylic fibers.*)
  4:221, 249

Red organic pigments
in enamels 12:185
identification 12:185
Redox reactions
of chlorine 9:333
of chlorine compounds 9:333
Redox titration
of aniline 5:445
of chromium 9:640
of cinchona alkaloids 4:606
of copper in copper(I) chloride 10:660
of oxalates 7:617
of sodium bismuthate 7:207
Red phosphorus. See Phosphorus, red.
Red pigments. (See also *Pigments*; and
individual compounds.) 17:192
Reduced specific viscosity 4:303
Reducing power
of cuprous oxide 17:195
Reducing substances
in beer 7:618
2,6-dichlorindophenol indicator time test
7:618
in formic acid 13:131
in nitrous oxide 5:366
in pyridine 17:439
Reducing sugars. (See also *Sugar.*)
achroic R value 7:556
in beer 7:563
in cocoa beans 9:558
in confectionery products 10:562
copper sulfate–carbonate–tartrate colorimetric
method 9:559
cupric tartrate colorimetric method 10:543
in ferrous gluconate 10:543
in green coffee 10:374
Hassid method 9:559
iodometric titration 9:560; 10:428
Lane and Eynon method 9:239; 10:562;
11:40
Munson-Walker method 7:563
in roasted coffee 10:375, 428
Schoorl copper reduction method 11:41
Shaffer–Somogyi method 9:560
in starch hydrolyzates 11:40
in sugar 18:348
Reducing value
of corn starch 11:27
ferricyanide volumetric method 11:27
Reducing water 4:468, 479
Reduction methods
for azo dyes 6:404
identification test for dyes 6:390, 391
for nitrate 19:138
for thiophene 18:544
Reductodehydrocholic acid
color tests 7:147

properties 7:143
Red violet 5RS. (See also *Triphenylmethane
and related dyes.*)
visible spectroscopy 19:167
Refining loss
of fats and fatty oils 12:492
Reflectance
of acrylic and modacrylic polymers and fibers
4:321
Reflectance, absolute
of coal and coke 10:222
Reflectance standards
for brightness testing 7:260
magnesium oxide 7:260, 262
Refractaloy. (See also Cobalt alloys.)
composition 10:286
Refractive index
of acetate and triacetate fibers 4:77
of acrylamide polymers 4:177
of acrylic and modacrylic fibers 4:257
of acrylonitrile 4:377
of acrylonitrile polymers 4:390
of almond oil, bitter 12:271
of aniline and its derivatives 5:454
of anise oil 12:271
of antifreezes 6:8
of barium nitrate 12:417
of bay leaves, oil of 12:271
of benzene–acetone–water mixtures 7:38
of benzene–carbon tetrachloride–acetic acid
mixtures 7:37
of benzene–carbon tetrachloride mixtures
7:37
of bergamot oil 12:271
of bois de rose oil 12:271
of camphor oil, white 12:271
of cananaga oil 12:271
of caraway oil 12:271
of cassia oil 12:271
of cedarwood oil 12:271
of cinnamon oil, Ceylon 12:271
of citronella oil 12:271
of clove oil 12:271
of coal and coke 10:224
of cocoa beans 9:548
of cocoa butter 9:576
of copaiba oil 12:271
of coriander oil 12:271
of diamonds, industrial 11:471
of dipropylene glycol 13:584
of ethanol 19:432
of ethylene and propylene polymers 12:382
of eucalyptus oil 12:271
of fats and fatty oils 12:475
of fennel oil 12:271
of fructose 18:343
of geranium oil 12:271

of glass and glass-ceramics 12:417; 13:473, 475
of glucose 18:343
of glycerol 13:530
of glycols 13:584
of hexylene glycol 13:584
of hydrocarbon resins 14:294
of hydrochloric acid 14:361
of invert sugar 18:343
of lactose hydrate 18:343
of lavender oil 12:271
of lead nitrate 12:417
of lemongrass oil 12:271
of lemon oil 12:271
of linaloe oil 12:271
of mace oil 12:271
of maltose hydrate 18:343
of mercaptans 15:571
of mineral oils 12:285
of mustard oil 12:271
of musts 19:403
of orange oil, sweet 12:271
of organo-chlorosilanes 18:63
of origanum oil 12:271
of patchouly oil 12:271
of peppermint oil 12:271
of petitgrain oil 12:271
of pimenta oil 12:271
of polyamide fibers 17:314
of propylene glycol 13:584
of resins 4:23
of rosemary oil 12:271
of sandalwood oil 12:271
of SBR 12:137
of spearmint oil 12:271
of spike oil 12:271
of styrene 18:311
of sucrose 18:343
of sugar solutions 18:343
of sweet birch 12:271
of terpenes 14:284
of thyme oil, red 12:271
of triethylene glycol 13:584
of turpentine 12:271
of vetiver oil 12:271
of wines 19:403
of ylang-ylang oil 12:271
Refractometry
of crotonaldehyde 11:179
for proofing 4:483
Refractories. (See also *Ceramics*; and Chrome
ores.) 9:201, 700
analysis 9:700
classification 9:201
composition 9:206
raw material composition 9:202
raw material properties 9:204
silicon content 18:30

Refractory boron compounds. See Boron
compounds, refractory.
Refractory carbides. (See also *Carbides*; and
individual compounds.) 8:130
nitrogen content 16:467
Refractory materials
in brake linings 7:447
carrier gas fusion methods 8:626, 683
oxygen in 4:15; 8:626, 631, 633, 683
quality control tests 4:25
Refractory minerals
decomposition 19:8
lithium metaborate fusion 15:32
thorium content 19:8
Refractory ores
potassium hydroxide fusion 19:236
sodium peroxide fusion 19:236
Refrigerator oil. (See also *Lubricants*.)
commercial grades and specifications 15:310
Reichert-Meissel number
for milk fat in chocolate 9:580
Reid vapor pressure test
for gasoline 13:341
of hydrocarbons 14:342
Reimer-Tiemann synthesis
of salicylaldehyde 6:602
Reinecke's salt
analysis 9:693
assay 9:693
colorimetric method for antihistamines 6:33
colorimetric method for choline and its
derivatives 9:622
colorimetric method for mepivacaine hydrochloride 5:397
gravimetric method for choline base 9:619
gravimetric method for gold 14:19
gravimetric method for piperidine 9:620
gravimetric method for pyridine 9:620
uses 9:693
Reinforcement
for grinding wheels 7:294
Relative potency
in bioassay 7:175, 176
Relative viscosity 4:303
Relaxin. (See also Progestational hormones.)
composition 14:186
occurrence 14:186
uses 14:186
Remalloy. (See also Cobalt alloys.) 10:285
René 41. (See also Cobalt alloys.)
composition 10:286
Renierite
occurrence 13:376
Rennet casein. See *Casein*.
Reodorants
in embalming fluids 12:167
Rescinnamine 4:589

analysis 4:604
Reserpine 4:589
  analysis 4:603
Residue by distillation
  of emulsified asphalts 6:325
Residue on evaporation
  acetic acid 4:100
  of acid anhydrides 8:568
  of amides 8:609
  of benzene 7:47
  of bergamot oil 12:273
  of brake fluids 7:431
  of bromine 8:5
  of carbon disulfide 8:275
  of carbon tetrachloride 8:327
  of emulsified asphalts 6:326
  of essential oils 12:273
  of fluorosilicic acid 13:50
  of formic acid 13:131
  of grapefruit oil 12:273
  of hydrofluoric acid 13:34
  of lemon oil 12:273
  of lime oil 12:273
  of mandarin oil 12:273
  of orange oil 12:273
  of pyridine 17:438
  of tangerine oil 12:273
Residue on ignition
  of acid anhydrides 8:568
  of acid halides 8:575
  of amino acids 8:442
  of ammonium acetate 5:327
  of ammonium chloride 5:338
  of ammonium iodide 14:604
  of ammonium nitrate 5:344
  of ammonium oxalate 5:345
  of ammonium persulfate 5:347
  of ammonium sulfate 5:352
  of ammonium thiocyanate 5:354
  of amobarbital 6:508
  of antibiotics 5:492
  of aprobarbital 6:512
  of arsenic chloride 6:234
  of arsenic oxide 6:232
  of benzoic acid 7:66
  of benzoyl chloride 7:77
  of benzyl alcohol 5:379
  of bithionol 11:547
  of caprolactam 8:126
  of choline and its derivatives 9:628
  of citric acid 10:93
  of gelatin 13:369
  of glycerol 13:521
  of hexachlorophene 11:547
  of hydrofluoric acid 13:34
  of hydroxylamine salts 14:454
  in mephobarbital 6:518

  of mercuric chloride 16:60
  of mercuric iodide 16:60
  of mercuric oxide 16:61
  of mercury 16:24
  of naphthalene 16:211
  in phenobarbital 6:523
  in secobarbital 6:525
  in vinbarbital 6:529
Residue, total
  of water 19:381
Resifilm. See Thiram.
Resin cure
  of resinoid grinding wheels 7:299
Resinoid-bonded products
  diamond reclaiming from 11:482
  diamonds, industrial in 11:468
Resinoid grinding wheels
  analysis 7:298
Resins. (See also Resins, synthetic.)
  analysis 7:444, 448
  in beer 7:604
  in brake linings 7:441, 444, 448
  in chrome protectors 6:354
  in enamels 12:170
  ethylene glycol content 13:596
  gas chromatography 7:444
  glycerol content 13:596
  glycol content 13:596
  in grinding wheels 7:299
  in hair sprays 11:90
  inclined plate flow test 7:444
  infrared spectroscopy 7:444
  in-process controls 7:448
  pentaerythritol content 13:596
  polyhydric alcohol content 13:596
  quality control tests 4:23
  in silicone resins 18:122
  solvent extraction 19:151
  in tobacco 19:151
Resins, dental. See Dental resins.
Resins, synthetic
  carbazole colorimetric method 10:282
  ferric chloride–ammonium hydroxide colorimetric
    method 10:282
  gel time 10:274
  viscosity 10:272
  volatile content 10:275
  water tolerance 10:274
Resins, thermosetting
  in brake linings 7:441
Resistance to oxidation
  of brake fluids 7:432
Resistivity
  apparatus 8:234
  of carbon black 8:193, 233
  of silicone fluids 18:166
Resistox 6:76

Resmethrin. (See also *Pesticides and insecticides.*)
  19:586
Resorcinol. (See also Dihydroxybenzenes.)
  4-aminoantipyrine colorimetric method 17:18
  analysis 17:46
  assay 17:46
  bromometric titration 17:46
  colorimetric method for captan 13:210
  colorimetric method for dexon 13:221
  colorimetric method for diethyl malonate
    15:431
  colorimetric method for ethylene glycol 13:570
  colorimetric method for folpet 13:210
  colorimetric method for phthalate plasticizers
    16:96
  colorimetric method for propylene glycol
    13:570
  commercial grades and specifications 17:15
  derivatives for identification 17:21
  ferric chloride colorimetric method 17:16
  in green tea leaf 18:459
  in hair dyes 11:103
  identification of formals 4:68
  iodine–pyrocatechol colorimetric method
    17:47
  manufacture 17:1, 10
  mercurous nitrate colorimetric method 17:20
  Millon's reagent color test 17:18
  occurrence 17:1
  paper chromatography 17:24
  in pharmaceuticals 11:548
  phthalic anhydride colorimetric method 17:19
  potassium titanium oxalate colorimetric method
    18:459
  properties 17:5
  solubility 17:5
  thin-layer chromatography 11:103; 17:24
  ultraviolet spectroscopy 11:548
  use in manufacture of azo dyes 6:403
Resorcinol–fuming hydrochloric acid
  colorimetric method for caramel 10:538
Resorcinol monobenzoate
  in photographic films 12:592
Resorcinol resin adhesives, evaluation 4:402
Resorcinol–sulfuric acid
  colorimetric method for nitroparaffins 16:535
  colorimetric method for 2-nitropropane 16:436
β-Resorcylaldehyde. See 2,4-Dihydroxybenz-
    aldehyde.
β-Resorcylic acid. (See also 2,4-Dihydroxybenzoic
    acid.)
  paper chromatography 8:379
Retentivity test
  for gas-phase carbons 8:144
Reticulate structure
  of carbon black 8:187, 189, 225
Returned trim

in brake linings 7:447
Rex alloys. (See also Cobalt alloys.)
  composition 10:287
Rhamnose
  in green tea leaf 18:478
  paper chromatography 18:478
Rhenium. (See also Rhenium alloys; Rhenium
    compounds; and Rhenium metal.)
  atomic absorption spectroscopy 17:534
  in chromic acid 9:684
  colorimetric determination 17:531, 534
  in columbite 17:529
  coprecipitation 17:525
  2,4-diphenylthiosemicarbazide colorimetric
    method 17:541
  distillation methods 17:522
  emission spectroscopy 17:530
  flame tests 6:589, 592
  α-furildioxime colorimetric method 17:540
  gravimetric determination 17:524, 530, 532
  history 17:515
  hydrogen sulfide gravimetric method 17:524,
    526, 530, 533
  infrared spectroscopy 17:540
  ion exchange chromatography 17:526, 533,
    537
  in iron 17:523
  isotopes 17:516
  in molybdenite 17:523, 529, 535
  in molybdenum 17:529
  nitron gravimetric method 17:531
  occurrence 17:515
  potassium chloride colorimetric method 17:531
  potassium iodide colorimetric method 17:531
  in pyrolusite 17:523, 538
  sample decomposition 17:521
  separation 17:522
  in sodium dichromate 9:684
  solvent extraction 17:528, 536, 540
  stannous chloride–thiocyanate colorimetric
    method 17:531, 535
  in steel 17:523
  tellurium gravimetric method 17:541
  tetraphenylarsonium chloride gravimetric
    method 17:525, 533
  tetraphenylarsonium perchlorate gravimetric
    method 17:525
  ultraviolet spectroscopy 17:540
Rhenium alloys. (See also Rhenium metal; and
    individual alloys.)
  analysis 17:521
  uses 17:516
Rhenium and technetium. (See also Rhenium;
    Technetium.) 17:515–543
Rhenium bromides
  properties 17:519
Rhenium chlorides

properties 17:519
Rhenium compounds. (See also individual
    compounds.)
  toxicity 17:519
Rhenium fluorides
  properties 17:519
Rhenium halides
  properties 17:519
  safety 17:520
Rhenium heptoxide
  manufacture 17:520
  properties 17:520, 523
  safety 17:520
Rhenium iodides
  properties 17:519
Rhenium metal
  analysis 17:521
  carrier gas fusion methods 8:682
  commercial grades and specifications 17:521
  hydrogen in 8:616, 618, 681
  manufacture 17:518, 520
  nitrogen in 8:616, 618
  oxygen in 8:616, 618, 621, 629, 638, 639,
    681, 682
  properties 17:517
  sampling 17:521
  toxicity 17:519
  uses 17:516
  vacuum fusion methods 8:616, 618, 621, 681
  vacuum hot extraction methods 8:638
Rhenium oxides
  manufacture 17:518
  properties 17:518
Rheology
  of bituminous materials 6:290
Rheometer test
  for caulking compounds 9:19
  of pigments 17:207
  temperature susceptibility 6:304
Rheometer torque curves
  of brake linings 7:448
Rhodamine. (See also *Triphenylmethane and
    related dyes.*)
  titanous chloride titration 19:177
Rhodamine B. (See also *Triphenylmethane and
    related dyes.*)
  colorimetric method for antimony 6:68, 225,
    231; 7:193; 10:294, 619; 11:141, 144; 15:
    197; 16:323
  colorimetric method for antimony trioxide
    13:431
  colorimetric method for gallium 13:256, 262
  colorimetric method for gold 12:570
  colorimetric method for tantalum 16:370;
    18:449
  colorimetric method for thallium 18:526, 529
  colorimetric method for uranium 19:242, 247

spectroscopic method for gold 14:11
structure 19:161
thin-layer chromatography 19:173
titanous chloride titration 19:181
Rhodamine B–stearate. See D & C Red No. 37.
Rhodamine C
  colorimetric method for tantalum 16:370
Rhodamine D. See D & C Red No. 19.
Rhodamine 6G. (See also *Pigments*; *Triphenyl-
    methane and related dyes.*)
  colorimetric method for tantalum 16:369;
    18:449
  structure 17:182
  thin-layer chromatography 19:173
Rhodamine lakes
  in enamels 12:183, 185
  identification 12:183, 185
Rhodamine S. (See also *Triphenylmethane
    and related dyes.*)
  thin-layer chromatography 19:173
Rhodamine WT. (See also *Triphenylmethane and
    related dyes.*)
  polarography 19:182
  voltammetry 19:182
Rhodanates. See Thiocyanates.
Rhodinyl acetate
  in essential oils 12:276
  molecular weight 12:276
Rhodium. (See also *Platinum group metals*;
    Rhodium compounds; and Rhodium metal.)
  bead test 6:588
  colorimetric determination 17:248, 262, 265
  detection 17:248
  emission spectroscopy 14:48; 17:246, 270
  flame photometry 17:247
  flame tests 6:589
  in gold 14:45
  gravimetric determination 17:238, 242
  history 17:215
  hydrochloric acid colorimetric method 17:248
  hydrogen sulfide gravimetric method 17:251
  identification 17:248
  isotopes 17:215
  2-mercapto-4,5-dimethylthiazole colorimetric
    method 17:265
  occurrence 17:214
  in ores 17:266
  oxidation numbers 17:220
  oxidation potentials 17:221
  paper chromatography 14:18
  separation 17:253, 261, 265
  sodium hypochlorite colorimetric method
    17:248
  sodium sulfide gravimetric method 17:253
  tin bromide colorimetric method 17:253
  tin chloride colorimetric method 17:248,
    262

tin chloride–potassium iodide colorimetric
    method 17:248
uses 17:217
Rhodium alloys. See Rhodium metal.
Rhodium chloride
    manufacture 17:225
Rhodium compounds
    manufacture 17:224
Rhodium hydrosulfide
    manufacture 17:225
    solubility 17:225
Rhodium hydroxide
    manufacture 17:224
    properties 17:224
Rhodium–iridium alloy
    uses 17:218
Rhodium metal
    carrier gas fusion methods 8:682
    hydrogen in 8:680
    oxygen in 8:680, 682
    properties 17:219, 220, 224
    uses 17:218
    vacuum fusion methods 8:680
Rhodium oxide
    manufacture 17:224
Rhodium sulfate
    manufacture 17:225
    properties 17:225
    uses 17:273
Rhodium trifluoride
    manufacture 17:225
    solubility 17:225
Rhodizite 7:315
Rhodochrosite 15:448
Rhodonite 9:278; 14:519; 15:448
Rhombic sulfur. (See also Sulfur, elemental.)
    18:364
Rhovyl fiber
    solubility 4:248
Ribitol
    liquid chromatography 13:503
Riboflavin
    assay 10:540
    in bakery products 6:455, 457
    in cereals 9:233, 245
    in chocolate and cocoa products 9:538
    in cocoa beans 9:540
    commercial specifications 10:532
    in eggs and egg products 12:66
    fluorometry 6:457
    impurities, determination of 10:544
    standard for bioassay of malt 7:159
    structure 10:539
    visible spectroscopy 10:540
Ribose
    in green tea leaf 18:478
    paper chromatography 18:478

Rice
    composition 9:233
    mercury content 16:40
    mineral content 9:233
    vitamin content 9:233
Rice vacuum method
    specific gravity of bituminous paving mixtures
    6:336
Ricinoleic acid. (See also Fatty acids.)
    properties 12:473
Ricinoleyl alcohol
    thin-layer chromatography 12:204
Rideal-Stewart titration
    of oxygen 16:526
Ridgway scale of hardness 4:32
Rifamycin 5:412, 495
    analysis 5:500, 505, 509
    bioautography 5:500
    colorimetric methods 5:509
    identification 5:500
    microbiological assay 5:505
    paper chromatography 5:500
    paper electrophoresis 5:500
    properties 5:493
    separation 5:500
    solubility 5:493, 494
    structure 5:462
    thin-layer chromatography 5:500
    uses 5:495
Rifamycin B 5:509
Rifamycin SV 5:505
    solubility 5:494
    structure 5:462
Ring and ball method
    apparatus 6:297
    for softening point of bituminous materials
    6:297
Rio Resin 6:76
Ristocetin 5:477, 599
    analysis 5:605, 610, 614
    column chromatography 5:605
    composition 5:477
    identification 5:605
    microbiological assay 5:610
    properties 5:600
    separation 5:605
    solubility 5:601, 602
    ultraviolet spectroscopy 5:614
    uses 5:599
RMDI
    gas chromatography 19:315
    in polyurethanes 19:315
Roasted coffee 10:375
    analysis 10:414
    aroma 10:436
    ash content 10:423
    caffeine content 10:409, 431

carbohydrate content 10:425
chlorine content 10:424
color 10:420
composition 10:375, 417
crude fiber content 10:425
impurities, determination of 10:434
lipid content 10:430
macroscopic examination 10:418
niacin content 10:434
organic acid content 10:430
particle size distribution 10:420
phosphorus content 10:424
reducing sugar content 10:428
sampling 10:416
soluble matter in 10:421
starch content 10:429
sucrose content 10:429
volume increase 10:419
water content 10:421
yield of roasting 10:414
Roberts and Smith method
for theaflavin 18:498
for thearubigens 18:498
Roboc 6:44, 76
Rock gypsum 14:88
Rocks
boron in 7:416
cobalt in 10:369
fluoride in 13:29
mercury in 16:44
niobium in 16:365, 376, 378
rubidium in 17:551
sulfur in 18:380, 385
thallium in 18:522, 524
thorium in 19:7, 14, 16
tin in 19:64
titanium in 19:96
Rock salt. See Halite; Sodium chloride.
Roese-Gottlieb test
for milkfat in milk 16:132
Roga index
of coal and coke 10:246
Rohan and Stewart method
for organic acids in cocoa beans 9:566
Roller analyzer
particle size distribution of adsorbents 4:442
Roll sulfur. (See also Sulfur, elemental.) 18:365
Roofing shingles, asbestos–cement 6:270
p-Rosaniline hydrochloride
colorimetric method for sulfite 18:405
colorimetric method for sulfur dioxide 7:644;
18:408
Roscoelite
composition 19:321
occurrence 19:321
Rosemary oil. (See also *Essential oils*.)
refractive index 12:271

Rosen method
color test for Carbowax 13:573
color test for diethylene glycol 13:573
color test for dipropylene glycol 13:573
color test for ethylene glycol 13:573
color test for polyethylene glycols 13:573
color test for polyglycols 13:573
color test for propylene glycol 13:573
color test for propylene oxide–ethylene oxide
polymers 13:573
Rosenmund and Kuhnhenn method
for allyl alcohol 5:82
for allyl compounds 5:82
for allyl esters 5:82, 96
for allyl ethers 5:99
for methallyl chloride 5:102
for olefins 5:82
Rosin
acid number 8:393
in alkyd resins 5:38
analysis 15:529
commercial grades and specifications 15:511
in fatty acids 18:186
manufacture 15:511
in matches 15:511
modifier for alkyd resins 5:24
2-naphthalenesulfonic acid volumetric method
18:186
particle size 15:529
solubility 13:539, 552
Rosin acids
in alkyd resins 5:38
Rosin color scale 14:294
p-Rosolic acid. See Aurin.
Rotating spindle method
for viscosity of epoxy resins 12:251
Roweite 7:315
Royalene 12:87
R salt. (See also 3-Hydroxy-2,7-naphthalenedi-
sulfonic acid.)
analysis 16:237
bromometric titration 16:234, 237, 239
colorimetric method for arylamines 6:408
commercial grades and specifications 16:234
ferric chloride colorimetric method 16:235
in G salt 16:239
impurities, determination of 16:237
manufacture 16:233
naphtholsulfonic acid content 16:238
in Schaeffer's salt 16:234
specification tests 16:238
RTV silicones. See Dental impression materials,
silicone-based.
Rubber. (See also *Elastomers, synthetic;*
Rubber, natural.)
analysis 7:447
aniline point, relation of 15:320

in brake linings 7:441, 447
carbon black in 8:197, 238
extraction of antidegradants from 6:108
in grinding wheels 7:299
properties 18:127
pyrolysis 8:238
solubility 13:539, 552
sulfur content 18:371, 374
Rubber-based adhesives, evaluation 4:403
Rubber, chlorinated
in alkyd resins 5:24
Rubber grinding wheels
analysis 7:298
Rubber-modified polystyrene. See impact
polystyrene.
Rubber, natural. (See also Rubber.)
bromination methods 12:122
castor oil in 12:145
effect of brake fluids on 7:429
gel permeation chromatography 12:119
identification 12:90
infrared spectroscopy 12:92, 98
intrinsic viscosity 12:119
manganese content 15:472
molecular weight distribution 12:119
pyrolysis–gas chromatography 12:102, 104
thermal analysis, differential 12:133
types 12:85
unsaturation, determination of 12:122
Rubber, synthetic. See *Elastomers, synthetic*;
Rubber.
Rubbing compounds 6:353
Rubeanic acid
colorimetric method for cobalt 11:140, 144
colorimetric method for copper 11:140, 144
Rubidium. (See also Rubidium alloys; Rubidium
compounds; and Rubidium metal.)
in alkali salts 15:267
ammonium sulfate gravimetric method 17:554
atomic absorption spectroscopy 17:552
bead test 6:588
in cements 9:131
in cesium 9:282, 284
chloroplatinic acid gravimetric method 17:553
detection 17:548
emission spectroscopy 17:550
flame emission spectroscopy 13:452; 17:551
flame photometry 9:284; 17:548, 555
flame tests 6:592
gravimetric determination 17:549, 553
half-life 17:545
history 17:543
iodometric method 17:555
ion exchange chromatography 17:366, 550
isotopes 17:544
occurrence 17:543
perchloric acid gravimetric method 15:267

9-phosphomolybdic acid gravimetric method
17:553
in rocks 17:551
safety 17:546
separation 17:549
silicomolybdic acid gravimetric method 17:549
sodium bismuth nitrite gravimetric method
17:549
in sodium dichromate 9:684
in water 18:221
x-ray spectroscopy 17:550
Rubidium alloys. (See also Rubidium metal.)
17:545
Rubidium alum. (See also Rubidium compounds.)
17:558
manufacture 17:558
solubility 17:558
uses 17:558
Rubidium aluminum sulfate. See Rubidium alum.
Rubidium and rubidium compounds. (See also
Rubidium; Rubidium compounds.) 17:543–
559
Rubidium bromide. (See also Rubidium compounds.) 17:557
manufacture 17:557
properties 17:557
solubility 17:557
Rubidium carbonate. (See also Rubidium
compounds.) 17:557
manufacture 17:547, 557
properties 17:557
solubility 17:557
uses 17:557
Rubidium chloride. (See also Rubidium compounds.) 17:558
manufacture 17:547, 558
properties 9:335; 17:558
solubility 9:336
uses 17:558
Rubidium chlorostannate. (See also Rubidium
compounds.)
manufacture 17:547
Rubidium compounds. (See also individual
compounds.) 17:557
analysis 17:555, 558
impurities, determination of 17:555, 558
sample preparation 17:555
sampling 17:548
solubility 17:546
toxicity 17:546
uses 17:545
Rubidium cyanide. (See also Cyanides.) 14:424
properties 14:424
uses 14:424
Rubidium fluoride. (See also Rubidium compounds.) 17:558
manufacture 17:558

properties 13:11; 17:558
solubility 17:558
Rubidium hydroxide. (See also Rubidium
    compounds.) 17:558
  commercial grades 17:558
  manufacture 17:546, 558
  properties 17:558
  safety 17:546
  uses 17:558
Rubidium iodide. (See also Rubidium compounds.)
    17:558
  manufacture 17:558
  properties 17:558
  uses 17:558
Rubidium metal
  analysis 17:548, 555
  commercial grades and specifications 17:548
  impurities, determination of 17:555
  manufacture 17:543, 546
  properties 17:545
  sample preparation 17:556
  sampling 17:548
  uses 17:545
Rubidium nitrate. (See also Rubidium com-
    pounds.) 17:558
  manufacture 17:558
  properties 17:558
  solubility 17:558
  uses 17:558
Rubidium oxide
  in alkali barium silicate glass 9:218
  emission spectroscopy 9:218
  in glass and glass-ceramics 13:415, 452
Rubidium sulfate. (See also Rubidium com-
    pounds.) 17:558
  manufacture 17:558
  properties 17:558
Rue oil. (See also *Essential oils.*)
  methyl nonyl ketone content 12:282
Rum 4:466
  analysis 4:480
  congeneric content 4:467
  quality control 4:470
Rupp method
  for mercury 16:33
Rust-prevention tests
  for turbine oil 15:335
Ruthenates
  manufacture 17:224
Ruthenium. (See also *Platinum group metals*;
    Ruthenium alloys; Ruthenium compounds;
    and Ruthenium metal.)
  in catalysts 17:272
  colorimetric determination 17:247, 263
  detection 17:247
  1,4-diphenylthiosemicarbazide colorimetric
    method 17:248

emission spectroscopy 14:48; 17:246
flame photometry 17:247
flame tests 6:589
in gold 14:45
gravimetric determination 17:237, 247, 256,
    264
history 17:215
hydrochloric acid colorimetric method 17:247
hydrogen sulfide gravimetric method 17:251
identification 17:247
isotopes 17:215
occurrence 17:214
oxidation numbers 17:220
oxidation potentials 17:221
paper chromatography 14:18
potassium thiocyanate colorimetric method
    17:247
ruthenium tetroxide volatilization 17:252,
    254, 263
separation 17:252, 254, 263
sodium hydroxide colorimetric method 17:247
sodium sulfide gravimetric method 17:256
thionalide gravimetric method 17:256, 264
thiourea colorimetric method 17:247, 263
uses 17:217
Ruthenium alloys. (See also Ruthenium metal.)
  uses 17:218
Ruthenium chloride
  manufacture 17:223
Ruthenium compounds. (See also individual
    compounds.)
  manufacture 17:223
Ruthenium disulfide
  manufacture 17:224
Ruthenium hydroxide
  manufacture 17:223
Ruthenium metal
  carrier gas fusion methods 8:682
  hydrogen in 8:680
  oxygen in 8:680, 682
  properties 17:219, 220, 223
  solubility 17:244
  uses 17:218
  vacuum fusion methods 8:680
Ruthenium oxide, hydrous
  manufacture 17:223
Ruthenium tetrachloride
  manufacture 17:224
Ruthenium tetroxide
  manufacture 17:223, 224
  properties 17:224
  toxicity 17:224, 228
  volatilization method for ruthenium 17:252,
    254
Ruthenium tribromide
  manufacture 17:224
Ruthenium trichloride

manufacture 17:224
Ruthenium triiodide
   manufacture 17:224
Rutile. (See also Titanium dioxide.) 19:81,
   123
   composition 10:104
   occurrence 19:81
   titanium content 19:104
Rutin
   Arnow's colorimetric method 19:146
   gas chromatography 19:146
   paper chromatography 19:146

solvent extraction 19:146
   in tobacco 19:146
Rye
   composition 9:233
   mineral content 9:233
   vitamin content 9:233
Rye, for distilled alcoholic beverages analysis
   4:469
   quality control 4:467
Rye whisky 4:464
   analysis 4:480
   quality control 4:470

S

Saare and Martens method
  for gel strength of corn starch 11:27
Sabinene
  gas chromatography 12:268
  in majoram, oil of 12:268
Saccharide content
  of starch hydrolysates 11:39
Saccharimeters 18:344
Saccharin
  in carbonated beverages 8:173
Safety precautions
  for explosives 12:405
Safety tests
  for antibiotics 5:493
  for penicillins 16:566
Safex. (See also Dithiocarbamate esters.)
  properties 11:589
  structure 11:589
  uses 11:589
Safflorite 10:284
Safflower oil
  in alkyd resins 5:23
Saffron
  identification 10:537
Saffron tagetes
  commercial specifications 10:532
  composition 10:532
Safranine T 6:381, 382, 383, 384
  application method 6:398
  manufacture 6:384
  visible spectroscopy 6:394
  volumetric method using titanous chloride
    6:395
Safrole
  analysis 5:91
  boiling point 5:90
  congealing point 12:286
  in manufacture of heliotropin 6:601
  in manufacture of vanillin 6:603, 604
  melting point 5:90
  ultraviolet spectroscopy 5:91
Sage oil, Dalmatian. (See also Essential oils.)
  thujone content 12:282
Sal ammoniac. See Ammonium chloride.
Salicylaldehyde (2-Hydroxybenzaldehyde) 6:597
  colorimetric method for cyclohexylamine
    11:227
  colorimetric method for deoxycholic acid
    7:147, 152
  colorimetric method for monoalkanolamines
    5:18

column chromatography 6:618
commercial specifications 6:610
derivatives for identification 6:612
fluorometry 6:626
hydroxylamine hydrochloride method 8:357
infrared spectroscopy 6:615, 617
manufacture 6:602
near-infrared spectroscopy 6:629
properties 6:598
ultraviolet spectroscopy 6:632
uses 6:606; 8:332
Salicylamide. (See also Salicylic acid derivatives.)
  6:124
  infrared spectrum 6:129
  ion exchange chromatography 6:182
  manufacture 6:126
  properties 6:124
Salicylanilides. (See also individual compounds.)
  analysis 11:551
  in disinfectants and antiseptics 11:550
  ion exchange chromatography 11:552
  nuclear magnetic resonance spectroscopy
    11:551
  thin-layer chromatography 11:552
  ultraviolet spectroscopy 11:552
Salicylate E 6:76
Salicylates
  ferric chloride colorimetric method 16:95
Salicylhydroxamic acid
  gravimetric method for niobium 16:367
  gravimetric method for tantalum 16:367
Salicylic acid. (See also Salicylic acid derivatives.)
  8:400
  in acetylsalicylic acid 6:138
  column chromatography 6:139, 150, 154
  commercial grades and specifications 6:126
  ferric ammonium sulfate colorimetric method
    10:166
  ferric chloride colorimetric method 19:484
  gas chromatography 8:388
  impurities, determination of 6:143
  ion exchange chromatography 6:144
  manufacture 6:124
  paper chromatography 8:400
  in phenprocoumon 10:166
  properties 6:124; 8:402
  solvent extraction 19:484
  ultraviolet spectroscopy 6:140, 144, 150,
    156, 159
  in wines 19:484
Salicylic acid–ammonium salicylate

gravimetric method for thorium 19:95
gravimetric method for zirconium 19:95
Salicylic acid derivatives. (See also Acetyl-
    salicylic acid; Salicylamide; Sodium
    salicylate.) 6:124
  analysis 6:127
  colorimetric method for boric acid 7:388
  commercial grades and specifications 6:126
  manufacture 6:124
  properties 6:124
N-Salicylidene-2-amino-3-hydroxyfluorene
  fluorometric method for aluminum 5:237, 241
Salicylidene-o-aminophenol
  fluorometric method for aluminum 5:237
Salicylideneamino-2-thiophenol
  colorimetric method for tin 19:64
Salicylsalicylic acid
  ion exchange chromatography 6:144
  in salicylic acid 6:143, 146
  thin-layer chromatography 6:146
  ultraviolet spectroscopy 6:146
Salmonella
  biochemical tests 9:603
  in cocoa products 9:601
  serological identification 9:605
Salol. See Phenyl salicylate.
Salt cake. (See also Sodium sulfate.) 18:244, 422
Salt, common. (See also Sodium chloride.)
  in bread 6:440
  in butter 16:138
  in cakes 6:446
  in cheese 16:142
  in chocolate 9:537
  iodide content 14:601
  in margarine 15:502
  silver nitrate titration 15:502
  Volhard titration 16:138, 142
Salting-out test
  for azo dyes 6:415
Salts, acid. See Acid salts.
Salt solutions
  caramel content 10:544
Salt test
  for triphenylmethane and related dyes 19:187
Samarium. (See also *Rare earths*; Samarium
    metal.)
  absorption spectroscopy 17:483
  in aluminum and aluminum alloys 5:152
  atomic absorption spectroscopy 17:494
  bead test 6:587
  emission spectroscopy 14:123; 15:153, 155;
    17:488, 501
  in europium oxide 17:502
  flame photometry 17:492
  in gadolinium oxide 17:502
  in hafnium 14:123
  in lanthanum oxide 15:153, 155; 17:501

in neodymium oxide 17:501
  polarography 17:493
  in praseodymium oxide 17:501
  x-ray fluorescence spectroscopy 17:490
  in ytterbium oxide 17:503
  in yttrium oxide 17:501
  in zirconium 14:123
Samarium fluoride
  carrier gas distillation fusion methods 8:636
  oxygen in 8:635, 636
  vacuum distillation fusion methods 8:635
Samarium metal. (See also Rare earth metals.)
  carrier gas fusion methods 8:627, 682
  hydrogen in 8:680
  oxygen in 8:623, 627, 680, 682
  properties 17:471
  uses 17:470
  vacuum fusion methods 8:623, 680
Samarium oxide. (See also Rare earth oxides.)
  emission spectroscopy 17:502
  impurities, determination of 17:502
  properties 17:497
  uses 17:509
Samarium perchlorate. (See also Rare earth
    compounds.)
  absorption spectrum 17:482
Samarskite 14:519
  composition 16:349
  decomposition 19:235
Sanbornite 6:533
SAN copolymers. See Styrene–acrylonitrile
    copolymers.
Sand
  commercial grades and specifications 15:513
  iron content 15:53
  in matches 15:513
  in match strikers 15:522
  particle size 15:529
  in set plaster 14:100
Sandalwood oil. (See also *Essential oils.*)
  refractive index 12:271
Sandstone 4:31, 33
  silicon content 18:27
Sanitizers
  definition 11:538
Santoflex 6:13, 76, 113
  analysis 6:114
  commercial grades and specifications 6:113
  properties 6:113
Santoflex AW 6:76, 111
  analysis 6:112
  commercial specifications 6:112
  gas chromatography 6:81
  properties 6:112
Santoflex B (DD, DPA, GP, 17, 36, 66, and 217)
    6:76
Santoflex BX (HP and 75) 6:76

colorimetric method based on formation of
  Wurster salts 6:100
Santomerse S
  colorimetric method for tin 10:304
Santonox 6:76
  ultraviolet spectrum 6:87
Santonox R 6:76
  column chromatography 6:87
Santovar A (and O) 6:76
  colorimetric method using diazotized *p*-
    nitroaniline 6:106
Santowhite C1 6:76
  colorimetric method based on formation of
    Wurster salts 6:101, 103
Santowhite CM (L, 54, and 54-S) 6:76
Santowhite Crystals (and Santowhite MK) 6:76
  colorimetric method using diazotized *p*-
    nitroaniline 6:106
Santowhite Powder 6:76
  colorimetric method using diazotized *p*-
    nitroaniline 6:106
  column chromatography 6:87
  ultraviolet spectrum 6:87
Saponification methods
  for acetyl content of starch derivatives 11:29
  for acrylates and methacrylates 4:196
  for acrylic ester content of polymers 4:276
  for acrylic ester monomers in polymers 4:216
  for adipic acid esters 4:426
  for allyl esters 5:94
  for amides 8:606
  for benzoic acid esters 7:84
  for benzyl cinnamate 10:67
  for cellulose acetate 9:70
  for cellulose acetate butyrate 9:74
  for cellulose acetate propionate 9:74
  for cinnamyl acetate 10:83
  for cinnamyl anthranilate 10:83
  for cinnamyl butyrate 10:83
  for cinnamyl formate 10:83
  for cinnamyl isobutyrate 10:83
  for cinnamyl isovalerate 10:83
  for cinnamyl propionate 10:83
  for citrate esters 10:98
  for diethylethoxymethylene malonate 15:438
  for diethylethylphenyl malonate 15:435
  for diethyl malonate 15:433
  for dimethyl malonate 15:433
  for dimethylmethoxymethylene malonate
    15:438
  for esters 4:484; 8:585
  for esters in essential oils 12:275
  for ethyl cinnamate 10:67
  for ethyl formate 13:132
  for formic acid esters 13:132
  for hydrocinnamyl acetate 10:83
  for hydrocinnamyl cinnamate 10:68, 83

  for hydrocinnamyl formate 10:83
  for isobutyl cinnamate 10:67
  for malonates, substituted 15:435
  for methyl cinnamate 10:67
  for methyl formate 13:132
  for neocinchophen 6:204
  for pentyl formate 13:132
  for phenethyl cinnamate 10:68
  for phthalate plasticizers 16:97
Saponification value
  of dimethyl terephthalate 17:165
  of fats and fatty oils 12:479
  of fatty acids 18:184
  of fatty alcohols 4:575
  of hydrocarbon resins 14:284
  of lubricants 15:321
Saponins
  infrared spectroscopy 12:577
Saponite. (See also *Clays and clay minerals.*)
  composition 10:102
Sapphirine 7:316
Saran
  infrared spectra 4:243
  microscopy 4:264
  solubility 4:250
Sarkomycin 5:463, 509
  analysis 5:510, 514, 517
  colorimetric method 5:517
  microbiological assay 5:514
  properties 5:509
  solubility 5:509
  structure 5:463
  uses 5:509
Sassafras oil
  solubility 13:540
Sassolite 7:315
Satin spar 14:88
Sauconite. (See also *Clays and clay minerals.*)
  composition 10:102
Saybolt color values 14:343
Saybolt Furol viscosity
  of bituminous materials 6:295
  of emulsified asphalts 6:326
  temperature susceptibility 6:304
Saybolt viscometer 6:294
Sborgite 7:314
SBR. (See also *Elastomers, synthetic;*
  Styrene–butadiene rubber.)
  bound rubber content 12:132
  colorimetric determination 12:91
  commercial grades and specifications 12:83,
    88
  composition 12:83
  effect of brake fluids on 7:429
  gel content 12:132
  gel permeation chromatography 12:119
  glass transition temperature 12:133

infrared spectroscopy 12:93, 96
intrinsic viscosity 12:119
molecular weight distribution 12:119
refractive index 12:137
spectroscopic method 12:138
styrene content 12:137
Scandates. (See also *Scandium*; Scandium
　　compounds.)
manufacture 17:562
properties 17:562
Scandium. (See also Scandium alloys; Scandium
　　compounds; and Scandium metal.) 17:560–
　　580
alizarin red S colorimetric method 17:571
alizarin sulfonic acid colorimetric method
　　17:574
in aluminum 17:577
in aluminum metal 5:141
ammonium tartrate gravimetric method 17:570
ammonium yttrium tartrate gravimetric method
　　17:568, 572
analysis 17:563
arsenazo colorimetric method 17:574
atomic absorption spectroscopy 15:390; 17:494,
　　577
benzilic acid gravimetric method 17:570
in cassiterite 17:577
in coal ash 17:574, 577
colorimetric determination 17:571
coprecipitation 17:568
detection 17:568
EDTA titration 13:445; 17:571
emission spectroscopy 17:568, 577
flame photometry 17:493
in glass and glass-ceramics 13:445
gravimetric determination 17:569
half-life 17:561
history 17:560
8-hydroxyquinoline gravimetric method 17:570
ion exchange chromatography 17:564, 575
isotopes 17:561
in magnesium 15:390
in magnesium alloys 15:390
neutron activation analysis 13:265; 17:577
occurrence 17:560
paper chromatography 17:568
properties 17:560
sample decomposition 17:564
sampling 17:564
separation 17:564
in silicate rocks 17:575, 577
sodium fluorosilicate gravimetric method 17:564
solvent extraction 17:566, 572, 574
in thortveitite 17:564, 569
toxicity 17:563
in uranium 19:240
in uranium compounds 17:574

volumetric determination 17:571
in wolframite 17:564
x-ray fluorescence 17:577
Scandium alloys. (See also Scandium metal.)
uses 17:561
Scandium compounds. See individual compounds.
Scandium fluoride
carrier gas distillation fusion methods 8:636
oxygen in 8:636
solubility 17:562
Scandium hydroxide
manufacture 17:562
properties 17:562
solubility 17:562
Scandium metal
carrier gas fusion methods 8:627, 682
commercial grades and specifications 17:563
hot extraction methods 8:684
hydrogen in 8:623, 638, 680, 684
manufacture 17:563
nitrogen in 8:623, 627
oxygen in 8:623, 627, 680, 682
properties 17:561
uses 17:561
vacuum fusion methods 8:623, 680
vacuum hot extraction methods 8:638
Scandium nitrate
colorimetric method for DTPA 12:28
colorimetric method for TTHA 12:28
Scandium oxalate
solubility 17:562
Scandium oxide
manufacture 17:562
properties 17:562
Scapolite 8:75
Scarlet 2R lake
in enamels 12:182, 185
identification 12:182, 185
Schaeffer's acid. See 2-Naphthol-6-sulfonic acid.
Schaeffer's salt
analysis 16:234
assay 16:234
bromometric titration 16:234, 237, 239
commercial grades and specifications 16:233
in G salt 16:239
impurities, determination of 16:234
in R salt 16:237
sodium nitrite colorimetric method 16:237
Scharer method
for pasteurization efficiency of milk 16:133
Scheelite 19:190
decomposition 19:197, 203
tungsten content 19:195, 203
Scheel's Green. See Copper arsenite.
Schiff base
preparation from benzaldehyde 6:595
Schiff's reagent

colorimetric method for acetals 4:70
colorimetric method for aldehydes 4:375, 484;
8:344
colorimetric method for formaldehyde 13:106
Schmall extractor 6:168
Schöniger method
for bromine 8:22
for sulfur content of acrylic polymers 4:280
Schoorl copper reduction
of reducing sugars 11:41
Schopper folding endurance tester 7:287
Schorlemmer method
for basicity of basic chromium salts 9:691
for basicity of chrome tanning compounds
9:691
Schultz-Blaschke equation
for intrinsic viscosity 17:343
Schütze reagent 8:640, 661, 673
Schütze-Unterzacher method
for oxygen in organic compounds 16:533
Schweizer's reagent
gravimetric method for tragacanth 12:582
Scopolamine 4:587
analysis 4:596
bioassay 7:181
gas chromatography 6:210, 211
Scopoletin. See 6-Methoxy-7-hydroxycoumarin.
gas chromatography 19:146
paper chromatography 19:146
solvent extraction 19:146
in tobacco 19:146
Scopolin
gas chromatography 19:146
paper chromatography 19:146
solvent extraction 19:146
in tobacco 19:146
Scoring-grading scales 17:639
Scotch whisky 4:465
analysis 4:480
congeneric content 4:467
quality control 4:470
Scouring soaps. (See also *Soaps.*)
composition 18:180
uses 18:180
Sealers
analysis 6:361
for cooling systems 6:352, 361, 375
performance tests 6:375
Seamanite 7:315
Searlesite 7:316
Seawater
nitrate content 16:505
Seawater product
composition 9:203
Sebacic acid 5:22
in alkyd resins 5:37
gas chromatography 5:25; 17:290; 19:294

thin-layer chromatography 17:292
Secobarbital 5:91; 6:468, 524
acid–base titration, nonaqueous 6:524
analysis 6:524
bromate–bromide titration 6:495, 497
commercial grades and specifications 6:524
derivatives for identification 6:477
gas chromatography 6:210, 211, 480, 481
impurities, determination of 6:525
infrared spectroscopy 6:496
paper chromatography 6:478, 479
properties 6:470
in sodium secobarbital 6:525
solubility 6:476
thin-layer chromatography 6:480, 481, 482
ultraviolet spectroscopy 6:485, 501
Sectioning techniques
for microscopy of fibers 4:265
Sediment
in beer 7:529, 532
mercury content 16:45, 46
in raw milk 16:126
in sugar 8:176
tin content 19:64
Sedimentation methods
for abrasive grains 7:303
for acrylamide polymers 4:174
for acrylic polymers 4:314
for adsorbents 4:442
for aerosols 4:459
apparatus 7:304
Sedimentation velocity
of polycarbonates 17:344
Seeds
dexon content 13:220
oil content 12:491
Segal test
for hydrochloric acid in gastric juices 14:367
Selenic acid. (See also Selenium compounds.)
properties 17:585
Selenide. (See also Selenium; Selenium compounds.)
in rare earth selenides 17:512
sulfur dioxide gravimetric method 17:512
Selenious acid. (See also Selenium compounds.)
gravimetric method for niobium 18:446
gravimetric method for tantalum 18:446
hydrogen sulfide gravimetric method 17:591
manufacture 17:584, 585
properties 17:584, 585
sodium thiosulfate titration 13:439
uses 17:584, 585
Selenious acid–sulfuric acid
colorimetric method for hydrocodone 16:246
Selenite 14:88
Selenium. (See also Selenium compounds;
Selenium, elemental.)
analysis 17:589

ascorbic acid colorimetric method 14:99
barium chloride gravimetric method 17:591
in cadmium–mercury reds 17:195
in chrome steels 9:662
in cobalt alloys 10:320
in cobalt metal 10:302
colorimetric determination 17:591, 599
in copper 17:593, 595
in copper alloys 10:645
in copper metal 10:625
in cotton 11:142, 154, 156
in cyclamates 11:231
detection 17:590
3,3′-diaminobenzidine colorimetric method
    10:302, 625; 17:599
diaminonaphthalene colorimetric method
    15:413
distillation methods 17:590, 593
flame tests 6:592
in food 17:598
in gallium arsenide 13:279
in gallium metal 13:266
in germanium 13:396
in glass 9:200
in glass and glass-ceramics 13:413, 415, 439
gravimetric determination 17:590, 592
in gypsum 14:99
history 17:580
p-hydrazinebenzenesulfonate colorimetric
    method 13:398
hydrazine gravimetric method 17:591, 592
hydroxylamine hydrochloride colorimetric
    method 17:597
hydroxylamine hydrochloride gravimetric
    method 17:591, 592, 596
hypophosphorous acid gravimetric method
    17:591
impurities, determination of 17:605
iodometric method 17:598, 600
ion exchange chromatography 14:17
in iron 15:59
isotopes 17:580
in magnesium sulfate 15:413
in metals 17:589
methylene blue colorimetric method 17:592
neutron activation analysis 13:265
occurrence 17:580
in ores 17:589
in organic materials 17:589
paper chromatography 14:18
phosphorous acid gravimetric method 17:591
in plants 17:596
potassium iodide gravimetric method 17:591
potassium permanganate titration 17:597
sample decomposition 17:589, 591
sampling 17:589
separation 17:590

silver foil gravimetric method 11:142
in smelter dust 17:593
stannous chloride gravimetric method 17:591
in steel 17:598
in sulfide ores 17:593
sulfur dioxide–hydrochloric acid gravimetric
    method 13:439; 17:195, 590, 592
sulfuric acid colorimetric method 17:591
thiocyanate gravimetric method 17:591
toxicity 17:580
volumetric determination 17:597
in water 17:599
x-ray fluorescence spectroscopy 11:156
zirconium hydroxide gravimetric method 17:590
zirconium sulfate gravimetric method 17:595
Selenium and tellurium. (See also Selenium;
    Tellurium.) 17:580–608
Selenium compounds. (See also individual com-
    pounds.) 17:606
toxicity 17:586
Selenium diethyldithiocarbamate. (See also
    Dithiocarbamic acid salts.)
properties 11:575
uses 11:576
Selenium dimethyldithiocarbamate. (See also
    Dithiocarbamic acid salts.)
uses 11:576
Selenium dioxide. (See also Selenium compounds.)
manufacture 17:585
properties 17:585
uses 17:607
Selenium disulfide. (See also Selenium compounds.)
uses 17:607
Selenium, elemental
assay 17:597
commercial grades and specifications 17:588
manufacture 17:586
properties 17:582
toxicity 17:586
uses 17:582
Selenium hexafluoride
toxicity 13:13
Selenium monochloride. (See also Selenium
    compounds.)
manufacture 17:586
Selenium oxychloride. (See also Selenium
    compounds.)
toxicity 17:586
uses 17:607
Selenium sulfide
colorimetric method for mercury 16:23
Selenium trioxide. (See also Selenium compounds.)
manufacture 17:585
properties 17:585
Selenous acid
colorimetric method for m-cresol 17:20
Self-processed material

in brake linings 7:447

Sellier tube 19:414

Semicarbazide
  gravimetric method for cinnamaldehyde
    10:45
  identification reagent for carbonyl compounds
    8:342
  reaction of carbonyl compounds 8:333

Semicarbazone derivative
  of acetone 4:110
  of acetophenone 4:124
  of benzaldehydes 6:595, 612, 613

Semiconductors. (See also *Ceramics*; and
  individual compounds.)
  analysis 14:557
  properties 9:210, 213
  uses 9:210, 213

Semisweet chocolate. See Chocolate.

Semisynthetic penicillins. (See also *Penicillins
  and related compounds.*)
  manufacture 16:553
  paper chromatography 16:558
  thin-layer chromatography 16:559

Senarmontite 6:39

Sensitizing ability
  of manganese dioxide 15:531
  of potassium dichromate 15:530

Sensory testing methods 17:608–644
  acceptability testing 17:610
  analysis of test results 17:621
  applications 17:610
  classification 17:609
  descriptive tests 17:609, 642
  difference testing 17:609
  directional triangle tests 17:627
  Ducan multiple range test 17:642
  duo-trio testing 17:627
  flavor profiling 17:643
  hedonic scale technique 17:639
  intensity testing 17:609
  objectives 17:609
  paired comparison test 17:629
  paired difference testing 17:630
  paired intensity testing 17:630
  paired preference test 17:633
  paired sensory tests 17:626
  panel indoctrination 17:613
  panel morale 17:614
  panel screening 17:612
  panel selection 17:611
  preference testing 17:610
  ranking test 17:633
  rating tests 17:633
  reporting of test results 17:621
  sample carriers 17:617
  sample coding 17:620
  sample handling 17:617

  sample presentation 17:617
  scoring-grading scales 17:639
  sequential triangle testing 17:627
  Smiley scale 17:639
  test design 17:609
  testing conditions 17:615
  test purposes 17:609
  triangle tests 17:619, 621

Sepiolite
  composition 10:103

Sequential triangle testing 17:627
  evaluation 17:628

Serendibite 7:316

Serine. (See also *Carboxylic acids, amino.*)
  8:413, 519
  analysis 8:523
  chromotropic acid colorimetric method 8:523
  in cocoa beans 9:556
  commercial specifications 8:522
  crystalline form 8:520
  gas chromatography 8:429
  infrared spectrum 8:521
  ion exchange chromatography 8:415, 421
  manufacture 8:522
  nuclear magnetic resonance spectrum 8:521
  paper chromatography 8:422
  periodate oxidation 8:523, 524
  potassium permanganate test 8:522
  properties 8:519
  in protein 8:414
  solubility 8:520
  in yeast 19:529

Serological identification
  of salmonella 9:605

Serpentine 6:254; 15:358
  in chromite 9:700

Serum
  insulin content 14:572
  iron content 15:32, 54, 57
  phenindione content 10:183

Seselin. (See also *Coumarins.*) 11:161
  gas chromatography 11:168, 170

Setocyanine. (See also *Triphenylmethane and
  related dyes.*)
  visible spectroscopy 19:167

Setocyanine supra. (See also *Triphenylmethane
  and related dyes.*)
  ceric sulfate titration 19:175

Setoglaucine. (See also *Triphenylmethane and
  related dyes.*)
  visible spectroscopy 19:167

Setoglaucine O. (See also *Triphenylmethane
  and related dyes.*)
  ceric sulfate titration 19:175

Set plaster. (See also *Gypsum and gypsum
  products.*)
  definition 14:89

sand content 14:100
Set point
    of naled 19:609
Setting expansion
    of dental amalgams 11:291
    of dental gypsum products 11:266
    of dental investments 11:269
Setting point
    of fats and fatty oils 12:476
    of fatty alcohols 4:570
Setting time
    of calcium hydroxide–salicylate dental products
        11:298
    of dental gypsum products 11:266
    of dental impression materials, alginate-based
        11:281
    of dental investments 11:269
    of dental silicate cements 11:259
    of dental zinc phosphate cements 11:263
    of zinc oxide–eugenol dental products 11:297
Settleable matter
    in water 19:381
Settlement test
    for emulsified asphalts 6:327
Sewage
    nitrile content 16:492
    oxygen content 16:524
    sulfide content 18:396
Sex hormones. (See also Androgens; Estrogens;
        Progestational hormones; and individual
        compounds.) 14:185
    analysis 14:220
    classification 14:185
    composition 14:185
    properties 14:199
    sources 14:199
    uses 14:199
Seybertite
    composition 10:103
Shaffer-Somogyi method
    for reducing sugars 9:560
    for sugar 18:356
Shale
    in concrete 9:142
Shampoo. (See also *Cosmetics*; and *Detergents*.)
        11:74, 325
    alcohol content 11:74
    alkanolamine content 11:82
    alkyl aryl polyether alcohol content 11:82
    alkyl sulfate content 11:83
    ammonium sulfate in 11:84
    analysis 11:74, 81, 82, 83
    ash content 11:75
    composition 11:74, 326
    fatty acid content 11:81
    gum content 11:75
    ion exchange chromatography 11:85, 86

lanolin content 11:82
    nonvolatile matter in 11:74, 75
    organic constituents 11:75
    polyhydroxy compounds in 11:82
    polyoxyethylene compounds in 11:82, 83
    protein content 11:75
    quaternary ammonium compounds in 11:82
    soap content 11:81
    surfactant content 11:77
    water content 11:74
Shape count
    of diamonds, industrial 11:473
Shattuck test
    for asphalts 6:307
Shear strength
    of graphite 14:80
Sheeter-type brake linings 7:442
Sheet machine
    for preparation of pulp handsheets 7:243, 244
Shellac
    solubility 13:539, 552
Shell aging index test
    for asphalts 6:307
Shell BR 11
    thermal analysis, differential 12:133
Shell content
    of cocoa products 9:573
Sherbet. (See also Frozen desserts; Ice cream.)
    commercial specifications 16:142
Shrinkage
    of caulking compounds 9:25
    of dental restorative materials, composite 11:314
    of felt 12:511
    of fibers 4:89
    of polyamide fibers 17:325
Siderite 14:519; 15:21
    composition 15:22
Siding, asbestos–cement 6:270
Siegenite 10:284
    composition 16:286
Sieve analysis
    of abrasive grain 7:303
    of adsorbents 4:440
    of aerosols 4:458
    of bituminous paving mixtures 6:332
    dry 4:440
    of emulsified asphalts 6:329
    wet 4:441
Sigma value method
    for foaming tendency of beer 7:536
Sigmoidal
    thin-layer chromatography 6:481
Silane coupling agents. See Organofunctional
    silanes.
Silane esters, organic. See Organosilane esters.
Silanes. (See also Silicon compounds.) 18:48
    toxicity 18:8

Silanic hydrogen
  hydrogen evolution method 18:67
  in organo-chlorosilanes 18:67
  in organosilicon copolymers 18:148
  in silicone resins 18:123
Silastic 12:87
Silazanes. (See also Silylation reagents; and
      individual compounds.) 18:89–94
  manufacture 18:91
  properties 18:89
  uses 18:89
Silica. (See also *Silicon*; Silicon dioxide.)
      18:1
  ablator 4:2
  in alkali barium silicate glass 9:218
  in alumina 5:116
  in aluminates 5:160
  in aluminum fluoride 5:180
  in aluminum oxide abrasives 4:35, 49
  in asbestos 6:259
  in asbestos–cement products 6:270
  in barite 6:556, 558, 563
  in barium sulfate 17:192
  in bauxite ore 4:37
  in beer 7:639
  bismuth in 7:216
  carrier gas fusion methods 8:626, 683
  in cellulose 9:49
  in cements 9:111, 126
  in ceramics 9:194
  in chrome cleaners 6:353
  in chrysotile 6:259, 267
  in clay 9:191
  in clays and clay minerals 10:113
  colorimetric method using ammonium molyb-
      date 7:430
  commercial specifications 12:174
  in diatomaceous earth 11:489
  in elastomers, synthetic 12:151
  emission spectroscopy 4:3; 9:218
  in enamels 9:195; 12:174
  ethylene oxide polymers gravimetric method
      18:28
  filler for charring ablators 4:16
  in fluorosilicic acid 13:49
  gelatin gravimetric method 18:28
  in glass and glass-ceramics 4:26; 13:413, 415,
      416
  in glazes 9:196
  gravimetric method 4:37; 6:259
  gravimetric method for silicon 18:17, 22, 28
  heteropoly blue colorimetric method 6:260
  in hydrated lime 9:191
  in hydraulic cements 9:96, 100
  in hydrofluoric acid 18:30
  hydrofluoric acid–sulfuric acid gravimetric
      method 9:111, 294, 704; 10:113; 13:416;

      14:139; 17:507; 18:45
  infrared spectrum 12:152
  ion exchange chromatography 18:15
  in lead silicochromate 9:707; 17:199
  in limestone 9:191
  microcosmic salt bead test 6:588
  molybdenum blue colorimetric method 4:38;
      10:114
  molybdic acid colorimetric method 9:126
  oxygen in 8:626, 683
  perchloric acid gravimetric method 13:417
  in pollucite 9:294
  properties 18:5
  quality control tests 4:25
  in quicklime 9:191
  quinoline silico molybdate gravimetric method
      12:155
  in rare earth fluorides 17:507
  in refractories 9:206
  separation 18:15
  in silicate glasses 9:198
  in silicates 18:44
  in silicate white lead 17:189
  silicomolybdic acid volumetric method 18:31
  in silicon carbide abrasives 4:57
  sodium carbonate fusion 13:417
  in sodium fluoaluminate 5:180
  sodium hydroxide volumetric method 18:30
  solubility 18:7
  in stannic oxide 19:71
  in titanium dioxide pigments 19:125
  x-ray emission spectroscopy 4:49
  in zircon 14:139
Silica–alumina
  chromium content 9:656
Silica brick. (See also Refractories.)
  acid dissolution 9:214
  composition 9:206
  fluxes for 9:216
Silica flour. See Quartz.
Silica gel
  as adsorbent 4:434
  manufacture 18:43
Silica glass. (See also *Glass and glass-ceramics*;
      Silicate glasses.) 9:197
  composition 9:199; 13:411
  manufacture 13:411
  properties 9:197, 204
  uses 9:199; 13:411
Silica refractories. (See also Refractories.)
  composition 9:208
  properties 9:208
Silica sand 4:31, 33
Silicate cements, dental. See Dental silicate
      cements.
Silicate glasses. (See also Glass.) 9:197
  commercial grades and specifications 9:198

composition 9:198
uses 9:198
Silicate minerals
  iron content 15:56
Silicate rocks
  chromium content 9:643, 646, 648, 653
  decomposition 19:210, 214
  impurities, determination of 17:71
  iron content 9:644
  molybdenum content 19:214
  nickel content 16:320
  potassium content 17:381
  sample decomposition 18:10
  sampling 18:9
  scandium content 17:575, 577
  silicon content 18:35
  sodium carbonate fusion 9:643
  sodium carbonate–sodium nitrate fusion 9:647
  sodium content 17:381
  sodium hydrogen sulfate fusion 9:644
  strontium content 18:273, 277, 280
  sulfur content 18:380
  thallium content 18:522, 531
  tin content 19:62, 64
  titanium content 19:109, 118, 120
  tungsten content 19:210, 214
  uranium content 19:250
  vanadium content 9:643; 19:109, 339, 342
Silicates. (See also individual compounds.) 18:1,
    42–45
  acid–base titration 18:44
  acid decomposition 18:14
  adhesives 4:403
  in alkaline flushes 6:359, 360
  alkalinity 18:44
  ammonium molybdate–benzidine colorimetric
    method 11:142
  analysis 17:361, 364; 18:44
  in antifreezes 6:13
  antirusts 6:351
  in barium sulfate 6:568
  in brake linings 7:447
  calcium fluoride–sulfuric acid gravimetric
    method 18:16
  composition, determination of 17:361
  cooling system sealer 6:353
  in cotton 11:142
  in detergents 11:379
  fluxes for 9:216
  hydrofluoric acid–sulfuric acid gravimetric
    method 11:379; 18:193
  hydrofluoric acid turbidimetric method 18:16
  ion exchange chromatography 17:364
  iron content 19:113
  lithium metaborate fusion 15:32
  manufacture 18:6
  occurrence 18:42

  in perchloric acid 9:364, 365
  in polyurethanes 19:312
  properties 18:5
  sample preparation 6:59, 236; 15:31; 17:361
  silicon content 18:37, 44
  Smith decomposition method 17:361
  in soaps 18:182, 193
  sodium carbonate–hydrochloric acid gravimetric
    method 18:16
  stannous chloride colorimetric method 9:365
  thallium content 18:524
  titanium content 19:113
Silicate slag
  sulfide content 18:397
Silicate white lead
  analysis 17:188
  commercial specifications 17:188
  lead content 17:188
  phosphorus content 17:188
  silica content 17:189
  uses 17:188
  water content 17:189
Silica W. (See also Silica.) 18:5
Siliceous aggregates
  in concrete 9:143
Siliceous bauxite
  composition 9:202
Silicic acid. (See also Silicon compounds.)
  in fluorosilicic acid 13:49
  gravimetric method for silicon 18:17
  ion exchange chromatography 18:15
  properties 18:6
  solubility 18:6
Silicides
  fluxes for 9:216
Silicium. See Silicon.
Silicofluoric acid. See Fluorosilicic acid.
Silicomanganese. (See also Manganese alloys;
    Silicon alloys.) 5:61
  commercial grades and specifications 15:453
  composition 5:64; 15:451; 18:2
  manganese content 15:460
  phosphorus content 17:86
  sample decomposition 18:24
  silicon content 18:24
Silico-12-molybdate. (See also Silicon compounds.)
  manufacture 18:31
  properties 18:31
  solubility 18:31
Silicomolybdic acid
  gravimetric method for cesium 17:549
  gravimetric method for rubidium 17:549
  volumetric method for silica 18:31
Silicon. (See also Silica; Silicon alloys; Silicon
    compounds; Silicon compounds, organic;
    and Silicon, elemental.) 18:1–48
  in alumina 5:116

in aluminates 5:160
in aluminum and aluminum alloys 5:141;
  18:34
in aluminum oxide 4:51; 5:187
in aluminum silicate 5:190; 18:29
ammonium molybdate–benzidine colorimetric
  method 18:16
ammonium molybdate colorimetric method
  13:281; 14:64; 15:274; 18:31
ammonium molybdate–oxine gravimetric method
  18:78
ammonium molybdate–stannous chloride
  colorimetric method 16:334
atomic absorption spectroscopy 9:346; 18:37
in bauxite 5:115; 18:37
bisulfate fusion 18:14
borax fusion 18:13
boric oxide fusion 18:13
boron content 18:13, 23
in brass 18:25
in bronze 18:25
calcium carbonate fusion 18:12
in calcium metal 8:76
in carbon 14:63, 74; 18:30, 37
in carbonate rocks 18:27
in carbonates 18:37
carrier gas fusion methods 8:625
in cement 18:38
in chrome ore 9:701, 703
in chrome steels 9:662
in chromic acid 9:684
in chromic oxide 9:692
in chromium 9:635
in chromium alloys 9:674
in clays and clay minerals 10:136
in cobalt alloys 10:320
in cobalt metal 10:293, 302
in cocoa products 9:595
colorimetric method 18:5, 16, 31
colorimetric method using molybdenum
  yellow 5:141
in copper 18:32
in copper alloys 10:645
in copper metal 10:625
in cotton 11:154, 157
in cutting alloys 10:289
in dental alloys 10:290
detection 18:16
in detergents 18:30
emission spectroscopy 8:76; 9:346; 13:277;
  14:48, 74, 122, 542; 15:155, 375; 16:160,
  387; 18:16
in enamels 12:184
in face powders 18:30
in ferrochromium 9:661; 15:80
in ferromanganese 15:78
in ferromolybdenum 18:24

in ferroniobium 16:387
in ferrosilicon 15:74; 18:26
in ferrotitanium 18:24
in ferrotungsten 18:24, 30
in ferrovanadium 18:24
in fluoride 18:32
fluoride content 18:22
in fluosilicate 18:32
free, in silicon carbide abrasives 4:55
in gallium arsenide 13:276
gallium content 13:262
in gallium metal 13:266
in gallium phosphide 13:281, 282
in glass 18:30
in gold 14:45
in graphite 14:63, 74; 18:30, 37
gravimetric method 18:5, 7, 16, 17
gravimetric method using hydrofluoric acid
  5:142
in hafnium 14:122, 135
in hard facing alloys 10:289
in high-temperature alloys 10:286
history 18:1
Hoffman and Lundell gravimetric method 18:22
hydrochloric acid gravimetric method 18:19
hydrofluoric acid–sulfuric acid gravimetric method
  9:674; 10:320, 645; 12:184; 15:63, 74; 16:184;
  18:7, 17, 87
hydrogen in 8:618, 680, 681
ion exchange chromatography 13:281; 18:15
in indium 14:542
in iron 15:59, 63; 18:26, 27, 38
isotopes 18:1
in lanthanum oxide 15:155
in lead phosphosilicate 18:30
in lead silico–chromate 18:30
in lithium 15:274
in magnesium 15:375, 390; 18:34
in magnesium alloys 15:375, 390
in magnesium silicate pigment 18:29
in manganese 15:479; 18:27
in manganese–silicon 18:24
microscopy 9:227
in molybdenum 16:161
molybdenum blue colorimetric method 10:302;
  18:34
in molybdenum trioxide 16:161
molybdenum yellow colorimetric method 15:391
molybdic acid–hydrofluoric acid colorimetric
  method 14:135
molybdisilicic acid colorimetric method 10:625
in molybdosilicic acid 16:184
in nickel 16:334; 18:37
in nickel alloys 16:334
in nickel–copper alloys 18:25
in niobium 16:387
occurrence 18:1

in organofunctional silanes 18:87
in organosilane esters 18:77
oxygen in 8:616, 618, 619, 620, 621, 623,
    625, 632, 680, 681
perchloric acid gravimetric method 15:391;
    18:21
in plants 18:28
potassium acid fluoride fusion 18:14
potassium carbonate fusion 18:12
potassium hexafluosilicate volumetric method
    18:30
potassium hydroxide fusion 18:12
pyrosulfate fusion 18:14
in quartz 18:27
in red lead 18:29
in refractories 18:30
sample decomposition 18:10
sampling 18:9
in sandstone 18:27
separation 18:15
silica gravimetric method 18:17, 22, 28
in silicate rocks 18:35
in silicates 18:37
silicic acid gravimetric method 18:17
in silicomanganese 18:24
in silicon carbide 8:136
in silicone fluids 18:112
in silicone resin adhesives 4:404
in soaps 18:30
sodium carbonate–calcium oxide fusion 18:12
sodium carbonate fusion 18:10
in sodium chloride 9:346
in sodium dichromate 9:684
sodium hydroxide fusion 18:12
sodium peroxide fusion 18:13
in spring steels 10:290
in steel 18:15, 35, 38
sulfuric acid gravimetric method 18:21
in surgical alloys 10:290
in titanium dioxide 18:30
in titanium metal 19:121
in tungsten 18:30
vacuum fusion methods 8:616, 618, 619, 620,
    621, 623, 680, 681
volumetric determination 18:5, 7, 30
in water 18:29
in wear-resistant alloys 10:289
x-ray fluorescence spectroscopy 10:136; 11:157
in zircaloy 14:135
in zirconium 14:122, 135; 18:37
Silicon alloys. (See also individual alloys.) 5:61,
    66
composition 18:2
nitrogen in 8:630
uses 18:2
Silicon bronze. (See also Silicon alloys.)
uses 18:2

Silicon–calcium–manganese alloys 5:61
composition 5:64
Silicon carbide. (See also *Carbides*; Silicon
    compounds.) 18:39
abrasive 7:294
analysis 7:295; 18:40
carbon content 8:134
formation 8:130
free carbon in 8:134
identification 7:295
impurities, determination of 18:40
manufacture 18:39
particle diameter values 7:305
properties 8:131, 132; 9:204; 18:39
silicon content 8:136
sodium carbonate fusion 18:12
sodium peroxide fusion 18:13
specific gravity 7:296
uses 8:132; 18:39
Silicon carbide abrasives 4:31, 33, 52
analysis 4:52
fusion of samples 4:56
impurities, determination of 4:55
Silicon carbide refractories. (See also Refractories.)
composition 9:207
Silicon cast iron. (See also Iron alloys.)
commercial specifications 15:73
composition 15:73
Silicon compounds. (See also *Silicon compounds,
    organic*; and individual compounds.) 18:39–
    45
safety 18:8
toxicity 18:8
Silicon compounds, organic. (See also Chloro-
    silanes; Organo-chlorosilanes; Organofunctional
    silanes; Organosilane esters; Organosilicon co-
    polymers; Silicone fluids; Silicone resins;
    Silicone rubbers; and Silylation reagents.) 18:
    48–155
Silicon dioxide. (See also Silica.)
in calcium oxide 8:94
carrier gas fusion methods 8:624, 626, 682
in clays and clay minerals 10:112
in cobalt(II) orthophosphate 10:343
in cobalt oxides 10:342
in cobalt(II) sulfate 10:344
in gypsum and gypsum products 14:94, 95
hydrofluoric acid–sulfuric acid gravimetric
    method 10:622; 14:95; 15:207
hydrogen in 8:616
in lime and limestone 15:207
nitrogen in 8:616
oxygen in 8:616, 624, 626, 682
structure 18:1
vacuum fusion methods 8:616
in vitrified grinding wheels 7:298
Silicone-based impression materials. See

Dental impression materials, silicone-based.
Silicone elastomers. See Silicone rubbers.
Silicone fluids. (See also individual compounds.)
  18:95–117
  absolute viscosity 18:113
  acid–base titration 18:114
  acid number 18:114
  analysis 18:104
  commercial grades and specifications 18:103
  composition, determination of 18:104
  desolvated viscosity 18:113
  dielectric constant 18:116
  dielectric strength 18:115
  dissipation factor 18:116
  emulsion stability 18:116
  gas chromatography 18:106
  gel permeation chromatography 18:110
  identification 18:104
  impurities, determination of 18:114
  infrared spectroscopy 18:104
  intrinsic viscosity 18:114
  kinematic viscosity 18:113
  manufacture 18:101
  molecular weight distribution 18:110
  pour point 18:115
  power factor 18:116
  properties 18:97
  properties, determination of 18:113
  refractive index 18:97
  resistivity 18:116
  sample decomposition 18:108
  silicon content 18:112
  specification tests 18:114
  uses 18:96
  viscosity 18:97, 113
  viscosity–temperature coefficient 18:114
  volatility 18:115
Silicone gums. (See also Silicones.)
  manufacture 18:103
Silicon, elemental
  analysis 18:9
  commercial grades and specifications 18:9
  impurities, determination of 18:39
  manufacture 18:8
  properties 18:4
  safety 18:8
  toxicity 18:8
  uses 18:1
Silicone oils. (See also Silicone fluids.)
  in automobile waxes 6:353
Silicone–organic copolymers. See Organosilicon
  copolymers.
Silicone–oxyalkylene copolymers. See Organosilicon copolymers.
Silicone resins 18:117–126
  ablator 4:23
  adhesives 4:403

  alkoxy content 18:123
  analysis 18:119
  commercial grades and specifications 18:119
  composition, determination of 18:122
  condensation 18:123
  corrosion test 18:124
  drying time 18:124
  electrical tests 18:124
  haze number 18:124
  impurities, determination of 18:122
  infrared spectroscopy 18:119
  manufacture 18:118
  nuclear magnetic resonance spectroscopy 18:121
  properties 18:117
  properties, determination of 18:124
  quality control tests 4:22
  resin content 18:122
  silanic hydrogen content 18:123
  types 18:118
  uses 18:117
  viscosity 18:124
  volatility 18:124
Silicone rubbers 18:126–141
  additives 18:129
  ageing tests 18:140
  analysis 18:131
  commercial grades and specifications 18:131
  composition 18:126
  compression set tests 18:131
  electrical tests 18:139
  elongation 18:131, 138
  hardness 18:131, 139
  manufacture 18:129
  oil immersion tests 18:140
  properties 18:126
  properties, determination of 18:131
  solubility 18:128
  tear strength 18:131, 139
  tensile strength 18:131, 138
  types 18:129
  uses 18:126
Silicones. (See also *Elastomers, synthetic*; Organosilicon copolymers; Silicone fluids; Silicone
  gums; Silicone resins; and Silicone rubbers.)
  in alkyd resins 5:45
  in automobile waxes and polishes 6:353
  commercial elastomers 12:87
  in creams and lotions 11:92
  density 12:90
  in hair sprays 11:90
  identification 12:90
  modifier for alkyd resins 5:24
  structure 18:95
  in waxes and polishes 6:363
Silicone surfactants. See Organosilicon copolymers.
Silicon halides 18:40
  properties 18:40

Silicon hydrides. See Silanes.
Silicon metal
  composition 18:2
  uses 18:2
Silicon oxychlorides
  properties 18:41
Silicon semiconductors. (See also *Ceramics.*)
  properties 9:213
  uses 9:213
Silicon tetrabromide. (See also Silicon compounds.)
  properties 18:40
Silicon tetrachloride. (See also Silicon compounds.)
  commercial grades and specifications 18:41
  manufacture 18:6, 41
  properties 9:335; 18:6, 40; 19:96
  solubility 9:336
  uses 18:41
Silicon tetrafluoride. (See also Silicon compounds.)
  distillation method for fluorine 15:226
  gas chromatography 13:64
  gasometric method for fluorides 13:15, 16
  manufacture 18:6
  properties 18:40
Silicon tetraiodide. (See also Silicon compounds.)
  properties 18:40
Silicophosphate
  manufacture 18:5
Silicothermic process
  for chromium 9:634
Silicotungstic acid
  gravimetric method for cesium 17:549, 553
  gravimetric method for polyoxyethylene 11:95
  gravimetric method for polyoxyethylene compounds 11:82
Silk fiber
  microscopy 4:265
  mixtures, determination in 4:353, 360
  solubility 4:250, 353, 360
Sillimanite
  in clay 9:191
  composition 9:203, 208
  properties 9:204, 208
  x-ray diffraction 9:222
Silomelane
  composition 15:22
Silt
  in concrete 9:142
Silver. (See also Silver alloys; Silver compounds; and Silver metal.) 18:155–178
  acid–base titration 11:289
  in aluminum 18:172
  in aluminum alloys 5:144
  ammonium hydroxide gravimetric method 18:164

ammonium thiocyanate volumetric method 15:392
in antimony 18:172
in antimony metal 6:44, 46
in arsenic metal 6:227
atomic absorption spectroscopy 6:46; 10:622; 14:49; 15:195, 196; 18:172
azide gravimetric method 18:164
bead test 6:588
1,2,3-benzotriazole gravimetric method 18:162, 166
in bismuth nitrate 7:203
in bismuth subcarbonate 7:196
in bullets 18:172
in cadmium 18:172
carbonate gravimetric method 18:164
catalytic methods 18:172
ceric ammonium nitrate color test 18:164
chromate gravimetric method 18:164
in chromium 9:635
in cobalt alloys 10:320
in cobalt metal 10:303
colorimetric determination 18:168
in copper alloys 10:645
copper gravimetric method 18:165
in copper metal 10:621
in cotton 11:156
cyanide gravimetric method 18:164
in dental amalgams 11:289
detection 18:163
5-(*p*-dimethylaminobenzylidine)rhodanine colorimetric method 10:303; 18:164, 168
dithizone colorimetric method 18:164, 170
dithizone titration method 6:44, 227
electrodeposition 18:163, 165, 171
emission spectroscopy 14:48, 538; 15:155, 187, 197; 17:270; 18:163, 172
in feed 18:172
ferricyanide gravimetric method 18:164
in ferrochromium 9:661
ferrocyanide gravimetric method 18:164
ferrous sulfate gravimetric method 18:165
fire assay 15:196
flame emission spectroscopy 18:172
flame tests 6:589
formaldehyde gravimetric method 18:165
in gallium metal 13:266
Gay-Lussac volumetric method 18:167
in glass and glass-ceramics 13:413
in gold 14:45, 49; 18:172
gravimetric determination 18:162, 165
half-life 18:155
history 18:155
in hydrochloric acid 14:374
hydrochloric acid gravimetric method 10:645; 18:165
hydrochloric acid turbidimetric method 18:171

hydrogen sulfide gravimetric method 18:164
in indium 14:538
in indium alloys 14:555
iodide gravimetric method 18:164
ion exchange methods 18:163
iron sulfate–EDTA gravimetric method 18:164
isotopes 18:155
in lanthanum oxide 15:155
in lead 15:195, 196; 18:171, 172
in lead alloys 15:196
in magnesium 15:392
in magnesium alloys 15:392
neutron activation analysis 13:265
nickel–dimethylglyoxime colorimetric method 18:164
occurrence 18:155
in ores 17:266; 18:172
oxidation potentials 18:158, 165
phosphate gravimetric method 18:164
in photographic films 12:586, 594
in photographic solutions 18:172
in plating solutions 18:172
in platinum group metals 17:229
polarography 12:567
potassium bromide–methylamino-4-phenyl sulfate colorimetric method 18:164
potassium chromate colorimetric method 18:164
potassium cyanide volumetric method 18:168
potassium ferricyanide–α,α'-dipyridyl colorimetric method 18:164
pyrogallol red colorimetric method 18:164
reduction methods for iron 15:36
in rocks 15:187
separation 18:162
in silver compounds 18:175
in silver iodide 12:567
sodium bromide gravimetric method 18:164
sodium bromide titration 14:555
sodium chloride gravimetric method 10:621; 18:164
in sodium dichromate 9:684
sodium hydroxide gravimetric method 18:164
solvent extraction 18:162
spot tests 18:164
stannous chloride gravimetric method 18:162, 164
in sulfide minerals 18:172
thioacetamide titration 18:168
thiocyanide gravimetric method 18:164
in tin metal 19:67
Volhard titration 18:167
in water 18:171, 172
x-ray fluorescence spectroscopy 11:156
zinc gravimetric method 18:162, 165
zolon red colorimetric method 18:164
Silver acetate. (See also Silver compounds.)

properties 18:173
solubility 18:160
uses 18:173
Silver alloys. (See also Silver metal; and individual alloys.)
assay 18:173
uses 18:156
Silver argentocyanide
gravimetric method for mercuric cyanide 16:23
Silver arsenate. (See also Silver compounds.)
properties 18:173
Silver azide
sodium nitrite titration 16:473
Silver beads
gold content 14:30
Silver brazing alloys. (See also Copper alloys; Silver alloys.) 10:631
composition 10:631; 18:156
uses 18:156
Silver bromate. (See also Silver compounds.)
solubility 18:160
Silver bromide. (See also Silver compounds.)
properties 18:173
solubility 18:160
uses 18:173
Silver carbonate. (See also Silver compounds.)
properties 18:173
solubility 18:160
uses 18:173
Silver chlorate. (See also Chlorates; Silver compounds.)
properties 9:358; 18:173
solubility 9:359
uses 18:173
Silver chloride. (See also Silver compounds.)
gravimetric method for chloride 9:341, 412
for lamination of infrared samples 4:236
manufacture 18:159
properties 9:335; 18:173
in silver sulfate 18:176
solubility 9:336; 18:160
turbidimetric method for chloride 9:419; 10:93
uses 18:159, 173
Silver chromate. (See also Silver compounds.)
properties 18:173
solubility 18:160
uses 18:173
Silver citrate. (See also Silver compounds.)
properties 18:173
uses 18:173
Silver coating weight
of photographic films 12:561
Silver compounds. (See also individual compounds.) 18:173
analysis 18:175
assay 18:175
properties 18:173

silver content 18:175
toxicity 18:160
uses 18:157, 173
Silver cyanate. (See also Silver compounds.)
properties 18:173
solubility 18:160
uses 18:173
Silver cyanide. (See also Silver compounds.)
color test for hydrogen sulfide 13:321; 15:255;
18:394
properties 18:173
solubility 18:160
uses 18:173
Silver dibutyldithiocarbamate. (See also Dithiocarbamic acid salts.)
properties 11:575
Silver diethyldithiocarbamate. (See also Dithiocarbamic acid salts.)
colorimetric method for arsenic 6:250; 10:95;
13:369; 14:373; 15:143, 223; 18:419
properties 11:575
Silver dimethyldithiocarbamate. (See also
Dithiocarbamic acid salts.)
properties 11:575
Silver dithizonate
colorimetric method for mercaptans 15:561
Silver ferricyanide. (See also Silver compounds.)
solubility 18:160
Silver fluoborate. (See also Silver compounds.)
properties 18:173
uses 18:173
Silver fluoride. (See also Silver compounds.)
properties 18:173
uses 18:173
Silver foil
gravimetric method for selenium 11:142
Silver–gold alloys
gold content 14:14
uses 14:8
Silver halides. (See also individual compounds.)
atomic absorption spectroscopy 12:564
halide content 12:566
neutron activation methods 12:568
in photographic films 12:561
polarographic method 12:563, 566
potassium cyanide titration 12:562, 566
thioacetamide titration 12:562
x-ray fluorescence spectroscopy 12:565, 567
Silver halide sol test
for gelatin 12:555
Silver iodate. (See also Silver compounds.)
properties 18:173
solubility 18:160
uses 18:173
Silver iodide. (See also Silver compounds.)
iodide content 12:566
properties 18:173

silver content 12:567
solubility 18:160
uses 18:173
Silver lactate. (See also Silver compounds.)
uses 18:173
Silver mercaptides
gravimetric determination of mercaptans 6:16
Silver–mercuric nitrate
gasometric method for olefins 15:244
gasometric method for propylene 15:244
Silver metal
analysis 18:161, 173
assay 18:173
carrier gas fusion methods 8:682
commercial grades and specifications 18:161
hydrogen in 8:680
impurities, determination of 18:173
manufacture 18:160
nitrogen in 8:617
oxygen in 8:616, 617, 623, 680, 682
properties 18:157
sampling 18:161
solubility in gold 14:9
spot test for sulfur 18:370
toxicity 18:160
uses 18:156, 174
vacuum fusion methods 8:616, 617, 623, 680
Silver molybdate. (See also Silver compounds.)
solubility 18:160
Silver nitrate. (See also Silver compounds.) 12:
540
absorption method for acetylenes 11:507
analysis 18:175
colorimetric method for acetylene 12:223, 333
color test for phosphine 17:61
commercial grades and specifications 12:541
commercial specifications 18:175
gravimetric method for bromide 8:17
gravimetric method for chloride 9:242; 10:672;
11:92; 15:282; 16:222
gravimetric method for chlorine 10:424, 442
gravimetric method for halides 13:69; 14:597
gravimetric method for iodides 14:596
gravimetric method for iodoform 14:614
gravimetric method for lead azide 12:419, 423
gravimetric method for lead thiocyanate 12:427,
430
gravimetric method for potassium chlorate 12:425,
427
gravimetric method for thiocyanate 12:423
gravimetric method for thiosulfate 18:431
gravimetric method for thiuram disulfides 11:595
identification method for bromide 8:15
impurities, determination of 12:540; 18:175
manufacture 18:158
properties 18:173
specification tests 18:175

spectroscopic methods for chloride 14:127
titration of barbiturates 6:493
titration of bromide 8:17, 21; 13:436
titration of chloride 6:434; 9:686; 11:47, 429;
 12:338; 13:436, 525; 14:96; 15:120, 122,
 279, 431; 18:88, 203; 19:364, 497, 531
titration of chlorine 9:52; 12:250; 19:154
titration of chlorosulfonic acid 9:530
titration of cyanides 13:178
titration of ethylene chlorohydrin 14:164
titration of halogens 14:513
titration of hydrogen cyanide 14:417
titration of hydrogen sulfide 19:504
titration of inorganic sulfides 8:276
titration of iodide 13:436
titration of mercaptans 8:276; 15:561, 562,
 566; 16:99
titration of phenindione 10:182
titration of salt in margarine 15:502
titration of sodium chloride 10:491; 11:355;
 12:66
titration of strontium chloride 18:275
titration of succinic acid 19:423
titration of thiols 17:38
titration of thiosulfate 12:550
turbidimetric method for bromide 14:592,
 604
turbidimetric method for chloride 8:348; 9:361,
 365; 10:329, 654, 656, 658; 12:337, 410, 542,
 544; 13:129, 525; 14:436, 592, 604; 15:142,
 275, 398, 401, 407, 411, 490; 16:61, 344, 498,
 536; 17:438; 18:175, 233, 283; 19:19, 222
turbidimetric method for chlorine compounds
 17:403
turbidimetric method for halides 14:610
turbidimetric method for methanol 16:113
uses 18:159, 160, 173
Silver nitrate–ammonium thiocyanate
volumetric method for phenylacetylene 18:308
Silver nitrate–formaldehyde
gravimetric method for chlorate 12:423
Silver nitrate–periodic acid
gravimetric method for glycols 13:572
Silver nitride. (See also Silver compounds.)
solubility 18:160
Silver nitrite. (See also Silver compounds.)
solubility 18:160
Silver orthophosphate
volumetric method for phosphoric acid 13:46
Silver oxalate
properties 18:173
uses 18:173
Silver oxide. (See also Silver compounds.)
carrier gas fusion methods 8:624
oxygen in 8:624
properties 18:173
solubility 18:160

uses 18:173
Silver perchlorate. (See also Perchlorates; Silver
 compounds.)
properties 9:363; 18:173
solubility 9:364
uses 18:173
Silver permanganate. (See also Silver compounds.)
properties 18:173
uses 18:173
Silver peroxide. (See also Silver compounds.)
properties 18:173
uses 18:173
Silver phosphate. (See also Silver compounds.)
properties 18:173
uses 18:173
Silver potassium cyanide. (See also Silver compounds.)
properties 18:173
uses 18:173
Silver pyrophosphate. (See also Silver compounds.)
properties 18:173
uses 18:173
Silver reducing substances
in glycerol 13:527
Silver–rhenium alloys. (See also Rhenium alloys.)
uses 17:517
Silver sulfate. (See also Silver compounds.)
analysis 18:176
commercial specifications 18:175
impurities, determination of 18:176
properties 18:173
solubility 18:160
specification tests 18:176
uses 18:173
Silver sulfide. (See also Silver compounds.)
methylene blue colorimetric method 12:586
in photographic films 12:586
properties 18:173
solubility 18:160
uses 18:173
Silver thiocyanide. (See also Silver carbonate.)
solubility 18:160
Silver toluenesulfonate distillation
of sodium acetate 10:492
Silylamides. (See also Silylation reagents; and
 individual compounds.) 18:89–94
properties 18:89
uses 18:89
Silylamines. (See also Silylation reagents; and
 individual compounds.) 18:89–94
manufacture 18:91
properties 18:89
uses 18:89
Silylation reagents. (See also individual compounds.) 18:89–94
analysis 18:92
commercial grades and specifications 18:92

gas chromatography 18:93
impurities, determination of 18:93
infrared spectroscopy 18:92
manufacture 18:91
nitrogen content 18:94
properties 18:89
safety 18:92
uses 18:89
Silyl ethers
gas chromatography 11:77
Simanal. (See also Silicon alloys.)
composition 18:2
Simpsonite
composition 18:441
Sinapic acid. (See also *Cinnamic acid and its derivatives.*) 10:64
Sintered alnico alloys. (See also Cobalt alloys.) 10:288
Sintered alumina
composition 9:203
Sintrom. See Acenocoumarol.
Si-O-flex 12:87
β-Sitosterol
in butter 16:138
gas chromatography 16:139; 19:150
in tobacco 19:150
Sixolite. (See also High explosives.)
melting point 12:461
properties 12:432
Sizing
in drycleaning solvents, used 11:609
Skala and Swanson method
for ash content of eggs 12:64
Skeletal density
of adsorbents 4:443
Skim milk. (See also Milk.)
analysis 16:131
manufacture 16:131
milkfat content 16:132
Skinkel technique
for finishing agents 11:126, 130
Skogbolite
composition 18:441
Skutterudite 10:284
Slag
in hydraulic cements 9:107
manganese content 15:468, 473
sample preparation 6:59, 236
sodium peroxide fusion 7:400
Slag cements. (See also Hydraulic cements.) 9:97
commercial grades and specifications 9:99, 100, 101
composition 9:96
hardening 9:98
impurities, determination of 9:110
Slaked lime. See Lime.
Slaking rate

of lime 15:230
Slaking, resistance to
of adsorbents 4:446
Sliding plate microviscometer method
for bituminous materials 6:301
Slip point
of fatty alcohols 4:571
Sludge
mercury content 16:44, 45, 46
radium content 17:464
Sludge density index 19:382
Sludge volume
of water 19:381
Slump test
for concrete 9:143
for dental gypsum products 11:264
Smallwood jet assembly
for emission spectrography 4:345
Smaltite 10:284
SMDC. See Sodium methyl dithiocarbamate.
Smelter dust
selenium content 17:593
Smiley scale 17:639
Smith and Rees spectroscopic method
for caffeine 10:407
Smith decomposition method
for silicates 17:361
Smithsonite 14:519
composition 19:545
Smoke point
of corn oil 11:58
Smoking machines
for cigarettes 10:11
SMZ alloy. (See also Silicon alloys.)
composition 18:2
Soap bars. (See also *Soaps.*)
composition 18:179
types 18:179
Soap emulsions
identification 11:93
Soap products. (See also *Soaps.*)
composition 18:179
uses 18:179
Soaps. (See also individual compounds.) 18:179–207
additives, determination of 18:183, 192, 194
analysis 18:180, 204
anhydrous soap content 18:188
combined alkali content 18:188
composition, determination of 18:184
cresol content 11:541
decomposition 18:188
definition 18:179
detergent content 11:363; 18:204
EDTA and related compounds in 12:43
in elastomers, synthetic 12:147
fatty acid content 18:184

in finishing agents 11:131
hexachlorophene content 11:550
history 18:179
identification 18:183
impurities, determination of 18:200
infrared spectroscopy 11:331
manufacture 18:179
in metal cleaners 6:355
pH 18:200
potassium content 18:190
quality control tests 18:204
sampling 18:180
separation 18:182
in shampoos 11:76, 81
silicon content 18:30
sodium content 18:190
solvent extraction 18:182
sulfuric acid gravimetric method 18:183
turbidimetric determination 11:131
types 18:179
uses 18:179
water content 18:191
Soda. (See also Sodium carbonate.) 18:207
Soda alum. See Aluminum sodium sulfate.
Soda ash. (See also Sodium carbonate.)
commercial grades and specifications 18:237
Soda bleach 9:352
Soda lime
absorbent for carbon dioxide 8:262
Soda lime–borosilicate glass. (See also Silicate glasses.)
composition 9:199
uses 9:199
Soda lime glass. (See also *Glass and glass-ceramics*; Silicate glasses.) 9:200
composition 9:198; 13:407, 409, 410
uses 9:198
Sodamide. (See also Sodium compounds.)
manufacture 18:212
uses 18:209
Soda soaps. See Sodium soaps.
Soda–titania–zirconium glass. (See also Silicate glasses.)
composition 9:199
uses 9:199
Soda–zinc glass. (See also Silicate glasses.)
composition 9:198
uses 9:198
Sodium. (See also Sodium alloys; Sodium compounds; and Sodium metal.) 18:207–260
acid salts 8:595
in acrylic fibers 4:349
in alkali salts 15:267
in alumina 5:116
in aluminum 18:218, 226
in aluminum alloys 5:144
in aluminum metal 5:144

in aluminum oxide 4:44; 5:186
in aluminum potassium sulfate 5:194
analysis 18:216
in asbestos 6:266, 267
atomic absorption spectroscopy 10:36, 131; 11:157; 14:125; 15:157, 393; 16:164; 18:220
in baking powder 17:388
in beer 7:652; 18:220
in blood serum 18:222
in calcium carbonate 8:84
in calcium metal 8:76
in cellulose 18:220
in cereals 9:233, 242
in cesium 9:282, 284
cesium–bismuth nitrite gravimetric method 18:229
in chloroplatinic acid 18:229
in chocolate and cocoa products 9:538, 595
in chromic acid 9:684, 699
in chrysotile 6:266, 267
in cigarette paper 10:36
in clays and clay minerals 10:130
in coal and coke 10:253
in cobalt compounds 10:327
in cobalt metal 10:303
colorimetric determination 18:224
in cotton 11:141, 157
in creams and lotions 11:92
in diatomaceous earth 11:492
in eggs and egg products 12:66
electrolytic method 10:327
emission spectroscopy 8:76; 11:493; 16:160; 18:217, 218
in fish 18:219, 220
flame emission spectroscopy 13:452
flame photometry 6:266; 7:652; 8:84, 595; 9:242, 284, 349, 362, 399; 10:303; 11:75, 226; 14:606; 15:272, 280, 399, 528; 17:381, 388, 400, 555; 18:218; 19:492
flame tests 6:583, 592
in fruit 18:219, 220
in fuel oil 18:219
fusion of nitrogen 16:454
in gas turbine fuel 18:222
in glass and glass-ceramics 13:452
gravimetric determination 18:217, 223
in hafnium 14:124
half-life 18:208
history 18:208
hydrogen chloride gravimetric method 15:267; 18:228
hydrolysis method for chlorocarbons 15:327
identification 18:217
iodometric method 18:229
ion exchange chromatography 16:188; 17:364, 550; 18:217
ion specific electrode techniques 18:222

isotopes 18:208
liquid chromatography 10:37
in lithium 15:272
in lithium carbonate 15:280
in lithium chloride 9:345
in lithium hydride 14:251
in lithium hydroxide 15:286
lithium tetra-*p*-tolylboron gravimetric method
  18:229
in lubricants 17:390
in lubricating oil 18:220, 226
in magnesium 15:393
in magnesium acetate 15:399
in magnesium alloys 15:393
in magnesium chloride 15:405
magnesium-1-naphthylamine-8-sulfonate
  octahydrate gravimetric method 18:229
magnesium-1-naphthylamine-8-sulfonate
  octahydrate volumetric method 18:229
in magnesium nitrate 15:408
in magnesium oxide 15:411
in magnesium sulfate 15:413
magnesium uranyl acetate colorimetric method
  18:227
magnesium uranyl acetate gravimetric method
  18:218, 223, 227
DL-methoxyphenylacetic acid gravimetric method
  18:218, 228
in milk 18:220
in molybdates 16:188
in molybdenum 16:161, 164
in molybdenum trioxide 16:161, 164
occurrence 18:207
in plants 17:384; 18:221, 227
in potassium bicarbonate 17:400
in potassium bromate 17:400
in potassium chlorate 9:359, 362; 15:528;
  17:400
in potassium chloride 9:339, 349
in potassium chromate 9:699
in potassium compounds 17:399
in potassium cyanide 17:400
in potassium dichromate 9:699; 17:400
in potassium hydrogen phthalate 17:403
in potassium iodate 14:609; 17:400
in potassium iodide 14:606
in potassium oxalate 17:400
potassium pyroantimonate gravimetric method
  18:229
in potassium-sodium chlorides 18:229
in potassium-sodium perchlorates 18:229
in potassium stannate 19:73
in qualitative test for alcohols 4:521
in rare earth chlorides 17:506
in rare earth fluorides 17:508
reduction of nitriles 16:399
in rubidium 17:555

sample decomposition 18:216
separation 17:364; 18:217
in shampoos 11:75
in silica ablators 4:3
in silicate rocks 17:381
silver nitrate volumetric method 16:189
in soaps 18:190
in sodium borohydride 14:251
in sodium cyclamate 11:226
in sodium hydride 14:251
in sodium thiopental 6:528
in soil 18:220
sulfuric acid gravimetric method 10:168
in terephthalic acid 17:163
in tobacco 19:152
in urine 18:222
uses 18:208
volumetric determination 14:251, 254; 18:224
in warfarin sodium 10:167, 168
in water 18:220, 221, 222
in wines 18:219, 221; 19:491, 492
Wurtz reaction 18:231
zinc uranyl acetate colorimetric method 11:141;
  18:225
zinc uranyl acetate gravimetric method 18:217,
  223
zinc uranyl acetate volumetric method 18:226
in zirconium 14:124
Sodium acetate. (See also Sodium compounds.)
  in certified colors 10:492
  chloride in 8:593
  commercial specifications 8:591; 18:235
  properties 8:591
  silver toluenesulfonate distillation 10:492
Sodium acetate-potassium dichromate
  turbidimetric method for barium 18:283
Sodium acetylide
  titration method 4:142
  xylene dispersions, analysis of 4:146
Sodium acid malonate
  acid-base titration, nonaqueous 15:431
  in malonic acid 15:431
Sodium acid pyrophosphate
  baking acid 6:426, 428, 434
  in baking powders 6:430, 434
  commercial specifications 6:428
  manufacture 17:126
Sodium adipate 4:429
Sodium alizarin sulfonate
  colorimetric method for zirconium 10:325
  identification reagent for aluminum 5:214, 233
Sodium alkyl sulfonates
  paper chromatography 12:575
Sodium alloys. (See also Sodium metal; and
  individual alloys.)
  uses 18:210
Sodium alum. See Aluminum sodium sulfate.

Sodium aluminates. (See also Aluminates.)
5:158
  analysis 5:159
  manufacture 5:159
  properties 5:158
  uses 5:159
Sodium aluminum hydride. (See also *Hydrides.*)
  analysis 14:247
  assay 14:247, 256
  commercial grades and specifications 14:247
  composition, determination of 14:247
  hydrogen content 14:247
  manufacture 14:246
  properties 14:243, 245
Sodium aluminum phosphate
  baking acid 6:426, 429
  in baking powders 6:430
  commercial specifications 6:429
Sodium aluminum sulfate
  baking acid 6:426, 429
Sodium amalgam. (See also Amalgams; Sodium
  alloys.)
  melting point 16:4
  uses 18:210
Sodium ammonium hydrogen phosphate
  gravimetric method for magnesium oxide 15:218
Sodium ammonium phosphate
  gravimetric method for aluminum 14:554
Sodium amobarbital 6:486, 502
  analysis 6:504, 510
  commercial grades and specifications 6:502
  impurities, determination of 6:510
  infrared spectrum 6:486
Sodium ampicillin. (See also Ampicillin.)
  impurities, determination of 16:566
  iodometric methods 16:562
Sodium arsenate 6:229
  analysis 6:234
  commercial specifications 6:230
  impurities, determination of 6:234
Sodium arsenite 6:229
  analysis 6:235
  commercial specifications 6:230
  impurities, determination of 6:235
  titration of chlorine 9:317, 353
  titration for manganese 15:465
Sodium azide. (See also Sodium compounds.)
  identification reagent for carbon disulfide
    8:273
  manufacture 16:472; 18:212
  properties 16:472
  sodium nitrite titration 16:473
Sodium azide–iodide
  volumetric method for oxygen 16:525, 528
Sodium barbital 6:486
  infrared spectrum 6:486
Sodium benzoate. (See also Benzoic

  acid salts.) 7:86
  acid–base titration 7:69; 8:63
  analysis 7:86
  in antifreezes 6:16
  antirusts 6:351
  in carbonated beverages 8:172
  commercial grades and specifications 7:86
  impurities, determination of 7:87
  in pharmaceuticals 8:63
  separation 7:62
  ultraviolet spectroscopy 7:70
Sodium benzylpenicillin. (See also Benzylpenicillin.)
  iodometric methods 16:562
  properties 16:556
  solubility 16:557
  structure 16:553, 567
Sodium bicarbonate. (See also Baking powder;
  Sodium compounds.) 6:425; 18:234
  acid–base titration 18:246
  in bakery chemicals 6:425, 436
  commercial grades and specifications 18:235
  gasometric determination of carbon dioxide
    6:436
  gravimetric method for chromium 9:647
  gravimetric method for vanadium 19:329
  in hypochlorite bleaches 9:355
  manufacture 18:234
  properties 18:234
  in sodium carbonate 18:246
  uses 18:234
Sodium bicarbonate–sodium chloride test
  for carboxyl content of cellulose 9:44
Sodium biphenyl
  hydrolysis of chlorocarbons 9:421
Sodium bismuthate. (See also Sodium compounds.) 7:206
  analysis 7:206
  chloride in 7:207
  colorimetric method for manganese 10:151;
    15:218, 457
  commercial specifications 7:206; 18:235
  manganese in 7:207
  redox titration 7:207
  volumetric method for manganese 10:314, 641;
    15:460
Sodium bismuthate–ferrous ammonium sulfate
  volumetric method for manganese 17:194
  volumetric method for manganous oxide 15:212
Sodium bismuth nitrite
  gravimetric method for cesium 17:549
  gravimetric method for rubidium 17:549
Sodium bisulfate. (See also Sodium compounds;
  Sulfates.)
  commercial specifications 18:235
  cooling system cleaners 6:351
  flux for ceramics 9:216

fusion of titanium dioxide pigments 19:125
manufacture 18:422
uses 18:422
Sodium bisulfite. (See also Sodium compounds.)
   colorimetric method for aldehydes 4:484
   commercial specifications 18:235
   identification reagent for carbonyl compounds
      8:342
   reaction of carbonyl compounds with 8:333
   volumetric method for acrylonitrile 16:409
   volumetric method for aldehydes 12:280
   volumetric method for furfural 13:237, 242
Sodium bisulfite addition method. (See also
   Bisulfite-addition method.)
   for acrylic monomers 4:195
   for acrylonitrile 4:341
   separation of volatile impurities in acrolein
      4:155
Sodium bisulfite–sodium hydroxide titration
   of cinnamaldehyde 10:52
Sodium bisulfite solubles
   in benzaldehydes 6:625
Sodium bitartrate
   gravimetric method for potassium 14:606
Sodium borate. (See also Sodium compounds.)
   in antifreezes 6:11
   commercial specifications 18:235
Sodium borate perhydrate
   active oxygen 7:224
Sodium borodeuteride 7:344
Sodium borohydride. (See also *Hydrides.*)
   7:312, 343, 344
   analysis 7:347; 14:247, 252
   argentimetric method 7:347
   assay 14:247, 252
   boron content 14:254
   commercial grades and specifications 14:246
   composition, determination of 14:247
   gas chromatography 14:253
   hydrogen content 14:247
   impurities, determination of 14:253
   iodometric titration 7:347; 14:253
   manufacture 14:245
   properties 14:243, 244
   sodium borohydride content 14:253
   sodium content 14:254
   uses 14:243
Sodium bromate
   analysis 8:8
   commercial grades and specifications 8:7
   properties 8:7
Sodium bromide. (See also Sodium compounds.)
   commercial grades and specifications 12:546
   commercial specifications 8:9, 10
   gravimetric method for silver 18:164
   manufacture 18:212
   properties 8:9

titration of hydroxylamine 14:452
titration of silver 14:555
Sodium butabarbital 6:512
   analysis 6:513
   commercial grades and specifications 6:513
   impurities, determination of 6:515
Sodium butyl phosphite extraction
   of sulfur from elastomers 12:131
Sodium camphorate
   gravimetric method for gallium 13:253, 258
Sodium carbide. (See also Sodium compounds.)
   manufacture 18:212
Sodium carbonate. (See also Sodium compounds.)
   18:235
   acid–base titration 6:359; 18:245
   in acid-type cooling system cleaners 6:352, 359
   alkalinity 18:245
   analysis 18:245
   assay 18:245
   commercial grades and specifications 18:235,
      237
   in embalming fluids 12:166
   fusion of boron and boron alloys 7:324
   fusion of ceramics 7:400
   fusion of glass 7:400
   fusion of lime and limestone 15:206
   fusion of nickel 16:300
   fusion of silica 13:417
   fusion of silicate rocks 9:643
   fusion of silicon 18:10
   fusion of silicon carbide 18:12
   fusion of stannic oxide 19:70
   fusion of strontium 18:274
   fusion of sulfur 15:223
   fusion of sulfur, inorganic 18:380
   fusion of tungsten 15:82; 16:338; 19:205, 206,
      207, 208, 214, 219
   fusion of tungsten minerals 19:197
   fusion of zirconium 16:173, 174
   gravimetric method for mercury 16:6, 7
   gravimetric method for nickel 16:292
   handling 18:245
   in hypochlorite bleaches 9:355
   impurities, determination of 18:246
   loss on heating 18:246
   manufacture 18:213, 235
   microchemical method for oxycodone 16:250
   microchemical method for phenazocine 16:250
   microchemical method for racemorphan 16:250
   microchemical method for thebaine 16:250
   properties 18:235
   specification tests 18:245
   uses 18:235, 238
   volumetric method for hydrogen chloride 14:
      396
Sodium carbonate–calcium oxide fusion
   of silicon 18:12

Sodium carbonate–hydrochloric acid
  gravimetric method for silicate 18:16
Sodium carbonate–potassium carbonate fusion
  of F D & C Blue No. 1 10:494
  of sulfur, inorganic 18:380
Sodium carbonate–potassium nitrate fusion
  of sulfur, inorganic 18:381
Sodium carbonate–sodium nitrate fusion
  of silicate rocks 9:647
Sodium carbonate–sodium peroxide fusion
  of sulfur, inorganic 18:381
Sodium carbonate–sodium sulfide
  colorimetric method for sulfur dyes 12:19
Sodium carbonyl. (See also Sodium compounds.)
  manufacture 18:213
Sodium carboxymethylcellulose. (See also
    Cellulose derivatives.) 9:87
  analysis 9:88
  commercial grades and specifications 9:88
  composition, determination of 9:89
  impurities, determination of 9:88
  manufacture 9:88
  properties 9:87
  viscosity 9:62, 90
Sodium cephalothin. (See also Cephalothin.)
  iodometric methods 16:562
  solubility 16:557
Sodium chlorate 9:357
  analysis 9:360
  ceric sulfate titration 9:353
  chlorate content 9:360
  commercial grades and specifications 9:359
  in hypochlorite bleaches 9:353
  impurities, determination of 9:360
  properties 9:358
  in sodium chlorite 9:355
  solubility 9:359
  uses 9:357
Sodium chlorate–nitric acid
  gravimetric method for manganese 16:297
Sodium chloride. (See also Salt, common;
    Sodium compounds.) 9:337; 18:239
  antifreeze 6:1
  in bakery chemicals 6:434
  in bread 6:440
  in cakes 6:446
  in certified colors 10:491
  colorimetric method for ethylene oxide 12:320
  colorimetric method for iridium 17:261
  commercial grades and specifications 9:337,
    338, 345, 347; 12:546
  commercial specifications 18:235
  in detergents 11:336, 355
  in eggs and egg products 12:66
  gravimetric method for mercury 16:29
  gravimetric method for silver 10:621; 18:164
  in hypochlorite bleaches 9:354

  impurities, determination of 9:343, 347
  manufacture 9:337; 18:212
  mercuric nitrate titration 9:360
  occurrence 18:239
  properties 9:335
  silver nitrate titration 6:434; 10:491; 11:355;
    12:66
  in soaps 18:182
  in sodium carbonate 18:247
  in sodium chlorate 9:360
  in sodium chlorite 9:355
  solubility 9:336
  uses 9:337; 18:239, 240
  Volhard titration 18:247
Sodium chloride–chlorine fusion
  of platinum group metals 17:238
Sodium chlorite 9:355
  analysis 9:356
  available chlorine 7:224, 227
  bleaching agents 7:222
  in chlorine dioxide bleaches 9:356
  colorimetric method for acetals 4:67
  colorimetric method for aniline and its derivatives
    5:452
  identification 9:356
  iodometric titration 7:226; 9:356
  properties 9:356
  uses 9:355
Sodium chromate. (See also Chromium compounds.)
    9:681
  analysis 9:683, 693
  assay 9:684, 693
  chromate content 9:684
  colorimetric method for hydrogen peroxide
    14:431
  commercial grades and specifications 9:683,
    695
  in cooling system flushes 6:352
  ferrous sulfate titration 9:685
  iodometric titration 9:684
  in metal cleaners 6:355, 368
  properties 9:682
  uses 9:681
Sodium citrate. (See also Citric acid and salts.)
  commercial grades and specifications 10:91
  uses 10:86
Sodium cloxacillin. (See also Cloxacillin.)
  iodometric methods 16:562
  properties 16:556
  solubility 16:557
Sodium cobalt(II) acetylacetonate. See Cobalt
    complexes.
Sodium cobaltic nitrate
  gravimetric method for potassium chlorate
    12:420
Sodium cobaltinitrite. (See also Sodium com-
    pounds.)

colorimetric method for potassium 17:377
commercial specifications 18:235
gravimetric method for potassium 17:368, 376
volumetric method for potassium 17:377
Sodium colistimethate. See Colistin sodium
   methanesulfonate.
Sodium compounds. (See also individual
   compounds.) 18:234–258
analysis 18:244
commercial specifications 18:234
impurities, determination of 18:245
sampling 18:216
specification tests 18:244
toxicity 18:214
Sodium copper chromate
commercial specifications 9:697
solubility 9:697
Sodium cyanide. (See also Cyanides; Sodium
   compounds.) 14:424
colorimetric method for hydroquinone 17:20
commercial specifications 14:424; 18:235
impurities in 14:425
manufacture 18:209
properties 14:424
solubility 14:424
uses 14:424
volumetric method for sulfur 18:376
Sodium cyclamate. (See also Cyclamates.)
   11:220
commercial grades and specifications 11:223
properties 11:221
sodium content 11:226
solubility 11:221
Sodium cyclohexylsulfamate. See Sodium
   cyclamate.
Sodium dehydrocholate 7:146
acid–base titration 7:149
Sodium dehydrocholic acid
commercial specifications 7:145, 146
uses 7:146
Sodium diborate. See Borax, anhydrous.
Sodium dibutyldithiocarbamate. (See also
   Dithiocarbamic acid salts.)
properties 11:575
uses 11:576
Sodium dichloroisocyanurate
available chlorine 7:224, 229
identification 7:225
iodometric titration 7:229
Sodium dichromate. (See also Chromium
   compounds.) 9:681
analysis 9:683, 693
assay 9:684, 693
chromium content 9:684
commercial grades and specifications 9:683,
   695
ferrous sulfate titration 9:685

iodometric titration 9:684
properties 9:682
in sodium chlorate 9:360
uses 9:681
Sodium dicloxacillin. (See also Dicloxacillin.)
iodometric methods 16:562
properties 16:556
solubility 16:557
Sodium diethyldithiocarbamate. (See also
   Dithiocarbamic acid salts; Sodium compounds.)
colorimetric method for copper 4:90; 9:51,
   241, 345; 10:95, 612; 11:45; 13:114; 18:176,
   202; 19:495
colorimetric method for nickel 10:128
colorimetric method for tellurium 10:304, 626
commercial specifications 18:235
extraction method for lead 15:178
gravimetric method for niobium 16:366
properties 11:575
ultraviolet spectroscopy 11:583
Sodium diisopropyldithiocarbamate. (See also
   Dithiocarbamic acid salts.)
ultraviolet spectroscopy 11:583
Sodium 2,6-dimethoxyphenylpenicillin. See
   Sodium methicillin.
Sodium dimethyldithiocarbamate. (See also
   Dithiocarbamic acid salts.)
properties 11:575
ultraviolet spectroscopy 11:583
uses 11:576
Sodium diphenyl
in decomposition of organic compounds 8:18
Sodium 2,2'-disulfostilbene-4,4'-diazobisphenol.
   See Brilliant yellow.
Sodium dithionate
antichlor 7:222
Sodium EDTA. (See also EDTA; *EDTA and
   related compounds.*)
etchant for ceramics 9:225
Sodium–ethanolamine
hydrolysis of chlorocarbons 9:419
Sodium ethylenediaminedithiocarbamate. (See
   also Dithiocarbamic acid salts.)
properties 11:575
Sodium ethylmercurithiosalicylate. See Thimerosal.
Sodium ferric pyrophosphate. (See also Iron
   compounds.)
uses 15:86
Sodium ferrocyanide. (See also Ferrocyanides.)
uses 15:86
Sodium fluoaluminate 5:177
analysis 5:178
assay methods 5:178
commercial grades and specifications 5:178
impurities, determination of 5:180
properties 5:177
Sodium fluoberyllate 7:122

Sodium fluoborate 7:337
Sodium fluoride. (See also Sodium compounds.)
    6:49
  carrier gas distillation fusion methods 8:636
  commercial grades and specifications 13:12
  commercial specifications 18:235
  fluorometric method for uranium 19:241, 242
  in glass 9:200
  impurities, determination of 13:33
  loss on drying 13:33
  manufacture 18:212
  properties 13:11
  in sodium fluoaluminate 5:180
  specification tests 13:33
  toxicity 13:13
  uses 13:12, 13
  volumetric method for aluminum 5:217
Sodium fluoroacetate
  toxicity 13:13
Sodium fluoroaluminate
  properties 13:11
Sodium fluoroborate
  properties 13:11
Sodium fluorosilicate
  gravimetric method for scandium 17:564
Sodium formaldehyde sulfoxalate
  color test for vat dyes 12:18
Sodium formate
  acid–base titration, nonaqueous 13:135
  analysis 13:135
  assay 13:135
  commercial specifications 13:134
  gravimetric method for palladium 17:258
  gravimetric method for platinum 17:260
  impurities, determination of 13:136
  iodometric titration 13:135
  manufacture 13:122
  reduction method for platinum group metals
    17:250
  uses 13:120, 134
Sodium fusidate
  solubility 5:509, 510
Sodium gluconate
  complexing agent for aluminum 5:219
Sodium glycochenodeoxycholate
  properties 7:143
Sodium glycocholate 7:142
  hydrolysis 7:141
  properties 7:143
Sodium glycodeoxycholate 7:142
  properties 7:143
Sodium glycolate
  in sodium carboxymethylcellulose 9:91
Sodium glycolithocholate
  properties 7:143
Sodium–gold alloy. (See also Sodium alloys.)
  uses 18:211

Sodium heparin 10:140
  analysis 10:140
  anthrone colorimetric method 10:149
  assay 10:143
  cetylpyridinum chloride turbidimetric method
    10:149
  electrophoresis 10:142
  enzyme methods 10:149
  ferric chloride colorimetric method 10:143
  identification 10:140
  impurities, determination of 10:150
  infrared spectroscopy 10:143
  ion exchange chromatography 10:140
  occurrence 10:140
  paper chromatography 10:141
  polarography 10:142
  potency tests 10:143
  properties 10:140
  protamine sulfate turbidimetric method 10:149
  separation 10:140
  solubility 10:150
  specification tests 10:150
  specific rotation 10:143
  streptomycin sulfate turbidimetric method
    10:149
  thin-layer chromatography 10:142
  toluidine blue colorimetric method 10:148
  Wilson technique 10:143
Sodium hexachloroiridate
  manufacture 17:226
  properties 17:226
Sodium hexahydroxyplatinate
  manufacture 17:227
  properties 17:227
Sodium hexanitritorhodate
  manufacture 17:225
  solubility 17:225
  uses 17:225
Sodium hexavanadate. (See also Vanadates.)
  manufacture 19:325
Sodium hexobarbital 5:369
  acid–base titration 5:370
  analysis 5:368
  anesthetic 5:368
  assay 5:369
  bromination method 5:369
  infrared spectrum 5:368
  properties 5:368
  ultraviolet spectra 5:370, 371
Sodium hydride. (See also *Hydrides*; Sodium
    compounds.)
  analysis 14:247, 250
  assay 14:247
  commercial grades and specifications 14:246
  composition, determination of 14:247, 251
  hydrogen content 14:247
  manufacture 14:245; 18:209, 212

properties 14:243, 244
sodium content 14:251
uses 14:242
Sodium hydrogen carbonate. See Sodium
    bicarbonate.
Sodium hydrogen sulfate fusion
    of silicate rocks 9:644
Sodium hydrosulfite. (See also Sodium
    compounds.)
    color test for indigoid dyes 14:503
    color test for vat dyes 12:14
    manufacture 18:213
    reducing agent for azo dyes 6:406
Sodium hydroxide. (See also Sodium compounds.)
    18:240
    absorption of hydrogen cyanide 14:412
    acid–base titration 14:253; 18:250
    alkalinity 18:250
    analysis 18:248
    assay 18:250
    colorimetric method for p-t-butylcatechol
        18:304
    colorimetric method for ethylene chlorohydrin
        14:155
    colorimetric method for phenindione 10:182,
        183
    colorimetric method for ruthenium 17:247
    colorimetric method for styphnate 12:415
    color test for natural coloring materials 10:537
    color test for reactive dyes 12:19
    commercial grades and specifications 18:235,
        241, 242
    distillation of ammonia 11:350
    etchant for ceramics 9:224
    flux for ceramics 9:216
    fusion of cements 9:125
    fusion of silicon 18:12
    fusion of titanium minerals 19:93
    fusion of tungsten minerals 19:198
    gravimetric method for chromium 9:688
    gravimetric method for magnesium 15:360
    gravimetric method for mercury 16:6, 7
    gravimetric method for nickel 16:297
    gravimetric method for rare earths 17:476
    gravimetric method for silver 18:164
    gravimetric method for tin 19:33
    gravimetric method for titanium 16:337; 19:94
    gravimetric method for uranium 19:232
    gravimetric method for vanadium 19:330
    handling 18:249
    hydrolysis of sulfamic acid 11:229
    in hypochlorite bleaches 9:354
    impurities, determination of 18:245, 252
    manufacture 18:212, 240
    mercury content 16:45, 46
    in methanol 16:115
    properties 18:240

safety 18:214, 249
sampling 18:249
in sodium borohydride 14:253
in sodium chlorite 9:355
in sodium metal 18:232
solubility 18:240
specification tests 18:250
spot test for polyurethane foams 19:282
uses 18:241
volumetric method for ammonium 13:136
volumetric method for ethylene chlorohydrin
    14:165
volumetric method for silica 18:30
in whitewall tire cleaners 6:354
Sodium hydroxide–cupric sulfate
    colorimetric method for protein 19:476
Sodium hydroxide–ferric chloride
    colorimetric method for nitroethane 16:434
    colorimetric method for nitroparaffins 16:434
    colorimetric method for 1-nitropropane 16:434
Sodium hydroxide–silver nitrate
    volumetric method for chloroformate groups
        17:340
Sodium 3-hydroxy-4-[(1-hydroxy-2-naphthyl)azo]-
    7-nitro-1-naphthalenesulfonate. See Eriochrome
    Black T.
Sodium hypobromite
    colorimetric method for chlorhexidine 11:570
Sodium hypochlorite. (See also Hypochlorites.)
    9:352
    available chlorine 7:224, 226
    bleaching agents 7:222
    in chlorine dioxide bleaches 9:356
    colorimetric method for rhodium 17:248
    identification reagent for amides 8:600
    iodometric titration 7:226
    oxidation agent for methyl ketones 8:338
    properties 9:352
    uses 9:352
    volumetric method for hypophosphite 17:115
    volumetric method for 1-nitrobutane 16:431
    volumetric method for 2-nitrobutane 16:431
    volumetric method for nitroethane 16:431
    volumetric method for nitromethane 16:431
    volumetric method for 1-nitro-2-methylpropane
        16:431
    volumetric method for nitroparaffins 16:428
    volumetric method for 1-nitropropane 16:431
    volumetric method for 2-nitropropane 16:431
    volumetric method for oxygen 16:527
    volumetric method for phosphite 17:115
Sodium hyponitrite. (See also Sodium com-
    pounds.)
    manufacture 18:213
Sodium hyposulfite
    color test for natural coloring materials
        10:537

Sodium indigotindisulfonate. See Indigo carmine.
Sodium iodate. (See also Iodates.) 14:607
  properties 14:607
  solubility 14:607
Sodium iodide. (See also Iodates.) 14:595
  assay 14:599
  in certified colors 10:493
  colorimetric method for hydroperoxides
    12:115
  commercial grades and specifications 12:546;
    14:595
  impurities, determination of 14:605
  liquid chromatography 10:493
  potassium iodate titration 14:599
  properties 14:595
  titration of mercury 16:37
  ultraviolet spectroscopy 10:493
  uses 14:595
Sodium lactate. (See also Lactic acid salts.)
  commercial grades and specifications 15:129
  manufacture 15:129
  properties 8:591; 15:129
  uses 15:129
Sodium lauryl sulfate
  benzethonium chloride titration 12:65
  in egg white 12:65
  properties 18:144
  upholstery cleaner 6:354
Sodium–lead alloys. (See also Sodium alloys.)
  uses 18:210
Sodium liothyronine. (See also Thyroid hormones.)
  analysis 14:207
  assay 14:207
  manufacture 14:190
  properties 14:190
  solubility 14:190
  uses 14:190
Sodium–liquid ammonia hydrolysis
  of chlorocarbons 9:418
Sodium–lithium fluoride
  fluorometric method for uranium 19:251, 252
  fusion of uranium 19:252
Sodium lithol
  in enamels 12:184
  identification 12:184
Sodium malonate
  infrared spectroscopy 15:426
  near-infrared spectroscopy 15:426
Sodium metabisulfite. (See also Sodium com-
    pounds; Sulfites.)
  commercial specifications 18:235
  gravimetric method for palladium 17:264
  gravimetric method for platinum 17:264
  iodometric method 18:395
  volumetric method for aldehydes 12:334
Sodium metaborate 7:378
  iodometric method 14:254, 255

  properties 7:378
  in sodium borohydride 14:254
  uses 7:378
Sodium metal
  analysis 18:216, 229
  commercial grades and specifications 18:216
  handling 18:214
  impurities, determination of 18:229
  manufacture 18:214
  mercury extraction 18:231
  oxygen in 8:635, 636
  properties 18:211
  safety 18:214
  sample preparation 18:233
  sampling 18:230
  solubility 18:212
  uses 18:208
  vacuum distillation fusion methods 8:635
Sodium metaniobate
  manufacture 16:355
  solubility 16:355
Sodium metaperiodate. (See also Periodates.)
    14:609
  assay 14:609
  commercial grades and specifications 14:609
  impurities, determination of 14:610
  potassium iodide–sodium thiosulfate volumetric
    method 14:610
  properties 14:609
  solubility 14:609
  uses 14:609
  volumetric method for glycols 13:591
Sodium metaphosphates 17:123
Sodium metasilicate. (See also Sodium silicates.)
    18:42
  commercial grades and specifications 18:44
  cooling system flushes 6:352
  sodium silicate content 18:45
Sodium metatantalate
  manufacture 16:355
  solubility 16:355
Sodium metavanadate. (See also Metavanadates.)
  colorimetric method for tartaric acid 19:420
  manufacture 19:325
  properties 19:345
Sodium methicillin. (See also Methicillin.)
  iodometric methods 16:562
  properties 16:556
  solubility 16:557
Sodium methohexital 5:369
  analysis 5:371
  assay 5:371
  properties 5:371
  ultraviolet spectroscopy 5:371
Sodium methylate titration
  of dimethyl terephthalate 17:165
Sodium methyldithiocarbamate. (See also

Dithiocarbamic acid salts.) 13:194
  manufacture 13:194
  properties 13:194
  solubility 13:195
  toxicity 13:195
  uses 11:576; 13:194
Sodium molybdate. (See also Molybdates.)
  analysis 16:184
  molybdenum content 16:184
  uses 16:184
Sodium molybdate–chlorostannous acid
  colorimetric method for phosphate 16:117
Sodium molybdate–citric acid
  gravimetric method for phosphorus 17:98
Sodium molybdate–hydrazine sulfate
  colorimetric method for phosphorus 15:284
Sodium molybdate–quinoline
  gravimetric method for phosphorus 17:68
Sodium molybdate–quinoline hydrochloride
  volumetric method for phosphorus 12:116
Sodium molybdate–stannous chloride
  colorimetric method for phosphorus 14:134;
    16:334
  colorimetric method for tantalum 18:449
Sodium monoxide. See Sodium oxide.
Sodium nafcillin. (See also Nafcillin.)
  iodometric methods 16:562
  properties 16:556
  solubility 16:557
  ultraviolet spectroscopy 16:560
Sodium 1,2-naphthoquinone-4-sulfonate
  colorimetric method for anesthetics 5:405, 407
  colorimetric method for cystine 8:467
  colorimetric method for morpholine 16:201
  colorimetric method for nitromethane 16:435
  colorimetric method for procaine hydrochloride
    5:407
Sodium nitrate. (See also Sodium compounds.)
  color test for leuco ester vat dyes 12:18
  commercial grades and specifications 16:497
  commercial specifications 18:235
  manufacture 18:213
  nitrate content 12:587
  in photographic films 12:587
Sodium nitride. (See also Sodium compounds.)
  manufacture 18:212
Sodium nitrite. (See also Sodium compounds.)
  analysis 16:491
  assay 16:491
  colorimetric method for 1-chlorobutane 9:388
  colorimetric method for 2-chlorobutane 9:388
  colorimetric method for 1-chlorodecane 9:388
  colorimetric method for 1-chloroheptane 9:388
  colorimetric method for 2-chloroheptane 9:388
  colorimetric method for 3-chloroheptane 9:388
  colorimetric method for 4-chloroheptane 9:388

  colorimetric method for 1-chlorohexane 9:388
  colorimetric method for 1-chlorooctadecane
    9:388
  colorimetric method for 1-chlorooctane 9:388
  colorimetric method for 2-chlorooctane 9:388
  colorimetric method for 1-chloropentane 9:388
  colorimetric method for 2-chloropentane 9:388
  colorimetric method for 1-chloropropane 9:388
  colorimetric method for 2-chloropropane 9:388
  colorimetric method for morphine 16:262
  colorimetric method for m-phenylenediamine
    11:446
  colorimetric method for Schaeffer's salt 16:237
  commercial grades and specifications 16:490
  commercial specifications 18:235
  diazotization method for acetaminophen 6:195
  diazotization method for acetanilide 6:194
  diazotization method for p-aminophenol 6:195
  diazotization method for phenacetin 6:194
  diazotization method for p-phenetidine 6:194
  gravimetric method for gold 14:19, 36
  gravimetric method for mercury 16:7
  hydrazine sulfate–iodine volumetric method
    16:492
  manufacture 18:213
  potassium chlorate–silver nitrate volumetric
    method 16:492
  potassium permanganate titration 16:491
  titration of aminonaphthalenesulfonic acids
    16:228
  titration of cyclamates 11:225
  titration of hydrazoic acid 16:472
  titration of lead azide 16:473
  titration of manganese 15:467
  titration of naphthylamines 16:227
  titration of palladium azide 16:473
  titration of phenacetin 6:190
  titration of silver azide 16:473
  titration of sodium azide 16:473
Sodium nitrite–Cellosolve
  colorimetric method for toluene diisocyanate
    19:313
Sodium nitrite–silver nitrite
  colorimetric method for mercury 16:19, 21
Sodium nitroferricyanide. (See also Sodium
    compounds.)
  commercial specifications 18:235
Sodium nitroprusside. (See also Iron compounds.)
  colorimetric method for acetone 4:111
  colorimetric method for p-cresol 17:20
  colorimetric method for lactic acid 19:422
  colorimetric method for methionine 8:505, 506
  colorimetric method for sulfide 11:142; 18:393
  uses 15:86
Sodium nitroprusside–diethanolamine
  colorimetric method for acetaldehyde 19:308

colorimetric method for propionaldehyde
19:308
Sodium nitroprusside dihydrate
microchemical method for hydromorphone
16:250
Sodium nitroprusside–morpholine
colorimetric method for acetaldehyde 12:237
Sodium orthoniobate
manufacture 16:355
Sodium orthophosphates. (See also individual
compounds.) 17:119
Sodium orthosilicate. (See also Sodium
silicates.) 18:42
Sodium orthotantalate
manufacture 16:355
Sodium oxacillin. (See also Oxacillin.)
iodometric methods 16:562
properties 16:556
solubility 16:557
Sodium oxalate. (See also Sodium compounds.)
commercial specifications 18:235
extraction of milk protein from chocolate
9:580
impurities, determination of 18:245
manufacture 18:213
properties 8:591
Sodium oxide. (See also Sodium compounds.)
acid–base titration 18:122, 231
in alkali barium silicate glass 9:218
in cements 9:120
in ceramics 9:194
in clays and clay minerals 10:123
in cobalt(II) orthophosphate 10:343
in cobalt oxides 10:342
in cobalt(II) sulfate 10:344
emission spectroscopy 9:218; 15:226
in enamels 9:195
flame photometry 9:121; 10:123
in glass and glass-ceramics 13:413, 415, 423
in glazes 9:196
in hydraulic cements 9:100
in lime and limestone 9:191; 15:226
manufacture 18:212
in refractories 9:206
in silicate glasses 9:198
in silicone resins 18:122
in sodium metal 18:231
in sodium silicate 18:44
zinc uranyl acetate gravimetric method 13:423
Sodium penicillin G. See Sodium benzylpenicillin.
Sodium penicillin O. (See also Penicillin O.)
iodometric methods 16:562
Sodium pentaborate 7:379
commercial specifications 7:379
properties 7:379
Sodium perborate 7:379

available oxygen content of 7:381
commercial grades and specifications 7:379
identification 7:381
oxidation of cobalt 10:370
properties 7:379
Sodium perchlorate. (See also Perchlorates.)
properties 9:363
solubility 9:364
Sodium periodate. (See also Sodium compounds.)
colorimetric method for ethylene glycol 13:571
colorimetric method for glycerol 13:571
colorimetric method for glycols 13:571
colorimetric method for propylene glycol 13:571
commercial specifications 18:235
volumetric method for glycerol 11:365; 13:514;
14:580
Sodium periodate–chromotropic acid
colorimetric method for glycols 13:598
Sodium permanganate. (See also Manganese
compounds.)
properties 15:450
Sodium peroxide. (See also Sodium compounds.)
18:242
active oxygen 7:224
analysis 18:258
assay 16:536; 18:258
commercial grades and specifications 16:535;
18:235, 243
flux for ceramics 9:216
fusion of alloys 7:400
fusion of ceric oxide 9:257
fusion of chrome ores 9:700
fusion of chromium 9:637, 653
fusion of ferrosilicon 18:13
fusion of iron ore 9:645
fusion of molybdenum disilicide 18:13
fusion of organosilane esters 18:78
fusion of phosphorus compounds, organic 17:68
fusion of refractory ores 19:236
fusion of slags 7:400
fusion of silicon 18:13
fusion of silicon carbide 18:13
fusion of sulfur compounds, organic 18:379
fusion of sulfur, inorganic 18:381
fusion of tin 19:60
fusion of tungsten minerals 19:197, 203
fusion of vanadium 19:328
fusion of zircon 14:140
gravimetric method for nickel 16:297
handling 18:243
impurities, determination of 16:536
manufacture 16:535; 18:212, 243
potassium permanganate titration 16:536;
18:258
properties 16:535; 18:242
specification tests 16:536
uses 16:535; 18:243

Sodium peroxide bomb method
  for chlorinated paraffins 9:498
  for chlorocarbons 9:414
  for sulfur compounds, organic 18:380
Sodium phosphate. (See also Sodium com-
    pounds.)
  commercial grades and specifications 18:235
  separation scheme for cobalt 10:352
  whitewall tire cleaners 6:354
Sodium phosphate dibasic
  commercial specifications 17:127
Sodium phosphate glasses 17:124
  manufacture 17:126
Sodium phosphate monobasic
  commercial specifications 17:127
Sodium phosphate tribasic
  commercial specifications 17:127
Sodium phosphide. (See also Sodium com-
    pounds.)
  manufacture 18:212, 213
Sodium phosphite. (See also Phosphites.)
  manufacture 17:112
Sodium piperidinedithiocarbamate
  gravimetric method for niobium 16:366
Sodium plumbite
  color test for sulfur 14:347
  color test for sulfur compounds 14:347
Sodium polyphosphates 17:123
Sodium-potassium alloy. (See also Sodium
    alloys.)
  impurities, determination of 18:231
  Wurtz reaction 18:231
Sodium potassium tartrate
  colorimetric method for nickel 9:345
Sodium pyroantimonate 6:52
Sodium pyrophosphate
  fusion of molybdenum 16:172
  properties 17:122
  volumetric method for manganese 15:458, 462
Sodium pyrosulfate
  flux for ceramics 9:216
  fusion of cobalt 16:327
  fusion of niobium ores 16:361
  fusion of tantalum ores 16:361
Sodium rhodizonate
  colorimetric method for barium 11:140, 144
  colorimetric method for calcium 11:140
  colorimetric method for lead 15:173
Sodium salicylate. (See also Salicylic acid deriva-
    tives.) 6:124
  commercial grades and specifications 6:127
  infrared spectrum 6:128
  manufacture 6:125, 126
  properties 6:124
  ultraviolet spectroscopy 6:130, 131, 132
Sodium secobarbital 6:524
  analysis 6:525

commercial grades and specifications 6:524
impurities, determination of 6:525
Sodium selenide. (See also Sodium compounds.)
  manufacture 18:212
Sodium selenite. (See also Selenium compounds.)
  uses 17:607
Sodium sesquicarbonate. (See also Sodium
    carbonate.) 18:239
  commercial specifications 18:239
  manufacture 18:239
  occurrence 18:239
  properties 18:239
  uses 18:239
Sodium sesquisilicate. (See also Sodium silicates.)
    18:42
  commercial grades and specifications 18:44
  sodium silicate content 18:45
Sodium silicates. (See also Silicates; and individual
    compounds.) 18:42
  acid-base titration 18:44
  adhesives 4:403
  analysis 18:44
  commercial grades and specifications 18:43
  manufacture 18:43
  properties 18:42
  in sodium metasilicate 18:45
  sodium oxide content 18:44
  in sodium sesquisilicate 18:45
  uses 18:43
  whitewall tire cleaners 6:354
Sodium silicide. (See also Sodium compounds.)
  manufacture 18:212
Sodium soaps. (See also Soaps.)
  manufacture 18:179
  uses 18:179, 180
Sodium stannate. (See also Stannates.)
  alkali content 19:72
  analysis 19:72
  commercial grades and specifications 19:72
  impurities, determination of 19:72
  manufacture 19:72
  tin content 19:72
  uses 19:72
Sodium sulfate. (See also Sodium compounds;
    Sulfates.) 18:243
  in ammonium dichromate 9:687
  barium chloranilate colorimetric method 11:84
  barium chloride gravimetric method 10:492;
    18:247
  barium chloride titration 10:492
  in car washes 354, 365
  in certified colors 10:491
  colorimetric method for ethinyl estradiol
    10:588
  commercial grades and specifications 18:235,
    244, 422
  manufacture 18:244, 422

phosphoric acid–hydrochloric acid colorimetric method 9:360
properties 18:243
in sodium carbonate 18:247
in sodium chlorate 9:360
solubility 18:243
uses 18:244, 422
Sodium sulfate–hydrogen peroxide
colorimetric method for titanium 11:75
Sodium sulfate–potassium permanganate
colorimetric method for barium 11:140
Sodium sulfide. (See also Sodium compounds; Sulfides.)
colorimetric method for lead 15:192
commercial specifications 18:235
gravimetric method for lead 10:95
gravimetric method for rhodium 17:253
gravimetric method for ruthenium 17:256
iodometric method 18:395
manufacture 18:212, 393
titration of thiuram disulfides 11:594
Sodium sulfite. (See also Sodium compounds; Sulfites.)
commercial specifications 18:235
iodometric method 18:395
mercuric thiocyanate spectroscopic method 18:408
in paper pulp 18:408
reagent for acetals 4:69
reagent for free formaldehyde 4:399
reagent for vinyl ethers 4:69
sodium thiosulfate content 18:432
uses 18:402
volumetric method for aldehydes 12:281
volumetric method for ethylene oxide 12:323
volumetric method for formaldehyde 12:168; 13:108; 17:301
volumetric method for ketones 12:281
volumetric method for sulfur 12:146; 18:375
Sodium tartrate. (See also Sodium compounds.)
commercial specifications 18:235
Sodium taurochenodeoxycholate
properties 7:143
Sodium taurocholate 7:142
hydrolysis 7:141
properties 7:143
Sodium taurodeoxycholate 7:142
properties 7:143
Sodium taurohyodeoxycholate 7:142
Sodium taurolithocholate
properties 7:143
Sodium telluride. (See also Sodium compounds.)
manufacture 18:212
Sodium tetraborate
gel test for locust bean gum 12:580
Sodium tetraborate decahydrate. See Borax.
Sodium tetraborate pentahydrate 7:375

commercial grades and specifications 7:376
properties 7:376
screen analysis 7:377
solubility 7:376
uses 7:376
Sodium tetraborates. (See also Borax; Kernite; Tincalconite.) 7:313, 368, 373
Sodium 1, 2, 3, 4-tetrahydro-2-methyl-1,4-dioxo-2-naphthalenesulfonate. See Menadione sodium bisulfite.
Sodium tetrametaphosphate
properties 17:123
Sodium tetraphenylborate. (See also Sodium compounds.)
commercial specifications 18:235
Sodium tetraphenylboron
colorimetric method for potassium 17:381
gravimetric method for benzethonium chloride 11:558
gravimetric method for cesium 9:298
gravimetric method for potassium 17:367, 378; 19:492
titration of benzethonium chloride 11:557
titration of cesium 9:298
titration of cetylpyridinium chloride 11:557
titration of potassium 17:379, 409
Sodium thiamylal 5:369; 6:526
analysis 5:372; 6:526
assay 5:372
commercial specifications 6:526
impurities, determination of 6:526
properties 5:372
Sodium thiocyanate. (See also Sodium compounds; Thiocyanates.)
colorimetric method for iron 15:49
colorimetric method for molybdenum 16:158
commercial grades and specifications 18:235, 434
properties 18:434
Sodium thiocyanate–stannous chloride
colorimetric method for molybdenum 16:382
Sodium thiopental 5:369; 6:527
analysis 5:372; 6:527
assay 5:372
commercial specifications 6:527
properties 5:372
sodium content of 6:528
ultraviolet spectroscopy 6:527
Sodium thiosulfate. (See also Sodium compounds; Thiosulfates.)
commercial grades and specifications 18:235, 431
iodometric titration 18:432
properties 18:430
in sodium sulfite 18:432
titration of antimony 6:49, 56, 66
titration of arsenic 6:243

titration of bromine 9:322, 344
titration of iodine 9:322, 344; 14:591
titration of maltose 12:607, 609
titration of selenious acid 13:439
titration of sugar 10:6
uses 18:430
Sodium thiosulfate–acetic acid
  volumetric method for epoxy compounds
    12:216
Sodium thiosulfate–Chloramine-T
  colorimetric method for bromide 19:496
Sodium thiosulfate–hydrochloric acid
  volumetric method for mercury fulminate
    12:409
Sodium thiosulfate–potassium iodide titration
  of bromate 9:285
Sodium thyroxine. See Thyroxine.
Sodium toluene sulfonate
  in detergents 11:366
  ultraviolet spectroscopy 11:366
Sodium triethylborohydride 7:344
Sodium trimetaphosphate
  properties 17:123
Sodium trimethoxyborohydride 7:344
Sodium tripolyphosphate
  impurities, determination of 17:137
  manufacture 17:126
  paper chromatography 17:137
Sodium tungstate. (See also Sodium compounds;
    Tungstates.)
  alkalinity 19:221
  analysis 19:221
  colorimetric method for vanadium 12:158
  commercial specifications 18:235; 19:221
  impurities, determination of 19:221
  manufacture 19:220
  specification tests 19:221
  uses 19:220
Sodium tungstate–sulfuric acid
  gravimetric method for gelatin 13:364
Sodium tungsten bronzes. (See also Tungsten
    compounds.)
  uses 19:220
Sodium vanadate. (See also Vanadates.)
  analysis 19:327
  dissolution 19:327
Sodium warfarin. (See also Coumarins.)
  infrared spectroscopy 11:162
Sodium–zinc alloy. (See also Sodium alloys.)
  uses 18:211
Soft drinks
  colors in 10:527
  cyclamate content 11:233
Softening point
  of bituminous materials 6:297
  of copper metal 10:627
  of epoxy resins 12:252, 257

of hydrocarbon resins 14:266, 281, 290
  temperature susceptibility 6:304
Soft magnetic alloys. (See also Cobalt alloys.)
    10:288
Soil
  analysis 17:363
  benomyl content 13:206
  boron in 7:407, 409, 411, 416, 417
  cadmium content 15:186
  calcium content 8:110; 18:220
  cobalt content 10:369
  copper content 10:614; 15:186
  dibromochloropropane content 13:184
  ethylene bromide content 13:184
  halogenated hydrocarbon content 13:183
  lead content 15:186
  magnesium content 15:371; 18:220
  mercury content 16:41
  methyl bromide content 13:184
  methyl isothiocyanate content 13:192
  nickel content 16:299
  potassium content 17:363, 384; 18:220
  radium content 17:453
  sample preparation 17:363
  sodium content 18:220
  strontium content 18:269
  tin content 19:64
  zinc content 15:186
Soil cement. See Cement-treated soils.
Soil removal tests
  for drycleaning detergents 11:613, 615
Solder. (See also Copper alloys; Tin alloys.)
  aluminum in 5:226
  composition 10:631; 19:24
  tin content 19:52
Solidification point
  of benzaldehydes 6:624
  of benzene 7:48
  of caprolactam 8:123
  of fats and fatty oils 12:476
  of high-explosive mixtures 12:468
  of naphthalene 16:211
  of phenol 17:40
  of phenols 17:25, 40
Solids content
  of amino resins 5:280
  of amino resin adhesives 4:399
  of condensed milk 16:145
  of distilled alcoholic beverages and intermediates
    4:485
  of drycleaning solvents, used 11:610
  of dry milk 16:145
  of eggs and egg products 12:59
  of epoxy resins 12:251
  of fermented milk products 16:136
  of frozen desserts 16:143
  of ice cream 16:143

of ion exchange resins 15:13
of musts 19:402
of oxygen 16:533
of polytetrafluoroethylene 13:82
of reducing water 4:479
of rubber-based adhesives 4:403
of undistilled intermediates for distilled
    alcoholic beverages 4:480
of water 19:380
of water for carbonated beverages 8:178
of wines 19:404
of yeast 19:522, 526
Solids, dissolved
in beer 7:485, 572
Solids mass spectroscopy
of ceramics 9:217
Solids-not-fat content
of raw milk 16:128
Solox 6:76
colorimetric method using diazotized
    p-nitroaniline 6:106
Solox NS 6:77
Solprene 12:87
Solubility
of dental porcelains 11:278
of dental restorative materials, composite
    11:314
for diatomites 11:494
of dry milk 16:146
of eggs and egg products 12:68
identification of acrylic and modacrylic fibers
    4:247
identification of textile finishes 4:329
of naphthalene 16:211
separation of opium alkaloids 4:597
of sodium heparin 10:150
Solubility temperatures
of ethylene and propylene polymers 12:348
Soluble ash content
of black tea 18:487
Soluble material
in barium chromate 9:698
in carbon black 17:202
in chromium pigments 9:707
in copper(I) oxide 10:672
in corn products 11:11
in cuprous oxide 17:196
in diatomaceous silica 17:192
in elastomers, synthetic 12:89
in iron blues 17:201
in leaded zinc oxide 17:189
in magnesium carbonate 15:402
in magnesium oxide 15:409
in pigments 17:177
in roasted coffee 10:417, 421, 435
in talc 10:544
in zinc sulfide pigments 17:190

Solution fractionation
of polycarbonates 17:346
Solution method
for infrared samples 4:239
Solution viscosity
of impact polystyrene 18:329
of polystyrene 18:323
Solvay process
manufacture of ammonium chloride 5:333
Solvent dyes 6:402
Solvent extraction
of caprolactam 17:296
of fat in corn products 11:7
of fatty acids in shampoos 11:81
Solvents
in enamels 12:170, 175, 188
mercaptan content 15:560
of methacrylate polymers 16:98
in penicillins 16:566
in pesticides and insecticides 19:614
in polycarbonates 17:349
in polyurethanes 19:310
water in 4:24
Solvents, residual
in acrylic and modacrylic fibers 4:343
Solvent test
for reactive dyes 12:19
Solway purple
colorimetric method for boric acid 7:408
Somatotropin. See Growth hormone.
Sonic grading test
for verified grinding wheels 7:302
Sonic velocity
of polyamide fibers 17:315
Sorbates
in wines 19:482
Sorbic acid
properties 8:397
separation 19:482
solvent extraction 19:482
steam distillation 19:482
2-thiobarbituric acid colorimetric method
    19:482
ultraviolet spectroscopy 19:482
in wines 19:416, 482
Sorbitan derivatives
in automobile polishes 6:353
Sorbitol 5:22
in embalming fluids 12:166
enzymatic method 19:451
gas chromatography 5:31; 19:451
ion exchange chromatography 19:450
paper chromatography 13:502
polarography 19:449
separation 19:444
structure 19:449
thin-layer chromatography 13:502; 19:450

in wines 19:425, 444, 449
Soreflon. See Polytetrafluoroethylene.
Sorel cement. (See also Hydraulic cements.)
  9:96
Sour cream. (See also Fermented milk products.)
  composition 16:135
  manufacture 16:135
Sour half and half. (See also Fermented milk
  products.)
  composition 16:135
  manufacture 16:135 ·
Soybean glues 4:400
Soybean oil. (See also *Fats and fatty oils.*)
  in alkyd resins 5:23
  composition 12:473
  gas chromatography 5:30
  properties 12:473
  solubility 13:539, 552
Soybean oil fatty acids
  gas chromatography 5:25
SPADNS
  colorimetric method for fluorine 11:493, 494
Spandex
  microscopy 4:264
  solubility 4:248, 250
Spark source spectroscopy
  of glass and glass-ceramics 13:456
Sparteine 4:590
  analysis 4:609
Spearmint oil. (See also *Essential oils.*)
  carvone content 12:282
  commercial specifications 12:262
  refractive index 12:271
Special glass. (See also *Glass and glass-ceramics.*)
  composition 13:407
Specific conductance 19:359
  of fluorosulfuric acid 13:54
  of potassium chloride 13:56
  of water 19:359
Specific gravity. (See also *Density.*)
  of abrasive grains 7:295
  of acetate and triacetate fibers 4:77
  of acetic acid–water mixtures 4:96
  of acetone 4:113
  of acrylic and modacrylic fibers 4:258
  of acrylonitrile polymers 4:390
  of alcohols 4:542
  of alkanolamines 5:14
  of alkyd resins 5:52
  of ammonium hydroxide 5:317
  of antifreezes 6:5, 17
  apparatus for ammonium hydroxide 5:319,
    322
  of beer 7:485
  of benzene 7:48
  of bituminous materials 6:318
  of bituminous paving mixtures 6:335

of *n*-butane 15:245
of cake batter 12:74
calculation from composition 13:329
of carbonyl compounds 8:358
of caulking compounds 9:19
centrifugal force method 13:330
of chlorophenols 9:527
of confectionery products 10:555
conversion to degrees Plato 7:487
of cresols 17:44
of cresylic acid 17:44
of crotonaldehyde 11:182
density gradient method 4:259
displacement balance method 13:330
of drycleaning detergents 11:611
of egg shell 12:77
of epoxy resins 12:253
of essential oils 12:269
of ethane 15:245
of ethanol 4:539; 12:272
of ethanol–water mixtures 7:544; 11:360
of ethyleneamines 11:429
of ethylenediamine 11:414
of fabrics 4:25
of fats and fatty oils 12:475
flotation method 4:258
of glycerol 13:528, 530
of glycols 13:602
of high-explosive mixtures 12:467
of hydrocarbons 14:341
of hydrochloric acid 14:361
of hydrogen cyanide 14:406, 416
hydrometer method 6:319; 15:246
of isobutane 15:245
of isopropyl alcohol 4:540
of isopropyl alcohol–water solutions 11:361
kinetic energy method 13:330
of lactic acid 15:147
of liquefied petroleum gas 15:245
of liquid resins 4:23
of lubricants 15:342
of match heads 15:539
of mercaptans 15:571
mercury displacement method 12:467
of methanol 4:538
of morpholine 16:199
of musts 19:402
of natural gas 13:329
of organo-chlorosilanes 18:63
of orthophosphoric acid 17:129
of phenol 17:44
of phenols 17:43
of polyethylenimine 14:491
of polyhydric alcohols 13:602
of polytetrafluoroethylene 13:83, 85
pressure balance method 13:330
for proof determination 4:481

of propane 15:245
of propellants in aerosols 11:97
of propylene 15:245
of pulp handsheets 7:256
pycnometer method 6:318; 15:247
of starch hydrolysates 11:42
of styrene 18:311
of sucrose solutions 18:338, 341
of sulfuric acid 18:411, 421
of tetracene 12:416
of titanium dioxide 15:532
weight in air and water method 6:319
Wharton and Forziati method 4:260
of wines 19:403
of xylenols 17:44
Specific gravity separation
of bakery chemicals 6:431
Specific heat
of beer 7:510
of diethylenetriamine 11:415
of ethylenediamine 11:415
of polypropylene 12:384
of tetraethylenepentamine 11:415
of triethylenetetraamine 11:415
Specific pore volume
of activated carbon 8:140
Specific refraction
of ethylene 12:382
of ethylene and propylene polymers 12:382
of ethylene–vinyl acetate copolymers 12:382
of vinyl acetate 12:382
Specific rotation
of amino acids 8:443
of medroxyprogesterone acetate 10:582
of sodium heparin 10:143
Specific surface area
of activated carbon 8:140
of polytetrafluoroethylene 13:88
Specific viscosity 4:303
of bituminous materials 6:296
of ethylene and propylene polymers 12:370
measurement of 4:306
of polyethylene 12:373
of polypropylene 12:373
Specific volume
of carbon black 8:227
of polyethylene 12:381
of polypropylene 12:381
of tobacco 19:158
Spectroscopic methods
for carbon 14:73
for chromium 9:641
for cinnamaldehyde 10:52
for graphite 14:73
for SBR 12:138
Speiss
sample preparation 6:59, 236

Spelters
aluminum in 5:228
Spergon. See Chloranil.
Sperm oil
solubility 13:539, 552
Sperrylite
composition 17:215
occurrence 17:216
Sphaerocobaltite 10:284, 331
Sphalerite 14:519; 15:166; 18:360
composition 19:544, 545, 572
occurrence 13:376
Sphene 8:75; 19:81
Sphondin. (See also *Coumarins*.) 11:160
gas chromatography 11:168, 170
Spiegeleisen. (See also Iron alloys; Manganese
alloys.) 5:61
commercial grades and specifications 15:72,
453
composition 5:64; 15:72, 451
manganese content 15:460
uses 15:452
Spinal anesthetics 5:373
Spinel
properties 9:204
Spinel ferrites. See *Ferrites*.
Spiramycins 5:464, 519
analysis 5:525, 529, 533
bioautography 5:524
identification 5:525
microbiological assay 5:529
paper chromatography 5:524
properties 5:518
separation 5:525
solubility 5:519, 520
thin-layer chromatography 5:524
uses 5:519
Spirit blends
analysis 4:480
quality control 4:470
Spirits 4:463
Splint strength
of matches 15:544
of paper matchboard 15:534
Splitting resistance
of felt 12:510
Splitting strength
of dental porcelains 11:274
Spodumene 9:278; 15:261
lithium content 15:270
Spontaneous ignition
of matches 15:549
Spot tests
for silver 18:164
for technetium 17:531
Spray tests
for aerosols 11:97

Spring alloys. (See also Cobalt alloys.) 10:290
    composition 10:290
Spurrite 8:75
Stabilite 6:77
    colorimetric method using diazotized *p*-
        nitroaniline 6:98
    color test 6:93
Stabilite Alba 6:77
    colorimetric method using diazotized *p*-
        nitroaniline 6:98
    color test 6:93
Stabilite FLX (and White) 6:77
Stabilite L 6:77
    colorimetric method using diazotized *p*-
        nitroaniline 6:98
Stabilite White Powder 6:77
    colorimetric method using diazotized *p*-
        nitroaniline 6:106
Stability. (See also Heat stability.)
    of acrylonitrile 4:377
    of bituminous paving mixtures 6:341
    of brake linings 7:455
    of copper(I) oxide 10:672
    of perchloroethylene 11:608
    of pigments 17:208
Stability, accelerated
    of 1,1,2,2-tetrachloroethane 9:448
Stability, dimensional
    of dental elastomers 11:286
Stability, gaseous
    of beer 7:538
Stability, physical
    of beer 7:532
Stability, structural
    of adsorbents 4:445
Stabilizers
    in confectionery products 10:551, 552
    in enamels 12:177
    in hydrogen peroxide 14:438
    in photographic films 12:579, 592
    solvent extraction 12:579
    in styrene polymers 18:321
Stabilizers, polymer 6:78
Stachydrine
    ion exchange chromatography 9:618
Stachyose content
    of green coffee 10:393
Stainierite 10:284
Staining characteristics
    of pesticides and insecticides 19:618
Staining tendency
    of antifreezes 6:9
    of automotive chemical specialties 6:357
Staining tests
    identification of acrylic and modacrylic fibers
        4:252
    of polyamide fibers 17:316

Stainless alloys
    composition 16:351
Stainless steel. (See also Chrome steel; *Steel.*)
        5:58
    cobalt content 10:369
    composition 5:60
Stalite 6:77
    color tests 6:90
Standard consistency
    of lime 15:230
Standard deviation 7:161
Standard distribution 7:161
Standard potentials
    of lead 15:163
Stannates. (See also Tin compounds; and
        individual compounds.) 19:71
    analysis 19:72
    manufacture 19:72
    solubility 19:72
    tin content 19:72
α-Stannic acid. See Stannic hydroxide.
β-Stannic acid. (See also Tin compounds.)
        19:27
    solubility 19:27
Stannic bromide. (See also Tin compounds.)
    properties 19:28, 34
Stannic chloride. (See also Tin compounds.)
        19:68
    analysis 19:70
    colorimetric method for amine antioxidants
        6:93
    commercial grades and specifications 19:69
    etchant for ceramics 9:224
    manufacture 19:68
    properties 19:28, 34, 68
    safety 19:68
    sample preparation 19:70
    solubility 19:28, 68
    uses 19:68
Stannic hydroxide. (See also Tin compounds.)
    properties 19:27
    solubility 19:27
Stannic oxide. (See also Cassiterite; Tin com-
        pounds.) 19:70
    analysis 19:70
    in cerium stannate 9:263
    impurities, determination of 19:70
    manufacture 19:70
    potassium iodate volumetric method 9:263
    properties 19:70
    sample decomposition 19:70
    sodium carbonate fusion 19:70
    solubility 19:27, 70
    uses 19:70
Stannic oxide, hydrated
    manufacture 19:26
Stannic sulfide. (See also Tin compounds.) 19:71

manufacture 19:71
solubility 19:28
uses 19:71
Stannite 19:22
Stannochlor. See Stannous chloride.
Stannous chloride. (See also Tin compounds.)
   19:73
  analysis 19:73
  colorimetric method for arsenic 11:140
  colorimetric method for 1,4-naphthoquinone
    17:157
  colorimetric method for phosphates 9:251
  colorimetric method for silicate 9:365
  colorimetric method for tellurium 17:605
  colorimetric method for tungsten 19:201
  commercial grades and specifications 19:69, 73
  gasometric method for arsenic 15:223
  gravimetric method for mercury fulminate
    12:419
  gravimetric method for selenium 17:591
  gravimetric method for silver 18:162, 164
  gravimetric method for tellurium 17:591
  impurities, determination of 19:74
  manufacture 19:73
  properties 9:335; 19:73
  solubility 9:336; 19:73, 74
  specification tests 19:74
  reduction method for iron 15:39, 42
  tin content 19:73
  uses 19:73
Stannous chloride–dimethylglyoxime
  colorimetric method for technetium 17:531
Stannous chloride–hydroquinone
  colorimetric method for niobium 16:333
Stannous chloride–lead acetate
  colorimetric method for sulfur dyes 12:19
Stannous chloride–potassium dichromate
  volumetric method for iron 16:330
Stannous chloride–potassium thiocyanate
  colorimetric method for technetium 17:531
  colorimetric method for tungsten 10:648
Stannous chloride–sulfosalicylic acid
  colorimetric method for technetium 17:531
Stannous chloride–thiocyanate
  colorimetric method for rhenium 17:531, 535
Stannous fluoborate. (See also Tin compounds.)
   19:74
  commercial grades 19:74
  manufacture 19:74
  uses 19:74
Stannous fluoride. (See also Tin compounds.)
   19:74
  analysis 19:75
  commercial specifications 19:69
  fluoride content 19:75
  impurities, determination of 19:75
  manufacture 19:74

properties 19:74
solubility 19:74
tin content 19:75
uses 19:75
Stannous hydroxide. (See also Tin compounds.)
  properties 19:27
  solubility 19:27
Stannous octoate. (See also Tin compounds.)
  manufacture 19:76
  uses 19:76
Stannous oxalate. (See also Tin compounds.)
  commercial specifications 19:69
Stannous oxide. (See also Tin compounds.)
   19:76
  commercial specifications 19:69
  manufacture 19:76
  properties 19:27, 76
  solubility 19:27
  uses 19:76
Stannous pyrophosphate. (See also Tin compounds.) 19:76
  commercial specifications 19:69
  manufacture 19:76
  properties 19:76
  solubility 19:76
  uses 19:76
Stannous sulfate. (See also Tin compounds.)
   19:76
  commercial specifications 19:69
  manufacture 19:76
  properties 19:76
  uses 19:76
Stannous sulfide. (See also Tin compounds.)
  solubility 19:28
Stanzolol. (See also Steroids.)
  manufacture 14:205
  properties 14:205
  solubility 14:205
  structure 14:205
  uses 14:205
Staphylomycin S
  structure 5:471
Staphylomycins 5:471, 551
  analysis 5:556, 560, 563
  bioautography 5:556
  colorimetric method 5:563
  column chromatography 5:556
  identification 5:556
  microbiological assay 5:560
  paper chromatography 5:556
  properties 5:550
  separation 5:556
  solubility 5:551, 552
  thin-layer chromatography 5:557
  uses 5:551
Starch
  acid hydrolysis 9:585

adhesives 4:404
in baking powders 6:429
in beer 7:555
in cereals 9:238
in chocolate liquor 9:585
in cocoa beans 9:548
color test for iodine 14:591
in confectionery products 10:550
cooling system sealer 6:353, 362
diastase method 9:586
in feed products 11:48
in fermentation of bread 6:442
in finishing agents 11:126, 132
in green coffee 10:374
in green tea leaf 18:477
hydrolysis 19:142
identification 6:362
iodine colorimetric method 7:555; 9:238;
    11:127, 132
iodine gravimetric method 19:142
Munson and Walker method 18:197
in photographic films 12:579
polarimetric methods 11:48
in roasted coffee 10:429
in soaps 18:182, 197
solubility 11:126
solvent extraction 19:142
in tobacco 19:142
Starch, damaged
α-amylase volumetric method 12:606
in flour 12:606
Starch derivatives. (See also *Corn products.*)
acetyl content 11:28
alkoxyl content 11:29
analysis 11:28
carboxyl content 11:34
hydroxyalkoxyl content 11:31
phosphorus content 11:35
saponification methods 11:29
sulfur content 11:37
Zeisel method 11:29, 32
Starch haze
in beer 7:520
Starch hydrolysates. (See also *Corn products.*)
analysis 11:38
assay 11:42
degrees Baumé 11:42
dextrose content 11:38
dextrose equivalent 11:40
impurities, determination of 11:44
reducing sugar content 11:40
saccharide content 11:39
specific gravity 11:42
visible spectroscopy 11:13
Starch–iodide
color test for sulfur dioxide 14:346
Starch–sulfuric acid

colorimetric method for iodate 12:545; 14:605
color test for iodide 14:609
Stark and Dean distillation
of water in toluenediamines 11:454
Statistics
in bioassay 7:162
STD 6:77
Steam distillation
of aniline and its derivatives 5:435
of black tea volatiles 18:512
for carboxylic acids 4:416
of fluoride 13:433
of hydrocarbons 18:195
of water in dry milk 16:145
of water in molybdenum disulfide 16:181
Steam emulsion number
of lubricants 15:337
Stearamide
derivatives for identification 8:601
in finishing agents 11:126, 130
gravimetric determination 11:126
properties 8:597
solubility 11:126, 130
Stearic acid. (See also Fatty acids.)
composition 8:281
extraction 6:169
gas chromatography 5:25; 8:382, 388; 18:325
in polystyrene 18:325
properties 12:473
thin-layer chromatography 12:205
Stearic acidamide
thin-layer chromatography 12:205
Stearic anhydride method
esterification of alcohols 4:537
Stearoyl-*p*-aminophenol 6:77
column chromatography 6:88
Stearoyl dilactoyl lactylate. (See also Lactic
    acid esters.)
gas chromatography 15:136
Stearoyl lactoyl lactylate. (See also Lactic
    acid esters.)
gas chromatography 15:136
Stearoyl lactylate. (See also Lactic acid esters.)
gas chromatography 15:136
Stearoyl trilactoyl lactylate. (See also Lactic
    acid esters.)
gas chromatography 15:136
Stearyl alcohol. (See also Fatty alcohols.)
commercial grades 4:555, 563
properties 4:561
specifications 4:564
thin-layer chromatography 12:204
uses 4:562
Stearyl isocyanate. (See also *Isocyanic acid
    esters.*)
infrared spectroscopy 15:106
Stearyl methacrylate. (See also

Acrylic and methacrylic acids and esters.)
commercial grades and specifications 4:187
polymer in 4:200
properties 4:182
saponification method 4:196
solubility 4:184
Steatite
properties 9:211
in whitewares 9:191
Steel. (See also Chrome steels; Iron alloys;
and Stainless steel.) 5:58; 18:261–263
aluminum in 5:227, 239, 246, 248
analysis 18:262
boron in 7:409, 411, 417, 418
butyl acetate extraction 9:645
carrier gas fusion methods 8:624, 626, 627,
682
carrier gas hot extraction methods 8:638
chromium content 9:645, 647, 648, 650,
653, 655; 19:207
classification 18:262
commercial specifications 18:261
composition 5:59, 60; 15:26; 18:261
copper content 10:617
cupferron–chloroform extraction 9:645
decomposition 19:235, 327
gallium content 13:261
hafnium content 14:116
hot extraction methods 8:684
hydrogen 8:613, 616, 617, 618, 619, 620,
624, 626, 638, 680, 684
impurities, determination of 17:78
iron content 15:69
manufacture 18:261
molybdenum content 16:382
niobium content 16:362, 364, 376, 380, 383
nitrogen in 8:616, 618, 619, 621, 623, 624,
626, 627, 628, 629, 630, 631, 634; 16:461
oxygen 8:616, 618, 619, 620, 621, 623, 624,
626, 627, 631, 665, 680, 682
properties 18:261
pyrohydrolysis 7:325
rare earth content 17:477
rhenium content 17:523
selenium content 17:598
silicon content 18:15, 35, 38
sulfur content 18:382, 385, 388
tantalum content 16:381
tin content 19:49, 64
titanium content 16:382; 19:111, 119, 120
tungsten content 16:382; 19:198, 206
uranium content 19:235
uses 18:262
vacuum fusion methods 8:616, 617, 618, 619,
620, 621, 623, 628, 680
vacuum hot extraction methods 8:638
vanadium content 9:645; 19:330, 342

zinc content 19:559
zirconium content 14:116
Steel blue enamel. (See also *Enamels.*)
composition 9:195
Steel corrosion test
for lubricants 15:336
Steepwater. See Feed products.
Stendomycin 5:477, 599
analysis 5:605, 611, 614
composition 5:477
countercurrent distribution 5:477, 605
identification 5:605
microbiological assay 5:611
paper chromatography 5:605
properties 5:600
separation 5:605
solubility 5:601, 602
uses 5:599
Stenediol. See Methandriol.
Stereon 12:87
Sterling silver. (See also Silver alloys.)
composition 18:156
uses 18:156
Steroid oxides. (See also *Epoxy compounds.*)
ultraviolet spectroscopy 12:203
Steroids. (See also Adrenal hormones; and
individual compounds.) 14:183
adsorption chromatography 14:208
analysis 14:207
blue tetrazolium colorimetric method 14:207
carbonyl compounds 8:354
colorimetric determination 14:207
fluorometric methods 14:207
gas chromatography 7:148; 14:213
infrared spectroscopy 14:207
liquid chromatography 14:209
paper chromatography 14:210
partition chromatography 14:208
polarographic methods 14:208
properties 14:191
sources 14:191
structure 14:191
thin-layer chromatography 14:211
in tobacco 19:149
ultraviolet spectroscopy 14:207
uses 14:191
Sterols
digitonin gravimetric method 19:149
in soaps 18:201
solvent extraction 19:149
in tobacco 19:149
in yeast 19:533
STH. See Growth hormone.
Stibiconite 6:39
Stibiotantalite
composition 18:441
Stibnite 6:39; 18:360

metallurgy 6:40
Stiffness
    of felt 12:513
Stigmasterol
    gas chromatography 19:150
    in tobacco 19:150
Stilbazo
    colorimetric method for tantalum 16:369
Stilbene-4,4'-bisazo-1-3,4-dihydroxybenzene-2,2-
        disulfonic acid diammonium salt. See
        Stilbazo.
Stilbene dyes 12:7
Stilbite 8:75
Stillage
    analysis 4:480
    quality control 4:468
Stillman-Reed
    volumetric method for aldehydes 12:283
    volumetric method for ketones 12:283
Stishovite. (See also Silica.) 18:5
Stoddard solvent. (See also *Hydrocarbons.*)
        11:604–607
    commercial grades and specifications 11:604;
        14:325
    corrosion test 11:606
    distillation range 11:606
    flash point 11:605
    impurities, determination of 11:606
    odor 11:605
    specification tests 11:605
Stokes' law 4:442; 9:156
Stolzite
    decomposition 19:203
    tungsten content 19:203
Stone coating test
    for emulsified asphalts 6:329
Stoneware. (See also *Ceramics.*) 9:191
    composition 9:194
Stop-leaks. See Sealers.
Storage quality
    of photographic films 12:594
Straight whisky 4:464
    analysis 4:480
Strain modulus
    of elastomers, synthetic 12:126
"Straits" tin. (See also Tin metal.) 19:30
Strasheim and Van Wamelen method
    for gold ores 14:10
Stream sediment
    copper content 10:614
    nickel content 16:299
Strecker degradation
    of amino acids in cocoa beans 9:542
Strecker reaction 8:481
Strength
    of adhesives 4:393
    of black tea 18:486

of egg shells 12:78, 79
of fibers 4:89
of paper 7:277
of pulp handsheets 7:254
Streptomycin 5:476, 582
    analysis 5:587, 593, 596
    bioautography 5:588
    colorimetric methods 5:598
    countercurrent distribution 5:587
    fluorometry 5:598
    identification 5:588
    ion exchange separation 5:588
    maltol assay 5:596
    microbiological assay 5:593
    paper chromatography 5:587
    properties 5:579
    separation 5:587
    solubility 5:580, 581
    structure 5:476
    thin-layer chromatography 5:588
    ultraviolet spectroscopy 5:597
    uses 5:582
Streptomycin sulfate
    infrared spectrum 5:588
    turbidimetric method for sodium heparin
        10:149
Stress relaxation of polymers 6:79
Stretch
    of paper 7:277
    of pulp handsheets 7:254
Stripping test
    for bituminous materials 6:345
Strobane. (See also *Pesticides and insecticides.*)
    commercial grades and specifications 19:593
    composition 19:581, 588, 597
    gas chromatography 19:389
    manufacture 19:597
    properties 19:588
    solubility 19:587
    toxicity 19:592
    in water 19:389
Stroehlein nitrogen–oxygen analyzer 8:665
Stroking test apparatus 7:435, 438
Strontianite. (See also Strontium carbonate.)
        18:264, 281
Strontium. (See also Strontium compounds;
        Strontium metal.) 18:264–285
    in acid salts 8:595
    in aluminum metal 5:152
    ammonium carbonate gravimetric method
        18:272, 274
    ammonium oxalate gravimetric method 18:272
    analysis 18:267
    atomic absorption spectroscopy 9:346; 18:270,
        280
    in barite 6:557, 560
    in barium 18:267, 273

in barium acetate 6:537
in barium carbonate 6:541, 544
in barium chloride 6:546
in barium hydroxide 6:550
in barium nitrate 6:553
in beer 7:652
in bones 18:269
in calcium 18:267, 273
in calcium carbonate 8:84
in chromic acid 9:684
in coal ash 18:280
colorimetric determination 18:270, 276
detection 18:270
DTPA titration 18:275
EDTA titration 13:445; 18:275
electrodeposition 18:270
emission spectroscopy 9:346; 11:493; 15:156, 187; 16:160; 18:270, 277, 281
flame emission photometry 18:277
flame photometry 7:652; 8:84, 595; 15:399
flame tests 6:583, 589, 592; 18:270
in food 18:269
gravimetric determination 18:267, 271
half-life 18:264
history 18:264
ion exchange chromatography 13:255; 18:268
isotope dilution techniques 18:281
isotopes 18:264
in lanthanum oxide 15:156
in magnesium acetate 15:399
in magnesium chloride 15:405
in magnesium nitrate 15:408
in magnesium oxide 15:411
in magnesium sulfate 15:413
methyl sulfate gravimetric method 18:268
in molybdenum 16:161
in molybdenum trioxide 16:161
murexide colorimetric method 18:276
neutron activation analysis 18:271, 281
nitric acid gravimetric method 18:274
occurrence 18:264
paper chromatography 18:270
in plants 18:280
in polymers 4:345
potassium rhodizonate gravimetric method 18:268
precipitation as nitrate 8:101
in rocks 15:187
sample decomposition 18:273
separation 18:267
in silicate rocks 18:273, 277, 280
sodium carbonate fusion 18:273
in sodium chloride 9:346
in soil 18:269
in strontium chromate 9:707
in strontium chromate pigments 18:272
in strontium sulfate 17:199

sulfonazo III colorimetric method 18:276
sulfuric acid gravimetric method 17:199; 18:272, 274
toxicity 18:266
volumetric determination 18:275
in water 18:221, 269, 275, 278, 280
x-ray fluorescence spectroscopy 18:270, 281
2,6-xylenolphthalein-α,α'-bis(imino)diacetic acid colorimetric method 18:276
Strontium acetate. (See also Strontium compounds.)
properties 18:282
uses 18:282
Strontium acetylide. (See also Strontium compounds.)
manufacture 18:265
uses 18:266
Strontium bromide. (See also Strontium compounds.)
properties 18:282
uses 18:266, 282
Strontium carbonate. (See also Strontium compounds.)
acid–base titration 18:275
manufacture 18:281
properties 18:282
solubility 18:267
uses 18:281, 282
Strontium chlorate. (See also Strontium compounds.)
properties 18:282
uses 18:282
Strontium chloride. (See also Strontium compounds.)
analysis 18:283
commercial specifications 18:282
impurities, determination of 18:283
properties 9:335; 18:282
silver nitrate titration 18:275
solubility 9:336
specification tests 18:283
uses 18:266, 282
Strontium chromate. (See also Chromium pigments; Strontium compounds.)
analysis 9:705; 17:199
chromium content 9:705; 17:199
commercial grades and specifications 9:706
commercial specifications 17:199
impurities, determination of 17:200
properties 18:282
solubility 18:267
strontium content 9:707; 17:199
uses 9:705; 17:199; 18:282
Strontium chromate pigments
strontium content 18:272
Strontium citrate complex
stability 10:89

Strontium complexes 8:98
Strontium compounds. (See also individual
    compounds.) 18:281
  properties 18:281
  toxicity 18:266
  uses 18:265, 281
Strontium dichromate. (See also Strontium
    compounds.)
  properties 18:282
  uses 18:282
Strontium fluoride. (See also Strontium
    compounds.)
  properties 13:11; 18:282
  uses 18:282
Strontium hydride. (See also Strontium
    compounds.)
  manufacture 18:265
Strontium hydroxide. (See also Strontium
    compounds.)
  properties 18:282
  uses 18:282
Strontium iodide. (See also Strontium com-
    pounds.)
  properties 18:282
  uses 18:282
Strontium lactate. (See also Strontium
    compounds.)
  uses 18:282
Strontium metal
  analysis 18:267
  commercial grades 18:267
  manufacture 18:264, 267
  properties 18:265
  uses 18:265
Strontium nitrate. (See also Strontium
    compounds.)
  analysis 18:283
  commercial specifications 18:282
  impurities, determination of 18:283
  potassium dichromate titration 18:275
  properties 18:282
  specification tests 18:283
  uses 18:282
Strontium nitride. (See also Strontium
    compounds.)
  manufacture 18:265
Strontium oxalate. (See also Strontium
    compounds.)
  potassium permanganate titration 18:275
  properties 18:282
  solubility 18:267
  uses 18:282
Strontium oxide. (See also Strontium
    compounds.)
  acid–base titration 18:275
  in alkali barium silicate glass 9:218
  emission spectroscopy 9:218

in glass and glass-ceramics 13:415, 445
in lime and limestone 15:217
properties 18:282
sulfuric acid gravimetric method 15:217
uses 18:282
Strontium perchlorate. (See also Strontium
    compounds.)
  uses 18:282
Strontium permanganate. (See also Strontium
    compounds.)
  properties 18:282
Strontium peroxide. (See also Strontium
    compounds.)
  properties 18:282
  uses 18:282
Strontium, radioactive
  in water 19:395
Strontium salicylate. (See also Strontium
    compounds.)
  uses 18:282
Strontium salt of 2-(2-hydroxy-1-naphthylazo)-
    1-naphthalenesulfonic acid. See D & C Red
    No. 13.
Strontium soap. (See also *Soaps.*)
  uses 18:180
Strontium stannate. (See also Strontium
    compounds.)
  uses 18:282
Strontium stearate. (See also Strontium
    compounds.)
  properties 18:282
  uses 18:282
Strontium sulfate. (See also Strontium
    compounds.)
  properties 18:282
  solubility 18:267
  uses 18:282
Strontium sulfide. (See also Strontium
    compounds.)
  manufacture 18:281
  properties 18:282
  uses 18:282
Strontium titanate. (See also Strontium
    compounds.)
  in ceramics 9:225
  microscopy 9:225
  properties 18:282
  uses 18:282
Strontium zirconate. (See also Strontium
    compounds.)
  properties 18:282
  uses 18:282
Structural adhesives
  testing of 4:392, 395
Structural stability of adsorbents 4:445
Strychnine 4:589
  analysis 4:592, 602

bioassay 7:181
gas chromatography 6:210
Strychnos alkaloids 4:589
analysis 4:602
Stucco. (See also *Gypsum and gypsum products.*)
uses 14:88
Styphen I 6:77
colorimetric method using diazotized *p*-nitroaniline 6:106
Styphnate
sodium hydroxide colorimetric method 12:415
in lead styphnate 12:415
Styphnic acid. (See also High explosives.)
acid–base titration 12:419
melting point 12:419, 461
properties 12:434
solvent extraction 12:419
Styrenated polyester
as ablator 4:19
Styrene. (See also *Hydrocarbons.*) 14:325; 18:285–313
in acrylonitrile–butadiene–styrene polymer 18:334
in air 18:311
in alkyd resins 5:43
analysis 18:294
assay 18:299
in 1,3-butadiene 11:504, 506
color 18:309
commercial grades and specifications 18:293
distillation 18:291
freezing point 18:299
gas chromatography 4:212, 214, 342; 7:7, 8; 11:506; 12:104; 18:297, 311, 327, 335
history 18:285
identification 18:294
impurities, determination of 18:297, 304-
induction time 18:311
infrared spectroscopy 18:294
inhibitors, determination of 18:303
liquid chromatography 7:13, 14, 15, 16
manufacture 18:288
mass spectrum 18:296
modifier for alkyd resins 5:24
nuclear magnetic resonance spectroscopy 18:294
occurrence 18:285
properties 18:285, 309
purification 18:289
Raman spectroscopy 18:294
refractive index 18:311
residual monomer in polymers 4:214, 341
safety 18:294
sampling 18:294
in SBR 12:137
separation 18:297
shipping 18:293

solvent extraction 18:335
specific gravity 18:311
storage 18:293
in styrenated fatty acids 5:48
in styrene–acrylonitrile copolymer 18:334
in styrene polymers 18:311, 325, 327
toxicity 18:288
ultraviolet spectroscopy 4:341; 5:48; 12:138; 18:296, 334
uses 14:325; 18:285
viscosity 18:311
Styrene–acrylonitrile–acrylic elastomer polymer. (See also *Styrene polymers.*)
properties 18:314
Styrene–acrylonitrile–butadiene polymers 4:383
Styrene–acrylonitrile copolymer
acrylonitrile content 4:335; 18:334
analysis 18:334
composition, determination of 18:334
impurities, determination of 18:335
infrared spectroscopy 18:319
properties 18:314
solubility 18:316
styrene content 18:334
Styrene–acrylonitrile polymers. (See also *Acrylic and modacrylic fibers; Acrylonitrile polymers; Styrene polymers.*) 4:383
infrared spectroscopy 4:386
polarography 4:335
Styrene–butadiene copolymers. (See also Styrene–butadiene rubber; *Styrene polymers.*)
infrared spectroscopy 18:332
properties 18:314
solubility 18:316
solvent extraction 18:332
Styrene–butadiene–methyl methacrylate terpolymer. (See also *Styrene polymers.*)
properties 18:314
Styrene–butadiene rubber. (See also *Elastomers, synthetic;* SBR.) 12:83
colorimetric determination 12:92
commercial elastomers 12:86
density 12:91
identification 12:90
in impact polystyrene 18:332
nuclear magnetic resonance spectroscopy 12:104
pyrolysis–gas chromatography 12:103, 104
uses 12:83
Styrene dimers
gas chromatography 18:309
in styrene 18:309
in styrene polymers 18:325
Styrene graft copolymer content
of impact polystyrene 18:330
Styrene–maleic anhydride copolymer. (See also *Styrene polymers.*)
properties 18:314

uses 18:314
Styrene–methyl methacrylate copolymers. (See
    also *Methacrylate polymers*; *Styrene
    polymers*.)
    infrared spectroscopy 16:74
    properties 18:314
Styrene–α-methylstyrene copolymer. (See
    also *Styrene polymers*.)
    properties 18:314
Styrene oxide. (See also *Epoxy compounds*.)
    12:228
    analysis 12:229
    assay 12:229
    hydrochlorination methods 12:208
    infrared spectroscopy 12:200
    magnesium sulfate–sodium thiosulfate volu-
        metric method 12:229
    manufacture 12:228
    near-infrared spectroscopy 12:217
    properties 12:228
    solubility 12:228
Styrene–polyethylene copolymers. (See also
    Ethylene and propylene polymers.) 12:345
    manufacture 12:345
    properties 12:345
Styrene polymers. (See also individual polymers.)
    18:314–336
    additives 18:315
    additives, separation of 18:320
    analysis 18:316, 323
    gas chromatography 4:286
    gel permeation chromatography 18:323
    identification 18:316
    infrared spectroscopy 18:316
    light scattering photometry 18:322
    manufacture 18:314
    methanol turbidimetric method 18:304
    molecular weight distribution 18:321
    properties 18:314, 316
    sample preparation 18:316
    solubility 18:316
    in styrene 18:291, 304
    styrene content 18:311
    uses 18:285, 314, 315
Styrene–polypropylene copolymers. (See also
    Ethylene and propylene polymers.) 12:345
    manufacture 12:345
Styrene resins. (See also *Hydrocarbon resins*;
    Polystyrene.) 14:263
    cloud point 14:292
    commercial grades and specifications 14:278,
        280
    density 14:267
    manufacture 14:271
    structure 14:267
Styrenes, substituted
    in styrene 18:291

Styrene trimers
    gas chromatography 18:309
    in styrene 18:309
    in styrene polymers 18:325
Styrol. See *Styrene*.
Styryl carbinol. See Cinnamyl alcohol.
4-Styrylpyridine. (See also *Pyridine and pyridine
    derivatives*.)
    ionization constants 17:416
Suberic acid
    gas chromatography 17:290
    properties 8:397
Sublimation test
    for indigo 14:504
Subsidiary colors
    in certified colors 10:511
Substances darkened by sulfuric acid
    in benzene 7:48
Substances reducing permanganate. See
    Permanganate time test.
Succinic acid 5:22; 8:399
    acid–base titration 19:424
    in adipic acid 4:415
    in alkyd resins 5:37
    colorimetric method 5:48
    derivatives for identification 8:376
    gas chromatography 5:25; 15:132; 17:290;
        19:296
    ion exchange chromatography 19:423
    liquid chromatography 15:133
    in musts 19:423
    paper chromatography 15:133; 19:418
    preparation 8:373
    properties 8:397, 399
    silver nitrate titration 19:423
    solvent extraction 19:423
    thin-layer chromatography 8:381; 19:419
    uses 8:399
    in wines 19:418, 423
Succinic anhydride
    infrared spectroscopy 8:562
    properties 8:559
Succinonitrile. (See also *Nitriles*.)
    gas chromatography 16:404
    properties 16:396
Succinyldicholine chloride. (See also *Choline
    and its derivatives*.)
    dipicrylamine colorimetric method 9:617
    Dragendorff's colorimetric method 9:617
    iodine colorimetric method 9:617
    iodoplatinate colorimetric method 9:617
    thin-layer chromatography 9:616
Succinylmonocholine iodide. (See also *Choline
    and its derivatives*.)
    dipicrylamine colorimetric method 9:617
    Dragendorff's colorimetric method 9:617
    iodine colorimetric method 9:617

iodoplatinate colorimetric method 9:617
thin-layer chromatography 9:616
Suconox 12 (and 18) 6:77
Sucrose. (See also *Sugar*.) 18:337
  assay 18:341
  in chocolate 9:550
  in cocoa beans 9:558
  in cocoa products 9:578, 590
  in confectionery products 10:561
  diphenylamine colorimetric method 19:411
  double polarization method 9:578
  enzymatic method 19:410
  in fermentation of bread 6:442
  in green coffee 10:374, 392
  in green tea leaf 18:478
  in molasses 18:341
  in musts 19:410
  paper chromatography 7:567; 18:478; 19:142
  polarimetry 18:344
  refractive index 18:343
  in roasted coffee 10:429
  specific gravity 18:338, 341
  in sugar solutions 18:343
  in syrups 18:341
  thin-layer chromatography 7:566; 13:502
  thiobarbituric acid colorimetric method 19:411
  in tobacco 19:141
  in wines 19:410, 411
Sugar. (See also Sucrose; and individual com-
    pounds.) 18:337–359
  acid hydrolysis 18:346
  alumina cream clarifying method 18:345
  analysis 8:175; 18:337
  as antifreeze 6:1
  ash content 8:175; 18:357
  assay 18:341
  in bakery products 6:452
  in bread 6:440
  in brake fluids 7:423
  in cakes 6:446
  in cereals 9:238
  in chocolate 9:540, 545
  in cocoa beans 9:558
  in cocoa products 9:578
  color 8:175; 10:558
  colorimetric determination 18:354, 358
  commercial grades and specifications 10:549
  in confectionery products 10:549, 555, 561
  density of syrup 8:177
  3,6-dinitrophthalic acid colorimetric method
    18:354
  in eggs and egg products 12:63
  Emerich method 18:354
  enzymatic method 19:409
  Fehling's solution gravimetric method 15:147
  in grapes 19:405
  gravimetric determination 18:351

  in green tea leaf 18:477
  impurities, determination of 18:357
  invertase hydrolysis 18:347
  iodometric method 18:353, 356
  ion exchange chromatography 19:409
  Knight and Allen method 18:355
  in lactic acid 15:146
  Lane and Eynon method 18:348; 19:408
  lead acetate clarifying method 18:345
  in mash and distillers beer 4:480
  Munson-Walker method 6:452; 18:196, 351
  in musts 19:402, 405
  occurrence 18:337
  Ofner method 18:353
  paper chromatography 18:477
  polarimetry 18:344
  potassium ferricyanide colorimetric method
    19:141
  reducing sugar content 18:348
  refractive index 18:343
  in roasted coffee 10:434
  sample preparation 18:337
  sediment 8:176
  Shaffer-Somogyi method 18:356
  in soaps 18:196
  sodium thiosulfate titration 10:6
  solvent extraction 18:477
  sucrose content 18:343
  in tobacco 19:141
  in tobacco for cigarettes 10:6
  types 18:337
  use in carbonated beverages 8:164, 173
  uses 18:337
  volumetric determination 18:348, 355; 19:408
  in wastes 18:358
  in wines 19:403, 405
Sugar beets
  carboxin content 13:213
  oxycarboxin content 13:213
Sugar derivatives
  paper chromatography 12:582
Sugar of lead. See Lead acetate trihydrate.
Sugar solutions
  caramel content 10:544
  degrees Plato 7:487
Sugars, reducing
  achroic R value 7:556
  in beer 7:563
  for invertase 19:534
  for lactase 19:534
  Munson-Walker method 7:563
Sugar syrup test
  for essential oils 12:264
Sugar test pills 19:410
Sulfamic acid
  in cyclamates 11:229
  etchant for ceramics 9:224

sodium hydroxide hydrolysis 11:229
Sulfamic acid–sodium nitrite
    volumetric method for nitrates 16:502
Sulfanilamide-N-(1-naphthyl)ethylenediamine
        dihydrochloride
    colorimetric method for nitrite 16:494
Sulfanilic acid
    4-amino-1,3-benzenesulfonic acid in 5:446
    azo dye colorimetric method 5:445, 451
    azo dye for identification 5:430
    bromination methods 5:444, 448
    colorimetric method for p-cresol 17:32
    color reactions for identification 5:431
    identification 5:432
    infrared spectrum 5:441
    manufacture 5:426
    in metanilic acid 5:448
    properties 5:423
    ultraviolet spectroscopy 5:433
    uses 5:428
Sulfanilic acid, diazotized
    colorimetric method for phenol 8:146
Sulfanilic acid–1-naphthylamine
    colorimetric method for nitrite 16:492
Sulfated alcohol
    barium chloride gravimetric method 11:126,
        131
    in finishing agents 11:126, 131
    solubility 11:126
Sulfated ash content
    of sugar 18:358
Sulfated monoglycerides
    infrared spectroscopy 11:330
Sulfated phenoxy ethers
    infrared spectroscopy 11:330
Sulfates. (See also individual compounds.)
        18:421–429
    in acid anhydrides 8:568
    in acid halides 8:575
    in acid salts 8:594
    in alumina 5:116
    in aluminum hydroxide 5:188
    in aluminum sulfate 5:193
    in amino acids 8:442
    in ammonium acetate 5:327
    in ammonium chloride 5:338
    in ammonium iodide 14:605
    in ammonium nitrate 5:344
    in ammonium oxalate 5:345
    in ammonium thiocyanate 5:354
    analysis 18:423
    in antimony pentachloride 6:52
    in antimony trichloride 6:51
    in antimony trioxide 6:53
    in arsenic(III) chloride 6:234
    in asbestos–cement products 6:270
    in bakery chemicals 6:437

    in barite 6:558, 561
    barium acetate gravimetric method 11:75
    barium carbonate–phenolphthalein color test
        18:423
    barium chloranilate colorimetric method 18:427
    barium chlorate titration 18:425
    barium chloride gravimetric method 9:117, 292,
        344, 348, 365, 686, 698; 10:252, 330, 333,
        340, 654, 661, 664, 672; 11:48, 228, 335, 381;
        12:186; 13:56; 14:146, 371, 604; 15:399, 404,
        407, 431, 480, 492; 16:239, 536; 17:188, 197,
        508; 18:184, 194, 258, 283, 386, 423, 424;
        19:74, 499
    barium chloride–methanol turbidimetric method
        10:94
    barium chloride titration 18:387, 425
    barium chloride turbidimetric method 9:129,
        350; 11:228; 13:33, 34, 129, 525; 14:371,
        437, 455, 605; 15:143; 16:62; 17:438; 18:176,
        233, 426; 19:19, 222, 364
    barium chromate titration 18:425
    barium iodate titration 18:425
    barium nitrate–sodium chloride turbidimetric
        method 18:426
    barium perchlorate titration 18:386
    benzidine hydrochloride titration 18:425
    in bismuth chloride 7:200
    in bismuth nitrate 7:203
    in bismuth oxide 7:205
    in bismuth subcarbonate 7:196
    in bismuth subnitrate 7:204
    in borax 7:383
    in boric acid 7:382
    in brake linings 7:447
    in cadmium bromide 8:38
    in cadmium carbonate 8:40
    in cadmium chloride 8:42
    in cadmium nitrate 8:45
    in calcium acetate 8:78
    in calcium carbonate 8:83
    in calcium chloride 8:88
    in calcium fluoride 8:90
    in calcium oxide 8:94
    in calcium phosphate 8:96
    in carbon disulfide 8:276
    in carbonyl compounds 8:363
    in cellulose acetate 9:72
    in cements 9:117, 129
    in ceric hydroxide 9:257
    in cerous chloride 9:260
    in cesium hydroxide 9:290
    in cesium sulfate 9:292
    in chromium compounds 9:686, 698
    in chromium plating solutions 18:427
    in citric acid and salts 10:94
    in clays and clay minerals 10:124
    in coal and coke 10:252

in cobalt(II) acetate 10:330
in cobalt carbonate 10:333
in cobalt chloride 10:336
in cobalt nitrate 10:340
colorimetric determination 18:425
commercial grades and specifications 18:421
in copper(II) acetate 10:654, 656
in copper(II) carbonate, basic 10:659
in copper(I) chloride 10:661
in copper(II) chloride 10:662
in copper(I) cyanide 10:664
in copper(II) hydroxide 10:666
in copper(I) oxide 10:672
copper sulfide volumetric method 12:146
in cotton 11:142
in creams and lotions 11:92
in cuprous oxide 17:196
in cyclamates 11:228
detection 18:423
in detergents 11:335, 381
EDTA titration 18:391
in elastomers, synthetic 12:156
in enamels 12:186
in fluorosilicic acid 13:50
in fluorosulfuric acid 13:56
in formaldehyde 13:114
in formic acid 13:129
in glass 9:200
in glycerol 13:525
gravimetric method 6:437; 18:424
gravimetric determination of barium 6:539,
    572, 575
gravimetric method for cesium 9:297
in hydraulic cements 9:100
in hydriodic acid 14:604
in hydrochloric acid 9:340, 350; 14:371
in hydrofluoric acid 13:34
in hydrogen peroxide 14:437
in hydroxylammonium chloride 14:455
identification 18:423
infrared spectroscopy 18:423
ion exchange chromatography 18:431
in lactic acid 15:143
in lead chromate 17:197
lead citrate colorimetric method 18:427
lead nitrate titration 18:425
in limestone 9:191
in magnesium acetate 15:399
in magnesium chloride 15:404
in magnesium nitrate 15:407
in magnesium perchlorate 9:364
in malonic acid 15:431
in manganese 15:480
in mercuric oxide 16:61
in nicotinic acid 17:439
in nitric acid 16:498
in perchloric acid 9:364, 365

o-phenanthroline colorimetric method 18:427
in polyurethanes 19:312
in potassium chlorate 9:359, 361
in potassium chloride 9:339; 12:545
in potassium hydroxide 18:258
in potassium iodate 14:609
in potassium iodide 14:606
in potassium permanganate 15:492
potassium permanganate–barium chloride
    colorimetric method 11:142; 18:423
in pyridine 17:438
in rare earth chlorides 17:506
in rare earth fluorides 17:508
in rare earth sulfates 17:512
in R salt 16:239
in shampoos 11:75
in silicate glasses 9:198
in silver nitrate 18:176
in soaps 18:182, 184, 194
in sodium arsenate 6:235
in sodium arsenite 6:235
in sodium chlorate 9:360
in sodium chloride 9:337, 338, 344, 348
in sodium fluoride 13:33
in sodium hydroxide 18:258
in sodium iodide 14:606
in sodium metal 18:233
in sodium peroxide 16:536
in sodium tungstate 19:222
in stannous chloride 19:74
in starch hydrolysates 11:48
in strontium chloride 18:283
in strontium chromate 9:707; 17:200
in strontium nitrate 18:283
in sulfate white lead 17:188
in sulfites 18:404
in thorium nitrate 19:19
in uranium 19:240
uses 18:421
in vinbarbital 6:529
volumetric determination 18:425
in water 18:427; 19:364
in wines 19:499
in zinc chromate 9:707; 17:199
in zirconium sulfate 14:146
Sulfates, acid-soluble
in asbestos 6:265
in chrysotile 6:265
Sulfate white lead
analysis 17:188
commercial specifications 17:188
impurities, determination of 17:188
lead content 17:188
sulfate content 17:188
zinc oxide content 17:188
Sulfatoceric acid. See Ceric sulfate.
Sulfhydryl group

in beer 7:620, 642
reduction to hydrogen sulfide 7:642
Sulfide minerals
　silver content 18:172
Sulfide ores
　sampling 18:378
　selenium content 17:593
　sulfur content 18:378
Sulfides. (See also individual compounds.)
　18:392–401
　acidimetric titration 18:397
　p-aminodimethylaniline–ferric chloride colori-
　　metric method 19:365
　ammoniacal cadmium chloride colorimetric
　　method 9:119
　in ammonium hydroxide 5:323
　analysis 18:393
　in arsenic(III) oxide 6:233
　in arsenic(V) oxide 6:233
　in barium carbonate 6:540, 544
　in barium hydroxide 6:550, 551
　in barium sulfate 6:568
　in carbon disulfide 8:274, 275
　carrier gas fusion methods 8:682
　colorimetric determination 18:398
　in cotton 11:142
　in depilatory powders 18:395
　detection 18:393
　in elastomers, synthetic 12:128, 156
　gas chromatography 8:274
　in hydraulic cements 9:100
　hydrofluoric acid gasometric method 13:438
　hydrogen sulfide volumetric method 15:480
　iodometric titration 9:119; 15:480; 18:390,
　　395; 19:365
　ion exchange chromatography 18:431
　in iron 15:71
　lead acetate color test 15:71
　lead nitrate titration 17:512
　in manganese 15:480
　manufacture 18:393
　oxygen in 8:682
　precipitation of antimony 6:62
　precipitation of arsenic 6:240
　properties 18:393
　in rare earth sulfides 17:512
　sample preparation 15:31
　separation 18:394
　in sewage 18:396
　in silicate slag 18:397
　sodium nitroprusside colorimetric method
　　11:142; 18:393
　sulfur content 18:380
　volumetric determination 18:395
　in water 18:396; 19:365
　zinc acetate gravimetric method 18:396
Sulfite pulp

bleaching 7:222
Sulfites. (See also individual compounds.)
　18:401–409
　analysis 18:402
　in barium carbonate 6:541
　barium chloride gravimetric method 18:402,
　　403
　in carbon disulfide 8:276
　cobalt–sodium azide color test 18:403
　in cotton 11:142
　detection 18:402
　in food additives 18:405
　in fruit 18:405
　fuchsine color test 18:402
　in hydrochloric acid 9:340, 350; 14:371
　in hydrofluoric acid 13:34
　iodometric method 13:34
　iodometric titration 9:350; 13:33; 14:371;
　　18:403, 404; 19:366
　ion exchange chromatography 18:431
　malachite green color test 18:402
　manufacture 18:401
　in nitrates 16:505
　potassium ferrocyanide–sodium nitroprusside
　　colorimetric method 11:142
　potassium permanganate color test 18:402
　properties 18:402
　p-rosaniline hydrochloride colorimetric method
　　18:405
　in sodium fluoride 13:33
　solubility 18:402
　sulfate content 18:404
　sulfuric acid gasometric method 18:402
　in water 18:404; 19:366
　x-ray diffraction 18:403
　zinc ferrocyanide colorimetric method 18:402
m-Sulfobenzaldehyde
　in F D & C Violet No. 1 10:498, 507
　liquid chromatography 10:507
　ultraviolet spectroscopy 10:507
Sulfobenzaldehydes
　in F D & C Blue No. 1 10:498, 506
　liquid chromatography 10:506
　ultraviolet spectroscopy 10:507
Sulfoborite 7:315
2-Sulfo-4-chlorophenol S
　colorimetric method for niobium 16:369, 374
Sulfocyanides. See Thiocyanates.
5-Sulfo-4'-diethylamino-2,2'-dihydroxyazobenzene
　colorimetric method for aluminum 5:235
　synthesis 5:235
Sulfonamides
　infrared spectroscopy 12:576
Sulfonaphthol–azoresorcinol
　fluorometric method for gallium 5:129
1-(4-Sulfo-1-naphthylazo)-2-naphthol-3-sulfonic
　acid

in F D & C Red No. 2 10:513
  visible spectroscopy 10:513
2-(4-Sulfo-1-naphthylazo)-1-naphthol-4-sulfonic
  acid
  in F D & C Red No. 4 10:516
  liquid chromatography 10:516
Sulfonated amides
  infrared spectroscopy 11:330
Sulfonated compounds
  bromide–bromate titration 10:519
  in D & C Blue No. 4 10:519
  in D & C Green No. 5 10:519
  in D & C Orange No. 4 10:519
  in D & C Yellow No. 10 10:524
  in F D & C Blue No. 2 10:512
  in F D & C Green No. 3 10:513
  in F D & C Violet No. 1 10:517
  in F D & C Yellow No. 5 10:517
  in F D & C Yellow No. 6 10:518
  liquid chromatography 10:512
Sulfonated esters
  infrared spectroscopy 11:330
Sulfonated oil
  in finishing agents 11:126, 131
  solubility 11:126
  turbidimetric determination 11:126, 131
Sulfonation method
  for allyl alcohol 5:86
Sulfonazo III
  colorimetric method for barium 18:276
  colorimetric method for strontium 18:276
  structure 18:276
Sulfonic acid
  cooling system cleaners 6:351
  gravimetric method for barium 6:576
2,4-Sulfophenol R
  colorimetric method for niobium 16:369
3-(4-Sulfophenylazo)chromotropic acid
  colorimetric method for fluorine 14:157
Sulfosalicylic acid
  colorimetric method for iron 15:439
  colorimetric method for niobium 16:375
  colorimetric method for titanium 14:137;
    19:107, 117
Sulfosuccinates
  infrared spectroscopy 12:576
5-Sulfo-2,2′,4′-trihydroxyazobenzene. See
  Superchrome Garnet Y.
2-(5-Sulfo-2,4-xylylazo)-1-naphthol
  in F D & C Red No. 4 10:515
  visible spectroscopy 10:516
2-(3-Sulfo-2,5-xylylazo)-1-naphthol-4-sulfonic
  acid
  liquid chromatography 10:516
2-(3-Sulfo-2,6-xylylazo)-1-naphthol-4-sulfonic
  acid
  in F D & C Red No. 4 10:516

2-(6-Sulfo-2,3-xylylazo)-1-naphthol-4-sulfonic
  acid
  in F D & C Red No. 4 10:516
  liquid chromatography 10:516
Sulfur. (See also Sulfur compounds; Sulfur,
  elemental.) 18:360–440
  in acetylene 4:134
  in acrylic polymers 4:280
  in alcohols 4:545
  in alkylnaphthalenes 16:219
  in aluminum 18:385
  in aluminum and aluminum alloys 5:144
  in aluminum oxide 5:187
  in aluminum oxide abrasives 4:46
  p-aminodimethylaniline–ferric sulfate colorimetric
    method 14:545
  in amino resins 5:287
  in ammonia 5:314
  in antimony 6:45; 18:382
  analysis 15:529
  barium chloride gravimetric method 13:437;
    14:72, 514; 15:62, 76, 223, 329; 16:179,
    213, 335; 18:386
  barium chloride titration 13:325
  barium chloride turbidimetric method 15:253,
    280, 283
  barium perchlorate titration 12:117
  barium sulfate gravimetric method 17:189,
    194, 195
  bomb method 15:328
  in brass 18:382, 390
  bromine–nitric acid oxidation 18:384
  in bronze 18:382, 390
  in 1,3-butadiene 11:504
  in cadmium–mercury reds 17:195
  in calcines 18:383
  in carbon 14:71
  in carbonate rocks 18:380
  in carbon black 8:236
  in chlorocarbons 9:409
  in chrome steels 9:662
  in chromic oxide 9:692
  in chromium 9:635
  in chromium alloys 9:674
  in coal and coke 10:240, 252
  in cobalt alloys 10:320
  in cobalt metal 10:293, 303
  in cobalt(II) orthophosphate 10:343
  in cobalt oxides 10:342
  in cobalt(II) sulfate 10:344
  in cocoa products 9:595
  colorimetric determination 18:391
  colorimetric method for cyclopropene groups
    12:484
  color test for fluorosulfuric acid 13:55
  combustion methods 13:324, 437; 18:308, 385
  commercial grades and specifications 15:511

in cotton 11:154, 157
in copper alloys 10:645
in copper metal 10:624, 625
coulometric methods 15:254
in cyclamates 11:230
in elastomers, synthetic 12:117, 128
Eschka gravimetric method 14:72
in ferrochromium 9:661
in ferromanganese 15:78; 18:382
in ferromolybdenum 18:382
in ferrosilicon 15:76; 18:382
in ferrotungsten 18:382, 385
in ferrovanadium 18:382
in fertilizers 18:384
in gallium arsenide 13:279
in gallium metal 13:266
in gallium phosphide 13:282
in gasoline 13:359
in gelatin 12:548
in glass and glass-ceramics 9:200; 13:415, 436
in graphite 4:14; 14:71
gravimetric determination 18:386
in hafnium 14:136
history 18:360
in hydrocarbons 14:346
hydrogenation method 15:253
hydrogen sulfide evolution method 18:384
hydrogen sulfide volumetric method 16:213
in indigoid dyes 14:514
in indium 14:542
infrared detector method 15:480
in inorganic materials 18:380
iodometric titration 13:437, 438; 15:480;
    16:118; 18:387, 395
iodometric titration of sulfur dioxide 10:303
in iron 15:59, 62; 18:382, 385
in iron oxide pigments 17:194
isotopes 18:361
lamp method 14:346; 16:213
lead acetate trihydrate colorimetric method
    12:545
in leaded zinc oxide 17:189
lead nitrate volumetric method 11:37
in lead sulfide 18:383
Leco volumetric method 14:71
in lime and limestone 15:223
in liquefied petroleum gas 15:253, 255
lithium aluminum hydride volumetric method
    12:129
in lithium carbonate 15:280
in lithium fluoride 15:283
in lubricants 15:328
in manganese 15:479
in manganese oxide pigments 17:194
in matches 15:511
mercury turbidimetric method 15:256
in metals 18:380

in methanol 16:118
methylene blue colorimetric method 12:560
in molybdenum 18:383
in molybdenum disulfide 16:178
in naphthalene 16:212
in natural gas 13:324
in nickel 16:335; 18:382
in nickel alloys 16:335
nitric acid–potassium chlorate–bromine oxidation
    18:383
nitric–hydrochloric acid–potassium chlorate
    oxidation 18:383
occurrence 18:360, 361
in ores 18:380
in organic materials 18:378
oxygen bomb method 14:71
oxy–hydrogen combustion method 15:253
Parr oxygen bomb method 16:213
particle size 15:529
in pesticides and insecticides 19:609
in petroleum products 18:378
in photographic paper 12:560
in plants 18:379
polarographic method 12:146, 560
in potassium bromide 12:545
potassium iodate volumetric method 9:675
propane–2-thiol–piperidine extraction 12:130
in pyrite 18:369
in rocks 18:380, 385
sample decomposition 18:378
in silicate rocks 18:380
sodium butyl phosphite extraction 12:131
sodium carbonate fusion 15:223
sodium plumbite color test 14:347
sodium sulfite volumetric method 12:146
in starch derivatives 11:37
in steel 18:382, 385, 388
in styrene 18:308
in sulfide ores 18:378
in sulfides 18:480
in sulfur compounds, organic 18:378, 391
sulfur dioxide volumetric method 14:136;
    15:63, 480
thin-layer chromatography 11:230
in thorium 18:383
in tin metal 19:66, 68
triphenyl phosphine extraction 12:128
tris(hydroxymethyl)aminomethane titration
    12:117
in tungsten 18:383
ultraviolet spectroscopy 11:231
in vanadium 18:383
volumetric determination 18:386
wet oxidation methods 18:381
x-ray fluorescence spectroscopy 11:157
x-ray spectrography 9:409
in zinc 18:383

in zinc oxide pigment 17:189
in zinc powder 17:204
in zirconium 14:136
λ-Sulfur. (See also Sulfur, elemental.) 18:365
μ-Sulfur. (See also Sulfur, elemental.) 18:365
π-Sulfur. (See also Sulfur, elemental.) 18:365
Sulfur acids 18:368
Sulfur bromide
  properties 18:366
Sulfur chloride
  properties 18:366
Sulfur compounds. (See also individual
    compounds.) 18:391–440
  in ammonium carbonate 5:333
  in ammonium phosphate, dibasic 5:350
  barium chloride turbidimetric method 15:401,
    410; 17:403
  in beer 7:640
  in benzene 7:48
  in benzoic acid 7:67
  in bromine 8:6
  in carbon disulfide 8:275
  in carbon tetrachloride 8:328
  in hydrocarbons 14:347
  in magnesium carbonate 15:401
  in magnesium oxide 15:410
  in natural gas 13:321
  in phenols 17:38
  in potassium hydrogen phthalate 17:403
  safety 18:367
  sodium plumbite color test 14:347
  toxicity 18:367
  uses 18:361
Sulfur compounds, inorganic
  calcined magnesium oxide–sodium carbonate
    fusion 18:380
  sodium carbonate fusion 18:380
  sodium carbonate–potassium carbonate fusion
    18:380
  sodium carbonate–potassium nitrate fusion
    18:381
  sodium carbonate–sodium peroxide fusion
    18:381
  sodium peroxide fusion 18:381
Sulfur compounds, organic
  Carius method 18:378
  catalytic hydrogenation method 18:379
  dry combustion method 18:380
  flask combustion method 18:378
  high-temperature combustion method 18:379
  lamp method 18:378
  magnesium nitrate fusion 18:379
  oxygen bomb method 18:378
  oxyhydrogen method 18:379
  sample decomposition 18:378
  sodium peroxide bomb method 18:380
  sodium peroxide fusion 18:379

  sulfur content 18:378, 391
Sulfur content, organic
  of certified colors 10:488
  peroxide decomposition 10:488
Sulfur crosslinkages
  in elastomers, synthetic 12:128
Sulfur dichloride
  properties 9:335
  solubility 9:336
Sulfur dioxide 18:401–409
  acid–base titration 19:480
  in air 18:408
  astrazone pink FG color test 18:403
  in beer 7:581, 643; 18:406
  in benzene 7:47
  bleaching agents 7:222
  in carbon dioxide 8:260
  colorimetric method using p-aminodimethyl-
    aniline sulfate 7:645
  colorimetric method using p-rosaniline hydro-
    chloride 7:644
  in corn products 11:9
  detection 18:403
  ferrozine colorimetric method 18:408
  in fluorosulfuric acid 13:57
  gas chromatography 14:392
  in gelatin 12:551; 13:371
  gravimetric method for gold 10:638; 14:15,
    19
  gravimetric method for selenium 17:195, 512,
    590, 592
  gravimetric method for tellurium 14:15; 17:590,
    601
  in hydrocarbons 14:346
  in hydrogen cyanide 14:422
  hydrogen peroxide volumetric method 18:408
  instrumental determination 18:408
  iodine–barium chloride gravimetric method
    13:371
  iodine–tetramethyldiaminodiphenylmethane
    color test 18:403
  iodometric titration 14:136, 422; 18:387,
    395, 408, 418; 19:479, 481
  lead perchlorate titration 18:391
  malachite green colorimetric method 12:551
  manufacture 18:365, 401
  Monier-Williams method 11:9
  occurrence 18:369, 401
  odor 18:403
  odor test for sulfur 18:370
  pH 19:478
  o-phenanthroline colorimetric method 18:408
  polarography 12:551; 13:57
  properties 18:401
  reduction of chlorate 9:287
  p-rosaniline colorimetric method 18:408
  starch–iodide color test 14:346

in sulfuric acid 18:418
toxicity 18:367
uses 18:401
volumetric method for sulfur 14:136; 15:63, 480
in water 18:406
West-Gaeke method 18:408
in wines 19:405, 411, 477
Sulfur dioxide–hydrochloric acid
gravimetric method for selenium 13:439
Sulfur dioxide–hydrazine sulfate
gravimetric method for telluride 17:512
Sulfur dyes 12:4
identification 12:19
sodium carbonate–sodium sulfide colorimetric method 12:19
stannous chloride–lead acetate colorimetric method 12:19
structure 12:9
Sulfur, elemental
acetone–carbon disulfide colorimetric method 18:371
in acetone extracts 18:371, 376
allotropes 18:363
analysis 18:370
barium chloride gravimetric method 18:374
benzoin–lead acetate colorimetric method 18:371
in butyl carbitol 18:377
colorimetric determination 18:371
commercial grades and specifications 18:370
copper(I) ammonium sulfate turbidimetric method 18:377
copper–cadmium acetate volumetric method 18:375
detection 18:370
in elastomers, synthetic 12:146
in fertilizers 18:373
in gasoline 18:374
impurities, determination of 18:377
manufacture 18:369
mercury turbidimetric method 18:374
occurrence 18:361, 369
in ores 18:372
in petroleum products 18:374
in plant sprays 18:377
potassium cyanide–ferric chloride colorimetric method 18:371
properties 18:363
in rubber 18:371, 374
silver spot test 18:370
sodium cyanide volumetric method 18:376
sodium sulfite volumetric method 18:375
solubility 18:364, 365, 370
solvent extraction 18:372
sulfur dioxide odor test 18:370
thallium(I) sulfide colorimetric method 18:371

toxicity 18:367
uses 18:361, 362
Sulfur fluorides. (See also individual compounds.) 13:10
properties 18:366
Sulfur halides
properties 18:366
safety 18:367
toxicity 18:367
Sulfur hexafluoride
gas chromatography 13:64
infrared spectroscopy 13:294
manufacture 13:10
in noble gases 13:294
properties 13:11
toxicity 13:12
Sulfuric acid 18:409–421
absorption of unsaturated hydrocarbons 11:606
acid–base titration 13:40; 16:499; 18:416
acidity, total 18:416
in acid mixtures 14:367; 16:499
analysis 18:416
assay 18:416
barium chloride gravimetric method 14:456
barium perchlorate–sodium sulfate volumetric method 14:367
Baumé gravity 18:411, 421
boiling point 18:410
in chlorine 9:309, 328
in chlorine dioxide absorption tower solution 7:227
colorimetric method for ethinyl estradiol 10:569, 571, 573, 585, 590
colorimetric method for mestranol 10:576, 585, 594
colorimetric method for narcotine 16:247
colorimetric method for selenium 17:591
colorimetric method for tellurium 17:591
color test for natural coloring materials 10:537
color test for nitrosodiphenylamines 11:458
color test for vat dyes 12:18
commercial grades and specifications 18:415
density 18:410
etchant for ceramics 9:224
in fluoroboric acid 13:40
in fluorosulfuric acid 13:55
freezing point 18:410
gasometric method for sulfite 18:402
gravimetric method for barium 19:126
gravimetric method for barium nitrate 12:427
gravimetric method for barium oxide 13:419
gravimetric method for calcium 15:402
gravimetric method for copper phthalocyanine blue 17:205
gravimetric method for gold 14:39
gravimetric method for lead 10:639; 15:172, 175, 178, 530; 17:189, 196, 197

gravimetric method for lead oxide 13:419
gravimetric method for lignins 19:147
gravimetric method for potassium 10:175
gravimetric method for silicon 18:21
gravimetric method for soaps 18:183
gravimetric method for sodium 10:168
gravimetric method for strontium 17:199;
    18:272, 274
gravimetric method for strontium oxide 15:217
in hydrogen cyanide 14:421
in hydroxylammonium acid sulfate 14:456
impurities, determination of 18:418
ionization constants 18:412
manufacture 18:365, 414
in methanol 16:115
in musts 19:412
odor test for chlormadinone acetate 10:576
properties 18:409
safety 18:367, 413
solubility 18:411
specific gravity 18:411, 421
thin-layer chromatography 8:381
toxicity 18:367, 413
turbidimetric method for barium 14:437
turbidimetric method for thiosulfate 14:606
uses 18:362, 409
volumetric method for mineral oils 12:285
in wines 19:412
Sulfuric acid–carbazole
colorimetric method for casein–formaldehyde
    resins 11:127, 133
colorimetric method for melamine–formaldehyde
    resins 11:133
colorimetric method for phenol–formaldehyde
    resins 11:127, 132
colorimetric method for urea–formaldehyde
    resins 11:133
Sulfuric acid–cupric sulfate
colorimetric method for vinyl ethers 12:314
Sulfuric acid digestion
of nitrogen, organic 10:489
Sulfuric acid insoluble matter
in iron 15:71
Sulfuric acid–persulfate digestion
of organic matter in water 19:369
Sulfuric acid–resorcinol
colorimetric method for 1,1-dichloro-1-
    nitroethane 13:189
Sulfuric acid–sodium thiosulfate
volumetric method for iodates 14:607
Sulfuric acid, substances darkened by
in benzene 7:48
in carbon tetrachloride 8:327
in chloroform 9:435
in citric acid 10:96
in ethyl ether 12:309
in glycerol 13:530

in lactic acid 15:141
in methanol 16:113
in 1,1,1-trichloroethane 9:444
Sulfuric acid, substances not precipitated by
in barium acetate 6:536
in barium carbonate 6:544
in barium chloride 6:546
in barium hydroxide 6:550
in barium nitrate 6:552
Sulfuric acid wash color
of aromatic hydrocarbons 7:42
of benzene 7:42
Sulfur monochloride
properties 9:335
solubility 9:336
Sulfur monofluoride. See Disulfur difluoride.
Sulfurous acid 18:401–409
in acetic acid 4:100
barium chloride gravimetric method 18:406
detection 18:403
in food additives 18:403, 405, 406
gravimetric method for niobium 16:380
gravimetric method for tantalum 16:380
gravimetric method for tungsten 16:380
hydrogen sulfide evolution method 18:403
iodometric titration 18:395, 407; 19:415
manufacture 18:365, 401
Monier-Williams method 18:406
in wines 19:414
Sulfur, oxidizable
bromometric titration 13:325
in natural gas 13:325
Sulfur pentafluoride. See Disulfur decafluoride.
Sulfur tetrafluoride
manufacture 13:11
properties 13:11
toxicity 13:13
uses 13:11
Sulfur trioxide. (See also Sulfate.)
barium chloride gravimetric method 14:96;
    15:222
in chlorosulfonic acid 9:531
in fluorosulfuric acid 13:55
in glass and glass-ceramics 13:413, 415
in gypsum and gypsum products 14:96
hydrogen chloride temperature-rise method
    9:531
in lime and limestone 15:222
safety 18:367
sulfur color test 13:55
Sulfuryl chloride
in chlorosulfonic acid 9:533
Sulfuryl fluoride 13:195
acid–base titration 13:195
in air 13:195
analysis 13:195
gas chromatography 13:196

manufacture 13:195
properties 13:195
solubility 13:195
toxicity 13:195
uses 13:195
Sulphenone
  gas chromatography 19:389
  in water 19:389
Sulphur. See *Sulfur.*
Sulvanite
  composition 19:321
  occurrence 19:321
Sunflower oil. (See also *Fats and fatty
    oils.*)
  properties 12:473
Sunscreens. (See also *Cosmetics.*) 11:90
  analysis 11:90
  composition, determination of 11:90
  effectiveness 11:91
  in hair sprays 11:90
  K value 11:91
  ultraviolet spectroscopy 11:91, 92
Sunset Yellow. See F D & C No. 6.
Superamides
  in shampoos 11:76
Superchrome Garnet Y
  use in polarography of aluminum 5:250, 251
Superinvar. (See also Cobalt alloys.) 10:290
  composition 10:290
Supermendur. (See also Cobalt alloys.) 10:288
Superoxides 16:537
  manufacture 16:537
Superphosphate. (See also Calcium phosphates.)
    17:141
  manufacture 17:143
Supersulfated cement. (See also Hydraulic
    cements.) 9:97
  composition 9:96
  hardening 9:98
Supertherm. (See also Cobalt alloys.)
  composition 10:286
Surface additives
  solvent extraction 18:320
  on styrene polymers 18:320
Surface area
  of carbonaceous materials 4:14
  of carbon black 8:186, 209, 214
  of cigarette filters 10:39
Surface area, internal
  of carbon black 8:187
Surface area, specific
  of activated carbon 8:140
Surface reactions
  of ethylene and propylene polymers
    12:345
Surface tension
  of beer 7:511, 595

of egg white 12:75
of detergents 11:386
of *n*-heptyl *p*-hydroxybenzoate solutions in
    beer 7:595
of preservatives in beer 7:597
Surfactants
  in carburetor cleaners 6:356
  charge, determination of 12:573
  colorimetric determination 12:573
  in cooling system cleaners 6:352, 359
  in detergents 11:335
  dye solubilization test 12:573
  in embalming fluids 12:167
  in engine degreasers 6:355
  infrared spectroscopy 12:575
  ion exchange chromatography 6:365; 11:80;
    12:574
  in metal cleaners 6:355
  methylene blue–catechol violet colorimetric
    method 12:573
  paper chromatography 12:575
  in photographic films 12:572
  separation 11:79
  in shampoos 11:76, 77
  solvent extraction 12:574
  ultraviolet spectroscopy 12:575
  in water 19:385
Surgical alloys. (See also Cobalt alloys.) 10:290
  composition 10:290
Suspended matter
  in acetone 4:115
  in alcohols 4:543
  in beer 7:529
  in ethylene oxide 12:223, 333
  in natural gas 13:328
  in water 19:381
Sussexite 7:315
Sweet chocolate. See Chocolate.
Sweeteners
  in carbonated beverages 8:164, 173
Swell
  of organic particles 7:447
Swelling
  of photographic films 12:593
Swelling value
  of epoxy resins 12:257
  of impact polystyrene 18:329
Swelling volume
  of elastomers, synthetic 12:125
Sylvanite 9:338; 14:1
  composition 17:357; 18:240
  uses 17:357; 18:240
Sylvite 9:304; 17:353
  uses 17:357
Synergists
  in confectionery products 10:552
Syngenite 17:353

Syngesterone. See Progesterone.
Synkayvite. See Menadiol sodium diphosphate.
Synpol 12:87
Synthetic diamonds 11:467
  commercial grades and specifications 11:467,
    468
  manufacture 11:468
  properties 11:467
  uses 11:467, 468
Synthetic felt. See *Felt.*
Synthetic rubber. See *Elastomers, synthetic.*
Synthetic salt cake 18:244

Syringaldehyde (3,5-Dimethoxy-4-hydroxybenz-
    aldehyde) 6:596
  column chromatography 6:618
  gas chromatography 11:167
  paper chromatography 6:620
  properties 6:598
  ultraviolet spectroscopy 6:633
Syrups
  assay 18:341
  potassium content 17:387
  sucrose content 18:341
Szaibelyite 7:315

T

Taberié formula 19:405
Tablets
  disintegration test 6:508
  weight variation test 6:510
Tack-free time
  of caulking compounds 9:24
Tacryl. (See also *Acrylic and modacrylic
    fibers*; *Acrylic fibers*.) 4:221
  cross section 4:223, 263
  properties 4:222
  solubility 4:249
  staining tests 4:255
Tails
  in alcoholic beverages 4:463
  analysis 4:480
  quality control 4:469
Takteen 12:87
Talbutal 5:91; 6:468
  properties 6:470
  solubility 6:476
  thin-layer chromatography 6:481
Talc 15:358
  in brake linings 7:441
  in clay 9:191
  commercial specifications 10:534
  composition 10:102, 534
  impurities, determination of 10:544
Tall oil 5:30
  solubility 13:539, 552
Tall oil fatty acids
  in alkyd resins 5:23
  gas chromatography 5:25
Tangerine oil. (See also *Essential oils*.)
  residue on evaporation 12:273
  ultraviolet spectroscopy 12:288
Tannic acid
  gravimetric method for animal glue 10:282
  test for caffeine 8:58
  turbidimetric method for gelatin 13:363
Tannin phenols
  cinchonine sulfate gravimetric method 19:466
  in wines 19:466
Tannins
  in beer 7:604, 646
  in black tea 18:505
  in cacao 9:536
  in carbonated beverages 8:174
  in chocolate 9:540
  in cocoa products 9:593
  colorimetric method using Folin Denis reagent
    8:174

gravimetric method for gallium 13:253, 258
gravimetric method for niobium 16:363, 366,
    370
gravimetric method for tantalum 16:363, 366
gravimetric method for titanium 19:95
gravimetric method for triphenylmethane and
    related dyes 19:165
gravimetric method for tungsten 16:363
reaction with dyes 6:389
solubility 19:145
solvent extraction 19:147
in tobacco 19:145, 147
ultraviolet spectroscopy 7:647
volumetric method 18:505
Tansy oil. (See also *Essential oils*.)
  thujone content 12:282
Tantalates
  acid dissolution 9:215
  manufacture 16:354
Tantalic acid
  manufacture 16:354
Tantalite
  composition 16:349; 18:441
Tantalocolumbite
  composition 16:349
Tantalum. (See also Tantalum alloys; Tantalum
    compounds; and Tantalum metal.) 18:440–
    454
  acid chrome dark blue colorimetric method
    16:369
  acridine gravimetric method 16:367
  alizarin red S colorimetric method 16:368
  aluminon colorimetric method 16:368
  in aluminum metal 5:152
  ammonium benzeneseleninate gravimetric
    method 16:367
  ammonium naphthaleneseleninate gravimetric
    method 16:367
  arsenazo I colorimetric method 16:369; 18:449
  arsenazo III colorimetric method 16:369
  atomic absorption spectroscopy 14:125
  auramine colorimetric method 16:370
  benzohydroxamic acid gravimetric method
    16:367
  N-benzoylphenylhydroxylamine gravimetric
    method 16:366
  boric acid gravimetric method 16:388
  brilliant green colorimetric method 16:370
  bromopyrogallol red colorimetric method
    16:368
  butylrhodamine B colorimetric method 16:369

in chromium 9:639
in chromium alloys 9:673
chromotropic acid colorimetric method 16:369;
  18:449
cinnamylhydroxamic acid gravimetric method
  16:367
in cobalt alloys 10:318
colorimetric method 16:365, 368; 18:447
crystal violet colorimetric method 16:370;
  18:449
cupferron gravimetric method 10:318; 16:366,
  378, 392; 18:445
detection 18:445
7,8-dihydroxy-4-methylcoumarin gravimetric
  method 16:366
dimethylfluorone colorimetric method 16:368;
  18:448
emission spectroscopy 14:122; 16:387; 18:445,
  450
Eriochrome black 100 colorimetric method
  16:368
eriochromecyanin colorimetric method 16:368
in ferrochromium 9:661
ferrodipyridyl gravimetric method 16:367
in ferroniobium 16:387, 389, 390
ferrophenanthroline gravimetric method 16:367
ferropyridyl gravimetric method 18:445
fuchsin colorimetric method 16:370
gallic acid gravimetric method 16:366
gravimetric determination 16:366; 18:445
in hafnium 14:122, 125
half-life 18:440
hematein colorimetric method 16:369
hematoxylin colorimetric method 16:369;
  18:449
in high-temperature alloys 10:286
history 18:440
hydrogen peroxide colorimetric method 18:449
hydrogen peroxide volumetric method 18:446
hydroquinone colorimetric method 16:368
3-hydroxy-5,7,3',4'-tetramethoxyquercetin
  colorimetric method 16:368
ion exchange chromatography 10:318; 14:113;
  16:153, 364, 377, 385, 390; 17:526; 18:445
isotopes 18:440
malachite green colorimetric method 16:370
methylthymol blue colorimetric method 16:369
methyl violet colorimetric method 16:370;
  18:449
morin colorimetric method 16:368; 18:449
morin gravimetric method 16:366
β-naphthoquinoline gravimetric method 16:367
neutron activation analysis 13:265; 18:452
in nickel 16:333
in nickel alloys 16:333
in niobium 16:387, 389
occurrence 16:349; 18:440

in ores 18:451
PAR colorimetric method 18:449
phenylacetylhydroxamic acid gravimetric
  method 16:367
phenylarsonic acid gravimetric method 16:363,
  366
phenylfluorone colorimetric method 16:368;
  18:449
potassium fluorotantalate gravimetric method
  18:445
potassium hydrogen sulfate fusion 16:393
potassium pyrogallosulfonate colorimetric
  method 16:368; 18:448
properties 16:353
n-propylarsonic acid gravimetric method 16:366
4-(2-pyridylazo)resorcinol colorimetric method
  16:369
pyrocatechol colorimetric method 16:368;
  18:448
pyrocatechol violet colorimetric method 16:368
pyrogallol colorimetric method 10:319; 16:368,
  392; 18:447
pyrogallol gravimetric method 16:367
quercetin colorimetric method 16:368; 18:449
quercetin gravimetric method 16:366
quinaldinehydroxamic acid gravimetric method
  16:367
quinalizarin colorimetric method 16:368
rhodamine C colorimetric method 16:370;
  18:449
rhodamine 6G colorimetric method 16:369;
  18:449
salicylhydroxamic acid gravimetric method
  16:367
sample preparation 18:451
selenious acid gravimetric method 18:446
separation 18:445, 447
sodium molybdate–stannous chloride colorimetric
  method 18:449
in steel 16:380
stilbazo colorimetric method 16:369
sulfurous acid gravimetric method 16:380
tannin gravimetric method 16:363, 366
in tantalum carbide 8:133, 136
tetraphenylarsonium chloride gravimetric
  method 16:366
thoron colorimetric method 16:369
tiron colorimetric method 16:368
in titanium 16:385
in titanium alloys 16:377
in titanium metal 19:121
in tungsten carbide 19:219
in uranium 16:385
urotropin–pyrogallol gravimetric method
  16:366
volumetric method 18:446
x-ray emission spectroscopy 18:451

x-ray fluorescence spectroscopy 16:389, 390; 18:452

in zircaloy 14:125

in zirconium 14:122, 125; 16:385

Tantalum alloys. (See also Tantalum metal.)
analysis 16:360
sample preparation 16:362
uses 18:443

Tantalum carbide. (See also *Carbides*; Tantalum compounds.) 18:452
carbon content 8:134
carrier gas fusion methods 8:624
formation 8:130
free carbon in 8:134
hydrogen in 8:624
nitrogen in 8:624
oxygen in 8:624
properties 8:131, 132; 18:452
solubility 18:452
tantalum content 8:133, 136
uses 8:132; 18:443, 452

Tantalum chloride. (See also Tantalum compounds.) 18:452
properties 18:452
uses 18:452

Tantalum compounds. (See also individual compounds.) 18:452
toxicity 18:444
uses 18:443

Tantalum metal
analysis 16:360; 18:445
carrier gas fusion methods 8:624, 625, 626, 627, 682
commercial grades and specifications 18:444
hot extraction methods 8:684
hydrogen in 8:617, 619, 622, 624, 638, 680, 684
impurities, determination of 18:452
manufacture 18:444
nitrogen in 8:619, 622, 623, 624, 626, 627, 638
oxygen in 8:617–627, 631–633, 638, 680, 682
properties 18:442, 444
sample preparation 16:362
solubility in gold 14:9
toxicity 18:444
uses 18:440, 442
vacuum fusion methods 8:617–623, 680
vacuum hot extraction methods 8:638

Tantalum ores
analysis 16:360
potassium pyrosulfate fusion 16:361
sample decomposition 16:361
sampling 16:360
sodium pyrosulfate fusion 16:361

Tantalum oxide. See Tantalum pentoxide.

Tantalum pentabromide

properties 16:355

Tantalum pentachloride
manufacture 16:355
properties 16:355; 19:96

Tantalum pentafluoride
manufacture 16:356
properties 16:355
solubility 16:356

Tantalum pentaiodide
properties 16:355

Tantalum pentoxide. (See also Tantalum compounds.) 18:452
commercial grades and specifications 18:452
manufacture 16:354; 18:447
properties 16:354; 18:452
solubility 16:354
uses 18:443, 452

Tantalum pentoxide hydrate. See Tantalic acid.

Tantalum tetrabromide
properties 16:355

Tantalum tetrachloride
manufacture 16:356
properties 16:356

Tantalum trichloride
manufacture 16:356

Tantenxenite
composition 18:441

Tapiolite
composition 18:441

Tar removers 6:354
analysis 6:365
performance tests 6:377

Tars. (See also Bituminous materials.) 6:274
commercial grades and specifications 6:283
composition 6:275
distillation test 6:315
manufacture 6:279
properties 6:275

Tartar emetic. See Antimony potassium tartrate.

Tartaric acid (Dihydroxysuccinic acid) 8:399
acid–base titration 19:420
calcium hydroxide turbidimetric method 15:145
in confectionery products 10:552
gravimetric determination 19:419
in green coffee 10:374
ion exchange chromatography 19:420
in lactic acid and its derivatives 15:145
in musts 19:411, 419
paper chromatography 15:133; 19:418
potassium chloride gravimetric method 19:419
properties 8:397, 399
in roasted coffee 10:375
sodium metavanadate colorimetric method 19:420
thin-layer chromatography 8:381; 19:418
use in polarimetric determination of boron 7:412

uses 8:399
in wines 19:411, 418, 419
Tartaric acid–hydrogen peroxide
colorimetric method for titanium 14:142
Tartrazine. (See also F D & C Yellow No. 5;
Triphenylmethane and related dyes.)
in enamels 12:184
identification 12:184
paper chromatography 19:172
titration of Victoria blue B 19:175
Taste perception. See *Sensory testing methods.*
Taurine 7:141
Taurocholic acid
liquid chromatography 7:148
Taurodeoxycholic acid
liquid chromatography 7:149
Taurolithocholic acid
liquid chromatography 7:149
Tax gallon 4:463
Taylor method
identification of acrylic and modacrylic
fibers by solubility 4:248
TBA
colorimetric method for furan aldehydes
9:565
colorimetric method for pyrrole aldehydes
9:565
TBC. See *p-t*-Butylcatechol.
TCNE. See Tetracyanoethylene.
TDA. See Triaminotriphenylmethane.
TDE
gas chromatography 19:389
in raw milk 16:129
in water 19:389
TDI. See Toluene diisocyanate.
TD nickel. (See also Nickel alloys.)
thorium content 16:335
Tea. (See also Black tea; Green Tea leaf; and
Instant tea.) 18:455–518
agronomy 18:455
manufacture 18:455
varieties 18:455
Tea, black. See Black tea.
Tea cream
composition 18:506
gravimetric method 18:506
separation 18:506
volumetric method 18:506
Tea, green. See Green tea; Green tea leaf.
Teallite 19:22
Tear factor
of pulp handsheets 7:256
Tearing resistance
of paper 7:282
Tearing strength
Elmendorf tester 7:283
of paper 7:282

of pulp handsheets 7:255
of silicone rubbers 18:131, 139
Technetium. (See also Technetium compounds;
Technetium metal.)
atomic absorption spectroscopy 17:534
colorimetric method 17:531, 534
coprecipitation 17:525
distillation methods 17:522
in fissium 17:541
gravimetric method 17:525
half-life 17:517
history 17:515
hydrogen sulfide gravimetric method 17:525,
526
infrared spectroscopy 17:540
ion exchange chromatography 17:526
isotopes 17:516
nitron gravimetric method 17:531
occurrence 17:515
potassium ferrocyanide colorimetric method
17:531
potassium thiocyanate colorimetric method
17:531, 538
potassium xanthate colorimetric method 17:531
properties 17:519
radioactivity 17:531
sample decomposition 17:521
separation 17:522
solvent extraction 17:528, 538
spot tests 17:531
stannous chloride–dimethylglyoxime colorimetric method 17:531
stannous chloride–potassium thiocyanate
colorimetric method 17:531
stannous chloride–sulfosalicylic acid colorimetric
method 17:531
tetraphenylarsonium chloride gravimetric
method 17:525, 531
tetraphenylarsonium perchlorate gravimetric
method 17:526
thiourea colorimetric method 17:531
toluene-3,4-dithiol colorimetric method 17:541
in uranium 17:529
Technetium chloride
properties 17:519
Technetium compounds. (See also individual
compounds.)
properties 17:518
Technetium fluoride
properties 17:519
Technetium metal
analysis 17:521
manufacture 17:521
properties 17:517
sampling 17:521
uses 17:517
Technetium oxides

properties 17:518
Technicon amino acid analyzer 8:415, 416
Technoflon 12:87
Tedlar. See Poly(vinyl fluoride).
Teepleite 7:315
Teflon. See Polytetrafluoroethylene.
Teflon FEP–fluorocarbon resins. See Tetra-
    fluoroethylene–hexafluoropropylene
    copolymer.
TEGN. (See also High explosives.)
  infrared spectrum 12:457
  manufacture 12:445
  properties 12:432
  solubility 12:451
  uses 12:445
TEL. See Tetraethyllead.
Telluric acid. (See also Tellurium compounds.)
  properties 17:585
  ultraviolet spectroscopy 17:606
Telluride. (See also Tellurium; Tellurium
    compounds.)
  in rare earth tellurides 17:512
  sulfur dioxide–hydrazine sulfate gravimetric
    method 17:512
Tellurium. (See also Tellurium alloys; Tellurium
    compounds; and Tellurium elemental.)
  aluminum amalgam gravimetric method 17:591
  in aluminum metal 5:152
  analysis 17:589
  barium chloride gravimetric method 17:591
  ceric sulfate volumetric method 17:602
  in cobalt alloys 10:320
  in cobalt metal 10:304
  colorimetric method 17:591, 605
  in copper 17:595, 601, 604
  in copper alloys 10:645
  in copper metal 10:626
  in copper–tellurium alloys 17:602
  in cotton 11:156
  detection 17:590
  diethyldithiocarbamate colorimetric method
    17:606
  distillation methods 17:590
  EDTA titration 13:445
  emission spectroscopy 14:48, 542; 16:387;
    17:592
  in ferroniobium 16:387
  flame tests 6:589, 592
  in gallium arsenide 13:279
  in gallium metal 13:266
  in gallium phosphide 13:282
  in glass 9:200
  in glass and glass-ceramics 13:413, 415, 445
  in gold 14:45
  gravimetric method 17:590, 601
  gravimetric method for rhenium 17:541
  history 17:580

hydroxylamine hydrochloride gravimetric
    method 17:591
hypophosphorous acid colorimetric method
    10:646
hypophosphorous acid gravimetric method
    17:591
in indium 14:542
ion exchange chromatography 14:17
in iron 15:59
isotopes 17:580
magnesium gravimetric method 17:591
in metals 17:589
neutron activation analysis 13:265
in niobium 16:387
occurrence 17:580
in ores 17:589
in organic materials 17:589
paper chromatography 14:18
potassium dichromate volumetric method 17:
    602
potassium iodide colorimetric method 17:606
potassium permanganate titration 17:603
sample decomposition 17:589, 591
sampling 17:589
in selenium 17:605
separation 17:590, 606
sodium diethyldithiocarbamate colorimetric
    method 10:304, 626
stannous chloride colorimetric method 17:605
stannous chloride gravimetric method 17:591
sulfur dioxide gravimetric method 14:15;
    17:590, 601
sulfuric acid colorimetric method 17:591
thiourea colorimetric method 17:606
ultraviolet spectroscopy 17:605
volumetric method 17:602
x-ray fluorescence spectroscopy 11:156
zinc gravimetric method 17:591
zirconium hydroxide gravimetric method 17:
    590
zirconium sulfate gravimetric method 17:595
Tellurium alloys. (See also Tellurium, elemental.)
    17:602
  uses 17:582
Tellurium compounds. (See also individual
    compounds.) 17:606
  toxicity 17:586
Tellurium diethyldithiocarbamate. (See also
    Dithiocarbamic acid salts.)
  uses 11:576
Tellurium dioxide. (See also Tellurium com-
    pounds.)
  manufacture 17:585
  properties 17:585
  toxicity 17:586
Tellurium, elemental
  assay 17:602

commercial grades and specifications 17:588
manufacture 17:586
properties 17:582
toxicity 17:586
uses 17:582
Tellurium hexafluoride
toxicity 13:13
Tellurium trioxide. (See also Tellurium compounds.)
manufacture 17:585
properties 17:585
Tellurous acid. (See also Tellurium compounds.)
hydrogen sulfide gravimetric method 17:591
manufacture 17:585
properties 17:585, 586
uses 17:585
Telone. See 1,3-Dichloropropene.
TEML. See Triethylmethyllead.
Temperature for zero birefringence (TZB) of
cellulose triacetate fiber 4:87
Temperature-rise method
for methyl methacrylate dental materials
11:305
moisture determination of magnesium perchlorate
9:366
reaction with triethylamine 4:103
reaction with water 4:103, 106
Temperature susceptibility
of bituminous materials 6:306
Tempered lead. (See also Sodium alloys.)
composition 18:210
uses 18:210
Temporary hardness
of water 19:362
Tenamene 2 6:77
colorimetric method based on formation of
Wurster salts 6:101, 103
Tenamene 3 6:77
Tenderness
of cakes 12:73
Tennantite
copper content 10:600
Tenorite
copper content 10:600
Tenox BHA (BHT and PC) 6:77
Tensile properties
of chromium 9:635
of polyamide fibers 17:309
Tensile strength
of cakes 12:74
of carbonaceous ablative materials 4:13
of cements 9:136
of chrome steels 9:664
of cigarette paper 10:34
of concrete 9:136
of dental cast gold alloys 11:293
of dental chromium–cobalt alloys 11:292

of dental wrought gold alloys 11:295
of felt 12:510
of graphite 14:77
of mortar 9:136
of paper 7:277
of pulp handsheets 7:254
of silicone rubbers 18:131, 138
of vinylidene fluoride–hexafluoropropylene
copolymer 13:99
Tensile testing
of polyamide fibers 17:323
TEPA. See Tetraethylenepentamine.
Terbium. (See also *Rare earths*; Terbium metal.)
absorption spectroscopy 17:483
atomic absorption spectroscopy 17:494
bead test 6:587
in dysprosium oxide 17:502
emission spectroscopy 15:153, 155; 17:501
in europium oxide 17:502
flame photometry 17:492
in gadolinium oxide 17:502
in lanthanum oxide 15:153, 155; 17:501
x-ray fluorescent spectroscopy 17:490
in yttrium oxide 17:501
Terbium fluoride
carrier gas distillation fusion methods 8:636
oxygen in 8:636
Terbium metal. (See also Rare earth metals.)
carrier gas fusion methods 8:627, 682
hot extraction methods 8:684
hydrogen in 8:623, 638, 680, 684
nitrogen in 8:623, 627
oxygen in 8:623, 627, 629, 637, 680, 682
properties 17:471
vacuum fusion methods 8:623, 680
vacuum hot extraction methods 8:638
Terbium oxide. (See also Rare earth oxides.)
emission spectroscopy 17:502
impurities, determination of 17:502
properties 17:497, 509
Terbium perchlorate. (See also Rare earth
compounds.)
absorption spectrum 17:484
Terephthalaldehydic acid
gas chromatography 8:385
Terephthalic acid 17:158–163
acid–base titration 17:158, 161
acid number 17:161
in alkyd resins 5:37
analysis 17:158, 160
ash content 17:162
assay 17:160
barium acetate gravimetric method 17:160
color tests 17:163
commercial grades and specifications 17:160
derivatives for identification 17:160
gas chromatography 8:385; 17:158

identification 17:160
impurities, determination of 17:161
in isophthalic acid 17:167
manufacture 17:158, 159
polarography 17:167
properties 17:159
solubility 17:159
thin-layer chromatography 19:293
uses 17:158
Terephthalonitrile. (See also *Nitriles.*)
polarography 16:398
Tergitol
properties 18:144
Terpene carbonyls
gas chromatography 8:354
Terpene polychlorinates. See Strobane.
Terpene resins. (See also *Hydrocarbon resins.*)
14:263
commercial grades and specifications 14:278,
280
density 14:267
infrared spectroscopy 14:268
manufacture 14:271
structure 14:267
uses 14:261, 263
Terpenes. (See also *Hydrocarbons*; and individual
compounds.) 14:266, 316
analysis 14:283
color 14:284
commercial specifications 14:316
copper corrosion test 14:347
flash point 14:284
gas chromatography 14:334
impurities, determination of 14:283
moisture content 14:345
occurrence 14:316
peroxide number 14:283
refractive index 14:284
uses 14:266, 316
Terphenyls 11:532–536
analysis 11:534
assay 11:534
ceric sulfate–nitric acid colorimetric method
11:526
commercial grades and specifications 11:533
in diphenyl 11:528
formaldehyde–sulfuric acid colorimetric
method 11:526
freezing point 11:534
gas chromatography 11:528, 535
identification 11:534
impurities, determination of 11:535
infrared spectroscopy 11:534
liquid chromatography 11:529, 536
manufacture 11:532
properties 11:532
solubility 11:532

tetracyanoethylene colorimetric method
11:527
thin-layer chromatography 11:530, 536
toxicity 11:532
ultraviolet spectroscopy 11:534
uses 11:524
α-Terpinene
gas chromatography 12:268
in majoram, oil of 12:268
γ-Terpinene
gas chromatography 12:268
in majoram, oil of 12:268
α-Terpineol
in black tea 18:511
gas chromatography 12:268; 18:511
in majoram, oil of 12:268
γ-Terpineol
gas chromatography 12:268
in majoram, oil of 12:268
Terpinyl acetate
in essential oils 12:276
molecular weight 12:276
Terraclor 13:214
analysis 13:215
in food crops 13:215
liquid chromatography 13:215
manufacture 13:214
1-naphthylamine colorimetric method 13:215
polarographic method 13:216
properties 13:214
solubility 13:214
toxicity 13:214
uses 13:214
Tersan. See Thiram.
Tertiary alcohols
acetyl chloride–dimethylaniline acylation
12:279
in essential oils 12:279
Tertschite 7:314
Tesla coil method
specimen preparation for electron microscopy
8:203
Test car method
for gasoline in natural gas 13:320
Testosterone. (See also Steroids.) 14:186
gas chromatography 14:219
infrared spectrum 14:203
liquid chromatography 14:210
manufacture 14:203
occurrence 14:186, 203
properties 14:203
solubility 14:203
structure 14:187
trade names 14:192
uses 14:186, 203
Testosterone propionate. (See also Steroids.)
uses 14:187

Testrone. See Testosterone.
Testryl. See Testosterone.
TETA. See Triethylentetraamine.
N,N,N',N'-Tetraallylmethylenediamine 5:87
Tetraborane(10) 7:340
  properties 7:343
Tetraboranes
  gas chromatography 7:353
Tetrabromochrysazin
  colorimetric method for boron 7:410
1,1,2,2-Tetrabromoethane. See Acetylene
  tetrabromide.
2,4,5,7-Tetrabromo-3,6-fluorandiol. See D & C
  Red No. 21.
2,4,5,7-Tetrabromofluorescein
  in D & C Red No. 21 10:523
  paper chromatography 10:520, 523
  visible spectroscopy 10:523
5,7,5',7'-Tetrabromoindigo. See CI Vat Blue 5.
2,4,5,7-Tetrabromo-12,13,14,15-tetrachloro-
  3,6-fluorandiol. See D & C Red No. 27.
Tetrabutylammonium hydroxide titration
  of acids in cigarette smoke 10:19
  of malononitrile 15:444
Tetrabutylammonium methoxide titration
  of cyanoacetamide 15:442
Tetrabutylthiuram disulfide. (See also Thiuram
  disulfides.)
  structure 11:592
  uses 11:592
Tetrabutyl titanate. (See also Organotitanium
  compounds.) 19:130
Tetracaine
  analysis 5:414
  anesthetic 5:414
  assay 5:414
  colorimetric method using bromothymol
   blue 5:417
  diazotization method 5:416
  gas chromatography 6:210, 211
  mixtures, determination in 5:385
  properties 5:414
  ultraviolet spectroscopy 5:417
Tetracaine hydrochloride 5:377
  analysis 5:414
  anesthetic 5:414
  gravimetric method 5:416
  infrared spectrum 5:416
  paper chromatography 5:418
  polarography 5:415, 418
  properties 5:414
  ultraviolet spectra 5:415, 416
  ultraviolet spectroscopy 5:414, 417, 419
Tetracalcium orthophosphate
  manufacture 17:121
Tetracene 12:415
  analysis 12:415

commercial specifications 12:415
explosion point 12:416
in explosive priming mixtures 12:417, 420,
  423, 429, 430
granulation 12:416
gravimetric method 12:420
impurities, determination of 12:415
manufacture 12:415
melting point 12:416
microscopy 12:417
nitrogen content 12:415
properties 12:406, 415
solubility 12:415, 420
solvent extraction 12:423, 429
specification tests 12:415
specific gravity 12:416
uses 12:415
x-ray diffraction 12:418
1,2,3,4-Tetrachlorobenzene. (See also Chlorinated
  benzenes.) 9:482
gas chromatography 9:378
infrared spectroscopy 9:394
mass spectroscopy 9:408
properties 9:482
1,2,3,5-Tetrachlorobenzene. (See also Chlorinated
  benzenes.) 9:482
infrared spectroscopy 9:394
mass spectroscopy 9:408
properties 9:482
1,2,4,5-Tetrachlorobenzene. (See also Chlorinated
  benzenes.) 9:482
composition 9:411
gas chromatography 9:378
infrared spectroscopy 9:394
manufacture 9:482
mass spectroscopy 9:399
properties 9:482
solubility 9:482
uses 9:482
2,3,5,6-Tetrachlorobenzoquinone. See Chloranil.
Tetrachlorobiphenyls. (See also Chlorinated
  biphenyls.)
properties 9:492
uses 15:302
Tetrachlorobisphenol A
  gas chromatography 17:338
  thin-layer chromatography 17:338
Tetrachlorobisphenol A polycarbonate. (See also
  Polycarbonates.)
  infrared spectrum 17:332
1,2,2,4-Tetrachlorobutane
  mass spectroscopy 9:403
1,2,3,4-Tetrachlorocyclo-1,3-pentadiene
  mass spectroscopy 9:406
1,1,1,2-Tetrachlorodifluoroethane. (See also
  Chlorofluorocarbons.)
  properties 13:61

1,1,2,2-Tetrachlorodifluoroethane. (See also
  Chlorofluorocarbons.)
  properties 13:61
1,1,1,2-Tetrachloroethane. (See also C_2
  Chlorinated hydrocarbons.)
  Fujiwara colorimetric method 9:386
  gas chromatography 9:377, 380
  mass spectroscopy 9:404
  properties 9:466
1,1,2,2-Tetrachloroethane. (See also C_2
  Chlorinated hydrocarbons.) 9:445
  analysis 9:447
  appearance 9:447
  azeotropes 9:446
  Beilstein test 9:382
  commercial grades and specifications 9:447
  composition 9:411
  distillation range 9:447
  Fujiwara colorimetric method 9:385
  gas chromatography 9:375, 377, 380, 381
  impurities, determination of 9:447
  infrared spectroscopy 9:394
  manufacture 9:447
  mass spectroscopy 7:31; 9:399
  properties 9:445, 466
  solubility 9:446
  specification tests 9:447
  stability, accelerated 9:448
  toxicity 9:446
  uses 9:445
sym-Tetrachloroethane. See 1,1,2,2-Tetrachloro-
  ethane.
1,1,2,2-Tetrachloroethene. See Tetrachloroethylene.
Tetrachloroethylene. (See also C_2 Chlorinated
  hydrocarbons; Perchloroethylene.) 9:459
  analysis 9:462
  azeotropes 9:461
  Beilstein test 9:382
  cloud point 9:463
  color 9:463
  commercial grades and specifications 9:462
  composition 9:411
  Fujiwara colorimetric method 9:386
  gas chromatography 9:375, 376, 377, 380,
    381, 442
  impurities, determination of 9:463
  infrared spectroscopy 9:390, 394
  manufacture 9:461
  mass spectroscopy 9:406
  odor 9:463
  properties 9:460
  pyridine colorimetric method 9:462
  solubility 9:460; 13:539, 552
  specification tests 9:463
  stability 9:465
  toxicity 9:460

in 1,1,1-trichloroethane 9:442
uses 9:459
Tetrachloroethylmercaptocyclohexenedi-
  carboximide. See Difolatan.
Tetrachloromethane. See *Carbon tetrachloride.*
1,1,3,3-Tetrachloro-2-methylpropane
  mass spectroscopy 9:403
Tetrachloronaphthalenes. (See also Chlorinated
  naphthalenes.)
  properties 9:490
  toxicity 9:490
1,1,1,5-Tetrachloropentane
  mass spectroscopy 9:401
Tetrachlorophenols. (See also Chlorophenols.)
  acid-base titration 9:514
  commercial grades and specifications 9:513
  gas chromatography 9:519
  infrared spectroscopy 9:515
  ion exchange chromatography 9:521
  manufacture 9:512
  properties 9:511
  ultraviolet spectroscopy 9:521
1,1,1,3-Tetrachloropropane
  mass spectroscopy 9:403
1,1,2,3-Tetrachloropropane
  mass spectroscopy 9:403
Tetrachloropropanes. (See also C_3 Chlorinated
  hydrocarbons.)
  properties 9:471
1,2,3,3-Tetrachloropropene
  mass spectroscopy 9:405
Tetrachloropropenes. (See also C_3 Chlorinated
  hydrocarbons.)
  properties 9:471
Tetrachlorosilane. (See also Chlorosilanes; Silicon
  tetrachloride.)
  commercial grades and specifications 18:49
  gas chromatography 18:51, 60, 64
  impurities, determination of 18:51
  manufacture 18:49
  in methyldichlorosilane 18:64
  properties 18:48
  safety 18:49
  in trimethylchlorosilane 18:64
  uses 18:49
Tetrachlorotetrabromofluorescein. See D & C
  Red No. 27.
4,7,4′,7′-Tetrachlorothioindigo. See CI Pigment
  Red 88.
o,α,α,α-Tetrachlorotoluene
  mass spectroscopy 9:407
Tetracyanoethylene
  colorimetric method for diphenyl 11:526
  colorimetric method for methyldiphenyls
    11:527
  colorimetric method for naphthalene
    11:527

colorimetric method for pyridine and its
    derivatives 17:425
colorimetric method for terphenyls 11:527
Tetracyclines 5:479
Tetradecafluorohexane. (See also *Fluorocarbons.*)
    properties 13:61
1,2-Tetradecanediol
    gas chromatography 13:577
Tetradecanol. (See also Alcohols, higher.)
    commercial grades 4:555
1-Tetradecanol. See Myristyl alcohol.
n-Tetradecanonitrile. (See also *Nitriles.*)
    derivatives for identification 16:402
    properties 16:396
1-Tetradecene. (See also Alpha olefins.)
    commercial specifications 14:314
n-Tetradecylisocyanate. (See also *Isocyanic
    acid esters.*)
    near-infrared spectroscopy 15:105
n-Tetradecyl methacrylate
    nuclear magnetic resonance spectroscopy
    16:86
Tetradifon. (See also *Pesticides and insecticides.*)
    commercial grades and specifications 19:595
    composition 19:591, 599
    manufacture 19:599
    properties 19:591
    solubility 19:587
    structure 19:585
    toxicity 19:592
Tetraerythritol
    gas chromatography 13:578
Tetraethoxysilane. (See also Organosilane
    esters.)
    gas chromatography 18:75
    properties 18:70
    uses 18:69
Tetraethylammonium borohydride 7:344
Tetraethylammonium bromide–perchloric acid
    volumetric method for epoxy resins 12:243
Tetraethylene glycol
    distillation range 13:602
    gas chromatography 13:577, 589
Tetraethylenepentamine. (See also Ethylene-
    amines.)
    amine value 11:425
    commercial specifications 11:418
    density 11:409
    gas chromatography 11:421
    infrared spectroscopy 11:420
    latent heat 11:417
    nitrogen content 11:427
    properties 11:408
    specific heat 11:415
    thermal conductivity 11:416
    vapor pressure 11:410
    viscosity 11:412, 413

Tetraethyllead. (See also Lead compounds.)
    in air 15:193
    atomic absorption spectroscopy 13:358
    bromine–nitric acid extraction method 15:194
    gas chromatography 13:358
    in gasoline 13:352; 15:184, 185, 194
    manufacture 15:167; 18:209
    properties 15:166
    x-ray emission spectroscopy 15:184
Tetraethyl rhodamine. See Rhodamine B.
Tetraethyl silicate
    uses 15:302
Tetraethylthiuram disulfide. (See also Thiuram
    disulfides.)
    polarography 11:598
    structure 11:592
    uses 11:592
Tetrafluoroboric acid. See Fluoroboric acid.
1,1,1,2-Tetrafluoroethane. (See also *Fluoro-
    carbons.*)
    properties 13:60
Tetrafluoroethylene. (See also *Fluorocarbons.*)
    properties 13:60
Tetrafluoroethylene–hexafluoropropylene co-
    polymer 13:89–92
    analysis 13:92
    commercial grades and specifications 13:91
    differential thermal analysis 13:92
    infrared spectrum 13:91
    manufacture 13:89
    melt flow 13:92
    properties 13:89
    solubility 13:89
    toxicity 13:89
    uses 13:89
Tetrafluoroethylene polymers. (See also Poly-
    tetrafluoroethylene; Tetrafluoroethylene-
    hexafluoropropylene copolymer.)
Tetrafluorohydrazine
    manufacture 13:9
    properties 13:9
Tetrahedrite 6:39; 18:155
    copper content 10:600
Tetrahydrocortisone
    gas chromatography 14:218
    liquid chromatography 14:210
Tetrahydrofuran 13:230
    analysis 13:231
    assay 13:231
    commercial grades and specifications 13:231
    in enamels 12:175
    in furan 13:229
    gas chromatography 13:229, 231; 16:109
    handling 13:230
    impurities, determination of 13:231
    infrared spectroscopy 12:199
    manufacture 13:231

in methanol 16:109
properties 12:175; 13:230
solubility 13:230
specific gravity 13:232
structure 13:228
toxicity 13:230
uses 13:230
Tetrahydrofurfuryl alcohol 13:242
analysis 13:243
assay 13:243
commercial grades and specifications 13:243
gas chromatography 13:243
impurities, determination of 13:244
manufacture 13:243
properties 13:242
solubility 13:242
structure 13:228
uses 13:243
1,2,3,4-Tetrahydronaphthalene. (See also
Hydrogenated naphthalenes.)
gas chromatography 13:352; 16:220
in gasoline 13:352, 355
liquid chromatography 7:13, 15
manufacture 16:219
properties 13:352; 16:219
solubility 16:219
uses 16:219
Tetrahydro-2*H*-1,4-oxazine. See *Morpholine.*
Tetrahydrophthalimide
gas chromatography 13:212
thin-layer chromatography 13:211
Tetrahydrophthalimide epoxide
gas chromatography 13:212
thin-layer chromatography 13:211
Tetrahydroxybenzenes. (See also *Phenols*; and
individual compounds.)
properties 17:5
solubility 17:5
2,2′,4,4′-Tetrahydroxybenzophenone
thin-layer chromatography 16:96
in sunscreens 11:92
ultraviolet spectroscopy 11:92; 16:95
Tetrahydroxy benzoquinone
color test for lead 15:193
1, 3, 4, 5-Tetrahydroxycyclohexanecarboxylic
acid. See Quinic acid.
1,3,4,5-Tetrahydroxycyclohexanecarboxylic acid
3-(3,4-dihydroxycinnamate). See Chlorogenic
acid.
Tetrahydroxyquinone
color test for lead 15:193
2,4,5,7-Tetraiodofluoresceins
in F D & C Red No. 3 10:513
liquid chromatography 10:514
paper chromatography 10:514
visible spectroscopy 10:514
Tetraisopropyl titanate. See

Titanium tetraisopropoxide.
Tetralin. See 1,2,3,4-Tetrahydronaphthalene.
Tetrametaphosphates
ion exchange chromatography 17:131
paper chromatography 17:135
Tetramethylammonium borohydride 7:344
1,2,4,5-Tetramethylbenzene. See Durene.
Tetramethylbenzenes. (See also *Hydrocarbons*;
and individual compounds.) 14:323
gas chromatography 7:8; 13:352
in gasoline 13:352, 355
properties 13:352
2,2,3,3-Tetramethylbutane
gas chromatography 14:330
N,N,N′,N′-Tetramethyl-1,3-butanediamine. (See
also *Diamines and higher amines, aliphatic.*)
properties 11:410
1,1,3,3-Tetramethylbutylamine
manufacture, methods of 5:262
properties 5:260
Tetramethylcyclopentane
gas chromatography 13:351
in gasoline 13:351
properties 13:351
Tetramethyl decynediol
commercial grades and specifications 4:139
properties 4:137
solubility 4:138
Tetramethyldichlorodisilane. (See also Organo-
chlorosilane.)
gas chromatography 18:61
1,1,3,3-Tetramethyldisilazane. (See also Silylation
reagents.)
properties 18:90
Tetramethylenediamine. (See also *Diamines and
higher amines, aliphatic.*)
gas chromatography 17:290
properties 11:410
Tetramethylene glycol. See 1,4-Butanediol.
Tetramethylene oxide. See *Epoxy compounds*;
Tetrahydrofuran.
Tetramethylhexanes
gas chromatography 13:351
in gasoline 13:351, 354
properties 13:351
Tetramethyllead. (See also Lead compounds.)
atomic absorption spectroscopy 13:358
gas chromatography 13:358
in gasoline 13:352; 15:185
manufacture 15:167
properties 15:166
Tetramethylolcyclohexanol pentanitrate. See
Sixolite.
Tetramethylolcyclopentanone tetranitrate. See
Fivonite.
2,3,4,6-Tetramethylphenol. (See also *Phenols.*)
derivatives for identification 17:21

Tetramethylsilane
 mass spectrum 18:59
Tetramethylthiuram disulfide. (See also
 Thiuram disulfides.)
 copper(II) sulfate titration 11:595
 in elastomers, synthetic 12:143
 polarography 11:598
 structure 11:592
 ultraviolet spectrum 12:143
 uses 11:592
Tetramethyl uric acid
 in black tea 18:488
Tetran. See Polytetrafluoroethylene.
2,3,4,6-Tetranitroaniline. (See also High
 explosives.) 12:435
 manufacture 12:435
 solubility 12:435
 uses 12:435
Tetranitroanisoles. (See also High explosives.)
 12:436
 uses 12:436
Tetranitrobenzene. (See also High explosives.)
 12:436
 properties 12:436
 uses 12:436
Tetranitrocarbazole. (See also High explosives.)
 12:436
 commercial specifications 12:437
 infrared spectrum 12:460
 manufacture 12:436
 properties 12:437
 uses 12:436
2,4,5,6-Tetranitro-*m*-cresol. (See also High
 explosives.)
 manufacture 12:437
 solubility 12:437
 uses 12:437
Tetranitrodiglycerin. (See also High explosives.)
 manufacture 12:444
 properties 12:444
 solubility 12:445
 uses 12:444
Tetranitromethane. (See also Nitroparaffins.)
 properties 16:414
Tetranitronaphthalene. (See also High explosives;
 Nitronaphthalenes.)
 manufacture 12:438
 melting point 16:224
 solubility 12:438
 uses 12:438
Tetraoxane 13:104
 properties 13:104
Tetraoxygen difluoride
 manufacture 13:10
 properties 13:10
Tetrapentaerythritol. (See also Polyhydric
 alcohols.)

 gas chromatography 13:594
Tetraphenylarsonium chloride
 colorimetric method for gold 14:25
 gravimetric method for hexafluorosilicic acid
 13:47
 gravimetric method for rhenium 17:525, 535
 gravimetric method for tantalum 16:366
 gravimetric method for technetium 17:525,
 531
Tetraphenylarsonium perchlorate
 gravimetric method for rhenium 17:525
 gravimetric method for technetium 17:526
Tetraphenylbenzidine
 colorimetric method for carbon monoxide
 16:532
Tetraphenylboron
 gravimetric method for choline and its derivatives
 9:623
 gravimetric method for potassium 16:166
 titration of procaine hydrochloride 5:407
Tetraphosphorus pentasulfide
 properties 17:117
Tetraphosphorus trisulfide. See Phosphorus
 sesquisulfide.
Tetrapolyphosphates
 ion exchange chromatography 17:131
Tetrapotassium pyrophosphate. (See also
 Potassium compounds.)
 properties 17:397
 solubility 17:397
Tetrasodium-2-methyl-1,4-naphthalenediol
 bis(dihydrogen phosphate). See Menadiol
 sodium diphosphate.
Tetrasodium pyrophosphate
 manufacture 17:125
 properties 17:123
 uses 17:123
Tetryl. (See also High explosives.) 12:442
 ammonium hydroxide colorimetric method
 12:421
 colorimetric method 12:421, 442
 commercial specifications 12:442
 in explosive priming mixtures 12:418, 421
 infrared spectrum 12:455
 manufacture 12:442
 melting point 12:461
 nitrogen content 12:442
 properties 12:432, 442
 solubility 12:421, 442
 uses 12:442
 x-ray diffraction 12:418
Tetrylol. (See also High-explosive mixtures.)
 analysis 12:450
 composition 12:450
 uses 12:450
Textiles
 bleaching 7:221

bleaching evaluation 7:232
manganese content 15:472
Textiles, bleached
properties, determination of 7:233
Texture
of pigments 17:207
Thallin–nitric acid
colorimetric method for thiophene 18:538
Thallium. (See also Thallium alloys; Thallium compounds; and Thallium metal.) 18:519–536
in air 18:524
in aluminum and aluminum alloys 5:152
ammonium dichromate gravimetric method 18:526
ammonium sulfide gravimetric method 18:522
analysis 18:522
atomic absorption spectroscopy 12:572; 18:534
in biological samples 18:524
brilliant green colorimetric method 18:521
in cadmium 18:524
chloroplatinic acid gravimetric method 18:526
colorimetric method 18:526, 529
detection 18:522, 525
EDTA titration 18:528
electrodeposition 18:526
emission spectroscopy 14:48, 538; 18:525
flame tests 6:589, 592; 18:525
in flue dust 18:524
fluorescence method 18:526
gallium content 13:262
in gold 14:45
gravimetric method 18:523, 526
half-life 18:519
history 18:519
hydrogen sulfide gravimetric method 18:523
in indium 14:538
iodometric methods 18:532
iodometric titration 18:529
ion exchange chromatography 13:254, 255; 18:524
isotopes 18:519
magnesium gravimetric method 18:523
in manganese 18:524
in minerals 18:522, 524
in nonferrous metals 18:524
occurrence 18:519
in ores 18:524
in organic material 18:523
phosphomolybdic acid gravimetric method 18:526
phosphomolybdotungstic acid colorimetric method 18:533
in photographic films 12:571
polarography 12:571; 18:534
potassium bromate titration 18:528

potassium chromate gravimetric method 18:523, 527
potassium ferricyanide–potassium hydroxide gravimetric method 18:528
potassium iodide gravimetric method 18:526, 527
in pyrite 18:522
radiochemical methods 18:534
Rhodamine B colorimetric method 18:526, 529
in rocks 18:522, 524
sample decomposition 18:522
separation 18:523
in silicate rocks 18:522, 531
in silicates 18:524
solvent extraction 18:523, 532
in thallium alloys 18:523
thallium(I) iodide gravimetric method 18:523
thionalide gravimetric method 18:523, 533
thiourea gravimetric method 18:523
trimethylphenylammonium bromide gravimetric method 18:527
in urine 18:524
volatilization methods 18:523
volumetric determination 18:528
in wine 18:524
zinc gravimetric method 18:523
Thallium acetate. (See also Thallium compounds.)
properties 18:534
Thallium-activated sodium iodide
uses 18:520
Thallium alloys. (See also Thallium metal.)
thallium content 18:523
uses 18:520
Thallium bromide. (See also Thallium compounds.)
properties 18:534
solvent extraction 18:524
uses 18:535
Thallium carbonate. (See also Thallium compounds.)
properties 18:534
Thallium chloride. (See also Thallium compounds.)
properties 18:534
solvent extraction 18:523
Thallium chromate. (See also Thallium compounds.)
properties 18:534
Thallium compounds. (See also individual compounds.) 18:534
solubility 18:521
toxicity 18:521
uses 18:520, 535
Thallium fluoride. (See also Thallium compounds.)
properties 13:11; 18:534
Thallium formate. (See also Thallium compounds.)
uses 18:535

Thallium hydroxide. (See also Thallium compounds.)
manufacture 18:521
Thallium iodide. (See also Thallium compounds.)
gravimetric method for thallium 18:523
Thallium metal
in alloys 18:523
analysis 18:522
commercial grades and specifications 18:521
manufacture 18:521
properties 18:520
toxicity 18:521
uses 18:520
Thallium nitrate. (See also Thallium compounds.)
uses 18:535
Thallium nitrate–hydrogen peroxide
gravimetric method for tungsten 19:203
Thallium nitrate–tin chloride
color test for platinum 17:249
Thallium oxide. (See also Thallium compounds.)
properties 18:534
Thallium oxysulfate. (See also Thallium compounds.)
uses 18:535
Thallium sulfate. (See also Thallium compounds.)
manufacture 18:534
in pesticides 18:535
potassium iodide gravimetric method 18:535
properties 18:534
in rodenticides 18:535
toxicity 18:520
uses 18:520
Thallium sulfide. (See also Thallium compounds.)
colorimetric method for sulfur 18:371
properties 18:534
uses 18:535
Thallous sulfate. See Thallium sulfate.
THAM. See Tris(hydroxymethyl)aminomethane.
THE
thin-layer chromatography 14:217
Theaflavin
analysis 18:497
in black tea 18:492
colorimetric method 18:498
flavognost colorimetric method 18:500
gas chromatography 18:503
liquid chromatography 18:497, 501
manufacture 18:497
Roberts and Smith method 18:498
solvent extraction 18:497, 498
structure 18:493
thin-layer chromatography 18:495, 496
Theaflavin-3,3'-digallate
in black tea 18:493
colorimetric method 18:500
gas chromatography 18:503
liquid chromatography 18:502

manufacture 18:497
structure 18:493
thin-layer chromatography 18:495, 496
Theaflavin gallate
in black tea 18:492
colorimetric method 18:499
gas chromatography 18:503
liquid chromatography 18:502
manufacture 18:497
structure 18:493
thin-layer chromatography 18:495, 496
Theanine
in green tea leaf 18:476
structure 18:476
Thearubigens
analysis 18:497
in black tea 18:492
colorimetric method 18:498
liquid chromatography 18:501
Roberts and Smith method 18:498
solvent extraction 18:498, 504
thin-layer chromatography 18:495, 496
Theaspirone
in black tea 18:508
structure 18:508
Thebaine. (See also *Narcotics.*) 4:588
ammonium molybdate colorimetric method
16:247
ammonium vanadate colorimetric method
16:247
analysis 4:597, 600
color tests 16:247
formaldehyde–sulfuric acid colorimetric method
16:247
gas chromatography 16:254
microchemical tests 16:250
paper chromatography 16:255
phosphorimetry 16:264
platinic chloride microchemical method 16:250
sodium carbonate microchemical method 16:250
structure 16:244
thin-layer chromatography 16:256
ultraviolet spectroscopy 16:253
Thenardite. (See also Sodium sulfate.) 18:207,
244
Thenard's blue. See Cobalt aluminate.
Thenoyltrifluoroacetone
colorimetric method for cerium 9:275
extraction method for cerium 9:269
gravimetric method for niobium 16:366
Thenyldiamine hydrochloride. (See also *Anti-histamines.*) 6:29
properties 6:32
ultraviolet spectroscopy 6:164
Theobroma oil. (See also Cocoa butter.) 9:545
commercial specifications 9:550
Theobromine 8:55

in black tea 18:488
in chocolate 9:540
in chocolate liquor 9:588
in cocoa beans 9:548
in cocoa powders 8:67
column chromatography 9:589
identification 8:58
liquid chromatography 8:61
paper chromatography 8:59
structure 18:488
sublimation 8:62
thin-layer chromatography 8:59; 18:492
ultraviolet spectroscopy 8:67
Theogallin
in green tea leaf 18:462, 475
liquid chromatography 18:476
structure 18:462
thin-layer chromatography 18:463, 475
Theonine
in cocoa beans 9:556
Theophylline 7:499; 8:55
in black tea 18:488
identification 8:58
liquid chromatography 8:61
paper chromatography 8:59
structure 18:488
sublimation 8:62
thin-layer chromatography 8:59; 18:492
Thermal black 8:179
commercial grades and specifications 8:196
composition 8:191
manufacture 8:194
Thermal conductivity
of argon 14:380
of carbon dioxide 14:380
of carbon monoxide 14:380
of diethylenetriamine 11:416
of ethane 14:380
of ethylenediamine 11:416
of helium 14:380
of hydrogen 14:380
measurement in carbon, hydrogen, and nitrogen
microanalysis 8:283, 294
of methane 14:380
of nitrogen 14:380
of oxygen 14:380
of tetraethylenepentamine 11:416
of triethylenetetraamine 11:416
of water vapor 14:380
Thermal decomposition
of hydrides 14:249
Thermal expansion
of dental investments 11:269
of dental porcelains 11:276
of dental restorative materials, composite
11:314
of dental waxes 11:320

of glass 13:404
of glass ceramics 13:404
of graphite 14:84
Thermal instability
of polytetrafluoroethylene 13:88
Thermal methods
for ethylene and propylene polymers 12:382
for polycarbonates 17:335
for polyurethanes 19:288
Thermal shock resistance
of dental porcelains 11:275
Thermal stability. (See also Heat stability.)
of diamonds, industrial 11:481
Thermofax microfilm reader printer 8:207
Thermoflex A 6:77
colorimetric method based on formation of
Wurster salts 6:101, 103
colorimetric method using diazotized *p*-
nitroaniline 6:99
Thermoflex C
colorimetric method using diazotized *p*-
nitroaniline 6:99
Thermogravimetry
of charring ablators 4:18
of graphite 4:8
Thermolastic 12:87
torsional modulus 12:136
Thermometers
specifications 6:506
Thermosetting resins
in brake linings 7:441
THF
thin-layer chromatography 14:217
Thiacril 12:87
Thiamine
in bakery products 6:445, 459
in cereals 9:233, 245
in chocolate and cocoa products 9:538
in cocoa beans 9:540
Thiamine hydrochloride
in bakery products 6:456
fluorometry 6:456
Thiamylal 6:468, 526
gas chromatography 6:483, 484
properties 6:470
in sodium thiamylal 6:526
thin-layer chromatography 6:481
Thianaphthalene
boiling point 16:209
Thiazine 6:381
Thiazine dyes. (See also *Azine and related
dyes.*) 6:381, 382; 12:8
analysis 6:388
identification 6:388
manufacture 6:383, 386
solubility 6:383
Thiazole dyes 12:8

Thiazolidomycin 5:463, 510
  analysis 5:510, 515, 517
  bioautography 5:511
  identification 5:510
  microbiological assay 5:515
  paper chromatography 5:510
  properties 5:509
  separation 5:510
  solubility 5:509, 510
  structure 5:463
  uses 5:510
Thickeners
  in confectionery products 10:552
  in enamels 12:176
Thickness
  of egg shells 12:78
  of felt 12:510
  of match strikers 15:547
  of pulp handsheets 7:254
7-(2-Thienyl)acetamido-3-(1-pyridylmethyl)-3-
    cephein-4-carboxylic acid betaine. See
    Cephaloridine.
Thimerosal. (See also Mercurials; Merthiolate.)
  ammonium thiocyanate volumetric method
    11:564
  assay 11:564
  commercial specifications 11:562
  dithizone colorimetric method 11:564
  polarography 11:565
  structure 11:562
  zinc reduction 11:564
Thin-layer chromatography
  of acenocoumarol 10:151
  of acetals 4:65
  of acetylcholine chloride 9:616
  of actinomycins 5:554
  of adipic acid 17:292
  of adrenocorticoids 14:216
  of alcohols 4:528
  of alkaloids 4:594
  of alkyd resins 5:49
  of alkyl iodides 16:93
  of allyl methacrylate 16:91
  of amines 19:140, 302, 305
  of aminocaproic acid 17:292
  of 2-amino-4,6-dinitrophenol 11:103
  of aminonitrophenols 11:103
  of aminophenols 11:103
  of anisidine 11:103
  of antihistamines 6:35
  of antimycin 5:613
  of antioxidants 6:89; 16:94; 19:313
  of apomorphine 16:256
  of ascorbic acid 19:425
  of bacitracin 5:538
  of barbiturates 6:479
  of benzaldehydes 6:620
  of benzylpenicillin 16:559
  of bethanechol chloride 9:616
  of bile acids and salts 7:148
  of bisepigallocatechin 18:495, 496
  of bishydroxycoumarin 10:151; 11:165
  of bisphenol A 17:338
  of brominated salicylanilides 11:552
  of 1,3-butanediol 13:582
  of 1,4-butanediol 13:582
  of 2,3-butanediol 13:502, 582
  of butter 10:545
  of butyl 3,5-dinitrobenzoate 16:94
  of *n*-butyl methacrylate 16:90
  of 4-*t*-butylphenyl salicylate 16:96
  of caffeine 8:59; 18:492
  of captan 13:211
  of captan epoxide 13:211
  of carbamylcholine chloride 9:616
  of carbohydrates 7:566
  of carbonyl compounds 8:348
  of carbonyls in cocoa beans 9:565
  of carboxylic acids 8:380
  of castor oil 12:145
  of catechin 18:463, 470, 495, 496
  of catechin gallate 18:463
  of catechol 11:103
  of cephalosporins 16:559
  of chloramphenicol 5:496
  of chlormadinone acetate 10:579
  of chlorogenic acids 18:463, 475
  of 5-chloro-2-hydroxybenzophenone 16:96
  of choline 9:616
  of choline chloride 9:616
  of cinnamaldehyde 10:48
  of cinnamaldehyde derivatives 10:49
  of citric acid 15:133; 19:419
  of cocaine 16:256
  of cocoa dialysates 9:593
  of codeine 16:256
  of Compound S 14:217
  of corticosteroids 14:213
  of corticosterone 14:217
  of cortisol 14:217
  of cortisone 14:217
  of coumarins 11:164
  of *p*-coumarylquinic acid 18:463, 475
  of cresols 17:24
  of crystal violet 19:173
  of cyclamates 11:235
  of cyclohexylamine 11:212
  of D & C Orange No. 4 11:69
  of D & C Orange No. 5 11:69
  of D & C Orange No. 17 11:69
  of D & C Red No. 7 11:69
  of D & C Red No. 9 11:69
  of D & C Red No. 10 11:69
  of D & C Red No. 13 11:69

of D & C Red No. 19 11:69
of D & C Red No. 21 11:69
of D & C Red No. 27 11:69
of D & C Red No. 30 11:69
of D & C Red No. 34 11:69
of D & C Red No. 36 11:69
of D & C Yellow No. 5 11:69
of decyl 3,5-dinitrobenzoate 16:94
of n-decyl methacrylate 16:90
of 2,4′-diamino-4-(4″-aminobenzyl)di-
  phenylmethane 19:306
of 2,4-diaminoanisole 11:103
of diaminodiphenylmethane 19:303, 306
of 2,4-diaminophenol 11:103
of diaminotoluenes 11:103
of 2,6-di-t-butyl-4-methylphenol 16:95
of dicarboxylic acids 19:293
of 9,10,12,13-diepoxystearic acid 12:205
of diethylamine 19:140
of difolatan 13:211
of difolatan epoxide 13:211
of digalloybisepigallocatechin 18:495, 496
of diglucosides 19:465
of dihydrocodeine 16:256
of dihydroergotamine 15:295
of 2,4-dihydroxybenzophenone 16:96
of 2,2′-dihydroxy-4-methoxybenzophenone
  16:96
of 9,10-dihydroxystearic acid 12:205
of dimethylamine 19:140
of 3,3′-dimethyl-4,4′-diaminodiphenylmethane
  19:306
of 2,3-dimethyl-1-hexyne-3,4-diol 13:582
of dimethylphenols 17:24
of 2,4-dinitrophenyl hydrazine derivatives of
  aldehydes 8:350
of diphenyl 11:526, 530; 13:225
of N,N'-diphenyl-p-phenylenediamine 12:145
of distearin 12:205
of dithiocarbamic acid salts 11:584
of dodecyl methacrylate 16:91
of dyes in lipsticks 11:68
of EDTA and related compounds 12:30
of epicatechin 18:463, 470, 495, 496
of epicatechin gallate 18:463, 470, 495, 496
of epigallocatechin 18:463, 470, 495, 496
of epigallocatechin gallate 18:463, 470, 495,
  496
of epoxy compounds 12:204
of epoxy resins 12:240
of 9,10-epoxystearic acid 12:205
of cis-9,10-epoxystearyl alcohol 12:204
of ergonovine 15:295
of ergotamine 15:295
of esters 8:585
of estrogens 14:213, 222
of ethinyl estradiol 10:590

of ethyl biscoumacetate 10:151
of ethyl 3,5-dinitrobenzoate 16:94
of ethylene glycol 13:502, 582
of ethylene glycol dimethacrylate 16:90
of ethylmorphine 16:256
of ethyl violet 19:173
of etryptamine 15:295
of fatty alcohols 12:204
of F D & C Blue No. 1 11:69; 19:173
of F D & C Greens 19:173
of F D & C Red No. 3 19:173
of F D & C Violet No. 1 19:173
of flavanols 18:470
of flavonol glycosides 18:474
of fluorene 11:532
of folpet 13:211
of formic acid 13:125
of fructose 13:502
of fuchsine 19:173
of fuchsine base 19:173
of fumaric acid 15:133
of fusidic acid 5:511
of gallic acid 18:475
of gallocatechin 18:463, 470, 495, 496
of gallocatechin gallate 18:463
of galloylbisepigallocatechin 18:495, 496
of glucose 13:502
of glycerol 12:587; 13:501, 582
of glycols 12:587; 13:582
of grape pigments 19:463
of hair dyes 11:102
of halides 8:16
of heroin 16:256
of hexamethylenediamine 19:305
of 1,6-hexamethylene diisocyanate 19:315
of hexyl 3,5-dinitrobenzoate 16:94
of n-hexyl methacrylate 16:90
of hydrocodone 16:256
of hydromorphone 16:256
of hydroquinone 11:103; 17:24
of p-hydroxyisophthalic acid 6:145
of 2-hydroxy-4-methoxybenzophenone 16:96
of 2-hydroxy-4-methoxy-4′-methyl-benzophenone
  16:96
of 2-hydroxy-4-octoxybenzophenone 16:95
of 12-hydroxystearic acid 12:205
of indole 15:295
of isophthalic acid 19:293
of isopropylphenols 17:24
of isotheaflavin 18:495, 496
of kaempferol-3-glucoside 18:463
of kaempferol-3-rhamnoglucoside 18:463
of lactic acid 15:133; 19:419
of leucoanthocyanins 18:463
of levorphanol 16:256
of LSD 15:294
of malachite green 19:173

of malic acid 15:133; 19:419
of malonic acid 15:428
of malononitrile 15:444
of malvidin diglucoside 19:465
of mannitol 19:450
of menadione 10:199
of meperidine 16:256
of mestranol 10:577
of methacholine chloride 9:616
of methacrylate polymers 16:90
of methacrylic acid 16:90
of methadone 16:256
of p-methylaminophenol 11:103
of β-methylcholine chloride 9:616
of methyl 3,5-dinitrobenzoate 16:94
of methylene blue 19:173
of methylergonovine 15:295
of methylethylamine 19:140
of 2-methyl-6-isopropylphenol 17:24
of methyl isothiocyanate 13:193
of methyl methacrylate 16:90
of methyl methacrylate dimer 16:90
of methyl methylergonovine 15:295
of methylnaphthalenes 11:532
of 2-methyl-2,4-pentanediol 13:582
of 2-methyl-1,2,3-pentanetriol 13:582
of N-methyl-n-propylamine 19:140
of methyl violet 19:173
of monostearin 12:205
of morphine 16:256
of morpholine 16:198
of myricetin-3-glucoside 18:463
of myricetin-3-rhamnoglucoside 18:463
of myristic aldehyde 12:205
of nalorphine 16:256
of naphthalene 11:532
of 1,5-naphthalenediol 11:103
of α-naphthol 11:103
of narceine 16:256
of narcotics 16:255
of narcotine 16:256
of neochlorogenic acid 18:475
of neomycins 5:586
of night blue 19:173
of nitrophenylenediamines 11:103
of norethindrone 10:583
of norethindrone acetate 10:584, 590
of normorphine 16:256
of nystatin 5:570
of octyl 3,5-dinitrobenzoate 16:94
of oleyl alcohol 12:204
of organotin compounds 19:312
of oxosteroids 14:213
of oxymorphone 16:256
of papaverine 16:256
of paromomycin 5:587
of penicillins 16:559

of pentaerythritol 13:582
of 1,5-pentanediol 13:582
of pesticides and insecticides 19:612
of phenazocine 16:256
of phenol 17:24
of phenolic acids 18:475
of phenolic steroids 14:213
of phenols 11:540; 17:25
of phenoxymethylpenicillin 16:559, 572
of phenprocoumon 10:151
of 1-(phenylazo)-2-naphthol 10:522
of phenylenediamine 11:103; 19:306
of phenyl naphthylamines 12:140
of phthalic acid 19:293
of phthalimide 13:211
of phloroglucinol 17:24
of plasticizers 12:592
of polyamides 17:291
of polycarbonates 17:338
of polyethylene glycols 13:582
of polyhydric alcohols 13:582
of polyphenols 18:462, 495, 496
of procaine hydrochloride 5:409
of progesterone 14:213
of propylene glycol 13:502, 582
of pyrocatechol 17:24
of pyrogallol 11:103; 17:24
of pyromelletic acid 8:405
of pyronine G 19:173
of quaterphenyls 11:531
of quercetin-3-glucoside 18:463
of quercetin-3-rhamnoglucoside 18:463
of quinidine 15:295
of quinine 15:295
of resorcinol 11:103; 17:24
of rhodamine B 19:173
of rhodamine 6G 19:173
of rhodamine S 19:173
of ricinoleyl alcohol 12:204
of rifamycins 5:500
of salicylanilides 11:552
of salicylsalicylic acid 6:146
of sebacic acid 17:292
of semisynthetic penicillins 16:559
of sodium heparin 10:142
of sorbitol 13:502; 19:450
of spiramycins 5:425
of staphylomycin 5:556
of stearic acid 12:205
of stearic acidamide 12:205
of stearyl alcohol 12:204
of steroids 14:211
of streptomycin 5:588
of succinic acid 19:419
of succinyldicholine chloride 9:616
of succinylmonocholine iodide 9:616
of sucrose 13:502

of sulfur 11:230
of tartaric acid 19:419
of terephthalic acid 19:293
of terphenyls 11:530, 536
of tetrachlorobisphenol A 17:338
of tetrahydrophthalimide 13:211
of tetrahydrophthalimide epoxide 13:211
of 2,2',4,4'-tetrahydroxybenzophenone
    16:96
of THE 14:217
of theaflavin 18:495, 496
of theaflavin-3,3'-digallate 18:495, 496
of theaflavin gallate 18:495, 496
of thearubigen 18:495, 496
of thebaine 16:256
of theobromine 18:492
of theogallin 18:463, 475
of theophylline 18:492
of THF 14:217
of thiophene 18:538
of thiophene derivatives 18:538
of thiostrepton 5:557
of toluenediamine 19:302
of *o*-toluidine 11:454
of triaminotriphenylmethane 19:303
of 1,3,5-trimethyl-2,4,6-tri(3,5-di-*t*-butyl-
    4-hydroxy benzyl)benzene 16:95
of triphenylmethane and related dyes 19:172
of tris(2-methyl-4-hydroxy-5-*t*-butylphenyl)-
    butane 16:95
of tristearin 12:205
of tryptamine 15:295
of tyrothricin 5:542
of ultraviolet absorbers 19:312
of urine 11:212
of Victoria blue 19:173
of warfarin 10:151; 11:165
Thin-section analysis
    of coal and coke 10:217
Thioacetamide separation
    of aluminum 5:207
Thioacetamide titration
    of silver 18:168
    of silver halides 12:562
2-Thiobarbituric acid. (See also TBA.)
    colorimetric method for margarine 15:501
    colorimetric method for sorbic acid 19:482
    colorimetric method for sucrose 19:411
4,4'-Thiobis(6-*t*-butyl-*m*-cresol) 6:76
    column chromatography 6:87
    thin-layer chromatography 6:89
    ultraviolet spectrum 6:87
4,4'-Thiobis(6-*t*-butyl-*o*-cresol.) (See also
    Phenols.)
    structure 17:3
4,4'-Thiobis(6-*t*-butyl-3-methylphenol)
    paper chromatography 6:85

Thiobis(di-*sec*-amylphenol) 6:76
4,4'-Thiobis(2,5-di-*t*-amylphenol) 6:76
    paper chromatography 6:85
4,4'-Thiobis(2,6-di-*t*-butylphenol) 6:74
2,2'-Thiobis(4,6-dichlorophenol). See Bithionol.
2,2'-Thiobis(4-methyl-6-*t*-butylphenol) 6:74
Thiocyanate method for iron 6:231, 233, 542,
    543; 7:194, 196, 200, 205, 382, 383
    in aluminum chloride 5:176
    in aluminum compounds 5:176
    in alums 5:194
    in ammonium acetate 5:326
    in ammonium bicarbonate 5:330
    in ammonium carbonate 5:330
    in ammonium hydroxide 5:321
    in ammonium oxalate 5:333
    in ammonium persulfate 5:347
    in ammonium sulfate 5:352
    in ammonium thiocyanate 5:354
Thiocyanate-stannous chloride
    colorimetric method for molybdenum 10:315
    colorimetric method for tungsten 10:323
Thiocyanates. (See also individual compounds.)
    18:433-437
    in ammonium thiocyanate 5:353
    analysis 18:435
    benzidine hydrochloride–pyridine colorimetric
        method 18:437
    in chlorides 9:340
    cobalt sulfate–acetone colorimetric method
        18:435
    colorimetric method 18:436
    colorimetric method for tungsten 19:208,
        209
    colorimetric method for uranium 19:247
    copper acetate–benzidine acetate colorimetric
        method 18:435
    copper(II) sulfate gravimetric method 18:435
    cyanogen bromide volumetric method 18:436
    detection 18:435
    ferric ammonium sulfate color test 12:428
    ferric nitrate colorimetric method 18:435,
        436
    gravimetric method 18:435
    gravimetric method for selenium 17:591
    iodometric methods 18:436
    in lead thiocyanate 12:421, 423
    in nitrates 16:505
    properties 18:433
    pyridine–pyrazolone colorimetric method
        14:420
    silver nitrate gravimetric method 12:423
    solubility 18:433
    titration of mercury 17:196
    toxicity 18:433
    uses 18:433
    volumetric method 18:436

Thiocyanation method
  for allyl alcohol 5:85
  for biallyl 5:104
  for olefinic compounds 5:85
Thiocyanic acid. (See also Thiocyanates.)
    18:433–437
  manufacture 18:433
  toxicity 18:433
Thiocyanide
  gravimetric method for silver 18:164
Thiocyanogen
  volumetric method for margarine 15:499
Thiocyanomercurate
  gravimetric method for zinc 19:560
Thiodiphenylamine
  chloramine-T colorimetric method 11:458
Thioglycolic acid. (See also *Mercaptans.*)
    8:398
  colorimetric method for nitrite 16:494
  N-ethylmaleimide colorimetric method 15:561
  gravimetric method for gold 14:19
  phospho-18-tungstic acid colorimetric method
    15:561
  properties 8:397, 398; 18:433
  uses 15:555
Thioglycolic acid–ammonium hydroxide
  colorimetric method for iron 11:429
Thioindigo. (See also Thioindigoid dyes.)
  history 14:498
  infrared spectrum 14:505
  properties 14:496
  structure 14:495
  uses 14:498
  visible spectroscopy 14:508
Thioindigo Bordeaux. (See also *Pigments.*)
  structure 17:184
Thioindigoid dyes. (See also *Indigoid dyes*; and
    individual compounds.)
  manufacture 14:500
  structure 14:495
  uses 14:498
Thioketone. See 4,4′-Bisdimethylaminothio-
    benzophenone.
Thiokol 12:87
Thiols
  in phenols 17:38
  silver nitrate titration 17:38
Thionalide
  gravimetric method for ruthenium 17:256,
    264
  gravimetric method for thallium 18:523, 533
Thionyl chloride–ammonium hydroxide
  gravimetric method for cinnamic acid 10:58
Thionyl chloride–aniline
  gravimetric method for cinnamic acid 10:58
Thionyl chloride-*p*-toluidine
  gravimetric method for cinnamic acid 10:58

Thiopental 6:468, 527
  bromide–bromate titration 6:497
  derivatives for identification 6:477
  gas chromatography 6:483, 484
  properties 6:470
  in sodium thiopental 6:528
  thin-layer chromatography 6:481
Thiophene 18:537–545
  alloxan colorimetric method 18:541
  analysis 18:537
  barium chloride gravimetric method 18:543
  in benzene 7:11, 48; 18:543
  benzoyl chloride gravimetric method 18:540
  bromometric titration 18:540
  in carbon disulfide 8:274, 276
  cerium(IV) nitrate colorimetric method 18:537
  colorimetric method 18:537, 540
  colorimetric method using isatin 7:48
  copper sulfate–lactic acid colorimetric method
    18:538
  p-dimethylaminobenzaldehyde colorimetric
    method 18:541
  gas chromatography 7:11; 8:274; 18:538
  gravimetric method 18:539
  history 18:537
  hydrogen sulfide evolution method 18:544
  identification 18:537
  infrared spectroscopy 18:542
  isatin colorimetric method 8:276; 18:537, 540
  Laubenheimer reaction 18:537
  Liebermann reaction 18:537
  mercurimetric titration 18:540
  mercury(II) acetate gravimetric method 18:539
  in motor oil 18:542
  ninhydrin spectroscopic method 18:542
  1-nitroso-2-naphthol colorimetric method 18:542
  oxidation methods 18:543
  phenanthraquinone colorimetric method 18:537
  polarography 18:542
  potassium nitrite colorimetric method 18:537
  properties 18:537
  reduction methods 18:544
  separation 18:538
  thallin–nitric acid colorimetric method 18:538
  thin-layer chromatography 18:538
  ultraviolet spectroscopy 18:543
  Van Urk reagent color test 18:541
  volumetric method 18:540
7-(Thiophene-2-acetamido)cephalosporanic acid.
    See Cephalothin.
Thiophene derivatives. (See also *Thiophene*; and
    individual compounds.)
  colorimetric methods 18:537
  gas chromatography 18:538
  thin-layer chromatography 18:538
Thiophenol
  gravimetric method for gold 14:19

Thiophosphates
  liquid chromatography 19:156
  solvent extraction 19:156
  in tobacco 19:156
Thiophosphoryl chloride. See Phosphorus
    sulfochloride.
Thiosalicylamide
  colorimetric method for mercury 16:42
Thiosalicylic acid 8:401
Thiosan. See Thiram.
Thiosemicarbazide
  identification method for carbonyl compounds
    8:342
  reaction of carbonyl compounds with 8:333
Thiostrepton 5:471, 551
  analysis 5:556, 561, 563
  bioautography 5:557
  identification 5:556
  microbiological assay 5:561
  paper chromatography 5:556
  properties 5:550
  separation 5:556
  solubility 5:551, 552
  thin-layer chromatography 5:557
  uses 5:551
Thiosulfates. (See also individual compounds.)
    18:429–433
  acid turbidimetric method 18:431
  ammonium molybdate–sulfuric acid colorimetric
    method 18:431
  analysis 18:431
  blue ring test 18:431
  colorimetric method for mercury fulminate
    12:420
  detection 18:431
  in gelatin 12:548; 13:371
  iodine–azide gasometric method 12:549
  iodometric titration 12:586; 18:432
  ion exchange chromatography 18:431
  manufacture 18:430
  mercuric chloride acidimetric method 18:431
  methylene blue colorimetric method 12:549,
    594; 13:371
  in photographic films 12:586, 594
  polarographic method 12:548, 586
  in potassium iodide 14:606
  properties 18:429
  separation 18:431
  silver nitrate gravimetric method 18:431
  silver nitrate titration 12:550
  in sodium iodide 14:606
  solubility 18:429
  sulfuric acid turbidimetric method 14:606
  uses 18:430
  volumetric method 18:432
  volumetric method for antimony 6:49, 56, 66
  volumetric method for arsenic(V) 6:243

Thiosulfuric acid. (See also Thiosulfates.)
  manufacture 18:429
  properties 18:429
Thiourea 5:276
  antichlor 7:222
  colorimetric method for bismuth 7:214;
    10:635; 15:198
  colorimetric method for osmium 17:248, 256,
    263
  colorimetric method for ruthenium 17:247,
    263
  colorimetric method for technetium 17:531
  colorimetric method for tellurium 17:606
  gravimetric method for thallium 18:523
  manufacture 18:434
Thiourea derivative
  of amines 5:266
Thiovanadates. (See also Vanadium compounds.)
    19:345
Thiram 13:204
  manufacture 13:204
  properties 13:204
  solubility 13:204
  uses 13:204
Thiurad. See Thiram.
Thiuram disulfides. (See also individual com-
    pounds.) 11:591–598; 13:202–205
  acid–base titration 11:595
  acid hydrolysis method 11:593
  analysis 11:593
  biological methods 11:598
  copper(I) dithiocarbamate colorimetric method
    11:594
  copper(I) iodide colorimetric method 11:593
  copper(II) sulfate–hydroquinone titration
    11:594
  infrared spectroscopy 11:596
  manufacture 11:592
  mass spectroscopy 11:597
  polarography 11:597
  potassium cyanide–silver nitrate titration 11:595
  properties 11:591
  reduction methods 11:593
  silver nitrate gravimetric method 11:595
  sodium sulfide titration 11:594
  ultraviolet spectroscopy 11:597
  uses 11:591
Thiuramyl. See Thiram.
Thixotropy
  of enamels 12:178
Thompson-Oakdale halogen apparatus 16:220
Thonzylamine
  gas chromatography 6:210, 211
Thonzylamine hydrochloride. (See also *Anti-
    histamines.*) 6:30
  column chromatography 6:150
  properties 6:32

ultraviolet spectroscopy 6:150
Thoreaulite
  composition 18:441
Thoria. (See also Thorium oxide.)
  composition 9:207
  in tungsten metal 19:218
Thorianite
  composition 19:2
  decomposition 19:8, 235
  thorium content 19:2, 8
Thorin
  colorimetric method for thorium 9:256;
    17:495; 19:13, 16
  structure 19:13
Thorite
  composition 19:2
  thorium content 19:2
Thorium. (See also Thorium alloys; Thorium
    compounds; and Thorium metal.) 19:1–21
  in aluminum and aluminum alloys 5:152
  ammonium hydroxide gravimetric method
    19:9
  ammonium oxalate gravimetric method 19:12
  arsenazo I colorimetric method 19:13
  arsenazo II colorimetric method 19:13
  arsenazo III colorimetric method 19:13, 14
  benzoic acid gravimetric method 19:9, 12
  benzoic acid–oxalic acid gravimetric method
    15:393
  in ceric oxide 9:256
  in cheralite 19:2
  colorimetric method 15:393; 19:13
  decomposition 19:7, 16
  detection 19:10
  diphenic acid gravimetric method 19:9
  DTPA titration 15:159
  EDTA titration 13:445; 19:13
  emission spectroscopy 9:257; 14:123;
    15:375; 17:488, 502; 19:10
  in europium oxide 12:403; 17:502
  fluorometric methods 19:17
  gas chromatography 19:17
  in glass and glass-ceramics 13:445
  gravimetric method 15:393; 19:8, 10, 12
  in hafnium 14:123
  half-life 19:1
  hexamine gravimetric method 17:495
  history 19:2
  in huttonite 19:2
  hydrofluoric acid gravimetric method 19:9
  hydrogen peroxide–nitric acid gravimetric method
    19:9
  8-hydroxyquinoline extraction 17:477
  iodate gravimetric method 19:9
  ion exchange chromatography 17:495, 575;
    19:10
  isotopes 19:1

in lanthanum glass 15:160
in lutetium oxide 17:503
in magnesium 15:375, 393
in magnesium alloys 15:375, 393; 19:12
mass spectroscopy 19:17
in monazite 19:6, 8, 12
neutron activation analysis 19:17
in nickel 16:335
in nickel alloys 16:335
occurrence 19:2
in ores 19:7, 16
oxalic acid gravimetric method 16:335; 19:9
in pilbarite 19:2
potassium hydroxide gravimetric method
  19:14
potassium iodate gravimetric method 19:14
pyrocatechol violet titration 19:13
radioactive decay 17:446
radioactivity methods 19:17
in rare earths 17:477, 495
in refractory minerals 19:8
in rocks 19:7, 14, 16
salicylic acid–ammonium salicylate gravimetric
  method 19:95
in samarium oxide 17:502
in scandium 17:573
separation 19:7
solvent extraction 19:10
in TD nickel 16:335
in thorianite 19:2, 8
thorin colorimetric method 17:495; 19:13, 16
in thorite 19:2
in thorium alloys 19:8
thorium dioxide gravimetric method 19:12
in thorogummite 19:2
in uranium 19:240
in uranothorianite 19:8
volumetric method 15:393; 19:13
in water 19:16
x-ray emission spectroscopy 9:257
x-ray fluorescent methods 19:17
x-ray spectroscopy 19:11
in ytterbium oxide 17:503
in zircon 19:16
in zirconium 14:123
Thorium alloys. (See also Thorium metal.)
  carrier gas fusion methods 8:682
  decomposition 19:8
  oxygen in 8:682
  thorium content 19:8
  uses 19:3
Thorium boride. (See also Thorium compounds.)
  manufacture 19:4
Thorium carbide. (See also Thorium compounds.)
  manufacture 19:4
Thorium carbonate. (See also Thorium compounds.)

solubility 19:5
Thorium compounds. (See also individual
   compounds.) 19:17–20
   aluminum in 5:250, 251
   analysis 19:18
   properties 19:4
   safety 19:6
   toxicity 19:6
   uses 19:3, 4
Thorium dioxide. (See also Thorium compounds.)
   19:18
   carrier gas fusion methods 8:624, 626
   gravimetric method for thorium 19:12
   manufacture 19:4, 18
   nitrogen in 8:624
   oxygen in 8:624, 626
   properties 9:204; 19:18
   in refractories 9:207
   safety 19:6
   solubility 19:18
   uses 19:7, 18
Thorium fluoride. (See also Thorium compounds.)
   carrier gas distillation fusion methods 8:636
   solubility 19:5
   uses 19:6
Thorium fluoride hydrate. (See also Thorium
   compounds.)
   manufacture 19:5
   properties 19:5
   uses 19:5
Thorium halides. (See also Thorium compounds.)
   manufacture 19:4
Thorium hydride. (See also Thorium compounds.)
   manufacture 19:4
Thorium hydrous hydrate. (See also Thorium
   compounds.)
   manufacture 19:18
   properties 19:18
   solubility 19:18
Thorium hydroxide. (See also Thorium compounds.)
   solubility 19:5
Thorium iodate. (See also Thorium compounds.)
   manufacture 19:5
   solubility 19:5
   uses 19:5
Thorium iodide. (See also Thorium compounds.)
   uses 19:7
Thorium–magnesium alloy. (See also Thorium
   alloys.)
   uses 19:3
Thorium metal
   analysis 19:7
   carrier gas fusion methods 8:624, 626, 627, 682
   commercial grades and specifications 19:7
   hot extraction methods 8:684
   hydrogen in 8:619, 622, 626, 638, 680, 684

manufacture 19:6
   nitrogen in 8:613, 619, 623, 624, 626, 627
   oxygen in 8:613, 619, 622, 623, 624, 626,
      627, 629, 630, 680, 682
   properties 19:4
   safety 19:6
   sulfur content 18:383
   toxicity 19:6
   uses 19:2, 3
   vacuum fusion methods 8:619, 622, 623,
      680
   vacuum hot extraction methods 8:638
Thorium molybdate. (See also Thorium compounds.)
   manufacture 19:5
   solubility 19:5
   uses 19:5
Thorium nitrate. (See also Thorium compounds.)
   19:17
   analysis 19:18
   commercial specifications 19:18
   gravimetric method for carrageenin 12:582
   gravimetric method for pectin 12:582
   impurities, determination of 19:18
   manufacture 19:17
   solubility 19:18
   specification tests 19:18
   titration of calcium chelates 12:50
   titration of cesium fluoride 9:289
   titration of fluoride 13:45, 434
   titration of fluorine 9:322; 17:507
   toxicity 19:6
   uses 19:18
Thorium nitrate–hexamethylenetetraamine
   colorimetric method for DCYTA 12:28
   colorimetric method for EDTA 12:28, 30
   colorimetric method for EGTA 12:28
   colorimetric method for NTA 12:28
   colorimetric method for PDTA 12:28
Thorium nitride. (See also Thorium compounds.)
   manufacture 19:4
Thorium oxalate. (See also Thorium compounds.)
   manufacture 19:5
   solubility 19:5
Thorium peroxide. (See also Thorium compounds.)
   manufacture 19:6
   solubility 19:6
   uses 19:6
Thorium phosphate. (See also Thorium compounds.)
   manufacture 19:5
   solubility 19:5
Thorium phosphide. (See also Thorium compounds.)
   manufacture 19:4
Thorium sulfide. (See also Thorium compounds.)
   manufacture 19:4

Thorogummite
  composition 19:2
  thorium content 19:2
Thoron. See Thorin.
  colorimetric method for lithium 15:269
  colorimetric method for tantalum 16:369
Thorone. See Thorin.
Thortveitite
  composition 17:560
  sample decomposition 17:564
  scandium content 17:564, 569
Threitol
  gas chromatography 13:507; 19:300
Threonine. (See also *Carboxylic acids, amino.*)
    8:413, 526
  analysis 8:522, 524, 528
  commercial specifications 8:528
  crystalline form 8:526
  dinitrobenzene identification test 8:528
  gas chromatography 8:429
  infrared spectrum 8:526
  ion exchange chromatography 8:415, 421
  manufacture 8:527
  microdiffusion method 8:528
  paper chromatography 8:422
  periodate oxidation 8:523
  potassium permanganate test 8:522
  properties 8:525
  in protein 8:414
  solubility 8:527
  structure 8:525
  in yeast 19:529
Threshold odor of activated carbon 8:148
Thucolite
  composition 19:225
  occurrence 19:225
α-Thujene
  gas chromatography 12:268
  in majoram, oil of 12:268
Thujone
  in cedar leaf oil 12:282
  hydroxylamine hydrochloride volumetric
    method 12:282
  in sage oil, Dalmatian 12:282
  in tansy oil 12:282
  in wormwood oil 12:282
Thulium. (See also *Rare earths*; Thulium metal.)
  atomic absorption spectroscopy 17:494
  bead test 6:587
  emission spectroscopy 15:153, 155; 17:501
  in erbium oxide 17:503
  flame photometry 17:493
  in lanthanum oxide 15:153, 155; 17:501
  in lutetium oxide 17:503
  x-ray fluorescent spectroscopy 17:490
  in ytterbium oxide 17:503
Thulium fluoride

  carrier gas distillation fusion methods 8:636
  oxygen in 8:635, 636
  vacuum distillation fusion methods 8:635
Thulium metal. (See also Rare earth metals.)
  carrier gas fusion methods 8:682
  hydrogen in 8:680
  oxygen in 8:680, 682
  properties 17:471
  vacuum fusion methods 8:680
Thulium oxide. (See also Rare earth oxides.)
  emission spectroscopy 17:503
  impurities, determination of 17:503
  properties 17:497
Thulium perchlorate. (See also Rare earth com-
    pounds.)
  absorption spectrum 17:485
Thylate. See Thiram.
Thyme oil, red. (See also *Essential oils.*)
  refractive index 12:271
Thymol. (See also 3-Methyl-6-isopropylphenol;
    Phenols.)
  colorimetric method for titanium 19:107, 118
  color test for ethylene glycol 13:570
  color test for propylene glycol 13:570
  commercial specifications 11:540
  phthalic anhydride colorimetric method 17:19
  properties 17:4
Thyroid hormones. (See also individual com-
    pounds.) 14:183
  analysis 14:206
  properties 14:189
  sources 14:189
  uses 14:189
Thyrotropic hormone. See Thyrotropin.
Thyrotropin. (See also Pituitary hormones.)
    14:180
  analysis 14:206
  assay 14:206
  manufacture 14:188
  occurrence 14:188
  uses 14:180, 188
Thyroxine. (See also Thyroid hormones.)
  analysis 14:207
  assay 14:207
  infrared spectrum 14:190
  manufacture 14:189
  occurrence 14:189
  properties 14:189
  solubility 14:189
  structure 14:183
Ticonal. (See also Cobalt alloys.) 10:288
  composition 10:288
Tiemannite 16:1
Tiglaldehyde
  absorption spectroscopy 9:564
  infrared spectroscopy 6:616
Tile

composition 9:194

Tilearex. See Terraclor.

Tilleyite 8:75

Time of setting
    of cements 9:134

Tin. (See also Tin alloys; Tin compounds; and
    Tin metal.) 19:21–80
in aluminum 19:53, 64
in aluminum and aluminum alloys 5:144
ammonium hydroxide gravimetric method
    19:33, 39
ammonium molybdate–ammonium thiocyanate
    colorimetric method 19:37
ammonium molybdate colorimetric method
    19:37
in antimony–lead alloys 19:64
in antimony metal 6:46
in arsenic metal 6:228
atomic absorption spectroscopy 14:125;
    15:196; 19:64
in beer 7:650, 662
Bendix volumetric method 19:55
in biological material 19:62
in brass 19:51
cacotheline colorimetric method 11:141, 144
in cadmium 19:64
in chromic acid 9:684
in chromium 9:635
Clarke weight-loss method 19:55
in cobalt alloys 10:320
in cobalt metal 10:304
colorimetric method 15:394; 19:37, 55
colorimetric method for titanium 12:186
in copper alloys 10:646; 19:65
in copper metal 10:626
in copper–nickel alloys 19:52
in copper–nickel–zinc alloys 19:52
in cotton 11:141, 154, 156
cupferron gravimetric method 19:34, 40
in dental amalgams 11:288
detection 19:37
distillation methods 19:35, 57
dithiol colorimetric method 11:141; 19:37, 55
emission spectroscopy 7:650; 14:34, 48, 122,
    538; 15:144, 156, 187, 197, 375; 16:160, 387;
    19:38, 64, 570
ferric chloride gravimetric method 16:336
in ferrochromium 9:661
in ferroniobium 16:387
in ferrotungsten 19:48
flame tests 6:589
in food 19:39, 53
in fruit 19:40, 61
in fruit juice 19:65
in gallium phosphide 13:282
in gallium metal 13:266
in gold 14:45

in gold alloys 14:34
gravimetric method 19:32, 38
in hafnium 14:122, 124
halogen volatilization methods 19:34, 71, 75
history 19:21
in hydrochloric acid 14:374
hydrogen sulfide gravimetric method 15:67;
    19:32, 39
4-hydroxy-3-nitrophenylarsonic acid colorimetric
    method 10:626
in indium 14:538
in indium alloys 14:553
iodometric titration 10:646; 14:136, 553;
    19:38, 42
ion exchange chromatography 19:37
in iron 15:59, 67
isotopes 19:21
in lactic acid and its derivatives 15:144
in lanthanum oxide 15:156
in lead 15:196; 19:64
in lead alloys 15:195
lead reduction method 15:68
in magnesium 15:375, 394; 19:53
in magnesium alloys 15:375, 394
mercuric chloride gravimetric method 19:37, 41
in metals 19:62
metastannic acid gravimetric method 19:33
in molybdenum 16:161
in molybdenum trioxide 16:161
neutron activation analysis 13:265
in nickel 16:336
in nickel alloys 16:336
in niobium 16:387
nitric acid gravimetric method 11:289; 19:33,
    41
6-nitro-2-naphthylamine-8-sulfonic acid ammon-
    ium salt fluorimetric method 19:38
occurrence 19:21
in organic compounds 19:54, 56, 77
phenylfluorone colorimetric method 6:46, 228;
    19:62
phosphomolybdic acid colorimetric method
    19:37
in photographic films 12:592
in plating solutions 19:64
polarography 7:662; 19:66
potassium iodate titration 15:68; 16:336
in potassium stannate 19:72
pyrocatechol violet colorimetric method 19:55,
    61
8-quinolinol colorimetric method 19:64
in rocks 15:187; 19:64
salicylideneamino-2-thiophenol colorimetric
    method 19:64
sample decomposition 19:31
sampling 19:31
Santomerse S colorimetric method 10:304

in sediment 19:64
separation 19:32
in silicate rocks 19:62, 64
in silver 18:173
in sodium dichromate 9:684
sodium hydroxide gravimetric method 19:33
sodium peroxide fusion 19:60
in sodium stannate 19:72
in soil 19:64
in solder 19:52
solvent extraction 19:36, 62
in stannates 19:72
in stannous chloride 19:73
in stannous fluoride 19:75
in steel 19:49, 64
in tin alloys 19:41, 47
in tin metal 19:66
in tinplate 19:54, 65
in titanium 19:52, 121
3,4',7-trihydroxyflavone fluorimetric method
    19:64
in tungsten 19:48
in type metal 19:64
volumetric determination 19:42
volumetric method using potassium iodate
    5:145
x-ray fluorescence spectroscopy 11:156; 19:65
in zinc 19:64, 570
in zirconium 14:122, 124, 136; 19:53
α-Tin. See Gray tin.
β-Tin. See White tin.
Tin alloys. (See also Tin metal.) 5:72
aluminum in 5:226
antimony in 6:60
bismuth in 7:214, 216
carrier gas fusion methods 8:682
commercial specifications 19:31
copper content 10:617
history 19:21
oxygen in 8:682
sample decomposition 19:31, 47
sample preparation 6:60
tin content 19:41, 47
uses 19:24
zinc content 19:568
Tin bromide
colorimetric method for rhodium 17:253
Tin bronze. See Stannic sulfide.
Tincal 7:314
Tincalconite. (See also Sodium tetraborate
    pentahydrate.) 7:313, 375
occurrence 7:313, 314
Tin chloride
colorimetric method for iridium 17:266
colorimetric method for platinum 17:249, 259
colorimetric method for rhodium 17:248, 262
gravimetric method for mercury 16:6, 7

Tin chloride–aniline
colorimetric method for mercury 16:19, 21
Tin chloride–cacotheline
colorimetric method for mercury 16:19, 21
Tin chloride–potassium iodide
colorimetric method for rhodium 17:248
Tin compounds. (See also individual compounds.)
    19:68–78
properties 19:27
sample decomposition 19:31
toxicity 19:28
uses 19:76
Tin compounds, organic. See Organotin com-
    pounds.
Tin concentrates
recovery of bismuth from 7:190
Tin crystals. See Stannous chloride.
Tinctorial strength
of organic dyes 15:532
of organic pigments 15:532
Tin–hydrochloric acid
gravimetric method for nitronaphthalenes
    16:225
Tin metal
allotropes 19:26
aluminum in 5:226
analysis 19:31, 66
carrier gas fusion methods 8:624, 682
commercial grades and specifications 19:29
impurities 19:29
impurities, determination of 19:66
hydrogen in 8:681
manufacture 19:28
nitrogen 8:617
oxygen in 8:617, 624, 681, 682
properties 19:25
reaction medium for fusion methods 8:629,
    632
sample decomposition 19:31, 66
sampling 19:66
solubility 19:25
tin content 19:66
toxicity 19:28
uses 19:23
vacuum fusion methods 8:617, 681
Tin–nickel alloy. (See also Nickel alloys.)
manufacture 16:287
uses 16:287
Tin oxide. See Stannic oxide.
in alkali barium silicate glass 9:218
carrier gas fusion methods 8:626
emission spectroscopy 9:218
in enamels 9:195
fluxes for 9:216
in glazes 9:196
oxygen in 8:626
Tinplate. (See also Tin metal.)

manufacture 19:24
tin content 19:54, 65
uses 19:24
Tin soap. (See also *Soaps.*)
uses 18:180
Tin tetrachloride. (See also Stannic chloride.)
properties 19:96
Tinting strength
of carbon black 8:186, 213
of pigments 17:206
Tinuvin P. (See also 2(2′-Hydroxy-5′-methyl-
phenyl)benzotriazole.) 6:77
Tiron
colorimetric method for iron 19:113
colorimetric method for niobium 16:368,
373
colorimetric method for tantalum 16:368
colorimetric method for titanium 19:99, 107,
112
colorimetric method for titanium dioxide
10:118; 13:427
Tishchenko reaction 8:580
Tissue
calcium in 8:110
insulin content 14:572
Titanated lithopone
analysis 17:190
commercial specifications 17:190
Titanates. (See also Titanium compounds; and
individual compounds.) 19:124
decomposition 19:92
fluxes for 9:216
manufacture 19:124
sampling 19:90
uses 19:124
Titania. (See also Titanium.)
in aluminum oxide abrasives 4:35, 49
decomposition 19:91
x-ray emission spectroscopy 4:49
Titania cover enamel. (See also *Enamels.*)
composition 9:195
Titaniferous magnetites 19:320, 325
composition 19:321
occurrence 19:321
vanadium content 19:325
Titanium. (See also Titanium alloys; Titanium
compounds; and Titanium metal.) 19:81–
132
absorption method for nitrogen 13:293
absorption spectroscopy 9:666
acetic acid gravimetric method 19:93
acetyl acetone colorimetric method 19:99
allotropes 19:84
in alumina 5:116
in aluminum 19:115, 119, 120
in aluminum alkyls 5:171
in aluminum alloys 5:145

in aluminum metal 5:145
in aluminum oxide 4:35, 40, 49; 5:188
ammonium hydroxide gravimetric method
16:174, 176; 19:94, 103, 111
ammonium phosphate gravimetric method
19:94
in amosite 6:267
in aplite 19:110
in asbestos 6:261, 267
ascorbic acid colorimetric method 19:107
atomic absorption spectroscopy 19:119
in barium titanate 19:126
in bauxite 5:116; 19:120
in biological materials 19:96
borax bead test 6:584
boron in 7:418
in carbon 14:67, 74
in carbonate rocks 19:120
in chromite 9:700
in chromium 9:639
in chromium alloys 9:663, 665
in chromium ores 9:701, 704
chromotropic acid colorimetric method 19:98,
107, 115
in chrysotile 6:261, 267
in clays and clay minerals 10:136
in cobalt alloys 10:320
colorimetric method 15:394; 19:97, 100, 107
colorimetric method using hydrogen peroxide
5:146
in copper alloys 10:646
in cotton 11:141, 154, 156
coulometric methods 19:119
in crocidolite 6:267
cupferron gravimetric method 9:665; 14:252;
15:83; 16:391; 19:95, 101, 111
DCyTA volumetric method 19:107
detection 19:97
diammonium hydrogen phosphate gravimetric
method 15:83
diantipyrylmethane colorimetric method
19:107, 118
in diatomaceous earth 11:490
3,6-dichlorochromotropic acid colorimetric
method 19:98
distillation methods 19:96
EDTA-8-hydroxyquinoline colorimetric method
19:97
EDTA titration 9:639
in elastomers, synthetic 12:151
in electrical resistance alloys 10:290
electrolytic method 10:321
emission spectroscopy 4:3; 11:493; 13:277;
14:48, 74, 122; 15:155, 187; 16:387; 18:445;
19:99, 119
in enamels 12:186
in face powder 19:106

ferric ammonium sulfate titration 10:321, 647;
19:104, 106
ferric chloride titration 8:136
ferric sulfate titration 19:105
in ferrochromium 9:661
in ferroniobium 16:387, 390
in ferrotitanium 15:82; 19:101, 105
ferrous ammonium sulfate titration 15:82;
19:127
flame emission spectroscopy 19:120
in gold 14:45
in graphite 14:67, 74
gravimetric method 19:93, 100, 101
in hafnium 14:122, 137
half-life 19:81
in high-temperature alloys 10:286
history 19:82, 88
in hydrochloric acid 14:375
hydrochloric acid gravimetric method 19:93
hydrogen evolution method 19:128
hydrogen peroxide colorimetric method 4:40;
10:321, 647; 11:141; 12:186; 14:67; 16:176,
337, 382, 391; 19:20, 97, 107
p-hydroxyphenylarsonic acid gravimetric method
19:96
8-hydroxyquinoline gravimetric method 9:664;
19:95
in ilmenite 19:104
ion exchange chromatography 13:255; 14:113,
140; 16:153, 385, 390; 17:526; 19:97
in iron 15:59; 19:111, 120
isotopes 19:81
Jones reductor volumetric method 17:187
in lanthanum oxide 15:155
in magnesium 15:394
in magnesium alloys 15:394
in meteorites 19:96
microcosmic salt bead test 6:587
in minerals 19:96
in molybdenum alloys 16:174
in molybdenum–titanium–zirconium alloys
16:176
morin colorimetric method 19:98
in nickel 16:337; 19:113, 120
in nickel alloys 16:337
in nickel–chromium alloys 19:112
in nickel–chromium–iron alloys 19:112
in niobium 16:387
nitrogen content 16:465
occurrence 19:81
in organic material 19:113
pertitanic acid colorimetric method 12:155
polarography 19:119
in polypropylene 19:120
potassium permanganate titration 19:103
potassium pyrosulfate fusion 16:176; 19:102,
111

rare earth content 17:477
in rocks 15:187; 19:96
in rutile 19:104
separation 19:93
in shampoos 11:75
in silica ablators 4:3
in silicate rocks 19:109, 118, 120
in silicates 19:113
in silicon carbide abrasives 4:56
sodium hydroxide gravimetric method 16:337;
19:94
sodium sulfate–hydrogen peroxide colorimetric
method 11:75
solvent extraction 19:95, 96, 110
standard solutions 19:102
in steel 16:382; 19:111, 119, 120
5-sulfosalicylic acid colorimetric method 14:137;
19:107, 117
tannin gravimetric method 19:95
tartaric acid–hydrogen peroxide colorimetric
method 14:142
in terephthalic acid 17:162
in thorium nitrate 19:20
thymol colorimetric method 19:107, 118
tin colorimetric method 12:186
tiron colorimetric method 19:99, 107, 112
in titanium bromide 7:362
in titanium carbide 8:133, 136
in titanium dioxide pigments 17:187; 19:104,
124
in titanium steel 19:111
in tungsten carbide 19:219
uses 19:82
valence state, determination of 19:126
volumetric determination 19:100, 104
x-ray emission spectroscopy 4:49
x-ray fluorescence spectroscopy 10:136;
11:156; 17:162; 19:99, 119
zinc colorimetric method 12:186
in zircon 14:139
in zirconium 14:122, 137; 19:117
α-Titanium. (See also Titanium metal.) 19:84
β-Titanium. (See also Titanium metal.) 19:84
Titanium alloys. (See also Titanium metal.)
5:61, 67
analysis 19:89, 120
boron in 7:410, 418
commercial grades and specifications 19:88
decomposition 19:90
hydrogen in 8:638, 680, 681
impurities, determination of 19:120
manganese content 15:466
niobium content 16:377
nitrogen in 8:631, 638
oxygen in 8:620, 631, 635, 636, 638, 680,
681
palladium content 17:271

properties 19:86
sampling 19:89
solubility 19:86
tantalum content 16:377
tin content 19:52
tungsten content 19:215
uses 19:83
vacuum distillation fusion methods 8:635
vacuum fusion methods 8:620, 680, 681
vacuum hot extraction methods 8:638
Titanium–barium pigment
analysis 17:187
barium sulfate content 17:187
commercial specifications 17:187
titanium dioxide content 17:187
Titanium boride. (See also Metal borides;
    Titanium compounds.) 7:358
compound impurities 7:366
manufacture 7:358
properties 7:358, 359; 19:124
titanium content of 7:362
uses 19:124
x-ray diffraction 7:366
Titanium bromide. (See also Titanium halides.)
sampling 19:90
Titanium butylate. See Tetrabutyl titanate.
Titanium–calcium pigment
analysis 17:187
calcium sulfate content 17:187; 19:125
commercial grades and specifications 17:187
titanium dioxide content 17:187
Titanium carbide. (See also *Carbides*; Titanium
    compounds.)
carbon content 8:134
formation 8:130
free carbon in 8:134
hydrogen in 8:681
manufacture 19:86, 124
oxygen in 8:618, 621, 638, 681
properties 8:131, 132; 19:124
solubility 19:124
in titanium boride 7:367
titanium content 8:133, 136
uses 8:132; 19:83, 124
vacuum fusion methods 8:618, 621, 681
vacuum hot extraction methods 8:638
Titanium carbide–molybdenum
in ceramics 9:225
microscopy 9:225
Titanium chelates. (See also Organotitanium
    compounds.)
uses 19:130
Titanium chloride
colorimetric method for phenolic hydroxyl
    groups 17:339
Titanium compounds. (See also individual com-
    pounds.) 19:121–130

analysis 19:90, 125
decomposition 19:92
manufacture 19:86
properties 19:87
sampling 19:90
solubility 19:86
toxicity 19:87
Titanium compounds, inorganic 19:121–129
Titanium compounds, organic. See Organo-
    titanium compounds.
Titanium dichloride. (See also Titanium halides.)
manufacture 19:122
properties 19:122
uses 19:122
Titanium dichloride–ferric ammonium sulfate
volumetric method for 1-nitronaphthalene
    16:225
Titanium dioxide. (See also Titanium compounds;
    Titanium dioxide pigments; and Titanium
    oxide.)
in alkali barium silicate glass 9:218
allotropes 19:123
analysis 15:531; 19:125
assay 10:541
in cements 9:122
in ceramics 9:194, 225
in cerium titanate 9:261
in clays and clay minerals 10:118
commercial grades and specifications 15:516
commercial specifications 10:533; 12:174;
    19:124
in creams and lotions 11:92
cupferron gravimetric method 13:426
delustering agent in fibers 4:329
emission spectroscopy 4:345; 9:218
in enamels 9:195; 12:174
ferric ammonium sulfate titration 9:261
in glass 9:200
in glass and glass-ceramics 13:413, 415, 425
in glazes 9:196
hydrogen in 8:616
hydrogen peroxide colorimetric method 9:123;
    10:118; 15:211; 17:302; 18:194
in lime and limestone 15:211
manufacture 19:86
in matches 15:516
microscopy 9:225
nitrogen in 8:616
occurrence 19:81
oxygen in 8:616
particle size 15:531
perchloric acid–hydrogen peroxide colorimetric
    method 13:426
in polyamides 17:302
properties 9:204; 19:123
in refractories 9:206
in silicate glasses 9:199

in silicon dioxide 18:30
in soaps 18:194
solubility 19:86
specific gravity 15:532
in styrene polymers 18:320
tiron colorimetric method 10:118; 13:427
in titanium–barium pigment 17:187
in titanium boride 7:367
in titanium–calcium pigment 17:187
toxicity 19:87
uses 19:82, 87, 123
vacuum fusion methods 8:616
vanadium content 19:343
wettability 15:531
zinc amalgam reduction method 10:541
Titanium dioxide pigments. (See also Titanium
    dioxide.)
analysis 17:187; 19:125
commercial grades and specifications 17:187; 19:
    123
impurities, determination of 19:125
manufacture 19:123
sodium bisulfate fusion 19:125
titanium content 17:187; 19:104, 124
Titanium ferricyanide. (See also Titanium com-
    pounds.)
solubility 19:86
Titanium fluoride. (See also Titanium halides.)
sampling 19:90
Titanium halides. (See also Titanium compounds;
    and individual compounds.) 19:122
analysis 19:127
decomposition 19:92
impurities, determination of 19:129
sampling 19:90
Titanium hydride. (See also *Hydrides*; Titanium
    compounds.) 19:121
analysis 14:249, 252
assay 14:249
commercial grades and specifications 14:246
composition 19:121
composition, determination of 14:249, 252
hydrogen content 14:249
manufacture 14:245; 19:121
properties 14:243, 244; 19:121
safety 19:122
titanium content 14:252
uses 14:242; 19:121
Titanium iodide. (See also Titanium halides.)
sampling 19:90
Titanium lactate. (See also Lactic acid salts.)
uses 15:130
Titanium metal
analysis 19:89, 120
carrier gas fusion methods 8:625, 626, 627,
    682
commercial grades and specifications 19:88

decomposition 19:90
hot extraction methods 8:684
hydrogen in 8:617, 618, 619, 621, 622, 623,
    638, 680, 684
impurities, determination of 19:120
manufacture 19:87
nitrogen in 8:618, 619, 622, 623, 627, 629,
    630
oxygen in 8:613, 616, 617, 619, 620, 621,
    622, 623, 625, 626, 627, 631, 632, 680, 682
properties 19:83
sampling 19:89
solubility 19:86
solubility in gold 14:9
tin content 19:52
in titanium halides 19:129
toxicity 19:87
tungsten content 19:215
uses 19:82, 84, 87
vacuum fusion methods 8:616–623, 680
vacuum hot extraction methods 8:638
Titanium minerals
acid fluoride fusion 19:93
borax fusion 19:93
decomposition 19:91
potassium pyrosulfate fusion 19:92
sampling 19:90
sodium hydroxide fusion 19:93
Titanium monoxide. (See also Titanium com-
    pounds.)
manufacture 19:86
solubility 19:86
Titanium nitride
in titanium boride 7:367
Titanium oxides. (See also Titanium compounds;
    and individual compounds.) 19:123
carrier gas fusion methods 8:624, 626, 682
decomposition 19:91
oxygen in 8:624, 626, 682
sampling 19:90
in soaps 18:182
Titanium peroxide. (See also Titanium compounds.)
uses 19:123
Titanium phosphate. (See also Titanium com-
    pounds.)
solubility 19:86, 100
Titanium refractory compounds. (See also
    individual compounds.) 19:124
properties 19:124
Titanium sesquioxide. (See also Titanium
    compounds.)
manufacture 19:86
properties 19:86
solubility 19:86
Titanium steel. (See also Titanium alloys.)
titanium content 19:111
Titanium sulfate. (See also Titanium compounds.)

colorimetric method for hydrogen peroxide
14:434
manufacture 19:124
Titanium tetrachloride. (See also Titanium
halides.)
colorimetric method for peroxides 12:308
manufacture 19:87, 122
properties 9:355; 19:96, 122
sampling 19:90
solubility 9:336; 19:122
toxicity 19:87
uses 19:87, 122
Titanium tetraisopropoxide. (See also Organotitanium compounds.)
manufacture 19:122, 130
uses 19:130
Titanium trichloride. (See also Titanium halides.)
manufacture 19:122
properties 9:335; 19:122
solubility 9:336
uses 19:122
Titanium–vanadium alloys. (See also Vanadium
alloys.)
decomposition 19:327
Titanium–zirconium alloys
iron content 15:57
Titanous chloride titration
of acid magenta 19:177
of azine dyes 6:394, 395
of certified colors 10:484
of crystal violet 19:181
of D & C Blue No. 4 and No. 6 10:485
of D & C Green No. 5 and No. 6 10:485
of D & C Orange No. 4 10:485
of D & C Reds 10:483, 485
of D & C Yellow No. 7 and No. 8 10:485
of eosin A 19:177
of Ext D & C Green No. 1 10:485
of Ext D & C Yellow No. 1 and No. 7 10:485
of F D & C Blue No. 1 and No. 2 10:485
of F D & C Green No. 2 10:485
of F D & C Red No. 2 and No. 4 10:485
of F D & C Violet No. 1 10:485
of F D & C Yellow No. 5 and No. 6 10:485
of F D & C Yellows 19:177
of ferric ammonium sulfate 19:180
of indigo 14:512
of indigo carmine 14:512
of magenta 19:177
of malachite green 19:177
of menadione 10:198
of menadione sodium bisulfite 10:202
of β-naphthol 10:503
of Orange B 10:485
of phenolphthalein 19:177
of rhodamine 19:177
of rhodamine B 19:181

Titanous chloride–ferric ammonium sulfate
volumetric method for nitrogen 12:467
Titanous chloride–monochlorobenzene
colorimetric method for gallium 14:555
Titanous sulfate. (See also Titanium compounds.)
properties 19:124
solubility 19:124
uses 19:124
Titan yellow
colorimetric method for magnesium 15:359,
367
Titanyl acetylacetonate. (See also Organotitanium compounds.)
manufacture 19:130
uses 19:130
Titanyl sulfate. (See also Titanium compounds.)
manufacture 19:87
uses 19:124
Titration. See Acid–base titration; Conductometric
titration; Potentiometric titration; and
Redox titration.
Titer test
for fats and fatty oils 12:476
for fatty acids 18:186
TMEL. See Trimethylethyllead.
TML. See Tetramethyllead.
TMPB. See Trimethylphenylammonium bromide.
TMSI. See N-Trimethylsilylimidazole.
TMTD. See Thiram.
TNT. (See also High explosives.) 12:439
ammonium hydroxide colorimetric method
12:421
colorimetric method 12:421
commercial specifications 12:440
in explosive priming mixtures 12:418, 421,
422, 426, 427
infrared spectrum 12:459
manufacture 12:439
melting point 12:461
properties 12:434
solubility 12:421, 439
solvent extraction 12:426, 427
uses 12:439
x-ray diffraction 12:418
Tobias acid. See 2-Naphthylamine-1-sulfonic acid.
Tobias B.O.N. maroon
in enamels 12:182, 186
identification 12:182, 186
Tobacco 19:133–160
alkaloid content 19:135
analysis 19:133
cellulose content 19:142
color 19:157
composition, determination of 19:135
crude fiber content 19:142
fire-holding capacity 19:158
impurities, determination of 19:154

inorganic content 19:152
lipid content 19:149
moisture content 19:135
nicotine content 4:610; 19:135
nitrogenous compound content 19:137
nornicotine content 19:135
organic acid content 19:148
pectin content 19:143
pH 19:158
phenolic content 19:145
pigment content 19:152
potassium content 17:390
properties, determination of 19:157
protein content 19:138
radioelement content 19:153
sample preparation 19:133
sampling 19:134
specific volume 19:158
starch content 19:142
sugar content 19:141
Tobacco humectants
gas chromatography 13:503
TODA. See 3,3'-Bitolylene-4,4'-diamine.
Toffees. (See also *Confectionery products.*)
    10:548
manufacture 10:553
o-Tolidine–arsenite
colorimetric method for ozone 16:549
o-Tolidine–manganese
colorimetric method for ozone 16:549
Tollens reagent
color test for aldehydes 8:344; 15:145
color test for formaldehyde 13:106
m-Tolualdehyde (3-Methylbenzaldehyde) 6:597
commercial specifications 6:611
derivatives for identification 6:612
infrared spectrum 6:615
near-infrared spectroscopy 6:629
properties 6:598
ultraviolet spectroscopy 6:632
o-Tolualdehyde (2-Methylbenzaldehyde) 6:597
commercial specifications 6:611
derivatives for identification 6:612
infrared spectroscopy 6:616
properties 6:598
ultraviolet spectroscopy 6:632
p-Tolualdehyde (4-Methylbenzaldehyde) 6:597
commercial grades and specifications 6:610,
    611
derivatives for identification 6:612
hydroxylamine hydrochloride method 6:620
manufacture 6:603
near-infrared spectroscopy 6:629
properties 6:598
uses 6:607
Toluamide
derivatives for identification 8:601

Toluene. (See also *Hydrocarbons.*) 14:317
azeotropes 14:318
in benzene 7:11
in cigarette smoke 10:31
in cocoa bean vapor 9:570
commercial grades and specifications 14:318
in enamels 12:175
in formic acid 13:128
gas chromatography 6:369; 7:7–11; 10:31;
    11:73; 13:128, 351; 14:332, 335–337;
    16:432; 18:75, 299
in gas line antifreezes 6:355, 369
in gasoline 13:351, 354
impurities, determination of 7:9
infrared spectroscopy 7:30
in isooctane 14:332
liquid chromatography 7:13, 14, 15, 16
manufacture 14:317
in manufacture of benzaldehyde 6:599
in manufacture of benzoic acid 7:58, 60
in manufacture of *p*-tolualdehyde 6:603
mass spectroscopy 7:32; 9:570
in methyltrimethoxysilane 18:75
in nail enamels 11:73
in paraffins 14:336
properties 12:175; 13:351; 14:317
solubility 13:539, 552; 14:317
spectroscopic method for nitrate 16:504
in styrene 18:299
toxicity 14:317
ultraviolet spectroscopy 7:39
uses 14:318
4-Toluene-azo-2-naphthol-3-carboxylic acid
in D & C Red No. 6 and No. 7 10:520
visible spectroscopy 10:521
2,4-Toluenediamine. (See also Toluenediamines.)
in air 11:455
Fast Scarlet R colorimetric method 11:455
Toluenediamines. (See also *Diamines and higher
    amines, aromatic*; Diaminotoluenes; and
    individual compounds.) 11:449–455
analysis 11:451
assay 11:451
commercial grades and specifications 11:450
derivatives for identification 11:438, 453
gas chromatography 11:451
identification 11:451
impurities, determination of 11:453
manufacture 11:449, 450
properties 11:434, 450
solubility 11:435
structure 19:304
thin-layer chromatography 19:302
toxicity 11:450
uses 11:449
volumetric determination 11:451
p-Toluene diazonium chloride

colorimetric method for phenols 17:29
colorimetric method for toluenediisocyanate
  19:313
Toluene diisocyanate. (See also Isocyanates.)
  assay 19:271
  composition, determination of 19:273
  gas chromatography 19:313
  infrared spectroscopy 19:273
  isomer content 19:273
  nuclear magnetic resonance spectroscopy
    19:302
  in polyurethanes 19:313
  properties 19:274
  sodium nitrite–Cellosolve colorimetric method
    19:313
  structure 19:270
  p-toluene diazonium chloride colorimetric
    method 19:313
  trade names 19:270
Toluene dilution test
  for cellulose nitrate 9:68
Toluene, 4,4-dimethyloxazolidine
  gas chromatography 5:6
Toluene-3,4-dithiol. See Dithiol.
o-(*p*-)Toluenesulfonamide 5:276
p-Toluenesulfonyl chloride
  properties 8:570
p-Toluenesulfonyl chloride-*p*-nitroso-*N*,*N*-
  dimethylaniline
  colorimetric method for cinnamyl alcohol
    10:80
m-Toluic acid
  analysis 8:405
  derivatives for identification 8:376
  paper chromatography 8:379
  properties 8:403, 405
o-Toluic acid
  analysis 8:405
  derivatives for identification 8:476
  gas chromatography 17:154
  paper chromatography 8:379
  in phthalic anhydride 17:154
  properties 8:403, 405
p-Toluic acid
  analysis 8:405
  derivatives for identification 8:376
  gas chromatography 8:385; 17:162
  paper chromatography 8:379
  properties 8:403, 405
  in terephthalic acid 17:162, 163
Toluic aldehyde
  derivatives for identification 8:343
  properties 8:334
p-Toluidide
  of adipic acid 4:412
  derivative for identification of benzoic acid
    7:63

derivative for identification of benzoic acid
  esters 7:82
derivative for identification of benzoic anhydride
  7:73
p-Toluidide derivative
  of carboxylic acids 8:374, 376
  of esters 8:583
m-Toluidine
  azeotropes 5:424
  azo dyes for identification 5:430
  colorimetric methods 5:451
  color reactions for identification 5:431
  derivatives for identification 5:429
  gas chromatography 5:440
  infrared spectroscopy 5:437, 453
  near-infrared spectroscopy 5:435, 453
  properties 5:423
  ultraviolet spectroscopy 5:432
  x-ray diffraction 5:435
o-Toluidine. (See also *Diamines and higher
    amines, aromatic.*)
  azeotropes 5:424
  azo dyes for identification 5:430
  bromination methods 5:448
  colorimetric method for gold 14:24
  colorimetric method using β-naphthol 6:409
  color reactions for identification 5:431
  derivatives for identification 5:429; 11:438
  gas chromatography 5:440
  hydrazine hydrate in 5:446
  infrared spectrum 5:437
  near-infrared spectroscopy 5:435
  picryl chloride method 5:447
  properties 5:423; 11:434
  solubility 11:435
  thin-layer chromatography 11:454
  in toluenediamines 11:454
  ultraviolet spectroscopy 5:432
  use in preparation of Safranine T 6:384
  use in preparation of Toluidine Blue O 6:387
  x-ray diffraction 5:435
p-Toluidine
  azeotropes 5:424
  azo dye colorimetric method 5:445, 450
  azo dyes for identification 5:430
  chronopotentiometry 5:455
  colorimetric method using β-naphthol 6:409
  colorimetric method for *m*-diamines 6:410,
    411
  color reactions for identification 5:431
  derivatives for identification 5:429
  gas chromatography 5:440
  infrared spectroscopy 5:437, 453
  near-infrared spectroscopy 5:435
  picryl chloride method 5:447
  properties 5:423
  separation 5:435

ultraviolet spectroscopy 5:432
x-ray diffraction 5:435
Toluidine blue
colorimetric method for sodium heparin 10:148
Toluidine Blue O 6:382, 383, 387
commercial specifications 6:387
manufacture 6:387
solubility 6:383
ultraviolet spectroscopy 6:392
volumetric method using titanous chloride
6:395
p-Toluidine hydrochloride
volumetric method for anionic detergents
11:338
Toluidine maroon
in enamels 12:186
identification 12:186
Toluidine red. (See also *Pigments.*)
commercial specifications 17:204
in enamels 12:182, 185
identification 12:182, 185; 17:205
potassium hydroxide color test 17:205
solubility 17:205
specification tests 17:205
structure 17:178
Toluidines 5:421
acid–base titration 5:443, 447
bromination methods 5:445, 448
manufacture 5:425
in toluenediamines 11:454
Toluidine toners
commercial specifications 12:174
composition 12:174
in enamels 12:174, 183
identification 12:183
m-Tolunitrile. (See also *Nitriles.*)
derivatives for identification 16:402
properties 16:396
o-Tolunitrile. (See also *Nitriles.*)
derivatives for identification 16:399
properties 16:396
p-Tolunitrile. (See also *Nitriles.*)
derivatives for identification 16:402
properties 16:396
o-Tolyl chloride. See *o*-Chlorotoluene.
p-Tolyl chloride. See *p*-Chlorotoluene.
Tolylene diisocyanates. (See also *Isocyanic acid
esters.*)
amine equivalent 15:112
commercial grades and specifications 15:104
dibutylamine volumetric method 15:112
dielectric constants 15:117
gas chromatography 15:110
impurities, determination of 15:118
infrared spectroscopy 15:113
isocyanate number 15:112
isomer ratio 15:113

near-infrared spectroscopy 15:105
properties 15:95
trade names 15:103
Tolylisocyanates. (See also *Isocyanic acid esters.*)
gas chromatography 15:110
infrared spectroscopy 15:106
near-infrared spectroscopy 15:105
p-Tolyl-2-naphthylamine. (See also *Diamines and
higher amines, aromatic.*)
derivatives for identification 11:438
N-*o*-Tolyl-*N'*-phenyl-*p*-phenylenediamine 6:77
p-(*p*-Tolysulfanilamido)diphenylamine. See
Aranox.
p-(*p*-Tolylsulfonylamino)diphenylamine
paper chromatography 6:85
p-(*p*-Tolylsulfonylamino)phenyl-*p*-tolylamine
paper chromatography 6:85
Tonox 6:77
Topaz 13:2
Top engine cleaners 6:356
analysis 6:371
Toraylon. (See also *Acrylic and modacrylic
fibers.*) 4:221, 249
Torbernite
composition 19:225
decomposition 19:235
occurrence 19:225
Tordan
structure 17:414
uses 17:414
Torsional modulus
of elastomers, synthetic 12:135
of Kraton 12:136
of poly(vinyl chloride) 12:136
of Thermolastic 12:136
Toughness
of abrasive grains 7:309
of enamels 12:180
Tourmaline 7:316
Toxaphene. (See also *Pesticides and insecticides.*)
commercial grades and specifications 19:593
composition 19:581, 588, 597
gas chromatography 19:389
manufacture 19:597
properties 19:588
solubility 19:587
toxicity 19:592
in water 19:389
TPAC. See Tetraphenylarsonium chloride.
TPAP. See Tetraphenylarsonium perchlorate.
TPTZ
colorimetric method for iron 11:497
Tragacanth
composition 12:583
in embalming fluids 12:166
ferric chloride gravimetric method 12:582
gravimetric method 12:581

identification 6:362
infrared spectroscopy 12:583
in matches 15:508
in photographic films 12:582
properties 15:508
Schweizer's reagent gravimetric method 12:582
Transesterification
  manufacture of acrylates and methacrylates 4:186
  of organosilane esters 18:77
Transition metals
  nitrogen in 8:631
  oxygen in 8:631
Transition point, second-order of acrylonitrile
    polymers 4:390
Transuranium elements. (See also individual ele-
    ments.) 19:258
  manufacture 19:258
  properties 19:260
  separation 19:260
  uses 19:260
Transverse deflection
  of methyl methacrylate dental materials
    11:303
Tremolite 8:75
Triacetate fiber. See *Acetate and triacetate
    fibers.*
Triacetin
  in cigarette filters 10:40
  Cundiff's method 10:40
Triacetyloleandomycin
  infrared spectrum 5:524
Trialkanolamines. (See also *Alkanolamines.*)
  manufacture 5:3
  mixtures, determination of 5:20
Trialkylaluminums. See Aluminum alkyls.
Trialkylbismuth compounds 7:207
Triallylamine
  gas chromatography 5:88
  paper chromatography 5:89
  properties 5:87, 260
Triallyl cyanurate 5:93
Triallyl glycerol
  halogenation methods 5:96, 99
Triallyl isocyanurate 5:93
Triamcinolone. (See also 9-Fluoro-16-hydroxy
    prednisolone.)
  infrared spectrum 14:198
  manufacture 14:198
  properties 14:198
  solubility 14:198
  structure 14:198
  uses 14:198
Triamcinolone acetonide. (See also Steroids.)
  infrared spectrum 14:199
  manufacture 14:198
  properties 14:198
  solubility 14:198

structure 14:198
uses 14:198
1,3,5-Triaminobenzene. (See also *Diamines and
    higher amines, aromatic.*) 11:433
Triaminotriphenylmethane
  in polyurethanes 19:304
  structure 19:304
  thin-layer chromatography 19:303
Triamylamines
  commercial grades and specifications 5:264
  properties 5:260
  uses 5:263
Triangle tests 17:621
  design 17:619, 624
  evaluation 17:624
  significance 17:622
Triarylbismuth compounds 7:207
Tri-*n*-butylphosphine borane 7:344
Triarylmethane dyes. See *Triphenylmethane
    and related dyes.*
Triazine–formaldehyde resin
  infrared spectrum 5:284
Triazones 5:276
Tribasic copper sulfate. See Basic copper sulfate.
Tribasic lead phosphosilicate. See Silicate white
    lead.
Tribasic lead sulfate
  in enamels 12:177
  properties 12:177
Tribromoethanol 5:369
  analysis 5:373
  anesthetic 5:368, 373
  assay 5:373
  infrared spectrum 5:373
  properties 5:373
Tribromoethylene
  Fujiwara colorimetric method 9:386
2,4,5,-Tribromofluorescein
  in D & C Red No. 21 10:523
  paper chromatography 10:520, 523
  visible spectroscopy 10:523
Tribromomethane. See Bromoform.
Tribromopyrogallol
  colorimetric method for niobium 16:368, 373
3,4′,5-Tribromosalicylanilide. (See also
    Salicylanilides.)
  nuclear magnetic resonance spectroscopy 11:551
  properties 11:551
  in soaps 18:200
  structure 11:551
  ultraviolet spectroscopy 11:549, 552; 18:200
Tribromosilane
  properties 18:40
Tribromothymol
  in embalming fluids 12:165
Tributylamine
  properties 5:260

uses 5:263
Tri-*sec*-butylchlorobenzene
  mass spectroscopy 9:407
Tributyl citrate. (See also Citrate esters.)
  properties 10:97
Tricalcium phosphate
  manufacture 17:121
  uses 17:121
Tricalcium silicate
  in hydraulic cements 9:107
  light microscopy 9:103
  in portland cement 9:103
Tricaprylmethylammonium borohydride 7:344
1,2,4-Tricarboxylic acid. See Trimellitic acid.
Trichlorfon. (See also *Pesticides and insecticides.*)
  commercial grades and specifications 19:595
  composition 19:590, 598
  manufacture 19:598
  properties 19:590
  solubility 19:587
  structure 19:584
  toxicity 19:592
Trichloroacetic acid 8:395
  in chloroform 9:434
  Fujiwara colorimetric method 9:386, 434
  gravimetric method for protein 19:475
  infrared spectrum 8:378
  properties 8:395, 397
  turbidimetric method for protein 10:150
  uses 8:395
2,4,5-Trichloroanisole
  gas chromatography 9:519
  ion exchange chromatography 9:523
  in 2,4,5-trichlorophenol 9:523
1,2,3-Trichlorobenzene. (See also Trichlorobenzenes.) 9:481
  gas chromatography 9:378
  infrared spectroscopy 9:394
  properties 9:482
1,2,4-Trichlorobenzene. (See also Trichlorobenzenes.) 9:481
  gas chromatography 9:378
  infrared spectroscopy 9:394
  mass spectroscopy 9:407
  properties 9:482
1,3,5-Trichlorobenzene. (See also Trichlorobenzenes.) 9:481
  mass spectroscopy 9:407
  properties 9:482
sym-Trichlorobenzene. See 1,3,5-Trichloro-
  benzene.
vic-Trichlorobenzene. See 1,2,3-Trichloro-
  benzene.
Trichlorobenzenes. (See also Chlorinated
  benzenes; and individual compounds.)
  9:481
  composition 9:411

manufacture 9:482
  solubility 9:481
  toxicity 9:481
  uses 9:482
Trichlorobiphenyls. (See also Chlorinated
  biphenyls.)
  properties 9:492
1,1,1-Trichloro-2,2-bis(*p*-chlorophenyl)ethane.
  See DDT.
1,1,1-Trichloro-2,2-bis(*p*-methoxyphenyl)ethane.
  See Methoxychlor.
Trichlorobutane
  mass spectroscopy 9:398
3,4,4'-Trichlorocarbanilide. (See also Carbanilides.)
  nuclear magnetic resonance spectroscopy
   11:551
  properties 11:551
  in soaps 18:200
  structure 11:551
  ultraviolet spectroscopy 18:200
2,4,6-Trichlorocumene
  mass spectroscopy 9:407
1,1,1-Trichloroethane. (See also C₂ Chlorinated
  hydrocarbons.) 9:440
  analysis 9:441
  appearance 9:443
  Beilstein test 9:382
  color 9:443
  commercial grades and specifications 9:441
  composition 9:411
  Fujiwara colorimetric method 9:386
  gas chromatography 9:374, 375, 376, 377,
   381, 442
  impurities, determination of 9:442, 443
  infrared spectroscopy 9:390
  inhibitors, determination of 9:442
  manufacture 9:441
  mass spectroscopy 9:401
  properties 9:440
  solubility 9:440
  specification tests 9:443
  specific gravity 9:443
  toxicity 9:441
  uses 9:440
1,1,2-Trichloroethane. (See also C₂ Chlorinated
  hydrocarbons.) 9:444
  azeotropes 9:445
  composition 9:411
  gas chromatography 9:375, 376, 377, 380,
   381; 16:92
  infrared spectroscopy 9:390, 394
  manufacture 9:445
  mass spectroscopy 9:400
  properties 9:444
  solubility 9:444
  toxicity 9:444

uses 9:444
1,1,2-Trichloroethane. See Trichloroethylene.
Trichloroethylene. (See also C$_2$ Chlorinated
    hydrocarbons.) 9:454
    analysis 5:366; 9:457
    antioxidants 9:456
    appearance 9:458
    azeotropes 9:455
    Beilstein method 9:382
    in carbon tetrachloride 8:325
    commercial grades and specifications 9:456
    composition 9:411
    copper corrosion test 9:458
    distillation range 9:457
    Fujiwara colorimetric method 9:385, 386
    gas chromatography 8:325; 9:374, 375, 377,
      380, 381, 442
    identification 5:367
    impurities, determination of 5:367; 9:457
    infrared spectroscopy 8:325; 9:390, 394
    manufacture 9:456
    mass spectroscopy 9:401
    α-naphthol colorimetric method 9:387
    odor 9:458
    oxidation, accelerated 9:458
    phenolphthalein colorimetric method 9:387
    properties 5:357, 366; 9:454
    residual stain test 9:458
    solubility 9:455
    specification tests 9:457
    toxicity 9:455
    in 1,1,1-trichloroethane 9:442
    uses 9:454
Trichlorofluoromethane. (See also Chlorofluoro-
    carbons; Trichloromonofluoromethane.)
    commercial specifications 13:63
    gas chromatography 11:89, 98; 13:64
    properties 13:61
    in hair sprays 11:89
Trichloroisocyanuric acid
    available chlorine 7:224, 229
    iodometric titration 7:229
    separation 7:225
Trichloromethane. See Chloroform.
1,1,3-Trichloro-2-methylpropene
    mass spectroscopy 9:403
3,3,3-Trichloro-2-methylpropene
    mass spectroscopy 9:403
N-Trichloromethylthio-4-cyclohexene-1,2-
    dicarboximide. See Captan.
N-Trichloromethylthiophthalimide. See Folpet.
Trichloromonofluoromethane. (See also Chloro-
    fluorocarbons; Trichlorofluoromethane.)
    in aerosols 11:95
    gas chromatography 12:324
    properties 11:95
Trichloronaphthalenes. (See also

Chlorinated naphthalenes.)
    properties 9:490
    toxicity 9:490
Trichloronitromethane. See Chloropicrin.
1,1,5-Trichloro-1-pentene
    mass spectroscopy 9:398
2,3,6-Trichlorophenol. (See also Trichloro-
    phenols.)
    gas chromatography 9:519
    ion exchange chromatography 9:523
    in 2,4,5-trichlorophenol 9:523
2,4,5-Trichlorophenol. (See also Trichlorophenols.)
    assay 9:524
    commercial grades and specifications 9:513
    gas chromatography 9:519
    impurities, determination of 9:523
    infrared spectroscopy 9:515
    ion exchange chromatography 9:525
Trichlorophenols. (See also *Chlorophenols*;
    and individual compounds.)
    ion exchange chromatography 9:521, 523
    manufacture 9:512
    properties 9:511
    ultraviolet spectroscopy 9:521
    uses 9:511
2,4,5-Trichlorophenoxyacetic acid (2,4,5-T) 8:406
    analysis 8:406
    properties 8:403
1,1,2-Trichloropropane. (See also C$_3$ Chlorinated
    hydrocarbons.)
    infrared spectroscopy 9:390
    mass spectroscopy 9:398
    properties 9:471
1,2,3-Trichloropropane. (See also Trichloro-
    propanes.) 9:465
    azeotropes 9:467
    composition 9:411
    in epichlorohydrin 14:175
    gas chromatography 9:381; 14:175
    infrared spectroscopy 9:390
    manufacture 9:468
    mass spectroscopy 9:399
    properties 9:467
    solubility 9:468
    toxicity 9:468
    uses 9:465
Trichloropropanes. (See also C$_3$ Chlorinated
    hydrocarbons; and individual compounds.)
    properties 9:471
1,1,2-Trichloropropene
    infrared spectroscopy 9:390
    mass spectroscopy 9:403
1,2,3-Trichloropropene
    infrared spectroscopy 9:390
    uses 9:465
3,3,3-Trichloropropene
    mass spectroscopy 9:402

Trichloropropenes. (See also C₃ Chlorinated hydrocarbons; and individual compounds.)
  properties 9:471
3′,4′,5′-Trichlorosalicylanilide. (See also Salicylanilides.)
  ultraviolet spectroscopy 11:549
Trichlorosilane. (See also Chlorosilanes.)
  commercial grades and specifications 18:49
  gas chromatography 18:51, 60
  impurities, determination of 18:52
  infrared spectroscopy 18:50
  manufacture 18:42, 49
  properties 18:40, 48
  safety 18:48, 49
  uses 18:42, 49
β,2,4-Trichlorostyrene
  mass spectroscopy 9:407
β,2,5-Trichlorostyrene
  mass spectroscopy 9:407
Trichlorotoluene
  mass spectroscopy 9:406
α,α,α-Trichlorotoluene. See Benzotrichloride.
α,2,4-Trichlorotoluene
  mass spectroscopy 9:406
α,3,4-Trichlorotoluene
  mass spectroscopy 9:406
1,1,2-Trichlorotrifluoroethane. (See also Chlorofluorocarbons.)
  commercial specifications 13:62, 63
  properties 13:61
1,1,2-Trichloro-1,2,2-trifluoromethane
  gas chromatography 11:98
α,α,α-Trichloro-m-xylene
  mass spectroscopy 9:406
α,α,α-Trichloro-p-xylene
  mass spectroscopy 9:406
Tricholine citrate. (See also Choline and its derivatives.)
  properties 9:613
  uses 9:612
Trichomycin 5:473, 565
  analysis 5:571, 576, 579
  column chromatography 5:571
  identification 5:571
  microbiological assay 5:576
  paper chromatography 5:571
  properties 5:565
  separation 5:571
  solubility 5:566, 567
  uses 5:565
Tricobalt tetroxide. See Cobalto–cobaltic oxide.
Tricresyl phosphate
  in enamels 12:176
  properties 12:176
  uses 15:302
Tridecanol. (See also Alcohols, higher; Fatty alcohols.)

properties 4:561
specifications 4:557
uses 4:556
Tridecanone
  absorption spectroscopy 9:564
  hydroxylamine hydrochloride method 8:357
  2-hydroxy-1-naphthaldehyde colorimetric method 11:204
n-Tridecanonitrile. (See also Nitriles.)
  properties 16:396
Tridymite. (See also Silica.) 18:5
Trierythritol
  gas chromatography 13:578
Triethanolamine. (See also Ethanolamines.)
  acid–base titration 12:292
  assay 12:292
  in brake fluids 7:425, 427
  colorimetric method for boric acid 7:389
  infrared spectrum 5:7
  manufacture 5:3
  properties 5:2; 12:291
  specifications 5:4
Triethanolamine adipate 4:429
Triethanolamine–soap emulsions
  identification 11:93
Triethanolamine soaps
  in automobile polishes 6:353
Triethylamine
  azeotropes 5:261
  commercial grades and specifications 5:264
  gas chromatography 11:245
  manufacture, methods of 5:262
  properties 5:260
  temperature-rise method 4:103
Triethylchlorobenzene
  mass spectroscopy 9:407
Triethyl citrate. (See also Citrate esters.)
  commercial specifications 10:97
  properties 10:97
Triethylenediamine. (See also Diamines and higher amines, aliphatic.)
  properties 11:410
Triethylene glycol. (See also Polyglycols.)
  5:31
  in air 13:599
  azeotropes 13:555
  colorimetric determination 13:599
  commercial grades and specifications 13:558, 560
  density 13:584
  derivatives for identification 13:568
  distillation range 13:602
  freezing point 13:535
  gas chromatography 13:576, 579, 589; 19:300
  manufacture 13:556
  paper chromatography 13:581
  potassium dichromate volumetric method 13:599

uses 13:565
Tripentaerythritol octanitrate. (See also High
    explosives.)
  manufacture 12:445
  properties 12:445
  uses 12:445
Triphenols. (See also individual compounds.)
  analysis 17:46
Triphenylamine. (See also *Diamines and
    higher amines, aromatic.*) 11:436
  properties 11:434
  solubility 11:435
Triphenyl indium. (See also Indium compounds.)
    14:560
  manufacture 14:560
Triphenylmethane
  gas chromatography 9:520
Triphenylmethane and related dyes. (See also
    individual compounds.) 6:606; 12:7;
    19:161-189
  analysis 19:164
  assay 19:174
  blowout method 19:164
  capillary test 19:164
  classification 19:165, 170
  cotton dyeing test 19:165
  dyeing test 19:164
  electrochemical methods 19:169
  identification 19:164
  impurities, determination of 19:185
  infrared spectroscopy 19:168
  leuco content 10:510
  liquid chromatography 10:477; 19:173
  manufacture 19:161
  mass spectroscopy 19:168
  in mixtures 19:164
  nuclear magnetic resonance spectroscopy
    19:168
  paper chromatography 12:587; 19:171
  in photographic films 12:587
  polarography 19:170, 182
  salt test 19:187
  separation 19:170
  solubility 19:165
  solvent extraction 19:170
  standards, purification of 19:187
  tannin gravimetric method 19:165
  thin-layer chromatography 19:172
  ultraviolet spectroscopy 19:168
  visible spectroscopy 19:166, 183
  volumetric determination 19:174
  zinc-acetic acid color test 19:165, 166
Triphenyl phosphate
  in enamels 12:176
  in photographic films 12:591
  properties 12:176, 591
Triphenylphosphine borane 7:344

Triphenyl phosphine extraction
  of sulfur in elastomers 12:128
Triphilite 15:261
Triple superphosphate
  manufacture 17:143
Triplite 17:71
Tripoli 4:31, 33
  in automobile polishes 6:353
Tripolyphosphates
  ion exchange chromatography 17:131
Tripotassium orthophosphate
  properties 17:120
  solubility 17:120
Triprolidine hydrochloride. (See also *Anti-
    histamines.*) 6:30
Tripropylamine
  properties 5:260
Tripropylene glycol
  distillation range 13:602
  gas chromatography 13:577, 589; 19:298
  in polyurethanes 19:297
Tripropylene glycol methyl ether
  in brake fluids 7:425
Tripropyl indium. (See also Indium compounds.)
    14:560
  manufacture 14:560
  properties 14:558
2,4,6-Tripyridyl-S-triazine. See TPTZ.
Trisaziridinyl phosphine oxide
  in cotton 11:138
  infrared spectroscopy 11:139
Tris(dimethyldithiocarbamate) iron. See Ferbam.
Tris(hydroxymethyl)aminomethane
  freezing point determination 5:14
  infrared spectrum 5:10
  manufacture 5:3
  melting point determination 5:14
  properties 5:2
  specifications 5:5
  standard for acid-base titration 5:12, 13
  titration of sulfur 12:117
Tris(hydroxymethyl)nitromethane. (See also
    Nitro alcohols.)
  chromotropic acid colorimetric method 16:444
  commercial grades and specifications 16:440
  gas chromatography 16:444
  infrared spectrum 16:443
  properties 16:439
  toxicity 16:440
Tris(2-methyl-4-hydroxy-5-t-butylphenyl)butane
  thin-layer chromatography 16:95
Tris(nonylphenyl)phosphite 6:77
  in impact polystyrene 18:332
  solvent extraction 18:332
  ultraviolet spectroscopy 18:332
Trisodium citrate dihydrate. (See also Sodium
    citrate.) 10:86

Trisodium citrate 5.5-hydrate. (See also Sodium citrate.) 10:86
Trisodium pentacyanoammino ferrate
  colorimetric method for aminopyridines 17:433
  colorimetric method for L-anabasine 17:433
  colorimetric method for cyanopyridines 17:433
  colorimetric method for ethyl isonicotinate 17:433
  colorimetric method for ethyl nicotinate 17:433
  colorimetric method for 4-ethylpyridine 17:433
  colorimetric method for hexyl nicotinate 17:433
  colorimetric method for 3-hydroxypyridine 17:433
  colorimetric method for isonicotinic acid 17:433
  colorimetric method for isonicotinic acid amide 17:433
  colorimetric method for isonicotinic hydrazide 17:433
  colorimetric method for methylpyridines 17:433
  colorimetric method for nicotine 17:433
  colorimetric method for nicotinic acid 17:433
  colorimetric method for nicotinic acid amide 17:433
  colorimetric method for pyridine and its derivatives 17:425, 433
  colorimetric method for pyridine-3-sodium sulfonate 17:433
  colorimetric method for pyridyl methyl ketones 17:433
  colorimetric method for pyruvic isonicotinyl hydrazone 17:433
Trisodium phosphate
  composition 17:120
  cooling system flushes 6:352
  manufacture 17:125
  uses 17:120
Trisodium phosphate, chlorinated. See Chlorinated trisodium phosphate.
Trisodium salt of 3-carboxy-5-hydroxy-1-*p*-sulfophenyl-4-*p*-sulfophenylazopyrazole. See F D & C Yellow No. 5.
Trisodium salt of 10-hydroxy-3,5,8-pyrenetri-sulfonic acid. See D & C Green No. 8.
Trisodium salt of 1-(4-sulfo-1-naphthylazo)-2-naphthol-3,6-disulfonic acid. See F D & C Red No. 2.
Trisodium salts of 1-(4-sulfophenylazo)-2-naphtholdisulfonic acids
  in F D & C Yellow No. 6 10:518
  liquid chromatography 10:518
Tris(pentaerythritol) diformal

gas chromatography 13:578, 594
Tristearin
  thin-layer chromatography 12:205
Tritisan. See Terraclor.
Tritium 14:377
  properties 14:378
  in water 19:395
Tritomite 7:316
Tritonal. (See also High-explosive mixtures.)
  analysis 12:450
  composition 12:450
  uses 12:450
Triuranium octoxide. (See also Uranium oxides.) 19:256
  analysis 19:258
  manufacture 19:256
  properties 19:256
  solubility 19:256
1,3,5-Trivinylbenzene trioxide. (See also *Epoxy compounds.*)
  infrared spectroscopy 12:200
Tromexan. See Ethyl biscoumacetate.
Trona. (See also Sodium carbonate.) 18:207, 237
Trootsite
  composition 19:545
Tropacocaine. (See also *Narcotics.*)
  structure 16:244
Tropane alkaloids 4:587
  analysis 4:595
Tropic acid esters
  analysis 4:596
Tropical bleach 7:221
  available chlorine 7:224, 225
  iodometric titration 7:225
Tropine
  gas chromatography 4:594
Tropinone
  gas chromatography 4:594
Troy ounce 18:155
Truck brake system road test 7:470
Tryptamine
  thin-layer chromatography 15:295
Tryptophan. (See also *Carboxylic acids, amino.*) 8:413, 509
  analysis 8:533
  colorimetric methods 8:533
  commercial specifications 8:531
  composition 8:413
  crystalline form 8:530
  p-dimethylaminobenzaldehyde colorimetric method 8:534
  gas chromatography 8:429
  in gramicidin 5:548
  indole in 8:533
  infrared spectrum 8:530
  iodometric titration 8:506
  ion exchange chromatography 8:421

manufacture 8:531
paper chromatography 8:422
polarography 8:512
properties 8:529
solubility 8:530
in tyrothricin 5:549
ultraviolet spectroscopy 8:533
xanthydrol colorimetric method 8:534
in yeast 19:529
Tseow-Chow resorcinol reaction
    identification of formals 4:68
TSH. See Thyrotropin.
TTA. See 2-Thenoyltrifluoroacetone.
TTHA. (See also *EDTA and related compounds.*)
    cobaltous nitrate–ferric chloride colorimetric
        method 12:28
    iodometric method 12:28
    scandium nitrate colorimetric method 12:28
    structure 12:25
    thorium nitrate titration 12:36
    zinc nitrate–lanthanum–DCYTA colorimetric
        method 12:30
    zinc nitrate titration 12:36
Tuads. See Thiram.
d-Tubocurarine chloride 4:592
Tubular centrifuges 9:176, 177
    characteristics 9:160
    settling curves 9:161
Tulisan. See Thiram.
Tumors
    insulin content 14:575
Tung oil. (See also *Fats and fatty oils.*)
    in alkyd resins 5:23
    in brake linings 7:441
    properties 12:473
    solubility 13:539, 552
Tungstates. (See also Tungsten compounds; and
        individual compounds.) 19:220
    in antifreezes 6:14
    impurities, determination of 16:190
    manufacture 19:220
    uses 19:220
Tungsten. (See also Tungsten alloys; Tungsten
        compounds; and Tungsten metal.) 19:190–
        224
    in aluminum and aluminum alloys 5:152
    ammonium hydroxide gravimetric method
        19:200, 219
    ammonium sulfide gravimetric method 19:205
    ammonium thiocyanate colorimetric method
        16:162
    atomic absorption spectroscopy 19:218
    benzidine chloride gravimetric method 19:202
    α-benzoinoxime gravimetric method 19:200
    in biological material 19:213, 217
    borax bead test 6:584
    in chromium 9:639

in chromium alloys 9:676
cinchonine gravimetric method 8:137; 10:322;
    16:338; 19:199, 203
cinchonine–oxime gravimetric method 15:81
in cobalt alloys 10:321
colorimetric method 19:201, 208
in copper alloys 10:648
in cotton 11:141, 156
in cutting alloys 10:289
decomposition 19:196
detection 19:201
diphenyline colorimetric method 11:141;
    19:202
dithiol colorimetric method 19:208, 213
emission spectroscopy 14:122; 15:156;
    16:160, 387
in ferroniobium 16:387, 389
in ferrotungsten 15:81; 19:206
flame emission spectroscopy 19:218
flame tests 6:589
fluoride content 13:27
gravimetric method 19:198, 202, 203
in hafnium 14:122
half-life 19:190
in hard facing alloys 10:289
in high-temperature alloys 10:286
history 19:190
hydrochloric acid gravimetric method 16:172
hydrochloric acid–nitric acid gravimetric method
    16:332, 338
hydrogen sulfide gravimetric method 19:199, 206
8-hydroxyquinoline oxide colorimetric method
    11:141
ion exchange chromatography 14:113; 16:153,
    162, 390; 17:526
in iron 15:59; 19:198, 206
isotopes 19:190
in lanthanum oxide 15:156
malachite green colorimetric method 19:202
microcosmic salt bead test 6:587
in molybdenum 16:161
in molybdenum alloys 16:172
molybdenum–α-benzoin oxime gravimetric
    method 10:323
in molybdenum trioxide 16:161
neutron activation analysis 13:265
in nickel 16:332, 337; 19:207
in nickel alloys 16:332, 337
in niobium 16:363, 387, 389
occurrence 19:191
oxine gravimetric method 19:201, 203
potassium carbonate fusion 19:200
potassium cyanide fusion 19:200
in scheelite 19:195, 203
separation 19:198, 213
in silicate rocks 19:210, 214
silicon content 18:30

sodium carbonate fusion 15:82; 16:338;
19:205, 206, 207, 208, 214, 219
solvent extraction 16:163; 19:200, 209, 211,
214
in spring steels 10:290
stannous chloride-potassium thiocyanate colorimetric method 10:323, 648; 19:201
in steel 16:382; 19:198, 206
in stolzite 19:203
sulfur content 18:383
sulfurous acid gravimetric method 16:380
tannin gravimetric method 16:363
in tantalum 16:363
thallium(I) nitrate–hydrogen peroxide gravimetric
method 19:203
thiocyanate colorimetric method 19:208, 209
tin content 19:48
in titanium metal 19:121, 215
in tungsten alloys 19:206
in tungsten carbide 8:133, 137; 19:219
in tungsten compounds 19:221
in tungsten metal 19:206
in tungsten minerals 19:203, 204
volatilization methods 19:199, 218
volumetric determination 19:208
in wear-resistant alloys 10:289
in wolframite 19:195, 203
x-ray fluorescence spectroscopy 11:156
in zirconium 14:122; 19:212
α-Tungsten. (See also Tungsten metal.) 19:193
β-Tungsten. (See also Tungsten metal.) 19:193
Tungsten alloys. (See also Tungsten metal; and
individual alloys.) 5:61, 67
analysis 19:198
commercial grades and specifications 19:196
decomposition 19:198
hydrogen in 8:617, 623
nitrogen in 8:617, 623
oxygen in 8:617, 623
properties 19:192
tungsten content 19:206
uses 19:192
vacuum fusion methods 8:617, 623
Tungsten bronzes. (See also Tungsten compounds.)
19:220
Tungsten carbide. (See also *Carbides*; Tungsten
compounds; and individual compounds.)
10:291; 19:221
as abrasive 4:31, 34
analysis 19:219
carbon content 8:134
formation 8:130
free carbon in 8:134
hydrogen in 8:681
impurities, determination of 19:219
manufacture 10:291; 19:221
oxygen in 8:618, 621, 638, 681

properties 8:131, 133; 19:192, 221
solubility 19:221
tungsten content of 8:133, 137; 19:219
uses 8:132, 133; 19:192, 221
vacuum fusion methods 8:618, 621, 681
vacuum hot extraction methods 8:638
Tungsten carbide–cobalt
in ceramics 9:225
microscopy 9:225
Tungsten carbide–copper
in ceramics 9:225
microscopy 9:225
Tungsten–cobalt steel. (See also Cobalt alloys.)
composition 10:289
Tungsten compounds. (See also individual
compounds.) 19:219–222
analysis 19:221
assay 19:221
toxicity 19:194
tungsten content 19:221
Tungsten disulfide. (See also Tungsten compounds.)
manufacture 19:221
properties 19:221
solubility 19:221
uses 19:221
Tungsten halides. (See also Tungsten compounds; and individual compounds.) 19:220
Tungsten hexacarbonyl. (See also Tungsten
compounds.)
properties 19:221
solubility 19:221
uses 19:221
Tungsten hexachloride. (See also Tungsten
compounds.)
manufacture 19:220
properties 19:220
solubility 19:220
uses 19:220
Tungsten hexafluoride. (See also Tungsten
compounds.)
manufacture 19:220
properties 19:220
solubility 19:220
uses 19:220
Tungsten metal
analysis 19:196, 218
carrier gas fusion methods 8:625, 626, 627,
682
commercial grades and specifications 19:196
decomposition 19:198
hydrogen in 8:617, 618, 620, 623, 638, 681
impurities, determination of 19:218
manufacture 19:194
nitrogen in 8:617, 618, 620, 623, 638
oxygen in 8:617–623, 625–627, 633, 638,
639, 681, 682
properties 19:190, 193

toxicity 19:194
tungsten content 19:206
uses 19:190, 192
vacuum fusion methods 8:617, 618, 619,
   620, 621, 622, 623, 681
vacuum hot extraction methods 8:638
Tungsten minerals
   decomposition 19:196, 203, 204
   pyrosulfate fusion 19:197
   sodium carbonate fusion 19:197
   sodium hydroxide fusion 19:198
   sodium peroxide fusion 19:197, 203
   tungsten content 19:203, 204
Tungsten–molybdenum–cobalt steel. (See also
   Cobalt alloys.)
   composition 10:289
Tungsten oxides. (See also Tungsten trioxide.)
   19:220
Tungsten pentabromide. (See also Tungsten
   compounds.)
   manufacture 19:220
   properties 19:220
Tungsten–rhenium alloy. (See also Rhenium
   alloys; Tungsten alloys.)
   uses 17:517; 19:193
Tungsten steel. (See also Tungsten alloys.)
   decomposition 19:206
   history 19:190
   properties 19:192
   uses 19:192
Tungsten trioxide. (See also Tungsten compounds.)
   manufacture 19:193, 220
   properties 19:194, 220
   uses 19:220
Tungstic oxide. See Tungsten trioxide.
   pyrolytic method for fluorine 11:494
Turbidimeters 19:358
Turbidimetric method
   for benzyl alcohol 7:100
   of microbiological assay 5:489
Turbidity
   of beer 7:519
   of formazin solutions 7:523
   of formic acid 13:131
   of gelatin 12:556
   units 7:523
   of water 19:358
Turbidity temperatures
   of ethylene and propylene polymers 12:348
Turbine oil. (See also Lubricants.)
   commercial grades and specifications 15:310
   oxidation characteristics 15:340
   rust-prevention tests 15:335
Turf-tect. See Terraclor.
Turkey red oil
   solubility 13:539, 552

Turmeric
   assay 10:541
   colorimetric method for borates 15:527
   colorimetric method for boric acid 7:388, 409
   composition 10:533
   identification 10:537
   visible spectroscopy 10:541
Turmeric oleoresin
   composition 10:533
d-Turocurarine chloride
   analysis 4:613
Turpentine
   refractive index 12:271
TV bulb glass. (See also Glass and glass-
   ceramics.)
   composition 13:409
Twist level
   of polyamide fibers 17:326
Tygon
   ethylene oxide in 12:331
Tylosin 5:464, 519
   analysis 5:525, 529, 534
   identification 5:525
   infrared spectrum 5:524
   microbiological assay 5:529
   properties 5:519
   separation 5:525
   solubility 5:519, 520
   ultraviolet spectroscopy 5:533
   uses 5:519
Type metal. (See also Tin alloys.) 5:65, 69
   composition 19:25
   tin content 19:64
Tyrocidine 5:467, 535
   analysis 5:542, 547, 549
   column chromatography 5:542
   countercurrent distribution 5:542
   identification 5:534
   microbiological assay 5:547
   properties 5:534
   separation 5:534
   solubility 5:535
   uses 5:535
Tyrocidine A
   structure 5:467
Tyrocidine B
   structure 5:467
Tyrocidine hydrochloride
   infrared spectrum 5:541
   ultraviolet spectrum 5:541
Tyrosine. (See also Carboxylic acids, amino.)
   8:413, 534
   acid–base titration, nonaqueous 8:538
   analysis 8:538
   in cocoa beans 9:556
   commercial specifications 8:538
   composition 8:413

crystalline form 8:535
enzymatic decarboxylation 8:539
fluorometry 8:539
gas chromatography 8:429
infrared spectrum 8:535
iodometric titration 8:507
ion exchange chromatography 8:415, 421
in leucine 8:495
manufacture 8:537
Millon's reagent color test 17:17
nuclear magnetic resonance spectrum
    8:536
optical rotation 8:444
paper chromatography 8:422
polarography 8:512
properties 8:535
solubility 8:536
structure 8:536
ultraviolet spectroscopy 8:538
in yeast 19:529
L-Tyrosine
  in green tea leaf 18:459

potassium titanium oxalate colorimetric
    method 18:459
Tyrosine unit
  for protease activity in beer 7:589
Tyrothricin 5:467, 535
  analysis 5:542, 547, 549
  colorimetric methods 5:549
  identification 5:542
  infrared spectrum 5:541
  microbiological assay 5:547
  properties 5:534
  separation 5:542
  solubility 5:535
  thin-layer chromatography 5:542
  ultraviolet spectrum 5:542
  uses 5:535
Tyuyamite
  composition 19:225
  occurrence 19:225
Tyuyamunite 19:321
TZM alloys. See Molybdenum–titanium–
    zirconium alloys.

U

Ubbelohde viscometer 4:308
Udimet alloys. (See also Cobalt alloys.)
  composition 10:287
Ulexite 7:313, 314
Ultracentrifugation techniques
  for polystyrene 18:323
Ultracentrifuges 9:179, 183
  analytical 9:182
  in biological research 9:180
  characteristics 9:160
  optical systems 9:185
  preparative 9:179
  settling curves 9:161
Ultramarine blue
  commercial specifications 10:533; 12:174;
    17:201
  composition 10:533; 17:201
  in enamels 12:174
  identification 17:202
Ultramicrobalance 8:282
Ultraphosphates 17:118
  properties 17:118
Ultrasonic methods
  for milkfat in milk 16:128
  specimen preparation for electron microscopy
    8:205
Ultrathane
  differential thermal analysis 12:133
Ultraviolet absorbers
  in methacrylate polymers 16:95
  in polyurethanes 19:312
  solvent extraction 19:312
  thin-layer chromatography 19:312
Ultraviolet discoloration
  of caulking compounds 9:26
Ultraviolet spectra
  of acetaminophen 6:183, 184
  of acetanilide 6:181
  of acetylsalicylic acid 6:130
  of anisomycin 5:604
  of antipyrine 6:208
  of bacitracin 5:537
  of benzene 7:39
  of benzoic acid 7:70
  of benzyl benzoate 7:85
  of BHT 6:87
  of N-butyl acetanilide 7:85
  of caffeine 6:133
  of carbomycin 5:522
  of cerium(III) perchlorate 9:273

  of chloramphenicol 5:498
  of cinchophen 6:200, 201
  of cinnamaldehyde 6:631
  of coumarin 6:636
  of N-cyclohexyl-2-benzothiazole sulfenamide
    12:143
  of di-2-benzothiazyl disulfide 12:143
  of 2,6-di-t-butyl-4-methyl phenol 12:142
  of dihydrostreptomycin sulfate 5:583
  of 2,4-dinitrobenzaldehyde 6:635
  of diphenylamine–acetone 12:141
  of diphenylguanidine 12:143
  of N,N'-diphenyl-p-phenylenediamine 12:141
  of erythromycin 5:523
  of 6-ethoxy-1,2-dihydro-2,2,4-trimethyl-
    quinoline 12:141
  of fumagillin 5:569
  of gramicidin D 5:539
  of n-heptyl p-hydroxybenzoate 7:603
  of mepivacaine hydrochloride 5:397
  of mercaptobenzimidazole 12:141
  of 2-mercaptobenzothiazole 12:143
  of 1,4-naphthoquinone 7:70
  of neocinchophen 6:202
  of neomycin sulfate 5:587
  of m-nitrobenzaldehyde 6:635
  of nitroethane 16:423
  of nitromethane 16:423
  of 2-nitropropane 16:423
  of novobiocin 5:499
  of nystatin 5:570
  of phenacaine hydrochloride 5:400, 401
  of phenacetin 6:132
  of phenyl naphthylamines 12:141
  of phthalic acid 7:70
  of procaine hydrochloride 5:404, 405
  of propoxycaine hydrochloride 5:410, 411
  of salicylamide 6:131, 132
  of Santonox R 6:87
  of Santowhite powder 6:87
  of secobarbital 6:485
  of sodium hexobarbital 5:370
  of sodium salicylate 6:130, 131
  of sorbic acid 7:603
  of streptomycin sulfate 5:583
  of tetracaine hydrochloride 5:415, 416
  of tetramethylthiuram disulfide 12:143
  of toluene 7:39
  of tyrocidine hydrochloride 5:541
  of tyrothricin 5:542

of vanillin 6:636
of viomycin sulfate 5:557
Ultraviolet spectroscopy 6:131, 132, 133, 159,
    182
of acenocoumarol 10:151, 153
of acetaldehyde 11:180
of acetals 4:71
of acetaminophen 6:156, 182, 196
of acetone 4:116
of acetylsalicylic acid 6:137, 140, 147, 149,
    150, 153, 156, 159, 160
of α-acids 7:605
of acrylamide monomer in polymers 4:176,
    337
of acrylonitrile 16:411
of adenine 12:553
of alcohols 4:545
of aldehydes 12:335
of alkyd resins 5:50
of alkyl aryl sulfates 12:575
of alkyl aryl sulfonates 11:343
of alkylbenzylammonium halides 12:575
of alphaprodine 16:251
of alphaprodine hydrochloride 16:251
of amino acids 12:553
of aminopyrine 6:212
of amolanone hydrochloride 5:378
of ampicillin 16:565, 576, 577
of amyl-p-N,N-dimethylaminobenzoate 11:92
of anileridine 16:251
of anileridine hydrochloride 16:251
of aniline 11:217; 14:239
of aniline and its derivatives 5:432, 449
of anisindione 10:176, 177
of anthocyanins 7:647
of antidegradants 12:141, 142
of antihistamines 6:31
of antioxidants 16:94
of antioxidants and antiozonants 6:97
of antipyrine 6:208, 209, 214
of apomorphine 16:251, 276
of apomorphine hydrochloride 16:251, 276
of barbiturates 6:483, 484, 500
of benoxinate hydrochloride 5:374
of benzalacetophenone oxides 12:203
of benzaldehyde 6:630; 7:99
of benzalkonium chloride 11:556
of benzene 7:39, 41
of benzethonium chloride 11:556
of benzoic acid 7:67, 69, 70
of benzo[a]pyrene 10:19
of benzyl alcohol 7:99
of benzyl benzoate 7:85
of benzyl cinnamate 11:92
of benzylpenicillenic acid 16:561, 566
of benzylpenicillic acid 16:561, 564
of benzylpenicillin 16:560, 564, 571, 572

of benzyl salicylate 11:92
of bishydroxycoumarin 10:151, 157, 158, 160
of bisphenol A 17:348
of bithionol 11:548, 549
of 4'-bromosalicylanilide 11:552
of 5-bromosalicylanilide 11:552
of 1,3-butadiene 11:505
of butethamine hydrochloride 5:382
of N-butyl acetanilide 7:85
of butylated hydroxytoluene 6:116
of t-butyl cinnamate 10:69
of 4-t-butylphenyl groups 17:339
of 4-t-butylphenyl salicylate 16:95
of caffeine 6:147, 149, 150, 154, 156, 159;
    8:67; 10:407, 408, 412, 433; 18:491
of candicidin 5:578
of carbon disulfide 7:45; 8:272, 278; 18:398
of carbonyl compounds 8:347
of carzinophilin 5:614
of catechin 18:465
of cationic surfactants 12:575
of cephaloridine 16:560
of cephalosporin C 16:579
of cephalosporins 16:560, 579
of cerium 9:273
of cetylpyridinium chloride 11:556
of chloramphenicol 5:507
of chlormadinone acetate 10:579, 594
of chlorogenic acid 10:69, 398, 399, 400
of 5-chloro-2-hydroxybenzophenone 16:95
of chlorophenols 9:521
of chlorprophenpyridamine maleate 6:150
of cinchona alkaloids 4:605
of cinnamaldehyde 10:47
of cinnamic acid 10:59; 11:92
of cinnamic anhydride 10:71
of cinnamoyl chloride 10:72
of N-cinnamoylimidazole 10:72
of cinnamyl acetate 10:80
of cinnamyl alcohol 10:79
of cinnamyl cinnamate 10:69, 80
of citrus oils 12:287
of cloxacillin 16:565
of coal and coke 10:213
of cocaine 5:383, 385; 16:251
of cocaine hydrochloride 16:251
of cocoa dialysates 9:593
of codeine 16:251, 270
of codeine phosphate 6:150; 16:251, 270
of codeine sulfate 6:150; 16:270
of copper(II) diethyldithiocarbamate 11:580
of cotarnine 16:251
of coumarins 11:163
of cresol 11:548
of crotonaldehyde 11:180
of cyclohexanol 11:196
of cyclohexanone 11:206

of cyclopentenylallyl barbituric acid 6:150
of cystosine 12:553
of diamines and higher amines, aromatic 11:
  439, 442
of 3,5-dibromosalicylanilide 11:552
of 4′,5-dibromosalicylanilide 11:552
of dibucaine 5:386, 387
of dibucaine hydrochloride 5:386
of dichlorophene 11:549
of dichlorophenols 9:521
of 4,4′-dichloro-3(trifluoromethyl) carbanilide
  18:200
of dicloxacillin 16:565
of diethylaminodiethyldithiocarbamate 11:583
of *m*-diethylaminophenol 10:510
of diethylethoxymethylene malonate 15:437
of dihydrocodeine 16:251
of dihydromorphine 16:251
of 2,4-dihydroxybenzophenone 16:95
of 2,2′-dihydroxy-4,4′-dimethoxybenzophenone
  16:95
of 2,2′-dihydroxy-4-methoxybenzophenone
  16:95
of dimethisoquin hydrochloride 5:387
of dimethisterone 10:580
of *p*-dimethylaminobenzaldehyde 10:507
of *p*-dimethylaminobenzoic acid 10:507
of dimethylgloxime 7:583
of 2,4-dinitroaniline 10:509
of dioxane 11:520
of diphenadione 10:178, 179
of diphenoxylate hydrochloride 16:251
of diphenyl 11:527
of diphenylamine 10:510
of dipropylene glycol salicylate 11:92
of dithiocarbamate esters 11:590
of dithiocarbamic acid salts 11:580, 583
of divinyl acetylene 4:376
of doxylamine succinate 6:150
of ecgonine 16:251
of egg volatiles 12:71
of epicatechin 18:465
of epicatechin gallate 18:465
of epigallocatechin 18:465
of epigallocatechin gallate 18:465
of epoxy compounds 12:201, 202
of erythromycin 5:532
of ethinyl estradiol 10:569, 571
of ethylbenzylaniline sulfonic acid 10:507
of ethyl biscoumacetate 10:151, 162
of ethyl cinnamate 10:69
of ethylene oxide 12:335, 575
of ethylmorphine 16:251
of *N*-ethyl-*N*-(3-sulfobenzyl)-sulfanilic acid
  10:507
of ethynodiol diacetate 10:581
of fats and fatty oils 12:489

of fentanyl 16:251
of filipin 5:578
of finishing agents 11:133
of flavanols 18:465
of formic acid 13:119
of fumagillin 5:578
of furfural 13:238
of furfuryl alcohol 13:241
of gallocatechin 18:465
of gelatin 12:553
of glutaraldehyde 12:585
of gramicidin 5:548
of grapefruit oil 12:288
of griseofulvin 5:507
of guanine 12:553
of *n*-heptyl *p*-hydroxybenzoate 7:602
of heroin 16:251, 271
of hexachlorophene 11:87, 548, 549; 18:199
of hexylcaine hydrochloride 5:391
of hydrocarbons 7:39, 40
of hydrocodone 16:252
of hydrocodone bitartrate 16:252
of hydrogen peroxide 14:435
of hydrogen sulfide 8:278; 18:398
of hydromorphone 16:252
of hydroquinone 4:205
of *p*-hydroxybenzoic acid 6:146
of *o*-(2-hydroxy-4-diethylaminobenzoyl)
  benzoic acid 10:510
of 2-hydroxy-4-methoxybenzophenone 16:95
of 2(2′-hydroxy-5′-methylphenyl)benzotriazole
  16:95
of 1-hydroxy-4-*p*-toluinoanthraquinone 10:510
of insulin 14:568
of iron oxide 13:464
of isatinsulfonic acid 10:512
of isobucaine hydrochloride 5:394
of isobutyl salicylate 11:92
of isocyanic acid esters 15:104
of isohumulones 7:604
of isomethadone 16:252
of isopropyl cinnamate 10:69
of Lake Red C amine 10:509
of lead 11:315
of lemon oil 12:288
of leuco base 10:511
of levomethorphan 16:252
of levorphanol 16:252
of levorphanol tartrate 16:252
of lime oil 12:288
of LSD 15:299
of malononitrile 15:443
of mandarin oil 12:288
of medroxyprogesterone acetate 10:582
of melamine–formaldehyde resin 11:134
of menadiol sodium diphosphate 10:204
of meperidine 16:252

of mepivacaine hydrochloride 5:396, 398
of meprylcaine hydrochloride 5:398
of mercaptans 15:559
of metabutethamine hydrochloride 5:399
of metabutoxycaine hydrochloride 5:400
of methadone 16:252
of methanol 16:107
of methapyrilene hydrochloride 6:150
of methicillin 16:560, 565, 574
of methyl *p*-acetoxycinnamate 10:69
of methyl *m*-bromocinnamate 10:69
of methyl *p*-chlorocinnamate 10:69
of methyl cinnamate 10:68; 11:92
of methyldiphenyls 11:527
of methylene chloride 9:430
of methyl-*p*-hydroxybenzoate 11:548
of methyl hydroxycinnamate 10:69
of methyl methoxycinnamate 10:69
of methyl *p*-methylcinnamate 10:69
of methyl salicylate 16:95
of methylvinylpyridine copolymers 4:267
of metopon 16:252
of mikamycins 5:563
of monochlorophenols 9:521
of morphine 16:252
of morphine sulfate 16:252, 261
of mucochloric acid 12:585
of nafcillin 16:578
of nalorphine 16:252
of nalorphine hydrochloride 16:252, 268
of 1,4-naphthoquinone 7:70
of narceine 16:252
of narcotics 16:253
of narcotine 16:252
of neocinchophen 6:204
of nicotine 10:16
of nicotinic acid 17:439
of nitric oxide 16:478
of *o*-nitrobenzalacetophenone oxides 12:203
of nitroparaffins 16:417
of norethindrone 10:584
of norethindrone acetate 10:584, 591
of norpethidine 16:252
of novobiocin 5:507
of nystatin 5:578
of orange oil 12:288
of oxacillin 16:565
of oxycodone 16:252
of oxymorphone 16:253
of ozone 16:538, 546, 548
of papaverine 16:253
of penicillenic acid 16:561, 566
of penicillins 16:560, 564
of pentachlorophenol 9:521
of pesticides and insecticides 19:610
of phenacaine hydrochloride 5:400
of phenacetin 6:133, 147, 149, 150,

154, 159, 182
of phenazocine 16:253
of phenazocine hydrobromide 16:253
of phenethicillin 16:565
of phenindamine tartrate 6:150
of phenindione 10:182
of phenobarbital 6:150, 501, 521
of phenol 9:521; 11:548
of phenolic antiozonants 6:97
of phenolic plasticizers 17:300
of phenols 17:22, 28
of β-(phenoxyethyl)dimethyldodecylammonium
bromide 11:556
of phenoxymethylpenicillin 16:560, 565, 572
of phenprocoumon 10:151, 164
of phenyl cinnamate 10:69
of phenyl groups 17:339
of phenylmercuric acetate 11:565
of *N*-phenyl-1-naphthylamine 11:442
of phenyl salicylate 16:95
of phthalic acid 7:70
of *o*-phthalic acid 5:35
of phthalic acid derivatives 10:505
of phytonadione 10:186, 187
of pimaricin 5:579
of piminodine 16:253
of piminodine ethanesulfonate 16:253
of polyacrylamide 4:171
of polyamides 17:298
of polymers 4:267
of polyunsaturated acids 12:489
of potassium 17:375
of pramoxine hydrochloride 5:402
of procaine hydrochloride 5:404, 408, 419
of progesterone 14:222
of propoxycaine hydrochloride 5:409, 411, 413
of pyrilamine maleate 6:150
of pyrocaine hydrochloride 5:413
of quaternary ammonium compounds 11:556
of quinine 16:271
of racemethorphan 16:253
of racemorphan 16:253
of resorcinol 11:548
of rhenium 17:540
of ristocetin 5:614
of safrole 5:91
of salicylamide 6:131, 132, 133, 159, 182
of salicylanilide 11:552
of salicylic acid 6:140, 150, 156, 159
of salicylsalicylic acid 6:146
of sodium benzoate 7:85
of sodium diethyldithiocarbamate 11:583
of sodium diisopropyldithiocarbamate 11:583
of sodium dimethyldithiocarbamate 11:583
of sodium hexobarbital 5:371
of sodium iodide 10:493
of sodium methohexital 5:371

of sodium nafcillin 16:560
of sodium salicylate 6:130, 132
of sodium thiopental 5:372; 6:527
of sodium toluene sulfonate 11:366
of sorbic acid 19:482
of steroid epoxides 12:203
of steroids 14:207
of streptomycin 5:597
of styrene 12:138; 18:297, 334
of *m*-sulfobenzaldehyde 10:507
of sulfobenzaldehydes 10:507
of sulfur 11:231
of sunscreens 11:91, 92
of surfactants 12:575
of tangerine oil 12:288
of tannins 7:647
of telluric acid 17:606
of tellurium 17:605
of terphenyls 11:534
of tetracaine 5:417
of tetracaine hydrochloride 5:414, 417
of tetrachlorophenols 9:521
of 2,2′,4,4′-tetrahydroxybenzophenone
    11:92; 16:95
of thebaine 16:253
of thenylpyramine hydrochloride 6:164
of theobromine 8:67
of thiophene 18:543
of thiuram disulfides 11:597
of thonzylamine hydrochloride 6:150
of 3,4′,5-tribromosalicylanilide 11:549, 552;
    18:200
of 3,4,4′-trichlorocarbanilide 18:200
of trichlorophenols 9:521
of 3′,4′,5′-trichlorosalicylanilide 11:549
of trigonelline 8:68; 10:412, 414
of triphenylmethane and related dyes 19:168
of tris(nonylated phenyl) phosphite 18:332
of tryptophan 8:533
of tylosin 5:533
of tyrosine 8:538
of Uvinul 490 11:92
of Uvinul N35 11:92
of Uvinul N539 11:92
of vanillin 6:621, 635; 19:349
of variotin 5:579
of Vegadex 11:590
of vinbarbital 6:529
of vinylpyridine monomers in polymers 4:341
of warfarin 10:151; 11:170, 171
of warfarin potassium 10:175
of warfarin sodium 10:170
of zinc diethyldithiocarbamate 11:583
of zinc phenolsulfonate 11:87
Umbelliferone. See 7-Hydroxycoumarin.
Uncertified colors 10:529-546
  analysis 10:535

assay 10:538
  in cosmetics 10:535, 536
  in drugs 10:534, 535
  in food 10:530, 535
  identification 10:535
  impurities, determination of 10:541
n-Undecane
  gas chromatography 13:352
  in gasoline 13:352, 355
  properties 13:352
1-Undecanol. (See also Fatty alcohols.)
  properties 4:561
  solubility 4:562
2-Undecanone
  absorption spectroscopy 9:564
  2-hydroxy-1-naphthaldehyde colorimetric
    method 11:204
n-Undecanonitrile. (See also *Nitriles.*)
  properties 16:396
Undecylic acid
  properties 8:558
Unsaponifiable matter
  in alkyd resins 5:52
  in fats and fatty oils 12:479
  in soaps 18:201
  solvent extraction 18:201
Unsaturated alcohols. See Acetylenic alcohols;
    Ethylenic alcohols.
Unsaturation, degree of
  of epoxy resins 12:251
  of hydrocarbon resins 14:285
  of hydrocarbons 14:336
  in margarine 15:498
Unsoundness tests
  of cement 9:140
UOP 80 6:77
UOP 88 (and UOP 288) 6:77
  colorimetric method based on formation of
    Wurster salts 6:101, 103
Upholstery cleaners 6:354
  analysis 6:366
Uranine. See D & C Yellow No. 8.
Uraninite 19:225
  composition 19:225, 255
  decomposition 19:235
  occurrence 19:225
Uranium. (See also Uranium alloys; Uranium
    compounds; and Uranium metal.) 19:224-
    258
  acid–base titration 14:366
  in aluminum metal 5:152
  ammonium hydroxide gravimetric method
    19:243
  ammonium thiocyanate colorimetric method
    19:248
  analysis 19:233
  arsenazo III colorimetric method 19:249

atomic absorption spectroscopy 19:255
in bauxite 5:115
borax bead test 6:584
in carnotite 19:325
ceric sulfate titration 19:246
cerium(IV) titration 13:274
colorimetric method 19:241, 243, 247
decomposition 19:234
detection 19:241
dibenzoylmethane colorimetric method 19:247, 249
EDTA titration 19:247
electrolytic methods 19:255
emission spectroscopy 19:241
fluorometry 14:138; 19:251
in glass 9:200
gravimetric method 19:230, 232, 234, 237, 243
in hafnium 14:138
hafnium in 14:116
half-life 19:226
history 19:224, 227
in hydrochloric acid 14:366
hydrogen peroxide colorimetric method 19:248
impurities 19:238, 244
ion exchange chromatography 13:255; 17:575; 19:231, 236, 237, 238, 250
isotopes 19:226
microcosmic salt bead test 6:587
niobium in 16:385
in nuclear fuel 19:255
occurrence 19:225
oxalate gravimetric method 19:244
oxidation potentials 19:229
oxine gravimetric method 19:242, 245
polarography 19:234, 255
potassium dichromate volumetric method 19:246
potassium ferrocyanide colorimetric method 19:242
pyrophosphate gravimetric method 19:245
radioactive decay 17:444
rare earth content 17:477
rhodamine B colorimetric method 19:242, 247
safety 19:230, 234
sampling 19:234
separation 19:236
in silicate rocks 19:250
sodium fluoride fluorometric method 19:241, 242
sodium hydroxide gravimetric method 19:232
sodium–lithium fluoride fluorometric method 19:251, 252
sodium–lithium fluoride fusion 19:252
solvent extraction 19:231, 234, 236, 254
in steel 19:235
tantalum in 16:385
technetium in 17:529

thiocyanate colorimetric method 19:247
toxicity 19:230, 234
in uranium–gallium alloys 13:273
uses 19:226
vacuum hot extraction methods 8:638
volumetric determination 19:234, 243, 245
in water 19:252
in zircaloy 14:138
in zirconium 14:138
zirconium in 14:116
Uranium acetate
color test for natural coloring materials 10:537
Uranium alloys. (See also Uranium metal.)
boron in 7:418
carrier gas fusion methods 8:682
decomposition 19:235
manufacture 19:228
oxygen in 8:682
uses 19:227
Uranium and the transuranium elements. (See also Transuranium elements; Uranium.) 19:224–262
Uranium bromides. (See also Uranium compounds.) 19:257
manufacture 19:229, 257
Uranium carbides. (See also Uranium compounds.) 19:256
carrier fusion methods 8:624, 626, 682
hydrogen in 8:621, 623, 680
manufacture 19:229, 256
nitrogen in 8:621, 623
oxygen in 8:617, 621, 623, 624, 626, 638, 680, 682
properties 19:256
uses 19:227, 256
vacuum fusion methods 8:617, 621, 623, 680
vacuum hot extraction methods 8:638
Uranium chlorides. See individual compounds.
Uranium compounds. (See also individual compounds.) 19:255–258
analysis 19:258
decomposition 19:234
impurities, determination of 19:258
properties 19:230
safety 19:230, 234
scandium in 17:574
solvent extraction 19:230
toxicity 19:230, 234
uses 19:226, 255
Uranium dioxide. (See also Uranium oxides.) 19:255
analysis 19:258
manufacture 19:255
occurrence 19:255
properties 19:255
solubility 19:256
uses 19:227, 255

Uranium dioxide–lanthanum oxide
  in ceramics 9:225
  microscopy 9:225
Uranium fluorides. See Uranium compounds;
    and individual compounds.
Uranium–gallium alloys 13:273
  analysis 13:273
  gallium in 13:274
  uranium in 13:273
Uranium hexachloride
  manufacture 19:257
  properties 19:257
  solubility 19:257
Uranium hexafluoride. (See also Uranium
    fluorides.)
  manufacture 19:232
  properties 19:256
  uses 19:256
Uranium hydride. (See also Uranium compounds.)
    19:256
  manufacture 19:256
  properties 19:256
  uses 19:256
Uranium iodides. (See also Uranium compounds.)
    19:257
  manufacture 19:229, 257
Uranium metal
  analysis 19:233
  carrier gas fusion methods 8:624, 626, 682
  commercial grades and specifications 19:233
  decomposition 19:234
  hot extraction methods 8:684
  hydrogen in 8:616, 619, 622, 626, 680
  manufacture 19:231
  nitrogen in 8:616, 619, 621, 622, 624, 626
  oxygen in 8:616, 619, 620, 621, 622, 624,
    626, 680, 682, 684
  properties 19:228
Uranium minerals
  decomposition 19:235
Uranium monoxide. (See also Uranium oxides.)
    19:256
  manufacture 19:256
Uranium nitrides. (See also Uranium compounds.)
    19:256
  carrier gas fusion methods 8:624, 626, 636, 682
  hydrogen in 8:621, 680
  manufacture 19:229, 256
  nitrogen in 8:621, 624, 626
  oxygen in 8:621, 624, 636, 680, 682
  properties 19:256
  vacuum fusion methods 8:621, 680
Uranium oxides. (See also Uranium compounds;
    and individual compounds.) 19:255
  carrier gas distillation fusion methods 8:636
  carrier gas fusion methods 8:624, 626, 682
  in ceramics 9:225

decomposition 19:235
in glass 9:200
manufacture 19:229
microscopy 9:225
oxygen in 8:624, 626, 636, 682
properties 9:204
pyrohydrolytic method for fluoride 13:435
solubility 19:235
uses 19:227
Uranium pentachloride
  manufacture 19:257
  properties 19:257
  solubility 19:257
Uranium peroxide. (See also Uranium oxides.)
    19:256
  manufacture 19:256
  properties 19:256
  uses 19:256
Uranium phosphide. (See also Uranium com-
    pounds.)
  carrier gas distillation fusion methods 8:636
  manufacture 19:229
Uranium sulfides. (See also Uranium compounds.)
  in ceramics 9:225
  manufacture 19:229
  microscopy 9:225
Uranium tetrachloride
  manufacture 19:257
  properties 19:257
  solubility 19:257
Uranium tetrafluoride
  manufacture 19:232, 256
  uses 19:256
Uranium trichloride
  manufacture 19:257
  properties 19:257
  solubility 19:257
Uranium trifluoride
  manufacture 19:256
  properties 19:257
Uranium trioxide. (See also Uranium oxides.)
    19:256
  manufacture 19:256
  properties 19:256
  uses 19:256
Uranophane
  composition 19:225
  occurrence 19:225
Uranothorianite
  decomposition 19:8
  thorium in 19:8
Urano–uranic oxide. See Triuranium octoxide.
Uranous oxide. (See also Uranium oxides.)
  solubility 19:235
Uranyl acetate. (See also Uranium compounds.)
  colorimetric method for cochineal 10:538
  commercial specifications 19:258

uses 19:258
Uranyl carbonates. (See also Uranium compounds.)
  manufacture 19:258
  uses 19:258
Uranyl nitrate. (See also Uranium compounds.)
    19:258
  commercial specifications 19:258
  manufacture 19:229, 258
  solubility 19:258
  uses 19:258
Uranyl sulfate. (See also Uranium compounds.)
  uses 19:258
Urbanyi and Rehm
  colorimetric method for ethinyl estradiol 10:573
Urea 5:276
  in alkyd resins by infrared spectroscopy 5:44
  in amino resins 5:285
  analysis 5:280, 285
  properties 8:597
  solubility 13:539, 552
Urea, butylated 5:276
Urea–formaldehyde resin adhesives. See amino
    resin adhesives.
Urea–formaldehyde resins
  analysis 15:536
  anthrone colorimetric method 11:134
  in cotton 11:134
  curing ability 15:536
  curing range 5:279
  in finishing agents 11:126, 133
  infrared spectrum 5:282, 283
  in match strikers 15:520, 536
  properties 5:277
  solubility 11:126, 133
  structure 5:278
  sulfuric acid–carbazole colorimetric method
    11:133
  viscosity 15:536
Urea–melamine resin
  infrared spectrum 5:284
Urea, methylated 5:276
Urea resins
  in alkyd resins 5:24
Urepan 12:87
Uretediones. See Diisocyanates.
Urethane
  infrared spectroscopy 15:94
  manufacture 15:94
  properties 8:597
Urethane elastomers. See Polyurethanes.

Urethane foam
  manufacture 18:142
  properties 18:141
  properties, determination of 18:149
  surfactants in 18:141
  types 18:142
  uses 18:143
Urethane polymers. (See also Polyurethanes.)
    19:262–319
Urine
  boron in 7:417
  calcium in 8:110; 18:222
  carbenicillin in 16:574
  cyclohexylamine in 11:212
  EDTA and related compounds in 12:48
  fluoride in 13:13, 26
  gold in 14:19, 21, 44
  hexachlorophene in 11:550
  iron in 15:32, 54
  lead in 15:185, 186, 189
  magnesium in 15:371; 18:222
  meperidine in 16:276
  methadone in 16:279
  morphine in 16:263, 266
  potassium in 18:222
  radium in 17:456, 468
  sodium in 18:222
  thallium in 18:524
  thin-layer chromatography 11:212
  zinc in 19:568
Uronic acids
  paper chromatography 12:582
Urotropine
  gravimetric method for indium 14:526,
    529
Urotropin-pyrogallol
  gravimetric method for niobium 16:366
  gravimetric method for tantalum 16:366
Uvanite 19:321
Uvinul 490
  in sunscreens 11:92
  ultraviolet spectroscopy 11:92
Uvinul D50. See 2,2′,4,4′-Tetrahydroxy-
    benzophenone.
Uvinul N35
  in sunscreens 11:92
  ultraviolet spectroscopy 11:92
Uvinul N539
  in sunscreens 11:92
  ultraviolet spectroscopy 11:92

V

Vaccines
  formaldehyde in 11:566
Vacuum distillation
  of plasticizers in photographic films 12:591
Vacuum extraction apparatus
  for hydrogen 5:131
Vacuum fusion methods. (See also *Carrier gas and vacuum fusion methods.*) 8:616, 619, 620, 621, 622, 860
  apparatus 8:654, 667
  applications 8:616, 680
  distillation 8:635
  hot extraction 8:638, 684
  for hydrogen in magnesium alloys 15:384
  for nitrogen 8:628; 10:301
  for nitrogen in metal borides 7:365
  for oxygen 10:301
  for oxygen in metal borides 7:365
  reaction media 8:629
  theory 8:614
Vacuum pipet 5:318
Valentinite 6:39
Valeraldehyde
  in black tea 18:510
  commercial specifications 8:340
  derivatives for identification 8:343
  gas chromatography 18:510
  properties 8:334
m-Valeramide
  derivatives for identification 8:601
Valeric acid
  in adipic acid 4:416
  in black tea 18:511
  gas chromatography 8:383; 18:511
  properties 8:397, 558
  thin-layer chromatography 8:381
Valeric anhydride
  properties 8:558
n-Valeronitrile. See *n*-Pentanonitrile.
Valeryl halides
  derivatives for identification 8:572
Valine. (See also *Carboxylic acids, amino.*) 8:413, 540
  analysis 8:544
  in black tea 18:507
  in cocoa beans 9:556
  commercial specifications 8:543
  composition 8:413
  crystalline form 8:541
  gas chromatography 8:429

infrared spectrum 8:541
  ion exchange chromatography 8:415, 421
  manufacture 8:543
  ninhydrin colorimetric method 8:544
  nuclear magnetic resonance spectrum 8:542
  paper chromatography 8:422
  properties 8:540
  in protein 8:414
  solubility 8:541
  structure 8:541
  in yeast 19:529
Valve freeze method
  for water in liquefied petroleum gas 15:251
Vanadates. (See also Vanadium compounds; and individual compounds.) 19:344
  manufacture 19:344
  properties 19:345
  solubility 19:345
Vanadic acid–sodium molybdate
  colorimetric method for phosphorus pentoxide 13:433
Vanadiferrous phosphate rock
  composition 19:321
  occurrence 19:321
Vanadinite 19:320
  composition 19:321
  occurrence 19:321
Vanadites. (See also Vanadium compounds.)
  manufacture 19:344
Vanadium. (See also Vanadium alloys; Vanadium compounds; and Vanadium metal.) 19:320–347
  in aluminum alloys 19:339, 342
  in aluminum and aluminum alloys 5:147
  in aluminum oxide abrasives 4:47
  aniline colorimetric method 19:331, 334
  atomic absorption spectroscopy 9:346; 10:132; 19:342
  in bauxite 5:116
  in beer 7:651
  benzidine colorimetric method 19:331
  α-benzoinoxime colorimetric method 19:331, 333
  N-benzoyl-*o*-tolylhydroxylamine colorimetric method 19:342
  in biological material 19:340, 342
  borax bead test 6:584
  cacotheline colorimetric method 19:331, 332
  in carbon 14:68, 74
  in carnotite 19:325

in cements 9:131
in chrome tanning compounds 9:691
in chromite 9:700
in chromium alloys 9:677; 19:339
in chromium compounds 9:687
in chromium ores 9:701, 704
in clay 19:343
in clays and clay minerals 10:129, 130
in cobalt alloys 10:323; 19:340
colorimetric determination 19:331, 339
in constant-modulus alloys 10:290
cupferron–chloroform extraction 9:645
cupferron gravimetric method 19:329, 335
detection 19:331
diaminobenzidine hydrochloride colorimetric
    method 19:334
in diatomaceous earth 11:491; 19:340
3,3'-dimethylnaphthidine colorimetric method
    14:68; 19:331, 333
α,α'-dipyridyl colorimetric method 19:331, 332
EDTA titration 13:445
in elastomers, synthetic 12:157
electrolysis 10:324; 19:331
emission spectroscopy 7:651; 9:346; 14:74,
    122; 15:155, 187; 16:160, 387; 19:334, 342
in ferrochromium 9:661
in ferroniobium 16:387
ferrous ammonium sulfate–potassium permanga-
    nate volumetric method 16:339
ferrous ammonium sulfate titration 15:81;
    19:102, 337
ferrous sulfate titration 8:137
in ferrovanadium 15:80; 19:338
in food 19:328
in fuel oil 19:339, 342
gallic acid–ammonium persulfate colorimetric
    method 19:341
in gas turbine fuels 19:339
in glass 9:200
in glass and glass-ceramics 13:445
in graphite 14:68, 74; 19:342
gravimetric determination 19:328, 334
in hafnium 14:122
half-life 19:320
in high-speed steels 10:289
in high-temperature alloys 10:286
history 19:320
hydrogen peroxide colorimetric method 10:324;
    16:339; 19:98, 108, 331, 340
hydrogen sulfide gravimetric method 19:329
8-hydroxyquinoline colorimetric method 11:491;
    19:331, 333, 340
8-hydroxyquinoline extraction 9:644; 10:130
ion exchange chromatography 14:113; 16:185,
    390; 19:331
in iron 9:645; 15:59
iron–dimethylglyoxime colorimetric

method 19:331, 332
iron–1,10-phenanthroline colorimetric method
    19:331, 332
isotopes 19:320
in lanthanum oxide 15:155
microcosmic salt bead test 6:587
molybdenum blue color test 19:334
in molybdovanadophosphoric acid and salts
    16:185
in nickel 16:338
in nickel alloys 16:338
in niobium 16:387
occurrence 19:320
in oils 19:328
pervanadic acid colorimetric method 9:687
phosphomolybdic acid colorimetric method
    19:331
phosphoric acid–sodium tungstate colorimetric
    method 19:339
phosphotungstovanadic acid colorimetric
    method 5:147; 9:677; 10:130
potassium permanganate–ferrous ammonium
    sulfate titration 10:324
potassium permanganate titration 19:335
in rocks 15:187
separation 19:328
in silicate rocks 9:643; 19:109, 339, 342
in silicon carbide abrasives 4:56
sodium bicarbonate gravimetric method 19:329
in sodium chloride 9:346
in sodium dichromate 9:684
sodium hydroxide gravimetric method 19:330
sodium peroxide fusion 19:328
sodium tungstate colorimetric method 12:158
solvent extraction 19:329, 330, 333, 340
in steel 9:645; 19:330, 342
in synthetic elastomers 19:339
in titaniferous magnetites 19:325
in titanium dioxide 19:343
in titanium metal 19:121
in uranium 19:240
uses 19:322
in vanadium–aluminum alloys 19:338
in vanadium carbide 8:133, 137; 19:338
in vanadium metal 19:327
in vanadium steel 19:336
vanadotungstic acid colorimetric method
    10:324
volumetric determination 19:335
volumetric method for glycerol 13:520
in water 19:341
x-ray fluorescence spectroscopy 19:342
in zirconium 14:122
Vanadium alcoholates. (See also Vanadium
    compounds.)
uses 19:343, 346
Vanadium alloys. (See also

Vanadium metal.) 5:61, 67
  analysis 19:326
  manufacture 19:325
  sampling 19:326
  uses 19:322
Vanadium–aluminum alloys. (See also Vanadium
  alloys.)
  analysis 19:325
  decomposition 19:327
  manufacture 19:325
  sampling 19:325
  vanadium in 19:338
Vanadium butyrate. (See also Vanadium compounds.)
  uses 19:343
Vanadium carbide. (See also *Carbides*; Vanadium
  compounds.)
  carbon in 8:134
  formation 8:130
  free carbon in 8:134, 137
  manufacture 19:324, 325
  properties 8:131, 133
  uses 8:133
  vanadium in 8:133, 137; 19:338
Vanadium chloride. (See also Vanadium compounds; and individual compounds.)
  sampling 19:326
  solubility 9:336
Vanadium compounds. (See also individual compounds.) 19:343–346
  analysis 19:346
  uses 19:322, 343
Vanadium dichloride. (See also Vanadium
  chloride.)
  properties 9:335
  uses 19:343
Vanadium metal
  analysis 19:326, 343
  assay 19:343
  carrier gas fusion methods 8:624, 625, 626,
    627, 682
  commercial grades and specifications 19:326
  dissolution 19:327
  hydrogen in 8:618, 622, 624, 626, 630, 680
  impurities, determination of 19:343
  manufacture 19:325
  nitrogen in 8:618, 622, 623, 624, 627, 630
  oxygen in 8:613, 614, 618, 622, 623, 624,
    625, 626, 627, 630, 632, 680, 682
  properties 19:323
  sampling 19:326
  solubility in gold 14:9
  sulfur in 18:383
  uses 19:322
  vacuum fusion methods 8:618, 622, 623, 680
Vanadium nitride. (See also Vanadium compounds.)

manufacture 19:324
Vanadium(II) oxide
  manufacture 19:344
  properties 19:344
  solubility 19:344
Vanadium(III) oxide
  manufacture 19:344
  properties 19:344
  solubility 19:344
Vanadium(IV) oxide
  manufacture 19:344
  properties 19:344
  solubility 19:344
  uses 19:343, 345
Vanadium(V) oxide
  analysis 19:327
  dissolution 19:327
  manufacture 19:325, 344
  properties 19:344
  solubility 19:344
  uses 19:322, 343, 345
Vanadium oxinate test
  for alcohols 4:521
Vanadium oxohalides. (See also Vanadium
  compounds; and individual compounds.)
  properties 19:324
  uses 19:345
Vanadium oxychlorides. (See also Vanadium
  compounds; and individual compounds.)
  sampling 19:326
Vanadium oxytrichloride. (See also Vanadium
  oxychlorides.)
  uses 19:343, 345
Vanadium pentoxide. See Vanadium(V) oxide.
Vanadium propylate. (See also Vanadium
  compounds.)
  uses 19:343
Vanadium steel. (See also Vanadium alloys.)
  decomposition 19:336
  vanadium in 19:336
Vanadium sulfate. (See also Vanadium compounds.)
  uses 19:343
Vanadium tetrachloride. (See also Vanadium
  chloride.)
  manufacture 19:324
  properties 19:96
  uses 19:343, 345
Vanadium tetroxide. See Vanadium(IV) oxide.
Vanadium trichloride. (See also Vanadium
  chloride.)
  uses 19:343, 345
Vanadotungstic acid
  colorimetric method for vanadium 10:324
Vancide F. See Ferbam.
Vancide M. See Maneb.
Vancomycin 5:478, 599

analysis 5:605, 612, 614
identification 5:605
microbiological assay 5:612
paper chromatography 5:603
properties 5:600
separation 5:605
solubility 5:601, 602
uses 5:599
Vancomycin sulfate
infrared spectrum 5:606
van der Meulen method
for bromine 8:19, 25
Van der Smissen burners 9:383
Van der Waals forces in adhesion 4:391
Van Dyke brown
composition 17:194
Van Hasselt PBN 6:77
Vanillic acid (3-Methoxy-4-hydroxybenzoic acid)
8:406
gas chromatography 8:389
paper chromatography 8:379, 406
properties 8:403
uses 8:406
Vanillin 6:597; 19:347–351
aldehyde in 19:351
analysis 19:348
assay 19:349
carbon and hydrogen analysis 8:285
in chocolate 9:537, 545
colorimetric method for catechins 19:467
column chromatography 6:618
commercial grades and specifications 6:610,
611; 19:348
derivatives for identification 6:612
fluorometry 6:627
gas chromatography 6:620; 8:355; 11:167;
19:349
history 19:347
hydroxylamine hydrochloride method 8:357
iodometric method 19:351
manufacture 6:603; 19:348
occurrence 19:348
paper chromatography 6:620
properties 6:598; 19:347
sampling 19:348
separation 19:348
solubility 13:540; 19:347
solvent extraction 19:348
structure 19:347
ultraviolet spectroscopy 6:621, 633, 635;
19:349
uses 6:607; 8:332; 19:347, 348
Vanilmandelic acid
gas chromatography 8:389
Van Slyke apparatus 5:15
monoalkanolamines, determination of 5:15
Van Urk reagent

color test for thiophene 18:541
Vapam. See Sodium methyl dithiocarbamate.
Vapor pressure
of *n*-butane 15:249
of chlorine 9:307
of diethylenetriamine 11:410
of ethane 15:249
of ethylenediamine 11:410, 411
of gasoline 13:340
of germanium tetrachloride 13:400
of graphite 4:11
of hydrocarbons 14:342
of isobutane 15:249
of liquefied petroleum gas 15:248
of pentaethylenehexamine 11:410
of polycarbonates 17:341
of propane 15:249
of propylene 15:249
Reid method 14:342
of tetraethylenepentamine 11:410
of triethylenetetraamine 11:410
Variamine Blue B hydrochloride
colorimetric method based on formation of
Wurster salts 6:101, 103
Variance 7:162
Variotin 5:473, 565
analysis 5:571, 577, 579
microbiological assay 5:577
properties 5:565
solubility 5:566, 567
structure 5:473
ultraviolet spectroscopy 5:579
uses 5:565
Vasopressin. (See also Pituitary hormones.)
14:181
analysis 14:206
commercial specifications 14:188
occurrence 14:188
structure 14:181
uses 14:181, 188
Vat blue 6. See Indanthrone blue.
Vat dyes 12:3
identification 12:14
nitric acid color test 12:18
separation 12:14
sodium formaldehyde sulfoxalate color test
12:18
sodium hydrosulfite color test 12:14
sulfuric acid color test 12:18
Vat orange 7. See Perinone orange.
Vat red 23. See Perylene maroon.
Vat red 29. See Perylene scarlet.
Vatting method
for indigoid dyes 14:509
Vat violet 1. See Isodibenzanthrone violet.
Vat yellow 1. See Flavonthrone yellow.
Veatchite 7:314

Vegadex. (See also Dithiocarbamate esters.)
  analysis 11:589
  properties 11:589
  structure 11:589
  ultraviolet spectroscopy 11:590
  uses 11:589
Vegetable albumen
  in confectionery products 10:551
Vegetable fat
  in butter 16:138
  gas chromatography 16:139
  liquid chromatography 16:138
Vegetable fibers
  cooling system sealers 6:352
Vegetable juice
  composition 10:533
Vegetable oils
  in sunscreens 11:91
Vegetables
  arsenic in 6:248
Vegetable waxes
  in automobile waxes 6:353
Vegetation
  copper in 10:614
  nickel in 16:299
Vehicle testing
  accuracy of results 7:472
  of brake linings 7:465
  terminology 7:471
Velon fiber. See Vinylidene chloride–vinyl
  chloride copolymers.
Venetian red
  commercial grades and specifications 17:193
  commercial specifications 12:174
  composition 12:174
  in enamels 12:174
  impurities, determination of 17:194
Veratraldehyde. (See also 3,4-Dimethoxybenzal-
  dehyde.)
  gas chromatography 8:355; 11:167
Veratric acid
  gas chromatography 8:389
  paper chromatography 8:379
Verdigrise. (See also Copper acetate.)
  manufacture 10:653
  solubility 10:653
  uses 10:653
Verel. (See also *Acrylic and modacrylic fibers*;
  Acrylonitrile polymers; Modacrylic fibers.)
  birefringence 4:258
  cross section 4:225, 263
  infrared spectrum 4:242
  melting point 4:262
  microscopy 4:264
  mixtures, determination in 4:359
  moisture regain 4:333
  producer 4:221

  properties 4:222
  solubility 4:247, 251, 359
  specific gravity 4:258
  staining tests 4:252
Vermiculite. (See also *Clays and clay minerals.*)
  composition 10:102
Veronal 5:91
Vesuvianite 7:316; 8:75
Vetiver oil. (See also *Essential oils.*)
  refractive index 12:271
Vetiveryl acetate
  in essential oils 12:276
  molecular weight 12:276
VGB 6:77
  diazotized *p*-nitroaniline colorimetric method 6:98
Vibrathane 12:87
Vicalloys. (See also Cobalt alloys.) 10:285
Vicara
  mixtures, determination in 4:357
  solubility 4:357
Vicat time-of-set test
  for cement 9:135
Victoria Blue. (See also *Triphenylmethane and
  related dyes.*)
  in enamels 12:187
  identification 12:187
  thin-layer chromatography 19:173
Victoria Blue B. (See also *Triphenylmethane and
  related dyes.*)
  paper chromatography 19:172
  tartrazine titration 19:175
  visible spectroscopy 19:167
Victoria Blue BO. (See also *Triphenylmethane
  and related dyes.*)
  impurities, determination of 19:187
  structure 19:161
  visible spectroscopy 19:184
Victoria Violet
  colorimetric method for boron 7:410
Vinbarbital 6:468, 529
  acid–base titration 6:529
  analysis 6:529
  commercial grades and specifications 6:528
  derivatives for identification 6:447
  gas chromatography 6:483, 484
  impurities, determination of 6:529
  properties 6:470
  thin-layer chromatography 6:481
  ultraviolet spectroscopy 6:529
Vinegar. (See also *Acetic acid.*) 4:93, 96
  analysis of 4:101
Vinsol
  analysis 15:529
  commercial grades and specifications 15:511
  manufacture 15:511
  in matches 15:511
  particle size 15:529

Vinyl acetate
  acetaldehyde in 8:589
  commercial specifications 8:581
  gas chromatography 4:212; 8:587; 11:181
  hydroquinone 8:589
  in polymers 4:274
  properties 8:577
  saponification 4:341
  specific refraction 12:382
Vinyl acetate–acrylonitrile–methylvinylpyridine
    polymers.
  precipitation fractionation methods 4:299
Vinyl acetate–acrylonitrile polymers. See
    Acrylonitrile–vinyl acetate polymers.
Vinyl acetate polymers
  gas chromatography 4:286
  solubility 4:330
Vinyl acetate–polypropylene copolymers. (See
    also *Ethylene and propylene polymers.*)
    12:345
  manufacture 12:345
Vinyl acetate resin. See Polyvinyl acetate.
Vinyl acetate–vinylidene dinitrile polymer. See
    Darvan.
Vinyl acetic acid
  sodium bisulfite addition 4:341
Vinylacetonitrile. (See also *Nitriles.*)
  azeotropes 16:403
  properties 16:396
Vinylbenzene. See *Styrene.*
Vinyl *n*-butyl ether. (See also Vinyl ethers.)
  properties 12:297
Vinylcarboxylic acids
  sodium sulfite addition 4:341
Vinyl chloride. (See also C$_2$ Chlorinated
    hydrocarbons.)
  in aerosols 11:95
  Beilstein test 9:382
  gas chromatography 9:374, 375, 377, 380,
    381; 11:98; 13:65
  mass spectroscopy 9:398
  properties 9:466; 11:95
Vinyl chloride–acrylonitrile polymers. See
    Acrylonitrile–vinyl chloride polymers; Dynel.
Vinyl chloride–polypropylene copolymers.
    (See also *Ethylene and propylene polymers.*)
    12:345
  manufacture 12:345
  properties 12:345
Vinyl chloride resin
  in finishing agents 11:126, 132
  solubility 11:126, 132
Vinyl chloride–vinyl acetate polymers
  chlorine determination 4:280
  infrared spectroscopy 4:244
Vinyl chloride–vinyl acetate resin
  in finishing agents 11:126, 132

solubility 11:126, 132
Vinylchloropropene
  mass spectroscopy 9:399
Vinyl compounds
  Wijs method 5:84
Vinyl cyanide. See *Acrylonitrile.*
Vinylcyclohexane oxide. (See also *Epoxy
    compounds.*)
  infrared spectroscopy 12:200
1-Vinyl-3-cyclohexene
  liquid chromatography 7:13, 14, 15, 16
4-Vinyl-1-cyclohexene 11:512
  assay 11:513
  azeotropes 11:512
  in 1,3-butadiene 11:504
  commercial specifications 11:513
  properties 11:512
  toxicity 11:512
Vinylcyclohexene dioxide. (See also *Epoxy
    compounds.*)
  properties 12:195
Vinylene groups
  infrared spectroscopy 12:357
  in polyethylene 12:357, 360
Vinyl esters
  manufacture 8:580
Vinyl ether. (See also Vinyl ethers.)
  analysis 5:367
  impurities, determination of 5:367
  properties 5:357, 367; 12:297
Vinyl ethers. (See also *Ethers*; and individual
    compounds.)
  analysis 12:311
  iodometric titration 12:311
  mercurimetric titration 12:312
  sulfite method 4:69
  sulfuric acid–cupric sulfate colorimetric method
    12:314
Vinyl ethyl ether. (See also Vinyl ethers.)
  properties 12:297
2-Vinyl-5-ethylpyridine
  residual monomer in polymers 4:341
Vinyl fluoride. (See also *Fluorocarbons.*)
  properties 13:60
Vinyl groups
  infrared spectroscopy 12:357
  in polyethylene 12:357, 360
Vinylidene chloride. See 1,1-Dichloroethylene.
Vinylidene chloride–vinyl chloride polymer. See
    Saran.
Vinylidene fluoride. (See also *Fluorocarbons.*)
  properties 13:60
Vinylidene fluoride–hexafluoropropylene
    copolymer 13:95
  analysis 13:95
  commercial grades and specifications 13:95
  compression set 13:98

elongation 13:99
hardness 13:98
physical tests 13:97
properties 13:95
sample preparation 13:97
tensile strength 13:99
Vinylidene groups
  infrared spectroscopy 12:357, 361
  in polyethylene 12:357, 360
  in polypropylene 12:361
Vinylidene resin
  in finishing agents 11:126, 133
  solubility 11:126
Vinyl methyl ether. (See also Vinyl ethers.)
  properties 12:297
3-Vinyl-6-methylpyridine. (See also *Pyridine and pyridine derivatives.*)
  ionization constants 17:416
Vinylon HH. See Vinyl chloride–vinyl acetate polymers.
Vinyl polymers
  flame test 12:588
  in photographic films 12:588
Vinyl propionate
  gas chromatography 11:181
Vinyl propionic acid
  sodium bisulfite addition 4:341
2-Vinylpyridine. (See also *Pyridine and pyridine derivatives.*)
  commercial specifications 17:423
  ionization constants 17:416
  uses 17:412
4-Vinylpyridine. (See also *Pyridine and pyridine derivatives.*)
  commercial specifications 17:423
  ionization constants 17:416
Vinylpyridine–acrylonitrile polymers. See Acrylonitrile–vinylpyridine polymers.
Vinylpyridine derivatives
  acid–base titration, nonaqueous 4:341
  residual monomer in polymers 4:341
Vinyl resin adhesives, evaluation 4:404
Vinylsulfonamide dyes
  identification 12:20
Vinylsulfone dyes
  identification 12:20
Vinyltoluene. (See also *Hydrocarbons.*)
  in alkyd resins 5:43
  gas chromatography 18:299
  modifier for alkyd resins 5:24
  properties 14:326
  in styrene 18:299
Vinyltriacetoxysilane. (See also Organofunctional silanes.)
  commercial specifications 18:82
  gas chromatography 18:85
  properties 18:80

Vinyl trichloride. See 1,1,2-Trichloroethane.
Vinyltrichlorosilane. (See also Organo–chlorosilanes; Organofunctional silanes.)
  commercial specifications 18:56, 82
  gas chromatography 18:62, 66
  impurities, determination of 18:66
  infrared spectrum 18:57
  properties 18:54, 80
Vinyltriethoxysilane. (See also Organofunctional silanes.)
  commercial specifications 18:82
  gas chromatography 18:85
  properties 18:80
Vinyltrimethoxysilane. (See also Organofunctional silanes.)
  commercial specifications 18:82
  gas chromatography 18:85
  properties 18:80
Vinyl-tris(2-methoxyethoxy)silane. (See also Organofunctional silanes.)
  commercial specifications 18:82
  gas chromatography 18:83
  properties 18:80
Violarite
  composition 16:286
Violaxanthin
  in green tea leaf 18:480
  thin-layer chromatography 18:480
Violet BNP. (See also *Triphenylmethane and related dyes.*)
  paper chromatography 19:172
Violet ninhydrin
  colorimetric method for propylene glycol 13:599
Viomycidine
  structure 5:471
Viomycin 5:471, 551
  analysis 5:557, 561, 563
  bioautography 5:557
  colorimetric methods 5:563
  gravimetric method using phosphotungstic acid 5:564
  identification 5:557
  microbiological assay 5:561
  paper chromatography 5:557
  properties 5:550
  separation 5:557
  solubility 5:551
  uses 5:551
Viomycin sulfate
  colorimetric methods 5:563
  infrared spectrum 5:557
  ultraviolet spectrum 5:557
Viridic aldehyde. See Chlorogenic acid.
Viscometers 4:307; 10:269, 271, 557; 11:20, 22, 23, 387; 13:97; 18:113
  cuam type 7:238

Ostwald-Cannon-Fenske type 7:235, 236, 267
Viscose fiber
  mixtures, determination in 4:357
  solubility 4:357
Viscose process performance tests
  for cellulose 9:55
Viscosity. (See also Fluidity; Intrinsic viscosity.)
  of acrylamide polymers 4:173
  of acrylic polymers 4:302
  of acrylonitrile polymers 4:389
  of adhesives 15:535
  of alkyd resins 5:52
  of amino resin adhesives 4:399
  of animal glues 4:399
  of beer 7:513
  of bituminous materials 6:282, 284, 295
  of brake fluids 7:431
  bubble method 12:252
  of casein 9:11
  of cellulose 9:34
  of cellulose acetate 9:71
  of cellulose acetate butyrate 9:76
  of cellulose acetate polymers 4:76
  of cellulose acetate propionate 9:76
  of cellulose derivatives 9:62
  of cellulose nitrate 9:66
  of chocolate 9:572
  of confectionery products 10:556
  of corn starch 11:24; 15:526
  of corn starch paste 11:17
  correlation between methods 7:264, 265
  definition of terms 4:303
  of detergents 11:386
  of diethylenetriamine 11:412, 413
  of eggs and egg products 12:74
  of egg white 12:75
  of egg yolk 12:75
  of enamels 12:178
  of epoxy resins 12:251
  of ethyleneamines 11:429
  of ethylenediamine 11:411
  of flour–water suspensions 12:604
  of fluorocarbons, polymers 13:96
  Gardner-Holdt bubble rise method 14:295
  of glass 13:404
  of glues of animal origin 13:491
  of gums 15:527
  Hoeppler viscometer method 7:515
  of hot melt adhesives 4:401
  of hydrocarbon resins 14:294
  of hydroxyethylcellulose and its derivatives
    9:84
  kinematic method 12:252
  of liquid resins 4:23
  of lubricants 15:350, 351
  of match heads 15:539
  of match strikers 15:540

  of organosilicon copolymers 18:148
  Ostwald viscometer method 7:516
  of pentaethylenehexamine 11:412
  of polyamides 17:294
  of polycarbonates 17:342
  of polyethylenimine 14:491
  of pulp dispersions 7:264
  rotating spindle method 12:251
  of silicone fluids 18:97, 113
  of silicone resins 18:124
  of sodium carboxymethylcellulose 9:90
  of styrene 18:311
  temperature relationships 6:282, 284
  temperature susceptibility factors 6:304
  of tetraethylenepentamine 11:412, 413
  of triethylenetetraamine 11:412, 413
  of urea–formaldehyde resins 15:536
Viscosity–average molecular weight
  of ethylene and propylene polymers 12:372
Viscosity–temperature coefficient
  of silicone fluids 18:114
Visible spectra
  of F D & C Blue No. 1 10:472
  of F D & C Blue No. 2 10:473
  of F D & C Green No. 3 10:474
  of F D & C Red No. 2 10:476
  of F D & C Red No. 3 10:477
  of F D & C Red No. 4 10:478
  of F D & C Violet No. 1 10:479
  of F D & C Yellow No. 5 10:480
  of F D & C Yellow No. 6 10:481
Visible spectroscopy
  of aminobenzene 10:509
  of annatto 10:539
  of anthocyanins 19:464
  of β-apo-8'-carotenal 10:540
  of azo dyes 6:404, 421
  of beer 7:518
  of bismuth solutions 7:216
  of brilliant green 19:167
  of bromine 8:21
  of carbon black 12:150
  3-carboxy-5-hydroxy-1-phenyl-4-p-sulfo-
    phenylazopyrazole, disodium salt of 10:517
  3-carboxy-5-hydroxy-1-p-sulfophenyl-4-
    phenylazopyrazole, disodium salt of 10:517
  of carmine 10:540
  of carotene 11:55; 19:152
  of β-carotene 10:539, 545
  of chromium 10:494
  of CI Pigment Red 88 14:508
  of Citrus Red No. 2 10:486
  of CI Vat Black 1 14:508
  of CI Vat Blue 5 14:508, 511
  of CI Vat Brown 5 14:508
  of CI Vat Orange 5 14:508, 511
  of CI Vat Red 1 14:508, 511

of CI Vat Violet 2 14:508
of coal and coke 10:213
of copper(II) diethyldithiocarbamate 11:580
of copper(II) dimethyldithiocarbamate 11:583
of corn oil 11:57
of corn products 11:13
of crystal violet 19:166, 167
of D & C Blues 10:486
of D & C Greens 10:486
of D & C Oranges 10:486; 11:70
of D & C Reds 10:486; 11:70
of D & C Violet No. 2 10:486
of D & C Yellows 10:486, 523
of diamines and higher amines, aromatic
    11:439, 442
of dibromofluoresceins 10:523
of diiodofluoresceins 10:514
of eriochrome cyanine R 19:167
of erioglaucine A 19:167
of ethyl violet 19:167
of Ext D & C Green No. 1 10:486
of Ext D & C Yellows 10:486
of F D & C Blues 10:486; 11:70
of F D & C Green No. 3 10:486
of F D & C Reds 10:486; 11:70
of F D & C Violet No. 1 10:486
of F D & C Yellows 10:486; 11:70
of finishing agents 11:133
of fluorescein 10:514, 523
of Fuchsine NJ 19:167
of glass and glass-ceramics 13:461
of 2-(2-hydroxy-1-naphthylazo)-6-and-8-
    naphthalenesulfonic acids 10:522
of indigo 14:508
of indigo carmine 14:511
of indigoid dyes 14:507, 510
of inhibitors in isoprene 11:511
of iodine 8:21
of iodofluoresceins 10:514
of iron chelates 12:52
of Magenta P 19:167
of Malachite Green 19:166, 167, 168
of malononitrile 15:443
of methanol 16:119
of methyl violet 19:167
of 4-monobromofluorescein 10:523
of neodymium 15:154
of nickel 16:300, 321
of p-nitroaniline in beer 7:592
of nitrobenzene 7:17
of nitrogen dioxide 16:485
of nitrotoluene 7:17
of norethynodrel 10:596
of Orange B 10:486
of Paper Blue R 19:184
of 1-(phenylazo)-2-naphthol-6-sulfonic acid,
    disodium salt of 10:518

of phenylenediamines 11:448
of pigments in tobacco 19:152
of praseodymium 15:154
of Red Violet 5RS 19:167
of riboflavin 10:540
of setocyanine 19:167
of setoglaucine 19:167
of starch hydrolyzates 11:13
of 1-(4-sulfo-1-naphthylazo)-2-naphthol-3-
    sulfonic acid 10:513
of 1-(4-sulfophenylazo)-2-naphthol-6-sulfonic
    acid, disodium salt of 10:518
of 2,4,5,7-tetrabromofluorescein 10:523
of 2,4,5,7-tetraiodofluorescein 10:514
of thioindigo 14:508
of 4-toluene-azo-2-naphthol-3-carboxylic acid
    10:521
of 2,4,5-tribromofluorescein 10:523
of triethyl rhodamine 10:523
of triiodofluoresceins 10:514
of triphenylmethane and related dyes 19:166,
    183
of turmeric 10:541
of Victoria Blue B 19:167
of Victoria Blue BO 19:184
of Wool Green S 19:167
of xanthophyll 11:55; 19:152
Vistalon 12:87
Vistane 12:87
Vitali-Morin reaction
    tropane alkaloids, analysis 4:595
Vitality
    of yeast 19:541
Vitallium. (See also Cobalt alloys.) 10:290
    composition 10:290
    uses 10:290
Vitamin A
    in chocolate and cocoa products 9:538
    in eggs 12:66
    glycerol dichlorohydrine colorimetric method
        12:66
Vitamin B_1. See Thiamine; Vitamins.
Vitamin B_2. See Riboflavin; Vitamins.
Vitamin B_3. See Nicotinamide.
Vitamin B_5. See Nicotinic acid; Vitamins.
Vitamin B_6. See Pyridoxine; Vitamins.
Vitamin B_{12}. (See also Vitamins.)
    bioassay 7:182
    in eggs and egg products 12:66
Vitamin C. See Ascorbic acid.
Vitamin E
    in eggs and egg products 12:66
Vitamin K_1. See Phytonadione.
Vitamin K_3. See Menadione.
Vitamins
    in bakery products 6:455, 456
    in cereals 9:233, 245

of cocoa beans 9:540
in musts 19:469
in wines 19:469
in yeast 19:533
Vitavax. See Carboxin.
Viton 12:87
Vitreous carbon. See Glassy Carbon.
Vitrified-bonded products
diamond reclaiming from 11:483
diamonds, industrial in 11:468
Vitrified grinding wheels
analysis 7:297
composition 7:298
grading tests 7:302
Vodka 4:465
analysis 4:480
quality control 4:470
Volatile acids
in roasted coffee 10:375
in wines 19:413
Volatile bases
in caprolactam 8:124
Volatile matter
in adhesives 10:275
in adsorbents 4:435
in bituminous paving mixtures 6:333
in blast-furnace slag cement 9:110
in brake linings 7:448
in carbon black 8:228, 235
in carbon materials 14:61
in chromium compounds 9:687
in chromium pigments 9:707
in clays and clay minerals 10:111
in coal and coke 10:235
in enamels 12:175
in fatty alcohols 4:575
in fibers 4:25
in fluorocarbons, polymers 13:96
in furfural 13:239
in furfuryl alcohol 13:242
by gas chromatography 8:229
in germanium dioxide 13:399
in graphite 14:61
in metal cleaners 6:368
in methacrylate polymers 16:100
in molding compounds 4:23
in polystyrene 18:325
in polytetramethylene ether gylcols 13:249
in portland cement 9:109
potassium dichromate volumetric method 10:437
potassium permanganate volumetric method 10:436
in roasted coffee 10:417, 436
in slag cement 9:110
in tetrahydrofuran 13:231
in tetrahydrofurfuryl alcohol 13:244

in water 19:381
in waxes and polishes 6:363
Volatility
of silicone fluids 18:115
of silicone resins 18:124
Volatilization methods
for platinum group metals 17:245
Volhard titration
of aluminum chloride 5:174
of bromide 8:17; 9:285
of chloride 6:265; 9:80, 341, 413, 418; 18:257
of epichlorohydrin 12:227
of hydriodic acid 14:598
of iodoform 14:614
of mercury 16:29
of monochloracetic acid 19:486
of salt 16:138, 142
of silver 18:167
of sodium chloride 18:247
Voltammetry
of bismuth 7:217
of crystal violet 19:169
for monomethyl ether of hydroquinone in
acrylic monomers 4:209
of Rhodamine WT 19:182
Voltzine
composition 19:545
Volume
of bottled carbonated beverage 8:175
Volume, bulk
of pulp handsheets 7:256
Volume, pore
of activated carbon 8:140
Volume, specific
of carbon black 8:227
Volumetric methods. See Acid–base titration;
Conductometric titration; Potentiometric
titration; and Redox titration.
Vonnel. (See also *Acrylic and modacrylic fibers*;
Acrylic fibers.) 4:221
solubility 4:249
properties 4:222
Vonsenite 7:315
Vorlex
composition 13:192
uses 13:192
VPM. See Sodium methyl dithiocarbamate.
Vrbaite 18:519
Vu-Chek antifreeze tester 6:18
Vulcaflex 6:77
Vulcanizates. See *Elastomers, synthetic.*
Vulcanization accelerators
in elastomers, synthetic 12:143
ultraviolet spectroscopy 12:143
Vulcanized fibre 10:278
Vulklar. See Chloranil.

W

Wall cleaners. (See also *Detergents.*) 11:325
  composition 11:325
Warfarin. (See also *Coumarins.*) 11:159
  assay 11:162, 169
  in biological systems 11:171
  gas chromatography 6:209, 211
  identification 10:151
  thin-layer chromatography 10:151; 11:165
  ultraviolet spectroscopy 10:151; 11:170, 171
  uses 11:159
Warfarin potassium 10:174
  analysis 10:174
  assay 10:175
  flame test 10:175
  identification 10:174
  iodoform test 10:175
  melting point 10:175
  potassium in 10:175
  properties 10:174
  solubility 10:174
  specification tests 10:175
  ultraviolet spectroscopy 10:175
Warfarin sodium 10:166
  analysis 10:167
  argentometry 10:169
  assay 10:168
  flame test 10:167
  fluorescence spectroscopy 10:168
  Folin-Ciocalteu colorimetric method 10:173
  gas chromatography 10:169
  identification 10:167
  impurities, determination of 10:174
  infrared spectrum 10:167
  iodoform test 10:167
  melting point 10:167
  nitric acid colorimetric method 10:173
  paper chromatography 10:167
  polarographic methods 10:168
  properties 10:166
  separation 10:167
  sodium in 10:167, 168
  solubility 10:166
  specification tests 10:174
  ultraviolet spectroscopy 10:170
Warpage
  of grinding wheels 7:300
Warwickite 7:315
Washing soda. (See also Sodium carbonate.)
  manufacture 18:239
  uses 18:239

Waspaloy. (See also Cobalt alloys.)
  composition 10:287
Water 19:352-396
  in acetic acid 4:100
  in acetone 4:115
  in acetylene 4:135
  in acetylenic alcohols and glycols 4:145
  in acid–base titration 19:360
  acidity 19:360
  in acrolein 4:154, 159
  in acrylamide 4:167
  in acrylonitrile 4:371
  in activated carbon 8:160
  in adhesives 10:275
  in adipic acid 4:419
  in adsorbents 4:435
  in air 16:485
  in alcohols 4:543
  in alkaline flushes 6:360
  alkalinity 8:177; 19:361
  in alkanolamines 5:14
  in aluminum hydroxide 5:189
  in aluminum oxide 5:189
  in aluminum sulfate 5:194
  in alums 5:194
  in amides 8:609
  in ammonia 5:305
  in ammonium chloride 5:337
  in ammonium nitrate 5:343
  in ammonium sulfate 5:352
  analysis 8:177; 19:353
  anions, determination of 19:363
  in antifreeze 6:17; 13:600
  antifreezes for 6:1
  in antirusts 6:358
  in asbestos 6:259
  azeotropic distillation 13:507
  bacteriologic examinations 19:390
  in bakery products 6:448
  in barite 6:556, 558
  in barium carbonate 6:539
  in benzene 7:51
  beta OH method 13:462
  in bituminous materials 6:316, 317
  in black tea 18:486
  boron in 7:417
  in bread 6:439
  in bromine 8:6
  in bromohydrocarbons 8:13
  in 1,3-butadiene 11:504

in butter 16:138
in calcium acetate 8:79
in calcium hypochlorite 7:230
calcium in 18:221
in caprolactam 8:124
in carbon black 8:192, 235
in carbon dioxide 8:255, 260
in carbon disulfide 8:277
in carbon materials 14:61
in carbon monoxide 8:310
in carbon tetrachloride 8:326
carbon tetrachloride distillation 12:467
in carbonyl compounds 8:359
in car washes 6:364
in casein 9:9
cations, determination of 19:370
in cellulose 9:47
in cellulose acetate 9:69
in cement 9:133
centrifuge test 15:330
in cereals 9:235
in cheese 16:141
in chlorinated paraffins 9:499
in chlorine 9:309, 321, 326
chlorine in 9:330
in chlorobenzene 9:478
in chloroisocyanuric acids and salts 7:232
in chlorophenols 9:527
in chocolate 9:540, 550
in chromium compounds 9:687
in chromium pigments 9:707
in chrysotile 6:259
in cigarette smoke 10:13, 14
in citrate esters 10:99
in clays and clay minerals 10:111
in coal and coke 10:234, 250
cobalt bromide color test 14:345; 15:251
in cobalt oxides 10:342
in cocoa beans 9:548, 552
in cocoa products 9:573, 590
color 19:353
commercial grades and specifications 19:353, 354
in confectionery products 10:558
copper in 10:614
in copper hydroxide 10:666
in copper(I) oxide 10:672
in corn products 11:2
in cottage cheese 16:136
coulometry 5:306
in creams and lotions 11:91
in crotonaldehyde 11:183
in crystal violet 19:187
in cyclamates 11:230
in cyclohexylamine 11:214
Dean and Starks method 11:454; 13:239
density 7:38

in detergents 11:382
dew point 13:303
dewpoint analyzers 16:532
dew point method 5:306; 8:257; 13:327; 14:345
in diamines and higher amines, aromatic 11:441
in diammonium phosphate 15:535
in diatomaceous earth 11:489
in dimethyl terephthalate 17:166
in dioxane 11:520
in diphenylamine 11:458
distillation method 6:360; 7:230; 15:330; 18:191
in drycleaning solvents, used 11:609
in dry milk 16:145
EDTA and related compounds in 12:47
effects on bituminous materials 6:309
electrical conductivity 19:359
electrolytic method 13:68; 15:252; 16:532
in emulsified asphalts 6:324
in epichlorohydrin 12:227; 14:175
in esters 8:588
etchant for ceramics 9:224
in ethanolamines 12:295
in ethyleneamines 11:428
in ethylene oxide 12:221, 333
in ethylenimine 14:480
ethylenimine in 14:478
in ethyl ether 12:310
evaporation method for gases 5:305
in fatty alcohols 4:575
in fibers 4:21, 25
in flour 12:602
fluoride in 13:24, 30
in fluorocarbons 13:67
formaldehyde in 16:119
in formic acid 13:128
Fourier transform spectra 13:172, 173
in furan 13:230
in furfural 13:239
in furfuryl alcohol 13:242
gas chromatography 4:154, 371; 6:5; 10:14; 11:215, 441; 12:339; 13:68, 129, 576; 14:171, 175, 237, 239, 385, 393, 415, 480; 15:252; 16:108, 119, 425, 485; 17:40; 19:458
in gases 5:305
in gelatin 12:547
in glass and glass-ceramics 13:459, 462
in glues of animal origin 13:493
in glycerol 13:507, 508, 526, 600
in glycols 13:600
gold in 14:43
in grains 4:469, 470
in graphite 14:61
gravimetric determination 13:326
in green coffee 10:383
in gypsum and gypsum products 14:93

in hair sprays 11:90
hardness 19:362
in hexachlorobenzene 9:483
in hexachloroethane 9:452
in hexafluorophosphoric acid 13:47
in high-explosive mixtures 12:467
in hydrated lime 9:191
in hydrazines 14:237, 239
hydrocarbon in 14:349
in hydrocarbons 14:345
in hydrogen 14:380, 385
in hydrogen chloride 14:397
in hydrogen cyanide 14:416, 422
hydrogen cyanide in 14:416, 417
hydrogen sulfide in 18:399
in hydroxylammonium acid sulfate 14:455
impurities, determination of 19:353
infrared spectroscopy 13:68, 459, 462;
    14:239
in instant coffee 10:440
iodide in 14:600, 602
in iodides 14:603
iodometric titration 12:221
iron in 15:54
in iron blues 17:201
in isophthalic acid 17:168
Karl Fischer titration 4:115, 159, 419, 460;
    5:337; 7:53; 8:588; 11:3, 214, 230; 12:310;
    13:67, 128, 249, 526; 14:422, 455, 490;
    15:252, 330; 16:437; 17:37, 161, 301, 438;
    19:187, 279
lead in 15:173, 186, 189
in lime and limestone 15:228
in limestone 9:191
in liquefied petroleum gas 15:251
liquid carrier distillation 14:345
lithium absorption method 13:293
in lithium chloride 9:345
lithium chloride electrical conductivity method
    13:326
lithium in 18:221
in lubricants 15:329
magnesium in 18:221
in magnesium perchlorate 9:366
magnesium perchlorate absorption 9:326
magnesium perchlorate gravimetric method
    14:397
manganese in 15:471, 474
in margarine 15:502
mercury in 16:22, 36, 44, 45, 46, 49
in methacrylate polymers 16:99, 100
in methanol 16:108, 116
methanol in 16:119
in methyl chloride 9:424
in methylene chloride 9:430
in mixed acids 16:500
in molybdates 16:190

in molybdenum disulfide 16:181
in monoammonium phosphate 15:535
morpholine in 16:200
in natural gas 13:326
near-infrared spectroscopy 13:527
nickel in 16:299
nitrite in 16:491, 492
in nitro alcohols 16:446
in nitroparaffins 16:437
in nitrous oxide 5:366
in noble gases 13:293, 303
odor 19:357
in organosilicon copolymers 18:148
by oven-drying method 5:357
in oxygen 16:532
oxygen in 16:524, 529
ozone in 16:548
in paper matchboard 15:534
in penicillins 16:566
in pesticides and insecticides 19:614
pH 19:360
in phenols 17:36, 39
phenols in 17:28
in phenylenediamines 11:448
phosphate in 17:102
phosphoric anhydride gravimetric method
    4:460
phosphorus in 17:60, 101
phosphorus pentoxide absorption method
    13:67
in phosphorus sesquisulfide 15:531
piezoelectric analyzers 16:532
in pigments 17:17
pollution characteristics 19:371
in polyacrylamides 4:177
in polyamide fibers 17:322
in polyamides 17:296, 301
in polycarbonates 17:349
in polyethylenimine 14:490
polyethylenimine in 14:489
in polyols 19:279
in polytetrafluoroethylene 13:88
in polytetramethylene ether glycols 13:248
in potassium chlorate 9:359
in potassium chloride 9:339
potassium in 17:388; 18:221
in propane 14:345
in propellants 4:460
properties 6:2
properties, determination of 19:353
in pulp handsheets 7:255
in pyridine 17:438
in quicklime 9:191
radioactivity measurements 19:393
radium in 17:456, 457
in rare earth oxides 17:510
in raw milk 16:124

in red phosphorus 15:536
refractive index 7:38
in resins 4:24
in roasted coffee 10:417, 421
in R salt 16:237
rubidium in 18:221
selenium in 17:599
in shampoos 11:74
in silicate white lead 17:189
silicon in 18:29
silver in 18:171, 172
in soaps 18:191
in sodium carbonate 18:247
in sodium carboxymethylcellulose 9:89
in sodium chlorate 9:359, 360, 361
in sodium chloride 9:338
in sodium chlorite 9:355
sodium in 17:388; 18:220, 221, 222
in sodium heparin 10:150
solids in 8:178
in solvents 4:24
specific conductance 19:359
steam distillation 16:145, 181
strontium in 18:221, 269, 275, 278, 280
in styrene 18:289, 309
in sugar 18:357
sulfate in 18:427
sulfide in 18:396
sulfite in 18:404
sulfur dioxide in 18:406
in sunscreens 11:90
taste 19:357
temperature-rise method 4:103, 106
temperature-rise test of magnesium perchlorate
    9:366
in terephthalic acid 17:161
in terpenes 14:283, 345
in 1,1,2,2-tetrachloroethane 9:447
in tetrahydrofuran 13:231
in textiles 4:332
thorium in 19:16
in tobacco 10:5; 19:135
in toluenediamines 11:454
toxicity 19:390
treatment processes 19:352
in 1,1,1-trichloroethane 9:443
in trichloroethylene 9:457
in tungstates 16:190
turbidity 19:358
uranium in 19:252
use in carbonated beverages 8:164, 177
valve freeze method 15:251
vanadium in 19:341
waste treatment processes 19:353
in yeast 19:522, 526
in zinc chromates 17:199
zinc in 19:565, 568

Water absorption
    of diatomaceous earth 15:529
    of felt 12:511
    of flour 12:613
    of photographic films 12:593
Water adsorption
    by carbon black 8:192
Water-beading test
    for automobile waxes and polishes 6:376
Water–boron trifluoride
    volumetric method for nitriles 16:406
Water, carbonated. See *Carbonated beverages.*
Water drop test
    of paper matchboard 15:535
Water fade 7:471
Water flavor test
    for essential oils 12:264
Water-flow calorimeter method
    for calorific value of natural gas 13:332
Water, for proof reduction of alcoholic
        beverages
    analysis 4:479
    quality control of 4:468
Water, free
    in phthalic anhydride 17:155
Water-gas tars. (See also Bituminous materials.)
    6:277
    properties 6:276
Water glass. See Colloidal silicates.
Water-holding capacity
    of drycleaning detergents 11:612
Water insolubles
    in acid salts 8:593
    in acrylamide 4:167
    in aluminum chloride 5:177
    in arsenic(V) oxides 6:233
    in barium acetate 6:536
    in barium chloride 6:546
    in barium nitrate 6:552
    in borax 7:383
    in cadmium chloride 8:42
    in caprolactam 8:126
    in potassium pyroantimonate 6:56
    in sodium arsenate 6:235
    in sodium arsenite 6:235
Water miscibility test
    for emulsified asphalts 6:329
Water, odor-free 8:148
Water regain
    of acetate and triacetate plastics and fibers
        4:76
    of acrylic and modacrylic fibers 4:332
    of carbon fibers 4:22
Water-soluble colors. (See also Certified colors;
        and individual compounds.)
    liquid chromatography 10:525
Water-soluble matter

in adsorbents 4:438
in barium chromate 6:548
in cocoa beans 9:549
in felt 12:495
Water solubility
    of acid halides 8:576
    of aluminum oxide 5:189
    identification of dyes 6:389
Water sorption test
    for dental restorative materials, composite
        11:314
    for methyl methacrylate dental materials
        11:302, 306
Water thickness swell
    of felt 12:511
Water tolerance
    of adhesives 10:274
    of brake fluids 7:433
Water vapor
    thermal conductivity 14:380
Waxes
    analysis 6:362
    automobile 6:353, 354, 362, 363, 375
    in automobile polishes 6:353, 363
    in brake linings 7:447
    in chrome protectors 6:354
    in creams and lotions 11:95
    in finishing agents 11:126, 130
    in lipsticks 11:67
    liquid chromatography 11:67
    performance tests 6:375
    in roasted coffee 10:435
    solubility 11:126, 130
    solvent extraction 19:151
    in styrene polymers 18:321
    in sunscreens 11:91
    in tobacco 19:151
Waxoline purple AS
    colorimetric method for boric acid 7:408
Wax vesta splints
    analysis 15:535
    composition 15:519, 535
    in matches 15:519, 535
Wear-resistant alloys. (See also Cobalt alloys.)
        10:289
    composition 10:289
Wear tests
    of brake linings 7:460, 468
Weathering test
    for liquefied petroleum gas 15:242
Weave distortion of fabrics 4:25
Weber-Fechner rule 7:160
West-Gaeke method
    for sulfur dioxide 18:408
Westphal balance 4:483
Wet ashing
    of bakery chemicals 6:431

of beer 7:553, 648
of boron compounds, organic 7:399
Wet decomposition. See Ashing procedures.
Wet density
    of ion exchange resins 15:10
Wet gluten method
    for protein in flour 12:603
Wet oxidation methods
    for sulfur 18:381
Wet peroxide method
    for chromium 9:689
Wet screening methods
    for cocoa products 9:569
Wettability
    of titanium dioxide 15:531
Wetting agents
    in de-icers 6:356
    in polytetrafluoroethylene 13:82
    in windshield washer solvents 6:354
Wetting characteristics
    of pesticides and insecticides 19:616
Wetting tests
    for detergents 11:394
Wharton and Forziati method
    specific gravity of acrylic and modacrylic fibers
        4:260
Wheat. (See also *Flour.*)
    classification 12:599
    composition 9:233
    mineral content 9:233
    uses 12:600
    vitamin content 9:233
Wheat flour. See *Flour.*
Wheat, for distilled alcoholic beverages
    analysis 4:469
    quality control 4:467
Whey protein content
    of dry milk 16:146
Whey protein nitrogen
    hydrochloric acid colorimetric method
        16:147
Whipping test
    for egg albumen 12:72
Whisky. See *Alcoholic beverages, distilled*; and
        specific types.
White casse
    in wines 19:494
White coat plaster. (See also *Gypsum and gypsum
        products.*)
    gauging plaster content 14:101
White flour. See *Flour.*
White gold
    composition 14:8
    uses 14:8
White iron pyrite 15:166
White metal 5:69
Whiteness 7:233

of acrylic and modacrylic fibers 4:324
of bleached fabrics 7:233
of papers 7:257
White opaque pigments. (See also *Pigments*;
   and individual compounds.)
  analysis 17:187
  commercial grades and specifications 17:187
White phosphorus. See Phosphorus, white.
Whiteside method
  for raw milk 16:125
White solder 5:72
White tin. (See also Tin metal.) 19:26
Whitewall tire cleaners 6:354
  analysis 6:367
  performance tests 6:378
Whiteware. (See also *Ceramics*.) 9:191
  acid dissolution 9:215
Whiting
  commercial specifications 12:174
  composition 12:174
  in elastomers, synthetic 12:152
  in enamels 12:174
  infrared spectrum 12:152
Wichers test
  for mercury metal 16:24
Wiikite 17:561
Wijs method
  for allyl alcohol 5:84
  for allyl esters 5:97, 99
  for iodine number of corn oil 11:55
  for iodine number of fats and fatty oils
    12:481
  for iodine number of fatty alcohols 4:569
  for iodine number of hydrocarbon resins
    14:286
Willemite
  composition 19:545
Willis-Hime test
  for mortars 9:148
Wilson technique
  for sodium heparin 10:143
Window cleaners. (See also *Detergents*.) 11:325
  composition 11:325
Windshield de-icers 6:356
  analysis 6:370
  performance tests 6:378
Windshield washer solvents 6:354
  analysis 6:366
  performance tests 6:378
Wine and must. (See also Grapes; Musts; and
  Wines.) 19:397–514
Wines
  acetic acid in 19:405, 412, 413
  acid–base titration 19:411
  acid in 19:411
  acidity 19:411
  alcohol in 19:404, 425

  analysis 19:397
  anion in 19:495
  ash in 19:488
  boron in 7:407
  carbon dioxide in 19:415
  carbonyl in 19:452
  cation in 19:490
  chemical additives 19:477
  color 19:504
  cyclamate in 11:233
  ethyl acetate in 19:413
  extract content 19:405
  fixed acidity 19:417
  fructose in 19:410
  gases, dissolved 19:500
  glucose in 19:409
  impurities, determination of 19:482
  iron in 15:32, 55
  lactic acid in 15:141
  legal ethanol limits 19:426
  nitrogen compound in 19:469
  pH 19:412
  phenol in 19:460
  pigment in 19:460
  potassium in 18:221
  proof 19:434
  refractive index 19:403
  sampling 19:401
  sodium in 18:219, 221
  soluble solid in 19:404
  specific gravity 19:403
  sucrose in 19:410, 411
  sugar in 19:403, 405
  sulfur dioxide in 19:405, 411
  thallium in 18:524
  toxic content 19:471
  volatile acidity 19:413
Wingstay 100 6:77
  colorimetric method based on formation of
    Wurster salts 6:101, 103
Wingstay S 6:77
  colorimetric method using diazotized *p*-nitro-
    aniline 6:106
  color tests 6:90
  ultraviolet spectroscopy 6:116
Wingstay T 6:77
  colorimetric method using diazotized *p*-nitro-
    aniline 6:106
Winkler titration
  of oxygen 16:524
Wisconsin mastitis test
  for raw milk 16:125
Wiserite 7:315
Witherite. (See also Barium carbonate.) 6:533,
  538
Wolfram. See *Tungsten*.
Wolframite 14:519; 17:561; 19:190

decomposition 19:197, 203
sample decomposition 17:564
scandium in 17:564
tungsten in 19:195, 203
Wollastonite 8:75
Wood
  in concrete 9:142
Wood alcohol. See *Methanol.*
Wood ash
  potassium content 17:406
Wood fiber
  in asbestos–cement products 6:273
  in wood-fibered plaster 14:100
Wood-fibered plaster. (See also *Gypsum and
  gypsum products.*)
  wood fiber content 14:100
Wood flour
  analysis 7:447
  in brake linings 7:441, 447
Wood's metal 5:65
Wood pulp
  bleaching 7:222
  viscosity 7:265
Wood splints
  analysis 15:535
  in matches 15:519, 535
Wood tars. (See also *Bituminous materials.*)
  6:276
Wool
  application of dyes to 6:397
  in felt 12:495
Wool Fast Blue BL 6:382
  manufacture 6:383
Wool Fast Blue GL 6:381, 382, 385
  application test 6:397
  manufacture 6:383
  solubility 6:383
  visible spectroscopy 6:395
Wool fat, hydrous
  solubility 13:539, 552

Wool felt. See *Felt.*
Wool fiber
  microscopy 4:265
  mixtures, determination in 4:353
  solubility 4:248, 250, 353
Wool green S. (See also *Triphenylmethane and
  related dyes.*)
  manufacture 19:163
  visible spectroscopy 19:167
Wool violet. See F D & C Violet No. 1.
Working time
  of dental elastomers 11:285
  of dental impression materials, alginate-based
  11:281
  of dental restorative materials, composite 11:313
Wormwood oil. (See also *Essential oils.*)
  thujone in 12:282
Wort. (See also Beer; *Brewery products.*)
  definition 7:476
  extracts 7:495
  sample preparation 7:483
WPE number
  of dental restorative materials, composite
  11:312
Wrought gold alloys, dental. See Dental wrought
  gold alloys.
Wrought iron. (See also Iron metal.)
  manufacture 15:27
Wulfenite. See Lead molybdate.
Wurster salt formation
  colorimetric method for amine antioxidants
  6:100
Wurtzite
  composition 19:544, 545
  properties 19:544
Wurtz reaction
  for sodium 18:231
  for sodium–potassium alloy 18:231
Wynestron. See Estrone.
Wytox BHT (and 312) 6:77

X

Xanthane hydride
  colorimetric method for phytonadione 10:193
Xanthate method
  for carbon disulfide 8:273
  for disodium ethylenebisdithiocarbamate
    11:580
  for dithiocarbamic acid salts 11:578
  for iron(III) dimethyldithiocarbamate 11:580
  for manganese(II) ethylenebisdithiocarbamate
    11:580
  for zinc(II) dimethyldithiocarbamate 11:580
  for zinc(II) ethylenebisdithiocarbamate
    11:580
Xanthates
  of alcohols 4:517, 521
Xanthene dyes 12:7
Xanthine
  in black tea 18:488
Xanthines
  identification 8:58
  ion exchange chromatography 8:62
  paper chromatography 8:59
Xanthocillin 5:461, 462, 495
  analysis 5:500, 505, 508
  colorimetric methods 5:508
  identification 5:500
  microbiological assay 5:505
  paper chromatography 5:500
  properties 5:493
  separation 5:500
  solubility 5:494
  structure 5:462
  uses 5:495
Xanthocillin X
  structure 5:462
Xanthone
  ion exchange chromatography 6:144
  in salicylic acid 6:143
Xanthophyll
  in egg yolk 12:58
  in feed products 11:52
  liquid chromatography 11:53
  solvent extraction 19:152
  in tobacco 19:152
  visible spectroscopy 11:55; 19:152
Xanthotoxin. (See also *Coumarins*.) 11:161
  gas chromatography 11:168, 170
Xanthydrol
  colorimetric method for tryptophan 8:534
Xanthylamides 8:601

Xanthyl derivatives
  of barbiturates 6:478
Xanthyletin. (See also *Coumarins*.) 11:161
  gas chromatography 11:170
Xenon. (See also *Gases, noble*.)
  commercial specifications 13:287
  gas chromatography 13:297, 308
  in helium 13:298
  impurities, determination of 13:301
  isotopic abundance 13:289
  in krypton 13:300
  manufacture 13:290
  mass spectroscopy 13:301, 305
  in neon 13:299
  occurrence 13:290
  in oxygen 16:531
  properties 13:288
  solubility 13:289
Xenotime 17:71, 470
X-ray crystallography
  of choline chloride 9:613
X-ray diffraction
  of alcohols 4:526
  of aluminum 12:418
  of aniline and its derivatives 5:435
  of antimony sulfide 12:418
  of barium nitrate 12:418
  of calcium silicide 12:418
  of carborundum 12:418
  of cellulose 9:53
  of cellulose nitrate 12:418
  of ceramics 9:219
  of clays and clay minerals 10:107
  of coal and coke 10:228
  of copper(II) bromide 10:652
  of copper(II) carbonate, basic 10:652
  of copper chloride 10:652
  of copper(I) cyanide 10:652
  of copper(II) hydroxide 10:652
  of copper metal 10:652
  of copper(II) nitrate 10:652
  of copper oxide 10:652
  of copper(II) sulfate 10:652
  of crystallinity of polymers 4:86
  of cyclonite 12:418
  of detergents 11:333
  of diamonds, industrial 11:471
  of diazodinitrophenol 12:418
  of glass 9:221
  of glass and glass-ceramics 9:221; 13:465

of graphite 14:76
of iron 15:34
of lead azide 12:418
of lead 2,4-dinitroresorcinol 12:418
of lead mononitroresorcinol 12:418
of lead nitrate 12:418
of lead styphnate 12:418
of lead thiocyanate 12:418
of mercury fulminate 12:418
of pentaerythritol tetranitrate 12:418
of polyamide fibers 17:319
of polycarbonates 17:334
of polyethylene 12:350
of portland cement 9:103
of potassium chlorate 12:418
of potassium dinitrobenzofuroxane 12:418
of α-quartz 12:418
of sillimanite refractory 9:222
of sulfites 18:403
of tetracene 12:418
of tetryl 12:418
of titanium boride 7:366
of triacetate fiber 4:77
of trinitrotoluene 12:418
X-ray emission spectroscopy
of asbestos–cement products 6:273
calcium in silica ablators 4:3
of cesium 9:300, 301
of glass and glass-ceramics 13:454
of gold 14:5
of impurities in aluminum oxide abrasives 4:49
of impurities in asbestos 6:267
of impurities in chrysotile 6:267
of lead 15:183
of lithium 15:272
of mercury 10:495
of niobium 16:362
of potassium 17:371
of tantalum 18:451
of tetraethyllead 15:184
of thorium 9:257
X-ray fluorescence spectroscopy
of aluminum 10:136; 11:157
of aluminum and aluminum alloys 5:117
of antimony 6:57; 11:156
of arsenic 6:243
of barite 6:564
of barium 6:578
of bromide 8:26
of bromine 11:156
of cadmium 11:156
of calcium 10:136; 11:154, 157
of ceramics 9:217
of cerium 17:490, 505
of chlorine 11:157
of chromium 11:154, 156; 17:162
of clays and clay minerals 10:134

of coal and coke 10:231
of cobalt 11:156
of copper 11:156
of cotton 11:146, 151
of dysprosium 17:490
of erbium 17:490
of europium 12:399; 17:490
of europium oxide–yttrium oxide phosphors
    12:399
of gadolinium 17:490
of gallium 13:264
of gold 14:31
of hafnium 14:108, 116
of holmium 17:490
of hydrogen chloride 14:395
of iron 10:136; 11:156
of lanthanum 15:158; 17:490
of lead 11:156
of lutetium 17:490
of magnesium 10:136; 15:367
of manganese 11:154, 157; 15:456, 477
of mercury 11:156
of molybdenum 17:162
of neodymium 17:490
of niobium 14:132; 18:452
of potassium 10:136; 11:154, 157
of praseodymium 17:490
of rare earth compounds 17:504
of rare earths 17:489
of samarium 17:490
of scandium 17:577
of selenium 11:156
of silicon 10:136; 11:157
of silver 11:156
of silver halides 12:565, 567
of strontium 18:270, 281
of sulfur 11:157
of tantalum 16:389, 390; 18:452
of tellurium 11:156
of terbium 17:490
of thorium 19:17
of thulium 17:490
of tin 11:156; 19:65
of titanium 10:136; 11:156; 17:162; 19:99,
    119
of tungsten 11:156
of ytterbium 17:490
of yttrium 17:490
of zinc 11:156
of zirconium 11:154, 156; 14:108, 116
X-ray powder diffraction
of morpholine derivatives 16:197
X-ray scattering, small angle
of acrylic fibers 4:319
X-ray spectroscopy
of cesium 9:281
of chloride 14:128

of chlorocarbons 9:409
of cobalt 10:348, 363
of copper 10:602, 617
of explosive priming mixtures 12:417
of gallium 14:556
of germanium 14:557
of iron 15:33
of nickel 16:300, 321
of rubidium 17:550
of thorium 19:11
m-Xylene. (See also Xylenes.)
  azeotropes 14:320
  commercial specifications 14:321
  gas chromatography 7:7, 8, 9, 10; 8:351;
    13:351; 14:333, 335, 336, 337; 16:92;
    18:298
  in gasoline 13:351, 354
  manufacture 14:319
  in paraffins 14:336
  properties 13:351; 14:319
  in styrene 18:298
  ultraviolet spectroscopy 7:40
  uses 14:319, 321
o-Xylene. (See also Xylenes.)
  azeotropes 14:320
  commercial specifications 14:321
  gas chromatography 7:7, 8, 9, 10, 71;
    13:183, 351; 14:333, 335, 336, 337;
    18:298
  in gasoline 13:351, 354
  manufacture 14:319
  in paraffins 14:336
  properties 13:351; 14:319
  in styrene 18:298
  ultraviolet spectroscopy 7:40
  uses 14:319, 320
p-Xylene. (See also Xylenes.)
  azeotropes 14:320
  in carbon tetrachloride 8:325
  commercial specifications 14:321
  gas chromatography 7:7, 8, 9, 10; 8:325, 351;
    13:351; 14:333, 335, 336, 337; 16:220;
    18:298
  in gasoline 13:351, 354
  manufacture 14:319
  in paraffins 14:336
  properties 13:351; 14:319
  in styrene 18:298
  ultraviolet spectroscopy 7:40
  uses 14:319, 320
Xylenecyanol FF. (See also *Triphenylmethane
    and related dyes.*)
  ceric sulfate titration 19:175
Xylenes. (See also *Hydrocarbons*; and individual
    compounds.) 14:319
  in carburetor cleaners 6:356
  commercial grades and specifications 14:319

in enamels 12:175
gas chromatography 6:369; 11:73; 16:432;
  17:154
in gas line antifreezes 6:355, 369
in nail enamels 11:73
in phthalic anhydride 17:154
properties 12:175
toxicity 14:320
2,3-Xylenol. (See also 2,3-Dimethylphenol;
    Phenols.)
  gas chromatography 17:33
  infrared spectroscopy 17:29
2,4-Xylenol. (See also 2,4-Dimethylphenol;
    Phenols.)
  gas chromatography 17:33
  infrared spectroscopy 17:29
  in 2,6-xylenol 17:33
2,5-Xylenol. (See also 2,5-Dimethylphenol;
    Phenols.)
  gas chromatography 17:33
  infrared spectroscopy 17:29
  in 2,6-xylenol 17:33
2,6-Xylenol. (See also 2,6-Dimethylphenol;
    Phenols.)
  colorimetric method for nitrite 16:494
  gas chromatography 17:33
  impurities, determination of 17:33
  infrared spectroscopy 17:29
  manufacture 17:10
3,4-Xylenol. (See also 3,4-Dimethylphenol;
    Phenols.)
  colorimetric method for nitrate 12:587
  gas chromatography 17:33
  infrared spectroscopy 17:29
3,5-Xylenol. (See also 3,5-Dimethylphenol;
    Phenols.)
  gas chromatography 17:33
  infrared spectroscopy 17:29
Xylenol orange
  colorimetric method for hafnium 14:117
  colorimetric method for niobium 16:369, 373
  colorimetric method for zirconium 14:117
2,6-Xylenolphthalein-α,α'-bis(imino)diacetic
    acid
  colorimetric method for strontium 18:276
Xylenols. (See also Dimethylphenols; and
    individual compounds.) 17:1
  commercial grades and specifications 17:13
  manufacture 17:9
  occurrence 17:1, 9
  solubility 17:3
  specific gravity 17:44
3,4-Xylic acid
  gas chromatography 8:385
2,3-Xylidine
  color reactions for identification 5:431
  manufacture 5:425

properties 5:423
  ultraviolet spectroscopy 5:432
2,4-Xylidine
  azo dyes for identification 5:430
  bromination methods 5:448
  color reactions for identification 5:431
  infrared spectrum 5:438
  manufacture 5:425
  properties 5:423
  ultraviolet spectroscopy 5:432
2,5-Xylidine
  color reactions for identification 5:431
  manufacture 5:425
  properties 5:423
  ultraviolet spectroscopy 5:432
2,6-Xylidine
  bromination methods 5:448
  color reactions for identification 5:431
  manufacture 5:425
  properties 5:423
  ultraviolet spectroscopy 5:432
3,4-Xylidine
  color reactions for identification 5:431
  properties 5:423
  ultraviolet spectroscopy 5:432

3,5-Xylidine
  manufacture 5:425
  near-infrared spectroscopy 5:435
  properties 5:423
  ultraviolet spectroscopy 5:432
Xylidines 5:421
  acid–base titration 5:443
  colorimetric methods 5:452
  gas chromatography 5:440
  manufacture 5:425
Xylitol
  liquid chromatography 13:503
Xylose
  gas chromatography 19:300
  in green tea leaf 18:478
  paper chromatography 18:478
2-(2,3-Xylylazo)-1-naphthol-4-sulfonic acid
  in F D & C Red No. 4 10:515
2-(2,4-Xylylazo)-1-naphthol-4-sulfonic acid
  visible spectroscopy 10:516
Xylylene diisocyanate. (See also Isocyanates;
    Isocyanic acid esters.)
  near-infrared spectroscopy 15:105
  structure 19:271
  trade names 19:271

Y

Yarns. (See also Fibers.)
  ash 4:91
  finish, spin 4:81
  fluffing 4:85
  pigment, inorganic 4:91
Yeast 19:514–543
  amino nitrogen in 19:529
  ammonium nitrogen in 19:529
  analysis 4:480; 19:520, 521
  ash content 19:530
  in bread 6:439
  in carbonated beverages 8:167
  cation content 19:530
  cell constituents 19:526
  cell count 19:522, 523
  chloride in 19:531
  in cocoa products 9:600
  composition, determination of 19:522, 526
  definition 19:514
  dough raising ability 19:536
  durability 19:540
  enzyme in 19:533
  fat in 19:531
  fermentation ability 19:534, 540
  fiber in 19:532
  live cell in 19:524
  microscopy 19:523
  moisture in 19:522, 526
  nitrogen in 19:528
  nitrogen-free extract content 19:533
  occurrence 19:515
  in penicillins 16:566
  phosphorus in 19:529
  protein in 19:528
  purity 19:541
  quality control 4:468
  quantity 19:522
  sampling 19:521
  separation 19:523
  solids in 19:522, 526
  species 19:515, 541
  sterol in 10:533
  uses 19:515
  vitality 19:541
  vitamins in 19:533
Yeast cultures. (See also *Yeast*.) 19:520
  analysis 19:520
  properties 19:521
  species 19:516
  uses 19:516, 520, 521

Yeast derivatives. (See also *Yeast*.) 19:518, 519
  composition 19:518
  manufacture 19:518
  species 19:516
  uses 19:516, 518
Yeast enzyme preparations. (See also *Yeast*.)
    19:521
  species 19:516
  uses 19:516
Yeast extract. (See also *Yeast*.) 19:518
  commercial specifications 19:519
  composition 19:518
  species 19:516
  uses 19:516, 518
Yeast fermentable extract
  in beer 7:501
Yeast inoculums. (See also *Yeast*.) 19:519
  manufacture 19:519
  species 19:516
  uses 19:516, 519
Yeast products. (See also *Yeast*.)
  analysis 19:520, 521
Yeast vitamin supplement. (See also *Yeast*.)
    19:517
  composition 19:517
  species 19:516, 518
  uses 19:516
Yellow cake 19:232
Yellow 2G. (See also *Triphenylmethane and
    related dyes*.)
  paper chromatography 19:172
Yellow organic pigments
  in enamels 12:184
  identification 12:184
Yellow phosphorus. See Phosphorus, white.
Yellow pigments. (See also *Pigments*; and
    individual compounds.) 17:196
Yellow RFS. (See also *Triphenylmethane and
    related dyes*.)
  paper chromatography 19:172
Yellow RY. (See also *Triphenylmethane and
    related dyes*.)
  paper chromatography 19:172
Yield strength
  of dental chromium–cobalt alloys 11:291
  of dental wrought gold alloys 11:295
Ylang-ylang oil. (See also *Essential oils*.)
  refractive index 12:271
Yogurt. (See also Fermented milk products.)
  manufacture 16:135

Yohimbine
  detection reagent for borax 7:391
Yolk index
  of eggs 12:75
Young's modulus, dynamic
  of ethylene–propylene copolymers 12:386, 388
  of polyethylene 12:387
  of polypropylene 12:385
Ytterbium. (See also *Rare earths*; Ytterbium
    metal.)
  absorption spectroscopy 17:483
  atomic absorption spectroscopy 17:494
  emission spectroscopy 15:153, 155; 17:503
  in erbium oxide 17:503
  flame photometry 17:493
  ion exchange chromatography 13:255
  in lanthanum oxide 15:153, 155
  in lutetium oxide 17:503
  polarography 17:493
  in thulium oxide 17:503
  x-ray fluorescence spectroscopy 17:490
Ytterbium fluoride
  carrier gas distillation fusion methods 8:636
  oxygen in 8:635, 636
  vacuum distillation fusion methods 8:635
Ytterbium metal. (See also Rare earth metals.)
  properties 17:471
Ytterbium oxide. (See also Rare earth oxides.)
  emission spectroscopy 17:503
  impurities, determination of 17:503
  properties 17:497
Ytterbium perchlorate. (See also Rare earth
    compounds.)
  absorption spectrum 17:486
Yttria. (See also *Rare earths.*) 17:470
Yttrium. (See also *Rare earths*; Yttrium metal.)
  atomic absorption spectroscopy 17:494
  in dysprosium oxide 17:502
  EDTA titration 13:445
  emission spectroscopy 15:155; 17:488, 501

  in erbium oxide 17:503
  in europium oxide 17:502
  flame photometry 17:493
  in gadolinium oxide 17:502
  in glass and glass-ceramics 13:445
  in holmium oxide 17:503
  in lanthanum oxide 15:155; 17:501
  in magnesium 15:394
  in magnesium alloys 15:394
  in samarium oxide 17:502
  in terbium oxide 17:502
  in thulium oxide 17:503
  x-ray fluorescence spectroscopy 17:490
  in ytterbium oxide 17:503
Yttrium earth group. (See also *Rare earths.*)
    17:470
Yttrium fluoride
  carrier gas distillation fusion methods 8:636
  oxygen in 8:630, 635, 636
  vacuum distillation fusion methods 8:635
Yttrium iron garnet
  in ceramics 9:225
  microscopy 9:225
Yttrium metal. (See also Rare earth metals.)
  carrier gas fusion methods 8:625, 627, 682
  hot extraction methods 8:684
  hydrogen in 8:621, 623, 638, 680, 684
  nitrogen in 8:621, 623, 627
  oxygen in 8:613, 621, 623, 625, 627, 630,
    680, 682
  properties 17:471
  vacuum fusion methods 8:621, 623, 680
  vacuum hot extraction methods 8:638
Yttrium oxide. (See also Rare earth oxides.)
  emission spectroscopy 17:501
  impurities, determination of 17:501
  properties 17:497
  uses 17:509
Yttrotantalite
  composition 18:441

Z

Zahm-Hartung volume meter
for carbon dioxide 7:574
Zahm sample bottle
for beer 7:570
Zalba (Special) 6:77
Zeaxanthin
in green tea leaf 18:480
thin-layer chromatography 18:480
Zefran. (See also *Acrylic and modacrylic fibers*;
Acrylic fibers.) 4:221
cross section 4:223, 263
infrared spectrum 4:241
microscopy 4:264
moisture regain 4:333
properties 4:222
solubility 4:247
staining tests 4:252
Zeisel method
for alkoxy groups in organosilane esters 18:75
for alkoxyl content of starch derivatives 11:29
for diethers 4:70
for ethoxy content 6:190
for hydroxyalkyloxyl content of starch
derivatives 11:32
for hydroxyethylcellulose 9:85
Zeiss particle size analyzer 8:208
Zerlate. See Ziram.
Zero strength time
of polychlorotrifluoroethylene 13:95
Zimmerman-Reinhardt method
volumetric method for iron oxide 17:194
Zinc. (See also *Zinc alloys*; *Zinc compounds*;
and *Zinc metal*.) 19:544–577
in acrylic fibers 4:349
in alumina 5:116
in aluminum 19:568
in aluminum alloys 5:148
in aluminum metal 5:148
ammonium hydroxide gravimetric method
14:552
ammonium sulfide gravimetric method 19:554
analysis 19:552
atomic absorption spectroscopy 7:663;
10:131; 11:157; 14:49; 15:157, 186, 195,
196, 395; 19:568
in bauxite 5:116
in beer 7:651, 663
in blood 19:568
borax bead test 6:584
in brass 10:630; 19:558

cadmium in 8:52
in cadmium chloride 8:42
in cadmium iodide 8:44
in cement 9:131
in clays and clay minerals 10:124, 130
in cobalt(II) acetate 10:331
in cobalt alloys 10:324
in cobalt carbonate 10:333
in cobalt compounds 10:328
in cobalt metal 10:305
in cobalt nitrate 10:341
in cobalt(II) sulfate 10:343, 344
in cocoa products 9:595, 598
colorimetric determination 15:395; 19:556,
563
colorimetric method for titanium 12:186
in copper alloys 10:649
in copper metal 10:627
in copper-nickel-zinc alloys 19:559
in cotton 11:141, 153, 156
in crystal violet 19:185
decomposition 19:553
in dental amalgams 11:288
detection 19:556
diammonium hydrogen phosphate gravimetric
method 19:559
disodium ethylenediaminetetraacetate titration
18:205
displacement of copper 10:604
dithizone colorimetric method 5:149; 7:664;
10:125, 305, 331; 11:141, 144; 13:268;
14:579; 15:491; 16:340; 19:566, 563
EDTA titration 12:153, 532; 15:395; 19:562
in elastomers, synthetic 12:151
electrochemical determination 19:566
electrolytic method 19:555
in electroplating solutions 19:559, 562, 566
emission spectroscopy 7:651; 11:153; 13:277;
14:48, 122, 538; 15:144, 187, 197, 375;
16:387; 19:558, 568, 569
in enamels 12:186
ferricyanide colorimetric method 19:556
in ferrites 12:532
in ferrochromium 9:661
in ferroniobium 16:387
flame tests 6:589, 592
in gallium arsenide 13:279
in gallium metal 13:266, 268
in gallium phosphide 13:282
in gelatin 12:554

in gold 14:45, 49
gravimetric determination 19:554, 558
gravimetric method using ammonium
     mercuric thiocyanate 5:148
gravimetric method for tellurium 17:591
in hafnium 14:122
half-life 19:544
history 19:544
in hydrochloric acid 14:374
hydrogen sulfide gravimetric method 12:151;
     16:298; 19:554, 558
impurities 19:545, 551
in indium 14:538
in indium alloys 14:552
in insulin 14:579
ion exchange chromatography 12:532;
     13:254; 19:555, 564
in iron 15:59; 19:559
isotopes 19:544
in lactic acid and its derivatives 15:144
in lead 15:195, 196; 19:568
in lead alloys 15:195
in leaded zinc oxide 17:189
in lubricating oils 19:559, 566
in magnesium 15:375, 394
in magnesium alloys 15:375, 394
in manganese sulfate 15:491
mercury thiocyanate gravimetric method
     19:557
microcosmic salt bead test 6:587
neutron activation analysis 13:265
in nickel 16:298, 339
in nickel alloys 16:339
in niobium 16:387
occurrence 19:544
oxine gravimetric method 19:560
pentacyano-p-nitrosodimethylanilinoferrate
     colorimetric method 19:557
in plants 18:221
polarography 5:150; 12:555; 19:185, 566
potassium ferrocyanide–diethylaniline colori-
     metric method 19:556
potassium ferrocyanide-3,3'-dimethyl-
     naphthidine chloride colorimetric method
     19:557
potassium ferrocyanide gravimetric method
     12:186
potassium ferrocyanide titration 10:649;
     14:552; 19:186, 561
reducing agent for axo dyes 6:404
reduction methods for iron 15:35
reduction method for platinum group metals
     17:250
in rocks 15:187
sampling 19:553
separation 19:554
in silver 18:173

in sodium dichromate 9:684
in soil 15:186
solvent extraction 15:491; 16:340; 19:555,
     564
in steel 19:559
thiocyanomercurate gravimetric method 19:560
in tin alloys 19:568
in tin metal 19:67
in tobacco 19:152
in urine 19:568
uses 19:545
volumetric determination 15:395; 19:561
in water 19:565, 568
x-ray fluorescence spectroscopy 11:156
in zinc alloys 19:566, 569
in zinc chromate 9:707; 17:199
in zinc metal 19:569
zincon colorimetric method 19:563
in zinc powder 17:204
in zinc soap 18:205
in zirconium 14:122
Zinc acetate. (See also Zinc compounds.)
   gravimetric method for sulfide 18:396
   uses 19:574
Zinc–acetic acid
   color test for triphenylmethane and related
        dyes 19:165, 166
Zinc-activated polyethylene thiuram disulfide.
     See Mitiram.
Zinc alloys 5:72
   aluminum in 5:246, 248, 253
   analysis 19:569
   cadmium in 8:52; 19:570
   composition, determination of 19:569
   copper in 10:617; 19:569
   decomposition 19:553
   iron in 19:570
   lead in 19:570
   magnesium in 19:570
   tin in 19:64, 570
   uses 19:545
   zinc in 19:566, 569
Zinc amalgam
   reduction method for titanium dioxide 10:541
Zinc ammonium chloride. (See also Zinc
     compounds.)
   uses 19:573
Zinc antimonide semiconductors. (See also
     Ceramics.)
   properties 9:213
   uses 9:213
Zinc blende 13:251
   composition 19:544, 545
Zinc bromide. (See also Zinc compounds.)
   properties 19:572
Zinc bromide-2,4-dinitrophenylhydrazine
   gravimetric method for camphene oxide 12:214

gravimetric method for 1,2-diisobutylene
oxide 12:214
gravimetric method for epoxy compounds
12:214
gravimetric method for α-methylstyrene oxide
12:214
gravimetric method for α-pinene oxide 12:214
Zinc chelates
EDTA content 12:49
Zinc chloride. (See also Zinc compounds.)
colorimetric method for hypochlorite 11:142
commercial grades and specifications 19:573
manufacture 19:573
microchemical method for papaverine 16:250
properties 9:335; 19:572
solubility 9:336; 19:572
titration of DTPA 12:33
titration of EDTA 12:33
titration of ferrocyanide 15:90
titration of HOEDTA 12:33
toxicity 19:550
uses 19:573
Zinc chloride–hydrochloric acid
colorimetric method for menadione 10:196
Zinc–chlorine fusion
of platinum group metals 17:244
Zinc chromate. (See also Chromium pigments.)
17:199
analysis 9:705; 17:199
chromium in 9:705
commercial grades and specifications 9:706;
17:199
commercial specifications 12:174
composition 17:199
in enamels 12:174
impurities, determination of 17:199
uses 9:705; 17:199
zinc in 9:107; 17:199
Zinc citrate complex
stability 10:89
Zinc compounds. (See also individual compounds.)
19:570–575
analysis 19:574
decomposition 19:553
impurities, determination of 19:574
properties 19:574
solubility 19:549
toxicity 19:550
zinc in 19:574
Zinc cyanide. (See also Zinc compounds.)
uses 19:573
Zinc dibutyldithiocarbamate. (See also
Dithiocarbamic acid salts; Zinc compounds.)
properties 11:575
uses 11:576; 19:574
Zinc diethyldithiocarbamate. (See also Dithiocarbamic acid salts; Zinc compounds.)

colorimetric method for copper 12:157
properties 11:575
ultraviolet spectroscopy 11:583
uses 11:576; 19:574
Zinc diethylenebisdithiocarbamate. (See also
Zinc compounds.)
uses 19:574
Zinc dimethyldithiocarbamate. (See also Dithiocarbamic acid salts; Zinc compounds.)
13:203
manufacture 13:204
properties 11:575; 13:204
solubility 13:204
toxicity 13:204
uses 11:576; 13:203; 19:574
xanthate method 11:580
Zinc–dithiozonate complex
colorimetric method for EDTA and related
compounds 12:47
Zinc ethylenebisdithiocarbamate. (See also
Dithiocarbamic acid salts.) 13:203
manufacture 13:203
properties 13:203
solubility 13:203
toxicity 13:203
uses 11:577; 13:203
xanthate method 11:580
Zinc ethylenediaminedithiocarbamate. (See also
Dithiocarbamic acid salts.)
properties 11:575
Zinc ferrocyanide
colorimetric method for sulfite 18:402
Zinc fluoborate. (See also Zinc compounds.)
uses 19:573
Zinc fluoride. (See also Zinc compounds.)
properties 13:11; 19:572
Zinc iodide. (See also Zinc compounds.)
properties 19:572
Zincite
composition 19:545, 571
occurrence 19:545, 571
Zinc metal
analysis 19:552, 569
carrier gas distillation fusion methods 8:636
carrier gas fusion methods 8:682
commercial grades and specifications 19:551
copper in 10:617
decomposition 19:553
ferric chloride volumetric method 17:204
ferric phenanthroline colorimetric method
19:575
fusion of platinum group metals 17:231
gravimetric method for silver 18:162, 165
gravimetric method for thallium 18:523
impurities, determination of 19:569
manufacture 19:550
oxygen in 8:636, 682

properties 19:548
sampling 19:553
solubility in gold 14:9
sulfur in 18:383
tin in 19:64
toxicity 19:550
uses 19:545, 549, 552
zinc in 19:569
in zinc oxide 19:575
Zinc naphthenate. (See also Zinc compounds.)
   in enamels 12:176
   uses 19:574
Zinc-α-naphthylamine
   colorimetric method for nitrate 19:366
Zinc-N-1-naphthylethylenediamine
   colorimetric method for parathion 19:611
Zinc-nickel-cobalt-copper ferrite. (See also
      Ferrites.)
   commercial specifications 12:520
   composition 12:520
Zinc-nickel-manganese-cobalt-copper ferrite.
      (See also Ferrites.)
   commercial specifications 12:520
   composition 12:520
Zinc nitrate. (See also Zinc compounds.)
   titration of DTPA 12:36
   titration of EDTA 12:36, 38
   titration of NTA 12:38
   titration of TTHA 12:36
   uses 19:574
Zinc nitrate-lanthanum-DCYTA
   colorimetric method for TTAA 12:30
Zinc-nitric acid
   color test for methacrylate polymers 16:70
Zinc nitride. (See also Zinc compounds.)
   manufacture 19:549
Zinc octoate. (See also Zinc compounds.)
   uses 19:574
Zincon
   colorimetric method for zinc 19:563
Zinc oxide. (See also Zinc compounds.) 19:571
   analysis 15:529; 19:574
   assay 15:529
   carrier gas fusion methods 8:626
   commercial grades and specifications 12:174;
      519; 15:514; 19:571
   composition 12:519
   in creams and lotions 11:93
   EDTA titration 13:445, 448; 15:530
   in elastomers, synthetic 12:156
   in enamels 9:195; 12:174
   ferrocyanide titration 17:188, 190
   in glass and glass-ceramics 13:413, 415, 418,
      445, 448
   in glazes 9:196
   gravimetric method for cobalt 10:350
   gravimetric method for nickel 16:297

hydrogen sulfide gravimetric method 13:419
impurities, determination of 15:529; 19:574
manufacture 15:514; 19:548, 571
in matches 15:514
occurrence 19:571
oxygen in 8:626
particle size 15:529
properties 19:571
pyrolytic method 12:156
in silicate glasses 9:198
solubility 19:571
in sulfate white lead 17:188
toxicity 19:550
uses 19:571
in zinc oxide pigment 17:189
in zinc sulfide pigments 17:190
Zinc oxide-eugenol dental pastes. See Zinc
      oxide-eugenol dental products.
Zinc oxide-eugenol dental products 11:296
   analysis 11:296
   commercial grades and specifications 11:257,
      296
   composition 11:296
   consistency 11:297
   gypsum compatibility 11:297
   hardness 11:297
   properties 11:295, 297
   setting time 11:297
   uses 11:295
Zinc oxide-methacrylate dental products 11:299
   analysis 11:299
   composition 11:299
   uses 11:296
Zinc oxide pigment
   analysis 17:189
   commercial grades and specifications 17:189
   impurities, determination of 17:189
   zinc oxide in 17:189
Zinc oxyphosphate cement. See Dental zinc
      phosphate cements.
Zinc N,N-pentamethylene dithiocarbamic acid
      salts. (See also Dithiocarbamic acid salts.)
   uses 11:577
Zinc phenolsulfonate
   in deodorants 11:87
   EDTA titration 11:87
   ultraviolet spectroscopy 11:87
Zinc phosphate cements, dental. See Dental
      zinc phosphate cements.
Zinc-potassium iodide
   microchemical method for narceine 16:250
   volumetric method for ammoniated mercury
      11:563
Zinc powder
   analysis 17:204
   commercial grades and specifications 17:204
   impurities, determination of 17:204

zinc in 17:204
Zinc reduction
  of merbromin 11:564
  of phenylmercuric acetate 11:564
  of phenylmercuric nitrate 11:564
  of thimerosal 11:564
Zinc silicate. (See also Silicates.)
  uses 18:44
Zinc soap. (See also *Soaps.*)
  analysis 18:205
  zinc in 18:205
Zinc spar
  composition 19:545
Zinc stearate. (See also Zinc compounds.)
  in enamels 12:176
  uses 19:574
Zinc sulfate. (See also Zinc compounds.)
    19:573
  commercial grades and specifications 19:573
  manufacture 19:573
  occurrence 19:573
  properties 19:573
  uses 19:573
  volumetric method for phosphate 18:194
Zinc sulfate–EDTA
  volumetric method for alumina 19:125
Zinc sulfide. (See also Zinc compounds.) 19:572
  commercial grades and specifications 19:572
  in glass 9:200
  manufacture 19:549, 572
  occurrence 13:376; 19:572
  properties 19:572
  uses 19:572
  in zinc sulfide pigments 17:190
Zinc sulfide–barium pigment
  analysis 17:190
  commercial specifications 17:189
Zinc sulfide–magnesium pigment
  analysis 17:190
  commercial specifications 17:189
Zinc sulfide pigments
  analysis 17:190
  commercial grades and specifications 17:189
  impurities, determination of 17:190
  zinc sulfide in 17:190
Zinc uranyl acetate
  colorimetric method for sodium 11:141;
    18:225
  gravimetric method for lithium oxide 13:424
  gravimetric method for sodium 18:217, 223
  gravimetric method for sodium oxide 13:423
  volumetric method for sodium 18:226
Zinc yellow. See Zinc chromate.
Zinc–Zincon complex
  colorimetric method for NTA 12:48
Zineb. See Zinc diethylenebisdithiocarbamate.
Zinnwaldite 15:261

Ziram. See Zinc dimethyldithiocarbamate.
Zircaloy
  carrier gas fusion methods 8:625
  copper in 14:129
  fluoride in 14:130
  hydrogen in 8:621, 622
  iron in 14:131
  nitrogen in 8:623
  oxygen in 8:621, 622, 625
  phosphorus in 14:135
  silicon in 14:135
  tantalum in 14:125
  uranium in 14:138
  vacuum fusion methods 8:621, 622, 623
  vacuum hot extraction methods 8:638
Zircon 14:103, 139
  aluminum in 14:139
  analysis 14:139
  composition 9:203; 14:104
  composition, determination of 14:139
  decomposition 19:235
  fluxes for 9:216
  hafnium in 14:139
  impurities, determination of 14:139
  iron in 14:139
  occurrence 14:104
  properties 9:204, 211; 14:139
  silica in 14:139
  sodium peroxide fusion 14:140
  thorium in 19:16
  titanium in 14:139
  uses 14:139
  in whiteware 9:191
  zirconium in 14:139
Zirconia. See Zirconium oxide.
Zirconium. (See also Zirconium alloys;
    Zirconium metal.)
  alizarin red S colorimetric method 14:110, 118;
    15:396
  in aluminum 14:120
  in aluminum alloys 5:151
  in aluminum oxide abrasives 4:43, 49
  ammonium hydroxide gravimetric method
    14:111; 16:173, 174
  ammonium phosphate gravimetric method
    14:109, 114; 16:173, 174
  in asbestos 6:267
  atomic absorption spectroscopy 14:108, 120,
    124
  in bauxite 5:116
  in cement 9:131
  chloromandelic acid gravimetric method 5:152
  in chromium alloys 9:663, 666
  in chrysotile 6:267
  in cobalt alloys 10:324
  colorimetric determination 14:110, 117;
    15:396

in copper alloys 10:649
in cotton 11:141, 154, 156
cupferron gravimetric method 9:665;
    14:111, 252; 16:340
diammonium hydrogen phosphate gravimetric
    method 9:666; 10:325, 649; 19:102
diammonium phosphate–hydrogen peroxide
    gravimetric method 16:341
EDTA titration 13:445; 14:115, 143; 15:396
electrolytic method 10:325
emission spectroscopy 4:3; 13:277; 14:107,
    119, 121; 15:155, 375; 16:387
in ferrochromium 9:661
in ferroniobium 16:387
in gallium arsenide 13:276
gravimetric determination 4:43; 14:108, 110,
    114; 15:396
in hafnium 14:116
hafnium in 14:114
in high-temperature alloys 10:286
history 14:103
in hydrochloric acid 14:375
8-hydroxyquinoline colorimetric method
    14:118
8-hydroxyquinoline gravimetric method
    9:664
identification 14:107
ion exchange chromatography 13:255; 14:113,
    114; 16:153, 385, 390; 17:526
in lanthanum 14:120
in lanthanum oxide 15:155
in magnesium 15:375, 396
in magnesium alloys 15:375, 396
mandelic acid gravimetric method 7:362;
    14:109, 115, 140
manganese in 15:456, 470
in molybdenum alloys 16:173
morin colorimetric method 11:141
in nickel 16:340
in nickel alloys 16:340
in niobium 14:116; 16:387
niobium in 16:385
nitrogen in 16:464
β-nitroso-α-naphthol colorimetric method
    11:141
occurrence 14:104
potassium pyrosulfate fusion 16:340
pyrocatechol violet colorimetric method
    14:110, 119
rare earths in 17:477
salicylic acid–ammonium salicylate gravimetric
    method 19:95
in scandium 17:573
separation 14:110
separation from aluminum 5:208
in silica ablators 4:3
sodium alizarin sulfonate colorimetric

method 10:325
sodium carbonate fusion 16:173, 174
solvent extraction 14:111, 114
in steel 14:116
tantalum in 16:385
in titanium metal 19:121
in uranium 14:116; 19:240
x-ray emission spectroscopy 4:49
x-ray fluorescence spectroscopy 11:154,
    156; 14:108, 116
xylenol orange colorimetric method 14:117
in zircon 14:139
in zirconium boride 7:362
in zirconium carbide 8:133, 137; 14:143
in zirconium diboride 14:143
in zirconium hydride 14:144
in zirconium nitride 14:143
in zirconium sulfate 14:146
in zirconyl acetate 14:147
in zirconyl ammonium carbonate 14:147
in zirconyl nitrate 14:148
Zirconium alloys. (See also Zirconium metal.)
    14:106
carrier gas fusion methods 8:682
hydrogen in 8:638, 681
oxygen in 8:620, 681, 682
tin in 19:53
titanium in 19:117
tungsten in 19:212
vacuum hot extraction methods 8:638
vacuum fusion methods 8:620, 681
Zirconium boride. (See also Metal borides.)
    7:358
manufacture 7:358
properties 7:359, 360
zirconium in 7:362
Zirconium carbide. (See also Carbides.) 14:142
analysis 14:142
carbon in 8:134; 14:143
formation 8:130
free carbon in 8:134
manufacture 14:142
properties 8:131, 133; 14:142
uses 8:133; 14:142
zirconium in 8:133, 137; 14:143
Zirconium compounds. (See also individual
    compounds.) 14:138
Zirconium–copper alloys. (See also Copper
    alloys.)
properties 10:631
Zirconium–copper–molybdenum alloys
copper in 14:126
molybdenum in 14:126
Zirconium diboride 14:142
analysis 14:142
boron in 14:144
manufacture 14:142

properties 14:142
uses 14:142
zirconium in 14:143
Zirconium dioxide. See Zirconium oxide.
Zirconium–eriochrome cyanine R
colorimetric method for fluoride 13:32
Zirconium hydride. (See also *Hydrides.*)
14:144
analysis 14:144, 249, 252
assay 14:249
commercial grades and specifications 14:246
composition, determination of 14:249, 252
hydrogen in 14:144, 249
manufacture 14:144, 245
properties 14:243, 244
uses 14:144, 242
zirconium in 14:144, 252
Zirconium hydroxide
gravimetric method for selenium 17:590
gravimetric method for tellurium 17:590
Zirconium lactate. (See also Lactic acid salts.)
uses 15:130
Zirconium metal. (See also Zirconium alloys.)
carrier gas fusion methods 8:625, 626, 627,
682
commercial grades and specifications 14:106
hot extraction methods 8:684
hydrogen in 8:619, 622, 626, 638, 680, 684
impurities, determination of 14:121
manufacture 14:105
nitrogen in 8:618, 619, 622, 623, 626, 627,
629, 630
oxygen in 8:619, 620, 621, 622, 623, 625,
626, 627, 631, 680, 682
properties 14:104
silicon in 18:37
tin in 19:53
titanium in 19:117
tungsten in 19:212
vacuum fusion methods 8:618, 619, 620, 621,
622, 623, 680
vacuum hot extraction methods 8:638
Zirconium nitride 14:142
analysis 14:142
carrier gas fusion methods 8:624
hydrogen in 8:624
manufacture 14:142
nitrogen in 8:624; 14:144
oxygen in 8:624
properties 14:142
uses 14:142
zirconium in 14:143
Zirconium oxide 14:145
in alkali barium silicate glass 9:218
analysis 14:145
carrier gas fusion methods 8:626
in ceramics 9:225

in cerium zirconate 9:262
composition 9:207
emission spectroscopy 9:218
fluxes for 9:216
in glass and glass-ceramics 13:413, 415, 427,
445
in glazes 9:196
mandelic acid gravimetric method 9:262;
13:427
manufacture 14:145
microscopy 9:225
oxygen in 8:626
properties 9:204; 14:145
pyrocatechol violet colorimetric method
13:428
in refractories 9:207
in silicate glasses 9:198
uses 14:145
Zirconium sulfate 14:145
analysis 14:146
gravimetric method for selenium 17:595
gravimetric method for tellurium 17:595
manufacture 14:145
solubility 14:146
sulfate in 14:146
uses 14:145
zirconium in 14:146
Zirconium tetrachloride 14:146
analysis 14:147
manufacture 14:146
properties 9:335; 14:146
sample preparation 14:147
solubility 9:336
uses 14:146
Zirconium–uranium alloys. (See also Uranium
alloys.)
decomposition 19:235
Zirconium–xylenol orange complex
colorimetric method for EDTA 12:38, 45
colorimetric method for NTA 12:38
Zircon refractories. (See also Refractories.)
composition 9:207, 208
properties 9:208
Zirconyl acetate 14:147
analysis 14:147
manufacture 14:147
solubility 14:147
uses 14:147
zirconium in 14:147
Zirconyl ammonium carbonate 14:147
analysis 14:147
carbon dioxide in 14:147
manufacture 14:147
uses 14:147
zirconium in 14:147
Zirconyl chloride 14:146
manufacture 14:146

solubility 14:146
uses 14:146
Zirconyl chloride–alizarin red S
  colorimetric method for fluorides 14:98
Zirconyl nitrate 14:147
  acid–base titration 14:148
  analysis 14:148
  manufacture 14:147
  nitrate in 14:148
  properties 14:147

uses 14:148
zirconium in 14:148
Zlatkis' reagent
  color test for cholesterol 12:64
Zoisite 8:75
Zolon red
  colorimetric method for silver
    18:164
Zymase
  in yeast 19:534